NEUROGLIA IN NEUROLOGIC AND PSYCHIATRIC DISORDERS, PART II

HANDBOOK OF CLINICAL NEUROLOGY

Series Editors

MICHAEL J. AMINOFF, FRANÇOIS BOLLER, AND DICK F. SWAAB

VOLUME 210

ELSEVIER

NEUROGLIA IN NEUROLOGIC AND PSYCHIATRIC DISORDERS, PART II

Series Editors

MICHAEL J. AMINOFF, FRANÇOIS BOLLER, AND DICK F. SWAAB

Volume Editors

ALEXEI VERKHRATSKY, LOT D. DE WITTE,
ELEONORA ARONICA, AND ELLY M. HOL

VOLUME 210

3rd Series

ELSEVIER

ELSEVIER
Radarweg 29, PO Box 211, 1000 AE Amsterdam, Netherlands
125 London Wall, London EC2Y 5AS, United Kingdom
50 Hampshire Street, 5th Floor, Cambridge, MA 02139, United States

ISBN: 978-0-443-19102-2

For information on all Elsevier publications
visit our website at https://www.elsevier.com/books-and-journals

Publisher: Stacy Masucci
Acquisitions Editor: Megan McManus
Editorial Project Manager: Kristi Anderson
Production Project Manager: Punithavathy Govindaradjane
Cover Designer: Christian Bilbow

Typeset by STRAIVE, India

Printed in the USA
Last Digit is the Print Number: 10 9 8 7 6 5 4 3 2

Handbook of Clinical Neurology 3rd Series

Available titles
Vol. 79, The human hypothalamus: Basic and clinical aspects, Part I, D.F. Swaab, ed. ISBN 9780444513571
Vol. 80, The human hypothalamus: Basic and clinical aspects, Part II, D.F. Swaab, ed. ISBN 9780444514905
Vol. 81, Pain, F. Cervero and T.S. Jensen, eds. ISBN 9780444519016
Vol. 82, Motor neurone disorders and related diseases, A.A. Eisen and P.J. Shaw, eds. ISBN 9780444518941
Vol. 83, Parkinson's disease and related disorders, Part I, W.C. Koller and E. Melamed, eds. ISBN 9780444519009
Vol. 84, Parkinson's disease and related disorders, Part II, W.C. Koller and E. Melamed, eds. ISBN 9780444528933
Vol. 85, HIV/AIDS and the nervous system, P. Portegies and J. Berger, eds. ISBN 9780444520104
Vol. 86, Myopathies, F.L. Mastaglia and D. Hilton Jones, eds. ISBN 9780444518996
Vol. 87, Malformations of the nervous system, H.B. Sarnat and P. Curatolo, eds. ISBN 9780444518965
Vol. 88, Neuropsychology and behavioural neurology, G. Goldenberg and B.C. Miller, eds. ISBN 9780444518972
Vol. 89, Dementias, C. Duyckaerts and I. Litvan, eds. ISBN 9780444518989
Vol. 90, Disorders of consciousness, G.B. Young and E.F.M. Wijdicks, eds. ISBN 9780444518958
Vol. 91, Neuromuscular junction disorders, A.G. Engel, ed. ISBN 9780444520081
Vol. 92, Stroke – Part I: Basic and epidemiological aspects, M. Fisher, ed. ISBN 9780444520036
Vol. 93, Stroke – Part II: Clinical manifestations and pathogenesis, M. Fisher, ed. ISBN 9780444520043
Vol. 94, Stroke – Part III: Investigations and management, M. Fisher, ed. ISBN 9780444520050
Vol. 95, History of neurology, S. Finger, F. Boller and K.L. Tyler, eds. ISBN 9780444520081
Vol. 96, Bacterial infections of the central nervous system, K.L. Roos and A.R. Tunkel, eds. ISBN 9780444520159
Vol. 97, Headache, G. Nappi and M.A. Moskowitz, eds. ISBN 9780444521392
Vol. 98, Sleep disorders Part I, P. Montagna and S. Chokroverty, eds. ISBN 9780444520067
Vol. 99, Sleep disorders Part II, P. Montagna and S. Chokroverty, eds. ISBN 9780444520074
Vol. 100, Hyperkinetic movement disorders, W.J. Weiner and E. Tolosa, eds. ISBN 9780444520142
Vol. 101, Muscular dystrophies, A. Amato and R.C. Griggs, eds. ISBN 9780080450315
Vol. 102, Neuro-ophthalmology, C. Kennard and R.J. Leigh, eds. ISBN 9780444529039
Vol. 103, Ataxic disorders, S.H. Subramony and A. Durr, eds. ISBN 9780444518927
Vol. 104, Neuro-oncology Part I, W. Grisold and R. Sofietti, eds. ISBN 9780444521385
Vol. 105, Neuro-oncology Part II, W. Grisold and R. Sofietti, eds. ISBN 9780444535023
Vol. 106, Neurobiology of psychiatric disorders, T. Schlaepfer and C.B. Nemeroff, eds. ISBN 9780444520029
Vol. 107, Epilepsy Part I, H. Stefan and W.H. Theodore, eds. ISBN 9780444528988
Vol. 108, Epilepsy Part II, H. Stefan and W.H. Theodore, eds. ISBN 9780444528995
Vol. 109, Spinal cord injury, J. Verhaagen and J.W. McDonald III, eds. ISBN 9780444521378
Vol. 110, Neurological rehabilitation, M. Barnes and D.C. Good, eds. ISBN 9780444529015
Vol. 111, Pediatric neurology Part I, O. Dulac, M. Lassonde and H.B. Sarnat, eds. ISBN 9780444528919
Vol. 112, Pediatric neurology Part II, O. Dulac, M. Lassonde and H.B. Sarnat, eds. ISBN 9780444529107
Vol. 113, Pediatric neurology Part III, O. Dulac, M. Lassonde and H.B. Sarnat, eds. ISBN 9780444595652
Vol. 114, Neuroparasitology and tropical neurology, H.H. Garcia, H.B. Tanowitz and O.H. Del Brutto, eds. ISBN 9780444534903
Vol. 115, Peripheral nerve disorders, G. Said and C. Krarup, eds. ISBN 9780444529022
Vol. 116, Brain stimulation, A.M. Lozano and M. Hallett, eds. ISBN 9780444534972
Vol. 117, Autonomic nervous system, R.M. Buijs and D.F. Swaab, eds. ISBN 9780444534910
Vol. 118, Ethical and legal issues in neurology, J.L. Bernat and H.R. Beresford, eds. ISBN 9780444535016
Vol. 119, Neurologic aspects of systemic disease Part I, J. Biller and J.M. Ferro, eds. ISBN 9780702040863
Vol. 120, Neurologic aspects of systemic disease Part II, J. Biller and J.M. Ferro, eds. ISBN 9780702040870
Vol. 121, Neurologic aspects of systemic disease Part III, J. Biller and J.M. Ferro, eds. ISBN 9780702040887
Vol. 122, Multiple sclerosis and related disorders, D.S. Goodin, ed. ISBN 9780444520012
Vol. 123, Neurovirology, A.C. Tselis and J. Booss, eds. ISBN 9780444534880
Vol. 124, Clinical neuroendocrinology, E. Fliers, M. Korbonits and J.A. Romijn, eds. ISBN 9780444596024
Vol. 125, Alcohol and the nervous system, E.V. Sullivan and A. Pfefferbaum, eds. ISBN 9780444626196
Vol. 126, Diabetes and the nervous system, D.W. Zochodne and R.A. Malik, eds. ISBN 9780444534804
Vol. 127, Traumatic brain injury Part I, J.H. Grafman and A.M. Salazar, eds. ISBN 9780444528926
Vol. 128, Traumatic brain injury Part II, J.H. Grafman and A.M. Salazar, eds. ISBN 9780444635211
Vol. 129, The human auditory system: Fundamental organization and clinical disorders, G.G. Celesia and G. Hickok, eds. ISBN 9780444626301

Vol. 130, Neurology of sexual and bladder disorders, D.B. Vodušek and F. Boller, eds. ISBN 9780444632470
Vol. 131, Occupational neurology, M. Lotti and M.L. Bleecker, eds. ISBN 9780444626271
Vol. 132, Neurocutaneous syndromes, M.P. Islam and E.S. Roach, eds. ISBN 9780444627025
Vol. 133, Autoimmune neurology, S.J. Pittock and A. Vincent, eds. ISBN 9780444634320
Vol. 134, Gliomas, M.S. Berger and M. Weller, eds. ISBN 9780128029978
Vol. 135, Neuroimaging Part I, J.C. Masdeu and R.G. González, eds. ISBN 9780444534859
Vol. 136, Neuroimaging Part II, J.C. Masdeu and R.G. González, eds. ISBN 9780444534866
Vol. 137, Neuro-otology, J.M. Furman and T. Lempert, eds. ISBN 9780444634375
Vol. 138, Neuroepidemiology, C. Rosano, M.A. Ikram and M. Ganguli, eds. ISBN 9780128029732
Vol. 139, Functional neurologic disorders, M. Hallett, J. Stone and A. Carson, eds. ISBN 9780128017722
Vol. 140, Critical care neurology Part I, E.F.M. Wijdicks and A.H. Kramer, eds. ISBN 9780444636003
Vol. 141, Critical care neurology Part II, E.F.M. Wijdicks and A.H. Kramer, eds. ISBN 9780444635990
Vol. 142, Wilson disease, A. Członkowska and M.L. Schilsky, eds. ISBN 9780444636003
Vol. 143, Arteriovenous and cavernous malformations, R.F. Spetzler, K. Moon and R.O. Almefty, eds. ISBN 9780444636409
Vol. 144, Huntington disease, A.S. Feigin and K.E. Anderson, eds. ISBN 9780128018934
Vol. 145, Neuropathology, G.G. Kovacs and I. Alafuzoff, eds. ISBN 9780128023952
Vol. 146, Cerebrospinal fluid in neurologic disorders, F. Deisenhammer, C.E. Teunissen and H. Tumani, eds. ISBN 9780128042793
Vol. 147, Neurogenetics Part I, D.H. Geschwind, H.L. Paulson and C. Klein, eds. ISBN 9780444632333
Vol. 148, Neurogenetics Part II, D.H. Geschwind, H.L. Paulson and C. Klein, eds. ISBN 9780444640765
Vol. 149, Metastatic diseases of the nervous system, D. Schiff and M.J. van den Bent, eds. ISBN 9780128111611
Vol. 150, Brain banking in neurologic and psychiatric diseases, I. Huitinga and M.J. Webster, eds. ISBN 9780444636393
Vol. 151, The parietal lobe, G. Vallar and H.B. Coslett, eds. ISBN 9780444636225
Vol. 152, The neurology of HIV infection, B.J. Brew, ed. ISBN 9780444638496
Vol. 153, Human prion diseases, M. Pocchiari and J.C. Manson, eds. ISBN 9780444639455
Vol. 154, The cerebellum: From embryology to diagnostic investigations, M. Manto and T.A.G.M. Huisman, eds. ISBN 9780444639561
Vol. 155, The cerebellum: Disorders and treatment, M. Manto and T.A.G.M. Huisman, eds. ISBN 9780444641892
Vol. 156, Thermoregulation: From basic neuroscience to clinical neurology Part I, A.A. Romanovsky, ed. ISBN 9780444639127
Vol. 157, Thermoregulation: From basic neuroscience to clinical neurology Part II, A.A. Romanovsky, ed. ISBN 9780444640741
Vol. 158, Sports neurology, B. Hainline and R.A. Stern, eds. ISBN 9780444639547
Vol. 159, Balance, gait, and falls, B.L. Day and S.R. Lord, eds. ISBN 9780444639165
Vol. 160, Clinical neurophysiology: Basis and technical aspects, K.H. Levin and P. Chauvel, eds. ISBN 9780444640321
Vol. 161, Clinical neurophysiology: Diseases and disorders, K.H. Levin and P. Chauvel, eds. ISBN 9780444641427
Vol. 162, Neonatal neurology, L.S. De Vries and H.C. Glass, eds. ISBN 9780444640291
Vol. 163, The frontal lobes, M. D'Esposito and J.H. Grafman, eds. ISBN 9780128042816
Vol. 164, Smell and taste, Richard L. Doty, ed. ISBN 9780444638557
Vol. 165, Psychopharmacology of neurologic disease, V.I. Reus and D. Lindqvist, eds. ISBN 9780444640123
Vol. 166, Cingulate cortex, B.A. Vogt, ed. ISBN 9780444641960
Vol. 167, Geriatric neurology, S.T. DeKosky and S. Asthana, eds. ISBN 9780128047668
Vol. 168, Brain-computer interfaces, N.F. Ramsey and J. del R. Millán, eds. ISBN 9780444639349
Vol. 169, Meningiomas, Part I, M.W. McDermott, ed. ISBN 9780128042809
Vol. 170, Meningiomas, Part II, M.W. McDermott, ed. ISBN 9780128221983
Vol. 171, Neurology and pregnancy: Pathophysiology and patient care, E.A.P. Steegers, M.J. Cipolla and E.C. Miller, eds. ISBN 9780444642394
Vol. 172, Neurology and pregnancy: Neuro-obstetric disorders, E.A.P. Steegers, M.J. Cipolla and E.C. Miller, eds. ISBN 9780444642400
Vol. 173, Neurocognitive development: Normative development, A. Gallagher, C. Bulteau, D. Cohen and J.L. Michaud, eds. ISBN 9780444641502
Vol. 174, Neurocognitive development: Disorders and disabilities, A. Gallagher, C. Bulteau, D. Cohen and J.L. Michaud, eds. ISBN 9780444641489
Vol. 175, Sex differences in neurology and psychiatry, R. Lanzenberger, G.S. Kranz, and I. Savic, eds. ISBN 9780444641236
Vol. 176, Interventional neuroradiology, S.W. Hetts and D.L. Cooke, eds. ISBN 9780444640345
Vol. 177, Heart and neurologic disease, J. Biller, ed. ISBN 9780128198148
Vol. 178, Neurology of vision and visual disorders, J.J.S. Barton and A. Leff, eds. ISBN 9780128213773
Vol. 179, The human hypothalamus: Anterior region, D.F. Swaab, F. Kreier, P.J. Lucassen, A. Salehi and R.M. Buijs, eds. ISBN 9780128199756
Vol. 180, The human hypothalamus: Middle and posterior region, D.F. Swaab, F. Kreier, P.J. Lucassen, A. Salehi and R.M. Buijs, eds. ISBN 9780128201077
Vol. 181, The human hypothalamus: Neuroendocrine disorders, D.F. Swaab, R.M. Buijs, P.J. Lucassen, A. Salehi and F. Kreier, eds. ISBN 9780128206836
Vol. 182, The human hypothalamus: Neuropsychiatric disorders, D.F. Swaab, R.M. Buijs, F. Kreier, P.J. Lucassen, and A. Salehi, eds. ISBN 9780128199732

All volumes in the 3rd Series of the **Handbook of Clinical Neurology** are published electronically, on Science Direct: http://www.sciencedirect.com/science/handbooks/00729752.

Foreword

We are delighted to present two volumes on neuroglia in the *Handbook of Clinical Neurology* series. For a long time, the focus of neuroscientists and clinicians has been on neurons in health and disease. Glia were considered simply as comparable to the connective tissue in other organs. At best, its function was described as "supportive" of neurons. Indeed, phylogenetically bigger brains have much more glia than smaller brains. An increased ratio between the number of glial cells and neurons reflects a more highly developed brain. Traditionally, the relative glial expansion in evolution was held to support its trophic and metabolic function. However, with our current knowledge on the crucial role of neuroglia in neurotransmission, the phylogenetic increase in the number of glial cells may also favor their growing participation in neurocomputation and neurotransmission. Neuroglial cells control the homeostasis of ions and neuroactive molecules, contribute to neurogenesis, and are involved in synaptic pruning and myelination, in regulating the blood–brain barrier, and in the function of the glymphatic system.

The first of these two new volumes covers neuroglia in relation to evolution, development, and aging, and its physiology in relation to neurotransmission, energy metabolism, the blood–brain barrier, the use of in vitro and animal models, omics, and imaging. The second volume is dedicated to the changes in neuroglia that occur in neurologic and psychiatric disorders. Chapters deal with glial alterations in Alzheimer, Parkinson, and Huntington diseases, amyotrophic lateral sclerosis and frontotemporal dementia, epilepsy, cerebrovascular diseases, traumatic brain and spinal cord injuries, multiple sclerosis and neuromyelitis optica, autoimmune encephalitis, leukodystrophies, glioma, encephalopathy, infectious diseases, and the major psychiatric disorders.

We are very fortunate to have four distinguished volume editors, namely, Alexei Verkhratsky from the University of Manchester, United Kingdom, Lot D. de Witte from the Icahn School of Medicine at Mount Sinai, New York, United States, Eleonora Aronica from the University of Amsterdam, the Netherlands, and Elly M. Hol from the University Medical Center Utrecht, the Netherlands. They have brought together an excellent set of authors and compiled an outstanding volume. We are delighted that it presents such scholarly and comprehensive accounts of the many aspects of this complex and novel topic.

In addition to the printed version, the volume will also be available electronically on Elsevier's ScienceDirect website, which is popular with readers and facilitates the book's accessibility. Indeed, all of the volumes in the present series of the *Handbook* are available electronically on this website.

As always, it is a pleasure to thank Elsevier, our publisher, and in particular, Nikki Levy and Kristi Anderson in San Diego, Megan McManus in Salt Lake City, and Punithavathy Govindaradjane at Elsevier Global Book Production in Chennai, for their assistance in the development and production of the *Handbook of Clinical Neurology*.

Michael J. Aminoff
François Boller
Dick F. Swaab

Preface

The 21st century is a century of the brain, both in our understanding of its ways of operation and its links to the computerized world and interactions with rapidly evolving technoecosystems, including "artificial intelligence." Despite this progress in knowledge on brain function, it is still challenging to develop effective new approaches that can cure brain diseases. The tremendous advances in medicine over the past century, which led to significant advances in the treatment of various diseases, from infections and trauma to cardiac, metabolic, and neoplastic, are in sharp contrast with the fact that the clinical success in treating many diseases of the brain remains feeble. Indeed, so far we have not been able to develop pathophysiology-based effective therapeutic management of diseases of the brain, especially those resulting in cognitive deficits; if anything, clinical neurology mainly resorts to symptomatic treatments.

The lack of effective treatments underscores the enormous complexity of the human brain and the difficulties researchers have in creating valid and translatable animal and cell models. In addition, the fundamental neurologic thinking of recent decades is highly neuronocentric, considering that neurons only are at the core of morbid changes. In the nervous tissue, it is the neuroglia that steer homeostasis, protection, and restoration. Neuroglial cells contribute to all neurologic disorders, and they participate in the cognitive reserve of the brain. Reactive gliosis is the main defensive response of the nervous tissue, which allows isolation of the lesioned area through the formation of a barrier, as well as a supporting fibrotic scar and regeneration of affected yet surviving nervous tissue. In aging as well as in many age-dependent chronic diseases, neuroglia undergo atrophy and loss of function; the decline in the protective capacity of neuroglial cells opens the gate to neurodegenerative and cognitive disorders. Revealing disease-specific changes in neuroglial cells paves the way to cell-specific therapeutic strategies.

In the second volume on neuroglia in the *Handbook of Clinical Neurology*, we assembled chapters overviewing the contribution of neuroglia to major neuropathologies, from trauma and stroke to neurodegeneration and neuropsychiatric diseases. We hope this volume may influence neurologists and psychiatrists to move away from neuronocentricity and accept a more inclusive view of the pathogenesis of brain disorders. We also hope that this volume may precipitate the development of novel, neuroglia-specific therapeutic strategies to address the challenges of preventing and curing brain diseases, thus offering hope for the future.

Lo duca e io per quel cammino ascoso
intrammo a ritornar nel chiaro mondo;
e sanza cura aver d'alcun riposo,

salimmo sù, el primo e io secondo,
tanto ch'i' vidi de le cose belle
che porta 'l ciel, per un pertugio tondo.
E quindi uscimmo a riveder le stelle.
Dante Alligieri, Inferno, canto 34

Alexei Verkhratsky
Lot D. de Witte
Eleonora Aronica
Elly M. Hol

Contributors

J. Albrecht
Department of Neurotoxicology, Mossakowski Medical Research Institute, Polish Academy of Sciences, Warsaw, Poland

E. Aronica
Department of (Neuro)Pathology, Amsterdam UMC, University of Amsterdam, Amsterdam Neuroscience, Amsterdam, The Netherlands

C. Ash
Division of Medical Sciences, University of Victoria, Victoria, BC, Canada

I. Awogbindin
Division of Medical Sciences; Institute on Aging and Lifelong Health (IALH), University of Victoria, Victoria, BC, Canada

C. Bendotti
Laboratory of Neurobiology and Preclinical Therapeutics, ALS Center, Department of Neuroscience, IRCCS-"Mario Negri" Institute for Pharmacological Research, Milano, Italy

D.K. Binder
Division of Biomedical Sciences, School of Medicine; Center for Glial-Neuronal Interactions, University of California, Riverside, Riverside, CA, United States

R.E. Bonomi
Department of Psychiatry, Yale University, New Haven, CT, United States

M. Bugiani
Department of Child Neurology and Pathology, Amsterdam Leukodystrophy Center, Amsterdam University Medical Center, Amsterdam, The Netherlands

A. Buonfiglioli
Department of Psychiatry, Icahn School of Medicine at Mount Sinai, New York, NY, United States

E.M. Castro
Department of Pharmaceutical Sciences, School of Pharmacy and Pharmaceutical Sciences, University of California, Irvine, Irvine, CA, United States

L.J. Castro-Vega
Genetics & Development of Brain Tumors Laboratory, ICM - Paris Brain Institute, Sorbonne University, UMR S 1127, Inserm U 1127, CNRS UMR 7225, F-75013, Paris, France

F. Cavaliere
The Basque Biomodels Platform for Human Research (BBioH), Achucarro Basque Center for Neuroscience & Fundación Biofisica Bizkaia, Leioa, Spain

J. Chevalier
Department of Child Neurology and Pathology, Amsterdam Leukodystrophy Center, Amsterdam University Medical Center, Amsterdam, The Netherlands

F. Correa-da-Silva
Department of Endocrinology and Metabolism, Amsterdam Gastroenterology Endocrinology and Metabolism; Laboratory of Endocrinology, Amsterdam University Medical Center (UMC), University of Amsterdam; Netherlands Institute for Neuroscience, Amsterdam, The Netherlands

F. Corvace
Department of Neuroanatomy and Molecular Brain Research, Ruhr University Bochum, Bochum, Germany

K.P. Cosgrove
Department of Psychiatry, Yale University, New Haven, CT, United States

E. Cuadrado
Department of Translational Neuroscience, University Medical Center Utrecht Brain Center, Utrecht University, Utrecht, The Netherlands

F. De Marchi
ALS Centre, Neurology Unit, Maggiore della Carità Hospital, University of Piemonte Orientale, Novara, Italy

L.D. de Witte
Department of Human Genetics, Radboud University Medical Center; Donders Institute for Brain, Cognition and Behaviour, Radboud University; Department of Psychiatry, Radboud University Medical Center, Nijmegen, The Netherlands

S. Dooves
Department of Child and Adolescence Psychiatry, Emma Center for Personalized Medicine, Amsterdam Neuroscience, Emma Children's Hospital, Amsterdam UMC Location Vrije Universiteit Amsterdam; Department of Complex Trait Genetics, Center for Neurogenomics and Cognitive Research, Vrije Universiteit Amsterdam, Amsterdam Neuroscience, Amsterdam, The Netherlands

W. Dykstra
Department of Translational Neuroscience, University Medical Center Utrecht Brain Center, Utrecht University, Utrecht, The Netherlands

P.M. Faustmann
Department of Neuroanatomy and Molecular Brain Research, Ruhr University Bochum, Bochum, Germany

T.J. Faustmann
Department of Psychiatry and Psychotherapy, Medical Faculty, Heinrich Heine University, Düsseldorf, Germany

L.R. Frick
Departments of Neurology and Medicine, Neuroscience Graduate Program, Jacobs School of Medicine & Biomedical Sciences, State University of New York at Buffalo, Clinical and Translational Research Center, Buffalo, NY, United States

M.R. Ghouli
Division of Biomedical Sciences, School of Medicine; Center for Glial-Neuronal Interactions, University of California, Riverside, Riverside, CA, United States

F.A.J. Gigase
Department of Psychiatry, Icahn School of Medicine at Mount Sinai, New York, NY, United States

A. Grosche
Department of Physiological Genomics, Ludwig-Maximilians-Universität München, München, Germany

J. Grosche
Effigos AG, Leipzig, Germany

D. Guneykaya
Department of Neurobiology, Harvard Medical School, Boston, MA, United States

V.M. Heine
Department of Child and Adolescence Psychiatry, Emma Center for Personalized Medicine, Amsterdam Neuroscience, Emma Children's Hospital, Amsterdam UMC Location Vrije Universiteit Amsterdam; Department of Complex Trait Genetics, Center for Neurogenomics and Cognitive Research, Vrije Universiteit Amsterdam, Amsterdam Neuroscience, Amsterdam, The Netherlands

R.J. Henry
Department of Pharmacology, School of Biomolecular and Biomedical Science, University College Dublin, Dublin, Ireland

E.M. Hol
Department of Translational Neuroscience, University Medical Center Utrecht Brain Center, Utrecht University, Utrecht, The Netherlands

G. Huberfeld
Université Paris Cité, Institute of Psychiatry and Neuroscience of Paris (IPNP), INSERM U1266, Neuronal Signaling in Epilepsy and Glioma; Department of Neurology, Hôpital Fondation Adolphe de Rothschild, Paris, France

F.S. Ismail
Department of Neurology, Klinikum Vest, Academic Teaching Hospital of the Ruhr University Bochum, Recklinghausen, Germany

A. Laurenge
Genetics & Development of Brain Tumors Laboratory, ICM - Paris Brain Institute, Sorbonne University, UMR S 1127, Inserm U 1127, CNRS UMR 7225; AP-HP, Groupe Hospitalier Pitié-Salpêtrière, Neuro-Oncology Department, F-75013, Paris, France

F.M. Leslie
Department of Pharmaceutical Sciences, School of Pharmacy and Pharmaceutical Sciences, University of California, Irvine, Irvine, CA, United States

D. Lim
Department of Pharmaceutical Sciences, Università del Piemonte Orientale "Amedeo Avogadro", Novara, Italy

D.J. Loane
School of Biochemistry and Immunology, Trinity Biomedical Sciences Institute, Trinity College Dublin, Dublin, Ireland

S. Lotfipour
Department of Pharmaceutical Sciences, School of Pharmacy and Pharmaceutical Sciences; Department of Emergency Medicine; Department of Pathology and Laboratory Medicine, School of Medicine, University of California, Irvine, Irvine, CA, United States

D. Mattei
Nash Family Department of Neuroscience, Icahn School of Medicine at Mount Sinai, Friedman Brain Institute, New York, NY, United States

C. Matute
Department of Neurosciences, University of the Basque Country UPV/EHU and CIBERNED, Leioa, Bizkaia, Spain

R. Pietrzak
Department of Psychiatry, Yale University, New Haven, CT, United States

M. Popek
Department of Neurotoxicology, Mossakowski Medical Research Institute, Polish Academy of Sciences, Warsaw, Poland

V.T.S. Rao
Departments of Paediatrics (Neurology), Pathology and Laboratory Medicine (Neuropathology), and Clinical Neurosciences, University of Calgary Cumming School of Medicine and Alberta Children's Hospital Research Institute (Owerko Centre), Calgary, AB, Canada; Medical Affairs Laboratory, Argenx US, Inc., Boston, MA, United States

H.B. Sarnat
Departments of Paediatrics (Neurology), Pathology and Laboratory Medicine (Neuropathology), and Clinical Neurosciences, University of Calgary Cumming School of Medicine and Alberta Children's Hospital Research Institute (Owerko Centre), Calgary, AB, Canada; Medical Affairs Laboratory, Argenx US, Inc., Boston, MA, United States

F.-D. Shi
Department of Neurology, Tianjin Neurological Institute, Tianjin Medical University General Hospital, Tianjin; Department of Neurology, China National Clinical Research Center for Neurological Diseases, Beijing Tiantan Hospital, Capital Medical University, Beijing, China

E. Šimončičová
Division of Medical Sciences; Neuroscience Graduate Program, University of Victoria, Victoria, BC, Canada

G.J.L.J. Snijders
Department of Psychiatry, Icahn School of Medicine at Mount Sinai, New York, NY, United States

M.V. Sofroniew
Department of Neurobiology, David Geffen School of Medicine, University of California, Los Angeles, CA, United States

E.G. Spinelli
Neurology Unit, Department of Neuroscience, IRCCS Ospedale San Raffaele; Vita-Salute San Raffaele University, Milano, Italy

D.F. Swaab
Neuropsychiatric Disorders Lab, Neuroimmunology Group, Netherlands Institute for Neuroscience, An Institute of the Royal Netherlands Academy of Arts and Sciences, Amsterdam, The Netherlands

M.-E. Tremblay
Division of Medical Sciences; Institute on Aging and Lifelong Health (IALH); Centre for Advanced Materials and Related Technology (CAMTEC), University of Victoria, Victoria; Department of Biochemistry and Molecular Biology, University of British Columbia, Vancouver, BC, Canada

B. Ugursu
Department of Psychoneuroimmunology, Max-Delbrück-Center for Molecular Medicine in the Helmholtz Association, Berlin, Germany

A. Verkhratsky
Faculty of Biology, Medicine and Health, University of Manchester, Manchester, United Kingdom; Department of Neurosciences, University of the Basque Country UPV/EHU and CIBERNED, Leioa, Bizkaia; IKERBASQUE, Basque Foundation for Science, Bilbao, Spain

V. Vidal
Division of Medical Sciences, University of Victoria, Victoria, BC, Canada; Science and Technology Department, University of Bordeaux, Bordeaux, France

C.-X. Yi
Department of Endocrinology and Metabolism, Amsterdam Gastroenterology Endocrinology and Metabolism; Laboratory of Endocrinology, Amsterdam University Medical Center (UMC), University of Amsterdam; Netherlands Institute for Neuroscience, Amsterdam, The Netherlands

L. Zhang
Neuropsychiatric Disorders Lab, Neuroimmunology Group, Netherlands Institute for Neuroscience, An Institute of the Royal Netherlands Academy of Arts and Sciences, Amsterdam, The Netherlands

L. Zhang
Department of Neurology, Tianjin Neurological Institute, Tianjin Medical University General Hospital, Tianjin, China

M. Zielińska
Department of Neurotoxicology, Mossakowski Medical Research Institute, Polish Academy of Sciences, Warsaw, Poland

Contents

Part I

Glia in neurological disorders

Handbook of Clinical Neurology, Vol. 210 (3rd series)
Neuroglia in Neurologic and Psychiatric Disorders, Part II
A. Verkhratsky, L.D. de Witte, E. Aronica, and E.M. Hol, Editors
https://doi.org/10.1016/B978-0-443-19102-2.00013-2

Chapter 1

General pathophysiology of neuroglia

ALEXEI VERKHRATSKY[1,2,3]*, ELLY M. HOL[4], LOT D. DE WITTE[5,6,7], AND ELEONORA ARONICA[8]

[1]*Faculty of Biology, Medicine and Health, University of Manchester, Manchester, United Kingdom*

[2]*Department of Neurosciences, University of the Basque Country UPV/EHU and CIBERNED, Leioa, Bizkaia, Spain*

[3]*IKERBASQUE, Basque Foundation for Science, Bilbao, Spain*

[4]*Department of Translational Neuroscience, University Medical Center Utrecht Brain Center, Utrecht University, Utrecht, The Netherlands*

[5]*Department of Human Genetics, Radboud University Medical Center, Nijmegen, The Netherlands*

[6]*Donders Institute for Brain, Cognition and Behaviour, Radboud University, Nijmegen, The Netherlands*

[7]*Department of Psychiatry, Radboud University Medical Center, Nijmegen, The Netherlands*

[8]*Department of (Neuro)Pathology, Amsterdam UMC, University of Amsterdam, Amsterdam Neuroscience, Amsterdam, The Netherlands*

Abstract

Neuroglia in the CNS, represented by astroglia, oligodendroglia, and microglia, are responsible for the homeostatic support and protection of the nervous tissue. Neuroglia are intimately involved in the pathogenesis of all neurologic diseases, and neuroglial changes to a large extent define the progression of these diseases and their neurologic outcome. In contrast to neurons, neuroglia are capable of mounting an evolutionary conserved response to pathology known as reactive gliosis. Reactive gliosis is initially protective and allostatic, and it is aimed at preserving the nervous tissue function and integrity. However, in many diseases, neuroglial cells undergo atrophy and functional asthenia, contributing to nervous tissue damage.

NEUROGLIA IN NEUROPATHOLOGY: HISTORIC PRELUDE

Rudolf Virchow, when introducing the concept of neuroglia in his seminal book *Die Cellularpathologie*, highlighted the fundamental importance of this "*connective tissue of the brain and spinal cord*" for neuropathology. Virchow stated, "*experience shows us that this very interstitial tissue of the brain and spinal marrow is one of the most frequent seats of morbid change*" (Virchow, 1858), quoted from the first English edition (Virchow, 1860), p. 317. The contribution of neuroglia to neurologic disorders was recognized by all neurologists and neuroanatomists of the late 19th and early 20th centuries. Numerous observations demonstrated that neuroglial cells change in the diseased brain and spinal cord, and these changes clearly signal pathophysiologic developments (Lugaro, 1907; Achucarro, 1910; Penfiled and Buckley, 1928). Alois Alzheimer described pathologic glial phenotypes in various diseases, from tertiary syphilis, to neurodegeneration and toxic encephalopathies (Alzheimer, 1910). The first specific glial type, the astrocyte, was defined in 1895 by Michal von Lenhossék (Lenhossék, 1895). Slightly earlier, the pathologic hypertrophy of astrocytes was considered a universal pathologic response. In 1893, Willem Andriezen noted that "*The protoplasmic glia elements are really the elements which exhibit a morbid hypertrophy in pathological conditions (alcoholism, G.P.), and which may show further morbid activities, in the last stage of which their protoplasm will deposit numerous organised fibrillae, in the act of doing which the protoplasm proper is used up, except a scanty remnant, which may persist, ghost-like, to mark the position of what was once a protoplasmic cell body.*" (Andriezen, 1893). In this statement, Andriezen also highlighted the highly dynamic nature of astrocytic pathologic changes, in which degeneration may follow hypertrophy. By the 1930s, the concept of reactive astrogliosis became universally acknowledged, and it became quite clear that "*the appearance of neuroglia serves as a delicate indicator of the action of noxious influences upon the central nervous system*" (del Río-Hortega and Penfield, 1927). Astrogliosis was characterized as "*the hyperplasia and hypertrophy of astrocytes associated with fibrous change in these cells. Swelling and direct division of cells is followed by hypertrophy and the formation of glia fibrils…*," while the palisades of reactive astrocytes were recognized as the "*glial splint*" that served to separate the wound and healthy nervous tissue (del Río-Hortega

*Correspondence to: Professor Alexei Verkhratsky, Faculty of Biology, Medicine and Health, The University of Manchester, Oxford Road, Manchester, M13 9PT, United Kingdom. Tel: +44-1612755414, E-mail: Alexej.Verkhratsky@manchester.ac.uk

and Penfield, 1927; Penfiled and Buckley, 1928). Distinct stages and dynamics of astrogliosis, which include an increase in astrocytic numbers indicative of proliferation, astrocyte accumulation around the wound, changes in their morphology, and formation of a border surrounding a fibrotic scar, were also documented.

In 1899, Franz Nissl discovered aberrant glial morphotypes in the postmortem tissues of patients with dementia and named these cells Stäbchenzellen (rod cells) (Nissl, 1899), which today are defined as rod microglia (Taylor et al., 2014). In 1896, George Marinesco reported that glial cells can phagocytose dying neurones (Marinesco, 1896). In 1919, Pio del Rio Hortega discovered, named, and characterized in detail the microglia, which he defined as the brain "*garbage collectors*" (Del Río-Hortega, 1919, 1920; Sierra et al., 2016). Del Rio-Hortega also described the sequence of morphologic changes associated with reactive microgliosis, when microglial cells change from the highly arborized cells in the healthy brain to amoeboid macrophages in pathologic nervous tissue (Del Rio-Hortega, 1932). Del Rio Hortega also characterized and defined the myelinating oligodendroglia (Del Río-Hortega, 1921, 1922; Perez-Cerda et al., 2015), the third main class of neuroglia of the central nervous system (CNS). The term myelin (from Greek *μυελός*, myelos-marrow) was coined by Rudolf Virchow in 1854 (Virchow, 1854), whereas pathologic damage to myelin as a fundamental part of multiple sclerosis was recognized by Jean-Martin Charcot (Charcot, 1868; Bourneville and Guérard, 1869).

The multiple pathophysiologic roles of neuroglia were, unfortunately, neglected in the second half of the 20th century, when neuronocentric views prevailed; consequently, therapeutic strategies were aimed solely at the modification of neuron-specific molecules. The recent decade, however, has witnessed the renaissance of neurogliopathology, which is now considered to be the central (and often the major) element in the pathogenesis of virtually all neurologic diseases (Graeber and Streit, 2010; Streit et al., 2014; Verkhratsky et al., 2017; Stadelmann et al., 2019; Skinnider et al., 2023; Verkhratsky et al., 2023; Sofroniew, 2024; Tremblay and Verkhratsky, 2024).

PRINCIPLES OF PATHOPHYSIOLOGY OF NEUROGLIA

Pathologic changes of neuroglia are complex, as they are determined by a multiplicity of factors, including age, sex, overall physique of the body, and life-long pathologic load; some glial changes are disease-specific, while others are common across various pathologic conditions. Changes in glia are dynamic and often are associated with disease stages; several pathologic phenotypes may coexist; they may develop sequentially or substitute each other. Such mutability and complexity of glial pathophysiology makes classifying exceedingly difficult, yet defining major pathologic trends is critically important to understand pathophysiologic mechanisms, which define disease-modifying or disease-preventing strategies.

Conceptually, the contribution of neuroglia to pathology can be primary (for example, due to genetic defects when aberrant neuroglia damage nervous tissue) or secondary, when glial cells either respond to pathology (reactive astrogliosis and barrier formation) or are damaged by pathology; in the latter case, loss of glial homeostatic support and neuroprotection injures nervous tissue (Verkhratsky and Butt, 2023). The severity and the localization of damage are also critical factors defining neuroglial responses. For example, acute traumatic lesions result in widespread and multiple tissue damage, including cessation of blood supply leading to ischemia, damage to the blood–brain barrier, and infiltration of blood–borne cells and molecules; ischemia in its turn impairs metabolism and energy production that translates into a loss of ionic and neurotransmitter homeostasis leading to excitotoxicity (Rossi et al., 2007). In chronic disorders, when pathologic changes evolve over time, the more subtle changes associated with progressive loss of glial function that develop simultaneously with reactive responses define the progression and outcome. Thus in neurodegenerative diseases, accumulation of aberrant or disordered proteins (α-synucleinopathies, β-amyloidosis, tauopathies, or accumulation of TAR DNA-binding protein TDP-43) trigger heterogeneous responses of neuroglia manifested by reactivity as well as various forms of atrophy; strained homeostatic systems often fail, leading to accumulation of toxic factors such as reactive oxygen species or excitotoxic glutamate (Heneka et al., 2015; Spires-Jones et al., 2017; Kovacs, 2019; Verkhratsky et al., 2021). Progression of the chronic pathology is associated with further decline of homeostatic support, compromised metabolism, and malfunction or disruption of organ-wide protective systems (such as the blood–brain barrier or glymphatic clearance), which exacerbate the damage and trigger maladaptive inflammation, ultimately precipitating neuronal death and cognitive decline.

Neuroglial role in all pathologies is primarily defensive; fundamentally, neuroglia protect cellular and tissue homeostasis and counteract the damage by reactive remodeling. Widespread views on neuroglia as potentially toxic elements, the dormant killers that, when "activated," destroy nervous tissue, are fundamentally incorrect. The primary mechanism underlying nervous tissue damage is the loss of neuroglial homeostatic support and neuroprotection. Similarly, simplified views on glial dichotomy, including the concepts of good and bad, resting and activated, neurotoxic and neuroprotective, pro- and anti-inflammatory, or A1/A2 and M1/M2 glia, are misleading (Escartin et al., 2021; Paolicelli et al., 2022); neuroglial phenotypes are highly heterogeneous in both the healthy and diseased nervous systems; and the outcome of pathologic remodeling of neuroglial cells is diverse. In certain contexts, neuroglial cells can precipitate and propagate the damage, which reflects dysregulation (for example, excessive secretion of cytokines (Sofroniew, 2014)) or loss (for example, failed antioxidative support leading to accumulation of peroxide produced by the urea cycle (Ju et al., 2022)) of neuroglial functions. Microglial cells in particular perform phagocytosis of the cellular debris, and when the control over microglia is diminished, it can mediate the injury (Prinz et al., 1999). Neuroglia are primary elements of the processes controlling post-lesion regeneration, supporting axonal growth, synaptogenesis, and restoration of blood–brain barrier, remyelination, and neurogenesis (Anderson et al., 2016; Rawji et al., 2020; Hemati-Gourabi et al., 2022).

Neuroglial defense is highly effective; neuroglia protection permits the brain to withstand many life-long environmental challenges and diseases. Neuroglia are also a fundamental part of neuroplasticity and secure compensatory regeneration. In particular, neuroglial cells are key elements of the cognitive reserve that defines the resilience of the nervous system to various diseases and aging (Verkhratsky and Zorec, 2024). In most of the cases, neuroglial defenses hold up, thus allowing the majority of humans to retain their cognitive abilities throughout the lifetime; moreover, regenerative plasticity of the nervous tissue can often compensate even for severe traumatic lesions affecting substantial portions of the brain. Aging, excessive exposure to environmental pollution, detrimental lifestyle, and systemic pathologies—these factors impact the body at large and on the nervous tissue in particular. A decrease in the neuroprotection associated, for example, with an accumulation of dystrophic glial cells in the elderly brain, opens the gate to maladaptive plasticity and disease.

The multiple pathologic forms acquired by neuroglial cells in pathology can be broadly classified into (i) reactive and allostatic gliosis; (ii) glial dystrophy and asthenia, with loss of function; (iii) glial degeneration and glial cell death; and (vi) gliopathies resulting in the emergence of diseased and aberrant glia, which together drive neuropathology (Fig. 1.1 and Pekny et al., 2016; Verkhratsky et al., 2023).

The most studied and best known pathologic response of neuroglia is represented by reactive gliosis, a term derived from Greek: γλια —glia and οσισ—condition or process. Thus, gliosis stands for "glial condition or process." Reactive gliosis is an evolutionary conserved protective response of glial cells to the damage of the nervous tissue. Reactive gliosis serves as a fundamental survivalistic mechanism aimed at localizing and sealing the damage, destroying pathogens, and assisting regeneration. As such, reactive gliosis is a part of inflammation, which is a wider tissue response to damage. Reactive gliosis is a characteristic feature of all types of glial cells and is subclassified into reactive astrogliosis, reactive microgliosis, and reactive remodeling of oligodendroglial cells, with both mature oligodendrocytes and oligodendroglial precursor cells undergoing such reactive changes. Conceptually, glial reactivity is a classical allostatic response (*αλλο*—variable and *στασισ* —standing still—that is "remaining stable by being variable" or "maintaining homeostasis through change" (Sterling and Eyer, 1988)), aimed at keeping the nervous tissue alive and functional. Glial allostatic responses are highly heterogeneous: they can be local and transient, or long-lasting and permanent; again, different degrees of allostatic mobilization can coexist in the same pathology.

Atrophy and functional asthenia with loss of function represent the second main class of pathologic changes in neuroglia. Astrocytes present a form of well-developed morphologic plasticity that accompanies physiologic processes such as, for example, neuronal plasticity (Henneberger et al., 2020). Retraction or extension of astrocytic terminal processes known as leaflets (Semyanov and Verkhratsky, 2021) modifies synaptic coverage and the overall homeostatic support of the nervous tissue. In aging, as well as in a wide variety of diseases, from major depression to epilepsy and neurodegeneration, astrocytes demonstrate morphologic atrophy, which limits spatial homeostatic support, reduces neuroprotection, and impairs synaptic transmission and plasticity (Banasr and Duman, 2008; Verkhratsky et al., 2019; Ramos-Gonzalez et al., 2021; Aten et al., 2023; Popov et al., 2023; Villanueva et al., 2023; Anders et al., 2024). Astrocytic asthenia is similarly fundamental for many neuropathologies, when loss of vital astrocyte functions, such as, for example, decreased glutamate clearance, are primary causes for neuronal damage and death (Hazell et al., 2009; Provenzano et al., 2023; Villanueva et al., 2023; Long et al., 2024). Similarly, dystrophic microglia, which accumulates in the aging brain, limits the defensive capabilities and opens the gate for age-dependent neurodegenerative diseases (Streit et al., 2020; Costa et al., 2021). A similar dystrophy and loss of function are observed in the oligodendroglial lineage, with oligodendrocytes degenerating and oligodendrocyte precursor cells (OPCs) losing their proliferative and remyelinating potential in aging (Peters, 2002; Ibañez et al., 2023; Yi et al., 2023).

In many disorders, neuroglial cells can be overstressed and damaged, leading to their degeneration and death. In particular, such cell death is well documented for astrocytes. In traumatic lesions, stroke, or neurodegenerative diseases, astrocytes undergo a specific form of death known as clasmatodendrosis (from Greek "κλάσμα," fragment, "δένδρον," tree, "ωσις," process; fragmentation of processes). This form of glial death was described by Alois Alzheimer (Alzheimer, 1910) and characterized by Ramon y Cajal, who regarded clasmatodendrosis as an early postmortem artifact (Ramón y Cajal, 1913). Subsequently, clasmatodendrosis was described in traumatic brain injury, cerebral ischemia, post-stroke dementia, status epilepticus, demyelinating diseases, cerebral edema, toxic encephalopathies, small-vessel disease, and neuroinfection (Balaban et al., 2021). Obviously, astrocytic degeneration and death add to an overall decline in homeostatic support and neuroprotection.

Finally, a distinct class of neurogliopathies is associated with the emergence of aberrant phenotypes, which drive neuropathologic progression. Such aberrant astrocytes can be found in

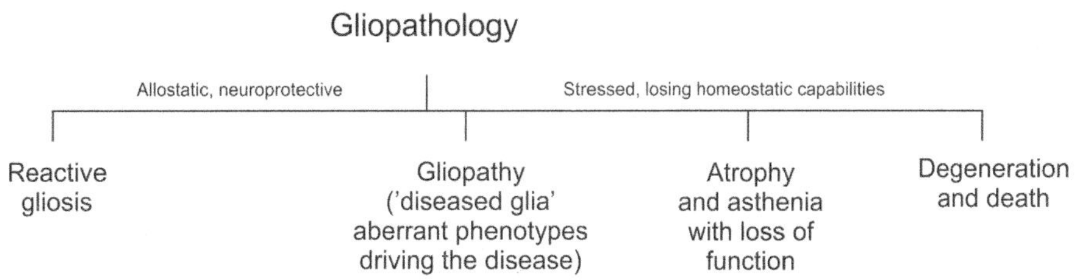

Fig. 1.1. Classification of pathophysiology of neuroglia. Modified from Verkhratsky A, Butt, A (2023). *Neuroglia: Function and pathology*. Academic Press, Elsevier. (Original Publication).

genetic disorders, of which the most known is the Alexander disease, a rare genetic neurologic disorder in which astrocytes express mutant GFAP that translates into profound leukomalakia (Messing et al., 2012). Several diseases in which astrocytes appear as the primary targets are also characterized by the appearance of aberrant glial cells. In amyotrophic lateral sclerosis, for example, the population of cells expressing markers of both astrocytes and microglia was described; these cells apparently contribute to neuronal degeneration (Wilhelmsson et al., 2017; Trias et al., 2018). In the Tuberous sclerosis complex, which is primary astrocytopathy, aberrant astrocytes are characterized by decreased homeostatic functions related to ion homeostasis, neurotransmitter metabolism, and calcium signaling (Zimmer et al., 2020; Luinenburg et al., 2023; Romagnolo et al., 2024).

CONCLUSIONS

Neuroglia are fundamental elements in all neurologic disorders and likely also contribute to all psychiatric disorders. When facing pathologic lesions, neuroglial cells undergo various metamorphoses mainly aimed at the preservation of nervous tissue homeostasis or isolating/bordering the damaged areas. At the same time, in many pathologic contexts, glial cells become atrophic and lose their homeostatic and protective functions, thereby contributing to the damage of the nervous tissue. The complex balance between neuroprotection and reactivity on one hand and atrophy and loss of function on the other define the progression and outcome of neuropathology.

While neuroglial cells preserve brain homeostasis and protect the brain against injury, it is now evident that these cells can also causally contribute to pathology. Genetic analyses have confirmed an important role of glial cells in neurodegenerative diseases, and this is further supported by preclinical studies. The neuronocentric view has broadened now, and the field has put a new focus on glial cells, their function in different pathologies, how this affects neurons, and whether and how glia can be targeted for the treatment of these diseases. Experts in their field are reviewing current advances for different pathologies in this book.

References

Achucarro N (1910). Some pathological findings in the neuroglia and in the ganglion cells of the cortex in senile conditions. Bull Gov Hosp Insane 2: 81–90.

Alzheimer A (1910). Beiträge zur Kenntnis der pathologischen Neuroglia und ihrer Beziehungen zu den Abbauvorgängen im Nervengewebe. In: F Nissl, A Alzheimer (Eds.), Histologische und histopathologische Arbeiten über die Grosshirnrinde mit besonderer Berücksichtigung der pathologischen Anatomie der Geisteskrankheiten. Gustav Fischer 3: 401–562.

Anders S, Breithausen B, Unichenko P et al. (2024). Epileptic activity triggers rapid ROCK1-dependent astrocyte morphology changes. Glia 72: 643–659.

Anderson MA, Burda JE, Ren Y et al. (2016). Astrocyte scar formation aids central nervous system axon regeneration. Nature 532: 195–200.

Andriezen WL (1893). The neuroglia elements of the brain. Br Med J 2: 227–230.

Aten S, Du Y, Taylor O et al. (2023). Chronic stress impairs the structure and function of astrocyte networks in an animal model of depression. Neurochem Res 48: 1191–1210.

Balaban D, Miyawaki EK, Bhattacharyya S et al. (2021). The phenomenon of clasmatodendrosis. Heliyon 7: e07605.

Banasr M, Duman RS (2008). Glial loss in the prefrontal cortex is sufficient to induce depressive-like behaviors. Biol Psychiatry 64: 863–870.

Bourneville DM, Guérard L (1869). De la Sclérose en Plaques disséminées, Adrien Delahaye. (Original Publication).

Charcot JM (1868). Histologie de la Sclérose en Plaques. In: Gazette des hopitaux civils et militaires, vol. 41. 554. 557-558, 566.

Costa J, Martins S, Ferreira PA et al. (2021). The old guard: age-related changes in microglia and their consequences. Mech Ageing Dev 197: 111512.

Del Río Hortega P (1922). Son homologables la glía de escasas radiaciones y las células de Schwann? Bol Soc Esp Biol 10: 25–28.

Del Río-Hortega P (1919). Poder fagocitario y movilidad de la microglia. Bol de la Soc esp de Biol 9: 154.

Del Rio-Hortega P (1920). La microglia y su transformacíon en células en bastoncito y cuerpos gránulo-adiposos. Trab del Lab de Invest Biol 18: 37.

Del Río-Hortega P (1921). Estudios sobre la neuroglia. La glia de escasas radiaciones oligodendroglia. Biol Soc Esp Biol 21: 64–92.

Del Rio-Hortega P (1932). Microglia. In: W Penfield (Ed.), Cytology and cellular pathology of the nervous system. Hoeber 2: 482–534.

del Río-Hortega P, Penfield WG (1927). Cerebral cicatrix: the reaction of neuroglia and microglia to brain wounds. Bull Johns Hopkins Hosp 41: 278–303.

Escartin C, Galea E, Lakatos A et al. (2021). Reactive astrocyte nomenclature, definitions, and future directions. Nat Neurosci 24: 312–325.

Graeber MB, Streit WJ (2010). Microglia: biology and pathology. Acta Neuropathol 119: 89–105.

Hazell AS, Sheedy D, Oanea R et al. (2009). Loss of astrocytic glutamate transporters in Wernicke encephalopathy. Glia 58: 148–156.

Hemati-Gourabi M, Cao T, Romprey MK et al. (2022). Capacity of astrocytes to promote axon growth in the injured mammalian central nervous system. Front Neurosci 16: 955598.

Heneka MT, Carson MJ, El Khoury J et al. (2015). Neuroinflammation in Alzheimer's disease. Lancet Neurol 14: 388–405.

Henneberger C, Bard L, Panatier A et al. (2020). LTP induction boosts glutamate spillover by driving withdrawal of Perisynaptic Astroglia. Neuron 108: 919–936.e911.

Ibañez S, Sengupta N, Luebke JI et al. (2023). Myelin dystrophy in the aging prefrontal cortex leads to impaired signal transmission and working memory decline: a multiscale computational study, Cold Spring Harbor Laboratory.

Ju YH, Bhalla M, Hyeon SJ et al. (2022). Astrocytic urea cycle detoxifies Abeta-derived ammonia while impairing memory in Alzheimer's disease. Cell Metab 34: 1104–1120.e1108.

Kovacs GG (2019). Molecular pathology of neurodegenerative diseases: principles and practice. J Clin Pathol 72: 725–735.

Lenhossék MV (1895). Der feinere Bau des Nervensystems im Lichte neuester Forschung, *second edn.* Fischer's Medicinische Buchhandlung H. Kornfield. Original Publication.

Long Y, Zhao Z, Xie W et al. (2024). Kallistatin leads to cognition impairment via downregulating glutamine synthetase. Pharmacol Res 202: 107145.

Lugaro E (1907). Sulle funzioni della nevroglia. Riv Patol Nerv Ment 12: 225–233.

Luinenburg MJ, Scheper M, Sorensen FNF et al. (2023). Loss of maturity and homeostatic functions in tuberous sclerosis complex-derived astrocytes. Front Cell Neurosci 17: 1284394.

Marinesco MG (1896). Lesions des centres nerveux produites par la toxine du *Bacillus Botulinus*. C R Soc Biol(Paris) 48: 989–991.

Messing A, Brenner M, Feany MB et al. (2012). Alexander disease. J Neurosci 32: 5017–5023.

Nissl F (1899). Über einige Beziehungen zwischen Nervenzellerkrankungen und gliösen Erscheinungen bei verschiedenen Psychosen. Arch Psychiatr 32: 1–21.

Paolicelli RC, Sierra A, Stevens B et al. (2022). Microglia states and nomenclature: a field at its crossroads. Neuron 110: 3458–3483.

Pekny M, Pekna M, Messing A et al. (2016). Astrocytes: a central element in neurological diseases. Acta Neuropathol 131: 323–345.

Penfiled W, Buckley RC (1928). Punctures of the brain. Arch Neurol Psychiatry 20: 1–13.

Perez-Cerda F, Sanchez-Gomez MV, Matute C (2015). Pio del Rio Hortega and the discovery of the oligodendrocytes. Front Neuroanat 9: 92.

Peters A (2002). The effects of normal aging on myelin and nerve fibers: a review. J Neurocytol 31: 581–593.

Popov A, Brazhe N, Morozova K et al. (2023). Mitochondrial malfunction and atrophy of astrocytes in the aged human cerebral cortex. Nat Commun 14: 8380.

Prinz M, Kann O, Draheim HJ et al. (1999). Microglial activation by components of gram-positive and -negative bacteria: distinct and common routes to the induction of ion channels and cytokines. J Neuropathol Exp Neurol 58: 1078–1089.

Provenzano F, Torazza C, Bonifacino T et al. (2023). The key role of astrocytes in amyotrophic lateral sclerosis and their commitment to glutamate excitotoxicity. Int J Mol Sci 24: 15430.

Ramón y Cajal S (1913). Contribucion al conocimiento de la neuroglia del cerebro humano. Trab Lab Invest Biol Univ Madrid 11: 255–315.

Ramos-Gonzalez P, Mato S, Chara JC et al. (2021). Astrocytic atrophy as a pathological feature of Parkinson's disease with LRRK2 mutation. NPJ Parkinsons Dis 7: 31.

Rawji KS, Gonzalez Martinez GA, Sharma A et al. (2020). The role of astrocytes in Remyelination. Trends Neurosci 43: 596–607.

Romagnolo A, Dematteis G, Scheper M et al. (2024). Astroglial calcium signaling and homeostasis in tuberous sclerosis complex. Acta Neuropathol 147: 48.

Rossi DJ, Brady JD, Mohr C (2007). Astrocyte metabolism and signaling during brain ischemia. Nat Neurosci 10: 1377–1386.

Semyanov A, Verkhratsky A (2021). Astrocytic processes: from tripartite synapses to the active milieu. Trends Neurosci 44: 781–792.

Sierra A, de Castro F, Del Rio-Hortega J et al. (2016). The "big-bang" for modern glial biology: translation and comments on Pio del Rio-Hortega 1919 series of papers on microglia. Glia 64: 1801–1840.

Skinnider MA, Gautier M, Teo AYY et al. (2023). The Tabulae Paralytica: multimodal single-cell and spatial atlases of spinal cord injury. bioRxiv. 2023.2006.2023.544348.

Sofroniew MV (2014). Multiple roles for astrocytes as effectors of cytokines and inflammatory mediators. Neuroscientist 20: 160–172.

Sofroniew MV (2024). Astrocyte cells in the brain have immune memory. Nature 627: 744–745.

Spires-Jones TL, Attems J, Thal DR (2017). Interactions of pathological proteins in neurodegenerative diseases. Acta Neuropathol 134: 187–205.

Stadelmann C, Timmler S, Barrantes-Freer A et al. (2019). Myelin in the central nervous system: structure, function, and pathology. Physiol Rev 99: 1381–1431.

Sterling P, Eyer J (1988). Allostasis: a new paradigm to explain arousal pathology. In: S Fisher, J Reason (Eds.), Handbook of life stress, cognition and health. John Wiley & Sons 629–649.

Streit WJ, Xue QS, Tischer J et al. (2014). Microglial pathology. Acta Neuropathol Commun 2: 142.

Streit WJ, Khoshbouei H, Bechmann I (2020). Dystrophic microglia in late-onset Alzheimer's disease. Glia 68: 845–854.

Taylor SE, Morganti-Kossmann C, Lifshitz J et al. (2014). Rod microglia: a morphological definition. PLoS One 9: e97096.

Tremblay M-E, Verkhratsky A (2024). Microglia: physiology, pathophysiology and therapeutic potential, Springer-Nature. Original Publication.

Trias E, Barbeito L, Yamanaka K (2018). Phenotypic heterogeneity of astrocytes in motor neuron disease. Clin Exp Neuroimmunol 9: 225–234.

Verkhratsky A, Butt A (2023). Neuroglia: function and pathology, Academic Press, Elsevier. Original Publication.

Verkhratsky A, Zorec R (2024). Neuroglia in cognitive reserve. MolPsychiatry. https://doi.org/10.1038/s41380-024-02644-z. E-pub ahead of print.

Verkhratsky A, Zorec R, Parpura V (2017). Stratification of astrocytes in healthy and diseased brain. Brain Pathol 27: 629–644.

Verkhratsky A, Rodrigues JJ, Pivoriunas A et al. (2019). Astroglial atrophy in Alzheimer's disease. Pflugers Arch 471: 1247–1261.

Verkhratsky A, Augusto-Oliveira M, Pivoriunas A et al. (2021). Astroglial asthenia and loss of function, rather than reactivity, contribute to the ageing of the brain. Pflugers Arch 473: 753–774.

Verkhratsky A, Butt A, Li B et al. (2023). Astrocytes in human central nervous system diseases: a frontier for new therapies. Signal Transduct Target Ther 8: 396.

Villanueva CB, Stephensen HJT, Mokso R et al. (2023). Astrocytic engagement of the corticostriatal synaptic cleft is disrupted in a mouse model of Huntington's disease. Proc Natl Acad Sci U S A 120: e2210719120.

Virchow R (1854). Ueber das ausgebreitete Vorkommen einer dem Nervenmark analogen Substanz in den thierischen Geweben. Virchows Arch 6: 562–572.

Virchow R (1858). Die Cellularpathologie in ihrer Begründung auf physiologische and pathologische Gewebelehre. Zwanzig Vorlesungen gehalten während der Monate Februar, März und April 1858 im pathologischen Institut zu Berlin. In: August Hirschwald, first edn. Original Publication.

Virchow RLK (1860). Cellular pathology, John Churchill. Original Publication.

Wilhelmsson U, Andersson D, de Pablo Y et al. (2017). Injury leads to the appearance of cells with characteristics of both microglia and astrocytes in mouse and human brain. Cereb Cortex 27: 3360–3377.

Yi C, Verkhratsky A, Niu J (2023). Pathological potential of oligodendrocyte precursor cells: terra incognita. Trends Neurosci 46: 581–596.

Zimmer TS, Broekaart DWM, Gruber VE et al. (2020). Tuberous sclerosis complex as disease model for investigating mTOR-related Gliopathy during Epileptogenesis. Front Neurol 11: 1028.

Handbook of Clinical Neurology, Vol. 210 (3rd series)
Neuroglia in Neurologic and Psychiatric Disorders, Part II
A. Verkhratsky, L.D. de Witte, E. Aronica, and E.M. Hol, Editors
https://doi.org/10.1016/B978-0-443-19102-2.00012-0

Chapter 2

Neuroglia in neurodegeneration: Alzheimer, Parkinson, and Huntington disease

DMITRY LIM[1*], CARLOS MATUTE[2], FABIO CAVALIERE[3], AND ALEXEI VERKHRATSKY[2,4,5]

[1]*Department of Pharmaceutical Sciences, Università del Piemonte Orientale "Amedeo Avogadro", Novara, Italy*

[2]*Department of Neurosciences, University of the Basque Country UPV/EHU and CIBERNED, Leioa, Bizkaia, Spain*

[3]*The Basque Biomodels Platform for Human Research (BBioH), Achucarro Basque Center for Neuroscience & Fundación Biofisica Bizkaia, Leioa, Spain*

[4]*Faculty of Biology, Medicine and Health, University of Manchester, Manchester, United Kingdom*

[5]*IKERBASQUE, Basque Foundation for Science, Bilbao, Spain*

Abstract

The conspicuous rise of chronic neurodegenerative diseases, including Alzheimer (AD), Parkinson (PD), and Huntington (HD) diseases, is currently without disease-modifying therapies and accompanied by an excessive rate of unsuccessful clinical trials. This reflects a profound lack of understanding of the pathogenesis of these diseases, indicating that the current paradigms guiding disease modeling and drug development are in need of reconsideration. The role of neuroglia, namely astrocytes, microglial cells, and oligodendrocytes, in the pathogenesis of neurodegenerative diseases emerged during the last decades. This chapter provides the state-of-the-art update on the changes of astrocytes, microglial cells, and oligodendrocytes in AD, PD, and HD. A growing body of evidence suggests that homeostatic and defensive functions of glial cells are compromised at different disease stages, leading to increased susceptibility of neurons to noxious stimuli, eventually resulting in their malfunction and degeneration. Investments are needed in the generation of novel preclinical models suitable for studying glial pathology, in "humanizing" research, and in-depth investigation of glial cell alterations to slow down and, possibly, halt and prevent the rise of neurodegenerative disease. Targeting glial cells opens new therapeutic avenues to treat AD, PD, and HD.

INTRODUCTION: THE RISE OF NEURODEGENERATIVE DISEASES

According to official estimates, age-dependent neurodegenerative diseases represent seventh cause of death, leading to 1.7 million deaths a year (https://www.who.int/news-room/fact-sheets/detail/the-top-10-causes-of-death). The prevalence of dementia in the European Union is near 8.9 million, which constitutes 1.73% of the population (https://www.alzheimer-europe.org/dementia/prevalence-dementia-europe), while in the United States, about 6.7 million (10.8%) of people aged over 65 live with dementia (https://www.alz.org/media/documents/alzheimers-facts-and-figures.pdf). In spite of enormous effort of biomedical research and financial expense to maintain neurodegenerative disease patients (more than $500 billion in 2019 in the United States), there is little progress in developing curative therapies for neurodegenerative diseases, including the diseases considered in this chapter, Alzheimer disease (AD), Parkinson disease (PD), and Huntington disease (HD). All in all, for AD, PD, and HD, the disease-modifying therapies do not exist, leaving patients with symptomatic (AD, HD) or replacement (PD) treatments. The hurdles in developing disease-modifying therapies for neurodegenerative diseases are best exemplified by AD, for which more than 500 clinical trials testing 98 unique drugs failed to reach the positive outcome in phase II or III (Kim et al., 2022). Numerous reasons for these failures, arguably, fall into three main categories, applicable also for other neurodegenerative diseases: (i) inadequate design of clinical trials, which have been extensively analyzed and

*Correspondence to: Dmitry Lim, Department of Pharmaceutical Sciences, Università del Piemonte Orientale "Amedeo Avogadro", Via Bovio 6, 28100, Novara, Italy. Tel: +39-0321-375822, Fax: +39-0321-375821, E-mail: dmitry.lim@uniupo.it

discussed elsewhere (Anderson et al., 2017; Mehta et al., 2017; Cummings, 2018); (ii) incomplete understanding of pathophysiology, which, in part, can be explained by the predominance of the neurocentric paradigm and the β-amyloid cascade hypothesis; and (iii) inadequate representation of human pathology by animal models used in preclinical research.

Discoveries of the last decades point to the key role of glial cells, especially astrocytes, in neurodegenerative diseases pathogenesis and mechanisms affecting nervous tissue independently of aggregation and deposition of proteins that represent archetypal hallmarks of neurodegenerative diseases and include β-amyloid in senile plaques of AD and α-synuclein in Lewy bodies in PD patients (Verkhratsky et al., 2010, 2019a; Hansen et al., 2018). This chapter provides the state-of-the-art update on the role of glial cells, astrocytes, microglial cells, and oligodendrocytes in AD, PD, and HD. After a brief introduction of classical pathophysiologic mechanisms of neurodegenerative diseases, we shall provide an overview of the models used in AD, PD, and HD research, and then focus on the alterations of glial cells in specific diseases.

NEUROGLIA IN ALZHEIMER DISEASE

Brief overview of AD pathogenesis and clinical presentation

Histopathology of AD is characterized by extracellular deposits of fibrillar β-amyloid (senile plaques), intracellular accumulation and aggregation of hyperphosphorylated tau proteins (neurofibrillary tangles, NFTs), neuronal damage and death, and reactive gliosis, which mostly develop around senile plaques (Braak and Braak, 1996b; Selkoe and Hardy, 2016; DeTure and Dickson, 2019). Familiar AD (FAD, $\leq$1% of cases) is caused by mutations in amyloid precursor protein (APP) and presenilins (PS1, 2). The etiology of sporadic AD (SAD, ~99% of cases) is less clear: 40% of SAD cases express ApoE4 allele of ApoE, while 70% of all SAD cases are associated with genetic mutations. The genetic landscape of SAD is represented by single nucleotide polymorphisms of more than 20 genetic loci. Some of these genes fall into three functional categories: cholesterol metabolism, immune response, and endocytosis/vesicle-mediated transport (Selkoe and Hardy, 2016; Carmona et al., 2018). Soluble, oligomeric β-amyloid species have been proposed to be toxic, although even high depositions of β-amyloid in genetic mouse models do not trigger massive neuronal death (Trujillo-Estrada et al., 2022).

Pathologic changes associated with AD progression start decades before the appearance of clinical symptoms, and disease evolution can be divided into (i) the early stage, in which there is ongoing cellular pathology but the disease is detectable neither by clinical nor by instrumental approaches; (ii) the preclinical stage, in which the ongoing pathophysiologic processes can be detected by biochemical biomarkers or by neuroimaging techniques; (iii) the emergence of prodromal clinical signs manifested by mild cognitive impairment (MCI) characterized with impairments of episodic memory and cognitive difficulties; and (iv) the dementia stage. The progression from moderate to severe dementia reflects the acceleration of tau pathology and neuronal death (Braak and Braak, 1991; Theofilas et al., 2018; Lim et al., 2023).

AD mouse models

Several approaches have been utilized to model AD in animals. Acute β-amyloid toxicity and lesion models reproduce neuronal dysfunction and death and trigger reactive gliosis, but are not recapitulating the mechanisms of the disease (Toledano and Alvarez, 2004; Balducci et al., 2017). Large animal models present difficulties in generating genetic models and a high cost of maintenance, which preclude their widespread use in preclinical studies. Insect and nematode models, useful to study some mechanistic, biochemical, and basic physiologic aspects of related pathologies, cannot recapitulate, even in the smallest part, the complexity of the human brain and, therefore, cannot be used in preclinical studies. Here we briefly discuss genetic mouse models of AD and their suitability to study glial pathobiology.

FAD-linked mutations recapitulate selected features of human pathology

Among mouse models, the most popular approach is to overexpress human APP, PS, or tau proteins bearing mutations associated with FAD (Table 2.1; Fig. 2.1). Because there are no known tau mutations associated with AD, tau pathology is modeled by either overexpression of human tau or by the mutations causing frontotemporal dementia with parkinsonism linked to chromosome 17 (FTDP-17) (Wszolek et al., 2006). Overexpression of human APP and PS mutants increased β-amyloid production, as well as the ratio between β-amyloid$_{42}$/β-amyloid$_{40}$ and extracellular deposition of β-amyloid aggregates. Tau mutants produced accumulation of hyperphosphorylated tau and formation of intracellular NFT-like structures. The mechanisms included increased phosphorylation and alteration of mRNA splicing with a higher level of 4R tau relative to 3R tau (Lee and Leugers, 2012; Götz et al., 2019). Reactive gliosis develops mostly around β-amyloid deposits. Single mutations resulted in a mild phenotype; therefore, a combination of mutations APP+APP, APP+PS APP+tau, and finally, APP+PS+tau was used to boost β-amyloid and tau pathology (Table 2.1). The race for the exacerbation of β-amyloid pathology and the recapitulation of both plaques and tangles led to the generation of the 3xTg-AD mouse in 2003 (Oddo et al., 2003) and the 5xFAD mouse in 2006 (Oakley et al., 2006). 3xTg-AD mice, overexpressing APP$^{K670N/M671L}$ (Swedish) and tau^{P301L} on the background of PS1^{M146V} knock-in, develop β-amyloid deposits and NFTs but fail to present neuronal death. 5xFAD, overexpressing APP$^{K670N/M671L}$ (Swedish), APP^{I716V} (Florida), APP^{V717I} (London), and PS1^{M146L} and PS1^{L286V} mutations, show intraneuronal accumulation of β-amyloid$_{42}$ from 1.5 months of age, and from 2 months of age demonstrate cerebral amyloid plaques and reactive gliosis. Memory impairment in the Y-maze is evident at 4–5 months of age, while significant neuronal loss is observed at 9 months of age. Histopathologically, however, this model does not develop NFTs (Oakley et al., 2006; Table 2.1).

Table 2.1

Genetic mouse models of AD.

Model name	Promoter	Proteins	Mutations	Plaques/ tangles	Pathology	References
Overexpression or knock-in of WT proteins						
NSE:b-APP751	Rat NSE	APP	No	Plaques	Extracellular β-amyloid immunoreactive deposits varying in size (<5–50 μm) and abundance. Age-dependent deficits in spatial learning in a water-maze task and in spontaneous alternation in a Y maze from 6 months.	Quon et al. (1991)
hTau40	Human Thy-1	hTau40 (2N4R)	No	PHF, pretangles		Götz et al. (1995)
HMG-Tau	Mouse HMG-CR	hTau23 (0N3R)	No	PHF, pretangles	Abundant tau in astrocytes	Brion et al. (1999)
T44	Mouse PrP	hTau23 (0N3R)	No	Pretangles	Insoluble, hyperphosphorylated tau; Axonal degeneration; reduced axonal transport; spinal cord gliosis and motor weakness.	Ishihara et al. (1999)
hTAU	Human tau	FL human tau	Murine tau KO	PHF	Pathologic tau accumulates in the cell bodies and dendrites of neurons in a spatiotemporally relevant distribution.	Andorfer et al. (2003), Duff et al. (2000)
Single mutations Tg and knock-in						
PDAPP	Human PDGF-b	APP-695; APP-751; APP-770	Tg APP^{IND}	Plaques	Extracellular thioflavin S-positive AP deposits, neuritic plaques, synaptic loss, astrocytosis, and microgliosis. No neuronal loss. Impaired LTP from 4 to 5 months. Behavioral alterations from 3 months.	Games et al. (1995), Irizarry et al. (1997), Larson et al. (1999)
Tg2576	Hamster PrP	APP-695	Tg APP^{SWE}	Plaques	Fivefold increased Aβ40 and 14-fold increased Ab42/43. At 9–10 months altered spatial reference and alternation memory. Memory loss from 6 months.	Hsiao et al. (1996), Westerman et al. (2002)
APP23	Mouse Thy-1.2	APP-751	Tg APP^{SWE}	Plaques	Sevenfold overexpression of APP. Widespread plaques from 6 months, plaques are immunoreactive for hyperphosphorylated tau, massive glial reaction.	Sturchler-Pierrat et al. (1997)
PS1-A246E; PS1-M146L; PS1dE9	Mouse PrP	PS1	Tg $PS1^{A246E}$, $PS1^{M146L}$, PS1dE9	No plaques	Higher PS1 accumulation (~40%–50%) in mutants than the fragments derived from wild-type PS1.	Lee et al. (1997)
PS1-KI	Endogenous PS1	PS1	$PS1^{M146V}$-KI	No plaques	No overt mutant phenotype, hypersensitive to seizure-induced synaptic degeneration and necrotic neuronal death in the hippocampus. Used in 3xTg-AD	Guo et al. (1999)
APP(NLh)	Endogenous APP	APP	APP^{SWE}-KI	No plaques	Normal APP levels in brain. Enhanced cleavage. Human Aβ ninefold higher than in normal human brain. Used in 2xKI	Reaume et al. (1996)
PS1(wt) or PS1(M146L)	Human PDGF-b	PS1	NO and Tg $PS1^{M146L}$	No plaques	Overexpression of mutant, but not wild-type, PS1 selectively increases brain Aβ42(43)	Duff et al. (1996)
JNPL3	Mouse PrP	hTau24 (0N4R)	Tg tau^{P301L}	Tangles	Motor and behavioral deficits in transgenic mice, with age- and gene-dose-dependent development of NFT from 6.5 months in hemizygous and 4.5 months in homozygous animals. 48% neuronal loss in spinal cord. Areas with NFT had reactive gliosis.	Lewis et al. (2000)
P301L	Mouse Thy1.2	hTau40 (2N4R)	Tg tau^{P301L}	Tangles	Accumulated tau was hyperphosphorylated and translocated from axonal to somatodendritic compartments and was accompanied by astrocytosis	Götz et al. (2001)
4R-P301S	Mouse Thy1.2	hTau24 (0N4R)	tau^{P301S}	PHF	Neurologic phenotype (paresis) from 5 to 6 months. A 49% reduction in the number of motor neurons in spinal cord.	Allen et al. (2002)
Tau-V337M Tg	Human PDGFb	hTau40 (2N4R)	tau^{V337M}	Tangles	Signs of neurodegeneration include loss of microtubules, accumulation of ribosomes, plasma and nuclear membrane ruffling, and swelling of the Golgi network.	Tanemura et al. (2002)
Tau-R406W Tg	Mouse CaMKII	hTau40 (2N4R)	tau^{R406W}	Tangles	Congophilic hyperphosphorylated tau inclusions in forebrain neurons. From 18 months of age. Impaired fear conditioning.	Tatebayashi et al. (2002)

Continued

Table 2.1

Continued

Model name	Promoter	Proteins	Mutations	Plaques/ tangles	Pathology	References
BRI-Ab	Mouse PrP	Aβ	BRI-Ab40, BRI-Ab42	Plaques	Aβ1–40 does not develop overt amyloid pathology. Aβ1–42 accumulates insoluble Aβ1–42 and develops compact amyloid plaques, congophilic amyloid angiopathy (CAA), and diffuse Abeta deposits.	McGowan et al. (2005)
Tg4510	TetO: Mouse PrP, TA: CaMKII	hTau24 (0N4R)	tau^{P301L}	Tangles	NFT progress from neocortex into the hippocampus and limbic structures with increasing age. Cognitive impairments from 4 months of age. Forebrain atrophy and a prominent neuronal loss in CA1.	Ramsden et al. (2005)
PS19	Mouse PrP	hTau34 (1N4R)	tau^{P301L}	Tangles	NFT progress from 6 months. Neuron loss, hippocampal and entorhinal cortical atrophy at 9–12 months. Hippocampal synapse loss, impaired synaptic function, microglial activation from 3 months.	Yoshiyama et al. (2007)
Combination of mutations Tg and knock-in						
APP-PS1	Mouse PrP	APP, PS1	Tg APP^{SWE}, Tg $PS1^{A246E}$	No plaques	Elevated ratio of Aβ1–42(43) to Aβ1–40 compared to APP^{SWE} alone or to PS1-WT and APP^{SWE}.	Borchelt et al. (1996)
HMG-APP(SDL)	Mouse HMG-CR	APP	APP^{SDL}	No plaques	Generation of C-terminal APP fragments and βA4. No signs of Alzheimer disease-related pathology or related behavioral deficits.	Czech et al. (1997)
PSAPP	Tg2576: Hamster PrP; PS1: humPDGF-b	APP-695, PS1	Tg APP^{SWE}; Tg $PS1^{M146L}$	Plaques	Amyloid deposition is accelerated compared with single Tg mice. Behavioral changes preceded Aβ pathology	Holcomb et al. (1998)
APP/Wt; APP/Sw; APP/Ld	Mouse Thy1	Wt APP-695; APP^{SWE}-770; APP_{LON}-695	Tg Wt APP; Tg APP^{SWE}; Tg APP^{LON}	Plaques	Cognitive impairment, LTP impairment	Moechars et al. (1999)
J20	Human PDGFb	APP	$APP^{SWE,IND}$	Plaques	High levels of Aβ1–42; age-dependent formation of amyloid plaques.	Mucke et al. (2000)
TgCRND8	Hamster PrP	APP-695	$APP^{SWE,IND}$	Plaques	3200–4600 pmol of Aβ42 per gram brain at age 6 months, with an excess of Aβ42 over Aβ40. Memory deficit from 3 months	Janus et al. (2000)
TgCRND8-PS1	Hamster PrP	APP-695, PS1	$APP^{SWE;IND}$; $PS1^{M146L;L286V}$	Plaques	Enhanced amyloid deposition compared with TgCRND8 mice from 1 month of age.	Chishti et al. (2001)
TAPP	Tg2576: Hamster PrP; JNPL3: Mouse PrP	APP, Tau	Tg APP^{SWE}; Tg tau^{P301L}	Plaques, Tangles	NFT pathology enhanced in the limbic system and olfactory cortex compared with JNPL3 mice. Behavioral alterations.	Lewis et al. (2001)
VLW	Mouse Thy1	hTau40 (2N4R)	Tg $tau^{G272V;P301L;R406W}$	Tangles	Ultrastructural abnormalities of lysosomes similar to those found in AD.	Lim et al. (2001)
APPSL, TASD-41	Mouse Thy1	APP-751	Tg $APP^{SWE,LON}$	Plaques	Plaques from 3 months in the frontal cortex; from 5 to 7 months—in the hippocampus, thalamus, and olfactory region.	Rockenstein et al. (2001)
Tg2576/PS-PS1KI	Tg2576: Hamster PrP; PS1: Endogenous	APP-695, PS1	Tg APP^{SWE}, $PS1^{P264L}$	Plaques	PS-1P264L mutation accelerated the onset of amyloid (A) deposition in a gene-dosage dependent manner	Flood et al. (2002)
DKI	PS1, APP: Endogenous	FL APP, PS1	$APP^{NLh/NLh}$; $PS1^{P264L/P264L}$	Plaques	Aβ deposition increased linearly over time.	Flood et al. (2002)
APP/PS1	APP: Mouse Thy1; PS1: Mouse HMG	APP-695, PS1	$APP^{SWE,LON}$; $PS1^{M146L}$	Plaques	Aβ deposits from 2.5 months of age. No neurodegeneration. Microglia alterations. Vascular alterations in MRT.	Blanchard et al. (2003), El Tayara et al. (2010), Jimenez et al. (2008)

3xTg-AD	PS1: Endogenous; APP, tau: Mouse Thy1.2	APP-695, PS1, hTau24 (0N4R)	APP^{SWE}; tau^{P301L}; $PS1^{M146V}$	Plaques, Tangles	Progressively develop plaques and tangles. Synaptic dysfunction, including LTP deficits, manifests in an age-related manner, but before plaque and tangle pathology. Deficits in long-term synaptic plasticity correlate with the accumulation of intraneuronal Aβ. Astrocytic atrophy from 1 month. Reactive astrocytes around plaques.	Oddo et al. (2003)
PS2APP	APP, PS2: Mouse Thy1.2	APP-751, PS2	APP^{SWE}; $PS2^{N141I}$	Plaques	Cognitive deficits from 8 months. Plaques from 5 months by TEM. Dentate gyrus posttetanic potentiation was significantly attenuated at 17 months.	Richards et al. (2003)
Tg-SweDI	Mouse Thy1.2	APP-770	APP^{SDI}	Plaques	APP levels below endogenous mouse APP. Perivascular/vascular Aβ deposits increased with age. Plaques from 3 months.	Davis et al. (2004)
APP/tau	Tg2576: Hamster PrP; Tau: Mouse Thy1	APP-695, hTau40 (2N4R)	Tg APPswe; Tg $tau^{G272V; P301L; R406W}$	Plaques, Tangles	Neuronal loss in limbic areas. NFTs at late age about 25 months.	Pérez et al. (2005), Ribé et al. (2005)
APP^{SWE}/PS1dE9	APP, PS1: Mouse PrP	APP-695, PS1	Tg APP^{SWE}; Tg PS1dE9	Plaques	Plaques from 6 months. At 18 months, amyloid burdens were higher in APP^{SWE}/PS1dE9. Cognitive impairment in APP^{SWE}/PS1dE9 mice stronger than in all other genotypes.	Savonenko et al. (2005)
tg-ArcSwe	Mouse Thy1.2	APP	Tg $APP^{SWE;ARC}$	Plaques	Intraneuronal Aβ and senile plaque deposition was markedly accelerated in tg-$APP^{SWE;ARC}$ mice, as compared to tg-APP^{SWE} mice. Plaques from 6–7 months. Cerebral amyloid angiopathy (CAA) in close proximity to astrocytic end feet.	Lord et al. (2006)
5xFAD	Mouse Thy1.2	APP-695, PS1	Tg APP^{SFL}; Tg $PS1^{M146L; L286V}$	Plaques	Intraneuronal Aβ42 from 1.5 months. Amyloid deposition and gliosis from 2 months. Large pyramidal neurons in cortical layer 5 and subiculum are lost. Memory deficit in elevated Y-maze at 4–5 months.	Oakley et al. (2006)
APPPS1	Mouse Thy1.2	APP, PS1	APP^{SWE}; $PS1^{L166P}$	Plaques	Amyloidosis from 6 to 8 weeks and the ratio of human amyloid Aβ42 to Aβ40 is 1.5 and 5 in predepositing and amyloid-depositing mice, respectively. Robust gliosis, from 1 to 8 months of age. Neuron loss from 8 months	Radde et al. (2006)
THY-Tau22	Mouse Thy1.2	hTau34 (1N4R)	Tg $tau^{G272V; P301S}$	PHF, Tangles	Hyperphosphorylation of tau on several Alzheimer disease-relevant tau epitopes (AT8, AT100, AT180, AT270, 12E8, taupSer396, and AP422), mild astrogliosis. Deficits in hippocampal synaptic transmission and impaired behavior characterized by increased anxiety, delayed learning from 3 months, and reduced spatial memory at 10 months.	Schindowski et al. (2006)
APP23 × B6/P301L	APP: Mouse Thy1.2; JNPL3: Mouse PrP	APP-751, hTau24 (0N4R)	Tg APP^{SWE}; Tg tau^{P301L}	Plaques, Tangles	Increased tau pathology compared to single B6/P301L tau transgenic mice predominately in areas with high Aβ plaque load.	Bolmont et al. (2007)
rTg3696AB	TetO: Mouse PrP; TA: CaMKII	APP-695, hTau24 (0N4R)	$APP^{SWE;LON}$; tau^{P301L}	Plaques, Tangles	Plaques, neurofibrillary tangles, and neurodegeneration	Paulson et al. (2008)
APP(NL/NL); APP(NL-F/NL-F); APP(NL-G-F/NL-G-F)	APP: Endogenous	FL mouse APP with humanized Ab sequence	KI $APP^{SWE; ARC; BEY/IBE}$	Plaques	Typical Aβ pathology, neuroinflammation and memory impairment in an age-dependent manner. The NL-G-F mice had a greater and more rapid accumulation of Ab deposits and glial responses. In NL-G-F mice, alterations in MWM and Fear Conditioning from 6 months; Aβ plaques from 6 to 9 months; GFAP from 6 months (CTX), 9 months (Hip); ChAT from 6 months.	Masuda et al. (2016), Mehla et al. (2019), Saito et al. (2014)
dual Tg	J20: Human PDGFb; Tg4510: TetO: Mouse PrP; TA: CaMKII		J20 (hAPP), Tg4510 (htau)	Plaques, Tangles	Spatial memory deficit from 3.5. Dual Tg mice demonstrated behavioral deficits earlier than those shown by the htau mice.	Lippi et al. (2018)

Continued

Table 2.1

Continued

Model name	Promoter	Proteins	Mutations	Plaques/ tangles	Pathology	References
Late onset AD and humanized models						
NSE-apoE3, NSE-apoE4	NSE	ApoE3, ApoE4	No	No plaques	Impaired spatial memory in NSE-apoE4 mice compared with NSE-apoE3.	Raber et al. (1998)
CRND8/E4	APP: Hamster PrP; hE4: Endogenous mouse E4	APP-695, human ApoE4	$APP^{SWE,IND}$	Plaques	E4 alleviates circadian rhythm and inflammation in CRND8 background.	Graybeal et al. (2015)
MAPT-KI	Human tau promoter and regulatory elements	Human tau gene locus	No	No tangles	No acceleration of neuroinflammation, neuronal death and brain atrophy in MAPT KI. Cross-breeding of MAPT KI with AppNL-G-F KI did not alter amyloid pathology, neuroinflammation, and neuronal death of AppNL-G-F KI.	Hashimoto et al. (2019)
AppNL-G-F/ hTau-KI	APP: Endogenous mouse; Tau: full length human	Human tau; humanized mouse APP	KI $APP^{SWE;\ ARC;\ BEY/IBE}$	Plaques, No tangles		Saito et al. (2019)
hAβ-KI	Endogenous mouse APP	Humanized Ab sequence	No	No plaques	Age-dependent impairments in cognition and synaptic plasticity, inflammatory alterations. Cre-mediated ablation of hAβ expression rescues cognition and formation of PAS granules.	Baglietto-Vargas et al. (2021)
LOAD1	Endogenous mouse	Human APOEϵ4, Trem2	$Trem2^{R47H}$	No plaques	Regional alterations in glycolytic metabolism and vascular perfusion on PET/MRI neuroimaging. Sex and age dependent transcriptional profile	Kotredes et al. (2021a,b)
LOAD2	Endogenous mouse	Human APOEϵ4, Trem2	$Trem2^{R47H}$	No plaques	No AD-related phenotype. Increased plasma cytokines and neuroinflammation from 12 months.	Kotredes et al. (2021a,b)

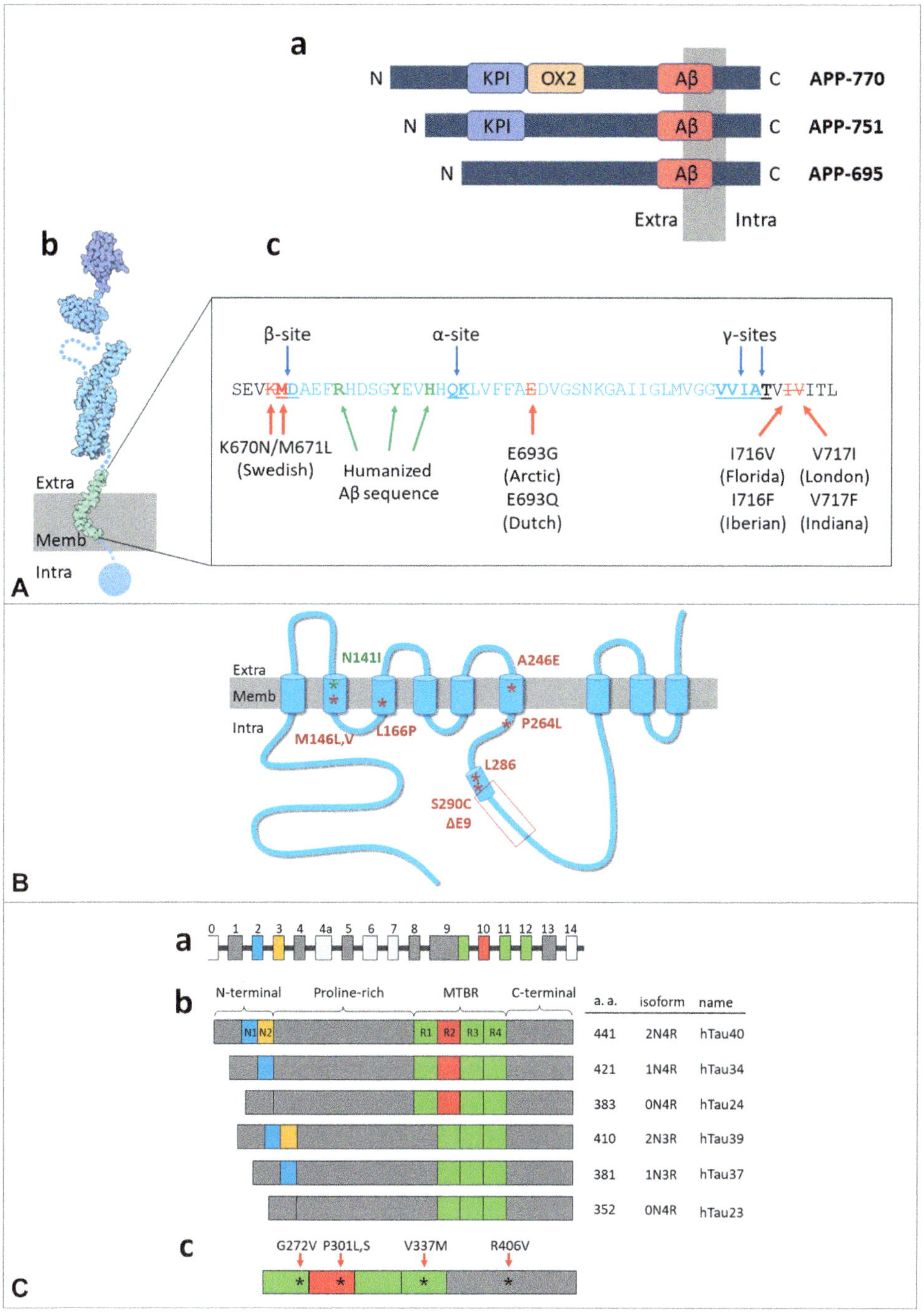

Fig. 2.1. APP, PS1, 2, and tau mutations used to generate transgenic and knock-in mouse models. (A) APP isoforms (a), membrane topology (b), and β-amyloid sequence (c). In (c), humanized amino acids are *underlined*. (B) Membrane topology of PS1 protein. Mutations of PS1 are shown in *dark-red*; mutations in PS2^{N141I} are shown in *green*. Exon 9, deleted in the S290C mutation is shown in *rectangle*. (C) Genomic exon organization (a), protein isoforms (b), and modeled mutations (c) of tau protein.

Alternative to overexpression, the knock-in strategy was developed to demonstrate that the increased total β-amyloid and β-amyloid$_{42}$/β-amyloid$_{40}$ ratio can be obtained without up-regulation of APP or PS. However, in spite of numerous alterations, including reactive gliosis, impaired neurogenesis, progressive memory impairment, oxidative damage, proteome, and phospholipid alterations (Siman et al., 2000; Zhang et al., 2007; Abdul et al., 2008; Bader Lange et al., 2010; Bruce-Keller et al., 2011; Sultana et al., 2011; Webster et al., 2013), double knock-in mice presented neither tau pathology nor neuronal loss (Malthankar-Phatak et al., 2012). The most recent models, harboring increasing number of point mutations recreating Swedish, Iberian, and Arctic mutations in a humanized β-amyloid sequence (APP$^{NL/NL}$, APP$^{NL-F/NL-F}$, and APP$^{NL-G-F/NL-G-F}$ mice) (Fig. 2.1A), developed β-amyloid deposition starting from 2 months of age, memory impairment from 6 months of age, and reactive gliosis from 9 months of age (Saito et al., 2014; Table 2.1). All in all, none of the existing animal models faithfully represents human pathology, and their relevance for understanding the pathophysiology of the disease remains rather limited.

Are existing mouse models useful to study neuroglial AD-related pathology?

Regarding neuroglial pathology, reactive astrocytes and microglia are present in many, if not all, APP, PS, and tau overexpressing mice. Moreover, reactive gliosis can be readily initiated by direct infusion of β-amyloid species into the brain (Toledano and Alvarez, 2004; Balducci et al., 2017). Therefore, reactive gliosis in these models likely represents a secondary reaction of neuroglia to β-amyloid and/or damage of other neural cells. To discriminate cell-autonomous from noncell-autonomous pathology, astrocytic expression of AD-related proteins is required. The use of neuron-specific promoters precludes β-amyloid and tau production by neuroglia, stipulating that the glial response in these models is noncell-autonomous. However, some promoters have weaker neuronal specificity. For example, the hamster prion protein (PrP) promoter, used to create popular Tg2576 mice bearing overexpressing human APP^{SWE} protein, was shown to be also active in astrocytes (Scott et al., 1989, 1992; Moser et al., 1995). Several reports detected increased hAPP and β-amyloid production from cultured astrocytes of Tg2576 mice (Zhao et al., 2011; Katsurabayashi et al., 2016; Heiland et al., 2019). Astrocytic expression of human APP was also detected in Tg2576 mice in vivo (Heiland et al., 2019). Unexpectedly, hAPP was present in astrocytes distant to β-amyloid deposits, while it was absent in reactive astrocytes surrounding senile plaques (Heiland et al., 2019).

Knock-in of AD-related mutation in mouse APP or PS genes allows expression from endogenous promoter and regulatory elements, allowing normal expression levels of mutant proteins. One such mouse model, 3xTg-AD mice, harboring knocked-in of $PS1^{M146V}$ mutation, has been extensively studied. In 3xTg-AD mice, astrocytic alterations appear as early as 1 month of age (Yeh et al., 2011), i.e., far earlier than β-amyloid depositions, synaptic/neuronal malfunctions, and detectable memory impairments (Clark et al., 2015; Stimmell et al., 2019). Numerous studies on primary and immortalized astrocytic cultures from 3xTg-AD mice provide further insights in cell-autonomous mechanisms of astrocytic alterations in AD (Grolla et al., 2013; Rocchio et al., 2019; Dematteis et al., 2020; Tapella et al., 2022; Gong et al., 2023; see also Section "Functional significance of astrocytic changes in AD: Homeostatic failure and glial paralysis"). It is unknown whether increased β-amyloid expression from the APP^{SWE} transgene contributes to early astrocytic pathology in the 3xTg-AD mice. Among existing models, recently generated knock-in models (APP^{NLh}; PS1-KI; DKI, $APP^{NL\text{-}F/NL\text{-}F}$, and $APP^{NL\text{-}G\text{-}F/NL\text{-}G\text{-}F}$ mice) may be useful to study neuroglial cell-autonomous pathology and the role of neuroglia in progression of AD (Table 2.1). However, to specifically address neuroglial contribution to AD pathogenesis, mice expressing APP/PS transgenic or knocked-in mutations under astroglial or microglial promoters need to be developed.

Human iPSC-derived astrocytes in AD research

Following the discovery of a reproducible method of reprogramming somatic cells into iPSCs, there is a renaissance of cellular models in the form of human-induced pluripotent stem cells (iPSC)-differentiated neural cells (iPSC-DNCs), including neurons, astrocytes, oligodendrocytes, and microglial cells. Being committed to terminal differentiation, iPSC-DNCs are suitable for the generation of complex, multicellular models, like organoids, to replicate the full spectrum of cell-to-cell interactions present in the brain (Lee et al., 2017a; Cerneckis et al., 2023).

Morphologic aspects were extensively studied in iPSC-derived FAD and SAD models. Both FAD ($PS1^{M146V}$ carrier) and SAD ($ApoE4^{+/+}$ carrier) astrocytes present reduction in cell volume and complexity, associated with decrease of astrocytic markers such as S100B (S100 Calcium Binding Protein B), EAAT1 (Excitotoxic Amino Acid Transporter 1; GLAST), and GS (Glutamine Synthase) (Jones et al., 2017). Alterations of metabolism, Ca^{2+} signaling, redox state, and cytokine production, as well as of lipid homeostasis, were also found (Kuijlaars et al., 2016; Oksanen et al., 2017; Zhao et al., 2017; Konttinen et al., 2019; Sienski et al., 2021; de Leeuw et al., 2022). iPSC-derived astrocytes can be grown in three-dimensional (3D) structures, commonly defined as cerebral organoids (Fan et al., 2019; Barak et al., 2022). In a multicellular environment, astrocytes showed dramatic improvement in development and maturation compared to 2D monolayer cultures, enabling the investigation of the effect of cell-specific AD-related mutations by the generation of chimeric cerebral organoids containing neurons and astrocytes with different genotypes (Kuijlaars et al., 2016; Zhang et al., 2016; Quadrato et al., 2017; Sloan et al., 2017; Huang et al., 2022). Astrocytes, derived from iPSCs, can also be grafted into the mouse brain where they develop into fully functional human astroglial cells, including primate-specific interlaminar astrocytes (Preman et al., 2021a,b). Furthermore, a number of reports prove that iPSC-derived astrocytes are suitable for personalized medicine (Konttinen et al., 2019; Shinohara et al., 2019; van der Kant et al., 2019; Arnst et al., 2021; Brookhouser et al., 2021; Jun et al., 2022).

Astrocytes in AD

Morphologic changes: Reactivity and atrophy

During AD pathogenesis, astrocytes undergo complex brain-region-specific and disease-stage-dependent alterations. Early changes are characterized by asthenia and atrophy, while the development of vascular amyloidosis and parenchymal β-amyloid deposits triggers reactive astrogliosis. Studies on 3xTg-AD mice bearing knock-in of $PS1^{M146V}$ and overexpressing human APP^{SWE} and tau^{P301L} mutants show that atrophic changes in astrocytes are detectable in the entorhinal cortex (EC) as early as at 1 month of age (Yeh et al., 2011). A significant (by about 50%) reduction of surface and volume of glial fibrillary acidic protein

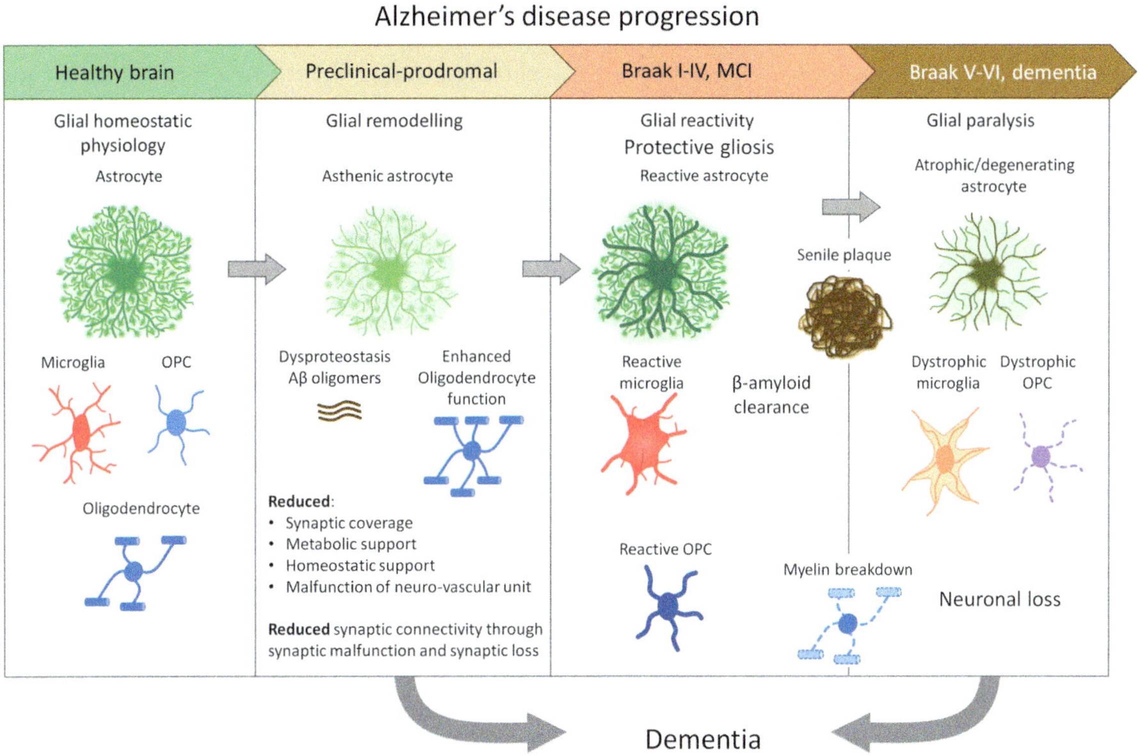

Fig. 2.2. Changes of neuroglia in AD.

(GFAP)-positive profiles indicate astrocytic atrophy. In the medial prefrontal cortex (PFC) the cytoskeletal atrophy is evident from 3 months of age, while in the hippocampal it starts from the dentate gyrus from 6 months, and in the CA1 region—from 18 months of age (Olabarria et al., 2010; Kulijewicz-Nawrot et al., 2012). Atrophic changes in astrocytic cytoskeleton fail to recover with the age persisting up to 18 months. Atrophic astrocytes in the hippocampus are distant from senile plaques; in the EC and PFC astrocytes generally do not associate with β-amyloid depositions (Olabarria et al., 2010; Yeh et al., 2011; Kulijewicz-Nawrot et al., 2012). In astrocytes from nontransgenic mice (non-Tg), the immunoreactivity against cytoskeletal proteins such as GFAP decreased with age, suggesting age-related atrophic processes (Verkhratsky et al., 2020a).

Normal aging of mice and humans was also accompanied by a general decrease in astrocytic morphologic complexity and territorial domains (Popov et al., 2021; Verkhratsky and Semyanov, 2023). Atrophic changes were not accompanied by the decrease in number of GFAP-positive astrocytes in 3xTg-AD mice (Verkhratsky et al., 2019a). Atrophic changes were found in the hippocampus of 5-month-old J20 mice overexpressing human APP with two mutations, Swedish and Indian (Beauquis et al., 2013). Astrocytic atrophy was also found in hiPSC-derived astrocytes from FAD and SAD patients (Jones et al., 2017) and in normal hiPSC-derived astrocytes grafted into the brain of an AD mouse model (Preman et al., 2021b).

The appearance of reactive astrocytes is most often judged by the increased immunoreactivity against GFAP and the intermediate filament of the astrocytic cytoskeleton (Potokar et al., 2020). In mice, increased GFAP staining is found mostly around parenchymal and perivascular Aβ deposits (Oddo et al., 2003; Lim et al., 2013). In human AD brains, both plaque-associated and diffuse reactive astrocytes were found in different regions (Simic et al., 2000; Porchet et al., 2003; Taipa et al., 2018). Although, at later Braak stages, no differences were found between demented and cognitively preserved individuals (Wharton et al., 2009), early detection of GFAP may have a prognostic value. Recently, reactive astrogliosis was detected in vivo in the brain of AD mouse models and AD patients using specific accumulation of ^{11}C-acetate by reactive astrocytes with consequent detection using positron emission tomography (PET) imaging (Nam et al., 2023). Several reports suggest that the presence of GFAP in plasma may be a reliable prognostic biomarker for AD as well as for other neurologic and neurodegenerative diseases (Chatterjee et al., 2021; Cicognola et al., 2021; Pereira et al., 2021; Verberk et al., 2021; Abdelhak et al., 2022; Shir et al., 2022). Finally, reactive astrocytes in AD do not lose their territorial domains, thus undergoing the isomorphic variant of reactive astrogliosis (Verkhratsky et al., 2019a; Fig. 2.2).

Functional significance of astrocytic changes in AD: Homeostatic failure and glial paralysis

Functional significance of astrocytic changes during AD development reflects their central role in CNS homeostasis and synaptic transmission (Verkhratsky and Nedergaard, 2018; Verkhratsky and Butt, 2023). Astrocytic atrophy and loss of their complexity in young 3xTg-AD mice leads to the reduction of synaptic coverage by perisynaptic astrocytic leaflets (Verkhratsky et al., 2015, 2019a). Decreased astrocyte presence in the active brain milieu diminishes homeostatic support, neurotransmitter uptake, supply of neurotransmitter precursors (glutamine and L-serine), and therefore the activity and plasticity of neuronal networks, which altogether contribute to early synaptic malfunction and cognitive decline (Verkhratsky et al., 2015, 2019a, 2023). This conclusion is supported by both in vivo and in vitro observations. In vivo, environmental enrichment and physical exercise reversed astrocytic atrophy and improved cognitive performance in mice (Beauquis et al., 2013; Rodríguez et al., 2013; Ryan and Kelly, 2016). There is growing appreciation that atrophic changes, with consequent homeostatic failure, may develop in astrocytes as cell-autonomous alterations mainly related to metabolic abnormalities, dysregulation of Ca^{2+} homeostasis, and cellular proteostasis (De Strooper and Karran, 2016; Verkhratsky et al., 2019a; Dematteis et al., 2020; Andersen et al., 2022; Tapella et al., 2022; Gong et al., 2023; Lim et al., 2023; Verkhratsky and Butt, 2023).

Reactive astrogliosis is one of the most studied phenomena of AD pathogenesis. In AD, reactive gliosis represents the defensive reaction to extracellular accumulation of β-amyloid. Functional outcome of reactive astrogliosis is less understood (Ben Haim et al., 2015a; Escartin et al., 2021). Reactive transformation likely includes a continuum of states of morphologic, molecular, and functional remodeling in response to injury, disease, or infection, which cannot be limited to a binarization on neurotoxic or neuroprotective effect (Escartin et al., 2021). Reports suggest a prevailing neuroprotective and anti-β-amyloid effect of reactive astrocytes, although the issue remains controversial (Ben Haim et al., 2015a; Escartin et al., 2021). Genetic ablation of GFAP or/and vimentin from AD mice either did not change or exacerbated AD pathology, arguing for the neuroprotective role of reactive astrocytes (Kraft et al., 2013; Kamphuis et al., 2015; Katsouri et al., 2020). However, in human brains, increased GFAP staining correlated with loss of high affinity glutamate transporters EAAT2 (GLT1), suggesting impaired glutamate uptake (Simpson et al., 2010). It was also shown that inhibition of the Janus Kinase 2 (JAK2)-signal transducer and activator of transcription 3 (STAT3) pathway in reactive astrocytes improved functional deficits in mouse models of Alzheimer disease (Ceyzériat et al., 2018; Guillemaud et al., 2020), suggesting that targeting specific aspects of pathologically remodeled reactive astrocytes might be beneficial.

The observation of the reduction of reactive astrogliosis, associated with the progression from MCI to cognitive impairment, found in vivo by the neuroimaging approach, suggests that a loss of protective properties, named "functional paralysis" may play a role in the acceleration of AD pathology (Verkhratsky et al., 2015; Fig. 2.2).

Calcium signaling in AD astrocytes

The calcium hypothesis of aging and neurodegenerative diseases, including AD (Toescu et al., 1998, 2004), shaped the paradigm for research and drug development, which culminated with the approval of memantine, an *N*-methyl-D-aspartate (NMDA) receptor antagonist, for AD treatment in 2003. Over the years the hypothesis has been mainly developed based on pathophysiology of neuronal cells, largely overlooking Ca^{2+} alterations in glial cells (Thibault et al., 2007; Berridge, 2010; Supnet and Bezprozvanny, 2010; Lim et al., 2014, 2016; Popugaeva et al., 2018; Verkhratsky, 2019). The advances in multiphoton fluorescence microscopy and the development of new generations of fluorescent probes for Ca^{2+} made possible assessment of astrocytic Ca^{2+} signaling not only in vitro but also in vivo highlighting the complexity of the phenomenon and the associated limitation of experimental protocols and interpretation (Volterra et al., 2014; Shigetomi et al., 2016; Semyanov et al., 2020).

Astrocytes, nonexcitable cells of the CNS, engage intracellular ionic signaling (which includes Ca^{2+}, Na^+, and Cl^- signals) as a substrate of their excitability (Rose and Verkhratsky, 2016; Verkhratsky and Nedergaard, 2018; Verkhratsky et al., 2019b; Lim et al., 2021; Untiet et al., 2023). Intracellular Ca^{2+} signals are generated by Ca^{2+} release from the ER through inositol-1,4,5-trisphosphate receptors ($InsP_3Rs$) and ryanodine receptors (RyR) receptors; $InsP_3R2$ is the predominant type localized in the soma and main astrocytic processes/branches, while in distant branches $InsP_3R1$ is active (Shigetomi et al., 2016; Lim et al., 2021). Store-operated channels contribute to prolongation of Ca^{2+} signals, while transient receptor potential (TRP) channels, ligand-gated Ca^{2+} channels, and Na^+/Ca^{2+} exchanger, operating in reverse mode, mediate Ca^{2+} signals in leaflets, which are devoid of the ER (Verkhratsky et al., 2020b; Lim et al., 2021). Astrocytic Ca^{2+} signaling modulates many processes from local (synaptic activity and vascular tone) to global—memory, sleep, active brain states, and cognition (Lim et al., 2021).

Exogenous application of β-amyloid to cultured astrocytes increased basal Ca^{2+} levels in the cytosol and sensitized astrocytes to an agonist-evoked Ca^{2+} increase through the mechanism of transcriptional remodeling of the Ca^{2+} signaling toolkit (Casley et al., 2009; Alberdi et al., 2013; Grolla et al., 2013; Lim et al., 2013; Delekate et al., 2014; Ronco et al., 2014; Tapella et al., 2018).

Acute β-amyloid Ca^{2+} toxicity for astrocytes has been tested in vivo in 2- to 4-month-old Tg-SweDI mice overexpressing Swedish and Dutch/Iowa APP mutations (Table 2.1), which at this age present rare (0.7 ± 0.1/min) spontaneous oscillations. Intravenous administration of β-amyloid$_{40}$ (0.4 mg/kg) resulted in a rapid increase in the frequency of Ca^{2+} oscillations (~3.7/min 50 min after injection) (Takano et al., 2007).

Contradictory data were published regarding the role of PS in cell Ca^{2+} homeostasis. It was suggested that in physiologic conditions PSs localized in the ER membrane act as Ca^{2+} permeable channels mediating passive leak of Ca^{2+} from the ER; FAD-linked mutations disrupt PS Ca^{2+} permeability, which leads to excessive accumulation of Ca^{2+} in the ER with consequent enhanced agonist-induced Ca^{2+} signals in the cytosol, which altogether contribute to chronic Ca^{2+} overload and cell damage (Supnet and Bezprozvanny, 2011). Conversely, it was suggested that PS2 FAD-linked mutants inhibit sarco-endoplasmic reticulum Ca^{2+} ATPase (SERCA), thereby depleting ER from Ca^{2+} (Brunello et al., 2009; Galla et al., 2020). It seems that PS1 and PS2 mutations produce differential effects on ER Ca^{2+} handling (Pizzo et al., 2020). In cultured astrocytes, reports indicate that enhanced $[Ca^{2+}]_i$ signals result from increased steady-state $[Ca^{2+}]_L$ levels (Rocchio et al., 2019; Dematteis et al., 2020). This was corroborated by observation of enhanced thapsigargin-induced Ca^{2+} release from the ER of hiPSC-derived astrocytes from FAD patients (Oksanen et al., 2017).

In vivo, increased spontaneous astrocytic Ca^{2+} activity was frequently reported. In $APP^{SWE}/PS1dE9$ Ca^{2+} signals started from astrocytes located close to β-amyloid plaques (~25 μm) and propagated to more distant astrocytes (Kuchibhotla et al., 2009). Consistent with this, in the APP/PS1 model overexpressing human $APP^{SWE}/PS1^{L166P}$ under murine Thy1.2 neuron-specific promoter, astrocytes located in the vicinity of plaques (≤50 μm) showed a higher frequency of spontaneous oscillation (Delekate et al., 2014). In $APP^{SWE}/PS1dE9$ mice, astrocytes located in areas with increased β-amyloid density showed higher spontaneous activity in peripheral astrocytic processes but not in soma (Lines et al., 2022). However, sensory stimulus-evoked Ca^{2+} responses were lower and slower in $APP^{SWE}/PS1dE9$ astrocytes (Lines et al., 2022). An increase in frequency of spontaneous Ca^{2+} activity was also detected in the cortex of young, preplaque 2- to 4-month-old Tg2576 (2.3 ± 0.6/min) and 3xTg-AD mice (3.5 ± 0.1/min) compared with control mice, which never presented a frequency higher than 0.4 ± 0.1/min (Takano et al., 2007).

Increased Ca^{2+} activity in plaque-associated astrocytes has been linked to astrocytic reactivity. Reactive astrocytes around β-amyloid deposits express increased levels of purinergic $P2Y_1$ receptor, which contribute to propagation of Ca^{2+} signal across astrocytic syncytium (Delekate et al., 2014); exogenous application of β-amyloid in vitro, in situ, and in vivo readily induces astrogliotic response (Alberdi et al., 2013; Ronco et al., 2014; Balducci et al., 2017), while inhibition of Ca^{2+} release from the ER suppressed astroglial reactivity (Alberdi et al., 2013; Kanemaru et al., 2013). Last, brain regional differences in the remodeling of astrocytic Ca^{2+} signaling toolkit were associated with ability to mount astrogliotic response (Grolla et al., 2013).

Recently, the hypothesis of the enhanced Ca^{2+} signaling in AD astrocytes has been challenged in the in vivo experiments. First, in the cingulate cortex astrocytes of 3-month-old APP^{NL-F} mice, spontaneous Ca^{2+} signals showed significantly lower frequency and amplitude compared with APP^{NL} control mice, which correlated with a decreased expression of IP_3R2 (Shah et al., 2022). Next, in awake 15-month-old Tg-ArcSwe mice, an attenuated astrocytic Ca^{2+} signaling was recorded during locomotion with no spatial correlation with β-amyloid plaques (Åbjørsbråten et al., 2022). Last, in a longitudinal study on PS2APP mice overexpressing APP^{SWE} and $PS2^{N141I}$ human proteins, no differences were found at 3 months, while reduced spontaneous Ca^{2+} signals in astrocytic microdomains were recorded in PS2APP mice at 6 months (Lia et al., 2023). The inhibition of Ca^{2+} signals was explained by the reduction of the ER Ca^{2+} content, which is characteristic for the $PS2^{N141I}$ mutation (Galla et al., 2020; Pizzo et al., 2020). Both suppressed Ca^{2+} signals in PS2APP astrocytes and correlated neuronal hyperactivity were corrected by lentivirus-mediated overexpression of STIM1 (Lia et al., 2023).

To recapitulate, (i) the response of Ca^{2+} homeostasis and signaling, both in neurons and astrocytes appears to be PS isoform and mutation-specific: PS1 mutations being associated with a potentiation, while PS2 mutations—with a suppression of ER-dependent Ca^{2+} signals; (ii) Ca^{2+} recordings in vivo, especially in awake and freely behaving animals are challenging (Khiroug and Verkhratsky, 2021); only few studies addressed alterations of Ca^{2+} signaling in AD models in vivo; (iii) discrepancies in the results from in vivo studies may arise from the differences in FAD-related genetic mutations (pathobiology of a specific mutations; neuron-specific transgenics vs knock-in), brain areas assessed (hippocampus, somatosensory cortex, visual cortex or cingulate cortex) and/or experimental protocols (slices, anesthetized or behaving mice). The shortages may also be due to a general poor understanding of the role of Ca^{2+} signals and Ca^{2+}-regulated processes in physiology of astrocytes (Lim et al., 2021). For example, efforts have been mainly directed to the description of Ca^{2+} signals themselves and their direct effects of synaptic transmission through regulation of Ca^{2+}-mediated gliotransmission (Zorec et al., 2012; Savtchouk and Volterra, 2018). However, little is known about how they are linked with altered protein turnover (protein synthesis, degradation, ER stress/unfolded protein response) and/or bioenergetic deficit, which are thought to be the earliest cellular alterations in AD (De Strooper and Karran, 2016; Lim et al., 2023).

Microglia in AD

Human tissues and preclinical AD models revealed multiple pathologic phenotypes of microglia in the context of AD (Schwabe et al., 2020). These phenotypes can coexist at different stages of the disease, reflecting a continuous process of remodeling and adaptation, with the balance of protective and detrimental traits defining the progression of the disease (Leyns et al., 2019; Zhang et al., 2021). Accumulation of fibrillar β-amyloid acts as an instigator of microglial pathologic remodeling (Sebastian Monasor et al., 2020). Similarly, APOE4 activates triggering receptor expressed on myeloid cells 2 (TREM2), an exclusively microglial AD risk factor (Gratuze et al., 2018), while β-amyloid stimulates the shedding of TREM2 from microglial membranes (Vilalta et al., 2021). Microglial TREM2 is needed for neuroprotection as it guides microglia toward senile plaques, where they, together with astrocytes, erect a protective barrier (von Saucken et al., 2020). This astrocyte-microglial barrier is instrumental for compacting and clearing β-amyloid depositions and in limiting the formation of toxic β-amyloid42 species (Meilandt et al., 2020). At the early stages, microglial TREM2, in particular, mediated phagocytosis of β-amyloid, thus counteracting the pathologic process; at the more advanced stages of AD, TREM2 is linked to instigating microgliosis and arguably potentiating tau pathology (Suárez-Calvet et al., 2019; Schwabe et al., 2020). At the advanced stages of the disease, microglia loses its phagocytotic capabilities (Lewcock et al., 2020), which may also be linked to TREM2. Incidentally, in patients with loss of function TREM2 mutations R47H or R62H, microglia show decreased defensive and neuroprotective capacity (Suárez-Calvet et al., 2019).

Oligodendrocytes in AD

Changes in white matter integrity emerge in the early stages of neurodegeneration (Lee et al., 2017b), which may provide a window of opportunity for new treatment strategies. For instance, in studies conducted on asymptomatic subjects with an increased risk of developing AD, a decline in white matter integrity has been associated with pathologic AD biomarker levels (Dean et al., 2017). Despite these discovered associations, the role of myelin and oligodendrocytes in neurodegenerative progress remains relatively poorly studied.

Myelin breakdown in AD

Myelin, a spirally wrapped and compacted glial membrane, enhances axonal conduction speed and allows oligodendrocytes to provide metabolic support to axons (Saab et al., 2016). Myelin and oligodendrocytes deteriorate with age (Bowley et al., 2010) and may well contribute to axonal functional impairment and demise in AD. Malfunction at both sides of the axon-myelin unit has been long associated with and modeled in aging and AD (Bartzokis et al., 2010). This model posits that because of the unique vulnerability of late-developing oligodendrocytes, myelin breakdown is at the core of the earliest changes involved in both brain aging and AD. Myelin breakdown disrupts brain functions that depend on highly synchronized timing of neuronal impulses and eventually results in functional 'disconnections' of associated cortical regions, with subsequent loss of neurons and progression to permanent deficits (Fig. 2.2). The loss of synchrony affects progressively more networks, and therefore results in an increasing group of devastating symptoms that we currently refer to as AD (Bartzokis et al., 2010).

On the contrary, the pattern of neurofibrillary changes that gradually develop in the course of AD bears a striking resemblance to the inverse sequence of cortical myelination (Braak and Braak, 1996a). Accordingly, it was hypothesized that factors released by oligodendrocytes exert a strong influence upon nerve cells and suppress disordered neuritic outgrowth, and that scarcity of such factors due to dysfunction of aging oligodendrocytes leads to the appearance of Alzheimer-type neurofibrillary changes (Braak and Braak, 1996a).

Macroscopic brain imaging studies show that cortical myelin damage occurs in the preclinical phase of AD (Araque Caballero et al., 2018; Wang et al., 2019). Indeed, recent evidence shows that genetic pathways of myelin malfunction and demyelinating injuries drive β-amyloid deposition in mouse models of AD (Depp et al., 2023). Thus, myelin malfunction causes the accumulation of the β-amyloid-producing machinery within axonal swellings and increases the cleavage of cortical β-amyloid precursor protein. Intriguingly, β-amyloid disease-associated microglia that usually corral and clear β-amyloid plaques are apparently distracted to nearby myelin damage. These data indicate that age-dependent structural defects of myelin promote β-amyloid plaque formation directly and indirectly and are therefore an upstream AD risk factor (Depp et al., 2023).

An important clinical implication of these findings is the possible coexistence of AD and multiple sclerosis (MS). This possibility is sustained by the increase in the β-amyloid pathology in young transgenic mice emulating familial AD (5xFAD) when subjected to experimental autoimmune encephalomyelitis (EAE) or cuprizone intoxication, two models reproducing features of the inflammatory and noninflammatory phases of MS (Depp et al., 2023). However, proper controls for EAE/cuprizone experiments were lacking using adult-aging mice, and therefore, it is premature to conclude that experimental MS increases β-amyloid pathology in 5xAD mice and the other way around. Moreover, the clinical evidence suggesting an increased rate of AD as a comorbidity of MS is scant and based in a single study indicating that patients with MS have a higher risk of receiving an AD or dementia diagnosis (Mahmoudi et al., 2022). A review on the coexistence of MS and AD in PubMed reports between 1976 and 2014 found only 24 patients having both diseases (Luczynski et al., 2019).

The findings in genetic models of AD now provide a possible causal link in line with the predominant β-amyloid hypothesis and the role of neuroinflammation in disease progression. Thus, cortical ensheathment by oligodendrocytes is thin and may lack the distinct myelinic channel architecture important for glial metabolic support found in subcortical tracts that appear resistant to AD pathology (Depp et al., 2023). However, there is no mechanistic appraisal of myelin deterioration in late-onset (sporadic) AD which represents the vast majority of cases. In here, it is plausible that the unique metabolic demands of producing and maintaining massive myelin sheaths and synthesizing the brain's cholesterol supply, makes oligodendrocytes especially susceptible to injury. Also, oligodendrocyte vulnerability increases with age as later-differentiating cells myelinate increasing numbers of axonal segments (Bartzokis et al., 2010). An intriguing possibility is that metabolic support to axons via monocarboxylate transporters (Saab et al., 2016), or other as yet unknown mechanisms, can cause a chronic power supply shortage that may ultimately cause axonal loss and neurodegeneration. Previous stages to this fatal demise may result in a slowly progressive disruption of neural impulse transmission that degrades the temporal synchrony of widely distributed neural networks underlying normal brain function. The resulting network 'disconnections' first impact functions that are most dependent on large-scale synchronization including higher cognitive functions and formation of new memories (Bartzokis et al., 2010).

Recent imaging studies link β-amyloid burden, intracortical myelin, and cortical synchronization in cognitively normal older individuals using ^{18}F-Florbetaben-PET imaging, cortical T1-weigthed/T2-weighted (T1w/T2w) ratio maps, and resting-state functional connectivity (rs-FC) (Fernandez-Alvarez et al., 2023). Collectively, these findings provide novel insights into the relationship between initial β-amyloid plaques and intracortical myelin before the onset of cognitive decline, which may contribute to monitor the efficacy of novel disease-modifying strategies in normal elderly individuals at risk for cognitive impairment. Notably, plasma markers of β-amyloid and neurodegeneration are related to deficits in intracortical myelin content and associated with altered patterns of resting-state functional connectivity (rs-FC). Thus, these data establish a link between plasma markers of β-amyloid/neurodegeneration and intracortical myelin content in cognitively normal older adults (Fernandez-Alvarez et al., 2022).

Oligodendroglia malfunction in AD

Oligodendroglia and myelin are profoundly damaged in AD, and their alterations contribute to the overall neurodegenerative process as they disrupt neuronal communication, impair synaptic function, and compromise the overall integrity of the brain's wiring. However, the extent of damage has not been fully assessed, nor have the mechanisms involved and whether they are causative or secondary to disease onset and progression. Compelling evidence connects changes of oligodendrocyte function with the onset of neurodegeneration in AD. However, the mechanisms controlling oligodendrocyte responses to β-amyloid are still elusive. Earlier results indicated that β-amyloid is toxic to oligodendroglia; however, it was not clear which species (monomers, oligomers, fibrils) were mediating the damage (Xu et al., 2001). In contrast, we found that β-amyloid promotes oligodendrocyte differentiation, maturation, and survival in isolated oligodendrocytes and in organotypic cerebellar slices (Quintela-López et al., 2019; Fig. 2.2). Mechanistically, β-amyloid peptides specifically induce local translation of 18.5-kDa myelin basic protein (MBP) isoform in distal cell processes, which is concomitant with an increase in process complexity of MBP-expressing oligodendrocytes that is mediated by the integrin β1 receptor, Src-family kinase Fyn, and Ca^{2+}/CaMKII as effectors to modulate MBP protein expression. The pharmacologic inhibition of Fyn kinase also attenuated oligodendrocyte differentiation and survival induced by β-amyloid oligomers. Similarly, using ex vivo organotypic cerebellar slices, β-amyloid promoted MBP upregulation through Fyn kinase and modulated oligodendrocyte population dynamics by inducing cell proliferation and differentiation. Importantly, application of β-amyloid to cerebellar organotypic slices enhanced remyelination and oligodendrocyte lineage recovery in lysolecithin (LPC)-induced demyelination (Quintela-López et al., 2019). These data reveal an important role of β-amyloid in oligodendrocyte lineage function and maturation, which may be relevant to AD pathogenesis.

White matter abnormalities

White matter changes are a key factor in lipid dyshomeostasis-associated neurodegenerative diseases and dementia (Bartzokis, 2011; Nasrabady et al., 2018). White matter changes and notable demyelination have been associated with neurodegenerative diseases such as AD, VaD, PD, HD, and dementia with Lewy bodies (Bartzokis et al., 2003; Ihara et al., 2010; Shaffer et al., 2017; Kavroulakis et al., 2018; de Schipper et al., 2019). White matter hyperintensities (WMH) seen in magnetic resonance imaging (MRI) or computed tomography scans are associated with cognitive impairment (Wardlaw et al., 2015). WMHs indicate the existence of myelin lesions that are traditionally considered to originate from vascular damage (Wardlaw et al., 2015). This perception could be attributed at least to some extent to 27-hydroxycholesterol (27-OH), as it plays a role in vascular pathologies such as atherosclerosis and the brain's renin–angiotensin system (Mateos et al., 2011; Gargiulo et al., 2018). Nevertheless, it is unlikely that these mechanisms solely underlie white matter changes, as in AD they do not occur randomly. Rather, demyelination follows the inverse order of myelogenesis (Braak and Braak, 1996a; Benitez et al., 2014), denoting a contradiction with a pure vascular etiology.

NEUROGLIA IN PD

Brief overview of PD pathogenesis and clinical picture

PD is a progressive neurodegenerative disease, characterized by selective degeneration of dopaminergic neurons in the substantia nigra pars compacta and widespread intracellular accumulation of Lewy bodies made of aggregated α-synuclein (Stefanis, 2012; Poewe et al., 2017). Symptomatology includes a core set of motor abnormalities such as slowness of movement, muscle rigidity, and tremor; nonmotor features of PD include constipation, anxiety, and progressive dementia. The genetic PD landscape includes about 20 loci with variable penetrance. The most penetrant genes include those causing monogenic PD: *SNCA1* (*PARK1*, coding for α-synuclein), *VPS35* (*PARK17*), phosphatase and tensin homolog (PTEN) Induced Kinase 1 (*PINK1*) (*PARK6*), parkinsonism-associated deglycase (*DJ-1*) (*PARK7*), and *PARKIN* (*PARK2*), among others (Day and Mullin, 2021). The etiology of PD is multifactorial, likely resulting from a combined effect of genetic and environmental factors such as exposure to toxicant chemicals and head injury (Simon et al., 2020). Cellular pathophysiology suggests involvement of mitochondrial dysfunction, oxidative stress, altered proteostasis, autophagy, and neuroinflammation (Abou-Sleiman et al., 2006; Wang et al., 2015; Park et al., 2018; Simon et al., 2020).

PD animal models

Chemical PD models

As outlined above, the epidemiologic context of PD is complex and involves a strong toxicologic component (Simon et al., 2020). Therefore, chemical models of PD are widely used until now in preclinical studies, especially in testing symptomatic drugs (Hamadjida et al., 2019; Ke et al., 2021; Thirugnanam and Santhakumar, 2022). Popular neurotoxins used to induce parkinsonism include 6-hydroxydopamine (6-OHDA), paraquat, rotenone, and 1-methyl-4-phenyl-1,2,3,6-tetrahydropyridine (MPTP), all of which target mitochondria by inhibiting mitochondrial respiration and inducing ROS-mediated cell death. 6-OHDA has increased affinity to the dopamine transporter, which contributes to dopamine uptake by DA neurons, while the specificity of other toxins for DA neurons is not well understood.

The loss of DA neurons is accompanied by reactive gliosis. Astroglial reactivity, assessed by the expression of GFAP, parallels degeneration of DA neurons, suggesting that astrogliosis follows neuronal death (Kohutnicka et al., 1998; Liberatore et al., 1999; Aoki et al., 2009). Astrocytic contribution may include multiple mechanisms: (i) reactive astrocytes produce elevated levels of ROS, contributing to oxidative neuronal damage; (ii) they release proinflammatory cytokines that contribute to microglial activation; (iii) reactive astrocytes reduce homeostatic support, leaving neurons to cope with neurotoxins effects.

While representing a validated model for preclinical screening of symptomatic drugs, chemical PD models do not provide insights in the mechanistic aspects of PD-associated mitochondrial damage and realistic disease pathophysiology.

Genetic models of PD

A rich genetic PD landscape provides ample room for researchers to use different PD-associated mutations to model the disease (Table 2.2). The discovery of the *SNCA* gene causing autosome-dominant PD variant and α-synuclein as major constituent of Lewy bodies prompted the generation of various models, which can be grouped in three main categories: (i) α-synuclein overexpression through animal transgenesis or viral expression; (ii) transgenic or knock-in expression of PD-associated mutations in the *SNCA* gene; and (iii) administration of preformed α-synuclein fibrils. Overexpression of normal or mutated α-synuclein (A53T, A30P, and E46K mutations) (Fig. 2.3A) only partially reproduces neuropathology and behavioral alterations characteristic of PD without overt DA neuron degeneration and shows mild motor deficit (Masliah et al., 2000; Richfield et al., 2002; Rockenstein et al., 2002; Cannon et al., 2013). Of note, α-synuclein inclusions are present in some models but absent in others (Gispert et al., 2003; Thiruchelvam et al., 2004). However, PD-like phenotypes can be enforced by neurotoxins in animals bearing α-synuclein mutations (Thiruchelvam et al., 2004; Cannon et al., 2013). In contrast to WT α-synuclein, truncated form induced degeneration of DA neurons (Wakamatsu et al., 2008; Table 2.2).

Viral (mostly lentiviral and recombinant adeno-associated virus (rAAV)) vectors have been widely used to increase α-synuclein expression in rodent and NHP brains (Van der Perren et al., 2015b; Fajardo-Serrano et al., 2021). In these models, high α-synuclein expression levels were achieved in parallel with overt degeneration of DA neurons, reduction of striatal DA, and behavioral alterations (Yang et al., 2015; Van der Perren et al., 2015a,b; Koprich et al., 2016; Sucunza et al., 2021). However, no behavioral phenotype was reported in these models. rAAV vectors are being tested in preclinical studies and clinical trials as a gene therapy approach to lower α-synuclein levels and prevent α-synuclein pathology propagation (Fajardo-Serrano et al., 2021; Sandoval et al., 2021). Lentivirus-mediated overexpression of α-synuclein specifically in microglial cells produced progressive degeneration of DA neurons accompanied by strong microglial reactivity with phagocytic exhaustion and excessive production of oxidative and proinflammatory molecules (Bido et al., 2021).

Mutations in the leucine rich repeat kinase 2 (LRRK2, PARK8) gene, causing most of the genetic PD cases, were extensively modeled in different transgenic, knock-in, and knock-out animal models (Crabtree and Zhang, 2012; Konnova and Swanberg, 2018; Dues and Moore, 2020; Fig. 2.3B). Overexpression or knock-in of LRRK2 along was not enough to induce PD-characteristic degeneration of DA neurons, although various degrees of motor phenotype and an increased sensitivity of mice to noxious stimuli were reported (Crabtree and Zhang, 2012). Combination of LRRK2 mutant mice with other PD-related models, especially with α-synuclein transgenics, has been used. However, deletion or mutation of LRRK2 does not seem to affect α-synuclein-related pathology (Dues and Moore, 2020; Banerjee et al., 2022). Finally, combination of LRRK2 and tau (P301S) overexpressing mice did not alter tau pathology (Mikhail et al., 2015; Nguyen et al., 2018; Table 2.2).

Loss-of-function mutations in Parkin, PINK1, and DJ-1 (coded by PARK2, PARK6, and PARK7, respectively) causing autosomal-recessive PD variants were modeled using the knock-out strategy (Fleming et al., 2005; Kitada et al., 2009; Konnova and Swanberg, 2018; Chia et al., 2020; Table 2.2). Although these models poorly represented PD-related neuropathology, they can be useful to screen symptomatic drugs and to test neuroprotective strategies (Fleming et al., 2005).

iPSC-derived astrocytes in PD research

The use of human biomodels has become a compelling urgency to complement animal models in studying the participation of astrocytes in the onset and propagation of PD. In the last years, many protocols to obtain a highly pure population of iPSC-derived astrocytes (hiA) were published (Barbar et al., 2020; Ramos-Gonzalez et al., 2021). Using human hiA is an advantage to study several factors of a chronic and multifactorial neurodegenerative disease, such as sex dimorphism, or to model the chronic neurodegenerative processes. However, depending on the specific protocol used, the hiA can manifest different phenotypes. Ramos-Gonzalez et al. (2021) demonstrated for the first time that hiA generated from patients with the G2019S mutation on LRRK2 showed a morphologic atrophy, similar to the astrocytic atrophy observed in other neurodegenerative pathologies like AD. This phenotype was also accompanied by mitochondrial respiration alteration and metabolism drop. This general astrocytic asthenia depicted a novel role for astrocytes in the propagation of PD, suggesting that they may contribute to neuronal death propagation through decreased homoeostatic support, elevated oxidative stress, and failed neuroprotection. In another study, the detrimental role of $LRRK2^{G2019S}$ hiA was confirmed upon coculturing PD astrocytes with control iPSC-derived dopaminergic neurons (hiDAN) (di Domenico et al., 2019). In fact, hiDAN displayed morphologic signs of neurodegeneration and abnormal, astrocyte-derived α-synuclein accumulation. Conversely, control hiA partially prevented the appearance of disease-related phenotypes in PD hiDAN. Studies conducted on hiA generated from other neurodegenerative diseases suggest that astrocytes can regulate not only the neuronal biology but also the physiology of other glial cells like oligodendrocytes in MS (Smith et al., 2022) or microglia in AD (Guttikonda et al., 2021).

More recently, 3D biomodels have been used to study the role of astrocytes in PD. de Rus et al. (2023) demonstrated that inflammatory astrocytes contributed to the blood–brain barrier (BBB) impairment in a 3D brain-chip model of PD ($LRRK2^{G2019S}$), and inhibition of mitogen-activated protein kinase 1/2 (MEK1/2) signaling in astrocytes contributed to rescue BBB integrity. hiA generated from LRRK2 patients showed an inflammatory and angiogenic profile altered with respect to the non-PD hiA;

Table 2.2

Genetic mouse models of PD.

Model name	Promoter	Protein	Mutation	Phenotype	References
α-synuclein models					
hPDGF-α-synuclein	Human PDGFb	FL human α-synuclein	No	Intranuclear deposits and cytoplasmic inclusions from 3 months. Loss of dopaminergic terminals in the basal ganglia and motor impairments at 12 months	Masliah et al. (2000)
Huα-Syn(A30P); Huα-Syn(A53T)	Mouse PrP	FL human α-synuclein	α-Syn^{A30P}; α-Syn^{A53T}	Adult-onset neurodegenerative disease with a progressive motor dysfunction leading to death in mice expressing the A53T human α-synuclein.	Lee et al. (2002)
hwα-SYN-5	Rat TH, 9 kb fragment	FL human α-synuclein	No	No locomotor alterations or dopamine content. Increased density of the DAT, altered locomotor activity responses to amphetamine, and increased sensitivity to MPTP	Richfield et al. (2002)
hm2α-SYN-39	Rat TH	FL human α-synuclein	α-Syn^{G88C}; α-Syn^{G209A};	No inclusions. Progressive locomotor impairment and age-related alterations in dopamine and metabolite levels. Decrease of TH+ neurons from 2.5 to 19 months.	Richfield et al. (2002)
mThy1-α-synuclein	Mouse Thy-1	FL human α-synuclein	No	Mouse Thy-1 Promoter induced higher α-Synuclein expression compared to human PDGFβ promoter (Masliah 2000). α-Synuclein expression in glial cells under human PDGFβ promoter.	Rockenstein et al. (2002)
Thy1αSNA53T	Mouse Thy-1	FL human α-synuclein	α-Syn^{A53T}	Progressive decline of motor performance from 3 weeks. Lewy-like pathology in neurons. Astrogliosis and microgliosis.	van der Putten et al. (2000)
PrPmt	Mouse PrP	FL human α-synuclein	α-Syn^{A53T}	Altered localization of α-synuclein. No aggregates or inclusions. Progressive motor phenotype	Gispert et al. (2003)
Syn130m	Rat TH, 9 kb fragment	C-terminal 130 a.a. fragment	α-Syn^{A53T}	Loss of dopaminergic neurons in the substantia nigra pars compacta with decrease in striatal dopamine content. Impairment of locomotor activity, reverted by L-DOPA.	Wakamatsu et al. (2008)
BAC-aSynE46K Rat	Human a-SYN	Human a-Syn Gene	α-Syn^{E46K}	No overt damage to the nigrostriatal dopamine system at 12 months. Alterations in striatal neurotransmitter metabolism, accumulation and aggregation of α-synuclein in nigral dopamine neurons, oxidative stress.	Cannon et al. (2013)
LRRK2 models					
LRRK2 R1441C-KI	Endogenous mouse	Endogenous mouse LRRK2	Point mutation, LRRK2^{R1441C}	No dopaminergic neurodegeneration. Reductions in amphetamine-induced locomotor activity.	Tong et al. (2009)
BAC-LRRK2-R1441G	Human LRRK2	Human LRRK2 Gene	LRRK2^{R1441G}	Neuritic inclusion pathology in the cortex and striatum. Age-dependent and levodopa-responsive slowness of movement associated with diminished dopamine release and axonal pathology of nigrostriatal dopaminergic projection.	Li et al. (2009)
tetO-LRRK2 tetOA53T	TA: CamKII; TetO: CMV-LRKK2, aSYN	FL human LRRK2; FL human α-synuclein	LRRK2^{G2019S}; α-Syn^{A53T}	Striatal and cortical neuronal loss. Elevated levels of insoluble α-synuclein. LRRK2 enhances α-synuclein-mediated cytotoxicity	Lin et al. (2009, p. 2)
R1441C-LRRK2; G2019S-LRRK2	CMV-enhanced human PDGFb	FL human LRRK2	LRRK2^{G2019S} or LRRK2^{1441G}	Age-dependent dopamine neuronal loss in G2019S mice. No motor phenotype at 6 and 15 months in G2019S mice; progressive impairment of locomotor activity in R1441C mice.	Ramonet et al. (2011)
(G2019S) LRRK2	CMV-enhanced human PDGFb	FL human LRRK2	LRRK2^{G2019S}	Progressive degeneration of substantia nigra pars compacta dopaminergic neurons from 8 to 16 months. Hypokinesia from 12 to 16 months.	Chen et al. (2012)

Continued

Table 2.2

Continued

Model name	Promoter	Protein	Mutation	Phenotype	References
Combined models					
LRRK2-Tg × Huα-Syn(A53T)	LRRK2: CMV-enhanced human PDGFb; a-SYN: Mouse PrP	FL human LRRK2	α-Syn^{A53T}	No influence of LRRK2-G2019S expression on aSYN-A53T phonotype except subtle increase in α-synuclein pathology in select brainstem regions.	Daher et al. (2012)
LRRK2 R1441G BAC × tau P301S (PS19)	Human LRRK2; Mouse PrP for tau	FL human LRRK2; 1N4R isoform tau	LRRK2^{G2019S}; tau^{P301S}	No increase of tau or dopaminergic neurons' pathology; no exacerbation of behavioral or motor phenotype; no shortening of lifespan compared with tau-P301S mice. No exacerbation of cell loss, microgliosis or astrogliosis in multiple brain regions.	Mikhail et al. (2015)
G2019S-LRRK2 × P301S-Tau (PS19)	Tau: mouse PrP; LRRK2: CMV-enhanced human PDGFb	FL human LRRK2; 1N4R isoform tau	LRRK2^{G2019S}; tau^{P301S}	Endogenous or pathogenic LRRK2 expression has minimal effects on the steady-state levels, solubility and abnormal phosphorylation of human P301S-tau throughout the mouse brain	Nguyen et al. (2018)
PARK genes mutations and knock-out models					
hUCHL1-I93M	Human PDGFb	Human UCH-L1 (PARK5)	UCHL1^{I93M}	Reduction in the dopaminergic neurons in the substantia nigra and dopamine content in the striatum at 20 months. Increased sensitivity to MPTP.	Setsuie et al. (2007)
parkin −/−		Endogenous mouse parkin (PARK2)	Knock-out	Reduced synaptic excitability. Locomotor deficit from 2 months. No detectable alterations in nigrostriatal system. Mitochondrial dysfunction and oxidative damage (Palacino et al., 2004).	Goldberg et al. (2003), Palacino et al. (2004)
DJ-1 −/−		Endogenous mouse DJ-1 (PARK7)	Knock-out	No neuropathologic alterations. Reduced evoked dopamine overflow and increased dopamine reuptake. Normal corticostriatal LTP, absent LTD. Locomotor deficit from 3 months.	Goldberg et al. (2005)
DJ-1 Null		Endogenous mouse DJ-1 (PARK7)	Knock-out	Age-dependent and task-dependent motor deficit from 5 months. No dopamine neuron loss from 6 to 11 months. Increased dopamine reuptake rates and elevated tissue dopamine content	Chen et al. (2005)
DJ-1 −/−		Endogenous mouse DJ-1 (PARK7)	Knock-out	Lesser body weight from 12 months. Progressive motor deficit from 6 months. No detectable alterations in nigrostriatal system	Chandran et al. (2008)
PINK1 −/−		Endogenous mouse PINK1 (PARK6)	Knock-out	No neuropathologic alterations. Locomotor deficit from 16 months. Reduced DA from 9 months. Deficit of mitochondrial protein import. Impaired mitochondrial bioenergetics.	Gispert et al. (2009)
Parkin/PINK1/DJ-1-TKO		Endogenous mouse Parkin (PARK2), PINK1 (PARK6), DJ-1 (PARK7)	Knock-out	No neuropathologic alterations up to 24 months. Increased striatal DA at 24 months	Kitada et al. (2009)
PINK1-KO rats		Endogenous mouse PINK1 (PARK6)	Knock-out	Early and progressive vocalization and oromotor deficits, reduced TH immunoreactivity in the locus coeruleus that correlates with vocal loudness and tongue force, and αSyn neuropathology in brain regions important for cranial sensorimotor control.	Grant et al. (2015)

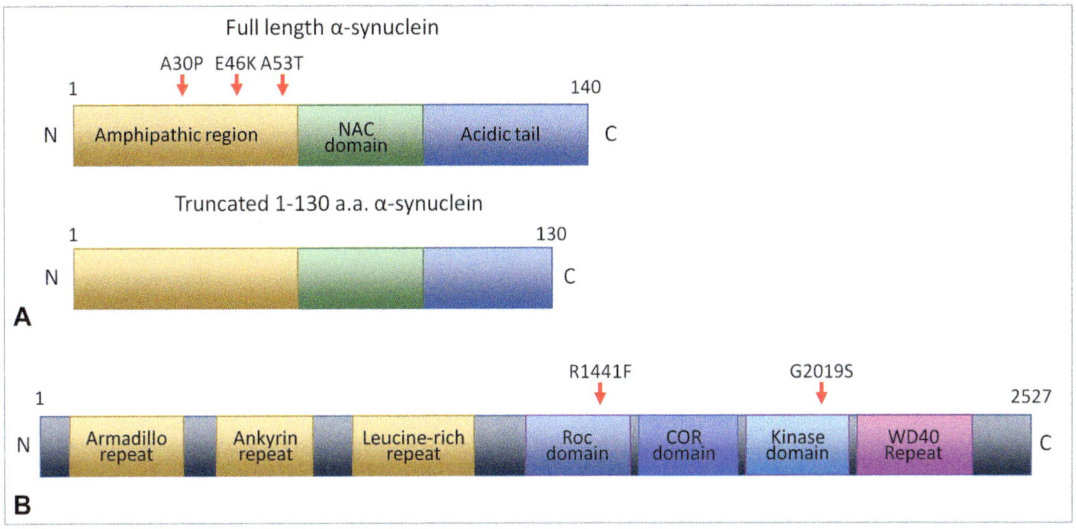

Fig. 2.3. Protein variants and mutations of α-synuclein and LRRK2 in popular mouse PD models. (A) Domain organization and mutations of human α-synuclein. (B) Domain organization and mutations of human LRRK2 protein.

moreover, PD hiA were unable to form the in vitro BBB on a chip. Vascular changes were also observed and confirmed in the human postmortem substantia nigra. If confirmed, the participation of reactive astrocytes in the BBB integrity can be considered as a valuable target to counteract the onset of neurodegeneration in PD.

Astrocytes in PD

As highly versatile cells, astrocytes retain their adaptive plasticity throughout the entire life of an individual, customizing their morphology and functional properties to changes in the surrounding environment. This feature explains the immense heterogeneity of astrocytes, displaying distinctive phenotypes according to regional specificities. Human astrocytes are particularly voluminous (four times higher than rodent astrocytes) and highly ramified, covering large territories that enable them to contact and modulate up to 2 million synapses. In certain PD-relevant regions such as the striatum, the density of excitatory synapses implies that a single astrocyte territory encompasses around 50,000 synaptic boutons (Khakh, 2019). In the substantia nigra pars compacta, astrocytic density is lower compared to other regions such as the substantia nigra pars reticulata or ventral tegmental area, and vulnerable dopaminergic neurons find less support to cope with oxidative stress and energetic deficit.

Astrocytes display numerous receptors that perceive the extracellular microenvironment, enabling them to tune homeostatic responses and modulate the composition of the extracellular milieu. This is achieved by several means, including neurotransmitter uptake alongside the perisynaptic sheath, regulation of water and ion content, and maintenance of the brain extracellular matrix. Astrocytic reactivity, along with microglial activation and infiltration of lymphocytes, are well-documented phenomena in post-mortem PD brains. Upon injury and disease, astrocytes rapidly become reactive and proliferate, changing their morphology and transcriptional profile.

Astroglial atrophy, asthenia, and loss of homeostatic and protective function have been described in aged brains and contribute to several neurodegenerative and psychiatric diseases. Likely, in PD, astrocytes become metabolically dysfunctional and lose their protective capabilities (Booth et al., 2017; Fig. 2.4). Metabolic dysfunction in astrocytes is mainly evidenced by the loss of function of cytoplasmic organelles. A recent study demonstrated that the treatment of $LRRK2^{G2019S}$ transgenic mice with α-synuclein increases the expression of ER stress proteins in astrocytes, thus affecting neurite length and neuronal viability, supporting the idea that ER stress in PD astrocytes can aggravate neuronal damage (Lee et al., 2019). Mitochondrial failure of nigro-striatal neurons is a key fact in PD. The large nigro-striatal axons are highly demanding energy, and they are normally supported by the homeostatic and metabolic support of astrocytes. The damaged mitochondria are accumulated in the synapses in the striatum, sequestered by autophagosomes, and retrogradely transported to the substantia nigra for their degradation. In PD, all these mechanisms are altered with the accumulation of altered mitochondria at the striatal neuronal terminals. In normal conditions, astrocytes may work as acceptors of damaged mitochondria for their further degradation. However, the loss of astrocytic homeostasis, may hinder the mitochondrial transfer to astrocytes and their degradation.

Microglia in PD

Microglial involvement into pathophysiology of PD is most likely secondary, in which microglial reactivity is triggered by either neuronal (or other cells) damage and death or by α-synuclein aggregates (Fig. 2.3). The latter are known to activate Toll-like receptors (TLR) TLR2 and TLR4, which trigger microglial reactivity and a wide range of changes, including metabolic, transcriptional, and physiologic functions (Harrison et al., 1998; Fellner et al., 2013).

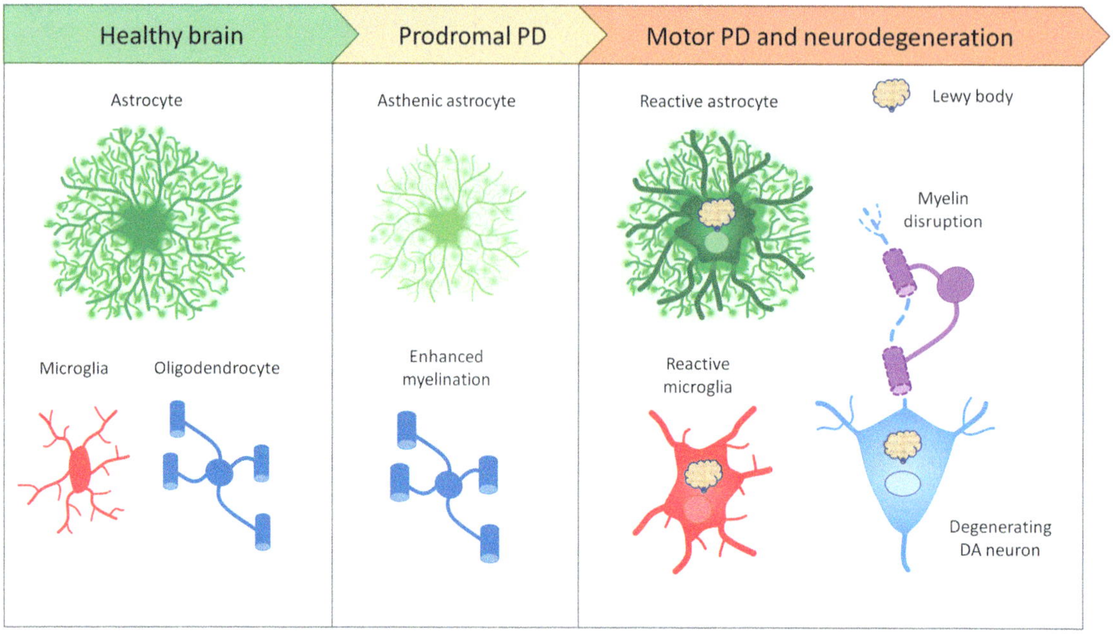

Fig. 2.4. Changes of glial cells in PD.

PET of PD patients using a peripheral benzodiazepine receptor ligand [^{11}C]-PK11195 revealed changes that were interpreted as indicative of microgliosis (Bartels et al., 2010), although the probe is far from being a reliable indicator of neither neuroinflammation nor microglial reactivity (Šimončičová et al., 2022). Microglial loss of function manifested in limited phagocytosis as well as microglial secretion of proinflammatory factors and release of reactive oxygen species were documented in postmortem tissues of PD patients (Imamura et al., 2003; Croisier et al., 2005).

Microglial cells in relevant PD brain areas were found to express major histocompatibility complex II (MHC II) (Imamura et al., 2003; Croisier et al., 2005). In addition, microglia in PD postmortem samples showed increased expression of genes linked to oxidative stress, cell proliferation, and ER stress (Smajić et al., 2022; Chatila et al., 2023). Microgliosis is regularly found in the animal models of PD, although the role of these reactive microglia seems to provide neuroprotection because inhibition of reactive microgliosis with genetic deletion of fractalkine receptor (CX3CR1) exacerbated neuronal damage and death in PD animal pathology (Cardona et al., 2006). Several microglia-specific genes, including TREM2 and its R47.H missense mutation, the human leukocyte antigen gene (HLA-DRA), and leucine-rich repeat kinase 2 (LRRK2), are regarded as PD-risk genes (Hamza et al., 2010; Rayaprolu et al., 2013). Patients with sporadic PD also have a high appearance of the p.R62H missense mutation of TREM2 (Nudelman et al., 2022).

Oligodendrocytes and myelin in PD

PD is a movement disorder characterized by the early loss of nigrostriatal dopaminergic pathways, producing significant network changes impacting motor coordination and long-distant connectivity, which involves myelinated axons. PD can be categorized into three motor stages: a silent period when nigrostriatal loss begins, a prodromal motor period with subtle focal manifestations, and clinical PD (Blesa et al., 2022). Interestingly, longitudinal diffusion tensor imaging before and after the development of prodromal motor PD showed higher fractional anisotropy in motor cortex and corticospinal tract compared to controls, indicating adaptive structural changes in motor networks in concert with nigrostriatal dopamine loss (Fu et al., 2022). Moreover, histologic analyses of the white matter underlying the motor cortex showed progressive disorientation of axons with segmental replacement of neurofilaments with α-synuclein, enlargement of myelinating oligodendrocytes, and increased density of their precursors, along with early cortical increases in myelin proteins in association with α-synuclein accumulation (Fu et al., 2022). These findings indicate that there are early profound alterations in the primary motor cortex myelin and oligodendroglia lineage and its output pathways that begin in the prodromal motor stage of PD and progress thereafter with disease aggravation.

These observations are at odds with the classical view of PD as a continued degeneration of the nigrostriatal system through the silent, prodromal, and motor stages, leaving the motor cortex and tracts largely spared (Blesa et al., 2022). In contrast, adaptive structural changes occur in neurons and oligodendrocytes of the motor cortex as well as in their axons and supporting oligodendroglia in the underlying white matter. Furthermore, such changes begin just prior to nigrostriatal degenerative changes and involve axon collateralization as well as increased myelination (Fu et al., 2022). Intriguingly, there is a different response of oligodendrocytes in the motor cortex vs white matter motor tracts, which may reflect the diverse caliber of myelinated axons: small, finely branched intracortical axons of inhibitory interneurons vs very large pyramidal projection neurons. This is consistent with the idea that oligodendrocytes produce an intrinsic length of myelin sheath (Bechler et al., 2015), with intracortical oligodendroglia increasing their myelin to wrap more small inhibitory axon collaterals in their vicinity, while their white matter counterparts rising their number and size to ensheath more large pyramidal axon collaterals in the white matter motor tracts.

These findings in humans add to experimental observations in transgenic mice over-expressing α-synuclein showing higher levels of myelin phospholipids preceding myelin loss (Grigoletto et al., 2017; Fig. 2.4).

NEUROGLIA IN HUNTINGTON DISEASE

Brief overview of HD pathogenesis and clinical picture

HD is a monogenic, dominantly inherited fatal neurodegenerative disorder with a prevalence of 10.6–13.7 individuals per 100,000 in western populations (Bates et al., 2015). HD is caused by the expansion of a CAG repeat coding for glutamine (polyglutamine tract or polyQ) in the first exon of the huntingtin gene HTT on chromosome 4. The polyglutamine tract length up to 35 amino acids is physiologic and does not cause disease; from 35 to 40 are at risk to develop HD, while the CAG repeat expansion over 40 residues results in a disease. Usually only one allele of HTT genes is affected. Symptoms typically include motor dysfunction (chorea), cognitive impairment (deficit of attention and emotion recognition, problems with memory), and psychiatric features (mood swings, depression, and irritability) (Bates et al., 2015). Neuropathology is characterized by preferential neurodegeneration of gamma aminobutyric acid (GABA)-ergic medium spiny neurons (MSNs) in the striatum. However, results from neuroimaging studies suggest that neurodegeneration occurs also in extrastriatal regions, such as the cortex, nucleus accumbens, globus pallidus, thalamus, and hypothalamus (Gray, 2019).

Neuropathologic classification considers five grades scale from grade 0 to grade 4. Grade 0 corresponds to no discernible neuropathologic abnormalities, with progressive neuronal loss from grade 1 (50% neuronal loss) to grade 4 (95% neuronal loss) (Vonsattel et al., 1985). Analysis of fluorescence-activated cell sorting (FACS)-separated CNS cells from mouse HD model brains showed that HTT was equally expressed by neurons and non-neuronal cells at the mRNA and protein levels (Orre et al., 2014; Jansen et al., 2017). However, in mice, nuclear HTT inclusions were mainly found in neurons (~50% neurons at late AD stage), while much fewer GFAP-positive astrocytes and oligodendrocytes were labeled (Jansen et al., 2017). In human brains of juvenile HD, ~4.5% of neurons had HTT nuclear inclusions, while positive glial cells, including astrocytes, oligodendrocytes, and microglial cells, were roughly 10 times less (Jansen et al., 2017).

HD animal models

Neurotoxin-mediated striatal lesions

The observation that intraparenchymal application of glutamate receptor agonists, such as glutamic and kainic acids, induces selective degeneration of the GABA-ergic projection neurons, which are the most affected neuronal type in HD, stimulated intensive in vitro and in vivo research in the HD field, suggesting excitotoxicity as a mechanism of neurodegeneration in HD (Alberch et al., 2002; Gil and Rego, 2008; André et al., 2010).

Accidental ingestion of some mitochondrial toxins, such as 3-nitropropionic acid (3-NPA), produces bilateral striatal lesions with preferential degeneration of MSNs and an HD-like phenotype (Ludolph et al., 1991). 3-NPA, along with another mitochondrial toxin malonate, has been widely used to replicate HD-related mitochondrial dysfunction in neurons both in vivo and in vitro. 3-NPA and malonate inhibit the respiratory chain, leading to energy depletion, implicating high sensitivity to bioenergetic deficit and metabolic alterations as the principal pathogenetic mechanisms of striatal MSNs' vulnerability in HD (Borlongan et al., 1997; Brouillet et al., 2005; Upadhayay et al., 2023). Further findings that intrastriatal delivery of mitochondrial toxins, such as amino-oxyacetate, 1-methyl-4-phenylpyridinium (MPP +), malonate, and rotenone, produce excitotoxic lesions resembling those of glutamate receptor agonists, suggested that mitochondrial dysfunction may cause vulnerability to excitotoxicity of MSNs (Brouillet et al., 2005). Mitochondrial contribution to HD has been extensively studied and was proposed for the development of anti-HD therapy (Lim et al., 2008; Costa et al., 2010; Damiano et al., 2010; Sawant et al., 2021).

Mouse HD models

Overexpressing a genomic fragment containing the promoter and exon 1 of the human HTT gene was used to create the first genetic HD mouse model, the R6/2 line bearing one transgene copy with 120 CAG repeats, which remain popular nowadays (Mangiarini et al., 1996). In the R6/2 line, the onset of motor symptoms occurs starting at 9 weeks, followed by death between 10 and 13 weeks. Motor phenotype includes choreiform-like movements, involuntary stereotypic movements, tremor, and epileptic seizures. Neuropathologic assessment revealed a lesser brain size and weight and the presence of intranuclear inclusions in R6/2 mice compared with nontransgenic control. However, no overt neuronal death or astrogliosis or microgliosis were found (Mangiarini et al., 1996; Yu et al., 2003). The R6/2 line has been extensively used to study excitation–inhibition imbalance and neuronal connectivity in striatal and extra-striatal neurons and cortico-striatal communication (Kaye et al., 2021).

A mouse bearing 171 amino acid N-terminal human HTT fragment with 82 CAG repeats (N171-82Q line) exhibits lower brain weight and premature death at about 15 weeks. This line showed progressive motor abnormalities and defects of neuronal connectivity. Although no overt neuronal death was observed in N171-82Q mice, apoptotic neurons positive to activated caspase-3 were evident (Yu et al., 2003). Astrocytes were positive to GFAP (Yu et al., 2003; Table 2.3; Fig. 2.5).

Models expressing the full-length human HTT gene were generated using yeast artificial chromosome (YAC) and bacterial artificial chromosome (BAC) plasmids. YAC72 and YAC128 mice show progressive motor and cognitive deficit, intranuclear

Table 2.3

Genetic mouse models of HD.

Name	Promoter	Protein	Phenotype	Reference
Overexpression of N-terminal fragment				
R6/2	Human HTT promoter 1 kb fragment	Human HTT 67 a.a.	Progressive neurologic phenotype including choreiform-like movements, involuntary stereotypic movements, tremor, and epileptic seizures. Lower brain size. No neuronal loss or gliosis, no neuroinflammation. Premature death at 10–13 weeks.	Mangiarini et al. (1996)
HD94	TetO: CMV; TA: CaMKII	Human HTT:mouse Htt exon 1, 67 a.a.	Neuronal inclusions and progressive motor dysfunction from 4 weeks. Increased GFAP staining.	Yamamoto et al. (2000)
N171-82Q	Mouse PrP	Human HTT 171 a.a.	Diffuse HTT nuclear labeling. Behavioral abnormalities, including loss of coordination, tremors, hypokinesis and abnormal gait, before dying prematurely at about 12–16 weeks	Schilling et al. (1999)
N118-82Q	Mouse PrP	Human HTT 118 a.a.	Inclusion pathology accompanied failure to gain weight, poor grooming, and hypoactivity. Premature death at about 20 weeks	Tebbenkamp et al. (2011)
N586-82Q	Mouse PrP	Human HTT 586 a.a. (Caspase cleavage product)	Failure to maintain weight, small brain weight and reductions in specific mRNAs in the striatum. Progressive movement disorder that includes dramatic deficits in motor performance on the rotarod and ataxia.	Tebbenkamp et al. (2011)
HTT84Q-GFP	Human Ubiquitin	Human HTT 67 a.a. GFP fused at C-term	Motor deficit from 4 months. Premature death at 6–12 weeks	Cheng et al. (2013)
HD150QG, HD190QG	Human HTT promoter 1 kb fragment	Human HTT 150 or 190 a. a. GFP fused at C-term	HTT aggregates in hypothalamus. Decrease of hypothalamic transcripts	Kotliarova et al. (2005)
Overexpression of full length human HTT				
HD48, HD89	CMV	FL human HTT	Progressive behavioral and motor dysfunction from 8 weeks, with neuron loss and gliosis in striatum, cerebral cortex, thalamus and hippocampus from 24 weeks.	Reddy et al. (1998)
YAC48, YAC72	Human HTT promoter and regulatory elements	FL human HTT	Electrophysiologic abnormalities from 6 months, Selective degeneration of medium spiny neurons in the lateral striatum at 12 months. Translocation of HTT fragments to the nucleus.	Hodgson et al. (1999)
YAC128	Human HTT promoter and regulatory elements	FL human HTT	Motor abnormalities from 6 months and age-dependent brain atrophy including cortical and striatal atrophy associated with striatal neuronal loss from 9 months.	Slow et al. (2003)
BACHD	Human HTT promoter and regulatory elements	FL human HTT	Progressive motor deficits from 2 months, neuronal synaptic dysfunction, and late-onset selective neuropathology at 12 months, which includes significant cortical and striatal atrophy and striatal dark neuron degeneration.	Gray et al. (2008)
Knock-in models				
HdhQ50	Endogenous mouse Htt	FL chimeric human HTT exon 1: mouse Htt	Heterozygous and homozygous mice do not show HD-related abnormalities	White et al. (1997)
HdhQ71, HdhQ94	Endogenous mouse Htt	FL chimeric human HTT exon 1: mouse Htt	No behavioral abnormalities or nuclear inclusions. HdhQ94 mice show increased sensitivity to AMPA and kainate	Levine et al. (1999)
CAG140	Endogenous mouse Htt	FL chimeric human HTT exon 1: mouse Htt	Motor abnormalities from 1 month. Nuclear HTT inclusions at 4 months	Menalled et al. (2003)
Hdh6/Q72, Hdh4/Q80	Endogenous mouse Htt	FL chimeric human HTT exon 1: mouse Htt	Abnormal social behavior in the absence of acute neurodegeneration	Shelbourne et al. (1999)
HdhQ92, HdhQ111	Endogenous mouse Htt	FL chimeric human HTT exon 1: mouse Htt	HTT nuclear inclusions and aggregates predominantly in MSNs. Late onset neurodegeneration in HdhQ111	Wheeler et al. (2000)
HdhQ150	Endogenous mouse Htt	FL chimeric human HTT exon 1: mouse Htt	Age-dependent, late-onset behavioral phenotype with motor abnormalities from 70 weeks of age. Loss of striatal dopamine D1 and D2 receptor binding sites at 70 and 100 weeks. Loss of striatal neurons 100 weeks.	Lin et al. (2001)
HdhQ200	Endogenous mouse Htt	FL chimeric human HTT exon 1: mouse Htt	Accelerated phenotype compared with HdhQ150 mice	Heng et al. (2010)
zQ175	Endogenous mouse Htt	FL chimeric human HTT exon 1: mouse Htt	Motor and grip strength abnormalities with an early onset (8 and 4 weeks of age, respectively), which were followed by deficits in rotarod and climbing activity at 30 weeks of age and by cognitive deficits at around 1 year of age	Menalled et al. (2012)

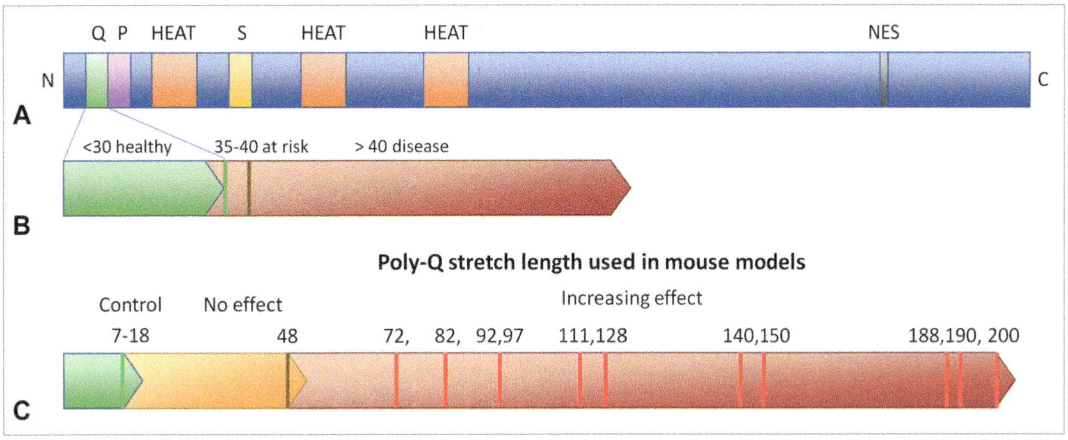

Fig. 2.5. Poly-glutamine stretch lengths used in popular mouse HD models. (A) Domain organization of human HTT protein. Q, poly-glutamine stretch; P, poly-proline domain; S, poly-serine domain; HEAT, HEAT repeat, acronym of (Huntingtin, elongation factor 3 (EF3), protein phosphatase 2A (PP2A), and the yeast kinase TOR1); NES, nuclear export sequence. (B) Poly-Q length related to HD disease: <30, healthy subjects; 35–40 repeats, at risk to develop HD; <40 repeats will develop HD. (C) lengths of poly-Q stretch modeled in mouse HD models.

inclusions, and neuronal loss in the striatum (Hodgson et al., 1999; Slow et al., 2003). Reactive gliosis was found in the cortex but not in the striatum starting from 15 months of age in YAC128 mice (Bayram-Weston et al., 2012). A BAC-based transgenic model expressing full-length HTT protein with 97 CAG and CAA repeats under endogenous regulatory machinery showed motor deficit stating from 3 months with late onset neuropathology at about 12 months (Gray et al., 2008; Table 2.3).

A number of knock-in HD mouse models has been generated (Pouladi et al., 2013; Kosior and Leavitt, 2018; Kaye et al., 2021). Mice bearing less than 80 CAG repeats, such as HdhQ50, HdhQ71, Hdh6/Q72, and Hdh4/Q80 (White et al., 1997; Levine et al., 1999; Lin et al., 2001; Fig. 2.5), showed no detectable phenotypic abnormalities or neuropathology (Kosior and Leavitt, 2018). Among mice expressing knocked-in polyQ tract longer than 80 CAG repeats, longer polyQ tract typically produces more severe phenotype. These models include HdhQ150, HdhQ111, CAG140, zQ175, and HdhQ200. CAG140 mice develop motor abnormalities by about 12 months. Nuclear inclusions appear at about 4–6 months without neuronal loss (Menalled et al., 2003). The zQ175 line showed motor dysfunction as early as 2 months, while cognitive deficit appeared at 12 months. Recently, zQ175 mice were crossed with FVB/N mice with consequent excision of Neo cassette (designated as Q175FDN) (Southwell et al., 2016). These manipulations resulted in a twofold increase in HTT expression with a significant acceleration of motor, cognitive, and neuropathologic phenotypes (Southwell et al., 2016; Table 2.3).

Astrocyte-specific expression of HTT in mice has also been pursued, resulting in HD-like phenotype and corroborating astrocytic contribution to HD pathogenesis (Bradford et al., 2009; Ben Haim et al., 2015b; Gray, 2019).

Astrocytes in HD

Morphologic changes of astrocytes in HD

Although reactive gliosis is a prominent feature of HD neuropathology in postmortem human brain samples, reports suggest that in HD patients, neuronal dysfunction, clinical symptoms, and neurodegeneration occur prior to the development of reactive gliosis. Astrogliosis, absent at grade 0, gradually increased from grade 1 to 4 (Vonsattel et al., 2008; Rüb et al., 2015). In animal HD models, assessment of astrocytic reactivity by GFAP staining found no significant differences between WT and transgenic mice (Mangiarini et al., 1996; Tong et al., 2014; Ben Haim et al., 2015b; Khakh and Goldman, 2023). However, in response to intrastriatal injection of lentiviral particles expressing Htt82Q, strong upregulation of GFAP and vimentin was observed (Faideau et al., 2010; Ben Haim et al., 2015b). Nonetheless, assessment of astrocytic morphology using either the expression of a cell surface fluorescent reporter or by filling astrocytes with a fluorescent dye lucifer yellow revealed a significant reduction of astrocyte volume, suggesting a withdrawal of perisynaptic astrocytic processes from excitatory synapses (Octeau et al., 2018; Benraiss et al., 2021; Fig. 2.6).

Alterations of neurotransmitter homeostasis and K^+ buffering

The excitotoxic hypothesis of MSN death in HD strongly implicates astrocytic insufficiency to maintain ionic and neurotransmitter homeostasis. HD grade-dependent downregulation of high affinity glutamate transporter GLT-1 was frequently reported alongside the reduction of glutamate transport activity (Gray, 2019; Khakh and Goldman, 2023). Astrocyte-specific expression of mutant HTT in mice led to reduced GLT1 expression, astrogliosis, and features of HD-like pathology (Bradford et al., 2010; Faideau et al., 2010).

Cultured astrocytes from BACHD mice (in which human HD expressed from BAC) produce more glutamate through de novo synthesis with consequent enhanced release of glutamate in medium compared with astrocytes from WT mice (Lee et al., 2013). Expression of mHTT in astrocytes induces death of cocultured neurons (Shin et al., 2005). These data are in line with reduced astrocytic volume and synaptic found in R6/2 mice (Tong et al., 2014) and strongly support the causative role of astrocytes in excitotoxic death of MSNs.

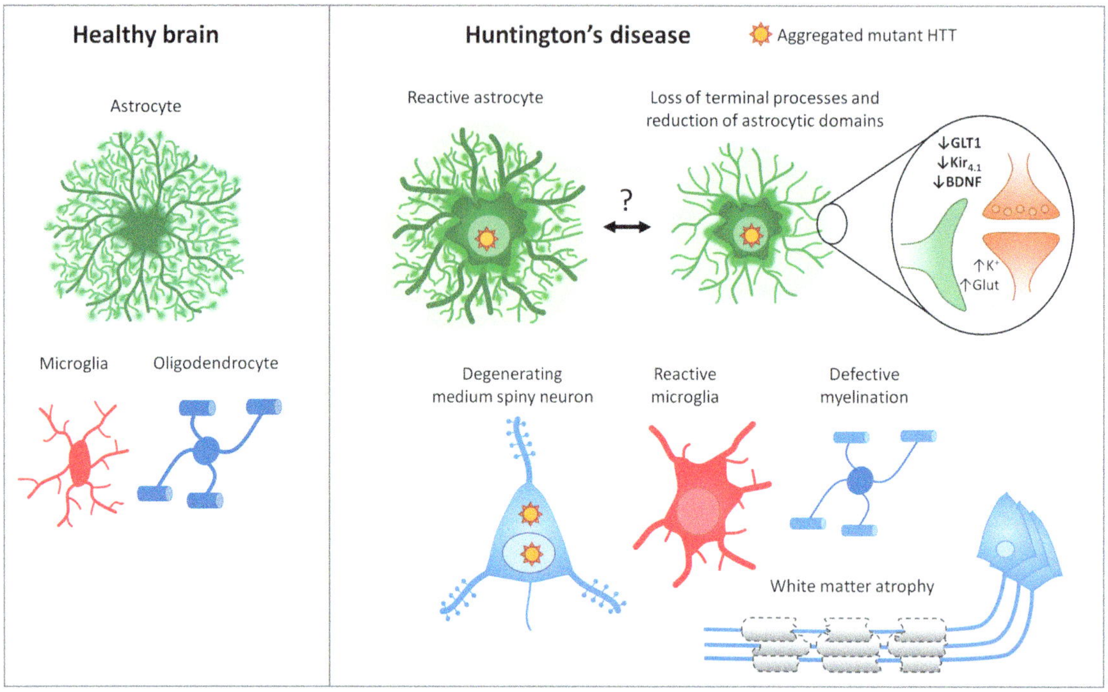

Fig. 2.6. Changes of glial cells in HD.

Another proposed mechanism through which astrocytes may be involved in HD-associated neuronal hyperexcitability implicates the reduction of $K_{ir}4.1$ inwardly-rectifying K^+ channels, found in 2- to 3-month-old R6/2 mice (Tong et al., 2014). Reduction of astrocytic $K_{ir}4.1$ expression was associated with doubling of extracellular K^+ concentration from ~1.5 to ~3 mM, depolarization of neuronal membrane, and hyperexcitability. AAV-mediated reintroduction of $K_{ir}4.1$ in astrocytes rescued motor dysfunction in 3-month-old R6/2 mice (Tong et al., 2014). However, to the best of our knowledge, no change of $K_{ir}4.1$ expression in HD human brains has been reported.

Reduced BDNF release by HD astrocytes

Brain-derived neurotrophic factor (BDNF) is reduced in the brains of HD patients and HD model mice (Zuccato et al., 2001; Zuccato and Cattaneo, 2007; Fig. 2.6). Although it is postulated that BDNF is produced in neurons, a growing body of evidence suggests that astrocytes can contribute to brain BDNF signaling through uptake recycling and release of BDNF (Bergami et al., 2008; Albini et al., 2023). Mutant HTT was shown to disrupt secretory function of astrocytes, reducing BDNF release in several HD model mice (Wang et al., 2012; Hong et al., 2016). These findings have been exploited by several groups to restore astrocytic BDNF and rescue BDNF deficits (Giralt et al., 2010; Arregui et al., 2011). Glatiramer acetate, a drug known to increase BDNF expression, increased production of BDNF in astrocytes in culture and in the brains of R6/2 and YAC128 mice, exerting a beneficial effect on the lifespan and motor performance of HD model mice (Reick et al., 2016). In addition, BDNF was shown to produce an autocrine effect on astrocytes, protecting them from 3-NPA-induced cell death, which, in turn, protected neurons (Saba et al., 2020).

Microglia in HD

Microglial reactivity in HD is rather widespread and is of the secondary nature being initiated by neuronal pathology (Yang et al., 2017). Reactive microglia have been observed in postmortem brains of HD patients, with some correlation between accumulation of reactive microglial phenotypes and the degree of neuronal loss (Singhrao et al., 1999; Sapp et al., 2001). Microglial reactivity was also deduced from PET imaging using [^{11}C]-PK11195 (Pavese et al., 2006), the accuracy of which remains questionable. Microglial cells in both human postmortem tissue and in animal HD models demonstrated reactive phenotypes with enlarged somata and fewer, thicker, and less motile processes (Franciosi et al., 2012; Savage et al., 2020). In vitro, in organotypic slices or in primary cocultures, neuronal expression of mHTT initiated microglia reactivity with some accumulation of proliferative microglia around neurites (Kraft et al., 2012).

Striatal microglia in the context of HD were also reported to accumulate ferritin while the disease progresses (Dexter et al., 1992). In an R6/2 polyglutamine repeat transgenic model of HD, microglial cells contained HTT and showed dystrophic appearance (Simmons et al., 2007).

Oligodendrocytes and myelin in HD

Clinical studies of HD patients and autopsy findings indicate that there are macroscopic changes in white matter, including atrophy preceding motor symptoms (Sun et al., 2022). MRI studies show that white matter volume decline in HD patients is age-dependent, and in the presymptomatic patient group, white matter volume decrease was not as significant as that in the late stage HD (Rosas et al., 2006; Fig. 2.6). In turn, asymptomatic mutation carriers have a significantly smaller ratio of white matter to intraventricular volume than healthy controls (Ciarmiello et al., 2006). Later on, voxel-based morphometry analysis determined that white matter in the frontal regions is damaged preferentially during the early stages of the disease, and that white matter loss progresses from the peristriatal, corpus callosum, and posterior white matter tracts to widespread loss throughout the brain (Tabrizi et al., 2009). However, these studies could not determine whether myelin changes are secondary to neuronal degeneration, despite the proposed hypothesis that HD pathogenesis begins with a deleterious effect of the mHTT on myelination (Bartzokis et al., 2007).

Mechanisms mediating myelin defects in HD include the effects of mHTT on the normal function of myelin regulatory factor (MYRF) and the downregulation of peroxisome-proliferator-activated receptor gamma coactivator 1 (PGC1; Sun et al., 2022). Thus, MYRF contributes to oligodendrocyte progenitor differentiation by regulating myelin protein expression, a signaling cascade that is abrogated by mHTT binding. In turn, PGC1 is a transcriptional coactivator that regulates myelin basic protein levels and rate-limiting enzyme mRNA for cholesterol synthesis, which upon mHTT binding contributes to demyelination in HD (for a review, see Sun et al., 2022). Consistent with those findings, HD animal models harboring this genetic mutation with different polyQ repeat lengths show variable myelin distortion phenotypes resulting in reduced myelination status defects (Sun et al., 2022).

ASTROTAUPATHIES

Astrotauopathies are a distinct class of neurodegenerative diseases characterized by tau accumulation exclusively in astrocytes (Kovacs et al., 2016; Kovacs, 2020), which are classified into: (i) astrocytic plaques, (ii) tufted astrocytes, (iii) ramified astrocytes, (iv) globular astroglial inclusions, (v) thorn-shaped astrocytes, and (vi) granular/fuzzy astrocytes. From the clinical perspective, astrotauopathies are primary elements in the pathophysiology of progressive supranuclear palsy, corticobasal degeneration, Picks disease, argyrophilic grain disease, globular glial tauopathies, and aging-related tau astrogliopathy (ARTAG). Tau pathology is also prominent in oligodendrocytes in progressive supranuclear palsy, corticobasal degeneration, and Pick disease (Ferrer, 2018).

Progressive supranuclear palsy is a rare neurodegenerative with progressive impairment of balance, walking, and eye movement, muscular rigidity, dysarthria, and dysphagia. Tufted astrocytes, characterized by phosphorylated tau immunoreactivity in the proximal segments of astrocytic processes and often localized near the blood vessels, are the histopathologic hallmark of this disease (Hauw et al., 1990; Yamada et al., 1992; Shibuya et al., 2011).

Corticobasal degeneration (manifested by impairment of movement, speech, memory, and swallowing) is associated with the early emergence of astrocytic plaques, which appear in the form of fuzzy short argyrophilic processes arranged annularly with fine collaterals at vertical or sharp angles (Hattori et al., 2003; Yoshida, 2014).

Ramified astrocytes, in which tau is accumulated in somata and processes, are a hallmark of the frontotemporal lobar degeneration known as Pick disease; pathophysiology of which is driven by early and profound apoptotic death and dystrophy of astrocytes, as well as some astrocytic reactivity, which correlates to the severity of dementia (Broe et al., 2004; Kersaitis et al., 2004; Kovacs et al., 2013).

Finally, ARTAG is a spectrum of age-dependent neurodegenerative diseases associated with the appearance of specific thorn-shaped and granular-fuzzy astrocytes in subpial and subependymal regions, in the depths of gyri, and the basal forebrain and brainstem, as well as in the amygdala (Kovacs et al., 2016).

CONCLUSIONS

The main breakthrough in studying chronic neurodegenerative diseases, arguably, has been achieved in two planes: first, the understanding that the pathogenetic process begins much earlier than clinical symptoms appear, while second regarded the role of nonneuronal cells, specifically considered in this chapter, astrocytes, microglial cells, and oligodendrocytes. In astrocytes, cell-autonomous alterations occur at very early stages of pathogenesis, and this may trigger and define progression of the disease. During pathogenesis, astrocytes undergo complex yet disease- and brain region-specific changes. However, a general pattern of alterations can be described in the following dynamics: in preclinical and prodromal stages, astrocytes become asthenic and/or atrophic with reduced homeostatic and protective capabilities; later, mainly upon deposition of aggregated protein, reactive astrocytes develop. Astrocytic reactivity is a protective reaction represented by a spectrum of morphologic and functional states. Exhaustion of protective capability marks progression to "glial paralysis" and atrophy, which accelerates neurodegeneration and progression to terminal stage. Pathologic remodeling of microglial cells, oligodendrocyte progenitor cells, and oligodendrocytes is complex and disease stage-specific. Nevertheless, microglial reactivity, myelin breakdown, and white matter abnormalities are common features of chronic neurodegenerative diseases.

Modeling human chronic neurodegenerative diseases is challenging, none model fully recapitulates human pathology and is suitable for testing disease-modifying drugs, while some models, including both chemical and genetic, may be used to test symptomatic drugs. Furthermore, modeling of glial pathology is strongly overlooked, although sporadic reports on glial cell-specific expression of mutant/pathogenic proteins suggest their key role in disease pathogenesis. Novel models, specifically addressing the pathology of glial cells, in-depth investigation of glial alterations at different disease stages, and correlation of these alterations with novel and available biomarkers are necessary for the development of therapies aiming at prevention and/or reversion of AD, PD, and HD.

CONFLICT OF INTEREST

The authors have no competing interests to declare.

FUNDING

This work had the following financial supports: grants FAR-2016 and FAR-2019 from The Università del Piemonte Orientale, and grant PNRR-MCNT2-2023-12377363 (NextGenerationEU, Italian Ministry of Health) to D.L.; grants PID2022-143020OB-I00 (Spanish MICINN), IT-1551-22 (Basque Govern), CB06/05/0076 (CIBERNED), and BIO22/ALZ/015 (Bioef) to C.M.; and grant 2022333055 (Basque Government-Health Dep.) to F.C.

References

Abdelhak A, Foschi M, Abu-Rumeileh S et al. (2022). Blood GFAP as an emerging biomarker in brain and spinal cord disorders. Nat Rev Neurol 18: 158–172. https://doi.org/10.1038/s41582-021-00616-3.

Abdul HM, Sultana R, St Clair DK et al. (2008). Oxidative damage in brain from human mutant APP/PS-1 double knock-in mice as a function of age. Free Radic Biol Med 45: 1420–1425. https://doi.org/10.1016/j.freeradbiomed.2008.08.012.

Åbjørsbråten KS, Skaaraas GHES, Cunen C et al. (2022). Impai red astrocytic Ca^{2+} signaling in awake-behaving Alzheimer's disease transgenic mice. eLife 11: e75055. https://doi.org/10.7554/eLife.75055.

Abou-Sleiman PM, Muqit MMK, Wood NW (2006). Expanding insights of mitochondrial dysfunction in Parkinson's disease. Nat Rev Neurosci 7: 207–219. https://doi.org/10.1038/nrn1868.

Alberch J, Pérez-Navarro E, Canals JM (2002). Neuroprotection by neurotrophins and GDNF family members in the excitotoxic model of Huntington's disease. Brain Res Bull 57: 817–822. https://doi.org/10.1016/s0361-9230(01)00775-4.

Alberdi E, Wyssenbach A, Alberdi M et al. (2013). Ca^{2+}-depende nt endoplasmic reticulum stress correlates with astrogliosis in oligomeric amyloid β-treated astrocytes and in a model of Alzheimer's disease. Aging Cell 12: 292–302. https://doi.org/10.1111/acel.12054.

Albini M, Krawczun-Rygmaczewska A, Cesca F (2023). Astrocytes and brain-derived neurotrophic factor (BDNF). Neurosci Res. https://doi.org/10.1016/j.neures.2023.02.001. S0168-0102(23)00035–4..

Allen B, Ingram E, Takao M et al. (2002). Abundant tau filaments and nonapoptotic neurodegeneration in transgenic mice expressing human P301S tau protein. J Neurosci 22: 9340–9351. https://doi.org/10.1523/JNEUROSCI.22-21-09340.2002.

Andersen JV, Schousboe A, Verkhratsky A (2022). Astrocyte energy and neurotransmitter metabolism in Alzheimer's disease: integration of the glutamate/GABA-glutamine cycle. Prog Neurobiol 217: 102331. https://doi.org/10.1016/j.pneurobio.2022.102331.

Anderson RM, Hadjichrysanthou C, Evans S et al. (2017). Why do so many clinical trials of therapies for Alzheimer's disease fail? Lancet 390: 2327–2329. https://doi.org/10.1016/S0140-6736(17)32399-1.

Andorfer C, Kress Y, Espinoza M et al. (2003). Hyperphosphorylation and aggregation of tau in mice expressing normal human tau isoforms. J Neurochem 86: 582–590. https://doi.org/10.1046/j.1471-4159.2003.01879.x.

André VM, Cepeda C, Levine MS (2010). Dopamine and glutamate in Huntington's disease: a balancing act. CNS Neurosci Ther 16: 163–178. https://doi.org/10.1111/j.1755-5949.2010.00134.x.

Aoki E, Yano R, Yokoyama H et al. (2009). Role of nuclear transcription factor kappa B (NF-kappaB) for MPTP (1-methyl-4-phenyl-1,2,3,6-tetrahyropyridine)-induced apoptosis in nigral neurons of mice. Exp Mol Pathol 86: 57–64. https://doi.org/10.1016/j.yexmp.2008.10.004.

Araque Caballero MÁ, Suárez-Calvet M, Duering M et al. (2018). White matter diffusion alterations precede symptom onset in autosomal dominant Alzheimer's disease. Brain 141: 3065–3080. https://doi.org/10.1093/brain/awy229.

Arnst N, Belio-Mairal P, García-González L et al. (2021). Deficiency in MT5-MMP supports branching of human iPSCs-derived neurons and reduces expression of GLAST/S100 in iPSCs-derived astrocytes. Cells 10: 1705. https://doi.org/10.3390/cells10071705.

Arregui L, Benítez JA, Razgado LF et al. (2011). Adenoviral astrocyte-specific expression of BDNF in the striata of mice transgenic for Huntington's disease delays the onset of the motor phenotype. Cell Mol Neurobiol 31: 1229–1243. https://doi.org/10.1007/s10571-011-9725-y.

Bader Lange ML, St Clair D, Markesbery WR et al. (2010). Age-related loss of phospholipid asymmetry in APP(NLh)/APP(NLh) x PS-1 (P264L)/PS-1(P264L) human double mutant knock-in mice: relevance to Alzheimer disease. Neurobiol Dis 38: 104–115. https://doi.org/10.1016/j.nbd.2010.01.004.

Baglietto-Vargas D, Forner S, Cai L et al. (2021). Generation of a humanized Aβ expressing mouse demonstrating aspects of Alzheimer's disease-like pathology. Nat Commun 12: 2421. https://doi.org/10.1038/s41467-021-22624-z.

Balducci C, Frasca A, Zotti M et al. (2017). Toll-like receptor 4-dependent glial cell activation mediates the impairment in memory establishment induced by β-amyloid oligomers in an acute mouse model of Alzheimer's disease. Brain Behav Immun 60: 188–197. https://doi.org/10.1016/j.bbi.2016.10.012.

Banerjee R, Rai A, Iyer SM et al. (2022). Animal models in the study of Alzheimer's disease and Parkinson's disease: a historical perspective. Animal Model Exp Med 5: 27–37. https://doi.org/10.1002/ame2.12209.

Barak M, Fedorova V, Pospisilova V et al. (2022). Human iPSC-derived neural models for studying Alzheimer's disease: from neural stem cells to cerebral organoids. Stem Cell Rev Rep 18: 792–820. https://doi.org/10.1007/s12015-021-10254-3.

Barbar L, Jain T, Zimmer M et al. (2020). CD49f is a novel marker of functional and reactive human iPSC-derived astrocytes. Neuron 107: 436–453.e12. https://doi.org/10.1016/j.neuron.2020.05.014.

Bartels AL, Willemsen ATM, Doorduin J et al. (2010). [11C]-PK11195 PET: quantification of neuroinflammation and a monitor of anti-inflammatory treatment in Parkinson's disease? Parkinsonism Relat Disord 16: 57–59. https://doi.org/10.1016/j.parkreldis.2009.05.005.

Bartzokis G (2011). Alzheimer's disease as homeostatic responses to age-related myelin breakdown. Neurobiol Aging 32: 1341–1371. https://doi.org/10.1016/j.neurobiolaging.2009.08.007.

Bartzokis G, Cummings JL, Sultzer D et al. (2003). White matter structural integrity in healthy aging adults and patients with Alzheimer disease: a magnetic resonance imaging study. Arch Neurol 60: 393–398. https://doi.org/10.1001/archneur.60.3.393.

Bartzokis G, Lu PH, Tishler TA et al. (2007). Myelin breakdown and iron changes in Huntington's disease: pathogenesis and treatment implications. Neurochem Res 32: 1655–1664. https://doi.org/10.1007/s11064-007-9352-7.

Bartzokis G, Lu PH, Tingus K et al. (2010). Lifespan trajectory of myelin integrity and maximum motor speed. Neurobiol Aging 31: 1554–1562. https://doi.org/10.1016/j.neurobiolaging.2008.08.015.

Bates GP, Dorsey R, Gusella JF et al. (2015). Huntington disease. Nat Rev Dis Primers 1: 15005. https://doi.org/10.1038/nrdp.2015.5.

Bayram-Weston Z, Jones L, Dunnett SB et al. (2012). Light and electron microscopic characterization of the evolution of cellular pathology in YAC128 Huntington's disease transgenic mice. Brain Res Bull 88: 137–147. https://doi.org/10.1016/j.brainresbull.2011.05.005.

Beauquis J, Pavía P, Pomilio C et al. (2013). Environmental enrichment prevents astroglial pathological changes in the hippocampus of APP transgenic mice, model of Alzheimer's disease. Exp Neurol 239: 28–37. https://doi.org/10.1016/j.expneurol.2012.09.009.

Bechler ME, Byrne L, Ffrench-Constant C (2015). CNS myelin sheath lengths are an intrinsic property of oligodendrocytes. Curr Biol 25: 2411–2416. https://doi.org/10.1016/j.cub.2015.07.056.

Ben Haim L, Carrillo-de Sauvage M-A, Ceyzériat K et al. (2015a). Elusive roles for reactive astrocytes in neurodegenerative diseases. Front Cell Neurosci 9: 278. https://doi.org/10.3389/fncel.2015.00278.

Ben Haim L, Ceyzériat K, Carrillo-de Sauvage MA et al. (2015b). The JAK/STAT3 pathway is a common inducer of astrocyte reactivity in Alzheimer's and Huntington's diseases. J Neurosci 35: 2817–2829. https://doi.org/10.1523/JNEUROSCI.3516-14.2015.

Benitez A, Fieremans E, Jensen JH et al. (2014). White matter tract integrity metrics reflect the vulnerability of late-myelinating tracts in Alzheimer's disease. Neuroimage Clin 4: 64–71. https://doi.org/10.1016/j.nicl.2013.11.001.

Benraiss A, Mariani JN, Osipovitch M et al. (2021). Cell-intrinsic glial pathology is conserved across human and murine models of Huntington's disease. Cell Rep 36: 109308. https://doi.org/10.1016/j.celrep.2021.109308.

Bergami M, Santi S, Formaggio E et al. (2008). Uptake and recycling of pro-BDNF for transmitter-induced secretion by cortical astrocytes. J Cell Biol 183: 213–221. https://doi.org/10.1083/jcb.200806137.

Berridge MJ (2010). Calcium hypothesis of Alzheimer's disease. Pflugers Arch 459: 441–449. https://doi.org/10.1007/s00424-009-0736-1.

Bido S, Muggeo S, Massimino L et al. (2021). Microglia-specific overexpression of α-synuclein leads to severe dopaminergic neurodegeneration by phagocytic exhaustion and oxidative toxicity. Nat Commun 12: 6237. https://doi.org/10.1038/s41467-021-26519-x.

Blanchard V, Moussaoui S, Czech C et al. (2003). Time sequence of maturation of dystrophic neurites associated with Abeta deposits in APP/PS1 transgenic mice. Exp Neurol 184: 247–263. https://doi.org/10.1016/s0014-4886(03)00252-8.

Blesa J, Foffani G, Dehay B et al. (2022). Motor and non-motor circuit disturbances in early Parkinson disease: which happens first? Nat Rev Neurosci 23: 115–128. https://doi.org/10.1038/s41583-021-00542-9.

Bolmont T, Clavaguera F, Meyer-Luehmann M et al. (2007). Induction of tau pathology by intracerebral infusion of amyloid-beta - containing brain extract and by amyloid-beta deposition in APP x Tau transgenic mice. Am J Pathol 171: 2012–2020. https://doi.org/10.2353/ajpath.2007.070403.

Booth HDE, Hirst WD, Wade-Martins R (2017). The role of astrocyte dysfunction in Parkinson's disease pathogenesis. Trends Neurosci 40: 358–370. https://doi.org/10.1016/j.tins.2017.04.001.

Borchelt DR, Thinakaran G, Eckman CB et al. (1996). Familial Alzheimer's disease-linked presenilin 1 variants elevate Abeta1-42/1-40 ratio in vitro and in vivo. Neuron 17: 1005–1013. https://doi.org/10.1016/s0896-6273(00)80230-5.

Borlongan CV, Koutouzis TK, Sanberg PR (1997). 3-Nitropropionic acid animal model and Huntington's disease. Neurosci Biobehav Rev 21: 289–293. https://doi.org/10.1016/s0149-7634(96)00027-9.

Bowley MP, Cabral H, Rosene DL et al. (2010). Age changes in myelinated nerve fibers of the cingulate bundle and corpus callosum in the rhesus monkey. J Comp Neurol 518: 3046–3064. https://doi.org/10.1002/cne.22379.

Braak H, Braak E (1991). Neuropathological staging of Alzheimer-related changes. Acta Neuropathol 82: 239–259. https://doi.org/10.1007/BF00308809.

Braak H, Braak E (1996a). Evolution of the neuropathology of Alzheimer's disease. Acta Neurol Scand Suppl 165: 3–12. https://doi.org/10.1111/j.1600-0404.1996.tb05866.x.

Braak H, Braak E (1996b). Development of Alzheimer-related neurofibrillary changes in the neocortex inversely recapitulates cortical myelogenesis. Acta Neuropathol 92: 197–201. https://doi.org/10.1007/s004010050508.

Bradford J, Shin J-Y, Roberts M et al. (2009). Expression of mutant huntingtin in mouse brain astrocytes causes age-dependent neurological symptoms. Proc Natl Acad Sci USA 106: 22480–22485. https://doi.org/10.1073/pnas.0911503106.

Bradford J, Shin J-Y, Roberts M et al. (2010). Mutant huntingtin in glial cells exacerbates neurological symptoms of Huntington disease mice. J Biol Chem 285: 10653–10661. https://doi.org/10.1074/jbc.M109.083287.

Brion JP, Tremp G, Octave JN (1999). Transgenic expression of the shortest human tau affects its compartmentalization and its phosphorylation as in the pretangle stage of Alzheimer's disease. Am J Pathol 154: 255–270. https://doi.org/10.1016/S0002-9440(10)65272-8.

Broe M, Kril J, Halliday GM (2004). Astrocytic degeneration relates to the severity of disease in frontotemporal dementia. Brain 127: 2214–2220. https://doi.org/10.1093/brain/awh250.

Brookhouser N, Raman S, Frisch C et al. (2021). APOE2 mitigates disease-related phenotypes in an isogenic hiPSC-based model of Alzheimer's disease. Mol Psychiatry 26: 5715–5732. https://doi.org/10.1038/s41380-021-01076-3.

Brouillet E, Jacquard C, Bizat N et al. (2005). 3-Nitropropionic acid: a mitochondrial toxin to uncover physiopathological mechanisms underlying striatal degeneration in Huntington's disease. J Neurochem 95: 1521–1540. https://doi.org/10.1111/j.1471-4159.2005.03515.x.

Bruce-Keller AJ, Gupta S, Knight AG et al. (2011). Cognitive impairment in humanized APP×PS1 mice is linked to Aβ(1-42) and NOX activation. Neurobiol Dis 44: 317–326. https://doi.org/10.1016/j.nbd.2011.07.012.

Brunello L, Zampese E, Florean C et al. (2009). Presenilin-2 d ampens intracellular Ca^{2+} stores by increasing Ca^{2+} leakage and reducing Ca^{2+} uptake. J Cell Mol Med 13: 3358–3369. https://doi.org/10.1111/j.1582-4934.2009.00755.x.

Cannon JR, Geghman KD, Tapias V et al. (2013). Expression of human E46K-mutated α-synuclein in BAC-transgenic rats replicates early-stage Parkinson's disease features and enhances vulnerability to mitochondrial impairment. Exp Neurol 240: 44–56. https://doi.org/10.1016/j.expneurol.2012.11.007.

Cardona AE, Pioro EP, Sasse ME et al. (2006). Control of microglial neurotoxicity by the fractalkine receptor. Nat Neurosci 9: 917–924. https://doi.org/10.1038/nn1715.

Carmona S, Hardy J, Guerreiro R (2018). The genetic landscape of Alzheimer disease. Handb Clin Neurol 148: 395–408. https://doi.org/10.1016/B978-0-444-64076-5.00026-0.

Casley CS, Lakics V, Lee H et al. (2009). Up-regulation of astrocyte metabotropic glutamate receptor 5 by amyloid-β peptide. Brain Res 1260: 65–75. https://doi.org/10.1016/j.brainres.2008.12.082.

Cerneckis J, Bu G, Shi Y (2023). Pushing the boundaries of brain organoids to study Alzheimer's disease. Trends Mol Med 29: 659–672. https://doi.org/10.1016/j.molmed.2023.05.007.

Ceyzériat K, Ben Haim L, Denizot A et al. (2018). Modulation of astrocyte reactivity improves functional deficits in mouse models of Alzheimer's disease. Acta Neuropathol Commun 6: 104. https://doi.org/10.1186/s40478-018-0606-1.

Chandran JS, Lin X, Zapata A et al. (2008). Progressive behavioral deficits in DJ-1-deficient mice are associated with normal nigrostriatal function. Neurobiol Dis 29: 505–514. https://doi.org/10.1016/j.nbd.2007.11.011.

Chatila ZK, Yadav A, Mares J et al. (2023). RNA- and ATAC-sequencing reveals a unique CD83+ microglial population focally depleted in Parkinson's disease. bioRxiv. https://doi.org/10.1101/2023.05.17.540842. 2023.05.17.540842.

Chatterjee P, Pedrini S, Stoops E et al. (2021). Plasma glial fibrillary acidic protein is elevated in cognitively normal older adults at risk of Alzheimer's disease. Transl Psychiatry 11: 27. https://doi.org/10.1038/s41398-020-01137-1.

Chen L, Cagniard B, Mathews T et al. (2005). Age-dependent motor deficits and dopaminergic dysfunction in DJ-1 null mice. J Biol Chem 280: 21418–21426. https://doi.org/10.1074/jbc.M413955200.

Chen C-Y, Weng Y-H, Chien K-Y et al. (2012). (G2019S) LRRK2 activates MKK4-JNK pathway and causes degeneration of SN dopaminergic neurons in a transgenic mouse model of PD. Cell Death Differ 19: 1623–1633. https://doi.org/10.1038/cdd.2012.42.

Cheng P-H, Li C-L, Her L-S et al. (2013). Significantly differential diffusion of neuropathological aggregates in the brain of transgenic mice carrying N-terminal mutant huntingtin fused with green fluorescent protein. Brain Struct Funct 218: 283–294. https://doi.org/10.1007/s00429-012-0401-x.

Chia SJ, Tan E-K, Chao Y-X (2020). Historical perspective: models of Parkinson's disease. Int J Mol Sci 21: 2464. https://doi.org/10.3390/ijms21072464.

Chishti MA, Yang DS, Janus C et al. (2001). Early-onset amyloid deposition and cognitive deficits in transgenic mice expressing a double mutant form of amyloid precursor protein 695. J Biol Chem 276: 21562–21570. https://doi.org/10.1074/jbc.M100710200.

Ciarmiello A, Cannella M, Lastoria S et al. (2006). Brain white-matter volume loss and glucose hypometabolism precede the clinical symptoms of Huntington's disease. J Nucl Med 47: 215–222.

Cicognola C, Janelidze S, Hertze J et al. (2021). Plasma glial fibrillary acidic protein detects Alzheimer pathology and predicts future conversion to Alzheimer dementia in patients with mild cognitive impairment. Alzheimers Res Ther 13: 68. https://doi.org/10.1186/s13195-021-00804-9.

Clark JK, Furgerson M, Crystal JD et al. (2015). Alterations in synaptic plasticity coincide with deficits in spatial working memory in presymptomatic 3xTg-AD mice. Neurobiol Learn Mem 125: 152–162. https://doi.org/10.1016/j.nlm.2015.09.003.

Costa V, Giacomello M, Hudec R et al. (2010). Mitochondrial fission and cristae disruption increase the response of cell models of Huntington's disease to apoptotic stimuli. EMBO Mol Med 2: 490–503. https://doi.org/10.1002/emmm.201000102.

Crabtree DM, Zhang J (2012). Genetically engineered mouse models of Parkinson's disease. Brain Res Bull 88: 13–32. https://doi.org/10.1016/j.brainresbull.2011.07.019.

Croisier E, Moran LB, Dexter DT et al. (2005). Microglial inflamma tion in the parkinsonian substantia nigra: relationship to α-synuclein deposition. J Neuroinflammation 2: 14. https://doi.org/10.1186/1742-2094-2-14.

Cummings J (2018). Lessons learned from Alzheimer disease: clinical trials with negative outcomes. Clin Transl Sci 11: 147–152. https://doi.org/10.1111/cts.12491.

Czech C, Delaère P, Macq AF et al. (1997). Proteolytical processing of mutated human amyloid precursor protein in transgenic mice. Brain Res Mol Brain Res 47: 108–116. https://doi.org/10.1016/s0169-328x(97)00039-9.

Daher JPL, Pletnikova O, Biskup S et al. (2012). Neurodegenerative phenotypes in an A53T α-synuclein transgenic mouse model are independent of LRRK2. Hum Mol Genet 21: 2420–2431. https://doi.org/10.1093/hmg/dds057.

Damiano M, Galvan L, Déglon N et al. (2010). Mitochondria in Huntington's disease. Biochim Biophys Acta 1802: 52–61. https://doi.org/10.1016/j.bbadis.2009.07.012.

Davis J, Xu F, Deane R et al. (2004). Early-onset and robust cerebral microvascular accumulation of amyloid beta-protein in transgenic mice expressing low levels of a vasculotropic Dutch/Iowa mutant form of amyloid beta-protein precursor. J Biol Chem 279: 20296–20306. https://doi.org/10.1074/jbc.M312946200.

Day JO, Mullin S (2021). The genetics of Parkinson's disease and implications for clinical practice. Genes (Basel) 12: 1006. https://doi.org/10.3390/genes12071006.

de Leeuw SM, Kirschner AWT, Lindner K et al. (2022). APOE2, E3, and E4 differentially modulate cellular homeostasis, cholesterol metabolism, and inflammatory response in isogenic iPSC-derived astrocytes. Stem Cell Reports 17: 110–126. https://doi.org/10.1016/j.stemcr.2021.11.007.

de Rus Jacquet A, Alpaugh M, Denis HL et al. (2023). The contribution of inflammatory astrocytes to BBB impairments in a brain-chip model of Parkinson's disease. Nat Commun 14: 3651. https://doi.org/10.1038/s41467-023-39038-8.

de Schipper LJ, Hafkemeijer A, Bouts MJRJ et al. (2019). Age- and disease-related cerebral white matter changes in patients with Parkinson's disease. Neurobiol Aging 80: 203–209. https://doi.org/10.1016/j.neurobiolaging.2019.05.004.

De Strooper B, Karran E (2016). The cellular phase of Alzheimer's disease. Cell 164: 603–615. https://doi.org/10.1016/j.cell.2015.12.056.

Dean DC, Hurley SA, Kecskemeti SR et al. (2017). Association of amyloid pathology with myelin alteration in preclinical Alzheimer disease. JAMA Neurol 74: 41–49. https://doi.org/10.1001/jamaneurol.2016.3232.

Delekate A, Füchtemeier M, Schumacher T et al. (2014). Metabotropic P2Y1 receptor signalling mediates astrocytic hyperactivity in vivo in an Alzheimer's disease mouse model. Nat Commun 5: 5422. https://doi.org/10.1038/ncomms6422.

Dematteis G, Vydmantaitė G, Ruffinatti FA et al. (2020). Proteomic analysis links alterations of bioenergetics, mitochondria-ER interactions and proteostasis in hippocampal astrocytes from 3xTg-AD mice. Cell Death Dis 11: 645. https://doi.org/10.1038/s41419-020-02911-1.

Depp C, Sun T, Sasmita AO et al. (2023). Myelin dysfunction drives amyloid-β deposition in models of Alzheimer's disease. Nature 618: 349–357. https://doi.org/10.1038/s41586-023-06120-6.

DeTure MA, Dickson DW (2019). The neuropathological diagnosis of Alzheimer's disease. Mol Neurodegener 14: 32. https://doi.org/10.1186/s13024-019-0333-5.

Dexter DT, Jenner P, Schapira AH et al. (1992). Alterations in levels of iron, ferritin, and other trace metals in neurodegenerative diseases affecting the basal ganglia. The Royal Kings and Queens Parkinson's Disease Research Group. Ann Neurol 32: S94–100. https://doi.org/10.1002/ana.410320716.

di Domenico A, Carola G, Calatayud C et al. (2019). Patient-specific iPSC-derived astrocytes contribute to non-cell-autonomous neurodegeneration in Parkinson's disease. Stem Cell Reports 12: 213–229. https://doi.org/10.1016/j.stemcr.2018.12.011.

Dues DJ, Moore DJ (2020). LRRK2 and protein aggregation in Parkinson's disease: insights from animal models. Front Neurosci 14: 719. https://doi.org/10.3389/fnins.2020.00719.

Duff K, Eckman C, Zehr C et al. (1996). Increased amyloid-beta42(43) in brains of mice expressing mutant presenilin 1. Nature 383: 710–713. https://doi.org/10.1038/383710a0.

Duff K, Knight H, Refolo LM et al. (2000). Characterization of pathology in transgenic mice over-expressing human genomic and cDNA tau transgenes. Neurobiol Dis 7: 87–98. https://doi.org/10.1006/nbdi.1999.0279.

El Tayara NET, Delatour B, Volk A et al. (2010). Detection of vascular alterations by in vivo magnetic resonance angiography and histology in APP/PS1 mouse model of Alzheimer's disease. MAGMA 23: 53–64. https://doi.org/10.1007/s10334-009-0194-y.

Escartin C, Galea E, Lakatos A et al. (2021). Reactive astrocyte nomenclature, definitions, and future directions. Nat Neurosci 24: 312–325. https://doi.org/10.1038/s41593-020-00783-4.

Faideau M, Kim J, Cormier K et al. (2010). In vivo expression of polyglutamine-expanded huntingtin by mouse striatal astrocytes impairs glutamate transport: a correlation with Huntington's disease subjects. Hum Mol Genet 19: 3053–3067. https://doi.org/10.1093/hmg/ddq212.

Fajardo-Serrano A, Rico AJ, Roda E et al. (2021). Adeno-associated viral vectors as versatile tools for Parkinson's research, both for disease modeling purposes and for therapeutic uses. Int J Mol Sci 22: 6389. https://doi.org/10.3390/ijms22126389.

Fan W, Sun Y, Shi Z et al. (2019). Mouse induced pluripotent stem cells-derived Alzheimer's disease cerebral organoid culture and neural differentiation disorders. Neurosci Lett 711: 134433. https://doi.org/10.1016/j.neulet.2019.134433.

Fellner L, Irschick R, Schanda K et al. (2013). Toll-like receptor 4 is required for α-synuclein dependent activation of microglia and astroglia. Glia 61: 349–360. https://doi.org/10.1002/glia.22437.

Fernandez-Alvarez M, Atienza M, Zallo F et al. (2022). Linking plasma amyloid beta and neurofilament light chain to intracortical myelin content in cognitively normal older adults. Front Aging Neurosci 14: 896848. https://doi.org/10.3389/fnagi.2022.896848.

Fernandez-Alvarez M, Atienza M, Cantero JL (2023). Cortical amyloid-beta burden is associated with changes in intracortical myelin in cognitively normal older adults. Transl Psychiatry 13: 115. https://doi.org/10.1038/s41398-023-02420-7.

Ferrer I (2018). Oligodendrogliopathy in neurodegenerative diseases with abnormal protein aggregates: the forgotten partner. Prog Neurobiol 169: 24–54. https://doi.org/10.1016/j.pneurobio.2018.07.004.

Fleming SM, Fernagut P-O, Chesselet M-F (2005). Genetic mouse models of parkinsonism: strengths and limitations. NeuroRx 2: 495–503. https://doi.org/10.1602/neurorx.2.3.495.

Flood DG, Reaume AG, Dorfman KS et al. (2002). FAD mutant PS-1 gene-targeted mice: increased a beta 42 and a beta deposition without APP overproduction. Neurobiol Aging 23: 335–348. https://doi.org/10.1016/s0197-4580(01)00330-x.

Franciosi S, Ryu JK, Shim Y et al. (2012). Age-dependent neurovascular abnormalities and altered microglial morphology in the YAC128 mouse model of Huntington disease. Neurobiol Dis 45: 438–449. https://doi.org/10.1016/j.nbd.2011.09.003.

Fu Y, Zhou L, Li H et al. (2022). Adaptive structural changes in the motor cortex and white matter in Parkinson's disease. Acta Neuropathol 144: 861–879. https://doi.org/10.1007/s00401-022-02488-3.

Galla L, Redolfi N, Pozzan T et al. (2020). Intracellular calcium dysregulation by the Alzheimer's disease-linked protein presenilin 2. Int J Mol Sci 21: E770. https://doi.org/10.3390/ijms21030770.

Games D, Adams D, Alessandrini R et al. (1995). Alzheimer-type neuropathology in transgenic mice overexpressing V717F beta-amyloid precursor protein. Nature 373: 523–527. https://doi.org/10.1038/373523a0.

Gargiulo S, Rossin D, Testa G et al. (2018). Up-regulation of COX-2 and mPGES-1 by 27-hydroxycholesterol and 4-hydroxynonenal: a crucial role in atherosclerotic plaque instability. Free Radic Biol Med 129: 354–363. https://doi.org/10.1016/j.freeradbiomed.2018.09.046.

Gil JM, Rego AC (2008). Mechanisms of neurodegeneration in Huntington's disease. Eur J Neurosci 27: 2803–2820. https://doi.org/10.1111/j.1460-9568.2008.06310.x.

Giralt A, Friedman HC, Caneda-Ferrón B et al. (2010). BDNF regulation under GFAP promoter provides engineered astrocytes as a new approach for long-term protection in Huntington's disease. Gene Ther 17: 1294–1308. https://doi.org/10.1038/gt.2010.71.

Gispert S, Del Turco D, Garrett L et al. (2003). Transgenic mice expressing mutant A53T human alpha-synuclein show neuronal dysfunction in the absence of aggregate formation. Mol Cell Neurosci 24: 419–429. https://doi.org/10.1016/s1044-7431(03)00198-2.

Gispert S, Ricciardi F, Kurz A et al. (2009). Parkinson phenotype in aged PINK1-deficient mice is accompanied by progressive mitochondrial dysfunction in absence of neurodegeneration. PLoS One 4: e5777. https://doi.org/10.1371/journal.pone.0005777.

Goldberg MS, Fleming SM, Palacino JJ et al. (2003). Parkin-deficient mice exhibit nigrostriatal deficits but not loss of dopaminergic neurons. J Biol Chem 278: 43628–43635. https://doi.org/10.1074/jbc.M308947200.

Goldberg MS, Pisani A, Haburcak M et al. (2005). Nigrostriatal dopaminergic deficits and hypokinesia caused by inactivation of the familial Parkinsonism-linked gene DJ-1. Neuron 45: 489–496. https://doi.org/10.1016/j.neuron.2005.01.041.

Gong C, Bonfili L, Zheng Y et al. (2023). Immortalized Alzheimer's disease astrocytes: characterization of their proteolytic systems. Mol Neurobiol. https://doi.org/10.1007/s12035-023-03231-z.

Götz J, Probst A, Spillantini MG et al. (1995). Somatodendritic localization and hyperphosphorylation of tau protein in transgenic mice expressing the longest human brain tau isoform. EMBO J 14: 1304–1313. https://doi.org/10.1002/j.1460-2075.1995.tb07116.x.

Götz J, Chen F, Barmettler R et al. (2001). Tau filament formation in transgenic mice expressing P301L tau. J Biol Chem 276: 529–534. https://doi.org/10.1074/jbc.M006531200.

Götz J, Halliday G, Nisbet RM (2019). Molecular pathogenesis of the tauopathies. Annu Rev Pathol 14: 239–261. https://doi.org/10.1146/annurev-pathmechdis-012418-012936.

Grant LM, Kelm-Nelson CA, Hilby BL et al. (2015). Evidence for early and progressive ultrasonic vocalization and oromotor deficits in a PINK1 gene knockout rat model of Parkinson's disease. J Neurosci Res 93: 1713–1727. https://doi.org/10.1002/jnr.23625.

Gratuze M, Leyns CEG, Holtzman DM (2018). New insights into the role of TREM2 in Alzheimer's disease. Mol Neurodegener 13: 66. https://doi.org/10.1186/s13024-018-0298-9.

Gray M (2019). Astrocytes in Huntington's disease. Adv Exp Med Biol 1175: 355–381. https://doi.org/10.1007/978-981-13-9913-8_14.

Gray M, Shirasaki DI, Cepeda C et al. (2008). Full-length human mutant huntingtin with a stable polyglutamine repeat can elicit progressive and selective neuropathogenesis in BACHD mice. J Neurosci 28: 6182–6195. https://doi.org/10.1523/JNEUROSCI.0857-08.2008.

Graybeal JJ, Bozzelli PL, Graybeal LL et al. (2015). Human ApoE ϵ4 alters circadian rhythm activity, IL-1β, and GFAP in CRND8 mice. J Alzheimers Dis 43: 823–834. https://doi.org/10.3233/JAD-132009.

Grigoletto J, Pukaß K, Gamliel A et al. (2017). Higher levels of myelin phospholipids in brains of neuronal α-Synuclein transgenic mice precede myelin loss. Acta Neuropathol Commun 5: 37. https://doi.org/10.1186/s40478-017-0439-3.

Grolla AA, Sim JA, Lim D et al. (2013). Amyloid-β and Alzheimer's disease type pathology differentially affects the calcium signalling toolkit in astrocytes from different brain regions. Cell Death Dis 4: e623. https://doi.org/10.1038/cddis.2013.145.

Guillemaud O, Ceyzériat K, Saint-Georges T et al. (2020). Complex roles for reactive astrocytes in the triple transgenic mouse model of Alzheimer disease. Neurobiol Aging 90: 135–146. https://doi.org/10.1016/j.neurobiolaging.2020.02.010.

Guo Q, Fu W, Sopher BL et al. (1999). Increased vulnerability of hippocampal neurons to excitotoxic necrosis in presenilin-1 mutant knock-in mice. Nat Med 5: 101–106. https://doi.org/10.1038/4789.

Guttikonda SR, Sikkema L, Tchieu J et al. (2021). Fully defined human pluripotent stem cell-derived microglia and tri-culture system model C3 production in Alzheimer's disease. Nat Neurosci 24: 343–354. https://doi.org/10.1038/s41593-020-00796-z.

Hamadjida A, Frouni I, Kwan C et al. (2019). Classic animal models of Parkinson's disease: a historical perspective. Behav Pharmacol 30: 291–310. https://doi.org/10.1097/FBP.0000000000000441.

Hamza TH, Zabetian CP, Tenesa A et al. (2010). Common genetic variation in the HLA region is associated with late-onset sporadic Parkinson's disease. Nat Genet 42: 781–785. https://doi.org/10.1038/ng.642.

Hansen DV, Hanson JE, Sheng M (2018). Microglia in Alzheimer's disease. J Cell Biol 217: 459–472. https://doi.org/10.1083/jcb.201709069.

Harrison JK, Jiang Y, Chen S et al. (1998). Role for neuronally derived fractalkine in mediating interactions between neurons and CX3CR1-expressing microglia. Proc Natl Acad Sci USA 95: 10896–10901. https://doi.org/10.1073/pnas.95.18.10896.

Hashimoto S, Matsuba Y, Kamano N et al. (2019). Tau binding protein CAPON induces tau aggregation and neurodegeneration. Nat Commun 10: 2394. https://doi.org/10.1038/s41467-019-10278-x.

Hattori M, Hashizume Y, Yoshida M et al. (2003). Distribution of astrocytic plaques in the corticobasal degeneration brain and comparison with tuft-shaped astrocytes in the progressive supranuclear palsy brain. Acta Neuropathol 106: 143–149. https://doi.org/10.1007/s00401-003-0711-4.

Hauw JJ, Verny M, Delaère P et al. (1990). Constant neurofibrillary changes in the neocortex in progressive supranuclear palsy. Basic differences with Alzheimer's disease and aging. Neurosci Lett 119: 182–186. https://doi.org/10.1016/0304-3940(90)90829-x.

Heiland T, Zeitschel U, Puchades MA et al. (2019). Defined astrocytic expression of human amyloid precursor protein in Tg2576 mouse brain. Glia 67: 393–403. https://doi.org/10.1002/glia.23550.

Heng MY, Duong DK, Albin RL et al. (2010). Early autophagic response in a novel knock-in model of Huntington disease. Hum Mol Genet 19: 3702–3720. https://doi.org/10.1093/hmg/ddq285.

Hodgson JG, Agopyan N, Gutekunst CA et al. (1999). A YAC mouse model for Huntington's disease with full-length mutant huntingtin, cytoplasmic toxicity, and selective striatal neurodegeneration. Neuron 23: 181–192. https://doi.org/10.1016/s0896-6273(00)80764-3.

Holcomb L, Gordon MN, McGowan E et al. (1998). Accelerated Alzheimer-type phenotype in transgenic mice carrying both mutant amyloid precursor protein and presenilin 1 transgenes. Nat Med 4: 97–100. https://doi.org/10.1038/nm0198-097.

Hong Y, Zhao T, Li X-J et al. (2016). Mutant huntingtin impairs BDNF release from astrocytes by disrupting conversion of Rab3a-GTP into Rab3a-GDP. J Neurosci 36: 8790–8801. https://doi.org/10.1523/JNEUROSCI.0168-16.2016.

Hsiao K, Chapman P, Nilsen S et al. (1996). Correlative memory deficits, Abeta elevation, and amyloid plaques in transgenic mice. Science 274: 99–102. https://doi.org/10.1126/science.274.5284.99.

Huang S, Zhang Z, Cao J et al. (2022). Chimeric cerebral organoids reveal the essentials of neuronal and astrocytic APOE4 for Alzheimer's tau pathology. Sig Transduct Target Ther 7: 1–10. https://doi.org/10.1038/s41392-022-01006-x.

Ihara M, Polvikoski TM, Hall R et al. (2010). Quantification of myelin loss in frontal lobe white matter in vascular dementia, Alzheimer's disease, and dementia with Lewy bodies. Acta Neuropathol 119: 579–589. https://doi.org/10.1007/s00401-009-0635-8.

Imamura K, Hishikawa N, Sawada M et al. (2003). Distribution of major histocompatibility complex class II-positive microglia and cytokine profile of Parkinson's disease brains. Acta Neuropathol 106: 518–526. https://doi.org/10.1007/s00401-003-0766-2.

Irizarry MC, Soriano F, McNamara M et al. (1997). Abeta deposition is associated with neuropil changes, but not with overt neuronal loss in the human amyloid precursor protein V717F (PDAPP) transgenic mouse. J Neurosci 17: 7053–7059. https://doi.org/10.1523/JNEUROSCI.17-18-07053.1997.

Ishihara T, Hong M, Zhang B et al. (1999). Age-dependent emergence and progression of a tauopathy in transgenic mice overexpressing the shortest human tau isoform. Neuron 24: 751–762. https://doi.org/10.1016/s0896-6273(00)81127-7.

Jansen AHP, van Hal M, Op den Kelder IC et al. (2017). Frequency of nuclear mutant huntingtin inclusion formation in neurons and glia is cell-type-specific. Glia 65: 50–61. https://doi.org/10.1002/glia.23050.

Janus C, Pearson J, McLaurin J et al. (2000). A beta peptide immunization reduces behavioural impairment and plaques in a model of Alzheimer's disease. Nature 408: 979–982. https://doi.org/10.1038/35050110.

Jimenez S, Baglietto-Vargas D, Caballero C et al. (2008). Inflammatory response in the hippocampus of PS1M146L/APP751SL mouse model of Alzheimer's disease: age-dependent switch in the microglial phenotype from alternative to classic. J Neurosci 28: 11650–11661. https://doi.org/10.1523/JNEUROSCI.3024-08.2008.

Jones VC, Atkinson-Dell R, Verkhratsky A et al. (2017). Aberrant iPSC-derived human astrocytes in Alzheimer's disease. Cell Death Dis 8: e2696. https://doi.org/10.1038/cddis.2017.89.

Jun GR, You Y, Zhu C et al. (2022). Protein phosphatase 2A and complement component 4 are linked to the protective effect of APOE ϵ2 for Alzheimer's disease. Alzheimers Dement. https://doi.org/10.1002/alz.12607.

Kamphuis W, Kooijman L, Orre M et al. (2015). GFAP and vimentin deficiency alters gene expression in astrocytes and microglia in wild-type mice and changes the transcriptional response of reactive glia in mouse model for Alzheimer's disease. Glia 63: 1036–1056. https://doi.org/10.1002/glia.22800.

Kanemaru K, Kubota J, Sekiya H et al. (2013). Calcium-dependent N-cadherin up-regulation mediates reactive astrogliosis and neuroprotection after brain injury. Proc Natl Acad Sci USA 110: 11612–11617. https://doi.org/10.1073/pnas.1300378110.

Katsouri L, Birch AM, Renziehausen AWJ et al. (2020). Ablation of reactive astrocytes exacerbates disease pathology in a model of Alzheimer's disease. Glia 68: 1017–1030. https://doi.org/10.1002/glia.23759.

Katsurabayashi S, Kawano H, Ii M et al. (2016). Overexpression of Swedish mutant APP in aged astrocytes attenuates excitatory synaptic transmission. Physiol Rep 4: e12665. https://doi.org/10.14814/phy2.12665.

Kavroulakis E, Simos PG, Kalaitzakis G et al. (2018). Myelin content changes in probable Alzheimer's disease and mild cognitive impairment: associations with age and severity of neuropsychiatric impairment. J Magn Reson Imaging 47: 1359–1372. https://doi.org/10.1002/jmri.25849.

Kaye J, Reisine T, Finkbeiner S (2021). Huntington's disease mouse models: unraveling the pathology caused by CAG repeat expansion. Fac Rev 10: 77. https://doi.org/10.12703/r/10-77.

Ke M, Chong C-M, Zhu Q et al. (2021). Comprehensive perspectives on experimental models for Parkinson's disease. Aging Dis 12: 223–246. https://doi.org/10.14336/AD.2020.0331.

Kersaitis C, Halliday GM, Kril JJ (2004). Regional and cellular pathology in frontotemporal dementia: relationship to stage of disease in cases with and without Pick bodies. Acta Neuropathol 108: 515–523. https://doi.org/10.1007/s00401-004-0917-0.

Khakh BS (2019). Astrocyte-neuron interactions in the striatum: insights on identity, form, and function. Trends Neurosci 42: 617–630. https://doi.org/10.1016/j.tins.2019.06.003.

Khakh BS, Goldman SA (2023). Astrocytic contributions to Huntington's disease pathophysiology. Ann N Y Acad Sci 1522: 42–59. https://doi.org/10.1111/nyas.14977.

Khiroug L, Verkhratsky A (2021). Coming full circle: in vivo Veritas, or expanding the neuroscience frontier. Cell Calcium 98: 102452. https://doi.org/10.1016/j.ceca.2021.102452.

Kim CK, Lee YR, Ong L et al. (2022). Alzheimer's disease: key insights from two decades of clinical trial failures. J Alzheimers Dis 87: 83–100. https://doi.org/10.3233/JAD-215699.

Kitada T, Tong Y, Gautier CA et al. (2009). Absence of nigral degeneration in aged parkin/DJ-1/PINK1 triple knockout mice. J Neurochem 111: 696–702. https://doi.org/10.1111/j.1471-4159.2009.06350.x.

Kohutnicka M, Lewandowska E, Kurkowska-Jastrzebska I et al. (1998). Microglial and astrocytic involvement in a murine model of Parkinson's disease induced by 1-methyl-4-phenyl-1,2,3,6-tetrahydropyridine (MPTP). Immunopharmacology 39: 167–180. https://doi.org/10.1016/s0162-3109(98)00022-8.

Konnova EA, Swanberg M (2018). Animal models of Parkinson's disease. In: TB Stoker, JC Greenland (Eds.), Parkinson's disease: pathogenesis and clinical aspects. Codon Publications, Brisbane, Australia.

Konttinen H, Gureviciene I, Oksanen M et al. (2019). PPARβ/δ-agonist GW0742 ameliorates dysfunction in fatty acid oxidation in PSEN1ΔE9 astrocytes. Glia 67: 146–159. https://doi.org/10.1002/glia.23534.

Koprich JB, Johnston TH, Reyes G et al. (2016). Towards a non-human primate model of alpha-synucleinopathy for development of therapeutics for Parkinson's disease: optimization of AAV1/2 delivery parameters to drive sustained expression of alpha synuclein and dopaminergic degeneration in macaque. PLoS One 11: e0167235. https://doi.org/10.1371/journal.pone.0167235.

Kosior N, Leavitt BR (2018). Murine models of Huntington's disease for evaluating therapeutics. Methods Mol Biol 1780: 179–207. https://doi.org/10.1007/978-1-4939-7825-0_10.

Kotliarova S, Jana NR, Sakamoto N et al. (2005). Decreased expression of hypothalamic neuropeptides in Huntington disease transgenic mice with expanded polyglutamine-EGFP fluorescent aggregates. J Neurochem 93: 641–653. https://doi.org/10.1111/j.1471-4159.2005.03035.x.

Kotredes KP, Oblak A, Pandey RS et al. (2021a). Uncovering disease mechanisms in a novel mouse model expressing humanized APOEε4 and Trem2*R47H. Front Aging Neurosci 13: 735524. https://doi.org/10.3389/fnagi.2021.735524.

Kotredes KP, Oblak AL, Preuss C et al. (2021b). LOAD2: a late-onset Alzheimer's disease mouse model expressing APOEε4, Trem2*R47H, and humanized amyloid-beta. Alzheimers Dement 17: e056017. https://doi.org/10.1002/alz.056017.

Kovacs GG (2020). Astroglia and tau: new perspectives. Front Aging Neurosci 12: 96. https://doi.org/10.3389/fnagi.2020.00096.

Kovacs GG, Rozemuller AJM, van Swieten JC et al. (2013). Neuropathology of the hippocampus in FTLD-Tau with Pick bodies: a study of the BrainNet Europe Consortium. Neuropathol Appl Neurobiol 39: 166–178. https://doi.org/10.1111/j.1365-2990.2012.01272.x.

Kovacs GG, Ferrer I, Grinberg LT et al. (2016). Aging-related tau astrogliopathy (ARTAG): harmonized evaluation strategy. Acta Neuropathol 131: 87–102. https://doi.org/10.1007/s00401-015-1509-x.

Kraft AD, Kaltenbach LS, Lo DC et al. (2012). Activated microglia proliferate at neurites of mutant huntingtin-expressing neurons. Neurobiol Aging 33. https://doi.org/10.1016/j.neurobiolaging.2011.02.015. 621.e17–33.

Kraft AW, Hu X, Yoon H et al. (2013). Attenuating astrocyte activation accelerates plaque pathogenesis in APP/PS1 mice. FASEB J 27: 187–198. https://doi.org/10.1096/fj.12-208660.

Kuchibhotla KV, Lattarulo CR, Hyman BT et al. (2009). Synchronous hyperactivity and intercellular calcium waves in astrocytes in Alzheimer mice. Science 323: 1211–1215. https://doi.org/10.1126/science.1169096.

Kuijlaars J, Oyelami T, Diels A et al. (2016). Sustained synchronized neuronal network activity in a human astrocyte co-culture system. Sci Rep 6: 36529. https://doi.org/10.1038/srep36529.

Kulijewicz-Nawrot M, Verkhratsky A, Chvátal A et al. (2012). Astrocytic cytoskeletal atrophy in the medial prefrontal cortex of a triple transgenic mouse model of Alzheimer's disease. J Anat 221: 252–262. https://doi.org/10.1111/j.1469-7580.2012.01536.x.

Larson J, Lynch G, Games D et al. (1999). Alterations in synaptic transmission and long-term potentiation in hippocampal slices from young and aged PDAPP mice. Brain Res 840: 23–35. https://doi.org/10.1016/s0006-8993(99)01698-4.

Lee G, Leugers CJ (2012). Tau and tauopathies. Prog Mol Biol Transl Sci 107: 263–293. https://doi.org/10.1016/B978-0-12-385883-2.00004-7.

Lee MK, Borchelt DR, Kim G et al. (1997). Hyperaccumulation of FAD-linked presenilin 1 variants in vivo. Nat Med 3: 756–760. https://doi.org/10.1038/nm0797-756.

Lee, M.K., Stirling, W., Xu, Y., Xu, X., Qui, D., Mandir, A.S., Dawson, T.M., Copeland, N.G., Jenkins, N.A., Price, D.L., 2002. Human alpha-synuclein-harboring familial Parkinson's disease-linked Ala-53 –> Thr mutation causes neurodegenerative disease with alpha-synuclein aggregation in transgenic mice. Proc Natl Acad Sci USA 99, 8968–8973. doi:https://doi.org/10.1073/pnas.132197599

Lee W, Reyes RC, Gottipati MK et al. (2013). Enhanced Ca(2+)-dependent glutamate release from astrocytes of the BACHD Huntington's disease mouse model. Neurobiol Dis 58: 192–199. https://doi.org/10.1016/j.nbd.2013.06.002.

Lee C-T, Bendriem RM, Wu WW et al. (2017a). 3D brain organoids derived from pluripotent stem cells: promising experimental models for brain development and neurodegenerative disorders. J Biomed Sci 24: 59. https://doi.org/10.1186/s12929-017-0362-8.

Lee SE, Sias AC, Mandelli ML et al. (2017b). Network degeneration and dysfunction in presymptomatic C9ORF72 expansion carriers. Neuroimage Clin 14: 286–297. https://doi.org/10.1016/j.nicl.2016.12.006.

Lee JH, Han J-H, Kim H et al. (2019). Parkinson's disease-associated LRRK2-G2019S mutant acts through regulation of SERCA activity to control ER stress in astrocytes. Acta Neuropathol Commun 7: 68. https://doi.org/10.1186/s40478-019-0716-4.

Levine MS, Klapstein GJ, Koppel A et al. (1999). Enhanced sensitivity to N-methyl-D-aspartate receptor activation in transgenic and knockin mouse models of Huntington's disease. J Neurosci Res 58: 515–532.

Lewcock JW, Schlepckow K, Di Paolo G et al. (2020). Emerging microglia biology defines novel therapeutic approaches for Alzheimer's disease. Neuron 108: 801–821. https://doi.org/10.1016/j.neuron.2020.09.029.

Lewis J, McGowan E, Rockwood J et al. (2000). Neurofibrillary tangles, amyotrophy and progressive motor disturbance in mice expressing mutant (P301L) tau protein. Nat Genet 25: 402–405. https://doi.org/10.1038/78078.

Lewis J, Dickson DW, Lin WL et al. (2001). Enhanced neurofibrillary degeneration in transgenic mice expressing mutant tau and APP. Science 293: 1487–1491. https://doi.org/10.1126/science.1058189.

Leyns CEG, Gratuze M, Narasimhan S et al. (2019). TREM2 function impedes tau seeding in neuritic plaques. Nat Neurosci 22: 1217–1222. https://doi.org/10.1038/s41593-019-0433-0.

Li Y, Liu W, Oo TF et al. (2009). Mutant LRRK2(R1441G) BAC transgenic mice recapitulate cardinal features of Parkinson's disease. Nat Neurosci 12: 826–828. https://doi.org/10.1038/nn.2349.

Lia A, Sansevero G, Chiavegato A et al. (2023). Rescue of astrocyte activity by the calcium sensor STIM1 restores long-term synaptic plasticity in female mice modelling Alzheimer's disease. Nat Commun 14: 1590. https://doi.org/10.1038/s41467-023-37240-2.

Liberatore GT, Jackson-Lewis V, Vukosavic S et al. (1999). Inducible nitric oxide synthase stimulates dopaminergic neurodegeneration in the MPTP model of Parkinson disease. Nat Med 5: 1403–1409. https://doi.org/10.1038/70978.

Lim F, Hernández F, Lucas JJ et al. (2001). FTDP-17 mutations in tau transgenic mice provoke lysosomal abnormalities and tau filaments in forebrain. Mol Cell Neurosci 18: 702–714. https://doi.org/10.1006/mcne.2001.1051.

Lim D, Fedrizzi L, Tartari M et al. (2008). Calcium homeostasis and mitochondrial dysfunction in striatal neurons of Huntington disease. J Biol Chem 283: 5780–5789. https://doi.org/10.1074/jbc.M704704200.

Lim D, Iyer A, Ronco V et al. (2013). Amyloid beta deregulates astroglial mGluR5-mediated calcium signaling via calcineurin and Nf-kB. Glia 61: 1134–1145. https://doi.org/10.1002/glia.22502.

Lim D, Ronco V, Grolla AA et al. (2014). Glial calcium signalling in Alzheimer's disease. Rev Physiol Biochem Pharmacol 167: 45–65. https://doi.org/10.1007/112_2014_19.

Lim D, Rodríguez-Arellano JJ, Parpura V et al. (2016). Calcium signalling toolkits in astrocytes and spatio-temporal progression of Alzheimer's disease. Curr Alzheimer Res 13: 359–369.

Lim D, Semyanov A, Genazzani A et al. (2021). Calcium signaling in neuroglia. Int Rev Cell Mol Biol 362: 1–53. https://doi.org/10.1016/bs.ircmb.2021.01.003.

Lim D, Tapella L, Dematteis G et al. (2023). The endoplasmic reticulum stress and unfolded protein response in Alzheimer's disease: a calcium dyshomeostasis perspective. Ageing Res Rev 87: 101914. https://doi.org/10.1016/j.arr.2023.101914.

Lin CH, Tallaksen-Greene S, Chien WM et al. (2001). Neurological abnormalities in a knock-in mouse model of Huntington's disease. Hum Mol Genet 10: 137–144. https://doi.org/10.1093/hmg/10.2.137.

Lin X, Parisiadou L, Gu X-L et al. (2009). Leucine-rich repeat kinase 2 regulates the progression of neuropathology induced by Parkinson's-disease-related mutant alpha-synuclein. Neuron 64: 807–827. https://doi.org/10.1016/j.neuron.2009.11.006.

Lines J, Baraibar AM, Fang C et al. (2022). Astrocyte-neuronal network interplay is disrupted in Alzheimer's disease mice. Glia 70: 368–378. https://doi.org/10.1002/glia.24112.

Lippi SLP, Smith ML, Flinn JM (2018). A novel hAPP/htau mouse model of Alzheimer's disease: inclusion of APP with tau exacerbates behavioral deficits and zinc administration heightens tangle pathology. Front Aging Neurosci 10: 382. https://doi.org/10.3389/fnagi.2018.00382.

Lord A, Kalimo H, Eckman C et al. (2006). The Arctic Alzheimer mutation facilitates early intraneuronal Abeta aggregation and senile plaque formation in transgenic mice. Neurobiol Aging 27: 67–77. https://doi.org/10.1016/j.neurobiolaging.2004.12.007.

Luczynski P, Laule C, Hsiung G-YR et al. (2019). Coexistence of multiple sclerosis and Alzheimer's disease: a review. Mult Scler Relat Disord 27: 232–238. https://doi.org/10.1016/j.msard.2018.10.109.

Ludolph AC, He F, Spencer PS et al. (1991). 3-Nitropropionic acid-exogenous animal neurotoxin and possible human striatal toxin. Can J Neurol Sci 18: 492–498. https://doi.org/10.1017/s0317167100032212.

Mahmoudi E, Sadaghiyani S, Lin P et al. (2022). Diagnosis of Alzheimer's disease and related dementia among people with multiple sclerosis: large cohort study, USA. Mult Scler Relat Disord 57: 103351. https://doi.org/10.1016/j.msard.2021.103351.

Malthankar-Phatak GH, Lin Y-G, Giovannone N et al. (2012). Amyloid deposition and advanced age fails to induce Alzheimer's type progression in a double knock-in mouse model. Aging Dis 3: 141–155.

Mangiarini L, Sathasivam K, Seller M et al. (1996). Exon 1 of the HD gene with an expanded CAG repeat is sufficient to cause a progressive neurological phenotype in transgenic mice. Cell 87: 493–506. https://doi.org/10.1016/s0092-8674(00)81369-0.

Masliah E, Rockenstein E, Veinbergs I et al. (2000). Dopaminergic loss and inclusion body formation in alpha-synuclein mice: implications for neurodegenerative disorders. Science 287: 1265–1269. https://doi.org/10.1126/science.287.5456.1265.

Masuda A, Kobayashi Y, Kogo N et al. (2016). Cognitive deficits in single App knock-in mouse models. Neurobiol Learn Mem 135: 73–82. https://doi.org/10.1016/j.nlm.2016.07.001.

Mateos L, Ismail M-A-M, Gil-Bea F-J et al. (2011). Upregulation of brain renin angiotensin system by 27-hydroxycholesterol in Alzheimer's disease. J Alzheimers Dis 24: 669–679. https://doi.org/10.3233/JAD-2011-101512.

McGowan E, Pickford F, Kim J et al. (2005). Abeta42 is essential for parenchymal and vascular amyloid deposition in mice. Neuron 47: 191–199. https://doi.org/10.1016/j.neuron.2005.06.030.

Mehla J, Lacoursiere SG, Lapointe V et al. (2019). Age-dependent behavioral and biochemical characterization of single APP knock-in mouse (APPNL-G-F/NL-G-F) model of Alzheimer's disease. Neurobiol Aging 75: 25–37. https://doi.org/10.1016/j.neurobiolaging.2018.10.026.

Mehta D, Jackson R, Paul G et al. (2017). Why do trials for Alzheimer's disease drugs keep failing? A discontinued drug perspective for 2010-2015. Expert Opin Investig Drugs 26: 735–739. https://doi.org/10.1080/13543784.2017.1323868.

Meilandt WJ, Ngu H, Gogineni A et al. (2020). Trem2 deletion reduces late-stage amyloid plaque accumulation, elevates the Aβ42:Aβ40 ratio, and exacerbates axonal dystrophy and dendritic spine loss in the PS2APP Alzheimer's mouse model. J Neurosci 40: 1956–1974. https://doi.org/10.1523/JNEUROSCI.1871-19.2019.

Menalled LB, Sison JD, Dragatsis I et al. (2003). Time course of early motor and neuropathological anomalies in a knock-in mouse model of Huntington's disease with 140 CAG repeats. J Comp Neurol 465: 11–26. https://doi.org/10.1002/cne.10776.

Menalled LB, Kudwa AE, Miller S et al. (2012). Comprehensive behavioral and molecular characterization of a new knock-in mouse model of Huntington's disease: zQ175. PLoS One 7: e49838. https://doi.org/10.1371/journal.pone.0049838.

Mikhail F, Calingasan N, Parolari L et al. (2015). Lack of exacerbation of neurodegeneration in a double transgenic mouse model of mutant LRRK2 and tau. Hum Mol Genet 24: 3545–3556. https://doi.org/10.1093/hmg/ddv105.

Moechars D, Dewachter I, Lorent K et al. (1999). Early phenotypic changes in transgenic mice that overexpress different mutants of amyloid precursor protein in brain. J Biol Chem 274: 6483–6492. https://doi.org/10.1074/jbc.274.10.6483.

Moser M, Colello RJ, Pott U et al. (1995). Developmental expression of the prion protein gene in glial cells. Neuron 14: 509–517. https://doi.org/10.1016/0896-6273(95)90307-0.

Mucke L, Masliah E, Yu GQ et al. (2000). High-level neuronal expression of abeta 1-42 in wild-type human amyloid protein precursor transgenic mice: synaptotoxicity without plaque formation. J Neurosci 20: 4050–4058. https://doi.org/10.1523/JNEUROSCI.20-11-04050.2000.

Nam M-H, Ko HY, Kim D et al. (2023). Visualizing reactive astrocyte-neuron interaction in Alzheimer's disease using 11C-acetate and 18F-FDG. Brain 146: 2957–2974. https://doi.org/10.1093/brain/awad037.

Nasrabady SE, Rizvi B, Goldman JE et al. (2018). White matter changes in Alzheimer's disease: a focus on myelin and oligodendrocytes. Acta Neuropathol Commun 6: 22. https://doi.org/10.1186/s40478-018-0515-3.

Nguyen APT, Daniel G, Valdés P et al. (2018). G2019S LRRK2 enhances the neuronal transmission of tau in the mouse brain. Hum Mol Genet 27: 120–134. https://doi.org/10.1093/hmg/ddx389.

Nudelman KNH, Brumm MC, Marek K et al. (2022). TREM2 variants in Parkinson's disease: results from the Parkinson's Progression Markers Initiative (PPMI) Study. Alzheimers Dement 18: e062316. https://doi.org/10.1002/alz.062316.

Oakley H, Cole SL, Logan S et al. (2006). Intraneuronal beta-amyloid aggregates, neurodegeneration, and neuron loss in transgenic mice with five familial Alzheimer's disease mutations: potential factors in amyloid plaque formation. J Neurosci 26: 10129–10140. https://doi.org/10.1523/JNEUROSCI.1202-06.2006.

Octeau JC, Chai H, Jiang R et al. (2018). An optical neuron-astrocyte proximity assay at synaptic distance scales. Neuron 98: 49–66.e9. https://doi.org/10.1016/j.neuron.2018.03.003.

Oddo S, Caccamo A, Shepherd JD et al. (2003). Triple-transgenic model of Alzheimer's disease with plaques and tangles: intracellular Abeta and synaptic dysfunction. Neuron 39: 409–421.

Oksanen M, Petersen AJ, Naumenko N et al. (2017). PSEN1 mutant iPSC-derived model reveals severe astrocyte pathology in Alzheimer's disease. Stem Cell Reports 9: 1885–1897. https://doi.org/10.1016/j.stemcr.2017.10.016.

Olabarria M, Noristani HN, Verkhratsky A et al. (2010). Concomitant astroglial atrophy and astrogliosis in a triple transgenic animal model of Alzheimer's disease. Glia 58: 831–838. https://doi.org/10.1002/glia.20967.

Orre M, Kamphuis W, Osborn LM et al. (2014). Acute isolation and transcriptome characterization of cortical astrocytes and microglia from young and aged mice. Neurobiol Aging 35: 1–14. https://doi.org/10.1016/j.neurobiolaging.2013.07.008.

Palacino JJ, Sagi D, Goldberg MS et al. (2004). Mitochondrial dysfunction and oxidative damage in parkin-deficient mice. J Biol Chem 279: 18614–18622. https://doi.org/10.1074/jbc.M401135200.

Park J-S, Davis RL, Sue CM (2018). Mitochondrial dysfunction in Parkinson's disease: new mechanistic insights and therapeutic perspectives. Curr Neurol Neurosci Rep 18: 21. https://doi.org/10.1007/s11910-018-0829-3.

Paulson JB, Ramsden M, Forster C et al. (2008). Amyloid plaque and neurofibrillary tangle pathology in a regulatable mouse model of Alzheimer's disease. Am J Pathol 173: 762–772. https://doi.org/10.2353/ajpath.2008.080175.

Pavese N, Gerhard A, Tai YF et al. (2006). Microglial activation correlates with severity in Huntington disease: a clinical and PET study. Neurology 66: 1638–1643. https://doi.org/10.1212/01.wnl.0000222734.56412.17.

Pereira JB, Janelidze S, Smith R et al. (2021). Plasma GFAP is an early marker of amyloid-β but not tau pathology in Alzheimer's disease. Brain 144: 3505–3516. https://doi.org/10.1093/brain/awab223.

Pérez M, Ribe E, Rubio A et al. (2005). Characterization of a double (amyloid precursor protein-tau) transgenic: tau phosphorylation and aggregation. Neuroscience 130: 339–347. https://doi.org/10.1016/j.neuroscience.2004.09.029.

Pizzo P, Basso E, Filadi R et al. (2020). Presenilin-2 and calcium handling: molecules, organelles, cells and brain networks. Cells 9: 2166. https://doi.org/10.3390/cells9102166.

Poewe W, Seppi K, Tanner CM et al. (2017). Parkinson disease. Nat Rev Dis Primers 3: 17013. https://doi.org/10.1038/nrdp.2017.13.

Popov A, Brazhe A, Denisov P et al. (2021). Astrocyte dystrophy in ageing brain parallels impaired synaptic plasticity. Aging Cell 20: e13334. https://doi.org/10.1111/acel.13334.

Popugaeva E, Pchitskaya E, Bezprozvanny I (2018). Dysregulation of intracellular calcium signaling in Alzheimer's disease. Antioxid Redox Signal 29: 1176–1188. https://doi.org/10.1089/ars.2018.7506.

Porchet R, Probst A, Bouras C et al. (2003). Analysis of glial acidic fibrillary protein in the human entorhinal cortex during aging and in Alzheimer's disease. Proteomics 3: 1476–1485. https://doi.org/10.1002/pmic.200300456.

Potokar M, Morita M, Wiche G et al. (2020). The diversity of intermediate filaments in astrocytes. Cells 9: 1604. https://doi.org/10.3390/cells9071604.

Pouladi MA, Morton AJ, Hayden MR (2013). Choosing an animal model for the study of Huntington's disease. Nat Rev Neurosci 14: 708–721. https://doi.org/10.1038/nrn3570.

Preman P, Alfonso-Triguero M, Alberdi E et al. (2021a). Astrocytes in Alzheimer's disease: pathological significance and molecular pathways. Cells 10: 540. https://doi.org/10.3390/cells10030540.

Preman P, Tcw J, Calafate S et al. (2021b). Human iPSC-derived astrocytes transplanted into the mouse brain undergo morphological changes in response to amyloid-β plaques. Mol Neurodegener 16: 68. https://doi.org/10.1186/s13024-021-00487-8.

Quadrato G, Nguyen T, Macosko EZ et al. (2017). Cell diversity and network dynamics in photosensitive human brain organoids. Nature 545: 48–53. https://doi.org/10.1038/nature22047.

Quintela-López T, Ortiz-Sanz C, Serrano-Regal MP et al. (2019). Aβ oligomers promote oligodendrocyte differentiation and maturation via integrin β1 and Fyn kinase signaling. Cell Death Dis 10: 445. https://doi.org/10.1038/s41419-019-1636-8.

Quon D, Wang Y, Catalano R et al. (1991). Formation of beta-amyloid protein deposits in brains of transgenic mice. Nature 352: 239–241. https://doi.org/10.1038/352239a0.

Raber J, Wong D, Buttini M et al. (1998). Isoform-specific effects of human apolipoprotein E on brain function revealed in ApoE knockout mice: increased susceptibility of females. Proc Natl Acad Sci USA 95: 10914–10919. https://doi.org/10.1073/pnas.95.18.10914.

Radde R, Bolmont T, Kaeser SA et al. (2006). Abeta42-driven cerebral amyloidosis in transgenic mice reveals early and robust pathology. EMBO Rep 7: 940–946. https://doi.org/10.1038/sj.embor.7400784.

Ramonet D, Daher JPL, Lin BM et al. (2011). Dopaminergic neuronal loss, reduced neurite complexity and autophagic abnormalities in transgenic mice expressing G2019S mutant LRRK2. PLoS One 6: e18568. https://doi.org/10.1371/journal.pone.0018568.

Ramos-Gonzalez P, Mato S, Chara JC et al. (2021). Astrocytic atrophy as a pathological feature of Parkinson's disease with LRRK2 mutation. NPJ Parkinsons Dis 7: 31. https://doi.org/10.1038/s41531-021-00175-w.

Ramsden M, Kotilinek L, Forster C et al. (2005). Age-dependent neurofibrillary tangle formation, neuron loss, and memory impairment in a mouse model of human tauopathy (P301L). J Neurosci 25: 10637–10647. https://doi.org/10.1523/JNEUROSCI.3279-05.2005.

Rayaprolu S, Mullen B, Baker M et al. (2013). TREM2 in neurodegeneration: evidence for association of the p.R47H variant with frontotemporal dementia and Parkinson's disease. Mol Neurodegener 8: 19. https://doi.org/10.1186/1750-1326-8-19.

Reaume AG, Howland DS, Trusko SP et al. (1996). Enhanced amyloidogenic processing of the beta-amyloid precursor protein in gene-targeted mice bearing the Swedish familial Alzheimer's disease mutations and a "humanized" Abeta sequence. J Biol Chem 271: 23380–23388. https://doi.org/10.1074/jbc.271.38.23380.

Reddy PH, Williams M, Charles V et al. (1998). Behavioural abnormalities and selective neuronal loss in HD transgenic mice expressing mutated full-length HD cDNA. Nat Genet 20: 198–202. https://doi.org/10.1038/2510.

Reick C, Ellrichmann G, Tsai T et al. (2016). Expression of brain-derived neurotrophic factor in astrocytes—beneficial effects of glatiramer acetate in the R6/2 and YAC128 mouse models of Huntington's disease. Exp Neurol 285: 12–23. https://doi.org/10.1016/j.expneurol.2016.08.012.

Ribé EM, Pérez M, Puig B et al. (2005). Accelerated amyloid deposition, neurofibrillary degeneration and neuronal loss in double mutant APP/tau transgenic mice. Neurobiol Dis 20: 814–822. https://doi.org/10.1016/j.nbd.2005.05.027.

Richards JG, Higgins GA, Ouagazzal A-M et al. (2003). PS2APP transgenic mice, coexpressing hPS2mut and hAPPswe, show age-related cognitive deficits associated with discrete brain amyloid deposition and inflammation. J Neurosci 23: 8989–9003. https://doi.org/10.1523/JNEUROSCI.23-26-08989.2003.

Richfield EK, Thiruchelvam MJ, Cory-Slechta DA et al. (2002). Behavioral and neurochemical effects of wild-type and mutated human alpha-synuclein in transgenic mice. Exp Neurol 175: 35–48. https://doi.org/10.1006/exnr.2002.7882.

Rocchio F, Tapella L, Manfredi M et al. (2019). Gene expression, proteome and calcium signaling alterations in immortalized hippocampal astrocytes from an Alzheimer's disease mouse model. Cell Death Dis 10: 24. https://doi.org/10.1038/s41419-018-1264-8.

Rockenstein E, Mallory M, Mante M et al. (2001). Early formation of mature amyloid-beta protein deposits in a mutant APP transgenic model depends on levels of Abeta(1-42). J Neurosci Res 66: 573–582. https://doi.org/10.1002/jnr.1247.

Rockenstein E, Mallory M, Hashimoto M et al. (2002). Differential neuropathological alterations in transgenic mice expressing alpha-synuclein from the platelet-derived growth factor and Thy-1 promoters. J Neurosci Res 68: 568–578. https://doi.org/10.1002/jnr.10231.

Rodríguez JJ, Terzieva S, Olabarria M et al. (2013). Enriched environment and physical activity reverse astrogliodegeneration in the hippocampus of AD transgenic mice. Cell Death Dis 4: e678. https://doi.org/10.1038/cddis.2013.194.

Ronco V, Grolla AA, Glasnov TN et al. (2014). Differential deregulation of astrocytic calcium signalling by amyloid-β, TNFα, IL-1β and LPS. Cell Calcium 55: 219–229. https://doi.org/10.1016/j.ceca.2014.02.016.

Rosas HD, Tuch DS, Hevelone ND et al. (2006). Diffusion tensor imaging in presymptomatic and early Huntington's disease: selective white matter pathology and its relationship to clinical measures. Mov Disord 21: 1317–1325. https://doi.org/10.1002/mds.20979.

Rose CR, Verkhratsky A (2016). Principles of sodium homeostasis and sodium signalling in astroglia. Glia 64: 1611–1627. https://doi.org/10.1002/glia.22964.

Rüb U, Vonsattel JPG, Heinsen H et al. (2015). The neuropathology of Huntington's disease: classical findings. In: Recent Developments and Correlation to Functional Neuroanatomy, Springer.

Ryan SM, Kelly ÁM (2016). Exercise as a pro-cognitive, pro-neurogenic and anti-inflammatory intervention in transgenic mouse models of Alzheimer's disease. Ageing Res Rev 27: 77–92. https://doi.org/10.1016/j.arr.2016.03.007.

Saab AS, Tzvetavona ID, Trevisiol A et al. (2016). Oligodendroglial NMDA receptors regulate glucose import and axonal energy metabolism. Neuron 91: 119–132. https://doi.org/10.1016/j.neuron.2016.05.016.

Saba J, López Couselo F, Turati J et al. (2020). Astrocytes from cortex and striatum show differential responses to mitochondrial toxin and BDNF: implications for protection of striatal neurons expressing mutant huntingtin. J Neuroinflammation 17: 290. https://doi.org/10.1186/s12974-020-01965-4.

Saito T, Matsuba Y, Mihira N et al. (2014). Single App knock-in mouse models of Alzheimer's disease. Nat Neurosci 17: 661–663. https://doi.org/10.1038/nn.3697.

Saito T, Mihira N, Matsuba Y et al. (2019). Humanization of the entire murine Mapt gene provides a murine model of pathological human tau propagation. J Biol Chem 294: 12754–12765. https://doi.org/10.1074/jbc.RA119.009487.

Sandoval IM, Marmion DJ, Meyers KT et al. (2021). Gene therapy to modulate alpha-synuclein in synucleinopathies. J Parkinsons Dis 11: S189–S197. https://doi.org/10.3233/JPD-212679.

Sapp E, Kegel KB, Aronin N et al. (2001). Early and progressive accumulation of reactive microglia in the Huntington disease brain. J Neuropathol Exp Neurol 60: 161–172. https://doi.org/10.1093/jnen/60.2.161.

Savage JC, St-Pierre M-K, Carrier M et al. (2020). Microglial physiological properties and interactions with synapses are altered at presymptomatic stages in a mouse model of Huntington's disease pathology. J Neuroinflammation 17: 98. https://doi.org/10.1186/s12974-020-01782-9.

Savonenko A, Xu GM, Melnikova T et al. (2005). Episodic-like memory deficits in the APPswe/PS1dE9 mouse model of Alzheimer's disease: relationships to beta-amyloid deposition and neurotransmitter abnormalities. Neurobiol Dis 18: 602–617. https://doi.org/10.1016/j.nbd.2004.10.022.

Savtchouk I, Volterra A (2018). Gliotransmission: beyond black-and-white. J Neurosci 38: 14–25. https://doi.org/10.1523/JNEUROSCI.0017-17.2017.

Sawant N, Morton H, Kshirsagar S et al. (2021). Mitochondrial abnormalities and synaptic damage in Huntington's disease: a focus on defective mitophagy and mitochondria-targeted therapeutics. Mol Neurobiol 58: 6350–6377. https://doi.org/10.1007/s12035-021-02556-x.

Schilling G, Becher MW, Sharp AH et al. (1999). Intranuclear inclusions and neuritic aggregates in transgenic mice expressing a mutant N-terminal fragment of huntingtin. Hum Mol Genet 8: 397–407. https://doi.org/10.1093/hmg/8.3.397.

Schindowski K, Bretteville A, Leroy K et al. (2006). Alzheimer's disease-like tau neuropathology leads to memory deficits and loss of functional synapses in a novel mutated tau transgenic mouse without any motor deficits. Am J Pathol 169: 599–616. https://doi.org/10.2353/ajpath.2006.060002.

Schwabe T, Srinivasan K, Rhinn H (2020). Shifting paradigms: the central role of microglia in Alzheimer's disease. Neurobiol Dis 143: 104962. https://doi.org/10.1016/j.nbd.2020.104962.

Scott M, Foster D, Mirenda C et al. (1989). Transgenic mice expressing hamster prion protein produce species-specific scrapie infectivity and amyloid plaques. Cell 59: 847–857. https://doi.org/10.1016/0092-8674(89)90608-9.

Scott MR, Köhler R, Foster D et al. (1992). Chimeric prion protein expression in cultured cells and transgenic mice. Protein Sci 1: 986–997. https://doi.org/10.1002/pro.5560010804.

Sebastian Monasor L, Müller SA, Colombo AV et al. (2020). Fibrillar Aβ triggers microglial proteome alterations and dysfunction in Alzheimer mouse models. eLife 9: e54083. https://doi.org/10.7554/eLife.54083.

Selkoe DJ, Hardy J (2016). The amyloid hypothesis of Alzheimer's disease at 25 years. EMBO Mol Med 8: 595–608. https://doi.org/10.15252/emmm.201606210.

Semyanov A, Henneberger C, Agarwal A (2020). Making sense of astrocytic calcium signals—from acquisition to interpretation. Nat Rev Neurosci 21: 551–564. https://doi.org/10.1038/s41583-020-0361-8.

Setsuie R, Wang Y-L, Mochizuki H et al. (2007). Dopaminergic neuronal loss in transgenic mice expressing the Parkinson's disease-associated UCH-L1 I93M mutant. Neurochem Int 50: 119–129. https://doi.org/10.1016/j.neuint.2006.07.015.

Shaffer JJ, Ghayoor A, Long JD et al. (2017). Longitudinal diffusion changes in prodromal and early HD: evidence of white-matter tract deterioration. Hum Brain Mapp 38: 1460–1477. https://doi.org/10.1002/hbm.23465.

Shah D, Gsell W, Wahis J et al. (2022). Astrocyte calcium dysfunction causes early network hyperactivity in Alzheimer's disease. Cell Rep 40: 111280. https://doi.org/10.1016/j.celrep.2022.111280.

Shelbourne PF, Killeen N, Hevner RF et al. (1999). A Huntington's disease CAG expansion at the murine Hdh locus is unstable and associated with behavioural abnormalities in mice. Hum Mol Genet 8: 763–774. https://doi.org/10.1093/hmg/8.5.763.

Shibuya K, Yagishita S, Nakamura A et al. (2011). Perivascular orientation of astrocytic plaques and tuft-shaped astrocytes. Brain Res 1404: 50–54. https://doi.org/10.1016/j.brainres.2011.06.014.

Shigetomi E, Patel S, Khakh BS (2016). Probing the complexities of astrocyte calcium signaling. Trends Cell Biol 26: 300–312. https://doi.org/10.1016/j.tcb.2016.01.003.

Shin J-Y, Fang Z-H, Yu Z-X et al. (2005). Expression of mutant huntingtin in glial cells contributes to neuronal excitotoxicity. J Cell Biol 171: 1001–1012. https://doi.org/10.1083/jcb.200508072.

Shinohara, Motoko, Shinohara, Mitsuru, Zhao, J., Fu, Y., Liu, C.-C., Kanekiyo, T., Bu, G., 2019.5-HT3 Antagonist ondansetron increases apoE secretion by modulating the LXR-ABCA1 pathway. Int J Mol Sci 20, E1488. doi:https://doi.org/10.3390/ijms20061488

Shir D, Graff-Radford J, Hofrenning EI et al. (2022). Association of plasma glial fibrillary acidic protein (GFAP) with neuroimaging of Alzheimer's disease and vascular pathology. Alzheimers Dement (Amst) 14: e12291. https://doi.org/10.1002/dad2.12291.

Sienski G, Narayan P, Bonner JM et al. (2021). APOE4 disrupts intracellular lipid homeostasis in human iPSC-derived glia. Sci Transl Med 13: eaaz4564. https://doi.org/10.1126/scitranslmed.aaz4564.

Siman R, Reaume AG, Savage MJ et al. (2000). Presenilin-1 P264L knock-in mutation: differential effects on abeta production, amyloid deposition, and neuronal vulnerability. J Neurosci 20: 8717–8726. https://doi.org/10.1523/JNEUROSCI.20-23-08717.2000.

Simic G, Lucassen PJ, Krsnik Z et al. (2000). nNOS expression in reactive astrocytes correlates with increased cell death related DNA damage in the hippocampus and entorhinal cortex in Alzheimer's disease. Exp Neurol 165: 12–26. https://doi.org/10.1006/exnr.2000.7448.

Simmons DA, Casale M, Alcon B et al. (2007). Ferritin accumulation in dystrophic microglia is an early event in the development of Huntington's disease. Glia 55: 1074–1084. https://doi.org/10.1002/glia.20526.

Simon DK, Tanner CM, Brundin P (2020). Parkinson disease epidemiology, pathology, genetics, and pathophysiology. Clin Geriatr Med 36: 1–12. https://doi.org/10.1016/j.cger.2019.08.002.

Šimončičová E, Gonçalves de Andrade E, Vecchiarelli HA et al. (2022). Present and future of microglial pharmacology. Trends Pharmacol Sci 43: 669–685. https://doi.org/10.1016/j.tips.2021.11.006.

Simpson JE, Ince PG, Lace G et al. (2010). Astrocyte phenotype in relation to Alzheimer-type pathology in the ageing brain. Neurobiol Aging 31: 578–590. https://doi.org/10.1016/j.neurobiolaging.2008.05.015.

Singhrao SK, Neal JW, Morgan BP et al. (1999). Increased complement biosynthesis by microglia and complement activation on neurons in Huntington's disease. Exp Neurol 159: 362–376. https://doi.org/10.1006/exnr.1999.7170.

Sloan SA, Darmanis S, Huber N et al. (2017). Human astrocyte maturation captured in 3D cerebral cortical spheroids derived from pluripotent stem cells. Neuron 95. https://doi.org/10.1016/j.neuron.2017.07.035. 779–790.e6.

Slow EJ, van Raamsdonk J, Rogers D et al. (2003). Selective striatal neuronal loss in a YAC128 mouse model of Huntington disease. Hum Mol Genet 12: 1555–1567. https://doi.org/10.1093/hmg/ddg169.

Smajić S, Prada-Medina CA, Landoulsi Z et al. (2022). Single-cell sequencing of human midbrain reveals glial activation and a Parkinson-specific neuronal state. Brain 145: 964–978. https://doi.org/10.1093/brain/awab446.

Smith MD, Chamling X, Gill AJ et al. (2022). Reactive astrocytes derived from human induced pluripotent stem cells suppress oligodendrocyte precursor cell differentiation. Front Mol Neurosci 15: 874299. https://doi.org/10.3389/fnmol.2022.874299.

Southwell AL, Smith-Dijak A, Kay C et al. (2016). An enhanced Q175 knock-in mouse model of Huntington disease with higher mutant huntingtin levels and accelerated disease phenotypes. Hum Mol Genet 25: 3654–3675. https://doi.org/10.1093/hmg/ddw212.

Stefanis L (2012). α-Synuclein in Parkinson's disease. Cold Spring Harb Perspect Med 2: a009399. https://doi.org/10.1101/cshperspect.a009399.

Stimmell AC, Baglietto-Vargas D, Moseley SC et al. (2019). Impaired spatial reorientation in the 3xTg-AD mouse model of Alzheimer's disease. Sci Rep 9: 1311. https://doi.org/10.1038/s41598-018-37151-z.

Sturchler-Pierrat C, Abramowski D, Duke M et al. (1997). Two amyloid precursor protein transgenic mouse models with Alzheimer disease-like pathology. Proc Natl Acad Sci USA 94: 13287–13292. https://doi.org/10.1073/pnas.94.24.13287.

Suárez-Calvet M, Morenas-Rodríguez E, Kleinberger G et al. (2019). Early increase of CSF sTREM2 in Alzheimer's disease is associated with tau related-neurodegeneration but not with amyloid-β pathology. Mol Neurodegener 14: 1. https://doi.org/10.1186/s13024-018-0301-5.

Sucunza D, Rico AJ, Roda E et al. (2021). Glucocerebrosidase gene therapy induces alpha-synuclein clearance and neuroprotection of midbrain dopaminergic neurons in mice and macaques. Int J Mol Sci 22: 4825. https://doi.org/10.3390/ijms22094825.

Sultana R, Robinson RAS, Di Domenico F et al. (2011). Proteomic identification of specifically carbonylated brain proteins in APP(NLh)/APP(NLh) × PS-1(P264L)/PS-1(P264L) human double mutant knock-in mice model of Alzheimer disease as a function of age. J Proteomics 74: 2430–2440. https://doi.org/10.1016/j.jprot.2011.06.015.

Sun Y, Tong H, Yang T et al. (2022). Insights into white matter defect in Huntington's disease. Cells 11: 3381. https://doi.org/10.3390/cells11213381.

Supnet C, Bezprozvanny I (2010). The dysregulation of intracellular calcium in Alzheimer disease. Cell Calcium 47: 183–189. https://doi.org/10.1016/j.ceca.2009.12.014.

Supnet C, Bezprozvanny I (2011). Presenilins function in ER calcium leak and Alzheimer's disease pathogenesis. Cell Calcium 50: 303–309. https://doi.org/10.1016/j.ceca.2011.05.013.

Tabrizi SJ, Langbehn DR, Leavitt BR et al. (2009). Biological and clinical manifestations of Huntington's disease in the longitudinal TRACK-HD study: cross-sectional analysis of baseline data. Lancet Neurol 8: 791–801. https://doi.org/10.1016/S1474-4422(09)70170-X.

Taipa R, Ferreira V, Brochado P et al. (2018). Inflammatory pathology markers (activated microglia and reactive astrocytes) in early and late onset Alzheimer disease: a post mortem study. Neuropathol Appl Neurobiol 44: 298–313. https://doi.org/10.1111/nan.12445.

Takano T, Han X, Deane R et al. (2007). Two-photon imaging of astrocytic Ca2+ signaling and the microvasculature in experimental mice models of Alzheimer's disease. Ann N Y Acad Sci 1097: 40–50. https://doi.org/10.1196/annals.1379.004.

Tanemura K, Murayama M, Akagi T et al. (2002). Neurodegeneration with tau accumulation in a transgenic mouse expressing V337M human tau. J Neurosci 22: 133–141. https://doi.org/10.1523/JNEUROSCI.22-01-00133.2002.

Tapella L, Cerruti M, Biocotino I et al. (2018). TGF-β2 and TGF-β3 from cultured β-amyloid-treated or 3xTg-AD-derived astrocytes may mediate astrocyte-neuron communication. Eur J Neurosci 47: 211–221. https://doi.org/10.1111/ejn.13819.

Tapella L, Dematteis G, Moro M et al. (2022). Protein synthesis inhibition and loss of homeostatic functions in astrocytes from an Alzheimer's disease mouse model: a role for ER-mitochondria interaction. Cell Death Dis 13: 878. https://doi.org/10.1038/s41419-022-05324-4.

Tatebayashi Y, Miyasaka T, Chui D-H et al. (2002). Tau filament formation and associative memory deficit in aged mice expressing mutant (R406W) human tau. Proc Natl Acad Sci USA 99: 13896–13901. https://doi.org/10.1073/pnas.202205599.

Tebbenkamp ATN, Swing D, Tessarollo L et al. (2011). Premature death and neurologic abnormalities in transgenic mice expressing a mutant huntingtin exon-2 fragment. Hum Mol Genet 20: 1633–1642. https://doi.org/10.1093/hmg/ddr040.

Theofilas P, Ehrenberg AJ, Nguy A et al. (2018). Probing the correlation of neuronal loss, neurofibrillary tangles, and cell death markers across the Alzheimer's disease Braak stages: a quantitative study in humans. Neurobiol Aging 61: 1–12. https://doi.org/10.1016/j.neurobiolaging.2017.09.007.

Thibault O, Gant JC, Landfield PW (2007). Expansion of the calcium hypothesis of brain aging and Alzheimer's disease: minding the store. Aging Cell 6: 307–317. https://doi.org/10.1111/j.1474-9726.2007.00295.x.

Thiruchelvam MJ, Powers JM, Cory-Slechta DA et al. (2004). Risk factors for dopaminergic neuron loss in human alpha-synuclein transgenic mice. Eur J Neurosci 19: 845–854. https://doi.org/10.1111/j.0953-816x.2004.03139.x.

Thirugnanam T, Santhakumar K (2022). Chemically induced models of Parkinson's disease. Comp Biochem Physiol C Toxicol Pharmacol 252: 109213. https://doi.org/10.1016/j.cbpc.2021.109213.

Toescu EC, Möller T, Kettenmann H et al. (1998). Long-term ac tivation of capacitative Ca^{2+} entry in mouse microglial cells. Neuroscience 86: 925–935. https://doi.org/10.1016/s0306-4522(98)00123-7.

Toescu EC, Verkhratsky A, Landfield PW (2004). Ca^{2+} regulation and gene expression in normal brain aging. Trends Neurosci 27: 614–620. https://doi.org/10.1016/j.tins.2004.07.010.

Toledano A, Alvarez MI (2004). Lesions and dysfunctions of the nucleus basalis as Alzheimer's disease models: general and critical overview and analysis of the long-term changes in several excitotoxic models. Curr Alzheimer Res 1: 189–214. https://doi.org/10.2174/1567205043332117.

Tong Y, Pisani A, Martella G et al. (2009). R1441C mutation in LRRK2 impairs dopaminergic neurotransmission in mice. Proc Natl Acad Sci USA 106: 14622–14627. https://doi.org/10.1073/pnas.0906334106.

Tong X, Ao Y, Faas GC et al. (2014). Astrocyte Kir4.1 ion channel deficits contribute to neuronal dysfunction in Huntington's disease model mice. Nat Neurosci 17: 694–703. https://doi.org/10.1038/nn.3691.

Trujillo-Estrada L, Sanchez-Mejias E, Sanchez-Varo R et al. (2022). Animal and cellular models of Alzheimer's disease: progress, promise, and future approaches. Neuroscientist 28: 572–593. https://doi.org/10.1177/10738584211001753.

Untiet V, Beinlich FRM, Kusk P et al. (2023). Astrocytic chloride is brain state dependent and modulates inhibitory neurotransmission in mice. Nat Commun 14: 1871. https://doi.org/10.1038/s41467-023-37433-9.

Upadhayay S, Yedke NG, Rahi V et al. (2023). An overview of the pathophysiological mechanisms of 3-nitropropionic acid (3-NPA) as a neurotoxin in a Huntington's disease model and its relevance to drug discovery and development. Neurochem Res 48: 1631–1647. https://doi.org/10.1007/s11064-023-03868-1.

van der Kant R, Langness VF, Herrera CM et al. (2019). Cholesterol metabolism is a druggable axis that independently regulates tau and amyloid-β in iPSC-derived Alzheimer's disease neurons. Cell Stem Cell 24. https://doi.org/10.1016/j.stem.2018.12.013. 363–375.e9.

Van der Perren A, Toelen J, Casteels C et al. (2015a). Longitudinal follow-up and characterization of a robust rat model for Parkinson's disease based on overexpression of alpha-synuclein with adeno-associated viral vectors. Neurobiol Aging 36: 1543–1558. https://doi.org/10.1016/j.neurobiolaging.2014.11.015.

Van der Perren A, Van den Haute C, Baekelandt V (2015b). Viral vector-based models of Parkinson's disease. Curr Top Behav Neurosci 22: 271–301. https://doi.org/10.1007/7854_2014_310.

van der Putten H, Wiederhold KH, Probst A et al. (2000). Neuropathology in mice expressing human alpha-synuclein. J Neurosci 20: 6021–6029. https://doi.org/10.1523/JNEUROSCI.20-16-06021.2000.

Verberk IMW, Laarhuis MB, van den Bosch KA et al. (2021). Serum markers glial fibrillary acidic protein and neurofilament light for prognosis and monitoring in cognitively normal older people: a prospective memory clinic-based cohort study. Lancet Healthy Longev 2: e87–e95. https://doi.org/10.1016/S2666-7568(20)30061-1.

Verkhratsky A (2019). Astroglial calcium signaling in aging and Alzheimer's disease. Cold Spring Harb Perspect Biol 11: a035188. https://doi.org/10.1101/cshperspect.a035188.

Verkhratsky A, Butt AM (2023). Neuroglia: function and pathology. Elsevier, New York, 730 pp.

Verkhratsky A, Nedergaard M (2018). Physiology of Astroglia. Physiol Rev98: 239–389. https://doi.org/10.1152/physrev.00042.2016.

Verkhratsky A, Semyanov A (2023). Astrocytes in ageing. Subcell Biochem103: 253–277. https://doi.org/10.1007/978-3-031-26576-1_11.

Verkhratsky A, Olabarria M, Noristani HN et al. (2010). Astrocytes in Alzheimer's disease. Neurotherapeutics 7: 399–412. https://doi.org/10.1016/j.nurt.2010.05.017.

Verkhratsky A, Marutle A, Rodríguez-Arellano JJ et al. (2015). Glial asthenia and functional paralysis: a new perspective on neurodegeneration and Alzheimer's disease. Neuroscientist 21: 552–568. https://doi.org/10.1177/1073858414547132.

Verkhratsky A, Rodrigues JJ, Pivoriunas A et al. (2019a). Astroglial atrophy in Alzheimer's disease. Pflugers Arch. https://doi.org/10.1007/s00424-019-02310-2.

Verkhratsky A, Untiet V, Rose CR (2019b). Ionic signalling in astroglia beyond calcium. J Physiol (Lond). https://doi.org/10.1113/JP277478.

Verkhratsky A, Augusto-Oliveira M, Pivoriūnas A et al. (2020a). Astroglial asthenia and loss of function, rather than reactivity, contribute to the ageing of the brain. Pflugers Arch. https://doi.org/10.1007/s00424-020-02465-3.

Verkhratsky A, Semyanov A, Zorec R (2020b). Physiology of astroglial excitability. Function (Oxf) 1: zqaa016. https://doi.org/10.1093/function/zqaa016.

Verkhratsky A, Butt A, Li B et al. (2023). Astrocytes in human central nervous system diseases: a frontier for new therapies. Signal Transduct Target Ther 8: 396.

Vilalta A, Zhou Y, Sevalle J et al. (2021). Wild-type sTREM2 blocks Aβ aggregation and neurotoxicity, but the Alzheimer's R47H mutant increases Aβ aggregation. J Biol Chem 296: 100631. https://doi.org/10.1016/j.jbc.2021.100631.

Volterra A, Liaudet N, Savtchouk I (2014). Astrocyte Ca^{2+} signalling: an unexpected complexity. Nat Rev Neurosci 15: 327–335. https://doi.org/10.1038/nrn3725.

von Saucken VE, Jay TR, Landreth GE (2020). The effect of amyloid on microglia-neuron interactions before plaque onset occurs independently of TREM2 in a mouse model of Alzheimer's disease. Neurobiol Dis 145: 105072. https://doi.org/10.1016/j.nbd.2020.105072.

Vonsattel JP, Myers RH, Stevens TJ et al. (1985). Neuropathological classification of Huntington's disease. J Neuropathol Exp Neurol 44: 559–577. https://doi.org/10.1097/00005072-198511000-00003.

Vonsattel JPG, Keller C, Del Pilar Amaya M (2008). Neuropathology of Huntington's disease. Handb Clin Neurol 89: 599–618. https://doi.org/10.1016/S0072-9752(07)01256-0.

Wakamatsu M, Ishii A, Iwata S et al. (2008). Selective loss of nigral dopamine neurons induced by overexpression of truncated human alpha-synuclein in mice. Neurobiol Aging 29: 574–585. https://doi.org/10.1016/j.neurobiolaging.2006.11.017.

Wang L, Lin F, Wang J et al. (2012). Expression of mutant N-terminal huntingtin fragment (htt552-100Q) in astrocytes suppresses the secretion of BDNF. Brain Res 1449: 69–82. https://doi.org/10.1016/j.brainres.2012.01.077.

Wang Q, Liu Y, Zhou J (2015). Neuroinflammation in Parkinson's disease and its potential as therapeutic target. Transl Neurodegener 4: 19. https://doi.org/10.1186/s40035-015-0042-0.

Wang Q, Wang Y, Liu J et al. (2019). Quantification of white matter cellularity and damage in preclinical and early symptomatic Alzheimer's disease. Neuroimage Clin 22: 101767. https://doi.org/10.1016/j.nicl.2019.101767.

Wardlaw JM, Valdés Hernández MC, Muñoz-Maniega S (2015). What are white matter hyperintensities made of? Relevance to vascular cognitive impairment. J Am Heart Assoc 4: 001140. https://doi.org/10.1161/JAHA.114.001140.

Webster SJ, Bachstetter AD, Van Eldik LJ (2013). Comprehensive behavioral characterization of an APP/PS-1 double knock-in mouse model of Alzheimer's disease. Alzheimers Res Ther 5: 28. https://doi.org/10.1186/alzrt182.

Westerman MA, Cooper-Blacketer D, Mariash A et al. (2002). The relationship between Abeta and memory in the Tg2576 mouse model of Alzheimer's disease. J Neurosci 22: 1858–1867. https://doi.org/10.1523/JNEUROSCI.22-05-01858.2002.

Wharton SB, O'Callaghan JP, Savva GM et al. (2009). Population variation in glial fibrillary acidic protein levels in brain ageing: relationship to Alzheimer-type pathology and dementia. Dement Geriatr Cogn Disord 27: 465–473. https://doi.org/10.1159/000217729.

Wheeler VC, White JK, Gutekunst CA et al. (2000). Long glutamine tracts cause nuclear localization of a novel form of huntingtin in medium spiny striatal neurons in HdhQ92 and HdhQ111 knock-in mice. Hum Mol Genet 9: 503–513. https://doi.org/10.1093/hmg/9.4.503.

White JK, Auerbach W, Duyao MP et al. (1997). Huntingtin is required for neurogenesis and is not impaired by the Huntington's disease CAG expansion. Nat Genet 17: 404–410. https://doi.org/10.1038/ng1297-404.

Wszolek ZK, Tsuboi Y, Ghetti B et al. (2006). Frontotemporal dementia and parkinsonism linked to chromosome 17 (FTDP-17). Orphanet J Rare Dis 1: 30. https://doi.org/10.1186/1750-1172-1-30.

Xu J, Chen S, Ahmed SH et al. (2001). Amyloid-beta peptides are cytotoxic to oligodendrocytes. J Neurosci 21: RC118. https://doi.org/10.1523/JNEUROSCI.21-01-j0001.2001.

Yamada T, McGeer PL, McGeer EG (1992). Appearance of paired nucleated, Tau-positive glia in patients with progressive supranuclear palsy brain tissue. Neurosci Lett 135: 99–102. https://doi.org/10.1016/0304-3940(92)90145-w.

Yamamoto A, Lucas JJ, Hen R (2000). Reversal of neuropathology and motor dysfunction in a conditional model of Huntington's disease. Cell 101: 57–66. https://doi.org/10.1016/S0092-8674(00)80623-6.

Yang W, Wang G, Wang C-E et al. (2015). Mutant alpha-synuclein causes age-dependent neuropathology in monkey brain. J Neurosci 35: 8345–8358. https://doi.org/10.1523/JNEUROSCI.0772-15.2015.

Yang H-M, Yang S, Huang S-S et al. (2017). Microglial activation in the pathogenesis of Huntington's disease. Front Aging Neurosci 9: 193. https://doi.org/10.3389/fnagi.2017.00193.

Yeh C-Y, Vadhwana B, Verkhratsky A et al. (2011). Early astrocytic atrophy in the entorhinal cortex of a triple transgenic animal model of Alzheimer's disease. ASN Neuro 3: 271–279. https://doi.org/10.1042/AN20110025.

Yoshida M (2014). Astrocytic inclusions in progressive supranuclear palsy and corticobasal degeneration. Neuropathology 34: 555–570. https://doi.org/10.1111/neup.12143.

Yoshiyama Y, Higuchi M, Zhang B et al. (2007). Synapse loss and microglial activation precede tangles in a P301S tauopathy mouse model. Neuron 53: 337–351. https://doi.org/10.1016/j.neuron.2007.01.010.

Yu Z-X, Li S-H, Evans J et al. (2003). Mutant huntingtin causes context-dependent neurodegeneration in mice with Huntington's disease. J Neurosci 23: 2193–2202. https://doi.org/10.1523/JNEUROSCI.23-06-02193.2003.

Zhang C, McNeil E, Dressler L et al. (2007). Long-lasting impairment in hippocampal neurogenesis associated with amyloid deposition in a - knock-in mouse model of familial Alzheimer's disease. Exp Neurol 204: 77–87. https://doi.org/10.1016/j.expneurol.2006.09.018.

Zhang Z-N, Freitas BC, Qian H et al. (2016). Layered hydrogels accelerate iPSC-derived neuronal maturation and reveal migration defects caused by MeCP2 dysfunction. Proc Natl Acad Sci USA 113: 3185–3190. https://doi.org/10.1073/pnas.1521255113.

Zhang G, Wang Z, Hu H et al. (2021). Microglia in Alzheimer's disease: a target for therapeutic intervention. Front Cell Neurosci 15: 749587. https://doi.org/10.3389/fncel.2021.749587.

Zhao J, O'Connor T, Vassar R (2011). The contribution of activated astrocytes to Aβ production: implications for Alzheimer's disease pathogenesis. J Neuroinflammation 8: 150. https://doi.org/10.1186/1742-2094-8-150.

Zhao J, Davis MD, Martens YA et al. (2017). APOE ϵ4/ϵ4 diminishes neurotrophic function of human iPSC-derived astrocytes. Hum Mol Genet 26: 2690–2700. https://doi.org/10.1093/hmg/ddx155.

Zorec R, Araque A, Carmignoto G et al. (2012). Astroglial excitability and gliotransmission: an appraisal of Ca2+ as a signalling route. ASN Neuro 4: e00080. https://doi.org/10.1042/AN20110061.

Zuccato C, Cattaneo E (2007). Role of brain-derived neurotrophic factor in Huntington's disease. Prog Neurobiol 81: 294–330. https://doi.org/10.1016/j.pneurobio.2007.01.003.

Zuccato C, Ciammola A, Rigamonti D et al. (2001). Loss of huntingtin-mediated BDNF gene transcription in Huntington's disease. Science 293: 493–498. https://doi.org/10.1126/science.1059581.

Handbook of Clinical Neurology, Vol. 210 (3rd series)
Neuroglia in Neurologic and Psychiatric Disorders, Part II
A. Verkhratsky, L.D. de Witte, E. Aronica, and E.M. Hol, Editors
https://doi.org/10.1016/B978-0-443-19102-2.00004-1

Chapter 3

Neuroglia in neurodegeneration: Amyotrophic lateral sclerosis and frontotemporal dementia

FABIOLA DE MARCHI[1,†], EDOARDO GIOELE SPINELLI[2,3,†], AND CATERINA BENDOTTI[4*]

[1]*ALS Centre, Neurology Unit, Maggiore della Carità Hospital, University of Piemonte Orientale, Novara, Italy*

[2]*Neurology Unit, Department of Neuroscience, IRCCS Ospedale San Raffaele, Milano, Italy*

[3]*Vita-Salute San Raffaele University, Milano, Italy*

[4]*Laboratory of Neurobiology and Preclinical Therapeutics, ALS Center, Department of Neuroscience, IRCCS-"Mario Negri" Institute for Pharmacological Research, Milano, Italy*

Abstract

Amyotrophic lateral sclerosis (ALS) and frontotemporal dementia (FTD) are devastating neurodegenerative diseases sharing significant pathologic and genetic overlap, leading to consider these diseases as a continuum in the spectrum of their pathologic features. Although FTD compromises only specific brain districts, while ALS involves both the nervous system and the skeletal muscles, several neurocentric mechanisms are in common between ALS and FTD. Also, recent research has revealed the significant involvement of nonneuronal cells, particularly glial cells such as astrocytes, oligodendrocytes, microglia, and peripheral immune cells, in disease pathology. This chapter aims to provide an extensive overview of the current understanding of the role of glia in the onset and advancement of ALS and FTD, highlighting the recent implications in terms of prognosis and future treatment options.

INTRODUCTION

Amyotrophic lateral sclerosis (ALS) and frontotemporal dementia (FTD) are devastating neurodegenerative diseases (NDDs) that share significant pathologic and genetic overlap. Approximately 10%–15% of ALS cases and as much as 40% of FTD cases are hereditary, typically characterized by dominant genetic traits. Except for mutations in *SOD1*, *MAPT*, or *GRN*, which are exclusively linked to either ALS or FTD, these disorders often manifest common gene mutations, such as those seen in *C9ORF72*, *TARDBP*, *UBQLN2*, *OPT*, and *TBK1*, leading to consider these diseases as a continuum in the spectrum of their pathologic features. Several common mechanisms contribute to the development of ALS and FTD, including impaired protein homeostasis, oxidative stress, excitotoxicity, aberrant RNA metabolism, impairment in axonal transport, and neuroinflammation. However, while FTD essentially compromises specific brain districts, in ALS, the impairment extends to the peripheral nervous system and to the skeletal muscles, which actively contribute to the pathogenesis through a "dying back" mechanism, thus increasing the complexity of the pathogenesis of this disease (Feldman et al., 2022).

Although the main emphasis in understanding the development of these disorders has traditionally been neurocentric, recent research has revealed the significant involvement of nonneuronal cells, particularly glial cells such as astrocytes, oligodendrocytes (OL), microglia, and peripheral immune cells, in a noncell autonomous scenario where the contribution of all these cells is closely interconnected in controlling the state of neighboring neurons. This chapter delves into the complex role of glial cells in the onset and advancement of ALS and FTD, underlining their influence on the progression of these diseases.

Amyotrophic lateral sclerosis

ALS is a complex, progressive, and fatal NDD of the central nervous system (CNS) (Feldman et al., 2022; De Marchi et al., 2023). Although considered a rare disorder, the lifetime risk of

†Co-first authors.

*Correspondence to: Caterina Bendotti, Laboratory of Neurobiology and Preclinical Therapeutics, ALS Center, Department of Neuroscience, IRCCS-"Mario Negri" Institute for Pharmacological Research, Milano, Italy. Tel: +39-0239014488, E-mail: caterina.bendotti@marionegri.it

developing ALS is roughly one in 350–400 persons, which means an incidence of 2–3/100,000 people (Chiò et al., 2013; Longinetti and Fang, 2019). ALS seems to be slightly more prevalent in males than females, with a ratio of 1.35:1. Also, compared to other NDDs, the prevalence is not notably higher than the incidence, considering the short disease duration (Erkkinen et al., 2018). The risk of disease development progressively increases with age, with a median age of onset at 58–60 years (Talbott et al., 2016), but juvenile and late-onset cases are not rare.

ALS is characterized by the progressive degeneration of motor neurons (MNs) in the brain and spinal cord, involving both upper (UMNs) and lower MNs (LMNs) (Zakharova and Abramova, 2022; Colombo et al., 2023). Clinically, this manifests as a variable phenotype, presenting signs and symptoms related to UMN and LMN (Chiò et al., 2020). There are two primary presentation phenotypes: the spinal one, accounting for approximately 70% of cases, and the bulbar one, comprising 30%. Within the classic spinal phenotype, various clinical nuances can be grouped, including predominant impairment of UMN or of LMN, flail arm, flail leg, and hemiplegic (Chiò et al., 2020). This phenotype, characterized by typical symptoms such as muscle weakness, hypotrophy, fasciculations, and cramps, is more frequently observed in males under 60. Conversely, the prevalent presentation in bulbar forms involves symptoms such as dysarthria, dysphagia, and sialorrhea. This phenotype, more common in women over 60, may also include rare cases with a respiratory onset. Currently, the El Escorial criteria (Brooks et al., 2000), modified in 2015 (Ludolph et al., 2015), are utilized for clinical diagnosis, relying on the distribution of MN involvement in various body segments, supported by active and chronic denervation activity observed in electromyographic examinations. Alongside the classical phenotypes of ALS, with documented UMN and LMN involvement, it is necessary to mention primary lateral sclerosis (PLS), a rare, slowly progressive disease characterized by the involvement of the UMN only (Paganoni et al., 2020a). Differentiating PLS from UMN-ALS is challenging, especially in the initial disease stages, with consequent prognostic and therapeutic implications (such as the inclusion in clinical trials) (Turner et al., 2020). Also, the cooccurrence of cases of PLS and ALS in the same families supports the idea of a possible common mechanism of motoneurons' damage in the PLS and ALS (De Boer et al., 2024).

In ALS, the prognosis depends on the rate of progression, and in a clinical setting, three indicators are commonly used to monitor progression: (1) the ALS Functional Rating Scale-Revised (ALSFRS-R) score, a functional scale (Cedarbaum et al., 1999; Gordon et al., 2004); (2) the pulmonary function tests, indicating the pulmonary involvement (Heiman-Patterson et al., 2021); and (3) the body mass index, considering the malnutrition a negative prognostic factor (Dardiotis et al., 2018).

It is now recognized that approximately 50% of cases exhibit concurrent cognitive involvement (Murphy et al., 2016; De Marchi et al., 2021a; Rusina et al., 2021), which primarily affects executive functions. Behavioral disturbances, including apathy, irritability, and alterations in eating habits, often accompany cognitive impairments. According to the Strong classification (Strong et al., 2017), patients are categorized into five groups based on the presence or absence of cognitive, behavioral, or both symptoms: ALS-normal, ALS-ci (with mild executive deficits), ALS-bi (with behavioral deficits), ALS-cbi (with both executive and behavioral deficits), and ALS-FTD (with severe cognitive and behavioral problems).

Recognizing cognitive symptoms in ALS patients has significant implications in pathophysiological, clinical-diagnostic, and ethical-prognostic terms. It is evident from the literature that cognitive deficits are associated with a shorter survival period (Chio et al., 2009; Westeneng et al., 2018; Goutman et al., 2022) and that executive deficits can complicate the end-of-life decisions that the disease necessitates throughout its progression (Connolly et al., 2015).

ALS is predominantly considered a sporadic disease (sALS). However, the percentage of familial cases (fALS) varies from 10% to 20%, depending on the case studies (Zou et al., 2017; Zhang et al., 2022). In most cases, fALS is monogenic, linked to single mutations in genes known to be causative of the disease. Over 40 genes have been discovered to be associated with ALS, although most forms are associated with mutations in *SOD1*, *C9ORF72*, *FUS*, *TARDBP*, and *TBK1* (Grassano et al., 2022; Megat et al., 2023). Notably, in the context of this chapter, repeat expansions in the promoter region of the *C9ORF72* gene are equally associated with exclusively motor forms, often with bulbar onset, exclusively cognitive forms, more frequently with behavioral onset, or mixed forms, involving both motor and cognitive symptoms, configuring ALS-FTD (Breevoort et al., 2022). The improvement of genetic knowledge is inevitably reflected in the development of new drugs acting in this regard. For example, Tofersen (also called BIIB067), a small string of DNA designed to bind to specific molecules of RNA and act as an antisense oligonucleotide (ASO), was recently approved for ALS patients' carriers of the *SOD1* mutation. This molecule explicitly targets the RNA produced from mutated *SOD1* genes, stopping toxic SOD1 proteins from being made. At the time of writing, the Food and Drug Administration recently approved Tofersen for patients with the *SOD1* mutation, and the concomitant compassionate use is permitted in Europe. This scenario mandates a genetic testing routine in clinical practice.

Currently, ALS remains an incurable disease, and no drugs can significantly modify the disease course (Johnson et al., 2022). For over 20 years, the only treatment approved for ALS was riluzole, an antiglutamatergic drug that can slow down the ALS progression by a few months (Bensimon et al., 1994; Lacomblez et al., 1996; Fang et al., 2018; Andrews et al., 2020). In recent years, two other drugs have been associated with riluzole (to date, approved only in the United States): edaravone (Abe et al., 2014), a free radical scavenger that suppresses the generation of hydroxyl radicals and peroxynitrite radicals, and sodium phenylbutyrate/taurursodiol (Relyvrio), a pan-histone deacetylase inhibitor, able to prevent nerve cell death by blocking stress signals acting on mitochondria and endoplasmic reticulum (Paganoni et al., 2020b; Wong et al., 2021) (see paragraph "Glial-based clinical trials for ALS/FTD" for new treatments and ongoing clinical trials).

Frontotemporal dementia

The term FTD stands for a diverse and challenging spectrum of NDDs characterized by progressive cognitive decline and/or behavioral alterations accompanied by damage to the brain frontal and temporal lobes. It is the most frequent neurodegenerative cause of dementia in people under the age of 65 (Erkkinen et al., 2018), and the understanding of distinct FTD subtypes is crucial for accurate diagnosis, prognosis, and the development of targeted therapeutic interventions.

The behavioral variant of FTD (bvFTD) represents the most common clinical presentation, making up around 40% of all cases (Logroscino et al., 2023). Patients present with prominent changes in behavior, personality, and social conduct (Boxer and Miller, 2005). Core symptoms include disinhibition, apathy, loss of empathy, and mostly executive cognitive dysfunction (Rascovsky et al., 2011). According to the current criteria (Rascovsky et al., 2011), a diagnosis of probable bvFTD is supported by typical neuroimaging findings, including atrophy or hypometabolism of the frontal and/or anterior temporal brain regions.

Among other FTD presentations, those primarily affecting linguistic functions are collectively known as primary progressive aphasia (PPA) syndromes. Even though other cognitive or behavioral impairments can emerge later in the disease course, language has to be the first and most impaired domain throughout the disease duration to prompt a diagnosis of PPA (Mesulam, 1982, 2003). The semantic variant of PPA (svPPA) primarily affects word comprehension and object recognition. Patients with svPPA demonstrate impaired naming and exhibit deficits in understanding the meaning of words and objects (i.e., loss of semantic knowledge) (Gorno-Tempini et al., 2011), accompanied by neuroanatomic damage typically affecting the temporal poles bilaterally, with a left-hemispheric prevalence (Rogalski et al., 2011). By contrast, the nonfluent/agrammatic variant of PPA (nfvPPA) is characterized by effortful speech due to difficulties in coordinating the motor movements required for speech production (i.e., apraxia of speech) and grammar difficulties (Gorno-Tempini et al., 2011). Neuroanatomic damage of nfvPPA patients is characteristically localized to the left posterior frontal and insular cortical regions (Gorno-Tempini et al., 2011; Rogalski et al., 2011).

While 10%–15% of ALS cases fulfill criteria for bvFTD or PPA variants, features of MN disease appear in approximately 15% of FTD patients during the disease progression (Ling et al., 2013). The clinical and neuropsychologic spectrum for ALS/FTD is shown in Fig. 3.1 (van Es et al., 2017; Cividini et al., 2022).

Neuroimaging studies in ALS have consistently reported extramotor cortical involvement in this disease, and similarities in patterns of regional cortical volume loss have also been identified in cognitively intact ALS patients and those with ALS-FTD, providing further evidence of a neuroanatomic "continuum" between these two disorders (Ling et al., 2013). Converging evidence comes from genetic studies as well. The hexanucleotide repeat expansion in the *C9ORF72* gene is the most common mutation in inherited forms of ALS (40%) and FTD (25%), and in a significant number of sALS (5%–20%) and FTD (6%) cases,

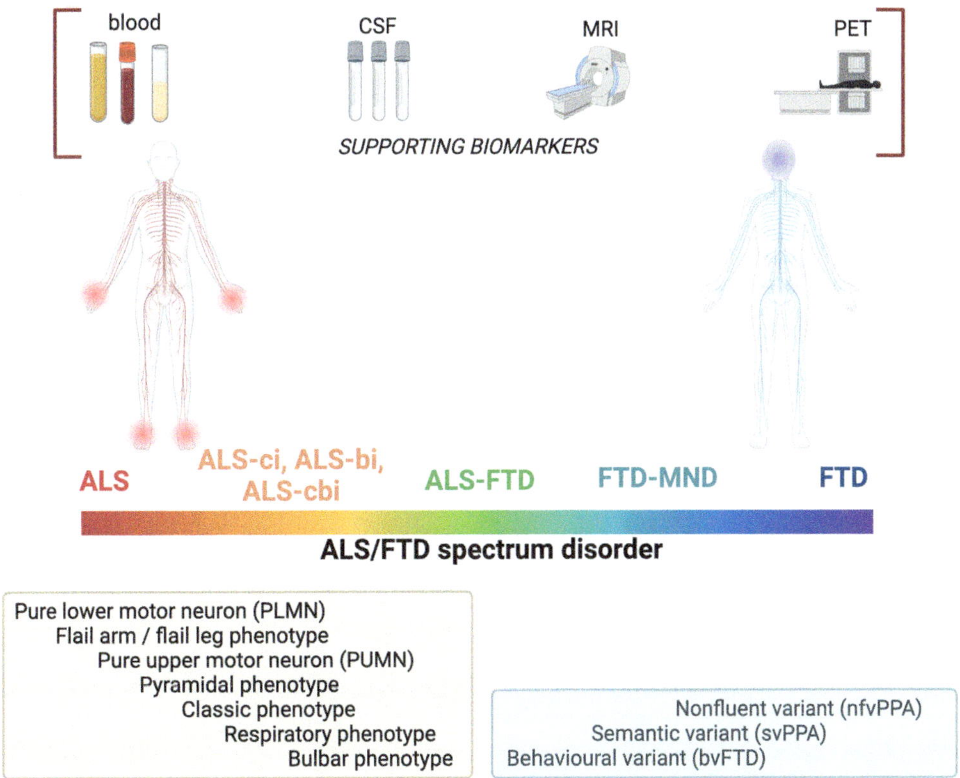

Fig. 3.1. The clinical and cognitive variability in the ALS/FTD spectrum disorder. The man in red represents a subject with pure motor ALS, while the man in blue represents a subject with pure FTD. Between the two clinical extremes, there are different nuances, both from a motor and cognitive point of view. In the top line, the supporting biomarkers (fluids and imaging) used for ALS/FTD diagnosis. Created with Biorender.com.

and explains the majority of the overlap of these two conditions (Majounie et al., 2012). Moreover, mutations in the *TARDBP* gene have been found to be causative for both ALS and FTD (Lattante et al., 2015). The overlap between ALS and FTD has been further confirmed by pathologic studies, where hyperphosphorylated TDP-43 inclusions have been observed in neurons and glial cells of nearly half of bvFTD cases as well as the majority of ALS patients (Ferrari et al., 2011; Brettschneider et al., 2014a,b; Prudlo et al., 2016; Hirsch-Reinshagen et al., 2023).

ANIMAL MODELS OF ALS AND FTD

The discovery of genetic mutations associated with ALS and FTD constituted a decisive turning point in expanding knowledge on these diseases, as it allowed the development of cellular and animal models to elucidate the mechanisms underlying the pathogenesis of these diseases. Among a plethora of models so far developed, those involving rodents hold particular significance. Indeed, by randomly inserting human mutant ALS or FTD genes into the mouse or rat genome, researchers can investigate these pathologies in a living mammalian organism, allowing for a closer recapitulation of the complexity of the human disease. In addition, rodent model paradigms have the advantage of allowing researchers to investigate tissue changes and molecular mechanisms that occur before the symptom's onset and at different stages during the progression of the disease. While a comprehensive discussion on all the available animal models of ALS/FTD can be appreciated in various exhaustive reviews (De Giorgio et al., 2019; Bonifacino et al., 2021; Todd and Petrucelli, 2022), here we briefly summarize those models carrying the most representative gene mutations related exclusively to ALS (*SOD1*) or both disorders (*TARDBP*, *C9ORF72*, *TBK1*, and *OPT*), providing clear evidence supporting the involvement of glia and immune cells in shaping their phenotype. For a list and description of currently available mouse models of ALS/FTD, see http://www.alzforum.org/research-models/search.

SOD1 rodent models of ALS

One year after the discovery of the first ALS-linked gene mutation related to Cu/Zn superoxide dismutase 1 (*SOD1*) in 1993, Gurney et al. (1994) developed the first transgenic mouse made by randomly inserting human mutant *SOD1* in the mouse genome, leading to a ubiquitous overexpression of the human protein with a glycine for alanine substitution in position 93 (SOD1G93A). These mice show progressive motor deficits, paralysis, and lethality due to progressive degeneration of spinal MNs and skeletal muscle atrophy recapitulating most of the ALS features (Bendotti and Carrì, 2004). So far, several rodent models, including rats, have been developed harboring different *SOD1* mutations (De Giorgio et al., 2019). Despite developing progressive MN degeneration, they differ in the onset and progression of the pathology and their lifespan depending on the type of mutation, the level of expression of the transgene, and the genetic background in a setting of heterogeneity that mimics that of SOD1-ALS patients (Bendotti et al., 2020). However, they share common traits in the pathogenic mechanisms of MN degeneration such as glutamate excitotoxicity, mitochondrial alterations, defects in axonal transport, protein aggregation, and neuroinflammation that recapitulate many of the mechanisms implicated also in human disease (Taylor et al., 2016). In this perspective, the mSOD1 model is viewed as an experimental paradigm for testing therapeutic approaches targeting specific disease mechanisms and, to date, remains the primary model for evaluating potential ALS therapeutic interventions as well as for the development of pharmacodynamic biomarkers that can be used in a translational approach. Notably, the SOD1G93A transgenic mouse has been essential to the preclinical studies for the main FDA-approved therapies for ALS (Gurney et al., 1998; Ryu et al., 2005; Ito et al., 2008) and for the novel gene therapy approach with the ASO targeted to *SOD1* (McCampbell et al., 2018). Regrettably, other treatments that effectively slowed down the disease progression and extended the survival of these mice did not produce the same positive results in ALS patients (Lee et al., 2023). However, the introduction of specific guidelines for using these mice (Ludolph et al., 2010) highlighted that the ability to replicate published results in this mouse model is also quite limited (Perrin, 2014). This raises concerns and suggests that caution is necessary when interpreting and extrapolating results from this mouse model to clinical applications. A disadvantage of these models is that SOD1-ALS accounts for only about 2% of human ALS, and SOD1-ALS tissue does not show the typical TDP-43 inclusions seen in most ALS cases. These limitations called the researcher to explore other models to enhance the translational relevance of their findings.

TDP-43 rodent models of ALS and FTD

Rodents carrying mutations in the *TARDBP* gene (encoding for TAR DNA-binding protein 43, or TDP-43) represent important models to investigate neurotoxic mechanisms in both ALS and FTD. Indeed, although *TARDBP* mutations account only for approximately 4% of fALS and FTD patients, mutated TDP-43 is the main component of cytoplasmic ubiquitinated inclusions in the CNS of most cases of fALS and sALS and in 50% of FTD. Various TDP-43 rodent models, with different mutations, promoters, and expression levels, have been generated to understand TDP-43-linked pathology (De Giorgio et al., 2019; Todd and Petrucelli, 2022). Disease severity in these models, whether carrying TDP-43 mutants or wild-type, is dosage-sensitive, with minimal overexpression causing RNA changes, neuronal toxicity, and a rather late-onset neuromuscular junction damage (Yang et al., 2022), while when the overexpression overcame a particular threshold, neurodegeneration and the pathologic phenotype can be very rapid (Gendron and Petrucelli, 2011). Mouse models overexpressing human TDP-43 with different mutations exhibit varying degrees of neurologic abnormalities, dependent on expression levels, with some showing premature lethality unrelated to neurodegeneration (De Giorgio et al., 2019; Todd and Petrucelli, 2022).

Models expressing mutant TDP-43 at endogenous physiological levels like those expressing a single-copy bacterial artificial chromosome (BAC) carrying TDP43-M337V (Gordon et al., 2019) or the mice bearing mutations in endogenous TDP-43 gene by knock-in approaches like the Q331K mutant TDP-43 (Arnold et al., 2013; White et al., 2018; Huang et al., 2020; Watanabe et al., 2020) develop a mild ALS-like phenotype at advanced ages, with or without neurodegeneration and without expected TDP43 pathology.

Cytoplasmic mislocalization of TDP-43, leading to a loss of its endogenous nuclear functions (Igaz et al., 2011), may be critical for initial neurodegeneration. However, TDP-43 mouse models, even with mislocalized cytoplasmic TDP-43, exhibit neuronal degeneration without typical cytoplasmic aggregates, suggesting that overt TDP-43 deposition may be critical for the late stage of the disease.

Hence, TDP-43 proteinopathy observed in postmortem specimens may be a marker of the disease end stage rather than contributing to pathogenesis. Loss of TDP43 in conditional knockout animals induces pronounced ALS/FTD-like defects (Wu et al., 2012, 2019).

In addition, MNs dysfunction may be influenced by the selective loss of function of glial TDP43, suggesting a noncell autonomous mechanism of neurodegeneration (Tong et al., 2013; Yang et al., 2014b; Peng et al., 2020).

In summary, the variability in the phenotype of the numerous rodent models is likely attributed to the tightly regulated levels of TDP-43 in cells. This insight underscores the importance of precise control over experimental conditions, including the manipulation of TDP-43 levels, to ensure the reliability and reproducibility of research outcomes.

Chromosome 9 open reading frame 72 (C9ORF72) mutant models of ALS and FTD

The presence of the G4C2 hexanucleotide repeat expansion (HRE) in the first intron of the human *C9ORF72* gene is currently the most commonly known genetic cause of both familial and sporadic ALS/FTD (Renton et al., 2011), sparking significant interest in developing mouse models with this genetic mutation. As the G4C2 repeat expansion is believed to cause toxicity through gain- and loss-of-function mechanisms, mouse models expressing the *C9ORF72* HRE and dipeptide repeat proteins (DPRs) or reducing endogenous C9orf72 levels have been developed (Balendra and Isaacs, 2018).

Transgenic C9orf72 models with BACs carrying the human *C9ORF72* gene, despite reproducing RNA foci and DPRs, do not exhibit ALS/FTD phenotypes (O'Rourke et al., 2015), except in one case (Liu et al., 2016; Nguyen et al., 2020), which was not reproduced (Mordes et al., 2020).

Adenoviral vector models expressing G4C2 repeats show ALS/FTD features but lack MN loss, indicating they may not fully represent ALS (Chew et al., 2015).

Tet-on inducible transgenic mice expressing pure G4C2 repeats exhibit myopathy and locomotor defects without RNA foci, pTDP43 inclusions, or neurodegeneration, highlighting systemic toxicity (Riemslagh et al., 2021)—an aspect that should not be overlooked given the high expression of C9orf72 in peripheral immune cells (O'Rourke et al., 2016) and the emerging evidence that alteration in immune cells may play a critical role in the distal axonopathy and skeletal muscle atrophy in ALS mouse models and patients (Nardo et al., 2016; Trias et al., 2018; Schreiber et al., 2019; Angelini et al., 2020; Trolese et al., 2022).

Indeed, C9orf72 knockout mice exhibit severe autoimmunity and lymphatic defects although, they fail to show neurodegeneration or ALS/FTD phenotypes (Sudria-Lopez et al., 2016).

However, reduction or absence of C9orf72 exacerbates gain-of-toxicity from G4C2 repeats, suggesting a synergy between gain and loss of function in *C9ORF72* gene mutation (Zhu et al., 2020).

As *C9ORF72* plays a physiological role in the autophagy-lysosome pathway, the reduction or loss of C9orf72 protein could prevent repeat-mediated autophagy induction, leading to increased accumulation of DPR proteins. The influence of C9orf72 on autophagy in the CNS raises the possibility of converging disease mechanisms for other ALS and FTD-linked genes that are directly or indirectly involved in autophagy-related protein clearance and membrane trafficking pathways (i.e., TBK1 and OPTN). It is also worth noting that these genes, as well as those for *TARDBP* and *C9ORF72*, are highly expressed in innate immune cells, and there is evidence linking their dysfunction to impaired immunity.

TBK1 mutant models of ALS and FTD

Tumor necrosis factor receptor-associated factor NF-kB activator (TANK)-binding kinase 1 (TBK1), a serine–threonine kinase that regulates type 1 interferon (IFN) production and autophagy, exhibits loss-of-function (LOF) gene mutations in ALS/FTD (Cirulli et al., 2015; Freischmidt et al., 2015; Ahmad et al., 2016). However, mice expressing ALS-specific *Tbk1* mutations or conditional and heterozygous *TBK1*-knockout mice do not show ALS-like phenotypes, while homozygous TBK1-knockout mice exhibit embryonic lethality due to severe degeneration of the fetal liver (Marchlik et al., 2010). In addition, conditional *TBK1* deletion in dendritic cells leads to splenomegaly, lymphadenopathy, and an increased inflammatory state (i.e., activated T-lymphocytes), associated with a hyperactive broader interferon response despite *TBK1*-deficient mice having attenuated type 1 interferon production (Xiao et al., 2017). Interestingly, LOF mutations in *TBK1* exacerbate ALS-like symptoms in SOD1G93A and TDP-43G298S transgenic mice (Gerbino et al., 2020; Sieverding et al., 2021). The effects of TBK1 loss could be differential in neurons vs glial cells, impacting protein clearance through autophagy dysfunction and exacerbating neuroinflammation, respectively.

OPTN mutant models of ALS and FTD

OPTN is another gene of which more than 20 missense variants, believed to result in a LOF, are associated with ALS/FTD (Markovinovic et al., 2017). Homozygous *OPTN* mutations induce ALS, while heterozygous mutations are seen as disease risk factors (Cirulli et al., 2015). OPT is a substrate of TBK1 and is crucial in various cellular pathways, including autophagy, vesicle trafficking, inflammatory signaling, and RIPK-1-mediated necroptosis, a programmed form of necrosis linked to inflammation (Ito et al., 2016).

OPTN-knockout mice fail to recapitulate the ALS phenotypes (Munitic et al., 2013), although conditional removal of *OPTN* from OL, microglia, or myeloid cells, but not from astrocytes or MNs, results in axonal pathology, myelin abnormalities, and NMJ defects, suggesting a noncell autonomous mechanism of axonopathy (Ito et al., 2016). Recently, other *OPTN*-knockout mice also developed progressive axonal degeneration, associated with partial MNs loss and the TDP-43 inclusions, and nuclear TDP-43 depletion in those that survived (Kurashige et al., 2021). However, none of these mouse models display motor defects, suggesting that these phenotypes are reminiscent of early disease stages during which possible compensatory mechanisms occur to maintain a proper neuromuscular function.

Consistent with the hypothesis that *OPTN* mutations lead to a loss of the protein's neuroprotective function in ALS, the intrathecal injection of adeno-associated virus 9 (AAV9)-OPTN into SOD1-G93A transgenic mice delayed the disease progression and the MN degeneration by increasing autophagy and protecting mice mitochondria (Wen et al., 2024).

GLIAL INVOLVEMENT IN ALS/FTD: IN HUMAN EVIDENCE

Glial histopathology in ALS and FTD: Postmortem evidence

The earliest data suggesting the glia role as the hallmark of ALS were obtained from *postmortem* neuropathological studies. In these studies, from the first report in 1996 to the most recent, no significant abnormalities are observed at a macroscopical level, except for limited brain atrophy at the precentral level. On the contrary, a reduction in the number of MNs in the cerebral cortex and spinal cord is described, associated with an increase in microglia activation and a process of reactive astrogliosis, as a critical aspect of glial neuropathology (Brownell et al., 1970; Ikeda et al., 2002). *Postmortem* investigations in ALS have unveiled neurodegenerative changes impacting both UMNs and LMNs, which are linked with neuroinflammatory responses and the activation of inflammatory and glial cells, encompassing microglia, astrocytes, OL, and lymphocytes. In the context of the role of glia in the ALS-FTD, multiple phenomena have been revealed, including astrogliosis leading to a proinflammatory environment, microgliosis, and neuroinflammation, as well as degeneration of glial cells and cellular damage associated with a genetic mutation in glia.

Histopathologic studies have shown evidence of astrogliosis in the brain and spinal cord of ALS patients surrounding MNs (Schiffer et al., 1996), while cortical astrogliosis might not be unique to ALS and could result from concurrent damage associated with prolonged hypoxic distress in the end-stage disease (Schiffer et al., 1996). It was also demonstrated that the reactive astrocytes are associated with an increased expression of proinflammatory markers, such as interleukin-1β (IL-1β) and tumor necrosis factor-alpha (TNF-α) in astrocytes, to higher immunoreactivity for glial fibrillary acidic protein (GFAP) and the calcium-binding protein S100beta and express oxidative inflammatory markers (i.e., nitric oxide) (Kushner et al., 1991; Migheli et al., 1999; Barbeito et al., 2004; Schiffer and Fiano, 2004), resulting in an inflammatory state associated with excitotoxic MN damage (Troost et al., 1990; Vargas and Johnson, 2010).

Microglial functions undergo alterations in ALS, as evidenced by microglial morphology and protein expression changes. It is well-known that these modifications in microglial function occur as a protective response against brain degeneration in the first disease phases but also have the potential to contribute to the neurodegeneration process. Evaluating a large cohort of autoptic cases, the group of Brettschneider demonstrated by microglia-specific markers (CD68 and Iba1) a more extensive microglial pathology in the corticospinal tract of ALS patients, which also correlates with UMN impairment and disease progression (Brettschneider et al., 2012). Lastly, regarding OL, *postmortem* analyses of ALS patients reported TDP-43 positive inclusions in OL that strongly correlated with MN loss and myelin alteration (Brettschneider et al., 2014a,b). TDP-43 and FUS-positive inclusions are likely implicated in OL degeneration either by altering their normal metabolic function or forming toxic aggregates responsible for the impairment of myelin binding protein mRNA translation and transport (Lorente Pons et al., 2020), thus suggesting that TDP-43 functional alterations in OL may directly promote the axonal damage rather than being a consequence of MN loss. Interestingly, in a comprehensive study of ALS cortex tissue sequencing, three distinct molecular subtypes were identified using machine learning algorithms (Tam et al., 2019). The majority of patient samples (60%) exhibited gene expression markers associated with oxidative and proteotoxic stress (ALS-Ox group). Another subgroup (20%) showed extensive glial involvement (ALS-Glial group), while the third (ALS-TE group) displayed highly elevated retrotransposon expression linked to TDP-43 dysfunction. The study suggests a disruption in retrotransposon silencing by TDP-43, potentially contributing to cellular toxicity. More recently, Eshima et al. (2023) analyzed how these gene expression patterns impact patient prognosis and survival. Noteworthy findings include the identification of an immunological eigengene correlated with age of onset and survival and the association of the ALS-Glia subtype with significantly shorter survival, supporting previous works linking inflammatory gene expression to accelerated disease progression in ALS patients. The transcriptional differences observed in postmortem cortex tissues were independent of genotype. On the other hand, another recent work by Rifai et al. (2023) evaluated potential differences in neuroinflammatory dysregulation and heterogeneity at the gene expression level in *postmortem* motor cortex tissue between *C9ORF72* and sALS patients. In the C9orf72 group, the analysis revealed distinct neuroinflammatory panels, including proinflammatory, axonal transport, and synaptic signaling pathways, in *C9ORF72* ALS patients compared to sALS. These findings highlight the importance of personalized treatment approaches in each ALS/FTD patient, considering also their genetic status, and emphasize the need to consider diverse pathogenic mechanisms in clinical trials. Further investigations into peripheral tissues may offer additional insights into potential markers for patient prognosis and aid in cohort selection for targeted therapies (Tam et al., 2019; Eshima et al., 2023; De Marchi et al., 2023).

In vivo imaging of glia in ALS and FTD

With the increasing interest in glial cell-specific targets in several NDDs, including the ALS/FTD spectrum, there is a growing need for neuroimaging tools to visually evaluate and track glial activation in vivo. The advancement of PET tracers and novel MRI techniques designed to target glial-cell-specific processes is crucial not only for elucidating the pathophysiological underpinnings of these human diseases but also for identifying potential responders to cell-specific drugs and facilitating the corresponding clinical translation (Cavaliere et al., 2020; Valori et al., 2023).

Several targets for PET imaging of glial cell functions have been identified. Monoamine oxidase type B (MAO-B) is the primary subtype in the human brain, accounting for approximately 70% of the total brain MAO activity (Meyer and Braga, 2022). In the CNS, this enzyme plays a significant role in metabolizing monoamines (including dopamine and norepinephrine) and is prominently present in the outer mitochondrial membrane of astrocytes. This characteristic has identified it as a notable PET imaging marker for astrocytes. ^{11}C-deuterium-L-deprenyl (^{11}C-DED) is an irreversible MAO-B radiobinder that has been mostly applied to the study of reactive astrogliosis in human PET research, although the recent development of ^{18}F-labeled radiotracers, such as ^{18}F-THK5351, benefits from their longer half-life (Liu et al., 2022). A first, pioneering study showed increased ^{11}C-DED binding in diffuse subcortical white matter and pontine regions of ALS patients compared to healthy controls (Johansson et al., 2007), supporting the presence of widespread astrogliosis in this disease. A recent case study performed in two patients (i.e., one with ALS and prevalent lower limb deficits, and the other with ALS and comorbid svPPA) demonstrated good correspondence between regions of increased ^{18}F-THK5351 binding and neurodegeneration (respectively, in the motor cortical leg area and in the left temporal pole) (Higashihara et al., 2021).

Another glial cell target studied in several PET studies is the translocator protein (TSPO) of the outer mitochondrial membrane. Although TSPO has been mostly considered a marker of activated microglia and reactive astrocytes, (Brown et al., 2007; Lavisse et al., 2012) although recently it has been suggested that this signal in human reflects the density of inflammatory cells rather than their activation (Nutma et al., 2023). ^{11}C-PK11195 was the first TSPO radiobinder to be extensively assessed in this context. More recently, second-generation TSPO ligands, such as ^{11}C-PBR28, have shown higher affinity and selectivity for TSPO over ^{11}C-PK11195, although this is highly influenced interindividual by a single nucleotide polymorphism (rs6971) in the gene encoding TSPO (Guo et al., 2014). Increased local uptake of TSPO tracers was consistently detected in the motor cortex, prefrontal cortex, pons, and thalami of sALS patients, colocalizing with structural abnormalities of both gray and white matter and showing significant correlations with clinical measures of UMN burden (Turner et al., 2004; Zürcher et al., 2015; Alshikho et al., 2016). Similar findings were shown in ALS patients with pathogenic genetic variations in the *SOD1* gene, even in the presymptomatic phases (Tondo et al., 2020). Increased local uptake of TSPO ligands colocalizing with regions of neurodegeneration was also found in FTD presentations (Cagnin et al., 2004), including presymptomatic carriers of *MAPT* mutations (Miyoshi et al., 2010).

Recent, exciting developments in the context of advanced MRI techniques have suggested possible alternative tools to describe glial alterations in NDDs. Several studies assessing iron-sensitive MRI sequences have indicated that signal alterations related to iron deposition within the motor white matter tracts and subcortical regions might correlate with astrogliosis and/or microglial activation in MN diseases (Kwan et al., 2012; Bhattarai et al., 2022; Zejlon et al., 2022). Ultra-high-field MRI, possibly combined with direct histopathologic evidence, might be used to further elucidate the relationship between iron deposition and underlying astrogliosis and microglial activation both in ALS (Northall et al., 2024) and FTD presentations (Tisdall et al., 2022). A recently developed neuroimaging technique, DTI-ALPS (diffusion tensor imaging along the perivascular space), has been proposed to assess in vivo the activity of the glymphatic system (Taoka et al., 2022). This could prove to be a valuable tool to assess astrocyte-specific dysfunction in patients with ALS/FTD in the near future.

THE ROLE OF GLIA IN ALS AND FTD

The conventional belief that the brain is an immune-privileged organ, largely due to the presence of the blood–brain barrier (BBB) and the absence of peripheral immune cells, has been challenged by growing evidence of the involvement of neuroinflammation in numerous neurologic disorders, including ALS and FTD. While resident glial cells in the CNS, such as microglia and astrocytes, are primarily engaged in protective functions, aiming to maintain CNS homeostasis and repair damaged tissue, their persistent activation following CNS damage can lead to excessive neuroinflammation with adverse consequences, resulting in neuronal dysfunction, synaptic degeneration, and neuronal death. In this scenario, also alterations of OL seem to take part in the neurotoxic effects either by impairing the metabolic support to the axon and/or by actively participating in immune-mediated processes (Nutma et al., 2020). Interestingly, a recent analysis of *postmortem* frontal and motor cortex transcriptomes derived from 208 patients identifies three subtype-specific disease pathways in this ALS cohort that correspond to glia activation being significantly enriched in patients with the shortest disease duration and median survival, underlying the relationship between elevated inflammatory gene expression and worse disease prognosis (Eshima et al., 2023). Thus, understanding how coordinated neuronal-glial signaling unfolds across various disease stages can provide valuable insights for developing potential therapeutic interventions to modulate inflammation and decelerate disease progression.

Astrocytes in ALS and FTD

Astrocytes are one of the most abundant cell types in the CNS. They possess a wide range of functions to maintain healthy CNS, including supplying energy molecules and neurotrophic factors to neurons, regulating the balance of ions in the extracellular environment, buffering the action of neurotransmitters, particularly the glutamate, influencing synaptic activity, and regulating the BBB function.

Reactive astrogliosis is a well-recognized aspect of neuroinflammation found in most NDDs, including ALS/FTD, and detected by the overexpression of GFAP and by the increased number and morphology of astrocytes in the brain and spinal cord (Haidet-Phillips et al., 2011; Qian et al., 2017; Guttenplan et al., 2020). However, degeneration of astrocytes has also been found in different cortical layers of FTLD cases as well as in the spinal cord of sALS patients (Martorana et al., 2012), suggesting that different astrocyte populations provide distinct contributions to disease pathogenesis (Sofroniew, 2020).

While from the *postmortem* samples of patients it is difficult to establish whether alteration of astrocyte is a trigger or consequence of neuronal damage, in cellular and animal models of ALS/FTD astrocytes have been shown to elicit both a cell-autonomous and noncell autonomous mechanism that may trigger and perpetuate the death of neurons in the brain and spinal cord. For instance, iPSC-derived astrocytes from patients with ALS/FTD-linked mutations exhibit transcriptional dysregulation, showing both cell-autonomous reactive transformation and susceptibility of astrocytes, as well as noncell autonomous mechanisms that lead to their dysfunction, reducing their trophic support to neurons (Taha et al., 2022; Ziff et al., 2022). In line with this, a partial loss of astrocytes surrounding spinal MNs was found to precede MN degeneration and late reactive astrogliosis in SOD1G93A mice (Rossi et al., 2008), suggesting that distinct astrocyte subpopulation might capture different disease stages (Sofroniew, 2020). The perturbed intracellular mechanisms in astrocytes include impaired calcium homeostasis (Martorana et al., 2012), proteostasis and autophagy (Serio et al., 2013; Szebényi et al., 2021), defects in lactate production and release (Ferraiuolo et al., 2011; Velebit et al., 2020), mitochondrial dysfunction, and loss of antioxidant secretion (Cassina et al., 2008; Birger et al., 2019). In addition, a meta-analysis of gene expression independent studies in astrocytes from mice along disease progression as well as of human astrocytes in monolayer cultures or from *postmortem* material showed that fALS astrocytes undertake a proinflammatory phenotype that shares similarities with the neurotoxic phenotype induced by TNF-α, IL1α, and C1q (Ziff et al., 2022). Indeed, reactive astrocytes contribute to the release of proinflammatory molecules, reactive oxygen species, nitric oxide, and cytokines, such as interleukin 6 (IL-6), IL1β, and TNF-α, that promote the degeneration of healthy neurons when directly cocultured with or receiving conditioned media from primary human or rodent *SOD1*-mutant astrocytes (Nagai et al., 2007; Di Giorgio et al., 2008). Noteworthy, abolishing these proinflammatory elements of astrocytes prolonged mouse survival (Guttenplan et al., 2020). Consistently, transplantation of astrocyte precursor cells with ALS mutations in mouse models has been shown to induce MN death and impair neuron function. Conversely, transplanting healthy astrocyte precursors can have a protective effect on MNs in ALS models in line with a noncell-autonomous role of astrocytes.

Another astrocyte-mediated noncell autonomous mechanism is the excessive extracellular glutamate resulting from the dysfunction of EAAT2, the selective astrocyte glutamate transporter. The presence of mutant *SOD1* or *C9ORF72* mutations has been reported to disrupt EAAT2 function, both in vitro and in vivo, supporting this hypothesis (Foran et al., 2011; Kwon et al., 2014; Rosenblum et al., 2017). Efforts to target astrocytes for therapy include gene expression modification and restoration of EAAT2. However, while preclinical studies have shown promise in mouse models of ALS (Rothstein et al., 2005) and FTLD-tau (Decker et al., 2016), no positive effect was obtained in ALS patients (Cudkowicz et al., 2014).

While the noncell autonomous toxicity of astrocytes carrying mutations of *SOD1* or *C9ORF72* is quite consolidated, other studies with selective mutant TDP-43 overexpression in rodent models' astrocytes or TDP-43-ALS (h)iPSC-derived astrocytes showed controversial results as concerns the pathologic effects on MNs (Haidet-Phillips et al., 2013; Serio et al., 2013; Tong et al., 2013; Yang et al., 2014a), underlying the complexity in deciphering TDP-43 linked pathology.

Besides the proinflammatory and glutamate-mediated toxicity, astrocytes in ALS/FTD patients may induce MN damage due to their metabolic dysfunction. Two approaches validated this concept: one demonstrated that by supplementing the coculture medium of mouse ALS astrocytes and MNs with lactate, a metabolite that the mouse ALS astrocytes fail to release, rescued the neuronal degeneration (Ferraiuolo et al., 2011); another showed a similar outcome when astrocytes carrying the human C9orf72-HRE were treated with inosine, which enhanced their metabolism and, consequently, lactate production and release (Allen et al., 2019). This observation aligns with a recent study showing that Cu-ATSM reduces elevated basal and ATP-linked mitochondrial respiration while increasing glycolysis in human astrocytes from ALS patients with consequent lactate production and release (Dennys et al., 2023). Interestingly, several preclinical studies are showing the efficacy of this compound in slowing the disease progression in ALS mouse models (Williams et al., 2016;

Vieira et al., 2017; Kuo et al., 2019; Lum et al., 2021), and different clinical trials are ongoing (NCT04082832, NCT03136809, NCT04313166).

Microglia in ALS and FTD

Microglia, the CNS's primary intrinsic immune effector cells, constitute a dynamic cell population actively surveying the microenvironment through motile cell processes. Microglia plays a pivotal role in maintaining brain homeostasis in response to invading pathogens or cell death, and its activation is a central component of neuroinflammation in the CNS (Kettenmann et al., 2011). The functional plasticity exhibited by activated microglia involves complex changes in morphology, number, expression of surface receptors, and the production of cytokines and growth factors. Their roles include initiating immune defenses, supporting tissue maintenance, remodeling synapses through phagocytic activity, and releasing chemokines, cytokines, and reactive oxygen species. In a healthy environment, microglia support neuronal survival by secreting BDNF and participate in synapse pruning through continuous remodeling of their cellular processes. Microglia secrete both pro- and antiinflammatory cytokines and chemokines, regulating adaptive immunity and neural homeostasis and function. Also, microglia possess tools for identifying and responding to changes in the brain parenchyma, including toll-like receptors, pattern recognition receptors, and triggering receptors expressed on myeloid cells 2 (TREM2), essential for pathogen recognition and binding and internalization of foreign or misfolded proteins. Overall, once microglia are immune-activated, they migrate to the site of the lesion and secrete immunomodulatory factors that can negatively influence the neuronal environment, developing a chronic inflammatory state (Kettenmann et al., 2011; Brites and Vaz, 2014). While it has long been considered that microglia could be represented as a spectrum between two phenotypes—neuroprotective (M2) and toxic (M1)—recent advances using transcriptional single-cell sorting showed that microglia can assume different activation states that correlate with either a damaging or protective phenotype depending on the context of their activation (Chiu et al., 2013; Ransohoff, 2016).

Microglia are unlikely to be the primary cell type responsible for the MN death in ALS, but they may still have a significant modulatory role. In mSOD1 mouse models, microglial activation has been observed since the disease onset (Chiu et al., 2009). However, Boillée et al. demonstrated that removal of CD11b-expressing microglia from mutant SOD1 mice extended lifespan without altering disease onset, implying that microglia is toxic in the later stages of the disease (Boillée et al., 2006). Although with a simplified vision, during the progression of ALS, activated microglia shift from a neuroprotective state, which promotes tissue repair by releasing protective molecules and antiinflammatory cytokines, to a toxic phenotype producing harmful molecules and proinflammatory cytokines that promote neuronal death (Liao et al., 2012). Factors primarily involved in microglial protection such as IL-4, IL-10, and G-CSF are altered in postmortem ALS/FTD tissues and mouse models. For example, Rossi et al. (2018) showed that delivering IL-4—through a lentiviral gene therapy in mSOD1 mice increased microglia proliferation, delaying disease onset and improving motor function. Likewise, inhibiting IL-10 with antibodies increased microglial activation and accelerated the motor deficits in mSOD1 mice (Gravel et al., 2016). Conversely, boosting IL-10 expression specifically in microglia through lentiviral upregulation delayed the disease onset and extended survival in the same mouse model (Gravel et al., 2016). Previous studies demonstrated that increased levels of IGF-1 were associated with reduced neurotoxic microglial activation, partially preventing MN death, slowing disease progression, and extending the lifespan of mSOD1 mice (Pollari et al., 2011).

Nevertheless, many factors released from microglia have been correlated with neurotoxicity in ALS/FTD. For example, administering the cytokine M-CSF, which led to increased microglial proliferation and proinflammatory signaling, accelerates the disease progression in mSOD1 mice (Gowing et al., 2009). Conversely, the IFN-γ inhibition extended lifespan and improved motor function in mSOD1 mice, shifting microglial gene expression from a neurotoxic to a neuroprotective pattern (Garofalo et al., 2020).

Oligodendrocytes in ALS and FTD

Besides astrocytes and microglia, OLs also contribute to the patho-dynamics of ALS/FTD through prevalent loss of function mechanisms. OLs are responsible for myelinating axons in the CNS and play a crucial role in supporting MNs by supplying them with the energy metabolite lactate and pyruvate via monocarboxylic acid transporters (MCT1 and MCT2) (Fünfschilling et al., 2012).

OL dysfunction, including abnormal morphology and decreased myelin basic protein levels, has been reported in ALS. In *SOD1* mutant mouse models, the OL degeneration appears to precede the death of MNs and symptoms onset, prompting a rapid increase in the population of OL progenitors' cells (OPCs) (Kang et al., 2013; Philips et al., 2013). However, these newly generated cells fail to mature, replace the degenerated OLs, and remyelinate axons (Kang et al., 2013). Interestingly, activation of the Notch signaling pathway, which inhibits the OL differentiation, has been reported in SOD1G93A mice (Liu et al., 2020b); however, the conditional deletion of this factor did not yield significant improvements in ALS mouse models (Eykens et al., 2018). In addition, in ALS mice, the mSOD1G93A disrupts the expression of MCT1 in OLs, mimicking the reduction in MCT1 accumulation observed in sALS (Lee et al., 2012). However, while reducing mSOD1 expression in OLs showed delayed disease onset and prolonged survival (Kang et al., 2013), restoring MCT1 levels in the spinal cord of mSOD1 mice did not result in an improvement in the pathology (Eykens et al., 2021). This is consistent with the in vitro studies showing that the neurotoxicity of OLs differentiated from sALS or fALS cases was not rescued by lactate addition while it was reversed by the *SOD1* knockdown in early progenitor OLs, except for the OLs associated with *C9ORF72* expansions (Ferraiuolo et al., 2016).

TDP-43 and FUS-positive inclusions were found in OLs of ALS patients, suggesting an impairment of autophagy in these cells (Kang et al., 2013; Philips et al., 2013).

TDP-43 seems to be essential for the proper function of mature OLs in a cell-autonomous manner. Indeed, mice with OL-specific TDP-43 knockout exhibited progressive motor deficits and defective myelination due to RIPK1-mediated necroptosis of oligodendroglia, although MNs were not directly affected (Wang et al., 2018). Impaired mRNA processing and alterations in the key regulators of OL differentiation further contribute to OL dysfunction in ALS/FTD.

In addition to the cell-autonomous mechanism, OLs significantly contribute to ALS/FTD pathogenesis through a cross-talk with microglia and astrocytes during the process of neuroinflammation (Peferoen et al., 2014; Nutma et al., 2020). This interaction is critical for myelin maintenance and repair. Microglia and astrocytes influence myelination by regulating the OPC differentiation, maintaining myelin homeostasis, and controlling axonal conduction speed (Lampron et al., 2015; Lloyd and Miron, 2019; Rawji et al., 2020). Astrocytes also support myelin health by providing nutrients and enhancing axonal conduction efficiency (Dutta et al., 2018; Lombardi et al., 2019; Raffaele et al., 2021).

In response to myelin injury, microglia can initially adopt a proresolving phenotype, promoting lesion repair. However, in the later stage of chronic disease, they shift to proinflammatory state triggering a detrimental astrocyte phenotype. This creates a hostile local environment that hampers myelin regeneration and accelerates the degeneration of OLs and MNs (Miron et al., 2013).

Glia-immunity cross talk

In ALS/FTD, damaged neurons and MNs engage with glial cells, both microglia and astrocytes, releasing specific cytokines and chemokines such as IL-1, IL-6, TNF, IFNγ, and MCP1. These, in turn, trigger the recruitment of both innate and adaptive immune cells into the CNS by increasing CNS barrier permeability and promoting immune cell trafficking, thereby fostering the neuroinflammation. The proinflammatory signals propagate from the central to peripheral immune systems and vice versa, thereby contributing to the overall inflammatory environment seen in ALS. Patients with ALS exhibit consistent changes in systemic inflammatory markers and immune cell populations compared to healthy individuals. These changes include differences in levels of CD4 and CD8 T lymphocytes, monocytes (including macrophages), neutrophils, natural killer (NK) cells, mast cells (MCs), and dendritic cells.

ALS patients show abnormalities in circulating cytokines, such as IFNγ, IL-2, IL-8, IL12p70, TNFα, IL-1b, CK, ferritin, IL-4, IL-5, IL-10, and IL-13, with increased ferritin and IL-2 correlating with poorer survival probability (De Marchi et al., 2021b). Similar alterations in cytokine levels are observed in patients with FTD, although further studies are needed for confirmation.

The presence of T lymphocytes in ALS is well-established (Sta et al., 2011; Rolfes et al., 2021). Chemokines and their receptors play a crucial role in the infiltration of these cells into the parenchyma. The chronic inflammatory environment leads to increased leukocyte cell adhesion on endothelial cell surfaces, facilitating their entry into the brain parenchyma (Larochelle et al., 2015; Engelhardt et al., 2016). T lymphocyte-derived TNF-α and IL-17 induce MM-9 secretion in immune cells and MNs, promoting T lymphocyte infiltration into the CNS (Song et al., 2016). Distinct T-cell subsets exhibit specific mechanisms for entering the ALS CNS (Yazdani et al., 2022).

CD8+ T cells display intense infiltration, leading to MN death through MHC-I expressed in activated microglia and injured MNs (Nardo et al., 2018; Coque et al., 2019; Liu et al., 2020a). Interestingly, the deletion of MHC-I and CD8+T cells reduced the inflammation in the spinal cord of SOD1G93A mice, as demonstrated by reduced levels of IFNγ and MCP1 (Nardo et al., 2018). Conversely, in the PNS, the interaction between MHC-I and CD8+ T cells is essential to preserve the quality of the connections of motor axons with skeletal muscles, which slows down the disease progression (Nardo et al., 2016, 2018).

Also, CD4+ T cells are recruited through CNS barriers by the release of chemokines like MCP-1 and other chemokines from the endothelial cell. Among these CD4+T cells, regulatory T lymphocytes (Treg), that are responsible for suppressing immune responses and reducing neuroinflammation, are found to be lower in ALS patients and rodent models compared to controls. In addition, Treg levels and function negatively correlate with disease progression rates (Henkel et al., 2013; Yazdani et al., 2022). A limited number of activated peripheral monocytes infiltrate the CNS, impacting neuroinflammation in ALS (Chiot et al., 2020). Previous studies indicate alterations in monocyte proportions in ALS (McGill et al., 2020). In rapidly progressing ALS patients, peripheral monocytes tend to be in a proinflammatory state (Zhao et al., 2017). Recent evidence confirms the infiltration of peripheral monocytes into the CNS, linked to improved MN survival in ALS, though infiltration may be restricted (Peake et al., 2017). Monocyte-derived macrophages are activated in ALS, exerting neuroprotective functions in protein clearance during the disease (Chiu et al., 2009; Shiraishi et al., 2021). Macrophages also show limited CNS infiltration, suggesting that accumulated monocytes in the CNS may result from the proliferation of infiltrated cells rather than the infiltration of circulating monocytes (Chiot et al., 2020).

Aberrant activation of the complement cascade, a key component of innate immunity, is thought to play a central role in synaptic degeneration across NDDs, including ALS/FTD (Bahia El Idrissi et al., 2016; McDonald et al., 2020; van der Ende et al., 2022). Elevated concentrations of complement activation products, such as C1q, C3b, C5a, and MAC (C5b-9), are detected in biological fluids and tissues of ALS patients, including serum, cerebrospinal fluid, spinal cord, and motor cortex (McDonald et al., 2020). The neuroinflammatory response in ALS involves the activation of the terminal complement pathway, particularly C5aR1 expression, in MNs, activated microglia, and macrophages. Postmortem tissue analysis of ALS donors demonstrates the deposition of complement activation products on motor endplates in the intercostal muscle of ALS patients, which precedes denervation, suggesting its potential role in the initiation of the disease (Bahia El Idrissi et al., 2016). A similar effect was found in both mutant *SOD1* and *TARDBP* ALS mouse models (Wang et al., 2017; Lee et al., 2017, 2018). Inhibition of C5aR1 formation extends survival and improves motor function in ALS animal models (Lee et al., 2017).

Limited studies have explored the roles of neutrophils and NK cells in neuroimmune crosstalk. Despite the significant correlation between increased neutrophil and NK cell numbers in peripheral blood and disease progression, their impact on CNS neuroinflammation still needs to be thoroughly investigated (Murdock et al., 2017; Grassano et al., 2023; Leone et al., 2022). For example, end-stage ALS mice exhibit a high frequency of NK cells in the spinal cord (Finkelstein et al., 2011), and NK cell-derived IFN-γ induces microglia toward an inflammatory phenotype, regulates the release of CCL2 from MNs, and impairs Treg cell migration (Garofalo et al., 2020).

MCs also appear to play an important role in the pathogenesis of ALS. Earlier studies suggested a role for these cells in early degeneration in the PNS, with subsequent impacts on neuronal damage (Trias et al., 2018; Angelini et al., 2020). Recent studies confirmed MC infiltration in the spinal cord of ALS patients (Fiala et al., 2010; Kovacs et al., 2021). IL-6, CCL5, and TNF-α released by activated microglia affect receptor expression on MCs, regulating MC activation and CNS recruitment (Jones et al., 2019). MCs release proteases to tight junctions and extracellular matrix components, influencing BBB permeability and integrity, thereby facilitating CNS invasion of MCs (Mattila et al., 2011; Jones et al., 2019). Lastly, genes such as *C9ORF72*, *TBK1*, *OPTN*, and *TARDBP* are highly expressed in innate immune cells, with evidence linking their dysfunction to altered immunity.

GLIAL BIOMARKERS IN CSF AND BLOOD

Alongside neuroimaging techniques, the identification of glial biomarkers in the CSF and blood is of paramount importance in the field of ALS/FTD research, not only to offer a window into the pathologic mechanisms underlying disease progression and the development of different clinical phenotypes but also for the design of future clinical trials targeting neuroinflammation.

Chitinase-3-like protein 1 (CHI3L1, or YKL-40) is a secreted astrocytic glycoprotein involved in tissue remodeling and repair processes, which has been related to the intensity of the inflammatory response in several NDDs, including AD and other tauopathies (Querol-Vilaseca et al., 2017; Pelkmans et al., 2024). Increased levels of YKL-40 have been demonstrated in the CSF of patients with ALS, correlating with the rate of functional decline (Thompson et al., 2019; Vu et al., 2020), the number of body regions clinically involved (Gille et al., 2019), and shorter survival time (Illán-Gala et al., 2018). YKL-40 levels were also found to discriminate ALS from ALS-mimic diseases (Thompson et al., 2019). An investigation assessing CSF levels of YKL-40 and the functionally related chitotriosidase 1 (CHIT1) protein showed similarly increased levels both in sALS and fALS, whereas presymptomatic individuals showed levels that were comparable to controls (Oeckl et al., 2019). Another recent study assessing YKL-40 and other chitinases in presymptomatic patients carrying ALS-related genetic mutations showed that YKL-40 remained relatively stable over time, although a slow elevation of CHIT1 protein was observed over time, suggesting that neuroinflammation mediated by astrocytic activation might represent a feature of the late presymptomatic to early symptomatic phase of ALS (Gray et al., 2020). When compared with pure FTD patients, those with cooccurrent ALS-FTD showed significantly higher CSF levels of YKL-40, suggesting also a possible role for differential diagnosis (Abu-Rumeileh et al., 2019). In fact, in recent multicentric studies, CSF YKL-40 and chitotriosidase levels were found to be only mildly elevated in patients with clinical FTD syndromes (Woollacott et al., 2020), particularly in symptomatic patients carrying a *GRN* mutation (Woollacott et al., 2022).

Another glial protein currently at the forefront of investigation as a potential biomarker is the GFAP, which is the most abundant cytoskeletal protein of astrocytes and is considered strongly indicative of the activation of this cell type (Middeldorp and Hol, 2011). In the context of ALS, an early study demonstrated moderately increased CSF GFAP levels, as compared to controls with other neurologic diseases (Benninger et al., 2016). However, this finding was not validated by a later investigation assessing both sALS and fALS (Oeckl et al., 2019). More recently, GFAP serum concentrations tested using SIMOA assays were found to be notably higher in individuals with cooccurrent cognitive impairment, also showing a negative correlation with performance at neuropsychologic testing assessing ALS-nonspecific functions, particularly memory (Falzone et al., 2022; Verde et al., 2023), thus suggesting a potential link between neuroinflammatory response and cognitive function in MN disease. When assessing FTD, increased plasma levels of GFAP in patients, either independently or in conjunction with the neurofilament light chain (NfL) protein, have shown possible use to distinguish between FTD and AD patients (Zhu et al., 2021). Moreover, a greater increase of plasma GFAP levels in FTLD-tau was found to be mirrored by a greater increase in NfL levels in FTLD-TDP, supporting the evaluation of the GFAP/NfL ratio as a marker to distinguish in vivo even among molecularly distinct FTLD subtypes (Cousins et al., 2022). Furthermore, GFAP levels in plasma have been found to correlate with a higher burden of cognitive impairment as FTD advances, suggesting its utility in tracking disease progression (Zhu et al., 2021). The prospect of utilizing GFAP as part of a biomarker panel for precise and differential diagnoses is a notable stride in pursuing targeted and personalized treatment strategies.

GLIAL-BASED CLINICAL TRIALS IN ALS AND FTD

The disease heterogeneity, the improvement of knowledge in pathogenetic mechanisms, and the new genetic discoveries draw attention to therapeutic trials that act on a broader spectrum and multiple mechanisms rather than against a single target. Researchers have tried to adapt to this change by promoting new experimental designs and betting on new therapeutic strategies. An example of a new trial design is the "Adaptive Platform trial," developed to investigate several interventions for a single disease continuously to rapidly identify novel treatments, biomarkers, and trial endpoints (Paganoni et al., 2022). Instead, regarding the new therapeutic strategies, gene and stem cell therapy are the most promising. Gene therapy considers various major strategies, including the ability to remove or inhibit abnormal transcribed RNA using microRNA or ASOs, inhibit the mutant proteins, or degrade the abnormal mRNA using RNA interference (Fang et al., 2022). In this regard, as already mentioned, the ASO BIIB067 has recently been approved for ALS-SOD1 patients; other clinical trials with ASOs targeting other autosomal dominant mutations (i.e., *C9ORF72*, *FUS*, and *ATXN2*) are underway (Breevoort et al., 2022). Stem cell therapy, including treatment with mesenchymal stem cells or neural stem cells, has the opportunity to strike multiple altered pathways and provide CNS neurotrophic support. Although the preclinical data is strong and the cells resulted are safe and well-tolerated clinical trials, few data on efficacy have been published (Sironi et al., 2023).

Another drug line in development concerns treatments able to act on glial and immune cells. In an ALS mouse model, various molecules with immune-modulatory properties have been described as able to act on the microglia and immune cells. However, to date, clinical results could be more encouraging. Studies with drugs acting on the immune system in ALS began more than 15 years ago, and among the first investigated, we should remember the failure of the minocycline trial (Gordon et al., 2007). Minocycline has antiapoptotic and antiinflammatory effects in vitro, affecting survival in ALS mouse models. In 2007, a phase 3 multicenter study involving a large cohort of patients tested minocycline in ALS patients, with harmful results (ALSFRS-R score deterioration was faster in the treated group than in the placebo group). Despite this, other compounds targeting neuroinflammation and microglia were tested in the following years. Ibudilast (MN-166), an inhibitor of macrophage migration inhibitory factor and phosphodiesterases 3, 4, 10, and 11, able to attenuate CNS microglial activation and secretion of proinflammatory cytokines, was recently investigated (Babu et al., 2021). The authors showed no significant reductions in motor cortical glial activation measured by PBR28-PET over time or CNS neuroaxonal loss, measured by serum NfL in treated ALS patients compared to controls. Similarly, the drug profile was not optimal, and the number of dropouts at the maximum dose was high. NP001, a macrophage activation regulator, slowed the disease progression in ALS patients with higher levels of the plasma inflammatory marker C-reactive protein in a phase 2A clinical trial (Miller et al., 2015). However, in a recently published phase 2B trial, also NP001 did not show significant differences compared to placebo concerning change in ALSFRS-R and vital capacity (Miller et al., 2022). A recent phase 2 trial with RNS60 achieved encouraging results: the molecule that in ALS mouse models reduces glial inflammation and increases Tregs in humans obtained positive effects on selected measures of respiratory and bulbar functions (Beghi et al., 2023), but a new trial using nebulized RNS60 is still ongoing (NCT02988297).

Another drug, masitinib, is still in study after showing promising results in phase 2. Masitinib is an oral tyrosine kinase inhibitor exerting neuroprotection via its immunomodulatory properties, mainly targeting microglia, macrophages, and mast cell activity in the CNS and PNS. In humans, in a phase 2/3 trial (Mora et al., 2020), this compound showed a significant benefit over the placebo in functional status, with a significant difference in ALSFRS-R progression between the treated group and controls (Mora et al., 2020). Also, in a long-term analysis, the drug prolonged survival of about 2 years. Phase 3 is ongoing and recruiting (NCT03127267). Other tested drugs acting on neuroinflammation, without clear clinical positive results, include several molecules such as 3K3A-APC, ANX005, AT-1501, caricitinib, sotuletinib, fasudil, IL-2, pegcetacoplan, rapamycin, and tocilizumab (Johnson et al., 2022).

Although astrocytes' dysfunction is not a core feature in the pathologic disease mechanism in the ALS/FTD, it can contribute to worsening the neurotoxic environment and, as such, become an excellent therapeutic target. Among the most promising, human embryonic stem cell-derived astrocytes were transplanted by an intrathecal injection in SOD1G93A and NSG animal models, resulting in safe and potentially sound ALS treatment (Izrael et al., 2018; Thomsen et al., 2018). Starting from positive preclinical assumptions, modified neural progenitor cells that differentiate into astrocytes and secrete GDNF (CNS10-NPC-GDNF) were produced and transplanted in ALS patients' spinal cord in a phase 1/2a study (Baloh et al., 2022). The study met the primary endpoint for safety at 1 year with no negative effect of the transplant on motor function. Also, *postmortem* analysis showed graft survival and GDNF production for up to 42 months posttransplantation. Another signaling pathway that raised considerable interest as a potential therapeutic target within astrocytes, is the antioxidant response mediated by the transcription factor nuclear factor erythroid 2-related factor 2 (NRF2) (Jiménez-Villegas et al., 2021). Preclinical studies convincingly showed that the selective rescue of functional deficit of astrocyte NRF2 is neuroprotective both in vitro (Diaz-Amarilla et al., 2016) and in animal models (Vargas et al., 2008), thus setting a strong rationale for astrocyte-targeted therapeutics. Interestingly, clinical studies with substances like curcumin (found in turmeric), a polyphenol that stimulates the NRF2 signaling pathway (Ashrafizadeh et al., 2020) are ongoing in ALS patients (NCT04499963; NCT04654689). Other therapeutic trials targeting astrocytes have not given encouraging results. Among them, we should cite the ceftriaxone trial, which increases EAAT2 promoter activity and

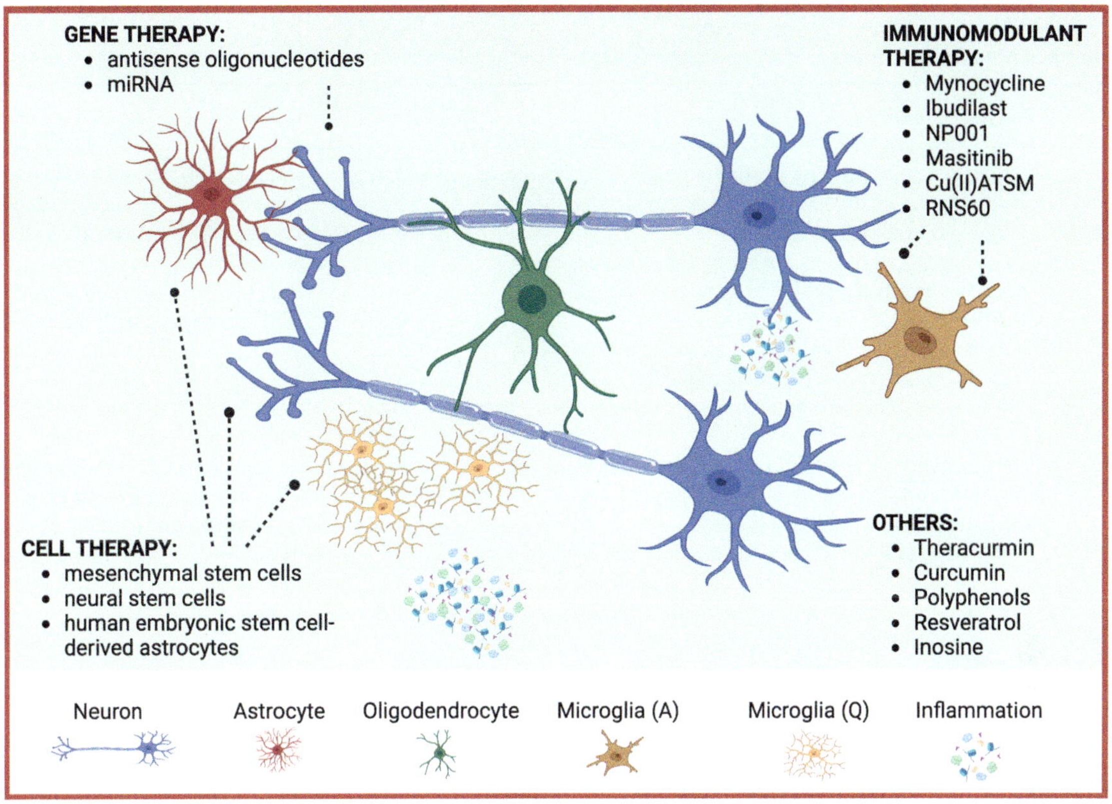

Fig. 3.2. Glial-based experimental therapies in ALS. A: activated; Q: quiescent. The NCT of clinical trials included in the figures and available on clinicaltrials.gov are listed here. *Gene therapy*: antisense oligonucleotides ➔ NCT04931862 (for C9orf72), NCT05683860 (for C9orf72), NCT02623699 (for SOD1), NCT04768972 (for FUS), NCT04494256 (for ATXN2); miRNA ➔ NCT06100276 (for SOD1). *Cell therapy*: mesenchymal stem cells ➔ NCT04821479, NCT01609283 and NCT01142856, NCT01051882 and NCT01777646 and NCT02017912 and NCT03280056 and NCT04681118, NCT04745299, NCT02290886; neural stem cells ➔ NCT01640067, NCT01730716 and NCT01348451; human neural progenitor cells transduced with GDNF ➔ NCT02943850 and NCT05306457. *Immunomodulant therapy*: NCT02238626 and NCT04057898 (ibudilast), NCT01281631 and NCT02794857 and NCT02794857 (NP001), NCT03127267 and NCT02588677 (masitinib), NCT02988297 and NCT02525471 and NCT03456882 (RNS60). *Others*: NCT04499963 (theracurmin), NCT04654689 (polyphenols), NCT02288091, and NCT03168711 (inosine). Figure created with Biorender.com.

protects MNs from excitotoxicity in human astrocyte cultures. Phase 3 of the study (Cudkowicz et al., 2014), published by a large collaborative group from the U.S., failed to show clinical efficacy. Also, due to the lack of pharmacodynamic markers, the potential biological effect of ceftriaxone on CNS EAAT2 mRNA or proteins was not determined. Always in this regard, also the inosine pathway was tested, starting from the description of a higher susceptibility to adenosine-induced toxicity in ALS patient-derived induced astrocytes (Allen et al., 2019). However, in 2018, the group of Paganoni (Nicholson et al., 2018) showed that inosine supplementation to ALS patients was safe, but it did not show efficacy in ALSFRS-R changes.

A summary of the main investigated treatments, and the related cellular target, is shown in Fig. 3.2.

CONCLUSIONS

Recent discoveries of ALS/FTD-associated genes indicate that, beyond inducing neuronal dysfunction, altered functioning of inflammatory cells in both the brain and periphery may directly contribute to disease pathogenesis (the main findings regarding glia and immune cell dysregulation are summarized in Table 3.1).

Extensive research has focused on astrocytes and microglia activation states, heterogeneity and communication between different cell types at various disease stages, and understanding how their interaction with neuron dysfunction could drive neurodegeneration. Examining ALS/FTD-associated genes, which are ubiquitously expressed, reveals that some genes, such as *GRN* and *C9ORF72*, exhibit higher expression in nonneuronal cells, including microglia. Others, such as *TBK1*, *OPTN*, and *SQSTM1*, are known for regulating innate immune cell function. This suggests a direct influence of ALS/FTD mutations on immune cell function, potentially acting in concert with their impact on neurons to drive disease progression.

Also, the key issue of the effect of modulating glial cells for therapeutic purposes has yet to be systematically addressed, considering the failure of several therapeutic trials in this regard. A translational approach encompassing both in vitro and in vivo models, genetic characterization, and in vivo biomarkers will provide further insight into the role of glia in the ALS/FTD spectrum disorder.

Table 3.1

Summary on glial and immune cell dysregulation in the amyotrophic lateral sclerosis—frontotemporal dementia spectrum disorders.

	Preclinical findings	Clinical findings
Astrocytes 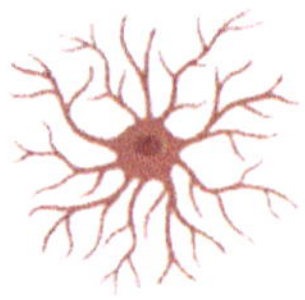	– Both degenerating or hyperproliferating aberrant astrocyte phenotypes are found in ALS rodent models promoting MN loss (Valori et al., 2023); – Astrocytes carrying ALS/FTD-linked mutations show reduced trophic support to MNs in a cell coculture system (Rossi et al., 2008); – Activated ALS/FTD mutant astrocytes overexpress GFAP and exert neurotoxic effects throughout: • Release of proinflammatory and neurotoxic molecules (Ziff et al., 2022) • Impaired calcium homeostasis (Martorana et al., 2012) • Impaired proteostasis and autophagy (Serio et al., 2013; Szebényi et al., 2021) • Metabolic dysfunction with defects in lactate production and release (Ferraiuolo et al., 2011; Velebit et al., 2020) • Mitochondrial dysfunction and loss of antioxidant secretion (Cassina et al., 2008; Birger et al., 2019) • Disruption of EAAT2 and excessive extracellular glutamate (Foran et al., 2011; Rosenblum et al., 2017)	– Astrocytes reactivity (GFAP and S100b overexpression) is a prominent hallmark in different cortical layers and spinal cord of postmortem ALS patients (Schiffer et al., 1996; Migheli et al., 1999); – Increased levels of YKL-40 and other chitinases are found in the ALS CSF, correlating with the rate of functional decline (Oeckl et al., 2019; Thompson et al., 2019; Vu et al., 2020); – Higher CSF levels of YKL-40 in ALS/FTD compared to pure FTD (Abu-Rumeileh et al., 2019); – Increased chitotriosidase CSF levels in symptomatic FTD carriers of GRN mutations (Woollacott et al., 2022); – Increased PET imaging markers for activated astrocytes (i.e., MAO-B radiobinders) are found in cortical and subcortical regions of ALS/FTD patients (Johansson et al., 2007; Higashihara et al., 2021)
Microglia 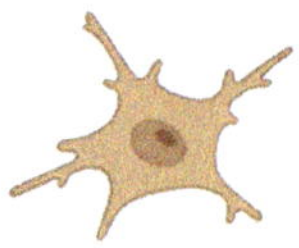	– Microglial expression profile changes during disease progression in mSOD1 mice, initially expressing an overall protective phenotype (M2) but undergoing a transformation to an overall toxic state (M1) at the later stage (Liao et al., 2012; Clarke and Patani, 2020) – Enhancing M2 microglia by the delivery of IL-4 or enhancing microglia-specific expression of IL-10 delays disease onset and improves motor dysfunction in mSOD1 mice (Gravel et al., 2016; Rossi et al., 2018); – IFN-γ inhibition shifts microglia from M1 to M2 and improves motor deficit and lifespan in mSOD1 mice (Garofalo et al., 2020); – IL-10 inhibition and M-CSF increase microglial proliferation and proinflammatory signaling, speeding up disease progression in mSOD1 mice (Gravel et al., 2016; Garofalo et al., 2020); – Proinflammatory microglia induce a detrimental astrocyte phenotype, hindering myelin regeneration and precipitating OLs and MNs degeneration (Rawji et al., 2020)	– Factors associated with microglia protection (IL-4, IL-10, G-CSF) are altered in postmortem tissues of ALS patients; – In PET imaging, increased local uptake of TSPO tracers was detected in the motor cortex, prefrontal cortex, pons, and thalami of sALS patients (Turner et al., 2004; Zürcher et al., 2015; Alshikho et al., 2016) and in SOD1-ALS patients (Tondo et al., 2020); – In FTD, increased local uptake of TSPO ligands in regions of neurodegeneration was reported (Cagnin et al., 2004), including presymptomatic carriers of MAPT mutations (Miyoshi et al., 2010)
Oligodendrocytes 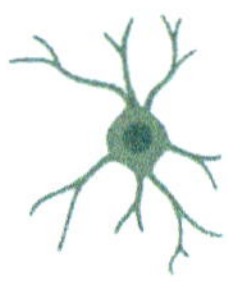	– OLs abnormal morphology and degeneration and myelin disruption precede the MN death in mSOD1 mice (Raffaele et al., 2021); – Defective OPCs regenerative response and axons remyelination in ALS rodent models (Kang et al., 2013); – MCT1 deficiency in mSOD1 OLs contributes to MNs loss (Lee et al., 2012); – OL-specific TDP-43 knockout mice exhibit progressive motor deficits and defective myelination without apparent toxicity on MNs (Wang et al., 2018); – OPC differentiation, myelin homeostasis, and axonal conduction speed during myelination are regulated by microglia and astrocytes (Lampron et al., 2015; Lloyd and Miron, 2019; Rawji et al., 2020)	– TDP-43 and FUS inclusions found in OLs of sporadic ALS patients (Brettschneider et al., 2014a; Valori and Neumann, 2021)

Table 3.1

Continued

	Preclinical findings	Clinical findings
Immune cells 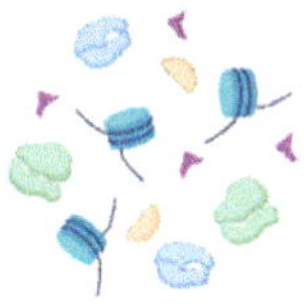	– Degenerating MNs and reactive microglia and astrocytes secrete molecules (i.e., MCP-1) that promote the infiltration of immune cells through CNS barriers (De Marchi et al., 2021b; Berriat et al., 2023)	– Increased inflammatory cells and complement activation in ALS spinal cord, motor cortex, and muscular endplate (Troost et al., 1990; Sta et al., 2011; Bahia El Idrissi et al., 2016) – Abnormalities in circulating cytokines, with increased ferritin and IL-2 correlating with poorer survival (De Marchi et al., 2021b)
Neutrophils 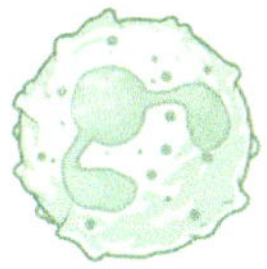	– Controversial role at the level of NMJ on disease progression in mSOD1 rodents (Trias et al., 2018; Trolese et al., 2022)	– NLR is considered a promising biomarker for ALS progression and prognosis (Murdock et al., 2017; Leone et al., 2022; Grassano et al., 2023)
Natural killer cells 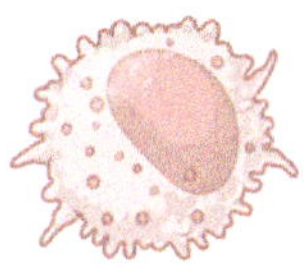	– ALS mice exhibit a high frequency of NK cells in the spinal cord (Finkelstein et al., 2011); – NK cell-derived IFN-γ induces microglia toward an inflammatory phenotype (Garofalo et al., 2020); – Longer survival is associated with decreased microglial activation and increased presence of Treg cells after NK cell depletion in mSOD1 mice (Garofalo et al., 2020)	– Controversial results on increased NK cell numbers in peripheral blood and correlation with disease progression (Murdock et al., 2017; Cui et al., 2022)
Monocytes/ macrophages 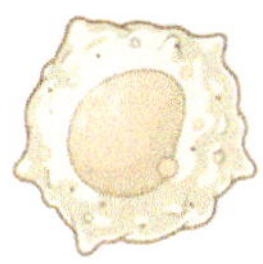	– Peripheral monocyte infiltration is limited in the CNS of rodent models (Berriat et al., 2023); – Modifying peripheral monocyte-derived macrophages suppresses proinflammatory microglia response and exerts neuroprotective functions in MNs (Chiot et al., 2020)	– ALS blood monocyte-derived macrophages produced more proinflammatory TNF than control macrophages, positively correlated with disease progression rate (Zhao et al., 2017; McGill et al., 2020)
Mast cells 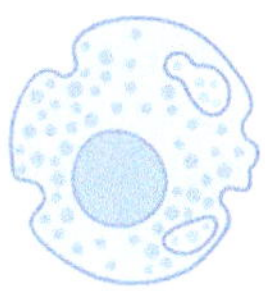	– Infiltration of mast cells correlates with denervation of NMJ, and their inhibition increased survival in mSOD1 rats (Trias et al., 2018)	– Mature mast cells are detected in the spinal cord, muscles, and nerves of individuals with ALS (Trias et al., 2018)
T cells—CD8 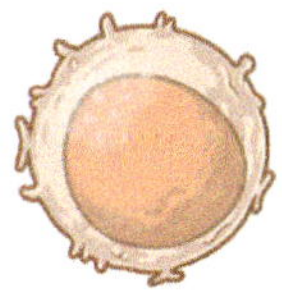	– Intense infiltration leads to MN death through MHC-I in activated microglia and injured MNs (Coque et al., 2019; Liu et al., 2020a); – Infiltrated CD8+ T cells are detrimental to MNs, while supporting peripheral axon regeneration, preserving MNJ in mSOD1 mice (Nardo et al., 2018)	– Blood CD8+ T cell levels are positively correlated with death risk (Cui et al., 2022)

Continued

Table 3.1

Continued

	Preclinical findings	Clinical findings
T cells—CD4 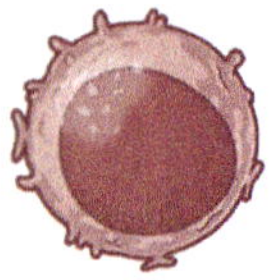	– Genetic deletion in mSOD1 mice reduces survival (Beers et al., 2008); – Tregs support microglial neuroprotection and slow disease progression in mSOD1 mice (Beers et al., 2008)	– CD4 effector T cells and Treg cells are differentially associated with ALS progression rate (Henkel et al., 2013; Yazdani et al., 2022)

ALS, amyotrophic lateral sclerosis; *CNS*, central nervous system; *EAAT2*, excitatory amino acid transporter; *FTD*, frontotemporal dementia; *G-CSF*, granulocyte colony-stimulating factor; *GFAP*, glial fibrillary acidic protein; *IFN*, interferon; *IL*, interleukin; *MCP-1*, monocyte chemoattractant protein-1; *MCT1*, monocarboxylate transporter 1; *MHC-I*, major histocompatibility complex class I; *MMP9*: matrix metallopeptidase 9; *MN*, motor neuron; *NLR*, NOD-like receptor; *NLRP3*, NLR family pyrin domain containing 3; *NMJ*, neuromuscular junction; *NK*, natural killer; *NLR*, neutrophil and lymphocyte ratio; *OLs*, oligodendrocytes; *OPC*, oligodendrocyte progenitor cells; *PET*, positron emission tomography; *S100b*, S100 calcium-binding protein B; *TDP-43*, TAR DNA-binding protein 43; *Tregs*: regulatory T cells; *YKL-40*, chitinase-3-like protein 1.

References

Abe K, Itoyama Y, Sobue G et al. (2014). Confirmatory double-blind, parallel-group, placebo-controlled study of efficacy and safety of edaravone (MCI-186) in amyotrophic lateral sclerosis patients. Amyotroph Lateral Scler Frontotemporal Degener 15: 610–617.

Abu-Rumeileh S, Steinacker P, Polischi B et al. (2019). CSF biomarkers of neuroinflammation in distinct forms and subtypes of neurodegenerative dementia. Alzheimers Res Ther 12: 2. https://doi.org/10.1186/s13195-019-0562-4.

Ahmad L, Zhang SY, Casanova JL, Sancho-Shimizu V (2016). Human TBK1: a gatekeeper of neuroinflammation. Trends Mol Med 22: 511–527.

Allen SP, Hall B, Castelli LM et al. (2019). Astrocyte adenosine deaminase loss increases motor neuron toxicity in amyotrophic lateral sclerosis. Brain 142: 586–605.

Alshikho MJ, Zürcher NR, Loggia ML et al. (2016). Glial activation colocalizes with structural abnormalities in amyotrophic lateral sclerosis. Neurology 87: 2554–2561.

Andrews JA, Jackson CE, Heiman-Patterson TD et al. (2020). Real-world evidence of riluzole effectiveness in treating amyotrophic lateral sclerosis. Amyotroph Lateral Scler Frontotemporal Degener 21: 509–518. https://doi.org/10.1080/21678421.2020.1771734.

Angelini DF, De Angelis F, Vacca V et al. (2020). Very early involvement of innate immunity in peripheral nerve degeneration in SOD1-G93A mice. Front Immunol 11: 575792.

Arnold ES, Ling SC, Huelga SC et al. (2013). ALS-linked TDP-43 mutations produce aberrant RNA splicing and adult-onset motor neuron disease without aggregation or loss of nuclear TDP-43. Proc Natl Acad Sci U S A 110: E736–E745.

Ashrafizadeh M, Ahmadi Z, Mohammadinejad R et al. (2020). Curcumin activates the Nrf2 pathway and induces cellular protection against oxidative injury. Curr Mol Med 20: 116–133.

Babu S, Hightower BG, Chan J et al. (2021). Ibudilast (MN-166) in amyotrophic lateral sclerosis-an open label, safety and pharmacodynamic trial. Neuroimage Clin 30: 102672.

Bahia El Idrissi N, Bosch S, Ramaglia V et al. (2016). Complement activation at the motor end-plates in amyotrophic lateral sclerosis. J Neuroinflammation 13: 72. https://doi.org/10.1186/s12974-016-0538-2.

Balendra R, Isaacs AM (2018). C9orf72-mediated ALS and FTD: multiple pathways to disease. Nat Rev Neurol 14: 544–558.

Baloh RH, Johnson JP, Avalos P et al. (2022). Transplantation of human neural progenitor cells secreting GDNF into the spinal cord of patients with ALS: a phase 1/2a trial. Nat Med 28: 1813–1822.

Barbeito LH, Pehar M, Cassina P et al. (2004). A role for astrocytes in motor neuron loss in amyotrophic lateral sclerosis. Brain Res Rev 47: 263–274.

Beers DR, Henkel JS, Zhao W et al. (2008). CD4+ T cells support glial neuroprotection, slow disease progression, and modify glial morphology in an animal model of inherited ALS. Proc Natl Acad Sci U S A 105: 15558–15563.

Beghi E, Pupillo E, Bianchi E et al. (2023). Effect of RNS60 in amyotrophic lateral sclerosis: a phase II multicentre, randomized, double-blind, placebo-controlled trial. Eur J Neurol 30: 69–86.

Bendotti C, Carrì MT (2004). Lessons from models of SOD1-linked familial ALS. Trends Mol Med 10: 393–400.

Bendotti C, Bonetto V, Pupillo E et al. (2020). Focus on the heterogeneity of amyotrophic lateral sclerosis. Amyotroph Lateral Scler Frontotemporal Degener 21: 485–495.

Benninger F, Glat MJ, Offen D et al. (2016). Glial fibrillary acidic protein as a marker of astrocytic activation in the cerebrospinal fluid of patients with amyotrophic lateral sclerosis. J Clin Neurosci 26: 75–78.

Bensimon G, Lacomblez L, Meininger V (1994). A controlled trial of riluzole in amyotrophic lateral sclerosis. N Engl J Med 330: 585–591.

Berriat F, Lobsiger CS, Boillée S (2023). The contribution of the peripheral immune system to neurodegeneration. Nat Neurosci 26: 942–954.

Bhattarai A, Egan GF, Talman P et al. (2022). Magnetic resonance iron imaging in amyotrophic lateral sclerosis. J Magn Reson Imaging 55: 1283–1300.

Birger A, Ben-Dor I, Ottolenghi M et al. (2019). Human iPSC-derived astrocytes from ALS patients with mutated C9ORF72 show increased oxidative stress and neurotoxicity. EBioMedicine 50: 274–289.

Boillée S, Yamanaka K, Lobsiger CS et al. (2006). Onset and progression in inherited ALS determined by motor neurons and microglia. Science 1979: 1389–1392.

Bonifacino T, Zerbo RA, Balbi M et al. (2021). Nearly 30 years of animal models to study amyotrophic lateral sclerosis: a historical overview and future perspectives. Int J Mol Sci 22: 12236.

Boxer AL, Miller BL (2005). Clinical features of frontotemporal dementia. Alzheimer Dis Assoc Disord 19: S3–S6.

Breevoort S, Gibson S, Figueroa K et al. (2022). Expanding clinical spectrum of C9ORF72-related disorders and promising therapeutic strategiesa review. Neurol Genet 8: e670.

Brettschneider J, Toledo JB, Van Deerlin VM et al. (2012). Microglial activation correlates with disease progression and upper motor neuron clinical symptoms in amyotrophic lateral sclerosis. PLoS One 7: e39216.

Brettschneider J, Arai K, Del Tredici K et al. (2014a). TDP-43 pathology and neuronal loss in amyotrophic lateral sclerosis spinal cord. Acta Neuropathol 128: 423–437.

Brettschneider J, Del Tredici K, Irwin DJ et al. (2014b). Sequential distribution of pTDP-43 pathology in behavioral variant frontotemporal dementia (bvFTD). Acta Neuropathol 127: 423–439.

Brites D, Vaz AR (2014). Microglia centered pathogenesis in ALS: insights in cell interconnectivity. Front Cell Neurosci 8: 117.

Brooks BR, Miller RG, Swash M et al. (2000). El Escorial revisited: revised criteria for the diagnosis of amyotrophic lateral sclerosis. Amyotroph Lateral Scler Other Motor Neuron Disord 1: 293–299.

Brown AK, Fujita M, Fujimura Y et al. (2007). Radiation dosimetry and biodistribution in monkey and man of 11C-PBR28: a PET radioligand to image inflammation. J Nucl Med 48: 2072–2079.

Brownell B, Oppenheimer DR, Hughes JT (1970). The central nervous system in motor neurone disease. J Neurol Neurosurg Psychiatry 33: 338–357.

Cagnin A, Rossor M, Sampson EL et al. (2004). In vivo detection of microglial activation in frontotemporal dementia. Ann Neurol 56: 894–897.

Cassina P, Cassina A, Pehar M et al. (2008). Mitochondrial dysfunction in SOD1G93A-bearing astrocytes promotes motor neuron degeneration: prevention by mitochondrial-targeted antioxidants. J Neurosci 28: 4115–4122.

Cavaliere C, Tramontano L, Fiorenza D et al. (2020). Gliosis and neurodegenerative diseases: the role of PET and MR imaging. Front Cell Neurosci 14: 75.

Cedarbaum JM, Stambler N, Malta E et al. (1999). The ALSFRS-R: a revised ALS functional rating scale that incorporates assessments of respiratory function. J Neurol Sci 169: 13–21.

Chew J, Gendron TF, Prudencio M et al. (2015). C9ORF72 repeat expansions in mice cause TDP-43 pathology, neuronal loss, and behavioral deficits. Science 1979: 1151–1154.

Chio A, Logroscino G, Hardiman O et al. (2009). Prognostic factors in ALS: a critical review. Amyotroph Lateral Scler 10: 310–323.

Chiò A, Logroscino G, Traynor BJ et al. (2013). Global epidemiology of amyotrophic lateral sclerosis: a systematic review of the published literature. Neuroepidemiology 41: 118–130.

Chiò A, Moglia C, Canosa A et al. (2020). ALS phenotype is influenced by age, sex, and genetics: a population-based study. Neurology 94: e802–e810.

Chiot A, Zaïdi S, Iltis C et al. (2020). Modifying macrophages at the periphery has the capacity to change microglial reactivity and to extend ALS survival. Nat Neurosci 23: 1339–1351. https://doi.org/10.1038/s41593-020-00718-z.

Chiu IM, Phatnani H, Kuligowski M et al. (2009). Activation of innate and humoral immunity in the peripheral nervous system of ALS transgenic mice. Proc Natl Acad Sci U S A 106: 20960–20965.

Chiu IM, Morimoto ETA, Goodarzi H et al. (2013). A neurodegeneration-specific gene-expression signature of acutely isolated microglia from an amyotrophic lateral sclerosis mouse model. Cell Rep 4: 385–401.

Cirulli ET, Lasseigne BN, Petrovski S et al. (2015). Exome sequencing in amyotrophic lateral sclerosis identifies risk genes and pathways. Science 1979: 1436–1441.

Cividini C, Basaia S, Spinelli EG et al. (2022). Amyotrophic lateral sclerosis–frontotemporal dementia: shared and divergent neural correlates across the clinical spectrum. Neurology 98: e402–e415.

Clarke BE, Patani R (2020). The microglial component of amyotrophic lateral sclerosis. Brain 143: 3526–3539.

Colombo E, Gentile F, Maranzano A et al. (2023). The impact of upper motor neuron involvement on clinical features, disease progression and prognosis in amyotrophic lateral sclerosis. Front Neurol 14: 1249429.

Connolly S, Galvin M, Hardiman O (2015). End-of-life management in patients with amyotrophic lateral sclerosis. Lancet Neurol 14: 435–442.

Coque E, Salsac C, Espinosa-Carrasco G et al. (2019). Cytotoxic CD8+ T lymphocytes expressing ALS-causing SOD1 mutant selectively trigger death of spinal motoneurons. Proc Natl Acad Sci U S A 116: 2312–2317.

Cousins KAQ, Shaw LM, Chen-Plotkin A et al. (2022). Distinguishing frontotemporal lobar degeneration tau from TDP-43 using plasma biomarkers. JAMA Neurol 79: 1155–1164.

Cudkowicz ME, Titus S, Kearney M et al. (2014). Safety and efficacy of ceftriaxone for amyotrophic lateral sclerosis: a multi-stage, randomised, double-blind, placebo-controlled trial. Lancet Neurol 13: 1083–1091.

Cui C, Ingre C, Yin L et al. (2022). Correlation between leukocyte phenotypes and prognosis of amyotrophic lateral sclerosis. eLife 11: e74065.

Dardiotis E, Siokas V, Sokratous M et al. (2018). Body mass index and survival from amyotrophic lateral sclerosis: a meta-analysis. Neurol Clin Pract 8: 437–444.

De Boer EMJ, Demaegd KC, De Bie CI et al. (2024). Familial motor neuron disease: co-occurrence of PLS and ALS (-FTD). Amyotroph Lateral Scler Frontotemporal Degener 25: 53–60.

De Giorgio F, Maduro C, Fisher EMC et al. (2019). Transgenic and physiological mouse models give insights into different aspects of amyotrophic lateral sclerosis. Dis Model Mech 12: dmm037424.

De Marchi F, Carrarini C, De Martino A et al. (2021a). Cognitive dysfunction in amyotrophic lateral sclerosis: can we predict it? Neurol Sci 42: 2211–2222.

De Marchi F, Munitic I, Amedei A et al. (2021b). Interplay between immunity and amyotrophic lateral sclerosis: clinical impact. Neurosci Biobehav Rev 958–978.

De Marchi F, Tondo G, Corrado L et al. (2023). Neuroinflammatory pathways in the ALS-FTD continuum: a focus on genetic variants. Genes (Basel) 14: 1658.

Decker JM, Krüger L, Sydow A et al. (2016). The Tau/A152T mutation, a risk factor for frontotemporal-spectrum disorders, leads to NR 2B receptor-mediated excitotoxicity. EMBO Rep 17: 552–569.

Dennys CN, Roussel F, Rodrigo R et al. (2023). CuATSM effectively ameliorates ALS patient astrocyte-mediated motor neuron toxicity in human in vitro models of amyotrophic lateral sclerosis. Glia 71: 350–365.

Di Giorgio FP, Boulting GL, Bobrowicz S et al. (2008). Human embryonic stem cell-derived motor neurons are sensitive to the toxic effect of glial cells carrying an ALS-causing mutation. Cell Stem Cell 3: 637–648.

Diaz-Amarilla P, Miquel E, Trostchansky A et al. (2016). Electrophilic nitro-fatty acids prevent astrocyte-mediated toxicity to motor neurons in a cell model of familial amyotrophic lateral sclerosis via nuclear factor erythroid 2-related factor activation. Free Radic Biol Med 95: 112–120.

Dutta DJ, Woo DH, Lee PR et al. (2018). Regulation of myelin structure and conduction velocity by perinodal astrocytes. Proc Natl Acad Sci U S A 115: 11832–11837.

Engelhardt B, Carare RO, Bechmann I et al. (2016). Vascular, glial, and lymphatic immune gateways of the central nervous system. Acta Neuropathol 132: 317–338.

Erkkinen MG, Kim M-O, Geschwind MD (2018). Clinical neurology and epidemiology of the major neurodegenerative diseases. Cold Spring Harb Perspect Biol 10: a033118.

Eshima J, O'Connor SA, Marschall E et al. (2023). Molecular subtypes of ALS are associated with differences in patient prognosis. Nat Commun 14: 95.

Eykens C, Nonneman A, Jensen C et al. (2018). Conditional deletion of Id2 or Notch1 in oligodendrocyte progenitor cells does not ameliorate disease outcome in $SOD1^{G93A}$ mice. Neurobiol Aging 68: 1–4.

Eykens C, Rossaert E, Duqué S et al. (2021). AAV9-mediated gene delivery of MCT1 to oligodendrocytes does not provide a therapeutic benefit in a mouse model of ALS. Mol Ther Methods Clin Dev 20: 508–519.

Falzone YM, Domi T, Mandelli A et al. (2022). Integrated evaluation of a panel of neurochemical biomarkers to optimize diagnosis and prognosis in amyotrophic lateral sclerosis. Eur J Neurol 29: 1930–1939.

Fang T, Al Khleifat A, Meurgey J-H et al. (2018). Stage at which riluzole treatment prolongs survival in patients with amyotrophic lateral sclerosis: a retrospective analysis of data from a dose-ranging study. Lancet Neurol 17: 416–422.

Fang T, Je G, Pacut P et al. (2022). Gene therapy in amyotrophic lateral sclerosis. Cells 11: 2066.

Feldman EL, Goutman SA, Petri S et al. (2022). Amyotrophic lateral sclerosis. Lancet 400: 1363–1380.

Ferraiuolo L, Higginbottom A, Heath PR et al. (2011). Dysregulation of astrocyte-motoneuron cross-talk in mutant superoxide dismutase 1-related amyotrophic lateral sclerosis. Brain 134: 2627–2641. https://doi.org/10.1093/brain/awr193.

Ferraiuolo L, Meyer K, Sherwood TW et al. (2016). Oligodendrocytes contribute to motor neuron death in ALS via SOD1-dependent mechanism. Proc Natl Acad Sci U S A 113: E6496–E6505.

Ferrari R, Kapogiannis D, Huey ED et al. (2011). FTD and ALS: a tale of two diseases. Curr Alzheimer Res 8: 273–294.

Fiala M, Chattopadhay M, La Cava A et al. (2010). IL-17A is increased in the serum and in spinal cord CD8 and mast cells of ALS patients. J Neuroinflammation 7: 1–14.

Finkelstein A, Kunis G, Seksenyan A et al. (2011). Abnormal changes in NKT cells, the IGF-1 axis, and liver pathology in an animal model of ALS. PLoS One 6: e22374.

Foran E, Bogush A, Goffredo M et al. (2011). Motor neuron impairment mediated by a sumoylated fragment of the glial glutamate transporter EAAT2. Glia 59: 1719–1731.

Freischmidt A, Wieland T, Richter B et al. (2015). Haploinsufficiency of TBK1 causes familial ALS and fronto-temporal dementia. Nat Neurosci 18: 631–636.

Fünfschilling U, Supplie LM, Mahad D et al. (2012). Glycolytic oligodendrocytes maintain myelin and long-term axonal integrity. Nature 485: 517–521.

Garofalo S, Cocozza G, Porzia A et al. (2020). Natural killer cells modulate motor neuron-immune cell cross talk in models of amyotrophic lateral sclerosis. Nat Commun 11: 1–16.

Gendron TF, Petrucelli L (2011). Rodent models of TDP-43 proteinopathy: investigating the mechanisms of TDP-43-mediated neurodegeneration. J Mol Neurosci 45: 486–499.

Gerbino V, Kaunga E, Ye J et al. (2020). The loss of TBK1 kinase activity in motor neurons or in all cell types differentially impacts ALS disease progression in SOD1 mice. Neuron 106: 789–805.e5.

Gille B, De Schaepdryver M, Dedeene L et al. (2019). Inflammatory markers in cerebrospinal fluid: independent prognostic biomarkers in amyotrophic lateral sclerosis? J Neurol Neurosurg Psychiatry 90: 1338–1346. https://doi.org/10.1136/jnnp-2018-319586.

Gordon PH, Miller RG, Moore DH (2004). ALSFRS-R. Amyotroph Lateral Scler Other Motor Neuron Disord 5: 90–93.

Gordon PH, Moore DH, Miller RG et al. (2007). Efficacy of minocycline in patients with amyotrophic lateral sclerosis: a phase III randomised trial. Lancet Neurol 6: 1045–1053.

Gordon D, Dafinca R, Scaber J et al. (2019). Single-copy expression of an amyotrophic lateral sclerosis-linked TDP-43 mutation (M337V) in BAC transgenic mice leads to altered stress granule dynamics and progressive motor dysfunction. Neurobiol Dis 121: 148–162.

Gorno-Tempini ML, Hillis AE, Weintraub S et al. (2011). Classification of primary progressive aphasia and its variants. Neurology 76: 1006–1014.

Goutman SA, Hardiman O, Al-Chalabi A et al. (2022). Recent advances in the diagnosis and prognosis of amyotrophic lateral sclerosis. Lancet Neurol 21: 480–493.

Gowing G, Lalancette-Hébert M, Audet J-N et al. (2009). Macrophage colony stimulating factor (M-CSF) exacerbates ALS disease in a mouse model through altered responses of microglia expressing mutant superoxide dismutase. Exp Neurol 220: 267–275.

Grassano M, Calvo A, Moglia C et al. (2022). Systematic evaluation of genetic mutations in ALS: a population-based study. J Neurol Neurosurg Psychiatry 93: 1190–1193.

Grassano M, Manera U, De Marchi F et al. (2023). The role of peripheral immunity in ALS: a population-based study. Ann Clin Transl Neurol 10: 1623–1632.

Gravel M, Béland L-C, Soucy G et al. (2016). IL-10 controls early microglial phenotypes and disease onset in ALS caused by misfolded superoxide dismutase 1. J Neurosci 36: 1031–1048.

Gray E, Thompson AG, Wuu J et al. (2020). CSF chitinases before and after symptom onset in amyotrophic lateral sclerosis. Ann Clin Transl Neurol 7: 1296–1306.

Guo Q, Owen DR, Rabiner EA et al. (2014). A graphical method to compare the in vivo binding potential of PET radioligands in the absence of a reference region: application to [11C] PBR28 and [18F] PBR111 for TSPO imaging. J Cereb Blood Flow Metab 34: 1162–1168.

Gurney ME, Pu H, Chiu AY et al. (1994). Motor neuron degeneration in mice that express a human Cu,Zn superoxide dismutase mutation. Science 264: 1772–1775.

Gurney ME, Fleck TJ, Himes CS et al. (1998). Riluzole preserves motor function in a transgenic model of familial amyotrophic lateral sclerosis. Neurology 50: 62–66.

Guttenplan KA, Weigel MK, Adler DI et al. (2020). Knockout of reactive astrocyte activating factors slows disease progression in an ALS mouse model. Nat Commun 11: 3753.

Haidet-Phillips AM, Hester ME, Miranda CJ et al. (2011). Astrocytes from familial and sporadic ALS patients are toxic to motor neurons. Nat Biotechnol 29: 824–828.

Haidet-Phillips AM, Gross SK, Williams T et al. (2013). Altered astrocytic expression of TDP-43 does not influence motor neuron survival. Exp Neurol 250: 250–259.

Heiman-Patterson TD, Khazaal O, Yu D et al. (2021). Pulmonary function decline in amyotrophic lateral sclerosis. Amyotroph Lateral Scler Frontotemporal Degener 22: 54–61.

Henkel JS, Beers DR, Wen S et al. (2013). Regulatory T-lymphocytes mediate amyotrophic lateral sclerosis progression and survival. EMBO Mol Med 5: 64–79.

Higashihara M, Ishibashi K, Tokumaru AM et al. (2021). 18F-THK5351 PET can identify core lesions in different amyotrophic lateral sclerosis phenotypes. Clin Nucl Med 46: e582–e583.

Hirsch-Reinshagen V, Hercher C, Vila-Rodriguez F et al. (2023). Psychotic symptoms in frontotemporal dementia with TDP-43 tend to be associated with type B pathology. Neuropathol Appl Neurobiol 49: e12921.

Huang S-L, Wu L-S, Lee M et al. (2020). A robust TDP-43 knock-in mouse model of ALS. Acta Neuropathol Commun 8: 1–19.

Igaz LM, Kwong LK, Lee EB et al. (2011). Dysregulation of the ALS-associated gene TDP-43 leads to neuronal death and degeneration in mice. J Clin Invest 121: 726–738.

Ikeda K, Akiyama H, Arai T et al. (2002). Morphometrical reappraisal of motor neuron system of Pick's disease and amyotrophic lateral sclerosis with dementia. Acta Neuropathol 104: 21–28.

Illán-Gala I, Alcolea D, Montal V et al. (2018). CSF sAPPβ, YKL-40, and NfL along the ALS-FTD spectrum. Neurology 91: E1619–E1628. https://doi.org/10.1212/WNL.0000000000006383.

Ito H, Wate R, Zhang J et al. (2008). Treatment with edaravone, initiated at symptom onset, slows motor decline and decreases SOD1 deposition in ALS mice. Exp Neurol 213: 448–455.

Ito Y, Ofengeim D, Najafov A et al. (2016). RIPK1 mediates axonal degeneration by promoting inflammation and necroptosis in ALS. Science 1979: 603–608.

Izrael M, Slutsky SG, Admoni T et al. (2018). Safety and efficacy of human embryonic stem cell-derived astrocytes following intrathecal transplantation in SOD1G93A and NSG animal models. Stem Cell Res Ther 9: 1–17.

Jiménez-Villegas J, Ferraiuolo L, Mead RJ et al. (2021). NRF2 as a therapeutic opportunity to impact in the molecular roadmap of ALS. Free Radic Biol Med 173: 125–141.

Johansson A, Engler H, Blomquist G et al. (2007). Evidence for astrocytosis in ALS demonstrated by [11C](L)-deprenyl-D2 PET. J Neurol Sci 255: 17–22.

Johnson SA, Fang T, De Marchi F et al. (2022). Pharmacotherapy for amyotrophic lateral sclerosis: a review of approved and upcoming agents. Drugs 82: 1367–1388.

Jones MK, Nair A, Gupta M (2019). Mast cells in neurodegenerative disease. Front Cell Neurosci 13: 171.

Kang SH, Li Y, Fukaya M et al. (2013). Degeneration and impaired regeneration of gray matter oligodendrocytes in amyotrophic lateral sclerosis. Nat Neurosci 16: 571–579.

Kettenmann H, Hanisch U-K, Noda M et al. (2011). Physiology of microglia. Physiol Rev 91: 461–553.

Kovacs M, Alamón C, Maciel C et al. (2021). The pathogenic role of c-Kit+ mast cells in the spinal motor neuron-vascular niche in ALS. Acta Neuropathol Commun 9: 1–18.

Kuo MTH, Beckman JS, Shaw CA (2019). Neuroprotective effect of CuATSM on neurotoxin-induced motor neuron loss in an ALS mouse model. Neurobiol Dis 130: 104495.

Kurashige T, Kuramochi M, Ohsawa R et al. (2021). Optineurin defects cause TDP43-pathology with autophagic vacuolar formation. Neurobiol Dis 148: 105215.

Kushner PD, Stephenson DT, Wright S (1991). Reactive astrogliosis is widespread in the subcortical white matter of amyotrophic lateral sclerosis brain. J Neuropathol Exp Neurol 50: 263–277.

Kwan JY, Jeong SY, Van Gelderen P et al. (2012). Iron accumulation in deep cortical layers accounts for MRI signal abnormalities in ALS: correlating 7 tesla MRI and pathology. PLoS One 7: e35241.

Kwon I, Xiang S, Kato M et al. (2014). Poly-dipeptides encoded by the C9orf72 repeats bind nucleoli, impede RNA biogenesis, and kill cells. Science 1979: 1139–1145.

Lacomblez L, Bensimon G, Leigh PN et al. (1996). Dose-ranging study of riluzole in amyotrophic lateral sclerosis. Amyotrophic lateral sclerosis/Riluzole Study Group II. Lancet 347: 1425–1431.

Lampron A, Larochelle A, Laflamme N et al. (2015). Inefficient clearance of myelin debris by microglia impairs remyelinating processes. J Exp Med 212: 481–495.

Larochelle C, Lécuyer MA, Alvarez JI et al. (2015). Melanoma cell adhesion molecule-positive CD8 T lymphocytes mediate central nervous system inflammation. Ann Neurol 78: 39–53.

Lattante S, Ciura S, Rouleau GA et al. (2015). Defining the genetic connection linking amyotrophic lateral sclerosis (ALS) with frontotemporal dementia (FTD). Trends Genet 31: 263–273.

Lavisse S, Guillermier M, Hérard A-S et al. (2012). Reactive astrocytes overexpress TSPO and are detected by TSPO positron emission tomography imaging. J Neurosci 32: 10809–10818.

Lee Y, Morrison BM, Li Y et al. (2012). Oligodendroglia metabolically support axons and contribute to neurodegeneration. Nature 487: 443–448.

Lee JD, Kumar V, Fung JNT et al. (2017). Pharmacological inhibition of complement C5a-C5a1 receptor signalling ameliorates disease pathology in the hSOD1G93A mouse model of amyotrophic lateral sclerosis. Br J Pharmacol 174: 689–699.

Lee JD, Levin SC, Willis EF et al. (2018). Complement components are upregulated and correlate with disease progression in the TDP-43Q331K mouse model of amyotrophic lateral sclerosis. J Neuroinflammation 15: 1–13.

Lee AJB, Kittel TE, Kim RB et al. (2023). Comparing therapeutic modulators of the SOD1 G93A amyotrophic lateral sclerosis mouse pathophysiology. Front Neurosci 16: 1111763.

Leone MA, Mandrioli J, Russo S et al. (2022). Neutrophils-to-lymphocyte ratio is associated with progression and overall survival in amyotrophic lateral sclerosis. Biomedicine 10: 354.

Liao B, Zhao W, Beers DR et al. (2012). Transformation from a neuroprotective to a neurotoxic microglial phenotype in a mouse model of ALS. Exp Neurol 237: 147–152.

Ling S-C, Polymenidou M, Cleveland DW (2013). Converging mechanisms in ALS and FTD: disrupted RNA and protein homeostasis. Neuron 79: 416–438.

Liu Y, Pattamatta A, Zu T et al. (2016). C9orf72 BAC mouse model with motor deficits and neurodegenerative features of ALS/FTD. Neuron 90: 521–534.

Liu Z, Cheng X, Zhong S et al. (2020a). Peripheral and central nervous system immune response crosstalk in amyotrophic lateral sclerosis. Front Neurosci 14: 575.

Liu C, Li D, Lv C et al. (2020b). Activation of the Notch signaling pathway and cellular localization of notch signaling molecules in the spinal cord of SOD1-G93A ALS model mice. Neuroscience 432: 84–93.

Liu Y, Jiang H, Qin X et al. (2022). PET imaging of reactive astrocytes in neurological disorders. Eur J Nucl Med Mol Imaging 49: 1275–1287.

Lloyd AF, Miron VE (2019). The pro-remyelination properties of microglia in the central nervous system. Nat Rev Neurol 15: 447–458.

Logroscino G, Piccininni M, Graff C et al. (2023). Incidence of syndromes associated with frontotemporal lobar degeneration in 9 European countries. JAMA Neurol 80: 279–286.

Lombardi M, Parolisi R, Scaroni F et al. (2019). Detrimental and protective action of microglial extracellular vesicles on myelin lesions: astrocyte involvement in remyelination failure. Acta Neuropathol 138: 987–1012.

Longinetti E, Fang F (2019). Epidemiology of amyotrophic lateral sclerosis: an update of recent literature. Curr Opin Neurol 32: 771.

Lorente Pons A, Higginbottom A, Cooper-Knock J et al. (2020). Oligodendrocyte pathology exceeds axonal pathology in white matter in human amyotrophic lateral sclerosis. J Pathol 251: 262–271.

Ludolph AC, Bendotti C, Blaugrund E et al. (2010). Guidelines for preclinical animal research in ALS/MND: a consensus meeting. Amyotroph Lateral Scler 11: 38–45.

Ludolph A, Drory V, Hardiman O et al. (2015). A revision of the El Escorial criteria—2015. Amyotroph Lateral Scler Frontotemporal Degener 16: 291–292.

Lum JS, Brown ML, Farrawell NE et al. (2021). CuATSM improves motor function and extends survival but is not tolerated at a high dose in SOD1 G93A mice with a C57BL/6 background. Sci Rep 11: 19392.

Majounie E, Renton AE, Mok K et al. (2012). Frequency of the C9orf72 hexanucleotide repeat expansion in patients with amyotrophic lateral sclerosis and frontotemporal dementia: a cross-sectional study. Lancet Neurol 11: 323–330.

Marchlik E, Thakker P, Carlson T et al. (2010). Mice lacking Tbk1 activity exhibit immune cell infiltrates in multiple tissues and increased susceptibility to LPS-induced lethality. J Leukoc Biol 88: 1171–1180.

Markovinovic A, Cimbro R, Ljutic T et al. (2017). Optineurin in amyotrophic lateral sclerosis: multifunctional adaptor protein at the crossroads of different neuroprotective mechanisms. Prog Neurobiol 154: 1–20.

Martorana F, Brambilla L, Valori CF et al. (2012). The BH4 domain of Bcl-XL rescues astrocyte degeneration in amyotrophic lateral sclerosis by modulating intracellular calcium signals. Hum Mol Genet 21: 826–840.

Mattila OS, Strbian D, Saksi J et al. (2011). Cerebral mast cells mediate blood-brain barrier disruption in acute experimental ischemic stroke through perivascular gelatinase activation. Stroke 42: 3600–3605.

McCampbell A, Cole T, Wegener AJ et al. (2018). Antisense oligonucleotides extend survival and reverse decrement in muscle response in ALS models. J Clin Invest 128: 3558–3567.

McDonald TS, McCombe PA, Woodruff TM et al. (2020). The potential interplay between energy metabolism and innate complement activation in amyotrophic lateral sclerosis. FASEB J 34: 7225–7233.

McGill RB, Steyn FJ, Ngo ST et al. (2020). Monocytes and neutrophils are associated with clinical features in amyotrophic lateral sclerosis. Brain Commun 2: fcaa013. https://doi.org/10.1093/braincomms/fcaa013.

Megat S, Mora N, Sanogo J et al. (2023). Integrative genetic analysis illuminates ALS heritability and identifies risk genes. Nat Commun 14: 342.

Mesulam M (1982). Slowly progressive aphasia without generalized dementia. Ann Neurol 11: 592–598.

Mesulam MM (2003). Primary progressive aphasia—a language-based dementia. N Engl J Med 349: 1535–1542.

Meyer JH, Braga J (2022). Development and clinical application of positron emission tomography imaging agents for monoamine oxidase B. Front Neurosci 15: 773404.

Middeldorp J, Hol EM (2011). GFAP in health and disease. Prog Neurobiol 93: 421–443.

Migheli A, Cordera S, Bendotti C et al. (1999). S-100beta protein is upregulated in astrocytes and motor neurons in the spinal cord of patients with amyotrophic lateral sclerosis. Neurosci Lett 261: 25–28.

Miller RG, Block G, Katz JS et al. (2015). Randomized phase 2 trial of NP001, a novel immune regulator: safety and early efficacy in ALS. Neurol Neuroimmunol Neuroinflamm 2: e100.

Miller RG, Zhang R, Bracci PM et al. (2022). Phase 2B randomized controlled trial of NP001 in amyotrophic lateral sclerosis: pre-specified and post hoc analyses. Muscle Nerve 66: 39–49.

Miron VE, Boyd A, Zhao J-W et al. (2013). M2 microglia and macrophages drive oligodendrocyte differentiation during CNS remyelination. Nat Neurosci 16: 1211–1218.

Miyoshi M, Shinotoh H, Wszolek ZK et al. (2010). In vivo detection of neuropathologic changes in presymptomatic MAPT mutation carriers: a PET and MRI study. Parkinsonism Relat Disord 16: 404–408.

Mora JS, Genge A, Chio A et al. (2020). Masitinib as an add-on therapy to riluzole in patients with amyotrophic lateral sclerosis: a randomized clinical trial. Amyotroph Lateral Scler Frontotemporal Degener 21: 5–14.

Mordes DA, Morrison BM, Ament XH et al. (2020). Absence of survival and motor deficits in 500 repeat C9ORF72 BAC mice. Neuron 108: 775–783.e4.

Munitic I, Giardino Torchia ML, Meena NP et al. (2013). Optineurin insufficiency impairs IRF3 but not NF-κB activation in immune cells. J Immunol 191: 6231–6240. https://doi.org/10.4049/jimmunol.1301696.

Murdock BJ, Zhou T, Kashlan SR et al. (2017). Correlation of peripheral immunity with rapid amyotrophic lateral sclerosis progression. JAMA Neurol 74: 1446–1454.

Murphy J, Factor-Litvak P, Goetz R et al. (2016). Cognitive-behavioral screening reveals prevalent impairment in a large multicenter ALS cohort. Neurology 86: 813–820.

Nagai M, Re DB, Nagata T et al. (2007). Astrocytes expressing ALS-linked mutated SOD1 release factors selectively toxic to motor neurons. Nat Neurosci 10: 615–622.

Nardo G, Trolese MC, de Vito G et al. (2016). Immune response in peripheral axons delays disease progression in SOD1 G93A mice. J Neuroinflammation 13: 1–16.

Nardo G, Trolese MC, Verderio M et al. (2018). Counteracting roles of MHCI and CD8(+) T cells in the peripheral and central nervous system of ALS SOD1(G93A) mice. Mol Neurodegener 13: 42. https://doi.org/10.1186/s13024-018-0271-7.

Nguyen L, Laboissonniere LA, Guo S et al. (2020). Survival and motor phenotypes in FVB C9-500 ALS/FTD BAC transgenic mice reproduced by multiple labs. Neuron 108: 784–796.e3.

Nicholson K, Chan J, Macklin EA et al. (2018). Pilot trial of inosine to elevate urate levels in amyotrophic lateral sclerosis. Ann Clin Transl Neurol 5: 1522–1533.

Northall A, Doehler J, Weber M et al. (2024). Multimodal layer modelling reveals in vivo pathology in amyotrophic lateral sclerosis. Brain 147: 1087–1099.

Nutma E, van Gent D, Amor S et al. (2020). Astrocyte and oligodendrocyte cross-talk in the central nervous system. Cells 9: 600.

Nutma E, Fancy N, Weinert M et al. (2023). Translocator protein is a marker of activated microglia in rodent models but not human neurodegenerative diseases. Nat Commun 14: 5247.

O'Rourke JG, Bogdanik L, Muhammad A et al. (2015). C9orf72 BAC transgenic mice display typical pathologic features of ALS/FTD. Neuron 88: 892–901.

O'Rourke JG, Bogdanik L, Yáñez A et al. (2016). C9orf72 is required for proper macrophage and microglial function in mice. Science 1979: 1324–1329.

Oeckl P, Weydt P, Steinacker P et al. (2019). Different neuroinflammatory profile in amyotrophic lateral sclerosis and frontotemporal dementia is linked to the clinical phase. J Neurol Neurosurg Psychiatry 90: 4–10.

Paganoni S, De Marchi F, Chan J et al. (2020a). The NEALS primary lateral sclerosis registry. Amyotroph Lateral Scler Frontotemporal Degener 21: 74–81.

Paganoni S, Macklin EA, Hendrix S et al. (2020b). Trial of sodium phenylbutyrate-taurursodiol for amyotrophic lateral sclerosis. N Engl J Med 383: 919–930. https://doi.org/10.1056/NEJMoa1916945.

Paganoni S, Berry JD, Quintana M et al. (2022). Adaptive platform trials to transform amyotrophic lateral sclerosis therapy development. Ann Neurol 91: 165–175.

Peake K, Manning J, Lewis C-A et al. (2017). Bone marrow-derived cell accumulation in the spinal cord is independent of peripheral mobilization in a mouse model of amyotrophic lateral sclerosis. Front Neurol 8: 75.

Peferoen L, Kipp M, van der Valk P et al. (2014). Oligodendrocyte-microglia cross-talk in the central nervous system. Immunology 141: 302–313.

Pelkmans W, Shekari M, Brugulat-Serrat A et al. (2024). Astrocyte biomarkers GFAP and YKL-40 mediate early Alzheimer's disease progression. Alzheimer's Dement 20: 483–493.

Peng AYT, Agrawal I, Ho WY et al. (2020). Loss of TDP-43 in astrocytes leads to motor deficits by triggering A1-like reactive phenotype and triglial dysfunction. Proc Natl Acad Sci U S A 117: 29101–29112.

Perrin S (2014). Preclinical research: make mouse studies work. Nature 507: 423–425.

Philips T, Bento-Abreu A, Nonneman A et al. (2013). Oligodendrocyte dysfunction in the pathogenesis of amyotrophic lateral sclerosis. Brain 136: 471–482.

Pollari E, Savchenko E, Jaronen M et al. (2011). Granulocyte colony stimulating factor attenuates inflammation in a mouse model of amyotrophic lateral sclerosis. J Neuroinflammation 8: 1–14.

Prudlo J, König J, Schuster C et al. (2016). TDP-43 pathology and cognition in ALS: a prospective clinicopathologic correlation study. Neurology 87: 1019–1023.

Qian K, Huang H, Peterson A et al. (2017). Sporadic ALS astrocytes induce neuronal degeneration in vivo. Stem Cell Rep 8: 843–855.

Querol-Vilaseca M, Colom-Cadena M, Pegueroles J et al. (2017). YKL-40 (Chitinase 3-like I) is expressed in a subset of astrocytes in Alzheimer's disease and other tauopathies. J Neuroinflammation 14: 1–10.

Raffaele S, Boccazzi M, Fumagalli M (2021). Oligodendrocyte dysfunction in amyotrophic lateral sclerosis: mechanisms and therapeutic perspectives. Cells 10: 565.

Ransohoff RM (2016). How neuroinflammation contributes to neurodegeneration. Science (1979) 353: 777–783.

Rascovsky K, Hodges JR, Knopman D et al. (2011). Sensitivity of revised diagnostic criteria for the behavioural variant of frontotemporal dementia. Brain 134: 2456–2477.

Rawji KS, Martinez GAG, Sharma A et al. (2020). The role of astrocytes in remyelination. Trends Neurosci 43: 596–607.

Renton AE, Majounie E, Waite A et al. (2011). A hexanucleotide repeat expansion in C9ORF72 is the cause of chromosome 9p21-linked ALS-FTD. Neuron 72: 257–268.

Riemslagh FW, van der Toorn EC, Verhagen RFM et al. (2021). Inducible expression of human C9ORF72 36x G4C2 hexanucleotide repeats is sufficient to cause RAN translation and rapid muscular atrophy in mice. Dis Model Mech 14: dmm044842.

Rifai O, O'shaughnessy J, Dando OR et al. (2023). Distinct neuroinflammatory signatures exist across genetic and sporadic amyotrophic lateral sclerosis cohorts. Brain 146: 5124–5138.

Rogalski E, Cobia D, Harrison TM et al. (2011). Anatomy of language impairments in primary progressive aphasia. J Neurosci 31: 3344–3350.

Rolfes L, Schulte-Mecklenbeck A, Schreiber S et al. (2021). Amyotrophic lateral sclerosis patients show increased peripheral and intrathecal T-cell activation. Brain Commun 3: fcab157.

Rosenblum LT, Shamamandri-Markandaiah S, Ghosh B et al. (2017). Mutation of the caspase-3 cleavage site in the astroglial glutamate transporter EAAT2 delays disease progression and extends lifespan in the SOD1-G93A mouse model of ALS. Exp Neurol 292: 145–153.

Rossi D, Brambilla L, Valori CF et al. (2008). Focal degeneration of astrocytes in amyotrophic lateral sclerosis. Cell Death Differ 15: 1691–1700.

Rossi C, Cusimano M, Zambito M et al. (2018). Interleukin 4 modulates microglia homeostasis and attenuates the early slowly progressive phase of amyotrophic lateral sclerosis. Cell Death Dis 9: 250.

Rothstein JD, Patel S, Regan MR et al. (2005). β-Lactam antibiotics offer neuroprotection by increasing glutamate transporter expression. Nature 433: 73–77.

Rusina R, Vandenberghe R, Bruffaerts R (2021). Cognitive and behavioral manifestations in ALS: beyond motor system involvement. Diagnostics 11: 624.

Ryu H, Smith K, Camelo SI et al. (2005). Sodium phenylbutyrate prolongs survival and regulates expression of anti-apoptotic genes in transgenic amyotrophic lateral sclerosis mice. J Neurochem 93: 1087–1098.

Schiffer D, Fiano V (2004). Astrogliosis in ALS: possible interpretations according to pathogenetic hypotheses. Amyotroph Lateral Scler Other Motor Neuron Disord 5: 22–25.

Schiffer D, Cordera S, Cavalla P et al. (1996). Reactive astrogliosis of the spinal cord in amyotrophic lateral sclerosis. J Neurol Sci 139: 27–33.

Schreiber S, Schreiber F, Garz C et al. (2019). Toward in vivo determination of peripheral nervous system immune activity in amyotrophic lateral sclerosis. Muscle Nerve 59: 567–576.

Serio A, Bilican B, Barmada SJ et al. (2013). Astrocyte pathology and the absence of non-cell autonomy in an induced pluripotent stem cell model of TDP-43 proteinopathy. Proc Natl Acad Sci U S A 110: 4697–4702.

Shiraishi W, Yamasaki R, Hashimoto Y et al. (2021). Clearance of peripheral nerve misfolded mutant protein by infiltrated macrophages correlates with motor neuron disease progression. Sci Rep 11: 16438.

Sieverding K, Ulmer J, Bruno C et al. (2021). Hemizygous deletion of Tbk1 worsens neuromuscular junction pathology in TDP-43^{G298S} transgenic mice. Exp Neurol 335: 113496.

Sironi F, De Marchi F, Mazzini L et al. (2023). Cell therapy in ALS: an update on preclinical and clinical studies. Brain Res Bull 194: 64–81.

Sofroniew MV (2020). Astrocyte reactivity: subtypes, states, and functions in CNS innate immunity. Trends Immunol 41: 758–770.

Song S, Miranda CJ, Braun L et al. (2016). Major histocompatibility complex class I molecules protect motor neurons from astrocyte-induced toxicity in amyotrophic lateral sclerosis. Nat Med 22: 397–403.

Sta M, Sylva-Steenland RM, Casula M et al. (2011). Innate and adaptive immunity in amyotrophic lateral sclerosis: evidence of complement activation. Neurobiol Dis 42: 211–220.

Strong MJ, Abrahams S, Goldstein LH et al. (2017). Amyotrophic lateral sclerosis-frontotemporal spectrum disorder (ALS-FTSD): revised diagnostic criteria. Amyotroph Lateral Scler Frontotemporal Degener 18: 153–174.

Sudria-Lopez E, Koppers M, de Wit M et al. (2016). Full ablation of C9orf72 in mice causes immune system-related pathology and neoplastic events but no motor neuron defects. Acta Neuropathol 132: 145–147.

Szebényi K, Wenger LMD, Sun Y et al. (2021). Human ALS/FTD brain organoid slice cultures display distinct early astrocyte and targetable neuronal pathology. Nat Neurosci 24: 1542–1554.

Taha DM, Clarke BE, Hall CE et al. (2022). Astrocytes display cell autonomous and diverse early reactive states in familial amyotrophic lateral sclerosis. Brain 145: 481–489.

Talbott EO, Malek AM, Lacomis D (2016). The epidemiology of amyotrophic lateral sclerosis. Handb Clin Neurol 138: 225–238.

Tam OH, Rozhkov NV, Shaw R et al. (2019). Postmortem cortex samples identify distinct molecular subtypes of ALS: retrotransposon activation, oxidative stress, and activated glia. Cell Rep 29: 1164–1177.

Taoka T, Ito R, Nakamichi R et al. (2022). Reproducibility of diffusion tensor image analysis along the perivascular space (DTI-ALPS) for evaluating interstitial fluid diffusivity and glymphatic function: CHanges in Alps index on multiple conditiON acquIsition eXperiment (CHAMONIX) study. Jpn J Radiol 40: 147–158.

Taylor JP, Brown Jr RH, Cleveland DW (2016). Decoding ALS: from genes to mechanism. Nature 539: 197–206.

Thompson AG, Gray E, Bampton A et al. (2019). CSF chitinase proteins in amyotrophic lateral sclerosis. J Neurol Neurosurg Psychiatry 90: 1215–1220. https://doi.org/10.1136/jnnp-2019-320442.

Thomsen GM, Avalos P, Ma AA et al. (2018). Transplantation of neural progenitor cells expressing glial cell line-derived neurotrophic factor into the motor cortex as a strategy to treat amyotrophic lateral sclerosis. Stem Cells 36: 1122–1131.

Tisdall MD, Ohm DT, Lobrovich R et al. (2022). Ex vivo MRI and histopathology detect novel iron-rich cortical inflammation in frontotemporal lobar degeneration with tau versus TDP-43 pathology. Neuroimage Clin 33: 102913.

Todd TW, Petrucelli L (2022). Modelling amyotrophic lateral sclerosis in rodents. Nat Rev Neurosci 23: 231–251.

Tondo G, Iaccarino L, Cerami C et al. (2020). 11C-PK11195 PET–based molecular study of microglia activation in SOD1 amyotrophic lateral sclerosis. Ann Clin Transl Neurol 7: 1513–1523.

Tong J, Huang C, Bi F et al. (2013). Expression of ALS-linked TDP-43 mutant in astrocytes causes non-cell-autonomous motor neuron death in rats. EMBO J 32: 1917–1926.

Trias E, King PH, Si Y et al. (2018). Mast cells and neutrophils mediate peripheral motor pathway degeneration in ALS. JCI Insight 3: e123249.

Trolese MC, Scarpa C, Melfi V et al. (2022). Boosting the peripheral immune response in the skeletal muscles improved motor function in ALS transgenic mice. Mol Ther 30: 2760–2784.

Troost D, Van den Oord JJ, Jong JMBVDE (1990). Immunohistochemical characterization of the inflammatory infiltrate in amyotrophic lateral sclerosis. Neuropathol Appl Neurobiol 16: 401–410.

Turner MR, Cagnin A, Turkheimer FE et al. (2004). Evidence of widespread cerebral microglial activation in amyotrophic lateral sclerosis: an [11C](R)-PK11195 positron emission tomography study. Neurobiol Dis 15: 601–609.

Turner MR, Barohn RJ, Corcia P et al. (2020). Primary lateral sclerosis: consensus diagnostic criteria. J Neurol Neurosurg Psychiatry 91: 373–377.

Valori CF, Neumann M (2021). Contribution of RNA/DNA binding protein dysfunction in oligodendrocytes in the pathogenesis of the amyotrophic lateral sclerosis/frontotemporal lobar degeneration spectrum diseases. Front Neurosci 15: 724891.

Valori CF, Sulmona C, Brambilla L et al. (2023). Astrocytes: dissecting their diverse roles in amyotrophic lateral sclerosis and frontotemporal dementia. Cells 12: 1450.

van der Ende EL, Heller C, Sogorb-Esteve A et al. (2022). Elevated CSF and plasma complement proteins in genetic frontotemporal dementia: results from the GENFI study. J Neuroinflammation 19: 217.

van Es MA, Hardiman O, Chio A et al. (2017). Amyotrophic lateral sclerosis. Lancet 390: 2084–2098.

Vargas MR, Johnson JA (2010). Astrogliosis in amyotrophic lateral sclerosis: role and therapeutic potential of astrocytes. Neurotherapeutics 7: 471–481.

Vargas MR, Johnson DA, Sirkis DW et al. (2008). Nrf2 activation in astrocytes protects against neurodegeneration in mouse models of familial amyotrophic lateral sclerosis. J Neurosci 28: 13574–13581.

Velebit J, Horvat A, Smolič T et al. (2020). Astrocytes with TDP-43 inclusions exhibit reduced noradrenergic cAMP and Ca2+ signaling and dysregulated cell metabolism. Sci Rep 10: 1–18.

Verde F, Milone I, Maranzano A et al. (2023). Serum levels of glial fibrillary acidic protein in patients with amyotrophic lateral sclerosis. Ann Clin Transl Neurol 10: 118–129.

Vieira FG, Hatzipetros T, Thompson K et al. (2017). CuATSM efficacy is independently replicated in a SOD1 mouse model of ALS while unmetallated ATSM therapy fails to reveal benefits. IBRO Rep 2: 47–53.

Vu L, An J, Kovalik T et al. (2020). Cross-sectional and longitudinal measures of chitinase proteins in amyotrophic lateral sclerosis and expression of CHI3L1 in activated astrocytes. J Neurol Neurosurg Psychiatry 91: 350–358.

Wang HA, Lee JD, Lee KM et al. (2017). Complement C5a-C5aR1 signalling drives skeletal muscle macrophage recruitment in the hSOD1(G93A) mouse model of amyotrophic lateral sclerosis. Skelet Muscle 7: 10. https://doi.org/10.1186/s13395-017-0128-8.

Wang J, Ho WY, Lim K et al. (2018). Cell-autonomous requirement of TDP-43, an ALS/FTD signature protein, for oligodendrocyte survival and myelination. Proc Natl Acad Sci U S A 115: E10941–E10950.

Watanabe S, Oiwa K, Murata Y et al. (2020). ALS-linked TDP-43 M337V knock-in mice exhibit splicing deregulation without neurodegeneration. Mol Brain 13: 1–4.

Wen D, Ji Y, Li Y et al. (2024).OPTN gene therapy increases autophagy and protects mitochondria in SOD1-G93A-expressing transgenic mice and cells. FEBS J 291: 795–813.

Westeneng H-J, Debray TPA, Visser AE et al. (2018). Prognosis for patients with amyotrophic lateral sclerosis: development and validation of a personalised prediction model. Lancet Neurol 17: 423–433.

White MA, Kim E, Duffy A et al. (2018). TDP-43 gains function due to perturbed autoregulation in a Tardbp knock-in mouse model of ALS-FTD. Nat Neurosci 21: 552–563.

Williams JR, Trias E, Beilby PR et al. (2016). Copper delivery to the CNS by CuATSM effectively treats motor neuron disease in SODG93A mice co-expressing the copper-chaperone-for-SOD. Neurobiol Dis 89: 1–9.

Wong C, Stavrou M, Elliott E et al. (2021). Clinical trials in amyotrophic lateral sclerosis: a systematic review and perspective. Brain Commun 3: fcab242.

Woollacott IOC, Nicholas JM, Heller C et al. (2020). Cerebrospinal fluid YKL-40 and chitotriosidase levels in frontotemporal dementia vary by clinical, genetic and pathological subtype. Dement Geriatr Cogn Disord 49: 56–76. https://doi.org/10.1159/000506282.

Woollacott IOC, Swift IJ, Sogorb-Esteve A et al. (2022). CSF glial markers are elevated in a subset of patients with genetic frontotemporal dementia. Ann Clin Transl Neurol 9: 1764–1777.

Wu LS, Cheng W-C, Shen C-KJ (2012). Targeted depletion of TDP-43 expression in the spinal cord motor neurons leads to the development of amyotrophic lateral sclerosis-like phenotypes in mice. J Biol Chem 287: 27335–27344.

Wu LS, Cheng W-C, Chen C-Y et al. (2019). Transcriptomopathies of pre-and post-symptomatic frontotemporal dementia-like mice with TDP-43 depletion in forebrain neurons. Acta Neuropathol Commun 7: 1–31.

Xiao Y, Zou Q, Xie X et al. (2017). The kinase TBK1 functions in dendritic cells to regulate T cell homeostasis, autoimmunity, and antitumor immunity. J Exp Med 214: 1493–1507.

Yang L, Gal J, Chen J et al. (2014a). Self-assembled FUS binds active chromatin and regulates gene transcription. Proc Natl Acad Sci U S A 111: 17809–17814.

Yang C, Wang H, Qiao T et al. (2014b). Partial loss of TDP-43 function causes phenotypes of amyotrophic lateral sclerosis. Proc Natl Acad Sci U S A 111: E1121–E1129.

Yang C, Qiao T, Yu J et al. (2022). Low-level overexpression of wild type TDP-43 causes late-onset, progressive neurodegeneration and paralysis in mice. PLoS One 17: e0255710.

Yazdani S, Seitz C, Cui C et al. (2022). T cell responses at diagnosis of amyotrophic lateral sclerosis predict disease progression. Nat Commun 13: 6733. https://doi.org/10.1038/s41467-022-34526-9.

Zakharova MN, Abramova AA (2022). Lower and upper motor neuron involvement and their impact on disease prognosis in amyotrophic lateral sclerosis. Neural Regen Res 17: 65.

Zejlon C, Nakhostin D, Winklhofer S et al. (2022). Structural magnetic resonance imaging findings and histopathological correlations in motor neuron diseases—a systematic review and meta-analysis. Front Neurol 13: 947347.

Zhang S, Cooper-Knock J, Weimer AK et al. (2022). Genome-wide identification of the genetic basis of amyotrophic lateral sclerosis. Neuron 110: 992-1008.e11.

Zhao W, Beers DR, Hooten KG et al. (2017). Characterization of gene expression phenotype in amyotrophic lateral sclerosis monocytes. JAMA Neurol 74: 677–685.

Zhu Q, Jiang J, Gendron TF et al. (2020). Reduced C9ORF72 function exacerbates gain of toxicity from ALS/FTD-causing repeat expansion in C9orf72. Nat Neurosci 23: 615–624.

Zhu N, Santos-Santos M, Illán-Gala I et al. (2021). Plasma glial fibrillary acidic protein and neurofilament light chain for the diagnostic and prognostic evaluation of frontotemporal dementia. Transl Neurodegener 10: 1–12.

Ziff OJ, Clarke BE, Taha DM et al. (2022). Meta-analysis of human and mouse ALS astrocytes reveals multiomic signatures of inflammatory reactive states. Genome Res 32: 71–84.

Zou Z-Y, Zhou Z-R, Che C-H et al. (2017). Genetic epidemiology of amyotrophic lateral sclerosis: a systematic review and meta-analysis. J Neurol Neurosurg Psychiatry 88: 540–549.

Zürcher NR, Loggia ML, Lawson R et al. (2015). Increased in vivo glial activation in patients with amyotrophic lateral sclerosis: assessed with [11C]-PBR28. Neuroimage Clin 7: 409–414.

Handbook of Clinical Neurology, Vol. 210 (3rd series)
Neuroglia in Neurologic and Psychiatric Disorders, Part II
A. Verkhratsky, L.D. de Witte, E. Aronica, and E.M. Hol, Editors
https://doi.org/10.1016/B978-0-443-19102-2.00016-8

Chapter 4

Neuroglia in epilepsy

MANOLIA R. GHOULI[1,2] AND DEVIN K. BINDER[1,2*]

[1]*Division of Biomedical Sciences, School of Medicine, University of California, Riverside, Riverside, CA, United States*
[2]*Center for Glial-Neuronal Interactions, University of California, Riverside, Riverside, CA, United States*

Abstract

Epilepsy is a group of neurologic diseases characterized by spontaneous, repetitive disruption to neuronal activity. Neurons have been at the core of epilepsy research efforts, and pharmacotherapies historically have been generated by targeting neuronal mechanisms. As a result, most currently available antiseizure drugs (ASDs) work to either decrease excitatory glutamatergic neurotransmission or to increase inhibitory GABAergic neurotransmission. However, ASDs may have undesirable side effects on cognition and also fail to control seizures in approximately 30% of epilepsy patients. In recent years, glia have surfaced as essential modulators of neuronal function in health and disease. The redirection of focus onto neuroglia provides new perspectives and opportunities to generate novel therapeutic targets that may treat refractory epilepsy and diminish the unwanted side effect profile of current treatments. In this chapter, we discuss the contribution of astroglia, oligodendroglia, and microglia to the genesis, development, and progression of epilepsy, and we highlight key enzymes, receptors, transporters, and channels that may be pursued as nonneuronal targets for novel ASDs.

EPILEPSY

The International League Against Epilepsy (ILAE) defines epilepsy as "a disorder of the brain characterized by an enduring predisposition to generate epileptic seizures and by the neurobiologic, cognitive, psychologic, and social consequences of this condition" (Fisher et al., 2014). In turn, an epileptic seizure is "a transient occurrence of signs and/or symptoms due to abnormal excessive or synchronous neuronal activity in the brain" (Fisher et al., 2005). An analysis of the burden of this disease reveals that as of 2016, 45.9 million patients have active epilepsy, and of these patients, 24.0 million suffer from idiopathic epilepsy (Beghi et al., 2019).

Our earliest account of seizures dates to an ancient Akkadian tablet that detailed a man whose "neck turns left, hands and feet are tense, and his eyes wide open, and from his mouth froth is flowing without him having any consciousness" (Kaculini et al., 2021). The symptoms described by the ancient Mesopotamians highlight cardinal physical manifestations of convulsive seizures caused by aberrant electrical activity in the brain. It was not until the advent of electroencephalogram technology by Hans Berger in 1924 that clinical seizures could be studied and characterized through the detection of various markers such as epileptiform spikes and discharges, focal interictal spikes, and 3-Hz spike–wave patterns (Louis et al., 2016).

Epilepsy arises from disturbances that tip the balance of the brain toward excitatory neurotransmission, leading to aberrant and synchronized neuronal firing (Patel et al., 2019). Such disturbances may arise from genetic predispositions, as 84 "epilepsy genes" have been identified that give rise to a primary epilepsy phenotype; or they may be acquired from, inter alia, brain tumors, brain trauma, cerebral infection, vascular disorders or damage, and immunologic disorders (Wang et al., 2017; Anwar et al., 2020). The classification of epilepsy requires (Fisher et al., 2014) the identification of a seizure as being epileptic according to the aforementioned ILAE conditions; (Fisher et al., 2005) characterization of the epilepsy type as being focal, generalized, combined, or unknown; and (Beghi et al., 2019) the diagnosis of a specific epilepsy syndrome (Scheffer et al., 2017).

Given that epilepsies are disorders of aberrant neuronal transmission, neuronal mechanisms have been assiduously pursued in the development of antiseizure drugs (ASDs). Most ASDs work (Fisher et al., 2014) to inhibit voltage-gated sodium channels to limit repetitive neuronal firing; (Fisher et al., 2005) to decrease excitatory glutamatergic neurotransmission; or (Beghi et al., 2019) to increase inhibitory GABAergic neurotransmission (Bittigau et al., 2002). Because they indiscriminately target neuronal populations throughout the central nervous system (CNS), ASDs may substantially diminish the cognitive capacity

*Correspondence to: Devin K. Binder, Division of Biomedical Sciences, School of Medicine, University of California-Riverside, Riverside, CA, United States. Tel: +1-951-827-2224, Fax: +1-951-827-5504, E-mail: devin.binder@ucr.edu

of patients to think, remember, and concentrate (Mutanana et al., 2020). Furthermore, approximately one-third of patients become pharmacoresistant to ASDs, leading to uncontrollable seizures and the detrimental socioeconomic effects associated therewith (Fattorusso et al., 2021). The high proportion of refractory epilepsy, the obscure cause of idiopathic epilepsies, and the often inadequate clinical effects of current ASDs all emphasize the urgent need for novel antiepileptogenic strategies. Recent research suggests that the targeting of glia may offer a promising approach in the treatment of epilepsy.

GLIA

The field of neuroscience—and by extension, the study of epilepsy—was revolutionized in 1858 when Rudolf Virchow delivered a lecture on the novel concept of "nervenkitt" (German for nerve cement), now known as "neuroglia" (Kettenmann and Verkhratsky, 2008). In his delivery, Virchow discussed a functionless "glue" that lies in between the neurons, providing them with basic trophic support. Originally, glia were viewed as only "passive support cells"; however, the past few decades have dramatically contributed to knowledge regarding the role of glia in health and disease. Neuroglia are now accepted to be key members and essential regulators of virtually all normal brain functions.

The primary neuroglial populations identified in the mature CNS are astrocytes, microglia, and oligodendrocytes (Purves et al., 2001). Among the diverse functions these distinct cells serve, glia work in concert to modulate synaptic transmission, maintain blood–brain barrier (BBB) permeability, mediate injury response and repair, and regulate axonal health and activity (Tran et al., 2021). Their abundance is a testament to their importance in the brain: an analysis of the glia/neuron ratio of the whole human brain revealed a conservative estimate of 0.99, suggesting there are as many glial cells as neurons in the human brain (Azevedo et al., 2009; Herculano-Houzel, 2014). Given their indispensable roles and their abundance in the CNS, the dysregulation or depletion of any glial population can challenge the delicate equilibrium they maintain, leading to aberrant neuronal signaling (Tran et al., 2021).

The role of neuroglia in pathology has been implicated in multiple neurologic diseases, including amyotrophic lateral sclerosis, Alzheimer disease, frontotemporal dementia, Parkinson disease, multiple sclerosis, and epilepsy (Stevenson et al., 2020). In other words, even in what are regarded to be "neuronal diseases," glial dysfunction is involved in the development of neuropathology. In the words of Rudolf Virchow: "it is extremely important to have a knowledge of that substance also which lies between the proper nervous parts, holds them together and gives the whole its form in a greater or less degree." Therefore, although current antiseizure pharmacotherapies target neuronal ion channels and receptors, advancing our understanding of the role of glia in epilepsy may lead to the development of novel treatment strategies.

In this chapter, we will investigate the contribution of astroglia, oligodendroglia, and microglia to epilepsy (Fig. 4.1). Historically, astroglia were studied to a greater extent as compared to other glia in the context of epilepsy; therefore, this review is necessarily "astrocentric." We performed a comprehensive search of the literature indexed in PubMed, ScienceDirect, Elsevier, Wiley Online Library, Oxford journals, and SAGE journals databases. The primary terms searched were "epilepsy and astroglia/ astrocytes," "seizures and astroglia/ astrocytes," "epilepsy and microglia," "seizures and microglia," "epilepsy and oligodendroglia/ oligodendrocytes" and "seizures and oligodendroglia/oligodendrocytes." Articles for this chapter were selected based on the significance of their contribution to the understanding of the pathophysiologic mechanisms induced by epilepsy. Articles not published in English were excluded. We apologize for any substantive omissions this strategy may have engendered.

ASTROGLIA

Astroglia are a heterogeneous group of star-shaped cells (*astro-* from the Greek word *ástron,* or "celestial body") that establish and maintain the homeostasis required for proper neuronal signaling (Allen and Barres, 2009). Two biochemically and spatially discrete populations of astroglia have been identified across species: the protoplasmic astroglia of the gray matter (also called type 1 astrocytes) and the fibrous astroglia of the white matter (type 2 astrocytes) (Miller and Raff, 1984). Protoplasmic astroglia exist in three-dimensional tessellating domains, acting as nonoverlapping synaptic islands that each contact and modulate 270,000–2,000,000 human tripartite synapses, depending on the brain region (Oberheim et al., 2012). In contrast, fibrous astroglia are longer, have fewer fine processes, are oriented in the direction of axon bundles, and share overlapping domains with other fibrous astrocytes (Oberheim et al., 2009).

In health, astroglia mediate a plethora of functions, including regulation of the BBB, modulation of synaptic transmission, spatial potassium buffering, and adenosine metabolism, and they provide metabolic trophic support to surrounding cells (Pascual et al., 2005; Eid et al., 2008a; Coulter and Steinhäuser, 2015; Boison and Steinhäuser, 2018). In neuropathology, however, under pathologic conditions, astroglia undergo a process referred to as gliosis (called "reactive astrogliosis"), which is a hallmark characteristic of various forms of epilepsy, including temporal lobe epilepsy (TLE) (Escartin et al., 2021). Astrogliosis is defined by the proliferation of astroglia, the morphologic and functional changes that astroglia undergo, and the consequential dysfunction of these cells to carry out their essential functions in the brain (Moulson et al., 2021). Even minute changes to astroglial morphology, gene expression, and metabolism can disrupt the homeostasis of the CNS and influence epileptogenesis (Binder and Steinhäuser, 2021). These various functions suggest that astroglia may serve as novel, nonneuronal therapeutic targets in the generation of future ASDs. Below, we consider studies that clarify the astroglial influence on the BBB in epilepsy, as well as the cellular and molecular changes of astrocytes in epilepsy.

Astroglia and the BBB in epilepsy

The blood–brain barrier

Although not the largest structure in the body, the brain is the most metabolically demanding organ, utilizing approximately 20% of all metabolic reserves even at rest. For oxygen, glucose, and other molecules to be delivered to the brain through the blood without

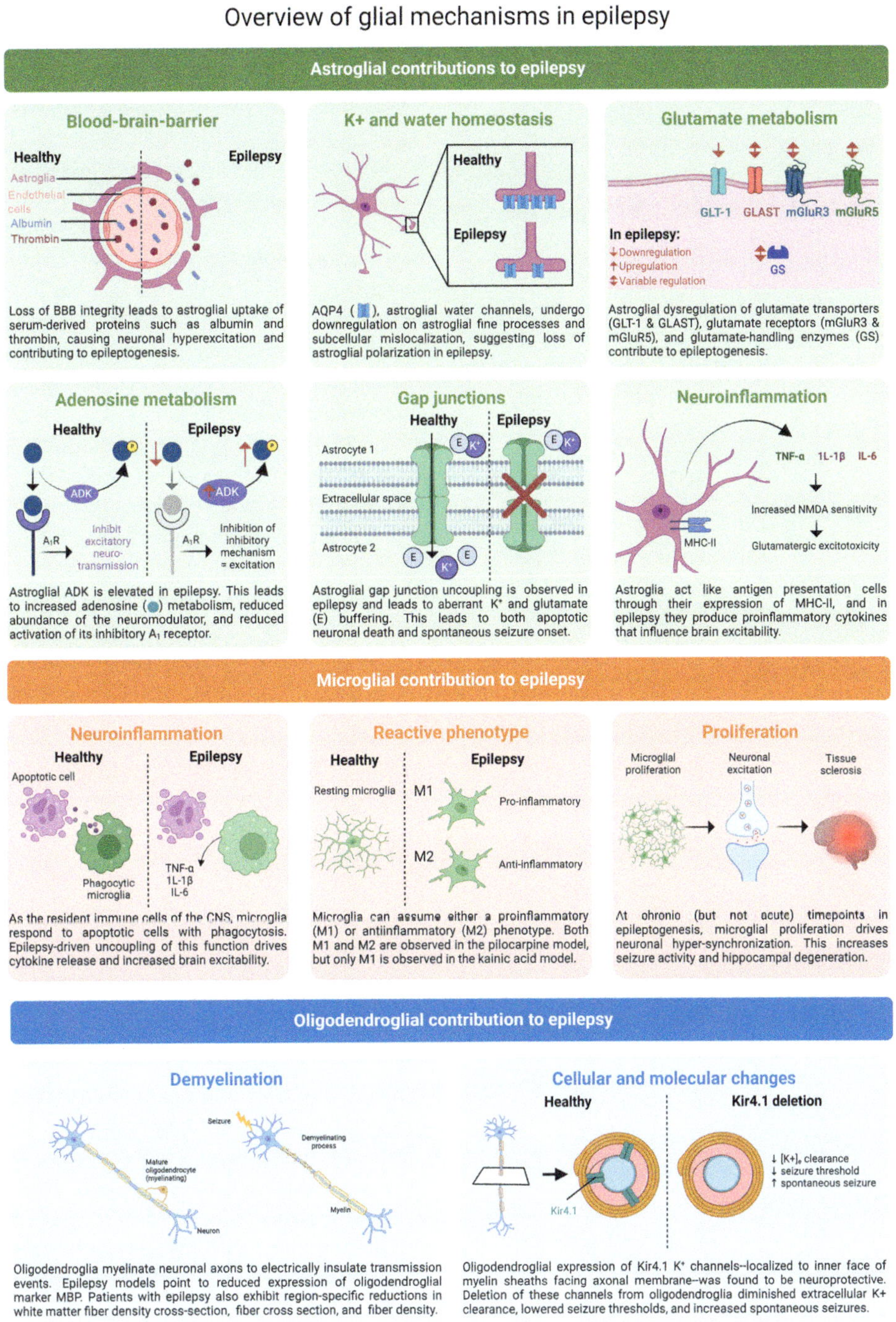

Fig. 4.1. Overview of glial mechanisms in epilepsy. Created by Biorender. Ghouli M (2024). https://BioRender.com/xyz.

compromising the homeostasis of the CNS microenvironment, the brain evolved to form the BBB (Watts et al., 2018).

The BBB is an interface that strictly regulates the movement of ions, molecules, and cells between the brain parenchyma and cerebrovasculature (Cabezas et al., 2014). Endothelial cells, pericytes, astroglia, and neurons come together to form a neurovascular unit that defines the permeability of the BBB (Muoio et al., 2014). Endothelial cells make up the capillary walls and express various junction proteins that regulate transcellular permeability (Stamatovic et al., 2016). Pericytes regulate BBB-specific gene expression in endothelial cells and induce astroglial end-feet polarization near the blood vessels (Armulik et al., 2010). Astroglia extend projections to engulf capillary walls and to ensheath neuronal synapses, thereby physically and functionally linking neurons and blood vessels together and allowing for synchronized neuronal activity and cerebral blood flow (Watts et al., 2018). This cluster of cells collaborates to ensure the delivery of trophic support to neurons, the elimination of waste, the protection of brain tissue from the invasion of foreign substances, and the regulation of the rheological properties of the blood–brain interface.

BBB perturbations

Disruption to the integrity of the BBB has been reported to cause neuronal hypersynchronization and epileptiform activity in both humans and animal studies. Acute seizure model studies show evidence of BBB breakdown approximately 5–30 min following the first seizure elicited by convulsants such as pentylenetetrazol, bicuculline, and hyperthermia (Nitsch and Klatzo, 1983). Interestingly, although BBB breakdown is observed bilaterally, each chemoconvulsive agent gives rise to spatially unique BBB pathology. Pentylenetetrazol primarily causes BBB leakage in the hypothalamus, whereas bicuculline perturbs BBB integrity in the pallidum (Nitsch and Klatzo, 1983). This suggests that chemoconvulsants give rise to region-specific breakdown of the BBB. As for nonchemical convulsants, in a cortical dysplasia hyperthermia-induced seizure model, the use of anticonvulsants such as topiramate and levetiracetam inhibited seizure activity and prevented BBB breakdown (Ahishali et al., 2010; Gürses et al., 2013). Chronic seizure models such as the kainic acid-, pilocarpine-, and electrical stimulation-induced models of TLE show evidence of extensive BBB leakage minutes after the development of *status epilepticus* (SE), which persists into the chronic phase of epilepsy (Ruth, 1984; van Vliet et al., 2007; Marchi et al., 2011; Zattoni et al., 2011; Michalak et al., 2013). These findings parallel human data that show increased BBB permeability following traumatic brain injuries, SE, and TLE—particularly in the periictal domain of TLE patients (van Vliet et al., 2007; Kimizu et al., 2018; Rüber et al., 2018).

BBB disruption is accompanied by infiltration of unwanted molecules, toxins, and cells, which may serve as prerequisites for neuronal hyperexcitation. One important molecule is albumin, the most abundant protein in blood (Sokrab et al., 1989). Humans with postencephalitic/encephalopathic epilepsy were found to exhibit higher levels of extravasated endogenous albumin in their CSF compared to disease controls (Kimizu et al., 2018). Moreover, resected hippocampal tissue from humans with TLE exhibits albumin immunoreactivity throughout the parenchyma and proximal to blood vessels (van Vliet et al., 2007). These findings align with animal studies that show evidence of albumin immunoreactivity in the rat hippocampus during the acute (1–2 days post-SE) and latent (1-week post-SE) phases of temporal lobe epileptogenesis (van Vliet et al., 2007).

Astroglial contribution to BBB perturbations

Extravasated albumin was found to be accumulated in the astroglia of resected epileptic human and rat hippocampi (van Vliet et al., 2007). In the kidney, albumin uptake is mediated by transforming growth factor beta receptors (TGF-βR), which are also expressed on astroglia and upregulated in the injured brain (Vivien et al., 1998; Morganti-Kossmann et al., 2002; Ivens et al., 2007). One study found that the astroglial uptake of albumin is inhibited by the TGF-βR1 kinase inhibitor, SB431542, suggesting that like in the kidney, astroglial albumin uptake is mediated by intracellular TGF-β signaling (Ivens et al., 2007). Because TGF-βR1 has been shown to downregulate astroglial Kir channels, it is hypothesized that astroglial Kir4.1 may be downregulated in this context—a hypothesis that aligns with the finding that K^+ buffering capacity is reduced following brain exposure to albumin in vivo (Perillan et al., 2002; Ivens et al., 2007). In turn, this leads to impaired astroglial K^+ buffering, the accumulation of extracellular K^+, and the facilitation of neuronal hypersynchronization (Ivens et al., 2007). Astroglial TGF-βR and TGF-βR1 kinase may serve as novel glial therapeutic targets for ASDs.

Other serum-derived proteins may similarly interact with astroglia and play a role in epileptogenesis. Thrombin, a blood coagulation serine protease, acts on the astroglial protease-activated receptor 1 (PAR1), a receptor that has been shown to significantly reorganize glutamatergic synapses (Sweeney et al., 2017). Through PAR1 activation, thrombin was determined to increase the reactivity of hippocampal CA1 pyramidal neurons and to lower the epileptic seizure threshold in the CA3 region of the hippocampus (Maggio et al., 2008). These findings demonstrate that following microvascular damage, the incursion of serum-born proteins such as albumin and thrombin through the breached BBB leads to epileptogenesis through the dysregulation of astroglia. Astroglial PAR1 kinase may serve as a novel glial therapeutic target for ASDs; however, more studies are required to clarify its role in epileptogenesis.

Cellular and molecular changes to astroglia in epilepsy

Astroglial membranes are decorated with various ion channels, receptors, and transporters that equip them to sample the extracellular milieu and actively regulate the excitability of neurons at the tripartite synapse. These cells are also endowed with intracellular enzymes and gap junction proteins that influence their function in health and disease. In the context of epilepsy, divergent regulation of astroglial machinery has been shown to contribute to pathogenesis.

Astroglial contribution to water and K^+ homeostasis

Water, the "solvent of life," is the major component of all living cells. A challenge that biologic organisms face is the movement of water across biologic barriers such as plasma membranes. Aquaporins are bidirectional passive water channels that mediate the exchange of fluid across biologic membranes in response to osmotic gradients. Aquaporin 4 (AQP4) is the most abundant water channel in the CNS (Mader and Brimberg, 2019). AQP4 is expressed in the endfeet of astroglia that contact major fluid compartments in the brain, such as the BBB and the brain-CSF border.

Animal studies have reported alterations to AQP4 expression and localization in epileptogenesis. Animals deficient in the astroglial AQP4 channel exhibit increased seizure threshold compared to wild-type animals when challenged with the chemoconvulsant pentylenetetrazol or with electrical stimulation (Binder et al., 2004, 2006a). However, AQP4 KO animals exhibit prolonged stimulation-evoked seizures with slowed potassium kinetics (Binder et al., 2006a,b). These studies suggest that although AQP4 deficiency increases the threshold required to induce an electrographic seizure, once the threshold has been surpassed, AQP4 deficiency worsens seizures. These findings are consistent with a posttraumatic epilepsy study that demonstrates that AQP4 KO animals exhibit a higher number of spontaneous seizures and increased seizure duration (Szu et al., 2020).

AQP4 expression was reduced on astroglial fine processes and endfeet following SE in a kainic acid model of TLE (Lee et al., 2012a). Additionally, this channel was found to be mislocalized preceding the onset of seizures in this model, suggesting that loss of astrocytic polarization and concomitant dysregulation of water homeostasis may be contributing factors to ictogenesis (Alvestad et al., 2013). One week following SE, a timepoint during which spontaneous seizures develop in this model, AQP4 KO animals displayed more seizure activity than wildtype controls (Lee et al., 2012a). Alternately, in a tuberous sclerosis complex genetic model of epilepsy, AQP4 was found to be upregulated in both human and rodent tissue (Short et al., 2019). These studies further emphasize that disturbances to water homeostasis due to AQP4 mislocalization and dysregulation may contribute to epileptogenesis. Astroglial AQP4 may serve as a novel glial therapeutic target for ASDs.

The subcellular distribution of AQP4 mirrors that of the inwardly rectifying potassium channel, Kir4.1. Kir4.1 channels mediate the astroglial clearance of extracellular K^+ following neuronal transmission events, a phenomenon known as "potassium buffering." The subsequent spike in intracellular K^+ osmotically drives water into the cell through AQP4 channels (Murphy et al., 2017). The spatial coupling of AQP4 and Kir4.1 suggests these channels work in concert to maintain water and potassium homeostasis in the brain (Nagelhus et al., 2004; Eid et al., 2005; Smith and Verkman, 2015). This homeostasis is disrupted with dysregulation of either channel. Kir4.1 has been found to be downregulated following SE in an animal model of TLE, following fluid percussion injury, following albumin exposure, and in genetic epilepsy models (Stewart et al., 2010; Frigerio et al., 2012; Harada et al., 2013). The downregulation of the channel and mutations to the gene encoding Kir4.1 (KCNJ10) were found to increase the concentration of K^+ and cause neuronal hypersynchronization (Bockenhauer et al., 2009; Scholl et al., 2009; Reichold et al., 2010). Similarly, astrocytes in human sclerotic tissue display reduced astroglial inward rectification compared to nonsclerotic equivalents, pointing to a relationship between Kir4.1 dysfunction and the development of hippocampal sclerosis in TLE (Hinterkeuser et al., 2000; Schröder et al., 2000). Astroglial Kir4.1 may serve as a novel glial therapeutic target for ASDs.

Astroglial contribution to glutamate metabolism

Glutamate is the most abundant excitatory neurotransmitter in the brain, and its clearance from the synaptic space is critical to the maintenance of neuronal homeostasis. Following neuronal stimulation, glutamate is released into the synaptic cleft quickly, reaching approximately 1 mM in concentration for only 1–2 ms (Clements et al., 1992). Because the synapse is continuous with the ECS, the diffusion of glutamate throughout the ECS, the binding of glutamate to its receptors, and the uptake of glutamate into nearby cells allow for the glutamate concentration to rapidly return to basal nanomolar levels (Bergles et al., 1999). If not for these mechanisms, persistently elevated glutamate concentrations in the ECS would lead to the tonic activation of nearby glutamatergic receptors, thereby increasing neuronal activity and driving network excitability (Romanos et al., 2019). In the epileptic human hippocampus, the concentration of glutamate was found to be higher than in the nonepileptic hippocampus at rest, and during seizure activity, the concentration of glutamate increased further and remained elevated throughout the course of the seizure (During and Spencer, 1993).

In health, astroglia facilitate the clearance of glutamate from the ECS. This is achieved through two astroglial Na^+-dependent high affinity glutamate transporters: glutamate transporter 1 (GLT-1; human homolog: excitatory amino acid transporter 2, EAAT2) and the glutamate aspartate transporter (GLAST; human homolog: excitatory amino acid transporter 1, EAAT1) (Peterson and Binder, 2020). Glutamate is then metabolized, used to synthesize other essential molecules, or otherwise shunted to neurons through the glutamate/GABA-glutamine cycle (Bak et al., 2006). In addition to glutamate transporters, astroglia also express metabotropic receptors 3 (mGluR3) and 5 (mGluR5). Dysregulation of both astroglial glutamate transporters as well as astroglial glutamate receptors has been found to contribute to epileptogenesis.

Glutamate transporters

GLT-1 is responsible for the clearance of approximately 90% of extracellular glutamate in the forebrain (Rimmele and Rosenberg, 2016). Global GLT-1 KO animals developed epilepsy and cell death, suggesting GLT-1 functions to maintain low extracellular glutamate concentrations (Tanaka et al., 1997). Although most of the GLT-1 in the brain is expressed in perisynaptic astroglial membranes, approximately 5%–10% of GLT-1 is expressed in neurons (Schreiner et al., 2014; Petr et al., 2015). To probe for the function of astroglial vs neuronal GLT-1, transgenic animals with cell-type-specific GLT-1 ablation were evaluated. These studies revealed that astroglial GLT-1 KO leads to the generation of lethal spontaneous seizures, whereas neuronal GLT-1 KO has no effect on mortality or seizures (Petr et al., 2015). Furthermore, transgenic animals expressing a 1.5- to 2-fold increase in astroglial GLT-1 exhibited reduced seizure frequency and mortality (Kong et al., 2012). Taken together, these studies point to the critical responsibility that astroglial GLT-1 bears in sustaining the excitatory-inhibitory balance of the CNS microenvironment and resisting epileptogenesis.

Animal models of TLE have revealed GLT-1 regulation following SE and throughout epileptogenesis. Ultrastructural imaging reveals that although astroglial ensheathment of the pre- and postsynaptic elements increases following SE, there is a marked reduction in the subcellular expression of GLT-1 at the perisynaptic astroglial membrane during the latency period (Clarkson et al., 2020). This persists into the early phase of epileptogenesis, as GLT-1 was found to be downregulated at the total protein level and in synaptosomes approximately 1 week post-SE, coinciding with the onset of spontaneous seizures in this model (Hubbard et al., 2016; Peterson and Binder, 2019). The AAV-mediated overexpression of GLT-1 following SE in an animal model of TLE was found to be neuroprotective, effectively reducing seizure frequency, total time spent in seizures, and eliminating large behavioral seizures (Peterson et al., 2021). These findings suggest that in the epileptic brain, downregulation of astroglial glutamate transporters leads to the accumulation of extracellular glutamate and the precipitation of neuronal hypersynchronization, and this can be attenuated through the restoration of normal GLT-1 levels.

Human data also demonstrate a reduction in GLT-1 expression in TLE patients (Sarac et al., 2009). Interestingly, GLT-1 levels in TLE patients with no hippocampal sclerosis were higher (in CA1 and CA2) than in nonepileptic patients, whereas GLT-1 levels were decreased in TLE patients with hippocampal sclerosis (Proper et al., 2002). In other words, it seems to be the case that epilepsy alone may lead to compensatory upregulation of astroglial GLT-1 to mediate the increased glutamatergic neurotransmission. However, in the sclerotic human hippocampus, chronic network alterations characterized by pyramidal neuron death, the reorganization of granule cells and interneurons, and rampant gliosis likely drive the aberrant regulation of GLT-1 expression (Thom, 2014).

GLAST KO animals are highly susceptible to developing seizures, suggesting that this transporter is also vital to the maintenance of neuronal homeostasis (Proper et al., 2002). Interestingly, GLAST KO animals and genetic controls expressed the same levels of GLT-1 in the hippocampus; however, neocortical GLT-1 expression was increased (210%), hippocampal mGluR1 and mGluR2 levels were increased (140% and 160%, respectively), cortical mGluR1, mGluR2, and mGluR3 were decreased (60%, 60%, and 70%, respectively), and cortical NMDAR levels were increased (~150%) in GLAST KO animals compared to genetic controls (Ueda et al., 2002). Thus, the seizure susceptibility of GLAST KO animals may be a product of the downstream dysregulation of various glutamate receptors and not a direct consequence of GLAST deficiency (Ueda et al., 2002). Animal models of TLE reveal that GLAST immunoreactivity was increased during SE and for a minor period thereafter, and synaptosomal protein levels of this transporter were increased 30 days following SE during the late epileptogenic period of this model. These findings suggest that astrocytic GLAST dysregulation could contribute to epileptogenesis. Human data reveals that GLAST was also found to be significantly downregulated in the hippocampus of patients with TLE (Sarac et al., 2009). The decrease in GLAST expression was most prominent in the CA4 region of the hippocampus and the polymorphic and supragranular layers of the dentate gyrus of TLE patients with hippocampal sclerosis but not in TLE patients without hippocampal sclerosis (Proper et al., 2002). Taken together, these studies suggest that glutamate transporter dysregulation causes accumulation of extracellular glutamate and contributes to epileptogenesis. Based on these studies, astroglial GLT-1 and GLAST may serve as novel nonneuronal therapeutic targets for ASDs.

Glutamate receptors

Metabotropic glutamate receptors (mGluRs) are G-protein-coupled receptors that modulate synaptic transmission and neuronal excitability in the CNS (Niswender and Conn, 2010). Where AMPA-type and NMDA-type ionotropic receptors mediate the cytotoxic effects of glutamate, mGluRs exert a neuroprotective force (Deng et al., 2003; Baskys et al., 2005; Jantzie et al., 2010). Astroglia express the mGluR3 and mGluR5 subtypes; however, the regulation of these receptors is variable in epilepsy.

Astroglial mGluR5 is predominantly expressed in the CNS during development in both humans and rodents, and the receptor is virtually undetectable in the rodent brain after postnatal week 3 (Sun et al., 2013). However, mGluR5 levels resurface in reactive astroglia following low-dose kainate-induced TLE (Umpierre et al., 2016). In a kindling model of TLE, mGluR5 levels were also found to be increased (Aronica et al., 2000). Acute stimulation of this receptor has been shown to have synergistic and modulatory effects on the uptake of extracellular K^+ and glutamate, suggesting its upregulation in the early phase of epilepsy is a compensatory mechanism to combat the increased neuronal activity (Devaraju et al., 2013).

Chronic activation of mGluR5, on the other hand, has been shown to downregulate both GLAST and GLT-1, leading to impaired clearance of extracellular glutamate (Aronica et al., 2003). In a pilocarpine model of TLE, stimulation of mGluR5 enhanced astrocytic calcium signaling, a phenomenon known to contribute to neuronal excitotoxicity after SE (Ding et al., 2007; Bazargani and Attwell, 2016). This aligns with the observation that mGluR5 stimulation in hippocampal slices increases ictal-like events (Bianchi et al., 2012). It also provides insight as to why increased hippocampal mGluR5 levels were found in TLE patients with and without hippocampal sclerosis (Notenboom et al., 2006; Kandratavicius et al., 2013).

mGluR3 is the major metabotropic glutamate receptor expressed by adult human cortical astrocytes (Sun et al., 2013). Unlike mGluR5, chronic activation of mGluR3 was found to upregulate both GLAST and GLT-1 and promote the clearance of extracellular glutamate (Aronica et al., 2003). It is hypothesized that mGluR3 exhibits neuroprotective effects also through its restoration of astroglial glutathione, a molecule that protects brain tissue against reactive oxygen species (Zhou et al., 2006). Furthermore, mGluR3 was found to protect against neuronal excitotoxicity and astroglial cell death (Bruno et al., 2017). In an electrical stimulation model of TLE, mGluR3 was increased in astroglia 1 week following SE, persisting up to 3 months following the initial insult (Aronica et al., 2000). Human TLE studies show variable regulation of mGluR3. Some studies have shown upregulation of mGluR2/3 in the hippocampus of TLE patients, whereas others suggest no changes are made to the expression of this receptor (Tang et al., 2004; Das et al., 2012). More studies must be done to tease apart these discrepancies. The activation of both astroglial mGluR3 and mGluR5 results in critical, differential modulation of glutamate transporter expression, thereby exerting variable modulation of neuronal activity. Thus, astroglial mGluR3 and mGluR5 may serve as novel nonneuronal therapeutic targets for ASDs.

Glutamine synthetase

Glutamine synthetase (GS) is an enzyme that converts the excitatory amino acid glutamate into the neutral amino acid glutamine. This enzyme is primarily expressed in the cytosol of astroglia; however, high concentrations are also found at the perivascular endfeet and in perisynaptic processes. Because GS is the only enzyme to produce glutamine—a molecule involved in the production of glutamate and GABA, in the synthesis of various proteins, and in osmoregulation—uncompensated disturbances to GS expression can challenge the excitatory-inhibitory balance of the CNS (Norenberg et al., 2007).

Astroglial GS dysregulation is implicated in several neurologic disorders such as Alzheimer disease, hepatic encephalopathy, and schizophrenia (Robinson, 2000; Bruneau et al., 2005; Brusilow et al., 2010; Olabarria et al., 2011). Due to its essential function in the glutamate/GABA-glutamine cycle, GS dysfunction and dysregulation inextricably correlate with perturbed neurotransmission. Case reports show that two unrelated newborns with congenital GS deficiency due to homozygous mutations to the GS gene (R324C and R341C) exhibited severe brain malformations, multiorgan failure, and neonatal death (Häberle et al., 2005). Similarly, animal models reveal that GS is essential to mouse embryogenesis and its deletion is lethal (He et al., 2007). Selective deletion of GS from the hippocampus and neocortex of mice gives rise to glial dysfunction, neurodegeneration, and epilepsy (Zhou et al., 2019). Furthermore, in a methionine sulfoximime-induced rat model of GS deficiency, >95% of animals developed recurrent seizures and exhibited neuropathologic features characteristic of TLE (Eid et al., 2008b).

In an IHKA animal model of TLE, GS was found to be upregulated in the epileptic hippocampus at an early timepoint (7 days post-IHKA), particularly in the CA1 and dentate gyrus domains (Ghouli et al., 2023). Alternately, in more chronic conditions, GS was found to be downregulated in the hippocampus of TLE patients (Eid et al., 2004). Interestingly, although astroglia normally have a high content of GS, downregulation of GS was most pronounced in regions of astrogliosis (Eid et al., 2004). Another study found that astroglial GS is specifically downregulated in the sclerotic hippocampus of TLE patients in areas with severe neuronal loss, namely CA1 and CA4 (van der Hel et al., 2005). Based on these findings, it is possible that the early upregulation of GS in TLE serves as a compensatory response to mediate the conversion of excess glutamate to glutamine; however, at later timepoints, the continued hyperexcitation of the hippocampus produces a pathological drop in GS levels. In addition to the loss of GS, the activity of GS was found to be reduced in TLE patients (Eid et al., 2004). This is likely due to the inhibitory phosphorylation of threonine residue 301 (T301) in the enzyme's active site (Huyghe et al., 2019). Enhanced T301 phosphorylation was observed in an animal model of TLE (Huyghe et al., 2019). Taken together, these studies show that astroglial GS is necessary for the maintenance of neuronal homeostasis, and this enzyme is temporally dysregulated throughout epileptogenesis. The reduction of its expression and activity results in the accumulation of glutamate, which drives the development of an epileptic phenotype. Thus, astroglial GS may serve as a novel nonneuronal therapeutic target for ASDs.

Astroglial contribution to adenosine metabolism

In the CNS, adenosine functions as an extracellular signaling molecule and a potent neuromodulator through its activation of the inhibitory A_1 and facilitatory A_{2A} adenosine receptors (Cunha, 2001; Studer et al., 2006; Layland et al., 2014). Inhibitory A_1 receptors are localized to the pre- and postsynaptic densities of hippocampal neurons that are vulnerable to seizures and cell death (Dunwiddie and Masino, 2001). In the context of epilepsy, adenosine acts on A_1 receptors, which then facilitate neuroprotection and inhibit excitatory neurotransmission (Aronica et al., 2013). Because it exerts antiepileptogenic effects, adenosine is sometimes called "the brain's endogenous anticonvulsant" (Ribeiro et al., 2003; Boison, 2005; Aronica et al., 2013). Corroborating this, the administration of A_1 receptor agonists was found to suppress seizure activity in rodent models of epilepsy (Dunwiddie, 1999; Dunwiddie and Masino, 2001; Boison, 2005). Alternately, deficient adenosinergic signaling can lead to ictogenesis (Boison, 2005).

Adenosine kinase (ADK) is the adenosine-metabolizing enzyme that controls ambient extracellular adenosine levels in the brain. ADK is primarily expressed by astroglia, and it regulates brain adenosine levels by phosphorylating adenosine to adenosine monophosphate (Studer et al., 2006; Aronica et al., 2013). In animal models of TLE, ADK was found to be significantly upregulated in reactive astrocytes following SE (Fedele et al., 2005; Aronica et al., 2011). The increased expression of astroglial ADK in this context was found to aggravate seizure activity (Fedele et al., 2005). These findings are consistent with human data that reveals increased ADK expression in the hippocampus and temporal cortex of TLE patients (Aronica et al., 2011). The upregulation of astroglial ADK suggests there is a functional increase in adenosine metabolism and a subsequent decrease in the abundance of this neuromodulator. In turn, this diminishes the activation of the inhibitory A_1 receptor, likely contributing to the development of seizures and epilepsy.

To dissect the molecular contribution of astrogliosis vs increased ADK to the development of seizures, one group transgenically uncoupled ADK expression from astrogliosis by replacing endogenous astroglial ADK with ectopic ADK expression in neurons (Li et al., 2008). They found that following SE, uncoupled animals developed astrogliosis without ADK upregulation or seizures, whereas wildtype animals developed astrogliosis with ADK upregulation and the onset of recurrent, spontaneous seizures (Li et al., 2008). This study highlighted the contribution of adenosine and ADK to ictogenesis independent from astrogliosis. Based on these findings, astroglial ADK may serve as a novel nonneuronal therapeutic target for ASDs.

Astroglial gap junctions

Glia communicate with one another through the intercellular diffusion of chemical messengers such as cytokines, chemokines, exosomes, and signaling molecules (Fields and Stevens-Graham, 2002; Nutma et al., 2020). Astroglia can also engage in interglial communication through syncytia that are created when the connexons of two adjacent cells dock together to form a gap junction, functionally fusing the cytoplasm of the cells (Zhong et al., 2023). Astroglial coupling enables the rapid diffusion of chemical messengers, metabolites, and Ca^{2+} signaling (Stephan et al., 2021).

Gap junction coupling can be studied in acute tissue slices through various methods, including paired recordings, determination of syncytial isopotentiality, tracer coupling followed by network topography analysis, and wide field imaging of ion-sensitive dyes (Stephan et al., 2021).

Some studies propose that epilepsy is caused by increased gap junction coupling, which drives neuronal hypersynchronization. One study determined that in TLE patients with hippocampal sclerosis, the expression of Cx32 decreases, Cx36 remains the same, and Cx43 increases (Collignon et al., 2006). This finding is corroborated in another study wherein Cx43 protein levels are significantly higher in epileptic tissue (Fonseca et al., 2002). Cx43 is found in great abundance in astroglia and is thought to be the main connexin involved in astroglial gap junctions; however, this connexin is also expressed in neurons (Fonseca et al., 2002). It is important to discern that an increase in Cx43 does not necessarily point to increased syncytial activity. In the case that it does correspond to increased gap junctional communication, the surge in extracellular K+ and glutamate mediated by injured neurons could trigger sweeping cytosolic waves of depression through coupled astroglia (Cotrina et al., 1998). In the context of epilepsy, this could lead to the spread of the lesion and epileptic focus.

Other studies propose that the *uncoupling* of glial gap junctions impairs the regulation of extracellular K^+, pH, and glutamate levels and therefore can cause ictogenesis (Bedner et al., 2015). TLE patients with hippocampal sclerosis were determined to have completely uncoupled astroglia, whereas nonsclerotic counterparts exhibited abundant astroglial coupling (Bedner et al., 2015). A lipopolysaccharide-induced animal model of gap junction uncoupling was found to hinder K^+ buffering, which is soon followed by both apoptotic neuronal death and the onset of spontaneous seizure (Bedner et al., 2015). Moreover, astroglia were found to undergo intracellular alkalization in response to epileptiform activity, followed by uncoupling (Onodera et al., 2021). When treated with a Na^+/HCO_3^- cotransporter blocker, intracellular alkalization, astroglial uncoupling, as well as neuronal hyperactivity, were prevented (Onodera et al., 2021). These findings point to the role that astroglial gap junction uncoupling may play in eliciting temporal lobe epileptogenesis. Thus, astroglial uncoupling may serve as a novel nonneuronal therapeutic target for ASDs; however, additional studies are required to clarify the role of gap junction uncoupling in epilepsy and tractable therapeutic targets.

Astroglial contribution to immune and inflammatory functions in the context of epilepsy

Inflammation is both a source and sequela of epilepsy. Neuroinflammation promotes brain hyperexcitation, and in return, seizures upregulate inflammatory mediators in glial cells. The immune system of the CNS reacts to a pathogen or insult through a general innate response, which then evolves into a specialized adaptive response (Chaplin, 2010). Astroglia have been shown to play a role in both the innate and adaptive immune responses in the CNS (Linnerbauer et al., 2020). Understanding the contribution of astroglia to the neurobiology of inflammation enables the identification of potential targets for antiepileptogenic therapies.

Astroglia act as antigen presenting cells in cerebral innate immunity

Innate immunity is the first line of resistance against invading pathogens and physical insults. Upon infection or insult, the innate immune system can either directly eliminate the pathogen through phagocytosis or it can activate the adaptive immune system through antigen presenting cells (APC). Professional APCs such as dendritic cells, macrophages, and B cells express toll-like receptors (TLR) that enable them to identify pathogen- and damage-associated molecular patterns (Gong et al., 2020). Professional APCs also express major histocompatibility class II (MHC-II), which allows them to present antigens to CD4+ helper T cells, which then initiates the adaptive immune response (Roche and Furuta, 2015).

Astroglia express various TLR and therefore act like APCs of the brain. Human astroglia express TLR1–5 and TLR9; however, mouse astroglia express TLR1–9 (Li et al., 2021). When challenged with lipopolysaccharide (LPS), a TLR agonist, astroglia are provoked to produce proinflammatory cytokines such as TNF-α, IL-1β, and IL-6 (Rodgers et al., 2020). This suggests that astroglia use their TLR to sample the extracellular milieu, and in the presence of pathogens or damage, they may elicit an inflammatory response. In addition to cytokine secretion, one study found that LPS-mediated TLR4 activation in postnatal astroglia also promotes the development of excitatory synapses (Shen et al., 2016). This suggests that inflammation-mediated activation of astroglial TLRs shapes the excitatory connectivity of the brain, and in the case of abnormal connectivity, may lead to neuronal hyperexcitability and seizures. In fact, one study found that human epileptogenic tissue exhibited increased levels of TLR4 (Maroso et al., 2010). It follows that TLR4-deficient mice were reported to be resistant to kainite-induced seizures (Maroso et al., 2010). These studies all point to astroglial TLRs as a potential target for future antineuroinflammatory and antiepileptogenic therapies.

In addition to TLR expression, astroglia also express MHC-II. In Parkinson disease (PD), one study found that astroglia colocalized with MHC-II and that these glial cells were surrounded by an infiltration of perivascular $CD4^+$ T cells (Rostami et al., 2020). Astroglial expression of MHC-II is likely induced by the proinflammatory cytokine gamma-interferon (IFN-γ), which is found to be expressed in high levels in PD brains (Hirsch et al., 1983; Vardjan et al., 2012; Wijdeven et al., 2018; Ferrari et al., 2021). Although few studies have looked at astroglia acting as APCs in epilepsy to trigger neuroinflammation, patients with epilepsy do exhibit increased levels of serum IFN-γ (Gao et al., 2017; Kocatürk and Kirmit, 2022). This may indicate that, like in the PD brain, IFN-γ in the epileptic brain may influence astroglial APC function through MHC-II expression and cytokine secretion. In turn, this may exacerbate the neuroinflammation-seizure spiral; however, further studies must be conducted to evaluate the biological role of astroglia acting as APCs in chronic seizure disorders.

Astroglia modulate cerebral adaptive immunity through cytokine and chemokine secretion

In addition to their capacity to identify pathogens and damage, astroglia also act as immune modulators through their secretion of various cytokines and chemokines. Nonreactive astroglia were found to express G-CSF, GM-CSF, GROα (CXCL1), IL-6, IL-8 (CXCL8), MCP-1 (CCL2), MIF, and Serpin E1 (Choi et al., 2014; Hyvärinen et al., 2019). Astroglia treated with TNF-α and 1L-1β to become reactive exhibited a distinct secretome, which includes IL-1β, IL-1ra, TNF-α, IP-10 (CXCL10), MIP-1α (CCL3), and RANTES (CCL5) (Choi et al., 2014). Following their secretion, proinflammatory cytokines and chemokines bind to their receptors and initiate downstream inflammatory cascades that feed into a neuroinflammation-hyperexcitation spiral, ultimately worsening seizure outcome in epileptic tissue.

For example, increased astroglial production of TNF-α can bind to astroglial TNF-α receptor 1 (TNFR1) and cause autocrine downregulation of astroglial GLT-1, thereby reducing glutamate reuptake from the synapse and increasing neuronal hyperexcitation (Fine et al., 1996; Carmen et al., 2009; Abd-El-Basset et al., 2021). TNF-α can also upregulate astroglial glutaminase, the enzyme responsible for glutamate production (Ye et al., 2013). Based on these findings, increased astroglial TNF-α can lead to excitotoxic glutamate accumulation and subsequent neuronal hyperexcitation and seizures. The inhibition of TNFR1 may therefore represent a novel strategy to prevent epileptogenesis or treat persistent epilepsy.

IL-6, another proinflammatory cytokine released by astroglia, was found to exhibit proepileptogenic effects. In a pentylenetetrazol seizure model, rats that were intranasally administered human IL-6 exhibited shorter seizure latency, increased seizure duration, and higher mortality (Kalueff et al., 2004). This may be because astroglial IL-6 expression increases sensitivity to *N*-methyl-D-aspartate (NMDA)-mediated seizures (Samland et al., 2003). Interestingly, IL-6 is involved in the formation of excitatory synapses. Studies show that transient maternal elevation of IL-6 leads to an aberrant, persistent increase in excitatory synapses in the offspring. This suggests that astroglial production of the proinflammatory mediator IL-6 can exert influence on the inhibitory-excitatory balance of the brain. Conversely, IL-6 is itself a byproduct of seizures in TLE, meaning it can be used as a diagnostic marker of new-onset refractory *status epilepticus* and may serve as a potential antiepileptogenic target (Alapirtti et al., 2018; Kwack and Kim, 2022). Together, these findings highlight the contribution of astroglia to the salient relationship between seizures and neuroinflammation.

Another proinflammatory cytokine produced by astroglia (and microglia) during seizures is IL-1β. The binding of IL-1β to its receptor, interleukin-1 receptor 1 (IL1R1), leads to several proepileptogenic phenomena. Notably, IL1R1 activation causes post-translational changes to NMDA receptors, leading to a persistent decrease in seizure threshold (Maroso et al., 2011). The underlying mechanism for this pathology involves the enhancement of NMDAR-mediated Ca^{2+} influx into neurons, resulting in increased depolarization, neuronal activation, and excitotoxicity (Zhand et al., 2018). IL-1β was found to be increased both in the sclerotic hippocampus of patients with TLE as well as in the rodent hippocampus in an IHKA model of TLE (Vezzani et al., 1999; Semple et al., 2017). In a rodent model of TBI-evoked epilepsy, pretreatment with an IL1R1 antagonist resulted in reduced seizure susceptibility and hippocampal astrogliosis at early timepoints, as well as improved spatial memory and fewer seizures at chronic timepoints (Semple et al., 2017). Taken together, these studies indicate that the IL-1β/IL1R1 pathway may serve as a potential target to combat epileptogenesis.

MICROGLIA

Microglia are the resident immune cells of the CNS that probe brain tissue for damage and signs of infection (Allen and Barres, 2009). These glial cells are highly sensitive to CNS dysfunction and quickly respond to threats through a multistage activation process (microgliosis) that converts "resting microglia" into "activated/reactive microglia" (Kettenmann et al., 2011). Similar to reactive astrocytes, reactive microglia also release a slew of cytokines and chemokines at various timepoints in disease progression. Cytokines are pro- or antiinflammatory molecules that are secreted and influence other cells, whereas chemokines are a subset of cytokines that specifically recruit other immune cells to the infected or inflamed site. Microglia use cytokines and chemokines as signaling proteins that facilitate the immune response in the body. In doing so, microglia can modulate neuronal circuits under physiological conditions as well as in pathological contexts such as epilepsy. As outlined below, there is a growing body of evidence that places microglia at the intersection between inflammation and epilepsy.

In health, microglia shape the adult hippocampal neurogenesis landscape through the phagocytosis of apoptotic newborn cells in the dentate gyrus (Sierra et al., 2010). Epileptic seizure activity has been shown to disrupt this function. In a kainic acid model of epilepsy, microglial phagocytosis was impaired, and this led to the collapse of the phagocytosis-apoptosis coupling in the epileptic hippocampus (Abiega et al., 2016). Disruption to phagocytosis-apoptosis coupling is also noted in the resected hippocampal tissue of TLE patients (Abiega et al., 2016). The loss of this coupling is associated with increased microglial expression of proinflammatory TNFα, IL-6, and IL-1β and decreased expression of antiinflammatory TGFβ (Abiega et al., 2016). Conversely, in the same model, microglial TGF-β1 mRNA expression was elevated in the epileptic hippocampus. For this reason, it is unclear whether the inflammatory response mediated by microglia is enhanced or diminished.

In animal models of epilepsy, rapidly activated microglia are reported to assume either a proinflammatory M1 phenotype or an antiinflammatory M2 phenotype. M1-microglia release proinflammatory cytokines that promote cellular infiltration into brain parenchyma, whereas M2-microglia carry out phagocytosis of debris and contribute to cellular repair processes (Benson et al., 2015). The pilocarpine model of epilepsy has been shown to lead to the expression of both M1- and M2-microglia, whereas the kainic acid model of epilepsy only generates M1-microglia, suggesting that the microglia-mediated inflammatory response to seizures is complex and varies depending on disease etiology (Benson et al., 2015). Furthermore, in the kainic acid model of epilepsy, the ablation of microglia worsened SE and neuronal degeneration and increased seizure frequency, interictal spiking, and seizure duration (Wu et al., 2020). However, in the pilocarpine model, the unilateral ablation of microglia from the dorsal hippocampus did not alter seizure dynamics. It was not until microglial ablation was coupled to LPS preconditioning that a proconvulsive effect was seen in the pilocarpine model (Mirrione et al., 2010). These studies highlight that the pro- or antiinflammatory properties of microglial activation are a function of the stimulus used to elicit seizures in epilepsy models.

There is also a temporal element to the variable action of reactive microglia. In the kainic acid model of epilepsy, microglial proliferation was blocked through the inhibition of the microglial CSF1 receptor (GW2580) in animals undergoing epileptogenesis (Di Nunzio et al., 2021). In the early phase of the disease, GW2580-treated animals underwent SE like their untreated counterparts. However, at chronic timepoints, GW2580-treated animals exhibited significant hippocampal neuroprotection and reduced seizures. This suggests that microglial proliferation has little effect on ictogenesis in the early phase of epilepsy; however, it plays a role in driving neuronal hypersynchronization in the chronic phase of epilepsy.

Neurons express the chemokine CX3CL1 that binds to the CX3CL1 receptor (CX3CR1) expressed on the cell membranes of microglia to prompt neuron–microglia communication (Wolf et al., 2013). TLE patients were found to have increased CX3CL1 expression in the hippocampus and temporal neocortex (Xu et al., 2012). Animal models corroborated this: in the pilocarpine model of epilepsy, CX3CL1 was acutely increased in the hippocampus (1–3 h after SE) (Yeo et al., 2011). At the 3-day timepoint, microglial CX3CR1 remains elevated; however, neuronal CX3CL1 decreases, correlating to the onset of neuronal cell death in this model (Yeo et al., 2011). These findings suggest that the increased neuronal-microglial communication at the onset of epileptogenesis may induce the neuronal death observed at the 3-day timepoint.

Astroglia and microglia both express the chemokine receptor CXCR4. When this receptor was treated with an antagonist (AMD3100) for 2 months prior to kainic acid injection, electrographic seizure number and duration were found to be reduced, and the aberrant neurogenesis observed in TLE was reversed (Song et al., 2016). The natural ligand of CXCR4 is CXCL12, a factor that induces microglial release of the proinflammatory molecule TNF-α and astrocytic release of glutamate, both of which can contribute to increased neuronal hyperexcitability (Bezzi et al., 2001; Luo et al., 2016). This suggests that CXCR4 may facilitate astroglia-microglia crosstalk that leads to the development of an epileptic phenotype.

Microglia also express ATP receptors, including P2X7 (Verkhratsky et al., 2012). This receptor promotes the processing of IL-1β and expression of TNFα—both of which aggravate neuronal hyperexcitability. However, the pro- and anticonvulsive profiles of P2X7 once again depend on the animal model in question. In the pilocarpine animal model of epilepsy, P2X7 exerts a proconvulsive effect, whereas it exerts an anticonvulsive effect in the kainic acid model of epilepsy (Kim and Kang, 2011; Jimenez-Pacheco et al., 2016).

It is clear that microglia facilitate intercellular communication and mitigate immune responses in the epileptic brain; however, inconsistent findings between different animal models of epilepsy pose a significant challenge in the interpretation of microglia data. With the pursuit of new studies that will clarify the role of microglia in epilepsy will come the ability to target and modulate microglial mechanisms underlying seizure disorders.

OLIGODENDROGLIA

Oligodendroglia are specialized cells that extend their membranes to envelop, myelinate, and electrically insulate the neurons of the CNS (Bradl and Lassmann, 2010). The myelin laid down by oligodendroglia enables fast conduction of electrical impulses along the axon, and it acts as a conduit through which oligodendroglia provide metabolic support to axons (Fünfschilling et al., 2012; Lee et al., 2012b). Oligodendrocytes also induce clustering of Na^+ channels at the node of Ranvier, a condition required for proper saltatory conduction (Kaplan et al., 1997; Kaplan et al., 2001). These cells are abundant throughout the CNS, with some reports claiming that oligodendroglia make up anywhere from 20% to 40% of cells, depending on the brain region (Valério-Gomes et al., 2018). A single oligodendroglial cell creates and maintains 20–60 myelin units across different neurons (Matthews and Duncan, 1971). Recently, it has been determined that oligodendroglia are heterogeneous with variable morphology and function; however, due to the recentness of this discovery, little consideration has been given to the physiological implications of this diversity in the context of neurological diseases such as epilepsy (Seeker and Williams, 2022). Although epilepsy is often regarded as a disease of gray matter, outlined below are studies that highlight the contribution of white matter to seizure pathogenesis.

Demyelination and epilepsy

Myelin is an insulating layer made up of fatty substances (~80%) and proteins (~20%) that enwraps axons in concentric lamellae (Guleria and Kelly, 2014). This sheath electrically insulates axons, facilitating axonal transmission and modulating network-level communication (de Curtis et al., 2021). An abundant protein in myelin is the myelin basic protein (MBP). Genetic suppression of the functional MBP gene was found to lead to hypomyelination, seizures, and premature death in rodents, suggesting that MBP maintains myelin integrity and can be used as one measure of myelination in the CNS (Delaney et al., 1995).

Animal models have been employed to assess whether epilepsy is associated with demyelination. In a kainic acid model of epilepsy, myelin changes were seen in the rat forebrain as early as 2 h following SE, and demyelination was found to be prominent 2 weeks following SE (Hopkins et al., 2000). In a pentylenetetrazol model of epilepsy, animals exhibited decreased MBP levels in the hippocampus and neocortex following SE, pointing to a loss of myelin in epileptic tissue (You et al., 2011). Similarly, in a pilocarpine model of epilepsy, MBP was reduced in the hippocampus, and this decrease in myelin protein was found to be associated with a decrease in the number of oligodendroglia and an increase in the number of oligodendroglia progenitors (OPCs) (Luo et al., 2015). These studies suggest that aberrant neuronal activity can lead to the dysregulation of oligodendroglia, which, in turn, may induce the loss of this glial population or the reduction of its capacity to produce myelin. As for the reciprocal question of whether demyelination can lead to epilepsy, pharmacological interventions were utilized. Most commonly, the cuprizone (CPZ) model of toxic demyelination has been used. CPZ-treated animals exhibit electrographic spike discharges and spontaneous, generalized tonic–clonic seizures (Hoffmann et al., 2008; Lapato et al., 2017).

Although demyelination has been shown to contribute to the development of debilitating neurological diseases, oligodendroglia that survive demyelination can remyelinate axons, and OPCs can generate new oligodendroglia that also remyelinate axons (Duncan et al., 2018; Yeung et al., 2019; Franklin et al., 2021; Neely et al., 2022). However, it is important to also note that aberrant neuronal activity characteristic of epilepsy can enhance maladaptive myelination and worsen the progression of pathogenesis (Knowles et al., 2022). Moreover, hypermyelination was also found to influence mental retardation and epilepsy (Sakuma et al., 2014). Therefore, not all remyelination is "good" myelination.

Limited human data exist on the relationship between oligodendroglia, myelination, and epilepsy. Patients with TLE and hippocampal sclerosis were found to have reduced axonal densities; however, no changes to oligodendroglial populations or myelination were reported. MRIs of these patients reveal blurring of gray matter and white matter (GM/WM blurring), a clinical finding associated with normal aging, pathological aging, and focal cortical dysplasia (FCD) (Blackmon et al., 2011). A comprehensive ENIGMA-epilepsy study was conducted to determine the microstructural differences across and within various adult epilepsies. In it, they determined that all epilepsies exhibit disrupted white matter tracts, particularly in the corpus callosum, cingulum, and external capsule (Hatton et al., 2020). TLE patients with hippocampal sclerosis were found to have pronounced white matter damage in the ipsilateral parahippocampal cingulum and external capsule (Hatton et al., 2020). TLE patients also exhibited disruptions to the corpus callosum, corona radiata, and external capsule (Hatton et al., 2020).

Tuberous sclerosis complex (TSC) is a monogenetic, multiorgan disease encoded by a mutation to either the TSC1 or TSC2 gene (Mühlebner et al., 2020). A primary manifestation of this disease is the formation of cortical tubers that highly correlate with the development of intractable epilepsy. TSC is also associated with developmental abnormalities in the form of autism spectrum disorder (ASD). Histological assessments of TSC brains show divergent amounts of MBP and amount of oligodendroglia in both GM and WM, and the degree of oligodendroglia depletion is positively correlated with ASD (Mühlebner et al., 2020).

MRI scans of TSC patients with ASD reveal disruptions to 9 out of 17 sampled white matter regions, including the arcuate fasciculus, cingulum, corpus callosum, anterior limbs of the internal capsule, and the sagittal stratum (Hatton et al., 2020). Importantly, TSC patients with ASD exhibited scored higher on the epilepsy severity score than TSC patients without ASD (Hatton et al., 2020). This suggests that the white matter tract changes observed via MRI may be a function of the higher incidence of epilepsy in this population. Another study corroborated these findings and showed that TSC patients exhibit reduced *fiber density cross-section* in the right superior longitudinal fasciculus (dorsal branch) and inferior longitudinal fasciculus (bilateral), reduced *fiber cross-section* in the right superior longitudinal fasciculus (ventral branch), and reduced *fiber density* in bilateral tapetum (Vanes et al., 2022).

Depletion of myelin and oligodendroglia was found in 62% of TSC cases and correlated strongly with mTOR upregulation (Scholl et al., 2017). mTOR upregulation, in turn, was positively associated with the development of epilepsy in TSC patients (Scholl et al., 2017). Another study showed that disturbances to mTOR signaling resulted in impairments to oligodendroglia maturation and myelin production (Gruber et al., 2021). These findings suggest that the abnormalities observed in the white matter of TSC patients (Fisher et al., 2014) may be a function of aberrant mTOR signaling; (Fisher et al., 2005) is associated with cognitive and developmental delays; and (Beghi et al., 2019) is associated with the genesis of epilepsy phenotypes. In other words, often times seizures are a symptom of demyelinating processes (de Curtis et al., 2021).

Alternately, resected tissue of pediatric patients with intractable epilepsy reveals increased numbers of oligodendroglia in GM, WM, and the gray-white matter junction (Sakuma et al., 2014). Another study found that surgically resected tissue from patients with FCD displays a reduced number of oligodendroglia and abnormal OPC proliferation and differentiation in the dysplastic cortex, as well as thinner myelin sheaths in cortical layer V axons (Donkels et al., 2020). Furthermore, a clinical case study of a patient with Mild Malformation of Cortical Development and Oligodendroglial Hyperplasia and Epilepsy reports that resected tissue from the epileptic focus reveals histological features of oligodendroglia hyperplasia (Verentzioti et al., 2019). The underlying cellular and network-level mechanism has yet to be resolved; however, this study—like the others—points to a relationship between aberrant oligodendroglial number/activity and epileptogenesis. In the face of seemingly contradictory findings, it is important to recall that demyelination rate is influenced temporally by neuronal hyperactivity. Therefore, it's possible that age and duration of pathogenesis impact the regulation of oligodendroglia and myelination in FCD.

Myelin oligodendrocyte glycoprotein (MOG) is a protein found on the surface of oligodendroglia. This protein is the target of a variety of autoimmune disorders, including optic neuritis and transverse myelitis (Chaudhuri et al., 2022). MOG antibody-associated disease (MOGAD) is an inflammatory disease of the CNS that is caused by immune-mediated demyelination. MOGAD patients often present with acute symptomatic seizures, mostly alongside encephalitis (Foiadelli et al., 2020; Montalvo et al., 2022). A small subset of patients that develop MOGAD-induced seizures go on to develop SE; however, the underlying cellular mechanism has yet to be resolved (Montalvo et al., 2022). Taken together, these studies emphasize the intimate, mutual relationship between oligodendroglia, myelination, and the modulation of neuronal transmission in the CNS.

Cellular and molecular changes to oligodendroglia in epilepsy

Like astroglia, oligodendroglia express Kir4.1 inwardly rectifying K^+ channels (Zhang et al., 2014). This channel localizes to the inner face of myelin sheaths that face the axon membrane (Schirmer et al., 2018). Cell-type-specific deletion of Kir4.1 from OPCs or mature oligodendroglia was shown to have no effect on oligodendroglia development or on neuronal myelination (Larson et al., 2018). However, oligodendroglia lacking Kir4.1 exhibited slower extracellular K^+ clearance, delayed axonal recovery following recurrent stimulation, lower seizure thresholds, and spontaneous seizure activity (Larson et al., 2018). Interestingly, these animals also displayed myelin vacuolization, a pathological finding indicative of myelin degeneration, suggesting that Kir4.1 deficiency may precipitate demyelination and thereby contribute to ictogenesis (Blackmon et al., 2011). Thus, oligodendroglial Kir4.1 may serve as a novel nonneuronal therapeutic target for ASDs. In fact, based on the dual expression of this potassium channel on both astroglia and oligodendroglia, targeting of Kir4.1 may have dual antiepileptogenic effects. Further studies are needed to determine the cell-type-specific vs global effect of Kir4.1 modulation as a treatment for seizures.

POTENTIAL GLIAL THERAPEUTIC TARGETS

Current ASDs primarily target neurons to increase inhibitory neurotransmission or to decrease excitatory neurotransmission. These therapies are ineffective at controlling seizures in greater than 30% of epilepsy patients. For the patients who are responsive to treatments, 88% experience adverse effects such as dizziness and sedation, as well as cognitive and neuropsychiatric symptoms (Laue-Gizzi, 2021). Moreover, for women of childbearing age, enzyme-inducing ASDs reduce the efficacy of oral contraceptives, which complicates family planning, and the teratogenicity of ASDs can be a great source of anxiety for those looking to conceive (Crawford, 2005; Perucca, 2005).

A fundamental paradigm shift is necessary in the generation of new ASDs. Epilepsy is not simply a neuronal disease; it is a network-level disorder. In this chapter, we considered the critical contributions of astroglia, oligodendroglia, and microglia to the genesis, development, and progression of epilepsy. Each glial cell type subserves distinct functions that inform its modulation of neuronal circuits. Seizures drive glial cells to undergo a gamut of morphologic and physiological changes that may causally attenuate or worsen seizure activity in the CNS.

Capitalizing on the recognition of epilepsy as a "gliopathy" leads to novel therapeutic targets (Fig. 4.2). Some clear target mechanisms based on the studies summarized above include: (Fisher et al., 2014) potassium homeostasis; (Fisher et al., 2005) water homeostasis; (Beghi et al., 2019) glutamate homeostasis; (Kaculini et al., 2021) adenosine metabolism; (Louis et al., 2016) gap junction coupling; (Patel et al., 2019) BBB mechanisms; and (Anwar et al., 2020) inflammatory mechanisms. It is also instructive to consider that epilepsy *comorbidities* also may be glial in nature. In other words, cognitive comorbidities in epilepsy may be astrocyte-based. For example, dysregulation of astrocyte adenosine metabolism may impair cognition (Boison and Aronica, 2015); and dysregulation of AQP4 may

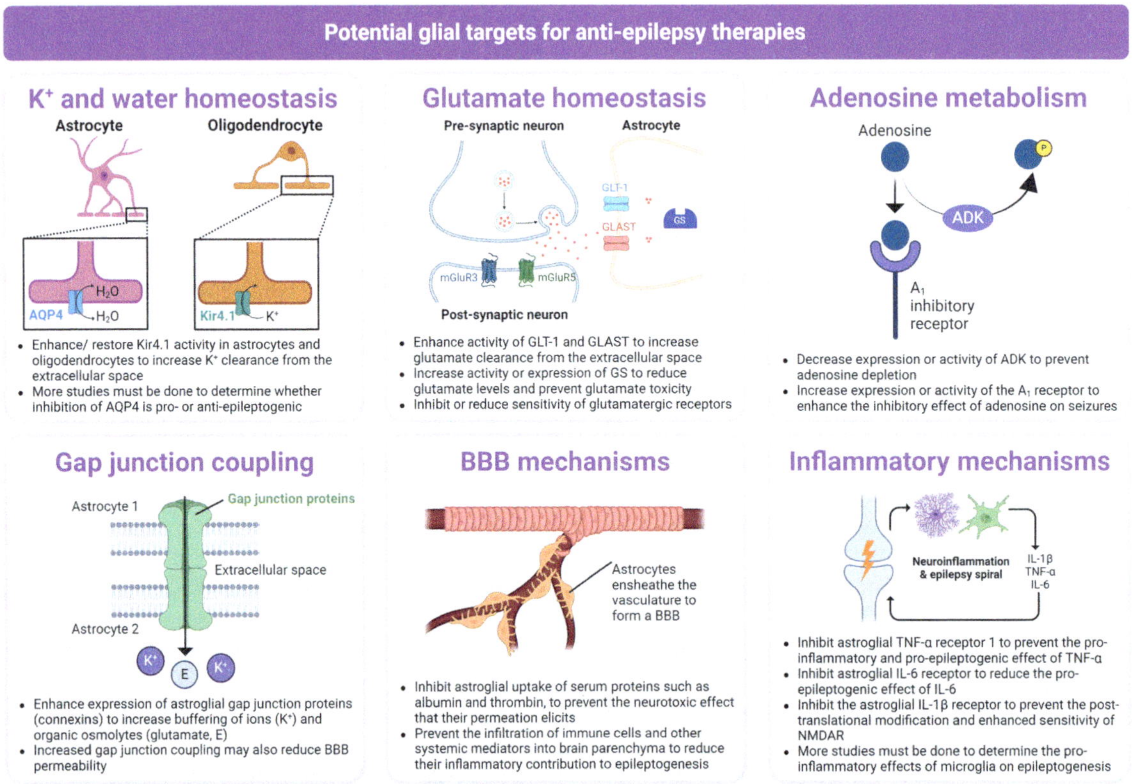

Fig. 4.2. Potential glial targets for antiepilepsy therapies. Created by Biorender. Ghouli M (2024). https://BioRender.com/xyz.

impair synaptic function and memory (Skucas et al., 2011; Szu and Binder, 2016). Thus, restoration of astrocyte homeostasis may not only be therapeutic in limiting seizures and epilepsy but also in ameliorating associated cognitive comorbidities.

Novel technologies also may help bring about a new era of seizure management that targets glial mechanisms. For example, microglial positron emission tomography (PET) imaging provides a noninvasive method through which the biological function of microglia can be measured in vivo and used to determine the reactivity, abundance, and location of these glial cells in the epileptic brain (Scott et al., 2017). If this technique could be combined with a drug delivery technique such as photoresponsive drug delivery (PDD) systems, one could envisage PET-guided localization of abnormal glial populations and PDD-mediated delivery of therapies targeting glial mechanisms (Singh et al., 2022).

In summary, ongoing advances in glioscience should allow specific targeting of gliopathic changes in various forms of epilepsy that would arrest the development of seizures and epilepsy and spare normal brain and cognitive function to the greatest extent possible, thereby approaching the primary goal of "no seizures, no side effects."

References

Abd-El-Basset EM, Rao MS, Alshawaf SM et al. (2021). Tumor necrosis factor (TNF) induces astrogliosis, microgliosis and promotes survival of cortical neurons. AIMS Neurosci 8: 558–584.

Abiega O et al. (2016). Correction: neuronal hyperactivity disturbs ATP microgradients, impairs microglial motility, and reduces phagocytic receptor expression triggering apoptosis/microglial phagocytosis uncoupling. PLoS Biol 14: e1002554.

Ahishali B et al. (2010). Effects of levetiracetam on blood-brain barrier disturbances following hyperthermia-induced seizures in rats with cortical dysplasia. Life Sci 87: 609–619.

Alapirtti T et al. (2018). The production of IL-6 in acute epileptic seizure: a video-EEG study. J Neuroimmunol 316: 50–55.

Allen NJ, Barres BA (2009). Glia—more than just brain glue. Nature 457: 675–677.

Alvestad S et al. (2013). Mislocalization of AQP4 precedes chronic seizures in the kainate model of temporal lobe epilepsy. Epilepsy Res 105: 30–41.

Anwar H et al. (2020). Epileptic seizures. Discoveries (Craiova) 8: e110.

Armulik A et al. (2010). Pericytes regulate the blood-brain barrier. Nature 468: 557–561.

Aronica E et al. (2000). Upregulation of metabotropic glutamate receptor subtype mGluR3 and mGluR5 in reactive astrocytes in a rat model of mesial temporal lobe epilepsy. Eur J Neurosci 12: 2333–2344.

Aronica E et al. (2003). Expression and functional role of mGluR3 and mGluR5 in human astrocytes and glioma cells: opposite regulation of glutamate transporter proteins. Eur J Neurosci 17: 2106–2118.

Aronica E et al. (2011). Upregulation of adenosine kinase in astrocytes in experimental and human temporal lobe epilepsy. Epilepsia 52: 1645–1655.

Aronica E, Sandau US, Iyer A et al. (2013). Glial adenosine kinase—a neuropathological marker of the epileptic brain. Neurochem Int 63: 688–695.

Azevedo FA et al. (2009). Equal numbers of neuronal and nonneuronal cells make the human brain an isometrically scaled-up primate brain. J Comp Neurol 513: 532–541.

Bak LK, Schousboe A, Waagepetersen HS (2006). The glutamate/GABA-glutamine cycle: aspects of transport, neurotransmitter homeostasis and ammonia transfer. J Neurochem 98: 641–653.

Baskys A et al. (2005). Group I metabotropic glutamate receptors reduce excitotoxic injury and may facilitate neurogenesis. Neuropharmacology 49: 146–156.

Bazargani N, Attwell D (2016). Astrocyte calcium signaling: the third wave. Nat Neurosci 19: 182–189.

Bedner P et al. (2015). Astrocyte uncoupling as a cause of human temporal lobe epilepsy. Brain 138: 1208–1222.

Beghi E et al. (2019). Global, regional, and national burden of epilepsy, 1990–2016: a systematic analysis for the Global Burden of Disease Study 2016. Lancet Neurol 18: 357–375.

Benson MJ, Manzanero S, Borges K (2015). Complex alterations in microglial M1/M2 markers during the development of epilepsy in two mouse models. Epilepsia 56: 895–905.

Bergles DE, Diamond JS, Jahr CE (1999). Clearance of glutamate inside the synapse and beyond. Curr Opin Neurobiol 9: 293–298.

Bezzi P et al. (2001). CXCR4-activated astrocyte glutamate release via TNFα: amplification by microglia triggers neurotoxicity. Nat Neurosci 4: 702–710.

Bianchi R, Wong RKS, Merlin LR (2012). GLutamate receptors in epilepsy: Group I mGluR-mediated epileptogenesis. In: JL Noebels, M Avoli, MA Rogawski, RW Olsen, AV Delgado-Escueta (Eds.), Jasper's basic mechanisms of the epilepsies. National Center for Biotechnology Information (US), Bethesda (MD). Copyright © 2012, Michael A Rogawski, Antonio V Delgado-Escueta, Jeffrey L Noebels, Massimo Avoli and Richard W Olsen.

Binder DK, Steinhäuser C (2021). Astrocytes and epilepsy. Neurochem Res 46: 2687–2695.

Binder DK, Oshio K, Ma T et al. (2004). Increased seizure threshold in mice lacking aquaporin-4 water channels. Neuroreport 15: 259–262.

Binder DK et al. (2006a). Increased seizure duration and slowed potassium kinetics in mice lacking aquaporin-4 water channels. Glia 53: 631–636.

Binder DK, Yao X, Verkman AS et al. (2006b). Increased seizure duration in mice lacking aquaporin-4 water channels. In: JT Hoff, RF Keep, G Xi, Y Hua (Eds.), Brain edema XIII. Springer Vienna, Vienna, pp. 389–392.

Bittigau P et al. (2002). Antiepileptic drugs and apoptotic neurodegeneration in the developing brain. Proc Natl Acad Sci U S A 99: 15089–15094.

Blackmon K et al. (2011). Individual differences in verbal abilities associated with regional blurring of the left gray and white matter boundary. J Neurosci 31: 15257–15263.

Bockenhauer D et al. (2009). Epilepsy, ataxia, sensorineural deafness, tubulopathy, and KCNJ10 mutations. N Engl J Med 360: 1960–1970.

Boison D (2005). Adenosine and epilepsy: from therapeutic rationale to new therapeutic strategies. Neuroscientist 11: 25–36.

Boison D, Aronica E (2015). Comorbidities in neurology: is adenosine the common link? Neuropharmacology 97: 18–34.

Boison D, Steinhäuser C (2018). Epilepsy and astrocyte energy metabolism. Glia 66: 1235–1243.

Bradl M, Lassmann H (2010). Oligodendrocytes: biology and pathology. Acta Neuropathol 119: 37–53.

Bruneau EG, McCullumsmith RE, Haroutunian V et al. (2005). Increased expression of glutaminase and glutamine synthetase mRNA in the thalamus in schizophrenia. Schizophr Res 75: 27–34.

Bruno V et al. (2017). The impact of metabotropic glutamate receptors into active neurodegenerative processes: a "dark side" in the development of new symptomatic treatments for neurologic and psychiatric disorders. Neuropharmacology 115: 180–192.

Brusilow SW, Koehler RC, Traystman RJ et al. (2010). Astrocyte glutamine synthetase: importance in hyperammonemic syndromes and potential target for therapy. Neurotherapeutics 7: 452–470.
Cabezas R et al. (2014). Astrocytic modulation of blood brain barrier: perspectives on Parkinson's disease. Front Cell Neurosci 8: 211.
Carmen J, Rothstein JD, Kerr DA (2009). Tumor necrosis factor-alpha modulates glutamate transport in the CNS and is a critical determinant of outcome from viral encephalomyelitis. Brain Res 1263: 143–154.
Chaplin DD (2010). Overview of the immune response. J Allergy Clin Immunol 125: S3–23.
Chaudhuri JR et al. (2022). Myelin oligodendrocyte glycoprotein antibody-associated disease presenting as intracranial hypertension. Case Rep 9: e200020.
Choi SS, Lee HJ, Lim I et al. (2014). Human astrocytes: secretome profiles of cytokines and chemokines. PLoS One 9: e92325.
Clarkson C et al. (2020). Ultrastructural and functional changes at the tripartite synapse during epileptogenesis in a model of temporal lobe epilepsy. Exp Neurol 326: 113196.
Clements JD, Lester RAJ, Tong G et al. (1992). The time course of glutamate in the synaptic cleft. Science 258: 1498–1501.
Collignon F et al. (2006). Altered expression of connexin subtypes in mesial temporal lobe epilepsy in humans. J Neurosurg 105: 77–87.
Cotrina ML et al. (1998). Astrocytic gap junctions remain open during ischemic conditions. J Neurosci 18: 2520–2537.
Coulter DA, Steinhäuser C (2015). Role of astrocytes in epilepsy. Cold Spring Harb Perspect Med 5: a022434.
Crawford P (2005). Best practice guidelines for the management of women with epilepsy. Epilepsia 46: 117–124.
Cunha RA (2001). Adenosine as a neuromodulator and as a homeostatic regulator in the nervous system: different roles, different sources and different receptors. Neurochem Int 38: 107–125.
Das A et al. (2012). Hippocampal tissue of patients with refractory temporal lobe epilepsy is associated with astrocyte activation, inflammation, and altered expression of channels and receptors. Neuroscience 220: 237–246.
de Curtis M, Garbelli R, Uva L (2021). A hypothesis for the role of axon demyelination in seizure generation. Epilepsia 62: 583–595.
Delaney KH et al. (1995). Familial dysmyelination in a long Evans rat mutant. Lab Anim Sci 45: 547–553.
Deng W, Rosenberg PA, Volpe JJ et al. (2003). Calcium-permeable AMPA/kainate receptors mediate toxicity and preconditioning by oxygen-glucose deprivation in oligodendrocyte precursors. Proc Natl Acad Sci U S A 100: 6801–6806.
Devaraju P, Sun MY, Myers TL et al. (2013). Astrocytic group I mGluR-dependent potentiation of astrocytic glutamate and potassium uptake. J Neurophysiol 109: 2404–2414.
Di Nunzio M et al. (2021). Microglia proliferation plays distinct roles in acquired epilepsy depending on disease stages. Epilepsia 62: 1931–1945.
Ding S et al. (2007). Enhanced astrocytic Ca2+ signals contribute to neuronal excitotoxicity after status epilepticus. J Neurosci 27: 10674–10684.
Donkels C et al. (2020). Oligodendrocyte lineage and myelination are compromised in the gray matter of focal cortical dysplasia type IIa. Epilepsia 61: 171–184.
Duncan ID et al. (2018). The adult oligodendrocyte can participate in remyelination. Proc Natl Acad Sci U S A 115: E11807–e11816.
Dunwiddie TV (1999). Adenosine and suppression of seizures. Adv Neurol 79: 1001–1010.
Dunwiddie TV, Masino SA (2001). The role and regulation of adenosine in the central nervous system. Annu Rev Neurosci 24: 31–55.
During MJ, Spencer DD (1993). Extracellular hippocampal glutamate and spontaneous seizure in the conscious human brain. Lancet 341: 1607–1610.
Eid T et al. (2004). Loss of glutamine synthetase in the human epileptogenic hippocampus: possible mechanism for raised extracellular glutamate in mesial temporal lobe epilepsy. Lancet 363: 28–37.
Eid T et al. (2005). Loss of perivascular aquaporin 4 may underlie deficient water and K+ homeostasis in the human epileptogenic hippocampus. Proc Natl Acad Sci U S A 102: 1193–1198.
Eid T, Williamson A, Lee T-SW et al. (2008a). Glutamate and astrocytes—key players in human mesial temporal lobe epilepsy? Epilepsia 49: 42–52.
Eid T et al. (2008b). Recurrent seizures and brain pathology after inhibition of glutamine synthetase in the hippocampus in rats. Brain 131: 2061–2070.
Escartin C et al. (2021). Reactive astrocyte nomenclature, definitions, and future directions. Nat Neurosci 24: 312–325.
Fattorusso A et al. (2021). The pharmacoresistant epilepsy: an overview on existant and new emerging therapies. Front Neurol 12: 674483.
Fedele DE et al. (2005). Astrogliosis in epilepsy leads to overexpression of adenosine kinase, resulting in seizure aggravation. Brain 128: 2383–2395.
Ferrari DP, Bortolanza M, Del Bel EA (2021). Interferon-γ involvement in the neuroinflammation associated with Parkinson's disease and L-DOPA-induced dyskinesia. Neurotox Res 39: 705–719.
Fields RD, Stevens-Graham B (2002). New insights into neuron-glia communication. Science 298: 556–562.
Fine SM et al. (1996). Tumor necrosis factor alpha inhibits glutamate uptake by primary human astrocytes. Implications for pathogenesis of HIV-1 dementia. J Biol Chem 271: 15303–15306.
Fisher RS et al. (2005). Epileptic seizures and epilepsy: definitions proposed by the international league against epilepsy (ILAE) and the International Bureau for Epilepsy (IBE). Epilepsia 46: 470–472.
Fisher RS et al. (2014). ILAE official report: a practical clinical definition of epilepsy. Epilepsia 55: 475–482.
Foiadelli T, Gastaldi M, Scaranzin S et al. (2020). Seizures and myelin oligodendrocyte glycoprotein (MOG) antibodies: two paradigmatic cases and a review of the literature. Mult Scler Relat Disord 41: 102011.
Fonseca CG, Green CR, Nicholson LFB (2002). Upregulation in astrocytic connexin 43 gap junction levels may exacerbate generalized seizures in mesial temporal lobe epilepsy. Brain Res 929: 105–116.
Franklin RJM, Frisén J, Lyons DA (2021). Revisiting remyelination: towards a consensus on the regeneration of CNS myelin. Semin Cell Dev Biol 116: 3–9.
Frigerio F et al. (2012). Long-lasting pro-ictogenic effects induced in vivo by rat brain exposure to serum albumin in the absence of concomitant pathology. Epilepsia 53: 1887–1897.
Fünfschilling U et al. (2012). Glycolytic oligodendrocytes maintain myelin and long-term axonal integrity. Nature 485: 517–521.
Gao F et al. (2017). Alteration of plasma cytokines in patients with active epilepsy. Acta Neurol Scand 135: 663–669.
Ghouli MR, Jonak CR, Sah R et al. (2023). Regulation of the volume-regulated anion channel pore-forming subunit LRRC8A in the intrahippocampal Kainic acid model of mesial temporal lobe epilepsy. ASN Neuro 15: 17590914231184072.
Gong T, Liu L, Jiang W et al. (2020). DAMP-sensing receptors in sterile inflammation and inflammatory diseases. Nat Rev Immunol 20: 95–112.
Gruber VE et al. (2021). Impaired myelin production due to an intrinsic failure of oligodendrocytes in mTORpathies. Neuropathol Appl Neurobiol 47: 812–825.

Guleria S, Kelly TG (2014). Myelin, myelination, and corresponding magnetic resonance imaging changes. Radiol Clin North Am 52: 227–239.

Gürses C et al. (2013). Topiramate reduces blood–brain barrier disruption and inhibits seizure activity in hyperthermia-induced seizures in rats with cortical dysplasia. Brain Res 1494: 91–100.

Häberle J et al. (2005). Congenital glutamine deficiency with glutamine synthetase mutations. N Engl J Med 353: 1926–1933.

Harada Y et al. (2013). Expressional analysis of inwardly rectifying Kir4.1 channels in Noda epileptic rat (NER). Brain Res 1517: 141–149.

Hatton SN et al. (2020). White matter abnormalities across different epilepsy syndromes in adults: an ENIGMA-epilepsy study. Brain 143: 2454–2473.

He Y, Hakvoort TB, Vermeulen JL et al. (2007). Glutamine synthetase is essential in early mouse embryogenesis. Dev Dyn 236: 1865–1875.

Herculano-Houzel S (2014). The glia/neuron ratio: how it varies uniformly across brain structures and species and what that means for brain physiology and evolution. Glia 62: 1377–1391.

Hinterkeuser S et al. (2000). Astrocytes in the hippocampus of patients with temporal lobe epilepsy display changes in potassium conductances. Eur J Neurosci 12: 2087–2096.

Hirsch MR, Wietzerbin J, Pierres M et al. (1983). Expression of Ia antigens by cultured astrocytes treated with gamma-interferon. Neurosci Lett 41: 199–204.

Hoffmann K, Lindner M, Gröticke I et al. (2008). Epileptic seizures and hippocampal damage after cuprizone-induced demyelination in C57BL/6 mice. Exp Neurol 210: 308–321.

Hopkins KJ, Wang G, Schmued LC (2000). Temporal progression of kainic acid induced neuronal and myelin degeneration in the rat forebrain. Brain Res 864: 69–80.

Hubbard JA, Szu JI, Yonan JM et al. (2016). Regulation of astrocyte glutamate transporter-1 (GLT1) and aquaporin-4 (AQP4) expression in a model of epilepsy. Exp Neurol 283: 85–96.

Huyghe D et al. (2019). Phosphorylation of glutamine synthetase on threonine 301 contributes to its inactivation during epilepsy. Front Mol Neurosci 12.

Hyvärinen T et al. (2019). Co-stimulation with IL-1β and TNF-α induces an inflammatory reactive astrocyte phenotype with neurosupportive characteristics in a human pluripotent stem cell model system. Sci Rep 9: 16944.

Ivens S et al. (2007). TGF-beta receptor-mediated albumin uptake into astrocytes is involved in neocortical epileptogenesis. Brain 130: 535–547.

Jantzie LL et al. (2010). Developmental regulation of group I metabotropic glutamate receptors in the premature brain and their protective role in a rodent model of periventricular leukomalacia. Neuron Glia Biol 6: 277–288.

Jimenez-Pacheco A et al. (2016). Transient P2X7 receptor antagonism produces lasting reductions in spontaneous seizures and gliosis in experimental temporal lobe epilepsy. J Neurosci 36: 5920–5932.

Kaculini CM, Tate-Looney AJ, Seifi A (2021). The history of epilepsy: from ancient mystery to modern misconception. Cureus 13: e13953.

Kalueff AV, Lehtimaki KA, Ylinen A et al. (2004). Intranasal administration of human IL-6 increases the severity of chemically induced seizures in rats. Neurosci Lett 365: 106–110.

Kandratavicius L et al. (2013). Distinct increased metabotropic glutamate receptor type 5 (mGluR5) in temporal lobe epilepsy with and without hippocampal sclerosis. Hippocampus 23: 1212–1230.

Kaplan MR et al. (1997). Induction of sodium channel clustering by oligodendrocytes. Nature 386: 724–728.

Kaplan MR et al. (2001). Differential control of clustering of the sodium channels Na(v)1.2 and Na(v)1.6 at developing CNS nodes of Ranvier. Neuron 30: 105–119.

Kettenmann H, Verkhratsky A (2008). Neuroglia: the 150 years after. Trends Neurosci 31: 653–659.

Kettenmann H, Hanisch UK, Noda M et al. (2011). Physiology of microglia. Physiol Rev 91: 461–553.

Kim JE, Kang TC (2011). The P2X7 receptor-pannexin-1 complex decreases muscarinic acetylcholine receptor-mediated seizure susceptibility in mice. J Clin Invest 121: 2037–2047.

Kimizu T et al. (2018). Chronic dysfunction of blood-brain barrier in patients with post-encephalitic/encephalopathic epilepsy. Seizure 63: 85–90.

Knowles JK et al. (2022). Maladaptive myelination promotes generalized epilepsy progression. Nat Neurosci 25: 596–606.

Kocatürk M, Kirmit A (2022). Evaluation of IL-10, IFN-γ, and thiol–disulfide homeostasis in patients with drug-resistant epilepsy. Neurol Sci 43: 485–492.

Kong Q et al. (2012). Increased glial glutamate transporter EAAT2 expression reduces epileptogenic processes following pilocarpine-induced status epilepticus. Neurobiol Dis 47: 145–154.

Kwack DW, Kim DW (2022). The increased interleukin-6 levels can be an early diagnostic marker for new-onset refractory status epilepticus. J Epilepsy Res 12: 78–81.

Lapato AS et al. (2017). Chronic demyelination-induced seizures. Neuroscience 346: 409–422.

Larson VA et al. (2018). Oligodendrocytes control potassium accumulation in white matter and seizure susceptibility. Elife 7: e34829.

Laue-Gizzi H (2021). Discontinuation of antiepileptic drugs in adults with epilepsy. Aust Prescr 44: 53–56.

Layland J, Carrick D, Lee M et al. (2014). Adenosine: physiology, pharmacology, and clinical applications. J Am Coll Cardiol Intv 7: 581–591.

Lee DJ, Hsu MS, Seldin MM et al. (2012a). Decreased expression of the glial water channel aquaporin-4 in the intrahippocampal kainic acid model of epileptogenesis. Exp Neurol 235: 246–255.

Lee Y et al. (2012b). Oligodendroglia metabolically support axons and contribute to neurodegeneration. Nature 487: 443–448.

Li T, Lan JQ, Boison D (2008). Uncoupling of astrogliosis from epileptogenesis in adenosine kinase (ADK) transgenic mice. Neuron Glia Biol 4: 91–99.

Li L, Acioglu C, Heary RF et al. (2021). Role of astroglial toll-like receptors (TLRs) in central nervous system infections, injury and neurodegenerative diseases. Brain Behav Immun 91: 740–755.

Linnerbauer M, Wheeler MA, Quintana FJ (2020). Astrocyte crosstalk in CNS inflammation. Neuron 108: 608–622.

Britton JW, Frey LC, Hopp JL et al. (2016). In: EK St. Louis, LC Frey (Eds.), Electroencephalography (EEG): an introductory text and atlas of normal and abnormal findings in adults, children, and infants [Internet]. American Epilepsy Society, Chicago.

Luo Y et al. (2015). Alterations in hippocampal myelin and oligodendrocyte precursor cells during epileptogenesis. Brain Res 1627: 154–164.

Luo X et al. (2016). Crosstalk between astrocytic CXCL12 and microglial CXCR4 contributes to the development of neuropathic pain. Mol Pain 12.

Mader S, Brimberg L (2019). Aquaporin-4 Water channel in the brain and its implication for health and disease. Cells 8: 90.

Maggio N, Shavit E, Chapman J et al. (2008). Thrombin induces long-term potentiation of reactivity to afferent stimulation and facilitates epileptic seizures in rat hippocampal slices: toward understanding

the functional consequences of cerebrovascular insults. J Neurosci 28: 732–736.
Marchi N et al. (2011). Modulation of peripheral cytotoxic cells and ictogenesis in a model of seizures. Epilepsia 52: 1627–1634.
Maroso M et al. (2010). Toll-like receptor 4 and high-mobility group box-1 are involved in ictogenesis and can be targeted to reduce seizures. Nat Med 16: 413–419.
Maroso M et al. (2011). Interleukin-1 type 1 receptor/Toll-like receptor signalling in epilepsy: the importance of IL-1beta and high-mobility group box 1. J Intern Med 270: 319–326.
Matthews MA, Duncan D (1971). A quantitative study of morphological changes accompanying the initiation and progress of myelin production in the dorsal funiculus of the rat spinal cord. J Comp Neurol 142: 1–22.
Michalak Z et al. (2013). Spatio-temporally restricted blood–brain barrier disruption after intra-amygdala kainic acid-induced status epilepticus in mice. Epilepsy Res 103: 167–179.
Miller RH, Raff MC (1984). Fibrous and protoplasmic astrocytes are biochemically and developmentally distinct. J Neurosci 4: 585–592.
Mirrione MM et al. (2010). Microglial ablation and lipopolysaccharide preconditioning affects pilocarpine-induced seizures in mice. Neurobiol Dis 39: 85–97.
Montalvo M et al. (2022). Acute symptomatic seizures secondary to myelin oligodendrocyte glycoprotein antibody-associated disease. Epilepsia 63: 3180–3191.
Morganti-Kossmann MC, Rancan M, Stahel PF et al. (2002). Inflammatory response in acute traumatic brain injury: a double-edged sword. Curr Opin Crit Care 8: 101–105.
Moulson AJ, Squair JW, Franklin RJM et al. (2021). Diversity of reactive astrogliosis in CNS pathology: heterogeneity or plasticity? Front Cell Neurosci 15: 703810.
Mühlebner A et al. (2020). Myelin pathology beyond white matter in tuberous sclerosis complex (TSC) cortical tubers. J Neuropathol Exp Neurol 79: 1054–1064.
Muoio V, Persson PB, Sendeski MM (2014). The neurovascular unit—concept review. Acta Physiol (Oxf) 210: 790–798.
Murphy TR et al. (2017). Hippocampal and cortical pyramidal neurons swell in parallel with astrocytes during acute hypoosmolar stress. Front Cell Neurosci 11: 275.
Mutanana N, Tsvere M, Chiweshe MK (2020). General side effects and challenges associated with anti-epilepsy medication: a review of related literature. Afr J Prim Health Care Fam Med 12: e1–e5.
Nagelhus EA, Mathiisen TM, Ottersen OP (2004). Aquaporin-4 in the central nervous system: cellular and subcellular distribution and coexpression with KIR4.1. Neuroscience 129: 905–913.
Neely SA et al. (2022). New oligodendrocytes exhibit more abundant and accurate myelin regeneration than those that survive demyelination. Nat Neurosci 25: 415–420.
Niswender CM, Conn PJ (2010). Metabotropic glutamate receptors: physiology, pharmacology, and disease. Annu Rev Pharmacol Toxicol 50: 295–322.
Nitsch C, Klatzo I (1983). Regional patterns of blood-brain barrier breakdown during epileptiform seizures induced by various convulsive agents. J Neurol Sci 59: 305–322.
Norenberg MD, Jayakumar AR, Rama Rao KV et al. (2007). New concepts in the mechanism of ammonia-induced astrocyte swelling. Metab Brain Dis 22: 219–234.
Notenboom RG et al. (2006). Up-regulation of hippocampal metabotropic glutamate receptor 5 in temporal lobe epilepsy patients. Brain 129: 96–107.
Nutma E, van Gent D, Amor S et al. (2020). Astrocyte and oligodendrocyte cross-talk in the central nervous system. Cells 9: 600.
Oberheim NA et al. (2009). Uniquely hominid features of adult human astrocytes. J Neurosci 29: 3276–3287.
Oberheim NA, Goldman SA, Nedergaard M (2012). Heterogeneity of astrocytic form and function. Methods Mol Biol 814: 23–45.
Olabarria M, Noristani HN, Verkhratsky A et al. (2011). Age-dependent decrease in glutamine synthetase expression in the hippocampal astroglia of the triple transgenic Alzheimer's disease mouse model: mechanism for deficient glutamatergic transmission? Mol Neurodegener 6: 55.
Onodera M et al. (2021). Exacerbation of epilepsy by astrocyte alkalization and gap junction uncoupling. J Neurosci 41: 2106–2118.
Pascual O et al. (2005). Astrocytic purinergic signaling coordinates synaptic networks. Science 310: 113–116.
Patel DC, Tewari BP, Chaunsali L et al. (2019). Neuron-glia interactions in the pathophysiology of epilepsy. Nat Rev Neurosci 20: 282–297.
Perillan PR, Chen M, Potts EA et al. (2002). Transforming growth factor-beta 1 regulates Kir2.3 inward rectifier K+ channels via phospholipase C and protein kinase C-delta in reactive astrocytes from adult rat brain. J Biol Chem 277: 1974–1980.
Perucca E (2005). An introduction to antiepileptic drugs. Epilepsia 46: 31–37.
Peterson AR, Binder DK (2019). Regulation of synaptosomal GLT-1 and GLAST during epileptogenesis. Neuroscience 411: 185–201.
Peterson AR, Binder DK (2020). Astrocyte glutamate uptake and signaling as novel targets for antiepileptogenic therapy. Front Neurol 11: 1006.
Peterson AR et al. (2021). Targeted overexpression of glutamate transporter-1 reduces seizures and attenuates pathological changes in a mouse model of epilepsy. Neurobiol Dis 157: 105443.
Petr GT et al. (2015). Conditional deletion of the glutamate transporter GLT-1 reveals that astrocytic GLT-1 protects against fatal epilepsy while neuronal GLT-1 contributes significantly to glutamate uptake into synaptosomes. J Neurosci 35: 5187–5201.
Proper EA et al. (2002). Distribution of glutamate transporters in the hippocampus of patients with pharmaco-resistant temporal lobe epilepsy. Brain 125: 32–43.
Purves D, Augustine G, Fitzpatrick D et al. (2001). Neuroscience. vol. 3: pp. 8–10.
Reichold M et al. (2010). KCNJ10 gene mutations causing EAST syndrome (epilepsy, ataxia, sensorineural deafness, and tubulopathy) disrupt channel function. Proc Natl Acad Sci U S A 107: 14490–14495.
Ribeiro JA, Lobo MG, Sebastião AM (2003). Endogenous adenosine modulation of 22Na uptake by rat brain synaptosomes. Neurochem Res 28: 1591–1595.
Rimmele TS, Rosenberg PA (2016). GLT-1: the elusive presynaptic glutamate transporter. Neurochem Int 98: 19–28.
Robinson SR (2000). Neuronal expression of glutamine synthetase in Alzheimer's disease indicates a profound impairment of metabolic interactions with astrocytes. Neurochem Int 36: 471–482.
Roche PA, Furuta K (2015). The ins and outs of MHC class II-mediated antigen processing and presentation. Nat Rev Immunol 15: 203–216.
Rodgers KR, Lin Y, Langan TJ et al. (2020). Innate immune functions of astrocytes are dependent upon tumor necrosis factor-alpha. Sci Rep 10: 7047.
Romanos J, Benke D, Saab AS et al. (2019). Differences in glutamate uptake between cortical regions impact neuronal NMDA receptor activation. Commun Biol 2: 127.
Rostami J et al. (2020). Astrocytes have the capacity to act as antigen-presenting cells in the Parkinson's disease brain. J Neuroinflammation 17: 119.
Rüber T et al. (2018). Evidence for peri-ictal blood-brain barrier dysfunction in patients with epilepsy. Brain 141: 2952–2965.

Ruth RE (1984). Increased cerebrovascular permeability to protein during systemic Kainic acid seizures. Epilepsia 25: 259–268.
Sakuma S, Halliday WC, Nomura R et al. (2014). Increased population of oligodendroglia-like cells in pediatric intractable epilepsy. Neurosci Lett 566: 188–193.
Samland H et al. (2003). Profound increase in sensitivity to glutamatergic- but not cholinergic agonist-induced seizures in transgenic mice with astrocyte production of IL-6. J Neurosci Res 73: 176–187.
Sarac S et al. (2009). Excitatory amino acid transporters EAAT-1 and EAAT-2 in temporal lobe and hippocampus in intractable temporal lobe epilepsy. Apmis 117: 291–301.
Scheffer IE et al. (2017). ILAE classification of the epilepsies: position paper of the ILAE Commission for Classification and Terminology. Epilepsia 58: 512–521.
Schirmer L et al. (2018). Oligodendrocyte-encoded Kir4.1 function is required for axonal integrity. Elife 7.
Scholl UI et al. (2009). Seizures, sensorineural deafness, ataxia, mental retardation, and electrolyte imbalance (SeSAME syndrome) caused by mutations in KCNJ10. Proc Natl Acad Sci U S A 106: 5842–5847.
Scholl T et al. (2017). Impaired oligodendroglial turnover is associated with myelin pathology in focal cortical dysplasia and tuberous sclerosis complex. Brain Pathol 27: 770–780.
Schreiner AE et al. (2014). Laminar and subcellular heterogeneity of GLAST and GLT-1 immunoreactivity in the developing postnatal mouse hippocampus. J Comp Neurol 522: 204–224.
Schröder W et al. (2000). Functional and molecular properties of human astrocytes in acute hippocampal slices obtained from patients with temporal lobe epilepsy. Epilepsia 41: S181–S184.
Scott G, Mahmud M, Owen DR et al. (2017). Microglial positron emission tomography (PET) imaging in epilepsy: applications, opportunities and pitfalls. Seizure 44: 42–47.
Seeker LA, Williams A (2022). Oligodendroglia heterogeneity in the human central nervous system. Acta Neuropathol 143: 143–157.
Semple BD et al. (2017). Interleukin-1 receptor in seizure susceptibility after traumatic injury to the pediatric brain. J Neurosci 37: 7864–7877.
Shen Y et al. (2016). Postnatal activation of TLR4 in astrocytes promotes excitatory synaptogenesis in hippocampal neurons. J Cell Biol 215: 719–734.
Short B et al. (2019). Cerebral aquaporin-4 expression is independent of seizures in tuberous sclerosis complex. Neurobiol Dis 129: 93–101.
Sierra A et al. (2010). Microglia shape adult hippocampal neurogenesis through apoptosis-coupled phagocytosis. Cell Stem Cell 7: 483–495.
Singh AK et al. (2022). Green light-activated single-component organic fluorescence-based Nano-drug delivery system for dual uncaging of anticancer drugs. ACS Appl Bio Mater 5: 1202–1209.
Skucas VA et al. (2011). Impairment of select forms of spatial memory and neurotrophin-dependent synaptic plasticity by deletion of glial aquaporin-4. J Neurosci 31: 6392–6397.
Smith AJ, Verkman AS (2015). Superresolution imaging of aquaporin-4 cluster size in antibody-stained paraffin brain sections. Biophys J 109: 2511–2522.
Sokrab TEO, Kalimo H, Johansson BB (1989). Endogenous serum albumin content in brain after short-lasting epileptic seizures. Brain Res 489: 231–236.
Song C et al. (2016). CXCR4 antagonist AMD3100 suppresses the long-term abnormal structural changes of newborn neurons in the intraventricular Kainic acid model of epilepsy. Mol Neurobiol 53: 1518–1532.
Stamatovic SM, Johnson AM, Keep RF et al. (2016). Junctional proteins of the blood-brain barrier: new insights into function and dysfunction. Tissue Barriers 4: e1154641.
Stephan J, Eitelmann S, Zhou M (2021). Approaches to study gap junctional coupling. Front Cell Neurosci 15.
Stevenson R, Samokhina E, Rossetti I et al. (2020). Neuromodulation of glial function during neurodegeneration. Front Cell Neurosci 14: 278. https://doi.org/10.3389/fncel.2020.00278.
Stewart TH et al. (2010). Chronic dysfunction of astrocytic inwardly rectifying K+ channels specific to the neocortical epileptic focus after fluid percussion injury in the rat. J Neurophysiol 104: 3345–3360.
Studer FE et al. (2006). Shift of adenosine kinase expression from neurons to astrocytes during postnatal development suggests dual functionality of the enzyme. Neuroscience 142: 125–137.
Sun W et al. (2013). Glutamate-dependent neuroglial calcium signaling differs between young and adult brain. Science 339: 197–200.
Sweeney AM et al. (2017). PAR1 activation induces rapid changes in glutamate uptake and astrocyte morphology. Sci Rep 7: 43606.
Szu JI, Binder DK (2016). The role of astrocytic Aquaporin-4 in synaptic plasticity and learning and memory. Front Integr Neurosci 10: 8.
Szu JI, Patel DD, Chaturvedi S et al. (2020). Modulation of post-traumatic epileptogenesis in aquaporin-4 knockout mice. Epilepsia 61: 1503–1514.
Tanaka K et al. (1997). Epilepsy and exacerbation of brain injury in mice lacking the glutamate transporter GLT-1. Science 276: 1699–1702.
Tang FR et al. (2004). Metabotropic glutamate receptor 2/3 in the hippocampus of patients with mesial temporal lobe epilepsy, and of rats and mice after pilocarpine-induced status epilepticus. Epilepsy Res 59: 167–180.
Thom M (2014). Review: hippocampal sclerosis in epilepsy: a neuropathology review. Neuropathol Appl Neurobiol 40: 520–543.
Tran A, Chen JTC, Belsham DD (2021). Glia-neuron communication: not a one-way street. In: JG Tasker, JS Bains, JA Chowen (Eds.), Glial-neuronal signaling in neuroendocrine systems. Springer International Publishing, Cham, pp. 155–180. https://doi.org/10.1007/978-3-030-62383-8_7.
Ueda Y, Doi T, Tsuru N et al. (2002). Expression of glutamate transporters and ionotropic glutamate receptors in GLAST knockout mice. Brain Res Mol Brain Res 104: 120–126.
Umpierre AD et al. (2016). Repeated low-dose kainate administration in C57BL/6J mice produces temporal lobe epilepsy pathology but infrequent spontaneous seizures. Exp Neurol 279: 116–126.
Valério-Gomes B, Guimarães DM, Szczupak D et al. (2018). The absolute number of oligodendrocytes in the adult mouse brain. Front Neuroanat 12: 90. https://doi.org/10.3389/fnana.2018.00090.
van der Hel WS et al. (2005). Reduced glutamine synthetase in hippocampal areas with neuron loss in temporal lobe epilepsy. Neurology 64: 326–333.
van Vliet EA et al. (2007). Blood-brain barrier leakage may lead to progression of temporal lobe epilepsy. Brain 130: 521–534.
Vanes LD et al. (2022). White matter disruptions related to inattention and autism spectrum symptoms in tuberous sclerosis complex. Neuroimage Clin 36: 103163.
Vardjan N et al. (2012). IFN-γ-induced increase in the mobility of MHC class II compartments in astrocytes depends on intermediate filaments. J Neuroinflammation 9: 144.
Verentzioti A et al. (2019). Epileptic patient with mild malformation of cortical development with oligodendroglial hyperplasia and epilepsy (MOGHE): a case report and review of the literature. Case Rep Neurol Med 2019: 9130780.

Verkhratsky A, Pankratov Y, Lalo U et al. (2012). P2X receptors in neuroglia. Wiley Interdiscip Rev Membr Transp Signal 1: e12. https://doi.org/10.1002/wmts.12.

Vezzani A et al. (1999). Interleukin-1beta immunoreactivity and microglia are enhanced in the rat hippocampus by focal kainate application: functional evidence for enhancement of electrographic seizures. J Neurosci 19: 5054–5065.

Vivien D et al. (1998). Evidence of type I and type II transforming growth factor-β receptors in central nervous tissues: changes induced by focal cerebral ischemia. J Neurochem 70: 2296–2304.

Wang J et al. (2017). Epilepsy-associated genes. Seizure 44: 11–20.

Watts ME, Pocock R, Claudianos C (2018). Brain energy and oxygen metabolism: emerging role in normal function and disease. Front Mol Neurosci 11: 216.

Wijdeven RH et al. (2018). Chemical and genetic control of IFNγ-induced MHCII expression. EMBO Rep 19: e45553.

Wolf Y, Yona S, Kim KW et al. (2013). Microglia, seen from the CX3CR1 angle. Front Cell Neurosci 7: 26.

Wu W et al. (2020). Microglial depletion aggravates the severity of acute and chronic seizures in mice. Brain Behav Immun 89: 245–255.

Xu Y et al. (2012). Altered expression of CX3CL1 in patients with epilepsy and in a rat model. Am J Pathol 180: 1950–1962.

Ye L et al. (2013). IL-1β and TNF-α induce neurotoxicity through glutamate production: a potential role for neuronal glutaminase. J Neurochem 125: 897–908.

Yeo SI et al. (2011). The roles of fractalkine/CX3CR1 system in neuronal death following pilocarpine-induced status epilepticus. J Neuroimmunol 234: 93–102.

Yeung MSY et al. (2019). Dynamics of oligodendrocyte generation in multiple sclerosis. Nature 566: 538–542.

You Y et al. (2011). Myelin damage of hippocampus and cerebral cortex in rat pentylenetetrazol model. Brain Res 1381: 208–216.

Zattoni M et al. (2011). Brain infiltration of leukocytes contributes to the pathophysiology of temporal lobe epilepsy. J Neurosci 31: 4037–4050.

Zhand A et al. (2018). Expression analysis of GRIN2B, BDNF, and IL-1β genes in the whole blood of epileptic patients. Neurol Sci 39: 1945–1953.

Zhang Y et al. (2014). An RNA-sequencing transcriptome and splicing database of glia, neurons, and vascular cells of the cerebral cortex. J Neurosci 34: 11929–11947.

Zhong S et al. (2023). Genesis of a functional astrocyte syncytium in the developing mouse hippocampus. Glia 71: 1081–1098.

Zhou F et al. (2006). Activation of group II/III metabotropic glutamate receptors attenuates LPS-induced astroglial neurotoxicity via promoting glutamate uptake. J Neurosci Res 84: 268–277.

Zhou Y et al. (2019). Selective deletion of glutamine synthetase in the mouse cerebral cortex induces glial dysfunction and vascular impairment that precede epilepsy and neurodegeneration. Neurochem Int 123: 22–33.

Handbook of Clinical Neurology, Vol. 210 (3rd series)
Neuroglia in Neurologic and Psychiatric Disorders, Part II
A. Verkhratsky, L.D. de Witte, E. Aronica, and E.M. Hol, Editors
https://doi.org/10.1016/B978-0-443-19102-2.00003-X

Chapter 5

Neuroglia pathology in genetic and epigenetic disorders of the central nervous system

HARVEY B. SARNAT[1,2*] AND VIJAYARAGHAVA T.S. RAO[1,2]

[1]*Departments of Paediatrics (Neurology), Pathology and Laboratory Medicine (Neuropathology), and Clinical Neurosciences, University of Calgary Cumming School of Medicine and Alberta Children's Hospital Research Institute (Owerko Centre), Calgary, AB, Canada*

[2]*Medical Affairs Laboratory, Argenx US, Inc., Boston, MA, United States*

Abstract

Glial cells are increasingly recognized for their important interactions with both developing and mature neurons, in particular for maintenance of dendritic ramifications and spines, synapses, and neurotransmitter uptake. MicroRNA abnormalities are demonstrated in individual astrocytes with alterations in neurological diseases. Alexander disease is a prototype astrocytic disease because of genetically altered glial fibrillary acidic protein (GFAP) filaments. Other genetic diseases are now recognized as involving glial cells in their pathogenesis: Rett, Fragile-X, Aicardi-Goutières, and Down syndromes, as well as epigenetic effects in the mechanism of fetal alcohol spectrum disorder. Many involve glial production of cytokines and neuroinflammation. Microglia also may contribute. The heat-shock protein α-B-crystallin is expressed in the Rosenthal fibers of Alexander disease, in which the molecular structure of GFAP is altered, in astrocytes secreting neurotoxic cytokines, and focally at or near epileptic foci. Satellite glial cells adherent to neuronal soma are frequent and diagnostically nonspecific but may contribute to neuronal degeneration, especially of hypermetabolic epileptogenic neurons. Glial cells have distorted size and morphology in mTOR malformations. Failure of glial apoptosis in the fetal lamina terminalis is the likely pathogenesis of callosal agenesis and of other cerebral dysgeneses.

INTRODUCTION

For more than a century, neuroglia never received the same attention as neurons. For many years, they were regarded as merely a filler in the neuropil space, perhaps with a function in the nutrition of neurons but not part of integrated metabolic pathways. Ramón y Cajal was aware of the presence of glial cells but did not focus his studies on these cells (Ramón y Cajal, 1995). Achúcarro was Ramón y Cajal's favorite student, and he had planned to make him his successor if Achúcarro had not died unexpectedly in 1918. Another student, of both Achúcarro and also of Cajal, Pío del Río Hortega, applied Achúcarro's Tannin silver carbonate method and then developed his own gold chloride impregnation technique in Cajal's laboratory that well demonstrated astrocytes and their complex processes that contacted capillaries, as well as microglia and neurons, published in 1916 (Del Rio, 1916). It was a parallel method to the rapid silver impregnations developed by Golgi and employed skillfully by Cajal to demonstrate neuronal processes, who then also interpreted these findings so brilliantly in developing the brain. Another of Cajal's younger trainees who contributed to defining glial morphogenesis and function was Lorente de Nó, though most of his work was in peripheral nerve. Later in the 20th century, it was discovered that astrocytes were important for neurotransmitter uptake at synapses (Bosworth and Allen, 2017). The immunohistochemical reactivity of glial fibrillary acidic protein (GFAP) as a monoclonal or polyclonal antibody is now the easiest and most efficient manner of demonstrating mature astrocytes and their processes and yields results similar to those of heavy metal impregnations (Fig. 5.1A), but vimentin and nestin demonstrate the transitory intermediate filaments of immature astrocytes or reactive astrocytes in a zone of ongoing gliosis (Fig. 5.1).

Astrocytes participate in the survival and function of neurons by maintaining extracellular ionic concentrations, promoting dendritic growth, synaptogenesis, and synaptic plasticity (Barres, 2008; Bosworth and Allen, 2017). The ontogenetic maturation of astrocytes is marked by an increased production

*Correspondence to: Dr. Harvey B. Sarnat, Alberta Children's Hospital Research Institute, Owerko Centre, CDC# 395, 2500 University Drive NW, Calgary, AB T2N 1N4 Canada. Tel: +1-403-441-8409, Fax: +1-403-955-7609. E-mail: harvey.sarnat@albertahealthservices.ca

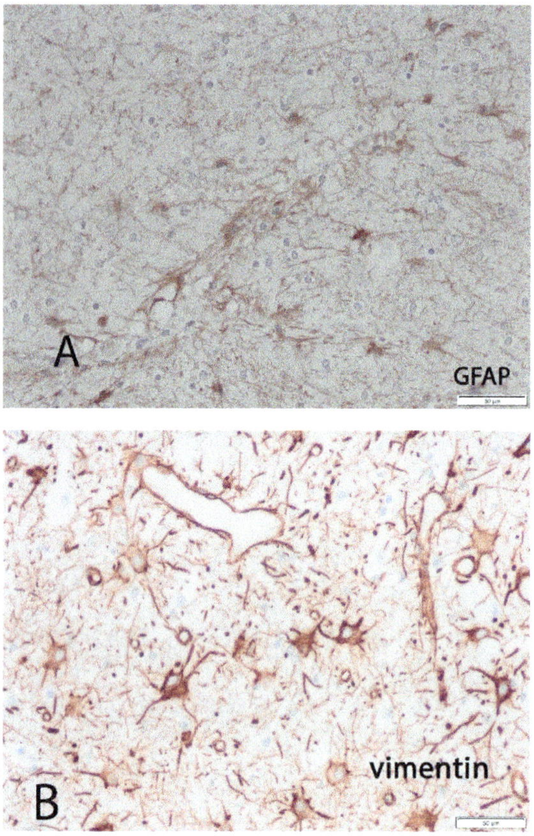

Fig. 5.1. Immunoreactivities of (A) glial fibrillary acidic protein, polyclonal antibody (GFAP; chronic), and (B) vimentin (reactive ongoing) gliosis. Astrocytic processes adhere to small blood vessels and also to neuronal somata in both. Same case as Fig. 5.4, an ultrastructural demonstration of satellitosis.

of microtubules, which coexist with intermediate filaments such as those of vimentin to form their processes, similar to the neuritic extensions of developing neurons (Schiweck et al., 2018). Immunocytochemistry used as neuronal and glial markers of maturation has greatly enhanced our understanding of differentiation of these cells during ontogenesis and regeneration (Sasaki et al., 1988; Honig et al., 1996; Yamaguchi, 2011).

Oligodendrocytes, both as myelinating and nonmyelinating cells, was a further important advance in the understanding of glia and the sequences of myelination in white matter tracts in the fetus and postnatally into adult life (Yakovlev and Lecours, 1967; Rorke and Riggs, 1969; Gilles, 1976; Kinney et al., 1988; Davidson and Rash, 2012). Communication between axons and oligodendrocytes is another factor regulating myelin formation (Hughes and Appel, 2016). Microglia help regulate myelin growth and integrity within the central nervous system (McNamara et al., 2023).

Some glial cells are capable of functions previously attributed only to neurons, for example, hippocampal astrocytes encoding reward locations in a spatial context in mice, a cognitive function previously attributed only to nerve cells (Kriegstein and Álvarez-Buyya, 2009). Glial-neuronal interactions are increasingly recognized as a critical part of synapses, particularly during epileptogenesis (Allen and Eroglu, 2017; Roy et al., 2020; Ammonthumkandy et al., 2022; Doron et al., 2022). A molecular switch for neuroprotection in the retina involves a balance between neurotoxic and neuroprotective astrocytes, regulated by discrete pools of compartmented cyclic adenosine monophosphate (Cameron et al., 2024). Evidence is now emerging that glial-neuronal synergy is also involved in long-term memory (Sun et al., 2024) and in certain psychiatric disorders such as perseverative behaviors (Ollivier et al., 2024). Neuronal-astrocytic integrated programs decline with normal aging and in schizophrenia (Ling et al., 2024).

Modulation of extracellular glutamate uptake by astrocytes via the Na^+-dependent glutamate transporters glutamate transporter-1 (GLT-1; EAAT2), glutamate aspartate transporter (GLAST; EAAT1), and metabotropic glutamate G-protein-coupled receptors-3 and -5 contribute to epileptogenesis in temporal lobe epilepsy in both humans and animal models and might offer targets for specific therapies (Peterson and Binder, 2020). Inhibitory input directs astrocyte morphogenesis through glial GABA-A receptors (Chang et al., 2023). Specialized astrocytes help mediate glutamatergic transmission directly (Ceglia et al., 2023). GLT-1 is responsible for 90% of glutamate uptake in the adult dorsal forebrain and crucial for maintaining low extracellular glutamate to enable efficient synaptic transmission (Chen et al., 2023). Glutamate-dependent neuroglial calcium signaling differs between immature and mature brains (Danbolt, 2001). Reactive astrocytes with cytoplasmic lipid storage also promote disease progression in epilepsy (Sun et al., 2013).

In mesial temporal lobe epilepsy with hippocampal sclerosis, long duration of epilepsy is associated with a decline in neuronogenesis but persistence and continued proliferation of astrogliosis (Kriegstein and Álvarez-Buyya, 2009). Whether glial "scars" after injury enhance or impair axonal regeneration remains controversial, but evidence is increasing that reactive astrocytes may promote regeneration (Anderson et al., 2016). Furthermore, reactive astrocytes as well as progenitor cells may at times be reprogrammed to differentiate as novel interneurons (Lentini et al., 2021). Abundant data now confirm a close relationship between the central nervous system and the immune system, including microglia, of major importance in the pathogenesis of some genetic diseases (Hayes et al., 2022; Lubarski et al., 2022; Rustenhoven and Kipnis, 2022).

MicroRNAs OF INDIVIDUAL GLIAL CELLS

MicroRNAs are small RNA molecules that are posttranscriptional gene expression regulators. They can offer spatio-temporal control of gene expression, which is relevant both during development and for immune regulation. These molecules can be key components of the cargo that are transferred via secreted vesicles from microglia to neurons under pathological conditions (Prada et al., 2018). The more recent demonstration of microRNA abnormalities in individual astrocytes with alterations in neurological diseases further highlights the importance of glial cells in normal and abnormal neural function (Rao et al., 2016; Ramírez et al., 2022). Many embryonic cells that serve a specific transitory function, such as radial glial cells that serve as a monorail-like transport for migratory neuroblasts from the subventricular zone to the cortical plate, also behave as resident stem cells capable of neuronal or glial differentiation after neuroblast migratory guidance is completed (Rao et al., 2016). Divergent transcriptional regulation of astrocyte reactivity is in part regulated by mRNAs (Burda et al., 2022). Some examples of important roles of microRNAs in the diseases that will be discussed in the following paragraphs are as follows:

An important microRNA that our earlier studies (Rao et al., 2016, 2019) highlighted, miR-146a, is the most commonly deregulated microRNA in developmental brain disorders. Its deficiency in mice has been found to "recapitulate several aspects reported in patients, including impaired neurogenesis, abnormal brain anatomy, and working and spatial memory deficits" (Fregeac et al., 2020). Issues of astrocytic microRNAs' biogeneses have been implicated in mislocalization of TAR DNA-binding protein (specifically TBP-43) in Alexander disease (Walker et al., 2014). Regional disturbances in microRNAs' expression can impact the transcriptional regulator MeCP2, a well-established pathophysiologic factor in Rett syndrome (Horvath et al., 2022). Dysregulated microRNAs' expressions have been reported in blood (Sotoudeh Anvari et al., 2022) and urine samples (Putkonen et al., 2020) from patients with fragile X syndrome. In Aicardi Goutières Syndrome, miR-219 may have a role in protecting astrocytes from potential neurotoxic action by lymphocytes (Pulliero et al., 2014). MicroRNAs suggest a link between astrocytic dysfunction, cognitive impairment, and neurodegenerative disease (Ramírez et al., 2022). A related technique, nucleic acid cytometry, can identify astrocytic regulators in single pathogenic glial cells to identify therapeutic targets (Clark et al., 2023).

Clearly, microRNAs shape the immune processes while also regulating developmental and/or repair mechanisms that are relevant physiologically and even more so under pathophysiologic developmental conditions. Not surprisingly, microRNAs can be valuable as therapeutic targets in many of these developmental disorders. However, there remain key gaps in our understanding of how these molecules fundamentally impact remote functions. Factors that regulate the synchrony of production, secretion, transportation, and the execution action of microRNAs are yet to be fully understood. MicroRNAs are certainly more readily beneficial as pathologic biomarkers.

EXAMPLES OF GENETIC/METABOLIC DISEASES IN WHICH GLIAL CELLS ARE ALTERED

Glial cells in diseases resulting from mutations in the mTOR signaling pathway and related genetic pathways (tuberous sclerosis complex; hemimegalencephaly; focal cortical dysplasia type II).

The mTOR genetic/metabolic pathway is complex and involves several other genetic pathways, including AKT, P13K, GATOR, RAS, PTEN, TSC1, and TSC2. The mTOR is central to these other pathways and follows two downstream paths, mTORC1 and mTORC2. Though our understanding of the physiological role of the mTOR pathway in normal neurogenesis is incomplete, it has importance in early cellular growth, lineage, and differentiation.

The histopathological hallmark of mTORopathies is the presence of megalocytic dysmorphic neurons in addition to dyslamination of the cortex and abnormal cerebral tissue architecture (Mühlebner et al., 2019). However, glial cells are also dysmorphic and enlarged, indicating that the growth and morphology of all cells of neuroepithelial origin are affected, regardless of their maturing lineage, and glial cells contribute to epileptogenesis (Crino, 2020). Astrocyte-specific *Tsc1*-null mice exhibit electroencephalographic-confirmed seizures from a month of age and die at 3–4 months. Their brains show increased numbers of megalocytic astrocytes throughout by age 3 weeks and abnormal neuronal lamination and architecture in the hippocampus and neocortex (Uhlmann et al., 2002). Balloon cells are large, globoid cells of mixed neuronal and glial lineage that express proteins of both and occur in FCD IIb, HME, and TSC (Yasin et al., 2010; Sarnat, 2023). In tuberous sclerosis complex, a somatic mTORopathy in addition to inherited germline mutations of TSC1 and TSC2 genes, the fetal and neonatal brain shows inflammatory markers and cells that also may contribute to epileptogenesis (Prabowo et al., 2013; Tsai et al., 2014).

Inhibitors of mTOR, such as everolimus and sirolimus, are effective in blocking overgrowth including both megalocytic neurons and glial cells in humans and murine models, with therapeutic potential (McDaniel and Wong, 2011; Overwater et al., 2016; Bevacqua et al., 2019; Xu et al., 2019; Huang et al., 2021). Moreover, administration of a combination of brain-permeable *Rapal. ink.1* and brain-impermeable *FKBP12* ligand RapaBlock enables brain-restricted inhibition of their respective kinase targets while mitigating adverse systemic effects of mTOR inhibitors (Zyang et al., 2022). Failure of astrocyte-dependent removal of extracellular debris and leakage of the blood–brain barrier at the contact of endothelial cells by astrocytic processes can result from impaired astrocytic functions in mTORopathies (Kielbinski et al., 2016).

Timing is of primordial importance in the initiation of mTOR activation and the extent of the lesion that results. The difference between type II focal cortical dysplasia type 2 (FCD II) is the timing of onset of expression within the approximately 33 mitotic cycles of neuroepithelial cells to exponentially produce all cerebral cortical neurons. Onset in one of the later cycles produces small lesions limited to one or two gyri as FCD II; earlier onset in one of the earlier cycles produces the more extensive lesions of hemimegalencephaly (Sarnat and Flores-Sarnat, 2014). In TSC, cortical tubers are rarely seen prenatally by ultrasound or even fetal MRI, though the somatic mutation is present from early fetal life. Tubers increase in number and size in the postnatal period over weeks, months, and years. A fundamental query for which the answer is still largely elusive is what factors determine the onset of genetic expression in tissues, in the mTOR pathways in particular.

Inflammation is another factor in the pathogenesis of some mTORopathies. In tuberous sclerosis, inflammatory markers and cells occur in the brain even during fetal development (McDaniel and Wong, 2011; Huang et al., 2021).

Agenesis of the corpus callosum

The callosal plate or bridge forms in the dorsal part of the lamina terminalis (a transitory embryonic structure that borders the

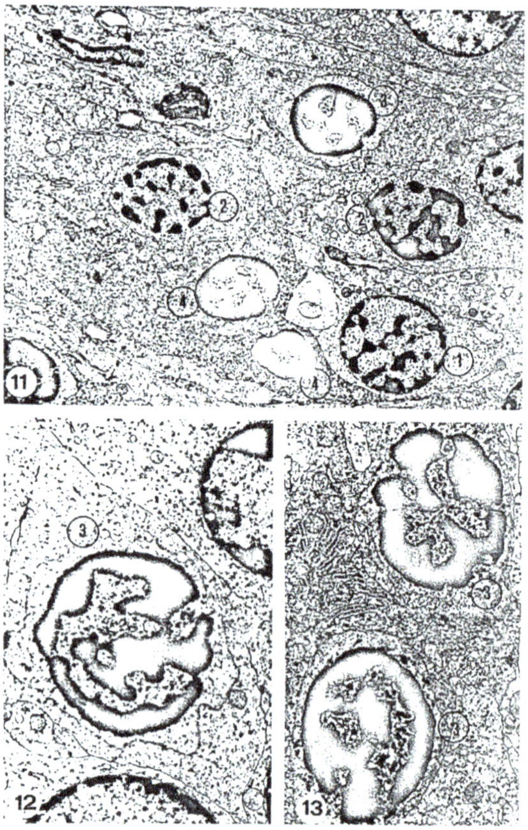

Fig. 5.2. Transmission electron micrographs of the early glial apoptosis for passage of callosal axons in the mouse. Stages of glial apoptosis are: (1) condensation of chromatin against the inner surface of the nucleus of an astrocyte; (2) progressive condensation of chromatin and separation of the chromatin from the inner surface of the nuclear membrane; (3) fusion of the aggregates of chromatin in the center of the nucleus and undulation of the still intact nuclear membrane; (4) condensation of the chromatin in an irregular mass and discontinuities in the nuclear membrane, swelling and disruption of microtubules, followed by disappearance of the cell. Reproduced from Zaki W (1985). Le processus dégénératif au cours du développement du corps calleux. Arch Anat Microsc Morphol Exp 74: 133–149.

rostral wall of the third ventricle and forms the boundary between the developing diencephalon and telencephalon after sagittal prosencephalic cleavage at 3–5 weeks gestation). Apoptotic degeneration of the glial wall of the lamina terminalis (Fig. 5.2) occurs at 42 days in the human fetus, and the first pioneer callosal axons from the subplate zone traverse the midline across this newly formed plate at 44 days. If these early fetal glial cells do not regress or disappear in that part of the lamina terminalis at the proper time, they form a barrier to the interhemispheric passage of axons, resulting in callosal agenesis. A glial barrier to passage of commissural axons of the corpus callosum occurs in a murine genetic model of agenesis of the corpus callosum in which there is failure of the physiological apoptosis of this wall in the dorsal part of the lamina terminalis to form a bridge shortly before the pioneer axons traverse the midline (Zaki, 1985). A similar process in humans probably explains the pathogenesis of callosal agenesis. Oligodendrocytes ensheath the immature axons of the corpus callosum, but myelin is not formed until the postnatal period.

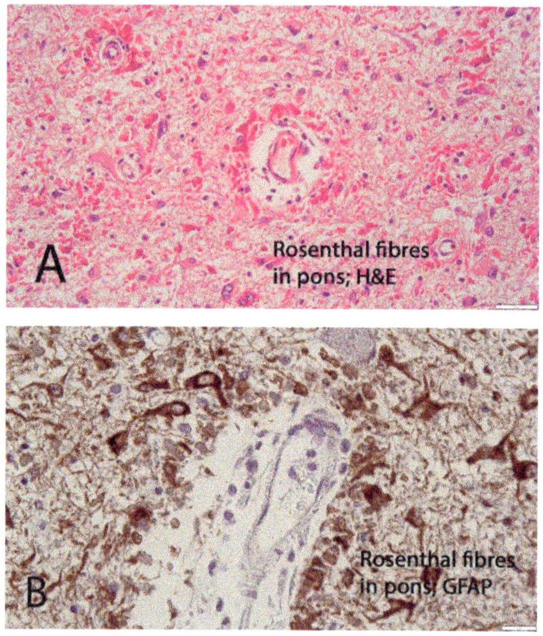

Fig. 5.3. Rosenthal fibers in the pons of a child dying with Alexander disease. (A) Hematoxylin-eosin (H&E) shows coarse glial processes that are strongly eosinophilic. (B) Glial fibrillary acidic protein (GFAP) confirms their origin from strongly immunoreactive astrocytes. These thick abnormal processes are especially abundant around blood vessels. Despite molecular abnormalities of the GFAP filaments, the antibody still recognizes them. Courtesy of Dr. E. Aronica of Amsterdam.

Alexander disease

Alexander disease is a prototype astrocytic disease because its etiology is an abnormal molecular structure of GFAP; hence, not only does it provide insight into the pathogenesis of this rare genetic/metabolic disease but also into physiological astrocytic functions in the normal developing CNS. Though *Rosenthal fibers* were first described histologically in 1898, their nature as thickened astrocytic processes filled with proliferated GFAP filaments in Alexander disease (Fig. 5.3) was revealed by immunocytochemical reactivity of GFAP and transmission electron microscopy in the mid-1980s (Eng, 1985; Choi, 1986).

This unique glial disease was described in 1949 by W. Stewart Alexander as a leukodystrophy, one of a group of progressive neurodegenerative metabolic white matter diseases (Alexander, 1949). The most severe form affects neonates, who exhibit failure-to-thrive, vomiting, gastro-esophageal reflux, and macrocephaly, leading to death in infancy (Lyon et al., 1996). More protracted juvenile and late adolescent forms also occur with slower progression, all characterized by developmental delay, spasticity, and epilepsy. A clinical trial of macrocephaly, seizures, and developmental delay in infancy and childhood is not universally present (Joung et al., 2023). Dysmyelination of white matter can be demonstrated by MRI to help confirm the diagnosis during life (Brenner et al., 2001; Wippold II et al., 2006; Srivastava et al., 2012; Kuhn and Cascella, 2022).

Because of the 15 mutations in the GFAP gene (Hsiao et al., 2005; Hagemann et al., 2006; Quinlan et al., 2007), the principal intermediate filament GFAP of mature astrocytes increases and results in dysmorphic astrocytes with coarse enlarged processes known as *Rosenthal fibers,* which are evident histologically and confirmed by immunocytochemistry for GFAP (Eliasson et al., 1999; Messing et al., 2012; Sosunov et al., 2018). These mutations primarily replace arginine with cysteine, resulting in increased cross-linkages of GFAP filaments and inter-mixture with other intermediate filaments, including vimentin within Rosenthal fibers (Chin and Goldman, 1996). Astrocytic end-feet that form the glia limitans on capillaries are disrupted, interfering with ion transport and water balance (Hsiao et al., 2005). Variable neuronal loss also occurs, indicating abnormal glial-neuronal interactions as well as activation of cell stress and inflammatory pathways, including the heat-shock protein α-B-crystallin (Eliasson et al., 1999; Sosunov et al., 2018). Astrocytic Rosenthal fibers are not pathognomonic of Alexander disease; however, because they also occur in old gliotic scars of cerebral infarcts, multiple sclerosis, cerebral contusions, and pilocytic astrocytomas (Chin and Goldman, 1996; Brenner et al., 2001; Sosunov et al., 2018) but they are not as extensive and prominent as in Alexander disease. Neither nestin nor vimentin, transitory intermediate filament proteins that precede GFAP in astrocytic maturation, are required for the formation of GFAP filaments (Eliasson et al., 1999).

GFAP is a type III intermediate filament protein transcribed from a 10 kb region on chromosome 17q21 in humans and chromosome 11 in the mouse (Hagemann et al., 2006). The genetic basis of Alexander disease by de novo mutations in four different GFAP residues was identified in 12 unrelated patients (Brenner et al., 2001; Quinlan et al., 2007). A murine model of Alexander disease mimics the human disease, including the formation of Rosenthal fibers and stress responses, with missense mutations homologous to two of the common mutations in human patients (Quinlan et al., 2007; Hagemann et al., 2018). Upregulation of α-B-crystallin occurs in Rosenthal fibers in both humans and mice (Rong et al., 2002; Der Perng et al., 2006; Tang et al., 2006; Quinlan et al., 2007; Messing et al., 2010; Hagemann et al., 2018; Messing, 2019; Viedma-Poyatos et al., 2022). GFAP mutants trigger the stress response, sometimes referred to as *GFAP toxicity*, and also show increased susceptibility to lipoxidation and mitochondrial dysfunction (Der Perng et al., 2006; Quinlan et al., 2007; Messing, 2019; Viedma-Poyatos et al., 2022). Gain of function due to a partial block of GFAP filament assembly can lead to the accumulation of an intermediate that participates in toxic interactions (Messing, 2019). GFAP astrocytopathy is also implicated as an important contributory cause of noninfectious meningoencephalomyelitis in children and adolescents (Uppal et al., 2023).

Until recently, treatment of patients with Alexander disease was largely symptomatic and palliative, including seizure control, but new metabolic strategies are devised to approach biochemical issues in at least slowing the progression of the disease, such as antisense suppression of GFAP (Rong et al., 2002; Hagemann et al., 2018; Messing, 2019).

Rett syndrome

Rett syndrome is a severe progressive neurodevelopmental disorder with an array of abnormal clinical neurological findings and a high risk of autism and epilepsy. The incidence of 1:15,000–22,000 makes it the second most frequent cause of severe cognitive disability in girls of all ethnicities (Banerjee et al., 2019; Vidal et al., 2019; Kwon, 2020). It occurs almost exclusively in girls because of a defect in the gene *MECP2* at chromosomal locus Xq28, though mutations in other genes, such as *CDKL5* and *FOXG1*, can also be associated and may modify *MECP2* expression. More than 500 distinct pathogenic mutations of the *MECP2* gene are described in Rett syndrome, mainly involving exons 3 and 4 and the C-terminal region (Gold et al., 2018; Banerjee et al., 2019; Sun et al., 2023). Epilepsy is a feature of the majority of patients with Rett syndrome, and several seizure types may occur, though generalized tonic–clonic seizures, often refractory to anti-seizure medications, are the most frequent (Cardoza et al., 2011). Various mutations in the *MECP2* gene have different influences on the epilepsy (Nisenkorn et al., 2015).

Disturbances in microRNA expression can impact the transcriptional regulator *MECP2* (Horvath et al., 2022). *MECP2* mutations in Rett syndrome impair genetic expression of human astrocytes and genes associated with astrocytic development. By negatively affecting their structural maturation and energy metabolism mediated by dysfunctional mitochondria, the important astrocytic/neuronal interaction for synaptic function is impaired (Medrihan et al., 2008).

A murine model is a transcriptional repressor methyl-CpG-binding protein 2 (*Mecp2*), also localized at Xq28, and is clinically expressed in the mouse by abnormal development and behavior (Medrihan et al., 2008; Banerjee et al., 2019; Vashi and Justice, 2019). GABAergic but not glycinergic synaptic transmission is strongly depressed in these mutant mice due to both decreased presynaptic GABA release and reduced levels of vesicular transmitter transporter and postsynaptic GABA(A)-receptor subunits, thus impairing inhibitory neurotransmission (Medrihan et al., 2008). Synaptic maturation at cortical projections to the lateral amygdala is impaired, as are GABAergic synapses in the brainstem; thus, the murine model of Rett syndrome involves synaptic circuitry at both cerebral cortical and subcortical levels and is described as a model synaptic disorder (Medrihan et al., 2008; Banerjee et al., 2019; Vashi and Justice, 2019). The murine model of Rett syndrome is not only exemplary of the value of animal models of human genetic diseases but also offers approaches to precision medicine in finding means of treatment by interfering with the pathogenesis (Gambino et al., 2010; Banerjee et al., 2019; Vashi and Justice, 2019).

Routine neuropathological studies of the brains of patients with Rett syndrome fail to demonstrate any major malformations, neuronal loss, or atrophy. However, brain weight is usually reduced to 60%–80% of age-matched controls (Kwon, 2020). Perturbed lipid metabolism is demonstrated in both humans and the murine model (Kyle et al., 2018). At a subcellular level, increased mitochondrial mass and cytosolic redox imbalance in hippocampal astrocytes are described in the murine Rett model (Bebensee et al., 2017). Cells of oligodendrocytic lineage may contribute to impaired myelin-related proteins in the murine model (Nguyen et al., 2013), but differentiating multipotential stem cells derived from Rett patients evolve in cell culture more as astrocytic lineage (Andoh-Noda et al., 2015). Similarly, astrocytic progenitor cells cultured with *Mecp2* display altered microtubule dynamics and growth rate, whereas neurons from the same subjects do not exhibit these impairments (Stobart and Anderson, 2013). Glial dysfunction in *MeCP2*-mutant mice (but not in cell cultures) is also documented to affect astrocytes, oligodendrocytes, and microglia (the latter are not glial cells from neuroepithelium) (Sharma et al., 2018; Kahanovitch et al., 2019). Microglia in Rett syndrome contribute to increased glutamate production and release in the hippocampus and damage dendrites and synapses in *MeCP2*-mutant mice (Maezawa and Jin, 2010). Microglia in both human patients and murine models of Rett syndrome are smaller and morphologically dysplastic (Armstrong, 2005; Kahanovitch et al., 2019). Astrocytic GFAP expression is not affected in Rett syndrome; however (Armstrong, 2005), though *Mecp2* directly binds several glial-specific genes, including GFAP and the S-100β protein (Kahanovitch et al., 2019).

Fragile-X syndrome

This monogenetic disorder is a cause of intellectual impairment and autistic spectrum disorder due to repetition of CGT triplets and their methylation that entails the silencing of the *FMR1* (fragile-X mental retardation protein-1) gene (Salcedo-Arellano et al., 2020). Absence or reduction of *FMR1* results in defects in neuronal differentiation and migration and in brain-derived neurotrophic factor (Castrén, 2016). Astrocytes are also affected by fragile-X syndrome, which impairs synaptogenesis and synaptic structure (Cheng et al., 2012; Jacobs et al., 2012, 2016). Abnormal dendritic morphology and altered synaptic protein expression occur; inflammatory cytokines, such as interleukin-6 (IL-6), are also produced secondary to abnormal astrocytic secretion of tenascin-C, an endogenous ligand of toll-like receptor 4 that induces proinflammatory cytokines (Kfasovska and Doering, 2018).

Aicardi-Goutières syndrome

The syndrome consists of a subacute but progressive encephalopathy beginning in early infancy that results in most, but not all, cases of severe intellectual and neurological disability (Sharma et al., 2018). Over time, about 40% develop chilblain cutaneous lesions on the fingers, toes, and ears, and a rarer subtype forms hepatosplenomegaly, elevated hepatic enzymes, and thrombocytopenia; thus, the clinical phenotype is broad (Crow, 2005). Neuropathologically and by neuroimaging, calcifications are demonstrated in the basal ganglia and white matter, and the white matter myelination is the pattern of dysmyelination or a leukodystrophy. Genetic and molecular studies have established that this disease is autosomal recessive and several genes are involved (Sase et al., 2018).

Both astrocytes and microglia are involved in pathogenesis (Cuadrado et al., 2013; Sase et al., 2018). Astrocytes are one of the major sources of the cytokine IFN (interferon-α) in the central nervous system and its downstream signaling cascades, and there is overproduction of this cytokine in Aicardi-Goutières syndrome and contribute to the pathogenetic mechanism of leukodystrophy and microangiopathy and downstream effects even after interferon is no longer active (Cuadrado et al., 2013; Sase et al., 2018). DNA damage is a major factor of neurotoxic inflammation in astrocytes in Aicardi-Goutières syndrome (Giordano et al., 2022). A murine model of this human disease without immune cell infiltration in the brain is available with similar neurodegenerative changes and high levels of distinct interferon-stimulated genes in the absence of interferon (Wiley et al., 2022). Brain microglial activation occurs and further contributes to the encephalopathy (La Maestra et al., 2018; West et al., 2019).

Down syndrome

Adults with Down syndrome (trisomy-21) undergo a neurodegenerative process that closely resembles Alzheimer disease. It is accompanied by severe glial as well as neuronal loss in the cerebral cortex, anterior thalamic nuclei, and basal ganglia (Karlsen and Pakkenberg, 2011; Perry et al., 2019). This glial loss contrasts with gliosis or astrocytic proliferation in the presence of neuronal loss in most other neurodegenerative conditions, including hippocampal sclerosis. Glial cell abnormalities in infants and children with Down syndrome are sparse, but in fetuses of 14 weeks gestation to term, the number of both GFAP-reactive astrocytes and Olig2-reactive oligodendrocytes in temporal lobe white matter was reduced in early and mid-gestation compared with controls, reflecting impaired differentiation of these glial cells (Kanaumi et al., 2013). Such findings merit closer examination, especially in the postnatal period.

Fetal alcohol spectrum disorder

Fetal alcohol spectrum disorder (FASD) is neither an inherited genetic disease nor a postzygotic somatic mutation of the mTOR signaling pathway, but rather is an epigenetic developmental disorder induced by fetal exposure to alcohol from maternal consumption in early pregnancy. Its clinical expression is mainly developmental delay, behavioral disorders, and a high risk of autism; epilepsy is only a mildly increased risk. The metabolic basis of FASD is abnormal DNA and histone methylation, particularly in imprinted regions of the genome, affecting transcription, which is further fine-tuned by altered microRNA expression (Ungerer et al., 2013; Kleiber et al., 2014).

Neuroanatomic lesions are demonstrated both by MRI and neuropathologically, involving not only the cerebral cortex but also the basal ganglia and cerebellum. The corpus callosum and anterior commissure may be hypoplastic. At the cellular level, both neurons and glial cells, both astrocytes and oligodendrocytes, exhibit delayed maturation, apoptosis, and degenerative features (Roebuck et al., 1998; Wilhelm and Guizzetti, 2016).

Other neuroembryologic and neuroanatomic results of FASD are delayed synaptogenesis and myelination, altered axonal growth and guidance, and defective neural crest migration, which explains the typical facial dysmorphism (Sarnat and Flores-Sarnat, 2014; Mathews et al., 2021). Effects on glia may affect their ability to modulate neuronal development, survival, and function (Guizzetti et al., 2014). Furthermore, ethanol also induces microglial activation and oxidative stress leading to neuronal cellular death, mediated by impairing extracellular vesicles and spreading apoptotic factors, including complement protein C1q (Guizzetti et al., 2014; Mukherjee et al., 2020).

Rodent models of FASD are created by administering ethanol prenatally to mice and rats. These experiments show decreased terminal dendritic ramifications and alterations or reduction in dendritic spines of pyramidal cells of both the neocortex and the CA1 sector of the hippocampus, using Golgi impregnations and transmission electron microscopy (Davies and Smith, 1981; Smith and Davies, 1990; Cul et al., 2010; Hamilton et al., 2010; Guizzetti et al., 2014). Astrocytes are also damaged, as shown in human cell cultures (Lokhorst and Druse, 1993; Pascual and Guerri, 2007).

OTHER NEUROPATHOLOGICAL FEATURES THAT INVOLVE GLIAL CELLS

Satellitosis

Satellite cells are glial cells that become tightly adherent to the somatic membrane of neurons, mostly in the cerebral cortex but also in Ammon's horn of the hippocampus and in heterotopic neurons in the U-fiber layer and deep subcortical white matter (Fig. 5.4A and B). The adhesion of glial cell bodies to neurons is very different from astrocytic processes' end-feet contacting neuronal somata. Satellitosis is prominent in brain resections of epileptic foci but also occurs in a variety of encephalopathies with or without seizures, in cerebral ischemic and inflammatory encephalopathies, in regions adjacent to cerebral tumors, in cerebral contusions, and in many other conditions. Because of this nonspecificity, many neuropathologists attribute little importance to it and do not even mention the phenomenon in their biopsy or autopsy reports.

Nevertheless, satellitosis can affect epileptogenesis of individual neurons because these glial cells occupy space on the neuronal soma and thus displace axosomatic synapses, thus shifting the ratio of excitation in axodendritic glutamatergic synapses and inhibition in the GABAergic axosomatic synapses toward excitation that enhances epileptic discharges. Their role in modulating metabolic activity in their adjacent neurons is largely unknown at this time. In satellitosis, keratan sulfate, an extracellular matrix proteoglycan that normally is adherent to neurons and influences the type of synapse that forms, envelops both the neuronal soma and adherent glial satellites (Sarnat HB, Yu W, Unpublished).

Furthermore, ultrastructural examination demonstrates that at epileptic foci, many well-preserved satellite glial cells are adherent to degenerating or dead neurons (Fig. 5.4C and D). Such satellite cells are usually astrocytes reactive for glial fibrillary acidic protein, but often also express vimentin or nestin, which indicates that they are immature or recently proliferated astrocytes. Oligodendrocytes also can contribute to satellitosis. The death of such neurons poses a question whether the satellites contribute to their removal from the cortex because they were epileptogenic neurons that were removed for the overall benefit of the functional cortex, or whether satellites are attempting to protect stressed neurons, sometimes without success: whether satellite cells are "guardian angels" or "angels of death" (Sarnat et al., 2010).

α-B-crystallin as a glial marker of epileptic foci

α-B-crystallin is a small heat-shock chaperone protein that normally is not expressed immunocytochemically in the CNS but becomes upregulated in regions of metabolic stress, such as in zones of chronic ischemia, cerebral contusions, near to cerebral neoplasms, and at epileptic foci (Becerra-Hernández et al., 2022; Tedesco et al., 2022). It normally is first expressed in the human fetal telencephalon at mid-gestation by a subset of progenitor cells (Kida et al., 2010). The overexpression of α-B-crystallin is demonstrated in glial cells within the gray matter of the cortex and in both acute and chronic glial cells in the subcortical white matter, often with a gradient of disappearance at 2–3 cm away from the epileptic focus (Sarnat and Flores-Sarnat, 2009). The upregulation of α-B-crystallin in the Rosenthal fibers of Alexander disease is well documented (Der Perng et al., 2006; Quinlan et al., 2007) and was discussed earlier. It also may be overexpressed throughout the brain in astrocytes secreting neurotoxic cytokines, as in several of the diseases discussed earlier.

Loss of subpial granular glial layer of Brun

The subpial granular glial layer of Brun (SGL) is a transitory layer of astrocytes overlying the marginal zone of the fetal cerebral cortex just beneath the pial surface between 13 and 24 weeks gestation (Fig. 5.5) (Sarnat, 1992, 2021). Though initially described briefly in 1874 by Boll (Boll, 1874) and in 1910 by Ranke (Ranke, 1910), this transitory cellular layer over the cerebral cortex was best defined by Brun in 1965 (Brun, 1965; Judas and Pletikos, 2010). Its cells originate from the basal peri-olfactory subventricular zone and migrate tangentially beneath the pia. Programmed cell death may, at least in part, contribute to the elimination of the SGL. During fetal life, the SGL probably contributes, together with the glia limitans (Sarnat, 1992) and pial membrane, to preventing overmigration. Subpial astrocytes have mature morphology, but unlike most astrocytes in gray or white matter, they retain expression of vimentin throughout fetal life, similar to radial glial fibers. The layer is deficient in brains with holoprosencephaly and lissencephaly, and glioneuronal heterotopia is frequent as nodules in the leptomeninges attached to the cortex by a thick or thin stalk (Sarnat, 1992; Sarnat and Flores-Sarnat, 2001).

Fetal astrocytic synthesis of keratan sulfate to guide axonal projections and insulate fascicles

Keratan sulfate is a proteoglycan of the extracellular space of the central nervous system that is secreted by fetal astrocytes and is an

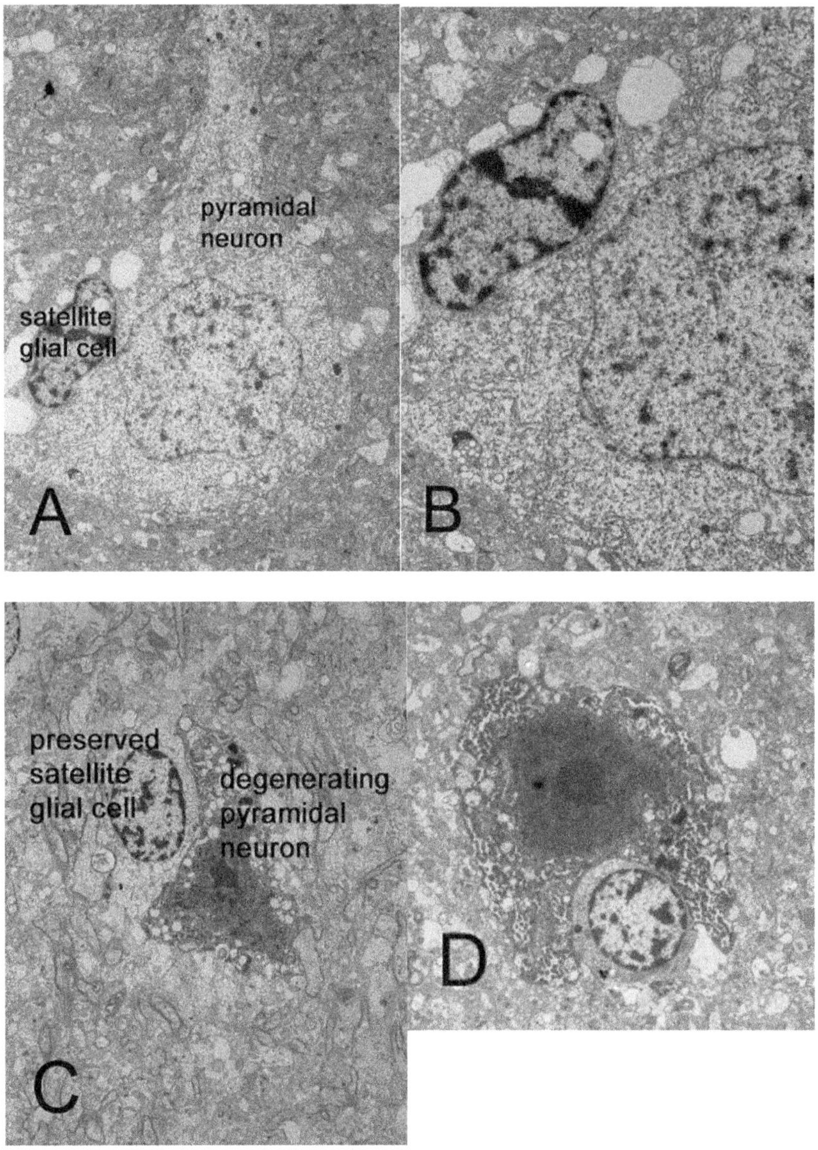

Fig. 5.4. Transmission electron microscopy (EM) of surgical neocortical resections in (A, B) parietal lobe of a 15-month-old infant and (C, D) temporal lobe of a 7-year-old boy, both with focal cortical dysplasia type Ia and refractory focal epilepsy. (A, B) Low and higher magnification EM showing satellitosis. A well-preserved glial cell body (not end-feet) is tightly adherent to a well-preserved normal pyramidal neuron of cortical layer 5, without space between their plasma membranes. Inhibition is diminished by displacing normal axosomatic synapses. (C, D) Two examples of well-preserved satellite glial cells adherent to dying neurons that exhibit dense coagulation of nuclear chromatin and of cytoplasmic organelles and also cytoplasmic vacuolation. The satellite cell in (D) indents the neuron for greater adhesion. Satellitosis of preserved neurons similar to (A, B) was also seen in this case (not illustrated). Neuronal degeneration poses the query of whether satellite cells contribute to neuronal death at epileptic foci or try to prevent apoptosis. Original magnifications (A, C, D) 2500; (B) 4000. Reproduced from Sarnat HB, Flores-Sarnat L (2021). Excitatory/inhibitory synaptic ratios in polymicrogyria and down syndrome help explain epileptogenesis in malformations. Pediatr Neurol 116: 41–54.

essential component of axonal pathfinding. It surrounds axonal fascicles to insulate them so that the fibers they contain are "pure" in the sense that they contain only axons of the same origin that secrete the same neurotransmitter and prevent them from exiting the fascicle before reaching their destination and also prevent axons from other sites *en route* from entering the fascicles (Sarnat, 2019). Keratan sulfate is closely related in molecular structure to another proteoglycan, chondroitin sulfate (aggrecan), mainly in cartilage, with cross-immunoreactivity that may confuse the identity of the molecules in tissue sections. In cerebral malformations, aberrant axonal projections are frequent that result in abnormal synaptic circuitry and sometimes epilepsy. In alobar holoprosencephaly, keratan sulfate not only surrounds axonal fascicles but also ensheaths individual axons, which may explain why 40% of infants and children with this severe cortical dysplasia do not have seizures (Sarnat et al., 2021). Keratan sulfate also repulses glutamatergic axons at the plasma membrane of the neuron and enhances GABAergic synapse formation; because this

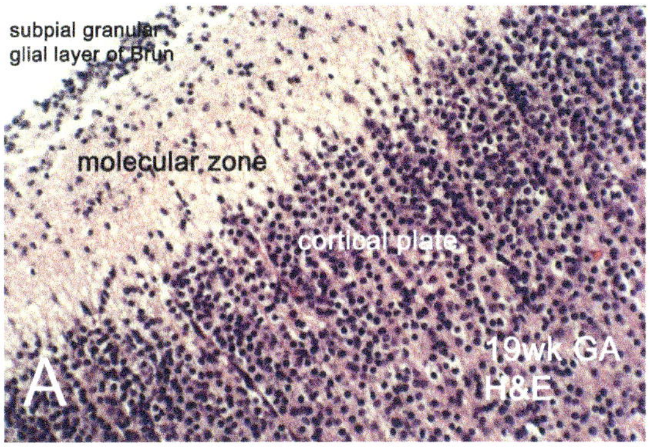

Fig. 5.5. Subpial granular glial layer over the surface of the cerebral cortex of a fetus of 19 weeks gestational age, a transitory structure of the fetal brain. The histological architecture of the cortical plate in the first half of gestation is radial microcolumnar rather than the more mature horizontal lamination of later gestation. Reproduced from reference Sarnat HB (1992). Cerebral dysgeneses: embryology and clinical expression, Oxford University Press, New York.

molecule surrounds the neuronal membrane except dendritic spines and terminal ramifications, it also explains why axo-dendritic synapses are excitatory and axo-somatic synapses are inhibitory in the cerebral cortex (Sarnat, 2019).

CONCLUSIONS

Glial cells play a major role in the pathogenesis of several genetic and epigenetic diseases by modulating dendritic and synaptic development and integrity and by secreting neurotoxic cytokines that have clinical neurological expression. Neuropathological investigation of affected patients should include immunoreactivities to intermediate filament proteins and inflammatory markers. Study of mRNAs and intensive genetic profiles are useful as well. Glial abnormalities contribute to some cerebral malformations. The purpose of the recognition of glial cells in developmental disorders of the nervous system is to enable personalized therapies that delay or arrest disease progression.

ACKNOWLEDGMENTS

We are grateful to Dr. Laura Flores-Sarnat (Calgary, Canada) and Dr. Violeta Medina Crespo (Mexico City, Mexico) for their valuable comments during the preparation of this manuscript.

ETHICS COMMITTEE APPROVAL

This study was approved by the Conjoint Health Research Ethics Committee of the University of Calgary and Alberta Health Services.

FUNDING

Alberta Children's Hospital Research Institute, Grant #60-28450, Project 10027949 (Dr. H.B. Sarnat).

DISCLOSURES

The authors have no financial disclosures or conflicts of interest to declare.

In our laboratories, fetal and postnatal autopsy tissues are always treated with respect.

References

Alexander WS (1949). Progressive fibrinoid degeneration of fibrillary astrocytes associated with mental retardation in a hydrocephalic infant. Brain 72: 373–381.

Allen NJ, Eroglu C (2017). Cell biology of astrocyte-synapse interactions. Neuron 96: 697–708.

Ammonthumkandy A, Ravina K, Wolseley V et al. (2022). Altered adult neurogenesis and gliogenesis in patients with mesial temporal lobe epilepsy. Nat Neurosci 25: 493–503.

Anderson MA, Burda JE, Ren Y et al. (2016). Astrocyte scar formation aids central nervous system axon regeneration. Nature 532: 195–200.

Andoh-Noda T, Akamatsu W, Miyake K et al. (2015). Differentiation of multipotent neural stem cells derived from Rett syndrome patients is biased toward the astrocytic lineage. Mol Brain 8: 31. https://doi.org/10.1186/s13041-015-0121-2.

Armstrong DD (2005). Neuropathology of Rett syndrome. J Child Neurol 20: 747–753.

Banerjee A, Miller MT, Li K et al. (2019). Toward a better diagnosis and treatment of Rett syndrome: a model synaptic disorder. Brain 142: 239–248.

Barres BA (2008). The mystery and magic of glia: a perspective on their roles in health and disease. Neuron 60: 430–440.

Bebensee DF, Can K, Müller M (2017). Increased mitochondrial mass and cytosolic redox imbalance in hippocampal astrocytes of a mouse model of Rett syndrome: subcellular changes revealed by ratiometric imaging of JC-1 and roGFP1 fluorescence. Oxid Met Cell Longev 3064016. https://doi.org/10.1155/2017/3064016.

Becerra-Hernández LV, Escobar-Betancourt MI, Pimienta-Jiménez HJ et al. (2022). Crystallin alpha-B overexpression as a possible marker of reactive astrogliosis in human cerebral contusions. Front Cell Neurosci 16: 838551.

Bevacqua M, Baldo F, Pastore S et al. (2019). Off-label use of sirolimus and everolimus in a pediatric center: a case series and review of the literature. Pediatr Drugs 21: 185–193.

Boll F (1874). Die Histologie und Histogenese der nervösen Centralorgane. Arch Psychiatr Nervenkr 4: 1–138.

Bosworth AP, Allen NJ (2017). The diverse actions of astrocytes during synaptic development. Curr Opin Neurobiol 47: 38–43.

Brenner M, Johnson AB, Boespflug-Tanguy O et al. (2001). Mutations in GFAP, encoding glial fibrillary acidic protein, are associated with Alexander disease. Nat Genet 27: 117–120.

Brun A (1965). The subpial granular layer of the fetal cerebral cortex in man. Acta Pathol Microbiol Scand Suppl 179: 1–71.

Burda JE, O'Shea TM, Ao Y et al. (2022). Divergent transcriptional regulation of astrocyte reactivity across disorders. Nature 606: 557–564.

Cameron EG, Nahmou M, Toth AB et al. (2024). A molecular switch for neuroprotective astrocyte reactivity. Nature 626: 574–782.

Cardoza B, Clarke A, Wilcox J et al. (2011). Epilepsy in Rett syndrome: association between phenotype and genotype, and implications for practice. Seizure 20: 646–649.

Castrén ML (2016). Cortical neuronogenesis in fragile X syndrome. Front Biosci 8: 160–168.

Ceglia R, Ledonne A, Litvin DG et al. (2023). Specialized astrocytes mediate glutamatergic gliotransmission in the CNS. Nature 622: 120–129.

Chang Y-T, Luna-Figueroa E, Woo J et al. (2023). Inhibitory input directs astrocyte morphogenesis through glial GABAABR. Nature 617: 369–376.

Chen ZP, Wang S, Zhao X et al. (2023). Lipid-accumulated reactive astrocytes promote disease progression in epilepsy. Nat Neurosci 26.

Cheng C, Sourial M, Doering LC (2012). Astrocytes and developmental plasticity in fragile X. Neural Plast 2012: 197491.

Chin SS, Goldman JE (1996). Glial inclusions in CNS degenerative diseases. J Neuropathol Exp Neurol 55: 499–508.

Choi BH (1986). Glial fibrillary acidic protein in radial glia of early human fetal cerebrum: a light and electron microscopic immunoperoxidase study. J Neuropathol Exp Neurol 45: 408–418.

Clark IC, Wheeler MA, Lee H-G et al. (2023). Identification of astrocyte regulators by nucleic acid cytometry. Nature 614: 326–333.

Crino PB (2020). Knockout of a tuberous sclerosis gene highlights role of glia in epileptogenesis. Epilepsy Curr 20: 139–141.

Crow YJ (2005). Aicardi-Goutières syndrome. GeneReviews updated 2016 Nov 22.

Cuadrado E, Jansen MH, Anink J et al. (2013). Chronic exposure of astrocytes to interferon-α reveals molecular changes related to Aicardi-Goutières syndrome. Brain 136: 245–258.

Cul ZJ, Zhao KB, Zhao HJ et al. (2010). Prenatal alcohol exposure induces long-term changes in dendritic spines and synapses in the mouse visual cortex. Alcohol 45: 312–319.

Danbolt NC (2001). Glutamate uptake. Prog Neurobiol 65: 1–105.

Davidson KGV, Rash JE (2012). Oligodendrocytes: gateway to the panglial syncytium. In: E Seemes, DC Spray (Eds.), Astrocytes wiring of the brain. CRC Press, Boca Raton, Florida p. 228.

Davies DL, Smith DE (1981). A Golgi study of mouse hippocampus CA1 pyramidal neurons following perinatal ethanol exposure. Neurosci Lett 26: 49–54.

Del Rio HP (1916). Estudios sobre el centrosoma de las células nerviosas y neuróglicas de los vertebrados, en sus formas normal y anormales. Trab Inst Cajal Invest Biol Univ Madrid 14: 117–153. (in Spanish).

Der Perng M, Su M, Wen SF et al. (2006). The Alexander disease-causing glial fibrillary acidic protein mutant, R416W, accumulates into Rosenthal fibers by a pathway that involves filament aggregation and the association of alpha-B-crystallin and HSP27. Am J Hum Genet 79: 197–213.

Doron A, Rubin A, Benmelech-Chovav A et al. (2022). Hippocampal astrocytes encode reward location. Nature 609: 772–778.

Eliasson C, Sahlgren C, Berthold CH et al. (1999). Intermediate filament protein partnership in astrocytes. J Biol Chem 274: 23996–24006.

Eng LF (1985). Glial fibrillary acidic protein (GFAP): the major protein of glial intermediate filaments in differentiating astrocytes. J Neuroimmunol 8: 203–214.

Fregeac J, Moriceau S, Poli A et al. (2020). Loss of the neurodevelopmental disease-associated gene miR-146a impairs neural progenitor differentiation and causes learning and memory deficits. Molecular. Autism 11 (1): 22.

Gambino F, Khelfaoui M, Poulain B et al. (2010). Synaptic maturation at cortical projections to the lateral amygdala in a mouse model of Rett syndrome. PloS One 5: e1399.

Gilles FH (1976). Myelination in the neonatal brain. Brain Pathol 7: 244–248.

Giordano AMS, Luciani M, Gatto F et al. (2022). DNA damage contributes to neurotoxic inflammation in Aicardi-Goutières syndrome astrocytes. J Exp Med 219: e20211121.

Gold WA, Krishnarajy R, Ellaway C et al. (2018). Rett syndrome: a genetic update and clinical review focussing on comorbidities. ACS Chem Nerosci 9: 167–176.

Guizzetti M, Zhang X, Goeke C et al. (2014). Glia and neurodevelopment: focus on fetal alcohol spectrum disorders. Front Pediatr 2: 123. https://doi.org/10.3389/fped.2014.00123. erratum. Front Pediatr 2015;3:27.

Hagemann TL, Connor JX, Messing A (2006). Alexander disease-associated glial fibrillary acidic protein mutations in mice induce Rosenthal fiber formation and a white matter stress response. J Neurosci 26: 11162–11173.

Hagemann TL, Powers B, Mazur C et al. (2018). Antisense suppression of glial fibrillary acidic protein as a treatment for Alexander disease. Ann Neurol 83: 27–39.

Hamilton GF, Whitcher LT, Klintsova AY (2010). Postnatal binge-like alcohol exposure decreases dendritic complexity while increasing the density of mature spines in mPFC layer II/III pyramidal neurons. Synapse 64: 127–135.

Hayes LN, An K, Carloni E et al. (2022). Prenatal immune stress blunts microglia reactivity, impairing neurocircuitry. Nature 610: 327–334.

Honig LS, Herrmann K, Shatz CJ (1996). Developmental changes revealed by immunohistochemical markers in human cerebral cortex. Cerebr Cortex 6: 794–806.

Horvath PM, Piazza MK, Kavalali ET et al. (2022). *MeCP2* loss-of-function dysregulates microRNAs regionally and disrupts excitatory/inhibitory synaptic transmission balance. Hippocampus 32: 610–623.

Hsiao VC, Tian R, Long H et al. (2005). Alexander disease mutation of GFAP causes filament disorganization and decreased solubility of GFAP. J Cell Sci 118: 2057–2065.

Huang XY, Hu QP, Shi HY et al. (2021). Everolimus inhibits PI3K/Akt/mTOR and NF-kB/IL-6 signaling and protects seizure-induced brain injury in rats. J Chem Neuroanat 114: 101960.

Hughes EG, Appel B (2016). The cell biology of CNS myelination. Curr Opin Neurobiol 39: 93–100.

Jacobs SD, Cheng C, Doering LC (2012). Probing astrocyte function in fragile-X syndrome. Results Probl Cell Differ 59: 15–31.

Jacobs SD, Cheng C, Doering LC (2016). Hippocampal neuronal subtypes develop abnormal dendritic arbors in the presence of fragile-X astrocytes. Neuroscience 324: 202–217.

Joung J, Gallison K, Sollee J et al. (2023). Acquisition and loss of developmental milestones and time to disease-related outcomes in cerebral Alexander disease. J Child Neurol 38: 672–678.

Judas M, Pletikos M (2010). The discovery of the subpial granular layer in the human cerebral cortex. Transl Neurosci 1: 255–260.

Kahanovitch U, Patterson KC, Hernández R et al. (2019). Glial dysfunction in *MeCP2* deficiency models: implications for Rett syndrome. Int J Mol Sci 20: 3813. https://doi.org/10.3390/ijms.20153813.

Kanaumi T, Milenkovic I, Adle-Biassette H et al. (2013). Non-neuronal cell responses differ between normal and down syndrome developing brains. Int J Dev Neurosci 31: 796–803.

Karlsen AS, Pakkenberg B (2011). Total numbers of neurons and glial cells in cortex and basal ganglia of aged brains with down syndrome—a stereological study. Cereb Cortex 21: 2519–2524.

Kfasovska V, Doering LC (2018). Regulation of IL-6 secretion by astrocytes via TLR4 in the fragile-X model. Front Mol Neurosci 11: 272.

Kida E, Wierzba-Bobrowicz T, Palminiello S et al. (2010). Molecular chaperone αB-crystallin is expressed in the human fetal telencephalon at midgestation by a subset of progenitor cells. J Neuropathol Exp Neurol 69: 745–759.

Kielbinski M, Gzielo K, Soltys Z (2016). Review: roles for astrocytes in epilepsy: insights from malformations of cortical development. Neuropathol Appl Neurobiol 42: 593–606.

Kinney HC, Brody BA, Kloman AS et al. (1988). Sequence of central nervous system myelination in human infancy. II. Patterns of myelination in autopsied infants. J Neuropathol Exp Neurol 47: 217–234.

Kleiber ML, Diehl EJ, Laufer BI et al. (2014). Long-term genomic and epigenomic dysregulation as a consequence of prenatal alcohol exposure: a model for fetal alcohol spectrum disorders. Front Genet 5: 161. eCollection 2014.

Kriegstein A, Álvarez-Buyya A (2009). The glial nature of embryonic and adult neural stem cells. Annu Rev Neurosci 32: 149–184.

Kuhn J, Cascella M (2022). Alexander disease, StatPearls (Internet).

Kwon JM (2020). Rett syndrome. In: RM Kliegman, JW St Geme III, NJ Blum, RC Tasker, SS Shah, KM Wilson, RE Behrman (Eds.), Nelson textbook of pediatrics, 21st edn. Elsevier, Philadelphia, pp. 3192–3194.

Kyle SM, Vashi N, Justice MJ (2018). Rett syndrome: a neurological disorder with metabolic components. Open Biol 8: 170216. https://doi.org/10.1098/rsob.170216.

La Maestra S, Frosina G, Micale RT et al. (2018). Brain microglía activation induced by intracranial administration of oligonucleotides and its pharmacological modulation. Drug Deliv Transl Res 8: 1345–1354.

Lentini C, d'Orange M, Marichal N et al. (2021). Reprogramming reactive glia into interneurons reduces chronic seizure activity in a mouse model of mesial temporal lobe epilepsy. Cell Stem Cell 28: 2104–2121.

Ling E, Nemesh J, Goldman M et al. (2024). A concerted neuron-astrocyte program declines in ageing and schizophrenia. Nature 627: 604–611.

Lokhorst DK, Druse MJ (1993). Effects of ethanol on cultured fetal astroglia. Alcohol Clin Exp Res 17: 810–815.

Lubarski K, Mania A, Malecki P et al. (2022). Inflammatory markers combined with metalloproteinase-9, neopterin, and S100B concentrations may indicate the pathogenesis of central nervous system diseases in children. J Child Neurol 37: 707–716.

Lyon G, Adams RD, Kolodny EH (1996). Neurology of hereditary metabolic diseases of children, second edn. McGraw-Hill, NY, pp. 74–76.

Maezawa I, Jin L-W (2010). Rett syndrome microglia damage dendrites and synapses by the elevated release of glutamate. J Neurosci 30: 5346–5356.

Mathews E, Dewees K, Diaz D et al. (2021). White matter abnormalities in fetal alcohol spectrum disorders: focus on axon growth and guidance. Exp Biol Med (Maywood) 246: 812–821.

McDaniel SS, Wong M (2011). Therapeutic role of mammalian target of rapamysin (mTOR) inhibition in preventing epileptogenesis. Neurosci Lett 497: 231–239.

McNamara NB, Munro DAD, Bestard-Cuche N et al. (2023). Microglia regulate central nervous system myelin growth and integrity. Nature 613: 120–129.

Medrihan L, Tantalaki E, Aramuni G et al. (2008). Early defects of GABAergic synapse in the brain stem of a MeCP2 mouse model of Rett syndrome. J Neurophysiol 99: 112–121.

Messing A (2019). Refining the concept of GFAP toxicity in Alexander disease. J Neurodev Disord 11: 27. https://doi.org/10.1186/s11689-019-9290-0.

Messing A, LaPash Daniels CM, Hagemann TL (2010). Strategies for treatment in Alexander disease. Neurotherapeutics 7: 507–515.

Messing A, Brenner M, Feany MB et al. (2012). Alexander disease. J Neurosci 32: 5017–5023.

Mühlebner A, Bongaarts A, Sarnat HB et al. (2019). New insights into a spectrum of developmental malformations related to mTOR dysregulations: challenges and perspectives. J Anat 235: 521–542.

Mukherjee S, Cabrera MA, Boyadjieva NI et al. (2020). Alcohol increases exosome release from microglia to promote complement C1q-induced cellular death of proopiomelanocortin neurons in the hypothalamus in a rat model of fetal alcohol spectrum disorders. J Neurosci 40: 7965–7979.

Nguyen MVC, Felice CA, Du F et al. (2013). Oligodendrocyte lineage cells contribute unique features to Rett syndrome neuropathology. J Neurosci 33: 18764–18774.

Nisenkorn A, Levy-Drummer R, Bondi O et al. (2015). Epilepsy in Rett syndrome: lessons from the Rett networked database. Epilepsia 56: 569–576.

Ollivier M, Soto JS, Linker KE et al. (2024). *Crym*-positive striatal astrocytes gate perseverative behaviour. Nature 627: 358–366.

Overwater IE, Rietman AB, Bindels-de Heus K et al. (2016). Sirolimus for epilepsy in children with tuberous sclerosis complex: a randomized trial. Neurology 87: 1011–1108.

Pascual M, Guerri C (2007). The peptide NAP promotes neuronal growth and differentiation through extracellular signal-regulated protein kinase and Akt pathways, and protects neurons co-cultured with astrocytes damaged by ethanol. J Neurochem 103: 557–568.

Perry JC, Pakkenberg B, Vann SD (2019). Striking reduction in neurons and glial cells in anterior thalamic nuclei of older patients with down syndrome. Neurobiol Aging 75: 54–61.

Peterson AR, Binder DK (2020). Astrocyte glutamate uptake and signaling as novel targets for antiepileptogenic therapy. Front Neurol 11: 1006. https://doi.org/10.3389/fneur.2020.01006.

Prabowo A, Anink J, Lammens M et al. (2013). Fetal brain lesions in tuberous sclerosis complex: TORC1 activation and inflammation. Brain Pathol 23: 45–59.

Prada I, Gabrielli M, Turola E et al. (2018). Glia-to-neuron transfer of miRNAs via extracellular vesicles: a new mechanism underlying inflammation-induced synaptic alterations. Acta Neuropathol 135: 529–550.

Pulliero A, Marengo B, Fenoglio D et al. (2014). Prevention of lymphocyte neurotoxic effects by microRNA delivery. MicroRNA 2: 187–193.

Putkonen N, Laiho A, Ethell D et al. (2020). Urine microRNA profiling displays miR-125a dysregulation in children with fragile X syndrome. Cells 9: 289.

Quinlan RA, Brenner M, Goldman JE et al. (2007). GFAP and its role in Alexander disease. Exp Cell Res 313: 2077–3087.

Ramírez AE, Gil-Jaramillo N, Tapias MA et al. (2022). MicroRNA: a linking between astrocyte dysfunction, mild cognitive impairment, and neurodegenerative diseases. Life 12: 1439. https://doi.org/10.3390/life12091439.

Ramón y Cajal S (1995). L'histologie du système nerveux de l'homme et des vértébrés. Paris: Masson. 1909-1911, Reprinted in English translation, histology of the nervous system of man and vertebrates (2 volumes) Oxford University Press, New York, Oxford.

Ranke O (1910). Beiträge zur Kenntnis der normalen und pathologischen Hirnindenbildung. Beitr Pathol Anat Allgern Pathol 47: 51–125.

Rao VTS, Ludwin SK, Fuh S-C et al. (2016). MiRNA expression in human astrocytes in relation to anatomical location and age. J Neuropathol Exp Neurol 75: 156–166.

Rao VTS, Fuh S-C, Karamchandani JR et al. (2019). Astrocytes in the pathogenesis of multiple sclerosis: an in situ MicroRNA study. J Neuropathol Exp Neurol 78: 1130–1146.
Roebuck TM, Matson SN, Riley EP (1998). A review of the neuroanatomical findings in children with fetal alcohol syndrome or prenatal exposure to alcohol. Alcohol Clin Exp Res 22: 339–344.
Rong L, Messing A, Goldman JE et al. (2002). GFAP mutations in Alexander disease. Int J Dev Neurosci 20: 259–268.
Rorke LB, Riggs HE (1969). Myelination in the brain of the newborn, Lippincott, Philadelphia, Toronto.
Roy CC, Meredith GH, John AW et al. (2020). Ultrastructural and functional changes at the tripartite synapse during epileptogenesis in a model of temporal lobe epilepsy. Exp Neurol 326: 113196. https://doi.org/10.1016/j.expneurol.2020.113196.
Rustenhoven J, Kipnis J (2022). Brain borders at the central stage of neuroimmunology. Nature 612: 417–429.
Salcedo-Arellano MJ, Hagerman RJ, Martínez-Cerdeño. (2020). Fragile-X syndrome: clinical presentation, pathology and treatment. Gac Med Mex 156: 60–66.
Sarnat HB (1992). Cerebral dysgeneses: embryology and clinical expression, Oxford University Press, New York.
Sarnat HB (2019). Proteoglycan (keratan sulfate) barrier in developing human telencephalon isolates cortical epileptic networks from deep heterotopia, insulates axonal fascicles, and explains why axosomatic synapses are inhibitory. J Neuropathol Exp Neurol 78: 1147–1159.
Sarnat HB (2021). Transitory and vestigial structures of the developing human nervous system. Pediatr Neurol 123: 86–101.
Sarnat HB (2023). New neuropathological concepts in focal cortical dysplasias (Chapter 18). In: E Wyllie (Ed.), Wyllie's treatment of epilepsy, eighth edn. Lippincott Williams & Wilkins, Philadelphia.
Sarnat HB, Flores-Sarnat L (2001). Neuropathological research strategies in holoprosencephaly. J Child Neurol 16: 918–931.
Sarnat HB, Flores-Sarnat L (2009). α-B-crystallin as a tissue marker of epileptic foci in paediatric resections. Can J Neurol Sci 36: 566–574.
Sarnat HB, Flores-Sarnat L (2014). Morphogenesis timing of genetically-programmed brain malformations in relation to epilepsy. Prog Brain Res 213: 181–198.
Sarnat HB, Flores-Sarnat L, Hader W (2010). Satellite glia in epilepsy: guardian angels or assassins of neurons? Proceedings of the XII international congress of neuropathology. Medimond, Rome, pp. 83–91.
Sarnat HB, Yu W, Flores-Sarnat L (2021). Keratan sulfate proteoglycan: axonal barrier in forebrain of fetuses with alobar/semi-lobar holoprosencephaly. Clin Neuropathol 40: 70–86.
Sasaki A, Hirato J, Nakazato Y et al. (1988). Immunohistochemical study of the early human fetal brain. Acta Neuropathol 76: 128–134.
Sase S, Takanohashi A, Venderver A et al. (2018). Astrocytes, an active player in Aicardi-Goutières syndrome. Brain Pathol 28: 399–407.
Schiweck J, Eickholt BJ, Murk K (2018). Important shapeshifter mechanisms allowing astrocytes to respond to the changing nervous system during development, injury and disease. Front Cell Neurosci 12: 261.
Sharma K, Singh J, Frost EE et al. (2018). *MeCP2* in central nervous system glial cells: current updates. Acta Neurobiol Exp 78: 30–40.
Smith DE, Davies DL (1990). Effect of perinatal administration of ethanol on the CA1 pyramidal cell of the hippocampus and Purkinje cell layer of the cerebellum: an ultrastructural survey. J Neurocytol 19: 708–717.
Sosunov A, Olabarria M, Goldman JE (2018). Alexander disease: an astrocytopathy that produces a leukodystrophy. Brain Pathol 28: 388–398.
Sotoudeh Anvari M, Vasei H, Najmabadi H et al. (2022). Identification of microRNAs associated with human fragile X syndrome using next-generation 15 sequencing. Sci Rep 12: 5011. [cited 2022 Jun 1].
Srivastava S, Waldman A, Naidu S et al. (2012). Alexander disease. GeneReviews 2002 updated Nov.
Stobart JL, Anderson CM (2013). Multifunctional role of astrocytes as gatekeepers of neural energy supply. Front Cell Neurosci 7: 38.
Sun W, McConnell E, Pare JF et al. (2013). Glutamate-dependent neuroglial calcium signaling differs between young and adult brain. Science 339: 197–200.
Sun J, Osenberg S, Irwin A et al. (2023). Mutations in the transcriptional regulator MeCP2 severely impact key cellular and molecular signatures of human astrocytes during maturation. Cell Rep 42: 111942.
Sun W, Liu Z, Jiang X et al. (2024). Spatial transcriptomics reveal neuron-astrocyte synergy in long-term memory. Nature 627: 374–381.
Tang G, Xu Z, Goldman JE (2006). Synergistic effects of the SAPK/JNK and the proteasome pathway on glial fibrillary acidic protein (GFAP) accumulation in Alexander disease. J Biol Chem 281: 38634–38643.
Tedesco B, Cristofani R, Cozzi M et al. (2022). Insights on human small heat shock proteins and their alterations in diseases. Front Mol Biosci 9: 842149. https://doi.org/10.3389/fmolb.2022.842149.
Tsai V, Parker W, Orlova KA et al. (2014). Fetal brain mTOR pathway activation in tuberous sclerosis complex. Cereb Cortex 24: 315–327.
Uhlmann EJ, Wong M, Baldwin RL et al. (2002). Astrocyte-specific *Tsc1* conditional knockout mice exhibit abnormal neuronal organization and seizures. Ann Neurol 52: 285–296.
Ungerer M, Knezovich J, Ramsay M (2013). In utero alcohol exposure, epigenetic changes, and their consequences. Alcohol Res 35: 37–46.
Uppal P, Smith J, Castillo-Pinto C et al. (2023). Glial fibrillary acidic protein (GFAP) astocytopathy: an emerging cause of meningoencephalomyelitis in children and adolescents. J Child Neurol 38: 659–664.
Vashi N, Justice MJ (2019). Treating Rett syndrome: from mouse models to human therapies. Mamm Genome 30: 90–110.
Vidal S, Xiol C, Pascual-Alonso A et al. (2019). Genetic landscape of Rett syndrome spectrum: improvements and challenges. Int J Mol Sci 20: 3925.
Viedma-Poyatos A, González-Jiménez P, Pajares MA et al. (2022). Alexander disease GFAP R239C mutant shows increased susceptibility to lipoxidation and elicits mitochondrial dysfunction and oxidative stress. Redox Biol 55: 102415. https://doi.org/10.1016/j.redox.2022.102415.
Walker AK, Daniels CML, Goldman JE et al. (2014). Astrocytic TDP-43 pathology in Alexander disease. J Neurosci 34: 6448–6458.
West PK, Viengkhou B, Campbell IL et al. (2019). Microglia responses to interleukin-6 and type 1 interferons in neuroinflammatory disease. Glia 67: 1821–1841.
Wiley CA, Steinman RA, Wang Q (2022). Innate immune activation without immune cell infiltration in brains of murine models of Aicardi-Goutières syndrome. Brain Pathol 33: e13118.
Wilhelm CJ, Guizzetti M (2016). Fetal alcohol spectrum disorders: an overview from the glia perspective. Front Integr Neurosci 9: 65. https://doi.org/10.3389/fnint.201`5.00065.
Wippold II FJ, Perry A, Lennerz J (2006). Neuropathology for the neuroradiologist. Rosenthal fibers. Am J Neuroradiol 27: 958–961.
Xu A, Uliel-Sibony S, Dunham C et al. (2019). mTOR inhibitors as a new therapeutic strategy in treatment resistant epilepsy in hemimegalencephaly: a case report. J Child Neurol 34: 132–138.

Yakovlev PI, Lecours A-R (1967). The myelination cycles of regional maturation of the brain. In: A Minkowsky (Ed.), Regional development of the brain in early life. FA Davis, Philadelphia, pp. 3–70.
Yamaguchi M (2011). Development and survival of adult-born olfactory neurons. In: T Seki, K Sawamoto, JM Parent, ER Álvarez-Buylla (Eds.), Neurogenesis in the adult brain I. Neurobiology. Springer, London, NY, pp. 357–369.
Yasin SA, Latak K, Becherini F et al. (2010). Balloon cells in human cortical dysplasia and tuberous sclerosis: isolation of a pathological progenitor-like cell. Acta Neuropathol 120: 85–96.
Zaki W (1985). Le processus dégénératif au cours du développement du corps calleux. Arch Anat Microsc Morphol Exp 74: 133–149.
Zyang Z, Fan Q, Luo X et al. (2022). Brain-restricted mTOR inhibition with binary pharmacology. Nature 609: 822–828.

Handbook of Clinical Neurology, Vol. 210 (3rd series)
Neuroglia in Neurologic and Psychiatric Disorders, Part II
A. Verkhratsky, L.D. de Witte, E. Aronica, and E.M. Hol, Editors
https://doi.org/10.1016/B978-0-443-19102-2.00009-0

Chapter 6

Neuroglia in stroke

ALEXEI VERKHRATSKY[1,2,3] AND MICHAEL V. SOFRONIEW[4*]

[1]*Faculty of Biology, Medicine and Health, University of Manchester, Manchester, United Kingdom*

[2]*Department of Neurosciences, University of the Basque Country UPV/EHU and CIBERNED, Leioa, Bizkaia, Spain*

[3]*IKERBASQUE, Basque Foundation for Science, Bilbao, Spain*

[4]*Department of Neurobiology, David Geffen School of Medicine, University of California, Los Angeles, CA, United States*

Abstract

Stroke, ischemic or hemorrhagic, triggers a complex and coordinated glial response, which, to a large extent, defines the progression and outcome of this focal damage of the nervous tissue. Massive cell death in the infarction core results in a release of damage-associated molecular patterns, which, together with blood-borne factors entering the brain through either ruptured vessels or through compromised blood–brain barrier, trigger reactive gliosis. Microglia are the first to migrate toward the lesion, proliferate, and phagocytose cellular debris in and around the infarct core. Reactive astrogliosis occurs around the margins of the infarct core and is characterized by astrocytic proliferation, morphologic remodeling with loss of territorial domain segregation, and transcriptional reprogramming into wound repair astrocytes that form a periinfarct border that protects the healthy tissue and assists postlesional regeneration.

HISTORIC PRELUDE: FROM HIPPOCRATIC APOPLEXY TO STROKE

It is impossible to cure a severe attack of apoplexy, and difficult to cure a mild one."

Aphorismes of Hippocrates, *quoted from E. Clarke, Bulletin of the History of Medicine, 1963, Vol. 37, p. 301–314.*

Stroke and epilepsy are probably the oldest neurological nosological forms, which were defined and described in the Hippocratic codex, written in the 5th and 4th centuries BC (Adams, 1849; Hippocrates, 1923-1931; Karimullah, 2020). Hippocratic codex placed stroke into the broader category of *apopliktiki* or *apoplexy;* the former related to the verb πλήττω—to strike down or to stun, whereas the latter to the noun πλήξη, meaning stroke. Both terms denoted a sudden, violent attack in which the patient falls down, loses senses, sensitivity, and faculty of movement. The loss of voluntary movements (paralysis) was the leading symptom; and it was pathogenetically linked to the stall of vital spirits (which were the driving force of brain—periphery signaling according to the pneumatic doctrine universally accepted in antiquity) in the brain, which made operation of the nervous system and muscles impossible. The terms apoplexy, paralysis, paraplegia, paraplexia, or paresis were used interchangeably in antique medical writings, all denoting the loss of body motions (Adams, 1856; Schiller, 1970). *Corpus Hippocraticum*; however, contains another nosological entry, the "Acute Brain Suffering" (Οξείες οδύνες του εγκεφάλου), which quite precisely describes acute stroke: "*there is also another disease, which appears with a concentration of fluids within the brain (edema or haemorrhage), acute strong pain in temporal and/or parietal lobe area (sudden headache with no cause), with a simultaneously appearance of shivering and fever in waves. The patient feels a kind of anguish pain in the area of his eyes (around orbital wall), his vision is afflicted, the eye pupil is divided and his thinks that he sees two objects instead of one. When he attempts to rise from his position (initially fallen to the ground, stroked down) he is conquered by vertigo (loss of balance, ataxia?), his ears bell, he vomits, while sometimes looses saliva, phlegm, and aliments*" (quoted from Tsoucalas et al., 2019). The pathophysiology of this disease was brain-centric and again connected to the stall of flow of vital spirits, which robbed the nervous system of the means of signaling and coordination. These ideas were further advanced by Galen, whose writings became a basis for the medicine of late antiquity, middle ages, and renaissance. The term apoplexy was in medical use till the end of the 19th century.

*Correspondence to: Michael Sofroniew, Department of Neurobiology, David Geffen School of Medicine, University of California, Los Angeles, CA, United States. Tel: +1-310-206-2262, Fax: +1-310-825-2224, E-mail: sofroniew@mednet.ucla.edu

The first autopsy, which described organic pathology associated with stroke in the postmortem human brain, was made in 1602 by Swiss physician Felix Plattner, who found that "*a phlegmatic humour is obstructing the inner passages of the brain*" (Platter, 1602), probably indicating that he found blocked arteries (Karenberg, 2020). The next conceptual step was made by Johann Jakob Wepfer, who emphasized the brain parenchymal structures (and not the ventricles) as sites of primary damage and found that the brain arteries were blocked by "*pituitous formations*" (most likely clots), which he directly associated with stroke; he also found that about half of strokes were linked to hemorrhage, whereas another half to artery obstruction (Wepfer, 1658).

The modern pathophysiology of stroke began in 1812 with the dissertation of Jean-André Rochoux, who linked the stroke lesion to hemorrhage: "*Apoplexy is a haemorrhage of the brain, by rupture, with more or less serious alteration of its substance*" (quoted from Karenberg, 2020) and described the appearance of the lesion in the postmortem brain (Rochoux, 1812). In 1820, Leon Rostan suggested that stroke may be linked not only to hemorrhage but also to ischemia, which causes softening of the brain (Karenberg, 2020). All these ideas culminated in the works of Rudolph Virchow, who realized the duality of stroke, which can be caused by either vessel rupture (hemorrhage) or arterial occlusion (ischemia); he also realized the link between thrombosis, clotting, embolism, and ischemic stroke. Virchow classified strokes into *sanguinea* and *ischaemica*, and contemplated that embolisms of the middle cerebral, the anterior cerebral, the carotid, and vertebral arteries were the main causes of stroke (Virchow, 1856; Schiller, 1970).

STROKE: GENERAL REMARKS

Stroke is precipitated by disruption of the blood flow in the brain, resulting in ischemia that triggers the death of neural cells with consequent neurological and cognitive deficits. The disruption of blood flow can be caused either by vessel occlusion due to thrombosis or embolism, a systemic decrease in the supply of blood (for example, due to heart failure), or by rupture of blood vessels resulting in hemorrhage. Global brain ischemia is a result of heart failure resulting in an immediate cessation of cerebral blood flow, which causes almost immediate loss of consciousness while electrical activity of the brain disappears in 30–40 s after the cardiac arrest. Global ischemia lasting for more than 10 min results in death of the human brain. Even short periods of global ischemia trigger severe disruptions in neuronal functions, including aberrant synaptic transmission, ion dyshomeostasis, and massive release of glutamate, which all result in neuronal death and cognitive abnormalities (Neumann et al., 2013).

The modern definition of stroke follows Virchow's concepts: strokes are classified into ischemic (which accounts for ~85% of all cases) and hemorrhagic (Fig. 6.1). Hemorrhagic stroke is further classified into subarachnoid hemorrhage (~5% of all cases), associated with rupture of vessels and blood escape to subarachnoid space, and intracerebral hemorrhage (~10% of all strokes) resulting from bursting of intraparenchymal blood vessels. The ischemic stroke has several classifications linked to the underlying mechanisms such as large-artery atherosclerosis, cardioembolism, small-vessel occlusion, and strokes of other etiologies (Adams Jr. et al., 1993; Amarenco et al., 2009). Focal brain ischemia triggers primary brain damage, which triggers a cascade of events that instigate wider tissue, organ, and systemic pathological consequences. The anatomical location of the vessel occlusion determines the site and the volume of the damage to the nervous tissue. The focal ischemia can be transient, with the blood flow being restored after the removal of the embolus, but the reperfusion may also add to the injury due to secondary oxidative stress and ion imbalances (Wu et al., 2018). The primary brain damage emerges almost immediately after stroke onset.

In the case of ischemic stroke, arterial occlusion affects the corresponding vascular territory and triggers death of all neuronal cells, thus forming an infarction core, the zone of pan-necrosis. This core is surrounded by a penumbra, where nervous tissues suffer from limited blood supply; the cells in penumbra are metabolically stressed and can either survive or die; this secondary death to a large extent defines the final size of the damage and neurological outcome (Paciaroni et al., 2009). The intercellular hemorrhage triggers similar consequence of events, which is further exacerbated by the formation of a hematoma, which adds to the structural damage through extravasation of blood-borne cells and proteins into the brain parenchyma. Both types of strokes trigger a multitude of cellular and molecular events such as loss of ion homeostasis, uncontrolled release of glutamate with associated excitotoxicity, and massive oxidative stress, to name but a few (Qin et al., 2022). Stroke also triggers an evolutionary conserved defensive response of the nervous tissue, generally known as neuroinflammation, which includes coordinated

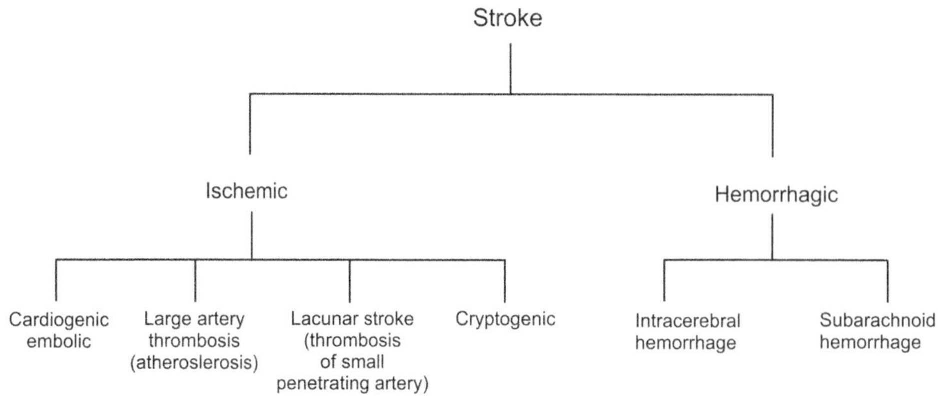

Fig. 6.1. Classification of stroke types.

and complex remodeling of neuroglial cells that undergo reactive gliosis. This response develops in stages and is aimed at erecting a protective border and poststroke regeneration, discussed below.

LOSS OF ION, WATER, AND METABOLIC HOMEOSTASIS IN THE ISCHEMIC CORE AND PENUMBRA

Either occlusion or rupture of arteries deprives nervous tissue of the blood flow delivering glucose and oxygen. In the ischemic core, ATP production is halted, which has immediate consequences on ion homeostasis, manifested by uncontrolled Na^+ entry and depolarization. Depolarization triggers opening of voltage-gated Ca^{2+} channels and Ca^{2+} entry, which in turn produces exocytotic release of glutamate that further depolarizes cells and amplifies Ca^{2+} entry. This positive feedback-based vicious circle underlies severe excitotoxicity and Ca^{2+} overload (Arundine and Tymianski, 2003; Bano and Ankarcrona, 2018), which ultimately results in the necrotic death of neurons and neuroglia with the core of infarction (Berliocchi et al., 2005; Zhivotovsky and Orrenius, 2011). In the gray matter subjected to focal ischemia, loss of ionic balance occurs within minutes after the cessation of blood flow: extracellular K^+ rises from 5 to 40–80 mM, extracellular Na^+ falls from 140 to 60 mM, while extracellular Ca^{2+} drops from 2.5 to 0.2–0.5 mM, indicating massive redistribution of ions between extracellular and intracellular space (Somjen, 2004; Rossi et al., 2007). Uncontrolled Ca^{2+} influx increases neuronal cytoplasmic Ca^{2+} to 30–60 μM (Silver and Erecinska, 1990), which triggers necrotic death cascades. Simultaneously, extracellular space acidifies with pH decreasing to ~6.5 (Rossi et al., 2007). Astrocytic ionic balance is also affected: cytoplasmic Na^+ concentration is increased to ~25 mM, whereas K^+ concentration falls to ~100 mM in response to short periods of ischemia (Eitelmann et al., 2022). Considering that astrocytic Na^+ regulates most of the homeostatic solute carrier transporters (including neurotransmitter transporters) and Na^+-K^+ ATPase (NKA) (Verkhratsky and Rose, 2020; Rose and Verkhratsky, 2024), these changes may substantially modify astrocyte-dependent control over interstitial ionostasis and neurotransmitters profiles. Loss of homeostatic control over the ionic composition of the interstitium, failure of NKA to maintain Na^+/K^+ balance, uncontrolled neurotransmitter release, edema, and mitochondrial failure all contribute to the generation of waves of spreading depolarization, which further exacerbate excitotoxicity and cell death (Luckl et al., 2018; Andrew et al., 2022). Cerebral edema is also prominent after stroke and is characterized by dysregulation of water transport by astrocytes. Deletion of the astrocyte water channel AQP4 reduces poststroke edema and improves neurological outcome (Manley et al., 2000).

In the penumbra, energy production decreases to about 50%–70%, which stresses neural cells but remains sufficient for their survival. Neuronal excitability is severely depressed, neuronal somata swell, and neuronal metabolism is suppressed. Transient ischemic depolarizations generated in the penumbra trigger new incidences of ion imbalance, which may lead to a secondary cellular death (Rossi et al., 2007). When cells in the penumbra die, the infarction core is expanded; cell death or survival defines the extent of stroke-associated damage. Cell death in penumbra occurs through several mechanisms; there are incidences of necrosis, but also cell die through apoptosis, autophagy, necroptosis, and ferroptosis (Garcia-Pupo et al., 2020). Necrosis dominates the ischemic core (Lipton, 1999); several necrosis routines characterized by distinct molecular pathways (for example, parthanatos, oxytosis, NETosis, and pyronecrosis (Vanden Berghe et al., 2014)) occur simultaneously or in sequence, depending on brain region, age, and gender (Puyal et al., 2013); in the penumbra various types of programmed cell death also take place (Graham and Chen, 2001; Vitale et al., 2023).

ASTROCYTES IN STROKE: NEUROPROTECTION AND REACTIVE ASTROGLIOSIS

Astrocytic protection of neurons

Astrocytes are the main cellular element of the homeostatic and protective system of the CNS, being responsible for virtually all homeostatic tasks required for proper function of the nervous tissue, from control over molecular composition of the interstitium to supporting the brain as an organ through regulation of the blood–brain interface, waste collection by the glymphatic system, and broad chemosensing contributing to the systemic homeostasis (Verkhratsky and Nedergaard, 2018; Verkhratsky and Butt, 2023). Astrocytes also contribute to all neuropathologies through highly complex, heterogeneous, mutable, age-, context-, and disease-specific changes, ranging from reactivity to atrophy, degeneration, and loss of function (Verkhratsky et al., 2023).

Astrocytes show more resilience to ischemia than neurons; for example, in primary neuronal-astrocytic cocultures, glucose and oxygen deprivation killed neurons within an hour, whereas astrocytes remained alive for 4–6 h (Almeida et al., 2002; Giffard and Swanson, 2005). In vivo, astrocytes are much more sensitive to ischemia and yet their survival rate is higher than that of neurons (Chen and Swanson, 2003a; Fern et al., 2014). In the stroke penumbra, astrocytes can remain functionally active due to a switch into glycolytic mode; astrocytes have some limited stores of glycogen. Astrocytes in penumbra are critical for neuronal survival through maintaining ion homeostasis, supplying neurons with lactate, and containing oxidative stress. Astrocytic resilience of ischemia is at least in part linked to glycogen, which, in the CNS, is found almost exclusively in astrocytes (Brown, 2004; Oz et al., 2015). Although the amount of glycogen in astrocytes is quite low (brain glycogen content is about 100 times lower than in the liver and 10 times lower that in the skeletal muscle (Brown and Ransom, 2007)), the glycogen breakdown contributes to supporting several vital astrocytic functions such as K^+ buffering, glutamate uptake, and synthesis of glutamine (Bak et al., 2018). Both ion homeostasis and glutamate uptake are critical for neuronal survival in the penumbra, and astrocytes play the leading role in the protection of stressed neurons.

In particular, astrocytic glutamate transport is central for containing excitotoxicity. Withdrawal of astrocytes from neuronal-astrocytic cocultures causes a 100 times increase in neuronal vulnerability to exogenous glutamate (Rosenberg and Aizenman, 1989). Likewise, ischemic damage is substantially exacerbated in the brains of mutant mice lacking glutamate transporter EAAT2/GLT-1/SLC1A2 (Mitani and Tanaka, 2003). Conversely, overexpression of EAAT1 reduced glutamate in poststroke tissue and decreased brain infarction in mice (Harvey et al., 2011). At the same time, evidence indicates that cerebral ischemia leads to a decreased expression and functionality of astrocytic EAAT2, linked to increased excitotoxicity and neuronal death (Hu et al., 2015; Zhang et al., 2019). Astrocytes also limit excitotoxicity through K^+ buffering (Larsen et al., 2014; MacAulay, 2020). Both glutamate uptake and K^+ clearance depend on NKA, which maintains the transmembrane Na^+ gradient and pumps K^+ into astrocytes, thus requiring substantial energy and constant ATP production through oxidative phosphorylation (Barros, 2022; Verkhratsky and Semyanov, 2022). Ischemic conditions therefore may reduce astrocytic neuroprotective capacity.

Astrocytic glycolysis not only generates ATP during ischemia thus supporting astrocyte survival, but also produces several glycolysis intermediates, and in particular lactate, that might be shuttled (Pellerin and Magistretti, 2012) to energy-deprived neurons (Tekkok et al., 2005). Lactate, however, is a double-edged sword—on the one hand, it is a neurone-preferred energy substrate (Smith et al., 2003); on the other hand, accumulation of extracellular lactate leads to acidosis. Astrocytes are quite sensitive to acidosis: lowering pH below 6.6 rapidly inhibits ATP production and kills these cells (Giffard et al., 1990). This has significant clinical implications as adding exogenous glucose exacerbates ischemic infarction by increasing acidosis and suppressing astrocytic neuroprotection (Zhao and Rempe, 2010).

Astrocytes also contribute to the protection of neurons through containing ROS accumulation in the ischemic and/or re-perfused tissue. Astroglial cells throughout the CNS are the major contributors to the antioxidant defense. In particular, astrocytes express high levels of the master regulator of the antioxidant system, nuclear factor-erythroid factor 2-related factor 2 or transcription factor Nrf2, which is scarcely present in neurons (Makar et al., 1994; Vargas et al., 2008). All in all, neurons contain relatively low quantities of ROS scavengers, and their antioxidant capacity is limited (Fernandez-Fernandez et al., 2012). In contrast, astrocytes have high levels of major ROS scavenger glutathione (Dringen et al., 2000). Astrocytes are generally responsible for the production and maintenance of glutathione and ascorbic acid, which represent the main ROS scavengers in the CNS (Fig. 6.2). The tripeptide glutathione fixes ROS either by nonenzymatic direct reaction or by acting as an electron donor for glutathione peroxidase (Sies, 1999). Synthesis of glutathione requires either cysteine or glutamylcysteine, which cannot be taken up by neurons; in contrast, astrocytes express high densities of glutamate-cystine exchanger SLC7A11/xCT/Sxc (Lewerenz et al., 2013), thus being able to accumulate cystine. After being taken up by astrocytes, cystine is reduced to cysteine or else converted to glutamylcysteine (CysGlu). Both of these molecules are used for

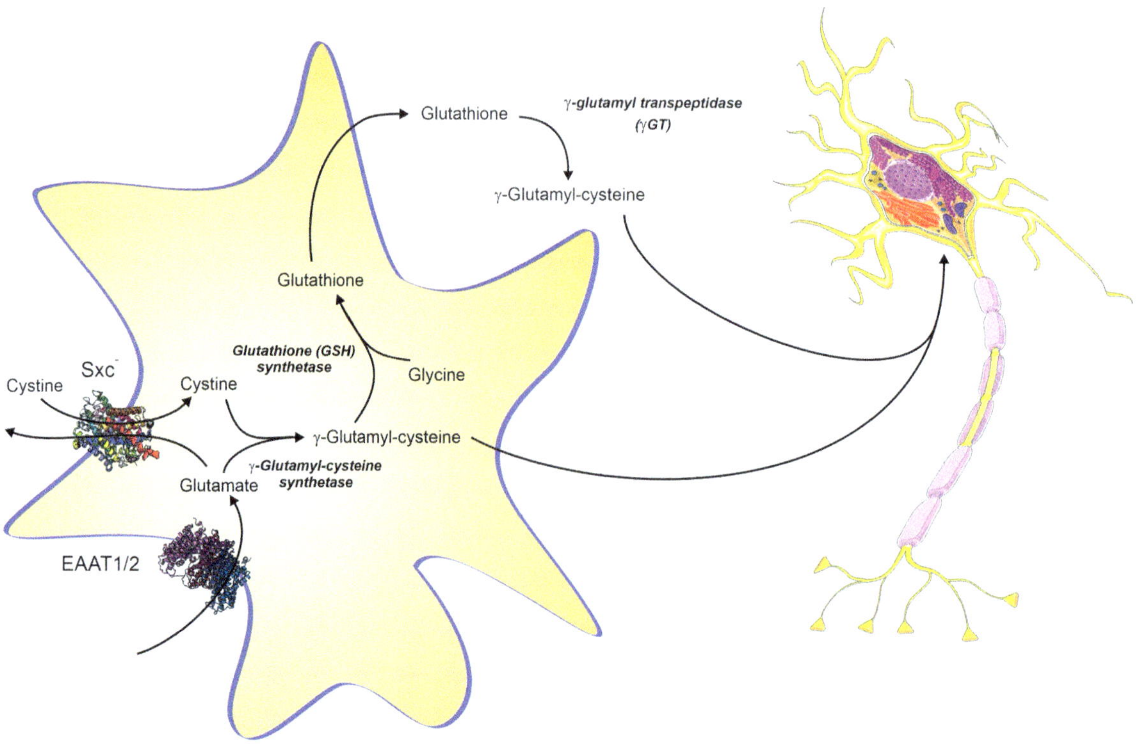

Fig. 6.2. Astrocytes and glutathione production. See text for further explanation. *EAAT1/2*, excitatory amino acid transporter 1/2; *Sxc*$^-$, glutamate/cystine exchanger.

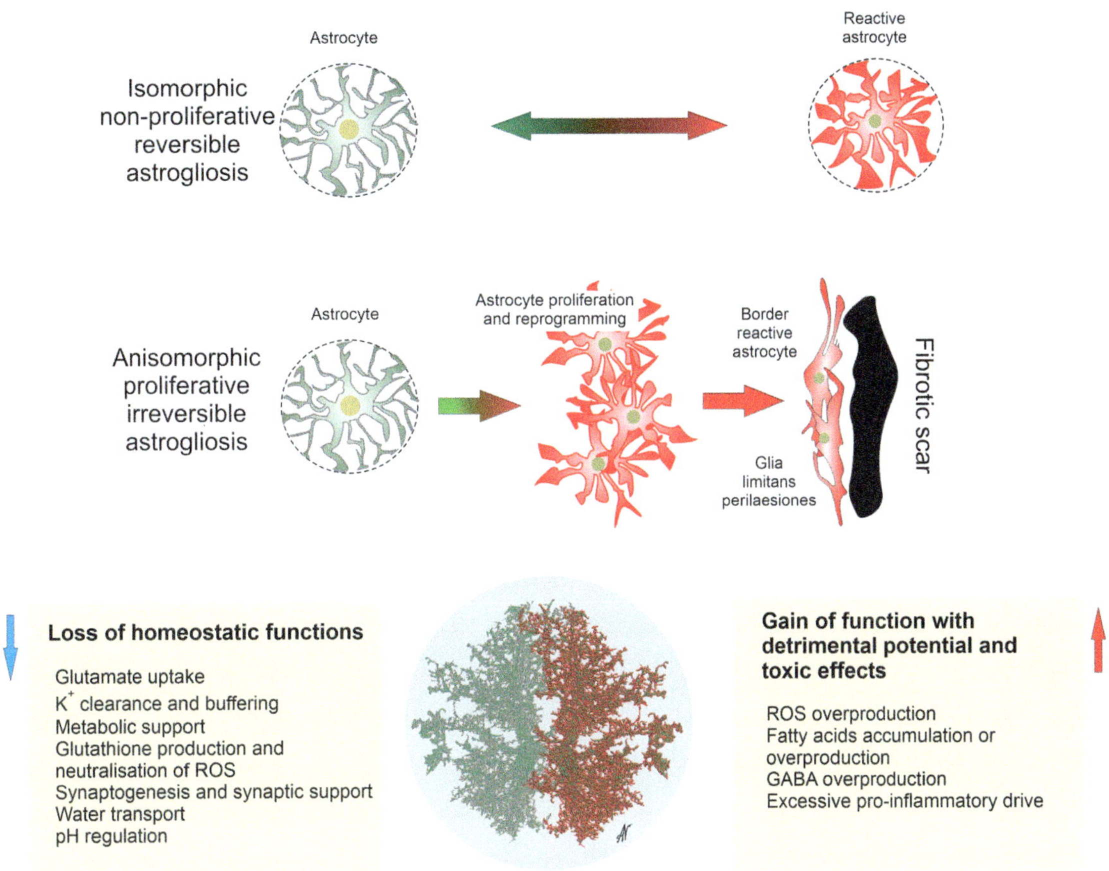

Fig. 6.3. Reactive astrogliosis. Top panel: Classification of reactive astrogliosis. Astrocyte reactivity phenotypes can at present be broadly classified into (i) nonproliferative astrogliosis, which is isomorphic with mainly preserved domain organization; and (ii) proliferative astrogliosis, which is anisomorphic with loss of domain organization, substantial structural reorganization and can either be diffuse or can result in the formation of new compact "limitans" borders around overt fibrotic tissue lesions. Bottom panel: Potential detrimental effects of diseased, atrophied, reactive, or aged astrocytes that can occur in specific contexts. See text for further explanation. Reproduced from Verkhratsky A, Butt A, Li B et al. (2023). Astrocytes in human central nervous system diseases: a frontier for new therapies. Signal Transduct Target Ther 8: 396.

astrocytic production of glutathione. In addition, cysteine and glutathione are released from astrocytes; glutathione is then converted to CysGlu by the ectoenzyme γ-glutamyl transpeptidase (γGT). Subsequently, CysGlu is moved to neurons by EAAT2/3 glutamate transporters, where it is used for the synthesis of glutathione (Dringen et al., 1999; Chen and Swanson, 2003b). Without astrocytes, neurons cannot tolerate oxidative stress. Experiments in neuronal-glial cocultures demonstrated that removal of astrocytes or pharmacological inhibition of γGT markedly increases neuronal vulnerability to oxidative stress (Dringen et al., 1999). In addition to glutathione, astrocytes are the main producers of ascorbic acid and provide for the recycling of vitamin C (Espinoza et al., 2020). Ascorbic acid oxidizes ROS and becomes dehydroascorbic acid in the process. Astrocytes take up this dehydroascorbic acid by the sodium-dependent vitamin C transporter 2 (SVCT2/SLC23A2) and the glucose transporter (GLUT1/SLC2A1); after entering astrocytes, dehydroascorbic acid is reduced to ascorbic acid, which is shuttled to neurons (Siushansian et al., 1996). An increase in neuronal activity stimulates the release of ascorbic acid from astrocytes (Wilson et al., 2000).

Astrocytic perilesional limitans border formation around fibrotic scar tissue

Every traumatic injury to the nervous tissue of the CNS (mechanical, ischemic, or infectious) associated with the breach of the blood–brain barrier and extravasations of blood-borne elements to the brain parenchyma triggers an innate inflammatory and glial response, aimed at isolation of the damaged area and fibrotic scar by a newly formed glia limitans perilesiones (Sofroniew, 2015; Wahane and Sofroniew, 2021; O'Shea et al., 2023; Verkhratsky et al., 2023). At the same time, various forms of reactive or asthenic astrocytes, with loss of homeostatic function are observed following brain damage; their emergence depends on the severity of the trauma, distance from the lesion site, and brain region (Munoz-Ballester and Robel, 2023).

The primary pathological element of the stroke-induced damage is associated with the breach of the blood–brain barrier with subsequent invasion of the brain with blood cells and various molecules. In hemorrhagic stroke this is further exacerbated by the formation of a hematoma. Blood-borne molecules (such as thrombin or fibronectin (Conforti et al., 2022)) act as danger-associated molecular patterns (DAMPs). Similarly, DAMPs are released from damaged neural cells and include, for example, ATP, heat shock proteins, peroxiredoxins, etc. (Garcia-Bonilla and Iadecola, 2012; Petrovic-Djergovic et al., 2016). DAMPs trigger reactive gliosis, represented by reactive astrogliosis, reactive microgliosis, and reactivity of oligodendroglial precursor cells and oligodendrocytes (Sofroniew, 2020; Escartin et al., 2021; Paolicelli et al., 2022; Yi et al., 2023). Reactive astrogliosis, characterized as a general pathological response of astrocytes to trauma in the 1920s ("*the glial splint*" (del Río-Hortega and Penfield, 1927)) is an evolutionary conserved, primarily defensive response of astrocytes to pathology, in which astrocytes instigate complex molecularly defined programs manifested by changes in gene expression, resulting in biochemical, morphologic, metabolic, and physiological remodeling, aimed at protection of the nervous system (Fig. 6.3) (Escartin et al., 2021).

The first cells to respond to a stroke are microglia, which start to migrate to the lesion site within ~30 min after the onset of the stroke. Oligodendrocyte precursors (OPC) are the next, with a response time of ~4 to 6 h, and finally, in about 24–48 h, reactive astrogliosis is initiated. The infarction core is rapidly filled with blood-borne macrophages, dendritic cells, and reactive microglia actively engaged in clearing the cellular debris. Reactive astrocytes, microglia, and OPCs proliferate and demarcate the lesion border with the glial barrier (Fig. 6.4). In contrast to microglia, astrocytes do not migrate toward the lesion site; their density around the infarction border increases due to symmetric division, with daughter cells staying together with their mother cells (Frik et al., 2018; Lange Canhos et al., 2021). The gradient of astrocytes, OPCs, and microglia in different reactive states is observed across penumbra and toward the healthy tissue (Fig. 6.4) (Choudhury and Ding, 2016; Sims and Yew, 2017). Reactive astrocytes surrounding the infarction core are highly heterogeneous; for example, the expression of the *Gfap* gene shows ~3000-fold difference between individual cells in the injured hippocampus (Wilhelmsson et al., 2017). Intermediate filaments, glial fibrillary acidic protein (GFAP), and vimentin are critical for mounting reactive astrogliosis; their genetic deletion ($Gfap^{-/-}Vim^{-/-}$ mice) prevents or attenuates reactive remodeling (Pekny et al., 1999; Wilhelmsson et al., 2017). Deletion of GFAP and vimentin increases the size of infarction following ischemic stroke induced by middle cerebral artery occlusion or transaction; in addition, these mice demonstrate impaired glutamate uptake and oxidative defense, thus indicating the protective role of astrogliosis in the context of acute ischemia (Li et al., 2008).

Over time, the lesion evolves: the stromal elements (fibroblasts and pericytes) populate the infarction zone and form the fibrotic scar, whereas reactive astrocytes erect a new glia limitans that fences the damaged tissue and assists poststroke regeneration (O'Shea et al., 2023; Verkhratsky et al., 2023). Over 90% of astrocytes in this new glia limitans border around infarcts derive from proliferation of local mature astrocytes, which dedifferentiate, proliferate, and become transcriptionally reprogrammed into wound repair cells. These new border-forming astrocytes downregulate their interactions with neurons and upregulate their interactions with stromal and inflammatory cells (O'Shea et al., 2023). Notably, in contrast with previous ideas that these border-forming astrocytes form "scars" that are detrimental to neural regeneration, these cells are able to support axon regeneration that has been stimulated by augmenting neuronal regrowth capacity and providing chemoattractive growth factors (Anderson et al., 2016, 2018). In line with these observations, it is now recommended to no longer refer to these structures as "glial scars" but instead refer to them as astrocyte borders, which form a new glia limitans perilesiones that reestablishes CNS integrity around tissue lesions by separating functioning neural tissue from stromal and immune cells in a manner similar to the perimeningeal glia limitans that surrounds the entire CNS, with which they share morphologic and molecular similarities (O'Shea et al., 2023).

NEUROGLIA AND BRAIN-WIDE POSTSTROKE PATHOLOGY

Stroke is a brain-wide and systemic pathology, and it is noteworthy that focal stroke results in wide-spread changes in glial cells across the brain (Shi et al., 2019). Pathological remodeling and functional changes of the brain structures in areas distant to the lesions, known as diaschisis (διασξισις—split throughout), are often observed in patients and in animal models (Carrera and Tononi, 2014). The concept of diaschisis was initially suggested by Charles-Édouard Brown-Séquard in 1875 (Brown-Séquard, 1875) and subsequently precisely defined and described by Constantin von Monakow (von Monakow, 1914). Reactive astrocytes characterized by aberrant GABA synthesis have been observed in cortical areas distant to subcortical photothrombotic stroke in mice (Nam et al., 2020). Similarly, microglia in various reactive states were found in brain regions distant to stroke or even throughout the whole brain in the extreme cases (Shi et al., 2019). Brain-wide disseminated reactive microgliosis was also supported by positron emission imaging of microglial markers (Zinnhardt et al., 2015). Global microglial reactivity usually peaks within a week after stroke and subsides within a month, yet sometimes it may persist for longer time (Shi et al., 2019). Reactive microglia can migrate from the site of the lesion (such as cortical microinfarcts) to distal brain regions (both ipsi- and contralesional) and instigate tissue remodeling, myelin damage, and degradation of axonal tracts (Lubart et al., 2021).

OLIGODENDROCYTES AND WHITE MATTER STROKE

White matter is frequently affected by focal ischemia, and white matter infarctions account for about 20% of all strokes, while massive cortical stroke often affects subcortical white matter (Wang et al., 2016). White matter damage also occurs in perinatal ischemia and stroke (Nelson, 2007). The vulnerability of white matter to ischemia stems from its high energy demand linked to nerve impulse propagation; even brief periods of insufficient blood flow may be quite damaging (Pantoni et al., 1996). Ischemic attack affects axons, as well as oligodendroglial cells, astrocytes, and microglia. Axonal death is mediated through massive

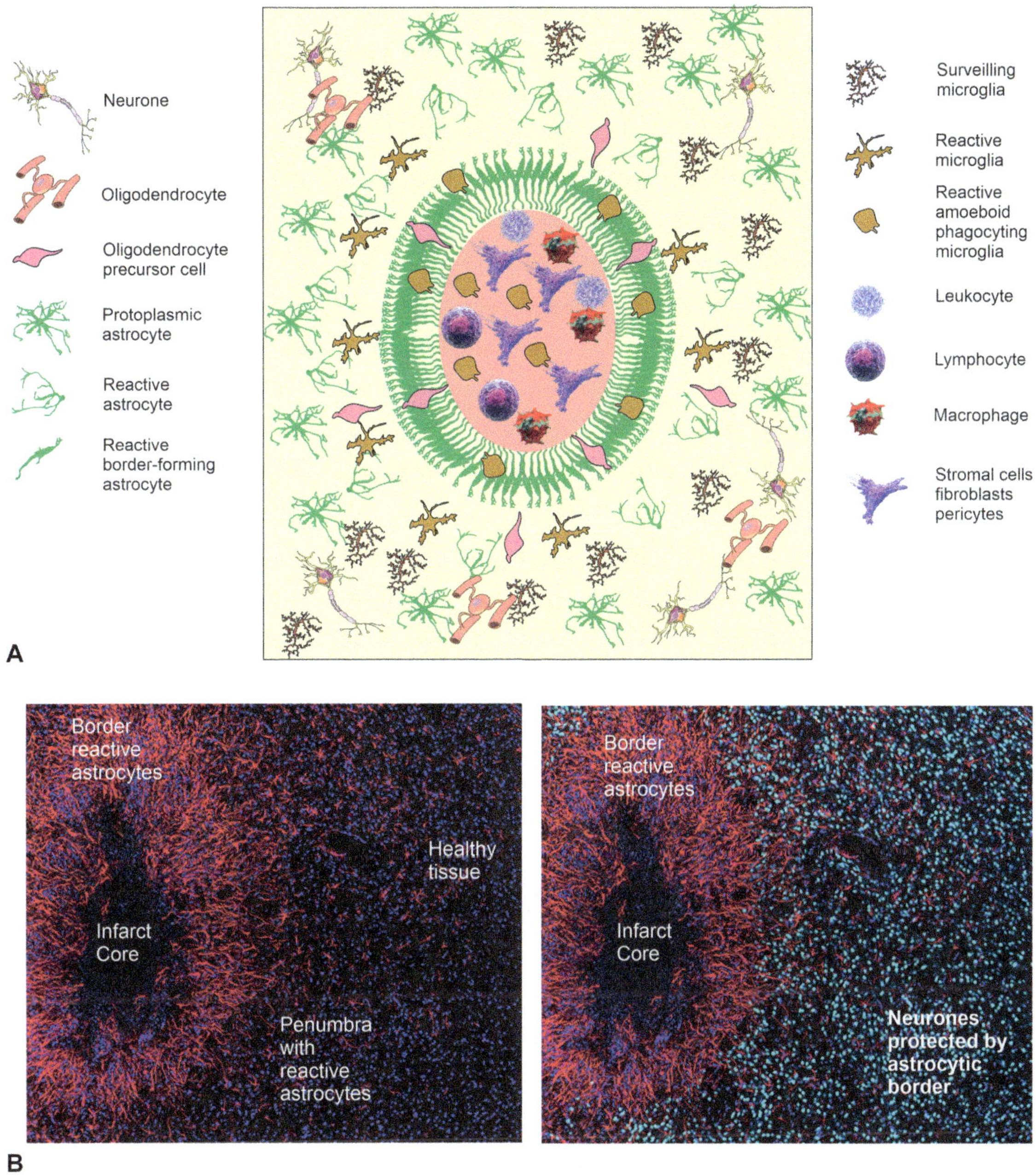

Fig. 6.4. Neuroglial border around the infarction core. See text for further explanation. (A) Schematic depiction of the infarction core, glial perilesional border, and surrounding with reactive gliocytes and nonreactive glial cells. (B) Reactive astrogliosis and protective astrocyte border formation in experimental stroke. Images show an ischemic infarct in mouse striatum at 14 days after injection of the vasoconstrictive agent, L-NIO (N^5-(1-iminoethyl)-L-ornithine) in a manner similar to that described previously (O'Shea et al., 2022). Immunohistochemistry shows GFAP-expressing astrocytes stained *red* and NeuN-expressing neurons stained *light blue*. The *left image* shows the border of newly-proliferated reactive astrocytes that surround the fibrotic tissue (unstained) of the infarct core and isolate it from adjacent neural tissue. The *right image* shows a normal density of healthy neurons immediately adjacent to the protective astrocyte border. Images are courtesy of Dr. Shinong Wang and Dr. Yan Ao. Reproduced from Verkhratsky A, Butt A, Li B et al. (2023). Astrocytes in human central nervous system diseases: a frontier for new therapies. Signal Transduct Target Ther 8: 396.

Ca^{2+} entry, mainly through reversed Na^+/Ca^{2+} exchanger (following a massive Na^+ influx), which is further amplified by Ca^{2+} release from the endoplasmic reticulum Ca^{2+} store (Fern et al., 2014). Axonal damage causes substantial release of glutamate and ATP (Back et al., 2007; Fields and Ni, 2010), which trigger secondary damage to axon-ensheathing oligodendrocytes and astrocytes through excessive stimulation of metabotropic and ionotropic (in particular $P2X_7$) glutamate and purinoceptors (Karadottir and Attwell, 2007; Matute et al., 2007; Hamilton et al., 2008). Of note, fibrous astrocytes residing in white matter are more vulnerable than protoplasmic astrocytes of gray matter. A brief 20-min period of oxygen glucose deprivation killed 82%

of astrocytes in the optic nerve, while only 3% in the hippocampus (Shannon et al., 2007).

Cells of the oligodendroglial lineage are also vulnerable to ischemia and glutamate excitotoxicity. Even brief exposure to ischemia affects oligodendroglial ion homeostasis, resulting in a considerable Ca^{2+} influx and cell death (Matute et al., 2006). The main route for Ca^{2+} entry is associated with AMPA/kainate and NMDA glutamate receptors (Karadottir et al., 2005; Micu et al., 2006). Expression of glutamate receptors is particularly high in OPCs, which consequently are highly sensitive to ischemia. Similarly, myelin sheaths contain many NMDA receptors, activation of which causes direct myelin damage: imaging experiments visualized NMDA-receptor-mediated Ca^{2+} entry into myelin following ischemic exposure, while antagonism of NMDA receptors prevented ischemic damage (Micu et al., 2006).

MICROGLIA IN STROKE

Microglial contributions to neuropathologies are, similar to other glial cells, complex and multifaceted, with numerous pathological states coexisting or substituting each other during the course of the disease (Paolicelli et al., 2022). Microglial cells seem to be more resistant to hypoxia/anoxia compared to neurons and oligodendrocytes (Yang et al., 2014), and yet severe ischemic attack can cause their demise. After middle cerebral artery occlusion, microglia die in the infarction core (Yang et al., 2014). Microglia death is mediated by ATP with a leading role for $P2X_7$ purinoceptors, which undergo pore dilation following excessive stimulation (Haas et al., 1996), thus instigating Ca^{2+} entry-dependent cell death (Eyo et al., 2013; He et al., 2017). In addition, $P2X_7$ receptors are coupled to the activation of inflammasomes, which is also linked to microglial apoptosis and pyroptosis (Poh et al., 2019).

Ischemia and stroke trigger reactive microgliosis, which, together with the gliotic response of macroglia, plays a central role in the pathophysiology of ischemic insults in the CNS (Planas, 2024). Microglial contributions in acute stroke phase are defensive, and ablation of microglia increases the size of infarction, potentiates neuronal death, and exacerbates poststroke inflammation (Lalancette-Hebert et al., 2007; Jin et al., 2017; Marino Lee et al., 2021). Similarly, depleting microglia in advance of experimental ischemia increased death of oligodendrocytes and OPCs (Xie et al., 2021), exacerbated stroke-induced neuronal malfunction neuronal activity (Szalay et al., 2016), and suppressed infiltration of protective $CD11c^+$ dendritic cells (Gallizioli et al., 2020). Microglial cells are capable of detecting decreases in local cerebellar blood flow in blood vessels located in their close positivity (Masuda et al., 2011). Stroke triggers overt morphologic remodeling of microglia toward reactive phenotypes, characterized by increases in the size of soma, reduced process ramification, with process shortening, thickening, and retraction, and sometimes the acquisition of an amoeboid, phagocyting phenotype (Kreutzberg, 1996; Jonas et al., 2012; Heindl et al., 2018; Otxoa-de-Amezaga et al., 2019). Simultaneously, stroke triggers widespread microglial proliferation (Denes et al., 2008; Otxoa-de-Amezaga et al., 2019). Reactive microglia also undergo functional remodeling and start to release cytokines within several hours after the stroke onset (Lambertsen et al., 2012).

Microglia are also instrumental for the poststoke recovery, and are particularly involved in supporting remyelination by inducing OPC proliferation and protecting myelin against damage (Pavic et al., 2021; Raffaele et al., 2021). Microglial cells also contribute to poststroke synaptic plasticity and restoration of neuronal ensembles by supporting synaptogenesis and performing synaptic pruning to shape synaptic connectivity (Szalay et al., 2016).

CONCLUSIONS

Stroke causes ischemic cell death, infiltration of blood-borne factors and leukocytes (secondary in the case of ischemic stroke and primary in the case of hemorrhage), edema, and an inflammatory tissue response. Astrocytes, oligodendrocytes, OPCs, and microglia die in the infarction core and become reactive in the surrounding penumbra. In the penumbra, astrocytes exert neuroprotective functions by maintaining ion, transmitter, energy, and water homeostasis. Over time after stroke, stromal-derived fibrotic scar tissue replaces the necrotic infarction core. Reactive microglia contribute to the cleaning of the cell debris, whereas reactive astrocytes create a protective perilesional barrier or glia limitans perilesiones. Glia contribute to the remodeling of nervous tissue surrounding the lesion, which defines the degree of functional compensation and neurological outcome.

References

Adams F (1849). The genuine works of Hippocrates, The Sydenham Society, London.

Adams F (1856). Aretaeus, the Cappadocean: the extant works, Sydenhwn Society, London.

Adams Jr HP, Bendixen BH, Kappelle LJ et al. (1993). Classification of subtype of acute ischemic stroke. Definitions for use in a multicenter clinical trial. Stroke 24: 35–41.

Almeida A, Delgado-Esteban M, Bolanos JP et al. (2002). Oxygen and glucose deprivation induces mitochondrial dysfunction and oxidative stress in neurones but not in astrocytes in primary culture. J Neurochem 81: 207–217.

Amarenco P, Bogousslavsky J, Caplan LR et al. (2009). Classification of stroke subtypes. Cerebrovasc Dis 27: 493–501.

Anderson MA, Burda JE, Ren Y et al. (2016). Astrocyte scar formation aids central nervous system axon regeneration. Nature 532: 195–200.

Anderson MA, O'Shea TM, Burda JE et al. (2018). Required growth facilitators propel axon regeneration across complete spinal cord injury. Nature 561: 396–400.

Andrew RD, Hartings JA, Ayata C et al. (2022). The critical role of spreading depolarizations in early brain injury: consensus and contention. Neurocrit Care 37: 83–101.

Arundine M, Tymianski M (2003). Molecular mechanisms of calcium-dependent neurodegeneration in excitotoxicity. Cell Calcium 34: 325–337.

Back SA, Craig A, Kayton RJ et al. (2007). Hypoxia-ischemia preferentially triggers glutamate depletion from oligodendroglia and axons in perinatal cerebral white matter. J Cereb Blood Flow Metab 27: 334–347.

Bak LK, Walls AB, Schousboe A et al. (2018). Astrocytic glycogen metabolism in the healthy and diseased brain. J Biol Chem 293: 7108–7116.

Bano D, Ankarcrona M (2018). Beyond the critical point: an overview of excitotoxicity, calcium overload and the downstream consequences. Neurosci Lett 663: 79–85.

Barros LF (2022). How expensive is the astrocyte? J Cereb Blood Flow Metab 42: 738–745.

Berliocchi L, Bano D, Nicotera P (2005). Ca^{2+} signals and death programmes in neurons. Philos Trans R Soc Lond B Biol Sci 360: 2255–2258.

Brown AM (2004). Brain glycogen re-awakened. J Neurochem 89: 537–552.

Brown AM, Ransom BR (2007). Astrocyte glycogen and brain energy metabolism. Glia 55: 1263–1271.

Brown-Séquard CE (1875). Seance du 18 decembre. C R Soc Biol 40: 424.

Carrera E, Tononi G (2014). Diaschisis: past, present, future. Brain 137: 2408–2422.

Chen Y, Swanson RA (2003a). Astrocytes and brain injury. J Cereb Blood Flow Metab 23: 137–149.

Chen Y, Swanson RA (2003b). The glutamate transporters EAAT2 and EAAT3 mediate cysteine uptake in cortical neuron cultures. J Neurochem 84: 1332–1339.

Choudhury GR, Ding S (2016). Reactive astrocytes and therapeutic potential in focal ischemic stroke. Neurobiol Dis 85: 234–244.

Conforti P, Mezey S, Nath S et al. (2022). Fibrinogen regulates lesion border-forming reactive astrocyte properties after vascular damage. Glia 70: 1251–1266.

del Río-Hortega P, Penfield WG (1927). Cerebral cicatrix: the reaction of neuroglia and microglia to brain wounds. Bull John Hopkins Hosp 41: 278–303.

Denes A, Ferenczi S, Halasz J et al. (2008). Role of CX3CR1 (fractalkine receptor) in brain damage and inflammation induced by focal cerebral ischemia in mouse. J Cereb Blood Flow Metab 28: 1707–1721.

Dringen R, Pfeiffer B, Hamprecht B (1999). Synthesis of the antioxidant glutathione in neurons: supply by astrocytes of CysGly as precursor for neuronal glutathione. J Neurosci 19: 562–569.

Dringen R, Gutterer JM, Hirrlinger J (2000). Glutathione metabolism in brain metabolic interaction between astrocytes and neurons in the defense against reactive oxygen species. Eur J Biochem 267: 4912–4916.

Eitelmann S, Stephan J, Everaerts K et al. (2022). Changes in Astroglial K^+ upon brief periods of energy deprivation in the mouse neocortex. Int J Mol Sci 23: 4836.

Escartin C, Galea E, Lakatos A et al. (2021). Reactive astrocyte nomenclature, definitions, and future directions. Nat Neurosci 24: 312–325.

Espinoza F, Magdalena R, Saldivia N et al. (2020). Vitamin C recycling regulates neurite growth in neurospheres differentiated in vitro. Antioxidants (Basel, Switzerland) 9: 1276.

Eyo UB, Miner SA, Ahlers KE et al. (2013). P2X7 receptor activation regulates microglial cell death during oxygen-glucose deprivation. Neuropharmacology 73: 311–319.

Fern RF, Matute C, Stys PK (2014). White matter injury: ischemic and nonischemic. Glia 62: 1780–1789.

Fernandez-Fernandez S, Almeida A, Bolanos JP (2012). Antioxidant and bioenergetic coupling between neurons and astrocytes. Biochem J 443: 3–11.

Fields RD, Ni Y (2010). Nonsynaptic communication through ATP release from volume-activated anion channels in axons. Sci Signal 3: ra73.

Frik J, Merl-Pham J, Plesnila N et al. (2018). Cross-talk between monocyte invasion and astrocyte proliferation regulates scarring in brain injury. EMBO Rep 19: e45294.

Gallizioli M, Miro-Mur F, Otxoa-de-Amezaga A et al. (2020). Dendritic cells and microglia have non-redundant functions in the inflamed brain with protective effects of type 1 cDCs. Cell Rep 33: 108291.

Garcia-Bonilla L, Iadecola C (2012). Peroxiredoxin sets the brain on fire after stroke. Nat Med 18: 858–859.

Garcia-Pupo L, Van San E, Delgado-Hernandez R et al. (2020). Emerging immune and cell death mechanisms in stroke: saponins as therapeutic candidates. Brain Behav Immun Health 9: 100152.

Giffard RG, Swanson RA (2005). Ischemia-induced programmed cell death in astrocytes. Glia 50: 299–306.

Giffard RG, Monyer H, Choi DW (1990). Selective vulnerability of cultured cortical glia to injury by extracellular acidosis. Brain Res 530: 138–141.

Graham SH, Chen J (2001). Programmed cell death in cerebral ischemia. J Cereb Blood Flow Metab 21: 99–109.

Haas S, Brockhaus J, Verkhratsky A et al. (1996). ATP-induced membrane currents in ameboid microglia acutely isolated from mouse brain slices. Neuroscience 75: 257–261.

Hamilton N, Vayro S, Kirchhoff F et al. (2008). Mechanisms of ATP- and glutamate-mediated calcium signaling in white matter astrocytes. Glia 56: 734–749.

Harvey BK, Airavaara M, Hinzman J et al. (2011). Targeted over-expression of glutamate transporter 1 (GLT-1) reduces ischemic brain injury in a rat model of stroke. PloS One 6: e22135.

He Y, Taylor N, Fourgeaud L et al. (2017). The role of microglial P2X7: modulation of cell death and cytokine release. J Neuroinflammation 14: 135.

Heindl S, Gesierich B, Benakis C et al. (2018). Automated morphological analysis of microglia after stroke. Front Cell Neurosci 12: 106.

Hippocrates (1923-1931). Hippocrates, *vols. I-IV*; with an English translation by W.H.S. Jones. William Heinemann New York : G.P. Putnam's Sons, London, New York.

Hu YY, Xu J, Zhang M et al. (2015). Ceftriaxone modulates uptake activity of glial glutamate transporter-1 against global brain ischemia in rats. J Neurochem 132: 194–205.

Jin WN, Shi SX, Li Z et al. (2017). Depletion of microglia exacerbates postischemic inflammation and brain injury. J Cereb Blood Flow Metab 37: 2224–2236.

Jonas RA, Yuan TF, Liang YX et al. (2012). The spider effect: morphological and orienting classification of microglia in response to stimuli in vivo. PloS One 7: e30763.

Karadottir R, Attwell D (2007). Neurotransmitter receptors in the life and death of oligodendrocytes. Neuroscience 145: 1426–1438.

Karadottir R, Cavelier P, Bergersen LH et al. (2005). NMDA receptors are expressed in oligodendrocytes and activated in ischaemia. Nature 438: 1162–1166.

Karenberg A (2020). Historic review: select chapters of a history of stroke. Neurol Res Pract 2: 34.

Karimullah KI (2020). Hippocrates transformed: crafting a Hippocratic discourse of medical semiotics in English, 1850–1930. Humanit Soc Sci Commun 7: 27.

Kreutzberg GW (1996). Microglia: a sensor for pathological events in the CNS. Trends Neurosci 19: 312–318.

Lalancette-Hebert M, Gowing G, Simard A et al. (2007). Selective ablation of proliferating microglial cells exacerbates ischemic injury in the brain. J Neurosci 27: 2596–2605.
Lambertsen KL, Biber K, Finsen B (2012). Inflammatory cytokines in experimental and human stroke. J Cereb Blood Flow Metab 32: 1677–1698.
Lange Canhos L, Chen M, Falk S et al. (2021). Repetitive injury and absence of monocytes promote astrocyte self-renewal and neurological recovery. Glia 69: 165–181.
Larsen BR, Assentoft M, Cotrina ML et al. (2014). Contributions of the Na^+/K^+-ATPase, NKCC1, and K_{ir}4.1 to hippocampal K^+ clearance and volume responses. Glia 62: 608–622.
Lewerenz J, Hewett SJ, Huang Y et al. (2013). The cystine/glutamate antiporter system xc^- in health and disease: from molecular mechanisms to novel therapeutic opportunities. Antioxid Redox Signal 18: 522–555.
Li L, Lundkvist A, Andersson D et al. (2008). Protective role of reactive astrocytes in brain ischemia. J Cereb Blood Flow Metab 28: 468–481.
Lipton P (1999). Ischemic cell death in brain neurons. Physiol Rev 79: 1431–1568.
Lubart A, Benbenishty A, Har-Gil H et al. (2021). Single cortical microinfarcts lead to widespread microglia/macrophage migration along the white matter. Cereb Cortex 31: 248–266.
Luckl J, Lemale CL, Kola V et al. (2018). The negative ultraslow potential, electrophysiological correlate of infarction in the human cortex. Brain 141: 1734–1752.
MacAulay N (2020). Molecular mechanisms of K^+ clearance and extracellular space shrinkage—glia cells as the stars. Glia 68: 2192–2211.
Makar TK, Nedergaard M, Preuss A et al. (1994). Vitamin E, ascorbate, glutathione, glutathione disulfide, and enzymes of glutathione metabolism in cultures of chick astrocytes and neurons: evidence that astrocytes play an important role in antioxidative processes in the brain. J Neurochem 62: 45–53.
Manley GT, Fujimura M, Ma T et al. (2000). Aquaporin-4 deletion in mice reduces brain edema after acute water intoxication and ischemic stroke. Nat Med 6: 159–163.
Marino Lee S, Hudobenko J, McCullough LD et al. (2021). Microglia depletion increase brain injury after acute ischemic stroke in aged mice. Exp Neurol 336: 113530.
Masuda T, Croom D, Hida H et al. (2011). Capillary blood flow around microglial somata determines dynamics of microglial processes in ischemic conditions. Glia 59: 1744–1753.
Matute C, Domercq M, Sanchez-Gomez MV (2006). Glutamate-mediated glial injury: mechanisms and clinical importance. Glia 53: 212–224.
Matute C, Torre I, Perez-Cerda F et al. (2007). P2X(7) receptor blockade prevents ATP excitotoxicity in oligodendrocytes and ameliorates experimental autoimmune encephalomyelitis. J Neurosci 27: 9525–9533.
Micu I, Jiang Q, Coderre E et al. (2006). NMDA receptors mediate calcium accumulation in myelin during chemical ischaemia. Nature 439: 988–992.
Mitani A, Tanaka K (2003). Functional changes of glial glutamate transporter GLT-1 during ischemia: an in vivo study in the hippocampal CA1 of normal mice and mutant mice lacking GLT-1. J Neurosci 23: 7176–7182.
Munoz-Ballester C, Robel S (2023). Astrocyte-mediated mechanisms contribute to traumatic brain injury pathology. WIREs Mech Dis 15: e1622.
Nam MH, Cho J, Kwon DH et al. (2020). Excessive astrocytic GABA causes cortical hypometabolism and impedes functional recovery after subcortical stroke. Cell Rep 32: 107861.
Nelson KB (2007). Perinatal ischemic stroke. Stroke 38: 742–745.
Neumann JT, Cohan CH, Dave KR et al. (2013). Global cerebral ischemia: synaptic and cognitive dysfunction. Curr Drug Targets 14: 20–35.
O'Shea TM, Ao Y, Wang S et al. (2022). Lesion environments direct transplanted neural progenitors towards a wound repair astroglial phenotype in mice. Nat Commun 13: 5702.
O'Shea T, Ao Y, Wang S et al. (2023). Border-forming wound repair astrocytes. bioRxiv. https://doi.org/10.1101/2023.1108.1125.554857. *preprint*.
Otxoa-de-Amezaga A, Miro-Mur F, Pedragosa J et al. (2019). Microglial cell loss after ischemic stroke favors brain neutrophil accumulation. Acta Neuropathol 137: 321–341.
Oz G, DiNuzzo M, Kumar A et al. (2015). Revisiting glycogen content in the human brain. Neurochem Res 40: 2473–2481.
Paciaroni M, Caso V, Agnelli G (2009). The concept of ischemic penumbra in acute stroke and therapeutic opportunities. Eur Neurol 61: 321–330.
Pantoni L, Garcia JH, Gutierrez JA (1996). Cerebral white matter is highly vulnerable to ischemia. Stroke 27: 1641–1646; discussion 1647.
Paolicelli RC, Sierra A, Stevens B et al. (2022). Microglia states and nomenclature: a field at its crossroads. Neuron 110: 3458–3483.
Pavic G, Petzsch P, Jansen R et al. (2021). Microglia contributes to remyelination in cerebral but not spinal cord ischemia. Glia 69: 2739–2751.
Pekny M, Johansson CB, Eliasson C et al. (1999). Abnormal reaction to central nervous system injury in mice lacking glial fibrillary acidic protein and vimentin. J Cell Biol 145: 503–514.
Pellerin L, Magistretti PJ (2012). Sweet sixteen for ANLS. J Cereb Blood Flow Metab 32: 1152–1166.
Petrovic-Djergovic D, Goonewardena SN, Pinsky DJ (2016). Inflammatory disequilibrium in stroke. Circ Res 119: 142–158.
Planas AM (2024). Role of microglia in stroke. Glia 72: 1016–1053.
Platter F (1602). Praxeos seu de cognoscendis, praedicendis, praecavendis curandisque affectibus homini incommodantibus tractatus tres, Translated into English under the title: a Golden practice of Physick. 1662. Peter Cole, London.
Poh L, Kang SW, Baik SH et al. (2019). Evidence that NLRC4 inflammasome mediates apoptotic and pyroptotic microglial death following ischemic stroke. Brain Behav Immun 75: 34–47.
Puyal J, Ginet V, Clarke PG (2013). Multiple interacting cell death mechanisms in the mediation of excitotoxicity and ischemic brain damage: a challenge for neuroprotection. Prog Neurobiol 105: 24–48.
Qin C, Yang S, Chu YH et al. (2022). Signaling pathways involved in ischemic stroke: molecular mechanisms and therapeutic interventions. Signal Transduct Target Ther 7: 215.
Raffaele S, Gelosa P, Bonfanti E et al. (2021). Microglial vesicles improve post-stroke recovery by preventing immune cell senescence and favoring oligodendrogenesis. Mol Ther 29: 1439–1458.
Rochoux J-A (1812). Propositions sur l'apoplexie, Thesis no. 76De l'Imprimerie de Didot jeune.
Rose CR, Verkhratsky A (2024). Sodium homeostasis and signalling: the core and the hub of astrocyte function. Cell Calcium 117: 102817.
Rosenberg PA, Aizenman E (1989). Hundred-fold increase in neuronal vulnerability to glutamate toxicity in astrocyte-poor cultures of rat cerebral cortex. Neurosci Lett 103: 162–168.

Rossi DJ, Brady JD, Mohr C (2007). Astrocyte metabolism and signaling during brain ischemia. Nat Neurosci 10: 1377–1386.

Schiller F (1970). Concepts of stroke before and after Virchow. Med Hist 14: 115–131.

Shannon C, Salter M, Fern R (2007). GFP imaging of live astrocytes: regional differences in the effects of ischaemia upon astrocytes. J Anat 210: 684–692.

Shi K, Tian DC, Li ZG et al. (2019). Global brain inflammation in stroke. Lancet Neurol 18: 1058–1066.

Sies H (1999). Glutathione and its role in cellular functions. Free Radic Biol Med 27: 916–921.

Silver IA, Erecinska M (1990). Intracellular and extracellular changes of [Ca^{2+}] in hypoxia and ischemia in rat brain in vivo. J Gen Physiol 95: 837–866.

Sims NR, Yew WP (2017). Reactive astrogliosis in stroke: contributions of astrocytes to recovery of neurological function. Neurochem Int 107: 88–103.

Siushansian R, Dixon SJ, Wilson JX (1996). Osmotic swelling stimulates ascorbate efflux from cerebral astrocytes. J Neurochem 66: 1227–1233.

Smith D, Pernet A, Hallett WA et al. (2003). Lactate: a preferred fuel for human brain metabolism in vivo. J Cereb Blood Flow Metab 23: 658–664.

Sofroniew MV (2015). Astrocyte barriers to neurotoxic inflammation. Nat Rev Neurosci 16: 249–263.

Sofroniew MV (2020). Astrocyte reactivity: subtypes, states, and functions in CNS innate immunity. Trends Immunol 41: 758–770.

Somjen GG (2004). Ions in the brain: normal function, seizures, and stroke, Oxford University Press, New York, NY.

Szalay G, Martinecz B, Lenart N et al. (2016). Microglia protect against brain injury and their selective elimination dysregulates neuronal network activity after stroke. Nat Commun 7: 11499.

Tekkok SB, Brown AM, Westenbroek R et al. (2005). Transfer of glycogen-derived lactate from astrocytes to axons via specific monocarboxylate transporters supports mouse optic nerve activity. J Neurosci Res 81: 644–652.

Tsoucalas G, Papaioannou TG, Karamanou M (2019). The Hippocratic doctrine of "the acute brain suffering" as the brain stroke. J Stroke Cerebrovasc Dis 28: 412–417.

Vanden Berghe T, Linkermann A, Jouan-Lanhouet S et al. (2014). Regulated necrosis: the expanding network of non-apoptotic cell death pathways. Nat Rev Mol Cell Biol 15: 135–147.

Vargas MR, Johnson DA, Sirkis DW et al. (2008). Nrf2 activation in astrocytes protects against neurodegeneration in mouse models of familial amyotrophic lateral sclerosis. J Neurosci 28: 13574–13581.

Verkhratsky A, Butt AM (2023). Neuroglia: function and pathology, Elsevier, New York.

Verkhratsky A, Nedergaard M (2018). Physiology of astroglia. Physiol Rev 98: 239–389.

Verkhratsky A, Rose CR (2020). Na^+-dependent transporters: the backbone of astroglial homeostatic function. Cell Calcium 85: 102136.

Verkhratsky A, Semyanov A (2022). The great astroglial metabolic revolution: mitochondria fuel astrocyte homeostatic support and neuroprotection. Cell Calcium 104: 102583.

Verkhratsky A, Butt A, Li B et al. (2023). Astrocytes in human central nervous system diseases: a frontier for new therapies. Signal Transduct Target Ther 8: 396.

Virchow R (1856). Gesammelte Abhandlungen zur Wissenschaftlichen Medicin, Verlag von Medinger & Comp.: Frankfurt A.M.

Vitale I, Pietrocola F, Guilbaud E et al. (2023). Apoptotic cell death in disease-current understanding of the NCCD 2023. Cell Death Differ 30: 1097–1154.

von Monakow C (1914). Die Localization im Grosshirn und der Abbau der Funktion durch korticale Herde, JF Bergmann, Wiesbaden.

Wahane S, Sofroniew MV (2021). Loss-of-functionmanipulations to identify roles of diverse glia and stromal cells during CNS scar formation. Cell Tissue Res387: 337–350.

Wang Y, Liu G, Hong D et al. (2016). White matter injury in ischemic stroke. Prog Neurobiol 141: 45–60.

Wepfer JJ (1658). Observationes anatomicae, ex cadaverum eorum, quos sustulit apoplexia. Cum exercitatione de eius loco affecto, Typis J.C. Suteri, Schafhusii.

Wilhelmsson U, Andersson D, de Pablo Y et al. (2017). Injury leads to the appearance of cells with characteristics of both microglia and astrocytes in mouse and human brain. Cereb Cortex 27: 3360–3377.

Wilson JX, Peters CE, Sitar SM et al. (2000). Glutamate stimulates ascorbate transport by astrocytes. Brain Res 858: 61–66.

Wu MY, Yiang GT, Liao WT et al. (2018). Current mechanistic concepts in ischemia and reperfusion injury. Cell Physiol Biochem 46: 1650–1667.

Xie D, Liu H, Xu F et al. (2021). IL33 (interleukin 33)/ST2 (interleukin 1 receptor-like 1) Axis drives protective microglial responses and promotes white matter integrity after stroke. Stroke 52: 2150–2161.

Yang Z, Zhao TZ, Zou YJ et al. (2014). Hypoxia induces autophagic cell death through hypoxia-inducible factor 1alpha in microglia. PLoS One 9: e96509.

Yi C, Verkhratsky A, Niu J (2023). Pathological potential of oligodendrocyte precursor cells: terra incognita. Trends Neurosci 46: 581–596.

Zhang LY, Hu YY, Zhao CC et al. (2019). The mechanism of GLT-1 mediating cerebral ischemic injury depends on the activation of p38 MAPK. Brain Res Bull 147: 1–13.

Zhao Y, Rempe DA (2010). Targeting astrocytes for stroke therapy. Neurotherapeutics 7: 439–451.

Zhivotovsky B, Orrenius S (2011). Calcium and cell death mechanisms: a perspective from the cell death community. Cell Calcium 50: 211–221.

Zinnhardt B, Viel T, Wachsmuth L et al. (2015). Multimodal imaging reveals temporal and spatial microglia and matrix metalloproteinase activity after experimental stroke. J Cereb Blood Flow Metab 35: 1711–1721.

Handbook of Clinical Neurology, Vol. 210 (3rd series)
Neuroglia in Neurologic and Psychiatric Disorders, Part II
A. Verkhratsky, L.D. de Witte, E. Aronica, and E.M. Hol, Editors
https://doi.org/10.1016/B978-0-443-19102-2.00015-6

Chapter 7

Unraveling the complexity of microglial responses in traumatic brain and spinal cord injury

REBECCA J. HENRY[1*] AND DAVID J. LOANE[2]

[1]*Department of Pharmacology, School of Biomolecular and Biomedical Science, University College Dublin, Dublin, Ireland*

[2]*School of Biochemistry and Immunology, Trinity Biomedical Sciences Institute, Trinity College Dublin, Dublin, Ireland*

Abstract

Microglia, the resident innate immune cells of the central nervous system (CNS), play an important role in neuroimmune signaling, neuroprotection, and neuroinflammation. In the healthy CNS, microglia adopt a surveillant and antiinflammatory phenotype characterized by a ramified scanning morphology that maintains CNS homeostasis. In response to acquired insults, such as traumatic brain injury (TBI) or spinal cord injury (SCI), microglia undergo a dramatic morphologic and functional switch to that of a reactive state. This microglial switch is initially protective and supports the return of the injured tissue to a physiologic homeostatic state. However, there is now a significant body of evidence that both TBI and SCI can result in a chronic state of microglial activation, which contributes to neurodegeneration and impairments in long-term neurologic outcomes in humans and animal models. In this review, we discuss the complex role of microglia in the pathophysiology of TBI and SCI, and recent advancements in knowledge of microglial phenotypic states in the injured CNS. Furthermore, we highlight novel therapeutic strategies targeting chronic microglial responses in experimental models and discuss how they may ultimately be translated to the clinic for human brain and SCI.

INTRODUCTION

Neuroinflammation is an important secondary injury mechanism in the pathophysiology of central nervous system (CNS) injury (SCI) (Alexander and Popovich, 2009; Johnson et al., 2013; Smith et al., 2013a; Faden and Loane, 2015; DiSabato et al., 2016). Microglia, the resident innate immune cells of the CNS, play essential roles in neuroimmune signaling, neuroprotection, and inflammation in the brain and spinal cord (Perry, 1998). Microglia maintain CNS homeostasis by supporting and protecting neuronal function (Perry, 1998; Perry et al., 2010). In the healthy CNS, microglia adopt a surveillant and antiinflammatory phenotype characterized by a ramified scanning morphology that maintains CNS homeostasis and contributes to learning-associated synaptic plasticity (Cornell et al., 2022). Following local tissue damage or viral/bacterial challenge, microglia undergo a morphologic transformation and dramatically alter their function. Initially, this phenotypic switch plays an important protective role against the initial pathologic insult, including clearance of cellular debris by phagocytosis that facilitates wound healing and tissue repair responses (Loane and Kumar, 2016). However, when microglial activation becomes chronic in nature, or dysregulated, it can have long-term detrimental effects on neuronal health and contribute to neurodegeneration (Loane et al., 2014; Faden et al., 2016; Loane and Kumar, 2016).

Based on morphologic characteristics and the presence of specific inflammatory markers, microglia were initially classified as either exhibiting a proinflammatory (activated) or antiinflammatory (resting) phenotypic state. However, this perspective is grossly oversimplified and fails to reflect the considerable molecular diversity of microglia in the healthy and diseased brain (Cherry et al., 2014; Ransohoff, 2016). Recent advancements in RNA-sequencing technologies have enabled unbiased transcriptional analyses of microglia phenotypic states that are related to functional responses (Keren-Shaul et al., 2017; Sousa et al., 2018; Masuda et al., 2020). Among others, disease-associated microglia (DAMs) have been identified as a unique microglial population present in several neurodegenerative diseases that are proposed to restrict neurodegenerative processes (Keren-Shaul et al., 2017). Thus, the identification of novel microglial phenotypic states in healthy and diseased states may have important implications for the development of novel, more selective

*Correspondence to: Rebecca J. Henry Ph.D., Department of Pharmacology, School of Biomolecular and Biomedical Science, University College Dublin, Dublin, Ireland. Tel: +353-1-896-1616, E-mail: rebecca.henry@ucd.ie

therapeutic approaches that target microglial responses in the injured CNS or during neurodegenerative disease.

In this review, we discuss the complex role of microglia in CNS injury and recent advancements in knowledge of microglial phenotypic states in the injured brain and spinal cord. Furthermore, we highlight novel therapeutic strategies targeting microglial responses in experimental models and discuss how they may ultimately be translated to the clinic for human CNS trauma.

ORIGIN AND FUNCTION OF MICROGLIA

Microglia are ubiquitously distributed throughout the CNS, accounting for 5%–20% of the total glial cell population within the CNS parenchyma (Perry, 1998; Ginhoux et al., 2013). High regional densities of microglia are reported in brain regions, such as the hippocampus and basal ganglia (Lawson et al., 1990; Mittelbronn et al., 2001). The Spanish Neuroscientist, Pío del Río-Hortega, was the first to clearly identify a small population of phagocytic, migratory cells in the brain, which he later described as *microglia*. He proposed that microglia arise from embryonic progenitors as meningeal macrophages infiltrating the brain during early embryonic development (del Rio-Hortega, 1932; Rio-Hortega, 1939). However, due to experimental misinterpretation, it took nearly 100 years for microglia to be recognized as a specialized and separate macrophage population within the CNS with distinct developmental origins (Ginhoux et al., 2010). Elegant fate mapping experiments have shown that the primitive or "early wave" of hematopoiesis in the yolk sac (YS) gives rise to adult microglia that are seeded in the brain rudiment at embryonic day 9.5 during murine embryonic development (Ginhoux et al., 2010; Schulz et al., 2012; Kierdorf et al., 2013). Notably, the YS origin of microglia is conserved across species (Herbomel et al., 1999). In humans, microglia-like cells can be detected in the fetus at gestational week 13 (Hutchins et al., 1990). Between gestational weeks 13 and 18, various morphologic features are evident across different brain regions including amoeboid and ramified cells (Hutchins et al., 1990). However, it appears that well-differentiated microglial populations are only evident in the developing human brain near the end of the gestational period (Esiri et al., 1991; Rezaie and Male, 2002; Rezaie et al., 2005; Verney et al., 2010).

Based on similar morphologic characteristics and phagocytic activities, there was initially strong support that microglia may eventually arise from blood monocytes that subsequentially invade the CNS (for review see Ginhoux et al., 2013). However, two pioneering studies definitively demonstrated that the maintenance and local expansion of microglia in the adult healthy brain depends exclusively on the self-renewal capacity of CNS-resident cells, with no requirement for an influx of circulating progenitors from the blood or bone marrow (Ajami et al., 2007; Mildner et al., 2007). Under pathologic conditions circulating monocytes are capable of infiltrating the CNS and contribute to disease/secondary injury (Mildner et al., 2007; Ajami et al., 2011; Butovsky et al., 2012; Jung and Schwartz, 2012; Prinz and Priller, 2014; Yamasaki et al., 2014; Jay et al., 2015; Theriault et al., 2015); however, they do not contribute to the resident microglial pool (Ajami et al., 2011). After colonizing the CNS, microglial progenitor cells quickly proliferate, and approximately 95% of microglia are generated during the first 2 weeks after birth (Alliot et al., 1999; Tambuyzer et al., 2009). Microglia maintain their population through rapid self-renewal, with several renewals occurring throughout the lifespan. This renewal was initially suggested to be largely derived from microglial progenitors expressing nestin in the brain (Elmore et al., 2014). However, this has been much debated, and a more recent study suggests that repopulated microglia are derived from the proliferation of the few remaining $<1\%$ nondepleted microglia (Huang et al., 2018).

Microglial development, differentiation, and maturation, are controlled by several factors, including cell–cell interactions with neurons and other cells (Wendimu and Hooks, 2022). Microglia activation is regulated by the interaction between microglial CX3-chemokine receptor 1 (CX3CR1) and its neuronal ligand, CX3CL1 (Lee et al., 2010; Wolf et al., 2013). Microglial proliferation and differentiation are dependent on the interaction between colony-stimulating factor 1 receptor (CSF1R), expressed on microglia, with its neuronal-derived ligands, colony-stimulating factor 1 (CSF-1) and interleukin (IL)-34 (Ginhoux et al., 2010; Wang et al., 2012; Elmore et al., 2014; Easley-Neal et al., 2019). Mutations in the CSF1R gene are associated with the development of CSF1R-related leukoencephalopathy (Konno et al., 2018). This results in progressive white matter neurodegeneration and is characterized by neuropsychiatric symptoms including progressive cognitive decline, and motor symptoms including Parkinsonism (Konno et al., 2018). CSF1R-deficient mice are largely devoid of microglia throughout their lifespan (Ginhoux et al., 2010; Elmore et al., 2014). However, and perhaps somewhat surprisingly, adult mice lacking microglia due to pharmacologic inhibition of CSF1R show no behavioral abnormalities (Elmore et al., 2014). In addition, while CSF1R deficient rats exhibit postnatal growth retardation (Pridans et al., 2018), genomic deletion of fms-intronic regulatory element (FIRE), a highly conserved super-enhancer located on the CSF1R locus, selectively deletes microglia independent of any behavioral abnormalities (Rojo et al., 2019).

In the uninjured brain and spinal cord, microglia display ramified morphologies characterized by a small cell body, thin and long highly branched processes. Initially, ramified microglia were incorrectly described as "resting." However, studies using transgenic GFP-labeled microglial reporter mice ($Cx3cr1^{GFP/+}$) and time-lapse two-photon imaging revealed that ramified microglia are never resting and are highly mobile, capable of extending and retracting their processes (Davalos et al., 2005; Nimmerjahn et al., 2005). They provide continual surveillance of the CNS microenvironment through the release of secreted mediators and/or through direct contact with neurons and other glial cells (Davalos et al., 2005; Nimmerjahn et al., 2005). Notably, following laser-induced injury, microglial processes rapidly converge to the site of focal injury without actual movement of the microglial cell body, thereby forming a barrier between healthy and injured tissue (Davalos et al., 2005). Furthermore, recent studies using spatial transcriptomics analysis have shed light on the transcriptional landscape of ramified microglia in the murine CNS and implicate them in synaptic integrity, neuronal maturation, and maintenance of homeostasis (Masuda et al., 2020). Thus, in addition to neuroprotection and maintenance of tissue homeostasis, ramified microglia play a critical role in neural development, synaptic plasticity, neurogenesis, and neuronal surveillance (Tremblay et al., 2011; Katsumoto et al., 2014; Salter and Beggs, 2014).

MICROGLIA IN THE INJURED CNS: PRIMED FOR IMMUNE DYSFUNCTION AND CHRONIC NEUROINFLAMMATION

Following traumatic brain injury (TBI), microglia transform from a ramified to an amoeboid morphology that is characterized by an enlarged cell body, short thick processes, and the presence of numerous cytoplasmic vacuoles. Phenotypic markers of amoeboid and proinflammatory microglia include cluster of differentiation (CD)16, CD32, CD68, major histocompatibility complex (MHC) II, inducible nitric oxide synthase (iNOS), tumor necrosis factor-α (TNFα) and interleukin (IL)-1β (Kaur et al., 1985; Nimmerjahn et al., 2005; Madore et al., 2013). In contrast, ramified microglia produce antiinflammatory cytokines and neurotrophins, including IL-10 and brain-derived neurotrophic factor (BDNF), that resolve inflammation and support a return to tissue homeostasis following injury (Loane and Kumar, 2016).

Microglia express immune pattern recognition receptors (PRRs), including toll-like receptors (TLRs), nucleotide-binding oligomerization domain (NOD)-like receptors (NLRs), scavenger receptors (SR), and purinergic receptors, that detect changes in the external environment (Jassam et al., 2017; Younger et al., 2019). In the immediate phase following TBI, pathogen-associated molecular patterns (PAMPs) and endogenous damage-associated molecular patterns (DAMPs) are released from the injured neurons and are recognized by PRRs on microglia. As a neuroprotective mechanism and to prevent ongoing neuronal damage, microglia transition to a proinflammatory state (Faden et al., 2016; Jassam et al., 2017). This initiates activation of intracellular signaling cascades, resulting in the NLR family pyrin domain containing 3 (NLRP3) inflammasome assembly that promotes the generation of mature IL-1β and IL-18 and the activation of nuclear factor kappa-light-chain-enhancer of activated B cells (NFκB). This in turn induces the production of proinflammatory cytokines, chemokines, and reactive oxygen species (ROS) (Jassam et al., 2017). A robust inflammatory response is required for the removal of injury-related debris and damaged neurons from the lesion microenvironment (Loane and Byrnes, 2010; Loane and Kumar, 2016). Following lesion clean-up, it would be expected that pro- and antiinflammatory microglia cooperate to resolve inflammation, thus allowing a return to homeostasis after TBI (Loane and Kumar, 2016). However, there is now a significant body of evidence that chronic nonresolving proinflammatory microglial activation occurs beyond 1-year postinjury in experimental models and also in severe TBI patients (Nagamoto-Combs et al., 2007; Ramlackhansingh et al., 2011; Acosta et al., 2013; Johnson et al., 2013; Smith et al., 2013b; Aungst et al., 2014; Loane et al., 2014; Mouzon et al., 2014), which may be a result of a self-perpetuating cycle of microglial-mediated neurotoxicity (Gao et al., 2003; Qin et al., 2004). Active microglial phagocytosis (i.e., phagoptosis) is also reported to contribute to delayed neuronal cell loss and long-term functional deficits following experimental stroke (Neher et al., 2013; Zhang et al., 2020). Similarly, in experimental TBI models, increased phagocytic activity, lysosomal burden, and lipid accumulation in lipofuscin-positive microglia are evident for 1-year postinjury; effects of which are mediated by phagocyte-mediated oxidative stress (Ritzel et al., 2023). Notably, pharmacologic depletion of lipofuscin-positive microglia in aged mice decreases chronic phagocytosis and inflammation, and results in reduced TBI-induced neurodegeneration and improved neurologic outcomes in aged TBI mice (Ritzel et al., 2023).

In contrast to the classic view of binary beneficial (antiinflammatory) vs detrimental (proinflammatory) "polarization" states in microglia, there is now robust evidence that microglia in the injured brain have heterogeneous and overlapping expression profiles of both pro- and antiinflammatory markers (Morganti et al., 2016; Kumar et al., 2016a; Lyu et al., 2021). In the experimental diffuse brain injury model, midline fluid percussion injury (mFPI), rod-like microglia with distinct morphologic characteristics, and inflammatory markers are evident (Taylor et al., 2014). These rod-like microglia form adjacent to hypertrophied microglia, cluster among layer V pyramidal neurons, and contribute to TBI-induced astrogliosis and cortical inflammation at 7 days postinjury (dpi) (Witcher et al., 2018). In addition, mFPI induces a more robust change in cortical inflammatory gene expression at subacute timepoints (7 dpi) when compared to the more focal injury model of lateral (l)FPI (Witcher et al., 2020). In the moderate–severe controlled cortical impact (CCI) experimental model of TBI, a peak in macrophage/microglia cortical proinflammatory gene expression is reported at 7 dpi (Kumar et al., 2016a; Henry et al., 2019). Notably, pharmacologic interventions targeting these inflammatory responses are associated with a decrease in neurodegenerative processes and improvements in motor recovery through 21 dpi (Henry et al., 2019). In the experimental model of mild closed head injury (CHI) model of TBI, microglial activation is brain region-specific with no evident activation in the cortex at 24 h postinjury or at 10 dpi (Mouzon et al., 2012). In contrast, evident microglia activation was observed in the corpus callosum at 12 months postinjury (Mouzon et al., 2014). Using $[^{18}F]$DPA-714 as a marker of activated microglia, another study showed that although the number of $Iba1^+$ microglia was increased following mild TBI, this was associated with a low uptake of $[^{18}F]$DPA-714 at 7 dpi (Israel et al., 2016). In an experimental model of repetitive mild TBI (CHI), activated $Iba1^+$ microglia were reported in the corpus callosum up to 12 months postinjury and this was associated with persistent cognitive deficits (Mouzon et al., 2014). In a similar manner, substantial neuronal loss, hippocampal microglial activation, and cognitive decline were reported up to 28 dpi in the repetitive mild lateral FPI injury model (Aungst et al., 2014). Overall, these studies suggest that several critical factors may influence microglial phenotypic states in TBI. These include the experimental model under investigation (focal contusion vs diffuse axonal injury), injury severity (mild vs moderate vs severe), and injury frequency (single vs repetitive). Additional factors include injury location within gray or white matter regions, sex, and aging (Loane and Kumar, 2016; Caplan et al., 2017; Doran et al., 2019; Ma et al., 2019; Rahimian et al., 2019; Green et al., 2021).

In a similar manner to TBI, microglia play an important role in the modulation of the inflammatory response to spinal cord injury (SCI). Following injury, microglia undergo activation, proliferation, and morphologic changes that are associated with either beneficial effects on repair processes or detrimental effects via excessive secretion of neurotoxic cytokines and reactive oxygen species (for review see (Kroner and Rosas, 2019; Xu et al., 2021; Deng et al., 2022). Notably, microglia are reported to be highly dynamic and undergo extensive proliferation during the first 2 weeks following exposure to moderate-level contusive SCI in mice, particularly at the lesion site where they play a crucial role in scar formation (Bellver-Landete et al., 2019). Microglia also majorly contribute to the early injury responses following SCI, via phagocytosis of damaged and degenerating tissue (Greenhalgh and David, 2014; Bellver-Landete et al., 2019). During the more chronic phases postinjury, peripherally derived macrophages are the primary contributors to phagocytosis, albeit less efficient at processing CNS debris (Greenhalgh and David, 2014). However, while beneficial effects of microglia responses are reported in the context of SCI, chronic and uncontrolled microglia activation can promote secondary injury responses, for review see (Courtine and Sofroniew, 2019; Venkatesh et al., 2019; Deng et al., 2022).

IDENTIFICATION OF NOVEL MICROGLIAL PHENOTYPIC STATES IN CNS INJURY STATES

The widespread use of RNA sequencing technologies has provided a deeper understanding of microglial phenotypic states in brain health and disease. Disease-associated microglia (DAMs) were first described in Alzheimer disease (AD) and have since been reported in other models of neurodegeneration (Keren-Shaul et al., 2017). Genes associated with DAM activation include Axl, apolipoprotein (Apoe), C-type lectin domain containing 7A (Clec7a), integrin subunit alpha X (Itgax), galectin-3 (Lgals3), and cystatin F (Cst7) (Keren-Shaul et al., 2017). Increased expression of DAM markers in microglia positively stained for intracellular β-amyloid has been reported in the brains of 5XFAD mice and in the postmortem brain of AD patients (Keren-Shaul et al., 2017). DAM activation is a two-step process, the first of which is triggering receptor expressed on myeloid cells 2 (TREM2)-independent, and involves downregulation of microglia checkpoints, while the second step from DAM1 to DAM2 is TREM2-dependent (Keren-Shaul et al., 2017). Although DAM activation is proposed to restrict neurodegenerative processes, further research is required to fully understand the role of DAM activation in neurodegenerative disease (Keren-Shaul et al., 2017; Deczkowska et al., 2018). Another unique microglial phenotype recently identified in neurodegenerative states is "interferon-response microglia (IRM)." Genes associated with IRM include interferon regulatory factor (Irf)7, interferon alpha (Ifna)4, Ifnb1, interferon alpha and beta receptor subunit 1 (Ifnar), interferon activated gene (Ifi)204, signal transducer and activator of transcription (Stat)1, Stat2, interferon-induced protein with tetratricopeptide repeats (Ifit)2, and C—C motif chemokine ligand (Ccl)5 (Jin et al., 2012; Witcher et al., 2018). Other unique microglial phenotypes have been identified in the diseased brain, including neurodegenerative microglia (MGnD) (Krasemann et al., 2017). The switch from homeostatic microglia to MGnD is induced via phagocytosis of apoptotic neurons (Krasemann et al., 2017). Furthermore, MGnD microglia hypersecrete phosphorylated tau extracellular vesicles (Clayton et al., 2021). The homeostatic to MGnD switch is mediated via the TREM2-Apoe pathway, which subsequently activates inflammatory genes, including Axl, Itgax, Clec7a, and Apoe, with a concomitant reduction in markers associated with homeostatic microglia, including transforming growth factor beta receptor (Tgfbr), hexosaminidase subunit beta (Hexb), purinergic receptor (P2ry12), and C-X3-C motif chemokine receptor (Cx3cr)1 (Krasemann et al., 2017). Like the DAM phenotype, the functional role of MGnD in neurodegeneration is not fully understood. Homeostatic microglia in the noninjured brain express phenotypic markers, including P2ry12, Fc receptor-like 2 (Fcrls), transmembrane protein (Tmem)119, olfactomedin (Olfml)3, spalt like transcription factor (Sall)1, transforming growth factor-b (TGF-β) signaling pathway (Tgfr1, Tgf-β, Smad3) and surface markers such as ectonucleoside triphosphate diphosphohydrolase (Entpd)1, Csfr1, and Trem2 (Sousa et al., 2018; Masuda et al., 2020). Lipid droplet-accumulating microglia (LDAM) in the aged brain express progranulin gene (GRN), solute carrier family 33 member 1 (SLC33A1), sorting nexin 17 (SNX17), and neuronal ceroid lipofuscinosis 3 (NCL3), among others (Marschallinger et al., 2020). Age-related LDAMs have a dysfunctional profile characterized by defective phagocytosis and the production of high levels of ROS and proinflammatory cytokines (TNF-α and IL-1β) (Afridi et al., 2020; Wei and Li, 2022). Several other microglial subtypes have been described, including "dark microglia" (St-Pierre et al., 2020), "proliferative-region-associated microglia (PAM)" (Li et al., 2019), white matter-associated microglia (WAMs) (Safaiyan et al., 2021), "activated response microglia (ARM)" (Friedman et al., 2018; Wei and Li, 2022), "necroptotic microglia (NM)" (Yang et al., 2017; Wei and Li, 2022), "late-response microglia (LRM)" (Mathys et al., 2017), and "dystrophic microglia" (Streit et al., 2009; Kenkhuis et al., 2021).

In diffuse and mixed focal/diffuse brain injury models, distinct microglial phenotypes associated with type I interferon signaling, inflammation, and damage/neuropathology have been identified at subacute timepoints postinjury (Todd et al., 2021; Witcher et al., 2021). In addition, significant overlap in genes altered following mild and severe midline FPI have been identified, including those related to type I interferon and DAM (Todd et al., 2021). In a similar manner, IFN-β and several IFN-related genes are reported to be upregulated in cortical tissue at 72h postmoderate level CCI (Barrett et al., 2020). Interestingly, expression of IFN-related genes including IFN-β, Ifi204, Irf7, and IFN-I signaling are enhanced in the injured cortex of aged mice following mild-level CCI at 24 h postinjury (Barrett et al., 2021). Subsequent findings demonstrated that genetic and/or pharmacologic targeting of IFN-I signaling was associated with protective effects on neurologic outcomes, albeit transient in the case of pharmacologic intervention in adult mice (Barrett et al., 2020).

Transcriptional analysis has also identified an increase in DAM signature genes up to 12 weeks following moderate-level CCI in the aged brain (Ritzel et al., 2022). This was characterized by a concurrent downregulation of homeostatic pathways and an upregulation of genes involved in lysosomal and phagocytic pathways (Ritzel et al., 2022). Notably, an overlap with fully activated DAM (stage 2), as well as stage 1 DAM, in the microglial transcriptomes is reported in mild and severe mFPI (Todd et al., 2021). However, the authors reported no significant overlap in either TBI model with the microglial checkpoint genes that are shown to be downregulated in DAM, nor was there an increase in microglial TREM2, a critical gene required for the full activation of DAM (Keren-Shaul et al., 2017). In contrast, TREM2 is reported to be increased in the chronically injured aged brain in a moderate-level CCI model (Ritzel et al., 2022). However, it is important to note that this analysis was carried out at a tissue level and may not be microglial-specific. Thus, it is plausible to suggest that TBI may induce a distinct microglial phenotypic response other than DAM and further emphasizes the need for microglia-specific transcriptional studies in the TBI setting (Todd et al., 2021).

In a severe contusion model of SCI, scRNAseq analysis demonstrated that baseline microglia undergo permanent transcriptional reprogramming to that of a DAM-like state (Hakim et al., 2021). Interestingly, DAM is reported to play an important role in locomotor recovery following exposure to SCI (Hakim et al., 2021). Deeper analysis by scRNAseq identified 11 major microglial subsets in a contusion SCI model (Brennan et al., 2022). Notably, homeostatic 1 (64.7%) and homeostatic 2 (26.3%) microglia predominate prior to induction of SCI and are replaced by proliferating, nonlipid phagocytosing, and lipid-processing 1 microglia at 7 dpi (Brennan et al., 2022). At 28 dpi, the predominate microglia subsets are lipid processing 1, patrolling, proliferating, and iron processing microglia. Consistently decreased genes at chronic timepoints include *P2ry12*, *HexB*, *Lgmn*, *Csf1r*, *Cx3cr1*, which play important roles in immune signaling, differentiation, and apoptosis (Brennan et al., 2022). In contrast, chronically upregulated genes included *Apoe*, *Spp1*, and *Fth1*, which enhance processes such as translation, transcription, and proliferation (Brennan et al., 2022). These data indicate that following SCI microglia upregulate molecular pathways involved in functional neuroimmune responses, including proliferation, innate immunity, immunometabolism, and iron and lipid processing (Brennan et al., 2022).

Examination of the microglial transcriptional profile up to 90 days following moderate-level CCI identified an initial upregulation followed by progressive downregulation of several genes involved in immune responses and tissue repair, including TNF superfamily member (Tnfsf)10, bone marrow stromal cell antigen (Bst)2, insulin growth factor (Igf)1, and CC motif chemokine receptor (Ccr)1 (Makinde et al., 2020). In contrast, genes associated with long-term regulation of synaptic potentiation, including protein tyrosine phosphatase nonreceptor type (Ptpn)5, SH3 and multiple ankyrin repeat domains (Shank)3, and sequestosome (Sqstm)1 were upregulated at 90-dpi when compared to 7 dpi (Makinde et al., 2020). Overall, these findings may provide novel insights into potential common molecular pathways altered in CNS injuries and other neurodegenerative disease states, as upregulated genes play important roles in cognitive impairments in AD (e.g., Ptpn5), schizophrenia (e.g., Shank3), and autism (e.g., Sqstm1) (Makinde et al., 2020).

Thus, the identification of these diverse microglial phenotypic states suggests that we need to move toward a more nuanced conception of microglia as having more heterogeneous phenotypes/genotypes depending on disease/injury context, CNS region under investigation, age, and sex (Fig. 7.1). Ultimately, the advanced understanding of microglial phenotypic states may lead to the development of novel, more selective therapeutic approaches to target microglial responses in CNS injuries and other related neurodegenerative diseases.

MICROGLIA AS A THERAPEUTIC TARGET IN TBI

Minocycline is a broad-spectrum tetracycline antibiotic that has antiinflammatory properties. It is often used as an inhibitor of microglial/macrophage function (Kobayashi et al., 2013; Han et al., 2019). In experimental TBI, differential effects of minocycline treatment are reported depending on injury model type and age at injury. Specifically, minocycline treatment exerts neuroprotective effects, including reductions in lesion volume and microglial activation at 24 h postinjury in a CHI model of mild TBI in adult male mice (Homsi et al., 2010). Moreover, improvements in TBI-induced locomotor hyperactivity are evident up to 8 weeks postinjury in minocycline-treated mice (Homsi et al., 2010). Similarly, neuroprotective effects of minocycline treatment attenuated memory deficits in male mice up to 13 weeks following exposure to CHI (Siopi et al., 2012). In contrast, in an experimental model of closed-head CCI in neonatal rats, although protective effects were evident during the acute phases (up to 3 dpi), minocycline treatment exacerbated TBI-induced cognitive deficits at more chronic timepoints (Hanlon et al., 2016). Furthermore, chronic minocycline treatment exacerbated microglial activation and neurodegeneration in neonatal rats exposed to CCI (Hanlon et al., 2017). Similarly, neither acute nor delayed minocycline treatment improved TBI-induced impulsivity and inattention in adult male rats following exposure to bilateral frontal CCI (Pechacek et al., 2022). Overall, these preclinical studies suggest that the injury model and age, as well as the treatment paradigm, may be important factors to consider when evaluating the therapeutic potential of minocycline. In human TBI, a study evaluated the therapeutic potential of minocycline treatment in 15 moderate–severe TBI patients, based on the Mayo classification

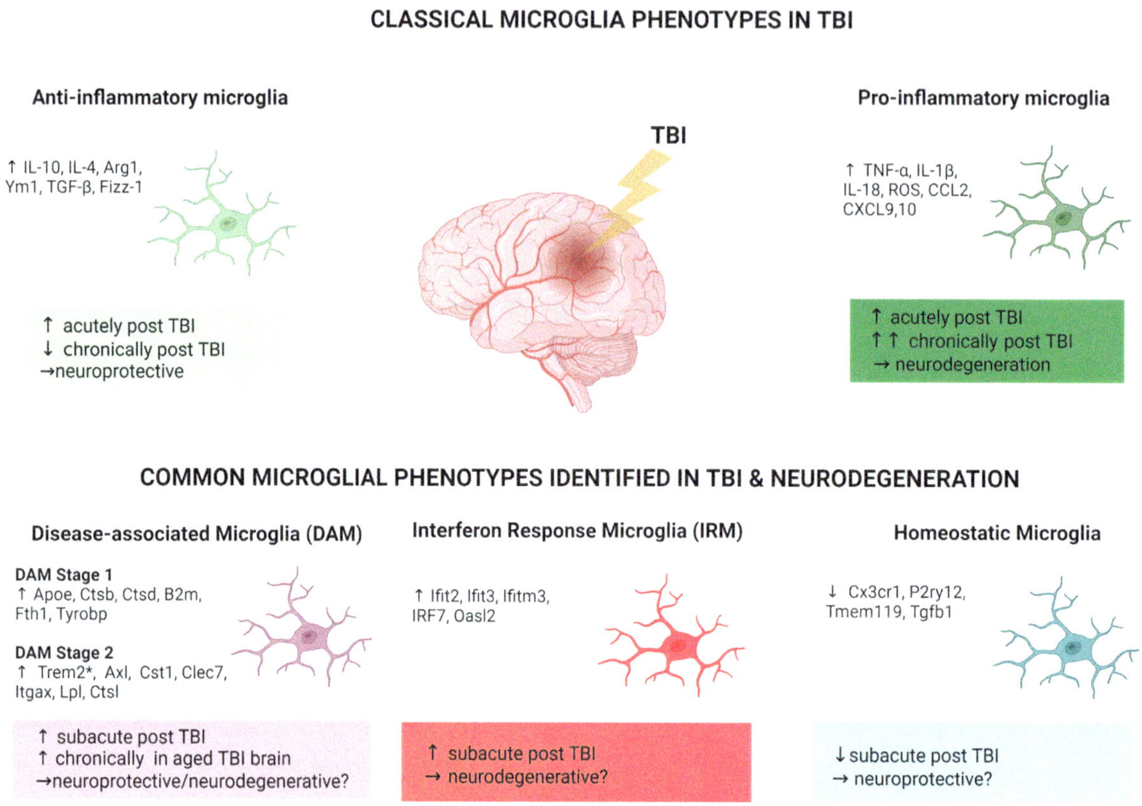

Fig. 7.1. Microglia phenotypes in the TBI brain. Traditionally, microglia were broadly classified as antiinflammatory (resting) or proinflammatory (activated) in the injured brain with distinct separation based on morphology and inflammatory markers. However, the advent of single-cell transcriptomic signatures has identified several microglia phenotypic states present in the normal and diseased brain including DAMs, IRM, and homeostatic. Although the specific role of these identified phenotypic states in mediating disease processes is not fully understood, it reveals that the microglial phenotypic states may be far more complex than originally suggested. This may have important implications in the development of novel therapeutics that target microglial responses in TBI. *Trem2 expression increased in the aged TBI brain only. *DAM*, disease-associated microglia; *IRM*, interferon response microglia; *TBI*, traumatic brain injury.

(Malec et al., 2007). Overall, minocycline treatment resulted in a decrease in 11C-PBR28 binding, a PET marker of microglia activation, at baseline in the cerebral white matter and thalamus. However, it also caused an increase in plasma neurofilament light levels and was associated with faster rates of brain atrophy (Scott et al., 2018). This clinical study suggests that chronic microglial activation may exert reparative effects during the chronic stages postinjury.

Another therapeutic target that has been extensively studied in preclinical models is phagocytic NADPH oxidase 2 (NOX2). NOX2 is a multisubunit enzyme complex responsible for producing extracellular and intracellular ROS in phagocytes, such as microglia and macrophages. Both in vitro and experimental (moderate-level CCI) TBI studies in adult male mice report that pharmacologic inhibition or constitutive NOX2 deletion is associated with a decrease in microglial proinflammatory gene expression, reduced neurodegeneration, and improvements in neurologic recovery (Zhang et al., 2012; Loane et al., 2013; Kumar et al., 2016a,b). Mechanistically, NOX2 deletion is associated with a decrease in microglial/macrophage signal transducer and activator of transcription (STAT)1 signaling and the proinflammatory phenotype in LPS/IL-4-stimulated bone marrow derived macrophages (BMDM) model (Barrett et al., 2017). Furthermore, NOX2 deletion promoted IL-10-mediated STAT3 signaling that upregulated an antiinflammatory molecular phenotype that resulted in inflammation resolution after moderate-level CCI in adult male mice (Barrett et al., 2017). Additional mechanistic studies demonstrate that deletion of NOX2 is associated with a reduction in the expression of components of the NLRP3 inflammasome in moderate-level CCI in adult male mice (Ma et al., 2017). Furthermore, delayed pharmacologic inhibition (1-day postinjury) of NOX2 and/or NOX2 deletion is associated with a marked downregulation of the NF-κB pathway in microglia/macrophages and reduced production of proinflammatory mediators following moderate-level CCI in adult male mice (Wang et al., 2017). Thus, inhibiting NOX2 may regulate microglial states that not only limit the toxic proinflammatory microglial phenotype but also actively promote an antiinflammatory and potentially neuroprotective phenotype (Barrett et al., 2017). Further evidence of microglial NOX2 in TBI pathophysiology was demonstrated when a metabotropic glutamate receptor 5 (mGluR5) agonist was administered centrally to injured adult male mice at 1 month after moderate-level CCI. This delayed treatment paradigm reduced NOX2/CD68-positive reactive microglia and associated neurodegenerative processes at 3 months postinjury (Byrnes et al., 2012).

Another potential therapeutic target that has gained significant attention in the TBI field is the NLRP3 inflammasome (O'Brien et al., 2020; Johnson et al., 2023). NLRP3 is expressed on microglia and astrocytes, and its activation results in oligomerization and assembly of the NLRP3 inflammasome and the subsequent activation of caspase 1 (Liu et al., 2013; Xu et al., 2018a). This initiates the processing and release of key proinflammatory cytokines, IL-1β and IL-18 (O'Brien et al., 2020; Blevins et al., 2022). Although activation of the NLRP3 inflammasome is important for host protection against invading pathogens, if left uncontrolled it can cause cell death through pyroptosis (Bergsbaken et al., 2009). Microglial NLRP3 expression is rapidly upregulated within hours of experimental TBI (Liu et al., 2013; Ismael et al., 2018; Xu et al., 2018b). Administration of MCC950, a highly selective NLRP3-specific small molecule inhibitor, during the acute phase following moderate-level CCI in adult mice resulted in a decrease in cerebral edema, improvements in neurologic function, and reduced cerebral edema up to 72 h postinjury (Ismael et al., 2018). Neuroprotective effects were associated with reduced proapoptotic markers and inflammation (Ismael et al., 2018). Another study demonstrated that the therapeutic window for MCC950 was up to 6 h following exposure to moderate-level CCI in adult male mice, whereby MCC950 treatment was associated with a decrease in TBI-induced neurologic deficits (motor and cognitive), reduced lesion volume, microglial activation, leukocyte recruitment, and proinflammatory cytokine production (Xu et al., 2018b). In a weight drop model of experiment TBI (moderate-level contusion), MCC950 treatment administered at 15 min prior to TBI and 3 h postinjury, resulted in an improvement in neurologic deficits up to 7 days postinjury in adult male mice. In addition, MCC950 attenuated TBI-induced mitochondrial dysfunction at 3 dpi (Chen et al., 2019). A novel NLRP3 inhibitor, JC124, was reported to exert neuroprotective effects in experimental TBI (Kuwar et al., 2019). Specifically, JC124 administered at 30 min postinjury and at 6, 24, and 30 h thereafter, resulted in a decrease in injury-induced degenerating neurons, reduced IL-1β focally at the level of the cerebral cortex, and a reduced lesion volume at 2 days following exposure to moderate-level CCI in adult male rats (Kuwar et al., 2019). Overall, these preclinical studies demonstrate the therapeutic potential of acute administration of NLRP3 inhibitors in alleviating the negative outcomes following exposure to TBI.

COULD MICROGLIAL ELIMINATION OR FORCED TURNOVER IN THE INJURED BRAIN BE NEUROPROTECTIVE FOR TBI?

CSF1R signaling via its ligands CSF1, and IL-34 is required for the development, survival, recruitment, and proliferation of microglia (Easley-Neal et al., 2019). Thus, the recent development of CSF1R inhibitors has greatly increased our ability to directly study microglial function under normal and disease states (Elmore et al., 2014; Spangenberg et al., 2019). Robust elimination of microglia in the adult mouse brain can be achieved (>95% depletion) following a 21-day feeding regime of a CSF1R inhibitor, PLX3397 (Elmore et al., 2015). Importantly, when PLX3397 chow is removed and mice are returned to a normal diet, microglia repopulate the brain and are restored in numbers to that of nondepleted mice (Elmore et al., 2015). Notably, repopulated microglia can mount a normal response to inflammatory stimuli (Elmore et al., 2015). A second-generation and more specific inhibitor of CSF1R, PLX5622, was developed and it has improved pharmacokinetic properties, including oral bioavailability with >20% brain penetration (Spangenberg et al., 2019). PLX5622 results in the elimination of microglia following 7 days of oral treatment and depletion can be sustained for many months when the second-generation CSF1R inhibitor is present in the diet (Spangenberg et al., 2019).

The first study to examine the effects of CSF1R depletion in CNS injury models reported differential effects on outcomes dependent on the timing of microglial depletion (Rice et al., 2015) (Table 7.1; Fig. 7.2). In the CaM/Tet-Diphtheria toxin A (DT_A)-induced model of the hippocampal lesion, PLX3397-mediated depletion of microglia during the recovery period resulted in improvements in cognitive- and anxiety-related behavior (Rice et al., 2015). This was associated with reduced neuroinflammation and lesion-induced alterations in synaptic markers, including synaptophysin and PSD-95, in a mix of adult male and female mice (Rice et al., 2015). In contrast, the depletion of microglia during lesioning resulted in increased neuronal loss (Rice et al., 2015). Subsequent studies showed that depletion of microglia in CNS injury models can have differential effects depending on the timing of microglia elimination and/or the type of injury model used. For example, prior depletion of microglia via PLX5622 treatment resulted in an increase in infarct volume up to 72 h following reperfusion in an experimental model of transient focal ischemia in aged male mice (Marino Lee et al., 2021). Similarly, prior microglial depletion (by PLX3397) resulted in an increase in infarct volume, dysregulated neuronal calcium responses, and increased neuronal death, at 24 h following reperfusion in middle cerebral artery occlusion (MCAo) in adult male mice (Szalay et al., 2016). Furthermore, microglia depletion led to an increase in brain inflammation, cell death, and leukocyte infiltration at 1 day following reperfusion (Jin et al., 2017). In an experimental model of permanent occlusion of the MCAo, depletion of microglia (PLX5622) prior to induction of ischemia led to an accumulation of neutrophils in the parenchyma and enlarged ischemic lesion acutely after ischemia (Otxoa-de-Amezaga et al., 2019). Overall, these preclinical studies suggest that acute activation of microglia may play an important protective role during the early stages following ischemia.

In contrast, PLX3397-induced microglial depletion prior to axonal injury, induced by lFPI in male mice, promoted neurite outgrowth, preserved dendritic spines, and reduced neuronal apoptosis acutely after TBI (Wang et al., 2020). Similarly, PLX5622-induced microglial depletion prior to mFPI in male

Table 7.1

Microglia depletion strategies in experimental CNS injury models.

Depletion model	Depletion strategy	Sex/species	Injury model	Time post-TBI	Main findings	Reference
PLX3397 (290 mg/kg)	Immediately postinduction of neuronal loss model	Adult male and female triple transgenic GFP-CaM/Tet mice	CaM/Tet-diphtheria toxin A (DT_A) mouse model of inducible neuronal loss	30 days postinjury	↑ Functional recovery ↓ Proinflammatory gene expression ↓ Lesion expansion	Rice et al. (2015)
PLX3397 (40 mg/kg)	21 days prior to ischemic injury	Adult male C57BL/6 and $Rag2^{-/-}\gamma c^{-/-}$ mice	Middle cerebral artery occlusion	1 day, 3 days following reperfusion	↑ Neurodeficits ↑ Brain infarction ↑ inflammatory mediators, leukocyte infiltration, cell death	Jin et al. (2017)
PLX3397 (290 mg/kg)	21 days prior to ischemic injury	C57BL/6J and *Cx3Cr1*$^{GFP/+}$ mice	Middle cerebral artery occlusion	1 day postreperfusion	↑ Brain infarction ↑ Neuronal death	Szalay et al. (2016)
PLX5622 (1200 mg/kg)	21 days prior to ischemic injury	Adult C57BL/6, CX3CR1$^{cre/ERT2}$Rosa26:tdT mice, Catchup^{IVM} CX3CR1$^{+/-}$ double reporter mice	Permanent middle cerebral artery occlusion	1 day, 4 days following ischemia	Dysregulated neuronal calcium responses ↑ Neutrophils in the parenchyma ↑ Enlarged ischemic lesion	Otxoa-de-Amezaga et al. (2019)
PLX5622 (1200 mg/kg)	21 days prior to ischemic injury	Adult and aged C57BL/6 mice	Middle cerebral artery occlusion	1 day, 3 days following reperfusion	↑ Infarct volume in aged mice ↑ Infiltrating monocytes in aged mice	Marino Lee et al. (2021)
PLX3397 (290 mg/kg)	21 days prior to TBI	Adult male thy-1 GFP mice	Lateral FPI	1 day, 3 days post-TBI	↑ Neurite outgrowth ↓ Neuronal apoptosis ↓ C/EBP-homologous protein and inositol-requiring kinase 1α	Wang et al. (2020)
PLX5622 (1200 mg/kg)	14 days prior to TBI	Adult male C57BL/6 mice	Midline FPI	7 days post-TBI	↓ Rod microglia ↓ Astrogliosis ↓ Inflammation	Witcher et al. (2018)
PLX5622 (1200 mg/kg)	14 days prior to TBI	Adult male C57BL/6 mice	Midline FPI	7 days, 30 days post-TBI	↓ Inflammation ↑ Cortical dendritic complexity (7 days) ↑ Neuronal connectivity (30 days) ↑ Cognitive function (30 days)	Witcher et al. (2021)
PLX5622 (1200 mg/kg)	14 days prior to TBI	Adult male C57BL/6 mice	CCI	3 days post-TBI	↓ Extracellular acidosis ↓ ROS production ↓ Inflammation	Ritzel et al. (2021)
PLX3397 (290 mg/kg)	Immediately post-TBI	Adult male and female C57BL/6 mice	CCI	5 days post-TBI	↓ Inflammation and apoptotic related genes	Wang et al. (2022)
PLX5622 (1200 mg/kg)	Immediately post-TBI for 14 days, followed by repopulation	Adult male and female C57BL/6 mice	CaM/DT_A mouse model of inducible neuronal loss	21 days postinjury	No change in TBI-induced lesion volume, impaired hematoma clearance ↓ Neuroinflammatory response ↑ Functional recovery ↑ PSD95 and synaptophysin	Rice et al. (2017)

PLX3397 (290 mg/kg)	21 days prior to ischemic injury, followed by repopulation	C57BL/6J and *Cx3Cr1*$^{GFP/+}$ mice	Middle cerebral artery occlusion	1 day postischemic injury	↓ Brain infarction	Szalay et al. (2016)
PLX5622 (1200 mg/kg)	7 days initiated at 1-month post-TBI, followed by repopulation	Adult male C57BL/6 mice	CCI	3-month post-TBI	↓ Lesion volume, hippocampal neuronal cell loss ↓ Neuroinflammation (NOX-2; NLRP3) ↑ Neurologic function	Henry et al. (2020)
PLX5622 (1200 mg/kg)	7 days initiated at 7 days post-TBI, followed by repopulation	Adult male C57BL/6 mice	Midline FPI	30 days post-TBI	↓ Neuropathology-related genes ↑ Neuronal connectivity Improvements in depressive-like and cognitive impairments	Bray et al. (2022)
PLX5622 (1200 mg/kg) and genetic depletion	Strategy I: continual depletion initiated at 21 days prior to TBI Strategy II: 21 days prior to TBI, followed by repopulation immediately following TBI Strategy III: 9 weeks post-TBI, followed by repopulation	Adult female C57BL/6 and $CX_3CR1^{creERT2}$ × iDTR × tdTomato mice	CCI	12 days, 30 days, 19 weeks post-TBI	Acute depletion, followed by repopulation (strategy II): ↓ Learning deficits ↑ Hippocampal neurogenesis Beneficial effects of repopulated microglia mediated by IL-6 trans signaling.	Willis et al. (2020)

Depletion model	Depletion strategy	Sex/species	Injury model	Time post-SCI	Main findings	Reference
PLX3397 (50 mg/kg)	Pregnant mothers received 290 mg/kg chow from E14 and newborn continued to receive 50 mg/kg sc on a daily basis	Newborn pups (P2/P7) and young adult mice	Neonatal crush injury	14 dpi	↓ Scar-free healing ↓ Axon regeneration	Li et al. (2020a)
PLX5622 (1200 ppm) or (200 ppm) PLX5622 (90 mg/kg)	Strategy I: 21 days prior to injury and continued through 35 dpi via chow Strategy II: 21 days up to injury; repopulation through 35 dpi via chow Strategy III: oral gavage for 7 dpi, followed by chow for 35 dpi Strategy IV: 3–35 dpi via chow	Adult C57Bl/6 mice Cx3cr1creER: Rosa26(R26)-TdT transgenic mice	Crush injury	35 dpi	Strategy I: ↓ locomotor function Strategy II: no effect on locomotor activity Strategy III: ↓ glial scar formation, ↑ parenchymal immune infiltration, ↓ neuronal and oligodendrocyte survival, ↓ locomotor function Strategy IV: no effect on locomotor activity	Bellver-Landete et al. (2019)
PLX3397 (290 mg/kg)	7 days prior to injury and continued through 7, 14 or 28 dpi	Adult female C57Bl/6 mice	Crush injury	7, 14, or 28 dpi	↓ Microglial proliferation Disrupted glial scar formation ↑ Immune cell infiltration ↓ Neuronal survival ↑ Locomotor deficits	Fu et al. (2020)

Continued

Table 7.1

Continued

Depletion model	Depletion strategy	Sex/species	Injury model	Time post-SCI	Main findings	Reference
PLX3397 (90 mg/kg)	14 days prior to injury and continued through 7, 14 or 28 dpi	Adult female C57Bl/6 mice	Crush injury	7, 14, or 28 dpi	Disrupted glial scar formation ↑ Inflammatory spillover ↑ Functional deficits ↑ Diffuse tissue damage	Zhou et al. (2023)
PLX5622 (1200 ppm)	Strategy I: 21 days prior to injury through 2 dpi Strategy II: Day 0 through 7 dpi Strategy III: Day 0 through 6 weeks postinjury	Adult male C57Bl/6 mice	Contusion injury	Up to 6 weeks postinjury	Preinjury treatment ↓ Microglia numbers, monocytes and neutrophils, ↓ ROS Postinjury treatment ↓ $CD45^{int}$ microglia at 7 dpi 6-week postinjury treatment ↓ neurologic deficits, ↑ Neuronal survival	Li et al. (2020b)
PLX5622 (1200 mg/kg) chow	Strategy I: 14 days prior to injury through 35 dpi (contusion) Strategy II: 14 days prior to injury through 7, 14, 21, or 28 dpi (crush injury)	Adult female C57Bl/6 mice	Contusion and crush injury	Up to 35 dpi	↑ Motor deficits ↑ Ectopic monocyte-derived macrophages (MDMs) in white matter ↓ Glial scar formation ↓ Astrocytes and $NG2^{+}BrdU^{+}$ cell proliferation and number	Brennan et al. (2022)
PLX3397 (290 mg/kg)	14 days prior to injury, followed by repopulation for 46 days	Adult C57Bl/6 mice	Transection	60 dpi	↓ $CD68^{+}$ reactive microglia/macrophages ↓ Proinflammatory markers ↑ $Tuj1^{+}$ neurons in lesioned area	Ma et al. (2020)
GW2580 (150 mg/kg)	Day 0 through 7 dpi	CX3CR1+/eGFP transgenic mice	Lateral hemisection	6 weeks postinjury	↑ Motor function ↑ Tissue preservation	Poulen et al. (2021a)
GW2580 (150 mg/kg)	Day 0 through 6 weeks postinjury	Female CX3CR1+/eGFP transgenic mice	Lateral hemisection	6 weeks postinjury	No effect on motor function or tissue reorganization	Poulen et al. (2021b)
GW2580 (0.1%)	4 weeks prior to injury followed by 6 weeks postinjury	CX3CR1+/eGFP transgenic mice	Lateral hemisection	6 weeks postinjury	↓ Proliferating microglia ↓ Gliosis Partial improvement in motor function	Gerber et al. (2018)

CCI, controlled cortical impact; *dpi*, days postinjury; *lFPI*, lateral fluid percussion injury; *mFPI*, midline fluid percussion injury; *SCI*, spinal cord injury; *TBI*, traumatic brain injury.

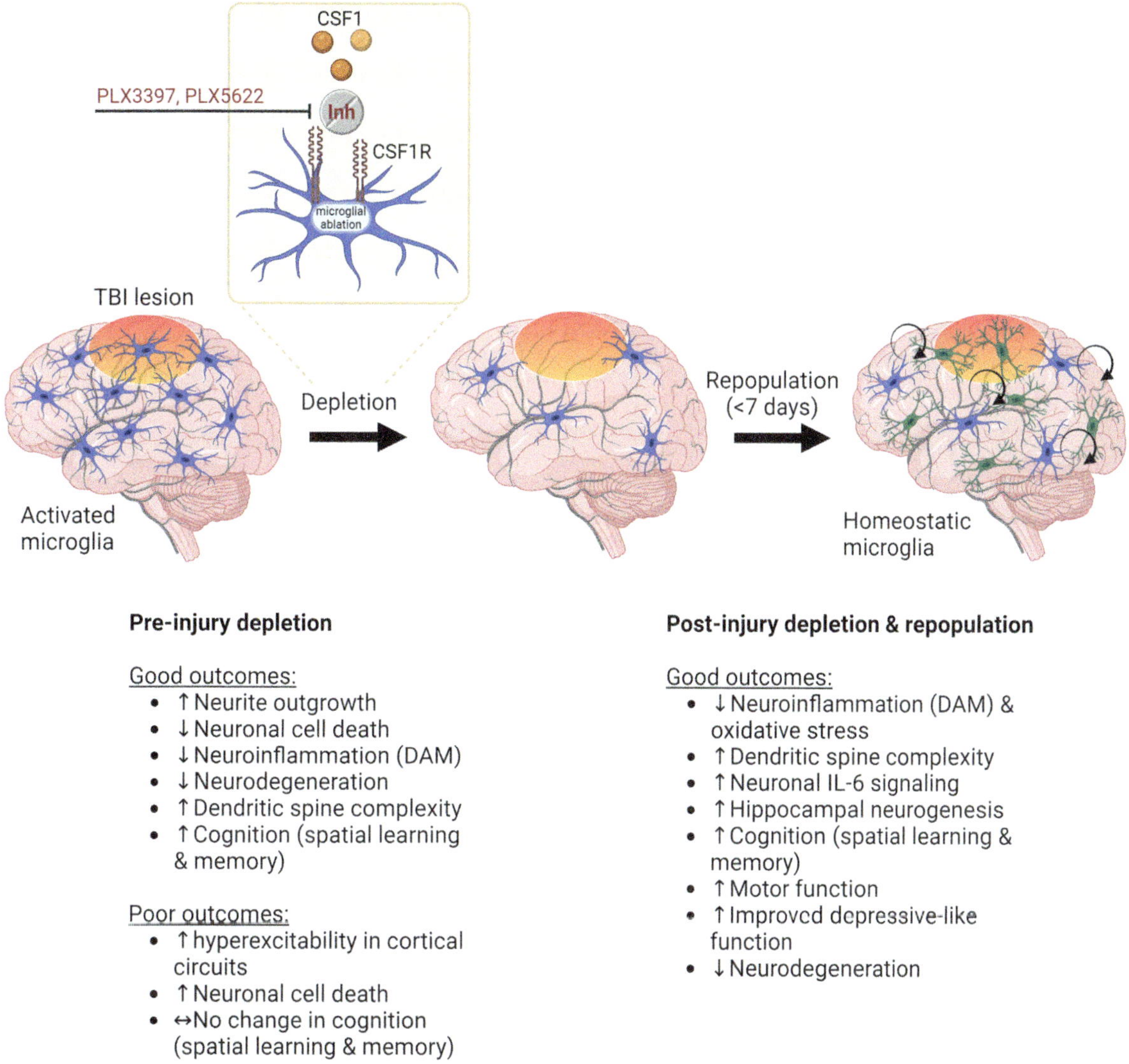

Fig. 7.2. Microglia depletion strategies in experimental TBI. Several studies have investigated the impact of microglia depletion on outcomes following exposure to experimental TBI. Both good and poor outcomes are reported following preinjury depletion strategies. Notably, depletion and subsequent forced turnover of microglia postinjury are reported to be associated with good outcomes in the context of experimental TBI. *CSF1*, colony-stimulating factor; *CSF1R*, colony-stimulating factor 1 receptor; *DAM*, disease associated microglia; *TBI*, traumatic brain injury.

mice resulted in reduced numbers of TBI-induced rod microglia, attenuated astrogliosis, and reduced expression of inflammatory genes associated with complement, cytokines, chemokines, and interferon signaling in the injured cortex (Witcher et al., 2018). Furthermore, TBI-induced cognitive impairments at 30 dpi were ameliorated when microglia were depleted by PLX5622 treatment (Witcher et al., 2021). In a similar manner, PLX5622-induced microglial depletion prior to a moderate-level CCI resulted in reduced extracellular acidosis, ROS production, and neuroinflammation at 3 dpi in male mice (Ritzel et al., 2021). Other studies have investigated the effects of microglial depletion during the acute phase postinjury. PLX3397-induced microglial depletion immediately following CCI resulted in decreased inflammatory- and apoptotic-related gene expression in both male and female mice at 5 dpi (Wang et al., 2022). However, acute depletion of microglia did not alter TBI-induced lesion volume or impair hematoma clearance (Wang et al., 2022). Furthermore, clodronate-induced depletion of microglia/macrophages immediately after closed-head injury in adolescent rats exacerbated acute neuronal injury responses in male and female rat pups (Hanlon et al., 2019). These conflicting studies suggest that the depletion of microglia in preclinical models may be time and injury-dependent (Table 7.1).

Additional studies have investigated the neuroprotective effects of transient microglial depletion and subsequent repopulation during acute, subacute, and chronic phases postinjury (Table 7.1), which is termed *forced turnover* of microglia. The studies performed to date show that transient depletion of microglia followed by repopulation, regardless of the timing of microglial forced turnover, results in broad neuroprotective effects. One of the first studies to examine the potential neuroprotective effects of microglial forced turnover was in experimental ischemic stroke. Transient depletion of microglia prior to cerebral ischemia, followed by repopulation beginning at 2 weeks following diet withdrawal resulted in a decreased infarct volume in injured mice (Szalay et al., 2016). In a DT_A model of inducible hippocampal neuronal loss, depletion of microglia for 14 days beginning at 7 days postlesion induction, followed by repopulation for 21 days decreased neuroinflammation, characterized by a reduced number of swollen microglia expressing CD68 and CD45, and reduced production of cytokines (Rice et al., 2017). In addition, the forced turnover of microglia improved long-term anxiety-like and cognitive functional recovery (Rice et al., 2017).

Other studies have reported differential outcomes dependent on the depletion/repopulation paradigm, injury model, and sex examined. Willis and colleagues demonstrated that both pharmacologic (PLX5622)- and genetic ($CX_3CR1^{creERT2} \times$ iDTR mice)-depletion of microglia prior to induction of moderate-level CCI, followed by repopulation during the acute phase postinjury, resulted in cognitive function recovery up to 9 months postinjury in female mice (Willis et al., 2020). Moreover, acute microglial turnover stimulated functional neurogenesis in the injured hippocampus. The neuroprotective effects of repopulated microglia were dependent on IL-6 trans-signaling via the soluble IL-6 receptor (Willis et al., 2020). Notably, neither continuous long-term microglial depletion nor transient delayed depletion starting at 9 weeks postinjury, followed by forced microglial turnover, led to improvements in TBI-induced cognitive deficits (Willis et al., 2020). This is in contrast with our findings, which demonstrated that delayed depletion of microglia initiated at 4 weeks postmoderate-level CCI via PLX5622 administration, followed by repopulation, resulted in a reduced cortical lesion volume, reduced hippocampal neuronal cell loss, and reduced NOX2- and NLRP3 inflammasome-associated neuroinflammation in PLX5622-treated forced turnover male mice (Henry et al., 2020). Importantly, these neuropathologic changes were associated with improvements in long-term motor and cognitive function up to 3 months postinjury (Henry et al., 2020). Pharmacologic depletion of microglia using PLX3397 immediately after moderate-level CCI, followed by repopulation at 5 dpi attenuated cortical lesion volume in both male and female mice at 30 dpi (Wang et al., 2022). Interestingly, microglial forced turnover by PLX3397 treatment decreased hippocampal atrophy and thalamic expression of $CD68^+$ microglia/macrophages in injured male, but not female mice (Wang et al., 2022). In addition, forced microglial turnover during the acute phase after CCI worsened neurologic function at more chronic timepoints postinjury in female CCI mice only (Wang et al., 2022).

Despite reported differential effects of microglial depletion strategies on injury outcomes, these studies provide important insights into the underlying mechanisms driving neuroprotective and/or detrimental effects of microglia signaling in the injured brain. While the ultimate goal is to translate these targeted depletion strategies into the clinical TBI setting, it is clear that additional experimental studies are required for researchers to fully understand the optimal depletion and/or forced turnover strategy. It will be important that future studies consider factors including the injury paradigm, injury severity, whether the therapeutic target is neuronal death or axonal injury, sex, and age on microglial responses. In addition, the animal model under investigation should be considered whether treatment paradigms are translatable across species.

UNDERSTANDING MICROGLIA RESPONSES IN LARGE ANIMAL MODELS OF TBI

Rodent TBI models offer important insights into the basic pathophysiologic mechanisms of TBI. However, they fail to fully recapitulate the biomechanics of human TBI. Crucially, there has been a universal failure to translate promising drug treatments developed at the bench in preclinical studies to the clinic for human TBI (Grovola et al., 2023). Due to a large brain mass and gyrencephalic convolutions similar to the human brain, large animal porcine and ferret TBI models may provide a more meaningful basis for clinical translation (Cullen et al., 2016; Lyu et al., 2021; Mayer et al., 2021; Grovola et al., 2023). To date, there have been several studies investigating microglial responses in large animal TBI models (Grovola et al., 2023). Using a porcine model of closed-head rotational velocity/acceleration-induced TBI, microglial reactivity was observed within 15 min of TBI exposure; effects of which were localized at proximal but not distal sites of injured neurons (Wofford et al., 2017). Following exposure to diffuse axonal injury (DAI) in micropigs using an FPI model, TBI resulted in significant axonal damage and increased $Iba1^+$ microglial process convergence at 6 h postinjury, effects of which were particularly localized within injured foci in the thalamus (Lafrenaye et al., 2015). Another study from the same research group demonstrated differential microglial responses between rodent (rat) and porcine DAI models. In micropigs, there was evidence of increased microglial-neuron interactions in the thalamus at 24 h after DAI (Gorse and Lafrenaye, 2018). In contrast, microglial-neuronal interactions were decreased by half at 6 h postinjury in rats (Gorse and Lafrenaye, 2018). These reported differences may be due to differences in the biomechanical response to central FPI in lissencephalic vs gyrencephalic brains, which may result in the induction of different injury severity across species (Gorse and Lafrenaye, 2018). Furthermore, in the chronic phase of TBI, there was evidence of neuropathology, including an increase in microglia density in hippocampal subregions, up to 1 year after closed-head rotational velocity/acceleration-induced mild TBI in pigs (Grovola et al., 2020). Overall, these studies demonstrate that a single mild TBI in a clinically relevant large animal model leads to long-term changes in microglial reactivity. However, how such changes in microglial morphology relate to microglial function and neuronal health requires additional mechanistic insight, including investigation of the expression profiles of genes associated with acute and chronic neuroinflammatory responses following mTBI in large animal models.

MICROGLIA IN HUMAN TBI: WHAT IS THE CURRENT EVIDENCE OF THEIR PHENOTYPIC AND FUNCTIONAL ROLES?

Humanized models of microglia have been developed to investigate human microglia at the single-cell level (Hasselmann and

Blurton-Jones, 2020; Zhang and Cui, 2021). Whole transcriptomic analysis reveals that human microglia-like cells (iMGLs) behave like cultured adult and fetal human microglia, can respond to inflammatory stimuli, and have key phagocytic activity (Abud et al., 2017). When iMGLs are transplanted into transgenic mice or human brain organoids they resemble microglia in vivo; thus, iMGLs may offer novel insights into microglia in TBI. A recent study demonstrated that stimulation of human-induced induced pluripotent stem cell (iPSC)-derived microglia with cytokines (TNF-α, IL-1β, IL-6, IL-4, and IL-10) at concentrations reported in human TBI resulted in time- and concentration-dependent effects on downstream inflammatory responses (Alam et al., 2023). Out of the five cytokines examined, TNF-α was the most potent inducer of downstream cytokine responses in human microglia (Alam et al., 2023). In addition, there are many other in vitro models of TBI, including stretch, compression, and blast injury models (Wu et al., 2021). Of these in vitro mechanical injuries, needle stretch, or transection has been proposed to mimic secondary injury responses, including glial activation (Morrison et al., 2011; Kumaria, 2017). A model of primary neuronal injury (needle stretch)-induced microglia activation demonstrated that this in vitro model closely mimics microglial proinflammatory activation and neurodegenerative responses described in vivo (Liu et al., 2023). A deeper analysis using this murine model determined an altered metabolic profile of proinflammatory microglia and the key role of glycolysis in driving microglial activation and subsequent neurodegeneration (Liu et al., 2023).

In addition to humanized in vitro models, the identification of TBI biomarkers will ultimately improve diagnostic assessment and therapeutic interventions for clinical TBI (Bogoslovsky et al., 2016; Zetterberg and Blennow, 2016). One such biomarker is translocator protein (TSPO), which is expressed on the outer mitochondrial membrane in multiple cell types, including microglia. TSPO is often used as a biomarker of neuroinflammation (Chen and Guilarte, 2008). TSPO positron emission tomography (PET) imaging, using the first-generation tracer ligand I-[^{11}C] PK11195, revealed persistent microglia activation up to 17 years following severe TBI (Ramlackhansingh et al., 2011). Microglia activation was present in subcortical regions including the thalamus, but not at the focal site of injury. Notably, [11C](R)-[^{11}C] PK11195 binding at the level of the thalamus was shown to correlate with worse cognitive outcomes (Ramlackhansingh et al., 2011). A subsequent study using a second-generation TSPO ligand, namely [(Loane et al., 2014)C]DPA-713, revealed increased binding in several brain regions including the supramarginal gyrus and amygdala, in retired National Football League (NFL) players (Coughlin et al., 2015). Significant hippocampal atrophy and varied performance in verbal learning and memory were observed in the former NFL players (Coughlin et al., 2015). Furthermore, using the [(Loane et al., 2014)C]DPA-713 PET ligand, there is evidence of increased binding primarily in medial temporal lobe regions and evidence of white matter changes, in a group of active or retired NFL players with a history of concussion (Coughlin et al., 2017). While these studies suggest a role for chronic microglial activation in neurodegenerative processes and neurologic decline, further longitudinal studies are required to understand the functional significance of neuroinflammation in human TBI and how it may be therapeutically targeted to slow disease progression.

Several clinical studies have attempted to correlate inflammatory cytokines profiles in blood with patient outcomes (Carpenter et al., 2015). Acute increases in serum levels of the prototypic proinflammatory cytokine IL-1β are reported to correlate with TBI severity (Tasci et al., 2003). Furthermore, increases in cerebrospinal fluid (CSF) levels of IL-1β are associated with poorer outcomes and increased intracranial pressure (ICP) following severe TBI (Hayakata et al., 2004; Shiozaki et al., 2005). Microdialysis revealed that IL-1β endogenous antagonist IL-1 receptor antagonist (IL-1Ra) may play a protective role in TBI (Hutchinson et al., 2007). A randomized Phase II controlled clinical trial found that recombinant human IL-1Ra (Anakinra) altered the neuroinflammatory response following severe TBI, although the sample size of 20 patients was not sufficiently powered to address potential therapeutic effects (Helmy et al., 2014). Both mRNA and protein expression of the pleiotropic cytokine TNF-α have been detected in the postmortem brain of severe TBI patients (Frugier et al., 2010). Moreover, increases in TNF-α levels in CSF are reported in TBI patients (Hayakata et al., 2004; Shiozaki et al., 2005; Stein et al., 2011). However, such increases do not correlate with poorer patient outcomes or increased ICP (Hayakata et al., 2004). For the inflammatory cytokine IL-6, discrepancies have been reported in relation to outcomes. Specifically, an increase in serum IL-6 levels during the acute phase after TBI correlates with increased ICP (Hergenroeder et al., 2010), and an increase of IL-6 within the CSF is associated with increased inflammation and poorer outcomes in the first year following TBI (Kumar et al., 2015). However, other studies have reported that increases in IL-6 are associated with improved overall outcomes (Carpenter et al., 2015). IL-10 is an antiinflammatory cytokine and higher serum levels of IL-10 are reported to be associated with hospital mortality in severely injured TBI patients (Schneider Soares et al., 2012). Overall, clinical studies investigating the potential use of cytokines as biomarkers in TBI are inconsistent and somewhat contradictory. There may be several reasons for this including the complexity of cytokine interactions, cytokine behavior within a given context, and the time course in which cytokines are released (Carpenter et al., 2015). Thus, a more nuanced approach may be required that considers the multiple possible interactions of cytokines in response to TBI (Helmy et al., 2012; Carpenter et al., 2015).

In addition to cytokines, extracellular vesicles (EVs) have generated a great amount of interest in recent years as potential diagnostic and therapeutic tools (Khan et al., 2022). In clinical TBI, increased concentrations of tau and IL-10 in neuronal-derived exosomes are reported in the blood of military personnel who previously suffered from mild TBI (Gill et al., 2018). Regression analysis demonstrated that postconcussive symptoms and posttraumatic stress disorder (PTSD) were associated with increases in exosomal tau and IL-10, respectively (Gill et al., 2018). In moderate–severe TBI, exosomal proteins were reported to be differentially increased in the blood acutely postinjury (Mondello et al., 2020). Differences in the dynamics and levels of exosomal proteins were dependent on injury type: diffuse injury patients displayed an increase in exosomal NFL and glial fibrillary acidic protein (GFAP), while focal injury resulted in an acute increase in exosomal levels of GFAP (48h postinjury) (Mondello et al., 2020). Ubiquitin carboxy-terminal hydrolase L1 (UCH-L1) decreased 24h after injury; however, a secondary peak was reported to be associated with early mortality (Mondello et al., 2020). A subsequent study from the same group examining exosomal miRNA expression from both mTBI and repetitive mTBI military personnel demonstrated dysregulation of 17 and 32 exomiRs in mTBI and repetitive mTBI samples (Devoto et al., 2020). IPA analysis revealed that dysregulated exomiRs correlated with pathways associated with inflammatory regulation, neurologic disease, etc. (Devoto et al., 2020). Overall, these studies suggest that analysis of injury exosomes, including inflammatory-related ones, may offer novel insights into TBI pathophysiology, including as a therapeutic target and/or treatment. Following experimental TBI, microglial-derived EVs with high concentrations of IL-1β and the proinflammatory microRNA (miR)-155 are discharged into the bloodstream at acute timepoints (24h postinjury). The enriched EVs from posttraumatic blood samples can activate microglial inflammatory responses in uninjured brains (Kumar et al., 2017). Another study reported a rapid increase in plasma EVs following experimental TBI, and systemic injection of enriched EVs from LPS-stimulated macrophages was associated with an increase in microglia activation and lesion volume following TBI (Hazelton et al., 2018). Interestingly, neutralization of EVs reduces proinflammatory mediators (Kumar et al., 2017) and confers significant neuroprotective effects (Bohman et al., 2016) in experimental TBI models.

THE THERAPEUTIC POTENTIAL OF TARGETING MICROGLIA RESPONSES IN SCI

Many preclinical studies have investigated the therapeutic potential of targeting microglial responses in experimental SCI models (Table 7.1; Fig. 7.3). Microglia depletion via PLX3397 administration, prior to induction of neonatal crush injury, led to a decrease in scar-free healing and reduced axonal regeneration (Li et al., 2020a). RNAseq analysis revealed that microglia contribute to scar-free healing following SCI via the secretion of fibronectin and the expression of extracellular and intracellular peptidase inhibitors (Li et al., 2020a). Another study reported that prior depletion (PLX5622) worsened motor function in adult mice following crush injury (Bellver-Landete et al., 2019). Notably, forced microglial turnover following injury did not alter SCI-induced locomotor deficits through 35dpi (Bellver-Landete et al., 2019). To investigate the beneficial vs detrimental role of microglia during the acute phase following injury, the authors depleted microglia acutely postinjury and reported a decrease in glial scar formation, an increase in peripheral immune cell infiltration, and an exacerbation of locomotor deficits (Bellver-Landete et al., 2019). Similarly, depletion of microglia (PLX3397) prior to induction of crush injury in adult female mice resulted in a decrease in microglial proliferation, disrupted glial scar formation, increased peripheral immune cell infiltration, decreases in neuronal survival, and worsened locomotor deficits, up to 28 dpi (Fu et al., 2020). Notably, restoring microglia-dependent signaling axes (CCL2 and TLR2) in microglia-depleted mice prevented secondary damage and promoted recovery (Brennan et al., 2022). Overall, these preclinical studies demonstrate the beneficial effects of microglia-dependent signaling in promoting homeostatic neuroinflammation and repair following SCI. However, it is important to note that other studies report beneficial effects of microglia depletion (Table 7.1; Fig. 7.3). Both prior (starting at 21 days prior to injury) and postinjury (day 0), microglia depletion (PLX5622) led to a decrease in microglia numbers, as well as a decrease in microglial ROS production at 2 dpi in adult male mice exposed to contusion injury (Li et al., 2020b). In addition, postinjury depletion was associated with a decrease in $CD45^{int}$ microglia at 7 dpi and an improvement in injury-induced neurologic recovery and neuronal survival up to 6 weeks postinjury (Li et al., 2020b). Similarly, prior depletion of microglia (PLX3397) resulted in reduced numbers of $CD68^{+}$ reactive microglia/macrophages and expression of proinflammatory markers at 60dpi in the transection model of SCI (Ma et al., 2020). In another study, pretreatment with the CSF1R antagonist, GW2580, decreased proliferating microglia and gliosis and resulted in a partial improvement in motor function recovery at 6 weeks following exposure to lateral hemisection in adult female mice (Gerber et al., 2018). In a similar manner, acute depletion of microglia via GW2580 treatment (up to 7 dpi) improved motor recovery and increased tissue preservation through 6 weeks postinjury (Poulen et al., 2021a). In contrast, continual depletion of microglia (GW2580) through 6 weeks postinjury had no effect on injury-induced motor deficits and tissue reorganization (Poulen et al., 2021b).

Thus, while there is strong evidence that microglia play an important role during the acute stages in SCI recovery, it is important to note that in a similar manner to experimental TBI, additional experimental studies are required to fully decipher the optimal depletion and/or forced turnover strategy. Once again, given the contradictory data and paucity of quality preclinical SCI studies, important factors to consider include the injury paradigm, injury severity, sex, and age on microglial responses. In addition, as previously mentioned regarding TBI, the animal model under investigation should be considered whether treatment paradigms are translatable across species.

In the clinical setting, a phase II double-blind, randomized, placebo-controlled study I demonstrated that intravenous administration of minocycline beginning at 12h and continuing for 7 days in patients with cervical SCI resulted in an average 14-point improvement in motor recovery compared to placebo controls (Casha et al., 2012). In contrast, there was no difference in motor score reported in minocycline-treated thoracic SCI patients compared to placebo controls (Casha et al., 2012). Using the same treatment paradigm, a phase III clinical was carried out

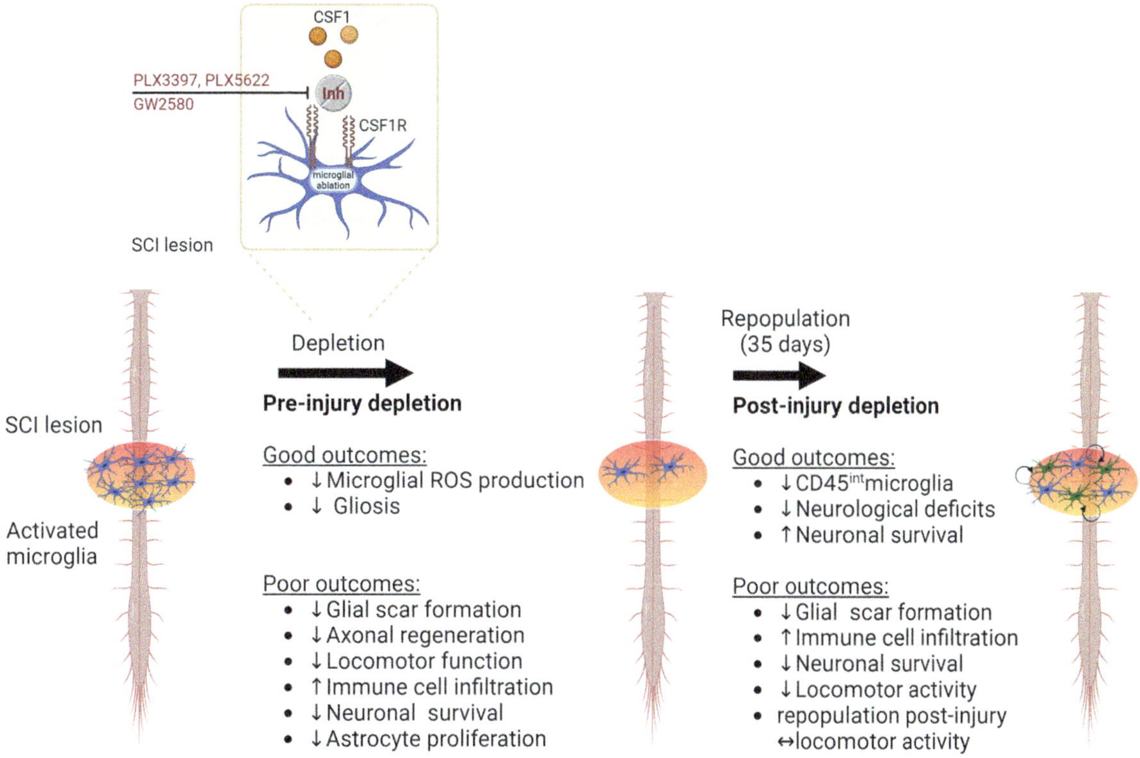

Fig. 7.3. Microglia depletion strategies in experimental SCI. Several studies have investigated the impact of microglia depletion on outcomes following exposure to experimental SCI. While primarily poor outcomes are reported in the context of preinjury depletion strategies; some good outcomes are evident. Both good and poor outcomes are reported in experimental models of SCI where microglia are depleted postinjury. Notably, forced microglial turnover has no impact on outcomes. *CSF1*, colony-stimulating factor; *CSF1R*, colony-stimulating factor 1 receptor; *ROS*, reactive oxygen species; *SCI*, spinal cord injury.

in patients with acute cervical SCI. However, to our knowledge, the results of this study are unknown (ClinicalTrials.gov ID: NCT01828203).

CONCLUSION

Research over the last decade has demonstrated that chronic uncontrolled microglial activation is a key driver of CNS pathophysiology. However, despite the identification of several potential therapeutic targets in the preclinical setting, these unfortunately have failed to translate to the clinic for human CNS injury. Several factors likely cause this lack of translation, including: (i) heterogeneity of CNS pathophysiology, (ii) lack of comparison of potential therapeutic targets across preclinical injury models and across different ages and sex, (iii) lack of investigation of appropriate therapeutic window to target neuroinflammatory/microglial responses, and (iv) inadequate data regarding pharmacokinetics and therapeutic brain concentrations (Loane et al., 2015; Faden et al., 2016). Recent advancements in microglial transcriptomics offer novel insights into microglial phenotypes (DAM, IRM, etc.) implicated in various neurodegenerative disease states, including CNS injury. Ultimately, understanding the role of specific microglia subpopulations in driving CNS injury pathophysiology will enable more targeted therapeutic approaches of implicated microglial pathways vs that of broad inhibition of microglial inflammatory responses. Furthermore, based on the heterogeneity of CNS injury, as well as the injury-severity-, sex-, and age-related differences in the microglial inflammatory response, it is vital that future preclinical studies consider to what extent potential new drug treatments are time-, sex-, and age-dependent, and whether they are effective across injury models and in large animal model species.

References

Abud EM, Ramirez RN, Martinez ES et al. (2017). iPSC-derived human microglia-like cells to study neurological diseases. Neuron 94: 278-93.e9.

Acosta SA, Tajiri N, Shinozuka K et al. (2013). Long-term upregulation of inflammation and suppression of cell proliferation in the brain of adult rats exposed to traumatic brain injury using the controlled cortical impact model. PLoS One 8: e53376.

Afridi R, Lee WH, Suk K (2020). Microglia gone awry: linking immunometabolism to neurodegeneration. Front Cell Neurosci 14: 246.

Ajami B, Bennett JL, Krieger C et al. (2007). Local self-renewal can sustain CNS microglia maintenance and function throughout adult life. Nat Neurosci 10: 1538–1543.

Ajami B, Bennett JL, Krieger C et al. (2011). Infiltrating monocytes trigger EAE progression, but do not contribute to the resident microglia pool. Nat Neurosci 14: 1142–1149.

Alam A, Singh T, Kayhanian S et al. (2023). Modeling the inflammatory response of traumatic brain injury using human induced pluripotent stem cell derived microglia. J Neurotrauma 40: 2164–2173.

Alexander JK, Popovich PG (2009). Neuroinflammation in spinal cord injury: therapeutic targets for neuroprotection and regeneration. Prog Brain Res 175: 125–137.

Alliot F, Godin I, Pessac B (1999). Microglia derive from progenitors, originating from the yolk sac, and which proliferate in the brain. Brain Res Dev Brain Res 117: 145–152.

Aungst SL, Kabadi SV, Thompson SM et al. (2014). Repeated mild traumatic brain injury causes chronic neuroinflammation, changes in hippocampal synaptic plasticity, and associated cognitive deficits. J Cereb Blood Flow Metab 34: 1223–1232.

Barrett JP, Henry RJ, Villapol S et al. (2017). NOX2 deficiency alters macrophage phenotype through an IL-10/STAT3 dependent mechanism: implications for traumatic brain injury. J Neuroinflammation 14: 65.

Barrett JP, Henry RJ, Shirey KA et al. (2020). Interferon-beta plays a detrimental role in experimental traumatic brain injury by enhancing neuroinflammation that drives chronic neurodegeneration. J Neurosci 40: 2357–2370.

Barrett JP, Knoblach SM, Bhattacharya S et al. (2021). Traumatic brain injury induces cGAS activation and type I interferon signaling in aged mice. Front Immunol 12: 710608.

Bellver-Landete V, Bretheau F, Mailhot B et al. (2019). Microglia are an essential component of the neuroprotective scar that forms after spinal cord injury. Nat Commun 10: 518.

Bergsbaken T, Fink SL, Cookson BT (2009). Pyroptosis: host cell death and inflammation. Nat Rev Microbiol 7: 99–109.

Blevins HM, Xu Y, Biby S et al. (2022). The NLRP3 inflammasome pathway: a review of mechanisms and inhibitors for the treatment of inflammatory diseases. Front Aging Neurosci 14: 879021.

Bogoslovsky T, Gill J, Jeromin A et al. (2016). Fluid biomarkers of traumatic brain injury and intended context of use. Diagnostics (Basel) 6: 37.

Bohman LE, Riley J, Milovanova TN et al. (2016). Microparticles impair hypotensive cerebrovasodilation and cause hippocampal neuronal cell injury after traumatic brain injury. J Neurotrauma 33: 168–174.

Bray CE, Witcher KG, Adekunle-Adegbite D et al. (2022). Chronic cortical inflammation, cognitive impairment, and immune reactivity associated with diffuse brain injury are ameliorated by forced turnover of microglia. J Neurosci 42: 4215–4228.

Brennan FH, Li Y, Wang C et al. (2022). Microglia coordinate cellular interactions during spinal cord repair in mice. Nat Commun 13: 4096.

Butovsky O, Siddiqui S, Gabriely G et al. (2012). Modulating inflammatory monocytes with a unique microRNA gene signature ameliorates murine ALS. J Clin Invest 122: 3063–3087.

Byrnes KR, Loane DJ, Stoica BA et al. (2012). Delayed mGluR5 activation limits neuroinflammation and neurodegeneration after traumatic brain injury. J Neuroinflammation 9: 43.

Caplan HW, Cox CS, Bedi SS (2017). Do microglia play a role in sex differences in TBI? J Neurosci Res 95: 509–517.

Carpenter KL, Czosnyka M, Jalloh I et al. (2015). Systemic, local, and imaging biomarkers of brain injury: more needed, and better use of those already established? Front Neurol 6: 26.

Casha S, Zygun D, McGowan MD et al. (2012). Results of a phase II placebo-controlled randomized trial of minocycline in acute spinal cord injury. Brain 135: 1224–1236.

Chen MK, Guilarte TR (2008). Translocator protein 18 kDa (TSPO): molecular sensor of brain injury and repair. Pharmacol Ther 118: 1–17.

Chen Y, Meng J, Xu Q et al. (2019). Rapamycin improves the neuroprotection effect of inhibition of NLRP3 inflammasome activation after TBI. Brain Res 1710: 163–172.

Cherry JD, Olschowka JA, O'Banion MK (2014). Neuroinflammation and M2 microglia: the good, the bad, and the inflamed. J Neuroinflammation 11: 98.

Clayton K, Delpech JC, Herron S et al. (2021). Plaque associated microglia hyper-secrete extracellular vesicles and accelerate tau propagation in a humanized APP mouse model. Mol Neurodegener 16: 18.

Cornell J, Salinas S, Huang HY et al. (2022). Microglia regulation of synaptic plasticity and learning and memory. Neural Regen Res 17: 705–716.

Coughlin JM, Wang Y, Munro CA et al. (2015). Neuroinflammation and brain atrophy in former NFL players: an in vivo multimodal imaging pilot study. Neurobiol Dis 74: 58–65.

Coughlin JM, Wang Y, Minn I et al. (2017). Imaging of glial cell activation and white matter integrity in brains of active and recently retired national football league players. JAMA Neurol 74: 67–74.

Courtine G, Sofroniew MV (2019). Spinal cord repair: advances in biology and technology. Nat Med 25: 898–908.

Cullen DK, Harris JP, Browne KD et al. (2016). A porcine model of traumatic brain injury via head rotational acceleration. Methods Mol Biol 1462: 289–324.

Davalos D, Grutzendler J, Yang G et al. (2005). ATP mediates rapid microglial response to local brain injury in vivo. Nat Neurosci 8: 752–758.

Deczkowska A, Keren-Shaul H, Weiner A et al. (2018). Disease-associated microglia: a universal immune sensor of neurodegeneration. Cell 173: 1073–1081.

del Rio-Hortega P (1932). Micorglia. In: W Penfield (Ed.), Cytology & cellular pathology of the nervous system. PB Hoeber, New York, pp. 483–534.

Deng J, Meng F, Zhang K et al. (2022). Emerging roles of microglia depletion in the treatment of spinal cord injury. Cells 11: 1871.

Devoto C, Lai C, Qu BX et al. (2020). Exosomal microRNAs in military personnel with mild traumatic brain injury: preliminary results from the chronic effects of neurotrauma consortium biomarker discovery project. J Neurotrauma 37: 2482–2492.

DiSabato DJ, Quan N, Godbout JP (2016). Neuroinflammation: the devil is in the details. J Neurochem 139: 136–153.

Doran SJ, Ritzel RM, Glaser EP et al. (2019). Sex differences in acute neuroinflammation after experimental traumatic brain injury are mediated by infiltrating myeloid cells. J Neurotrauma 36: 1040–1053.

Easley-Neal C, Foreman O, Sharma N et al. (2019). CSF1R ligands IL-34 and CSF1 are differentially required for microglia development and maintenance in white and gray matter brain regions. Front Immunol 10: 2199.

Elmore MR, Najafi AR, Koike MA et al. (2014). Colony-stimulating factor 1 receptor signaling is necessary for microglia viability, unmasking a microglia progenitor cell in the adult brain. Neuron 82: 380–397.

Elmore MR, Lee RJ, West BL et al. (2015). Characterizing newly repopulated microglia in the adult mouse: impacts on animal behavior, cell morphology, and neuroinflammation. PLoS One 10: e0122912.

Esiri MM, al Izzi MS, Reading MC (1991). Macrophages, microglial cells, and HLA-DR antigens in fetal and infant brain. J Clin Pathol 44: 102–106.

Faden AI, Loane DJ (2015). Chronic neurodegeneration after traumatic brain injury: Alzheimer disease, chronic traumatic encephalopathy, or persistent neuroinflammation? Neurotherapeutics 12: 143–150.

Faden AI, Wu J, Stoica BA et al. (2016). Progressive inflammation-mediated neurodegeneration after traumatic brain or spinal cord injury. Br J Pharmacol 173: 681–691.

Friedman BA, Srinivasan K, Ayalon G et al. (2018). Diverse brain myeloid expression profiles reveal distinct microglial activation states and aspects of Alzheimer's disease not evident in mouse models. Cell Rep 22: 832–847.

Frugier T, Morganti-Kossmann MC, O'Reilly D et al. (2010). In situ detection of inflammatory mediators in post mortem human brain tissue after traumatic injury. J Neurotrauma 27: 497–507.

Fu H, Zhao Y, Hu D et al. (2020). Depletion of microglia exacerbates injury and impairs function recovery after spinal cord injury in mice. Cell Death Dis 11: 528.

Gao HM, Liu B, Hong JS (2003). Critical role for microglial NADPH oxidase in rotenone-induced degeneration of dopaminergic neurons. J Neurosci 23: 6181–6187.

Gerber YN, Saint-Martin GP, Bringuier CM et al. (2018). CSF1R inhibition reduces microglia proliferation, promotes tissue preservation and improves motor recovery after spinal cord injury. Front Cell Neurosci 12: 368.

Gill J, Mustapic M, Diaz-Arrastia R et al. (2018). Higher exosomal tau, amyloid-beta 42 and IL-10 are associated with mild TBIs and chronic symptoms in military personnel. Brain Inj 32: 1277–1284.

Ginhoux F, Greter M, Leboeuf M et al. (2010). Fate mapping analysis reveals that adult microglia derive from primitive macrophages. Science 330: 841–845.

Ginhoux F, Lim S, Hoeffel G et al. (2013). Origin and differentiation of microglia. Front Cell Neurosci 7: 45.

Gorse KM, Lafrenaye AD (2018). The importance of inter-species variation in traumatic brain injury-induced alterations of microglial-axonal interactions. Front Neurol 9: 778.

Green TRF, Murphy SM, Ortiz JB et al. (2021). Age-at-injury influences the glial response to traumatic brain injury in the cortex of male juvenile rats. Front Neurol 12: 804139.

Greenhalgh AD, David S (2014). Differences in the phagocytic response of microglia and peripheral macrophages after spinal cord injury and its effects on cell death. J Neurosci 34: 6316–6322.

Grovola MR, Paleologos N, Wofford KL et al. (2020). Mossy cell hypertrophy and synaptic changes in the hilus following mild diffuse traumatic brain injury in pigs. J Neuroinflammation 17: 44.

Grovola MR, von Reyn C, Loane DJ et al. (2023). Understanding microglial responses in large animal models of traumatic brain injury: an underutilized resource for preclinical and translational research. J Neuroinflammation 20: 67.

Hakim R, Zachariadis V, Sankavaram SR et al. (2021). Spinal cord injury induces permanent reprogramming of microglia into a disease-associated state which contributes to functional recovery. J Neurosci 41: 8441–8459.

Han Y, Zhang L, Wang Q et al. (2019). Minocycline inhibits microglial activation and alleviates depressive-like behaviors in male adolescent mice subjected to maternal separation. Psychoneuroendocrinology 107: 37–45.

Hanlon LA, Huh JW, Raghupathi R (2016). Minocycline transiently reduces microglia/macrophage activation but exacerbates cognitive deficits following repetitive traumatic brain injury in the neonatal rat. J Neuropathol Exp Neurol 75: 214–226.

Hanlon LA, Raghupathi R, Huh JW (2017). Differential effects of minocycline on microglial activation and neurodegeneration following closed head injury in the neonate rat. Exp Neurol 290: 1–14.

Hanlon LA, Raghupathi R, Huh JW (2019). Depletion of microglia immediately following traumatic brain injury in the pediatric rat: implications for cellular and behavioral pathology. Exp Neurol 316: 39–51.

Hasselmann J, Blurton-Jones M (2020). Human iPSC-derived microglia: a growing toolset to study the brain's innate immune cells. Glia 68: 721–739.

Hayakata T, Shiozaki T, Tasaki O et al. (2004). Changes in CSF S100B and cytokine concentrations in early-phase severe traumatic brain injury. Shock 22: 102–107.

Hazelton I, Yates A, Dale A et al. (2018). Exacerbation of acute traumatic brain injury by circulating extracellular vesicles. J Neurotrauma 35: 639–651.

Helmy A, Antoniades CA, Guilfoyle MR et al. (2012). Principal component analysis of the cytokine and chemokine response to human traumatic brain injury. PLoS One 7: e39677.

Helmy A, Guilfoyle MR, Carpenter KL et al. (2014). Recombinant human interleukin-1 receptor antagonist in severe traumatic brain injury: a phase II randomized control trial. J Cereb Blood Flow Metab 34: 845–851.

Henry RJ, Doran SJ, Barrett JP et al. (2019). Inhibition of miR-155 limits neuroinflammation and improves functional recovery after experimental traumatic brain injury in mice. Neurotherapeutics 16: 216–230.

Henry RJ, Ritzel RM, Barrett JP et al. (2020). Microglial depletion with CSF1R inhibitor during chronic phase of experimental traumatic brain injury reduces neurodegeneration and neurological deficits. J Neurosci 40: 2960–2974.

Herbomel P, Thisse B, Thisse C (1999). Ontogeny and behaviour of early macrophages in the zebrafish embryo. Development 126: 3735–3745.

Hergenroeder GW, Moore AN, McCoy Jr JP et al. (2010). Serum IL-6: a candidate biomarker for intracranial pressure elevation following isolated traumatic brain injury. J Neuroinflammation 7: 19.

Homsi S, Piaggio T, Croci N et al. (2010). Blockade of acute microglial activation by minocycline promotes neuroprotection and reduces locomotor hyperactivity after closed head injury in mice: a twelve-week follow-up study. J Neurotrauma 27: 911–921.

Huang Y, Xu Z, Xiong S et al. (2018). Repopulated microglia are solely derived from the proliferation of residual microglia after acute depletion. Nat Neurosci 21: 530–540.

Hutchins KD, Dickson DW, Rashbaum WK et al. (1990). Localization of morphologically distinct microglial populations in the developing human fetal brain: implications for ontogeny. Brain Res Dev Brain Res 55: 95–102.

Hutchinson PJ, O'Connell MT, Rothwell NJ et al. (2007). Inflammation in human brain injury: intracerebral concentrations of IL-1alpha, IL-1beta, and their endogenous inhibitor IL-1ra. J Neurotrauma 24: 1545–1557.

Ismael S, Nasoohi S, Ishrat T (2018). MCC950, the selective inhibitor of nucleotide oligomerization domain-like receptor protein-3 inflammasome, protects mice against traumatic brain injury. J Neurotrauma 35: 1294–1303.

Israel I, Ohsiek A, Al-Momani E et al. (2016). Combined [(18)F]DPA-714 micro-positron emission tomography and autoradiography imaging of microglia activation after closed head injury in mice. J Neuroinflammation 13: 140.

Jassam YN, Izzy S, Whalen M et al. (2017). Neuroimmunology of traumatic brain injury: time for a paradigm shift. Neuron 95: 1246–1265.

Jay TR, Miller CM, Cheng PJ et al. (2015). TREM2 deficiency eliminates TREM2+ inflammatory macrophages and ameliorates pathology in Alzheimer's disease mouse models. J Exp Med 212: 287–295.

Jin X, Ishii H, Bai Z et al. (2012). Temporal changes in cell marker expression and cellular infiltration in a controlled cortical impact model in adult male C57BL/6 mice. PLoS One 7: e41892.

Jin WN, Shi SX, Li Z et al. (2017). Depletion of microglia exacerbates postischemic inflammation and brain injury. J Cereb Blood Flow Metab 37: 2224–2236.

Johnson VE, Stewart JE, Begbie FD et al. (2013). Inflammation and white matter degeneration persist for years after a single traumatic brain injury. Brain 136: 28–42.

Johnson NH, de Rivero Vaccari JP, Bramlett HM et al. (2023). Inflammasome activation in traumatic brain injury and Alzheimer's disease. Transl Res 254: 1–12.

Jung S, Schwartz M (2012). Non-identical twins—microglia and monocyte-derived macrophages in acute injury and autoimmune inflammation. Front Immunol 3: 89.

Katsumoto A, Lu H, Miranda AS et al. (2014). Ontogeny and functions of central nervous system macrophages. J Immunol 193: 2615–2621.

Kaur C, Ling EA, Wong WC (1985). Transformation of amoeboid microglial cells into microglia in the corpus callosum of the postnatal rat brain. An electron microscopical study. Arch Histol Jpn 48: 17–25.

Kenkhuis B, Somarakis A, de Haan L et al. (2021). Iron loading is a prominent feature of activated microglia in Alzheimer's disease patients. Acta Neuropathol Commun 9: 27.

Keren-Shaul H, Spinrad A, Weiner A et al. (2017). A unique microglia type associated with restricting development of Alzheimer's disease. Cell 169: 1276-90.e17.

Khan NA, Asim M, El-Menyar A et al. (2022). The evolving role of extracellular vesicles (exosomes) as biomarkers in traumatic brain injury: clinical perspectives and therapeutic implications. Front Aging Neurosci 14: 933434.

Kierdorf K, Erny D, Goldmann T et al. (2013). Microglia emerge from erythromyeloid precursors via Pu.1- and Irf8-dependent pathways. Nat Neurosci 16: 273–280.

Kobayashi K, Imagama S, Ohgomori T et al. (2013). Minocycline selectively inhibits M1 polarization of microglia. Cell Death Dis 4: e525.

Konno T, Kasanuki K, Ikeuchi T et al. (2018). CSF1R-related leukoencephalopathy: a major player in primary microgliopathies. Neurology 91: 1092–1104.

Krasemann S, Madore C, Cialic R et al. (2017). The TREM2-APOE pathway drives the transcriptional phenotype of dysfunctional microglia in neurodegenerative diseases. Immunity 47: 566-81.e9.

Kroner A, Rosas AJ (2019). Role of microglia in spinal cord injury. Neurosci Lett 709: 134370.

Kumar RG, Diamond ML, Boles JA et al. (2015). Acute CSF interleukin-6 trajectories after TBI: associations with neuroinflammation, polytrauma, and outcome. Brain Behav Immun 45: 253–262.

Kumar A, Alvarez-Croda DM, Stoica BA et al. (2016a). Microglial/macrophage polarization dynamics following traumatic brain injury. J Neurotrauma 33: 1732–1750.

Kumar A, Barrett JP, Alvarez-Croda DM et al. (2016b). NOX2 drives M1-like microglial/macrophage activation and neurodegeneration following experimental traumatic brain injury. Brain Behav Immun 58: 291–309.

Kumar A, Stoica BA, Loane DJ et al. (2017). Microglial-derived microparticles mediate neuroinflammation after traumatic brain injury. J Neuroinflammation 14: 47.

Kumaria A (2017). In vitro models as a platform to investigate traumatic brain injury. Altern Lab Anim 45: 201–211.

Kuwar R, Rolfe A, Di L et al. (2019). A novel small molecular NLRP3 inflammasome inhibitor alleviates neuroinflammatory response following traumatic brain injury. J Neuroinflammation 16: 81.

Lafrenaye AD, Todani M, Walker SA et al. (2015). Microglia processes associate with diffusely injured axons following mild traumatic brain injury in the micro pig. J Neuroinflammation 12: 186.

Lawson LJ, Perry VH, Dri P et al. (1990). Heterogeneity in the distribution and morphology of microglia in the normal adult mouse brain. Neuroscience 39: 151–170.

Lee S, Varvel NH, Konerth ME et al. (2010). CX3CR1 deficiency alters microglial activation and reduces beta-amyloid deposition in two Alzheimer's disease mouse models. Am J Pathol 177: 2549–2562.

Li Q, Cheng Z, Zhou L et al. (2019). Developmental heterogeneity of microglia and brain myeloid cells revealed by deep single-cell RNA sequencing. Neuron 101: 207-23.e10.

Li Y, He X, Kawaguchi R et al. (2020a). Microglia-organized scar-free spinal cord repair in neonatal mice. Nature 587: 613–618.

Li Y, Ritzel RM, Khan N et al. (2020b). Delayed microglial depletion after spinal cord injury reduces chronic inflammation and neurodegeneration in the brain and improves neurological recovery in male mice. Theranostics 10: 11376–11403.

Liu HD, Li W, Chen ZR et al. (2013). Expression of the NLRP3 inflammasome in cerebral cortex after traumatic brain injury in a rat model. Neurochem Res 38: 2072–2083.

Liu N, Li Y, Jiang Y et al. (2023). Establishment and application of a novel in vitro model of microglial activation in traumatic brain injury. J Neurosci 43: 319–332.

Loane DJ, Byrnes KR (2010). Role of microglia in neurotrauma. Neurotherapeutics 7: 366–377.

Loane DJ, Kumar A (2016). Microglia in the TBI brain: the good, the bad, and the dysregulated. Exp Neurol 275: 316–327.

Loane DJ, Stoica BA, Byrnes KR et al. (2013). Activation of mGluR5 and inhibition of NADPH oxidase improves functional recovery after traumatic brain injury. J Neurotrauma 30: 403–412.

Loane DJ, Kumar A, Stoica BA et al. (2014). Progressive neurodegeneration after experimental brain trauma: association with chronic microglial activation. J Neuropathol Exp Neurol 73: 14–29.

Loane DJ, Stoica BA, Faden AI (2015). Neuroprotection for traumatic brain injury. Handb Clin Neurol 127: 343–366.

Lyu J, Jiang X, Leak RK et al. (2021). Microglial responses to brain injury and disease: functional diversity and new opportunities. Transl Stroke Res 12: 474–495.

Ma MW, Wang J, Dhandapani KM et al. (2017). NADPH oxidase 2 regulates NLRP3 inflammasome activation in the brain after traumatic brain injury. Oxid Med Cell Longev 2017: 6057609.

Ma C, Wu X, Shen X et al. (2019). Sex differences in traumatic brain injury: a multi-dimensional exploration in genes, hormones, cells, individuals, and society. Chin Neurosurg J 5: 24.

Ma D, Zhao Y, Huang L et al. (2020). A novel hydrogel-based treatment for complete transection spinal cord injury repair is driven by microglia/macrophages repopulation. Biomaterials 237: 119830.

Madore C, Joffre C, Delpech JC et al. (2013). Early morphofunctional plasticity of microglia in response to acute lipopolysaccharide. Brain Behav Immun 34: 151–158.

Makinde HM, Just TB, Gadhvi GT et al. (2020). Microglia adopt longitudinal transcriptional changes after traumatic brain injury. J Surg Res 246: 113–122.

Malec JF, Brown AW, Leibson CL et al. (2007). The mayo classification system for traumatic brain injury severity. J Neurotrauma 24: 1417–1424.
Marino Lee S, Hudobenko J, McCullough LD et al. (2021). Microglia depletion increase brain injury after acute ischemic stroke in aged mice. Exp Neurol 336: 113530.
Marschallinger J, Iram T, Zardeneta M et al. (2020). Lipid-droplet-accumulating microglia represent a dysfunctional and proinflammatory state in the aging brain. Nat Neurosci 23: 194–208.
Masuda T, Sankowski R, Staszewski O et al. (2020). Microglia heterogeneity in the single-cell Era. Cell Rep 30: 1271–1281.
Mathys H, Adaikkan C, Gao F et al. (2017). Temporal tracking of microglia activation in neurodegeneration at single-cell resolution. Cell Rep 21: 366–380.
Mayer AR, Ling JM, Dodd AB et al. (2021). Reproducibility and characterization of head kinematics during a large animal acceleration model of traumatic brain injury. Front Neurol 12: 658461.
Mildner A, Schmidt H, Nitsche M et al. (2007). Microglia in the adult brain arise from Ly-6ChiCCR2+ monocytes only under defined host conditions. Nat Neurosci 10: 1544–1553.
Mittelbronn M, Dietz K, Schluesener HJ et al. (2001). Local distribution of microglia in the normal adult human central nervous system differs by up to one order of magnitude. Acta Neuropathol 101: 249–255.
Mondello S, Guedes VA, Lai C et al. (2020). Circulating brain injury exosomal proteins following moderate-to-severe traumatic brain injury: temporal profile, outcome prediction and therapy implications. Cells 9: 977.
Morganti JM, Riparip LK, Rosi S (2016). Call off the Dog(ma): M1/M2 polarization is concurrent following traumatic brain injury. PLoS One 11: e0148001.
Morrison 3rd B, Elkin BS, Dolle JP et al. (2011). In vitro models of traumatic brain injury. Annu Rev Biomed Eng 13: 91–126.
Mouzon B, Chaytow H, Crynen G et al. (2012). Repetitive mild traumatic brain injury in a mouse model produces learning and memory deficits accompanied by histological changes. J Neurotrauma 29: 2761–2773.
Mouzon BC, Bachmeier C, Ferro A et al. (2014). Chronic neuropathological and neurobehavioral changes in a repetitive mild traumatic brain injury model. Ann Neurol 75: 241–254.
Nagamoto-Combs K, McNeal DW, Morecraft RJ et al. (2007). Prolonged microgliosis in the rhesus monkey central nervous system after traumatic brain injury. J Neurotrauma 24: 1719–1742.
Neher JJ, Emmrich JV, Fricker M et al. (2013). Phagocytosis executes delayed neuronal death after focal brain ischemia. Proc Natl Acad Sci U S A 110: E4098–E4107.
Nimmerjahn A, Kirchhoff F, Helmchen F (2005). Resting microglial cells are highly dynamic surveillants of brain parenchyma in vivo. Science 308: 1314–1318.
O'Brien WT, Pham L, Symons GF et al. (2020). The NLRP3 inflammasome in traumatic brain injury: potential as a biomarker and therapeutic target. J Neuroinflammation 17: 104.
Otxoa-de-Amezaga A, Miro-Mur F, Pedragosa J et al. (2019). Microglial cell loss after ischemic stroke favors brain neutrophil accumulation. Acta Neuropathol 137: 321–341.
Pechacek KM, Reck AM, Frankot MA et al. (2022). Minocycline fails to treat chronic traumatic brain injury-induced impulsivity and attention deficits. Exp Neurol 348: 113924.
Perry VH (1998). A revised view of the central nervous system microenvironment and major histocompatibility complex class II antigen presentation. J Neuroimmunol 90: 113–121.
Perry VH, Nicoll JA, Holmes C (2010). Microglia in neurodegenerative disease. Nat Rev Neurol 6: 193–201.
Poulen G, Aloy E, Bringuier CM et al. (2021a). Inhibiting microglia proliferation after spinal cord injury improves recovery in mice and nonhuman primates. Theranostics 11: 8640–8659.
Poulen G, Bartolami S, Noristani HN et al. (2021b). Unlike brief inhibition of microglia proliferation after spinal cord injury, long-term treatment does not improve motor recovery. Brain Sci 11: 1643.
Pridans C, Raper A, Davis GM et al. (2018). Pleiotropic impacts of macrophage and microglial deficiency on development in rats with targeted mutation of the Csf1r locus. J Immunol 201: 2683–2699.
Prinz M, Priller J (2014). Microglia and brain macrophages in the molecular age: from origin to neuropsychiatric disease. Nat Rev Neurosci 15: 300–312.
Qin L, Liu Y, Wang T et al. (2004). NADPH oxidase mediates lipopolysaccharide-induced neurotoxicity and proinflammatory gene expression in activated microglia. J Biol Chem 279: 1415–1421.
Rahimian R, Cordeau Jr P, Kriz J (2019). Brain response to injuries: when microglia go sexist. Neuroscience 405: 14–23.
Ramlackhansingh AF, Brooks DJ, Greenwood RJ et al. (2011). Inflammation after trauma: microglial activation and traumatic brain injury. Ann Neurol 70: 374–383.
Ransohoff RM (2016). A polarizing question: do M1 and M2 microglia exist? Nat Neurosci 19: 987–991.
Rezaie P, Male D (2002). Mesoglia & microglia—a historical review of the concept of mononuclear phagocytes within the central nervous system. J Hist Neurosci 11: 325–374.
Rezaie P, Dean A, Male D et al. (2005). Microglia in the cerebral wall of the human telencephalon at second trimester. Cereb Cortex 15: 938–949.
Rice RA, Spangenberg EE, Yamate-Morgan H et al. (2015). Elimination of microglia improves functional outcomes following extensive neuronal loss in the hippocampus. J Neurosci 35: 9977–9989.
Rice RA, Pham J, Lee RJ et al. (2017). Microglial repopulation resolves inflammation and promotes brain recovery after injury. Glia 65: 931–944.
Rio-Hortega D (1939). The microglia. Lancet 233: 1023–1026.
Ritzel RM, He J, Li Y et al. (2021). Proton extrusion during oxidative burst in microglia exacerbates pathological acidosis following traumatic brain injury. Glia 69: 746–764.
Ritzel RM, Li Y, Lei Z et al. (2022). Functional and transcriptional profiling of microglial activation during the chronic phase of TBI identifies an age-related driver of poor outcome in old mice. Geroscience 44: 1407–1440.
Ritzel RM, Li Y, Jiao Y et al. (2023). Brain injury accelerates the onset of a reversible age-related microglial phenotype associated with inflammatory neurodegeneration. Sci Adv 9: eadd1101.
Rojo R, Raper A, Ozdemir DD et al. (2019). Deletion of a Csf1r enhancer selectively impacts CSF1R expression and development of tissue macrophage populations. Nat Commun 10: 3215.
Safaiyan S, Besson-Girard S, Kaya T et al. (2021). White matter aging drives microglial diversity. Neuron 109: 1100-17.e10.
Salter MW, Beggs S (2014). Sublime microglia: expanding roles for the guardians of the CNS. Cell 158: 15–24.
Schneider Soares FM, Menezes de Souza N, Liborio Schwarzbold M et al. (2012). Interleukin-10 is an independent biomarker of severe traumatic brain injury prognosis. Neuroimmunomodulation 19: 377–385.
Schulz C, Gomez Perdiguero E, Chorro L et al. (2012). A lineage of myeloid cells independent of Myb and hematopoietic stem cells. Science 336: 86–90.
Scott G, Zetterberg H, Jolly A et al. (2018). Minocycline reduces chronic microglial activation after brain trauma but increases neurodegeneration. Brain 141: 459–471.

Shiozaki T, Hayakata T, Tasaki O et al. (2005). Cerebrospinal fluid concentrations of anti-inflammatory mediators in early-phase severe traumatic brain injury. Shock 23: 406–410.

Siopi E, Llufriu-Daben G, Fanucchi F et al. (2012). Evaluation of late cognitive impairment and anxiety states following traumatic brain injury in mice: the effect of minocycline. Neurosci Lett 511: 110–115.

Smith C, Gentleman SM, Leclercq PD et al. (2013a). The neuroinflammatory response in humans after traumatic brain injury. Neuropathol Appl Neurobiol 39: 654–666.

Smith DH, Johnson VE, Stewart W (2013b). Chronic neuropathologies of single and repetitive TBI: substrates of dementia? Nat Rev Neurol 9: 211–221.

Sousa C, Golebiewska A, Poovathingal SK et al. (2018). Single-cell transcriptomics reveals distinct inflammation-induced microglia signatures. EMBO Rep 19:e46171.

Spangenberg E, Severson PL, Hohsfield LA et al. (2019). Sustained microglial depletion with CSF1R inhibitor impairs parenchymal plaque development in an Alzheimer's disease model. Nat Commun 10: 3758.

Stein DM, Lindell A, Murdock KR et al. (2011). Relationship of serum and cerebrospinal fluid biomarkers with intracranial hypertension and cerebral hypoperfusion after severe traumatic brain injury. J Trauma 70: 1096–1103.

St-Pierre MK, Simoncicova E, Bogi E et al. (2020). Shedding light on the dark side of the microglia. ASN Neuro 12: 1759091420925335.

Streit WJ, Braak H, Xue QS et al. (2009). Dystrophic (senescent) rather than activated microglial cells are associated with tau pathology and likely precede neurodegeneration in Alzheimer's disease. Acta Neuropathol 118: 475–485.

Szalay G, Martinecz B, Lenart N et al. (2016). Microglia protect against brain injury and their selective elimination dysregulates neuronal network activity after stroke. Nat Commun 7: 11499.

Tambuyzer BR, Ponsaerts P, Nouwen EJ (2009). Microglia: gatekeepers of central nervous system immunology. J Leukoc Biol 85: 352–370.

Tasci A, Okay O, Gezici AR et al. (2003). Prognostic value of interleukin-1 beta levels after acute brain injury. Neurol Res 25: 871–874.

Taylor SE, Morganti-Kossmann C, Lifshitz J et al. (2014). Rod microglia: a morphological definition. PLoS One 9: e97096.

Theriault P, ElAli A, Rivest S (2015). The dynamics of monocytes and microglia in Alzheimer's disease. Alzheimers Res Ther 7: 41.

Todd BP, Chimenti MS, Luo Z et al. (2021). Traumatic brain injury results in unique microglial and astrocyte transcriptomes enriched for type I interferon response. J Neuroinflammation 18: 151.

Tremblay ME, Stevens B, Sierra A et al. (2011). The role of microglia in the healthy brain. J Neurosci 31: 16064–16069.

Venkatesh K, Ghosh SK, Mullick M et al. (2019). Spinal cord injury: pathophysiology, treatment strategies, associated challenges, and future implications. Cell Tissue Res 377: 125–151.

Verney C, Monier A, Fallet-Bianco C et al. (2010). Early microglial colonization of the human forebrain and possible involvement in periventricular white-matter injury of preterm infants. J Anat 217: 436–448.

Wang Y, Szretter KJ, Vermi W et al. (2012). IL-34 is a tissue-restricted ligand of CSF1R required for the development of Langerhans cells and microglia. Nat Immunol 13: 753–760.

Wang J, Ma MW, Dhandapani KM et al. (2017). Regulatory role of NADPH oxidase 2 in the polarization dynamics and neurotoxicity of microglia/macrophages after traumatic brain injury. Free Radic Biol Med 113: 119–131.

Wang CF, Zhao CC, Liu WL et al. (2020). Depletion of microglia attenuates dendritic spine loss and neuronal apoptosis in the acute stage of moderate traumatic brain injury in mice. J Neurotrauma 37: 43–54.

Wang Y, Wernersbach I, Strehle J et al. (2022). Early posttraumatic CSF1R inhibition via PLX3397 leads to time- and sex-dependent effects on inflammation and neuronal maintenance after traumatic brain injury in mice. Brain Behav Immun 106: 49–66.

Wei Y, Li X (2022). Different phenotypes of microglia in animal models of Alzheimer disease. Immun Ageing 19: 44.

Wendimu MY, Hooks SB (2022). Microglia phenotypes in aging and neurodegenerative diseases. Cells 11: 2091.

Willis EF, MacDonald KPA, Nguyen QH et al. (2020). Repopulating microglia promote brain repair in an IL-6-dependent manner. Cell 180: 833-46.e16.

Witcher KG, Bray CE, Dziabis JE et al. (2018). Traumatic brain injury-induced neuronal damage in the somatosensory cortex causes formation of rod-shaped microglia that promote astrogliosis and persistent neuroinflammation. Glia 66: 2719–2736.

Witcher KG, Dziabis JE, Bray CE et al. (2020). Comparison between midline and lateral fluid percussion injury in mice reveals prolonged but divergent cortical neuroinflammation. Brain Res 1746: 146987.

Witcher KG, Bray CE, Chunchai T et al. (2021). Traumatic brain injury causes chronic cortical inflammation and neuronal dysfunction mediated by microglia. J Neurosci 41: 1597–1616.

Wofford KL, Harris JP, Browne KD et al. (2017). Rapid neuroinflammatory response localized to injured neurons after diffuse traumatic brain injury in swine. Exp Neurol 290: 85–94.

Wolf Y, Yona S, Kim KW et al. (2013). Microglia, seen from the CX3CR1 angle. Front Cell Neurosci 7: 26.

Wu YH, Rosset S, Lee TR et al. (2021). In vitro models of traumatic brain injury: a systematic review. J Neurotrauma 38: 2336–2372.

Xu KY, Wu CY, Tong S et al. (2018a). The selective Nlrp3 inflammasome inhibitor Mcc950 attenuates lung ischemia-reperfusion injury. Biochem Biophys Res Commun 503: 3031–3037.

Xu X, Yin D, Ren H et al. (2018b). Selective NLRP3 inflammasome inhibitor reduces neuroinflammation and improves long-term neurological outcomes in a murine model of traumatic brain injury. Neurobiol Dis 117: 15–27.

Xu L, Wang J, Ding Y et al. (2021). Current knowledge of microglia in traumatic spinal cord injury. Front Neurol 12: 796704.

Yamasaki R, Lu H, Butovsky O et al. (2014). Differential roles of microglia and monocytes in the inflamed central nervous system. J Exp Med 211: 1533–1549.

Yang SH, Lee DK, Shin J et al. (2017). Nec-1 alleviates cognitive impairment with reduction of Abeta and tau abnormalities in APP/PS1 mice. EMBO Mol Med 9: 61–77.

Younger D, Murugan M, Rama Rao KV et al. (2019). Microglia receptors in animal models of traumatic brain injury. Mol Neurobiol 56: 5202–5228.

Zetterberg H, Blennow K (2016). Fluid biomarkers for mild traumatic brain injury and related conditions. Nat Rev Neurol 12: 563–574.

Zhang Y, Cui D (2021). Evolving models and tools for microglial studies in the central nervous system. Neurosci Bull 37: 1218–1233.

Zhang QG, Laird MD, Han D et al. (2012). Critical role of NADPH oxidase in neuronal oxidative damage and microglia activation following traumatic brain injury. PLoS One 7: e34504.

Zhang Y, Li H, Li X et al. (2020). TMEM16F aggravates neuronal loss by mediating microglial phagocytosis of neurons in a rat experimental cerebral ischemia and reperfusion model. Front Immunol 11: 1144.

Zhou ZL, Xie H, Tian XB et al. (2023). Microglial depletion impairs glial scar formation and aggravates inflammation partly by inhibiting STAT3 phosphorylation in astrocytes after spinal cord injury. Neural Regen Res 18: 1325–1331.

Handbook of Clinical Neurology, Vol. 210 (3rd series)
Neuroglia in Neurologic and Psychiatric Disorders, Part II
A. Verkhratsky, L.D. de Witte, E. Aronica, and E.M. Hol, Editors
https://doi.org/10.1016/B978-0-443-19102-2.00001-6

Chapter 8

Astrocytes and microglia in multiple sclerosis and neuromyelitis optica

LINJIE ZHANG[1], ALEXEI VERKHRATSKY[2,3,4], AND FU-DONG SHI[1,5*]

[1]*Department of Neurology, Tianjin Neurological Institute, Tianjin Medical University General Hospital, Tianjin, China*

[2]*Faculty of Biology, Medicine and Health, University of Manchester, Manchester, United Kingdom*

[3]*Department of Neurosciences, University of the Basque Country UPV/EHU and CIBERNED, Leioa, Bizkaia, Spain*

[4]*IKERBASQUE, Basque Foundation for Science, Bilbao, Spain*

[5]*Department of Neurology, China National Clinical Research Center for Neurological Diseases, Beijing Tiantan Hospital, Capital Medical University, Beijing, China*

Abstract

Multiple sclerosis and neuromyelitis optica are autoimmune neurodegenerative diseases primarily targeting myelin sheath and neuroglia. In multiple sclerosis, autoantibodies destroy oligodendrocytes and myelin, which underlies primary neurologic symptoms. Focal damage to myelin triggers reactive astrogliosis and microgliosis, which contribute to and to a large extent define the disease progression. In neuromyelitis optica, autoantibodies against water channel aquaporin 4 (AQP4), which are localized at astrocytic endfeet mediate damage of the glia limitans thus facilitating infiltration of blood-borne molecules and cells that propagate the damage to nerves and neurons. This primary astrocytopathy recruits microglia, which contribute to the neuroinflammatory response. Neuroglial cells therefore are potential targets for cell-specific therapies.

HISTORIC PRELUDE

The history of multiple sclerosis began (according to a generally accepted legend—see comprehensive account of the disease history by Murray, 2005) in the 14th century AD. This very first case was the Saint Lidwina the Virgin from Schiedam, Holland (1380–1433), who developed a peculiar, relapsing, and progressive neurologic condition starting from 15 years old. The onset was manifested by the fall while skating (Fig. 8.1), and the disease was characterized by miscoordination, blindness in one eye, pain, and weakness. Later, she developed various visions and became a well-recognized mystic, a healer, and a holy woman. Lidwina was beatified in 1890 (Albers, 1910) and became a patron of multiple sclerosis patients. The firsthand detailed description of multiple sclerosis from the patient's perspective (including visual and motor symptoms) can be found in the diaries of Agustas d'Este, an illegitimate grandson of King George III of Great Britain (Firth, 1941).

The first clinical description of multiple sclerosis with an early (a 17-year-old man) onset belongs to Charles-Prosper Ollivier d'Angers (Ollivier, 1824). The first histopathologic description of the disseminated plaques, which are the hallmark of multiple sclerosis, was made by Robert Carswell in 1838 (Carswell, 1838). A decade later in 1849 Friedrich von Frerichs, provided the clinical and histopathologic description of the *Hirnsklerose* (brain sclerosis), which highlighted motor and visual symptoms and their course, and linked them to multiple, sclerotic lesions present through the brain and spinal cord (Frerich, 1849). Finally, in the 1860s, Jean-Martin Charcot and Edme Felix Alfred Vulpian produced a classical description of the disease, which they named *Sclérose en plaque disseminé* (Vulpian) or *Sclérose en plaque* (Charcot) (Vulpian, 1866; Charcot, 1877; Bogousslavsky et al., 2011; Zalc, 2018).

Neuromyelitis optica acuta was described by Eugène Devic in 1894 (Devic, 1895) and hence sometimes referred to as Devic's syndrome, characterized by acute myelitis and optica neuritis. Devic's student Fernand Gault published his doctoral thesis, entitled *De la neuro-myélite optique aiguë*, which consisted of a review of the previous medical literature and a clinicopathological

*Correspondence to: Fu-Dong Shi, Department of Neurology, Tianjin Neurological Institute, Tianjin Medical University General Hospital, Tianjin 300052, China; Department of Neurology, China National Clinical Research Center for Neurological Diseases, Beijing Tiantan Hospital, Capital Medical University, Beijing 100070, China. Tel: +86-02260814606, E-mail: fshi@tmu.edu.cn

Fig. 8.1. Saint Lidwina the Virgin, the patron of multiple sclerosis patients. "Lidwina's Fall on the Ice," printed by Peterezs Otgier Nachtegael, first edition 1498 (https://inpress.lib.uiowa.edu/feminae/DetailsPage.aspx?Feminae_ID=30904).

analysis of Devic's case (Gault, 1894). Of note, clinical descriptions of paresis, spinal pain, and amaurosis have been described earlier; in particular, such descriptions were made by Antoine Portal, the first physician to Louis XVIII in 1804, and Giovanni Battista Pescetto in 1844 (Jarius and Wildemann, 2012; Jarius and Wildemann, 2013).

The groundbreaking discovery of a novel, pathogenic autoantibody (termed NMO-IgG or AQP4-Ab) in a subset of patients by Vanda Lennon and colleagues in 2004 has led to a substantial increase in the interest in NMO (Lennon et al., 2004; Lennon et al., 2005). This discovery led to the recognition of "IgG-positive NMO" as an immunopathologically defined disease entity in its own right distinct from classical MS. The term "NMO spectrum disorder" was intended to be used to refer to all major clinical manifestations that had by then been reported in association with NMO-IgG (Jarius et al., 2013). According to the pathogenesis of AQP4-IgG-positive cases, the term "autoimmune AQP4 channelopathy" was proposed (Wingerchuk, 2007).

OVERVIEW OF MULTIPLE SCLEROSIS AND NEUROMYELITIS OPTICA

Multiple sclerosis (MS) is an autoimmune, inflammatory, and degenerative disease of the central nervous system (CNS), in which oligodendrocytes, axons, and neurons are lost in both white and gray matter leading to the accumulation of neurologic and cognitive deficits. The progression of MS is broadly classified into a relapsing–remitting MS (RRMS) phase, in which neurologic deficits manifest and recede, a secondary progressive MS (SPMS; when the progressive phase follows RRMS), and primary progressive MS (PPMS), wherein neurologic disability accumulates with minimal or absent periods of recovery. Multiple focal areas of myelin loss within the CNS are the main pathohistological hallmark of MS. Demyelination is accompanied by inflammation, reactive gliosis, and relative axonal preservation in the early stage of MS. The latter stages of the disease are accompanied by compartmentalized inflammation, contributing to continuous inflammatory and degenerative changes in the CNS, hence driving disease progression (Thompson et al., 2018; Haase and Linker, 2021). MS lesions are broadly segregated into gray matter lesions, where the cortical lesions are often associated with meningeal inflammation, and white matter lesions. The latter are further subclassified into (i) active lesions populated with immune cells (both microglia and invading macrophages); (ii) mixed active–inactive lesions characterized by an actively demyelinating rim associated with macrophage/microglia and a demyelinated immunologically inactive core; and (iii) inactive demyelinated lesions. Active and mixed active–inactive lesions are further subdivided into demyelinating and postdemyelinating ones depending on the presence or absence of myelin degradation products in macrophages or microglia (Kuhlmann et al., 2017). Advances in imaging, biomarker analysis, and genomics suggest that microglia and astrocytes have central roles in the disease progression. An in-depth understanding of inflammation driven by neuroglia is required to identify novel potential therapeutic opportunities.

Neuromyelitis optica (NMO), which belongs to a group of Neuromyelitis optica spectrum disorders (NMOSD), is a rare autoimmune disease. Clinical presentations of NMOSD are mainly represented by immune-mediated damage to the myelin sheath and axons in the optic nerve and the spinal cord, together with area postrema syndrome, brainstem syndrome, and diencephalic syndrome (Wingerchuk et al., 2015). Aetiopathologically, the NMOSD are subclassified into (i) diseases with identifiable antibodies against aquaporin-4 AQP4-IgG (NMOSD-AQP4); (ii) NMOSD without antiaquaporin-4 immunoglobulins (AQP4-IgG); (iii) with unknown AQP4-IgG status; and (iv) NMOSD with identified antibodies against myelin oligodendrocyte glycoprotein (NMOSD-MOG) (Wingerchuk et al., 2015; Dos Passos et al., 2018). About 70% of all NMSOD cases are associated with the generation of AQP4-IgG, which attack AQP4 that, in the CNS, are localized primarily at the endfeet of astrocytes. The NMO is therefore a primary astrocytopathy. The onset is associated with an autoimmune attack mediated by AQP4-IgG on the optic nerves and the spinal cord, often leading to blindness and paralysis. Seropositivity for aquaporin-4 IgG in the context of an inflammatory attack on the CNS, especially within the optic nerve or spinal cord, is all that is required to confirm a diagnosis of NMOSD, on the basis of the newest clinical criteria (Carnero Contentti and Correale, 2021; Wright et al., 2021). Histopathology studies of early NMO lesions demonstrated prominent reactive microgliosis. Significant microglial reactivity is also observed in NMO animal models induced by injecting AQP4-IgG into the CNS. The main therapeutic target is currently focused on blocking or reducing the production of autoantibodies and suppressing the antibody-mediated complement-mediated inflammatory response, to reduce the primary damage of the astrocytes. With further study of the role of microglia in the inflammatory response, the treatment of this aspect may be worth exploring (Chen et al., 2021) (Fig. 8.2).

OVERVIEW OF NEUROGLIA IN THE CONTEXT OF MS AND NMO

Neuroglia, which are classified into astroglia (parenchymal astrocytes, radial astrocytes, and ependymoglia), oligodendroglia (oligodendrocytes and oligodendrocyte precursor cells, OPCs), and microglia in the CNS (Verkhratsky and Butt, 2023), provide physical and functional support of the nervous tissue and form the active milieu of the CNS (Semyanov and Verkhratsky, 2021). Neuroglia contribute to the pathogenesis of all neurologic diseases; pathophysiology of neuroglia is complex, mutable, and highly heterogeneous (Verkhratsky et al., 2023); neuroglial cells are intimately involved in the pathogenesis of both MS and NMO. First and foremost, these disorders lead to massive myelin damage; demyelination leads to the disruption of action potential propagation, generation of myelin debris, and secondary inflammation. New myelin formation at the sites of CNS demyelination occurs through the differentiation of OPCs followed by remyelination. Remyelination is necessary to restore myelin structure, provide metabolic support to axons, and restore rapid action potential propagation along axons. Remyelination is a tightly controlled process (regulated by several transcription factors) that depends upon the functional state, proliferation, migration maturation, and age of OPCs (Yi et al., 2023). When OPC recruitment and maturation are impeded, progressive demyelination and chronic neuroinflammation prevail. Thus, a decrease in the damage to oligodendrocytes and the promotion of regeneration

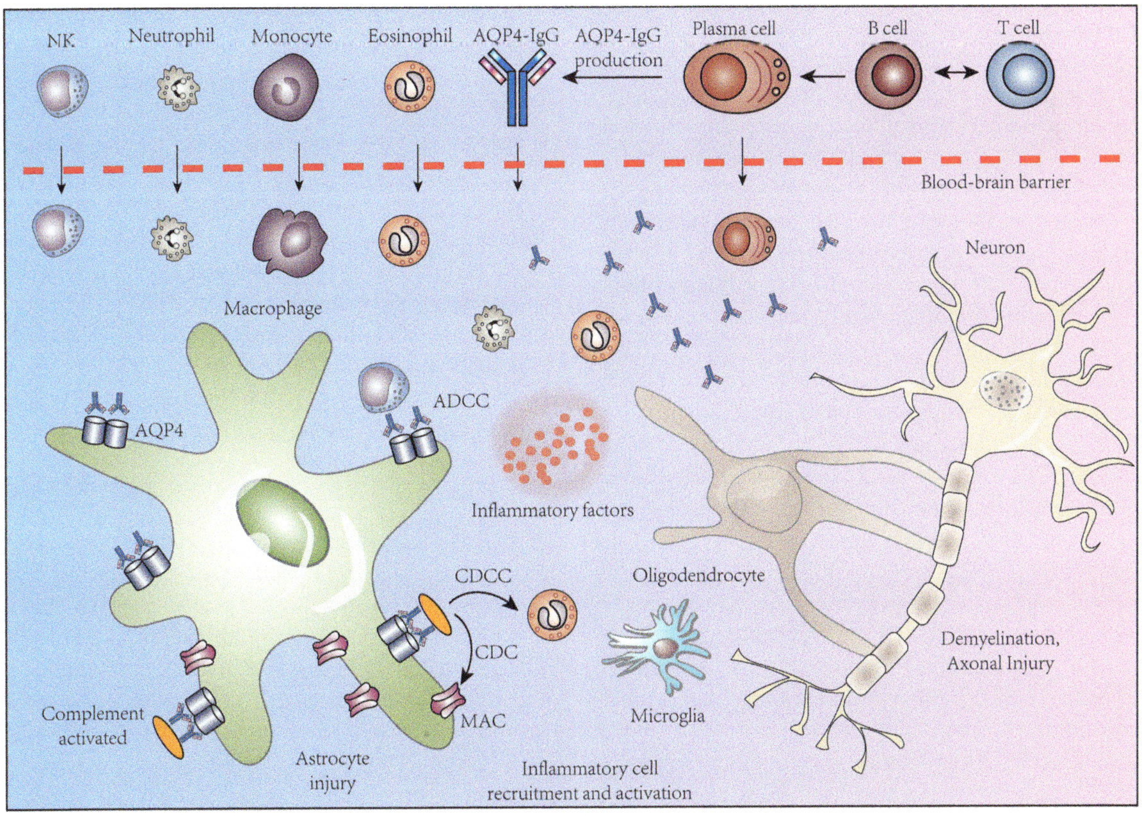

Fig. 8.2. Pathogenesis of neuromyelitis optica. Based on Papadopoulos MC, Bennett JL, Verkman AS (2014). Treatment of neuromyelitis optica: state-of-the-art and emerging therapies. Nat Rev Neurol 10: 493–506.

are major treatment directions (Kalafatakis and Karagogeos, 2021; Tepavčević and Lubetzki, 2022). Microglia are the resident, macrophages of the CNS, which are of myeloid origin; microglia derive from fetal macrophages that enter the neural tube early in development (Ginhoux and Prinz, 2015). In the CNS, microglia constitute 5%–10% of the total glial cell population; microglia perform numerous vital physiologic functions and respond to damage with graded reactive microgliosis (Kettenmann et al., 2011; Paolicelli et al., 2022). The class of astroglia covers a highly heterogeneous population of neural cells responsible for the homeostasis and defense of the nervous tissue (Verkhratsky and Nedergaard, 2018). In response to lesions to the nervous tissue (such as trauma, stroke, infection, and autoimmune attack) astrocytes undergo a complex program of reactive astrogliosis generality aimed at lesion localization and regeneration (Escartin et al., 2021).

In this chapter, we narrate the contribution of glial cells to the pathophysiology of MS and NMO and the potential for glio-centric therapies. In particular, we focus on the role of microglia and astrocytes, which define pathogenetic steps in both diseases.

MICROGLIA IN MULTIPLE SCLEROSIS

Microglial phenotypes

Microglia are multifunctional cells with a number of physiologic functions in the embryonic, early postnatal, and adult brain (Kettenmann et al., 2013; Stratoulias et al., 2019). Phagocytosis, surveillance, and capacity for releasing soluble factors are core properties through which microglia contribute to key biologic functions (Paolicelli et al., 2022). Pathologic insults instigate an evolutionary conserved defensive microglial response known as reactive microgliosis. Reactive microglia function as proinflammatory or antiinflammatory elements, maintaining tissue homeostasis and adaptation according to the environmental challenge. Various diseases are characterized by the emergence of multiple microglial phenotypes and states. Microglia support neuroinflammation through secreting proinflammatory mediators and oxidative compounds, including IL-1, IL-12, IL-23, Iba-1, CD68, IL-1β, TNF-α, and nitric oxide (NO). In addition, in the inflammatory environment, microglia express MHC class II, costimulatory molecules, Fc receptors, and integrins (Thompson et al., 2018; Wang et al., 2019; Liu et al., 2020b). At the same time, microglia are involved in rehabilitation and repair through secreting IL-4, IL-10, and IL-13 and upregulate the expression of the CD206 and arginase 1 (Arg1), promoting tissue regeneration (Colton, 2009). Besides, microglia can secrete growth factors and neurotrophic factors, such as IGF-I, CSF1, nerve growth factor (NGF), BDNF, neurotrophin (NT)4/5, and glial cell-derived neurotrophic factor (GDNF); all of which promote regeneration of the nervous tissue (Pocock and Kettenmann, 2007; Colonna and Butovsky, 2017). In the experimental autoimmune encephalomyelitis (EAE) model of MS, in the early demyelination stage, microglial proinflammatory properties predominate, while antiinflammatory regenerative microglia are more abundant in the late stage. These regenerative microglia promote OPC differentiation and remyelination (Baaklini et al., 2019).

Microglia contribution to pathogenesis of multiple sclerosis

Focal white matter demyelinating lesions characteristic of MS are distinguished by a compromised blood–brain barrier (BBB), deposition of blood plasma proteins, infiltration of B cells and (predominantly CD8+) T cells, infiltration of myeloid cells, degeneration of myelin, oligodendrocytes loss, and axonal damage (Filippi et al., 2018; Healy et al., 2022). In particular, the number of microglial cells at the lesion sites and their features are used to date MS lesions. In the active lesions, macrophages and microglia are present throughout the lesion area. The mixed active–inactive lesion features a hypocellular lesion centre, whereas macrophages and microglia are confined to the lesion border. In inactive lesions, macrophages and microglia are almost completely absent (Healy et al., 2022). In the healthy and periplaque white matter, diffuse microglial reactivity is more prominent with a longer disease duration and is more extensive in the progressive MS than in the RRMS (Healy et al., 2022; Kessler et al., 2023) (Fig. 8.3).

Although myelin is known to be critical for the support of CNS axons, the mechanisms sustaining myelin health remain unclear. Resident microglia are crucial for the maintenance of myelin health in adulthood in both mice and humans (McNamara et al., 2023). Microglia are indispensable for the development of myelin axonal coverage (Santos and Fields, 2021). Loss of myelin health due to the absence of microglia is associated with the appearance of myelinating oligodendrocytes with altered lipid metabolism; this aberrant state is linked to a disrupted tumor growth factor β1 and its receptor (TGFβ1-TGFβR1) axis. This highlights microglia as a therapeutic target for restoration of myelin growth and integrity in particular in aging and neurodegeneration (McNamara et al., 2023). Microglia (as well as infiltrating macrophages) from acute lesions in SPMS are characterized by iron accumulation (Tham et al., 2021). This iron accumulation occurs after phagocytotic uptake of myelin debris, dead oligodendrocytes, and extravasated red blood cells (Williams et al., 2012; Dal-Bianco et al., 2017; Gillen et al., 2018). Iron-rich microglia and macrophages are mainly present in the rims of focal MS lesions. These iron-rim lesions are substantially more destructive than noniron-rim lesions (Dal-Bianco et al., 2021).

ASTROCYTES IN MULTIPLE SCLEROSIS

Active MS lesions are characterized by the prevalence of reactive astrocytes, which hypertrophies morphology and high expression levels of glial fibrillary acidic protein (GFAP); they also upregulate the expression of proinflammatory cytokines, chemokines, and remyelination signaling molecules (Williams et al., 2007; Franklin and Goldman, 2015). Astrocytes in the center of inactive lesions typically have small cell bodies, are thought to be transcriptionally inactive, and have long processes rich in filaments, forming perilesional barriers (which is, often, erroneously referred to as an astrocytic scar (Rawji et al., 2020), of note, all scars are fibrotic (Verkhratsky et al., 2023). Like microglia,

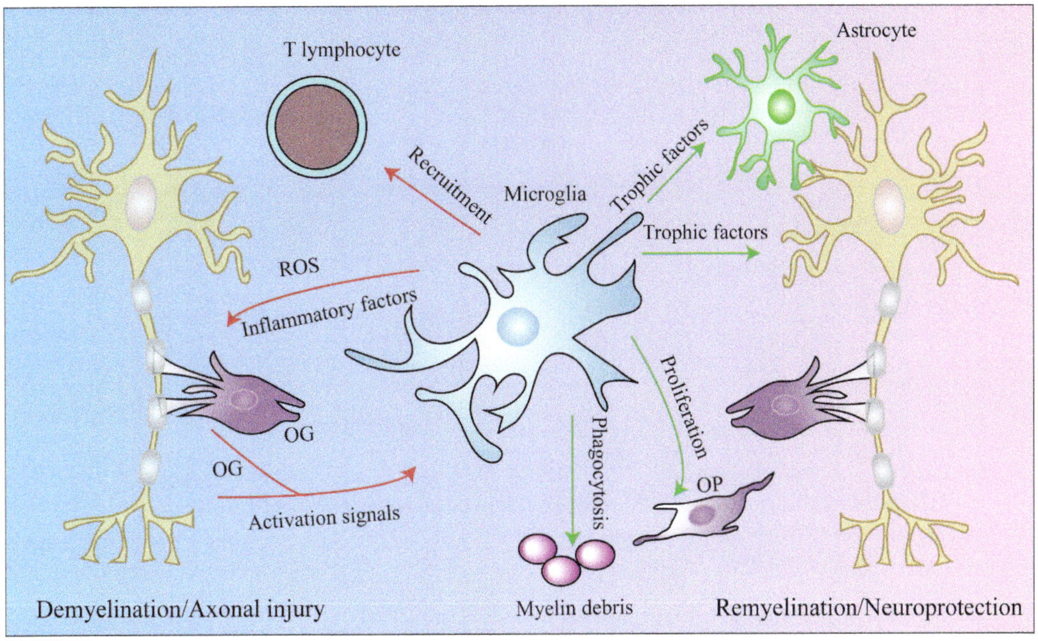

Fig. 8.3. Microglia in multiple sclerosis. Based on Espinosa-Parrilla J, Pugliese M et al. (2016). Neuroprotection: a new therapeutic approach of relapsing remitting multiple sclerosis [internet]. Trending topics in multiple sclerosis. InTech.E.

astrocytes also accumulate iron, predominantly in the inactive gliotic centers of slowly expanding lesions, and can phagocytose myelin-like macrophages and microglia (Ponath et al., 2017). Inflammatory stimulation boosts astrocytic reactivity and increases their capacity to sequester iron (Rathore et al., 2012). A better understanding of the functional consequences of iron accumulation in microglia and astrocytes is needed to accurately interpret the data available and target treatments. Astrocytic hyperplasia is marked in the periplaque area but is less in the normal-appearing white matter (Healy et al., 2022). Changes of the microglial reactivity and astrocytic hyperplasia may indicate the sites of the Wallerian degeneration or future lesion (Singh et al., 2013; Maggi et al., 2014; Healy et al., 2022). Serum GFAP levels correlate with and may reflect reactive astrogliosis and/or astrocytic damage. Serum GFAP is higher in the progressive MS compared to the RRMS. Serum GFAP levels also correlate with clinical severity scores, white matter lesion load and white matter and gray matter volume loss in progressive MS (Abdelhak et al., 2018; Ayrignac et al., 2020).

CROSS TALK BETWEEN ASTROCYTES AND MICROGLIA

All neuroglial cells contribute to the MS progression; intercellular interactions as well as neural cells' cross talk to the immune system must be considered; in particular, a lymphocyte-microglia-astrocyte axis was suggested to play a defining role in the chronic active MS (Absinta et al., 2021). Both microglia and astrocytes are disease modifiers being part of the innate immune system based on their ability to produce immunomodulators and expression of receptors associated with innate immunity, such as complement receptors or toll-like receptors (TLRs) (Matejuk and Ransohoff, 2020; Garland et al., 2022).

Cross talk between astrocytes and microglia is neurochemical in nature, being mediated by growth factors, neurotransmitters, cytokines, chemokines, innate-immunity mediators, and damage-associated molecular patterns (with ATP being an archetypal DAMP), mitogenic factors, NO, reactive oxygen species, and metabolic mediators linked to cell metabolism (Matejuk and Ransohoff, 2020). In addition, astrocytes, microglia, and neurons communicate by releasing and perceiving extracellular vesicles, which mediate long-distance signaling by carrying various cargo including mRNA and miRNA capable of modulating gene expression in distant cells (Matejuk and Ransohoff, 2020). In addition, reactive glia express and secrete canonical cytokines such as IL-1β, IL-6, TNF-α, IL-18, TGF-β, and IL-10.

Glial cells regulate and control each other's function, migration, and reactivity. In neurologic diseases, there is a dialogue between astrocytes and microglia. Current therapies for neurologic disorders and clinical trials based on blocking inflammatory responses are significantly inadequate. It is useful to be aware that the CNS environment involves programs and functions of astrocytes and microglia that are not understood in a typical immune response. The language of microglia and astrocytes is a key to understanding this complex system, which consists of unimaginable diversity and plasticity and a multitude of unknown functions (Fig. 8.4).

NEURODEGENERATION AS A FEATURE OF MS: ROLE FOR NEUROGLIA

Neurodegeneration is one of the intrinsic components of the pathophysiology of MS. Progression without activity, cognitive impairment, depression, fatigue, and even optic neuropathy are all linked to neurodegeneration (Sandi et al., 2021). Understanding the relationship between inflammation and neurodegeneration

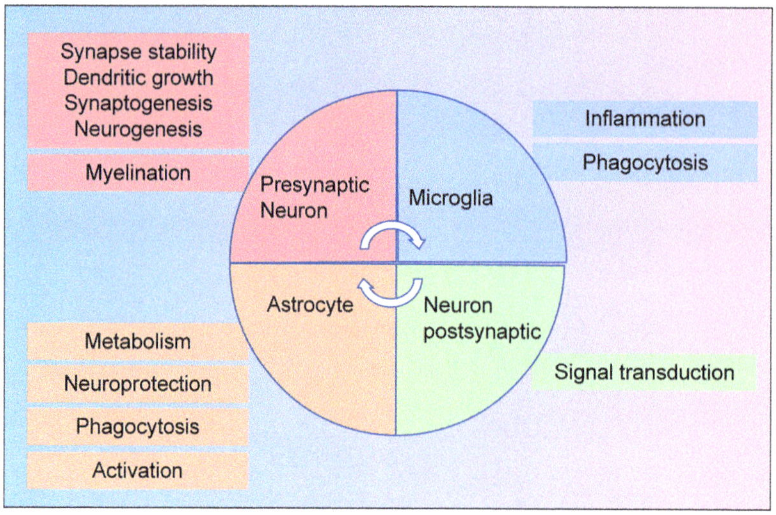

Fig. 8.4. Interactions of astrocytes, microglia, and neurons in MS and NMO. Based on Matejuk A, Ransohoff RM (2020). Crosstalk between astrocytes and microglia: an overview. Front Immunol 11: 1416.

is of key importance for developing future therapeutic strategies in multiple sclerosis. Neuroglia contribute to pathophysiology of neurodegeneration through many mechanisms. Multidimensional data revealing how immune cells interact with glia, and how these cells induce or regulate inflammation and downstream neurodegeneration, are key to identifying therapeutic targets for treating progressive MS (Ramaglia et al., 2021).

NEUROGLIA AS A THERAPEUTIC TARGET IN MULTIPLE SCLEROSIS

The major therapeutic goal of MS is to prevent the accumulation of disability not associated with acute inflammation, as it is not sufficiently controlled by currently available treatments. Disability progression can result from acute relapses as well as from CNS parenchymal disintegration and neurodegeneration without new CNS lesion formation (Husseini et al., 2024).

Microglia and astrocytes contribute to axonal and myelin injury, being thus targets for therapeutic intervention as demonstrated by in vitro and in vivo studies of MS (Brambilla, 2019; Voet et al., 2019; Healy et al., 2022). Both cell types mediate injury by producing and releasing soluble molecules, including proinflammatory cytokines, excitotoxins (glutamate), and free radicals (reactive oxygen and nitrogen species) (Sheng et al., 2013; Lassmann and van Horssen, 2016). Microglia are a source of complement components and can mediate effects that depend on cell contact, such as synaptic stripping and antibody-dependent cell cytotoxicity (Ulvestad et al., 1994; Voet et al., 2019). Astrocytes can also produce complement proteins, actively recruit peripheral lymphocytes, and prevent repair (Farina et al., 2007; Lian et al., 2015).

At the same time, astrocytes and microglia protect nervous tissue by producing neurotrophic factors and can promote myelin regeneration (Nociti and Romozzi, 2023; Kent and Miron, 2024). These protective effects provide additional pathways and molecular targets for therapeutic development, but also increase the complexity of drug intervention, as the harmful activity of glial cells needs to be targeted while maintaining their repair properties. Other considerations are the heterogeneity of glia and whether there are discrete subpopulations of damage- and repair-inducing cells, or whether all cells acquire these properties at different times. The impact of current multiple sclerosis treatments on glial cell function, and future strategies targeting astrocytes and microglia for multiple sclerosis therapy are discussed here.

Targeting microglia

Indirect and direct effects on microglia are presented by the currently available disease-modifying therapies (DMTs) for MS. IFNβ and glatiramer acetate, have generalized antiinflammatory effects on microglia. These agents are unlikely to access the CNS; hence, the antiinflammatory effects are probably mediated by reduced infiltration of T cells into the CNS or modulation of their immune regulatory properties (Prod'homme and Zamvil, 2019). However, other DMTs are thought to gain access to the brain parenchyma and they directly affect microglia. Dimethyl fumarate, for example, modulates microglial phenotypes through activation of the hydroxycarboxylic acid receptor 2, culminating in reduced production of proinflammatory molecules mediated by NF-κB (Parodi et al., 2015). Microglia express high levels of Bruton's tyrosine kinase (BTK) while BTK inhibitors can access the brain raising the possibility of targeting two independent cell types with a single compound. However, a clearer understanding of the functional effects of BTK inhibition on microglia is needed (Krämer et al., 2023; Geladaris et al., 2024). Depletion of microglia with colony-stimulation factor 1 receptor antagonists, such as PLX5622, is also being investigated in the context of MS treatment.

Targeting astrocytes

Sphingolipid metabolism was identified as a potential therapeutic target in CNS inflammation. Miglustat, a drug used to treat

Gaucher and Niemann-Pick disease (Venier and Igdoura, 2012), suppresses astrocyte pathogenetic activities and ameliorates EAE. In preclinical studies, Miglustat suppressed the progression of EAE in nonobese diabetic mice as indicated by a reduction in clinical scores, axonal loss, demyelination, and the recruitment of proinflammatory monocytes to the CNS (Chao et al., 2019). Miglustat's therapeutic mechanisms involve the suppression of cytosolic phospholipase A2 (cPLA2)—mitochondrial antiviral signaling protein signaling in astrocytes (Chao et al., 2019).

Astrocytes clearly have a role at all stages of the MS disease course, but successful therapeutic targeting of these cells requires an in-depth understanding of many pathophysiologic aspects, including the contribution of astrocytes to the overall genetic architecture of the MS, the roe remodeling and failure of homeostatic functions of astrocytes, the true extent of astrocyte heterogeneity in a pathologic context, and astrocytic aging. Novel, pathologically relevant, and humanized models are of particular importance for developing astrocyte-targeting therapeutic strategies.

NEUROGLIA IN NMO

Microglia

Microglia, as the resident immune sentinels of the CNS, play key roles in immunologic regulation in neuroinflammatory diseases. In the early NMO lesions within AQP4-rich CNS regions, prominent reactive microgliosis is evidenced by histopathologic analysis (Lucchinetti et al., 2014). Microglia are the major mediators of inflammatory response in CNS and are arguably the first to respond to signals deriving from astrocytes attached and damaged by AQP4-IgG. Complement activation and C5b-9 play a vital role in the pathogenesis of CNS disorders (Rus and Niculescu, 2001). C3aR and C5aR are highly expressed in myeloid cells, including macrophages and macroglia, and also expressed in nonmyeloid cells, including neurons and reactive astrocytes (Asavapanumas et al., 2021). Reactive astrocytes produce complement fragment C3 that is subsequently cleaved into fragments C3a and C3b. The complement system triggers, though binding to the C3a receptor, reactive microgliosis (Chen et al., 2020).

Reactive microglia recruit macrophages and neutrophilic granulocytes to the NMO lesions, and produce and secrete IL-1β, IL-6, IL-17, and high-mobility group box 1 protein (HMGB1), all contributing to the pathogenesis of NMO (Bradl et al., 2014). Microglia (and not astrocytes, oligodendrocytes, or neurons) express IL-6 receptor which is activated by IL-6 released from astrocytes attacked by AQP4-IgG. This cascade impairs the integrity of BBB and promotes diapedesis of leucocytes and autoantibodies entry (Takeshita et al., 2021).

In the NMO animal model experiment, anti-AQP4 antibodies cause the loss of AQP4 in astrocyte endfeet regardless of the presence or absence of microglia, but the motor impairment only occurred in the presence of microglia (Chen et al., 2020). Microglia contain abundant progranulin, a secreted glycosylated protein that regulates immune response and cancer growth (Yin et al., 2010). The CSF levels of progranulin positively correlate with CSF cell counts, CSF protein levels, and the degree of the spinal lesion (Kimura et al., 2017). Reactive microgliocytes increase secretion of C1q, proinflammatory cytokines (such as IL-1β, IL-6, and TNF-α), and progranulin, which promote microglial reactivity in an autocrine manner and result in damage of neurons, astrocytes, and oligodendrocytes in a paracrine manner (Li et al., 2021). Microglia can express and secret type I interferon (IFNI) along with neurons, astrocytes, and endothelial cells, microglia respond to IFNI by reactive remodeling more significantly than astrocytes and neurons. The IFNI-triggered reactive microglia play pathologic roles in NMO (Wlodarczyk et al., 2021).

Astrocytes

As alluded to before, the NMO is a primary autoimmune astrocytopathy mediated by AQP4-IgG that damages glia limitans, which translates into the breaching BBB, recruitments of granulocytes and macrophages to instigate inflammation, and astrocytic death through complement-dependent cellular cytotoxicity, all these cascades leading to an ultimate demyelination and axonal damage (Rash et al., 1998; Verkman, 2012). The primary NMO target, the AQP4, is expressed in astrocytes in the brain, spinal cord, and optic nerve; AQP4 channels are localized to glia limitans at the pia mater, to perivascular astrocytic endfeet, and to ependymocytes (Nagelhus and Ottersen, 2013). The presence of AQP4-IgG, loss of astrocytes, deposition of complement, and infiltration of macrophages in NMO lesions, all point out toward a specific immune lesion of astrocytes expressing AQP4 (Herwerth et al., 2016). Astrocytes contribute to the local innate immune response to all CNS insults, and express receptors involved in innate immunity, including TLRs, nucleotide-binding oligomerization domains, double-stranded RNA-dependent protein kinase, scavenger receptors, mannose receptors and components of the complement system (Farina et al., 2007). Astrocytes are key players in CNS inflammation, being involved in both pro- and antiinflammatory modulation through secretion of various cytokines, chemokines, and neuromodulatory molecules (Lee et al., 2022).

The binding of AQP4-IgG to AQP4 instigates the internalization of AQP4 with subsequent upregulation on the secretion of cytokines, chemokines, and complement components (Chen et al., 2020). Astrocytes in particular release complement C3, which is cleaved into C3a and C3b; C3a binds to C3a receptors in microglia and promotes the production of C1q. The latter is produced by microglia, astrocytes, and neurons, and contributes to the activation of the complement cascade, which, in the NMO context, amplifies astrocytic damage and injures neurons and oligodendrocytes (Chen et al., 2020; Moinfar and Zamvil, 2020). The C1q binds to the Fc region of AQP4-bound AQP4-IgG and activates the downstream proteins of membrane attack complex (MAC), astrocyte lysis, and tissue necrosis. Inhibiting C1q in the NMO mouse model attenuates the damage to astrocytes, oligodendrocytes, and neurones (Phuan et al., 2013). Astrocyte lysis induced by AQP4-IgG and complement contributes to the pronounced inflammation and necrotic tissue loss with axonal and oligodendrocyte damage (Roemer et al., 2007). Astrocytes also produce tumor growth factor-β (TGF-β), which amplifies the release of complement system from neurons and induces reactive microgliosis (Luo, 2022). C3a, C4a, and C5a, produced in complement activation, have proinflammatory potential and play a pivotal role in NMO (Kuroda et al., 2013). The C5a levels in CSF were elevated significantly in NMO patients and correlated with the severity of exacerbation (Kuroda et al., 2013).

Circulating AQP4-IgG may also trigger specific clinical presentations manifested by nausea and vomiting resulting from the pathologic remodeling of the area postrema without astrocyte loss or tissue destruction (Popescu et al., 2011). In this pathology, the levels of AQP4 in area postrema are severely depleted although there is no overt astrocytic damage, nor astrocyte loss, but rather increased reactive astrogliosis (Popescu et al., 2011). The BBB of the area postrema is somewhat leaky even in physiologic conditions so that some large molecules can enter CNS parenchyma (Broadwell and Sofroniew, 1993).

Astrocyte-derived extracellular vesicles (ADEVs) contain multiple proteins, metabolite, and signaling molecules, which can contribute to both physiologic and pathologic signaling (Taylor et al., 2007; Proia et al., 2008), being a nonsynaptic form of glial-neuronal communication (Chun et al., 2021; Rouillard et al., 2021). Incidentally, ADEVs increase the survival rate and normalize the functional activity of neurons exposed to oxidative stress (Chun et al., 2021; Rouillard et al., 2021). In addition, ADEVs are involved in the remyelination and CNS repair by modulating microglia (Kakimura et al., 2002; Wilson et al., 2015). The number of ADEVs is increased in plasma from patients with NMOSDs, whereas apolipoprotein E (APOE) content of these ADEVs was high (Jiang et al., 2024). Intracerebral injection of the APOE-mimetic peptide APOE130–149 attenuated microglial reactivity, neuroinflammation, and brain lesions in a mouse model of NMO (Jiang et al., 2024). The protective effect of APOE in NMO was further corroborated by the observations in APOE-deficient mice, which showed exacerbated brain lesions in response to intracerebral injection of AQP4-IgG; this can be rescued by exogenous APOE administration (Jiang et al., 2024). Genetic knockdown of the APOE lipoprotein receptor-related protein 1 (LRP1) inhibited the beneficial effects of APOE130–149 (Jiang et al., 2024). The transfusion of ADEVs derived from patients with NMO and healthy controls similarly alleviated astrocyte loss, reactive microgliosis, and demyelination in NMO model mice. Thus, APOE from ADEVs could mediate disease-modifying astrocyte-microglia cross talk in NMO (Jiang et al., 2024).

Interaction of microglia and astrocytes with immune cells

In physiologic conditions, astrocytes limit (working in concert with other cellular elements of the BBB) the peripheral immune cells' entry to the CNS. In pathologic conditions, astrocytes are involved in innate immune reactions and adaptive immune responses (Farina et al., 2007; Ransohoff and Brown, 2012). Both in physiologic and pathologic conditions, microglia supervise their surrounding environment and quickly respond to any change in the cerebral nerve system (Nimmerjahn et al., 2005; Block and Hong, 2007). Specific stimuli lead to the activation of microglia and astrocytes, followed by a breakdown of the BBB, infiltration of peripheral leukocytes and inflammation factors, then neuroinflammation is induced (Colombo and Farina, 2016). Microglia and CNS-resident innate immune cells producing complement contribute to the clinical manifestations of NMO. The AQP4-IgG belongs to IgG1, a T cell-dependent antibody, which activates the complement system thus combining cellular and humoral immunity in the pathogenesis of NMO (Varrin-Doyer et al., 2012).

In the peripheral blood of NMO patients, the plasma cells differentiated from AQP4-specific B lymphocytes produce anti-AQP4 antibodies and activate T-cell responses by antigen presentation or cytokine secretion (Wilson et al., 2018; Afzali et al., 2024). Myeloid antigen-presenting cells (APCs) process AQP4 and present antigen to CD4-T cells, and subsequently activated AQP4-specific T cells damage the BBB (Bennett et al., 2015). Specific T cells interact with B cells and dendritic cells and differentiate into Th17 cells driven by IL-6, IL-23, and TGF-β. The compromised BBB exacerbates local inflammation by allowing extravasation of inflammatory cells and proinflammatory factors, leading to more AQP4-IgG in the CNS parenchyma and recruitment of neutrophils. Blood-borne immune cells, such as eosinophils, neutrophils, basophils, mast cells, NK cells, and macrophages, are recruited following complement activation and astrocyte damage, promoting the release of cytokines (IL-17, IL-8), proteases and radicals, and leading to the damage of vasculature and parenchyma (Wu et al., 2019). In addition, these inflammatory cells produce complement-independent damage to astrocytes by antibody-dependent cell-mediated cytotoxicity (ADCC) or degranulation (Carnero Contentti and Correale, 2021). Cytotoxicity mediated by ADCC and complement leads to internalization of glutamate transporter EAAT-2, leading to the failure of astrocytic glutamate clearance and consequent excitotoxicity (Hinson et al., 2010). Compromised astrocytes fail to support oligodendrocytes, leading to demyelination and axonal loss (Verkman et al., 2013).

BTK, a crucial protein of B-cell receptor was upregulated both in the blood and cerebrospinal fluid of NMOSD patients. BTK activity is enhanced both in B cells and microglia. Blockade of BTK with zanubrutinib, a highly specific BTK inhibitor, mitigated the activation and maturation of B cells and reduced the production of causal AQP4 autoantibodies. BTK inhibition contributes to the amelioration of NMOSD pathology and could be a viable therapeutic target (Liu et al., 2023).

Interaction of microglia and astrocytes with brain resident cells

The interaction between microglia and astrocytes significantly influences the cerebral pathophysiology. Microglia/astrocytes are a part of the neurovascular unit, which is composed of microglia, neurons, the BBB, and the extracellular matrix. They achieve immune "optimization" through mutual communication and cooperation in the neurovascular unit and build a cascaded immune amplification network (Liu et al., 2020a). Astrocytes have a close relationship with neurons and blood vessels, supervise synapses, and modulate synaptic transmission (Wang et al., 2006; Santello et al., 2012; Verkhratsky and Butt, 2023). Astrocytes produce TGF-β and IL-33 to promote the synapse pruning of microglia in development, and upregulate lipocalin-2 and orosomucoid to trigger microglial reactivity under pathologic conditions to promote neuroinflammation (Norden et al., 2014; Jo et al., 2017; Kim et al., 2017; Vainchtein et al., 2018).

Microglia and astrocytes are the major resident phagocytic cells in the brain, and they participate in myelination by devouring unwanted myelin and the clearance of degenerated myelin debris for accelerating remyelination, or engulfing healthy myelin sheath for inhibiting remyelination (Xu et al., 2023). The interaction between microglia and astrocytes plays a vital role in their phagocytotic ability and the disease progression in the cerebral nerve system (Smith and Hoerner, 2000). Neurons and astrocytes reciprocally orchestrate microglial homeostasis (Baxter et al., 2021).

NEUROGLIA AS A THERAPEUTIC TARGET IN NMO

Microglia

Microglia, being a critical element of NMO pathogenesis represents a promising therapeutic target. Microglia, as well as IgG dose and AQP4, contribute to motor impairment and sublytic NMO-compatible immunopathology (Chen et al., 2020). The cross talk between astrocytes and microglia participates in the early activation of intrinsic complement components. Microglial C3a receptor signaling is one of the critical drivers of the precytolytic phase in the evolving NMO lesion. Insulin-like growth factor 1 (IGF1) gene therapy is a feasible approach to target brain cells and IGF1 therapy leads to a region-specific modification of the aged microglia population (Falomir-Lockhart et al., 2019). Thus, reactive microgliosis gene therapy might be a novel approach to modulating the progression of NMO (Li et al., 2021). In addition, microglia express C5 and C5a (Martin et al., 2007), which play vital roles in the activation of cytolytic complement. C5a is a strong chemoattractant for polymorph nuclear cells and mediates eosinophil and neutrophil degranulation (Giembycz and Lindsay, 1999; Saadoun et al., 2012). Therefore, inhibiting complement C5 may modulate microglia function in NMO and become a promising choice. Eculizumab, a monoclonal IgG neutralizing the complement C5, has effectively reduced the relapse frequency of NMO in the clinical trial (Pittock et al., 2019).

Minocycline, a microglial inhibitor, was shown to reduce reactive microgliosis in several CNS diseases (Manso et al., 2018; Park et al., 2020). Colony-stimulating factor 1 receptor antagonists, which deplete the microglia population, may be potentially effective, although their deployment needs more investigation because of side effects on the myeloid and lymphoid compartments (Lei et al., 2020). The binding of C3a and C3aR is a critical signal for reactive microglia response. The C3aR inhibitor and C3a neutralizer may represent another potential treatment (Chen et al., 2020; Zhang et al., 2020). Identification of targeted microglia-specific gene products is similarly promising and may be relevant not only for treating NMO but also other inflammatory and neurodegenerative CNS diseases associated with microglial reactivity (Moinfar and Zamvil, 2020).

Astrocytes

The production of AQP4-IgG occurs outside the CNS; hence, the targeted peripheral therapy such as B cell depletion could prevent damage to astrocytes. In an animal model astrocyte and microglia responses to B cell depleting treatments demonstrated that B cell depleting markedly reduced disease severity and myelin damage accompanied by reduced microglia and astrocyte reactivity in the animal models (Ahn et al., 2023). The drug bortezomib binds to the catalytic site of the 26S proteasome with high affinity and specificity to activate the unfolded terminal protein response to eliminate plasmablasts and plasma cells. Furthermore, bortezomib may protect astrocytes from NF-κB-dependent inflammatory damage in early events in NMO pathogenesis (Zhang et al., 2017). In an NMO mice model, the C1q inhibitors protected astrocytes, oligodendrocytes, and neurons from damage (Phuan et al., 2013). Synthetic nanoparticles may be used for astrocyte-specific targeting therapy (Zhang et al., 2016). Intraspinally administered poly-(lactic-co-glycolic acid) nanoparticles are taken up by astrocytes and remain for a long time in the SCI rat model (Wang et al., 2008). As the inhibitor of the terminal C5b-9 MAC, CD59, a glycophosphoinositol-anchored membrane protein, is an important modulator of the immune response in NMO (Wang et al., 2017). CD59 inhibition increases the pathology of NMO in the animal model (Zhang and Verkman, 2014) and CD59 plays a major protective role outside of the CNS in seropositive NMO (Yao and Verkman, 2017).

CONCLUSIONS

Multiple sclerosis and NMO are autoimmune primary neurogliopathies, the clinical progression of which, to a large extent, is defined by the adaptive response of all types of glial cells, oligodendrocytes, OPCs, astrocytes, and microglia. In the MS oligodendrocytes and myelin are primary targets; their damage, however, triggers reactive microgliosis, which localizes the lesions and slows down disease progression. In parallel remyelination supported by OPCs underlies remissions. In NMO the AQP4-IgG attack astrocytic endfeet, which when collapsed, compromise BBB and allow massive entry of autoantibodies and immune cells. In parallel damage to astrocytes triggers reactive microgliosis and astrogliosis, which, like MS, define the disease progression. Neuroglial cells therefore are legitimate targets for therapeutic interventions, which are yet to be developed.

References

Abdelhak A, Huss A, Kassubek J et al. (2018). Serum GFAP as a biomarker for disease severity in multiple sclerosis. Sci Rep 8: 14798.

Absinta M, Maric D, Gharagozloo M et al. (2021). A lymphocyte-microglia-astrocyte axis in chronic active multiple sclerosis. Nature 597: 709–714.

Afzali AM, Nirschl L, Sie C et al. (2024). B cells orchestrate tolerance to the neuromyelitis optica autoantigen AQP4. Nature 627: 407–415.

Ahn JJ, Islam Y, Clarkson-Paredes C et al. (2023). B cell depletion modulates glial responses and enhances blood vessel integrity in a model of multiple sclerosis. Neurobiol Dis 187: 106290.

Albers PH (1910). St.Lidwina the Catholic encyclopedia, vol. 9, Robert Appleton Company, New York, pp. 233.

Asavapanumas N, Tradtrantip L, Verkman AS (2021). Targeting the complement system in neuromyelitis optica spectrum disorder. Expert Opin Biol Ther 21: 1073–1086.

Ayrignac X, Le Bars E, Duflos C et al. (2020). Serum GFAP in multiple sclerosis: correlation with disease type and MRI markers of disease severity. Sci Rep 10: 10923.

Baaklini CS, Rawji KS, Duncan GJ et al. (2019). Central nervous system remyelination: roles of glia and innate immune cells. Front Mol Neurosci 12: 225.

Baxter PS, Dando O, Emelianova K et al. (2021). Microglial identity and inflammatory responses are controlled by the combined effects of neurons and astrocytes. Cell Rep 34: 108882.

Bennett JL, O'Connor KC, Bar-Or A et al. (2015). B lymphocytes in neuromyelitis optica. Neurol Neuroimmunol Neuroinflamm 2: e104.

Block ML, Hong JS (2007). Chronic microglial activation and progressive dopaminergic neurotoxicity. Biochem Soc Trans 35: 1127–1132.

Bogousslavsky J, Walusinski O, Moulin T (2011). Alfred Vulpian and Jean-Martin Charcot in each other's shadow? From Castor and Pollux at La Salpêtrière to neurology forever. Eur Neurol 65: 215–222.

Bradl M, Kanamori Y, Nakashima I et al. (2014). Pain in neuromyelitis optica–prevalence, pathogenesis and therapy. Nat Rev Neurol 10: 529–536.

Brambilla R (2019). The contribution of astrocytes to the neuroinflammatory response in multiple sclerosis and experimental autoimmune encephalomyelitis. Acta Neuropathol 137: 757–783.

Broadwell RD, Sofroniew MV (1993). Serum proteins bypass the blood-brain fluid barriers for extracellular entry to the central nervous system. Exp Neurol 120: 245–263.

Carnero Contentti E, Correale J (2021). Neuromyelitis optica spectrum disorders: from pathophysiology to therapeutic strategies. J Neuroinflammation 18: 208.

Carswell R (1838). Pathological anatomy: illustrations of the elementary forms of disease, Longman, Orme, Brown, Green and Longman, London.

Charcot JM (1877). Lectures on diseases of the nervous system, The New Sydenham Society, pp. 158–222.

Chao CC, Gutiérrez-Vázquez C, Rothhammer V et al. (2019). Metabolic control of astrocyte pathogenic activity via cPLA2-MAVS. Cell 179: 1483–1498.e1422.

Chen T, Bosco DB, Ying Y et al. (2021). The emerging role of microglia in neuromyelitis optica. Front Immunol 12: 616301.

Chen T, Lennon VA, Liu YU et al. (2020). Astrocyte-microglia interaction drives evolving neuromyelitis optica lesion. J Clin Invest 130: 4025–4038.

Chun C, Smith AST, Kim H et al. (2021). Astrocyte-derived extracellular vesicles enhance the survival and electrophysiological function of human cortical neurons in vitro. Biomaterials 271: 120700.

Colombo E, Farina C (2016). Astrocytes: key regulators of neuroinflammation. Trends Immunol 37: 608–620.

Colonna M, Butovsky O (2017). Microglia function in the central nervous system during health and neurodegeneration. Annu Rev Immunol 35: 441–468.

Colton CA (2009). Heterogeneity of microglial activation in the innate immune response in the brain. J Neuroimmune Pharmacol 4: 399–418.

Dal-Bianco A, Grabner G, Kronnerwetter C et al. (2017). Slow expansion of multiple sclerosis iron rim lesions: pathology and 7 T magnetic resonance imaging. Acta Neuropathol 133: 25–42.

Dal-Bianco A, Grabner G, Kronnerwetter C et al. (2021). Long-term evolution of multiple sclerosis iron rim lesions in 7 T MRI. Brain 144: 833–847.

Devic E (1895). Myélite aiguë dorso-lombaire avec névrite optique—Autopsie. Congrès français de medicine, Asselin et Houzeau, Louis Savy, Paris, Lyon 1895: 434–439.

Dos Passos GR, Oliveira LM, da Costa BK et al. (2018). MOG-IgG-associated optic neuritis, encephalitis, and myelitis: lessons learned from neuromyelitis optica spectrum disorder. Front Neurol 9: 217.

Escartin C, Galea E, Lakatos A et al. (2021). Reactive astrocyte nomenclature, definitions, and future directions. Nat Neurosci 24: 312–325.

Falomir-Lockhart E, Dolcetti FJC, García-Segura LM et al. (2019). IGF1 gene therapy modifies microglia in the striatum of senile rats. Front Aging Neurosci 11: 48.

Farina C, Aloisi F, Meinl E (2007). Astrocytes are active players in cerebral innate immunity. Trends Immunol 28: 138–145.

Filippi M, Bar-Or A, Piehl F et al. (2018). Multiple sclerosis. Nat Rev Dis Primers 4: 43.

Firth D (1941). The case of Augustus d'Este (1794-1848): the first account of disseminated sclerosis: (section of the history of medicine). Proc R Soc Med 34: 381–384.

Franklin RJ, Goldman SA (2015). Glia disease and repair-remyelination. Cold Spring Harb Perspect Biol 7: a020594.

Frerich FT (1849). Ueber Hirnsklerose. Arch Ges Med 10: 334–350.

Garland EF, Hartnell IJ, Boche D (2022). Microglia and astrocyte function and communication: what do we know in humans? Front Neurosci 16: 824888.

Gault F (1894). De la neuromyélite optique aiguë, Thése: Faculté de Médecine et de Pharmacie.

Geladaris A, Torke S, Saberi D et al. (2024). BTK inhibition limits microglia-perpetuated CNS inflammation and promotes myelin repair. Acta Neuropathol 147: 75.

Giembycz MA, Lindsay MA (1999). Pharmacology of the eosinophil. Pharmacol Rev 51: 213–340.

Gillen KM, Mubarak M, Nguyen TD et al. (2018). Significance and in vivo detection of iron-laden microglia in white matter multiple sclerosis lesions. Front Immunol 9: 255.

Ginhoux F, Prinz M (2015). Origin of microglia: current concepts and past controversies. Cold Spring Harb Perspect Biol 7: a020537.

Haase S, Linker RA (2021). Inflammation in multiple sclerosis. Ther Adv Neurol Disord 14: 17562864211007687.

Healy LM, Stratton JA, Kuhlmann T et al. (2022). The role of glial cells in multiple sclerosis disease progression. Nat Rev Neurol 18: 237–248.

Herwerth M, Kalluri SR, Srivastava R et al. (2016). In vivo imaging reveals rapid astrocyte depletion and axon damage in a model of neuromyelitis optica-related pathology. Ann Neurol 79: 794–805.

Hinson SR, McKeon A, Lennon VA (2010). Neurological autoimmunity targeting aquaporin-4. Neuroscience 168: 1009–1018.

Husseini L, Geladaris A, Weber MS (2024). Toward identifying key mechanisms of progression in multiple sclerosis. Trends Neurosci 47: 58–70.

Jarius S, Lauda F, Wildemann B et al. (2013). Steroid-responsive hearing impairment in NMO-IgG/aquaporin-4-antibody-positive neuromyelitis optica. J Neurol 260: 663–664.

Jarius S, Wildemann B (2012). The case of the Marquis de Causan (1804): an early account of visual loss associated with spinal cord inflammation. J Neurol 259: 1354–1357.

Jarius S, Wildemann B (2013). The history of neuromyelitis optica. J Neuroinflammation 10: 8.
Jiang S, Li X, Li Y et al. (2024). APOE from patient-derived astrocytic extracellular vesicles alleviates neuromyelitis optica spectrum disorder in a mouse model. Sci Transl Med 16: eadg5116.
Jo M, Kim JH, Song GJ et al. (2017). Astrocytic orosomucoid-2 modulates microglial activation and neuroinflammation. J Neurosci 37: 2878–2894.
Kakimura J, Kitamura Y, Takata K et al. (2002). Microglial activation and amyloid-beta clearance induced by exogenous heat-shock proteins. FASEB J 16: 601–603.
Kalafatakis I, Karagogeos D (2021). Oligodendrocytes and microglia: key players in myelin development, damage and repair. Biomolecules 11: 1058.
Kent SA, Miron VE (2024). Microglia regulation of central nervous system myelin health and regeneration. Nat Rev Immunol 24: 49–63.
Kessler W, Thomas C, Kuhlmann T (2023). Microglia activation in periplaque white matter in multiple sclerosis depends on age and lesion type, but does not correlate with oligodendroglial loss. Acta Neuropathol 146: 817–828.
Kettenmann H, Hanisch UK, Noda M et al. (2011). Physiology of microglia. Physiol Rev 91: 461–553.
Kettenmann H, Kirchhoff F, Verkhratsky A (2013). Microglia: new roles for the synaptic stripper. Neuron 77: 10–18.
Kim JH, Ko PW, Lee HW et al. (2017). Astrocyte-derived lipocalin-2 mediates hippocampal damage and cognitive deficits in experimental models of vascular dementia. Glia 65: 1471–1490.
Kimura A, Takemura M, Saito K et al. (2017). Increased cerebrospinal fluid progranulin correlates with interleukin-6 in the acute phase of neuromyelitis optica spectrum disorder. J Neuroimmunol 305: 175–181.
Krämer J, Bar-Or A, Turner TJ et al. (2023). Bruton tyrosine kinase inhibitors for multiple sclerosis. Nat Rev Neurol 19: 289–304.
Kuhlmann T, Ludwin S, Prat A et al. (2017). An updated histological classification system for multiple sclerosis lesions. Acta Neuropathol 133: 13–24.
Kuroda H, Fujihara K, Takano R et al. (2013). Increase of complement fragment C5a in cerebrospinal fluid during exacerbation of neuromyelitis optica. J Neuroimmunol 254: 178–182.
Lassmann H, van Horssen J (2016). Oxidative stress and its impact on neurons and glia in multiple sclerosis lesions. Biochim Biophys Acta 1862: 506–510.
Lee HG, Wheeler MA, Quintana FJ (2022). Function and therapeutic value of astrocytes in neurological diseases. Nat Rev Drug Discov 21: 339–358.
Lei F, Cui N, Zhou C et al. (2020). CSF1R inhibition by a small-molecule inhibitor is not microglia specific; affecting hematopoiesis and the function of macrophages. Proc Natl Acad Sci U S A 117: 23336–23338.
Lennon VA, Kryzer TJ, Pittock SJ et al. (2005). IgG marker of optic-spinal multiple sclerosis binds to the aquaporin-4 water channel. J Exp Med 202: 473–477.
Lennon VA, Wingerchuk DM, Kryzer TJ et al. (2004). A serum autoantibody marker of neuromyelitis optica: distinction from multiple sclerosis. Lancet 364: 2106–2112.
Li W, Liu J, Tan W et al. (2021). The role and mechanisms of microglia in neuromyelitis optica spectrum disorders. Int J Med Sci 18: 3059–3065.
Lian H, Yang L, Cole A et al. (2015). NFκB-activated astroglial release of complement C3 compromises neuronal morphology and function associated with Alzheimer's disease. Neuron 85: 101–115.
Liu LR, Liu JC, Bao JS et al. (2020a). Interaction of microglia and astrocytes in the neurovascular unit. Front Immunol 11: 1024.
Liu R, Xu NG, Yi W et al. (2020b). Electroacupuncture attenuates inflammation after ischemic stroke by inhibiting NF-κB-mediated activation of microglia. Evid Based Complement Alternat Med 2020: 8163052.
Liu Y, Huang Z, Zhang TX et al. (2023). Bruton's tyrosine kinase-bearing B cells and microglia in neuromyelitis optica spectrum disorder. J Neuroinflammation 20: 309.
Lucchinetti CF, Guo Y, Popescu BF et al. (2014). The pathology of an autoimmune astrocytopathy: lessons learned from neuromyelitis optica. Brain Pathol 24: 83–97.
Luo J (2022). TGF-β as a key modulator of astrocyte reactivitydisease relevance and therapeutic implications. Biomedicine 10: 1206.
Maggi P, Macri SM, Gaitán MI et al. (2014). The formation of inflammatory demyelinated lesions in cerebral white matter. Ann Neurol 76: 594–608.
Manso Y, Holland PR, Kitamura A et al. (2018). Minocycline reduces microgliosis and improves subcortical white matter function in a model of cerebral vascular disease. Glia 66: 34–46.
Martin CB, Ingersoll SA, Martin BK (2007). Regulation of the C5a receptor promoter in glial cells: minimal dependence upon the CCAAT element in astrocytes. Mol Immunol 44: 713–721.
Matejuk A, Ransohoff RM (2020). Crosstalk between astrocytes and microglia: an overview. Front Immunol 11: 1416.
McNamara NB, Munro DAD, Bestard-Cuche N et al. (2023). Microglia regulate central nervous system myelin growth and integrity. Nature 613: 120–129.
Moinfar Z, Zamvil SS (2020). Microglia complement astrocytes in neuromyelitis optica. J Clin Invest 130: 3961–3964.
Nagelhus EA, Ottersen OP (2013). Physiological roles of aquaporin-4 in brain. Physiol Rev 93: 1543–1562.
Murray TJ (2005). Multiple sclerosis: the history of a disease, Springer Publishing Company.
Nimmerjahn A, Kirchhoff F, Helmchen F (2005). Resting microglial cells are highly dynamic surveillants of brain parenchyma in vivo. Science 308: 1314–1318.
Nociti V, Romozzi M (2023). The role of BDNF in multiple sclerosis neuroinflammation. Int J Mol Sci 24: 8447.
Norden DM, Fenn AM, Dugan A et al. (2014). TGFβ produced by IL-10 redirected astrocytes attenuates microglial activation. Glia 62: 881–895.
Ollivier CP (1824). De La Moelle Epiniére et de ses maladies, Crevot, Paris.
Paolicelli RC, Sierra A, Stevens B et al. (2022). Microglia states and nomenclature: a field at its crossroads. Neuron 110: 3458–3483.
Park D, Kim S, Kim H et al. (2020). Seizure progression triggered by IQSEC3 loss is mitigated by reducing activated microglia in mice. Glia 68: 2661–2673.
Parodi B, Rossi S, Morando S et al. (2015). Fumarates modulate microglia activation through a novel HCAR2 signaling pathway and rescue synaptic dysregulation in inflamed CNS. Acta Neuropathol 130: 279–295.
Phuan PW, Zhang H, Asavapanumas N et al. (2013). C1q-targeted monoclonal antibody prevents complement-dependent cytotoxicity and neuropathology in in vitro and mouse models of neuromyelitis optica. Acta Neuropathol 125: 829–840.
Pittock SJ, Berthele A, Fujihara K et al. (2019). Eculizumab in aquaporin-4-positive neuromyelitis optica spectrum disorder. N Engl J Med 381: 614–625.
Pocock JM, Kettenmann H (2007). Neurotransmitter receptors on microglia. Trends Neurosci 30: 527–535.

Ponath G, Ramanan S, Mubarak M et al. (2017). Myelin phagocytosis by astrocytes after myelin damage promotes lesion pathology. Brain 140: 399–413.
Popescu BF, Lennon VA, Parisi JE et al. (2011). Neuromyelitis optica unique area postrema lesions: nausea, vomiting, and pathogenic implications. Neurology 76: 1229–1237.
Prod'homme T, Zamvil SS (2019). The evolving mechanisms of action of glatiramer acetate. Cold Spring Harb Perspect Med 9: a029249.
Proia P, Schiera G, Mineo M et al. (2008). Astrocytes shed extracellular vesicles that contain fibroblast growth factor-2 and vascular endothelial growth factor. Int J Mol Med 21: 63–67.
Ramaglia V, Rojas O, Naouar I et al. (2021). The ins and outs of central nervous system inflammation-lessons learned from multiple sclerosis. Annu Rev Immunol 39: 199–226.
Ransohoff RM, Brown MA (2012). Innate immunity in the central nervous system. J Clin Invest 122: 1164–1171.
Rash JE, Yasumura T, Hudson CS et al. (1998). Direct immunogold labeling of aquaporin-4 in square arrays of astrocyte and ependymocyte plasma membranes in rat brain and spinal cord. Proc Natl Acad Sci U S A 95: 11981–11986.
Rathore KI, Redensek A, David S (2012). Iron homeostasis in astrocytes and microglia is differentially regulated by TNF-α and TGF-β1. Glia 60: 738–750.
Rawji KS, Gonzalez Martinez GA, Sharma A et al. (2020). The role of astrocytes in remyelination. Trends Neurosci 43: 596–607.
Roemer SF, Parisi JE, Lennon VA et al. (2007). Pattern-specific loss of aquaporin-4 immunoreactivity distinguishes neuromyelitis optica from multiple sclerosis. Brain 130: 1194–1205.
Rouillard ME, Sutter PA, Durham OR et al. (2021). Astrocyte-derived extracellular vesicles (ADEVs): deciphering their influences in aging. Aging Dis 12: 1462–1475.
Rus H, Niculescu F (2001). The complement system in central nervous system diseases. Immunol Res 24: 79–86.
Saadoun S, Waters P, MacDonald C et al. (2012). Neutrophil protease inhibition reduces neuromyelitis optica-immunoglobulin G-induced damage in mouse brain. Ann Neurol 71: 323–333.
Sandi D, Fricska-Nagy Z, Bencsik K et al. (2021). Neurodegeneration in multiple sclerosis: symptoms of silent progression, biomarkers and neuroprotective therapy-kynurenines are important players. Molecules 26: 3423.
Santello M, Calì C, Bezzi P (2012). Gliotransmission and the tripartite synapse. Adv Exp Med Biol 970: 307–331.
Santos EN, Fields RD (2021). Regulation of myelination by microglia. Sci Adv 7: eabk1131.
Semyanov A, Verkhratsky A (2021). Astrocytic processes: from tripartite synapses to the active milieu. Trends Neurosci 44: 781–792.
Sheng WS, Hu S, Feng A et al. (2013). Reactive oxygen species from human astrocytes induced functional impairment and oxidative damage. Neurochem Res 38: 2148–2159.
Singh S, Metz I, Amor S et al. (2013). Microglial nodules in early multiple sclerosis white matter are associated with degenerating axons. Acta Neuropathol 125: 595–608.
Smith ME, Hoerner MT (2000). Astrocytes modulate macrophage phagocytosis of myelin in vitro. J Neuroimmunol 102: 154–162.
Stratoulias V, Venero JL, Tremblay M et al. (2019). Microglial subtypes: diversity within the microglial community. EMBO J 38: e101997.
Takeshita Y, Fujikawa S, Serizawa K et al. (2021). New BBB model reveals that IL-6 blockade suppressed the BBB disorder, Preventing Onset of NMOSD. Neurol Neuroimmunol Neuroinflamm 8: e1076.
Taylor AR, Robinson MB, Gifondorwa DJ et al. (2007). Regulation of heat shock protein 70 release in astrocytes: role of signaling kinases. Dev Neurobiol 67: 1815–1829.
Tepavčević V, Lubetzki C (2022). Oligodendrocyte progenitor cell recruitment and remyelination in multiple sclerosis: the more, the merrier? Brain 145: 4178–4192.
Tham M, Frischer JM, Weigand SD et al. (2021). Iron heterogeneity in early active multiple sclerosis lesions. Ann Neurol 89: 498–510.
Thompson AJ, Banwell BL, Barkhof F et al. (2018). Diagnosis of multiple sclerosis: 2017 revisions of the McDonald criteria. Lancet Neurol 17: 162–173.
Ulvestad E, Williams K, Matre R et al. (1994). Fc receptors for IgG on cultured human microglia mediate cytotoxicity and phagocytosis of antibody-coated targets. J Neuropathol Exp Neurol 53: 27–36.
Vainchtein ID, Chin G, Cho FS et al. (2018). Astrocyte-derived interleukin-33 promotes microglial synapse engulfment and neural circuit development. Science 359: 1269–1273.
Varrin-Doyer M, Spencer CM, Schulze-Topphoff U et al. (2012). Aquaporin 4-specific T cells in neuromyelitis optica exhibit a Th17 bias and recognize Clostridium ABC transporter. Ann Neurol 72: 53–64.
Venier RE, Igdoura SA (2012). Miglustat as a therapeutic agent: prospects and caveats. J Med Genet 49: 591–597.
Verkhratsky A, Butt AM (2023). Neuroglia: function and pathology, Elsevier, New York.
Verkhratsky A, Butt A, Li B et al. (2023). Astrocytes in human central nervous system diseases: a frontier for new therapies. Signal Transduct Target Ther 8: 396.
Verkhratsky A, Nedergaard M (2018). Physiology of astroglia. Physiol Rev 98: 239–389.
Verkman AS (2012). Aquaporins in clinical medicine. Annu Rev Med 63: 303–316.
Verkman AS, Phuan PW, Asavapanumas N et al. (2013). Biology of AQP4 and anti-AQP4 antibody: therapeutic implications for NMO. Brain Pathol 23: 684–695.
Voet S, Prinz M, van Loo G (2019). Microglia in central nervous system inflammation and multiple sclerosis pathology. Trends Mol Med 25: 112–123.
Vulpian A (1866). Note sur la sclérose en plaques de la moelle épinière. Communication à la Société Médicale des Hôpitaux le 9 Mai 1866. Union médicale 3: 75–100.
Wang AT, Lee SS, Sigman M et al. (2006). Developmental changes in the neural basis of interpreting communicative intent. Soc Cogn Affect Neurosci 1: 107–121.
Wang J, Wang J, Wang J et al. (2019). Targeting microglia and macrophages: a potential treatment strategy for multiple sclerosis. Front Pharmacol 10: 286.
Wang YC, Wu YT, Huang HY et al. (2008). Sustained intraspinal delivery of neurotrophic factor encapsulated in biodegradable nanoparticles following contusive spinal cord injury. Biomaterials 29: 4546–4553.
Wang Z, Guo W, Liu Y et al. (2017). Low expression of complement inhibitory protein CD59 contributes to humoral autoimmunity against astrocytes. Brain Behav Immun 65: 173–182.
Williams A, Piaton G, Lubetzki C (2007). Astrocytes–friends or foes in multiple sclerosis? Glia 55: 1300–1312.
Williams R, Buchheit CL, Berman NE et al. (2012). Pathogenic implications of iron accumulation in multiple sclerosis. J Neurochem 120: 7–25.

Wilson R, Makuch M, Kienzler AK et al. (2018). Condition-dependent generation of aquaporin-4 antibodies from circulating B cells in neuromyelitis optica. Brain 141: 1063–1074.
Wilson SM, Lam D, Babiak MC et al. (2015). Transient aphasias after left hemisphere resective surgery. J Neurosurg 123: 581–593.
Wingerchuk DM (2007). Neuromyelitis optica: new findings on pathogenesis. Int Rev Neurobiol 79: 665–688.
Wingerchuk DM, Banwell B, Bennett JL et al. (2015). International consensus diagnostic criteria for neuromyelitis optica spectrum disorders. Neurology 85: 177–189.
Wlodarczyk A, Khorooshi R, Marczynska J et al. (2021). Type I interferon-activated microglia are critical for neuromyelitis optica pathology. Glia 69: 943–953.
Wright SK, Wassmer E, Vincent A (2021). Pathogenic antibodies to AQP4: neuromyelitis optica spectrum disorder (NMOSD). Biochim Biophys Acta Biomembr 1863: 183772.
Wu Y, Zhong L, Geng J (2019). Neuromyelitis optica spectrum disorder: pathogenesis, treatment, and experimental models. Mult Scler Relat Disord 27: 412–418.
Xu T, Liu C, Deng S et al. (2023). The roles of microglia and astrocytes in myelin phagocytosis in the central nervous system. J Cereb Blood Flow Metab 43: 325–340.
Yao X, Verkman AS (2017). Complement regulator CD59 prevents peripheral organ injury in rats made seropositive for neuromyelitis optica immunoglobulin G. Acta Neuropathol Commun 5: 57.
Yi C, Verkhratsky A, Niu J (2023). Pathological potential of oligodendrocyte precursor cells: terra incognita. Trends Neurosci 46: 581–596.
Yin F, Banerjee R, Thomas B et al. (2010). Exaggerated inflammation, impaired host defense, and neuropathology in progranulin-deficient mice. J Exp Med 207: 117–128.
Zalc B (2018). One hundred and fifty years ago Charcot reported multiple sclerosis as a new neurological disease. Brain 141: 3482–3488.
Zhang C, Tian DC, Yang CS et al. (2017). Safety and efficacy of bortezomib in patients with highly relapsing neuromyelitis optica spectrum disorder. JAMA Neurol 74: 1010–1012.
Zhang F, Lin YA, Kannan S et al. (2016). Targeting specific cells in the brain with nanomedicines for CNS therapies. J Control Release 240: 212–226.
Zhang H, Verkman AS (2014). Longitudinally extensive NMO spinal cord pathology produced by passive transfer of NMO-IgG in mice lacking complement inhibitor CD59. J Autoimmun 53: 67–77.
Zhang LY, Pan J, Mamtilahun M et al. (2020). Microglia exacerbate white matter injury via complement C3/C3aR pathway after hypoperfusion. Theranostics 10: 74–90.

Handbook of Clinical Neurology, Vol. 210 (3rd series)
Neuroglia in Neurologic and Psychiatric Disorders, Part II
A. Verkhratsky, L.D. de Witte, E. Aronica, and E.M. Hol, Editors
https://doi.org/10.1016/B978-0-443-19102-2.00002-8

Chapter 9

Neuroglia in autoimmune encephalitis

FATME SEVAL ISMAIL[1], PEDRO M. FAUSTMANN[2*], FRANCO CORVACE[2,†], AND TIMO JENDRIK FAUSTMANN[3,†]

[1]*Department of Neurology, Klinikum Vest, Academic Teaching Hospital of the Ruhr University Bochum, Recklinghausen, Germany*

[2]*Department of Neuroanatomy and Molecular Brain Research, Ruhr University Bochum, Bochum, Germany*

[3]*Department of Psychiatry and Psychotherapy, Medical Faculty, Heinrich Heine University, Düsseldorf, Germany*

Abstract

Neuroglial cells play a crucial role in central nervous system (CNS) health and disease. Antibody-associated autoimmune encephalitis (AE) represents a group of inflammatory brain diseases with antibodies (Abs) against neuronal cell surface (e.g., anti-*N*-methyl-D-aspartate receptor (NMDAR), anti-leucine-rich glioma-inactivated 1 (LGI1), γ-aminobutyric acid (GABA) type A or B receptor ($GABA_{A/B}R$)) or intracellular neuronal proteins. AE with Abs against glial antigens, e.g., myelin oligodendrocyte glycoprotein (MOG), glial fibrillary acidic protein (GFAP) are also described. Besides the known pathomechanisms with direct pathogenic effects of primary neuronal Abs and activation of innate (dendritic cells) and adaptive (B and T cells) immune systems, research findings suggest the involvement of glial cells including astrocytes, microglia, oligodendrocytes in the pathogenesis of Ab-associated AE, but only a limited number of studies is available. Neuropathologic findings showed reactive astrogliosis and microgliosis with microglial activation/proliferation, e.g., in anti-NMDAR and anti-LGI1 encephalitis. Direct effects of the $GABA_AR$ and NMDAR Abs on astrocytic receptors are discussed. Because of the primary involvement of B and T cells in the pathogenesis of Ab-associated AE it can be assumed that astrocytic and microglial activation is largely a response to the primary changes, but additional direct effects of Abs on astrocytic receptors are possible. Further research in this field is required to explore the exact role of glial cells in Ab-associated AE.

INTRODUCTION

Antibody-associated autoimmune encephalitis (AE) represents a group of inflammatory brain diseases with antibodies (Abs) against neuronal cell surface (incl. ion channels, receptors) or intracellular neuronal proteins (Dalmau et al., 2017; Dalmau and Graus, 2018). Seizures, memory impairment, psychiatric symptoms such as psychosis, behavioral changes, abnormal movements, dysautonomia, and disturbance of consciousness are counted among common clinical symptoms of AE (Graus et al., 2016). Additionally, presentation as a separate clinical entity in terms of autoimmune psychosis is possible (Pollak et al., 2020). Several Abs associated with AE were reported, which are directed either against neuronal cell-surface proteins such as anti-*N*-methyl-D-aspartate receptor (NMDAR), anti-leucine-rich glioma-inactivated 1 (LGI1), α-amino-3-hydroxy-5-methyl-4-isoxazolepropionic acid receptor (AMPAR), γ-aminobutyric acid (GABA) type A or B receptor ($GABA_{A/B}R$), contactin-associated protein-like 2 (CASPR2), dipeptidyl-peptidase-like protein 6 (DPPX), metabotropic glutamate receptor 5 (mGluR5), dopamine 2 receptor (D2R), neurexin-3α, or intracellular neuronal proteins such as Hu, Ma2, glutamic acid decarboxylase 65 (GAD65) (Irani et al., 2011; Dalmau et al., 2017; Dalmau and Graus, 2018). In AE associated with Abs against neuronal cell-surface proteins, the Abs have direct effects on the target antigens and can potentially alter their structure and function. In contrast, Abs against intracellular (onco-neuronal) antigens did not have access to these antigens and cytotoxic T cell are predominantly involved in the pathogenesis. Recently, the dendritic cells (professional antigen-presenting cells) and their interactions have been discussed as key modulators in Ab-associated AE (Ismail et al., 2021b). In addition, diagnostic criteria for AE without any identifiable pathogenic Ab, the so-called antibody-negative AE, have been proposed (Lee et al., 2022; Dalmau and Graus, 2023).

[†]Franco Corvace and Timo Jendrik Faustmann have equally contributed as senior authors.

[*]Correspondence to: Pedro M. Faustmann, Department of Neuroanatomy and Molecular Brain Research, Ruhr University Bochum, Bochum, Germany. Tel: +49-2343224405, Fax: +49-2343214089, E-mail: pedro.faustmann@ruhr-uni-bochum.de

In human neuropathologic studies of anti-NMDAR encephalitis, the most common form of AE, moderate brain inflammatory infiltrates along with B and plasma cells (especially in perivascular space), deposits of IgG, and microglial proliferation have been found, but without evidence of T-cell-mediated neuronal loss (Dalmau et al., 2008; Bauer and Bien, 2016). Otherwise, for example, in patients with anti-Hu paraneoplastic encephalitis, inflammatory infiltrates with extensive T-cell-mediated (predominant $CD8^+$ T cells in parenchyma, $CD4^+$ T cells in perivascular space) destruction of neurons have been reported (Bauer and Bien, 2016). In addition, AE associated with Abs against glial antigens, for example, myelin oligodendrocyte glycoprotein (MOG), glial fibrillary acidic protein (GFAP) has been described (Lucchinetti et al., 2014; Shan et al., 2018; Marignier et al., 2021). Besides the known pathomechanisms with direct pathogenic effects of Abs and activation of innate (dendritic cells) and adaptive (B and T cells) immune systems, research findings suggest the involvement of glial cells including astrocytes, microglia, and oligodendrocytes in the pathogenesis of Ab-associated AE (Fig. 9.1) (Lucchinetti et al., 2014; Bauer and Bien, 2016; Shan et al., 2018; Ismail and Faustmann, 2020, 2023; Marignier et al., 2021). Neuroglial cells are not only key players in physiology, metabolism, and development of the brain and spinal cord, but are also involved in the pathogenesis of central nervous system (CNS) diseases (Verkhratsky et al., 2023). Microglia and astrocytes are the main cell types of the innate immune system in the CNS, which undergo complex reactive remodeling in response to various brain pathologies (Escartin et al., 2021; Paolicelli et al., 2022). In AE, the glia-mediated neuroinflammation may contribute directly to seizure generation, supporting the role of glia also in the pathogenesis of epilepsy (Ismail and Faustmann, 2023). Moreover, pericytes as part of the blood–brain barrier (BBB) could be involved in AE (Dalkara et al., 2011; Gong et al., 2023). Pericytes are located in capillaries and postcapillary venules subendothelially within the basal membrane. During neuroinflammation pericytes regulate CNS vascular immune homeostasis (Török et al., 2021). In recent studies using in vivo two-photon microscopy, a subpopulation of microglia was found to be associated with pericytes at all levels of the capillary tree. These pericyte-associated microglia play an important role in controlling cerebral blood flow (Morris et al., 2023). Only a few studies on BBB disruption, e.g., in anti-NMDAR encephalitis have been reported, confirming the importance of BBB breakdown, but the exact involved mechanisms remain unclear (Gong et al., 2023). Further research is needed to understand the association between the elements of the BBB and NMDARs, including astrocytes, microglia, pericytes, and endothelial cells (Gong et al., 2023).

In the following chapters, we focus on the role of astrocytes, microglia, and oligodendrocytes in Ab-associated AE.

ASTROCYTES

Astrocytes are the ubiquitously distributed glial cells in the CNS and are of ectodermal, neuroepithelial origin. They form complex connections with neurons, blood vessels, and other glial cells, contributing to the maintenance of normal brain functions. Astrocytes are highly heterogeneous in form and function. Astrocytes contribute to the CNS homeostasis including (1) molecular homeostasis by transporting ions (K^+, Cl^-, Ca^{2+}) and protons, by releasing scavengers of reactive oxygen species and by regulation of water transport and neurotransmitter balance (glutamate, GABA, adenosine, monoamines); (2) sustaining neurotransmission by supplying neurons with neurotransmitter precursors and regulation of cellular and network homeostasis through embryonic neurogenesis and adult neurogenesis, effects on synaptogenesis, synaptic maintenance/elimination and synaptic plasticity; (3)

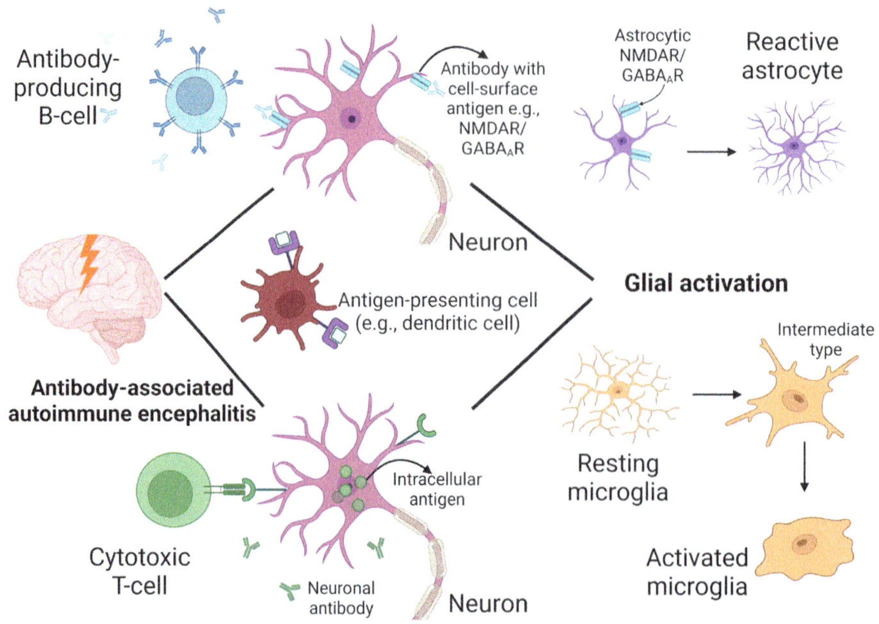

Fig. 9.1. An overview of cellular (B cell, T cell, antigen-presenting cell/dendritic cell, astrocyte, microglia) and humoral immunity involved in the pathogenesis of antibody-associated AE. $GABA_AR$, γ-aminobutyric acid type A receptor; NMDAR, anti-*N*-methyl-D-aspartate receptor. Created with BioRender.com.

metabolic homeostasis via glycogen synthesis and storage, supply of neurons with energy substrates, defining the cytoarchitecture of the gray matter, forming neuro-glia-vascular unit and glial-vascular interface for regulation of local blood flow; (4) systemic homeostasis (regulation of energy balance, blood pH and Na+ concentration) by control of BBB and acting as chemosensors; (5) forming together with microglia the main defensive system of the CNS (Verkhratsky and Nedergaard, 2018).

Astrocytes respond to damage and disease in the CNS through a complex process with physicochemical and physiologic changes, called reactive astrogliosis (Sofroniew, 2014, 2020). A recent evidence showed that reactive astrogliosis and perilesional glial barrier formation play important role in regulating CNS inflammation, because reactive astrocytes can produce different molecules with either pro- or antiinflammatory features in response to different kinds of stimulation and reactive astrocytes can exert both pro- and antiinflammatory effects on microglia (Sofroniew and Vinters, 2010; Verkhratsky et al., 2023). Severe proliferative reactive astrogliosis is characterized by pronounced upregulation of expression of GFAP and other genes, pronounced hypertrophy of cell body and processes, as well as astrocyte proliferation, resulting in long-lasting reorganization of tissue architecture and intercellular network.

Different types of intercellular signaling molecules, released by all cell types in CNS, including neurons, microglia, oligodendrocyte lineage cells, pericytes, endothelia, and other astrocytes, in response to CNS pathology, can trigger reactive astrogliosis or regulate specific aspects of reactive astrogliosis. Such molecular mediators include large polypeptide growth factors and cytokines such as interleukin 6 (IL6), tumor necrosis factor-α (TNF-α), and transforming growth factor-β1 (TGF-β1), mediators of innate immunity such as lipopolysaccharide (LPS) and other toll-like receptor ligands (Sofroniew, 2009). Also neurotransmitters, e.g., glutamate and noradrenalin, purines such as ATP, reactive oxygen species including nitric oxide (NO), hypoxia and glucose deprivation, products associated with neurodegeneration (e.g., β-amyloid), molecules associated with systemic metabolic toxicity such as $NH4^+$ and regulators of cell proliferation, e.g., endothelin-1, can trigger or regulate reactive astrogliosis (Sofroniew, 2009). Moreover, different intracellular signaling pathways associated with STAT3, NF-κB, SOCS3, Nrf2, cAMP, and Olig2 are involved in mediating different aspects or different degrees of reactive astrogliosis such as GFAP upregulation, cell hypertrophy, proliferation, and pro- or antiinflammatory effects (Sofroniew and Vinters, 2010).

In histopathologic studies of Ab-associated AE, astrogliosis and reactive astrocytes were detected (Vincent et al., 2004; Dunstan and Winer, 2006; Dauvilliers et al., 2013; Bauer and Bien, 2016; Kuehn et al., 2020). This was the case in LGI1-associated AE (previously described as voltage-gated potassium channel (VGKC) Abs) with postmortem demonstration of severe neuronal loss with multiple reactive astrocytes, macrophages, and scattered T-cells in the amygdaloid nucleus and adjacent hippocampus (Dunstan and Winer, 2006). In another case with LGI1/VGKC Abs, a stereotactic biopsy of the amygdala demonstrated perivascular and parenchymal lymphocytes, astrogliosis, and microglial activity (Vincent et al., 2004). In a further case of LGI1 Abs, the cellularity adjacent to the hippocampus was increased by astrocytes with large cytoplasm and accompanying macrophages and lymphocytes, and the immunostaining showed numerous GFAP-positive cells with isomorphic, reactive rather than pleomorphic fibrillary morphology (Kuehn et al., 2020).

In a case with intracellular Ma2-Abs and diencephalitis, severe neuronal loss and gliosis with extensive $CD8^+$ T-cell perivascular and parenchymal inflammatory infiltrates were found, particularly within the hypothalamus. Within the lesions, there was a profound astrocytic gliosis with increased expression of aquaporin 4 (AQP4) around the inflamed vessels (Dauvilliers et al., 2013). Following this, neuropathologic findings with regard to specific astrocytic changes in Ab-associated AE are very limited, because most forms of AE can be treated effectively, biopsies are no longer required and most patients survive.

We hypothesized and discussed the potential role of astroglial receptors in the pathogenesis of Ab-associated AE (Ismail and Faustmann, 2020, 2023). There is evidence that astrocytes, in addition to neurons, express functional NMDA receptors including the GluN1 subunit in different brain regions such as the cortex, hippocampus, and brainstem (Verkhratsky and Kirchhoff, 2007; Skowrońska et al., 2019). It is conceivable that the pathologic Abs against the NMDAR in AE affect not only the neuronal receptors but also the NMDARs on astrocytes (Ismail and Faustmann, 2020). Following this, NMDAR Abs can lead to cross-linking and internalization of the target receptors in astrocytes similar to neurons. The astrocytic NMDARs affect neuronal-glial signal transmission and regulate Ca^{2+} influx into the cell, therefore, the lack of NMDAR can lead to disruption of the local depolarization and the Ca^{2+} release within the astrocytes and consequently block the release of inhibitory gliotransmitters (e.g., ATP metabolized to adenosine or endocannabinoids, or GABA), resulting in reduced inhibition on excitatory synapses and additional impairment of synaptic dysfunction, in addition to direct effect of Abs on neurons. In conclusion, astrocytic NMDARs can codetermine the severity of clinical disease manifestation and recovery in NMDAR-associated AE, especially in view of the fact that NMDARs were found in cortical, hippocampal and brainstem astrocytes (Ismail and Faustmann, 2020). Functional NMDARs were also found in brain microglia and oligodendrocytes, especially in the myelin sheath (Verkhratsky and Kirchhoff, 2007). The role of microglial NMDARs in NMDAR encephalitis is unclear. In another study, NMDAR Abs alter the function of NMDARs in oligodendrocytes, causing a decrease of expression of glucose transporter GLUT1 (Matute et al., 2020).

Moreover, Abs against α1, β3, and γ2 subunits of the $GABA_A$ receptor was associated with a severe form of AE with characteristic clinical-radiologic patterns including encephalopathy and refractory seizures with multifocal brain lesions, affecting both gray and white matter (Petit-Pedrol et al., 2014; Pettingill et al., 2015; O'Connor et al., 2019). $GABA_A$ receptors ($GABA_AR$) mediate the inhibitory synaptic transmission in the CNS. Experimental studies showed multiple mechanisms and direct functional effects of $GABA_AR$ Abs on neurons with decreased GABAergic synaptic transmission and increased neuronal excitability, contributing to seizure generation in $GABA_AR$ AE (Spatola et al., 2017; Brändle et al., 2021; Menke et al., 2022;

Noviello et al., 2022; van Casteren et al., 2022). Previous studies confirmed the expression of $GABA_AR$ on astrocytes, but extensive studies about the effects of autoimmune $GABA_AR$ Abs on astrocytic $GABA_AR$ are missing (Backus et al., 1988; Fraser et al., 1994, 1995; Mederos and Perea, 2019; Liu et al., 2022). The main $GABA_AR$ subunits expressed in rodent astrocytes are α1, α2, β1, β3, and γ1 and can be found in both white and gray matter. To date, $GABA_AR$ subunits expressed in human astrocytes include α2, β1, and γ1. Following this, we assumed overlapping binding of $GABA_AR$ Abs to neuronal and astroglial $GABA_AR$ with additionally blocking astrocytic $GABA_AR$, resulting in impaired Ca^{2+} homeostasis/spreading, astrocytic Cl^- imbalance, dysfunction of astrocyte-mediated gliotransmission (e.g., decreased adenosine levels) and accumulation of excitatory neurotransmission, all this contributing to seizures, variable clinical/MRI presentations, and severity (Ismail and Faustmann, 2023; Untiet et al., 2023). In vitro and in vivo animal models can be used to test the effects of $GABA_AR$ Abs on glia.

Taken together, autoimmune disorders are complex and multiple mechanisms including glia could contribute to the pathogenesis of $GABA_AR$ and NMDAR encephalitis (Ismail and Faustmann, 2020, 2023).

Autoimmune GFAP astrocytopathy

There is increasing evidence for the concept of astrocytopathies in which a disruption of normal astrocyte functions represents the primary cause of neurologic dysfunction and disease (Verkhratsky et al., 2023). Autoimmune GFAP astrocytopathy and neuromyelitis optica spectrum disorders (NMOSD) are examples of autoimmune astrocytopathies (Lucchinetti et al., 2014; Fang et al., 2016; Shan et al., 2018; Kunchok et al., 2019). The GFAP astrocytopathy has typical symptoms and signs of meningoencephalomyelitis or a limited form thereof, is steroid-responsive, sometimes with relapsing course (Fang et al., 2016; Flanagan et al., 2017; Kunchok et al., 2019). The disorder is defined by the detection and confirmation of immunoglobulin G (IgG) Abs against the GFAP, which is an intracellular astrocytic intermediate filament and an important component of the cytoskeleton, involved in multiple astrocyte functions (Kunchok et al., 2019). The intracellular antigen location suggests that GFAP-specific cytotoxic T cells may be involved in the pathogenesis of the disorder rather than direct effects of the Abs. Neuropathologic findings from different studies with GFAP astrocytopathy demonstrated inflammatory changes with mononuclear infiltration by macrophages/microglia, T cells ($CD8^+$, $CD3^+$), B cells ($CD20^+$) (Shan et al., 2018; Kunchok et al., 2019). In an autopsy case, diffuse inflammation in the brain parenchyma, perivascular spaces, and leptomeninges with predominant T cells, macrophages, and activated microglia were reported, but there was no astrocyte involvement with change in GFAP immunostaining (Yamakawa et al., 2021). In other studies, immunopathology of the lesions from patients revealed absent or diminished GFAP and AQP4 staining (Shan et al., 2018; Kunchok et al., 2019). In a further study, pathologic examination of patients with GFAP astrocytopathy revealed astrocyte damage and elevated GFAP levels in the cerebrospinal fluid (CSF) (Yang et al., 2019). The main mechanisms involved in astrocyte injury are unknown, but it has been suggested that T-cell immune responses may cause astrocyte damage. It has also been discussed that other components of the immune system such as microglia, macrophages, cytokines, and chemokines could contribute to disease pathogenesis. Immune-mediated astrocyte dysfunction could cause chemokine release that further recruits inflammatory cells, enabling the spread of the immune attack to healthy nervous system tissue. Another hypothesis is that GFAP autoimmunity can occur as a secondary phenomenon as a result of undiscovered plasma membrane protein-directed IgGs initiating a primary autoimmune event with disruption of astrocytic function (Kunchok et al., 2019). Further research is required to uncover and prove the involved pathophysiologic mechanisms.

Moreover, the overlapping occurrence of GFAP-Abs in NMDAR encephalitis was described (Martinez-Hernandez et al., 2020). The presence of GFAP-Abs was usually clinically unsuspected and their pathogenic role in patients with NMDAR encephalitis remains unclear.

OLIGODENDROCYTES

Oligodendrocytes are the myelinating glial cells of the CNS, building the sheath of neuronal axons by wrapping a fatty layer around them (Bradl and Lassmann, 2010). Compared to the peripheral nervous system (PNS) (where only one single Schwann cell wraps one single neuronal axon) usually multiple neuronal axons can be wrapped by a single oligodendrocyte in the CNS.

Oligodendrocytes originate from specialized oligodendrocyte precursor cells in the ventral and dorsal spinal cord, from the anterior entopeduncular area and the medial ganglionic eminence of the ventral forebrain, the lateral and/or caudal ganglionic eminences and even from the postnatal cortex (Bradl and Lassmann, 2010).

The morphology of individual oligodendrocytes varies throughout the CNS depending on brain regions and surrounding axons and can be separated into Type I (small round shape cell body with many processes in gray and white matter), Type II (polygonal shaped cell body with fewer processes only found in the white matter), Type III (bulky cell body forming a few myelin sheaths on large caliber axons in the white matter), and Type IV (schwannoid oligodendrocytes, myelinating a single large caliber axon) (Pérez-Cerdá et al., 2015; Osanai et al., 2022).

Mechanisms regulating myelination in the CNS depend on neuronal differentiation and not on an intrinsic oligodendrocyte differentiation program, and are independent of the neuregulin-1/ErbB pathway, which could be found in the process of myelination in the PNS (Brinkmann et al., 2008). In this vein, one important mechanism seems to be the neuronal electric activity, e.g., light-induced myelination in the optic nerve or increasing neuronal activity of the motor cortex, which promoted oligodendrogenesis, increased myelination, and improved motor function in mice (Gibson et al., 2014). The myelin basic protein (MBP), the proteolipid protein (PLP), and galactosylceramide (GalC) are important proteins for the assembly of myelin. MBP here together with neuronal signals seems to cluster lipid bilayers, leading to an accumulation of membrane components within the myelin (Fitzner et al., 2006). In oligodendrocytes, this leads to an elaboration about three times its weight in membrane per day, and the increase in the membrane of oligodendrocytes could be a later risk factor under pathologic conditions (Bradl and Lassmann, 2010).

Proper myelination is important in supporting the formation of sodium channels at the node of Ranvier, saltatory nerve conduction, axonal transport, and an increase in the axon diameter via the building of compact myelin consisting of cytoplasmatic channels (inner and outer tongues). Next to the node, the paranodes can be found consisting of paranodal cytoplasmatic channels connected to inner and outer tongues (Kaplan et al., 2001; Osanai et al., 2022). Neurophysiological measuring the resistance of internodal processes in rats show different conduction speeds due to oligodendroglial myelination. The conduction speed for cerebellar axons is twice that for corpus callosal axons involving shorter internodes in corpus callosum compared to the cerebellum but having a similar size of somata. These findings most probably underline the faster conduction speed needed for the motor output in the cerebellum compared to that in the corpus callosum (Bakiri et al., 2011).

Oligodendrocytes are involved in the pathophysiology of different neurologic disorders, but only a few experimental data were described in the context of their role in Ab-associated AE (Matute et al., 2020). Similar to neurons, oligodendrocytes express NMDAR, which supports axons metabolically with lactate to maintain the proper propagation of action potentials (Saab et al., 2016). Stimulation of oligodendroglial NMDARs by glutamate released from axons mobilized glucose transporter GLUT1 into the oligodendrocyte plasma membrane and myelin compartment, enhancing glucose uptake and glycolytic support of fast-spiking axons (Saab et al., 2016). Recently, CSF of patients with NMDAR encephalitis and a human NMDAR-specific monoclonal antibody (SSM5) derived from plasma cells of a patient was studied (Matute et al., 2020). NMDAR agonist responses were reduced after preincubation of oligodendrocytes with patients' CSF or SSM5. Patients' CSF also led to a reduction of oligodendrocyte expression of glucose transporter GLUT1 induced by NMDAR activity. In conclusion, Abs from patients with NMDAR encephalitis altered the function of NMDARs in oligodendrocytes, causing a decrease in the expression of GLUT1. There is a possible link between Ab-mediated dysfunction of NMDARs in oligodendrocytes and the white matter alterations described in patients with this disorder, as normal GLUT1 expression in oligodendrocytes and myelin is required to metabolically support axonal function (Matute et al., 2020).

In addition, oligodendrocytes are also involved in the pathophysiology of MOG Ab-associated encephalitis and several overlapping entities have been described.

MOG ab-associated AE

The MOG is a characteristic protein expressed on oligodendrocytes and myelin in the CNS (Ramanathan et al., 2016). MOG is located at the extracellular surface of myelin sheaths and mature oligodendrocytes. Multiple functions were proposed as a cellular adhesive molecule, a regulator of oligodendrocyte microtubule stability, and a mediator of interactions between myelin and the immune system (Johns and Bernard, 1999; Corbali and Chitnis, 2023).

MOG Ab-associated AE, occurring in children and adults, is a distinct disease entity that can be immunopathogenetically distinguished from both classic multiple sclerosis (MS) and AQP4-positive NMOSD (Jarius et al., 2018). Clinical phenotypes encompass cerebral cortical encephalitis, limbic encephalitis, and acute disseminated encephalomyelitis (ADEM) (Lee et al., 2023). Further clinical presentations include relapsing/bilateral optic neuritis, brainstem encephalitis, and transverse myelitis.

Pathophysiologically, initial activation of $CD4^+$ T cells occurs in the periphery and is followed by reactivation in the subarachnoid/perivascular spaces by MOG-loaded antigen-presenting cells and inflammatory CSF milieu, contributing to infiltration of T cells and MOG Abs in CNS parenchyma. $CD4^+$ T cells are described as the dominant T-cell type in lesion histology (Takai et al., 2020). Granulocytes, macrophages/microglia, and activated complement are also detected in the lesions. B cells are part of the disease pathogenesis due to the production of MOG Abs. Potential MOG Ab-associated pathologies include MOG opsonization, complement activation, and antibody-dependent cellular cytotoxicity (Corbali and Chitnis, 2023).

MOG-associated pathology is dominated by the coexistence of both perivenous (around small veins) and confluent white matter demyelination (Höftberger et al., 2020). Cortical lesions also show perivenous confluent patterns and intracortical demyelination.

In addition to $CD4^+$ T-cell dominance, astrocytes are spared in MOG Ab-associated encephalitis, whereas pronouncedly reduced in NMOSD, and activated and proinflammatory in MS (Lucchinetti et al., 2014; Höftberger et al., 2020). The astrocyte involvement is different in these three diseases, e.g., GFAP elevation was detected in NMOSD patients but was absent in MOG-associated diseases (Kaneko et al., 2016). Conversely, oligodendrocytes are variably damaged and lost in all three diseases; for example, in MOG Ab-associated encephalitis progenitor cells are preserved, because they do not yet express MOG (Marignier et al., 2010; Lucchinetti et al., 2014; Höftberger et al., 2020). Another study demonstrated a complement-independent effect of NMOSD-IgG/AQP4 Abs on astrocytes, with secondary damage to oligodendrocytes (Marignier et al., 2010).

In addition, overlapping cases of MOG- and AQP4-associated syndromes with NMDAR encephalitis were described (Titulaer et al., 2014; Ramberger et al., 2015). The NMDARs can be found on the myelin sheath built by oligodendrocytes (Lipton, 2006; Saab et al., 2016). So, it is possible that NMDAR-Abs can also cause demyelination under an ongoing immune response. In another study, MOG-Abs rarely coexisted with Abs against LGI1, CASPR2, and $GABA_AR$ (Kunchok et al., 2021). Additionally, a case of overlapping encephalitis with coexistent MOG-, NMDAR-, and mGluR5-Abs was reported and initially presented with fever, headache, difficulty in language expression, lymphocytic-predominant pleocytosis, and cortical edema followed by seizures (Fu et al., 2022). Studies about exact pathomechanisms of overlapping syndromes are not available yet; however, due to the involvement of oligodendrocytes in MOG Ab-associated encephalitis and anti-NMDAR encephalitis additional demyelinating effects could lead to severe pathologic effects.

MICROGLIA

Microglia were first described in the 20th century and detected in the brain by a silver carbonate-based staining method developed by the Spanish neuroscientist Pío del Río-Hortega (Perry and Gordon, 1991; Prinz et al., 2019). Microglia belong to the mononuclear phagocytes and are correspondingly descendants of the monocyte–macrophage system. Microglial progenitors arise in the yolk sac and migrate into the brain rudiment at very early stages of embryonic development, persisting in the CNS into adulthood (Ginhoux and Prinz, 2015).

During brain ontogeny, migrated microglia phagocytose apoptotic neurons, trigger programmed cell death through activation of the neurotrophin receptor p57, and limit the growth and differentiation of axons and neurons (Streit, 2001; Mallat et al., 2005). After reaching their final position in the brain, reduction of the cell body, and formation of long thin projections, the microglia are no longer capable of migration and are referred to as surveilling microglia with highly motile processes, which is the primary existing phenotype under physiologic conditions (Kettenmann et al., 2011). Pathologic insults lead to reactive microgliosis and proliferation, whereby they transform from the surveilling ramified state through an intermediate form into the activated state (round phagocytic type, RPT) phenotypically, characterized by a round soma without ramified processes (Faustmann et al., 2003; Block et al., 2007; Kettenmann et al., 2011). This reactive remodeling of microglia occurs in several steps and is mediated by incoming stimuli such as ATP, adenosine, complement factors, cytokines, chemokines, and changes in potassium concentration. As a result of reactivity, increased mobility of astrocytes, and a chemotactic gradient, microglia actively migrate to the pathologic focus and exert curative or destructive effects depending on various factors (Walter and Neumann, 2009; Kettenmann et al., 2011; Paolicelli et al., 2022). Studies show that reactive microglia can express various neurotrophic factors such as neurotrophin 3 (NT3), growth factor BDNF, and nerve growth factor (NGF), but also proinflammatory molecules such as IL1, IL6, TNF-α, TGF-β, interferon-γ (IFN-γ), prostaglandins, ROS, oxygen radicals, proteinases, and nitric oxide (Kettenmann et al., 2011; Nayak et al., 2014; Wolf et al., 2017; Borst et al., 2021). The effect of microglia seems to depend in particular on stimulation and interaction with other molecules, for example, activation of microglia by interleukin IL4 and IL13 leads to antiinflammatory features promoting tissue remodeling and repair through the release of neurotrophic (TGF-β, insulin-like growth factor 1 (IGF-1)) and angiogenic factors (VEGF), while microglia activated by proinflammatory cytokines (e.g., IFN-γ) or LPS produce oxidative metabolites and proinflammatory cytokines such as IL1β, TNF-α, IL6 (Block et al., 2007; Czeh et al., 2011; Borst et al., 2021). Microglial antigen presentation in the context of MHC class (I and II) and the expression of stimulating proteins such as CD11a, CD40, CD54 leads to the induction of CD8$^+$ cytotoxic cells and CD4$^+$ T-helper cells, whereby microglia have a direct influence on the humoral and cellular immune response. The increased expression of proinflammatory cytokines leads overall to impaired glial cell-neuron communication and astrocyte syncytium through decreased expression of connexin 43, increased permeability of the BBB, ionic imbalance including impaired glutamate uptake (Faustmann et al., 2003; Devinsky et al., 2013). After neuronal lesions or generally under pathologic conditions, activation of inflammatory macrophages occurs in the CNS. Microglial activation is an indicator of an incipient pathologic process and at the same time the earliest response of the brain toward protection and regeneration.

Study findings about the involvement of microglia in the pathogenesis of Ab-associated AE are limited. As described earlier, the activation of microglia is an indicator of an incipient pathologic process and could therefore also represent one of the changes that can be observed in the context of Ab-associated AE. Indeed, patients with NMDAR encephalitis show increased microglial proliferation, activation, and formation of microglial nodules (Dalmau et al., 2007, 2008; Tüzün et al., 2009; Bauer and Bien, 2016; Dalmau and Graus, 2018). According to further studies, all areas of the CNS of patients with anti-NMDAR encephalitis showed increased reactive microglia. The extent of microgliosis was highest in the hippocampus, basal forebrain, basal ganglia, and spinal cord (Tüzün et al., 2009). Immunopathology of brain tissue (biopsy or autopsy material) of 17 cases with AE and Abs to either intracellular (Hu, Ma2, GAD) or surface-antigen targets (VGKC or NMDAR) showed numerous CD68$^+$ cells in the brain parenchyma and the majority of the CD68$^+$ cells were microglia (Bien et al., 2012). A translational study investigated activation of microglia and other innate immune cells in the mouse and human brain, visualized by targeting the 18-kDa translocator protein (TSPO) with second-generation positron emission tomography (PET) tracers such as [18F]DPA-714, in the context of neuron-directed CD8$^+$ T-cell-mediated autoimmune limbic encephalitis (ALE), a subtype of AE (Gallus et al., 2023). TSPO, formerly known as the peripheral benzodiazepine receptor, is present in the mitochondrial outer membrane, in particular, in microglia and in low levels in neurons and astrocytes. TSPO reactivity is greatly increased in activated microglia (Cosenza-Nashat et al., 2009). In humans, specimens from patients with ALE associated with anti-Hu Abs or anti-GAD65 Abs, strongly increased TSPO reactivity overlapped with Iba-1 positive activated microglia and reactive astrocytes (Gallus et al., 2023). A similar pattern was detected in a patient with Ab-negative ALE. These findings were confirmed in mice with CD8$^+$ T-cell-mediated ALE, in which [18F]DPA-714-PET-MRI showed strong parenchymal innate immune activation (Gallus et al., 2023). In a previous study, CD8$^+$ T-cell-initiated attack of distinct hippocampal neurons in mice was able to induce ALE and caused segmental apoptotic neurodegeneration, astrogliosis, and microglial activation (Pitsch et al., 2021).

Despite the crucial contribution that the discovery of Abs has made to the understanding of the pathophysiology of these diseases, the exact mediators and signaling pathways are not yet fully understood. However, cytokines have recently been shown to act as signaling molecules in the pathogenesis of AE (Ciano-Petersen et al., 2022). Cytokines are regulatory peptides that are predominantly expressed not only by cells of the reticuloendothelial system such as monocytes, endothelial cells, fibroblasts, or lymphocytes, but also by neurons, astrocytes, and microglia under pathologic conditions, controlling humoral immune responses. Cytokines are divided into five main groups based on their function: interferons, interleukins, the chemokine family, growth factors, the tumor necrosis factor family, and adipokines (Dinarello, 2007). An imbalance in the expression of various cytokines is a characteristic of many autoimmune diseases of the CNS. Despite various approaches to studying the cytokine profile of AE, only the cytokine profile of anti-NMDAR encephalitis has been studied comprehensively to date (Leypoldt et al., 2015; Byun et al., 2016; Kothur et al., 2016; Liba et al., 2016; Ciano-Petersen et al., 2021, 2022). In the CSF of several patients with anti-NMDAR encephalitis, various cytokines involved in the activation of B and T cells were found. These include IFN, TNF-α, IL1β and IL6 (Ciano-Petersen et al., 2022). A process in which microglia, among others, could be involved. The 17 kDa pleiotropic glycoprotein IL1β mediates its activity via two receptors type 1 (IL1R1) as well as type 2 (IL1R2) and is able to induce the proliferation and activation of T and B lymphocytes, the synthesis of COX-2, prostaglandin, inducible nitric oxide (NO−) synthase (iNOS) and chemokines. Under physiologic conditions, a low concentration of IL1β is constitutively expressed and stored by neurons and glial cells. Under pathological conditions, there is an additional predominant microglial expression of IL1β and, in the presence of the cytokine, increased expression of IL6 and TNF-α (Basu et al., 2004). The IL6 mediates its activity after dimerization via two specific receptor types, a membrane-bound IL6 receptor (IL6R) and a soluble IL6 receptor (sIL6R) with subsequent binding to the glycoprotein gp130, activation of Janus tyrosine kinases and phosphorylation of tyrosine residues of STAT proteins (signal transducer and activator of transcription), an important transcription factor of specific genes important for differentiation, proliferation and cell viability (Darnell, 1997). The IL6-mediated phosphorylation of STAT-3, for example, leads to the production of acute phase proteins and induction of B-cell proliferation (Levy and Lee, 2002). Increased expression of IL6 leads to lower IL1 and TNF-α concentrations in the sense of a negative feedback loop through an expression of an interleukin 1 receptor antagonist (IL1Ra) and soluble TNF-α-decoy receptors, whereby IL6 has an antiinflammatory as well as a proinflammatory effect (Borish and Steinke, 2003).

Furthermore, elevated levels of B-cell-activating factor (BAFF) and proliferation-inducing ligand (APRIL) were found in 40 patients with anti-NMDAR encephalitis in the acute phase (Deng et al., 2017; Ciano-Petersen et al., 2022). These two molecules play an important role in the survival, differentiation, and maturation of B cells, as well as in Ab synthesis, and can be associated with astrocytes and microglia, respectively (Thangarajh et al., 2007; Li et al., 2017). For example, Li et al. (2017) were able to show that under ischemic conditions BAFF was strongly upregulated in microglia and that this upregulation can at least be attributed to the JAK–STAT signaling pathway, which in turn is activated by IFN-γ and IL10. It should be said, however, that no elevated levels of BAFF or APRIL could be detected in long-term studies of anti-NMDAR encephalitis (Kimura et al., 2015; Kothur et al., 2016; Liba et al., 2016; Ciano-Petersen et al., 2022).

Unfortunately, relatively few data on microglia and cytokine expression in other Ab-mediated AE are currently studied. In an autopsy case with LGI1-associated AE (previously described as VGKC Abs), mild focal perivascular T-cell lymphocytic cuffing and parenchymal infiltration including also mild microglial proliferation was reported involving the leptomeninges, parenchyma of the limbic system (cingulate gyrus, hippocampus, and amygdala) and midbrain (Park et al., 2007). In another case with LGI1/VGKC Abs, a stereotactic biopsy of the amygdala demonstrated also microglial activity (Vincent et al., 2004).

In conclusion, the results show that the expression of cytokines and microglia, as major producers of proinflammatory cytokines in the CNS, may have an important role in the pathogenesis of Ab-mediated AE that needs to be explored.

CONCLUSIONS AND FUTURE PERSPECTIVES

Neuroglial cells including astrocytes, microglia, and oligodendrocytes play a crucial role in CNS health and disease. Findings about their involvement in Ab-associated AE are limited. Neuropathologic findings showed reactive astrogliosis and microgliosis, e.g., in anti-NMDAR encephalitis (Bauer and Bien, 2016). Direct effects of the $GABA_AR$ and NMDAR Abs on astrocytic receptors were discussed (Ismail and Faustmann, 2020, 2023). In addition, AE associated with Abs against glial antigens, e.g., MOG or GFAP, was described. Because of the primary involvement of B and T cells in the pathogenesis of Ab-associated AE it can be assumed that astrocytic and microglial activation is largely a response to the primary changes, but additional direct effects of Abs on astrocytic receptors are possible (Fig. 9.1).

Further research is necessary to explore the exact role of neuroglia in Ab-associated AE. For this purpose, in vitro and in vivo models can be used; for example, the astrocyte-microglia coculture model of inflammation is not only a well-established in vitro model for investigation of drug effects on glial cells but also offers the potential for studying Ab effects on glial cells (Faustmann et al., 2003; Ismail et al., 2021a).

References

Backus KH, Kettenmann H, Schachner M (1988). Effect of benzodiazepines and pentobarbital on the GABA-induced depolarization in cultured astrocytes. Glia 1: 132–140. https://doi.org/10.1002/GLIA.440010205.

Bakiri Y, Káradóttir R, Cossell L et al. (2011). Morphological and electrical properties of oligodendrocytes in the white matter of the corpus callosum and cerebellum. J Physiol 589: 559–573. https://doi.org/10.1113/JPHYSIOL.2010.201376.

Basu A, Krady JK, Levison SW (2004). Interleukin-1: a master regulator of neuroinflammation. J Neurosci Res 78: 151–156. https://doi.org/10.1002/JNR.20266.

Bauer J, Bien CG (2016). Neuropathology of autoimmune encephalitides. Handb Clin Neurol 133: 107–120. https://doi.org/10.1016/B978-0-444-63432-0.00007-4.

Bien CG, Vincent A, Barnett MH et al. (2012). Immunopathology of autoantibody-associated encephalitides: clues for pathogenesis. Brain 135: 1622–1638. https://doi.org/10.1093/brain/aws082.

Block ML, Zecca L, Hong JS (2007). Microglia-mediated neurotoxicity: uncovering the molecular mechanisms. Nat Rev Neurosci 8: 57–69. https://doi.org/10.1038/NRN2038.

Borish LC, Steinke JW (2003). 2. Cytokines and chemokines. J Allergy Clin Immunol 111. https://doi.org/10.1067/mai.2003.108.

Borst K, Dumas AA, Prinz M (2021). Microglia: immune and non-immune functions. Immunity 54: 2194–2208. https://doi.org/10.1016/J.IMMUNI.2021.09.014.

Bradl M, Lassmann H (2010). Oligodendrocytes: biology and pathology. Acta Neuropathol 119: 37–53. https://doi.org/10.1007/S00401-009-0601-5.

Brändle SM, Cerina M, Weber S et al. (2021). Cross-reactivity of a pathogenic autoantibody to a tumor antigen in GABAA receptor encephalitis. Proc Natl Acad Sci U S A 118. https://doi.org/10.1073/PNAS.1916337118/-/DCSUPPLEMENTAL.

Brinkmann BG, Agarwal A, Sereda MW et al. (2008). Neuregulin-1/ErbB signaling serves distinct functions in myelination of the peripheral and central nervous system. Neuron 59: 581–595. https://doi.org/10.1016/J.NEURON.2008.06.028.

Byun JI, Lee ST, Moon J et al. (2016). Distinct intrathecal interleukin-17/interleukin-6 activation in anti-N-methyl-d-aspartate receptor encephalitis. J Neuroimmunol 297: 141–147. https://doi.org/10.1016/J.JNEUROIM.2016.05.023.

Ciano-Petersen NL, Cabezudo-García P, Muñiz-Castrillo S et al. (2021). Current status of biomarkers in anti-N-methyl-D-aspartate receptor encephalitis. Int J Mol Sci 22: 13127. https://doi.org/10.3390/IJMS222313127.

Ciano-Petersen NL, Muñiz-Castrillo S, Birzu C et al. (2022). Cytokine dynamics and targeted immunotherapies in autoimmune encephalitis. Brain Commun 4. https://doi.org/10.1093/BRAINCOMMS/FCAC196.

Corbali O, Chitnis T (2023). Pathophysiology of myelin oligodendrocyte glycoprotein antibody disease. Front Neurol 14. https://doi.org/10.3389/FNEUR.2023.1137998.

Cosenza-Nashat M, Zhao ML, Suh HS et al. (2009). Expression of the translocator protein of 18 kDa by microglia, macrophages and astrocytes based on immunohistochemical localization in abnormal human brain. Neuropathol Appl Neurobiol 35: 306. https://doi.org/10.1111/J.1365-2990.2008.01006.X.

Czeh M, Gressens P, Kaindl AM (2011). The yin and yang of microglia. Dev Neurosci 33: 199–209. https://doi.org/10.1159/000328989.

Dalkara T, Gursoy-Ozdemir Y, Yemisci M (2011). Brain microvascular pericytes in health and disease. Acta Neuropathol 122: 1–9. https://doi.org/10.1007/S00401-011-0847-6.

Dalmau J, Graus F (2018). Antibody-mediated encephalitis. N Engl J Med 378: 840–851. https://doi.org/10.1056/NEJMRA1708712.

Dalmau J, Graus F (2023). Diagnostic criteria for autoimmune encephalitis: utility and pitfalls for antibody-negative disease. Lancet Neurol 22: 529–540. https://doi.org/10.1016/S1474-4422(23)00083-2.

Dalmau J, Tüzün E, Wu HY et al. (2007). Paraneoplastic anti-N-methyl-D-aspartate receptor encephalitis associated with ovarian teratoma. Ann Neurol. https://doi.org/10.1002/ana.21050.

Dalmau J, Gleichman AJ, Hughes EG et al. (2008). Anti-NMDA-receptor encephalitis: case series and analysis of the effects of antibodies. Lancet Neurol 7: 1091–1098. https://doi.org/10.1016/S1474-4422(08)70224-2.

Dalmau J, Geis C, Graus F (2017). Autoantibodies to synaptic receptors and neuronal cell surface proteins in autoimmune diseases of the central nervous system. Physiol Rev 97: 839–887. https://doi.org/10.1152/PHYSREV.00010.2016.

Darnell JE (1997). STATs and gene regulation. Science 277: 1630–1635. https://doi.org/10.1126/SCIENCE.277.5332.1630.

Dauvilliers Y, Bauer J, Rigau V et al. (2013). Hypothalamic immunopathology in anti-ma–associated diencephalitis with narcolepsy-cataplexy. JAMA Neurol 70: 1305–1310. https://doi.org/10.1001/JAMANEUROL.2013.2831.

Deng B, Liu XN, Li X et al. (2017). Raised cerebrospinal fluid BAFF and APRIL levels in anti-N-methyl-d-aspartate receptor encephalitis: correlation with clinical outcome. J Neuroimmunol 305: 84–91. https://doi.org/10.1016/J.JNEUROIM.2017.01.012.

Devinsky O, Vezzani A, Najjar S et al. (2013). Glia and epilepsy: excitability and inflammation. Trends Neurosci 36: 174–184. https://doi.org/10.1016/J.TINS.2012.11.008.

Dinarello CA (2007). Historical insights into cytokines. Eur J Immunol 37: 1. https://doi.org/10.1002/EJI.200737772.

Dunstan EJ, Winer JB (2006). Autoimmune limbic encephalitis causing fits, rapidly progressive confusion and hyponatraemia. Age Ageing 35: 536–537. https://doi.org/10.1093/AGEING/AFL045.

Escartin C, Galea E, Lakatos A et al. (2021). Reactive astrocyte nomenclature, definitions, and future directions. Nat Neurosci 24: 312–325. https://doi.org/10.1038/S41593-020-00783-4.

Fang B, McKeon A, Hinson SR et al. (2016). Autoimmune glial fibrillary acidic protein astrocytopathy: a novel meningoencephalomyelitis. JAMA Neurol. https://doi.org/10.1001/jamaneurol.2016.2549.

Faustmann PM, Haase CG, Romberg S et al. (2003). Microglia activation influences dye coupling and Cx43 expression of the astrocytic network. Glia 42: 101–108. https://doi.org/10.1002/GLIA.10141.

Fitzner D, Schneider A, Kippert A et al. (2006). Myelin basic protein-dependent plasma membrane reorganization in the formation of myelin. EMBO J 25: 5037–5048. https://doi.org/10.1038/SJ.EMBOJ.7601376.

Flanagan EP, Hinson SR, Lennon VA et al. (2017). Glial fibrillary acidic protein immunoglobulin G as biomarker of autoimmune astrocytopathy: analysis of 102 patients. Ann Neurol 81: 298–309. https://doi.org/10.1002/ANA.24881.

Fraser DD, Mudrick-Donnon LA, Macvicar BA (1994). Astrocytic GABA receptors. Glia 11: 83–93. https://doi.org/10.1002/GLIA.440110203.

Fraser DD, Duffy S, Angelides KJ et al. (1995). GABAA/benzodiazepine receptors in acutely isolated hippocampal astrocytes. J Neurosci 15: 2720–2732. https://doi.org/10.1523/JNEUROSCI.15-04-02720.1995.

Fu J, Peng L, Yang Y et al. (2022). Case report: overlapping syndrome mimicking infectious meningoencephalitis in a patient with coexistent MOG, NMDAR, mGluR5 antibody positivity. Front Immunol 13. https://doi.org/10.3389/FIMMU.2022.919125.

Gallus M, Roll W, Dik A et al. (2023). Translational imaging of TSPO reveals pronounced innate inflammation in human and murine CD8 T cell–mediated limbic encephalitis. Sci Adv 9. https://doi.org/10.1126/SCIADV.ABQ7595.

Gibson EM, Purger D, Mount CW et al. (2014). Neuronal activity promotes oligodendrogenesis and adaptive myelination in the mammalian brain. Science 344. https://doi.org/10.1126/SCIENCE.1252304.

Ginhoux F, Prinz M (2015). Origin of microglia: current concepts and past controversies. Cold Spring Harb Perspect Biol 7. https://doi.org/10.1101/CSHPERSPECT.A020537.

Gong X, Wang N, Zhu H et al. (2023). Anti-NMDAR antibodies, the blood–brain barrier, and anti-NMDAR encephalitis. Front Neurol 14. https://doi.org/10.3389/FNEUR.2023.1283511.

Graus F, Titulaer MJ, Balu R et al. (2016). A clinical approach to diagnosis of autoimmune encephalitis. Lancet Neurol 15: 391–404. https://doi.org/10.1016/S1474-4422(15)00401-9.

Höftberger R, Guo Y, Flanagan EP et al. (2020). The pathology of central nervous system inflammatory demyelinating disease accompanying myelin oligodendrocyte glycoprotein autoantibody. Acta Neuropathol 139: 875. https://doi.org/10.1007/S00401-020-02132-Y.

Irani SR, Michell AW, Lang B et al. (2011). Faciobrachial dystonic seizures precede Lgi1 antibody limbic encephalitis. Ann Neurol 69: 892–900. https://doi.org/10.1002/ANA.22307.

Ismail FS, Faustmann PM (2020). Astrocytes and their potential role in anti-NMDA receptor encephalitis. Med Hypotheses 139. https://doi.org/10.1016/j.mehy.2020.109612.

Ismail FS, Faustmann PM (2023). The potential role of astroglial GABAA receptors in autoimmune encephalitis associated with GABAA receptor antibodies and seizures. Epilepsia Open 8. https://doi.org/10.1002/EPI4.12750.

Ismail FS, Corvace F, Faustmann PM et al. (2021a). Pharmacological investigations in glia culture model of inflammation. Front Cell Neurosci 15. https://doi.org/10.3389/FNCEL.2021.805755.

Ismail FS, Meuth SG, Melzer N (2021b). The role of dendritic cells and their interactions in the pathogenesis of antibody-associated autoimmune encephalitis. J Neuroinflammation 18. https://doi.org/10.1186/S12974-021-02310-Z.

Jarius S, Paul F, Aktas O et al. (2018). MOG encephalomyelitis: international recommendations on diagnosis and antibody testing. J Neuroinflammation 15. https://doi.org/10.1186/S12974-018-1144-2.

Johns TG, Bernard CCA (1999). The structure and function of myelin oligodendrocyte glycoprotein. J Neurochem 72: 1–9. https://doi.org/10.1046/J.1471-4159.1999.0720001.X.

Kaneko K, Sato DK, Nakashima I et al. (2016). Myelin injury without astrocytopathy in neuroinflammatory disorders with MOG antibodies. J Neurol Neurosurg Psychiatry 87: 1257–1259. https://doi.org/10.1136/JNNP-2015-312676.

Kaplan MR, Cho MH, Ullian EM et al. (2001). Differential control of clustering of the sodium channels Na(v)1.2 and Na(v)1.6 at developing CNS nodes of Ranvier. Neuron 30: 105–119. https://doi.org/10.1016/S0896-6273(01)00266-5.

Kettenmann H, Hanisch UK, Noda M et al. (2011). Physiology of microglia. Physiol Rev 91: 461–553. https://doi.org/10.1152/PHYSREV.00011.2010.

Kimura A, Yoshikura N, Koumura A et al. (2015). B-cell-activating factor belonging to the tumor necrosis factor family (BAFF) and a proliferation-inducing ligand (APRIL) levels in cerebrospinal fluid of patients with meningoencephalitis. J Neurol Sci 352: 79–83. https://doi.org/10.1016/J.JNS.2015.03.036.

Kothur K, Wienholt L, Mohammad SS et al. (2016). Utility of CSF cytokine/chemokines as markers of active intrathecal inflammation: comparison of demyelinating, anti-NMDAR and enteroviral encephalitis. PLoS One 11. https://doi.org/10.1371/JOURNAL.PONE.0161656.

Kuehn JC, Scheuerle A, Bauer J et al. (2020). A 64-year-old patient with a mesiotemporal mass and symptomatic epilepsy. Brain Pathol 30: 413. https://doi.org/10.1111/BPA.12818.

Kunchok A, Zekeridou A, McKeon A (2019). Autoimmune glial fibrillary acidic protein astrocytopathy. Curr Opin Neurol 32: 452. https://doi.org/10.1097/WCO.0000000000000676.

Kunchok A, Flanagan EP, Krecke KN et al. (2021). MOG-IgG1 and coexistence of neuronal autoantibodies. Mult Scler 27: 1175. https://doi.org/10.1177/1352458520951046.

Lee WJ, Lee HS, Kim DY et al. (2022). Seronegative autoimmune encephalitis: clinical characteristics and factors associated with outcomes. Brain 145: 3509–3521. https://doi.org/10.1093/BRAIN/AWAC166.

Lee WJ, Kwon YN, Kim B et al. (2023). MOG antibody-associated encephalitis in adult: clinical phenotypes and outcomes. J Neurol Neurosurg Psychiatry 94: 102–112. https://doi.org/10.1136/JNNP-2022-330074.

Levy DE, Lee C (2002). What does Stat3 do? J Clin Invest 109: 1143. https://doi.org/10.1172/JCI15650.

Leypoldt F, Höftberger R, Titulaer MJ et al. (2015). Investigations on CXCL13 in anti–N-methyl-D-aspartate receptor encephalitis: a potential biomarker of treatment response. JAMA Neurol 72: 180. https://doi.org/10.1001/JAMANEUROL.2014.2956.

Li K, Yu W, Cao R et al. (2017). Microglia-mediated BAFF-BAFFR ligation promotes neuronal survival in brain ischemia injury. Neuroscience 363: 87–96. https://doi.org/10.1016/J.NEUROSCIENCE.2017.09.007.

Liba Z, Kayserova J, Elisak M et al. (2016). Anti-N-methyl-D-aspartate receptor encephalitis: the clinical course in light of the chemokine and cytokine levels in cerebrospinal fluid. J Neuroinflammation 13. https://doi.org/10.1186/S12974-016-0507-9.

Lipton SA (2006). NMDA receptors, glial cells, and clinical medicine. Neuron 50: 9–11. https://doi.org/10.1016/J.NEURON.2006.03.026.

Liu J, Feng X, Wang Y et al. (2022). Astrocytes: GABAceptive and GABAergic cells in the brain. Front Cell Neurosci 16. https://doi.org/10.3389/FNCEL.2022.892497.

Lucchinetti CF, Guo Y, Popescu BF\G et al. (2014). The pathology of an autoimmune astrocytopathy: lessons learned from neuromyelitis optica, in: brain pathology. https://doi.org/10.1111/bpa.12099.

Mallat M, Marín-Teva JL, Chéret C (2005). Phagocytosis in the developing CNS: more than clearing the corpses. Curr Opin Neurobiol 15: 101–107. https://doi.org/10.1016/J.CONB.2005.01.006.

Marignier R, Nicolle A, Watrin C et al. (2010). Oligodendrocytes are damaged by neuromyelitis optica immunoglobulin G via astrocyte injury. Brain 133: 2578–2591. https://doi.org/10.1093/BRAIN/AWQ177.

Marignier R, Hacohen Y, Cobo-Calvo A et al. (2021). Myelin-oligodendrocyte glycoprotein antibody-associated disease. Lancet Neurol 20: 762–772. https://doi.org/10.1016/S1474-4422(21)00218-0.

Martinez-Hernandez E, Guasp M, García-Serra A et al. (2020). Clinical significance of anti-NMDAR concurrent with glial or neuronal surface antibodies. Neurology 94: E2302–E2310. https://doi.org/10.1212/WNL.0000000000009239.

Matute C, Palma A, Serrano-Regal MP et al. (2020). N-methyl-D-aspartate receptor antibodies in autoimmune encephalopathy Alter oligodendrocyte function. Ann Neurol 87: 670–676. https://doi.org/10.1002/ANA.25699.

Mederos S, Perea G (2019). GABAergic-astrocyte signaling: a refinement of inhibitory brain networks. Glia 67: 1842–1851. https://doi.org/10.1002/GLIA.23644.

Menke AF, Ismail FS, Dornmair K et al. (2022). GABA a receptor autoantibodies decrease GABAergic synaptic transmission in the hippocampal CA3 network. Int J Mol Sci 23: 3707. https://doi.org/10.3390/IJMS23073707.

Morris GP, Foster CG, Courtney JM et al. (2023). Microglia directly associate with pericytes in the central nervous system. Glia 71: 1847–1869. https://doi.org/10.1002/GLIA.24371.

Nayak D, Roth TL, McGavern DB (2014). Microglia development and function. Annu Rev Immunol 32: 367–402. https://doi.org/10.1146/ANNUREV-IMMUNOL-032713-120240.

Noviello CM, Kreye J, Teng J et al. (2022). Structural mechanisms of GABAA receptor autoimmune encephalitis. Cell 185: 2469–2477. e13. https://doi.org/10.1016/J.CELL.2022.06.025.

O'Connor K, Waters P, Komorowski L et al. (2019). GABAA receptor autoimmunity: a multicenter experience. Neurol Neuroimmunol Neuroinflamm 6. https://doi.org/10.1212/NXI.0000000000000552.

Osanai Y, Yamazaki R, Shinohara Y et al. (2022). Heterogeneity and regulation of oligodendrocyte morphology. Front Cell Dev Biol 10. https://doi.org/10.3389/FCELL.2022.1030486.

Paolicelli RC, Sierra A, Stevens B et al. (2022). Microglia states and nomenclature: a field at its crossroads. Neuron 110: 3458–3483. https://doi.org/10.1016/J.NEURON.2022.10.020.

Park DC, Murman DL, Perry KD et al. (2007). An autopsy case of limbic encephalitis with voltage-gated potassium channel antibodies. Eur J Neurol 14. https://doi.org/10.1111/J.1468-1331.2007.01924.X.

Pérez-Cerdá F, Sánchez-Gómez MV, Matute C (2015). Pío del Río Hortega and the discovery of the oligodendrocytes. Front Neuroanat 9. https://doi.org/10.3389/FNANA.2015.00092.

Perry VH, Gordon S (1991). Macrophages and the nervous system. Int Rev Cytol 125: 203–244. https://doi.org/10.1016/S0074-7696(08)61220-6.

Petit-Pedrol M, Armangue T, Peng X et al. (2014). Encephalitis with refractory seizures, status epilepticus, and antibodies to the GABAA receptor: a case series, characterisation of the antigen, and analysis of the effects of antibodies. Lancet Neurol 13: 276–286. https://doi.org/10.1016/S1474-4422(13)70299-0.

Pettingill P, Kramer HB, Coebergh JA et al. (2015). Antibodies to GABAA receptor $\alpha 1$ and $\gamma 2$ subunits: clinical and serologic characterization. Neurology 84: 1233–1241. https://doi.org/10.1212/WNL.0000000000001326.

Pitsch J, van Loo KMJ, Gallus M et al. (2021). CD8+ T-lymphocyte-driven limbic encephalitis results in temporal lobe epilepsy. Ann Neurol 89: 666–685. https://doi.org/10.1002/ANA.26000.

Pollak TA, Lennox BR, Müller S et al. (2020). Autoimmune psychosis: an international consensus on an approach to the diagnosis and management of psychosis of suspected autoimmune origin. Lancet Psychiatry 7: 93–108. https://doi.org/10.1016/S2215-0366(19)30290-1.

Prinz M, Jung S, Priller J (2019). Microglia biology: one century of evolving concepts. Cell 179: 292–311. https://doi.org/10.1016/J.CELL.2019.08.053.

Ramanathan S, Dale RC, Brilot F (2016). Anti-MOG antibody: the history, clinical phenotype, and pathogenicity of a serum biomarker for demyelination. Autoimmun Rev 15: 307–324. https://doi.org/10.1016/J.AUTREV.2015.12.004.

Ramberger M, Bsteh G, Schanda K et al. (2015). NMDA receptor antibodies: a rare association in inflammatory demyelinating diseases. Neurol Neuroimmunol Neuroinflamm 2: e141. https://doi.org/10.1212/NXI.0000000000000141.

Saab AS, Tzvetavona ID, Trevisiol A et al. (2016). Oligodendroglial NMDA receptors regulate glucose import and axonal energy metabolism. Neuron 91: 119–132. https://doi.org/10.1016/J.NEURON.2016.05.016.

Shan F, Long Y, Qiu W (2018). Autoimmune glial fibrillary acidic protein Astrocytopathy: a review of the literature. Front Immunol 9. https://doi.org/10.3389/FIMMU.2018.02802.

Skowrońska K, Obara-Michlewska M, Zielińska M et al. (2019). NMDA receptors in astrocytes: in search for roles in neurotransmission and astrocytic homeostasis. Int J Mol Sci 20: 309. https://doi.org/10.3390/IJMS20020309.

Sofroniew MV (2009). Molecular dissection of reactive astrogliosis and glial scar formation. Trends Neurosci. https://doi.org/10.1016/j.tins.2009.08.002.

Sofroniew MV (2014). Astrogliosis. Cold Spring Harb Perspect Biol 7. https://doi.org/10.1101/CSHPERSPECT.A020420.

Sofroniew MV (2020). Astrocyte reactivity: subtypes, states, and functions in CNS innate immunity. Trends Immunol 41: 758–770. https://doi.org/10.1016/J.IT.2020.07.004.

Sofroniew MV, Vinters HV (2010). Astrocytes: biology and pathology. Acta Neuropathol 119: 7–35. https://doi.org/10.1007/S00401-009-0619-8.

Spatola M, Petit-Pedrol M, Simabukuro MM et al. (2017). Investigations in GABAA receptor antibody-associated encephalitis. Neurology 88: 1012–1020. https://doi.org/10.1212/WNL.0000000000003713.

Streit WJ (2001). Microglia and macrophages in the developing CNS. Neurotoxicology 22: 619–624. https://doi.org/10.1016/S0161-813X(01)00033-X.

Takai Y, Misu T, Kaneko K et al. (2020). Myelin oligodendrocyte glycoprotein antibody-associated disease: an immunopathological study. Brain 143: 1431–1446. https://doi.org/10.1093/BRAIN/AWAA102.

Thangarajh M, Masterman T, Hillert J et al. (2007). A proliferation-inducing ligand (APRIL) is expressed by astrocytes and is increased in multiple sclerosis. Scand J Immunol 65: 92–98. https://doi.org/10.1111/J.1365-3083.2006.01867.X.

Titulaer MJ, Höftberger R, Iizuka T et al. (2014). Overlapping demyelinating syndromes and anti–N-methyl-D-aspartate receptor encephalitis. Ann Neurol 75: 411–428. https://doi.org/10.1002/ANA.24117.

Török O, Schreiner B, Schaffenrath J et al. (2021). Pericytes regulate vascular immune homeostasis in the CNS. Proc Natl Acad Sci U S A 118. https://doi.org/10.1073/PNAS.2016587118/-/DCSUPPLEMENTAL.

Tüzün E, Zhou L, Baehring JM et al. (2009). Evidence for antibody-mediated pathogenesis in anti-NMDAR encephalitis associated with ovarian teratoma. Acta Neuropathol 118: 737–743. https://doi.org/10.1007/S00401-009-0582-4.

Untiet V, Beinlich FRM, Kusk P et al. (2023). Astrocytic chloride is brain state dependent and modulates inhibitory neurotransmission in mice. Nat Commun 14: 1871. https://doi.org/10.1038/S41467-023-37433-9.

van Casteren ACM, Ackermann F, Rahman KA et al. (2022). Differential modes of action of $\alpha 1$- and $\alpha 1\gamma 2$-autoantibodies derived from patients with GABAAR encephalitis. eNeuro 9. https://doi.org/10.1523/ENEURO.0369-22.2022.

Verkhratsky A, Kirchhoff F (2007). NMDA receptors in glia. Neuroscientist 13: 28–37. https://doi.org/10.1177/1073858406294270.

Verkhratsky A, Nedergaard M (2018). Physiology of astroglia. Physiol Rev 98: 239–389. https://doi.org/10.1152/PHYSREV.00042.2016.

Verkhratsky A, Butt A, Li B et al. (2023). Astrocytes in human central nervous system diseases: a frontier for new therapies. Signal Transduct Target Ther 8. https://doi.org/10.1038/S41392-023-01628-9.

Vincent A, Buckley C, Schott JM et al. (2004). Potassium channel antibody-associated encephalopathy: a potentially immunotherapy-responsive form of limbic encephalitis. Brain 127: 701–712. https://doi.org/10.1093/BRAIN/AWH077.

Walter L, Neumann H (2009). Role of microglia in neuronal degeneration and regeneration. Semin Immunopathol 31: 513–525. https://doi.org/10.1007/S00281-009-0180-5.

Wolf SA, Boddeke HWGM, Kettenmann H (2017). Microglia in physiology and disease. Annu Rev Physiol 79: 619–643. https://doi.org/10.1146/ANNUREV-PHYSIOL-022516-034406.

Yamakawa M, Hogan KO, Leever J et al. (2021). Autopsy case of meningoencephalomyelitis associated with glial fibrillary acidic protein antibody. Neurol Neuroimmunol Neuroinflamm 8. https://doi.org/10.1212/NXI.0000000000001081.

Yang X, Huang Q, Yang H et al. (2019). Astrocytic damage in glial fibrillary acidic protein astrocytopathy during initial attack. Mult Scler Relat Disord 29: 94–99. https://doi.org/10.1016/J.MSARD.2019.01.036.

Handbook of Clinical Neurology, Vol. 210 (3rd series)
Neuroglia in Neurologic and Psychiatric Disorders, Part II
A. Verkhratsky, L.D. de Witte, E. Aronica, and E.M. Hol, Editors
https://doi.org/10.1016/B978-0-443-19102-2.00032-6

Chapter 10

Neuroglia in leukodystrophies

ELLY M. HOL[1], WERNER DYKSTRA[1], JULIETTE CHEVALIER[2], ELOY CUADRADO[1], MARIANNA BUGIANI[2], ELEONORA ARONICA[3], AND ALEXEI VERKHRATSKY[4,5,6*]

[1]*Department of Translational Neuroscience, University Medical Center Utrecht Brain Center, Utrecht University, Utrecht, The Netherlands*

[2]*Department of Child Neurology and Pathology, Amsterdam Leukodystrophy Center, Amsterdam University Medical Center, Amsterdam, The Netherlands*

[3]*Department of (Neuro)Pathology, Amsterdam UMC, University of Amsterdam, Amsterdam Neuroscience, Amsterdam, The Netherlands*

[4]*Faculty of Biology, Medicine and Health, University of Manchester, Manchester, United Kingdom*

[5]*Department of Neurosciences, University of the Basque Country UPV/EHU and CIBERNED, Leioa, Bizkaia, Spain*

[6]*IKERBASQUE, Basque Foundation for Science, Bilbao, Spain*

Abstract

Leukodystrophies are a heterogeneous group of rare genetic neurologic disorders characterized by white matter degeneration resulting from mutations affecting glial cells. This review focuses on the primary subtypes—astroglial, oligodendroglial, and microglial leukodystrophies—offering a detailed description of their neuropathologic features and clinical manifestations. It delves into key aspects of the pathogenesis, emphasizing the distinct cellular mechanisms that drive white matter damage. Advances in disease modeling, including the development of animal models with pathologic gene expressions and patient-derived iPS-cell models, have significantly enhanced our understanding of these rare disorders. Insights into the roles of different glial cell types highlight the complexity of leukodystrophies and provide a foundation for the development of targeted therapeutic strategies.

INTRODUCTION

Leukodystrophies are a heterogeneous class of inherited diseases defined by the degeneration of white matter and myelin. The majority of leukodystrophies are associated with gliopathies, with oligodendrocytes, astrocytes, or microglia being the primary drivers of the pathogenesis (van der Knaap and Bugiani, 2017). The first clinical descriptions of inherited white matter pathology (defined as "a diffuse sclerosis with myelin lesions and white matter sclerosis") were provided by Friedrich Christoph Pelizaeus (Pelizaeus, 1898) and Ludwig Merzbacher (Merzbacher, 1910). The term "leukodystrophy" (from Greek λευκός—white and δυστροφία—*abnormal nourishing*) was introduced by Max Bielschowsky and Richard Henneberg in 1928 (Bielschowsky and Henneberg, 1928), who described familial diffuse sclerosis, naming it *Leukodystrophia Cerebri Progressiva Hereditaria*.

Leukodystrophies were traditionally defined as genetic and clinically progressive white matter disorders characterized by deficient myelin deposition or myelin loss, caused by defects in either myelin composition or in myelin-forming cells, the oligodendrocytes (Morell, 1984; Seitelberger, 1984). When this definition was formulated in the 1980s, the genetic causes of most leukodystrophies were unknown, and magnetic resonance imaging (MRI) was not yet a part of the routine clinical practice. These leukodystrophies were classified into hypomyelinating leukodystrophies with persistent lack of myelin, demyelinating disorders with loss of previously normally deposited myelin, and demyelinating diseases with myelin vacuolation. With the introduction of MRI into clinical practice at the end of the 1980s, it became obvious that MRI was highly sensitive to white matter signal changes, and a pattern recognition approach was developed using this imaging technique (van der Knaap et al., 1999). This approach allowed grouping patients with unknown leukodystrophies based on MRI patterns, accelerating the genetic testing procedure.

In the last decade, next generation sequencing has allowed the identification of an increasing number of genetically characterized leukodystrophies. It also demonstrated that most leukodystrophies are caused by pathogenic variants in genes encoding proteins

*Correspondence to: Professor Alexei Verkhratsky, Faculty of Biology, Medicine and Health, The University of Manchester, Oxford Road, Manchester, M13 9PT, United Kingdom. Tel: +44-1612755414, E-mail: Alexej.Verkhratsky@manchester.ac.uk

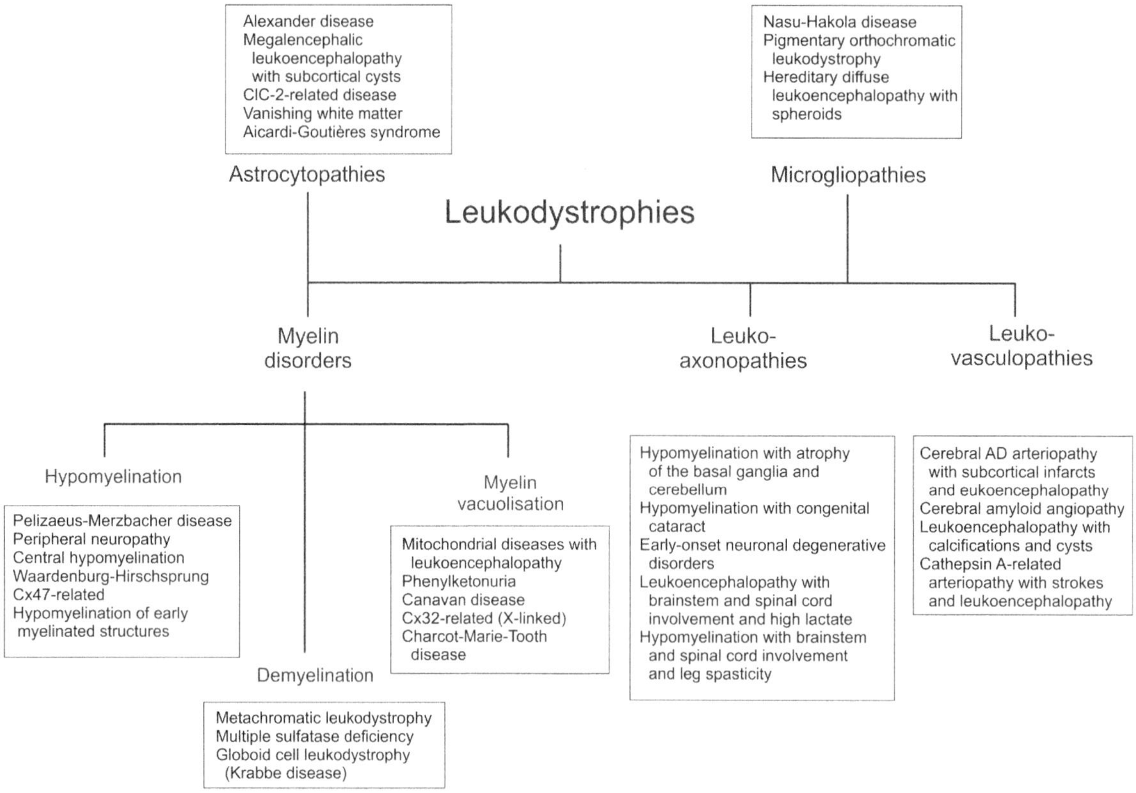

Fig. 10.1. Classification of leukodystrophies. Reproduced, with permission from Verkhratsky A, Butt AM (2023). Neuroglia: function and pathology, Elsevier, New York.

specific to cells other than the oligodendrocytes (Kevelam et al., 2016), making the traditional definition of leukodystrophies outdated. A new definition was introduced: leukodystrophies are genetically determined white matter disorders with onset at any age and a variable clinical course due to defects in any of the structural components of the white matter, including myelin and oligodendrocytes, as well as astrocytes, microglia, axons, and blood vessels (Kevelam et al., 2016). This definition is also upheld by basic research of the past three decades that revealed that developmental myelination, myelin maintenance, and remyelination are not solely functions of oligodendrocytes but are considerably influenced by the interactions between all components of white matter (van der Knaap and Bugiani, 2017). As a consequence, a new classification of leukodystrophies was proposed by van der Knaap and Bugiani (2017). In addition to myelin and oligodendrocyte-related disorders, this classification distinguishes astrocytic leukodystrophies, microgliopathic leukodystrophies, leuko-axonopathies, and leuko-vasculopathies, based on the driving cellular pathology and recognizing the possibility that a single disease might fall into more than one category (Fig. 10.1).

This chapter describes several major leukodystrophies, with particular emphasis on astroglial, oligodendroglial, and microglial leukodystrophies.

ASTROGLIAL LEUKODYSTROPHIES

Astrocytes are the central elements of pathogenesis of astroglial leukodystrophies. All these are rare disorders caused by mutations in either genes highly expressed in astrocytes (*GFAP*, *HEPACAM/GlialCAM*, and *MLC1*), more general genes (*EIF2B1*, *EIF2B2*, *EIF2B3*, *EIF2B4*, and *EIF2B5*), or genes involved in the immune response (*TREX1*, *RNASEH2B*, *RNASEH2C*, *RNASEH2A*, *SAMHD1*, *ADAR*, *IFIH1*, *LSM11*, and *RNU7–1*). The main nosological forms are represented by Alexander disease (AxD) caused by mutations in glial fibrillary acidic protein (*GFAP*) (Messing, 2018), vanishing white matter (VWM) disease linked to mutations in elongation initiation factor genes *EIF2B1*, *EIF2B2*, *EIF2B3*, *EIF2B4*, and *EIF2B5* (van der Knaap et al., 2006a), megalencephalic leukoencephalopathy with subcortical cysts (MLC) associated with mutations in *MLC1* or *GlialCAM* (van der Knaap et al., 2012), and Aicardi–Goutières syndrome (AGS) triggered by mutations in *TREX1*, *RNASEH2B*, *RNASEH2C*, *RNASEH2A*, *SAMHD1*, *ADAR*, *IFIH1*, *LSM11*, and *RNU7–1* (Livingston and Crow, 2016).

Alexander disease

Alexander disease (AxD) is a fatal primary astrocytopathy whereby dominant missense mutations in *GFAP*, encoding GFAP, account for nearly all cases (Brenner et al., 2001). The disorder was first described in 1949 by William Stewart Alexander, who published a report about a child with progressive fibrinoid degeneration of fibrillary astrocytes associated with mental retardation, megalencephaly, and hydrocephaly (Alexander, 1949). The condition was characterized by a lack or loss of myelin, and astrocytic eosinophilic deposits concentrating in the perivascular tissue and the pial surface of the brain. These hallmark protein aggregates were recognized as Rosenthal fibers (RFs), named after Werner Rosenthal who described them in 1898 (Rosenthal, 1898) in spinal cord tumors. The association of these fibers with astrocytes

Table 10.1

Classification of Alexander disease (AxD).

Age-based classification			
Neonatal (1%)	Infantile (42%)	Juvenile (22%)	Adult (33%)
Seizures	Seizures	Seizures	Spasticity
Severe motor and intellectual disability	Progressive psychomotor retardation	Gradual loss of intellectual function	Bulbar/pseudobulbar signs (palatal myoclonus, dysphagia, dysphonia, dysarthria)
		Ataxia	
Hydrocephalus	Loss of developmental milestones	Bulbar/pseudobulbar signs	Ataxia
	Megalencephaly	Breathing difficulties	Gait disturbance
	Frontal bossing		Cerebellar signs
			Sleep apnea

Phenotype-based classification	
Type I AxD	Type II AxD
Early age at onset (often before 4 years)	Manifests across the lifespan
Seizures	Autonomic dysfunction
Macrocephaly	Bulbar symptoms
Encephalopathy	Ocular movement abnormalities
Paroxysmal deterioration	Palatal myoclonus
Failures to thrive	Often negative for neurocognitive or developmental deficits
Developmental delay	
Classic radiologic features	Atypical radiologic features
Frontal predominance	Hindbrain predominance
Mean survival of 14.5 years	Mean survival of 25 years

and AxD was revealed in 1969 based on electron microscope studies (Seil et al., 1968). Appearance of white matter RFs correlated with myelin loss, thus defining AxD as a leukodystrophy (Wohlwill et al., 1959).

In 1964, this disease was named after Alexander (Friede, 1964) and, in 1998, it was found that mice overexpressing wild-type GFAP exhibited RFs (Messing et al., 1998). This toxic gain-of-function of GFAP implicated *GFAP* as a possible cause of the disease and led to the discovery that nearly all patients had heterozygous mutations in GFAP (Brenner et al., 2001). Many mutations, spanning almost the entire *GFAP* gene, have been reported since (Messing et al., 2012). Rare cases as a result of alternative splicing have also been described (Flint et al., 2012; Graff-Radford et al., 2014; Green et al., 2018; Helman et al., 2020). Even though most mutations are de novo, resulting in early-onset AxD, late-onset patients develop symptoms after their childbearing years and thus have a 50% chance of passing on a *GFAP* mutation. Since GFAP is primarily expressed in astrocytes, AxD became the first known primary genetic astrocytopathy. Moreover, *GFAP* mutations result in reactive astrogliosis and a marked inflammatory environment (Hagemann et al., 2005), which is a similar glial response occurring in many neurologic diseases (Pekny et al., 2016; Escartin et al., 2021).

AxD patients typically present with white matter abnormalities, developmental delay, and/or progressive motor, intellectual disability, and seizures, eventually culminating in death (Messing et al., 2012). To gain a better understanding of the disease variability between patients, a classification of AxD has been proposed (Table 10.1). Using the age-based classification, AxD is categorized by the proportion of patients, who have different phenotypes during a particular time of their life, thereby distinguishing between neonatal, infantile, juvenile, and adult cases (Russo et al., 1976; Srivastava et al., 1993). Alternatively, AxD can be classified by symptoms and MRI, which is called the phenotype-based description (Prust et al., 2011), whereby type I and type II are distinguished. Type I is characterized by early onset, developmental delay, seizures, frontal predominance, and a mean survival of 14.5 years. Type II is characterized by onset at any age, autonomic malfunction, hindbrain predominance, and a mean survival of 25 years. A third classification has been proposed that is also phenotype-based, but distinguishes a third "intermediate" form in addition to the prefrontal or hindbrain predominant forms (Yoshida et al., 2011).

To establish reliable biomarkers for AxD, collecting cerebrospinal fluid (CSF) by lumbar punctures and blood collection are used to gain insight into whether higher levels of GFAP correlate with disease progression, whether GFAP is always elevated, whether GFAP levels change over time, whether elevated GFAP correlates with clinical endpoints, and whether GFAP predicts disability. Elevated levels of GFAP have been found in the CSF of early-onset and late-onset AxD patients (Kyllerman et al., 2005; Jany et al., 2015). In blood samples of infantile and juvenile patients, but not in adult patients, GFAP levels were also elevated. This could indicate that the degree of GFAP increase correlates with the severity of the disease. Moreover, elevated levels of GFAP were observed in postmortem tissue of AxD patients (Tang et al., 2008; Walker et al., 2014). The mechanism by which GFAP is secreted from astrocytes into the CSF and blood is still unclear. One possible way of GFAP secretion could be through exosomes, since in glioma patients, GFAP-positive exosomes were found in the serum after survivin vaccination (Galbo et al., 2017). Another possible way could be through the glymphatic system, since biomarkers of traumatic brain injury were shown to be transported from the brain to the blood by glymphatic flow (Plog et al., 2015). This is relevant for AxD, because RFs are concentrated perivascularly, which could indicate that the brain is using the glymphatic system to clear the excess GFAP. It should be noted, however, that it is still unclear whether these mechanisms also contribute to GFAP secretion in AxD. The establishment of reliable biomarkers can also be used to assess the effect of drugs and other treatments such as antisense oligonucleotides (ASOs) that were recently successfully utilized in experiments with AxD mice and rats (Hagemann et al., 2018). In AxD mice, ASO treatment reduced RFs, stress, and immune markers, activated GFAP promotor, and normalized body weight (Hagemann et al., 2018). In AxD rats, the hallmark white matter pathology was present and could be prevented and even rescued with ASO treatment, clearly indicating that reducing the amounts of toxic GFAP is beneficial (Hagemann et al., 2021). Currently, phase 1–3 clinical trials for treating AxD patients with ASOs are in progress (https://clinicaltrials.gov/study/NCT04849741).

Neuropathologically, AxD is characterized by an upregulation of *GFAP*, an accumulation of GFAP in excessive amounts, the formation of RFs (Alexander, 1949), reactive astrogliosis and microgliosis, and myelin abnormalities (van der Knaap et al., 2001). RFs are cytoplasmic inclusions inside astrocytes. Their size varies between 4 and 20 μm width and can be as long as 100 μm. They accumulate throughout the cerebral cortex, white matter, brainstem, and spinal cord being specifically concentrated near subependymal, subpial, and perivascular regions (Sosunov et al., 2018). Using a light microscope, RFs were first described as eosinophilic deposits inside astrocytes (Alexander, 1949). Electron microscopy revealed that, ultrastructurally, RFs appear as cytosolic, dense, amorphous structures surrounded by intermediate filaments (Herndon et al., 1970; Borrett and Becker, 1985). In addition to GFAP, RFs include αB-crystallin, small heat shock protein 27 (Hsp27), ubiquitin, vimentin, plectin, c-Jun, the 20 S proteasome, synemin, and cyclin D2 (Iwaki et al., 1989, 1993; Tian et al., 2006; Pekny et al., 2014; Heaven et al., 2016). The small heat shock proteins αB-crystallin and Hsp27 are protein chaperones that interact with GFAP. αB-crystallin regulates GFAP filament assembly in mice overexpressing GFAP, as well as in AxD patient brains (Eng et al., 1998). The exact process of RF formation and their pathologic role remain elusive, although certain insights have emerged to be discussed below.

The degree to which myelin is affected varies significantly between AxD patients. In early-onset cases, the white matter pathology is more pronounced compared to late-onset patients. Affected children exhibit severe bilateral cerebral white matter degeneration with brainstem involvement in combination with many RFs and occasionally mass-like lesions as well as altered basal ganglia and thalamic structures (van der Knaap et al., 2001; Tavasoli et al., 2017). The subpial astrocytes are enlarged and their morphology presents a reactive phenotype, as characterized by hypertrophy of the processes. However, these astrocyte changes seem to be region-specific, as AxD astrocytes in the isocortex remain relatively unaltered. This variability between AxD astrocytes is likely explained by the difference in GFAP expression between astrocyte subtypes. Protoplasmic cortical astrocytes, for instance, express low levels of GFAP compared to subpial or white matter astrocytes (Sosunov et al., 2018). Other pathologic features include multinucleated astrocytes, macrophages in the white matter, and reactive microglia, resembling a marked inflammatory environment (Olabarria et al., 2015). Very rarely, infantile AxD patients present with white matter cavitation (Klein and Anzil, 1994).

Late-onset patients mainly exhibit brainstem and cerebellar white matter pathology with medullar and spinal cord involvement (van der Knaap et al., 2005, 2006b; Pareyson et al., 2008) where GFAP levels are high in comparison to other areas. Cerebral white matter pathology is also observed in late-onset patients, but this seems to be confined to periventricular regions. The deep white matter shows RFs, but the degree of myelination is normal. White matter astrocytes of the subcortical U fibers appear normal as well. Occasionally, white matter focal lesions are mistaken for gliomas (Barreau et al., 2011; Tavasoli et al., 2017). The variability within AxD patients in terms of white matter and astrocyte abnormalities is likely related to differences in GFAP expression in different cell types over time. How GFAP mutations lead to subsequent pathology remains largely unknown. As an important hub in the astrocytic cytoskeletal network, GFAP mutations are likely to distort a variety of intermediate filament functions, which eventually leads to cellular malfunction.

Despite significant progress in unraveling disease mechanisms, it remains unclear how AxD astrocytes accumulate high amounts of GFAP, what triggers the formation of RFs, whether RFs are protective or toxic, and what molecular processes underlie the subsequent white matter pathology. Mutations in GFAP initiate a cascade of events that causes astrocytes to adopt a reactive state. Aberrantly phosphorylated GFAP is cleaved by caspase and deposited in RFs (Battaglia et al., 2019) that also contain small heat shock proteins (Der Perng et al., 2006), p62 (Hagemann et al., 2012), and the intermediate filament proteins nestin, vimentin, and synemin (Mignot et al., 2007; Tanaka et al., 2007; Pekny et al., 2014). Lysosomes and $LC3^+$ autophagosomes surround these structures (Tang et al., 2008). The subsequent reactive astrogliosis, characterized by the upregulation of antioxidant response element (ARE), Nrf2, P53, P38/MAPK, MLK3/JNK, Akt, STAT3, and Nf-κB pathways represents a stress response (Sosunov et al., 2018). AxD astrocytes gain expression of CD44 and endothelial B receptors, show altered expression of inward rectifying $K_{ir}4.1$ channels, reduced expression of the ATPase ATP1B2, and downregulation of EAAT2 glutamate transporters (Hagemann et al., 2006; Tian et al., 2010; Sosunov et al., 2013). Potassium buffering and glutamate clearance are compromised, resulting in hyperexcitable neurons (Tian et al., 2010; Bellot-Saez et al., 2017). Structurally, AxD astrocytes exhibit hypertrophic processes and decreased gap junctional coupling through connexins and altered expression of cadherins (Hagemann et al., 2006; Sosunov et al., 2013, 2018; Kondo et al., 2016). mTOR signaling is increased and alterations in mechanotransduction include altered plectin levels, upregulation of phosphorylated focal adhesion kinase (p-FAK), increased Lamin A expression, increased yes-associated protein (YAP) expression, and decreased signaling of the Hippo pathway (Tian et al., 2006; Tang et al., 2008; Cho and Messing, 2009; Sosunov et al., 2013; Kondo et al., 2016; Wang et al., 2018). Organelle malfunction, such as endoplasmic reticulum (ER) stress, swollen ERs, displaced Golgi stacks, nuclear localization of lysosomes and mitochondria, toxic levels of mitochondrial reactive oxygen species, and impaired intercellular mitochondrial transfer further constitute the astrocytic pathology (Hagemann et al., 2006; Wang et al., 2015, 2016; Jones et al., 2018; Gao et al., 2019). Other observations include multinucleated and deformed nuclei, proteasome inhibition, cytoplasmic localization of TDP-43, iron accumulation, and a reduced capacity to release adenosine triphosphate (ATP), but an increase in intracellular Ca^{2+} in response to ATP (Hagemann et al., 2006; Tang et al., 2010; Hagemann et al., 2012; Sosunov et al., 2013; Walker et al., 2014; Canals et al., 2018; Hagemann et al., 2018; Jones et al., 2018; Battaglia et al., 2019). The overall result is that AxD astrocytes are less capable of communicating through Ca^{2+} waves and affect other cells, such as microglia and oligodendrocytes (Olabarria et al., 2015; Li et al., 2018; Hagemann et al., 2021; Saito et al., 2024), through the release of molecules such as lipocalin-2 (LCN2), tumor necrosis factor alpha (TNF-α), CXCL10, and interleukins (ILs) (Olabarria et al., 2015; Kondo et al., 2016; Saito et al., 2018). CHI3L1 inhibits oligodendrocyte precursor cell (OPC) proliferation and causes myelination defects (Li et al., 2018). Recent data show that neural cell differentiation is altered in AxD patient-derived induced pluripotent stem (iPS)-cell models (Matusova et al., 2024). AxD astrocytes are more susceptible to the extracellular matrix (ECM) molecule hyaluronan, which exerts a toxic effect in excessive amounts (Bugiani et al., 2013; Sosunov et al., 2013). Neurons in AxD are hyperexcitable, have swollen axons, are hampered by demyelination, and are affected by the increased astrocytic release of nitric oxide (NO) (Wang et al., 2015; Hagemann et al., 2021). All of the molecular changes and potentially involved mechanisms that are involved in AxD pathogenesis are summarized in Fig. 10.2.

New cellular and animal models for AxD have been developed, such as a rat model (Hagemann et al., 2021) in which a novel antisense therapy reverses AxD pathology and reprogramming of AxD patient fibroblasts into iPS cells to generate neural cell models that help to reveal insights in the pathogenesis of AxD (Jones et al., 2018; Li et al., 2018; Battaglia et al., 2019; Gao et al., 2019; Matusova et al., 2024).

Vanishing white matter disease

VWM, also referred to as "childhood ataxia with central nervous system hypomyelination" (CACH), is the most prevalent leukodystrophy affecting around 2.5% of all cases (van der Knaap et al., 1993, 2024). It can occur at any age and is characterized by mild cognitive decline, occasional epileptic seizures, and motor symptoms such as spasticity and cerebellar ataxia (Hamilton et al., 2018). In female patients, ovarian dysgenesis or premature ovarian failure can also occur and may even precede the neurologic symptoms (Mathis et al., 2008). VWM can occur at any age, typically presenting in childhood (ages 2–6), with approximately 60% of patients showing symptoms before the age of four (van der Knaap et al., 1993, 2024). Although VWM is most frequent in Caucasians, a disease form with onset between 3 and 9 months is exclusive to the Cree Indians in Quebec (Fogli et al., 2002; Hamilton et al., 2018). VWM can also occur in adolescence or adulthood and then it also features psychiatric symptoms (Denier et al., 2007). Disease severity is typically inversely correlated with age of onset. VWM runs a chronic progressive course, upon which stressful events such as head traumas and febrile infections may lead to episodes of rapid deterioration characterized by motor decline, hypotonia, loss of consciousness, or death (Hamilton et al., 2018).

VWM is an autosomal recessive polygenic disorder, caused by pathogenic variants in any of the five genes (*EIF2B1, EIF2B2, EIF2B3, EIF2B4*, and *EIF2B5*) that encode the subunits of the eukaryotic initiation translation factor 2B (eIF2B). eIF2B, a guanine nucleotide exchange factor (GEF), is responsible for initiating the translation of messenger RNA (mRNA) into protein (van der Knaap et al., 1993, 2024; Norris et al., 2021). Under normal conditions, eIF2B exchanges GDP for GTP on the translation initiation factor eIF2. This allows eIF2 to first bind the methionine-charged tRNA (met-tRNAi) and subsequently bind the small ribosomal subunit, which is required to initiate protein synthesis (Jennings and Pavitt, 2014; Norris et al., 2021). Under stress conditions, eIF2B is involved in the integrated stress response (ISR), a cellular stress response, which inhibits protein synthesis to preserve cellular energy and prevent cellular death (Kantor et al., 2005; Asundi et al., 2024). Stressors such as viral infections, amino acid deficiency, or protein misfolding activate the ISR causing eIF2 phosphorylation, which sequesters eIF2B (Marintchev and Ito, 2020; Schoof et al., 2021). Reduced activity of eIF2B results in decreased bulk protein synthesis. Similar to ISR activation, most VWM-causing gene variants reduce eIF2B activity leading to lower translation initiation rates and protein synthesis. In the context of VWM, the ISR is constitutively dysregulated (Abbink et al., 2019; Keefe et al., 2020).

Neuropathologic features of VWM include rarefaction and cavitation or cystic degeneration of cerebral white matter with absent or abnormally thin myelin sheaths, variable axonal loss and, despite extensive white matter damage, mild astrogliosis, and microgliosis (Man et al., 2022; van der Knaap et al., 2024). VWM selectively affects the oligodendrocytes and the astrocytes. Mature oligodendrocytes are reduced in numbers, whereas OPC numbers are increased correlating to the observed myelin pathology (Herrero et al., 2019; Man et al., 2023). The number of astrocytes is also reduced. White matter astrocytes are dysmorphic and atrophic, with blunt, coarse cell processes. Astrocytes in gray matter structures are less affected, and Bergmann glia, the astrocytes in the cerebellar cortex, are typically mislocalized to the molecular layer (Dooves et al., 2016; Bugiani et al., 2018).

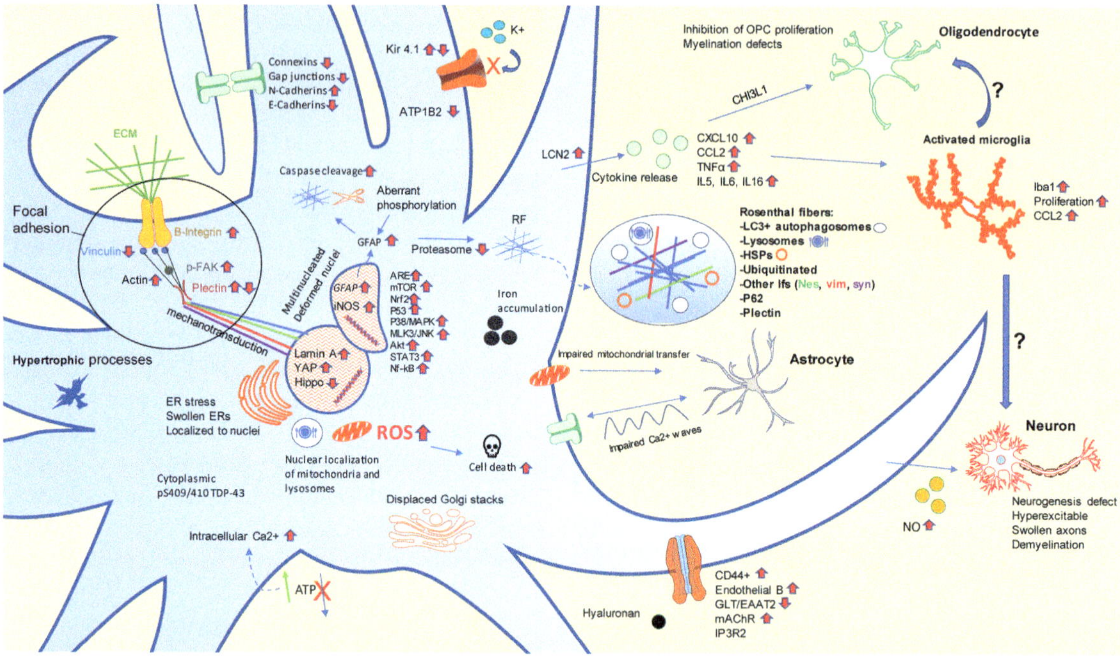

Fig. 10.2. Alexander disease (AxD) molecular pathogenesis in astrocytes. GFAP mutations trigger reactive astrogliosis, leading to the formation of Rosenthal fibers and activation of stress response pathways. Astrocytic dysfunction includes impaired K^+ buffering, glutamate clearance, organelle stress, and altered intercellular communication, affecting neurons and oligodendrocytes. Abbreviations: *Akt*, protein kinase B; *ARE*, antioxidant response element; *ATP*, adenosine triphosphate; *ATP1B2*, ATPase Na^+/K^+ transporting subunit beta 2; *CCL2*, C—C motif chemokine ligand 2; *CD44*$^+$, cluster of differentiation 44 immunopositive; *CHI3L1*, chitinase-3-like protein 1; *CXCL10*, C-X-C motif chemokine ligand 10; *ECM*, extracellular matrix; *ER*, endoplasmic reticulum; *GFAP*, glial fibrillary acidic protein; *GLT1/EAAT2*, glutamate transporter 1/excitatory amino acid transporter 2; *HSPs*, heat shock proteins; *Iba1*, ionized calcium-binding adapter molecule 1; *IL5*, interleukin-5; *IL6*, nterleukin-6; *IL16*, interleukin-16; *iNOS*, inducible nitric oxide synthase; *IP3R2*, inositol 1,4,5-trisphosphate receptor type 2; *JNK*, c-Jun N-terminal kinase; $K_{ir}4.1$, K^+ inwardly rectifying channel 4.1; *LCN2*, lipocalin 2; *LC3*$^+$, light chain 3 immunopositive; *mAChR*, muscarinic acetylcholine receptor; *MLK3*, mixed-lineage kinase 3, upstream regulator of JNK; *mTOR*, mammalian target of rapamycin; *Nes*, nestin; *Nf-κB*, nuclear factor kappa-light-chain-enhancer of activated B cells; *NO*, nitric oxide; *Nrf2*, nuclear factor erythroid 2-related factor 2; *OPC*, oligodendrocyte precursor cell; *P38/MAPK*, mitogen-activated protein kinase; *P53*, tumor suppressor protein encoded by the TP53 gene; *P62*, also known as sequestosome; *p-FAK*, phosphorylated focal adhesion kinase; *pS409/410 TDP*, TAR DNA-binding protein, phosphorylated on serine 409/410; *RF*, Rosenthal fiber; *STAT3*, signal transducer and activator of transcription 3; *TNF-α*, tumor necrosis factor-α; *Syn*, synemin; *Vim*, vimentin; *YAP*, yes-associated protein.

Histopathologic features of VWM astrocytes

Astrocytes and oligodendrocytes are specifically affected in VWM. Several studies revealed astrocytic malfunction as the major pathologic mechanism of VWM (Dooves et al., 2016; Bugiani et al., 2018). siRNA targeting of different EIF2B5 mRNA regions in human glial progenitor cells inhibits the generation of GFAP-positive astrocytes. VWM-causing variants cause loss of physiologic astrocytic functions and prevent astrocytic maturation. VWM astrocytes typically express the immature markers nestin and CD44, and fail to express S100β, a marker of end-stage astrocytic development (Bugiani et al., 2011). In VWM mouse models, nestin-positive astrocytes appear before other histopathologic abnormalities and clinical symptoms. GFAP splicing is also affected in VWM. In particular, the relative ratio between GFAPα, involved in filament formation and stability, and GFAPδ is abnormal (Bugiani et al., 2011, 2018). GFAPδ is unable to form filaments, and its overexpression results in collapse of the GFAP network (de Reus et al., 2024). White matter astrocytes from VWM patients are intensely GFAPδ-immunopositive while overexpression of GFAPδ occurs at the expense of GFAPα, arguing against proper reactive gliosis and resulting in an elevated GFAPδ/GFAPα ratio (Bugiani et al., 2011, 2018). During astrocyte differentiation and maturation, the GFAPδ/GFAPα ratio progressively decreases, suggesting that alternative splicing of GFAP is important for normal astrocyte development. In VWM astrocytes, the GFAPδ/GFAPα ratio remains elevated, indicating a persistent immature astrocytic phenotype (Bugiani et al., 2011, 2018). Several lines of evidence suggest that scarce and disproportionate gliosis, along with the lack of scar formation, relate to astrocytic immaturity (Bugiani et al., 2011, 2018).

Overexpression of GFAPδ may also relate to the abnormal morphology of VWM astrocytes. Overrepresentation of GFAPδ in GFAP networks increases the binding of αB-crystallin and the tendency of intermediate filaments to aggregate (Perng et al., 2008). Similarly, VWM astrocytes also accumulate αB-crystallin, possibly in response to GFAPδ overexpression. The blunt and coarse cellular processes typical of VWM atrophic astrocytes may also result from increased filament aggregation following GFAPδ overexpression (Bugiani et al., 2011, 2018).

Relating astrocytic malfunction to the pathologies of other cells

Impaired cross talk between astrocytes and oligodendrocytes may explain the deficient myelination in VWM. In cocultures with VWM astrocytes and wild-type oligodendrocyte progenitors, the number of myelinating oligodendrocytes is decreased as well as the mRNA levels of mature myelin proteins (de Waard and Bugiani, 2020). This suggests that VWM astrocytes inhibit wild-type oligodendrocyte progenitors to differentiate into myelinating cells, thus hampering myelination. In line with this, VWM OPCs undergo a normal differentiation process if they are cocultured with wild-type astrocytes, indicating that the VWM oligodendrocyte defects are not cell-autonomous but rather a consequence of astrocytic dysfunction (Bugiani et al., 2011, 2018). Pathologic remodeling of OPCs can be mediated by factors secreted in the ECM. In VWM, the expression of one such astrocyte-derived factor, hyaluronan, is increased together with that of its surface receptor CD44 (Bugiani et al., 2013). Hyaluronan inhibits oligodendrocyte maturation, and its level in VWM correlates with regional disease severity: hyaluronan expression is highest in the most severely affected frontal white matter and is much less expressed in the relatively spared cerebellar white matter, in agreement with regional vulnerability shown by imaging studies (Bugiani et al., 2013; Stellingwerff et al., 2021). This suggests a link between astrocyte-secreted hyaluronan levels, degrees of oligodendrocyte progenitor maturation and, in general, severity of affected areas. Together, these results highlight that astrocytic dysfunction is primary and that oligodendrocytes are the cells secondarily affected in this disorder.

VWM astrocytes also impact axonal function. Reduced axonal thickness in VWM white matter correlates with the degree of severity of tissue damage, suggesting impairments in the velocity of nerve-impulse propagation and consequently, in synaptic activity (Klok et al., 2018). In VWM mice, abnormalities in axonal diameter, myelin thickness, and axon-myelin G ratio coexist with an increase in the number of unmyelinated axons. Cocultures of wild-type neurons and VWM astrocytes show increased axonal density, whereas cocultures of VWM neurons with wild-type astrocytes show normal axonal density (Klok et al., 2018). New cellular and animal models for VWM are now emerging. Reprogramming of VWM patient fibroblasts into iPS cells to generate neurons, oligodendrocytes, and astrocytes has resulted in only astrocytes presenting with the same morphologic abnormalities as postmortem brain tissue astrocytes. Conditional mouse models are also being developed. These new models may give us better insights into VWM pathogenesis (Abbink et al., 2019; Zhou et al., 2019).

Megalencephalic leukoencephalopathy with subcortical cysts

Megalencephalic leukoencephalopathy with subcortical cysts (MLC) is an autosomal polygenic inherited disorder with infantile onset; it is characterized by chronic white matter edema, cerebral swelling, formation of subcortical cysts, and delayed white matter atrophy. The classic progressive form of MLC is caused by recessive pathogenic variants in the *MLC1* or *GlialCAM* genes (MLC2A form). Patients exhibit increased macrocephaly during the first year of life, followed by progressive cerebellar ataxia, spasticity, mild cognitive decline, and epilepsy (van der Knaap and Bugiani, 2017). A remitting phenotype has also been described, caused by dominant variants in *GlialCAM* (MLC2B form) (van der Knaap et al., 2010). Strikingly, although patients with MLC2B also present with macrocephaly during the first year of life, marked improvement and normalization are observed after the initial MRI, including a decrease in size or disappearance of subcortical cysts (van der Knaap and Bugiani, 2017; Passchier et al., 2024). Years after onset, these patients still develop ataxia, spasticity, mild cognitive decline, and sometimes seizures.

Irrespective of the causative gene, the main features of MLC include diffuse MRI white matter signal changes with swelling and, eventually, atrophy of the cerebral white matter; intramyelinic edema as documented at diffusion imaging; and the presence of subcortical cysts mainly in the anterior temporal, frontal, and parietal regions. Histopathology shows vacuoles within the outer part of myelin sheaths and inside end-feet of perivascular astrocytes. Fibrillary astrogliosis is observed in the white matter, but there is little, if any, reactive microgliosis. To date, only one brain autopsy has been performed on an MLC patient. The patient carried a homozygous missense variant (S69L) in *MLC1* and neuropathology showed reduced number of astrocytes and oligodendrocytes accompanied by axonal loss (Lopez-Hernandez et al., 2011).

Loss of function of astrocytes in MLC

MLC1 encodes a highly hydrophobic membrane protein similar, in structure, to ion channels and transporters (Brignone et al., 2014, 2015); *GlialCAM* encodes an adhesion-like molecule of the immunoglobulin superfamily that acts as an MLC1-chaperone and as a secondary subunit of the ClC-2 chloride channel (Jeworutzki et al., 2012). In the brain, MLC1 is expressed at inter-astrocytic junctions, astrocyte end-feet contacting the vasculature and pia mater (glia limitans), and in cerebellar Bergmann glia (Teijido et al., 2007). GlialCAM colocalizes with MLC1 at perivascular astrocytic end-feet and inter-astrocytic junctions (Hoegg-Beiler et al., 2014; Alonso-Gardon et al., 2021). Absence of GlialCAM leads to MLC1 mislocalization. As an adhesion molecule, GlialCAM acts as a guiding molecule directing MLC1 to cell junctions and enriching its expression at the plasma membrane (Jeworutzki et al., 2012; Hoegg-Beiler et al., 2014). The absence of MLC1 also impacts the localization of GlialCAM in astrocytes and oligodendrocytes. In astrocyte cultures prepared from $MLC1^{-/-}$ mice, GlialCAM has an impaired cellular localization only under elevated K^+ conditions, suggesting that correct GlialCAM localization in the absence of MLC1 depends on the extracellular K^+ concentrations (Sirisi et al., 2014). GlialCAM and MLC1 form a functional unit accounting for the reason why mutations in either of these genes lead to the same disease (Hoegg-Beiler et al., 2014; Sirisi et al., 2017; Perez-Rius et al., 2019). MLC patients develop white matter edema, suggesting a role in ion–water homeostasis for GlialCAM and MLC1 (van der Knaap et al., 2012).

Effects of MLC1/GlialCAM mutations on astrocyte function

MLC1 interacts with multiple different proteins, including ion transporters such as Na^+–K^+-ATPase (NKA), volume-regulated anion channels (VRAC), the transient receptor potential cation channel V, member 4 (TRPV4), $K_{ir}4.1$, and the water channel aquaporin-4 (AQP4) (Boor et al., 2007). It presumably plays a role in different cell signaling pathways, among which are the EGFR/ERK and PLCγ1 pathways that are involved in cellular processes as proliferation, maturation, reactive astrogliosis, and apoptosis (Lanciotti et al., 2016). The dysfunctional interaction between MLC1, GlialCAM, and all the above-mentioned proteins contributes to white matter malfunction and, thus, to the alteration of cerebral homeostasis (Brignone et al., 2015).

Under physiologic conditions, astrocytes buffer extracellular K^+ released after neuronal activity. In MLC, astrocyte impairment lead to an impaired extracellular K^+ homeostasis due to the abnormal interaction of GlialCAM with Na^+–K^+ ATPase, connexins, $K_{ir}4.1$, and AQP4 (Brignone et al., 2015; Min and van der Knaap, 2018). Aberrant K^+ buffering and deficient glutamate clearance also contribute to neuronal hyperexcitability and spreading depolarization leading to epileptic seizures (Dubey et al., 2018). In addition, ClC-2 protein levels are also reduced in both MLC1 and GlialCAM mouse models (Jeworutzki et al., 2012). Although mutations in *CLCN2* have not been found in MLC, several lines of evidence suggest that the activation of CLC-2 is dependent on interaction with MLC1 and GlialCAM during depolarization (Hoegg-Beiler et al., 2014). The formation of this tertiary complex may be important in reducing neuronal hyperexcitability through Cl^--dependent enhancement of inhibition (Untiet and Verkhratsky, 2024).

Potassium uptake by astrocytes is coupled to water influx, resulting in astrocyte swelling. MLC1 is implicated in the regulation of cell volume and osmotic balance in astrocytes (Brignone et al., 2015). In response to the disturbance of ion–water homeostasis or to oxidative stress, astrocyte swelling and reactivity occur. The return to osmotic balance is achieved through the activation of the volume-regulated anion channel and the calcium-activated $K_{Ca}3.1$ channel, and through the EGFR/ERK and PLCγ1 pathways, which modulate the activation of the regulatory volume decrease (RVD) (Okada, 2024). VRAC activity is fundamental for controlling astrocyte ionic concentration gradients by regulating ion and water flux. However, in MLC, astrocyte swelling is sustained and irreversible leading to permanent white matter edema (Min and van der Knaap, 2018; Brignone et al., 2019). In addition, underexpressed MLC1 in astrocytes, together with reduced expression of GlialCAM, results in aberrant chloride currents, hindering VRAC activity and, therefore, the adjustment of ion–water homeostasis in both the intracellular and extracellular environments. Thus, RVD is impaired, which contributes to the long-term maintenance of swollen astrocytes (Lafrenaye and Simard, 2019). A possible causal relation exists between disturbed ion–water homeostasis and vacuolization in astrocyte end-feet. Using $Clcn2^-$ mice, it has been shown that ClC-2 is involved in an astrocytic vacuolation phenotype similar to that of MLC (Jeworutzki et al., 2012; Hoegg-Beiler et al., 2014). In agreement with this, downregulation of MLC1 to mimic the expression levels in the diseased brain causes the appearance of vacuoles in astrocytes (Duarri et al., 2011). Increase in both MLC1 and GlialCAM expression reverses this phenotype (Bosch and Estevez, 2020). These data suggest a correlation between MLC1 and vacuolization in astrocytes, the former being a possible cause for the latter.

Calcium-dependent $K_{Ca}3.1$ channels are located together with MLC1 in perivascular astrocytic end-feet. An increase in intracellular Ca^{2+} concentration is mediated by TRPV4, further supporting a cross talk between different ion channels and MLC1. Based on this, it has been suggested that MLC1 and $K_{Ca}3.1$ may also cooperate to regulate BBB permeability, Na^+ influx, and ion–water homeostasis (Brignone et al., 2015; Lattier et al., 2020).

Wild-type MLC1 regulates astrocyte proliferation by downregulating various signaling pathways such as ERK1/2 and PLCγ1, blocking $K_{Ca}3.1$ channel, and promoting epidermal growth factor receptor (EGFR) degradation, all of which promote astrocyte growth (Yu et al., 2014; Lanciotti et al., 2016). By contrast, pathologic variants in *MLC1* favor astrocyte proliferation. This is corroborated by evidence showing that MLC1 expression increases when cell proliferation is inhibited (Lanciotti et al., 2016). The impairment of all the astrocytic functions described above drives the course of the disease, further suggesting that astrocytes play a central cell-autonomous role in MLC pathogenesis.

Role of astrocyte malfunction in oligodendrocytes and myelin pathology in MLC

MLC1 is present in all myelin-producing vertebrate species, while *MLC1* orthologs are absent in species that lack myelin. The highest MLC1 expression in mice and humans is observed during active myelination (Schmitt et al., 2003; Bosch and Estevez, 2020). Incidentally, in the context of MLC, white matter swelling is most severe during active myelination (van der Knaap et al., 2012; van der Knaap and Bugiani, 2017). Altogether, this points toward MLC1 playing a critical role in temporal regulation of the myelination processes. Myelin abnormalities in MLC1 may be caused by oligodendrocyte malfunction. GlialCAM and ClC-2 are expressed on oligodendrocyte processes and at astrocyte-oligodendrocyte contact sites (Jeworutzki et al., 2012). Although MLC1 is not expressed in oligodendrocytes, $MLC1^{-/-}$ mice show mislocalization of GlialCAM and ClC-2 in oligodendrocytes (Hoegg-Beiler et al., 2014; Perez-Rius et al., 2019). The absence of MLC1 may cause astrocytic GlialCAM instability, destabilizing the GlialCAM/ClC-2 complex, and potentially resulting in oligodendrocyte damage.

Astrocytes are integrated, through gap junctions, into functional syncytia, which allow exchange of ions, metabolites, and other small molecules. Astrocytes form gap connexin-based junctions with oligodendrocytes. Astrocytes mainly express Cx43, Cx30, and Cx26, while oligodendrocytes express Cx29, Cx32, and Cx47 with Cx29 and Cx32 being found deeper in myelin sheaths. GlialCAM associates with Cx43 and stabilizes it at the plasma membrane at cell–cell junctions. Overexpression of mutant MLC1 in U251 astrocytoma cells leads to decreased Cx43 stability at gap junctions (Lanciotti et al., 2020). Deletion of the astrocytic connexins Cx43 and Cx30 in mice leads to intramyelinic edema and vacuolated oligodendrocytes. These double-knockout mice have reduced MBP amounts and a significant decrease in the number of mature oligodendrocytes (Lutz et al., 2009; Nualart-Marti et al., 2013). Therefore, the loss of inter-astrocytic junctions hinders oligodendrocyte maturation and myelin production. Mice with mutations in both oligodendrocyte and astrocyte connexins develop white matter vacuolation accompanied by progressive loss of astrocytes (Magnotti et al., 2011; Li et al., 2016). Thus, in the context of MLC, *MLC1* and *GlialCAM* mutations could also alter the function of cell–cell junctions leading to the observed oligodendrocyte and myelin abnormalities.

Aicardi–Goutières syndrome

Aicardi–Goutières syndrome (AGS) is a severe, usually autosomal recessive familial leukoencephalopathy. The first cases, eight infants from five families, were described by Jean Aicardi and Françoise Goutières almost 30 years ago. These children suffered from an early-onset familial encephalopathy with chronic CSF lymphocytosis and basal ganglia calcifications, mimicking an intrauterine infection negative for the classical TORCH parameters (perinatal infection; e.g., due to toxoplasmosis, HIV, rubella virus, cytomegalovirus, and/or herpes simplex virus) (Aicardi and Goutières, 1984). Since these first published cases, substantial progress in understanding the molecular mechanisms underlying AGS was achieved. In particular, the discovery of causative mutations in six causative genes (*TREX1*, *RNASEH2A*, *RNASEH2B*, *RNASEH2C*, *ADAR1*, and SAMHD1 (Crow et al., 2006a,b; Rice et al., 2009, 2012)) together with improved understanding of their molecular function helped elucidating AGS pathophysiology. Studies have now uncovered that AGS-related genes are also tightly linked to a predisposition to other autoimmune disorders, such as familial chilblain lupus and systemic lupus erythematosus (Ramantani et al., 2010). Furthermore, evidence is emerging that cerebral vascular alterations are another factor that profoundly contributes to disease progression and mortality. The early onset and recurrence of ischemic and hemorrhagic strokes in *SAMHD1*-mutated individuals, secondary to cerebral vasculopathy, is associated with poor outcomes in these patients (Xin et al., 2011).

The clinical picture of AGS patients is relatively homogeneous despite its genetic heterogeneity. The onset is variable, but usually occurs in the first years of life (Goutières, 2005; Rice et al., 2007b). When presenting at early age, the disease is often marked by irritability, inconsolable crying, dystonia, hypotonia, seizures, and progressive microcephaly, with severe developmental delay and early death (between 1 and 17 years) in approximately 25% of cases (Barth et al., 1999; Crow and Rehwinkel, 2009).

Neuropathologically, AGS is characterized by cerebral atrophy, white matter abnormalities, and brain calcification (Lanzi et al., 2002; Goutières, 2005; Crow and Livingston, 2008; Orcesi et al., 2009). Brain atrophy is evident from enlarged ventricles and sulci. The atrophy is widespread as cortical, brain-stem, and cerebellar atrophy are present in many of the affected children (Rice et al., 2007b). Using MRI, diffuse white matter signal abnormalities can be seen. Hypointense signals on T1 and hyperintense signal on T2 can be detected as an indication of severe demyelination (Aicardi, 2002; Goutières, 2005). In severely affected patients, an extensive frontotemporal leukodystrophy with temporal cystic lesions is commonly observed (Rice et al., 2007b). The third typical finding in AGS patients is the presence of brain calcifications, best visualized using CT scans. The calcifications are typically bilateral and may be present in the basal ganglia, the dentate nuclei of the cerebellum, the subcortical white matter, and the deep white matter (Lanzi et al., 2002; Goutières, 2005; Rice et al., 2007b; Orcesi et al., 2009), though rare patients without calcifications may be seen.

One of the clinical hallmarks of the disease is the presence of raised levels of interferon-α (IFN-α) in the CSF and serum (Lebon et al., 1988). The IFN-α level is typically higher in the CSF than in the serum, suggesting a predominant intrathecal production (Lebon et al., 1988; Goutières et al., 1998; Abdel-Salam et al., 2004). Other cytokines such as C-X-C motif chemokine 10 (CXCL10) (van Heteren et al., 2008) and pterins (Blau et al., 2003) are also elevated in the CSF of AGS patients. The levels of IFN-α and CXCL10 are normally higher at the time of onset and tend to decrease with age (Rice et al., 2007b; van Heteren et al., 2008). The presence of white blood cells in the CSF was considered to be a typical finding (Aicardi and Goutières, 1984), although the number of cells tends to fall with age (Rice et al., 2007b) and leucocytes are absent in a considerable number of affected individuals (du Moulin et al., 2011).

Even though the organ most affected is the brain, extraneurologic signs are frequently observed in AGS. The most common finding is the presence of chilblain skin lesions in at least 43% of patients (Rice et al., 2007b). These ulcerating skin lesions typically appear in feet, toes, hands, ears, and nose (Rice et al., 2007b; Ramantani et al., 2010; Ramesh et al., 2010). Other far more rare features observed are hepatosplenomegaly, joint effusions and contractures, thrombocytopenia, glaucoma, and the presence of autoantibodies in the serum (Rice et al., 2007b; Ramantani et al., 2010).

GENETIC ETIOLOGY, MOLECULAR PATHOGENESIS, AND GENOTYPE–PHENOTYPE CORRELATION

Since the first discovery of an AGS causing genetic mutation in 2006, we have seen a great leap forward in our understanding of the mechanisms in disease in AGS. Presently, mutations in six genes, which lead to the disease, have been described. These mutations, which account for more than 90% of AGS cases, were discovered by Crow and collaborators, in genes coding for important proteins in nucleic acid metabolism. These genes encode the DNA exonuclease 1 (*TREX1*) (Crow et al., 2006a), the three subunits of the endoribonuclease H2 complex (*RNASEH2B*, *RNASEH2C*, and *RNASEH2A*) (Crow et al., 2006b), the deoxynucleoside triphosphate triphosphohydrolase (*SAMHD*) (Rice et al., 2009; Goldstone et al., 2011; Powell et al., 2011), and the recently discovered adenoside deaminase-1 (*ADAR1*) (Rice et al., 2012). AGS is normally inherited in an autosomal recessive manner (Rice et al., 2007b), but some de novo dominant mutations have been reported for *TREX1* (Rice et al., 2007a; Ramantani et al., 2010) and *ADAR1* (Rice et al., 2012). Even though mutations in these genes cover most of the patients suspected for AGS, there are still patients, who do not present with mutations in these genes, suggesting that mutations in at least one or more genes remain to be discovered.

The current genetic evidence has led to the consensus hypothesis that mutations, in all the previously described genes, result in nucleic acid accumulation, which triggers an immune response. In particular, deficiency in Trex1 activity results in an accumulation of single-stranded DNA (ssDNA) polynucleotides (Yang et al., 2007). This activates the interferon-stimulatory DNA response (Stetson et al., 2008), through the STING-dependent antiviral response pathway (Gall et al., 2012). Recent findings established RNase-H2 as a key mammalian genome surveillance enzyme required to maintain genome integrity (Hiller et al., 2012; Reijns et al., 2012). RNase-H2 represents the major RNase-H activity in eukaryotic cells (Cerritelli and Crouch, 2009) and the absence of RNase-H2 results in the accumulation of ribonucleoside monophosphates (Lazzaro et al., 2012). SAMHD1 is a dGTP-stimulated triphosphohydrolase that converts deoxynucleoside triphosphates to the constituent deoxynucleoside and inorganic triphosphate. It is hypothesized that loss of SAMHD1 function would result in a buildup of cytosolic DNA (deoxynucleotide triphosphates) and trigger an inappropriate interferon response (Goldstone et al., 2011). ADAR1 participates in one specific type of RNA editing that involves the conversion of adenosine into inosine residues in double-stranded RNA (dsRNA) (Wang et al., 2004; Lazzaro et al., 2012). Although not proven yet, it has been proposed that ADAR1 deficiency may lead to a cytoplasmic accumulation of dsRNA generated from genomic repetitive elements (Rice et al., 2012).

It is assumed that, for all of these mutations, accumulated immune stimulatory nucleic acids trigger primary intrathecal synthesis of IFN-α and CXCL10 in the central nervous system by astrocytes (van Heteren et al., 2008). Transgenic mice with an astrocyte-specific chronic overproduction of IFN-α develop a progressive inflammatory encephalopathy, with calcium deposits, gliosis, angiopathy, and neurodegeneration (Akwa et al., 1998; Campbell et al., 1999). These latter findings are remarkably similar to those observed in AGS patients, and thus implicate a critical pathogenic role for IFN-α in the cerebral pathology of AGS.

Taken together, AGS is a genetically heterogeneous disorder with a relatively uniform clinical presentation. Within this phenotype, there is a certain spectrum of disease burden; as a matter of fact, there is a certain genotype–phenotype correlation. Patients carrying mutations in *TREX1*, *RNASEH2A*, and *RNASEH2C* are more severely affected and have a higher mortality rate compared to patients with mutations in *RNASEH2B* (Rice et al., 2007b). In line with these observations, a remarkably different CSF cell gene expression signature was described in patients with different genotypes (Izzotti et al., 2012). Notably, patients with mutations in *TREX1* were those with the most severe and earliest onset of symptoms. Furthermore, it has been described that, in patients with mutations in *SAMHD1*, the onset of the disease occurs later and the clinical phenotype is milder (Ramesh et al., 2010; Thiele et al., 2010; Xin et al., 2011). Of note, severe cerebral vasculopathy is a typical feature of AGS patients associated with *SAMHD1* mutations. Several preclinical models are available and have recently expanded with patient-derived iPS-cell-derived models (Hofer et al., 2024).

OLIGODENDROGLIAL LEUKODYSTROPHIES

Oligodendrocytes are the primary elements in several leukodystrophies including Krabbe disease, adult-onset autosomal dominant leukodystrophy, metachromatic leukodystrophy (MLD), Canavan disease, Pelizaeus–Merzbacher disease, and hypomyelination with atrophy of basal ganglia and cerebellum; all these pathologies are discussed in detail below.

Krabbe disease or globoid cell leukodystrophy

Krabbe disease is an autosomal recessive demyelinating myelinopathy, with severe white matter loss and presenting globoid phagocytes (van der Knaap and Bugiani, 2017). Diffuse brain sclerosis accompanied with globoid cells in the brain (the histopathologic hallmark of the disease) was discovered by Knud Haraldsen Krabbe in 1916 (Krabbe, 1916). The causal role for the genetic deficiency of the β-galactocerebrosidase was revealed in 1970 (Suzuki and Suzuki, 1970). The loss of function of the responsible gene GALC leads to the insufficiency of this enzyme, which catalyzes cleaving the glycosylic linkage of galactosylceramide and galactosylsphingosine (also known as psychosine), which are highly toxic lipids (Hawkins-Salsbury et al., 2013). According to the clinical presentation, Krabbe disease is classified into infantile, juvenile, or adult (Lynch et al., 2019). Out of these, the infantile form, which is clinically manifested from 3 to 6 months of age, is the most severe with fatal outcome at 2 years of age. Both juvenile and adult forms have a relatively mild but heterogeneous clinical presentation with a slow progression rate.

More than 100 mutations of the human *GALC* gene were identified; and all of them are linked to a significant reduction of the functionally active β-galactocerebroside; the activity of the latter can be used as a diagnostic test (Jalal et al., 2012). The most widely used animal model of Krabbe disease is the twitcher mouse harboring naturally occurring *GALC* mutation and replicating human pathology (Suzuki and Suzuki, 1983). In this model, the development of oligodendrocytes is impaired leading to demyelination (Inamura et al., 2018). Cellular pathology includes the appearance of multinucleated globoid phagocytes with lysosomes; these cells are most likely of microglial origin and they emerge prior to demyelination and inflammation (Suzuki, 1998; Wenger et al., 2000; Nicaise et al., 2016). Oligodendrocyte demise is directly linked to an abnormal accumulation of psychosine, which triggers cell death and demyelination (Giri et al., 2006). Transplantation of bone marrow in the twitcher mice decreased the number of globoid phagocytes, reduced the level of psychosine, and rescued demyelination (Hoogerbrugge et al., 1988a,b). Transplantation of hematopoietic stem cells is also considered to be a treatment for the infantile form of Krabbe disease (Page et al., 2019).

Adult-onset autosomal dominant leukodystrophy

Autosomal dominant leukodystrophy, described in 1964 (Zerbin-Rudin and Peiffer, 1964), is manifested by demyelination and white matter atrophy results from aberrant peroxisomal fatty acid β-oxidation (Melberg et al., 2006; Sundblom et al., 2009). This is caused by an extra copy of the gene *LMNB1* encoding the nuclear lamina protein lamin B1 (Padiath et al., 2006). The family of intermediate proteins, lamins, is associated with nucleus as well as with the inner nuclear membrane; they interact with chromatin-associated proteins, which in turn, regulate DNA replication, RNA transcription, nuclear and chromatin organization, cell cycle regulation, cell development and differentiation, nuclear migration, and apoptosis (Gruenbaum et al., 2003). Although adult-onset autosomal dominant leukodystrophy (ADLD) clinical presentation is somewhat similar to progressive multiple sclerosis, the underlying neuropathology is distinct, with no loss of oligodendrocytes, no prominent reactive astrogliosis, and no increase of lymphocytes or other phagocytic cells (Coffeen et al., 2000). Oligodendrocyte-specific overexpression of LMNB1 in mice triggered pathology resembling human disease; in contrast, overexpression of LMNB1 in astrocytes or neurons did result in any overt pathologic changes, highlighting the primary pathophysiologic role of oligodendrocytes (Heng et al., 2013). Underlying molecular mechanisms precipitating oligodendrocyte pathology remain obscure; they might be related to epigenetic factors (Lu et al., 2019), and aberrant lipid and cholesterol synthesis (Rolyan et al., 2015; Yattah et al., 2020). Of note, astrocytes in human ADLD undergo substantial morphologic atrophy, which may limit their homeostatic and neuroprotective support further potentiating demyelination (Coffeen et al., 2000). In a mouse model of ADLD with Lamin B1 overexpressing oligodendrocytes, however, both astrogliosis and microgliosis were noted (Rolyan et al., 2015).

Metachromatic leukodystrophy

MLD is a lysosomal disorder associated with mutations in the *ARSA* gene encoding for the enzyme arylsulfatase A (ASA, also known as cerebroside sulfatase) or by mutations in the *PSAP* gene encoding the saponin precursor prosaponin (Meyer et al., 2014; van der Knaap and Bugiani, 2017). Deficiency in the ASA enzyme (which converts cerebroside 3-sulfate into cerebroside and sulfate) translates into the accumulation of fatty sulfatides and demyelination of axons in the CNS and in the periphery. Saponins, in their turn, degrade galactocerebrosides, GM1 ganglioside, globotriaosylceramide, and glycerolipids, and enhance hydrolysis of sphingomyelin by sphingomyelin phosphodiesterase (Meyer et al., 2014). ASA is mostly expressed in oligodendrocytes, although some expression is also detected in astrocytes, microglia, and neurons (Ceravolo et al., 2024). Clinically, MLD is characterized by demyelination and brain atrophy, associated with loss of oligodendrocytes and reactive astrogliosis, with metachromatic deposits of lysosomal accumulation of sulfatides in astrocytes, microglia, and neurons (van der Voorn et al., 2005; Eichler et al., 2008).

Canavan disease

Canavan disease was described by Myrtelle Canavan in 1931 (Canavan, 1931). Cytopathologically, this disease is characterized by widespread myelin vacuolation; this pathology is caused by mutations in the *ASPA* gene, which encodes aspartoacylase, a metabolic enzyme that catalyzes the deacetylation of *N*-acetylaspartic acid to produce acetate and L-aspartate (Gordon, 2001; Kumar et al., 2006). The *N*-acetylaspartic acid is, after glutamate, the second most abundant amino acid derivative in the brain; it is synthesized and secreted by neurons, and subsequently accumulated by oligodendrocytes, where it is broken down by ASPA to release acetate that is used for myelin lipid synthesis, being thus critical for myelination (Chakraborty et al., 2001; Francis et al., 2016).

Pelizaeus–Merzbacher disease

Pelizaeus–Merzbacher disease, named so after its discoverers (Pelizaeus, 1898; Merzbacher, 1910), is a hypomyelinating disease. It is caused by mutations in the *PLP1* gene encoding for proteolipid protein (PLP) and its alternatively spliced form DM20 (van der Knaap and Bugiani, 2017). The main form of Pelizaeus–Merzbacher disease is associated with *PLP1* duplications; this form of the disease is driven by massive reduction in the number of myelinating oligodendrocytes, which translates into diffuse severe hypomyelination and thinning of white matter (Koeppen and Robitaille, 2002). Hypomyelination is also accompanied by reactive astrogliosis and microgliosis (Garcia et al., 2020). The underlying mechanism is linked to the toxic gain of function of PLP1 point mutation, duplication, and overexpression, all of which lead to the misfolding and aggregation of mutated PLP1 and oligodendrocyte death (Inoue, 2005; Torii et al., 2014).

Hypomyelination with atrophy of basal ganglia and cerebellum

Hypomyelination with atrophy of the basal ganglia and cerebellum, abbreviated as H-ABC, is linked to the mutations of the *TUBB4A* gene encoding the IVA isoform of β-tubulin (Garcia et al., 2020). Tubulins, both α- and β-, create microtubules critical for the development and function of all neural cells including oligodendrocytes (Curiel et al., 2017). A spontaneously occurring *TUBB4A* mutation has been found in taiep (myelin mutant) rats, which present an accumulation of microtubules in oligodendrocytes together with demyelination; with smaller oligodendrocytes being more susceptible to pathologic changes (Song et al., 2001). Arguably aberrant microtubules affect the growth and development of oligodendroglia, thereby causing loss of myelin with subsequent axonopathy.

MICROGLIAL LEUKODYSTROPHIES

Polycystic lipomembranous osteodysplasia with sclerosing leukoencephalopathy or Nasu–Hakola disease

The first two cases of polycystic lipomembranous osteodysplasia with sclerosing leukoencephalopathy were described independently in 1970 by Nasu and Hakola (Nasu et al., 1970; Song et al., 2001). Nasu–Hakola disease is an autosomal recessive disorder characterized by progressive presenile dementia and bone cysts (Bianchin et al., 2004). This disease is also characterized by unusual psychiatric manifestations with numerous emotional abnormalities, including euphoric, facetious and unrestrained behavior together with social inhibition. Profound uncontrolled reactive microgliosis in the white matter constitutes the main pathophysiologic feature of Nasu–Hakola disease. Aberrant microglial reactivity results in demyelination and accumulation of axonal spheroids mainly in the white matter of the frontal and temporal lobes (Bianchin et al., 2010). Microglial toxicity was proposed (although not confirmed) to trigger secondary astrogliosis and promote neuronal death (Bianchin et al., 2010). At the same time, microglia in Nasu–Hakola disease are missing triggering receptor expressed on myeloid cells 2 (TREM2), which is fundamental for neuroprotection; and thus, loss of TREM2 function may be responsible for neuronal death (Satoh et al., 2011; Errichiello et al., 2019).

Hereditary adult-onset leukoencephalopathy with axonal spheroids and pigmented glia

Hereditary diffuse leukoencephalopathy with axonal spheroids (HDLS) is an autosomal dominant disordered linked to the mutations in the *CSF1R* gene (Lynch et al., 2016); this disease was initially described in a Western Swedish family in 1984 (Lin et al., 2010). The *CSF1R* gene encodes microglia specific colony stimulating factor 1 receptor (Konno et al., 2018); this receptor mediates cytokine-dependent activation of mononuclear phagocytic cells, and hence reactivity of microglia. The histopathology of HDLS is characterized by vacuolated myelin in the periventricular, deep and subcortical cerebral white matter, together with enlargement of the lateral ventricles and thinning of the corpus callosum. The specific sign of HDLS is represented by diffuse white matter lesions of swollen axons and axonal spheroids revealed by the immunoreactivity of phosphorylated neurofilaments and amyloid precursor protein (Konno et al., 2018). Underlying pathogenetic mechanisms remain unknown; genetic deletion of the *CSF1R* gene in mice leads to an almost complete ablation of microglia and decreased density of oligodendrocytes (Hagemeyer et al., 2017).

CONCLUSIONS

Neuroglial cells, i.e., astrocytes, oligodendrocytes, and microglia, are primary drivers of white matter pathology in heritable leukodystrophies. The pathophysiologic mechanisms are defined by the nature of on the cell primarily affected by the genetic mutation. In particular, oligodendrocytes drive MLD and Krabbe disease, whereas astrocytes drive AxD and VWM disease. Numerous animal models, expressing pathologic genes, were developed, and more recently patient-derived iPS-cell models are being developed, thus advancing our understanding of the pathogenesis of these rare diseases.

ACKNOWLEDGMENT

We would like to thank M.S. Jorge for help with the manuscript.

References

Abbink TEM, Wisse LE, Jaku E et al. (2019). Vanishing white matter: deregulated integrated stress response as therapy target. Ann Clin Transl Neurol 6: 1407–1422.

Abdel-Salam GM, Zaki MS, Lebon P et al. (2004). Aicardi-Goutières syndrome: clinical and neuroradiological findings of 10 new cases. Acta Paediatr 93: 929–936.

Aicardi J (2002). Aicardi-Goutières syndrome: special type early-onset encephalopathy. Eur J Paediatr Neurol 6: A1–A7. discussion A23–A25, A77–A86.

Aicardi J, Goutières F (1984). A progressive familial encephalopathy in infancy with calcifications of the basal ganglia and chronic cerebrospinal fluid lymphocytosis. Ann Neurol 15: 49–54.

Akwa Y, Hassett DE, Eloranta ML et al. (1998). Transgenic expression of IFN-alpha in the central nervous system of mice protects against lethal neurotropic viral infection but induces inflammation and neurodegeneration. J Immunol 161: 5016–5026.

Alexander WS (1949). Progressive fibrinoid degeneration of fibrillary astrocytes associated with mental retardation in a hydrocephalic infant. Brain 72: 373–381. 373 pl.

Alonso-Gardon M, Elorza-Vidal X, Castellanos A et al. (2021). Identification of the GlialCAM interactome: the G protein-coupled receptors GPRC5B and GPR37L1 modulate megalencephalic leukoencephalopathy proteins. Hum Mol Genet 30: 1649–1665.

Asundi J, Zhang C, Donnelly-Roberts D et al. (2024). GDF15 is a dynamic biomarker of the integrated stress response in the central nervous system. CNS Neurosci Ther 30: e14600.

Barreau P, Prust MJ, Crane J et al. (2011). Focal central white matter lesions in Alexander disease. J Child Neurol 26: 1422–1424.

Barth PG, Walter A, van Gelderen I (1999). Aicardi-Goutières syndrome: a genetic microangiopathy? Acta Neuropathol 98: 212–216.

Battaglia RA, Beltran AS, Delic S et al. (2019). Site-specific phosphorylation and caspase cleavage of GFAP are new markers of Alexander disease severity. Elife 8: e47789.

Bellot-Saez A, Kekesi O, Morley JW et al. (2017). Astrocytic modulation of neuronal excitability through K(+) spatial buffering. Neurosci Biobehav Rev 77: 87–97.

Bianchin MM, Capella HM, Chaves DL et al. (2004). Nasu-Hakola disease (polycystic lipomembranous osteodysplasia with sclerosing leukoencephalopathy—PLOSL): a dementia associated with bone cystic lesions. From clinical to genetic and molecular aspects. Cell Mol Neurobiol 24: 1–24.

Bianchin MM, Martin KC, de Souza AC et al. (2010). Nasu-Hakola disease and primary microglial dysfunction. Nat Rev Neurol 6: 2 p following 523.

Bielschowsky M, Henneberg R (1928). Über familiäre diffuse Sklerose (Leukodystrophia cerebri progressiva hereditaria). J Psychol Neurol 36: 131–181.

Blau N, Bonafe L, Krageloh-Mann I et al. (2003). Cerebrospinal fluid pterins and folates in Aicardi-Goutières syndrome: a new phenotype. Neurology 61: 642–647.

Boor I, Nagtegaal M, Kamphorst W et al. (2007). MLC1 is associated with the dystrophin-glycoprotein complex at astrocytic endfeet. Acta Neuropathol 114: 403–410.

Borrett D, Becker LE (1985). Alexander's disease. A disease of astrocytes. Brain 108: 367–385.

Bosch A, Estevez R (2020). Megalencephalic leukoencephalopathy: insights into pathophysiology and perspectives for therapy. Front Cell Neurosci 14: 627887.

Brenner M, Johnson AB, Boespflug-Tanguy O et al. (2001). Mutations in GFAP, encoding glial fibrillary acidic protein, are associated with Alexander disease. Nat Genet 27: 117–120.

Brignone MS, Lanciotti A, Visentin S et al. (2014). Megalencephalic leukoencephalopathy with subcortical cysts protein-1 modulates endosomal pH and protein trafficking in astrocytes: relevance to MLC disease pathogenesis. Neurobiol Dis 66: 1–18.

Brignone MS, Lanciotti A, Camerini S et al. (2015). MLC1 protein: a likely link between leukodystrophies and brain channelopathies. Front Cell Neurosci 9: 66.

Brignone MS, Lanciotti A, Serafini B et al. (2019). Megalencephalic leukoencephalopathy with subcortical cysts protein-1 (MLC1) counteracts astrocyte activation in response to inflammatory signals. Mol Neurobiol 56: 8237–8254.

Bugiani M, Boor I, van Kollenburg B et al. (2011). Defective glial maturation in vanishing white matter disease. J Neuropathol Exp Neurol 70: 69–82.

Bugiani M, Postma N, Polder E et al. (2013). Hyaluronan accumulation and arrested oligodendrocyte progenitor maturation in vanishing white matter disease. Brain 136: 209–222.

Bugiani M, Vuong C, Breur M et al. (2018). Vanishing white matter: a leukodystrophy due to astrocytic dysfunction. Brain Pathol 28: 408–421.

Campbell IL, Krucker T, Steffensen S et al. (1999). Structural and functional neuropathology in transgenic mice with CNS expression of IFN-alpha. Brain Res 835: 46–61.

Canals I, Ginisty A, Quist E et al. (2018). Rapid and efficient induction of functional astrocytes from human pluripotent stem cells. Nat Methods 15: 693–696.

Canavan MM (1931). Schilder's encephalitis periaxalis diffusa: report of a case in a child aged sixteen and one-half months. Arch Neurol Psychiatry 25: 299–308.

Ceravolo G, Zhelcheska K, Squadrito V et al. (2024). Update on leukodystrophies and developing trials. J Neurol 271: 593–605.

Cerritelli SM, Crouch RJ (2009). Ribonuclease H: the enzymes in eukaryotes. FEBS J 276: 1494–1505.

Chakraborty G, Mekala P, Yahya D et al. (2001). Intraneuronal N-acetylaspartate supplies acetyl groups for myelin lipid synthesis: evidence for myelin-associated aspartoacylase. J Neurochem 78: 736–745.

Cho W, Messing A (2009). Properties of astrocytes cultured from GFAP over-expressing and GFAP mutant mice. Exp Cell Res 315: 1260–1272.

Coffeen CM, McKenna CE, Koeppen AH et al. (2000). Genetic localization of an autosomal dominant leukodystrophy mimicking chronic progressive multiple sclerosis to chromosome 5q31. Hum Mol Genet 9: 787–793.

Crow YJ, Livingston JH (2008). Aicardi-Goutières syndrome: an important Mendelian mimic of congenital infection. Dev Med Child Neurol 50: 410–416.

Crow YJ, Rehwinkel J (2009). Aicardi-Goutières syndrome and related phenotypes: linking nucleic acid metabolism with autoimmunity. Hum Mol Genet 18: R130–R136.

Crow YJ, Hayward BE, Parmar R et al. (2006a). Mutations in the gene encoding the 3'-5' DNA exonuclease TREX1 cause Aicardi-Goutières syndrome at the AGS1 locus. Nat Genet 38: 917–920.

Crow YJ, Leitch A, Hayward BE et al. (2006b). Mutations in genes encoding ribonuclease H2 subunits cause Aicardi-Goutières syndrome and mimic congenital viral brain infection. Nat Genet 38: 910–916.

Curiel J, Rodriguez Bey G, Takanohashi A et al. (2017). TUBB4A mutations result in specific neuronal and oligodendrocytic defects that closely match clinically distinct phenotypes. Hum Mol Genet 26: 4506–4518.

de Reus AJEM, Basak O, Dykstra W et al. (2024). GFAP-isoforms in the nervous system: understanding the need for diversity. Curr Opin Cell Biol 87: 102340.

de Waard DM, Bugiani M (2020). Astrocyte-oligodendrocyte-microglia crosstalk in astrocytopathies. Front Cell Neurosci 14: 608073.

Denier C, Orgibet A, Roffi F et al. (2007). Adult-onset vanishing white matter leukoencephalopathy presenting as psychosis. Neurology 68: 1538–1539.

Der Perng M, Su M, Wen SF et al. (2006). The Alexander disease-causing glial fibrillary acidic protein mutant, R416W, accumulates into Rosenthal fibers by a pathway that involves filament aggregation and the association of alpha B-crystallin and HSP27. Am J Hum Genet 79: 197–213.

Dooves S, Bugiani M, Postma NL et al. (2016). Astrocytes are central in the pathomechanisms of vanishing white matter. J Clin Invest 126: 1512–1524.

du Moulin M, Nurnberg P, Crow YJ et al. (2011). Cerebral vasculopathy is a common feature in Aicardi-Goutières syndrome associated with SAMHD1 mutations. Proc Natl Acad Sci U S A 108: E232. author reply E233.

Duarri A, Lopez de Heredia M, Capdevila-Nortes X et al. (2011). Knockdown of MLC1 in primary astrocytes causes cell vacuolation: a MLC disease cell model. Neurobiol Dis 43: 228–238.

Dubey M, Brouwers E, Hamilton EMC et al. (2018). Seizures and disturbed brain potassium dynamics in the leukodystrophy megalencephalic leukoencephalopathy with subcortical cysts. Ann Neurol 83: 636–649.

Eichler FS, Ren JQ, Cossoy M et al. (2008). Is microglial apoptosis an early pathogenic change in cerebral X-linked adrenoleukodystrophy? Ann Neurol 63: 729–742.

Eng LF, Lee YL, Kwan H et al. (1998). Astrocytes cultured from transgenic mice carrying the added human glial fibrillary acidic protein gene contain Rosenthal fibers. J Neurosci Res 53: 353–360.

Errichiello E, Dardiotis E, Mannino F et al. (2019). Phenotypic expansion in Nasu-Hakola disease: immunological findings in three patients and proposal of a unifying pathogenic hypothesis. Front Immunol 10: 1685.

Escartin C, Galea E, Lakatos A et al. (2021). Reactive astrocyte nomenclature, definitions, and future directions. Nat Neurosci 24: 312–325.

Flint D, Li R, Webster LS et al. (2012). Splice site, frameshift, and chimeric GFAP mutations in Alexander disease. Hum Mutat 33: 1141–1148.

Fogli A, Wong K, Eymard-Pierre E et al. (2002). Cree leukoencephalopathy and CACH/VWM disease are allelic at the EIF2B5 locus. Ann Neurol 52: 506–510.

Francis JS, Wojtas I, Markov V et al. (2016). N-acetylaspartate supports the energetic demands of developmental myelination via oligodendroglial aspartoacylase. Neurobiol Dis 96: 323–334.

Friede RL (1964). Alexander's disease. Arch Neurol 11: 414–422.
Galbo Jr PM, Ciesielski MJ, Figel S et al. (2017). Circulating CD9+/GFAP+/survivin+ exosomes in malignant glioma patients following survivin vaccination. Oncotarget 8: 114722–114735.
Gall A, Treuting P, Elkon KB et al. (2012). Autoimmunity initiates in nonhematopoietic cells and progresses via lymphocytes in an interferon-dependent autoimmune disease. Immunity 36: 120–131.
Gao L, Zhang Z, Lu J et al. (2019). Mitochondria are dynamically transferring between human neural cells and Alexander disease-associated GFAP mutations impair the astrocytic transfer. Front Cell Neurosci 13: 316.
Garcia LM, Hacker JL, Sase S et al. (2020). Glial cells in the driver seat of leukodystrophy pathogenesis. Neurobiol Dis 146: 105087.
Giri S, Khan M, Rattan R et al. (2006). Krabbe disease: psychosine-mediated activation of phospholipase A2 in oligodendrocyte cell death. J Lipid Res 47: 1478–1492.
Goldstone DC, Ennis-Adeniran V, Hedden JJ et al. (2011). HIV-1 restriction factor SAMHD1 is a deoxynucleoside triphosphate triphosphohydrolase. Nature 480: 379–382.
Gordon N (2001). Canavan disease: a review of recent developments. Eur J Paediatr Neurol 5: 65–69.
Goutières F (2005). Aicardi-Goutières syndrome. Brain Dev 27: 201–206.
Goutières F, Aicardi J, Barth PG et al. (1998). Aicardi-Goutières syndrome: an update and results of interferon-alpha studies. Ann Neurol 44: 900–907.
Graff-Radford J, Schwartz K, Gavrilova RH et al. (2014). Neuroimaging and clinical features in type II (late-onset) Alexander disease. Neurology 82: 49–56.
Green L, Berry IR, Childs AM et al. (2018). Whole exon deletion in the GFAP gene is a novel molecular mechanism causing Alexander disease. Neuropediatrics 49: 118–122.
Gruenbaum Y, Goldman RD, Meyuhas R et al. (2003). The nuclear lamina and its functions in the nucleus. Int Rev Cytol 226: 1–62.
Hagemann TL, Gaeta SA, Smith MA et al. (2005). Gene expression analysis in mice with elevated glial fibrillary acidic protein and Rosenthal fibers reveals a stress response followed by glial activation and neuronal dysfunction. Hum Mol Genet 14: 2443–2458.
Hagemann TL, Connor JX, Messing A (2006). Alexander disease-associated glial fibrillary acidic protein mutations in mice induce Rosenthal fiber formation and a white matter stress response. J Neurosci 26: 11162–11173.
Hagemann TL, Jobe EM, Messing A (2012). Genetic ablation of Nrf2/antioxidant response pathway in Alexander disease mice reduces hippocampal gliosis but does not impact survival. PLoS One 7: e37304.
Hagemann TL, Powers B, Mazur C et al. (2018). Antisense suppression of glial fibrillary acidic protein as a treatment for Alexander disease. Ann Neurol 83: 27–39.
Hagemann TL, Powers B, Lin NH et al. (2021). Antisense therapy in a rat model of Alexander disease reverses GFAP pathology, white matter deficits, and motor impairment. Sci Transl Med 13: eabg4711.
Hagemeyer N, Hanft KM, Akriditou MA et al. (2017). Microglia contribute to normal myelinogenesis and to oligodendrocyte progenitor maintenance during adulthood. Acta Neuropathol 134: 441–458.
Hamilton EMC, van der Lei HDW, Vermeulen G et al. (2018). Natural history of vanishing white matter. Ann Neurol 84: 274–288.
Hawkins-Salsbury JA, Parameswar AR, Jiang X et al. (2013). Psychosine, the cytotoxic sphingolipid that accumulates in globoid cell leukodystrophy, alters membrane architecture. J Lipid Res 54: 3303–3311.
Heaven MR, Flint D, Randall SM et al. (2016). Composition of Rosenthal fibers, the protein aggregate hallmark of Alexander disease. J Proteome Res 15: 2265–2282.
Helman G, Takanohashi A, Hagemann TL et al. (2020). Type II Alexander disease caused by splicing errors and aberrant overexpression of an uncharacterized GFAP isoform. Hum Mutat 41: 1131–1137.
Heng MY, Lin ST, Verret L et al. (2013). Lamin B1 mediates cell-autonomous neuropathology in a leukodystrophy mouse model. J Clin Invest 123: 2719–2729.
Herndon RM, Rubinstein LJ, Freeman JM et al. (1970). Light and electron microscopic observations on Rosenthal fibers in Alexander's disease and in multiple sclerosis. J Neuropathol Exp Neurol 29: 524–551.
Herrero M, Mandelboum S, Elroy-Stein O (2019). eIF2B mutations cause mitochondrial malfunction in oligodendrocytes. Neuromolecular Med 21: 303–313.
Hiller B, Achleitner M, Glage S et al. (2012). Mammalian RNase H2 removes ribonucleotides from DNA to maintain genome integrity. J Exp Med 209: 1419–1426.
Hoegg-Beiler MB, Sirisi S, Orozco IJ et al. (2014). Disrupting MLC1 and GlialCAM and ClC-2 interactions in leukodystrophy entails glial chloride channel dysfunction. Nat Commun 5: 3475.
Hofer MJ, Modesti N, Coufal NG et al. (2024). The prototypical interferonopathy: Aicardi-Goutières syndrome from bedside to bench. Immunol Rev 327: 83–99.
Hoogerbrugge PM, Poorthuis BJ, Romme AE et al. (1988a). Effect of bone marrow transplantation on enzyme levels and clinical course in the neurologically affected twitcher mouse. J Clin Invest 81: 1790–1794.
Hoogerbrugge PM, Suzuki K, Suzuki K et al. (1988b). Donor-derived cells in the central nervous system of twitcher mice after bone marrow transplantation. Science 239: 1035–1038.
Inamura N, Kito M, Go S et al. (2018). Developmental defects and aberrant accumulation of endogenous psychosine in oligodendrocytes in a murine model of Krabbe disease. Neurobiol Dis 120: 51–62.
Inoue K (2005). PLP1-related inherited dysmyelinating disorders: Pelizaeus-Merzbacher disease and spastic paraplegia type 2. Neurogenetics 6: 1–16.
Iwaki T, Kume-Iwaki A, Liem RK et al. (1989). Alpha B-crystallin is expressed in non-lenticular tissues and accumulates in Alexander's disease brain. Cell 57: 71–78.
Iwaki T, Iwaki A, Tateishi J et al. (1993). Alpha B-crystallin and 27-kd heat shock protein are regulated by stress conditions in the central nervous system and accumulate in Rosenthal fibers. Am J Pathol 143: 487–495.
Izzotti A, Longobardi M, Cartiglia C et al. (2012). Different mutations in three prime repair exonuclease 1 and ribonuclease H2 genes affect clinical features in Aicardi-Goutières syndrome. J Child Neurol 27: 51–60.
Jalal K, Carter R, Yan L et al. (2012). Does galactocerebrosidase activity predict Krabbe phenotype? Pediatr Neurol 47: 324–329.
Jany PL, Agosta GE, Benko WS et al. (2015). CSF and blood levels of GFAP in Alexander disease. eNeuro 2: e0080-15.
Jennings MD, Pavitt GD (2014). A new function and complexity for protein translation initiation factor eIF2B. Cell Cycle 13: 2660–2665.
Jeworutzki E, Lopez-Hernandez T, Capdevila-Nortes X et al. (2012). GlialCAM, a protein defective in a leukodystrophy, serves as a ClC-2 Cl(−) channel auxiliary subunit. Neuron 73: 951–961.
Jones JR, Kong L, Hanna MG et al. (2018). Mutations in GFAP disrupt the distribution and function of organelles in human astrocytes. Cell Rep 25: 947–958.e4.

Kantor L, Harding HP, Ron D et al. (2005). Heightened stress response in primary fibroblasts expressing mutant eIF2B genes from CACH/VWM leukodystrophy patients. Hum Genet 118: 99–106.

Keefe MD, Soderholm HE, Shih HY et al. (2020). Vanishing white matter disease expression of truncated EIF2B5 activates induced stress response. Elife 9: e56319.

Kevelam SH, Steenweg ME, Srivastava S et al. (2016). Update on leukodystrophies: a historical perspective and adapted definition. Neuropediatrics 47: 349–354.

Klein EA, Anzil AP (1994). Prominent white matter cavitation in an infant with Alexander's disease. Clin Neuropathol 13: 31–38.

Klok MD, Bugiani M, de Vries SI et al. (2018). Axonal abnormalities in vanishing white matter. Ann Clin Transl Neurol 5: 429–444.

Koeppen AH, Robitaille Y (2002). Pelizaeus-Merzbacher disease. J Neuropathol Exp Neurol 61: 747–759.

Kondo T, Funayama M, Miyake M et al. (2016). Modeling Alexander disease with patient iPSCs reveals cellular and molecular pathology of astrocytes. Acta Neuropathol Commun 4: 69.

Konno T, Kasanuki K, Ikeuchi T et al. (2018). CSF1R-related leukoencephalopathy: a major player in primary microgliopathies. Neurology 91: 1092–1104.

Krabbe KA (1916). New familial, infantile form of diffuse brain-sclerosis. Brain 39: 74–114.

Kumar S, Mattan NS, de Vellis J (2006). Canavan disease: a white matter disorder. Ment Retard Dev Disabil Res Rev 12: 157–165.

Kyllerman M, Rosengren L, Wiklund LM et al. (2005). Increased levels of GFAP in the cerebrospinal fluid in three subtypes of genetically confirmed Alexander disease. Neuropediatrics 36: 319–323.

Lafrenaye AD, Simard JM (2019). Bursting at the seams: molecular mechanisms mediating astrocyte swelling. Int J Mol Sci 20: 330.

Lanciotti A, Brignone MS, Visentin S et al. (2016). Megalencephalic leukoencephalopathy with subcortical cysts protein-1 regulates epidermal growth factor receptor signaling in astrocytes. Hum Mol Genet 25: 1543–1558.

Lanciotti A, Brignone MS, Belfiore M et al. (2020). Megalencephalic leukoencephalopathy with subcortical cysts disease-linked MLC1 protein favors gap-junction intercellular communication by regulating connexin 43 trafficking in astrocytes. Cells 9: 1425.

Lanzi G, Fazzi E, D'Arrigo S (2002). Aicardi-Goutières syndrome: a description of 21 new cases and a comparison with the literature. Eur J Paediatr Neurol 6: A9–22. discussion A23–A25, A77–A86.

Lattier JM, De A, Chen Z et al. (2020). Megalencephalic leukoencephalopathy with subcortical cysts 1 (MLC1) promotes glioblastoma cell invasion in the brain microenvironment. Oncogene 39: 7253–7264.

Lazzaro F, Novarina D, Amara F et al. (2012). RNase H and postreplication repair protect cells from ribonucleotides incorporated in DNA. Mol Cell 45: 99–110.

Lebon P, Badoual J, Ponsot G et al. (1988). Intrathecal synthesis of interferon-alpha in infants with progressive familial encephalopathy. J Neurol Sci 84: 201–208.

Li J, Zhang L, Chu Y et al. (2016). Astrocytes in oligodendrocyte lineage development and white matter pathology. Front Cell Neurosci 10: 119.

Li L, Tian E, Chen X et al. (2018). GFAP mutations in astrocytes impair oligodendrocyte progenitor proliferation and myelination in an hiPSC model of Alexander disease. Cell Stem Cell 23: 239–251.e6.

Lin WL, Wszolek ZK, Dickson DW (2010). Hereditary diffuse leukoencephalopathy with spheroids: ultrastructural and immunoelectron microscopic studies. Int J Clin Exp Pathol 3: 665–674.

Livingston JH, Crow YJ (2016). Neurologic phenotypes associated with mutations in TREX1, RNASEH2A, RNASEH2B, RNASEH2C, SAMHD1, ADAR1, and IFIH1: Aicardi-Goutières syndrome and beyond. Neuropediatrics 47: 355–360.

Lopez-Hernandez T, Sirisi S, Capdevila-Nortes X et al. (2011). Molecular mechanisms of MLC1 and GLIALCAM mutations in megalencephalic leukoencephalopathy with subcortical cysts. Hum Mol Genet 20: 3266–3277.

Lu G, Zhang M, Wang J et al. (2019). Epigenetic regulation of myelination in health and disease. Eur J Neurosci 49: 1371–1387.

Lutz SE, Zhao Y, Gulinello M et al. (2009). Deletion of astrocyte connexins 43 and 30 leads to a dysmyelinating phenotype and hippocampal CA1 vacuolation. J Neurosci 29: 7743–7752.

Lynch DS, Jaunmuktane Z, Sheerin UM et al. (2016). Hereditary leukoencephalopathy with axonal spheroids: a spectrum of phenotypes from CNS vasculitis to parkinsonism in an adult onset leukodystrophy series. J Neurol Neurosurg Psychiatry 87: 512–519.

Lynch DS, Wade C, Paiva ARB et al. (2019). Practical approach to the diagnosis of adult-onset leukodystrophies: an updated guide in the genomic era. J Neurol Neurosurg Psychiatry 90: 543–554.

Magnotti LM, Goodenough DA, Paul DL (2011). Deletion of oligodendrocyte Cx32 and astrocyte Cx43 causes white matter vacuolation, astrocyte loss and early mortality. Glia 59: 1064–1074.

Man JHK, van Gelder C, Breur M et al. (2022). Cortical pathology in vanishing white matter. Cells 11: 3581.

Man JHK, van Gelder C, Breur M et al. (2023). Regional vulnerability of brain white matter in vanishing white matter. Acta Neuropathol Commun 11: 103.

Marintchev A, Ito T (2020). eIF2B and the integrated stress response: a structural and mechanistic view. Biochemistry 59: 1299–1308.

Mathis S, Scheper GC, Baumann N et al. (2008). The ovarioleukodystrophy. Clin Neurol Neurosurg 110: 1035–1037.

Matusova Z, Dykstra W, de Pablo Y et al. (2024). Aberrant neurodevelopment in human iPS cell-derived models of Alexander disease. Glia. https://doi.org/10.1002/glia.24618.

Melberg A, Hallberg L, Kalimo H et al. (2006). MR characteristics and neuropathology in adult-onset autosomal dominant leukodystrophy with autonomic symptoms. AJNR Am J Neuroradiol 27: 904–911.

Merzbacher L (1910). Eine eigenartige familiärhereditäre Erkrankungform (aplasia axialis extra-corticalis congenita). Z Gesamte Neurol Psychiat 3: 1–138.

Messing A (2018). Alexander disease. Handb Clin Neurol 148: 693–700.

Messing A, Head MW, Galles K et al. (1998). Fatal encephalopathy with astrocyte inclusions in GFAP transgenic mice. Am J Pathol 152: 391–398.

Messing A, Brenner M, Feany MB et al. (2012). Alexander disease. J Neurosci 32: 5017–5023.

Meyer RC, Giddens MM, Coleman BM et al. (2014). The protective role of prosaposin and its receptors in the nervous system. Brain Res 1585: 1–12.

Mignot C, Delarasse C, Escaich S et al. (2007). Dynamics of mutated GFAP aggregates revealed by real-time imaging of an astrocyte model of Alexander disease. Exp Cell Res 313: 2766–2779.

Min R, van der Knaap MS (2018). Genetic defects disrupting glial ion and water homeostasis in the brain. Brain Pathol 28: 372–387.

Morell P (1984). A correlative synopsis of the leukodystrophies. Neuropediatrics 15: 62–65.

Nasu T, Tsukahara Y, Terayama K et al. (1970). An autopsy case of "membranous lipodystrophy" with myeloosteopathy of long bones and leucodystrophy of the brain. The 59th Tokyo Meeting of Pathology, Tokyo Byori Shudankai Kiji, Tokyo, pp. 10–13.

Nicaise AM, Bongarzone ER, Crocker SJ (2016). A microglial hypothesis of globoid cell leukodystrophy pathology. J Neurosci Res 94: 1049–1061.

Norris K, Hodgson RE, Dornelles T et al. (2021). Mutational analysis of the alpha subunit of eIF2B provides insights into the role of eIF2B bodies in translational control and VWM disease. J Biol Chem 296: 100207.

Nualart-Marti A, Solsona C, Fields RD (2013). Gap junction communication in myelinating glia. Biochim Biophys Acta 1828: 69–78.

Okada Y (2024). Physiology of the volume-sensitive/regulatory anion channel VSOR/VRAC. Part 1: from its discovery and phenotype characterization to the molecular entity identification. J Physiol Sci 74: 3.

Olabarria M, Putilina M, Riemer EC et al. (2015). Astrocyte pathology in Alexander disease causes a marked inflammatory environment. Acta Neuropathol 130: 469–486.

Orcesi S, La Piana R, Fazzi E (2009). Aicardi-Goutières syndrome. Br Med Bull 89: 183–201.

Padiath QS, Saigoh K, Schiffmann R et al. (2006). Lamin B1 duplications cause autosomal dominant leukodystrophy. Nat Genet 38: 1114–1123.

Page KM, Stenger EO, Connelly JA et al. (2019). Hematopoietic stem cell transplantation to treat leukodystrophies: clinical practice guidelines from the Hunter's Hope leukodystrophy care network. Biol Blood Marrow Transplant 25: e363–e374.

Pareyson D, Fancellu R, Mariotti C et al. (2008). Adult-onset Alexander disease: a series of eleven unrelated cases with review of the literature. Brain 131: 2321–2331.

Passchier EMJ, Bisseling Q, Helman G et al. (2024). Megalencephalic leukoencephalopathy with subcortical cysts: a variant update and review of the literature. Front Genet 15: 1352947.

Pekny T, Faiz M, Wilhelmsson U et al. (2014). Synemin is expressed in reactive astrocytes and Rosenthal fibers in Alexander disease. APMIS 122: 76–80.

Pekny M, Pekna M, Messing A et al. (2016). Astrocytes: a central element in neurological diseases. Acta Neuropathol 131: 323–345.

Pelizaeus F (1898). Ueber eine eigenartige familiäre Entwickelungshemmung vornehmlich auf motorischem Gebiet. Arch Psychiatr Nervenkr 31: 101–104.

Perez-Rius C, Folgueira M, Elorza-Vidal X et al. (2019). Comparison of zebrafish and mice knockouts for megalencephalic leukoencephalopathy proteins indicates that GlialCAM/MLC1 forms a functional unit. Orphanet J Rare Dis 14: 268.

Perng MD, Wen SF, Gibbon T et al. (2008). Glial fibrillary acidic protein filaments can tolerate the incorporation of assembly-compromised GFAP-delta, but with consequences for filament organization and alphaB-crystallin association. Mol Biol Cell 19: 4521–4533.

Plog BA, Dashnaw ML, Hitomi E et al. (2015). Biomarkers of traumatic injury are transported from brain to blood via the glymphatic system. J Neurosci 35: 518–526.

Powell RD, Holland PJ, Hollis T et al. (2011). Aicardi-Goutières syndrome gene and HIV-1 restriction factor SAMHD1 is a dGTP-regulated deoxynucleotide triphosphohydrolase. J Biol Chem 286: 43596–43600.

Prust M, Wang J, Morizono H et al. (2011). GFAP mutations, age at onset, and clinical subtypes in Alexander disease. Neurology 77: 1287–1294.

Ramantani G, Kohlhase J, Hertzberg C et al. (2010). Expanding the phenotypic spectrum of lupus erythematosus in Aicardi-Goutières syndrome. Arthritis Rheum 62: 1469–1477.

Ramesh V, Bernardi B, Stafa A et al. (2010). Intracerebral large artery disease in Aicardi-Goutières syndrome implicates SAMHD1 in vascular homeostasis. Dev Med Child Neurol 52: 725–732.

Reijns MA, Rabe B, Rigby RE et al. (2012). Enzymatic removal of ribonucleotides from DNA is essential for mammalian genome integrity and development. Cell 149: 1008–1022.

Rice G, Newman WG, Dean J et al. (2007a). Heterozygous mutations in TREX1 cause familial chilblain lupus and dominant Aicardi-Goutières syndrome. Am J Hum Genet 80: 811–815.

Rice G, Patrick T, Parmar R et al. (2007b). Clinical and molecular phenotype of Aicardi-Goutières syndrome. Am J Hum Genet 81: 713–725.

Rice GI, Bond J, Asipu A et al. (2009). Mutations involved in Aicardi-Goutières syndrome implicate SAMHD1 as regulator of the innate immune response. Nat Genet 41: 829–832.

Rice GI, Kasher PR, Forte GM et al. (2012). Mutations in ADAR1 cause Aicardi-Goutières syndrome associated with a type I interferon signature. Nat Genet 44: 1243–1248.

Rolyan H, Tyurina YY, Hernandez M et al. (2015). Defects of lipid synthesis are linked to the age-dependent demyelination caused by Lamin B1 overexpression. J Neurosci 35: 12002–12017.

Rosenthal W (1898). Über eine eigenthümliche, mit syringomyelie complicirte geschwulst des rückenmarks. Beitr Pathol Anat 23: 111–143.

Russo Jr LS, Aron A, Anderson PJ (1976). Alexander's disease: a report and reappraisal. Neurology 26: 607–614.

Saito K, Shigetomi E, Yasuda R et al. (2018). Aberrant astrocyte Ca(2+) signals "AxCa signals" exacerbate pathological alterations in an Alexander disease model. Glia 66: 1053–1067.

Saito K, Shigetomi E, Shinozaki Y et al. (2024). Microglia sense astrocyte dysfunction and prevent disease progression in an Alexander disease model. Brain 147: 698–716.

Satoh J, Tabunoki H, Ishida T et al. (2011). Immunohistochemical characterization of microglia in Nasu-Hakola disease brains. Neuropathology 31: 363–375.

Schmitt A, Gofferje V, Weber M et al. (2003). The brain-specific protein MLC1 implicated in megalencephalic leukoencephalopathy with subcortical cysts is expressed in glial cells in the murine brain. Glia 44: 283–295.

Schoof M, Wang L, Cogan JZ et al. (2021). Viral evasion of the integrated stress response through antagonism of eIF2-P binding to eIF2B. Nat Commun 12: 7103.

Seil FJ, Schochet Jr SS, Earle KM (1968). Alexander's disease in an adult. Report of a case. Arch Neurol 19: 494–502.

Seitelberger F (1984). Structural manifestations of leukodystrophies. Neuropediatrics 15: 53–61.

Sirisi S, Folgueira M, Lopez-Hernandez T et al. (2014). Megalencephalic leukoencephalopathy with subcortical cysts protein 1 regulates glial surface localization of GLIALCAM from fish to humans. Hum Mol Genet 23: 5069–5086.

Sirisi S, Elorza-Vidal X, Arnedo T et al. (2017). Depolarization causes the formation of a ternary complex between GlialCAM, MLC1 and ClC-2 in astrocytes: implications in megalencephalic leukoencephalopathy. Hum Mol Genet 26: 2436–2450.

Song J, Goetz BD, Kirvell SL et al. (2001). Selective myelin defects in the anterior medullary velum of the taiep mutant rat. Glia 33: 1–11.

Sosunov AA, Guilfoyle E, Wu X et al. (2013). Phenotypic conversions of "protoplasmic" to "reactive" astrocytes in Alexander disease. J Neurosci 33: 7439–7450.

Sosunov A, Olabarria M, Goldman JE (2018). Alexander disease: an astrocytopathy that produces a leukodystrophy. Brain Pathol 28: 388–398.

Srivastava S, Waldman A, Naidu S (1993). Alexander disease. In: MP Adam, J Feldman, GM Mirzaa et al. (Eds.), GeneReviews(R). University of Washington, Seattle, WA.

Stellingwerff MD, Al-Saady ML, van de Brug T et al. (2021). MRI natural history of the leukodystrophy vanishing white matter. Radiology 300: 671–680.

Stetson DB, Ko JS, Heidmann T et al. (2008). Trex1 prevents cell-intrinsic initiation of autoimmunity. Cell 134: 587–598.

Sundblom J, Melberg A, Kalimo H et al. (2009). MR imaging characteristics and neuropathology of the spinal cord in adult-onset autosomal dominant leukodystrophy with autonomic symptoms. AJNR Am J Neuroradiol 30: 328–335.

Suzuki K (1998). Twenty five years of the "psychosine hypothesis": a personal perspective of its history and present status. Neurochem Res 23: 251–259.

Suzuki K, Suzuki Y (1970). Globoid cell leucodystrophy (Krabbe's disease): deficiency of galactocerebroside beta-galactosidase. Proc Natl Acad Sci U S A 66: 302–309.

Suzuki K, Suzuki K (1983). The twitcher mouse. A model of human globoid cell leukodystrophy (Krabbe's disease). Am J Pathol 111: 394–397.

Tanaka KF, Takebayashi H, Yamazaki Y et al. (2007). Murine model of Alexander disease: analysis of GFAP aggregate formation and its pathological significance. Glia 55: 617–631.

Tang G, Yue Z, Talloczy Z et al. (2008). Autophagy induced by Alexander disease-mutant GFAP accumulation is regulated by p38/MAPK and mTOR signaling pathways. Hum Mol Genet 17: 1540–1555.

Tang G, Perng MD, Wilk S et al. (2010). Oligomers of mutant glial fibrillary acidic protein (GFAP) inhibit the proteasome system in alexander disease astrocytes, and the small heat shock protein alphaB-crystallin reverses the inhibition. J Biol Chem 285: 10527–10537.

Tavasoli A, Armangue T, Ho CY et al. (2017). Alexander disease. J Child Neurol 32: 184–187.

Teijido O, Casaroli-Marano R, Kharkovets T et al. (2007). Expression patterns of MLC1 protein in the central and peripheral nervous systems. Neurobiol Dis 26: 532–545.

Thiele H, du Moulin M, Barczyk K et al. (2010). Cerebral arterial stenoses and stroke: novel features of Aicardi-Goutières syndrome caused by the Arg164X mutation in SAMHD1 are associated with altered cytokine expression. Hum Mutat 31: E1836–E1850.

Tian R, Gregor M, Wiche G et al. (2006). Plectin regulates the organization of glial fibrillary acidic protein in Alexander disease. Am J Pathol 168: 888–897.

Tian R, Wu X, Hagemann TL et al. (2010). Alexander disease mutant glial fibrillary acidic protein compromises glutamate transport in astrocytes. J Neuropathol Exp Neurol 69: 335–345.

Torii T, Miyamoto Y, Yamauchi J et al. (2014). Pelizaeus-Merzbacher disease: cellular pathogenesis and pharmacologic therapy. Pediatr Int 56: 659–666.

Untiet V, Verkhratsky A (2024). How astrocytic chloride modulates brain states. Bioessays 46: e2400004.

van der Knaap MS, Bugiani M (2017). Leukodystrophies: a proposed classification system based on pathological changes and pathogenetic mechanisms. Acta Neuropathol 134: 351–382.

van der Knaap MS, Breiter SN, Naidu S et al. (1999). Defining and categorizing leukoencephalopathies of unknown origin: MR imaging approach. Radiology 213: 121–133.

van der Knaap MS, Naidu S, Breiter SN et al. (2001). Alexander disease: diagnosis with MR imaging. AJNR Am J Neuroradiol 22: 541–552.

van der Knaap MS, Salomons GS, Li R et al. (2005). Unusual variants of Alexander's disease. Ann Neurol 57: 327–338.

van der Knaap MS, Pronk JC, Scheper GC (2006a). Vanishing white matter disease. Lancet Neurol 5: 413–423.

van der Knaap MS, Ramesh V, Schiffmann R et al. (2006b). Alexander disease: ventricular garlands and abnormalities of the medulla and spinal cord. Neurology 66: 494–498.

van der Knaap MS, Fogli A, Boespflug-Tanguy O et al. (1993). Childhood ataxia with central nervous system hypomyelination/vanishing white matter. In: MP Adam, J Feldman, GM Mirzaa et al. (Eds.), GeneReviews(R). University of Washington, Seattle, WA.

van der Knaap MS, Lai V, Kohler W et al. (2010). Megalencephalic leukoencephalopathy with cysts without MLC1 defect. Ann Neurol 67: 834–837.

van der Knaap MS, Boor I, Estevez R (2012). Megalencephalic leukoencephalopathy with subcortical cysts: chronic white matter oedema due to a defect in brain ion and water homoeostasis. Lancet Neurol 11: 973–985.

van der Knaap MS, Bugiani M, Abbink TEM (2024). Vanishing white matter. Handb Clin Neurol 204: 77–94.

van der Voorn JP, Pouwels PJ, Kamphorst W et al. (2005). Histopathologic correlates of radial stripes on MR images in lysosomal storage disorders. AJNR Am J Neuroradiol 26: 442–446.

van Heteren JT, Rozenberg F, Aronica E et al. (2008). Astrocytes produce interferon-alpha and CXCL10, but not IL-6 or CXCL8, in Aicardi-Goutières syndrome. Glia 56: 568–578.

Walker AK, Daniels CM, Goldman JE et al. (2014). Astrocytic TDP-43 pathology in Alexander disease. J Neurosci 34: 6448–6458.

Wang Q, Miyakoda M, Yang W et al. (2004). Stress-induced apoptosis associated with null mutation of ADAR1 RNA editing deaminase gene. J Biol Chem 279: 4952–4961.

Wang L, Hagemann TL, Kalwa H et al. (2015). Nitric oxide mediates glial-induced neurodegeneration in Alexander disease. Nat Commun 6: 8966.

Wang L, Hagemann TL, Messing A et al. (2016). An in vivo pharmacological screen identifies cholinergic signaling as a therapeutic target in glial-based nervous system disease. J Neurosci 36: 1445–1455.

Wang L, Xia J, Li J et al. (2018). Tissue and cellular rigidity and mechanosensitive signaling activation in Alexander disease. Nat Commun 9: 1899.

Wenger DA, Rafi MA, Luzi P et al. (2000). Krabbe disease: genetic aspects and progress toward therapy. Mol Genet Metab 70: 1–9.

Wohlwill FJ, Bernstein J, Yakovlev PI (1959). Dysmyelinogenic leukodystrophy; report of a case of a new, presumably familial type of leukodystrophy with megalobarencephaly. J Neuropathol Exp Neurol 18: 359–383.

Xin B, Jones S, Puffenberger EG et al. (2011). Homozygous mutation in SAMHD1 gene causes cerebral vasculopathy and early onset stroke. Proc Natl Acad Sci U S A 108: 5372–5377.

Yang YG, Lindahl T, Barnes DE (2007). Trex1 exonuclease degrades ssDNA to prevent chronic checkpoint activation and autoimmune disease. Cell 131: 873–886.

Yattah C, Hernandez M, Huang D et al. (2020). Dynamic Lamin B1-gene association during oligodendrocyte progenitor differentiation. Neurochem Res 45: 606–619.

Yoshida T, Sasaki M, Yoshida M et al. (2011). Nationwide survey of Alexander disease in Japan and proposed new guidelines for diagnosis. J Neurol 258: 1998–2008.

Yu Z, Yu P, Chen H et al. (2014). Targeted inhibition of KCa3.1 attenuates TGF-beta-induced reactive astrogliosis through the Smad2/3 signaling pathway. J Neurochem 130: 41–49.

Zerbin-Rudin E, Peiffer J (1964). Genetic contribution to the problem of the late form of Pelizaeus-Merzbacher disease. Humangenetik 1: 107–122.

Zhou L, Li P, Chen N et al. (2019). Modeling vanishing white matter disease with patient-derived induced pluripotent stem cells reveals astrocytic dysfunction. CNS Neurosci Ther 25: 759–771.

Handbook of Clinical Neurology, Vol. 210 (3rd series)
Neuroglia in Neurologic and Psychiatric Disorders, Part II
A. Verkhratsky, L.D. de Witte, E. Aronica, and E.M. Hol, Editors
https://doi.org/10.1016/B978-0-443-19102-2.00007-7

Chapter 11

Reciprocal interactions between glioma and tissue-resident cells fueling tumor progression

ALICE LAURENGE[1,2], LUIS JAIME CASTRO-VEGA[1], AND GILLES HUBERFELD[3,4*]

[1]*Genetics & Development of Brain Tumors Laboratory, ICM - Paris Brain Institute, Sorbonne University, UMR S 1127, Inserm U 1127, CNRS UMR 7225, F-75013, Paris, France*

[2]*AP-HP, Groupe Hospitalier Pitié-Salpêtrière, Neuro-Oncology Department, F-75013, Paris, France*

[3]*Université Paris Cité, Institute of Psychiatry and Neuroscience of Paris (IPNP), INSERM U1266, Neuronal Signaling in Epilepsy and Glioma, Paris, France*

[4]*Department of Neurology, Hôpital Fondation Adolphe de Rothschild, Paris, France*

Abstract

Gliomas are the most frequent primary brain tumor and are essentially incurable. While nondiffuse gliomas are circumscribed, diffuse gliomas display an aggressive behavior characterized by tumor cell migration over large distances into the brain parenchyma, thereby precluding curative surgical resection. Almost all diffuse gliomas progress and recur as higher grades and become resistant to standard-of-care treatments. It is being increasingly recognized that glioma cells establish functional interactions with cells residing in the tumor microenvironment. Of these, tumor-associated microglia and macrophages (TAMs) play critical roles in immunosuppression through modulation of the extracellular matrix, and the secretion of molecules such as cytokines, neurotrophic factors, and micro-RNAs (miRNAs). Conversely, glioma cell signals influence cell states and drive the metabolic reprogramming of TAMs. Similarly, emergent evidence indicates that neuronal activity influences glioma by released factors and by establishing functional synapses with glioma cells to promote tumor growth and invasion. Glioma cells also affect local neuronal activities and maintain connections through microtube gap junctions to amplify local effects. Here, we discuss the molecular mechanisms underlying bidirectional interactions between glioma cells and TAMs, as well as between glioma cells and neurons. A better understanding of these cellular cross talks is crucial for the development of novel therapeutic strategies for diffuse gliomas.

TAM-GLIOMA CELL INTERACTIONS

Compelling evidence from patient biopsies and in vivo murine models has established that the glioma mass comprises up to 50% myeloid cells (Roggendorf et al., 1996; Olah et al., 2012; Garcia et al., 2014; Huszthy et al., 2015; Zhang et al., 2015; Annovazzi et al., 2018). These tumor-associated microglia and macrophages (TAMs), henceforth referred to as TAMs, actively modulate the biology of the tumor, rather than merely serving as reactive cells. In this regard, they have long been known to establish an immunosuppressive tumor microenvironment (TME). Conversely, microenvironmental signals influence the phenotype and immune response of TAMs. In the following section, we will delve into the molecular mechanisms underlying TAM-glioma cell interactions, as well as the therapeutic implications (Fig. 11.1).

Soluble Factors Mediating TAM-Glioma Cross Talk

CHEMOKINES

Many chemokine receptors are highly expressed by TAMs and are important for their recruitment and activation (Hattermann et al., 2014; Laudati et al., 2017). For example, the monocyte chemoattractant protein-1 or chemokine ligand-2 (MCP-1/CCL2), secreted by tumor cells, is known to recruit macrophages and microglia into

*Correspondence to: Gilles Huberfeld, Université Paris Cité, Institute of Psychiatry and Neuroscience of Paris (IPNP), INSERM U1266, Neuronal Signaling in Epilepsy and Glioma, 75014 Paris, France; Department of Neurology, Hopital Fondation Adolphe de Rothschild, 75019 Paris, France. Tel: +33-1-40-78-92-87, Fax: +33-1-40-78-92-88, E-mail: gilles.huberfeld@inserm.fr

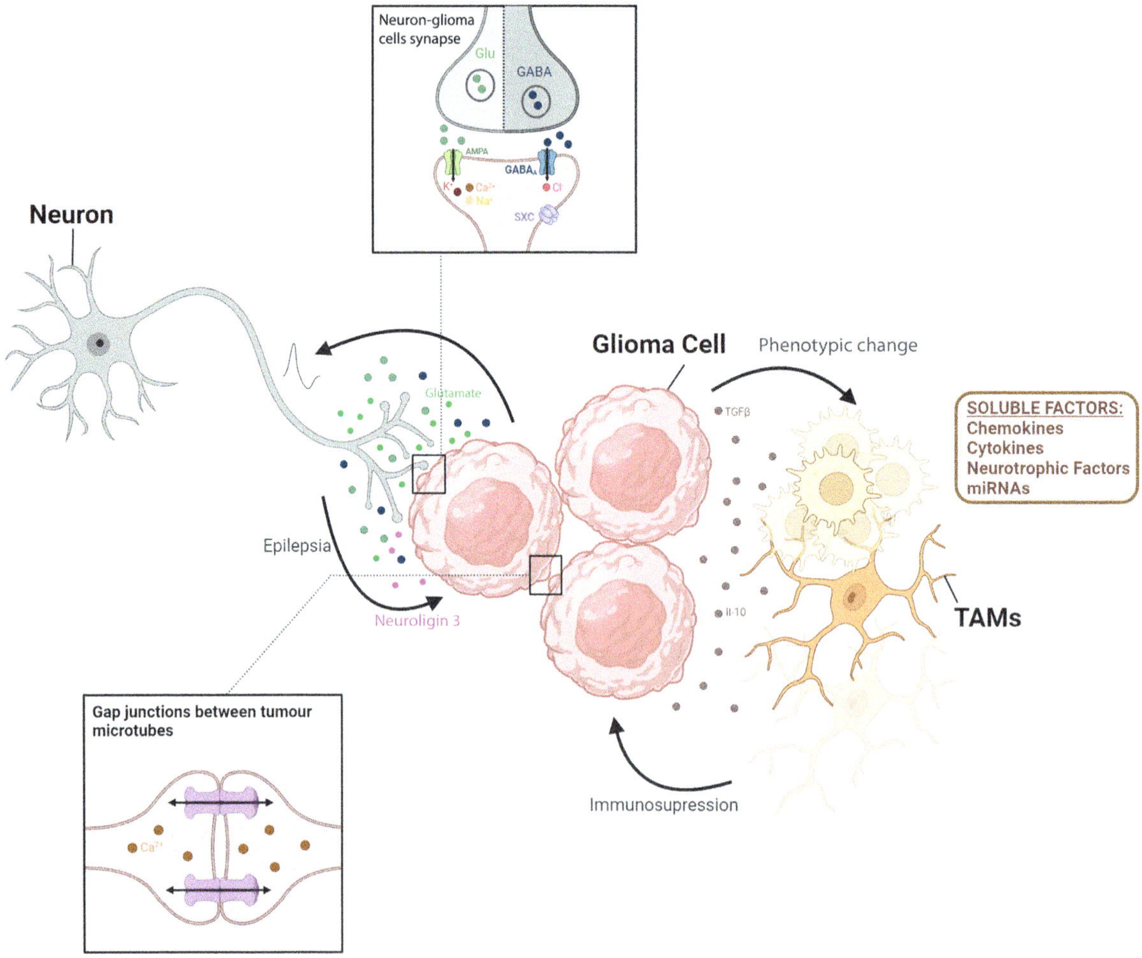

Fig. 11.1. A scheme on glioma cells, TAMs, and neuron interactions. Credit: Yvette Hayat.

tumor tissues (Platten et al., 2003; Jung et al., 2018; Felsenstein et al., 2020). Additionally, monocyte chemoattractant protein-3 or chemokine ligand-7 (MCP-3/CCL7) could play a crucial role in this recruitment (Okada et al., 2009). CCL2 also exerts other effects on TAMs: it notably increases the release of InterLeukine-6 (IL-6) by microglia in coculture systems in vitro, which leads to higher invasiveness of glioma cells (Zhang et al., 2012). Similarly to CCL2, macrophage colony-stimulating factor (M-CSF), encoded by the CSF1 gene and secreted by glioma cells, participates in the attraction of TAMs, with a reduction in TAM infiltration by colony-stimulating factor-1 receptor (CSF-1R) antagonism (Coniglio et al., 2012), particularly in demyelinated areas (Sielska et al., 2013). The chemokine (C-X-C motif) ligand 1 (CX3CL1), expressed by glioma cells, also leads to TAM recruitment via the chemokine (C-X-C motif) ligand-receptor CX3CR1 and increases the expression of matrix metalloproteinases (MMPs) 2, 9, and 14 by TAMs, thereby promoting tumor invasion, at least in vitro (Held-Feindt et al., 2010). Other TAM-recruiting chemokines secreted by glioma cells described recently include periostin (Zhou et al., 2015) and osteopontin, which also maintains the so-called M2 gene signature and phenotype (Wei et al., 2019).

Cytokines

Glioma cells and TAMs interact bidirectionally through cytokine secretion. For instance, transforming growth factor beta (TGF-β) and InterLeukine-10 (IL-10) have long been known to be highly expressed in glioblastoma (Huettner et al., 1997; Platten et al., 2001). TGF-β is secreted by both glioma cells (Sasaki et al., 1995) and microglia (Wesolowska et al., 2008). Glioma cells also stimulate the production of TGF-β by microglia; the concentration of TGF-β is indeed higher in microglia treated with glioma-conditioned medium (Zhu et al., 2016). TGF-β participates in the tumorigenicity of glioma cells; knockdown of the TGF-beta type II receptor in glioblastoma cells using short hairpin RNA strongly reduces the tumorigenicity of cells both in vivo and in coculture systems with microglia, resulting in a loss of microglial invasion-promoting activity (Wesolowska et al., 2008). TGF-β signaling also contributes to the stemness phenotype of glioma-initiating cells via the TGF-β-SRY-box Transcription factor 4-SRY-box Transcription factor 2 (TGF-β-Sox4-Sox2) pathway; inhibition of this TGF-β pathway significantly decreases the tumorigenicity of glioma-initiating cells by promoting their differentiation and shows a prognostic impact (Ikushima et al., 2009). TGF-β is also well-known for its immunosuppressive effects; it leads to the inhibition of the proliferation of mature T lymphocytes, limitation of their cytotoxic activity, or even induction of apoptosis (Platten et al., 2001). The cooperation between glioma cells and TAMs, with the former stimulating the secretion of TGF-β by the latter, thus contributes to the modulation of the lymphocytic compartment.

IL-10 and TGF-β are antiinflammatory cytokines enriched in the tumoral tissue, which are produced mainly by TAMs (Wagner et al., 1999; Wu et al., 2010). Other antiinflammatory characteristics include the downregulation of proinflammatory cytokine tumor necrosis factor (TNF) (Lang et al., 2002; Chuang et al., 2016), and reduction in the antigen-presentation capacity via downregulation of the major histocompatibility class (MHC) II and costimulatory molecules such as cluster of differentiation 86 (CD86) (Frei et al., 1994; Mittal et al., 2015). This occurs, at least in part, via activation of the signal transducer and activator of transcription 3 (STAT3) pathway (Riley et al., 1999). Glioma cells also release interleukin-4 (IL-4), whose receptors are highly expressed in TAMs, especially after IL-10 stimulation (Lang et al., 2002; Biswas et al., 2011), suggesting synergy between both cytokines. In this way, in vitro stimulation of macrophages by a combination of IL-10 and IL-4 leads to higher expression of antiinflammatory markers compared with single-cytokine treatment (Makita et al., 2015). Glioma cells are responsive to IL-10 released by TAMs, as they express the IL-10 receptors IL10RA and IL10RB (Akil et al., 2015; Donson et al., 2017), which activate the Janus kinase signal transducer and activator of transcription-3 (JAK STAT3) pathway. Thereby, this pathway participates in the regulation of glioma cell proliferation, migration, and invasion (Qi et al., 2016; Zhang et al., 2019). IL-10 could also indirectly increase glioma invasion via upregulation of the nuclear transporter KaryoPheriN subunit alpha 2 (KPNA2) (Zhang et al., 2019). Overall, some of these cytokines represent promising therapeutic targets currently at the preclinical stage (Lanza et al., 2021).

Neurotrophic factors

Neurotrophic factors are biomolecules that also participate in the interactions between TAMs and glioma cells. For instance, glioma cells secrete glial-derived neurotrophic factor (GDNF) (Ku et al., 2013), whose receptors GDNF family receptors GFRa-1 and GFRa-2 are expressed by TAMs (Honda et al., 1999; Rickert et al., 2014), allowing this neurotrophic factor to act as a chemoattractant, both in vitro and in vivo (Ku et al., 2013). Glioma cells also produce vascular endothelial growth factor (VEGF), known for its role in angiogenesis, which also contributes to the modulation of TAMs' activity (Turkowski et al., 2018). In a glioblastoma murine model, high levels of VEGF lead not only to a modification of the tumor's vascularization but also to a reduced infiltration of TAMs, enhancing the rate of apoptosis (Turkowski et al., 2018). Furthermore, the authors reported an increased expression of Major Histocompatibility Complexes MHCI and MHCII by TAMs in gliomas with high levels of VEGF, together with a downregulation of proinflammatory and suppressive cytokines (Turkowski et al., 2018). On the other hand, it has been reported that TAMs can also release VEGF (Chen et al., 2014; Brandenburg et al., 2016; Osterberg et al., 2016). Epidermal growth factor (EGF) is also involved in TAMs-glioma cell interactions. The use of an EGF receptor EGFR-specific inhibitor reduced microglia-stimulated glioblastoma invasion in an in vitro model, without impacting basal invasion (Coniglio et al., 2012). Expression of EGF by unpermeabilized microglia was reported in immunofluorescence assays, suggesting that EGF is associated with the plasma membrane (Coniglio et al., 2012) and stimulates glioma invasion via direct interactions with EGFR on glioma cells.

miRNAs

Micro-RNAs (miRNAs) are small noncoding RNAs (19–25-nucleotides) that regulate gene expression by posttranscriptional repression of messenger RNA (mRNA) targets. These molecules can be secreted in protein-bound and protein-free forms, as well as within extracellular vesicles. Trafficking of mRNAs in the TME has been reported between TAMs and glioma cells (Rooj et al., 2016).

Van der Vos et al. exposed primary mouse microglia to extracellular vesicles containing miRNAs released by primary human glioblastoma cells in vitro (van der Vos et al., 2016). After the uptake of vesicles containing high levels of miRNA-451 and miRNA-21 by microglia, increased proliferation and a shift in the cytokine profile toward immune suppression were observed. These effects were associated with a decrease in the mRNA levels of the miRNA-451 and miRNA-21 target, c-Myc. The authors confirmed the uptake of extracellular vesicles by TAMs in an in vivo murine model using intravital 2-photon microscopy and reported increased levels of miRNA-21, along with reduced levels of c-Myc mRNA.

Xu et al. reported that hypoxic glioma-derived exosomes, which contain a high level of IL-6 and miR-155-3p, facilitate autophagy and M2-like TAM polarization compared to normoxic glioma-derived exosomes. They demonstrated that this effect occurs through an IL-6-pSTAT3-miR-155-3p-autophagy-pSTAT3 positive feedback loop (Xu et al., 2021). Additionally, Azambuja et al. reported the release of Arginase-1+ extracellular vesicles by TAMs that promote glioma growth (Azambuja et al., 2020).

The cross talk between TAMs and glioma cells through miRNA could also occur without the use of extracellular vesicles. In particular, miRNA-146a was shown to be downregulated in microglia treated with a glioma-conditioned medium (Karthikeyan et al., 2018). Overexpression of miRNA-146a resulted in suppression of microglial migration linked to decreased expression of mothers against decapentaplegic homolog 4 (SMAD4), a transcription factor involved in TGF-β signal transduction in microglia. Glioma cells, on their part, exhibited decreased viability when treated with a conditioned medium derived from SMAD4 knockdown or miRNA-146a overexpressed microglia. Interestingly, a polymorphism in the pre-miRNA-146a gene was reported to be associated with susceptibility and outcome in glioma (Permuth-Wey et al., 2011). Overall, these results suggest that miRNA-146a-mediated participation in the cross talk between microglia and glioma promotes microglial migration and glioma cell viability (Guo et al., 2019b).

Moreover, miRNA-142-3p was shown to be the most downregulated miRNA between human glioblastoma-infiltrating macrophages and matched peripheral monocytes in a study using a miRNA microarray approach (Xu et al., 2014). In an in vivo murine model, administration of this miRNA led to a decrease in infiltrating macrophages, inhibition of glioma growth, and an increase in the median survival rate (Xu et al., 2014).

Oncometabolite D-2HG

A high proportion of gliomas harbor a mutation in the isocitrate dehydrogenase 1/2 (IDH) genes. Recent studies suggested that the IDH mutation shapes the cellular composition of the TME (Klemm et al., 2020; Friedrich et al., 2021). In particular, monocyte-derived macrophages (MDMs) predominate in gliomas without mutation, while microglia are abundant in IDH-mutant gliomas. Moreover, it has been shown that immune-related signaling pathways and chemotaxis factors are downregulated in IDH-mutant compared with IDH-wildtype gliomas (Amankulor et al., 2017). Recent evidence indicates that (D)-2-hydroxyglutarate (D-2HG), the oncometabolite released by IDH-mutant cells into the TME, may play immunomodulatory roles. For instance, D-2HG alters tryptophan metabolism in macrophages via activation of the aryl hydrocarbon receptor (AHR), driving an immunosuppressive state (Friedrich et al., 2021). Moreover, treatment of murine immortalized microglia with D-2HG precludes lipopolysaccharide (LPS)-mediated activation via the inhibition of adenosine triphosphate (ATP) synthase, which impairs phosphorylation in nuclear factor kappa B (NF-κB) and mammalian target of rapamycin (mTOR) signaling pathways (Han et al., 2019).

Modulation of the extracellular matrix

The extracellular matrix (ECM) is a dynamic and precisely regulated component of tissues that influences cell adhesion, migration, and proliferation, and participates in intercellular communication (Lu et al., 2012). TAMs and glioma cells can notably interact via the modulation of the ECM, which plays a role that goes far beyond simple mechanical support. Glioma cells notably migrate through identifiable anatomic pathways such as myelinated fiber tracts and blood vessels. In addition to these purely physical aspects, specific components of the ECM have been implicated in the regulation of glioma cell/TAMs invasion or activation state of TAMs.

TAMs may affect glioma cell migration by secreting matrix metalloproteases (MMPs) and by promoting the activation of pro-MMPs secreted by glioma cells themselves via membrane-type MMPs. In an organotypic brain slice model of glioma invasion, MMP-2 activity was notably higher in the presence of microglia; furthermore, in vitro data demonstrated that soluble factors released from glioma cells stimulate MMP-2 activity in microglia (Markovic et al., 2005). Similarly, glioma cells release soluble factors that induce the expression of membrane type 1 matrix metalloprotease (MT1-MMP) via activation of toll-like receptor 2 (TLR2), which promotes glioma invasion (Markovic et al., 2009; Vinnakota et al., 2013). For instance, mouse xenograft studies revealed that versican (or chondroitin sulfate proteoglycan 2 (CSPG2), a member of large chondroitin sulfate proteoglycans) acts as a ligand for TLRs on TAMs (Hu et al., 2015).

Tumor cells can also modulate the composition of the ECM to influence the state of microglial activation. For example, secretome proteomics has revealed that glioblastoma cells produce osteopontin and lactadherin. Additionally, a processed form of osteopontin specifically expressed by tumor cells attenuates the proinflammatory capacity of microglia while retaining their M2 phenotype (Ellert-Miklaszewska et al., 2016). Anatomopathologic studies of human high-grade diffuse glioma showed a correlation between tenascin expression and TAM density, suggesting that this substrate may favor TAM migration (Kulla et al., 2000). Xia et al. demonstrated that the loss of function of tenascin-C produced by glioblastoma cells influences the morphology of microglial cells toward an amoeboid-like phenotype (Xia et al., 2016). Furthermore, it was reported that tenascin-C stimulates the release of proinflammatory factors (including TNFα) by TAMs (Ma et al., 2019). More recently, Zhan et al. demonstrated that the inhibition of uridine diphosphate-glucose 6-dehydrogenase (UGDH), which participates in the synthesis of unbranched polysaccharides with repetitive units of disaccharide called UDP-glucuronic acid, in glioblastoma cells leads to an increase in the phagocytic capacity of TAMs (Zhan et al., 2022).

Contributions of TAM-Glioma Interactions for Tumor Behavior

Impact of glioma cells on TAM spatial organization

Glioma cells actively attract and recruit TAMs, to establish particular tumor niches. For instance, analysis of brain slices from a rat model of glioma using two-color time-lapse fluorescence microscopy showed the induction of microglia motility in the presence of glioma cells (Juliano et al., 2018). Different mechanisms responsible for this active recruitment were mentioned earlier. However, the relative contribution of those factors to the attraction of resident microglia or peripheral MDMs to the tumor remains unclear (Liu et al., 2022). Microglia and macrophages colonize different regions of diffuse gliomas. For instance, in a GL261 mouse model of glioblastoma, using chemokine (C-X-C motif) ligand receptor 1–green fluorescent protein, chemokine (C—C motif) receptor 2 wild type (CX3CR1GFP/WT; CCR2RFP/WT) double knock-in mice, Chen et al. demonstrated that macrophages/monocytes localize preferentially to perivascular areas, whereas resident microglia are prevalent in peritumoral regions (Chen et al., 2017b). Moreover, the recruitment of TAMs varies according to the type of diffuse glioma. Notably, by comparing TAMs from syngeneic pair mouse models of IDH1-mutant and IDH-wildtype gliomas; it was reported a reduced macrophage and microglia infiltration in mutant gliomas associated with lower levels of chemokine expression (Amankulor et al., 2017). Moreover, the TAMs content differs between astrocytomas and oligodendrogliomas, with higher myeloid infiltration in astrocytomas (Zhao et al., 2022), which correlates with the report of significant overexpression of chemotaxis factors colony stimulating factor 1 (CSF1) and

FMS-like tyrosine kinase 3 ligand (FLT3LG) in alpha thalassemia/mental retardation X linked (ATRX)-mutated astrocytomas (scRNA-seq and scATAC-seq data) (Babikir et al., 2021). Differences in TAM recruitment were also reported in low-grade gliomas. For example, Guo et al. established a series of Nf1 mutant optic glioma mouse strains harboring the coexistence of different additional genomic alterations, and they observed different patterns of chemokine expression, which differentially attract microglia (Guo et al., 2019a). These examples illustrate that glioma cells attract and recruit TAMs heterogeneously according to their molecular alterations and the grade of the glioma. This heterogeneity of cancer cells thus contributes to the observed heterogeneity of tumor microenvironments between different gliomas, but also between different areas of the same glioma in a given patient.

Impact of glioma cells on TAM polarization

TAM diversity is associated not only with different ontogenetic origins or localization within the TME but also with different activation states. Indeed, once recruited to the glioma microenvironment, TAMs acquire new molecular characteristics. The functions of TAMs are modulated by glioma-derived factors such as chemokines, neurotrophic factors, miRNA, or ECM components, as mentioned earlier. The particular transcriptional programs of TAMs resulting from these interactions have been well studied using glioma mouse models. For example, Szulzewsky et al. compared the expression profiles of TAMs from GL261-implanted mouse brains with those of naive control cells (Szulzewsky et al., 2015). They reported a specific TAM phenotype, for both microglia and invading macrophages, that does not correspond to the M1/M2 classification scheme, but instead includes a mixture of M1 and M2a,b,c-specific genes (Szulzewsky et al., 2015). They notably reported the increased expression of GPNMB and Spp1 by microglia from mouse and human gliomas, which was associated with poor prognosis in human glioblastoma (Szulzewsky et al., 2015).

The polarization of TAMs differs according to the tumor type. For example, TAMs from the mesenchymal-like glioblastoma subtype also display a mesenchymal-like phenotype. It has been suggested that this is achieved via ligands produced by glioblastoma cells that bind cognate receptors expressed by TAMs (Hara et al., 2021). The oncometabolite D-2HG produced in IDH-mutant glioma cells leads to a reduced expression of proinflammatory chemokines by murine microglia in vitro (Han et al., 2019). Additionally, the upregulation of CD163, a marker of immunosuppressive myeloid cells, is observed in ATRX-mutant astrocytomas compared with oligodendrogliomas (Babikir et al., 2021). The ATRX loss of function was confirmed in an ATRX-knockout mouse model (Babikir et al., 2021). Transcriptional analyses using single-cell RNA-seq approaches revealed not only high TAM heterogeneity between gliomas but also strong spatial heterogeneity within the same tumor for both glioma and myeloid cells. These analyses also identified ligand–receptor pairs for glioma-myeloid and myeloid-glioma signaling across all gliomas (Abdelfattah et al., 2022).

Moreover, the polarization of microglia by glioma is underpinned by distinct histone modifications. Maleszewska et al. showed that cultured microglia preexposed to glioma-conditioned medium present a reduced inflammatory response after rechallenging with lipopolysaccharide (Maleszewska et al., 2021). They observed that a glioma-conditioned medium leads to an increased expression and enzymatic activity of histone deacetylases, which leads to the erasure of histone acetylation at tested genes. The reduced expression of later inflammatory genes is associated with repressive histone marks such as H3K27 trimethylation. They showed that histone deacetylase inhibitors block the histone modifications induced by glioma-conditioned medium, which restores the inflammatory potential of microglia (Maleszewska et al., 2021). Altogether, these examples illustrate how glioma cells drive the epigenetic reprogramming of TAMs toward an immunosuppressive phenotype, ultimately favoring tumor progression.

Glioma cells drive the metabolic reprogramming of TAMs

It is well-known that TAMs preferentially utilize glycolysis, fatty acid synthesis, and glutamine-glutamate metabolism, as opposed to mitochondrial oxidative phosphorylation in normal cells, influencing their polarization (Mazzone et al., 2018; Puthenveetil and Dubey, 2020). For instance, Carroll et al. demonstrated that fatty acid synthase, involved in the synthesis of long-chain fatty acids, is essential for toll-like receptor activation of macrophages after stimulation by lipopolysaccharide (Carroll et al., 2018). TAMs adapt to the available nutrients in the tumor microenvironment, which are commonly limited in glucose and oxygen but contain high levels of fatty acids. Additionally, they are directly remodeled by metabolites released from cancer cells, acting as metabolic signals. While the influence of glucose, tryptophan, L-arginine, cysteine, glutamine, lactate, adenosine, or itaconic acid on TAM functions has been well described in many cancers, there is less data in the field of glioma (Won et al., 2019).

Metabolic switches in microglial cells in inflammatory environments have been reported in other pathologies such as Alzheimer disease. For instance, Baik et al. demonstrated using metabolic profiling that exposure of murine microglia to amyloid-β in vitro induces acute microglial inflammation, characterized by altered morphology, enhanced phagocytic activity, and increased expression of proinflammatory cytokines, alongside metabolic reprogramming from oxidative phosphorylation to aerobic glycolysis (Baik et al., 2019). Once activated, microglia exhibit altered energy metabolism, leading to a chronic tolerant phase with diminished immune responses (Baik et al., 2019). In Parkinson disease, it has been hypothesized that hyperactivity of glucose-6-phosphate dehydrogenase in microglia, resulting in increased nicotinamide adenine dinucleotide phosphate (NADPH) production and subsequent radical oxygen species (ROS) production, could lead to dopaminergic neurotoxicity (Tu et al., 2019). Metabolic changes similar to those observed in neurodegenerative diseases could play an important role in the context of chronic inflammation in gliomas.

Impact of TAMs on tumor growth

Markovic et al. effectively demonstrated the role of TAMs in promoting glioma growth. First, they provided in situ evidence of this role: the growth of gliomas was reduced in organotypic brain slices when microglia were depleted using liposomes filled with clodronate (Markovic et al., 2005). Second, they utilized an in vivo approach: microglia depletion attenuated glioma expansion in a mouse model where the herpes simplex virus thymidine kinase gene was expressed under the control of the CD11b promoter (Markovic et al., 2009). The mechanisms underlying this effect have been at least partially described, notably through the production of cytokines by TAMs, such as TGF-b, IL-6, IL-1b, or EGF. Additionally, microglia release the cochaperone stress-inducible protein 1, which increases the proliferation of glioblastoma cells both in vitro and in vivo (Carvalho da Fonseca et al., 2014).

TAMs also contribute to glioma growth by interacting with glioma stem cells. In vitro studies have shown that naive microglia can reduce the sphere-forming capacity of glioma patient-derived glioma stem cells, whereas TAMs from glioma patients do not have such an effect (Sarkar et al., 2014). Similarly, Shi et al. reported abundant secretion of pleiotrophin by TAMs, whose receptor PTPRZ1 is expressed by glioma stem cells (Shi et al., 2017). They demonstrated in vivo that the coimplantation of M2-like macrophages promotes glioma stem cell-driven tumor growth, which is reduced when pleiotrophin expression in macrophages is silenced. Moreover, using shRNA or anti-PTPRZ1 antibody to block pleiotrophin protein tyrosine phosphatase receptor type Z1 (PTPRZ1) signaling, suppressed glioblastoma tumor growth and resulted in prolonged animal survival (Shi et al., 2017). Gangoso et al. conducted serial transplantation of glioma stem cells into immunocompetent hosts and demonstrated that these cells have the capability to establish a favorable tumor microenvironment. They reported stimulation by TAMs of transcriptional changes responsible for immune evasion in glioma stem cells in coculture experiments (Gangoso et al., 2021).

There is also evidence of the role of TAMs in the growth of lower-grade gliomas. For instance, Pong et al. showed that optic glioma formation is delayed in mice with reduced expression of CX3CR1, which is involved in microglia migration (Pong et al., 2013). Similarly, in vivo optic glioma growth is reduced after administration of the microglial inhibitor minocycline (Daginakatte and Gutmann, 2007), or a neutralizing antibody against chemokine (C—C motif) ligand 5 (CCL5), which is a mitogen for glioma cells released by microglia (Solga et al., 2015).

Impact of TAMs on tumor migration and invasion

Modulation of the extracellular matrix composition by TAMs influences the migration capability of glioma cells. Using an organotypic brain slice model of glioma, Markovic et al. demonstrated that the invasiveness of tumors is dependent on the presence of microglia (Markovic et al., 2005). Further in vitro assays have shown that microglia stimulate the invasion of glioblastoma cells (Coniglio et al., 2012), in an STI1-dependent manner (da Fonseca et al., 2012). Similarly, Liu et al. showed an increase in the migration ability of glioma cells when cocultured with TAMs, and further linked this effect to the TGF-β pathway (Liu et al., 2018). More recently, Zhang et al. reported an increase in pseudopodia formation in a hydromatrix 3D cell culture model of glioma cells in the presence of CCL8, which is highly expressed by TAMs (Zhang et al., 2020). Consistent results were observed in a murine model. After 12 days of implantation of tumor cells, more tumor cells were present and multiplied at a distance from the initial tumor site in the presence of CCL8 (Zhang et al., 2020).

TAMs and glioma cells interact bidirectionally via many different mechanisms, and these interactions have a huge impact on the development and evolution of the tumor. The influence of these interactions on the survival of patients with different molecular subtypes of glioma and the resistance to classical therapies remain to be elucidated. Nevertheless, the targeting of these bidirectional interactions could open up promising avenues for treatments.

NEURON-GLIOMA INTERACTIONS

The brain's nervous tissue is not a passive host invaded by autonomous and self-determined glioma cells. Accumulating data indicate that while tumors grow within an excitable cortex, they generate a neuronal response, such as epilepsy. Furthermore, the neuronal environment affects the proliferation, migration, and invasive capacities of glioma cells. Neuronal activity drives glioma growth via released activity-dependent mitogens or by synaptic signaling. Neurons also directly establish functional synapses with glioma cells to promote tumor growth through glutamatergic pathways, a feature reminiscent of the glial or neuronal origin of glioma cells (oligodendrocyte precursor cells or neural precursor cells). Moreover, glioma cells establish connections through microtube gap junctions so that the local effects of neuronal activity may spread, making the developing glioma a singular network within the brain. In the following section, we will survey the mechanistic underpinnings establishing neuron-glioma interactions (Fig. 11.1).

Neuronal activity promotes tumor growth while glioma cells affect local neuronal network activities

Using optogenetics in animal models xenografted with human glioblastoma cells was instrumental in uncovering that increased neuronal activity from peritumoral areas directly influences tumor growth (Venkatesh et al., 2015, 2019; Tantillo et al., 2020). Conversely, reducing or dampening neuronal activity through anesthesia decreases both the proliferation and migration of glioma cells (Venkataramani et al., 2019). Additionally, it has been shown that optogenetic over-activation of inhibitory interneurons reduces the number of tumor cells undergoing division (Tantillo et al., 2020). The effects of neuronal activity and synaptic signaling on glioma dynamics have also been demonstrated in a more "natural" setting by grafting the tumor into the visual cortex and controlling light exposure, yielding similar findings (Tantillo et al., 2020). Nevertheless, translation to humans remains to be established. Indeed, epileptic discharges represent transient extreme neuronal activities and theoretically could promote glioma growth.

Notably, patients with low-grade gliomas and glioblastomas who experience epilepsy often have longer survival rates (Lote et al., 1998; Pallud et al., 2014a, 2024), which represent a paradoxical situation. The presence and severity of epilepsy may be positively correlated with tumor growth features (Vecht et al., 2014), and oncological treatments such as surgery, chemotherapy, and radiotherapy reduce seizure frequency along with their antitumor effects (Huberfeld and Vecht, 2016). Additionally, tumors associated with epilepsy tend to exhibit limited volume and slower progression (Berendsen et al., 2016). IDH mutations in both low and high-grade gliomas contribute to prolonged survival but also increased epileptogenicity (Chen et al., 2017a). Conversely, epileptic and oncogenic processes share several mechanisms (Huberfeld and Vecht, 2016), and in murine models of glioma, the onset of epilepsy and tumor progression are linked to the emergence of specific subtypes of astrocyte-derived subpopulations, which promote synaptogenesis (John Lin et al., 2017).

Neuronal activities in the gamma range recorded by ElectroCorticoGraphy in cortices surrounding gliomas and infiltrated by tumor cells have recently been reported to be enhanced as compared to those recorded from noninfiltrated areas (Krishna et al., 2023). However, analysis across all frequency bands suggests that peritumoral cortices display specific ElectroCorticoGraphy features at rest (Díaz-Fernández, in progress). Peritumoral areas also exhibit higher connectivity, which correlates with severity. Highly active/connected regions show overexpression of several genes, including axon pathfinding genes, synapse-associated genes, and synaptogenic factors. Among them, thrombospondin-1 is upregulated in certain populations of glioma cells, contributing to increased synapse formation and neuronal connectivity. Moreover, thrombospondin-1 expression promotes tumor cell division, invasion, and the extension of connected microtube-based networks. Patients with higher peritumoral connectivity were characterized by reduced survival and increased thrombospondin-1 expression. These findings clearly highlight the bidirectional interactions between glioma growth, the expression of specific genes, neuronal activity, connectivity, and synaptogenesis.

Neuronal activity-driven glioma growth is also mediated by secreted factors, as the medium in which optogenetically-activated neurons are grown exerts activity-dependent progrowth effects (Venkatesh et al., 2015). Screening for potent excreted mitogens has highlighted the role of BDNF (brain-derived neurotrophic factor) as well as the synaptic adhesion molecule Neuroligin-3. Neuroligin-3 is cleaved by the protease ADAM Metallopeptidase Model 10 (ADAM10), and its secretion by neurons is activity-dependent. Neuroligin-3 acts on glioma cells through both paracrine and autocrine effects, triggering its own secretion by glioma cells and favoring cell division via the phosphoinositide-3-kinase-protein kinase B (Akt)-mammalian target of rapamycin (PI3K-Akt-mTOR) pathway. Additionally, neuroligin-3 upregulates synapse-associated genes, the BDNF receptor gene neurotrophic receptor tyrosine kinase 2 (NTRK2), and tweety homolog-1 (TTHY1), which is involved in establishing glioma cell microtube networks (Venkatesh et al., 2017). Neuroligin-3 serves as a prognostic marker in glioblastoma, as patients expressing Neuroligin-3 below median levels survive on average 20.8 months (95% CI, 17.1–24.4), while patients with expression levels higher than the median survive 15.2 months (95% CI, 13.1–17.2). Finally, Neuroligin-3 appears to be crucial in oncogenesis, as xenografted glioblastoma cells fail to grow in knockout mice.

Remote neuronal activity may also promote tumor progression, necessitating consideration of global brain activity as a macroenvironment. In a mouse model of glioblastoma where tumor cells are grafted in one hemisphere, activation of contralateral neurons via designer receptor exclusively activated by designer drugs (DREADD) resulted in increased proliferation indexes and infiltration (Huang-Hobbs et al., 2023), emphasizing the role of callosal projections. In this model, a subpopulation of tumor cells was identified at the invasion front, characterized by their dependence on neuronal activity. Semaphorin-4F overexpression was proposed as an explanation for this behavior, as it mediates both activity-dependent infiltration and increased neuronal activities bilaterally.

Interactions between invasive glioma cells, other glioma cells, and neurons and glia are complex and highly dynamic. Migrating tumor cells become disconnected from other tumor cells but receive synaptic inputs from local neurons. They exhibit features of neuronal and neural-progenitor-like tumor cell states, consistent with neuronal migration behavior (Venkataramani et al., 2022). A key characteristic of their migration pattern relies on local environment probing, facilitated by the emission of new microtubes driven by glutamatergic AMPA synapses. This invasion pattern of glioma cells has been suggested to "hijack" neuronal properties.

However, the network organization of interconnected glioma cells in the tumor core also plays a key role in pathophysiology (Hausmann et al., 2023). Tumor cells are linked via microtubes, at the extremity of which gap junctions allow interconnections and propagation of calcium waves within the network. Some of these tumor cells serve as key components of the network and drive activities by displaying periodic calcium oscillations. These hub cells have been characterized and express the calcium-activated potassium channel KCa3.1 (also known as IK1, SK4, or KCNN4). Blockade of this conductance abolishes these intrinsic autonomous glioma cell network oscillations and prolongs survival. In human patients, high tumoral expression of KCa3.1 is associated with reduced survival, confirming data from animal models.

Synaptic influences on glioma cells

Apart from the effects of neuronal activity on glioma cells behavior—mediated by various activity-dependent mitogens—, glioma cells express functional receptors for several neurotransmitters, which constitutes another level of exposure to the neuronal environment. As a proof of concept, treatment with tricyclic antidepressants is associated with a reduced risk of developing a glioma in humans (Walker et al., 2011) and reduces glioma development in mice (Shchors et al., 2015). Among multiple signaling systems, glioblastoma cells express fully functional dopamine receptors, including the D2 (Li et al., 2014) and D4 subtypes (Dolma et al., 2016), whose blockade affects glioblastoma growth. The role of serotonin 5-HT7 receptors, also upregulated (Mahé et al., 2004), is less obvious since treatment with selective serotonin reuptake inhibitors does not affect survival (Otto-Meyer et al., 2020). However, two widespread systems have been shown to play a key role in glioma pathophysiology: GABAergic and glutamatergic signaling.

Glutamatergic signaling and glioma excitatory drive

Glutamatergic excitatory synaptic signaling is now recognized as a major microenvironmental driver of glioma growth. The extracellular glutamate concentration can increase up to 100-fold in both gliomas and the surrounding tissue in humans (Marcus et al., 2010) as well as in mouse models (Buckingham et al., 2011). However, periglioma glutamate concentration (as well as other markers of glutamate metabolism) measured from postoperative tissues, although a predictor of epilepsy, did not correlate with overall survival (Neal et al., 2016). A major provider for extracellular glutamate accumulation in glioblastoma is the cystine/glutamate antiporter Xc system (SXC-SLC7A11) (Ye and Sontheimer, 1999; Savaskan et al., 2008; Buckingham et al., 2011; Takano et al., 2012). SXC is an antiport that loads cells with cystine against glutamate and is primarily expressed by astrocytes. Intracellular cystine is converted to cysteine and then into the antiradical species glutathione, with a benefit for glioma cell survival (Kandil et al., 2010). SXC is upregulated in glioblastoma cells (Ye and Sontheimer, 1999; Savaskan et al., 2008, 2011), under the control of the EGF (Tsuchihashi et al., 2016) and largely contributes to the increase in extracellular glutamate (Lyons et al., 2007). About half of postoperative tissues from glioblastoma patients show an overexpression of SXC (Robert et al., 2015), which is associated with a significant 9-month survival reduction. In an animal model of glioblastoma, overexpression of SXC in xenografted tumor cells is associated with increased epileptogenicity and reduced survival (Robert et al., 2015), while its blockade by sulfasalazine or RNA interference (Lyons et al., 2007; Savaskan et al., 2008) reduces tumor growth, improves survival (Lyons et al., 2007; Robert et al., 2015; Sleire et al., 2015) and reduces tumor epileptogenicity (Buckingham et al., 2011). In humans, a single dose of sulfasalazine reduces brain glutamate concentration (measured in magnetic resonance spectroscopy) by 20%–50%, in correlation with SXC expression (Robert et al., 2015). In addition, extracellular glutamate uptake by astrocytes is impaired in glioma tissues due to the downregulation of the transporters excitatory aminoacid transporters EAAT1 (Glt-1) & EAAT2 (Glast) (Ye et al., 1999). As a proof of concept of its effects in glioma, EAAT2 upregulation by peroxisome proliferator-activated receptor gamma (PPARγ) reduces tumor growth in animals (de Groot et al., 2005). Furthermore, in a retrospective study of a cohort of diabetic patients, treatment by an antidiabetic compound with PPARγ activating effects, prolonged survival from 6 to 19 months in case of glioblastoma occurrence (Grommes et al., 2010).

More recently, direct "bona fide" synapses from neurons to glioma cells have been evidenced both in animal models and human tissue, especially in astrocytomas and glioblastomas (Venkataramani et al., 2019; Venkatesh et al., 2019), providing a new source for glutamatergic drive for gliomas. These synapses, connecting the axons of pyramidal cells with the microtubules of glioma cells, were observed in 10%–16% of glioma cells. They primarily consisted of postsynaptic glutamatergic AMPA receptors, with a high prevalence of edited GluR2 subunits, which allow for increased Ca^{++} permeability. These synapses were shown to be functional by recording postsynaptic glioma cells both by patch-clamp and Ca^{++} imaging. AMPA-like excitatory postsynaptic potentials were recorded together with K^+ currents. They displayed Ca^{++} waves that propagated through the glioma cells network connected by microtubes. The formation of AMPAergic synapses on glioma cells depends on the BDNF/tropomyosin-related kinase B (TrkB)/Ca^{2+}/calmodulin-dependent protein kinase II (CAMKII) pathway (Taylor et al., 2023) and depolarizing synaptic inputs correlate with tumor aggressiveness.

Glutamatergic signaling, either mediated by accumulating ambient or synaptically-released glutamate, leads to the activation of several pathways depending on intracellular Ca^{++} concentration, including focal adhesion kinase, mitogen-activated protein kinase (MAPK), Akt, leading to both division and migration of tumor cells (Ishiuchi et al., 2002; Venkataramani et al., 2019; Venkatesh et al., 2019). This pathway can be inhibited by silencing AMPA receptors or using antagonists, including the approved antiepileptic drug Perampanel. Moreover, in humans, the noncompetitive AMPA receptor blocker, Talampanel, has shown favorable effects on glioblastoma patients' survival (Grossman et al., 2009). Glutamatergic signaling in peritumoral areas, both through AMPA and NMDA signaling, is also involved in the genesis of epileptic activities (Buckingham et al., 2011; Pallud et al., 2014b; Huberfeld and Vecht, 2016; Augustin et al., 2018) and in excitotoxic effects (Ye and Sontheimer, 1999; Takano et al., 2001; Savaskan et al., 2011) that will deeply affect the microenvironment.

GABAergic signaling and chloride regulation in gliomas

Gamma aminobutyric acid (GABA) is known to regulate glioma cell proliferation (Young and Bordey, 2009). Glioma cells express ionotropic $GABA_A$ receptors with a different status according to their state of malignancy, functional receptors being identified on low-grade glioma cells while nonfunctional receptors mostly formed by Theta subunits are expressed by glioblastoma cells (Labrakakis et al., 1998; Smits et al., 2012). Although ambient GABA concentration reports are controversial (Beaumont and Whittle, 2000; Bianchi et al., 2004; Shamji et al., 2009), inhibitory interneurons are globally reduced and reshaped in the tumoral environment (Haglund et al., 1992; Wolf et al., 1996; Marco et al., 1997).

GABA effects on glioma cells are mediated by chloride. Ion fluxes, responsible for variations in cellular osmolarity and volume, are involved in mitosis and growth (Watkins and Sontheimer, 2011) among which chloride fluxes are key both for proliferation and migration (Habela et al., 2008). During the intermitotic period, Chloride accumulates in glioma cells due to the effects of the sodium/potassium/chloride cotransporter NKCC1 (Ernest et al., 2005; Habela et al., 2008; Haas and Sontheimer, 2010; Garzon-Muvdi et al., 2012) whose expression is correlated with the histologic grade (Garzon-Muvdi et al., 2012). Both mitosis and migration are characterized by a reduction in cellular volume under the osmotic pressure induced by chloride efflux through ClC and KCC channels (Ernest et al., 2005; Habela et al., 2008; Cuddapah et al., 2012) and K^+ efflux, triggered by intracellular Ca^{++} increase related to glutamatergic AMPA and bradykinin pathways (Lyons et al., 2007). Strategies including blockade of NKCC1-mediated chloride load by bumetanide (Haas and Sontheimer, 2010) or by upstream regulation of with-no-lysin WNK3 kinase (Haas et al., 2011) or blockade of ClC3-mediated chloride efflux (Lyons et al., 2002; Lui et al., 2010) reduce tumor growth. In such processes, GABA plays a key role by disrupting chloride accumulation.

Similar dysregulation of neuronal chloride regulation (Conti et al., 2011; Pallud et al., 2014b) have been implicated in the depolarizing effects of GABA, resulting in epilepsy, linking at another level the neuronal microenvironment and glioma growth.

CONCLUDING REMARKS

Bilateral and dynamic interactions between glioma and cells residing in the TME contribute to tumor progression. TAMs are highly abundant cells in gliomas and contribute to maintaining an immunosuppressive environment. In addition, glioma cells integrate the brain environment through neuron-glioma interactions, which are reminiscent of the glial or neuronal origin of glioma cells (oligodendrocyte precursor cells or neural precursor cells). It is important to consider the whole brain activity, including epilepsy, in the management of these aggressive tumors still with a dismal prognosis. It is expected that this knowledge will pave the way for the development of new therapies targeting such interactions.

ACKNOWLEDGMENTS

This work was supported by grants from the European Research Council to G.H. (consolidator grant #865592) and the French National Cancer Institute (INCa-PLBIO22-243). Work in the Genetics & Development of Bain Tumors Laboratory is supported by the grants: Fondation Bristol Myers Squibb pour la Recherche en Immuno-Oncologie (BMS 2104009NA) and Entreprises contre le Cancer Paris-Île-de-France (GEFLUC R20202DD). The laboratory is also supported by La Ligue Nationale contre le Cancer (Équipe Labellisée) and by the Site de Recherche Intégré sur le Cancer (SiRIC CURAMUS). A.L. is supported by the Fondation Recherche Médicale (FRM) scholarship.

References

Abdelfattah N, Kumar P, Wang C et al. (2022). Single-cell analysis of human glioma and immune cells identifies S100A4 as an immunotherapy target. Nat Commun 13: 767. https://doi.org/10.1038/s41467-022-28372-y.

Akil H, Abbaci A, Lalloué F et al. (2015). IL22/IL-22R pathway induces cell survival in human glioblastoma cells. PLoS One 10: e0119872. https://doi.org/10.1371/journal.pone.0119872.

Amankulor NM, Kim Y, Arora S et al. (2017). Mutant IDH1 regulates the tumor-associated immune system in gliomas. Genes Dev 31: 774–786. https://doi.org/10.1101/gad.294991.116.

Annovazzi L, Mellai M, Bovio E et al. (2018). Microglia immunophenotyping in gliomas. Oncol Lett 15: 998–1006. https://doi.org/10.3892/ol.2017.7386.

Augustin K, Williams S, Cunningham M et al. (2018). Perampanel and decanoic acid show synergistic action against AMPA receptors and seizures. Epilepsia 59: e172–e178. https://doi.org/10.1111/epi.14578.

Azambuja JH, Ludwig N, Yerneni SS et al. (2020). Arginase-1+ exosomes from reprogrammed macrophages promote glioblastoma progression. Int J Mol Sci 21: 3990. https://doi.org/10.3390/ijms21113990.

Babikir H, Wang L, Shamardani K et al. (2021). ATRX regulates glial identity and the tumor microenvironment in IDH-mutant glioma. Genome Biol 22: 311. https://doi.org/10.1186/s13059-021-02535-4.

Baik SH, Kang S, Lee W et al. (2019). A breakdown in metabolic reprogramming causes microglia dysfunction in Alzheimer's disease. Cell Metab 30: 493–507.e6. https://doi.org/10.1016/j.cmet.2019.06.005.

Beaumont A, Whittle IR (2000). The pathogenesis of tumour associated epilepsy. Acta Neurochir 142: 1–15. https://doi.org/10.1007/s007010050001.

Berendsen S, Varkila M, Kroonen J et al. (2016). Prognostic relevance of epilepsy at presentation in glioblastoma patients. Neuro Oncol 18: 700–706. https://doi.org/10.1093/neuonc/nov238.

Bianchi L, De Micheli E, Bricolo A et al. (2004). Extracellular levels of amino acids and choline in human high grade gliomas: an intraoperative microdialysis study. Neurochem Res 29: 325–334. https://doi.org/10.1023/b:nere.0000010462.72557.6d.

Biswas A, Bhattacharya A, Kar S et al. (2011). Expression of IL-10-triggered STAT3-dependent IL-4Rα is required for induction of arginase 1 in visceral leishmaniasis. Eur J Immunol 41: 992–1003. https://doi.org/10.1002/eji.201040940.

Brandenburg S, Müller A, Turkowski K et al. (2016). Resident microglia rather than peripheral macrophages promote vascularization in brain tumors and are source of alternative pro-angiogenic factors. Acta Neuropathol 131: 365–378. https://doi.org/10.1007/s00401-015-1529-6.

Buckingham SC, Campbell SL, Haas BR et al. (2011). Glutamate release by primary brain tumors induces epileptic activity. Nat Med 17: 1269–1274. https://doi.org/10.1038/nm.2453.

Carroll RG, Zasłona Z, Galván-Peña S et al. (2018). An unexpected link between fatty acid synthase and cholesterol synthesis in proinflammatory macrophage activation. J Biol Chem 293: 5509–5521. https://doi.org/10.1074/jbc.RA118.001921.

Carvalho da Fonseca AC, Wang H, Fan H et al. (2014). Increased expression of stress inducible protein 1 in glioma-associated microglia/macrophages. J Neuroimmunol 274: 71–77. https://doi.org/10.1016/j.jneuroim.2014.06.021.

Chen X, Zhang L, Zhang IY et al. (2014). RAGE expression in tumor-associated macrophages promotes angiogenesis in glioma. Cancer Res 74: 7285–7297. https://doi.org/10.1158/0008-5472.CAN-14-1240.

Chen H, Judkins J, Thomas C et al. (2017a). Mutant IDH1 and seizures in patients with glioma. Neurology 88: 1805–1813. https://doi.org/10.1212/WNL.0000000000003911.

Chen Z, Feng X, Herting CJ et al. (2017b). Cellular and molecular identity of tumor-associated macrophages in glioblastoma. Cancer Res 77: 2266–2278. https://doi.org/10.1158/0008-5472.CAN-16-2310.

Chuang Y, Hung ME, Cangelose BK et al. (2016). Regulation of the IL-10-driven macrophage phenotype under incoherent stimuli. Innate Immun 22: 647–657. https://doi.org/10.1177/1753425916668243.

Coniglio SJ, Eugenin E, Dobrenis K et al. (2012). Microglial stimulation of glioblastoma invasion involves epidermal growth factor receptor (EGFR) and colony stimulating factor 1 receptor (CSF-1R) signaling. Mol Med 18: 519–527. https://doi.org/10.2119/molmed.2011.00217.

Conti L, Palma E, Roseti C et al. (2011). Anomalous levels of cl- transporters cause a decrease of GABAergic inhibition in human peritumoral epileptic cortex. Epilepsia 52: 1635–1644. https://doi.org/10.1111/j.1528-1167.2011.03111.x.

Cuddapah VA, Habela CW, Watkins S et al. (2012). Kinase activation of ClC-3 accelerates cytoplasmic condensation during mitotic cell rounding. Am J Physiol Cell Physiol 302: C527–C538. https://doi.org/10.1152/ajpcell.00248.2011.

da Fonseca ACC, Romão L, Amaral RF et al. (2012). Microglial stress inducible protein 1 promotes proliferation and migration in human glioblastoma cells. Neuroscience 200: 130–141. https://doi.org/10.1016/j.neuroscience.2011.10.025.

Daginakatte GC, Gutmann DH (2007). Neurofibromatosis-1 (Nf1) heterozygous brain microglia elaborate paracrine factors that promote Nf1-deficient astrocyte and glioma growth. Hum Mol Genet 16: 1098–1112. https://doi.org/10.1093/hmg/ddm059.

de Groot JF, Liu TJ, Fuller G et al. (2005). The excitatory amino acid transporter-2 induces apoptosis and decreases glioma growth in vitro and in vivo. Cancer Res 65: 1934–1940. https://doi.org/10.1158/0008-5472.CAN-04-3626.

Dolma S, Selvadurai HJ, Lan X et al. (2016). Inhibition of dopamine receptor D4 impedes autophagic flux, proliferation, and survival of glioblastoma stem cells. Cancer Cell 29: 859–873. https://doi.org/10.1016/j.ccell.2016.05.002.

Donson AM, Apps J, Griesinger AM et al. (2017). Molecular analyses reveal inflammatory mediators in the solid component and cyst fluid of human Adamantinomatous Craniopharyngioma. J Neuropathol Exp Neurol 76: 779–788. https://doi.org/10.1093/jnen/nlx061.

Ellert-Miklaszewska A, Wisniewski P, Kijewska M et al. (2016). Tumour-processed osteopontin and lactadherin drive the protumorigenic reprogramming of microglia and glioma progression. Oncogene 35: 6366–6377. https://doi.org/10.1038/onc.2016.55.

Ernest NJ, Weaver AK, Van Duyn LB et al. (2005). Relative contribution of chloride channels and transporters to regulatory volume decrease in human glioma cells. Am J Physiol Cell Physiol 288: C1451–C1460. https://doi.org/10.1152/ajpcell.00503.2004.

Felsenstein M, Blank A, Bungert AD et al. (2020). CCR2 of tumor microenvironmental cells is a relevant modulator of glioma biology. Cancers (Basel) 12: 1882. https://doi.org/10.3390/cancers12071882.

Frei K, Lins H, Schwerdel C et al. (1994). Antigen presentation in the central nervous system. The inhibitory effect of IL-10 on MHC class II expression and production of cytokines depends on the inducing signals and the type of cell analyzed. J Immunol 152: 2720–2728.

Friedrich M, Sankowski R, Bunse L et al. (2021). Tryptophan metabolism drives dynamic immunosuppressive myeloid states in IDH-mutant gliomas. Nat Cancer 2: 723–740. https://doi.org/10.1038/s43018-021-00201-z.

Gangoso E, Southgate B, Bradley L et al. (2021). Glioblastomas acquire myeloid-affiliated transcriptional programs via epigenetic immunoediting to elicit immune evasion. Cell 184: 2454–2470.e26. https://doi.org/10.1016/j.cell.2021.03.023.

Garcia C, Dubois LG, Xavier AL et al. (2014). The orthotopic xenotransplant of human glioblastoma successfully recapitulates glioblastoma-microenvironment interactions in a non-immunosuppressed mouse model. BMC Cancer 14: 923. https://doi.org/10.1186/1471-2407-14-923.

Garzon-Muvdi T, Schiapparelli P, ap Rhys C et al. (2012). Regulation ofbrain tumor dispersal by NKCC1 through a novel role in focal adhesion regulation. PLoS Biol 10: e1001320. https://doi.org/10.1371/journal.pbio.1001320.

Grommes C, Conway DS, Alshekhlee A et al. (2010). Inverse association of PPARγ agonists use and high grade glioma development. J Neurooncol 100: 233–239. https://doi.org/10.1007/s11060-010-0185-x.

Grossman SA, Ye X, Chamberlain M et al. (2009). Talampanel with standard radiation and temozolomide in patients with newly diagnosed glioblastoma: a multicenter phase II trial. J Clin Oncol 27: 4155–4161. https://doi.org/10.1200/JCO.2008.21.6895.

Guo X, Pan Y, Gutmann DH (2019a). Genetic and genomic alterations differentially dictate low-grade glioma growth through cancer stem cell-specific chemokine recruitment of T cells and microglia. Neuro Oncol 21: 1250–1262. https://doi.org/10.1093/neuonc/noz080.

Guo Y, Hong W, Wang X et al. (2019b). MicroRNAs in microglia: how do microRNAs affect activation, inflammation, polarization of microglia and mediate the interaction between microglia and glioma? Front Mol Neurosci 12: 125. https://doi.org/10.3389/fnmol.2019.00125.

Haas BR, Sontheimer H (2010). Inhibition of the sodium-potassium-chloride cotransporter Isoform-1 reduces glioma invasion. Cancer Res 70: 5597–5606. https://doi.org/10.1158/0008-5472.CAN-09-4666.

Haas BR, Cuddapah VA, Watkins S et al. (2011). With-no-lysine kinase 3 (WNK3) stimulates glioma invasion by regulating cell volume. Am J Physiol Cell Physiol 301: C1150–C1160. https://doi.org/10.1152/ajpcell.00203.2011.

Habela CW, Olsen ML, Sontheimer H (2008). ClC3 is a critical regulator of the cell cycle in Normal and malignant glial cells. J Neurosci 28: 9205–9217. https://doi.org/10.1523/JNEUROSCI.1897-08.2008.

Haglund MM, Berger MS, Kunkel DD et al. (1992). Changes in gamma-aminobutyric acid and somatostatin in epileptic cortex associated with low-grade gliomas. J Neurosurg 77: 209–216. https://doi.org/10.3171/jns.1992.77.2.0209.

Han C-J, Zheng J-Y, Sun L et al. (2019). The oncometabolite 2-hydroxyglutarate inhibits microglial activation via the AMPK/mTOR/NF-κB pathway. Acta Pharmacol Sin 40: 1292–1302. https://doi.org/10.1038/s41401-019-0225-9.

Hara T, Chanoch-Myers R, Mathewson ND et al. (2021). Interactions between cancer cells and immune cells drive transitions to mesenchymal-like states in glioblastoma. Cancer Cell 39: 779–792.e11. https://doi.org/10.1016/j.ccell.2021.05.002.

Hattermann K, Sebens S, Helm O et al. (2014). Chemokine expression profile of freshly isolated human glioblastoma-associated macrophages/microglia. Oncol Rep 32: 270–276. https://doi.org/10.3892/or.2014.3214.

Hausmann D, Hoffmann DC, Venkataramani V et al. (2023). Autonomous rhythmic activity in glioma networks drives brain tumour growth. Nature 613: 179–186. https://doi.org/10.1038/s41586-022-05520-4.

Held-Feindt J, Hattermann K, Müerköster SS et al. (2010). CX3CR1 promotes recruitment of human glioma-infiltrating microglia/macrophages (GIMs). Exp Cell Res 316: 1553–1566. https://doi.org/10.1016/j.yexcr.2010.02.018.

Honda S, Nakajima K, Nakamura Y et al. (1999). Rat primary cultured microglia express glial cell line-derived neurotrophic factor receptors. Neurosci Lett 275: 203–206. https://doi.org/10.1016/s0304-3940(99)00769-7.

Hu F, Dzaye OD, Hahn A et al. (2015). Glioma-derived versican promotes tumor expansion via glioma-associated microglial/macrophages toll-like receptor 2 signaling. Neuro Oncol 17: 200–210. https://doi.org/10.1093/neuonc/nou324.

Huang-Hobbs E, Cheng Y-T, Ko Y et al. (2023). Remote neuronal activity drives glioma progression through SEMA4F. Nature 619: 844–850. https://doi.org/10.1038/s41586-023-06267-2.

Huberfeld G, Vecht CJ (2016). Seizures and gliomas—towards a single therapeutic approach. Nat Rev Neurol 12: 204–216. https://doi.org/10.1038/nrneurol.2016.26.

Huettner C, Czub S, Kerkau S et al. (1997). Interleukin 10 is expressed in human gliomas in vivo and increases glioma cell proliferation and motility in vitro. Anticancer Res 17: 3217–3224.

Huszthy PC, Sakariassen PØ, Espedal H et al. (2015). Engraftment of human glioblastoma cells in immunocompetent rats through acquired immunosuppression. PLoS One 10: e0136089. https://doi.org/10.1371/journal.pone.0136089.

Ikushima H, Todo T, Ino Y et al. (2009). Autocrine TGF-beta signaling maintains tumorigenicity of glioma-initiating cells through Sry-related HMG-box factors. Cell Stem Cell 5: 504–514. https://doi.org/10.1016/j.stem.2009.08.018.

Ishiuchi S, Tsuzuki K, Yoshida Y et al. (2002). Blockage of Ca2+−permeable AMPA receptors suppresses migration and induces apoptosis in human glioblastoma cells. Nat Med 8: 971–978. https://doi.org/10.1038/nm746.

John Lin C-C, Yu K, Hatcher A et al. (2017). Identification of diverse astrocyte populations and their malignant analogs. Nat Neurosci 20: 396–405. https://doi.org/10.1038/nn.4493.

Juliano J, Gil O, Hawkins-Daarud A et al. (2018). Comparative dynamics of microglial and glioma cell motility at the infiltrative margin of brain tumours. J R Soc Interface 15: 20170582. https://doi.org/10.1098/rsif.2017.0582.

Jung Y, Ahn S-H, Park H et al. (2018). MCP-1 and MIP-3α secreted from necrotic cell-treated glioblastoma cells promote migration/infiltration of microglia. Cell Physiol Biochem 48: 1332–1346. https://doi.org/10.1159/000492092.

Kandil S, Brennan L, McBean GJ (2010). Glutathione depletion causes a JNK and p38MAPK-mediated increase in expression of cystathionine-γ-lyase and upregulation of the transsulfuration pathway in C6 glioma cells. Neurochem Int 56: 611–619. https://doi.org/10.1016/j.neuint.2010.01.004.

Karthikeyan A, Gupta N, Tang C et al. (2018). Microglial SMAD4 regulated by microRNA-146a promotes migration of microglia which support tumor progression in a glioma environment. Oncotarget 9: 24950–24969. https://doi.org/10.18632/oncotarget.25116.

Klemm F, Maas RR, Bowman RL et al. (2020). Interrogation of the microenvironmental landscape in brain tumors reveals disease-specific alterations of immune cells. Cell 181: 1643–1660.e17. https://doi.org/10.1016/j.cell.2020.05.007.

Krishna S, Choudhury A, Keough MB et al. (2023). Glioblastoma remodelling of human neural circuits decreases survival. Nature 617: 599–607. https://doi.org/10.1038/s41586-023-06036-1.

Ku M-C, Wolf SA, Respondek D et al. (2013). GDNF mediates glioblastoma-induced microglia attraction but not astrogliosis. Acta Neuropathol 125: 609–620. https://doi.org/10.1007/s00401-013-1079-8.

Kulla A, Liigant A, Piirsoo A et al. (2000). Tenascin expression patterns and cells of monocyte lineage: relationship in human gliomas. Mod Pathol 13: 56–67. https://doi.org/10.1038/modpathol.3880010.

Labrakakis C, Patt S, Hartmann J et al. (1998). Functional GABA(a) receptors on human glioma cells. Eur J Neurosci 10: 231–238. https://doi.org/10.1046/j.1460-9568.1998.00036.x.

Lang R, Patel D, Morris JJ et al. (2002). Shaping gene expression in activated and resting primary macrophages by IL-10. J Immunol 169: 2253–2263. https://doi.org/10.4049/jimmunol.169.5.2253.

Lanza M, Casili G, Campolo M et al. (2021). Immunomodulatory effect of microglia-released cytokines in gliomas. Brain Sci 11: 466. https://doi.org/10.3390/brainsci11040466.

Laudati E, Currò D, Navarra P et al. (2017). Blockade of CCR5 receptor prevents M2 microglia phenotype in a microglia-glioma paradigm. Neurochem Int 108: 100–108. https://doi.org/10.1016/j.neuint.2017.03.002.

Li J, Zhu S, Kozono D et al. (2014). Genome-wide shRNA screen revealed integrated mitogenic signaling between dopamine receptor D2 (DRD2) and epidermal growth factor receptor (EGFR) in glioblastoma. Oncotarget 5: 882–893. https://doi.org/10.18632/oncotarget.1801.

Liu Z, Kuang W, Zhou Q et al. (2018). TGF-β1 secreted by M2 phenotype macrophages enhances the stemness and migration of glioma cells via the SMAD2/3 signalling pathway. Int J Mol Med 42: 3395–3403. https://doi.org/10.3892/ijmm.2018.3923.

Liu X, Liu Y, Qi Y et al. (2022). Signal pathways involved in the interaction between tumor-associated macrophages/TAMs and glioblastoma cells. Front Oncol 12: 822085. https://doi.org/10.3389/fonc.2022.822085.

Lote K, Stenwig AE, Skullerud K et al. (1998). Prevalence and prognostic significance of epilepsy in patients with gliomas. Eur J Cancer 34: 98–102. https://doi.org/10.1016/s0959-8049(97)00374-2.

Lu P, Weaver VM, Werb Z (2012). The extracellular matrix: a dynamic niche in cancer progression. J Cell Biol 196: 395–406. https://doi.org/10.1083/jcb.201102147.

Lui VCH, Lung SSS, Pu JKS et al. (2010). Invasion of human glioma cells is regulated by multiple chloride channels including ClC-3. Anticancer Res 30: 4515–4524.

Lyons SA, O'Neal J, Sontheimer H (2002). Chlorotoxin, a scorpion-derived peptide, specifically binds to gliomas and tumors of neuroectodermal origin. Glia 39: 162–173. https://doi.org/10.1002/glia.10083.

Lyons SA, Chung WJ, Weaver AK et al. (2007). Autocrine glutamate signaling promotes glioma cell invasion. Cancer Res 67: 9463–9471. https://doi.org/10.1158/0008-5472.CAN-07-2034.

Ma D, Liu S, Lal B et al. (2019). Extracellular matrix protein tenascin C increases phagocytosis mediated by CD47 loss of function in glioblastoma. Cancer Res 79: 2697–2708. https://doi.org/10.1158/0008-5472.CAN-18-3125.

Mahé C, Bernhard M, Bobirnac I et al. (2004). Functional expression of the serotonin 5-HT $_7$ receptor in human glioblastoma cell lines. Br J Pharmacol 143: 404–410. https://doi.org/10.1038/sj.bjp.0705936.

Makita N, Hizukuri Y, Yamashiro K et al. (2015). IL-10 enhances the phenotype of M2 macrophages induced by IL-4 and confers the ability to increase eosinophil migration. Int Immunol 27: 131–141. https://doi.org/10.1093/intimm/dxu090.

Maleszewska M, Steranka A, Smiech M et al. (2021). Sequential changes in histone modifications shape transcriptional responses underlying microglia polarization by glioma. Glia 69: 109–123. https://doi.org/10.1002/glia.23887.

Marco P, Sola RG, Ramón y Cajal S et al. (1997). Loss of inhibitory synapses on the soma and axon initial segment of pyramidal cells in human epileptic peritumoural neocortex: implications for epilepsy. Brain Res Bull 44: 47–66. https://doi.org/10.1016/s0361-9230(97)00090-7.

Marcus HJ, Carpenter KLH, Price SJ et al. (2010). In vivo assessment of high-grade glioma biochemistry using microdialysis: a study of energy-related molecules, growth factors and cytokines. J Neurooncol 97: 11–23. https://doi.org/10.1007/s11060-009-9990-5.

Markovic DS, Glass R, Synowitz M et al. (2005). Microglia stimulate the invasiveness of glioma cells by increasing the activity of metalloprotease-2. J Neuropathol Exp Neurol 64: 754–762. https://doi.org/10.1097/01.jnen.0000178445.33972.a9.

Markovic DS, Vinnakota K, Chirasani S et al. (2009). Gliomas induce and exploit microglial MT1-MMP expression for tumor expansion.

Proc Natl Acad Sci U S A 106: 12530–12535. https://doi.org/10.1073/pnas.0804273106.

Mazzone M, Menga A, Castegna A (2018). Metabolism and TAM functions-it takes two to tango. FEBS J 285: 700–716. https://doi.org/10.1111/febs.14295.

Mittal SK, Cho K-J, Ishido S et al. (2015). Interleukin 10 (IL-10)-mediated immunosuppression: MARCH-I INDUCTION REGULATES ANTIGEN PRESENTATION BY MACROPHAGES BUT NOT DENDRITIC CELLS. J Biol Chem 290: 27158–27167. https://doi.org/10.1074/jbc.M115.682708.

Neal A, Yuen T, Bjorksten AR et al. (2016). Peritumoural glutamate correlates with post-operative seizures in supratentorial gliomas. J Neurooncol 129: 259–267. https://doi.org/10.1007/s11060-016-2169-y.

Okada M, Saio M, Kito Y et al. (2009). Tumor-associated macrophage/microglia infiltration in human gliomas is correlated with MCP-3, but not MCP-1. Int J Oncol 34: 1621–1627. https://doi.org/10.3892/ijo_00000292.

Olah M, Raj D, Brouwer N et al. (2012). An optimized protocol for the acute isolation of human microglia from autopsy brain samples. Glia 60: 96–111. https://doi.org/10.1002/glia.21251.

Osterberg N, Ferrara N, Vacher J et al. (2016). Decrease of VEGF-A in myeloid cells attenuates glioma progression and prolongs survival in an experimental glioma model. Neuro Oncol 18: 939–949. https://doi.org/10.1093/neuonc/now005.

Otto-Meyer S, DeFaccio R, Dussold C et al. (2020). A retrospective survival analysis of glioblastoma patients treated with selective serotonin reuptake inhibitors. Brain Behav Immun Health 2: 100025. https://doi.org/10.1016/j.bbih.2019.100025.

Pallud J, Audureau E, Blonski M et al. (2014a). Epileptic seizures in diffuse low-grade gliomas in adults. Brain 137: 449–462. https://doi.org/10.1093/brain/awt345.

Pallud J, Le Van QM, Bielle F et al. (2014b). Cortical GABAergic excitation contributes to epileptic activities around human glioma. Sci Transl Med 6: . https://doi.org/10.1126/scitranslmed.3008065.

Pallud J, Roux A, Moiraghi A et al. (2024). Characteristics and prognosis of tumor-related epilepsy during tumor evolution in patients with IDH wild-type glioblastoma. Neurology 102: e207902. https://doi.org/10.1212/WNL.0000000000207902.

Permuth-Wey J, Thompson RC, Burton Nabors L et al. (2011). A functional polymorphism in the pre-miR-146a gene is associated with risk and prognosis in adult glioma. J Neurooncol 105: 639–646. https://doi.org/10.1007/s11060-011-0634-1.

Platten M, Wick W, Weller M (2001). Malignant glioma biology: role for TGF-beta in growth, motility, angiogenesis, and immune escape. Microsc Res Tech 52: 401–410. https://doi.org/10.1002/1097-0029(20010215)52:4<401::AID-JEMT1025>3.0.CO;2-C.

Platten M, Kretz A, Naumann U et al. (2003). Monocyte chemoattractant protein-1 increases microglial infiltration and aggressiveness of gliomas. Ann Neurol 54: 388–392. https://doi.org/10.1002/ana.10679.

Pong WW, Higer SB, Gianino SM et al. (2013). Reduced microglial CX3CR1 expression delays neurofibromatosis-1 glioma formation. Ann Neurol 73: 303–308. https://doi.org/10.1002/ana.23813.

Puthenveetil A, Dubey S (2020). Metabolic reprograming of tumor-associated macrophages. Ann Transl Med 8: 1030. https://doi.org/10.21037/atm-20-2037.

Qi L, Yu H, Zhang Y et al. (2016). IL-10 secreted by M2 macrophage promoted tumorigenesis through interaction with JAK2 in glioma. Oncotarget 7: 71673–71685. https://doi.org/10.18632/oncotarget.12317.

Rickert U, Grampp S, Wilms H et al. (2014). Glial cell line-derived neurotrophic factor family members reduce microglial activation via inhibiting p38MAPKs-mediated inflammatory responses. J Neurodegener Dis 2014: 369468. https://doi.org/10.1155/2014/369468.

Riley JK, Takeda K, Akira S et al. (1999). Interleukin-10 receptor signaling through the JAK-STAT pathway. Requirement for two distinct receptor-derived signals for anti-inflammatory action. J Biol Chem 274: 16513–16521. https://doi.org/10.1074/jbc.274.23.16513.

Robert SM, Buckingham SC, Campbell SL et al. (2015). SLC7A11 expression is associated with seizures and predicts poor survival in patients with malignant glioma. Sci Transl Med 7. https://doi.org/10.1126/scitranslmed.aaa8103.

Roggendorf W, Strupp S, Paulus W (1996). Distribution and characterization of microglia/macrophages in human brain tumors. Acta Neuropathol 92: 288–293. https://doi.org/10.1007/s004010050520.

Rooj AK, Mineo M, Godlewski J (2016). MicroRNA and extracellular vesicles in glioblastoma: small but powerful. Brain Tumor Pathol 33: 77–88. https://doi.org/10.1007/s10014-016-0259-3.

Sarkar S, Döring A, Zemp FJ et al. (2014). Therapeutic activation of macrophages and microglia to suppress brain tumor-initiating cells. Nat Neurosci 17: 46–55. https://doi.org/10.1038/nn.3597.

Sasaki A, Naganuma H, Satoh E et al. (1995). Secretion of transforming growth factor-beta 1 and -beta 2 by malignant glioma cells. Neurol Med Chir (Tokyo) 35: 423–430. https://doi.org/10.2176/nmc.35.423.

Savaskan NE, Heckel A, Hahnen E et al. (2008). Small interfering RNA–mediated xCT silencing in gliomas inhibits neurodegeneration and alleviates brain edema. Nat Med 14: 629–632. https://doi.org/10.1038/nm1772.

Savaskan NE, Seufert S, Hauke J et al. (2011). Dissection of mitogenic and neurodegenerative actions of cystine and glutamate in malignant gliomas. Oncogene 30: 43–53. https://doi.org/10.1038/onc.2010.391.

Shamji MF, Fric-Shamji EC, Benoit BG (2009). Brain tumors and epilepsy: pathophysiology of peritumoral changes. Neurosurg Rev 32: 275–284. discussion 284–286. https://doi.org/10.1007/s10143-009-0191-7.

Shchors K, Massaras A, Hanahan D (2015). Dual targeting of the Autophagic regulatory circuitry in gliomas with repurposed drugs elicits cell-lethal autophagy and therapeutic benefit. Cancer Cell28: 456–471. https://doi.org/10.1016/j.ccell.2015.08.012.

Shi Y, Ping Y-F, Zhou W et al. (2017). Tumour-associated macrophages secrete pleiotrophin to promote PTPRZ1 signalling in glioblastoma stem cells for tumour growth. Nat Commun 8: 15080. https://doi.org/10.1038/ncomms15080.

Sielska M, Przanowski P, Wylot B et al. (2013). Distinct roles of CSF family cytokines in macrophage infiltration and activation in glioma progression and injury response. J Pathol 230: 310–321. https://doi.org/10.1002/path.4192.

Sleire L, Skeie BS, Netland IA et al. (2015). Drug repurposing: sulfasalazine sensitizes gliomas to gamma knife radiosurgery by blocking cystine uptake through system Xc−, leading to glutathione depletion. Oncogene 34: 5951–5959. https://doi.org/10.1038/onc.2015.60.

Smits A, Jin Z, Elsir T et al. (2012). GABA-A Channel subunit expression in human glioma correlates with tumor histology and clinical outcome. PLoS One 7: e37041. https://doi.org/10.1371/journal.pone.0037041.

Solga AC, Pong WW, Kim K-Y et al. (2015). RNA sequencing of tumor-associated microglia reveals Ccl5 as a stromal chemokine critical for Neurofibromatosis-1 glioma growth. Neoplasia 17: 776–788. https://doi.org/10.1016/j.neo.2015.10.002.

Szulzewsky F, Pelz A, Feng X et al. (2015). Glioma-associated microglia/macrophages display an expression profile different from M1 and M2 polarization and highly express Gpnmb and Spp1. PLoS One 10: e0116644. https://doi.org/10.1371/journal.pone.0116644.

Takano T, Lin JH-C, Arcuino G et al. (2001). Glutamate release promotes growth of malignant gliomas. Nat Med 7: 1010–1015. https://doi.org/10.1038/nm0901-1010.

Takano H, McCartney M, Ortinski PI et al. (2012). Deterministic and stochastic neuronal contributions to distinct synchronous CA3 network bursts. J Neurosci 32: 4743–4754. https://doi.org/10.1523/JNEUROSCI.4277-11.2012.

Tantillo E, Vannini E, Cerri C et al. (2020). Differential roles of pyramidal and fast-spiking, GABAergic neurons in the control of glioma cell proliferation. Neurobiol Dis 141: 104942. https://doi.org/10.1016/j.nbd.2020.104942.

Taylor KR, Barron T, Hui A et al. (2023). Glioma synapses recruit mechanisms of adaptive plasticity. Nature 623: 366–374. https://doi.org/10.1038/s41586-023-06678-1.

Tsuchihashi K, Okazaki S, Ohmura M et al. (2016). The EGF receptor promotes the malignant potential of glioma by regulating amino acid transport system xc(−). Cancer Res 76: 2954–2963. https://doi.org/10.1158/0008-5472.CAN-15-2121.

Tu D, Gao Y, Yang R et al. (2019). The pentose phosphate pathway regulates chronic neuroinflammation and dopaminergic neurodegeneration. J Neuroinflammation 16: 255. https://doi.org/10.1186/s12974-019-1659-1.

Turkowski K, Brandenburg S, Mueller A et al. (2018). VEGF as a modulator of the innate immune response in glioblastoma. Glia 66: 161–174. https://doi.org/10.1002/glia.23234.

van der Vos KE, Abels ER, Zhang X et al. (2016). Directly visualized glioblastoma-derived extracellular vesicles transfer RNA to microglia/macrophages in the brain. Neuro Oncol 18: 58–69. https://doi.org/10.1093/neuonc/nov244.

Vecht CJ, Kerkhof M, Duran-Pena A (2014). Seizure prognosis in brain tumors: new insights and evidence-based management. Oncologist 19: 751–759. https://doi.org/10.1634/theoncologist.2014-0060.

Venkataramani V, Tanev DI, Strahle C et al. (2019). Glutamatergic synaptic input to glioma cells drives brain tumour progression. Nature 573: 532–538. https://doi.org/10.1038/s41586-019-1564-x.

Venkataramani V, Yang Y, Schubert MC et al. (2022). Glioblastoma hijacks neuronal mechanisms for brain invasion. Cell 185: 2899–2917.e31. https://doi.org/10.1016/j.cell.2022.06.054.

Venkatesh HS, Johung TB, Caretti V et al. (2015). Neuronal activity promotes glioma growth through Neuroligin-3 secretion. Cell 161: 803–816. https://doi.org/10.1016/j.cell.2015.04.012.

Venkatesh HS, Tam LT, Woo PJ et al. (2017). Targeting neuronal activity-regulated neuroligin-3 dependency in high-grade glioma. Nature 549: 533–537. https://doi.org/10.1038/nature24014.

Venkatesh HS, Morishita W, Geraghty AC et al. (2019). Electrical and synaptic integration of glioma into neural circuits. Nature 573: 539–545. https://doi.org/10.1038/s41586-019-1563-y.

Vinnakota K, Hu F, Ku M-C et al. (2013). Toll-like receptor 2 mediates microglia/brain macrophage MT1-MMP expression and glioma expansion. Neuro Oncol 15: 1457–1468. https://doi.org/10.1093/neuonc/not115.

Wagner S, Czub S, Greif M et al. (1999). Microglial/macrophage expression of interleukin 10 in human glioblastomas. Int J Cancer 82: 12–16. https://doi.org/10.1002/(sici)1097-0215(19990702)82:1<12::aid-ijc3->3.0.co;2-o.

Walker AJ, Card T, Bates TE et al. (2011). Tricyclic antidepressants and the incidence of certain cancers: a study using the GPRD. Br J Cancer 104: 193–197. https://doi.org/10.1038/sj.bjc.6605996.

Watkins S, Sontheimer H (2011). Hydrodynamic cellular volume changes enable glioma cell invasion. J Neurosci 31: 17250–17259. https://doi.org/10.1523/JNEUROSCI.3938-11.2011.

Wei J, Marisetty A, Schrand B et al. (2019). Osteopontin mediates glioblastoma-associated macrophage infiltration and is a potential therapeutic target. J Clin Invest 129: 137–149. https://doi.org/10.1172/JCI121266.

Wesolowska A, Kwiatkowska A, Slomnicki L et al. (2008). Microglia-derived TGF-beta as an important regulator of glioblastoma invasion—an inhibition of TGF-beta-dependent effects by shRNA against human TGF-beta type II receptor. Oncogene 27: 918–930. https://doi.org/10.1038/sj.onc.1210683.

Wolf HK, Roos D, Blümcke I et al. (1996). Perilesional neurochemical changes in focal epilepsies. Acta Neuropathol 91: 376–384. https://doi.org/10.1007/s004010050439.

Won W-J, Deshane JS, Leavenworth JW et al. (2019). Metabolic and functional reprogramming of myeloid-derived suppressor cells and their therapeutic control in glioblastoma. Cell Stress 3: 47–65. https://doi.org/10.15698/cst2019.02.176.

Wu A, Wei J, Kong L-Y et al. (2010). Glioma cancer stem cells induce immunosuppressive macrophages/microglia. Neuro Oncol 12: 1113–1125. https://doi.org/10.1093/neuonc/noq082.

Xia S, Lal B, Tung B et al. (2016). Tumor microenvironment tenascin-C promotes glioblastoma invasion and negatively regulates tumor proliferation. Neuro Oncol 18: 507–517. https://doi.org/10.1093/neuonc/nov171.

Xu S, Wei J, Wang F et al. (2014). Effect of miR-142-3p on the M2 macrophage and therapeutic efficacy against murine glioblastoma. J Natl Cancer Inst 106: dju162. https://doi.org/10.1093/jnci/dju162.

Xu J, Zhang J, Zhang Z et al. (2021). Hypoxic glioma-derived exosomes promote M2-like macrophage polarization by enhancing autophagy induction. Cell Death Dis 12: 373. https://doi.org/10.1038/s41419-021-03664-1.

Ye ZC, Sontheimer H (1999). Glioma cells release excitotoxic concentrations of glutamate. Cancer Res 59: 4383–4391.

Ye ZC, Rothstein JD, Sontheimer H (1999). Compromised glutamate transport in human glioma cells: reduction-mislocalization of sodium-dependent glutamate transporters and enhanced activity of cystine-glutamate exchange. J Neurosci 19: 10767–10777. https://doi.org/10.1523/JNEUROSCI.19-24-10767.1999.

Young SZ, Bordey A (2009). GABA's control of stem and Cancer cell proliferation in adult neural and peripheral niches. Phys Ther 24: 171–185. https://doi.org/10.1152/physiol.00002.2009.

Zhan D, Yalcin F, Ma D et al. (2022). Targeting UDP-α-d-glucose 6-dehydrogenase alters the CNS tumor immune microenvironment and inhibits glioblastoma growth. Genes Dis 9: 717–730. https://doi.org/10.1016/j.gendis.2021.08.008.

Zhang J, Sarkar S, Cua R et al. (2012). A dialog between glioma and microglia that promotes tumor invasiveness through the CCL2/CCR2/interleukin-6 axis. Carcinogenesis 33: 312–319. https://doi.org/10.1093/carcin/bgr289.

Zhang Z-M, Yang Z, Zhang Z (2015). Distribution and characterization of tumor-associated macrophages/microglia in rat C6 glioma. Oncol Lett 10: 2442–2446. https://doi.org/10.3892/ol.2015.3533.

Zhang Z, Huang X, Li J et al. (2019). Interleukin 10 promotes growth and invasion of glioma cells by up-regulating KPNA 2 in vitro. J Cancer Res Ther 15: 927–932. https://doi.org/10.4103/jcrt.JCRT_284_19.

Zhang X, Chen L, Dang W-Q et al. (2020). CCL8 secreted by tumor-associated macrophages promotes invasion and stemness of glioblastoma cells via ERK1/2 signaling. Lab Invest 100: 619–629. https://doi.org/10.1038/s41374-019-0345-3.

Zhao B, Xia Y, Yang F et al. (2022). Molecular landscape of IDH-mutant astrocytoma and oligodendroglioma grade 2 indicate tumor purity as an underlying genomic factor. Mol Med 28: 34. https://doi.org/10.1186/s10020-022-00454-z.

Zhou W, Ke SQ, Huang Z et al. (2015). Periostin secreted by glioblastoma stem cells recruits M2 tumour-associated macrophages and promotes malignant growth. Nat Cell Biol 17: 170–182. https://doi.org/10.1038/ncb3090.

Zhu W, Carney KE, Pigott VM et al. (2016). Glioma-mediated microglial activation promotes glioma proliferation and migration: roles of Na+/H+ exchanger isoform 1. Carcinogenesis 37: 839–851. https://doi.org/10.1093/carcin/bgw068.

Handbook of Clinical Neurology, Vol. 210 (3rd series)
Neuroglia in Neurologic and Psychiatric Disorders, Part II
A. Verkhratsky, L.D. de Witte, E. Aronica, and E.M. Hol, Editors
https://doi.org/10.1016/B978-0-443-19102-2.00011-9

Chapter 12

Neuroglia in hepatic encephalopathy

MAGDALENA ZIELIŃSKA*, MARIUSZ POPEK, AND JAN ALBRECHT

Department of Neurotoxicology, Mossakowski Medical Research Institute, Polish Academy of Sciences, Warsaw, Poland

Abstract

Neuroglia contribute to the pathophysiology of hepatic encephalopathy (HE) either beneficially or detrimentally. Pathogenesis of HE is linked to damage triggered by blood-derived toxins, with ammonia being the main causative factor. Neuroglial cells, especially astrocytes and microglia, respond to HE-associated systemic and central signals and undergo complex and variable changes in their metabolism, morphology, and function, which include ion and water dyshomeostasis in conjunction with neurotransmission imbalance and neuroinflammation. HE-induced alterations of astrocytes are defined as astrocytopathy, with aberrant astrocytes resulting in either gain or loss of functions. In the chronic HE, the presence of Alzheimer type II cells is a histologic hallmark, with asthenic astrocytes emerging as a newcomer. In acute HE, rapid swelling of astrocytes is a primary cause of cerebral edema and mortality. This chapter reviews the dominant role of astrocytes in the pathogenesis of HE resulting from acute and chronic liver failure, mainly in experimental models. The focus is on the loss of homeostatic function bearing upon the functioning of the glymphatic system, aberrant neurotransmission as a consequence of astrocyte-neuron miscommunication, and the concordant neuroinflammatory response of astrocytes and microglia. The chapter concludes with a delineation of concepts for future research.

HEPATIC ENCEPHALOPATHY: BASIC INFORMATION

Hepatic encephalopathy (HE) is a complex neuropsychiatric disorder that results from impaired liver function. Liver malfunction compromises clearance from the blood of toxins, which eventually enter the brain. While more than 20 different compounds occur in circulation in the progression of HE (Zieve, 1987), excessive accumulation of ammonia appears to be the most critical factor.

TYPES OF HE

A few HE classifications have been proposed in recent years (Rose et al., 2020; Häussinger et al., 2022). These classifications are considerably overlapping, and the high complexity of HE adds subjectivity to the formulated statements. An overall consensus classifies HE into four types:

Acute hepatic encephalopathy (AHE) (Type A) is associated with acute liver failure (ALF) resulting from severe inflammatory and/or necrotic liver disease of rapid onset. Progressive brain edema followed by increased intracranial pressure (ICP) and herniation is the ultimate cause of death in 50%–80% of type A HE patients (Ostapowicz, 2002; O'Grady, 2007).

Portal systemic encephalopathy (PSE) (Type B) results from portocaval shunting in the absence of parenchymal liver disease. A portocaval shunt is a treatment for high blood pressure in the liver and is also employed as a second-line therapeutic measure in patients with gastroesophageal varices (Garcia-Tsao et al., 2007). An anastomosis is made between the portal vein, which supplies 75% of the liver's blood, and the inferior vena cava that drains blood from the lower two-thirds of the body. This permits toxins of intestinal origin (ammonia, bile acids, mercaptans, manganese, etc.) to bypass the liver into the systemic circulation.

Chronic hepatic encephalopathy (CHE) (Type C) accompanies chronic liver failure (cirrhosis). Factors involved in the etiology of cirrhosis include excessive alcohol consumption, viral infection, biliary obstruction, but also exposure to drugs or toxins. Portal hypertension arises as a result of cirrhosis; the high portal pressure stimulates the opening of embryonic venous channels. Since those venous channels are considered "natural" portal-systemic shunts, in the past, many authors used the terms PSE and CHE interchangeably (Jones and Weissenborn, 1997; Albrecht and Jones, 1999; Vilstrup et al., 2014).

Acute upon chronic liver failure (Type ACLF) reflects rapid deterioration of a chronic balanced condition. This is a frequently

*Correspondence to: Magdalena Zielińska, Department of Neurotoxicology, Mossakowski Medical Research Institute, Polish Academy of Sciences, Warsaw, Poland. Tel: +48-22-6086470, Fax: +48-22-6685532, E-mail: mzielinska@imdik.pan.pl

Table 12.1

Symptoms and classification of HE, including minimal HE (MHE).

Manifestation	Stage	Description
Covert	MHE	Neuropsychologic/psychomotor alternations without clinical manifestations
	I	Mild confusion, discrete euphoria or irritability, decreased attention and concentration, decelerated ability to perform mental tasks, slurred speech, reversal of sleep rhythm
Overt	II	Lethargy or apathy, difficult to perform mental tasks, personality changes, inappropriate behavior, disorientation for time, lack of sphincter control
	III	Somnolent to semistupor, responsive to stimuli, unable to perform mental tasks, persistent disorientation with respect to time and/or place, delusions, bizarre behavior speech present but incoherent
	IV	Coma, no response to painful stimuli, wide pupils without reactions

Based on Albrecht J, Jones EA (1999). Hepatic encephalopathy: molecular mechanisms underlying the clinical syndrome. J Neurol Sci 170: 138–146. https://doi.org/10.1016/S0022-510X(99)00169-0 and Vilstrup H, Amodio P, Bajaj J et al. (2014). Hepatic encephalopathy in chronic liver disease: 2014 practice guideline by the American Association for the Study of Liver Diseases and the European Association for the Study of the liver. Hepatology 60: 715–735. https://doi.org/10.1002/hep.27210.

encountered HE variant that often leads to death in a sequence of events identical to that in type A HE patients.

CAUSES

Paracetamol overdose appears to be the most common cause of ALF and the ensuing type A HE. The most common causes of liver cirrhosis include alcohol abuse and chronic hepatitis B or C. Less frequently, liver cirrhosis is associated with biliary obstruction, nonalcoholic fatty liver disease leading to obesity, and hemochromatosis (for recent review and references, see Wazir et al., 2023).

SYMPTOMS

HE clinical evolution is classified into minimal (nonsymptomatic) HE and four symptomatic stages of HE (Table 12.1). Because of its multisymptomatic character, demarcation lines between stages I, II, III, and IV are difficult to draw; hence, details of symptom selection and assignment to a particular stage differ from one report to another (Albrecht and Jones, 1999; Vilstrup et al., 2014). In most general terms, the progression of HE through stages I to IV may be viewed as a progressing imbalance between inhibitory and excitatory neurotransmission in favor of the former. HE is considered to be to a great degree reversible; spontaneous liver recovery or successful liver transplantation will ameliorate HE symptoms. Severe brain edema in stage IV is often fatal (see Section "Astrocyte swelling and brain edema").

EXPERIMENTAL MODELS

The types of HE occurring in humans or selected aspects of those types have been more or less faithfully reproduced in animal models or simulated in the *in vitro* systems (Butterworth et al., 2009; DeMorrow et al., 2021). Type A HE has been studied in anhepatic or toxic models. The former includes liver devascularization and total hepatectomy; the latter treatment of animals with hepatotoxins: galactosamine (GLC) (rabbits and rats), thioacetamide (TAA) (rats and mice), azoxymethane (AZM) (mice), or acetaminophen (dogs). Of these, the TAA and AZM models are the most frequently used and will be overwhelmingly referred to in the further sections of this review. Type B is modeled in rats by constructing end-to-side portocaval anastomosis (PCA). Type C animal models include bile duct ligation (BDL) (rat) or long-term administration of carbon tetrachloride (rat). To induce type ACLF, BDL or PCA rats are fed with ammonium salts.

ALZHEIMER II ASTROCYTES: THE RESPONSE OF ASTROCYTES TO CHRONIC HE

By neurophysiologic criteria, HE reflects neuronal malfunction and, hence, is a disease of neurons. However, light and electron microscopic studies have unambiguously documented that astrocytes undergo complex morphologic alterations with the progression of HE. At the same time, neurons remain relatively intact or suffer limited damage (Albrecht, 1986). This contrasting perception of the two facets of HE has come to light long before it was realized that astrocytes are active partners of neurons in neurotransmission. Analyses of brain sections of cirrhotic patients and animals subjected to chronic HE models invariably demonstrated the presence of Alzheimer type II astrocytes (AIIA). AIIA have large lobulated nuclei with sparse marginal chromatin (Fig. 12.1A) and are often Periodic acid–Schiff staining (P.A.S.)—positive, indicating the presence of excess glycogen (Fig. 12.1B).

The presence of AIIA in a CHE patient was documented for the first time by Adams and Foley (1953) and in a PSE patient by Victor et al. (1965). Hepatic retinopathy, another complication of liver cirrhosis, is specifically associated with the occurrence in the eye of astrocytes and Müller cells with enlarged nuclei, strikingly resembling AIIA (Reichenbach et al., 1995; Albrecht et al., 1998). Kimura and Budka (1986) characterized in considerable detail the immunocytochemical staining patterns of the astrocytic proteins in the autopsy material of CHE patients. The major finding was that AIIA shows prominent upregulation of S-100β protein, whereas the reactive astroglia marker, glial fibrillary acidic protein (GFAP) is almost absent in these cells. The authors interpreted the above-described immunocytochemical profile of AIIA as reflecting "peculiar types of reactive astrocytes retaining characteristics of protoplasmic astrocytes" (Kimura and Budka, 1986). The absence or low GFAP immunoreactivity in AIIA was later confirmed by other authors in

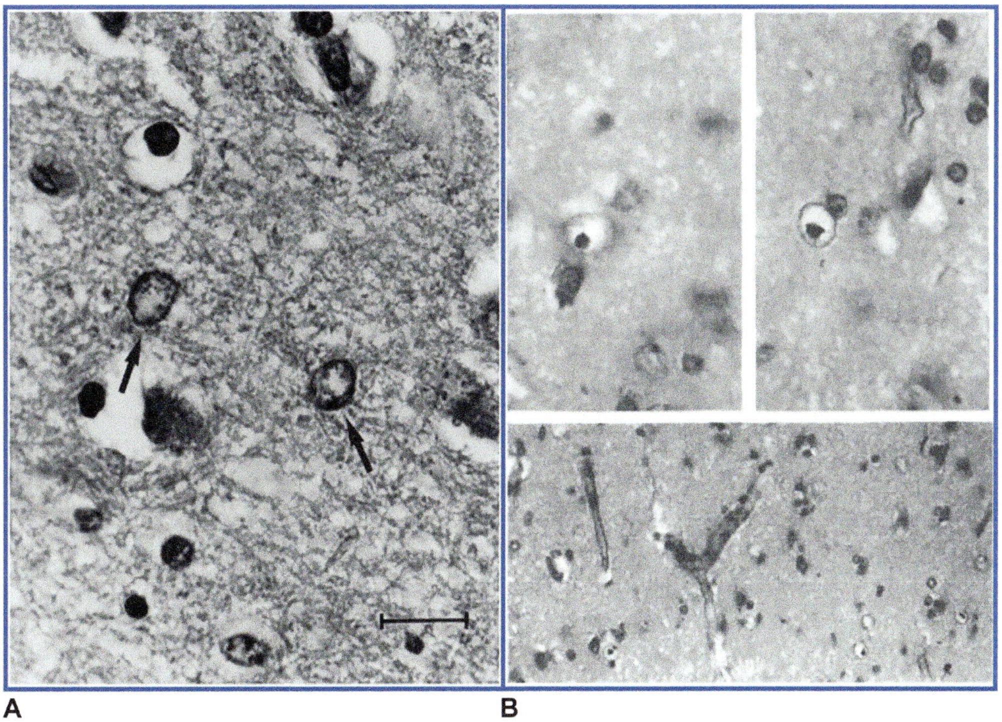

Fig. 12.1. Representative Alzheimer II astrocytes. (A) Alzheimer type II astrocyte (arrows), with no discernible cytoplasm and spare marginal chromatin in nuclei in a patient who died 6 months after the first signs of a hepatoma. (B) P.A.S, positive inclusions in astrocytic nuclei. Panel (A) Reproduced from Albrecht J, Jones EA (1999). Hepatic encephalopathy: molecular mechanisms underlying the clinical syndrome. J Neurol Sci 170: 138–146. https://doi.org/10.1016/S0022-510X(99)00169-0; picture courtesy of Dr. Arnulf H. Koeppen. Panel (B) Reproduced from Victor M, Adams RD, Cole M (1965). The acquired (non-WILSONIAN) type of chronic HEPATOCEREBRAL degeneration. Medicine 44: 345–396. https://doi.org/10.1097/00005792-196509000-00001.

experimental models: in PSE rats (Ma et al., 1991) and in cultured astrocytes subjected to prolonged ammonia treatment (Norenberg et al., 1990; Neary et al., 1994). Martin et al. (1987) performed a morphometric evaluation of astrocytic nuclei of autopsy material from CHE patients, which revealed a significant increase in the average nuclear area of astrocytes, a change largely contributed by AIIA, which is proportional to the advancement of the *premortem*-recorded HE symptoms (Martin et al., 1987).

Formation of AIIA was reproduced in chronic HE models, including PSE in rats or pigs (Diemer and Laursen, 1977; Diemer and Tønnesen, 1977), urease-induced hyperammonemia in rats (Diemer and Laursen, 1977), and in PSE rats in which (i) HE status was aggravated by clamping of the hepatic artery (Diemer and Laursen, 1977), or (ii) in animals brought to coma by feeding with ammoniated resin (Norenberg, 1976). These experimental studies provided information about the dynamics of astrocyte transformation and AIIA accumulation, including brain regional variations of the response. For instance, in PSE rats, AIIA became first apparent not earlier than 4 days after the operation and tended to reverse after ~35 days, being relatively most intense in the striatum and negligible in the cerebellum (Diemer, 1977). Norenberg (1976) reported AIIA to be reversible only in rats without ammonia-induced coma. Also in PSE rats, the increase in several AIIA was potentiated by hepatic ligation (Jover et al., 2006). Notably, ammonia may not be the sole AIIA-promoting factor in HE, as AIIA was not detected in the cerebral biopsy of a patient with severe Reye's syndrome, a condition associated with hyperammonemia (Partin et al., 1975).

It needs to be emphasized that AIIA is not pathognomonic to HE. It was originally visualized by von Hoesslin and Alzheimer (1912) in a patient with pseudosclerosis, identified by Spielmeyer (1920) as typical of Wilsonian hepatocellular degeneration, and ever since reported in multiple neurodegenerative conditions (for references, see Ma, 2001). Disappointingly, an elaborate description of the morphologic properties of AIIA provided no clues as to their functional role in the pathogenesis of CHE. However, as observed in PSE rats, AIIA presents a propensity to be transformed into reactive astrocytes upon direct mechanical insult (Ma et al., 1991). Moreover, their large nucleus has been speculated to reflect their present or past metabolic remodeling (Norenberg, 1976; Kimura and Budka, 1986). The above observations and their interpretation, in conjunction with the growing knowledge of the neuromodulatory function of astrocytes, prompted a hypothesis that metabolically activated astrocytes may play a role in the onset or progression of initial stages of HE. Early attempts at verifying this hypothesis are described in the next section.

METABOLIC REMODELING OF ASTROCYTES AT THE ONSET OF ALF

Early studies revealed that induction of full-blown acute HE in the TAA model requires three administrations of hepatotoxin at 24-h intervals (Albrecht and Hilgier, 1984). The advanced stage of HE in this model is preceded by a distinct stage after the second administration of TAA, characterized by hyperammonemia and hyperphenolemia (Albrecht and Hilgier, 1984; Hilgier et al., 1985), with subtle behavioral symptoms and weak biochemical changes. Specifically, rats at this stage show a moderate reduction in motor activity, together with an increased frequency and slightly decreased amplitude of electrocorticogram (Pluta and Albrecht, 1984), no changes in brain ammonia and 2-oxoglutarate content, and no reduction of the cerebral metabolic rate of oxygen (Pluta and Albrecht, 1986). The stage of HE after second TAA administration, further referred to as the onset stage, is characterized by the above control activity of some metabolic steps and changes in astrocytic transporters. At the whole-brain level, this stage is characterized by an increase in the activity of carbonic anhydrase (CA) (Albrecht and Hilgier, 1984), the CO_2-fixating enzyme predominantly located in astrocytes (Kimelberg et al., 1982). CA activation is a transient phenomenon, not observed in the advanced stage in this model (Albrecht and Hilgier, 1984). Studies with astrocyte-enriched fractions bulk isolated from TAA-treated rats identified the onset stage-specific activation of several processes critical for maintaining ion and neurotransmitter homeostasis, including Na^+-K^+ ATPase activity (Albrecht et al., 1985), chloride influx, and GABA uptake (Albrecht et al., 1985; Dabrowiecki and Albrecht, 1985). The above manifestations of astroglial response are considered an adaptive response of astrocytes toward improving ion and neurotransmitter homeostasis. All the above-mentioned parameters drop below the control level after the third administration of TAA (reviewed in Albrecht, 1986), with oxygen consumption being specifically reduced in astrocytes, but not in neurons (Albrecht et al., 1987). Hence, the beneficial responses of astrocytes at the onset of HE appear to be too short-lived to prevent further development of HE.

ASTROCYTIC IMPACT ON THE EXCITATION/INHIBITION BALANCE

In the most general terms, the progression of HE symptoms reflects a gradual decline of the brain inhibition and excitation balance in favor of inhibition. The onset and early phase of HE following ALF is accompanied by overexcitation and seizures (Vogels et al., 1997; Olivero et al., 2022), events mainly related to increased glutamatergic neurotransmission and associated with overactivation of ionotropic glutamate receptors (Marcaida et al., 1992; Cauli et al., 2008; Rodrigo et al., 2009). Acute HE of longer duration and/or progression toward chronic forms leads to glutamate receptor downregulation or loss (Peterson et al., 1990; Saransaari et al., 1997), contributing to an increased inhibition. Glutamatergic transmission deficiency is coincident with and augmented by an increased GABAergic tone (Bassett et al., 1990). While the magnitude and, at times, brain region specificity of the changes in the two neurotransmitter systems show considerable variations, the mechanisms underlying the interference of HE-affected astrocytes with the glutamatergic or GABAergic transmission have a common background.

Astrocytic contribution to alterations of glutamate and GABA synthesis in HE

Glutamate and GABA synthesis in the CNS is linked to glutamine-to-glutamate conversion that occurs in the framework of the glutamate/GABA/glutamine shuttle, molecular components of which are compartmentalized between astrocytes and neurons (Fig. 12.2). Neuronal glutamate and GABA derive from glutamine, which is produced from glutamate and ammonia in a reaction catalyzed by an astrocyte-specific enzyme glutamine synthetase (GS). GS activity is increased in hyperammonemia and has multiple implications for the pathogenesis of HE (Suárez et al., 2002). However, activation of GS is not directly translated to increased synthesis of glutamate or GABA. Studies using ^{13}C NMR spectroscopy have shown that a portion of the amine moiety of the glutamate needed for glutamine synthesis must have been derived from reductive amination of α-ketoglutarate catalyzed by GDH (reviewed by Cooper, 2012). Glutamate synthesis from α-KG production relies to a great degree on the anaplerotic generation of oxaloacetate from pyruvate, a reaction catalyzed by pyruvate carboxylase (PC), which is another enzyme specifically located in astrocytes (Shank and Campbell, 1983; Yu et al., 1983; Faff-Michalak and Albrecht, 1991). Acute hyperammonemia conditions increase the PC-mediated anaplerotic flux of pyruvate to glutamate and subsequent synthesis of glutamine (Lapidot and Gopher, 1997; Zwingmann et al., 1998, 2003; Kanamatsu and Tsukada, 1999; Leke et al., 2011). This increase likewise contributes to the excitotoxic, glutamatergic transmission-related aspect of HE. By contrast, chronic HE in the PSE model is associated with decreased anaplerosis in the frontal cortex but not in the thalamus (Zwingmann, 2007), and a decrease in PC activity in a fraction enriched in mitochondria of astrocytic origin was observed in a hepatotoxic and simple hyperammonemic model of acute HE (Faff-Michalak and Albrecht, 1991). Conversion of TCA cycle constituents to GABA appears unchanged by HE (Zwingmann et al., 1998). It must be noted, however, that the net effect of HE on the neurotransmitter pools of glutamate and GABA depends not only on glutamine synthesis in astrocytes but also on its glutaminase (PAG)-mediated conversion to glutamate and glutamate decarboxylase (GAD)-dependent conversion of glutamate to GABA (Fig. 12.2). Variable effects of HE were documented in different models on the expression/activity of GAD or PAG (Thomas et al., 1988; Leke et al., 2011; Probst et al., 2020; for a review see Cooper and Jeitner, 2016).

Astrocytic interference with the glutamatergic/GABAergic neurotransmission balance at steps beyond neurotransmitter synthesis

The glutamatergic transmission

Disturbances of the glutamatergic tone in HE are related to excessive glutamate accumulation in the interstitium. Increased extracellular glutamate, identified using the brain microdialysis technique, is a feature common to acute HE in human patients (Tofteng, 2002; Tofteng et al., 2002) and experimental animals (Bosman et al., 1992; De Knegt et al., 1994; Michalak et al., 1996;

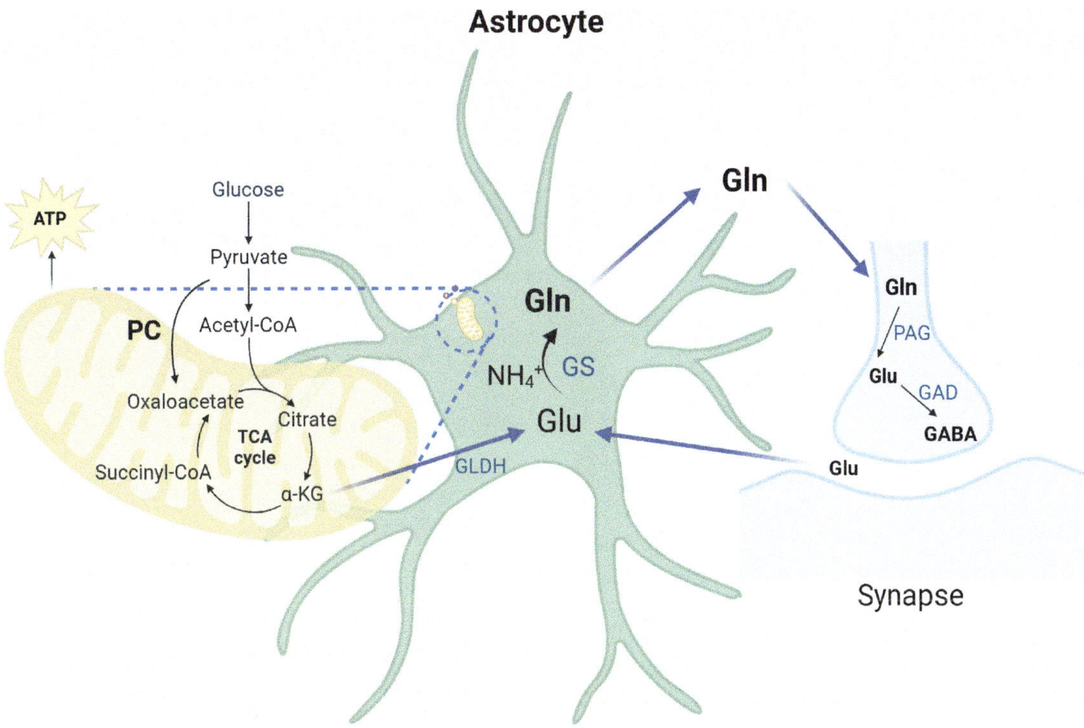

Fig. 12.2. Scheme of glutamate/GABA/glutamine astrocyte-neuronal shuttle and neurons. In astrocytic cytosol glutamate (Glu) is converted to glutamine (Gln) by glutamine synthetase (GS), and Gln is transported to extracellular space. After transfer to neurons, Gln is converted into Glu by phosphate-activated glutaminase (PAG). In inhibitory synapses, Glu is decarboxylated to GABA by the glutamate decarboxylase enzyme (GAD). In both astrocytes and neurons, part of Glu is utilized for the synthesis of alpha-ketoglutaric acid (α-KG), a substrate for the tricarboxylic acid (TCA) cycle, by oxidative deamination mediated by glutamate dehydrogenase (GLDH). Pyruvate carboxylase (PC) catalyzes oxaloacetate generation from pyruvate. Created with BioRender.com.

Hilgier et al., 1999) and to chronic HE in animal models (Therrien and Butterworth, 1991). Glutamate, measured by microdialysis, is believed to represent the pool released by exocytosis from neurons and a pool arising from nonvesicular glutamate release from intracellular compartments (Vaquero and Butterworth, 2006).

There is a consensus that in HE glutamate emerges in excess in the extracellular space mainly (albeit not exclusively, see the following section) because of its inefficient uptake to astrocytes. The expression of astroglia-specific glutamate transporter isoforms EAAT1 (GLAST) and EAAT2 (GLT1) was consistently found to be downregulated at the activity and/or expression level in HE models (Knecht et al., 1997; Norenberg et al., 1997; Chan and Butterworth, 1999; Zhou and Norenberg, 1999; Chan et al., 2000; Cabrera-Pastor et al., 2019) in ammonia-exposed astrocytes (Albrecht et al., 1988; Bender and Norenberg, 1996), and in EAAT1- and EAAT2-transfected oocytes upon prolonged ammonia treatment (Søgaard et al., 2012).

Impaired astrocytic glutamate uptake is not the only cause of its excessive extracellular accumulation. Ammonia directly depolarizes neuronal membranes (Szerb and Butterworth, 1992), and by this mechanism triggers calcium-dependent synaptic glutamate release as shown both *in vitro* (Hilgier et al., 1991) and *in vivo* (Moroni et al., 1983). However, the overwhelming role of inefficient glutamate uptake to astrocytes is highlighted when comparing the pertinent events in the TAA model of ALF. In this model, glutamate accumulation in microdialysates (Hilgier et al., 2014) is coincident with the loss of astrocytic glutamate transporters (Norenberg et al., 1997) and decreased glutamate uptake (Albrecht et al., 1988), whereas, at the same time, potassium-stimulated glutamate release measured *ex vivo* was substantially reduced (Hilgier et al., 1991). Ammonia has also been shown to stimulate calcium-dependent glutamate release from cultured astrocytes by a mechanism involving cytosolic alkalinization (Rose et al., 2005); however, the significance of this process *in vivo* remains to be documented.

Glutamatergic tonus is modulated by an astroglia-derived neuromodulatory compound, kynurenic acid (KYNA). KYNA is the only, hitherto, known endogenous antagonist of ionotropic *N*-methyl-D-aspartate (NMDA) receptor glutamate receptors, showing a strong affinity toward the glycine site of the latter (Schwarcz and Pellicciari, 2002; Rae et al., 2006). KYNA is synthesized from a tryptophan metabolite kynurenine in a reaction catalyzed by kynurenine aminotransferase II (KAT II), an enzyme predominantly located in astrocytes. Induction of ALF in rats in the TAA model led to an increase in KATII activity (Saran et al., 2004) and expression (Hilgier et al., 2014) in rat cerebral cortical homogenates, which was accompanied by increased KYNA accumulation in the extracellular space (Hilgier et al., 2014). Hence, KYNA is likely to contribute to the attenuation of the excitatory activity of glutamate. Increased KYNA synthesis and extracellular levels correlated with the comatose development in animals, despite the maintenance of increased extrasynaptic glutamate (Hilgier et al., 2014). The above again indicates that glutamate accumulation in the extracellular space may not necessarily lead to overexcitation.

The GABAergic transmission

The GABA content in cerebral cortical microdialysates is of one order of magnitude lower than that of glutamate and is not significantly altered in HE (Bosman et al., 1992). The increased concentration of GABA has so far only been reported in the microdialysates of the cerebellum of rats with chronic hyperammonemia (Gonzalez-Usano et al., 2014) and appeared to reflect a neuroinflammatory, mainly microglial response (Section "Neuroinflammation: A consequence of systemic inflammation and reactive microgliosis") (Dadsetan et al., 2016). Increased GABAergic transmission associated with HE is largely attributed to the $GABA_A$-benzodiazepine receptor complex ($GABA_A$-BDZR complex) activation at the synapse (Basile et al., 1989, 1990). However, ammonia also interacts with the peripheral BDZRs (PBDZRs, more recently renamed as translocator protein 18A (Papadopoulos et al., 2006), which are specifically localized in astrocytes (Anholt et al., 1986; Itzhak et al., 1993) and whose levels are increased in the brains of HE patients (Lavoie et al., 1990), chronic HE animals (Giguère et al., 1992; Desjardins et al., 1997), and cultured astrocytes exposed to ammonia (Ducis et al., 1990). Increased PBDZR activity upregulates synthesis of prognenelone, derivatives of which are strong positive modulators of the $GABA_A$-BDZR complex located at the synapse (Itzhak et al., 1993). In the brains of mice in the TAA model of ALF and ammonium acetate-induced hyperammonemia, increased PBDZR binding to cell membrane fractions and prognenelone synthesis were found (Itzhak et al., 1995). Enhanced release of GABA as observed in astrocytes isolated from rats with TAA-induced ALF may likewise contribute to enhanced GABAergic tone (Albrecht and Rafałowska, 1987); however, the latter observation remains to be verified in the *in vivo* setting.

As outlined in Section "Intracellular osmotic imbalance," taurine is released from astrocytes in response to swelling, and this capacity serves as an osmosensor, osmoregulator, and neuroprotectant. However, taurine is an agonist of $GABA_A$ receptors (review and references in Albrecht and Schousboe, 2005). Therefore, under HE conditions, its increased availability in the extracellular space is likely to promote inhibitory GABAergic transmission. Still, the significance of the interaction of endogenous taurine with $GABA_A$ receptors *in vivo* remains to be proven. It remains to be seen whether even under conditions of massive release from astrocytes, extracellular taurine concentrations in the CNS ever reach an effective threshold level. For more details regarding inhibitory transmission in HE, the reader is referred to an excellent review by Sergeeva (2013).

MALFUNCTIONAL ASTROCYTE AND IMPAIRED ION FLUXES

Functional impairments of astrocytes are manifested by the loss of homeostatic function (Verkhratsky et al., 2023). In the context of HE pathogenesis and hyperammonemia, Na^+, K^+, and Cl^- ionostasis, control of pH, and cell volume are the main players. The term "ammonia" refers to the sum of uncharged ammonia base (NH_3) and ammonium ions (NH_4^+). The conventional knowledge is that NH_3 as a lipid-soluble molecule may diffuse freely across cell membranes and thus pass the blood–brain barrier (BBB) with a net flux linearly correlated to its arterial concentration (Sørensen and Keiding, 2007); the fraction, however, is rather small at physiologic pH levels (Lockwood et al., 1980). At physiologic pH, more than 98% is in the NH_4^+ form and passes across cell membranes through K^+ channels (Cooper and Plum, 1987; Sugimoto et al., 1997; Nagaraja and Brookes, 1998). In addition, human nonerythroid Rhesus (Rh) glycoprotein B and nonerythroid Rh glycoprotein C were identified as mammalian ammonium/ammonia transporters (Weiner, 2006; Grishin et al., 2020).

Ammonium ions at cell membranes compete with K^+ at the Na^+-K^+ATPase and Na-K-2Cl cotransporter1 (NKCC1) that cotransports Na^+, K^+, $2Cl^-$, and water in an electro-neutral manner (Aickin et al., 1982; Epstein and Silva, 1985; Moser, 1987). Increased NKCC1 activity was first demonstrated in ammonia-exposed cultured astrocytes, where delayed cell swelling was associated with a significant increase in the uptake of the K^+ analog ^{86}Rb (Jayakumar et al., 2008) and increase in cytosolic Na^+ concentration (Kelly et al., 2009). The impairment of NKCC1 was confirmed by *ex vivo* experiments on hippocampal slices exposed to 5 mM ammonia and bumetanide block (Kelly and Rose, 2010). Accordingly, increased bumetanide-inhibited swelling and enhanced NKCC1 expression were seen in slices from the TAA-injected rats (Zielińska et al., 1999). Of note, the identification of K^+ clearance failure, sensitive to NKCC1 inhibition was described as a crucial mechanism in ammonia neurotoxicity in ornithine transcarbamylase (OTC) deficient or $Otc^{spf\text{-}ash}$ mice subjected to an ammonium salt infusion (Rangroo Thrane et al., 2013); however, both astrocyte swelling and brain edema were not detected in this experiment. High ammonia downregulates the expression of inward rectifying K^+ channel $K_{ir}4.1$ localized on the astrocyte membrane by a mechanism putatively involving NMDA receptors (Obara-Michlewska et al., 2015; Skowrońska et al., 2019). Downregulation of $K_{ir}4.1$ channel level impairs astrocytic K^+ clearance, thus affecting neuronal excitability (Ohno et al., 2021). Furthermore, the $K_{ir}4.1$ expression decrease coincides with astrocyte swelling in edema-related diseases (Milton and Smith, 2018). However, astrocyte swelling not involving $K_{ir}4.1$ was also reported (Larsen et al., 2014). Downregulation of these channels, which benefits the brain in terms of ammonia absorption, interferes with water homeostasis, potentially leading to cerebral edema resulting from water retention and extensive neuronal depolarization induced by increased extracellular K^+ levels (Lichter-Konecki et al., 2008). Impaired K^+ clearance with reduced glutamate turnover causes changes in synaptic transmission, which is responsible for neurologic manifestations of hyperammonaemia and HE (Chepkova et al., 2017; Popek et al., 2018).

A transient rise in intracellular Ca^{2+} was one of the earliest events noted in cultured astrocytes exposed to ammonia (Schliess et al., 2002; Rama Rao et al., 2003b; Rose, 2006) that further induced activation of certain Ca^{2+}-dependent enzymes (e.g., cNOS, NOX, and cPLA2) (Jayakumar et al., 2009). Increase of intracellular Ca^{2+} paralleled intracellular alkalinization, and glutamate release was observed upon acute treatment of astrocytes with ammonia (Rose et al., 2005; Liang et al., 2014). Prolonged exposure of cultured astrocytes to ammonia induces Ca^{2+} release from the endoplasmic reticulum store triggered by the activation of metabotropic receptors (Rose et al., 2005; Liang et al., 2014). Ammonia also modifies Ca^{2+} signals by increasing the expression of Ca^{2+}-permeable TRPC1 channels, upregulating the expression of $Ca_V1.2$ voltage-gated Ca^{2+} channels, and facilitating Ca^{2+} release from the intercellular stores (Wang et al., 2015, 2017).

Aberrant Ca^{2+} signals trigger astrocytic glutamate secretion that exacerbates excitotoxic damage of the nervous tissue (Hertz et al., 1989; Liang et al., 2014). Furthermore, excessive ammonium impairs the astrocytic transport of Na^+ and H^+, contributing to aberrant Na^+ signaling that further affects glutamate uptake and pH homeostasis in the brain tissue (Kelly et al., 2009). Astroglial Na^+ signals (Rose and Verkhratsky, 2024) directly control the activity of Na^+-K^+ ATPase and cotransporter NKCC1, thus regulating K^+ buffering. Moreover, Na^+ astrocyte levels control the supply of neurons with metabolic energy substrate lactate, supported by the astrocyte-neuron lactate shuttle stimulated by neuronal activity (Magistretti and Allaman, 2018). Notably, hyperammonemia impairs connexin hemichannel function and lactate release, which implies a limited energy supply to the already compromised neurons (Hadjihambi et al., 2017).

The homeostatic functions of astrocytes include pH regulation, spatial buffering of extracellular K^+ ions, energy substrate lactate supply to neurons, and ATP production, to name but a few (Allaman et al., 2011; Verkhratsky and Nedergaard, 2018; Verkhratsky et al., 2020). Homeostatic responses of astrocytes are controlled by dynamic changes in the cytoplasmic concentration of two cations, Ca^{2+} and Na^+, the intracellular concentration of which is controlled by several transporters and can be rapidly affected by ammonia-induced activation of respective fluxes through ionic channels or ion exchangers (Verkhratsky and Butt, 2023).

The glymphatic system integrates the homeostatic functions of astrocytes at the organ level. Specifically, it contributes to the CNS microenvironment maintenance by providing waste clearance; interstitial fluid from the parenchymal CNS drains into the perivascular space, the meningeal lymphatics, and reaches the systemic circulation (Rasmussen et al., 2018). Astrocytes do not possess an energy-dependent water pump; thus, water movement across the BBB and lipid membranes is driven by osmotic forces across cell membranes with a significant contribution of localized mainly at perivascular endfeet water channels aquaporin 4 (AQP4) (Nagelhus and Ottersen, 2013; Stokum et al., 2023). The glymphatic system is strictly dependent on the astrocytic AQP4, as has been demonstrated with the development of brain neurodegenerative changes linked with amyloid β accumulation (Iliff et al., 2014). An impaired brain fluid transport in AQP4 knockout mice resulted in reduced glymphatic clearance, leading to fluid stagnation in the brain interstitial space (Gomolka et al., 2023). The fact that AQP4 expression is altered in HE (Section "Astrocyte swelling and brain edema") prompted interest in its possible interference in the HE pathogenesis. Increased accumulation of the fluorescent CSF tracer in the brains of mice with AZM-induced ALF was interpreted as interstitial fluid retention and glymphatic function inhibition (Obara-Michlewska et al., 2018). Malfunction of the glymphatic system was reported in the olfactory bulb, hippocampus, and prefrontal cortex of BDL rats with chronic HE (Hadjihambi et al., 2019). This study also demonstrated an association of altered AQP4 expression with reduced cerebral fluid flow and suggested accumulation of interstitial toxins such as bile acids that result in interstitial fluid increase, resembling vasogenic edema.

ASTROCYTE SWELLING AND BRAIN EDEMA

Brain edema is defined as the excessive accumulation of intracellular and/or extracellular fluid in brain tissue and is a fatal clinical complication of patients with HE. Cerebral edema occurs in the majority of patients (50%–80%) with acute HE (Blei, 2000; Bernal et al., 2010), evolving to intracranial hypertension (Blei, 2007; Scott, 2013). In chronic HE magnetic resonance imaging (MRI), imaging revealed that low-grade brain edema occurs as a significant pathologic feature in some cases, although intracranial hypertension rarely develops (Cudalbu and Taylor-Robinson, 2019). Despite the fact that brain edema remains a leading cause of death in acute HE, no specific treatment other than whole-body cooling is available (Blei, 2007).

Characteristic features of cerebral edema, including swollen astrocytic endfeet along with abnormal mitochondria morphology, have been observed in patients with acute HE (Fig. 12.3) (Norenberg, 1977; Kato et al., 1992). These findings are compliant with the results of MRI imaging studies and computed tomography of ALF patients, corroborating the presence of cytotoxic edema, as evidenced by a decrease in the apparent diffusion coefficient (Detry, 2006). The occurrence of a cytotoxic mechanism of cerebral edema in acute HE was also documented in ALF animal models by morphologic (Kato et al., 1989) and MRI analysis of brain tissue (Obara-Michlewska et al., 2018; Hamdani et al., 2021). Astrogliopathy in acute HE includes a cytotoxic component resulting from

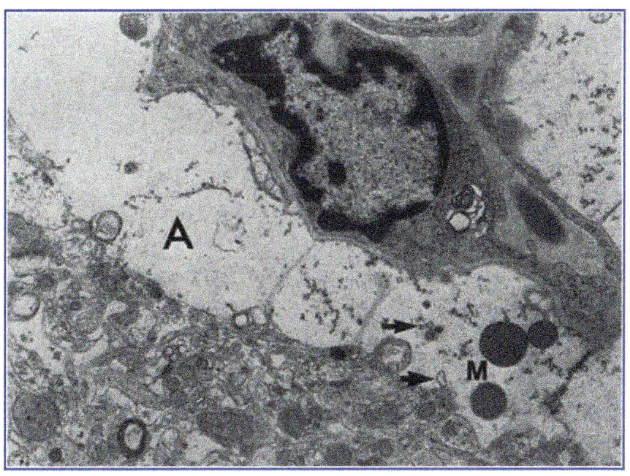

Fig. 12.3. Representation of swollen perivascular astrocytes (A) with swollen mitochondria (M) and dilated endoplasmic reticulum (*arrows*) in an ALF patient. Reproduced from Kato M, Hughes RD, Keays RT et al. (1992). Electron microscopic study of brain capillaries in cerebral edema from fulminant hepatic failure. Hepatology 15: 1060–1066. https://doi.org/10.1002/hep.1840150615.

intracellular metabolic and ion imbalance, followed by intracellular accumulation of water (Haussinger and Schliess, 2008; Albrecht et al., 2010). The multiplicity of factors involved in the formation of brain edema in acute HE renders its mechanism complex; however, the role of hyperammonemia is evident. Patients with genetic disorders of urea cycle enzymes who do not have other alterations of liver function also present brain edema (Brusilow, 1985). Furthermore, exposure to ammonia results in brain edema *in vivo* (Takahashi et al., 1991; Blei et al., 1994) and astrocytic swelling *in vitro* (Jayakumar et al., 2006). Moreover, in some patients with ALF, cerebral herniation occurred only if plasma ammonia levels exceeded 150 μM (Clemmesen et al., 1999).

Ammonia accumulation is tightly linked to increased synthesis of glutamine and its subsequent intra- and extraastrocytic accumulation (Desjardins et al., 2012; Jayakumar and Norenberg, 2018; Zielińska et al., 2022). This link has been verified in ALF patients, where increased intracranial pressure positively correlates with arterial and cerebral ammonia (Bernal et al., 2007) and cerebral glutamine content (Binesh et al., 2005; Jayakumar and Norenberg, 2018). Fixation of ammonia into glutamine in the brain and the ensuing increase of cerebral glutamine concentrations (Bjerring et al., 2008) were detected in patients with minimal HE (Binesh et al., 2005). Indirect evidence for the pathogenic role of glutamine was also provided by studies showing that brain glutamine level positively correlates with the severity of clinical symptoms in patients with acute upon chronic liver failure (Laubenberger et al., 1997; Tofteng et al., 2006). More explicit evidence of the role of high glutamine concentrations in the development of the cytotoxic component of cerebral edema derives from *in vivo* studies in a hyperammonemic rat model, where inhibition of GS with methionine sulfoximine led to amelioration of astrocyte swelling and, consequently, cerebral edema (Takahashi et al., 1991; Tanigami et al., 2005). As outlined in the following subsections, the mechanism of brain edema in ALF does not appear to simply reflect glutamine accumulation, but its interference with (i) intracellular osmotic imbalance and (ii) normal mitochondrial function, the latter related to the intramitochondrial release of ammonia, and oxidative/nitrosative stress, as described in the Trojan horse hypothesis (Albrecht and Norenberg, 2006; Zielińska et al., 2022).

Intracellular osmotic imbalance

Due to the lack of a complete urea cycle in the brain, excess blood-derived ammonia is converted by an astrocytic GS to glutamine (Norenberg and Martinez-Hernandez, 1979; Sonnewald et al., 1996), resulting in increased CSF and brain glutamine in animal models and humans alike (Hourani et al., 1971; Record et al., 1976; Cascino et al., 1982; Watanabe et al., 1984; McConnell et al., 1995; Czarnecka et al., 2018; Claeys et al., 2022). Restricted transfer of glutamine out of the cell associated with an increased synthesis within the cell could explain the cell swelling. While many glutamine transporters are localized in the astrocytic membrane, only transporters belonging to system N and the y^+LAT2 transporter have been so far shown to contribute to brain edema (Zielińska et al., 2022). It was suggested that accumulation of glutamine in the brain microdialysates caused by continuous infusion of [^{15}N]ammonium acetate reflects inhibition of glutamine efflux from astrocytes by reversed operation of system N transport (Kanamori and Ross, 2005). Toward this end, a decrease in the level of the SN1 transporter in the cerebral cortex of TAA rats (Zielińska et al., 2014) and AZM mice model of ALF correlated with the brain edema; similarly, local knockout of the SN1 transporter in the brain resulted in the swelling of nervous tissue (Hamdani et al., 2021).

Accumulation of glutamine in astrocytes is a likely cause of the increased intracranial pressure, as it generates an outward glutamine concentration gradient (Master et al., 1999; Tofteng et al., 2006). Compensatory osmoregulation through the removal from the brain of other osmotically active compounds, e.g., myo-inositol, choline, or taurine (Heins and Zwingmann, 2010; Jiménez et al., 2017), is effective in chronic HE (Cudalbu and Taylor-Robinson, 2019), but is mostly insufficient to compensate for fast and excessive glutamine accumulation in acute HE. Alterations in osmotically active substances in the brain may precede the onset of clinical symptoms (Larsen, 2002).

Some, but not all, ALF animal model studies confirmed the involvement of AQP4 protein as a driving force for the fluid flow and removal of molecular waste products (Bjerring et al., 2011; Sepehrinezhad et al., 2021). Indeed, alterations in AQP4 content were demonstrated in cultured astrocytes treated with ammonia (Rama Rao et al., 2003a) and in the brain tissue of animals with ALF and chronic HE, albeit in CHE, endfeet polarization of AQP4 was maintained (Wright et al., 2010; Rama Rao et al., 2010a). As to functional implications, AQP4 knockout mice injected with hepatotoxin TAA presented lower brain water content and neurologic deficits than TAA wild-type mice (Rama Rao et al., 2014). AQP4 immunoreactivity increase was likewise observed in the glial endfeet surrounding microvessels in *postmortem* brain tissue samples of ALF patients (Thumburu et al., 2014).

Most importantly for the translatability of the experimental studies, elevated cerebral glutamine and increased arterial ammonia concentration, and correlation of both increases with elevated intracranial pressure and clinical signs of cerebral edema frequently emerging in patients with acute HE (Tofteng et al., 2006). Of note, the severity of liver failure better correlates with elevated plasma glutamine levels, cerebral edema, and brain herniation syndromes in the acute and ACLF groups than in the chronic group (Helling et al., 2016; Gupta et al., 2017).

The oxidative/nitrosative stress and mitochondrial dysfunction

One of the key causative factors in the induction of astrocytic swelling is oxidative/nitrosative stress. Generation of free radicals is induced by high concentrations of ammonia (Kosenko et al., 1997; Murthy et al., 2001), and there is strong documentation confirming the occurrence of oxidative/nitrosative stress markers in the *postmortem* brain tissue of patients with chronic HE. The markers of this phenomenon include protein tyrosine nitration, RNA oxidation, increased heat shock protein 27, and upregulation of oxidative/nitrosative stress-related genes (Görg et al., 2010b, 2013). It was also found that oxidative/nitrosative stress induction is strongly associated with receptor NMDA-dependent elevated intracellular calcium ion concentration in astrocytes (Schliess et al., 2002; Görg, 2003; Suárez et al., 2006). Different calcium-related signaling cascades merge to stimulate nitric oxide synthase (NOS) and/or nicotinamide adenine dinucleotide phosphate (NADPH) oxidase (NOX)-dependent reactive nitrogen oxide species formation. *In vitro* and *in vivo* studies revealed that modulation of the Nrf2/hem oxygenase 1 (HO1) signaling pathway is a tool by which ammonia-induced oxidative/nitrosative stress and cytotoxicity may be antagonized (Wang et al., 2019; Jin et al., 2020; Khalil et al., 2021). Of note, NADPH oxidase inhibitors alleviate both reactive nitrogen oxide species-induced swelling and the downregulation of myo-inositol and taurine transporters mRNA level (Lachmann et al., 2013; Oenarto et al., 2014); the latter two effects may also interfere with swelling-induced osmolyte trafficking in astrocytes.

The y^+LAT2/SLC7A6 transporter is another potential contributor to oxidative/nitrosative stress in hyperammonemia. Pharmacologic manipulation with y^+LAT2 substrates glutamine, leucine, and isoleucine affected the amount of NO synthesized (Hilgier et al., 2009). Furthermore, induction of iNOS in astrocytes and the stimulation of NO synthesis by ammonia are associated with increased arginine uptake, processes documented to be promoted by a selective increase in the expression and activity of y^+LAT2 (Zielińska et al., 2011, 2015).

According to the Trojan horse hypothesis (Albrecht and Norenberg, 2006), glutamine accumulated in the cytoplasm enters mitochondria, opens the mitochondrial permeability transition pore (mPTP), and leads to the formation of ROS, further contributing to the astrocyte swelling (Ziemińska et al., 2000; Albrecht and Norenberg, 2006; Rama Rao and Norenberg, 2012). The postulated mechanism is that upon entry to mitochondria, subsequent deleterious effects are a result of glutamine degradation by PAG to ammonia and glutamate. Glutamine enters astrocytic mitochondria by active transport stimulated by pathophysiologic concentration of glutamine, rather than by diffusion (Dolińska et al., 1996). In isolated mitochondria and cultured astrocytes, glutamine induced intraastrocytic accumulation of ROS and astrocytic swelling (Ziemińska et al., 2000; Rama Rao et al., 2003b). Histidine, the inhibitor of glutamine uptake to the mitochondria, attenuated opening of mPTP, oxidative stress, and/or mitochondrial swelling in isolated astrocytic mitochondria (Ziemińska et al., 2000), cultured astrocytes (Pichili et al., 2007), and brain edema in ALF rats (Rama Rao et al., 2010b).

Neurogliovascular unit and perivascular edema

Astrocytes link the vasculature to neurons through neurovascular coupling, contributing to regulation of cerebral blood flow, which supports the brain energy demands in response to local neuronal needs (Attwell and Iadecola, 2002; Petzold and Murthy, 2011). A global decrease in the brain energy metabolism is one of the primary events associated with the pathogenesis of HE (Sections "Metabolic remodelling of astrocytes at the onset of ALF" and "Malfunctional astrocyte and impaired ion fluxes"). In general, the cerebral blood flow values reported in HE are variable (Ede and Williams, 1986; Almdal et al., 1989; Aggarwal et al., 1994; Wendon et al., 1994; Jalan et al., 2004; Iversen et al., 2009). Astrocytes regulate local cerebral blood flow through the secretion of vasoactive metabolites of arachidonic acid from astrocyte endfeet onto blood vessels (Filosa et al., 2006; Attwell et al., 2010). Moreover, K^+ release from astrocyte endfeet may also contribute to vasodilation, and so does oxygen (MacVicar and Newman, 2015; Sweeney et al., 2019). Astrocytes modulate the properties of the BBB-composing endothelial cells through vessels lining the endfeet (Abbott et al., 2006; Sofroniew and Vinters, 2010; Sweeney et al., 2019). While a complete breakdown of the BBB is unlikely to occur in HE, decreased functionality of tight junctions (TJ) and the subsequent uncontrolled influx of neurotoxic substances, including ammonia (Lockwood et al., 1991; Weiss et al., 2016), from the blood are apparent. Experimentally, significant damage to the BBB in the acute preclinical toxic (GLC, AZM, and TAA) animal models of ALF with increased levels of BBB permeability markers and TJ alterations was reported, the latter often preceding BBB permeability changes (Nguyen et al., 2006; Cauli et al., 2011; Nguyen, 2012; Grant et al., 2018; Obara-Michlewska et al., 2018; Orzeł-Gajowik et al., 2023). In turn, the rat BDL model of chronic HE often, albeit inconsistently, presented brain water content increase, BBB leakage, and loss of TJ proteins (Quinn et al., 2014; Dhanda and Sandhir, 2018). The variability of the BBB damage may to some degree reflect the diverse response of astrocytes in different HE models.

THE ASTHENIC ASTROCYTE

As outlined in Section "The oxidative/nitrosative stress and mitochondrial dysfunction," oxidative/nitrosative stress is a critical contributor to the pathogenesis of HE (Kosenko et al., 1999, 2017; Murthy et al., 2001; Schliess et al., 2002; Görg, 2003; Görg et al., 2008; Kruczek et al., 2011; Qvartskhava et al., 2015). In addition to a variety of intracellular consequences of ROS generation described in Section "The oxidative/nitrosative stress and mitochondrial dysfunction," markers of the senescence process were demonstrated in astrocytes (Schliess et al., 2004, 2006; Görg et al., 2013, 2015, 2018, 2019; Oenarto et al., 2016). In some brain pathologies (Chinta et al., 2015; Bussian et al., 2018), senescence markers were detected in degenerated, asthenic astrocytes. Notably, divergent pieces of evidence demonstrated features of senescence in ammonia-treated astrocytes, such as cell growth inhibition, upregulation of senescence-associated β-D-galactosidase (SA-β-Gal) (Kuilman et al., 2010; Görg et al., 2015), nuclear accumulation of a master regulator p53 (Görg et al., 2015), and its phosphorylation of at serine 392 by mitogen-activated protein kinases (Panickar et al., 2009) in line with enhanced transcription and nuclear accumulation of cell cycle inhibitory factor p21 (Görg et al., 2015, 2018).

The mechanisms underlying the appearance of senescence markers in astrocytes involve ammonia-induced upregulation of ROS-generating NADPH oxidase 4 (NOX4) isoform and of HO1, which is thought to be a protective response to oxidative stress (Warskulat et al., 2002; Görg et al., 2013). HO1 and NOX4 increase at mRNA level were detected in *postmortem* brain tissue from HE patients (Görg et al., 2015). In asthenic astrocytes, a functional link between senescent markers and ammonia-induced synthesis of glutamine and fructose-6-phosphate to glucosamine-6-phosphate

(GlcN-6-P) by rate-controlling enzyme was demonstrated. The synthesis of GlcN-6-P promoted activation of sugar nucleotides, utilized by O-GlcNAc transferase for the O-GlcNAcylation of proteins (Copeland et al., 2013). Interestingly, similar to, e.g., phosphorylation, O-GlcNAcylation can modulate the function of proteins in specific ways (Copeland et al., 2013; Saunders et al., 2023). Consequently, upon HE, O-GlcNAcylation of individual proteins was documented in *postmortem* brain samples from patients with liver cirrhosis and HE (Görg et al., 2019). The possible functional consequence of ammonia-induced astrocyte senescence can be related to synaptic contact loss, as documented in astrocyte-neuron cocultures treated with ammonia and in brains from HE animal models (Jayakumar et al., 2014; Galland et al., 2017). However, the latter link remains to be established in HE patients. Interestingly, the involvement of HO1-targeting miR326-3p downregulation makes the mechanism of ammonia-induced astrocyte progressive loss of function even more puzzling (Oenarto et al., 2016).

NEUROINFLAMMATION: A CONSEQUENCE OF SYSTEMIC INFLAMMATION AND REACTIVE MICROGLIOSIS

Systemic inflammation or systemic inflammatory response syndrome (SIRS) is associated with infection, sepsis, hepatocyte cell death, hyponatremia, gastrointestinal hemorrhage, or renal failure, and is mediated by proinflammatory cytokines and mediators release. Systemic inflammation is a precipitating factor both in ALF and chronic liver disease, contributing to mental status changes and motor deficits (Papadopoulos et al., 2000; Shawcross et al., 2004; Iacobone et al., 2009; Butterworth, 2011, 2013). Moreover, systemic infection correlates with the clinical phenotype in HE and ACLF and predicts clinical course (Clària et al., 2016; Trebicka et al., 2020). Synergistic effects of systemic infection and ammonia in HE have been initially suggested since circulating levels of the proinflammatory cytokines including tumor necrosis factor α (TNFα) positively correlated with the disease severity (Odeh et al., 2005; Odeh, 2007; Wright et al., 2007). Additionally, TNFα gene polymorphisms influence the status of ALF patients (Bernal et al., 1998). Some studies documented that the degree of systemic infection in HE patients correlates with HE severity, even irrespective of ammonia levels (Shawcross et al., 2011).

The biologic nature of the signaling between pathologic liver and neuroinflammation involves humoral and neural clues. Mechanisms that initiate the processes in the brain encompass direct active transport of cytokines and cytokine interaction with receptors (e.g., TNFα receptor 1) residing in circumventricular organs lacking the BBB or with afferent neurons of the vagus nerve (McCoy and Tansey, 2008). A significant correlation was found between C-reactive protein levels and inflammatory gene markers in the cortex of patients with HE, suggesting a systemic-brain link (Görg et al., 2013). Restoration of behavior and microglial morphology in BDL and hyperammonemic rats by the antiinflammatory agent ibuprofen added credence to this link (Rodrigo et al., 2010). Inhibition of TNFα signaling by systemic treatment with etanercept also reduced systemic inflammation, attenuated the neurologic decline, and prevented microglial activation in the cerebral cortex (Chastre et al., 2012). Moreover, systemic inflammatory signals altered the BBB permeability to cytokines (Varatharaj and Galea, 2017). The cerebrovascular endothelial cell activation by peripherally administered TNFα stimulated microglia to produce monocyte chemotactic protein 1 and mediated the recruitment of monocytes into the brain with subsequent in situ synthesis of TNFα in an animal model of biliary cirrhosis (D'Mello et al., 2009).

Liver-derived toxins other than cytokines also play a role in neuroinflammation. Sole or cumulative action of ammonia, lactate, glutamate, manganese, and neurosteroids is likely to trigger the transformation from the resting to the reactive brain microglia (Van Rossum and Hanisch, 2004; Claeys et al., 2021).

Neuroinflammation is associated with reactive astrogliosis and microgliosis (Conductier et al., 2010; Shulyatnikova and Verkhratsky, 2020; Escartin et al., 2021; Paolicelli et al., 2022). Reactive microgliosis, associated with increased synthesis or release of proinflammatory signaling molecules, such as cytokines (IL-1α, IL-1β, and TNFα) and chemokines (chemokine ligand 2, CCL2; C-X3-C ligand 1, CX3CL1), was demonstrated in HE animal models (Chastre et al., 2012; McMillin et al., 2014), and increased microglia motility was monitored using an *in vivo* two-photon imaging technique (Rangroo Thrane et al., 2012). The direct evidence of neuroinflammation, documenting reactive microgliosis with a concurrent increase of TNFα, IL-1β, and IL-6 levels in the brain, was demonstrated in an animal model of ALF resulting from hepatic devascularization (Jiang et al., 2006). The involvement of additional factors, ROS, and prostanoids, was documented in the ammonia-induced transition from resting to reactive microglia (Zemtsova et al., 2011).

Reactive microgliosis was likewise assessed by quantifications of ^{11}C-PK11195, a positive emission tomography ligand for translocator protein (Cui et al., 2009). Using this technique, neuroinflammation has been described by ^{11}C-PK1195 and ^{18}F-DPA-714 binding to translocator protein in an acute HE model of TAA-injected rats (Luo et al., 2018). However, the limited selectivity of the ^{11}C-PK11195 ligand implies its roles beyond sole inflammatory process indicator (Notter et al., 2021). Deletion of translocator protein from astrocytes was associated with ammonia-induced mitochondria permeability transition increase and elevated cell volume of the mitochondria, indicating, beyond neuroinflammation, its multifunctional character (Panickar et al., 2007) (see also Section "The oxidative/nitrosative stress and mitochondrial dysfunction"). Consequently, experimental studies demonstrated attenuation of disease progression by antiinflammatory strategies such as minocycline, proinflammatory cytokine gene deletion, etanercept, infliximab, and minocycline (Henry et al., 2008; Jiang et al., 2009; Bémeur et al., 2010; Dadsetan et al., 2016). While evidence in favor of the role of neuroinflammation in chronic HE models is less consistent (Cauli et al., 2007; D'Mello et al., 2009; Rodrigo et al., 2010; Brück et al., 2011), neuroinflammation was documented in both acute and chronic HE patients. In patients suffering from ALF resulting from viral hepatitis, reactive microgliosis was detected by the increase of immunostaining for human leukocyte antigen DR (CR3/43) (Butterworth, 2011). In turn, upregulation of the microglia marker ionized calcium-binding adaptor molecule 1 (Iba1) was demonstrated in *postmortem* brain tissue samples of HE patients with cirrhosis (Fig. 12.4) (Zemtsova et al., 2011; Dennis et al., 2014).

Astrocytes contribute to neuroinflammation in HE through the release of glutamate, prostanoids, and reactive oxygen/nitrogen species due to ammonia-induced upregulation of iNOS and NADPH-oxidase activation (Schliess et al., 2002; Reinehr et al., 2007; Görg et al., 2010a). Inflammatory cytokines such as TNFα aggravated ammonia-induced astrocyte swelling in an oxidative stress-dependent manner (Lachmann et al., 2013). Additive (or synergic) effects of proinflammatory cytokines (TNFα, IL-1, and IL-6) in astrocyte swelling induction suggested astrocyte inflammation and its contribution to HE pathomechanism (Rama Rao et al., 2010b). Inflammatory cytokines also induce protein tyrosine nitration in rat astrocytes (Görg et al., 2006), demonstrating the interdependence of different aspects of causative factors in HE.

The study of *postmortem* brain sections of HE patients showed GFAP decline in the cerebral cortex, hypothalamus, basal ganglia, and subcortical white matter (Sobel et al., 1981), a finding later confirmed and elaborated on, in *in vitro* in astrocytes treated with ammonia (Norenberg et al., 1990), and in acute HE animal models (Rubio et al., 1993; Bélanger et al., 2002). Of note, astrocyte remodeling, classically assessed through increased GFAP immunoreactivity was detected in rats with resected liver (Guazzelli et al., 2020). In addition, 3D reconstruction of GFAP-positive astrocytes in BDL mice revealed a subtle increase in arborization, documenting morphologic changes associated with glial cell dysfunction (Claeys et al., 2022).

CONCLUDING REMARKS

Advances in the knowledge of the complex and multiple roles of glia in the pathology of HE are spectacular; however, our understanding of details is far from satisfying. While the role of microglia in the pathogenesis of HE appears to be confined to neuroinflammation, that of astrocytes encompasses a multitude

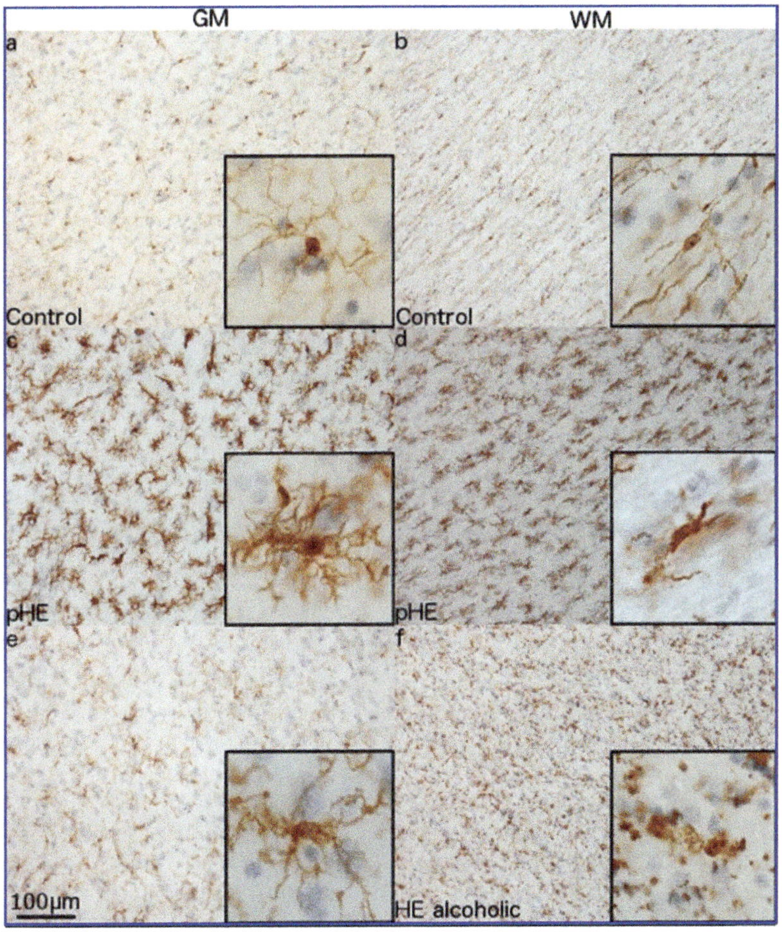

Fig. 12.4. Morphology of Iba1-stained microglia in cortical gray matter (GM; a, c, f) and white matter (WM; b, d, f) of patients with alcoholic HE. Distinction between proliferative (pHE; c–d) and nonproliferative cells (HE alcoholic; e–f). Reproduced from Dennis CV, Sheahan PJ, Graeber MB et al. (2014). Microglial proliferation in the brain of chronic alcoholics with hepatic encephalopathy. Metab Brain Dis 29: 1027–1039. https://doi.org/10.1007/s11011-013-9469-0.

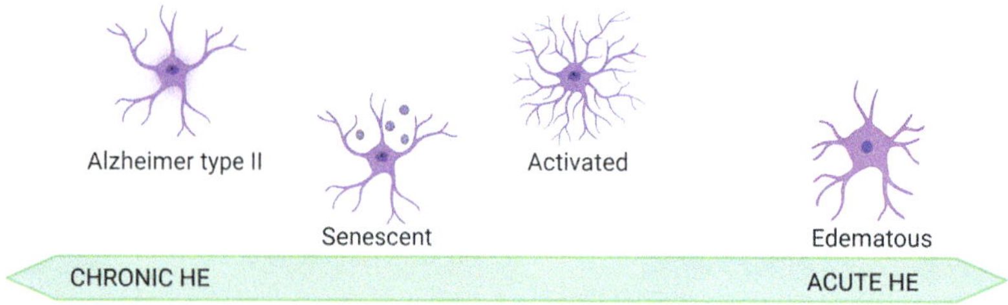

Fig. 12.5. Visualization of astrocytes manifested in HE. Created with BioRender.com.

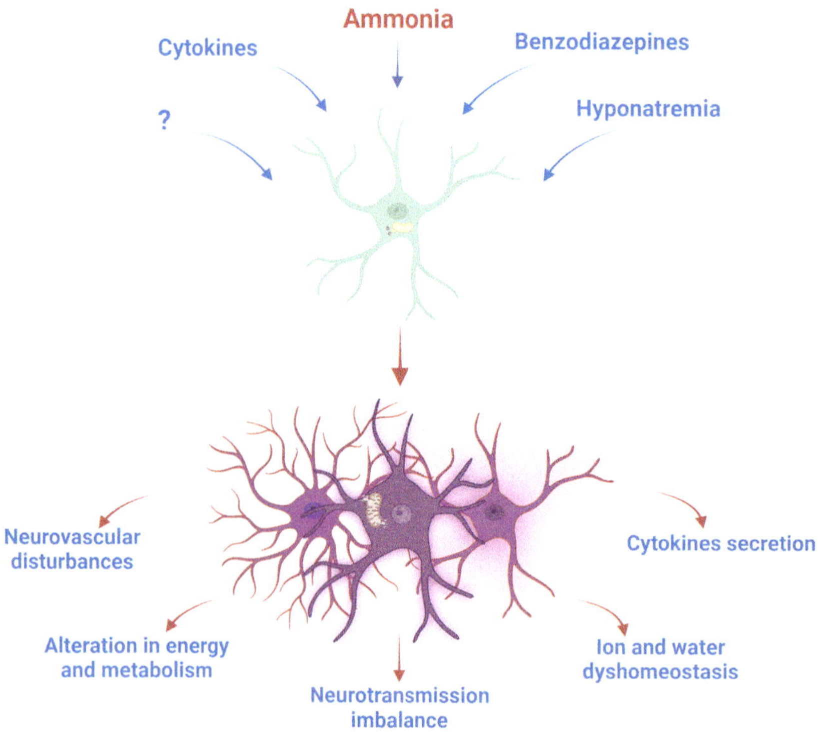

Fig. 12.6. HE-related factors targeting astrocytes (*upper*) and astrocytic contribution to brain dysfunction in HE (*bottom*). Created with BioRender.com.

of events. Excessive entry of ammonia into the brain primarily impairs the morphology and metabolism of astrocytes (Fig. 12.5), changing their energy status and impairing their ability to regulate transmission. In addition, the neuroimmune response, which largely advances the chain of pathogenic events, involves a concerted action of astrocytes and microglia. Functions of astrocytes in synaptic plasticity and the control of recruited immune cells have placed astrocytes at the center of the CNS functions; however, in the context of HE, details of the mechanisms are missing. Astrocytes receive and convey information to all neural cell types in a coordinated effort to respond to HE-relevant factors such as ammonia, benzodiazepines, and proinflammatory cytokines, initiating changes in molecular components and/or some cellular processes. The state-of-the-art new technology will make it possible to further verify the complexity of astrocytes. Single-cell transcriptional profiling techniques tool used specifically to reveal distinct astrocyte subtypes with regional distribution should allow to identify glia subpopulation, as well as their contribution to HE-pertinent changes in intracellular signaling networks (for basic data see Zeisel et al., 2018).

In conclusion, while an abundance of information is available on the role of glia in HE (Fig. 12.6), it mainly originates from experimental studies, with relatively few reports providing mechanistic insight into the HE-affected human brain. Clearly, this state of events falls short of drawing glia-targeting diagnostic and therapeutic perspectives and thus cannot satisfy even the minimum demands of clinical scientists. It is to be believed that future advancement of techniques will allow filling this gap. Then, time will come to examine how other biologic variables, including age and sex, influence astrocytes, microglia, neurons, and the perhaps unjustly bypassed status of oligodendrocytes within the HE brain and/or periphery.

ACKNOWLEDGMENTS

Funds from the authors' research unit were provided by statutory funds No. 13 from Mossakowski Medical Research Institute, Polish Academy of Sciences, Warsaw, Poland.

REFERENCES

Abbott NJ, Rönnbäck L, Hansson E (2006). Astrocyte-endothelial interactions at the blood-brain barrier. Nat Rev Neurosci 7: 41–53. https://doi.org/10.1038/nrn1824.

Adams RD, Foley JM (1953). The neurological disorder associated with liver disease. Res Publ Assoc Res Nerv Ment Dis 32: 198–237.

Aggarwal S, Kramer D, Yonas H et al. (1994). Cerebral hemodynamic and metabolic changes in fulminant hepatic failure: a retrospective study. Hepatology 19: 80–87.

Aickin CC, Deisz RA, Lux HD (1982). Ammonium action on post-synaptic inhibition in crayfish neurones: implications for the mechanism of chloride extrusion. J Physiol 329: 319–339. https://doi.org/10.1113/jphysiol.1982.sp014305.

Albrecht J (1986). Early metabolic changes in astrocytes in experimental hepatogenic encephalopathy. In: T Grisar et al. (Eds.), Dynamic properties of glial cells. II. Cellular and molecular aspects. Pergamon Press, Oxford, pp. 363–370.

Albrecht J, Hilgier W (1984). Brain carbonic anhydrase activity in rats in experimental hepatogenic encephalopathy. Neurosci Lett 45: 7–10. https://doi.org/10.1016/0304-3940(84)90320-3.

Albrecht J, Jones EA (1999). Hepatic encephalopathy: molecular mechanisms underlying the clinical syndrome. J Neurol Sci 170: 138–146. https://doi.org/10.1016/S0022-510X(99)00169-0.

Albrecht J, Norenberg MD (2006). Glutamine: a Trojan horse in ammonia neurotoxicity. Hepatology 44: 788–794. https://doi.org/10.1002/hep.21357.

Albrecht J, Rafałowska U (1987). Enhanced potassium-stimulated gamma-aminobutyric acid release by astrocytes derived from rats with early hepatogenic encephalopathy. J Neurochem 49: 9–11. https://doi.org/10.1111/j.1471-4159.1987.tb03385.x.

Albrecht J, Schousboe A (2005). Taurine interaction with neurotransmitter receptors in the CNS: an update. Neurochem Res 30: 1615–1621. https://doi.org/10.1007/s11064-005-8986-6.

Albrecht J, Wysmyk-Cybula U, Rafalowska U (1985). Na+/K+-ATPase activity and GABA uptake in astroglial cell-enriched fractions and synaptosomes derived from rats in the early stage of experimental hepatogenic encephalopathy. Acta Neurol Scand 72: 317–320. https://doi.org/10.1111/j.1600-0404.1985.tb00877.x.

Albrecht J, Wysmyk-Cybula U, Rafałowska U (1987). Cerebral oxygen consumption in experimental hepatic encephalopathy: different responses in astrocytes, neurons, and synaptosomes. Exp Neurol 97: 418–422. https://doi.org/10.1016/0014-4886(87)90102-6.

Albrecht J, Hilgier W, Łazarewicz J et al. (1988). Astrocytes in acute hepatic encephalopathy: metabolic properties and transport functions. In: MD Norenberg et al. (Eds.), Biochemical pathology of astrocytes. R. Liss, New York, pp. 465–476.

Albrecht J, Gadamski R, Kuhrt H et al. (1998). Retinal gliopathy accompanying thioacetamide-induced liver insufficiency: light and electron microscopic observations. Acta Neuropathol 96: 57–66. https://doi.org/10.1007/s004010050860.

Albrecht J, Zielińska M, Norenberg MD (2010). Glutamine as a mediator of ammonia neurotoxicity: a critical appraisal. Biochem Pharmacol 80: 1303–1308. https://doi.org/10.1016/j.bcp.2010.07.024.

Allaman I, Bélanger M, Magistretti PJ (2011). Astrocyte-neuron metabolic relationships: for better and for worse. Trends Neurosci 34: 76–87. https://doi.org/10.1016/j.tins.2010.12.001.

Almdal T, Schroeder T, Ranek L (1989). Cerebral blood flow and liver function in patients with encephalopathy due to acute and chronic liver diseases. Scand J Gastroenterol 24: 299–303. https://doi.org/10.3109/00365528909093050.

Anholt RR, Pedersen PL, De Souza EB et al. (1986). The peripheral-type benzodiazepine receptor. Localization to the mitochondrial outer membrane. J Biol Chem 261: 576–583.

Attwell D, Iadecola C (2002). The neural basis of functional brain imaging signals. Trends Neurosci 25: 621–625. https://doi.org/10.1016/s0166-2236(02)02264-6.

Attwell D, Buchan AM, Charpak S et al. (2010). Glial and neuronal control of brain blood flow. Nature 468: 232–243. https://doi.org/10.1038/nature09613.

Basile AS, Gammal SH, Jones EA et al. (1989). GABAA receptor complex in an experimental model of hepatic encephalopathy: evidence for elevated levels of an endogenous benzodiazepine receptor ligand. J Neurochem 53: 1057–1063. https://doi.org/10.1111/j.1471-4159.1989.tb07395.x.

Basile AS, Ostrowski NL, Gammal SH et al. (1990). The GABAA receptor complex in hepatic encephalopathy. Autoradiographic evidence for the presence of elevated levels of a benzodiazepine receptor ligand. Neuropsychopharmacology 3: 61–71.

Bassett ML, Mullen KD, Scholz B et al. (1990). Increased brain uptake of gamma-aminobutyric acid in a rabbit model of hepatic encephalopathy. Gastroenterology 98: 747–757. https://doi.org/10.1016/0016-5085(90)90298-f.

Bélanger M, Desjardins P, Chatauret N et al. (2002). Loss of expression of glial fibrillary acidic protein in acute hyperammonemia. Neurochem Int 41: 155–160. https://doi.org/10.1016/S0197-0186(02)00037-2.

Bémeur C, Qu H, Desjardins P et al. (2010). IL-1 or TNF receptor gene deletion delays onset of encephalopathy and attenuates brain edema in experimental acute liver failure. Neurochem Int 56: 213–215. https://doi.org/10.1016/j.neuint.2009.11.010.

Bender AS, Norenberg MD (1996). Effects of ammonia on L-glutamate uptake in cultured astrocytes. Neurochem Res 21: 567–573. https://doi.org/10.1007/BF02527755.

Bernal W, Donaldson P, Underhill J et al. (1998). Tumor necrosis factor genomic polymorphism and outcome of acetaminophen (paracetamol)-induced acute liver failure. J Hepatol 29: 53–59. https://doi.org/10.1016/S0168-8278(98)80178-5.

Bernal W, Hall C, Karvellas CJ et al. (2007). Arterial ammonia and clinical risk factors for encephalopathy and intracranial hypertension in acute liver failure. Hepatology 46: 1844–1852. https://doi.org/10.1002/hep.21838.

Bernal W, Auzinger G, Dhawan A et al. (2010). Acute liver failure. Lancet 376: 190–201. https://doi.org/10.1016/S0140-6736(10)60274-7.

Binesh N, Huda A, Bugbee M et al. (2005). Adding another spectral dimension to ^{1}H magnetic resonance spectroscopy of hepatic encephalopathy. Magn Reson Imaging 21: 398–405. https://doi.org/10.1002/jmri.20291.

Bjerring PN, Hauerberg J, Frederiksen H-J et al. (2008). Cerebral glutamine concentration and lactate–pyruvate ratio in patients with acute liver failure. Neurocrit Care 9: 3–7. https://doi.org/10.1007/s12028-008-9060-4.

Bjerring PN, Eefsen M, Larsen FS et al. (2011). Hypermagnesemia does not prevent intracranial hypertension and aggravates cerebral hyperperfusion in a rat model of acute hyperammonemia. Hepatology 53: 1986–1994. https://doi.org/10.1002/hep.24274.

Blei AT (2000). Medical therapy of brain edema in fulminant hepatic failure. Hepatology 32: 666–669. https://doi.org/10.1053/jhep.2000.17923.

Blei AT (2007). Brain edema in acute liver failure: can it be prevented? Can it be treated? J Hepatol 46: 564–569. https://doi.org/10.1016/j.jhep.2007.01.011.

Blei AT, Olafsson S, Therrien G et al. (1994). Ammonia-induced brain edema and intracranial hypertension in rats after portacaval anastomosis. Hepatology 19: 1437–1444. https://doi.org/10.1002/hep.1840190619.

Bosman DK, Deutz NE, Maas MA et al. (1992). Amino acid release from cerebral cortex in experimental acute liver failure, studied by in vivo cerebral cortex microdialysis. J Neurochem 59: 591–599. https://doi.org/10.1111/j.1471-4159.1992.tb09410.x.

Brück J, Görg B, Bidmon H-J et al. (2011). Locomotor impairment and cerebrocortical oxidative stress in portal vein ligated rats in vivo. J Hepatol 54: 251–257. https://doi.org/10.1016/j.jhep.2010.06.035.

Brusilow SW (1985). Inborn errors of urea synthesis. In: JK Lloyd, CR pp ? Scriver (Eds.), Genetic and pp. metabolic disease in pediatrics. Butterworth-Heinemann, pp. 140–165. https://doi.org/10.1016/B978-0-407-02312-3.50012-0.

Bussian TJ, Aziz A, Meyer CF et al. (2018). Clearance of senescent glial cells prevents tau-dependent pathology and cognitive decline. Nature 562: 578–582. https://doi.org/10.1038/s41586-018-0543-y.

Butterworth RF (2011). Hepatic encephalopathy: a central neuroinflammatory disorder? Hepatology 53: 1372–1376. https://doi.org/10.1002/hep.24228.

Butterworth RF (2013). The liver–brain axis in liver failure: neuroinflammation and encephalopathy. Nat Rev Gastroenterol Hepatol 10: 522–528. https://doi.org/10.1038/nrgastro.2013.99.

Butterworth RF, Norenberg MD, Felipo V et al. (2009). Experimental models of hepatic encephalopathy: ISHEN guidelines. Liver Int 29: 783–788. https://doi.org/10.1111/j.1478-3231.2009.02034.x.

Cabrera-Pastor A, Arenas YM, Taoro-Gonzalez L et al. (2019). Chronic hyperammonemia alters extracellular glutamate, glutamine and GABA and membrane expression of their transporters in rat cerebellum. Modulation by extracellular cGMP. Neuropharmacology 161: 107496. https://doi.org/10.1016/j.neuropharm.2019.01.011.

Cascino A, Cangiano C, Fiaccadori F et al. (1982). Plasma and cerebrospinal fluid amino acid patterns in hepatic encephalopathy. Dig Dis Sci 27: 828–832. https://doi.org/10.1007/BF01391377.

Cauli O, Rodrigo R, Piedrafita B et al. (2007). Inflammation and hepatic encephalopathy: ibuprofen restores learning ability in rats with portacaval shunts. Hepatology 46: 514–519. https://doi.org/10.1002/hep.21734.

Cauli O, Rodrigo R, Boix J et al. (2008). Acute liver failure-induced death of rats is delayed or prevented by blocking NMDA receptors in brain. Am J Physiol Gastrointest Liver Physiol 295: G503–G511. https://doi.org/10.1152/ajpgi.00076.2008.

Cauli O, López-Larrubia P, Rodrigo R et al. (2011). Brain region-selective mechanisms contribute to the progression of cerebral alterations in acute liver failure in rats. Gastroenterology 140: 638–645. https://doi.org/10.1053/j.gastro.2010.10.043.

Chan H, Butterworth RF (1999). Evidence for an astrocytic glutamate transporter deficit in hepatic encephalopathy. Neurochem Res 24: 1397–1401. https://doi.org/10.1023/a:1022532623281.

Chan H, Hazell AS, Des jardins P et al. (2000). Effects of ammonia on glutamate transporter (GLAST) protein and mRNA in cultured rat cortical astrocytes. Neuroc h em Int 37: 243–248. https://doi.org/10.1016/S0197-0186(00)00026-7.

Chastre A, Bélanger M, Beauchesne E et al. (2012). Inflammatory cascades driven by tumor necrosis factor-alpha play a major role in the progression of acute liver failure and its neurological complications. PLoS One 7: e49670. https://doi.org/10.1371/journal.pone.0049670.

Chepkova AN, Sergeeva OA, Görg B et al. (2017). Impaired novelty acquisition and synaptic plasticity in congenital hyperammonemia caused by hepatic glutamine synthetase deficiency. Sci Rep 7: 40190. https://doi.org/10.1038/srep40190.

Chinta SJ, Woods G, Rane A et al. (2015). Cellular senescence and the aging brain. Exp Gerontol 68: 3–7. https://doi.org/10.1016/j.exger.2014.09.018.

Claeys W, Van Hoecke L, Lefere S et al. (2021). The neurogliovascular unit in hepatic encephalopathy. JHEP Rep 3: 100352. https://doi.org/10.1016/j.jhepr.2021.100352.

Claeys W, Van Hoecke L, Geerts A et al. (2022). A mouse model of hepatic encephalopathy: bile duct ligation induces brain ammonia overload, glial cell activation and neuroinflammation. Sci Rep 12: 17558. https://doi.org/10.1038/s41598-022-22423-6.

Clària J, Stauber RE, Coenraad MJ et al. (2016). Systemic inflammation in decompensated cirrhosis: characterization and role in acute-on-chronic liver failure. Hepatology 64: 1249–1264. https://doi.org/10.1002/hep.28740.

Clemmesen JO, Larsen FS, Kondrup J et al. (1999). Cerebral herniation in patients with acute E liver failure is correlated with arterial ammonia concentration. Hepatology 29: 648–653. https://doi.org/10.1002/hep.510290309.

Conductier G, Blondeau N, Guyon A et al. (2010). The role of monocyte chemoattractant protein MCP1/CCL2 in neuroinflammatory diseases. J Neuroimmunol 224: 93–100. https://doi.org/10.1016/j.jneuroim.2010.05.010.

Cooper AJL (2012). The role of glutamine synthetase and glutamate dehydrogenase in cerebral ammonia homeostasis. Neurochem Res 37: 2439–2455. https://doi.org/10.1007/s11064-012-0803-4.

Cooper AJL, Jeitner TM (2016). Central role of glutamate metabolism in the maintenance of nitrogen homeostasis in normal and hyperammonemic brain. Biomolecules 6: 16. https://doi.org/10.3390/biom6020016.

Cooper AJ, Plum F (1987). Biochemistry and physiology of brain ammonia. Physiol Rev 67: 440–519. https://doi.org/10.1152/physrev.1987.67.2.440.

Copeland RJ, Han G, Hart GW (2013). O-GlcNAcomics—Revealing roles of O-GlcNAcylation in disease mechanisms and development of potential diagnostics. Proteomics Clin Appl 7: 597–606. https://doi.org/10.1002/prca.201300001.

Cudalbu C, Taylor-Robinson SD (2019). Brain edema in chronic hepatic encephalopathy. J Clin Exp Hepatol 9: 362–382. https://doi.org/10.1016/j.jceh.2019.02.003.

Cui Y, Takashima T, Takashima-Hirano M et al. (2009). ^{11}C-PK11195 PET for the in vivo evaluation of neuroinflammation in the rat brain after cortical spreading depression. J Nucl Med 50: 1904–1911. https://doi.org/10.2967/jnumed.109.066498.

Czarnecka A, Aleksandrowicz M, Jasiński K et al. (2018). Cerebrovascular reactivity and cerebral perfusion of rats with acute liver failure: role of L-glutamine and asymmetric dimethylarginine in L-arginine-induced response. J Neurochem 147: 692–704. https://doi.org/10.1111/jnc.14578.

D'Mello C, Le T, Swain MG (2009). Cerebral microglia recruit monocytes into the brain in response to tumor necrosis Factorα signaling during peripheral organ inflammation. J Neurosci 29: 2089–2102. https://doi.org/10.1523/JNEUROSCI.3567-08.2009.

Dabrowiecki Z, Albrecht J (1985). The utility of astroglia cell enriched fraction from rat brain for studying cell membrane potentials: changes induced by hepatogenic encephalopathy. Neurochem Res 10: 315–318. https://doi.org/10.1007/BF00964601.

Dadsetan S, Balzano T, Forteza J et al. (2016). Infliximab reduces peripheral inflammation, neuroinflammation, and extracellular GABA in the cerebellum and improves learning and motor coordination in rats with hepatic encephalopathy. J Neuroinflammation 13: 245. https://doi.org/10.1186/s12974-016-0710-8.

De Knegt RJ, Schalm SW, Van Der Rijt CCD et al. (1994). Extracellular brain glutamate during acute liver failure and during acute hyperammonemia simulating acute liver failure. J Hepatol 20: 19–26. https://doi.org/10.1016/S0168-8278(05)80462-3.

DeMorrow S, Cudalbu C, Davies N et al. (2021). 2021 ISHEN guidelines on animal models of hepatic encephalopathy. Liver Int 41: 1474–1488. https://doi.org/10.1111/liv.14911.

Dennis CV, Sheahan PJ, Graeber MB et al. (2014). Microglial proliferation in the brain of chronic alcoholics with hepatic encephalopathy. Metab Brain Dis 29: 1027–1039. https://doi.org/10.1007/s11011-013-9469-0.

Desjardins P, Bandeira P, Raghavendra Rao VL et al. (1997). Increased expression of the peripheral-type benzodiazepine receptor-isoquinoline carboxamide binding protein mRNA in brain following portacaval anastomosis. Brain Res 758: 255–258. https://doi.org/10.1016/s0006-8993(97)00339-9.

Desjardins P, Du T, Jiang W et al. (2012). Pathogenesis of hepatic encephalopathy and brain edema in acute liver failure: role of glutamine redefined. Neurochem Int 60: 690–696. https://doi.org/10.1016/j.neuint.2012.02.001.

Detry O (2006). Brain edema and intracranial hypertension in fulminant hepatic failure: pathophysiology and management. WJG 12: 7405. https://doi.org/10.3748/wjg.v12.i46.7405.

Dhanda S, Sandhir R (2018). Blood-brain barrier permeability is exacerbated in experimental model of hepatic encephalopathy via MMP-9 activation and downregulation of tight junction proteins. Mol Neurobiol 55: 3642–3659. https://doi.org/10.1007/s12035-017-0521-7.

Diemer NH (1977). Number of purkinje cells and bergmann astrocytes in rats with CCl_4-induced liver disease. Acta Neurol Scand 55: 1–15. https://doi.org/10.1111/j.1600-0404.1977.tb05623.x.

Diemer NH, Laursen H (1977). Glial cell reactions in rats with HYPERAMMONIEMIA induced by urease or PORTO-CAVAL anastomosis. Acta Neurol Scand 55: 425–442. https://doi.org/10.1111/j.1600-0404.1977.tb07623.x.

Diemer NH, Tønnesen K (1977). Glial changes in pigs with PORTO-CAVAL anastomosis and temporary or total hepatic artery clamping. Acta Pathol Microbiol Scand Sect A Pathol 85A: 721–730. https://doi.org/10.1111/j.1699-0463.1977.tb00464.x.

Dolińska M, Hilgier W, Albrecht J (1996). Ammonia stimulates glutamine uptake to the cerebral non-synaptic mitochondria of the rat. Neurosci Lett 213: 45–48. https://doi.org/10.1016/0304-3940(96)12827-5.

Ducis I, Norenberg LO, Norenberg MD (1990). Effect of phenol and sodium octanoate on the astrocyte benzodiazepine receptor. Brain Res 514: 349–351. https://doi.org/10.1016/0006-8993(90)91431-f.

Ede RJ, Williams RW (1986). Hepatic encephalopathy and cerebral edema. Semin Liver Dis 6: 107–118. https://doi.org/10.1055/s-2008-1040594.

Epstein FH, Silva P (1985). Na-K-cl cotransport in chloride-transporting epithelia[a]. Ann N Y Acad Sci 456: 187–197. https://doi.org/10.1111/j.1749-6632.1985.tb14864.x.

Escartin C, Galea E, Lakatos A et al. (2021). Reactive astrocyte nomenclature, definitions, and future directions. Nat Neurosci 24: 312–325. https://doi.org/10.1038/s41593-020-00783-4.

Faff-Michalak L, Albrecht J (1991). Aspartate aminotransferase, malate dehydrogenase, and pyruvate carboxylase activities in rat cerebral synaptic and nonsynaptic mitochondria: effects of in vitro treatment with ammonia, hyperammonemia and hepatic encephalopathy. Metab Brain Dis 6: 187–197. https://doi.org/10.1007/BF00996918.

Filosa JA, Bonev AD, Straub SV et al. (2006). Local potassium signaling couples neuronal activity to vasodilation in the brain. Nat Neurosci 9: 1397–1403. https://doi.org/10.1038/nn1779.

Galland F, Negri E, Da Ré C et al. (2017). Hyperammonemia compromises glutamate metabolism and reduces BDNF in the rat hippocampus. Neurotoxicology 62: 46–55. https://doi.org/10.1016/j.neuro.2017.05.006.

Garcia-Tsao G, Sanyal AJ, Grace ND et al. (2007). Prevention and management of gastroesophageal varices and variceal hemorrhage in cirrhosis. Hepatology 46: 922–938. https://doi.org/10.1002/hep.21907.

Giguère JF, Hamel E, Butterworth RF (1992). Increased densities of binding sites for the "peripheral-type" benzodiazepine receptor ligand [3H]PK 11195 in rat brain following portacaval anastomosis. Brain Res 585: 295–298. https://doi.org/10.1016/0006-8993(92)91222-z.

Gomolka RS, Hablitz LM, Mestre H et al. (2023). Loss of aquaporin-4 results in glymphatic system dysfunction via brain-wide interstitial fluid stagnation. Elife 12: e82232. https://doi.org/10.7554/eLife.82232.

Gonzalez-Usano A, Cauli O, Agusti A et al. (2014). Pregnenolone sulfate restores the glutamate-nitric-oxide-cGMP pathway and extracellular GABA in cerebellum and learning and motor coordination in hyperammonemic rats. ACS Chem Nerosci 5: 100–105. https://doi.org/10.1021/cn400168y.

Görg B (2003). Benzodiazepine-induced protein tyrosine nitration in rat astrocytes. Hepatology 37: 334–342. https://doi.org/10.1053/jhep.2003.50061.

Görg B, Bidmon HJ, Keitel V et al. (2006). Inflammatory cytokines induce protein tyrosine nitration in rat astrocytes. Arch Biochem Biophys 449: 104–114. https://doi.org/10.1016/j.abb.2006.02.012.

Görg B, Qvartskhava N, Keitel V et al. (2008). Ammonia induces RNA oxidation in cultured astrocytes and brain in vivo. Hepatology 48: 567–579. https://doi.org/10.1002/hep.22345.

Görg B, Morwinsky A, Keitel V et al. (2010a). Ammonia triggers exocytotic release of L-glutamate from cultured rat astrocytes. Glia 58: 691–705. https://doi.org/10.1002/glia.20955.

Görg B, Qvartskhava N, Bidmon H-J et al. (2010b). Oxidative stress markers in the brain of patients with cirrhosis and hepatic encephalopathy. Hepatology 52: 256–265. https://doi.org/10.1002/hep.23656.

Görg B, Bidmon H-J, Häussinger D (2013). Gene expression profiling in the cerebral cortex of patients with cirrhosis with and without hepatic encephalopathy. Hepatology 57: 2436–2447. https://doi.org/10.1002/hep.26265.

Görg B, Karababa A, Shafigullina A et al. (2015). Ammonia-induced senescence in cultured rat astrocytes and in human cerebral cortex in hepatic encephalopathy. Glia 63: 37–50. https://doi.org/10.1002/glia.22731.

Görg B, Karababa A, Häussinger D (2018). Hepatic encephalopathy and astrocyte senescence. J Clin Exp Hepatol 8: 294–300. https://doi.org/10.1016/j.jceh.2018.05.003.

Görg B, Karababa A, Schütz E et al. (2019). O-GlcNAcylation-dependent upregulation of HO1 triggers ammonia-induced oxidative stress and senescence in hepatic encephalopathy. J Hepatol 71: 930–941. https://doi.org/10.1016/j.jhep.2019.06.020.

Grant S, McMillin M, Frampton G et al. (2018). Direct comparison of the thioacetamide and azoxymethane models of type a hepatic encephalopathy in mice. Gene Expr 18: 171–185. https://doi.org/10.3727/105221618X15287315176503.

Grishin DV, Kasap EY, Izotov AA et al. (2020). Multifaceted ammonia transporters. All Life 13: 486–497. https://doi.org/10.1080/26895293.2020.1812443.

Guazzelli PA, Cittolin-Santos GF, Meira-Martins LA et al. (2020). Acute liver failure induces glial reactivity, oxidative stress and impairs brain energy metabolism in rats. Front Mol Neurosci 12: 327. https://doi.org/10.3389/fnmol.2019.00327.

Gupta T, Dhiman RK, Ahuja CK et al. (2017). Characterization of cerebral edema in acute-on-chronic liver failure. J Clin Exp Hepatol 7: 190–197. https://doi.org/10.1016/j.jceh.2017.04.001.

Hadjihambi A, De Chiara F, Hosford PS et al. (2017). Ammonia mediates cortical hemichannel dysfunction in rodent models of chronic liver disease. Hepatology 65: 1306–1318. https://doi.org/10.1002/hep.29031.

Hadjihambi A, Harrison IF, Costas-Rodríguez M et al. (2019). Impaired brain glymphatic flow in experimental hepatic encephalopathy. J Hepatol 70: 40–49. https://doi.org/10.1016/j.jhep.2018.08.021.

Hamdani EH, Popek M, Frontczak-Baniewicz M et al. (2021). Perturbation of astroglial Slc38 glutamine transporters by NH_4^+ contributes to neurophysiologic manifestations in acute liver failure. FASEB J 35. https://doi.org/10.1096/fj.202001712RR.

Haussinger D, Schliess F (2008). Pathogenetic mechanisms of hepatic encephalopathy. Gut 57: 1156–1165. https://doi.org/10.1136/gut.2007.122176.

Häussinger D, Dhiman RK, Felipo V et al. (2022). Hepatic encephalopathy. Nat Rev Dis Primers 8: 43. https://doi.org/10.1038/s41572-022-00366-6.

Heins J, Zwingmann C (2010). Organic osmolytes in hyponatremia and ammonia toxicity. Metab Brain Dis 25: 81–89. https://doi.org/10.1007/s11011-010-9170-5.

Helling G, Wahlin S, Smedberg M et al. (2016). Plasma glutamine concentrations in liver failure. PLoS One 11: e0150440. https://doi.org/10.1371/journal.pone.0150440.

Henry CJ, Huang Y, Wynne A et al. (2008). Minocycline attenuates lipopolysaccharide (LPS)-induced neuroinflammation, sickness behavior, and anhedonia. J Neuroinflammation 5: 15. https://doi.org/10.1186/1742-2094-5-15.

Hertz L, Bender AS, Woodbury DM et al. (1989). Potassium-stimulated calcium uptake in astrocytes and its potent inhibition by nimodipine. J Neurosci Res 22: 209–215. https://doi.org/10.1002/jnr.490220215.

Hilgier W, Zitting A, Albrecht J (1985). The brain octopamine and phenylethanolamine content in rats in thioacetamide-induced hepatogenic encephalopathy. Acta Neurol Scand 71: 195–198. https://doi.org/10.1111/j.1600-0404.1985.tb03188.x.

Hilgier W, Haugvicova R, Albrecht J (1991). Decreased potassium-stimulated release of [3H]D-aspartate from hippocampal slices distinguishes encephalopathy related to acute liver failure from that induced by simple hyperammonemia. Brain Res 567: 165–168. https://doi.org/10.1016/0006-8993(91)91451-6.

Hilgier W, Zielińska M, Borkowska HD et al. (1999). Changes in the extracellular profiles of neuroactive amino acids in the rat striatum at the asymptomatic stage of hepatic failure. J Neurosci Res 56: 76–84. https://doi.org/10.1002/(SICI)1097-4547(19990401)56:1<76::AID-JNR10>3.0.CO;2-Y.

Hilgier W, Freśko I, Klemenska E et al. (2009). Glutamine inhibits ammonia-induced accumulation of cGMP in rat striatum limiting arginine supply for NO synthesis. Neurobiol Dis 35: 75–81. https://doi.org/10.1016/j.nbd.2009.04.004.

Hilgier W, Kocki T, Obara-Michlewska M et al. (2014). Modulation by kynurenine of extracellular kynurenate and glutamate in cerebral cortex of rats with acute liver failure. Pharmacol Rep 66: 466–470. https://doi.org/10.1016/j.pharep.2013.11.007.

Hourani BT, Hamlin EM, Reynolds TB (1971). Cerebrospinal fluid glutamine as a measure of hepatic encephalopathy. Arch Intern Med 127: 1033–1036.

Iacobone E, Bailly-Salin J, Polito A et al. (2009). Sepsis-associated encephalopathy and its differential diagnosis. Crit Care Med 37: S331–S336. https://doi.org/10.1097/CCM.0b013e3181b6ed58.

Iliff JJ, Chen MJ, Plog BA et al. (2014). Impairment of glymphatic pathway function promotes tau pathology after traumatic brain injury. J Neurosci 34: 16180–16193. https://doi.org/10.1523/JNEUROSCI.3020-14.2014.

Itzhak Y, Baker L, Norenberg MD (1993). Characterization of the peripheral-type benzodiazepine receptors in cultured astrocytes: evidence for multiplicity. Glia 9: 211–218. https://doi.org/10.1002/glia.440090306.

Itzhak Y, Roig-Cantisano A, Dombro RS et al. (1995). Acute liver failure and hyperammonemia increase peripheral-type benzodiazepine receptor binding and pregnenolone synthesis in mouse brain. Brain Res 705: 345–348. https://doi.org/10.1016/0006-8993(95)01244-3.

Iversen P, Sørensen M, Bak LK et al. (2009). Low cerebral oxygen consumption and blood flow in patients with cirrhosis and an acute episode of hepatic encephalopathy. Gastroenterology 136: 863–871. https://doi.org/10.1053/j.gastro.2008.10.057.

Jalan R, Olde Damink SWM, Hayes PC et al. (2004). Pathogenesis of intracranial hypertension in acute liver failure: inflammation, ammonia and cerebral blood flow. J Hepatol 41: 613–620. https://doi.org/10.1016/j.jhep.2004.06.011.

Jayakumar AR, Norenberg MD (2018). Hyperammonemia in hepatic encephalopathy. J Clin Exp Hepatol 8: 272–280. https://doi.org/10.1016/j.jceh.2018.06.007.

Jayakumar AR, Rao KVR, Murthy CRK et al. (2006). Glutamine in the mechanism of ammonia-induced astrocyte swelling. Neurochem Int 48: 623–628. https://doi.org/10.1016/j.neuint.2005.11.017.

Jayakumar AR, Liu M, Moriyama M et al. (2008). Na-K-cl Cotransporter-1 in the mechanism of ammonia-induced astrocyte swelling. J Biol Chem 283: 33874–33882. https://doi.org/10.1074/jbc.M804016200.

Jayakumar AR, Rama Rao KV, Tong XY et al. (2009). Calcium in the mechanism of ammonia-induced astrocyte swelling. J Neurochem 109: 252–257. https://doi.org/10.1111/j.1471-4159.2009.05842.x.

Jayakumar AR, Tong XY, Curtis KM et al. (2014). Decreased astrocytic thrombospondin-1 secretion after chronic ammonia treatment reduces the level of synaptic proteins: *in vitro* and *in vivo* studies. J Neurochem 131: 333–347. https://doi.org/10.1111/jnc.12810.

Jiang W, Qu H, Desjardins P et al. (2006). Unequivocal evidence for cytokine accumulation in brain in experimental acute liver failure. Hepatology 44: 366A.

Jiang W, Desjardins P, Butterworth RF (2009). Cerebral inflammation contributes to encephalopathy and brain edema in acute liver failure: protective effect of minocycline. J Neurochem 109: 485–493. https://doi.org/10.1111/j.1471-4159.2009.05981.x.

Jiménez JV, Carrillo-Pérez DL, Rosado-Canto R et al. (2017). Electrolyte and acid–base disturbances in end-stage liver disease: a physiopathological approach. Dig Dis Sci 62: 1855–1871. https://doi.org/10.1007/s10620-017-4597-8.

Jin X, Chen D, Wu F et al. (2020). Hydrogen sulfide protects against ammonia-induced neurotoxicity through activation of Nrf2/ARE signaling in astrocytic model of hepatic encephalopathy. Front Cell Neurosci 14: 573422. https://doi.org/10.3389/fncel.2020.573422.

Jones EA, Weissenborn K (1997). Neurology and the liver. J Neurol Neurosurg Psychiatry 63: 279–293. https://doi.org/10.1136/jnnp.63.3.279.

Jover R, Rodrigo R, Felipo V et al. (2006). Brain edema and inflammatory activation in bile duct ligated rats with diet-induced hyperammonemia: a model of hepatic encephalopathy in cirrhosis. Hepatology 43: 1257–1266. https://doi.org/10.1002/hep.21180.

Kanamatsu T, Tsukada Y (1999). Effects of ammonia on the anaplerotic pathway and amino acid metabolism in the brain: an ex vivo 13C NMR spectroscopic study of rats after administering [2-13C] glucose with or without ammonium acetate. Brain Res 841: 11–19. https://doi.org/10.1016/s0006-8993(99)01772-2.

Kanamori K, Ross BD (2005). Suppression of glial glutamine release to the extracellular fluid studied *in vivo* by NMR and microdialysis in hyperammonemic rat brain. J Neurochem 94: 74–85. https://doi.org/10.1111/j.1471-4159.2005.03170.x.

Kato M, Sugihara J, Nakamura T et al. (1989). Electron microscopic study of the blood-brain barrier in rats with brain edema and encephalopathy due to acute hepatic failure. Gastroenterol Jpn 24: 135–142. https://doi.org/10.1007/BF02774187.

Kato M, Hughes RD, Keays RT et al. (1992). Electron microscopic study of brain capillaries in cerebral edema from fulminant hepatic failure. Hepatology 15: 1060–1066. https://doi.org/10.1002/hep.1840150615.

Kelly T, Rose CR (2010). Ammonium influx pathways into astrocytes and neurones of hippocampal slices. J Neurochem 115: 1123–1136. https://doi.org/10.1111/j.1471-4159.2010.07009.x.

Kelly T, Kafitz KW, Roderigo C et al. (2009). Ammonium-evoked alterations in intracellular sodium and pH reduce glial glutamate transport activity. Glia 57: 921–934. https://doi.org/10.1002/glia.20817.

Khalil HMA, Eliwa HA, El-Shiekh RA et al. (2021). Ashwagandha (*Withania somnifera*) root extract attenuates hepatic and cognitive deficits in thioacetamide-induced rat model of hepatic encephalopathy via induction of Nrf2/HO-1 and mitigation of NF-κB/MAPK signaling pathways. J Ethnopharmacol 277: 114141. https://doi.org/10.1016/j.jep.2021.114141.

Kimelberg HK, Stieg PE, Mazurkiewicz JE (1982). Immunocytochemical and biochemical analysis of carbonic anhydrase in primary astrocyte cultures from rat brain. J Neurochem 39: 734–742. https://doi.org/10.1111/j.1471-4159.1982.tb07954.x.

Kimura T, Budka H (1986). Glial fibrillary acidic protein and S-100 protein in human hepatic encephalopathy: Immunocytochemical demonstration of dissociation of two glia-associated proteins. Acta Neuropathol 70: 17–21. https://doi.org/10.1007/BF00689509.

Knecht K, Michalak A, Rose C et al. (1997). Decreased glutamate transporter (GLT-1) expression in frontal cortex of rats with acute liver failure. Neurosci Lett 229: 201–203. https://doi.org/10.1016/S0304-3940(97)00444-8.

Kosenko E, Kaminsky Y, Kaminsky A et al. (1997). Superoxide production and antioxidant enzymes in ammonia intoxication in rats. Free Radic Res 27: 637–644. https://doi.org/10.3109/10715769709097867.

Kosenko E, Kaminski Y, Lopata O et al. (1999). Blocking NMDA receptors prevents the oxidative stress induced by acute ammonia intoxication. Free Radic Biol Med 26: 1369–1374. https://doi.org/10.1016/s0891-5849(98)00339-6.

Kosenko EA, Tikhonova LA, Alilova GA et al. (2017). Portacaval shunting causes differential mitochondrial superoxide production in brain regions. Free Radic Biol Med 113: 109–118. https://doi.org/10.1016/j.freeradbiomed.2017.09.023.

Kruczek C, Görg B, Keitel V et al. (2011). Ammonia increases nitric oxide, free Zn^{2+}, and metallothionein mRNA expression in cultured rat astrocytes. Biol Chem 392: 1155–1165. https://doi.org/10.1515/BC.2011.199.

Kuilman T, Michaloglou C, Mooi WJ et al. (2010). The essence of senescence. Genes Dev 24: 2463–2479. https://doi.org/10.1101/gad.1971610.

Lachmann V, Görg B, Bidmon HJ et al. (2013). Precipitants of hepatic encephalopathy induce rapid astrocyte swelling in an oxidative stress dependent manner. Arch Biochem Biophys 536: 143–151. https://doi.org/10.1016/j.abb.2013.05.004.

Lapidot A, Gopher A (1997). Quantitation of metabolic compartmentation in hyperammonemic brain by natural abundance 13C-NMR detection of 13C-15N coupling patterns and isotopic shifts. Eur J Biochem 243: 597–604. https://doi.org/10.1111/j.1432-1033.1997.00597.x.

Larsen FS (2002). Cerebral blood flow in hyperammonemia: heterogeneity and starling forces in capillaries. Metab Brain Dis 17: 229–235. https://doi.org/10.1023/a:1021941414605.

Larsen BR, Assentoft M, Cotrina ML et al. (2014). Contributions of the Na^+/K^+-ATPase, NKCC1, and Kir4.1 to hippocampal K^+ clearance and volume responses. Glia 62: 608–622. https://doi.org/10.1002/glia.22629.

Laubenberger J, Haussinger D, Bayer S et al. (1997). Proton magnetic resonance spectroscopy of the brain in symptomatic and asymptomatic patients with liver cirrhosis. Gastroenterology 112: 1610–1616. https://doi.org/10.1016/S0016-5085(97)70043-X.

Lavoie J, Layrargues GP, Butterworth RF (1990). Increased densities of peripheral-type benzodiazepine receptors in brain autopsy samples from cirrhotic patients with hepatic encephalopathy. Hepatology 11: 874–878. https://doi.org/10.1002/hep.1840110524.

Leke R, Bak LK, Iversen P et al. (2011). Synthesis of neurotransmitter GABA via the neuronal tricarboxylic acid cycle is elevated in rats with liver cirrhosis consistent with a high GABAergic tone in chronic hepatic encephalopathy. J Neurochem 117: 824–832. https://doi.org/10.1111/j.1471-4159.2011.07244.x.

Liang C, Du T, Zhou J et al. (2014). Ammonium increases ca(2+) signalling and up-regulates expression of TRPC1 gene in astrocytes in primary cultures and in the in vivo brain. Neurochem Res 39: 2127–2135. https://doi.org/10.1007/s11064-014-1406-z.

Lichter-Konecki U, Mangin JM, Gordish-Dressman H et al. (2008). Gene expression profiling of astrocytes from hyperammonemic mice reveals altered pathways for water and potassium homeostasis in vivo. Glia 56: 365–377. https://doi.org/10.1002/glia.20624.

Lockwood AH, Finn RD, Campbell JA et al. (1980). Factors that affect the uptake of ammonia by the brain: the blood-brain pH gradient. Brain Res 181: 259–266. https://doi.org/10.1016/0006-8993(80)90611-3.

Lockwood AH, Yap EW, Rhoades HM et al. (1991). Altered cerebral blood flow and glucose metabolism in patients with liver disease and minimal encephalopathy. J Cereb Blood Flow Metab 11: 331–336. https://doi.org/10.1038/jcbfm.1991.66.

Luo S, Kong X, Wu JR et al. (2018). Neuroinflammation in acute hepatic encephalopathy rats: imaging and therapeutic effectiveness evaluation using 11C-PK11195 and 18F-DPA-714 micro-positron emission tomography. Metab Brain Dis 33: 1733–1742. https://doi.org/10.1007/s11011-018-0282-7.

Ma KC (2001). Alzheimer-type I astrogliopathy (AIA) and its implications for dynamic plasticity of astroglia: a historical review of the significance of AIA. J Neuropathol Exp Neurol 60: 121–131. https://doi.org/10.1093/jnen/60.2.121.

Ma KC, Chang ZH, Shih H et al. (1991). The compensatory "rebound" of reactive astrogliosis: glial fibrillary acidic protein immunohistochemical analysis of reactive astrogliosis after a puncture wound to the brain of rats with portocaval anastomosis. Acta Neuropathol 82: 72–77. https://doi.org/10.1007/BF00310926.

MacVicar BA, Newman EA (2015). Astrocyte regulation of blood flow in the brain. Cold Spring Harb Perspect Biol 7: a020388. https://doi.org/10.1101/cshperspect.a020388.

Magistretti PJ, Allaman I (2018). Lactate in the brain: from metabolic end-product to signalling molecule. Nat Rev Neurosci 19: 235–249. https://doi.org/10.1038/nrn.2018.19.

Marcaida G, Felipo V, Hermenegildo C et al. (1992). Acute ammonia toxicity is mediated by the NMDA type of glutamate receptors. FEBS Lett 296: 67–68. https://doi.org/10.1016/0014-5793(92)80404-5.

Martin H, Voss K, Hufnagl P et al. (1987). Morphometric and densitometric investigations of protoplasmic astrocytes and neurons in human hepatic encephalopathy. Exp Pathol 32: 241–250. https://doi.org/10.1016/s0232-1513(87)80035-x.

Master S, Gottstein J, Blei AT (1999). Cerebral blood flow and the development of ammonia-induced brain edema in rats after portacaval anastomosis. Hepatology 30: 876–880. https://doi.org/10.1002/hep.510300428.

McConnell JR, Antonson DL, Ong CS et al. (1995). Proton spectroscopy of brain glutamine in acute liver failure. Hepatology 22: 69–74.

McCoy MK, Tansey MG (2008). TNF signaling inhibition in the CNS: implications for normal brain function and neurodegenerative disease. J Neuroinflammation 5: 45. https://doi.org/10.1186/1742-2094-5-45.

McMillin M, Frampton G, Thompson M et al. (2014). Neuronal CCL2 is upregulated during hepatic encephalopathy and contributes to microglia activation and neurological decline. J Neuroinflammation 11: 121. https://doi.org/10.1186/1742-2094-11-121.

Michalak A, Rose C, Butterworth J et al. (1996). Neuroactive amino acids and glutamate (NMDA) receptors in frontal cortex of rats with experimental acute liver failure. Hepatology 24: 908–913. https://doi.org/10.1002/hep.510240425.

Milton M, Smith PD (2018). It's all about timing: the involvement of Kir4.1 channel regulation in acute ischemic stroke pathology. Front Cell Neurosci 12: 36. https://doi.org/10.3389/fncel.2018.00036.

Moroni F, Lombardi G, Moneti G et al. (1983). The release and neosynthesis of glutamic acid are increased in experimental models of hepatic encephalopathy. J Neurochem 40: 850–854. https://doi.org/10.1111/j.1471-4159.1983.tb08057.x.

Moser H (1987). Electrophysiological evidence for ammonium as a substitute for potassium in activating the sodium pump in a crayfish sensory neuron. Can J Physiol Pharmacol 65: 141–145. https://doi.org/10.1139/y87-028.

Murthy CR, Rama Rao KV, Bai G et al. (2001). Ammonia-induced production of free radicals in primary cultures of rat astrocytes. J Neurosci Res 66: 282–288. https://doi.org/10.1002/jnr.1222.

Nagaraja TN, Brookes N (1998). Intracellular acidification induced by passive and active transport of ammonium ions in astrocytes. Am J Physiol Cell Physiol 274: C883–C891. https://doi.org/10.1152/ajpcell.1998.274.4.C883.

Nagelhus EA, Ottersen OP (2013). Physiological roles of aquaporin-4 in brain. Physiol Rev 93: 1543–1562. https://doi.org/10.1152/physrev.00011.2013.

Neary JT, Whittemore SR, Zhu Q et al. (1994). Destabilization of glial fibrillary acidic protein mRNA in astrocytes by ammonia and protection by extracellular ATP. J Neurochem 63: 2021–2027. https://doi.org/10.1046/j.1471-4159.1994.63062021.x.

Nguyen JH (2012). Blood-brain barrier in acute liver failure. Neurochem Int 60: 676–683. https://doi.org/10.1016/j.neuint.2011.10.012.

Nguyen JH, Yamamoto S, Steers J et al. (2006). Matrix metalloproteinase-9 contributes to brain extravasation and edema in fulminant hepatic failure mice. J Hepatol 44: 1105–1114. https://doi.org/10.1016/j.jhep.2005.09.019.

Norenberg MD (1976). Histochemical studies in experimental portal-systemic encephalopathy: I. Glutamic dehydrogenase. Arch Neurol 33: 265. https://doi.org/10.1001/archneur.1976.00500040049007.

Norenberg MD (1977). A light and electron microscopic study of experimental portal-systemic (ammonia) encephalopathy. Progression and reversal of the disorder. Lab Invest 36: 618–627.

Norenberg MD, Martinez-Hernandez A (1979). Fine structural localization of glutamine synthetase in astrocytes of rat brain. Brain Res 161: 303–310. https://doi.org/10.1016/0006-8993(79)90071-4.

Norenberg MD, Neary JT, Norenberg LO et al. (1990). Ammonia induced decrease in glial fibrillary acidic protein in cultured astrocytes. J Neuropathol Exp Neurol 49: 399–405. https://doi.org/10.1097/00005072-199007000-00004.

Norenberg MD, Huo Z, Neary JT et al. (1997). The glial glutamate transporter in hyperammonemia and hepatic encephalopathy: relation to energy metabolism and glutamatergic neurotransmission. Glia 21: 124–133.

Notter T, Schalbetter SM, Clifton NE et al. (2021). Neuronal activity increases translocator protein (TSPO) levels. Mol Psychiatry 26: 2025–2037. https://doi.org/10.1038/s41380-020-0745-1.

O'Grady J (2007). Modern management of acute liver failure. Clin Liver Dis 11: 291–303. https://doi.org/10.1016/j.cld.2007.04.011.

Obara-Michlewska M, Ruszkiewicz J, Zielińska M et al. (2015). Astroglial NMDA receptors inhibit expression of Kir4.1 channels in glutamate-overexposed astrocytes in vitro and in the brain of rats with acute liver failure. Neurochem Int 88: 20–25. https://doi.org/10.1016/j.neuint.2014.10.006.

Obara-Michlewska M, Ding F, Popek M et al. (2018). Interstitial ion homeostasis and acid-base balance are maintained in oedematous brain of mice with acute toxic liver failure. Neurochem Int 118: 286–291. https://doi.org/10.1016/j.neuint.2018.05.007.

Odeh M (2007). Pathogenesis of hepatic encephalopathy: the tumour necrosis factor-α theory. Eur J Clin Invest 37: 291–304. https://doi.org/10.1111/j.1365-2362.2007.01778.x.

Odeh M, Sabo E, Srugo I et al. (2005). Relationship between tumor necrosis factor-alpha and ammonia in patients with hepatic encephalopathy due to chronic liver failure. Ann Med 37: 603–612. https://doi.org/10.1080/07853890500317414.

Oenarto J, Görg B, Moos M et al. (2014). Expression of organic osmolyte transporters in cultured rat astrocytes and rat and human cerebral cortex. Arch Biochem Biophys 560: 59–72. https://doi.org/10.1016/j.abb.2014.06.024.

Oenarto J, Karababa A, Castoldi M et al. (2016). Ammonia-induced miRNA expression changes in cultured rat astrocytes. Sci Rep 6: 18493. https://doi.org/10.1038/srep18493.

Ohno Y, Kunisawa N, Shimizu S (2021). Emerging roles of astrocyte Kir4.1 channels in the pathogenesis and treatment of brain diseases. Int J Mol Sci 22: 10236. https://doi.org/10.3390/ijms221910236.

Olivero M, Gagliardi D, Costamagna G et al. (2022). Newly diagnosed hepatic encephalopathy presenting as non-convulsive status epilepticus: a case report and literature review. Front Neurol 13: 880068. https://doi.org/10.3389/fneur.2022.880068.

Orzeł-Gajowik K, Milewski K, Zielińska M (2023). miRNA-ome plasma analysis unveils changes in blood-brain barrier integrity associated with acute liver failure in rats. Fluids Barriers CNS 20: 92. https://doi.org/10.1186/s12987-023-00484-7.

Ostapowicz G (2002). Results of a prospective study of acute liver failure at 17 tertiary care centers in the United States. Ann Intern Med 137: 947. https://doi.org/10.7326/0003-4819-137-12-200212170-00007.

Panickar KS, Jayakumar AR, Rama Rao KV et al. (2007). Downregulation of the 18-kDa translocator protein: effects on the ammonia-induced mitochondrial permeability transition and cell swelling in cultured astrocytes. Glia 55: 1720–1727. https://doi.org/10.1002/glia.20584.

Panickar KS, Jayakumar AR, Rao KVR et al. (2009). Ammonia-induced activation of p53 in cultured astrocytes: role in cell swelling and glutamate uptake. Neurochem Int 55: 98–105. https://doi.org/10.1016/j.neuint.2008.12.022.

Paolicelli RC, Sierra A, Stevens B et al. (2022). Microglia states and nomenclature: a field at its crossroads. Neuron 110: 3458–3483. https://doi.org/10.1016/j.neuron.2022.10.020.

Papadopoulos MC, Davies DC, Moss RF et al. (2000). Pathophysiology of septic encephalopathy: a review. Crit Care Med 28: 3019–3024. https://doi.org/10.1097/00003246-200008000-00057.

Papadopoulos V, Baraldi M, Guilarte TR et al. (2006). Translocator protein (18kDa): new nomenclature for the peripheral-type benzodiazepine receptor based on its structure and molecular function. Trends Pharmacol Sci 27: 402–409. https://doi.org/10.1016/j.tips.2006.06.005.

Partin JC, Partin JS, Schubert WK et al. (1975). Brain ultrastructure in Reye's syndrome: (excephalopathy and fatty alteration of the viscera). J Neuropathol Exp Neurol 34: 425–444. https://doi.org/10.1097/00005072-197509000-00005.

Peterson C, Giguere JF, Cotman CW et al. (1990). Selective loss of N-methyl-D-aspartate-sensitive L-[3H]glutamate binding sites in rat brain following portacaval anastomosis. J Neurochem 55: 386–390. https://doi.org/10.1111/j.1471-4159.1990.tb04149.x.

Petzold GC, Murthy VN (2011). Role of astrocytes in neurovascular coupling. Neuron 71: 782–797. https://doi.org/10.1016/j.neuron.2011.08.009.

Pichili VBR, Rao KVR, Jayakumar AR et al. (2007). Inhibition of glutamine transport into mitochondria protects astrocytes from ammonia toxicity. Glia 55: 801–809. https://doi.org/10.1002/glia.20499.

Pluta R, Albrecht J (1984). Thioacetamide-induced hepatic encephalopathy in the rat. Clinical observations. Neuropatol Pol 22: 379–385.

Pluta R, Albrecht J (1986). Changes in arterial and cerebral venous blood gases, cerebral blood flow and cerebral oxygen consumption at different stages of thioacetamide-induced hepatogenic encephalopathy in rat. Resuscitation 14: 135–139. https://doi.org/10.1016/0300-9572(86)90117-6.

Popek M, Bobula B, Sowa J et al. (2018). Cortical synaptic transmission and plasticity in acute liver failure are decreased by presynaptic events. Mol Neurobiol 55: 1244–1258. https://doi.org/10.1007/s12035-016-0367-4.

Probst J, Kölker S, Okun JG et al. (2020). Chronic hyperammonemia causes a hypoglutamatergic and hyperGABAergic metabolic state associated with neurobehavioral abnormalities in zebrafish larvae. Exp Neurol 331: 113330. https://doi.org/10.1016/j.expneurol.2020.113330.

Quinn M, McMillin M, Galindo C et al. (2014). Bile acids permeabilize the blood brain barrier after bile duct ligation in rats via Rac1-dependent mechanisms. Dig Liver Dis 46: 527–534. https://doi.org/10.1016/j.dld.2014.01.159.

Qvartskhava N, Lang PA, Görg B et al. (2015). Hyperammonemia in gene-targeted mice lacking functional hepatic glutamine synthetase. Proc Natl Acad Sci U S A 112: 5521–5526. https://doi.org/10.1073/pnas.1423968112.

Rae C, Moussa CE-H, Griffin JL et al. (2006). A metabolomic approach to ionotropic glutamate receptor subtype function: a nuclear magnetic resonance *in vitro* investigation. J Cereb Blood Flow Metab 26: 1005–1017. https://doi.org/10.1038/sj.jcbfm.9600257.

Rama Rao KV, Norenberg MD (2012). Brain energy metabolism and mitochondrial dysfunction in acute and chronic hepatic encephalopathy. Neurochem Int 60: 697–706. https://doi.org/10.1016/j.neuint.2011.09.007.

Rama Rao KV, Chen M, Simard JM et al. (2003a). Increased aquaporin-4 expression in ammonia-treated cultured astrocytes. Neuroreport 14: 2379–2382. https://doi.org/10.1097/00001756-200312190-00018.

Rama Rao KV, Jayakumar AR, Norenberg DM (2003b). Ammonia neurotoxicity: role of the mitochondrial permeability transition. Metab Brain Dis 18: 113–127. https://doi.org/10.1023/a:1023858902184.

Rama Rao KV, Jayakumar AR, Tong X et al. (2010a). Brain aquaporin-4 in experimental acute liver failure. J Neuropathol Exp Neurol 69: 869–879. https://doi.org/10.1097/NEN.0b013e3181ebe581.

Rama Rao KV, Reddy PVB, Tong X et al. (2010b). Brain edema in acute liver failure. Am J Pathol 176: 1400–1408. https://doi.org/10.2353/ajpath.2010.090756.

Rama Rao KV, Verkman AS, Curtis KM et al. (2014). Aquaporin-4 deletion in mice reduces encephalopathy and brain edema in experimental acute liver failure. Neurobiol Dis 63: 222–228. https://doi.org/10.1016/j.nbd.2013.11.018.

Rangroo Thrane V, Thrane AS, Chang J et al. (2012). Real-time analysis of microglial activation and motility in hepatic and hyperammonemic encephalopathy. Neuroscience 220: 247–255. https://doi.org/10.1016/j.neuroscience.2012.06.022.

Rangroo Thrane V, Thrane AS, Wang F et al. (2013). Ammonia triggers neuronal disinhibition and seizures by impairing astrocyte potassium buffering. Nat Med 19: 1643–1648. https://doi.org/10.1038/nm.3400.

Rasmussen MK, Mestre H, Nedergaard M (2018). The glymphatic pathway in neurological disorders. Lancet Neurol 17: 1016–1024. https://doi.org/10.1016/S1474-4422(18)30318-1.

Record CO, Buxton B, Chase RA et al. (1976). Plasma and brain amino acids in fulminant hepatic failure and their relationship to hepatic encephalopathy. Eur J Clin Invest 6: 387–394. https://doi.org/10.1111/j.1365-2362.1976.tb00533.x.

Reichenbach A, Fuchs U, Kasper M et al. (1995). Hepatic retinopathy: morphological features of retinal glial (Müller) cells accompanying hepatic failure. Acta Neuropathol 90: 273–281. https://doi.org/10.1007/BF00296511.

Reinehr R, Görg B, Becker S et al. (2007). Hypoosmotic swelling and ammonia increase oxidative stress by NADPH oxidase in cultured astrocytes and vital brain slices. Glia 55: 758–771. https://doi.org/10.1002/glia.20504.

Rodrigo R, Cauli O, Boix J et al. (2009). Role of NMDA receptors in acute liver failure and ammonia toxicity: therapeutical implications. Neurochem Int 55: 113–118. https://doi.org/10.1016/j.neuint.2009.01.007.

Rodrigo R, Cauli O, Gomez-Pinedo U et al. (2010). Hyperammonemia induces neuroinflammation that contributes to cognitive impairment in rats with hepatic encephalopathy. Gastroenterology 139: 675–684. https://doi.org/10.1053/j.gastro.2010.03.040.

Rose C (2006). Effect of ammonia on astrocytic glutamate uptake/release mechanisms. J Neurochem 97: 11–15. https://doi.org/10.1111/j.1471-4159.2006.03796.x.

Rose CR, Verkhratsky A (2024). Sodium homeostasis and signalling: the core and the hub of astrocyte function. Cell Calcium 117: 102817. https://doi.org/10.1016/j.ceca.2023.102817.

Rose C, Kresse W, Kettenmann H (2005). Acute insult of ammonia leads to calcium-dependent glutamate release from cultured astrocytes, an effect of pH. J Biol Chem 280: 20937–20944. https://doi.org/10.1074/jbc.M412448200.

Rose CF, Amodio P, Bajaj JS et al. (2020). Hepatic encephalopathy: novel insights into classification, pathophysiology and therapy. J Hepatol 73: 1526–1547. https://doi.org/10.1016/j.jhep.2020.07.013.

Rubio M, Bodega G, Suárez I et al. (1993). Effects of experimentally induced hyperammonemia on glial fibrillary acidic protein (GFAP) in the rhombencephalon of goldfish (*Carassius auratus* L.). Neurosci Res 16: 59–64. https://doi.org/10.1016/0168-0102(93)90009-f.

Saran T, Hilgier W, Urbańska EM et al. (2004). Kynurenic acid synthesis in cerebral cortical slices of rats with progressing symptoms of thioacetamide-induced hepatic encephalopathy. J Neurosci Res 75: 436–440. https://doi.org/10.1002/jnr.10862.

Saransaari P, Oja SS, Borkowska HD et al. (1997). Effects of thioacetamide-induced hepatic failure on the N-methyl-D-aspartate receptor complex in the rat cerebral cortex, striatum, and hippocampus. Binding of different ligands and expression of receptor subunit mRNAs. Mol Chem Neuropathol 32: 179–193. https://doi.org/10.1007/BF02815175.

Saunders H, Dias WB, Slawson C (2023). Growing and dividing: how O-GlcNAcylation leads the way. J Biol Chem 299: 105330. https://doi.org/10.1016/j.jbc.2023.105330.

Schliess F, Görg B, Fischer R et al. (2002). Ammonia induces MK-801-sensitive nitration and phosphorylation of protein tyrosine residues in rat astrocytes. FASEB J 16: 739–741. https://doi.org/10.1096/fj.01-0862fje.

Schliess F, Foster N, Görg B et al. (2004). Hypoosmotic swelling increases protein tyrosine nitration in cultured rat astrocytes. Glia 47: 21–29. https://doi.org/10.1002/glia.20019.

Schliess F, Görg B, Häussinger D (2006). Pathogenetic interplay between osmotic and oxidative stress: the hepatic encephalopathy paradigm. Biol Chem 387. https://doi.org/10.1515/BC.2006.171.

Schwarcz R, Pellicciari R (2002). Manipulation of brain kynurenines: glial targets, neuronal effects, and clinical opportunities. J Pharmacol Exp Ther 303: 1–10. https://doi.org/10.1124/jpet.102.034439.

Scott TR (2013). Pathophysiology of cerebral oedema in acute liver failure. WJG 19: 9240. https://doi.org/10.3748/wjg.v19.i48.9240.

Sepehrinezhad A, Shahbazi A, Sahab Negah S et al. (2021). Drug-induced-acute liver failure: a critical appraisal of the thioacetamide model for the study of hepatic encephalopathy. Toxicol Rep 8: 962–970. https://doi.org/10.1016/j.toxrep.2021.04.011.

Sergeeva OA (2013). GABAergic transmission in hepatic encephalopathy. Arch Biochem Biophys 536: 122–130. https://doi.org/10.1016/j.abb.2013.04.005.

Shank RP, Campbell GL (1983). Ornithine as a precursor of glutamate and GABA: uptake and metabolism by neuronal and glial enriched cellular material. J Neurosci Res 9: 47–57. https://doi.org/10.1002/jnr.490090107.

Shawcross DL, Davies NA, Williams R et al. (2004). Systemic inflammatory response exacerbates the neuropsychological effects of induced hyperammonemia in cirrhosis. J Hepatol 40: 247–254. https://doi.org/10.1016/j.jhep.2003.10.016.

Shawcross DL, Sharifi Y, Canavan JB et al. (2011). Infection and systemic inflammation, not ammonia, are associated with grade 3/4 hepatic encephalopathy, but not mortality in cirrhosis. J Hepatol 54: 640–649. https://doi.org/10.1016/j.jhep.2010.07.045.

Shulyatnikova T, Verkhratsky A (2020). Astroglia in sepsis associated encephalopathy. Neurochem Res 45: 83–99. https://doi.org/10.1007/s11064-019-02743-2.

Skowrońska K, Obara-Michlewska M, Zielińska M et al. (2019). NMDA receptors in astrocytes: in search for roles in neurotransmission and astrocytic homeostasis. Int J Mol Sci 20: 309. https://doi.org/10.3390/ijms20020309.

Sobel RA, DeArmond SJ, Forno LS et al. (1981). Glial fibrillary acidic protein in hepatic encephalopathy: an immunohistochemical study. J Neuropathol Exp Neurol 40: 625–632. https://doi.org/10.1097/00005072-198111000-00004.

Sofroniew MV, Vinters HV (2010). Astrocytes: biology and pathology. Acta Neuropathol 119: 7–35. https://doi.org/10.1007/s00401-009-0619-8.

Søgaard R, Novak I, MacAulay N (2012). Elevated ammonium levels: differential acute effects on three glutamate transporter isoforms. Am J Physiol Cell Physiol 302: C880–C891. https://doi.org/10.1152/ajpcell.00238.2011.

Sonnewald U, Westergaard N, Jones P et al. (1996). Metabolism of [U- ^{13} C $_5$]glutamine in cultured astrocytes studied by NMR spectroscopy: first evidence of astrocytic pyruvate recycling. J Neurochem 67: 2566–2572. https://doi.org/10.1046/j.1471-4159.1996.67062566.x.

Sørensen M, Keiding S (2007). New findings on cerebral ammonia uptake in HE using functional 13N-ammonia PET. Metab Brain Dis 22: 277–284. https://doi.org/10.1007/s11011-007-9066-1.

Spielmeyer W (1920). Die histopathologische Zusammengehörigkeit der Wilsonschen Krankheit und der Pseudosklerose. Z f d g Neur u Psych 57: 312–351.

Stokum JA, Shim B, Negoita S et al. (2023). Cation flux through SUR1-TRPM4 and NCX1 in astrocyte endfeet induces water influx through AQP4 and brain swelling after ischemic stroke. Sci Signal 16: eadd6364. https://doi.org/10.1126/scisignal.add6364.

Suárez I, Bodega G, Fernández B (2002). Glutamine synthetase in brain: effect of ammonia. Neurochem Int 41: 123–142. https://doi.org/10.1016/S0197-0186(02)00033-5.

Suárez I, Bodega G, Arilla E et al. (2006). The expression of nNOS, iNOS and nitrotyrosine is increased in the rat cerebral cortex in experimental hepatic encephalopathy. Neuropathol Appl Neurobiol 32: 594–604. https://doi.org/10.1111/j.1365-2990.2006.00768.x.

Sugimoto H, Koehler RC, Wilson DA et al. (1997). Methionine sulfoximine, a glutamine synthetase inhibitor, attenuates increased extracellular potassium activity during acute hyperammonemia. J Cereb Blood Flow Metab 17: 44–49. https://doi.org/10.1097/00004647-199701000-00006.

Sweeney MD, Zhao Z, Montagne A et al. (2019). Blood-brain barrier: from physiology to disease and back. Physiol Rev 99: 21–78. https://doi.org/10.1152/physrev.00050.2017.

Szerb JC, Butterworth RF (1992). Effect of ammonium ions on synaptic transmission in the mammalian central nervous system. Prog Neurobiol 39: 135–153. https://doi.org/10.1016/0301-0082(92)90008-3.

Takahashi H, Koehler RC, Brusilow SW et al. (1991). Inhibition of brain glutamine accumulation prevents cerebral edema in hyperammonemic rats. Am J Physiol Heart Circ Physiol 261: H825–H829. https://doi.org/10.1152/ajpheart.1991.261.3.H825.

Tanigami H, Rebel A, Martin LJ et al. (2005). Effect of glutamine synthetase inhibition on astrocyte swelling and altered astroglial protein expression during hyperammonemia in rats. Neuroscience 131: 437–449. https://doi.org/10.1016/j.neuroscience.2004.10.045.

Therrien G, Butterworth RF (1991). Cerebrospinal fluid amino acids in relation to neurological status in experimental portal-systemic encephalopathy. Metab Brain Dis 6: 65–74. https://doi.org/10.1007/BF00999904.

Thomas JW, Banner C, Whitman J et al. (1988). Changes in glutamate-cycle enzyme mRNA levels in a rat model of hepatic encephalopathy. Metab Brain Dis 3: 81–90. https://doi.org/10.1007/BF01001011.

Thumburu KK, Dhiman RK, Vasishta RK et al. (2014). Expression of astrocytic genes coding for proteins implicated in neural excitation and brain edema is altered after acute liver failure. J Neurochem 128: 617–627. https://doi.org/10.1111/jnc.12511.

Tofteng F (2002). Monitoring extracellular concentrations of lactate, glutamate, and glycerol by in vivo microdialysis in the brain during liver transplantation in acute liver failure. Liver Transpl 8: 302–305. https://doi.org/10.1053/jlts.2002.32283.

Tofteng F, Jorgensen L, Hansen BA et al. (2002). Cerebral microdialysis in patients with fulminant hepatic failure: cerebral microdialysis in patients with fulminant hepatic failure. Hepatology 36: 1333–1340. https://doi.org/10.1002/hep.1840360607.

Tofteng F, Hauerberg J, Hansen BA et al. (2006). Persistent arterial Hyperammonemia increases the concentration of glutamine and alanine in the brain and correlates with intracranial pressure in patients with fulminant hepatic failure. J Cereb Blood Flow Metab 26: 21–27. https://doi.org/10.1038/sj.jcbfm.9600168.

Trebicka J, Fernandez J, Papp M et al. (2020). The PREDICT study uncovers three clinical courses of acutely decompensated cirrhosis that have distinct pathophysiology. J Hepatol 73: 842–854. https://doi.org/10.1016/j.jhep.2020.06.013.

Van Rossum D, Hanisch U-K (2004). Microglia. Metab Brain Dis 19: 393–411. https://doi.org/10.1023/B:MEBR.0000043984.73063.d8.

Vaquero J, Butterworth RF (2006). The brain glutamate system in liver failure. J Neurochem 98: 661–669. https://doi.org/10.1111/j.1471-4159.2006.03918.x.

Varatharaj A, Galea I (2017). The blood-brain barrier in systemic inflammation. Brain Behav Immun 60: 1–12. https://doi.org/10.1016/j.bbi.2016.03.010.

Verkhratsky AN, Butt A (2023). Neuroglia: function and pathology, Academic Press, London, United Kingdom.

Verkhratsky A, Nedergaard M (2018). Physiology of astroglia. Physiol Rev 98: 239–389. https://doi.org/10.1152/physrev.00042.2016.

Verkhratsky A, Untiet V, Rose CR (2020). Ionic signalling in astroglia beyond calcium. J Physiol 598: 1655–1670. https://doi.org/10.1113/JP277478.

Verkhratsky A, Butt A, Li B et al. (2023). Astrocytes in human central nervous system diseases: a frontier for new therapies. Signal Transduct Target Ther 8: 396. https://doi.org/10.1038/s41392-023-01628-9.

Victor M, Adams RD, Cole M (1965). The acquired (non-WILSONIAN) type of chronic HEPATOCEREBRAL degeneration. Medicine 44: 345–396. https://doi.org/10.1097/00005792-196509000-00001.

Vilstrup H, Amodio P, Bajaj J et al. (2014). Hepatic encephalopathy in chronic liver disease: 2014 practice guideline by the American Association for the Study of Liver Diseases and the European Association for the Study of the liver. Hepatology 60: 715–735. https://doi.org/10.1002/hep.27210.

Vogels BA, van Steynen B, Maas MA et al. (1997). The effects of ammonia and portal-systemic shunting on brain metabolism, neurotransmission and intracranial hypertension in hyperammonaemia-induced encephalopathy. J Hepatol 26: 387–395. https://doi.org/10.1016/s0168-8278(97)80057-8.

von Hoesslin C, Alzheimer AV (1912). Ein Beitrag zur Klinik und pathologischen Anatomie der Westphal-Strümpellschen Pseudosklerose. Z f d g Neur u Psych 8: 183–209.

Wang F, Du T, Liang C et al. (2015). Ammonium increases ca(2+) signalling and upregulates expression of Cav1.2 gene in astrocytes in primary cultures and in the in vivo brain. Acta Physiol (Oxf) 214: 261–274. https://doi.org/10.1111/apha.12500.

Wang W, Gu L, Verkhratsky A et al. (2017). Ammonium increases TRPC1 expression via Cav-1/PTEN/AKT/GSK3β pathway. Neurochem Res 42: 762–776. https://doi.org/10.1007/s11064-016-2004-z.

Wang T, Suzuki K, Kakisaka K et al. (2019). L-carnitine prevents ammonia-induced cytotoxicity and disturbances in intracellular amino acid levels in human astrocytes. J Gastroenterol Hepatol 34: 1249–1255. https://doi.org/10.1111/jgh.14497.

Warskulat U, Görg B, Bidmon H-J et al. (2002). Ammonia-induced heme oxygenase-1 expression in cultured rat astrocytes and rat brain in vivo. Glia 40: 324–336. https://doi.org/10.1002/glia.10128.

Watanabe A, Takei N, Higashi T et al. (1984). Glutamic acid and glutamine levels in serum and cerebrospinal fluid in hepatic encephalopathy. Biochem Med 32: 225–231. https://doi.org/10.1016/0006-2944(84)90076-0.

Wazir H, Abid M, Essani B et al. (2023). Diagnosis and treatment of liver disease: current trends and future directions. Cureus. https://doi.org/10.7759/cureus.49920.

Weiner ID (2006). Expression of the non-erythroid Rh glycoproteins in mammalian tissues. Transfus Clin Biol 13: 159–163. https://doi.org/10.1016/j.tracli.2006.02.002.

Weiss N, Barbier Saint Hilaire P, Colsch B et al. (2016). Cerebrospinal fluid metabolomics highlights dysregulation of energy metabolism in overt hepatic encephalopathy. J Hepatol 65: 1120–1130. https://doi.org/10.1016/j.jhep.2016.07.046.

Wendon JA, Harrison PM, Keays R et al. (1994). Cerebral blood flow and metabolism in fulminant liver failure. Hepatology 19: 1407–1413.

Wright G, Shawcross D, Olde Damink SWM et al. (2007). Brain cytokine flux in acute liver failure and its relationship with intracranial hypertension. Metab Brain Dis 22: 375–388. https://doi.org/10.1007/s11011-007-9071-4.

Wright G, Soper R, Brooks HF et al. (2010). Role of aquaporin-4 in the development of brain oedema in liver failure. J Hepatol 53: 91–97. https://doi.org/10.1016/j.jhep.2010.02.020.

Yu AC, Drejer J, Hertz L et al. (1983). Pyruvate carboxylase activity in primary cultures of astrocytes and neurons. J Neurochem 41: 1484–1487. https://doi.org/10.1111/j.1471-4159.1983.tb00849.x.

Zeisel A, Hochgerner H, Lönnerberg P et al. (2018). Molecular architecture of the mouse nervous system. Cell 174: 999–1014.e22. https://doi.org/10.1016/j.cell.2018.06.021.

Zemtsova I, Görg B, Keitel V et al. (2011). Microglia activation in hepatic encephalopathy in rats and humans. Hepatology 54: 204–215. https://doi.org/10.1002/hep.24326.

Zhou B-G, Norenberg MD (1999). Ammonia downregulates GLAST mRNA glutamate transporter in rat astrocyte cultures. Neurosci Lett 276: 145–148. https://doi.org/10.1016/S0304-3940(99)00816-2.

Zielińska M, Hilgier W, Law RO et al. (1999). Effects of ammonia in vitro on endogenous taurine efflux and cell volume in rat cerebrocortical minislices: influence of inhibitors of volume-sensitive amino acid transport. Neuroscience 91: 631–638. https://doi.org/10.1016/S0306-4522(98)00639-3.

Zielińska M, Ruszkiewicz J, Hilgier W et al. (2011). Hyperammonemia increases the expression and activity of the glutamine/arginine transporter y+ LAT2 in rat cerebral cortex: implications for the nitric oxide/cGMP pathway. Neurochem Int 58: 190–195. https://doi.org/10.1016/j.neuint.2010.11.015.

Zielińska M, Popek M, Albrecht J (2014). Roles of changes in active glutamine transport in brain edema development during hepatic encephalopathy: an emerging concept. Neurochem Res 39: 599–604. https://doi.org/10.1007/s11064-013-1141-x.

Zielińska M, Milewski K, Skowrońska M et al. (2015). Induction of inducible nitric oxide synthase expression in ammonia-exposed cultured astrocytes is coupled to increased arginine transport by upregulated y(+)LAT2 transporter. J Neurochem 135: 1272–1281. https://doi.org/10.1111/jnc.13387.

Zielińska M, Albrecht J, Popek M (2022). Dysregulation of astrocytic glutamine transport in acute hyperammonemic brain edema. Front Neurosci 16: 874750. https://doi.org/10.3389/fnins.2022.874750.

Ziemińska E, Dolińska M, Lazarewicz JW et al. (2000). Induction of permeability transition and swelling of rat brain mitochondria by glutamine. Neurotoxicology 21: 295–300.

Zieve L (1987). Pathogenesis of hepatic encephalopathy. Metab Brain Dis 2: 147–165. https://doi.org/10.1007/BF00999607.

Zwingmann C (2007). The anaplerotic flux and ammonia detoxification in hepatic encephalopathy. Metab Brain Dis 22: 235–249. https://doi.org/10.1007/s11011-007-9069-y.

Zwingmann C, Brand A, Richter-Landsberg C et al. (1998). Multinuclear NMR spectroscopy studies on NH4Cl-induced metabolic alterations and detoxification processes in primary astrocytes and glioma cells. Dev Neurosci 20: 417–426. https://doi.org/10.1159/000017339.

Zwingmann C, Chatauret N, Leibfritz D et al. (2003). Selective increase of brain lactate synthesis in experimental acute liver failure: results of a [H-C] nuclear magnetic resonance study. Hepatology 37: 420–428. https://doi.org/10.1053/jhep.2003.50052.

Handbook of Clinical Neurology, Vol. 210 (3rd series)
Neuroglia in Neurologic and Psychiatric Disorders, Part II
A. Verkhratsky, L.D. de Witte, E. Aronica, and E.M. Hol, Editors
https://doi.org/10.1016/B978-0-443-19102-2.00027-2

Chapter 13

Neuroglial responses to bacterial, viral, and fungal neuroinfections

IFEOLUWA AWOGBINDIN[1,2,†], EVA ŠIMONČIČOVÁ[1,3,†], VIRGINIE VIDAL[1,4], CHANTAILLE ASH[1], AND MARIE-EVE TREMBLAY[1,2,5,6*]

[1]*Division of Medical Sciences, University of Victoria, Victoria, BC, Canada*

[2]*Institute on Aging and Lifelong Health (IALH), University of Victoria, Victoria, BC, Canada*

[3]*Neuroscience Graduate Program, University of Victoria, Victoria, BC, Canada*

[4]*Science and Technology Department, University of Bordeaux, Bordeaux, France*

[5]*Department of Biochemistry and Molecular Biology, University of British Columbia, Vancouver, BC, Canada*

[6]*Centre for Advanced Materials and Related Technology (CAMTEC), University of Victoria, Victoria, BC, Canada*

Abstract

Evidence regarding the host's response to peripheral pathogens in humans abound, whereas studies on the pathogenesis of central nervous system-penetrating infections are relatively scarce. However, given the spate of epidemic and pandemic neuroinfections in the 21st century, the field has experienced a renewed interest lately. This chapter discusses a timely and exciting topic on the roles of glial cells, mainly microglia and astrocytes, in neuroinvasive infections. This chapter considered fungal, viral, and bacterial neuroinfections, X-raying their neuroinvasiveness, neurotropism, and neurovirulence before focusing on specific examples notable for each category, including *Escherichia coli*, *Cryptococcus neoformans,* and SARS-CoV-2. These infections are renowned worldwide for a high case-fatality rate, leaving many survivors with life-long morbidity and others with a bleak future neurologic prognosis. Importantly, the chapter discusses possible ways microglia and astrocytes are culpable in these infections and provides approaches by which they can be manipulated for therapeutic purposes, identifying viable research gaps in the process. Additionally, it offers a synopsis of ongoing works considering microglial selective targeting to attenuate the pathology, morbidity, and mortality associated with these neuroinfections. Considering that microglia and astrocytes are first responders in the central nervous system, targeting these glial cells could be the game changer in managing existing and emerging neuroinvasive infections.

NEUROINFECTIONS AND THE NERVOUS SYSTEM

The immune response and its potential dysregulation have been placed at the core of the mechanisms underlying infectious disease outcomes. Baseline immune mediator signaling is constitutively maintained for an optimal course of homeostatic physiologic functions, including the regulation of immunocompetent cell populations and neural/synaptic plasticity (Dantzer et al., 2008; Boulanger, 2009; Dantzer, 2018). In response to an invading pathogen, peripheral immunologic response, mediated first by innate (i.e., monocytes/macrophages, neutrophiles) and then adaptive immune cells (B- and T-lymphocytes, natural killer [NK]-cells), most notably elevates the baseline levels of pro- and anti-inflammatory cytokines and chemokines (Chaplin, 2010). If prolonged or exaggerated, however, the initially beneficial immune response may become dysregulated, leading to a severe, potentially detrimental inflammation (Colonna and Butovsky, 2017). The peripheral immune response can affect the central nervous system (CNS) that was classically regarded as an immune-privileged organ, with microglia, tissue-resident macrophages of embryonic yolk-sac origin (Ginhoux et al., 2013; Utz et al., 2020), representing its

[†]Co-first authors

*Correspondence to: Marie-Ève Tremblay, Division of Medical Sciences, University of Victoria, Victoria, BC, Canada; Institute on Aging and Lifelong Health (IALH), University of Victoria, Victoria, BC, Canada; Department of Biochemistry and Molecular Biology, The University of British Columbia, Vancouver, BC, Canada; Centre for Advanced Materials and Related Technology (CAMTEC), University of Victoria, Victoria, BC, Canada. Tel: +1-250-721-8739, Fax: +1-250-721-8955, E-mail: evetremblay@uvic.ca

primary provider of an immune response alongside other neuroglial cells including astrocytes (Colonna and Butovsky, 2017). While a principal involvement of brain neuroglia with neuroinvasive infections is well established, there is a need for a deeper understanding of their action and reaction mechanisms. Therefore, interest in research characterizing drug targets on microglia and other glial cells to treat and eliminate pathogens in the CNS is increasing.

The chapter presents an overview of the current understanding of neuroglial interaction with bacterial, viral, and fungal pathogens. Specifically, it discusses the involvement of microglia during CNS infections caused by human immunodeficiency virus (HIV), severe acute respiratory syndrome-associated coronavirus 2 (SARS-CoV-2), *E. coli*, and *C. neoformans*. Additionally, it introduces experimental studies investigating targets for the modulation of glial functions in these conditions and explores the potential of some already-approved drugs for therapeutic targeting.

NEUROINVASIVENESS, NEUROTROPISM, AND NEUROVIRULENCE OF PATHOGENS: AN OVERVIEW

Pathogens, including those responsible for epidemics and pandemics, have commonly developed unprecedented abilities to access and disrupt the homeostasis of the CNS, leading to acute and post-acute devastating consequences (Dando et al., 2014; Govic et al., 2022). Pathogens enter the CNS through direct or indirect routes (Fig. 13.1A–G). Obligate neurotropic infectious agents exploit peripheral nerve endings (e.g., *Burkholderia pseudomallei,* herpes simplex virus [HSV], rabies virus) or neuromuscular junctions (NMJs; e.g., poliovirus, rabies virus; (Fig. 13.1G), including the trigeminal (Fig. 13.1F), olfactory (rabies virus, HSV, influenza A virus [IAV]; Fig. 13.1E), and vagal nerves, innervating peripheral tissues to access the CNS. Pathogens travel from the terminals or NMJs through the axons to soma and synaptic terminals, spreading to adjoining nerves, interneurons, glial cells, and CNS regions (Dando et al., 2014; Swanson and McGavern, 2015). Alternatively, neuroinfectious agents can attack the periphery-CNS barriers (Fig. 13.1B–D)—the hematogenous route—to achieve neuroinvasion after causing a burden in the blood. In this case, they compromise intercellular junctions or infect cellular components of the blood–brain barrier (BBB) (Fig. 13.1B), blood-meningeal (Fig. 13.1C), and blood-cerebrospinal fluid barriers (BCSFBs) (Fig. 13.1D), including the associated glial cells. Of note, pathogens and their strains may show a preference for a specific hematogenous portal. *Mycobacterium tuberculosis* compromised the structural and molecular integrity of the BCSFB but not the BBB to cause tuberculous meningitis (Sánchez-Garibay et al., 2022). Pathogens also gain CNS entry through the hematogenous route either by extravasating through cells constituting the barrier under pathogen-mediated neurotoxic conditions such as inflammation (e.g., *Haemophilus influenzae*) or as a decoy–Trojan horse–inside infected blood cells (e.g., *C. neoformans*) transiting through leaky CNS barriers (Fig. 13.1B) (Dando et al., 2014; Swanson and McGavern, 2015; Govic et al., 2022).

The response of multiple CNS components, particularly neuroglia, is crucial to determining the course, severity, and neuropathology associated with neurovirulent pathogens (Fig. 13.1H–J). Infection of glial cells may precipitate myriads of complications due to the impairment of a plethora of functions they perform, leading to aberrant synaptic sculpting and inflammation (Geyer et al., 2019; Awogbindin et al., 2021). Also, the trophic support roles of neuroglial cells to neurons can be dysregulated during infection, contributing to cell death (Quincozes-Santos et al., 2021). Reactivity of neuroglia, especially microglia, mounts CNS immune response in collaboration with peripheral immune cells to determine the outcome, i.e., whether the infection will be resolved, retained, or multiply and spread to other cell types and brain regions (Geyer et al., 2019; Awogbindin et al., 2021; Singh et al., 2021). Astrocytes and microglia are recruited to local sites of infection, with their numbers and intensity of response determining both the immediate, medium, and long-term sequelae of pathogen-mediated neuropathology (Tremblay et al., 2020; Singh et al., 2021). Therefore, neuropathology mediated by an infection is often not only a result of the direct pathogenic effect but also of (in)actions of microglia and astrocytes. Also, neuroglial cells are vulnerable to neurotoxic molecules, including cytokines, chemokines, antibodies, and peripheral cells infiltrating through nerve endings, and compromised periphery-CNS barriers, shaping the course and severity of pathogen-mediated CNS pathology. Chronic systemic infections also disrupt gut microbiome-brain endocrine, metabolic, and immunologic communications, resulting in imbalanced levels of neurochemicals that modulate glial functions (Gorlé et al., 2020). A gut lacking a beneficial microbiome composition and short-chain fatty acids products such as butyrate promotes BBB breakdown (Braniste et al., 2014). Infiltration of cytokines and neuropeptides (substance P [SP], melanocortin, kynurenic acid (Lun et al., 2023), opioid β-endorphin [β-END], leptin) can modulate neuroglial functions, while peripheral antibodies may self-react against myelin, culminating into inflammatory demyelinating disease-like complications (Marrodan et al., 2019). Similarly, the loss of function of primary peripheral tissues directly infected by microbes (e.g., pulmonary immunopathology arising from SARS-CoV-2 infection) may lead to secondary sequelae in the CNS premised by glial dysfunction and affecting neuronal health (Dheer et al., 2018).

The preferred route a pathogen uses to access the CNS determines the most vulnerable regions to infection, causing meningitis, the infection of the meningeal subarachnoid space and cerebrospinal fluid (CSF), or encephalitis, the inflammation of CNS parenchyma (Fig. 13.1H). Neuroinvasion may manifest immediately as fever, headache, seizures, and muscle stiffness with neurologic sequelae ranging from mild, moderate, to severe neurocognitive defects, myelitis, movement disorders, and neuropsychiatric diseases (Fig. 13.1I, J) (Swanson and McGavern, 2015). The spectrum of symptoms can also set the ground for future neurologic sequelae of significant public health concerns such as Alzheimer disease (AD) and Parkinson disease (Lotz et al., 2021). Often, neurotropic microbes use multiple portals simultaneously, resulting in tropism across several brain regions (Dando et al., 2014; Swanson and McGavern, 2015). Pathogens can demonstrate cell- or region-specific tropism infecting and spreading only in spinal cord regions or particular neuronal or

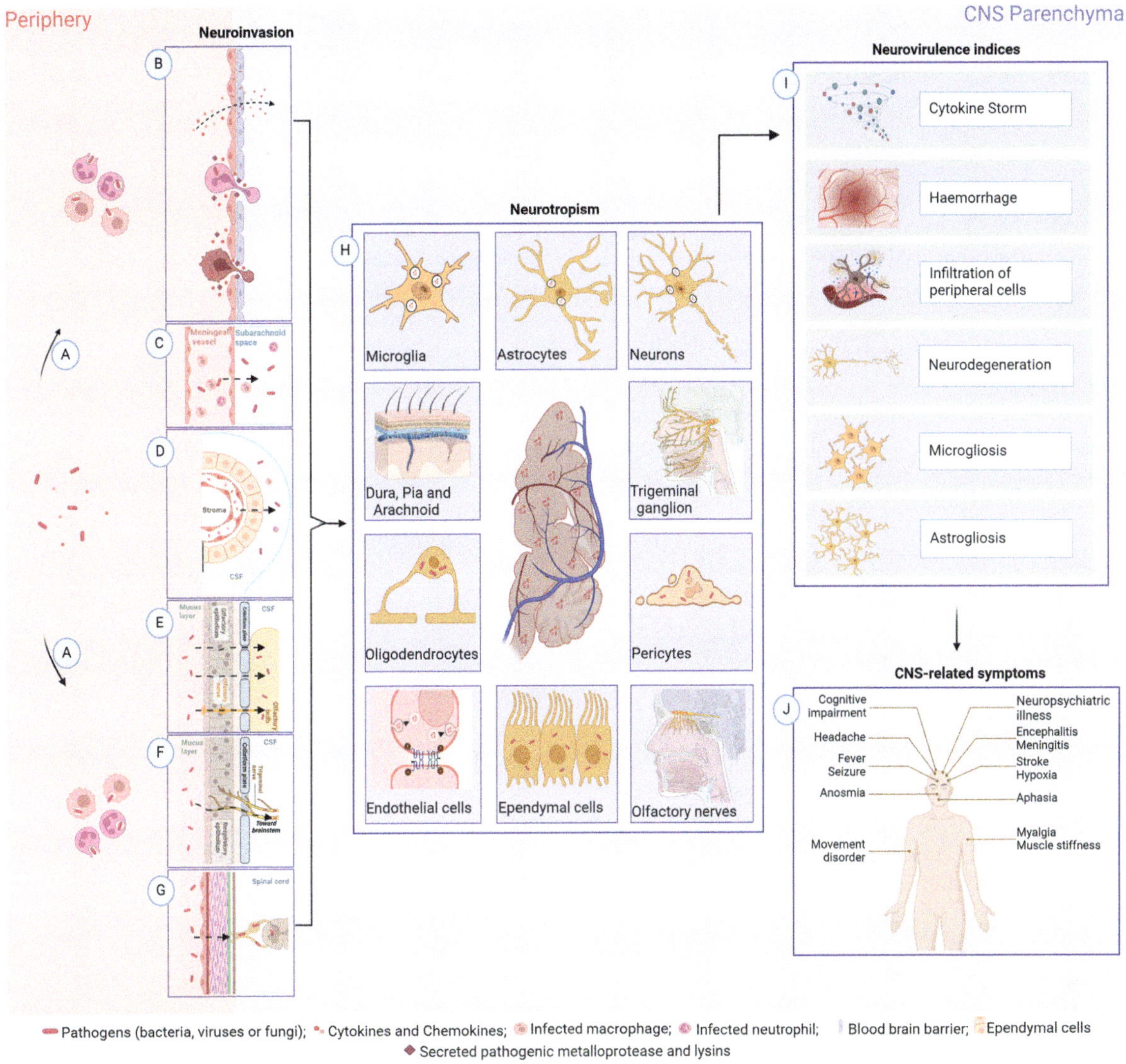

Fig. 13.1. Neuroinvasion, neurotropism, and neurovirulence of central nervous system (CNS)-penetrating infections. Neurovirulent pathogens have developed unprecedented abilities to penetrate the CNS from the (A) air or periphery, where they can infect blood cells, through different means. (B) First, peripheral infections can attain a high level of blood poisoning, eliciting cytokine storms that can penetrate the blood–brain barrier (BBB), rendering it leaky in the process. This paves the way for pathogens to access the CNS through leaky walls or via the infection of the cells constituting the barrier before emigrating to the CNS by transcytosis. In addition, the infected peripheral cells such as macrophages and neutrophiles can squeeze through the leaky BBB as Trojans, ferrying pathogens into the CNS parenchyma. A few pathogens secrete lysins (protein metalloproteinases), which digest BBB constituents to create a thorough path for pathogens into the CNS. Similarly, pathogens access the CNS by penetrating the meningeal vessel into (C) the arachnoid space or (D) using the fenestrated capillaries lining the blood-cerebrospinal fluid (CSF) barrier. Airborne pathogens may bypass CNS dissemination through the lung-blood route and follow the olfactory path wherein infection of the olfactory epithelium leads to the transfer of pathogens to the deeply penetrating adjoining nerves of the (E) olfactory or (F) trigeminal. (G) Likewise, some infections preferentially usurp afferent and efferent nerves of the neuromuscular junctions to reach the spinal cord. (H) In the CNS, while many pathogens leverage the neural (inter)connections to spread parenchyma-wide, a significant few prefer regional niches where they proliferate or hibernate while waiting for environmental stimuli such as psychologic stress for reactivation. In the same vein, neuroinfections may be selective for neuronal or non-neuronal cells, including components of the blood-periphery barriers or vasculature, causing significant damage to cellular and structural elements of the CNS (I), resulting in cytokine storm, hemorrhage, glial nodules, and neurodegeneration. (J) Since different CNS regions perform varying functions, the manifestation of CNS neuroinfections is multisymptomatic and often dynamic as the infection traverses through CNS regions and cells.

non-neuronal cells, possibly because of enrichment in receptors, nutrient, processing, and/or fusion proteins required for infection and survival (Fig. 13.1H) (Schweighardt and Atwood, 2001; Swanson and McGavern, 2015). They also consider the weak presence or absence of immune response and proclivity to mount successful evasive mechanisms when considering a target cell or brain region (Geyer et al., 2019; Nath and Johnson, 2022; Reddy et al., 2022). Direct infection of (non)neuronal cells during CNS infection, multiplication, or spread may result in a cytopathic effect, which culminates in cellular death or loss of function, manifesting as neurologic complications (Fig. 13.1I) (Bauer et al., 2022). These symptoms could be temporary, depending on the dynamic nature of pathogenic infections transiting through CNS sections, or persisting when the infection finally settles and disseminates within a specific cell type in a confined brain region. Under an immunocompetent condition, the infection may resolve or spread without any definite harm in the CNS and subsequently establish asymptomatic latent/quiescent infection in preferred CNS regions, waiting for a favorable stressful or immunocompromised state to reactivate (Koyuncu et al., 2013). The leakage of CNS barriers and consequent infiltration of peripheral cells can also cause focal hemorrhage, initiating further microglial and astrocytic reactivity.

Therefore, understanding the mechanisms by which glial cells react to CNS-invading microorganisms, whether they are infected themselves or not, is crucial to developing effective treatment and preventive measures against infections and associated neuropathologic changes (Awogbindin et al., 2021; Lotz et al., 2021; Govic et al., 2022).

NEUROGLIAL ROLES IN BACTERIAL INFECTIONS: A SPOTLIGHT ON MENINGOENCEPHALITIS-CAUSING *E. COLI* K1

Bacterial CNS infection

Neurotropic bacterial infections are notable global public health concerns for neonates, children, adults, and animal models (Govic et al., 2022; Hasbun, 2022). For instance, between 25% and 33% of the world population is infected with *M. tuberculosis,* with about 5%–10% manifesting in symptoms of pulmonary tuberculosis (TB) and 1% in CNS TB (WHO, 2019; Dian et al., 2021). Neurovirulent bacteria are largely prevalent worldwide, albeit with regional and continental differences (Barichello et al., 2023). For instance, *M. tuberculosis* is endemic in Africa/Asia (Sánchez-Garibay et al., 2022), whereas *Borrelia burgdorferi* causing neuroborreliosis abounds in North America/Eurasia (Govic et al., 2022). Some bacteria, including *Streptococcus agalactiae,* Group B *Streptococcus*, *E. coli*, and *Listeria monocytogenes*, pose a neurodevelopmental threat due to vertical and transplacental transmission, while airborne *Streptococcus pneumoniae* and *Neisseria meningitis* infect immunocompromised newborns and children. Others may elicit neurologic complications in adults (and immunocompromised children) through hematogenous routes following sexual transmission (*Treponema pallidum*), rodent bite (*Leptospira* spp.), inhalation (*M. tuberculosis*), or exposure to tick, an arthropod (*B. burgdorferi*) (Dando et al., 2014).

At the blood–CNS interface, *Streptococcus suis*, *L. monocytogenes*, *E. coli*, *M. tuberculosis*, *S. pneumoniae*, and *N. meningitis* access the CNS through host–pathogen receptor–ligand interaction or pinocytosis (Fig. 13.1) (Williams and Blakemore, 1990; Dando et al., 2014). This transcellular mechanism enables neuroinvasive bacteria to spread through transcytosis to adjoining cells. Bacteria can invade host cells via multiple receptors that are present on different cell types across CNS barriers, which equips them with a capacity to infect other cells constituting the BBB and BCSFB with complementing receptors. However, some bacteria have limited paracellular ability due to a few or lack of binding receptors. *L. monocytogenes*, for example, uses inflammation to weaken intercellular junctions through which pathogens and infected leukocytes traffic to the CNS. *E. coli*, *N. meningitis*, *M. tuberculosis*, *H. influenza*, and *Streptococcus* spp. can disrupt intercellular junctions between barrier's endothelial, epithelial cells and (sub)arachnoid membranes and spaces (Dando et al., 2014; Sánchez-Garibay et al., 2022). Neuroinvasive bacteria can additionally colonize the CNS through the diapedesis of infected peripheral monocytes (*L. monocytogenes*, *S. suis*, *M. tuberculosis*) and CD11b-positive myeloid cells (*B. pseudomallei*) (Williams and Blakemore, 1990; Liu et al., 2013; van Leeuwen et al., 2014; Sánchez-Garibay et al., 2022).

Overall, bacterial CNS invasion causes morbidity and mortality in neonates, children, and adults, and even despite the efficiency of many clinically available bactericidal drugs, leaving survivors with severe neurologic complications. Reasons for this are attributed to the evasive nature of bacteria developing resistance to drugs, escaping peripheral immune/complement response (Tseng et al., 2012; Yang et al., 2023), and achieving sufficient bacteremia required for CNS neurotropism, the ability to use multiple receptors and mechanisms for CNS entry (Huang and Jong, 2001), inaccessibility of CNS to antibiotics due to their large size of molecules, and/or insufficient insight into the pathophysiology and pathogenesis following CNS infection.

Microglia and astrocytes in meningoencephalitis-causing *E. coli* K1 infection

E. coli meningoencephalitis is most common in neonates and adults with an immunocompromised status resulting from age, disease conditions including cancers, acquired immunodeficiency diseases (AIDS), type I/II diabetes, and treatments such as chemotherapy (Kim, 2016). *E. coli* with polysaccharide capsular K1 antigen accounts for 80% of associated meningitis cases, causing 40% of mortality in neonates and life-long complications in half of survivors, without a vaccine currently available (Kim, 2016). Mediating its poor antigenicity, the K1 antigen usurps its molecular mimicry to polysialosylglycopeptides on fetal neurons and other host tissues, leading to high bacteremia in newborns and infants (Steenbergen et al., 2006). In addition, *E.coli* employs K1 to evade phagocytosis by the complement system's alternative pathway (Fig. 13.2A) (Moxon and Kroll, 1990). Peripheral and CNS *E. coli* infection was recapitulated in experimental post-adolescent young, aging, and immunocompromised rodents, albeit data reflecting the pathophysiology of a neonatal infection are scarce (Barrientos et al., 2009; Ribes et al., 2020). In the CNS, single peripheral *E. coli* infection resulted in late but sustained hippocampal inflammation and impaired long-term memory lasting up to 8 days in aged rats—despite receiving lower doses—that did not occur in young adult rats (Fig. 13.2B, C) (Barrientos et al., 2006; Tanaka et al., 2018). *E. coli* infection exacerbated hepatic and splenic cytokine storm in aged, compared to young adult rats, with no effect on the bacterial load (Barrientos et al., 2015). Although *E. coli* was not detected in the hippocampus of young adult and aged rats, inflammatory response persisted

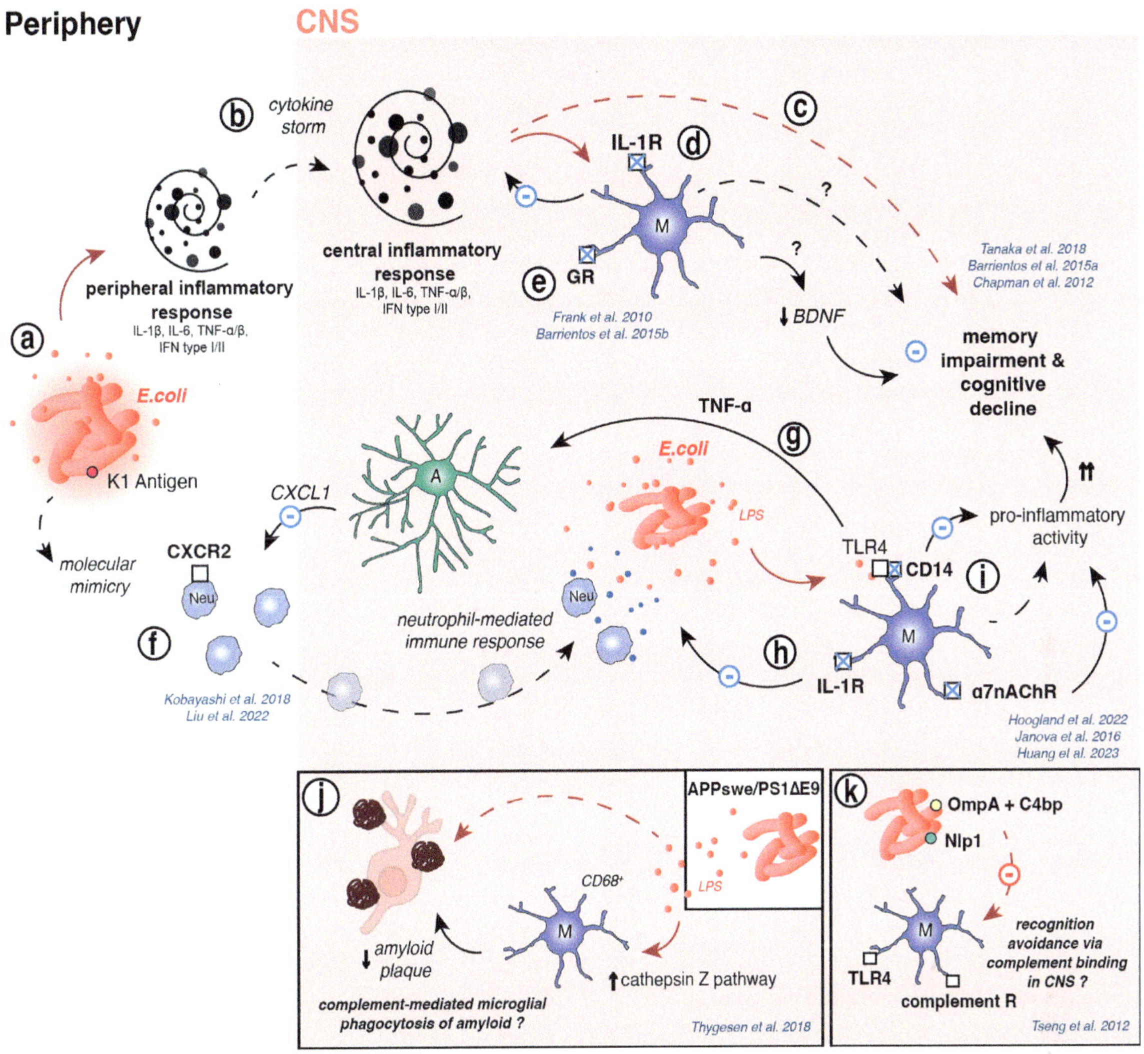

Fig. 13.2. Putative mechanisms of interaction between bacterial infection and glia. (A) *Escherichia coli* (*E. coli*), as an example of a bacterial pathogen, employs molecular mechanisms that allow it to be protected from a complement-mediated immune reaction (e.g., K1 antigen). (B) Regardless of its poor antigenicity, it was able to induce a peripheral inflammatory response that can initiate a central inflammatory response, which both affects microglial response and is further regulated by these cells. (C) By an upregulated proinflammatory signaling and tentatively decreased neurotrophic support, microglia may be contributing toward memory impairment and cognitive decline, as observed in rodent models. (D) Modulation of microglial IL-1 or (E) GR may attenuate their inflammatory response and prevent pathological outcomes. (F, G) Direct brain tissue bacterial invasion was associated with neutrophile infiltration to aid with a local immune response. Microglial TNF-α stimulates astrocytes to release CXCL1 that is recognized by peripheral neutrophiles, attracting them to the affected site. Rather than a downstream effect of microglial engulfment of *E. coli* particle, microglia recognize *E. coli* membrane particles (e.g., lipopolysaccharide) via pattern-recognition receptor (e.g., TLR4), which induces TNF-α production and inflammatory activity. (H) Agonists of CD14, TLR4 co-receptor, or α7nAChR receptors were found to modulate microglial proinflammatory activity while (I) IL-1R was recognized as a potential regulator of the microglia-astrocyte TNF-α signaling and neutrophile infiltration; if exaggerated, both processes can have pathological outcomes. (J) In the context of experimental amyloid beta pathology ($APP_{swe}/PS1_{\triangle E9}$ mouse model), LPS treatment reduced amyloid plaque load, potentially via stimulation of complement-mediated microglial phagocytosis. (K) Peripherally, *E. coli* proteins Nlp1 and OmpA aid in avoidance of bacterial recognition by the immune system via complement binding; however, whether similar mechanism applies in CNS and how it affects complement-stimulated microglial phagocytosis remain to be answered. *α7nAchR*, alpha7-nicotinic receptor; *A*, astrocytes; *BDNF*, brain-derived neurotrophic factor, *C4bp*, complement binding protein 4, *CD14*, cluster of differentiation 14; *CNS*, central nervous system; *CXCL1*, chemokine (C-X-C motif) ligand 1; *CXCR2*, chemokine (C-X-C motif) ligand receptor 2; *GR*, glucocorticoid receptor; *IFN*, interferon; *IL*, interleukin, *LPS*, lipopolysaccharide; *M*, microglia; *Neu*, neutrophile, *Nlp1*, new lipoprotein I; *OmpA*, outer membrane protein A; *TLR*, toll-like receptor, *TNF*, tumor necrosis factor.

in aged animals, resulting in memory impairment (Barrientos et al., 2015). The CSF injection of interleukine-1 receptor (IL-1R) antibody concurrently with a peripheral *E. coli* inoculation blocked cytokine (IL-1b and IL-6) surge after 24 h, and subsequent hippocampal-dependent cognitive decline, as well as its molecular drivers, at day 4 (Fig. 13.2D) (Frank et al., 2010). This suggests that a CNS-exclusive approach may be sufficient to prevent *E. coli*-mediated CNS sequelae. The infection also significantly impaired the hippocampal long-term potentiation, a brain-derived neurotrophic factor (BDNF)-dependent process of synapse strengthening, in aged rats (Fig. 13.2A) (Tanaka et al., 2018). It also reduced the level of BDNF in hippocampal synaptoneurosomes of *E. coli*-infected aged rats and intra-cisterna magna IL-1R antibody treatment (Cortese et al., 2011; Chapman et al., 2012). Since microglial BDNF is vital for memory-associated synaptic plasticity (Ferrini and De Koninck, 2013; Parkhurst et al., 2013), *E. coli* might be disrupting microglial response, including the secretion of BDNF at synaptic sites (Fig. 13.2C). Of note, CSF administration of mifepristone, an antagonist of a glucocorticoid receptor expressed on microglia, reversed the *E. coli*-associated cognitive defect, hippocampal inflammation, and elevated mRNA expression of ionized calcium-binding adapter molecule 1 (IBA1), a microglia/macrophage marker (Fig. 13.2E) (Barrientos et al., 2015). However, depletion of peripheral macrophages with clodronate-encapsulated liposomes—administered 1 day after *E. coli* exposure—blunted the proinflammatory response in the liver and spleen, but not the hippocampus, and did not prevent associated memory decline in aged rats. This suggests that peripheral cells other than macrophages underlie the CNS impact of *E. coli* (Barrientos et al., 2015). Taken together, *E. coli* may exploit the host's immune status to establish CNS infection through a defective immune response via an unknown bidirectional interaction between the periphery and microglia.

Similar to bacterial infection of peripheral tissues, invasion of CNS with *E. coli* resulted in the infiltration of neutrophiles, a frontline defense against bacterial infection (Fig. 13.2F) (Ribes et al., 2013). Recognition of surface bacterial antigens (e.g., lipopolysaccharide [LPS]) by the humoral complement system culminates into the recruitment of neutrophiles and phagocytes through opsonic tagging, chemotaxis, and inflammation (Kobayashi et al., 2018). Neutrophiles possess robust broad-spectrum innate mechanisms to eliminate invading bacteria, including oxidative burst, neutrophil extracellular traps, phagocytosis, and intracellular killing (Kobayashi et al., 2018). A study in mice showed that an intercellular communication between microglia and astrocytes through tumor necrosis factor alpha (TNF-α) resulted in C-X-C chemokine ligand 1 (CXCL1)-dependent neutrophil recruitment to the CNS during *E. coli* infection (Fig. 13.2G) (Liu et al., 2022). During an intravenous *E. coli* K1 infection in mice, the authors observed that microglial TNF-α stimulated astrocytic secretion of CXCL1, which invited peripheral neutrophiles via CXCL1 receptor (CXCR2). Ablation of microglia using colony-stimulating factor-1 receptor (CSF-1R) antagonist PLX3397 or *E. coli* K1 infection in mice lacking TNF-α resulted in an impaired neutrophil infiltration and exacerbated CNS bacterial load (Liu et al., 2022). Primary microglial cells isolated from *E. coli* K1-infected mice also responded to *E. coli* K1 LPS with a TNF-α secretion more robust than astrocytes', indicating that microglia may be the first responders during its CNS infection (Liu et al., 2022). Nevertheless, *E. coli* was not detected in microglia, suggesting the bacteria may evade direct microglial phagocytosis through a yet-to-be deciphered mechanism. In the absence of a direct microglia–*E. coli* K1 interaction, microglia can respond to secreted pathogen-associated molecular patterns (PAMPs) from the bacteria, including LPS-enriched outer membrane vesicles, using toll-like receptor (TLR) 4 (Fig. 13.2G) (Liu et al., 2022). Together, these studies demonstrated that early and timely microglial proinflammatory response is required for bacterial clearance through regulated microglial-astrocytic neutrophil recruitment. In pulmonary viral infection, unabated neutrophil infiltration contributes to immunopathology (Ayegbusi et al., 2019; Liu et al., 2022). Whether microglia-astrocytic neutrophil recruitment to the CNS parenchyma is self-resolving or contributes to CNS immunopathology and its underlying mechanisms require more investigation. For instance, IL-1RA antagonist blocked the microglial IL-1RA-astrocytic CXCL1 exaggerated neutrophil recruitment, which contributed to an acute phase-associated microvessel blockade and blood flow stagnation following an ischemic brain injury in mice (Fig. 13.2H) (Huang et al., 2023). Early evidence shows that CD14 (TLR4 co-receptor) and alpha-7 nicotinic acetylcholine receptor (α7nAChR) moderate microglial proinflammatory response to limit damaging effects during CNS *E. coli* infection (Fig. 13.2I) (Janova et al., 2016; Hoogland et al., 2022).

Meanwhile, during a transcellular infection of endothelial cells via receptor-mediated endocytosis, vacuoles containing *E. coli* K1 avoid fusion with lysosomes containing the degradation enzyme cathepsin D to achieve CNS entry (Kim et al., 2003). In aged $APP_{SWE}/PS1_{\Delta E9}$ transgenic mice with AD amyloid pathology, weekly intraperitoneal *E. coli* LPS injection activated the complement system. This effect was accompanied by a reduced amyloid pathology, probably indicating enhanced clearance via phagocytosis (Thygesen et al., 2018). The authors also observed elevated levels of lysosomal protease cathepsin Z pathways, which co-expressed with CD11b (complement 3 receptor) and CD68 (phagocytic receptor) on microglia (Fig. 13.2J) (Thygesen et al., 2018). This indicates that microglia respond to LPS via the complement system, enhancing phagocytosis. In the periphery, *E. coli* inhibits the complement system to avoid recognition, fostering septicemia and CNS infection. *E. coli* new lipoprotein I (Nlpl) recruits complement inhibitor C4b-binding protein (C4bp) to the bacterial outer membrane protein A (OmpA), shielding it from recognition by the peripheral classical complement system (Fig. 13.2K) (Tseng et al., 2012). Since the complement system is a mechanism by which microglia initiate phagocytosis, it is yet uncertain whether microglia competently use lysosomal degradative functions leading to killing of phagocytosed *E. coli* K1 or whether bacteria developed a sophisticated strategy to evade microglial recognition and engulfment in the CNS.

In neutropenic mice with low neutrophil levels, Ribes and colleagues showed that activating a similar nonspecific antiviral innate response in mice, mediated by NK cells and microglia, can prevent bacterial CNS infection from degenerating (Ribes et al., 2020). In the study, they used young adult mice that were depleted of $CD11b^{+}Ly\text{-}6G^{+}Ly\text{-}6C^{int}$ neutrophiles with anti-Ly6G monoclonal antibody. Intraperitoneal injection of the TLR3-activating viral mimic polyinosinic:polycytidylic acid (poly(I:C)) 72 h before direct injection of high dose *E. coli* K1 inoculum (which killed immunocompetent young mice) into the frontal neocortex resulted in significant survival (above 75% relative) 14 days post-infection (dpi). Authors observed reduced bacterial burden in the brain (and distant peripheral spleen) shortly (30 h) after bacterial inoculation with increased IBA1$^+$ cellular density, soma area, and thicker processes suggestive of reactive hypertrophic microglia at the inoculation site (Ribes et al., 2020). These changes were accompanied by elevated interferon gamma levels (IFN-γ) and increased number of NK-cells in the brain, believed to be recruited from the spleen (Ribes et al., 2020). The microglial morphology resembled an amoeboid state with enhanced cytokine and nitric oxide secretion, phagocytosis, and intracellular pathogenic killing in previous studies in which primary microglial cells were primed with poly (I:C) via TLR3 before exposure to *E. coli* K1 (Ribes et al., 2009, 2010). Similarly, primary microglial cells from immunocompromised neonate mice born to diet-induced vitamin D-deficient parents were unable to secrete the required TNF-α and IL-6 needed to achieve effective *E. coli* K1 phagocytosis and intracellular killing, despite TLR-3, -4 and -9 priming with poly (I:C), LPS or CpG oligodeoxynucleotide, respectively (Djukic et al., 2014). These observations further buttress the relevance of microglia-orchestrated early cellular innate response against CNS bacterial infection. If these and other mechanisms of pathogenesis and the response of non-neuronal cells are known, they may usher novel therapeutic approach to reduce morbidity and neurologic complications in the survivors of bacterial infections. Thus, microglia-targeted therapy for *E. coli* could be the game changer.

MICROGLIAL ROLES IN VIRAL INFECTIONS

An acute neurologic response consisting of an evolutionary conserved suite of behaviors—fever, pain, fatigue, social withdrawal, and impaired sleep—collectively termed as "sickness," follows viral exposure (Dantzer et al., 2008). Extensive research of past decades has further shown that severe, repetitive, or chronic viral infections may alter cognition and lead or contribute to the development of long-term, sometimes irreversible, neurologic outcomes, including a range of neuropsychiatric symptoms and disorders (Benros et al., 2011, 2013; Nersesjan et al., 2022; Granholm, 2023), accelerated aging and neurodegenerative diseases (Garden, 2002; Field et al., 2010; Awogbindin et al., 2021; Filgueira et al., 2021). This knowledge has existed since the late 19th to early 20th century with the first reports of encephalitis described after outbreaks of influenza or Spanish flu pandemics (Stefano, 2021). However, a major interest in this field was re-initiated in the past two decades with the identification of neurologic consequences of Zika virus (ZIKV) or coronaviruses exposure, pressing the need to understand their medium and long-term impact. Beyond that, individuals with HSV or rabies have been diagnosed with encephalitis and Guillain-Barré syndrome (GBS), a neuromuscular disorder. At the same time, myopathy was found to be present after infection with flavi- or coronaviruses (Conrady et al., 2013; Filippone et al., 2020; Granholm, 2023). While only a brief display of a few examples, it is importantly showing a similarity in neurologic outcomes across a range of viral infections of different origins, a tentative indication of a shared mechanism of action that is yet-to-be more deeply described. However, further specificity of action or symptomatic manifestation may be dependent on individual traits such as age, sex, medical history (i.e., previously encountered infections), in addition to genetic- or lifestyle-acquired predispositions, increasing the difficulty in uncovering shared traits.

Viral infiltration of the CNS is associated with synaptic loss, neuronal damage, and apoptosis, which can occur as an indirect consequence of a severe or dysregulated immune response (e.g., for non-neurotropic viruses) or a direct infection (neurotropic virus) (Jurgens et al., 2012). Particularly, post-viral encephalitis was, similarly to neurodegenerative diseases, associated with hippocampal synaptic or neuronal loss (Hong et al., 2016; Vasek et al., 2016; Garber et al., 2019). Microglial ability to exert swift and directed responses is represented not only by a dynamic cytoskeletal remodeling of their processes (Nimmerjahn et al., 2005; Franco-Bocanegra et al., 2019) but also by a vast expression of various sensor receptors. For both the homeostatic maintenance and response to disturbances, microglia present a range of PAMPs and damage-associated molecular patterns (DAMPs) receptors (Schafer et al., 2012; Colonna and Butovsky, 2017).

Non-neurotropic viral infection and the CNS

Being of a neurotropic character, meaning showing an ability to directly infect and replicate within the CNS, does not appear entirely essential for the onset of neurologic sequalae. It is to note that, while non-neurotropic viruses, such as H3N2 IAV, may have different preferred target tissues (e.g., lungs or epithelia), they can still impact the CNS (Hosseini et al., 2018; Hosseini and Korte, 2023). Instead, the most conserved mechanism is seemingly via induction of peripheral inflammatory response across many viral infectious conditions (Dantzer et al., 2008; Jurgens et al., 2012). This was observed, for instance, in a case of murine HSV-1 after exposure via various routes, such as intracerebral, intraperitoneal, or peripheral white adipose tissue injection, further highlighting the relevance of non-neurotropic viral challenges or even other severe peripheral inflammatory stimuli *per se* (Fekete et al., 2018). Similarly, peripheral H3N2 IAV infection exposure in $APP_{SWE}/PS1_{\Delta E9}$ mice worsened the course of the disease, potentially via reduced microglial clearance of amyloid beta and dysregulated microglial inflammatory signaling (Hosseini et al., 2021).

If severe enough, viral infections may impair the BBB endothelium through alteration of tight junction protein expression, allowing for both production of damage signals (cytokines, reactive oxygen species (ROS) or reactive nitrogen species (RNS)), as well as entry of peripheral inflammatory mediators, infected cells, antigen-presenting or reactive myeloid cells into the brain parenchyma, inducing a central immune response (Barbosa-Silva et al., 2018; Alquisiras-Burgos et al., 2021; Düsedau et al., 2023; Raber et al., 2023). Following this peripheral response, hippocampal inflammation (mediated by IL-1β, IL-6, TNF-α, IFN-α), as well as number and area of IBA1$^+$ microglia were increased within 7

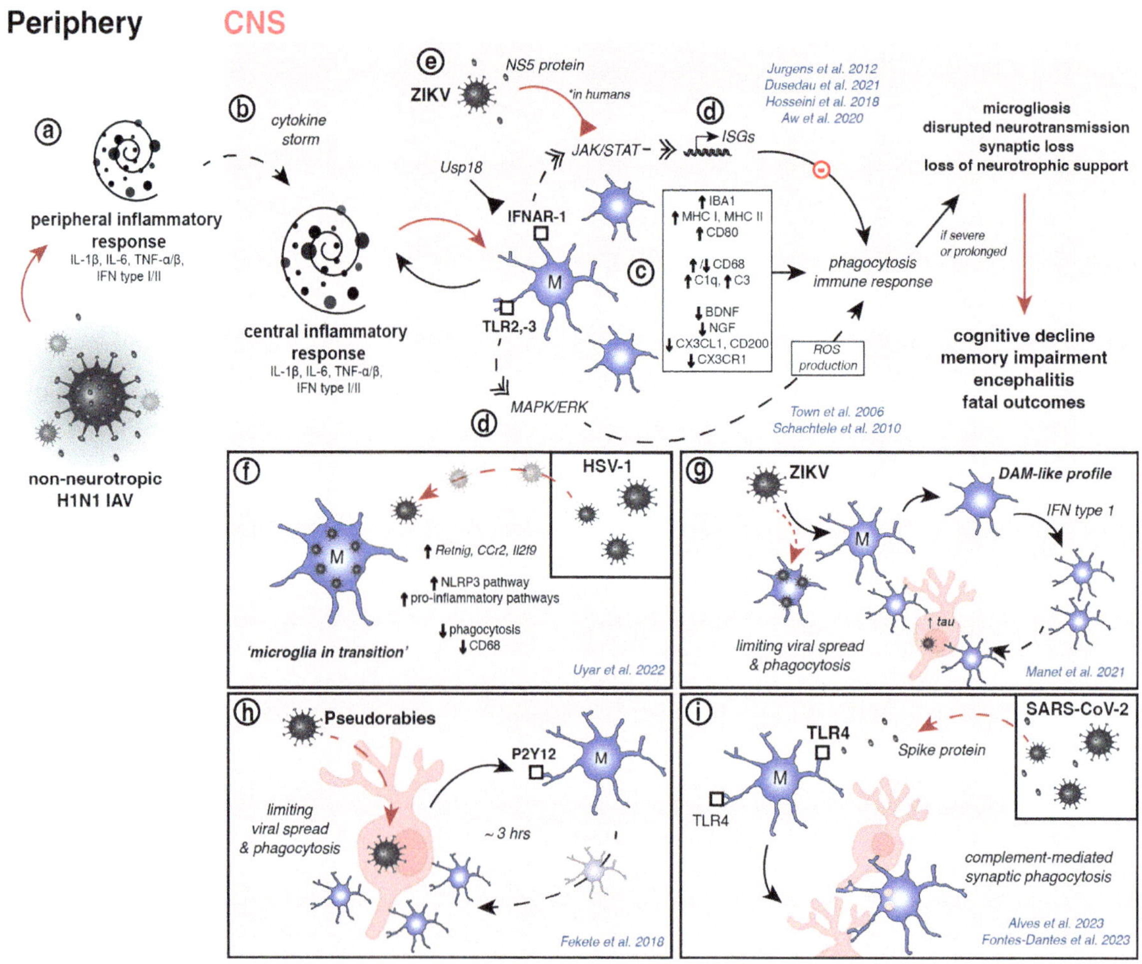

Fig. 13.3. Microglial features across various viral infection states. (A–C) In response to virus-induced peripheral inflammation, such as in case of a non-neurotropic H1N1 IAV, central inflammatory response is accompanied by an increase in number and area of IBA1-immunopositive (+) microglia and phase-dependent changes in microglial regulatory (CX3CL1, CD200), response-associated (MHC I and II, CD80, F4/80, CD68, CX3CR1), and neurotrophic (BDNF, NGF) factors, as well as changes in complement mediators (C1q, C3). These changes that underlie the early microglial inflammatory and phagocytic response, however, may be pathologic if prolonged or exaggerated. (D) Microglial IFNAR receptors mediate the homeostatic and immune response associated IFN signaling. Usp18 acts as one of the microglial regulators of this pathway in the homeostatic white matter. Several TLRs, including TLR-2 and -3, were implicated in microglial viral response, i.e., by activating the MAPK/ERK pathway and ROS production. (E) In humans, ZIKV proteins were reported to intervene with the IFNAR downstream signaling, the JAK/STAT pathway, preventing the expression of ISGs and inflammatory response. (F) A unique microglial named "*microglia in transition*" was reported in a model of rodent intranasal HSV-1 infection, showing upregulated NLRP3 inflammasome and proinflammatory signaling, albeit decreased phagocytic properties. (G) While holding a potential to become infected and serving as a viral reservoir, uninfected microglia were reported to adopt a disease-associated microglia (DAM)-like profile. Via IFN type 1, these cells signal to other microglia, stimulating the infected cell ensheathment and phagocytosis, limiting the viral spread. (H) Microglial purinergic receptors, i.e., P2Y12, participate in the early response of microglia to a virus infection, initiating chemotaxis, phagocytosis, and proinflammatory mediator release even prior to significant viral replication as shown in a model of pseudorabies. (I) As another pattern recognition receptor, TLR4 was found to recognize viral particles (SARS-CoV-2 spike protein) without associated viral genetic information, stimulating downstream pathways leading to microglial phagocytosis of synaptic connections via complement system. *BDNF*, brain derived neurotrophic factor; *C1q*, complement component 1q; *C3*, complement component 3, *Ccr2*, chemoattractant protein-1 receptor; *CD*, cluster of differentiation; *CNS*, central nervous system; *CX3CL1*, fractalkine; *CX3CR1*, fractalkine receptor; *DAM*, disease-associated microglia; *ERK*, *extracellular signal-regulated kinase 1/2*; *H1N1 IAV*, H1N1 strain of Influenza A; *HSV-1*, herpes simplex virus 1; *IBA1*, ionized calcium-binding adapter molecule 1; *IFN*, interferon; *IFNAR*, interferon receptor; *IL*, interleukin; *ISGs*, interferon stimulated genes; *JAK/STAT*, Janus kinase-signal transducer and activator of transcription pathway; *M*, microglia; *MAPK*, mitogen-activated protein kinase; *MHC II*, major histocompatibility complex II; *NGF*, neural growth factor; *NLRP3*, NLR family pyrin domain containing 3 protein complex; *P2Y12*, purinergic chemoreceptor; *SARS-CoV-2*, severe acute respiratory syndrome coronavirus 2; *TLR*, toll-like receptor; *TNF*, tumor necrosis factor; *Usp18*, ubiquitin-specific protease 18; *ZIKV*, Zika virus.

dpi in adult male mice with IAV H1N1 (Fig. 13.3A–C) (Jurgens et al., 2012). In parallel, factors with a modulatory impact on microglia (CD200, CX3CL1) or alluding to their neurotrophic activity (BDNF, NGF) were decreased (Fig. 13.3C) (Jurgens et al., 2012). Similar results, including an upregulated expression of major histocompatibility complexes I and II (MHC I and II), CD80 and F4/80, and downregulated microglial fractalkine receptor (CX3CR1) were observed on $CD45^{low}/Cd11b^{+}$

microglial cells isolated from IAV H1N1-exposed adult female mice at 7 and partially at 14 dpi (Fig. 13.3C) (Düsedau et al., 2023). Between 10 and 14 dpi, IAV-infected mice showed increased levels of phagocytosis-associated receptors (i.e., CD36 or CD68) and complement factors (C1q, C3) in the cerebral cortex and hippocampus, suggestive of increased synaptic pruning (Fig. 13.3C). A decrease in presynaptic vesicular glutamate transporter 1 (VGLUT1) expression was also observed, using a flow cytometry method to analyze synapses, indicative of a temporary impact on glutamatergic transmission potentially resulting from impaired neurotrophic support (Düsedau et al., 2023). A density and morphologic analysis did not reveal changes in IBA1^{+}/transmembrane protein 119 (TMEM119)$^{+}$ cells; however, there was an increased peripheral immune cell invasion that authors related to a decreased expression of tight junction protein genes (*Cldn5*, *Tjp1*), suggesting a temporary BBB impairment associated with elevated peripheral IFN-γ levels (Düsedau et al., 2023). Without providing direct evidence of a long-term neurologic impact of viral infection, this observation is important considering that a potential for re-activation has been described for viruses (Traylen et al., 2011), an effect that may act in a contributing manner. Expanding on these observations, Hosseini et al. (2018) compared the impact of the H3N2 (non-neurotropic) and H7N7 (neurotropic) strains in female adult mice between 30 and 120 dpi. In a similar manner, CNS inflammation, microgliosis (increase in microglial numbers), and dendritic loss in hippocampus were observed for both viral strains at 30 dpi contrary to 120 dpi, tentatively hinting at a lack of a more prolonged outcome (Hosseini et al., 2018). Occurrence of these effects beyond the initial sickness phase, however, was proposed to contribute to microglial priming, a phenomenon describing a developed sensitivity or decreased tolerance of microglia toward a secondary stimulus (Barbosa-Silva et al., 2018; Düsedau et al., 2023), hence serving as a risk factor for susceptibility to subsequent factors.

Toll-like receptors and IFN type 1-mediated response

The recognition of viral particles by macrophages and microglia is most notably mediated via transmembrane protein receptors as TLRs—so far most well characterized in this context—retinoic acid-inducible gene I (RIG)- or nucleotide oligomerization domain (NOD)-like receptors (Kawasaki and Kawai, 2014). Upon recruitment of adaptor proteins (e.g., Myd88, TRIF), TLRs act by activating mitogen-activated protein kinases (MAPKs), nuclear factor kappa B (NF-κB), and interferon regulatory factors (IRFs), leading to the expression of cytokines and IFN type I (Fig. 13.3D) (IFN-α, IFN-β) (Kawasaki and Kawai, 2014). Particularly, the release of IFNs and enhanced expression of interferon-stimulated genes (ISGs) have been widely described to hold a range of functions in infection pathologies (Fig. 13.3D). Early experiments with primary microglial cells obtained from wild-type mice showed that HSV-1 exposure initiates a pronounced cytokine release by microglia, followed by apoptosis. Knockout (KO) in the TLR2 was shown to downregulate the scale of microglial apoptosis when measured at 24-h timepoint (Aravalli et al., 2005, 2007).

After an acute peripheral recombinant IFN-α exposure, RNA sequencing of isolated mouse microglial cells showed an upregulation of ISGs (e.g., *Mx1, Rsad2*) and complement factors (e.g., C4b), which were also associated with an increased microglia-mediated presynaptic engulfment in vivo in the cerebellum (but not hippocampus or cerebral cortex) (Aw et al., 2020). IFN-α-responding microglia shifted into a state with increased soma size and a tendency toward an elevated ramification, but at the same time, expressed lower levels of CD68 and CD45 as indicated by whole-region flow cytometry, suggesting a unique microglial state (Aw et al., 2020). Given microglial ability to express IFN type I receptors (IFNAR 1 or 2) and release IFNs, apart from paracrine signaling to other microglia, this allows for an autocrine self-regulatory role (Aw et al., 2020). IFNAR stimulation also leads to transcription of ISGs via the Janus kinase (JAK)-signal transducer and activation of transcription (STAT) pathway (Fig. 13.3D) (see Hayashida et al., 2019). Fine balance of IFN type 1 levels, which are constitutively expressed for maintenance of homeostasis, also ensures cellular response to IFN type 2 (IFN-γ) signals (see Goldmann et al., 2016). Usp18, a ubiquitin-specific protease 18, was identified as a regulator of microglial homeostasis via negative regulation of IFNAR signaling in white matter mouse microglia during steady state (Fig. 13.3E) (Goldmann et al., 2015). Usp18 deficiency was associated with enhanced IBA1^{+} microglial density and, in case of cerebral subcortical white matter microglia, with expression of MHC II and phagocytic markers (e.g., Mac3) (Goldmann et al., 2015). Primary microglial cells from Usp18$^{flox/flox}$ mice displayed a prolonged IFN type 1 gene expression after stimulation with IFN-β underscored by an enhanced phosphorylation of STAT1 (Goldmann et al., 2015). This evidence of constitutive IFN type 1 signaling in the homeostatic brain reveals the detrimental outcomes that a dysregulated IFN type 1 signaling could incite in the brain. In case of ZIKV, its NS5 protein binds to STAT2, degrading it, and hence suppressing the IFN type 1 response in a human cell line (Fig. 13.3E) (Grant et al., 2016; Hayashida et al., 2019). This mechanism of IFN antagonism was not replicated in experiments with murine fibroblasts, suggesting a species-associated specificity and tentatively in line with reports of a lack of lethal outcomes in wild-type mice (Kumar et al., 2016; Rossi et al., 2016). Further, dysregulation of IFN 1 signaling was reported with aging and found to contribute to the impairment of hippocampal neurogenesis in aged mice, an effect that was partially rescued via administration of IFNAR-1 neutralizing antibodies (Baruch et al., 2014).

Rodent exposure to poly(I:C), a synthetic analog of double-stranded RNA viruses, was further associated with increased expression of IFN-α, IFN-β, and proinflammatory genes in the CNS (Costello and Lynch, 2013). Mediated through TLR3 activation, it caused the upregulation of microglial (CD11b, F4/80) and astrocytic (glial fibrillary acidic protein [GFAP]) cell surface markers in the cerebral cortex of wild-type mice, as well as the adoption of an amoeboid-like shape in mouse primary microglial cells (Town et al., 2006). This effect was attenuated with TLR3 knockout, potentially due to a slower onset of MAPK pathway activation, preventing microglial response upon its intracerebroventricular administration (Town et al., 2006). These TLR3-deficient mice showed more resistance to an intraperitoneal administration of West Nile virus than wild-type mice by a decreased brain viral load, microgliosis, and peripheral macrophage infiltration, suggesting an important role for TLR3 in viral entry to the CNS and permeability of BBB (Wang et al., 2004). Electrophysiologic recordings did not find differences in long-term potentiation after intraperitoneal injection of poly(I:C); however, changes in spontaneous neuronal activity and glutamate receptors, suggesting an impact on hippocampal excitability, were observed (Costello and Lynch, 2013). An immunostimulatory character of poly(I:C) pretreatment was previously proposed to serve as a priming factor for macrophages via TLR3-mediated IFN signaling. As shown in models of SARS-CoV-2 or IAV strain PR8, when poly(I:C) was intranasally applied to aged mice, it decreased viral replication and enhanced mouse survival (Zhao et al., 2012). This is seemingly very phase-dependent, as poly(I:C) administration during a state of disease (ME7 prion mouse model) had a more synergistic effect—increasing sickness response, neurologic impairment, and disease progression at the same time as expression of ISGs with antiviral properties and activation of IFN-regulated proapoptotic pathways (Field et al., 2010).

The downstream activated pathways for TLR signaling include activation of MAPK-extracellular signal regulated kinase (ERK) and ROS production (Fig. 13.3D). This oxidative response to HSV-1 was attenuated in primary microglial cells from TLR2 KO mice and was associated with a decreased cytopathogenic character of HSV-infected TLR2$^{-/-}$ microglial cells in a mixed neuronal culture (Schachtele et al., 2010). While ROS are important for microglia-mediated phagocytosis, an exacerbated ROS response detrimentally causes brain damage (Marques et al., 2008; Schachtele et al., 2010). Uyar et al. (2020, 2022) defined microglia with a transcriptomic signature enhanced in several genes (e.g., *Retnig, Cccr2, Il2f9*) as in "*transition*" in ventral thalamic regions of a HSV-1 infection mouse model (Fig. 13.3F) (Uyar et al., 2020, 2022). These microglia displayed elevated enrichment in nucleotide binding, leucine-rich-containing family pyrin domain containing 3 (NLRP3)-inflammasome and proinflammatory pathway, as well as presented a reduced phagocytic ability, possibly due to an increased intracellular number of viral particles (Uyar et al., 2022). Most of these TMEM119^{+} cells had an amoeboid shape and were hypertrophic, a feature that was reduced at 6 dpi. In coincidence, there was a lack of signal for CD68, a phagolysosomal marker, on these microglia in contact with infected material, while the infected thalamus showed increases in this marker, possibly indicating a compensation mediated by other neuroglial cells (including TMEM119-downregulating microglial states). As these TMEM19^{+}/CD68^{-} microglia contained HSV-1 particles, a hypothesis of intracellular antiviral response that may be affecting their functional capability was proposed (Uyar et al., 2022).

Inducing an encephalitis with lethal outcomes, ZIKV was associated with reduced levels of human IFN type 1 response (supressing effect) (Grant et al., 2016). Displaying a species-specific mechanism, peripheral ZIKV is seemingly unable to evade IFN signaling pathway and hence manifests as a nonlethal form with a systemic viremia in adult mice (Grant et al., 2016). Genetically induced deficiency in TLR3 and RIG-1 adaptor proteins (i.e., TRIF, IPS-1) in young mice elevated CNS viremia—without a clinical manifestation—early after systemic exposure to ZIKV strain PRVABC59 (Piret et al., 2018; Enlow et al., 2021). Density of IBA1^{+}/TMEM119^{+} microglia in infected mouse hippocampi did not change; however, these cells displayed an increased cell soma and arbour area across dorsal hippocampus CA1 layers (Enlow et al., 2021). Furthermore, ultrastructural analysis of these cells was suggestive of elevated phagocytosis and extracellular digestion. With PLX5662-mediated microglial depletion, both enhanced viral replication in neurons and astrocytes—based on the ZIKV immunogold labeling—and a potential for compensatory astrocyte-mediated phagocytosis of viral particles were additionally reported (Enlow et al., 2021). In addition, intracranial ZIKV administration induced 100% mortality even in immunocompetent mice with a more rapid development in case of a deficient IFNAR1 signaling (Hayashida et al., 2019). Experiments with human brain cell preparations introduced microglia as one of the primary ZIKV targets in vitro, altering expression of metabolites involved in neuronal differentiation, apoptosis, viral survival, or production of inflammatory mediators (IL-6, TNF-α, IL-1β, monocyte chemoattractant protein [MCP-1]) (Lum et al., 2017). Experiments showed that uninfected microglia attained a disease-associated microglia (DAM)-like profile after an intracerebroventricular administration of ZIKV (Fig. 13.3G) (Manet et al., 2022). Via IFN type 1 release, these responding microglia further stimulated other nonexposed/noninfected microglia. As another parallel to AD pathology and tauopathies, mature neurons were found to enhance phosphorylation, followed by accumulation of tau after exposure to ZIKV in this model; and this incited their encirclement and occasional engulfment by microglia (Manet et al., 2022).

Purinergic signaling as a driver of microglial chemotaxis in viral infection immune response

Paracrine signaling of damaged and/or infected neurons by a release of nucleotides (e.g., adenosine diphosphate [ADP], and adenosine triphosphate [ATP]) serves as a chemotactic signal for microglial migration toward the site of damage/infection, proinflammatory mediator release, and phagocytosis (Chen et al., 2019). A P2Y12 purinergic receptor was reported to be crucial for expansion of IBA1^{+} microglial population and stimulation of infected cell clearance in adult mice after pseudorabies virus exposure (Fig. 13.3H) (Fekete et al., 2018). This activity was stimulated within few hours, generally before neurons displayed signs of advanced viral replication or chromatin alterations (Fekete et al., 2018). Moreover, processes of on average 1–3 migrated microglia adopted a physical barrier-like structure,

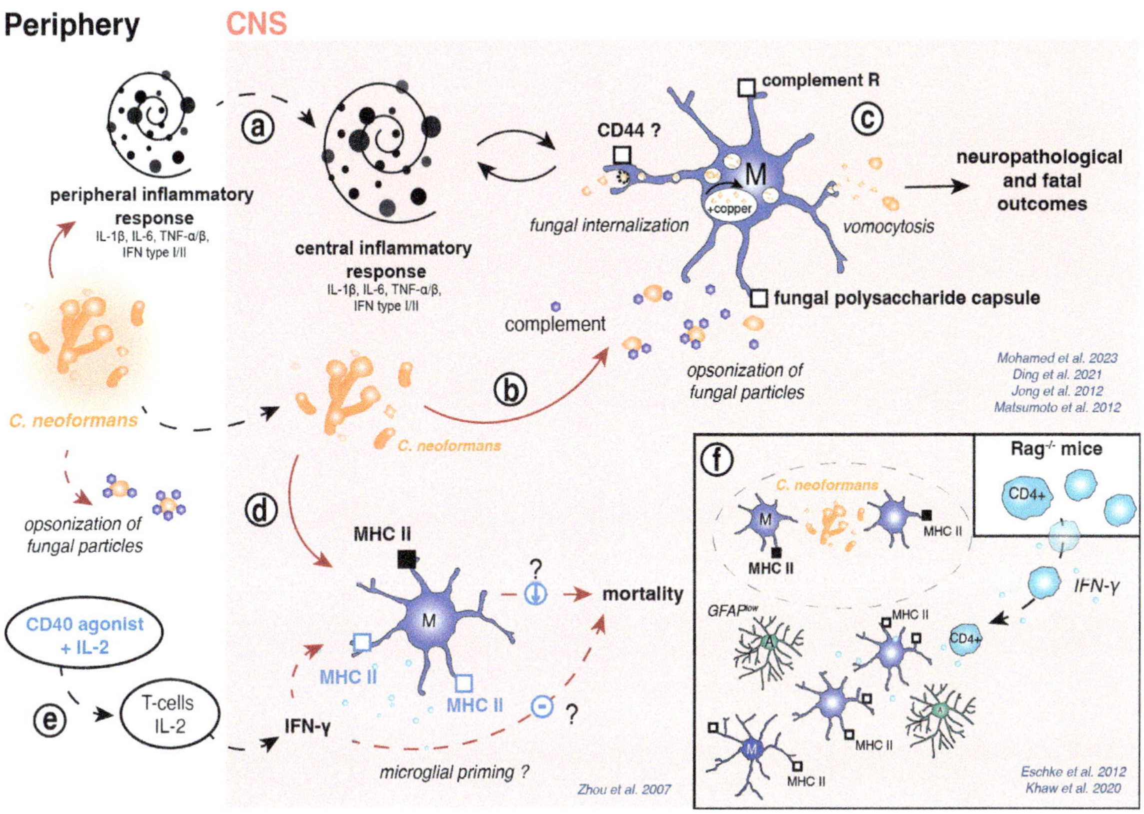

Fig. 13.4. Glial interaction with a neuroinvasive *Cryptococcus neoformans* infection. (A) After bacterial and viral pathogen exposure, fungal infection with *C. neoformans* activates both the peripheral and central immune system, eliciting microglial inflammatory response. (B) Mirroring the peripheral immune system response, microglia can opsonize *C. neoformans* particles as part of the immune response. (C) Following microglial phagocytosis (or other alternative pathway, e.g., CD44-mediated internalization), *C. neoformans* particles are hypothesized to replicate within the phagolysosomal system of microglia, bypassing the intracellular processing, and to use intracellular copper as a nutritional source. Mature fungal particles are then freely released into the extracellular space in a process of vomocytosis, ultimately leading to neuronal toxicity, neuropathologic consequences, or even fatal outcomes (if untreated). (D) Via yet unrevealed mechanism *C. neoformans* was observed to prevent induction of MHC II expression, a classical part of a microglial response to pathogens. (E) Peripheral immunotherapy with CD40 agonist and IL-2 cytokine extended survival and delayed mortality onset in *C. neoformans*-exposed rodents, presumably via IFN-γ-induced microglial priming and expression of MHC II. Role of MHC II in microglia-mediated fungal antigen clearance is unclear, considering that these mice still died due to accumulation of fungal particles in the brain. (F) Similarly, in Rag knockout ($Rag^{-/-}$) mice, an immunocompromised rodent model showing only a modest neuropathology after *C. neoformans* infection, microglia at the neuroinvasion site were shown to adopt an increased soma size but lacked MHC II expression. After CD4-immunopositive ($^{+}$) T-cell brain infiltration with IFN signaling, proximal and distal microglia increased in density and presented morphologic change toward a more reactive state, as well as MHC II expression. *A*, astrocyte; *M*, microglia; *CD*, cluster of differentiation; *CNS*, central nervous system, *GFAP*, glial fibrillary acidic protein; *IFN*, interferon; *IL*, interleukin; *MHC II*, major histocompatibility complex II; *Rag*, recombination activating gene 2 protein; *TNF*, tumor necrosis factor.

tightly wrapping around infected neuronal soma. Along with CD68 expression, these cells assisted in prevention of viral spread after neuronal apoptosis in these mice; corroborating the findings from human postmortem HSV brain samples (Fekete et al., 2018).

GLIAL ROLES IN FUNGAL INFECTIONS

Fungi with a notable history of CNS invasion include yeasts (*C. neoformans* and *Candida* spp.), molds (*Aspergillus* spp.), and dimorphic fungi (*Histoplasma capsulatum* and *Coccidioides* spp.), causing meningoencephalitis, ventriculitis, myelitis, hydrocephalus, meningitis, stroke/infarction, aneurysm, brain abscess, or granuloma. While *Cryptococcus* and *Aspergillus* are ubiquitous air contaminants globally, others are mostly found in moist soil (*Blastomyces*) or enriched in bird droppings (*Histoplasma*) within certain geographic regions (Murthy and Sundaram, 2014). Although CNS mycoses have been reported in healthy individuals (Panackal et al., 2015), more frequently, the immunocompromised patients with genetic disorders (e.g., caspase recruitment domain-containing protein [CARD] 9 deficiency) or chronic primary health condition (cancer, AIDS, or blood disorders, e.g., granulomatous disease and leukemia), occasioned by age (infants and premature babies), immunosuppressive therapy, organ/stem cell transplant, or metabolic disorders (diabetic ketosis and iron overload), are the most vulnerable to opportunistic fungal neuroinfections with fatal outcome (Davis et al., 2009; Dufresne et al., 2017; Góralska et al., 2018). For instance, in Canada, invasive candidemia (with 50% of patients needing intensive care) and aspergillosis, identified by blood and CSF culture, are estimated at a rate of 3.7 and 1.59 per 100,000 persons, respectively (Dufresne et al., 2017). In the United States, invasive candidiasis and cryptococcosis ranged from 72 to 228 and 30 to 66 cases per million persons, respectively, with up to 84% of cryptococcosis and 64% of candidiasis cases developing CNS mycoses (Góralska et al., 2018). Fungal CNS infection is further complicated by its multifaceted and unspecific clinical presentation, limited and challenging diagnostic approaches, as well as poor CNS and CSF pharmacokinetic profile, toxicity, and resistance to available treatment regimens, hence requiring a multidisciplinary team for global consensus (Schwartz et al., 2018; Ashley, 2019; Takoutsing et al., 2023). Survivors of CNS fungal infections present cognitive decline, vision impairment, motor imbalance, and speech difficulty (Góralska et al., 2018).

CNS mycoses typically result from a direct seeding of the CNS via the nasal and paranasal sinuses or by the dissemination of pulmonary infection to the CNS by hematogenous routes and neutrophilic Trojan horses after inhalation (Fig. 13.1A–H). These routes are modestly explained for *Candida* spp. and *C. neoformans* but not for others (Colombo and Rodrigues, 2015; Mohamed et al., 2022). The expression of surface hyaluronic acid and extracellular expression of laccase, urease, and phospholipase B1 aids *C. neoformans* CNS invasion (to break down host cell membrane) (Olszewski et al., 2004; Jong et al., 2007; Maruvada et al., 2012; Qiu et al., 2012). A direct endothelial internalization of *C. neoformans* was further described (Chen et al., 2021). Also, the transcellular CNS infiltration of *C. neoformans* and *Candida albicans*-associated mortality is strongly linked to a significant expression of secreted proteins, such as fungalysin metalloprotease (CnMpr1) and secreted aspartic proteinases 2 (Saps2) that primarily target the BBB by facilitating the attachment of pathogens to the endothelium, destroying tight junctions in the process (Vu et al., 2014; Wu et al., 2023). Of note, peripheral injection of *C. albicans* deficient in Saps2 did not result into CNS infiltration in young adult mice at 4 dpi, despite significant fungal burden observed in the lungs and liver (Wu et al., 2023). In addition, C-C chemokine receptor 2 (CCR2) ligand, CCL2, orchestrates significant infiltration of peripheral immune cells, including monocytes, macrophages, IFN-γ-secreting T cells, to CNS regions infected with *C. neoformans*, hence contributing to driving neuroimmunopathology (Xu et al., 2021). Morbidity and mortality associated with fungal CNS pathology are attributed to an uncontrolled neurotropism in immunocompetent and immunocompromised patients due to impaired T-cell response and fungal clearance. In immunocompetent individuals or HIV-positive individuals newly initiated on antiretroviral therapies (ARTs), clinical symptoms or neurologic deterioration do not abate despite weaning fungal titres post-treatment or following immune reconstitution with ART treatment (da Cunha Colombo et al., 2011; Panackal et al., 2015, 2017). A model of *C. neoformans* infection in immunocompromised $Rag^{-/-}$ mice with adoptively transferred $CD4^{+}$ T-cells, as well as in human cases, suggests the contribution of immune cells characterized by IFN-γ and $CD4^{+}$ cells-coordinated neuroimmunopathology, infiltration of peripheral cells, and elevated CSF levels of complement C3d receptor and neurofilament light chains (NFL) (Eschke et al., 2015; Panackal et al., 2015, 2017; Neal et al., 2017; Khaw et al., 2020). The CNS-damaging overzealous and paradoxical immune response (and cytokine storm) to antifungal or ART treatment is termed (post-infectious) immune reconstitution inflammatory syndrome (Brienze et al., 2021; Shi et al., 2022). In *Caenorhabditis elegans*, infection with *C. neoformans* resulted in selective degeneration of dopaminergic neurons and reduced lifespan (Kitisin et al., 2022). Of note, movement defects including ataxia and balance impairment are important components of the murine coma and behavioral scale (MCBS) adapted to measure the severity of cryptococcal meningoencephalitis in mice (Xu et al., 2023). The worsening neuropathology during immune reconstitution inflammatory syndrome is sensitive to adjunctive corticosteroid therapy (Nakajima et al., 2015), implying CNS inflammation plays a pivotal role in its pathogenesis. Meanwhile, the contributions of CNS-resident glial cells to the pathogen- and T-cells-centered fungal CNS pathologies are less discussed (Hole and Wormley, 2016). Therefore, the chapter provides information on the response and reactivity of glial cells during CNS-invasive fungal infections in the following section.

Microglia and astrocytes in neuroinvasive *C. neoformans* infection

Zhou et al. (2007) demonstrated that microglial response may be dysfunctional during *C. neoformans* CNS infection and microglial early priming may confer protection during the infection (Fig. 13.4A–D) (Zhou et al., 2007). Using a direct intracerebral *C. neoformans* injection, they showed that $CD45^{low}/CD11b^{+}$ cells representing the microglial population failed to express MHC II (Fig. 13.4D). However, peripheral injection of agonist antibody to CD40 (a receptor on antigen presenting cells that recruits effective and specific T cell response through CD154) and IL-2 (IFN-γ-secreting cytokine) elicited MHC II expression on $CD45^{low}/CD11b^{+}$ cells (Fig. 13.4E) (Zhou et al., 2007). The immunotherapy yielded similar results in the absence of *C. neoformans* and when $CD4^{+}$ T-cells were depleted, indicating its microglial priming effect (Zhou et al., 2007). It further delayed the mortality onset of *C. neoformans-infected* mice by 8 days. It extended their survival by 18 days in wild-type mice but not in mice deficient in IFN-γ signaling, which cannot upregulate MHC II on $CD45^{low}CD11b^{+}$ cells. The study did not show whether microglia upregulating MHC II phagocytosed or displayed fungicidal activity. Nonetheless, all infected mice died due to elevated *C. neoformans* burden after a prolonged delay (Zhou et al., 2007). While it is unknown whether microglial priming by the immunotherapy had a fungicidal effect in the direct model at any time point, as previously suggested (Blasi et al., 1995), the authors demonstrated reduced brain cryptococcal burden in another disseminated model when *C. neoformans* was injected peripherally with the immunotherapy (Fig. 13.4E) (Zhou et al., 2006). Together, this suggests that microglia may be equipped to orchestrate IFN-γ-inducible T-cell interaction in the CNS (Aguirre and Miller, 2002), and this is compromised by *C. neoformans* infection.

Evidence posits that opsonized *C. neoformans* can infect macrophages, including microglia, in mice and humans through phagocytosis but evade their antimicrobial process to fuel its proliferation after internalization (Johnston and May, 2013). Using a green fluorescent protein (GFP) reporter fungal strain, Mohamed et al. (2023) confirmed that *C. neoformans* directly infected microglia in young adult mice following CNS entry, and multiplied therein (Fig. 13.4B, C) (Mohamed et al., 2023). In human fetal microglial cell culture, antibody-labeled *C. neoformans* usurped the phagolysosomal pathway for entry and accumulated inside mature phagosomes, where they replicated before releasing the progeny extracellularly (Lee et al., 1995; Mohamed et al., 2022). The intra-phagolysosome replication resulted in an elevated number of the host's phagolysosome and fungal polysaccharide capsules, a pattern recognition receptor (Fig. 13.4B, C). In contrast to viral transcytosis, the release of fungal progeny for subsequent propagation through phagosomal extrusion (also called vomocytosis) did not result in the lysis and death of infected microglia (Alvarez and Casadevall, 2006; Johnston and May, 2013). Evidence suggests that microglia provide a nutrient rich haven to fungi following their entry (Mohamed et al., 2022, 2023). Particularly, in the CNS, brain (non)neuronal cells are enriched with copper at about threefold higher than the extracellular concentration (Gaier et al., 2013). Copper availability is vital to the survival and virulence of *C. neoformans*, which upregulate a copper importer to access the micronutrient in the host under limited copper concentrations (Ding et al., 2013; Hofer, 2013). Using a *C. neoformans* with GFP-labeled copper importer under the control of CTR4 promoter, the GFP expression was significantly reduced in $Ly6G^{-}Ly6C^{-}CD45^{int}CX3CR1^{hi}$ microglia, compared to extracellular GFP expression ($CD45^{-}$) indicating that the copper concentration in microglial phagosome sufficiently supports *C. neoformans* without a limited need for the importer (Fig. 13.4C) (Mohamed et al., 2022, 2023).

While the fungal receptor Clec7a linked to phagocytosis is robustly expressed on microglia, it has a limited role in the host response, internalization, or outcomes of *C. neoformans* infection (Nakamura et al., 2007; Walsh et al., 2017). Evidence from RAW264.7 macrophage-like cells derived from BALB/c mice suggested the interaction between *C. neoformans* hyaluronic acid and its receptor CD44 initiated the parasitic internalization process by recruiting the host autophagy-initiation complex related proteins and subsequent increase in cytosolic calcium concentration (Fig. 13.4C) (Ding et al., 2021). Of note, the internalization of *C. neoformans* was impaired in CD44 KO mice (Jong et al., 2012). In mouse brains, the expression of CD44 is absent or very low in healthy parenchyma but significantly induced during pathology, including models of ischemic stroke and amyotrophic lateral sclerosis (ALS) (Matsumoto et al., 2012; Sawada et al., 2020; Augusto-Oliveira et al., 2022). Although controversy exists on whether CD44 is exclusively expressed by peripheral macrophages (Jurga et al., 2020), microglial and astrocytic cell primary cultures from P1 wild-type mice robustly expressed CD44 following stimulation with LPS or IFN-γ (Matsumoto et al., 2012). In addition, reactive $IBA1^{+}$ microglia/macrophages upregulated and sustained CD44 expression within and around ischemic areas in young adult mice with middle cerebral artery occlusion (Sawada et al., 2020). It is currently unknown whether CD44 in microglia or astrocytes plays a specific role in the etiology and severity of CNS cryptococcosis and associated pathology.

$IBA1^{+}$ cells in *C. neoformans*-infected immunocompromised $Rag\text{-}1^{-/-}$ mice, which present with moderate morbidity and neuropathology, but low to no mortality, showed increased soma size only at peri-cryptococci lesion and without MHC II expression (Eschke et al., 2015). However, following $CD4^{+}$ reconstitution and brain $CD4^{+}$ infiltration to model immune reconstitution inflammatory syndrome in mice, proximal and distant $IBA1^{+}$ cells increased in density and adopted a reactive morphology with a significant MHC II expression (Fig. 13.4F) (Eschke et al., 2015; Khaw et al., 2020). However, astrocytes minimally expressed intense GFAP staining at the cryptococci lesion border in *C. neoformans*-infected $Rag\text{-}1^{-/-}$ mice, which was not exacerbated after receiving $CD4^{+}$ cells adoptively (Fig. 13.4F) (Eschke et al., 2015). The study additionally supports that peripheral IFN-γ, which diffuses freely through the BBB, is a major requirement for MHC II expression on microglia, as the transfer of $CD4^{+}$ cells that cannot secrete IFN-γ did not induce microglial expression of MHC II but was sufficient to trigger mortality and neuropathology comparable to IFN-γ-secreting $CD4^{+}$ T cells-recipient *C. neoformans*-infected $Rag\text{-}1^{-/-}$ mice. Meanwhile, *C. neoformans* infection during genetic (*Cx3cr1*-Cre^{ER}-$iDTR^{flox}$) and pharmacologic (PLX5622) depletion of microglia resulted in a marked reduction of fungal load in the murine brain (Mohamed et al., 2022, 2023). Using a microglia-specific *Sall1*-Cre^{ER}-$Csf1r^{flox}$ transgenic model with CSF1R deletion, because of the PLX5622 limitations, and *Cx3cr1*Cre^{ER}-$iDTR^{flox}$ in targeting microglia (Lei et al., 2020; Masuda et al., 2022), the brain fungal load significantly diminished in the reduced presence of microglia (Mohamed et al., 2022, 2023). Infiltration of peripheral cells also persisted in the reduced presence of microglia.

NEUROGLIA IN SARS-COV-2 AND COVID-19: A SPOTLIGHT

First identification of the SARS-CoV-2 causing an acute respiratory disease, coronavirus disease 2019 (COVID-19), occurred in late 2019 (Hu et al., 2021). Following its rapid expansion on a global scale with an increasing infectiousness and mortality rates, the World Health Organization declared COVID-19 a global pandemic by March 2020 (Lukiw et al., 2022; WHO, 2022). COVID-19 manifests as an array of symptoms, ranging from an asymptomatic and mild sickness characterized by any combination of fever, fatigue, dry cough, nasal congestion, headache, myalgia, and dyspnea, to severe intra- and extrapulmonary sickness such as encephalitis, anosmia, stroke, with symptoms so severe that a ventilator is required for assistance with breathing, or that the infection becomes fatal (Murta et al., 2020; Awogbindin et al., 2021; Tavčar et al., 2021; Valenza et al., 2021; Lukiw et al., 2022). The presence of comorbidities has demonstrated a significant impact on the symptom severity (Nanda et al., 2020). Studies on the outcomes of infections on the brain and with a focus on the role of non-neuronal cells, including glial cells, are growing as their relevance for immunologic response is being uncovered (Gonçalves de Andrade et al., 2021).

SARS-CoV-2 pathway to the brain

As anosmia, the loss of smell, is a reported COVID-19 symptom, the consensus for the primary viral particle entry to the brain was hypothesized to mainly involve the olfactory nerve pathway (Murta et al., 2020; Gonçalves de Andrade et al., 2021; Knyazev et al., 2021). However, the olfactory nerve lacks a significant expression of the two previously identified receptors for SARS-CoV-2 transmission: angiotensin converting enzyme 2 (ACE2) and transmembrane serine protease 2 (TMPRSS2) (Bilinska et al., 2021). In comparison, as it expresses ACE2 and TMPRSS2, the terminal nerve, a cranial nerve that bridges the olfactory epithelium with the hypothalamus and with projections to the forebrain (Bilinska et al., 2021; Butowt and von Bartheld, 2022), may serve as a more dominant route of entry for SARS-CoV-2 into the brain. In this context, it is important to call attention to brain endothelial cells to which SARS-CoV-2 envelope (S2E) protein was reported to bind, allowing for viral transmission across the BBB, presumably by transcytosis, inducing inflammatory responses in astrocytes (Knyazev et al., 2021; Ju et al., 2022). A small postmortem high-resolution magnetic resonance imaging (MRI) human study reported presence of microvascular injury and fibrinogen leakage in the olfactory bulb and brainstem after COVID-19 (Lee et al., 2021). Histologically, authors observed presence of $IBA1^+$ microglia and hypertrophic astrocytes around blood vessels with infiltrating macrophages and T-cells, potentially contributing to the vascular damage. In part of the human cohort, $IBA1^+$ and $CD68^+$ microglia were identified surrounding neurons, suggestive of neuronal phagocytosis (Lee et al., 2021; Tavčar et al., 2021). During the inflammatory response, reactive microglia can provoke neurotoxic astrocyte reactivity, inducing secretion of toxic lipids and inducing apoptosis of cells of the oligodendrocyte lineage; however, this is yet to be proven in human during SARS-CoV-2 infection, as it was so far shown only in mice (Monje and Iwasaki, 2022). If the BBB is a crucial entry point of SARS-CoV-2 to the brain (Erickson et al., 2021; Knyazev et al., 2021; Rhea et al., 2021; Ju et al., 2022), astrocytes and microglia could be key players that work together to provide an active defense against the virus through their homeostatic and plasticity functions.

SARS-CoV-2 infection affected myelin plasticity across animal models and human cases, potentially due to a reduced capability of oligodendrocytic cells to respond to neuronal activity with myelin changes (Monje and Iwasaki, 2022). Using an adult female mouse model of mild SARS-CoV-2 infection, loss of myelin, and reduced number of oligodendrocyte precursor cells in the hippocampus and oligodendrocytes in subcortical regions was also found (Fernández-Castañeda et al., 2022). In coincidence, these mice presented with increased CSF levels of CCL2 chemokine, which was previously connected to negative regulation of adult neurogenesis (Villeda et al., 2011), as well reported to be elevated in patients with long COVID-19, a post-COVID condition lasting >12 weeks of symptoms after infection (Fernández-Castañeda et al., 2022). This dysregulation of myelin turnover and/or oligodendrocyte lineage cell functions may be connected to reports indicating demyelinating diseases, such as GBS or multiple sclerosis, in patients following SARS-CoV-2 infection (Arnaud et al., 2020; Khan et al., 2021) or other viruses (Fazakerley and Walker, 2003). The results from a human study obtained with MRI also indicated changes in the extracellular water content and mean diffusivity of cerebral white matter that persist after SARS-CoV-2 infection in unvaccinated patients (Petersen et al., 2022). As this cohort did not report neuropsychologic impairments within up to a year post recovery, the impact of these persisting white matter alterations remains to be elucidated.

Neurodegenerative susceptibility following SARS-CoV-2 infection

A 2020 review reported the case of a patient with COVID-19 and associated encephalopathy, who had increased levels GFAP and of triggering receptor expressed on myeloid cells (TREM)2 in the CSF (Pilotto and Padovani, 2020). TREM2 is an innate pattern-recognition immune receptor that serves as an essential regulator of microglial function and whose polymorphic variants have been widely discussed in the context of AD (Gratuze et al., 2018). CSF tau and hyperphosphorylated tau (ptau) are used as efficient biomarkers of AD, with their presence positively correlating with AD genetic variants, cognitive impairment, or neuropathologic load in the brain parenchyma of humans (Cruchaga et al., 2013). A large-cohort genome-wide association study identified a single nucleotide polymorphism locus for CSF tau/ptau with a risk for developing AD within the TREM gene cluster (Cruchaga et al., 2013). Patients with COVID-19 report both acute and long-term neurologic complication, including cognitive decline and neurobehavioral impairment, similarly to other viral infections (see Microglia Roles and Viral Infections section earlier).

Considering microglial involvement with synaptic plasticity regulation, a hypothetic pipeline of downstream consequences for microglial activity during a severe/prolonged COVID-19 state was proposed to include altered microglia-mediated synaptic phagocytosis, contributing to impacting synaptic transmission (Monje and Iwasaki, 2022). This characteristic has been reported in a human brain organoid model with induced microglia-like cells (iMGs), in which SARS-CoV-2 exposure induced synaptic loss associated with a threefold enhanced presence of postsynaptic marker staining (PSD-95) within less-ramified $IBA1^+$ microglial cells (Samudyata et al., 2022). While warranting in situ and in vivo validation, these findings echo the potential for a dysregulated microglia-mediated synaptic plasticity induced by

SARS-CoV-2 exposure to contribute to mediating cognitive impairment in affected patients (Awogbindin et al., 2021; Poloni et al., 2021).

Upregulated expression of a range of purinergic receptors (P2X7, P2Y1, P2Y6, P2Y12) and regulator endonucleases was reported in a BV-2 microglia-like cell line exposed to a SARS-CoV-2 spike protein, while these results were later also majorly repeated in the hippocampus of spike protein-infused adult male mice (Alves et al., 2023). The spike protein (or its cleaved antigens) can reach the brain parenchyma, independently of the viral genetic information itself (Rhea et al., 2021), and was shown to be sustained in the circulation of patients diagnosed with long COVID-19 even months after infection (Swank et al., 2023). Intracerebroventricular delivery of the spike protein to the adult male mouse brain further induced microglia-complement mediated synaptic engulfment which manifested in a cognitive deficit (Fig. 13.3I) (Fontes-Dantas et al., 2023). This process was found to be TLR-4-dependent as its depletion via either genetic or pharmacologic means had a protective impact; an association that was further strengthened by the identification of a TLR-4 genetic polymorphism (genotype GG TLR-4-2604G>A (*rs10759931*)) in patients with a poor cognitive functioning post-COVID-19 syndrome (Fontes-Dantas et al., 2023).

In addition, myalgic encephalomyelitis/chronic fatigue syndrome (ME/CFS) has been reported after acute viral infections, such as the IAV, Ebola virus, and SARS coronavirus (see Renz-Polster et al., 2022). Long COVID-19 (>12 weeks of symptoms) manifests as a lack of recovery from COVID-19 or development of new neurologic symptoms or mental health conditions, resembling ME/CFS as both are characterized by lingering presence of symptoms of fatigue and cognitive deficiency (Paul et al., 2021). Sex differences are apparent across ME/CFS with females found to be three times more likely to be diagnosed with the disease (Renz-Polster et al., 2022). Conversationally, males are tentatively prospected to exhibit a higher susceptibility to SARS-CoV-2 infection and mortality compared to females due to varying biologic factors between sexes, i.e., immune system differences, hormonal impact, ACE-2 expression, risk comorbidities rate, or social factors, such as compliance with health measures and medical reporting (Taslem Mourosi et al., 2022). Additionally, it was suggested that females may be more predisposed toward the development of long COVID-19 (Taslem Mourosi et al., 2022); however, the underlying mechanism of this effect is yet unclear, with changes along the female hormonal cycles being a focus of investigation. Apart from the impact on physiology, the COVID-19 pandemic has been lengthily associated with an increased psychologic load (anxiety and stress) for both infected and noninfected parts of the population globally. A preclinical study in mice showed a sex-specific impact of chronic unpredictable mild stress (CUMS) exposure on the susceptibility to a subsequent SARS-CoV-2 infection (Yan et al., 2021). Accompanying signs of a depressive-like phenotype after CUMS, flow cytometry-sorted $Cd45^{low}/Cd11b^{+}$ microglial cells from female mice upregulated MHC II compared to male mice who showed more anxiety-like behaviors. Authors further speculated that this priming of microglia prior to infection implies an elevated susceptibility of the immune system to infection, particularly given the reports of an increased gene expression of SARS-CoV-2 receptors (*Ace2a*, *Bsg*) in the prefrontal cortex of female, but not male, mice (Yan et al., 2021). In a clinical context, this finding may provide more understanding to a previously observed higher rates of female susceptibility to stress-induced psychologic problems during the COVID-19 pandemic (Hou et al., 2020).

MACROPHAGES AND MICROGLIA AS VIRAL RESERVOIRS IN HIV: A SPOTLIGHT

Neurotropic viruses target neuronal cells for replication, inadvertently causing death of the host cells, neurons, and/or nonmyeloid glia. Macrophages, and in that respect microglia, display a specialized ability to accumulate some viral strains, but without any resulting cytopathogenicity (Suh et al., 2009; Campbell et al., 2019). Seemingly nondetrimental, functionally, these latent macrophage reservoirs display a viral replication "quiescence" after a later virus reactivation under specific contexts (e.g., stress) (Castellano et al., 2017). Due to a lack of success in using strategies such as microglial depletion in viral infection contexts (i.e., majority of experiments proving a necessity for microglia in driving an immunologic response), as well as resistance of these infected cells, and henceforth the virus, to ART (Castellano et al., 2017; Hatton and Duncan, 2019), prevention of microglia/macrophage infection or development of elimination strategies to target these cells upon infection may be crucial for success of any treatment. The HIV infection, upon which patients display a range of mild to severe permanent neurologic and memory deficits, collectively termed as HIV-associated neurocognitive disorder (HAND) (Saylor et al., 2016), will serve as an example of this effect.

In vitro experiments using human fetal microglia/macrophages revealed acute HIV-induced apoptosis. However, a small subpopulation of infected cells survived, maintaining their macrophage-like expression pattern and phagocytic properties. These cells exhibited a silenced viral replication, which could be re-activated by a subsequent stimulus (Castellano et al., 2017). Authors identified no alterations in proteins of various apoptotic pathways (Apaf-1, CytC, caspase-3) after HIV exposure, in which case, they proposed an early-stage HIV-mediated blockade of apoptosis through an induced retainment of these proteins within mitochondria (Castellano et al., 2017). A yet inexplicable finding in such conditions was an upregulation of Bim, a proapoptotic protein, that accumulated in mitochondria of these latently infected macrophages and was thus speculated to serve an alternative function (Castellano et al., 2017). Early on, the role of TREM1 was identified in the context of regulation of myeloid cell inflammatory responses to viral signaling via TLRs (Bleharski et al., 2003). Experiments with human macrophages (Yuan et al., 2017) or human monocyte-derived microglial cells (Campbell et al., 2023) showed that exposure to HIV, or its proteins (e.g., Tat, gp120), is associated with an increase in TREM1 expression, an effect that seems to be closely tied to apoptotic and T-cell-mediated cytopathogenic resistance in these cells. After miRNA silencing of TREM1, these human microglial cells showed an increased susceptibility toward HIV Tat-induced apoptosis but without initiation of overt inflammatory signaling (Campbell et al., 2023). A role for TLR-3 and -4 ligands (e.g., LPS, poly(I:C)) in suppression of HIV was also reported in human primary microglial cells via an activation of a downstream signaling pathway and anti-HIV activity depending on the type of infection (Suh et al., 2009). Using a combination of gene silencing and inhibitor administration in vitro, Campbell et al. (2023) further proposed a mechanism underlying Tat-induced TLR4 activation causing a downstream dose-dependent activation of cyclooxygenases (COX1, COX2) and prostaglandin E2 (PGE2) synthesis (Okorji and Olajide, 2014; Campbell et al., 2023). COX-2 is important for the homeostatic brain maintenance and fundamental cognitive functions, and additionally, it was proposed to be involved in proinflammatory signaling and be enhanced across neurodegenerative pathologies (Griffin et al., 1994; Chaudhry and Dore, 2009; Yagami et al., 2016; Campbell et al., 2023). Serving as a major modulator of inflammation, PGE2 pathway signaling through its receptor (EP2) was previously proposed to reduce glucose influx and mitochondrial oxidative phosphorylation, hence driving energy deficit and a maladaptive proinflammatory response in aged macrophages/microglia (Minhas et al., 2021). It is yet unclear if these pathways share more commonalities, albeit a significant metabolic transformation seems to underlie the phenotypic switches across states of the heterogeneous microglial spectrum. While the silencing of NF-κB pathway, a major regulator of cell's growth and apoptosis, does not seem to directly affect COX-PGE2 signaling in vitro (Campbell et al., 2023), it appears to enhance caspase 3 levels and decrease levels of Bcl-2, an antiapoptotic protein, leading to a proapoptotic effect in human monocyte-derived macrophages exposed to HIV proteins (Yuan et al., 2017). On that note, increased TREM1 seems to be a commonly present downstream inflammatory pathway across a range of viral infections, including Ebola, hepatitis C, West Nile virus, and ZIKV (see Campbell et al., 2019, 2023), endowing this target with a therapeutic potential (Yuan et al., 2017).

Another protein released by HIV-1-infected cells, Vpr, which has been detected in plasma and CSF of HIV-infected individuals, contributes to mediating IFN type 1-mediated cellular apoptosis and is necessary for viral replication (Walsh et al., 2014; Mamik et al., 2017). Soluble Vpr infection of human primary microglial cells was reported to activate the NLRP3-inflammasome pathway, a cytosolic protein complex that senses PAMPs and DAMPs, leading to induction of IL-1β transcription and release, as well as caspase 1 expression (Mamik et al., 2017). The latter was proposed to contribute to cell death of macrophage-like/microglial cells, which may support the hypothesis that a percentage of their population does not survive after acute expose to act as a reservoir (Castellano et al., 2017; Mamik et al., 2017). Further intraperitoneal treatment of Vpr transgenic mice with a caspase 1 inhibitor, VX-765, reduced caspase 1 and Il-1β transcription, improved anxiety-like symptoms, and attenuated the $IBA1^+$ microglial response to induced systemic inflammation via complete Freund's adjuvant exposure (Mamik et al., 2017). Conversationally, in mouse primary microglial cells, HIV-1 Tat exposure was also associated with fatty acyl-coenzyme A synthetase (ACSL4) expression and ferroptosis (Kannan et al., 2023), which could be pointing to a dysregulation of iron metabolism and which could have a harmful impact on the brain (Hare et al., 2013).

NEUROGLIA AS PHARMACOLOGIC TARGETS DURING NEUROINVASIVE INFECTIONS

Neuroglia as therapeutic target in HIV

HIV was first isolated in 1983 (Barré-Sinoussi et al., 1983), leading to development of the current therapy called combination ART (cART) (Hammer et al., 1997). cART targets the replication mechanism of HIV in cell reservoirs found in the periphery (Schnittman et al., 1990), as well as in the CNS, mainly in microglia and astrocytes (Watkins et al., 1990; Ranki et al., 1995; Takahashi et al., 1996; Li et al., 2015, 2016). While major benefits of this treatment strategy include a decline in morbidity and mortality caused by AIDS (Palella et al., 1998), as well as decreased sexual transmission of this disease (Cohen et al., 2011), the prevalence of cognitive impairment is still high (Wang et al., 2020). One of the contributing reasons is the low BBB penetration of the majority of antiretrovirals, thus limiting cART efficacy in the CNS (Yilmaz et al., 2012). Thereby, research efforts have been focusing on the development of nanoformulation strategies to target HIV directly in microglia and astrocytes. The first approaches with lone elvitegravir (EVG) or combined ritonavir (RTV)/atazanavir (ATV) successfully encapsulated this integrase inhibitor or the combination of protease inhibitors, respectively, within nanoparticles (NPs). First results were shown in in vitro models, including a BBB model (a culture of brain endothelial cells and astrocytes) and human primary microglial cells (Singh et al., 2020; Gong et al., 2020a,b). Although these experiments are still in an early stage (Cuadros et al., 2022; Kawakita et al., 2022), they are promising in sense of a preliminary display of nontoxicity of these NP-bound antiretrovirals on cultured cells, improved drug delivery across the BBB, and ability of elevating the intracellular drug uptake, leading to enhanced viral suppression. More favorable in vivo findings suggest that NPs containing several combined antiretrovirals could enhance drug distribution in the mouse brain (adult female Balb/c mice), as well as bioavailability in microglia and astrocytes (HIV-infected adult male C57BL/6 mice), thereby in line with the previously mentioned in vitro results (Surnar et al., 2021). On the other hand, many experiments have shown an increase in inflammatory and oxidative response in glial cells caused by cART (Jensen et al., 2015, 2019; Lisi et al., 2017; Gonzalez et al., 2021). A co-treatment including anti-inflammatory and antioxidative agents might make cART combined with NPs a highly effective treatment (Surnar et al., 2021). A role for negative factor (Nef) of HIV was shown in HIV's ability for avoidance of immune surveillance or viral dissemination. Found in exosomes released from HIV-infected cells, it was also associated with T-cell depletion and impairment of

BBB integrity and permeability in a mixed-cell in vitro BBB model (Raymond et al., 2016). A Nef-transfected microglial cell line was also able to release these Nef$^+$ exosomes-in comparison with transfected human primary astrocytes, which were able to stimulate cytokine release in a microglia-like cell line. To prevent the incorporation of Nef in exosomes (and thus reduce the impact on the BBB), a study investigated the delivery of magnetic NP-bound Nef peptides in a mixed-cell in vitro BBB model that would regulate the exosome content and Nef concentration via blockade of Nef myrisolation (Raymond et al., 2016), a lipid modification that promotes protein binding to cell membranes. Another target linked to HAND is glutamate metabolism (Ernst et al., 2010; Vázquez-Santiago et al., 2014; Saylor et al., 2016). Both experimental and clinical studies suggest that HIV particles contribute to dysregulating this mechanism, increasing extracellular levels of glutamate with a detrimental impact on excitatory neurotransmission (see Nedelcovych et al., 2017), a process to which HIV-infected glial cells may be particularly contributing (Potter et al., 2013). JHU-083, a prodrug of 6-diazo-5-oxo-L-norleucine, a glutaminase antagonist that normalizes glutamate production, showed an ability to reverse cognitive impairment in EcoHIV-infected mice (genetically modified HIV to infect nontransgenic mice and induce HAND) (Nedelcovych et al., 2017, 2019). Further approaches to regulate HIV infection, tested on HIV cell reservoirs—microglia and astrocytes—in vitro, included recombinant nuclear receptor related protein 1 (Ye et al., 2022), P27SJ protein (Darbinian-Sarkissian et al., 2006; Darbinian et al., 2008), or Crispr/Cas9 ribonucleoprotein complex (Huang and Nair, 2017; Campbell et al., 2019). Although the development of pharmacologic targets for an improved HIV therapy is on the rise, these are still majorly in the experimental phase, requiring a more advanced understanding of the mechanisms of HIV-associated disorders, such as HAND.

Neuroglia as therapeutic targets in nematode infection

Another neuroinvasive infection in which microglia have been explored as a potential therapeutic target is neural angiostrongylosis (NA), a disease prevalent in Southeast Asia and induced by *Angiostrongylus cantonensis*, a nematode whose larvae are able to cross the BBB and cause eosinophilic meningitis in the human brain (Marie and Petri, 2022). Intraperitoneal injection of *A. cantonensis* induced the expression of a proinflammatory signature represented by IL-6, IL-1β, and TNF-α increases, and demyelination in the brain of young adult mice at 21 dpi (Lin et al., 2010; Feng et al., 2014, 2019). Of note, inducible nitric oxide synthase (iNOS) was prominently upregulated on these microglia, suggesting an increased reactivity (Lin et al., 2010; Feng et al., 2014, 2019). Albendazole, an anthelmintic combined with corticosteroids, is the advised treatment for killing larvae and modulating inflammatory reactions in patients with NA (Ansdell et al., 2021; Jacob et al., 2021). Studies investigating neuropathologies presenting similar symptoms such as *A. cantonensis*-induced optic neuritis and cerebral ischemic reperfusion injury found that tanshinone IIA (TSIIA), a constituent of the *Salvia miltiorrhiza* root, is effective in alleviating CNS inflammation and demyelination in young adult Balb/c mice and adult male Sprague Dawley rats, respectively (Feng et al., 2015; Yan et al., 2016; Song et al., 2021). Indeed, the albendazole and TSIIA treatment of *A. cantonensis*-infected mice showed a decrease of demyelination, an increase in oligodendrocyte proliferation, as well as a reduced expression of proinflammatory cytokines and microglial markers associated with inflammation (CD86, iNOS, Cd11b) accompanied by an increase in IBA1$^+$/Arginase 1 (Arg-1)$^+$ cells in the cerebral cortex (Feng et al., 2019). These results, supported by further research, suggest that TSIIA modulates microglial inflammatory activity via a negative regulation of the NF-κB pathway (Pires et al., 2018; Song et al., 2021). TSIIA could be the next step in the treatment of NA and associated symptoms.

Bacterial CNS infection: Therapeutic insight

Strategies to prevent bacterial CNS invasion and complications are efforts of continuous research. A few vaccines have been developed and deployed for some notable bacterial infections, especially for neonates in endemic regions. Multimodal treatment approaches taking into consideration the use of anti-inflammatory drugs (dexamethasone), agents inhibiting the release of virulent bacterial molecules (rifampicin and daptomycin) and limiting host factors contributing to BBB permeability (e.g., metalloproteinase inhibitors) in combination with antibiotics, are being proposed and regularly updated (Woehrl et al., 2011; Hasbun, 2022). Barring direct CNS invasion through the olfactory system, effective antibiotic therapy is valuable, at least to stem bacteremia. However, the development of bacterial resistance due to the frequent clinical use of antibiotics is a major threat (Hasbun, 2022). Advances through well-designed proof-of-concept mechanistic studies have shown that targeting host-bacterial interaction contributing to BBB permeability is an innovative way of preventing bacterial CNS invasion (Yang et al., 2023). Nevertheless, multiple bacterial strains that can use a plethora of host receptors is a drawback for this approach. Future treatment options that will significantly alter the course and outcome of bacterial CNS infection will emerge out of the significant understanding of the pathophysiology of each microbe in the brain, including the response of different types of neuronal and non-neuronal cells. Furthermore, the evasive mechanisms observed in the periphery could provide insight into similar escape principles at play in the CNS. One of the approaches being developed aims to enhance the intra-CNS detection and intracellular killing of bacteria having crossed the barriers by enhancing microglial phagocytic response (Diesselberg et al., 2018; Thygesen et al., 2018). Also, a finding demonstrated that dual blocking of CD14 and complement component 5 (C5) significantly attenuated cytokine storm in the periphery following *E. coli* infection, a mechanism that warrants further investigation in the CNS although both mechanisms are triggered following *E. coli* invasion (Bjerkhaug et al., 2023). In addition, *L. monocytogenes* ligand internalin (InlB) binds to the globular head of complement C1q receptor (Braun et al., 2000) with yet-to-be deciphered implications in modulating glial cells such as microglia in the CNS.

CONCLUDING REMARKS

Compared to the peripheral immune response to pathogens, the pathogenesis of neuroinvasive infections and the reactions of first responders, microglia (and astrocytes), are less understood. This chapter started off by discussing some of the currently known routes for pathogens to impact the CNS and initiate the sickness response. In selected examples of widely abound bacterial, viral, and fungal pathogens, it further reviewed the current depth of knowledge on the neuroinvasiveness of these microbes while acknowledging the role of the peripheral system in the immune response and its impact on CNS functioning. Importantly, the chapter X-rayed the complex and intricate roles of microglia, together with astrocytes, during *E. coli*, *C. neoformans*, HIV, and SARS-CoV-2 infections in the CNS. Summarily, neuroinfections deploy several mechanisms to maneuver various microglial defense functions to flourish in the CNS. Maintaining a finely tuned response with subtle to massive immune events occurring at a speed of milliseconds is critical for the immune system, in the CNS majorly represented by microglia. Furthermore, as key components of neuroplasticity and homeostasis maintenance regulatory mechanisms and dynamic sensory hubs, microglia and astrocytes can be considered small but a mighty part of the CNS. With their activity tightly tailored to the needs of their microenvironment, they can become the best support (e.g., early stage of infection) or the greatest foe (e.g., in case of prolonged, severe, or recurring infection). It has become evident that the latter can have far-reaching implications for mental and neurologic health, especially in those belonging to immunocompromised and/or elderly populations, hence strongly calling for further investigation. As described across different sections of this chapter, there has been a lot of progress in recognition of receptors and mechanisms these glial cells use for pathogen recognition and response, which creates an opportunity for various modulatory interventions. With a foundation formed by in vitro and clinical case studies, the past two decades of infection research have been swiftly moving forward and building up its translational value notably by using novel rodent models and conducting larger-scale clinical observations; a research sector that has at present significantly expanded mainly due to funding opportunities created to address the global COVID-19 pandemic. To bypass the limitation of the BBB for pharmacologic targeting, development of drug-delivery methods such as NPs is also underway, firmly engaged in designing nontoxic and more efficient transfer molecules. Considering the seemingly limitless potential of functional states with variable sensitivity that the glia cells can adopt in various health and disease contexts, ranging within a spectrum of positively beneficial, nonbeneficial to detrimental, single cell/state/subpopulation targeting represents one of the main directions of current glial research. Such an approach is of particular interest for addressing in infection contexts the risk presented by macrophage/microglia virus reservoirs. Glial cells are inherently first responders during CNS infections. Understanding their interactions with other peripheral and central components is an essential step for further progress, paving the way for multimodal therapeutic strategies for preventing or attenuating immediate-, short-, and long-term infection-induced outcomes.

References

Aguirre K, Miller S (2002). MHC class II-positive perivascular microglial cells mediate resistance to Cryptococcus neoformans brain infection. Glia 39: 184–188. https://doi.org/10.1002/glia.10093.

Alquisiras-Burgos I, Peralta-Arrieta I, Alonso-Palomares LA et al. (2021). Neurological complications associated with the blood-brain barrier damage induced by the inflammatory response during SARS-CoV-2 infection. Mol Neurobiol 58: 520–535. https://doi.org/10.1007/s12035-020-02134-7.

Alvarez M, Casadevall A (2006). Phagosome extrusion and host-cell survival after Cryptococcus neoformans phagocytosis by macrophages. Curr Biol 16: 2161–2165. https://doi.org/10.1016/j.cub.2006.09.061.

Alves VS, Santos SACS, Leite-Aguiar R et al. (2023). SARS-CoV-2 spike protein alters microglial purinergic signaling. Front Immunol 14: 1158460. https://doi.org/10.3389/fimmu.2023.1158460.

Ansdell V, Kramer KJ, McMillan JK et al. (2021). Guidelines for the diagnosis and treatment of neuroangiostrongyliasis: updated recommendations. Parasitology 148: 227–233. https://doi.org/10.1017/S0031182020001262.

Aravalli RN, Hu S, Rowen TN et al. (2005). Cutting edge: TLR2-mediated proinflammatory cytokine and chemokine production by microglial cells in response to herpes simplex virus1. J Immunol 175: 4189–4193. https://doi.org/10.4049/jimmunol.175.7.4189.

Aravalli RN, Hu S, Lokensgard JR (2007). Toll-like receptor 2 signaling is a mediator of apoptosis in herpes simplex virus-infected microglia. J Neuroinflammation 4: 11. https://doi.org/10.1186/1742-2094-4-11.

Arnaud S, Budowski C, Tin NW et al. (2020). Post SARS-CoV-2 Guillain-Barré syndrome. Clin Neurophysiol 131: 1652–1654. https://doi.org/10.1016/j.clinph.2020.05.003.

Ashley ED (2019). Antifungal drugs: special problems treating central nervous system infections. J Fungu 5: 97. https://doi.org/10.3390/jof5040097.

Augusto-Oliveira M, Arrifano GP, Delage CI et al. (2022). Plasticity of microglia. Biol Rev Camb Philos Soc 97: 217–250. https://doi.org/10.1111/brv.12797.

Aw E, Zhang Y, Carroll M (2020). Microglial responses to peripheral type 1 interferon. J Neuroinflammation 17: 340. https://doi.org/10.1186/s12974-020-02003-z.

Awogbindin IO, Ben-Azu B, Olusola BA et al. (2021). Microglial implicat ions in SARS-CoV-2 infection and COVID-19lessons from viral RNA neurotropism and possible relevance to Parkinson's disease. Front Cell Neurosci 15: 670298. https://doi.org/10.3389/fncel.2021.670298.

Ayegbusi OT, Ajagbe OA, Afowowe TO et al. (2019). Virus genes and host correlates of pathology are markedly reduced during respiratory syncytial and influenza virus co-infection in BALB/c mice. Heliyon 5: e01094. https://doi.org/10.1016/j.heliyon.2018.e01094.

Barbosa-Silva MC, Santos LE, Rangel B (2018). The impact of non-neurotropic influenza strains on the brain: a role for microglial priming? J Neurosci 38: 7758–7760. https://doi.org/10.1523/JNEUROSCI.1368-18.2018. Society for Neuroscience.

Barichello T, Rocha Catalão CH, Rohlwink UK et al. (2023). Bacterial meningitis in Africa. Front Neurol 14: 822575. https://doi.org/10.3389/fneur.2023.822575.

Barré-Sinoussi F, Chermann JC, Rey F et al. (1983). Isolation of a T-Lymphotropic retrovirus from a patient at risk for acquired immune deficiency syndrome (AIDS). Science 220: 868–871. https://doi.org/10.1126/science.6189183. American Association for the Advancement of Science.

Barrientos RM, Higgins EA, Biedenkapp JC et al. (2006). Peripheral infection and aging interact to impair hippocampal memory consolidation. Neurobiol Aging 27: 723–732. https://doi.org/10.1016/j.neurobiolaging.2005.03.010.

Barrientos RM, Watkins LR, Rudy JW et al. (2009). Characterization of the sickness response in young and aging rats following E. coli infection. Brain Behav Immun 23: 450–454. https://doi.org/10.1016/j.bbi.2009.01.016.

Barrientos RM, Thompson VM, Arnold TH et al. (2015). The role of hepatic and splenic macrophages in E. coli-induced memory impairments in aged rats. Brain Behav Immun 43: 60–67. https://doi.org/10.1016/j.bbi.2014.07.004.

Baruch K, Deczkowska A, David E et al. (2014). Aging-induced type I interferon signaling at the choroid plexus negatively affects brain function. Science (New York, NY) 346: 89–93. https://doi.org/10.1126/science.1252945.

Bauer L, Laksono BM, de Vrij FMS et al. (2022). The neuroinvasiveness, neurotropism, and neurovirulence of SARS-CoV-2. Trends Neurosci 45: 358–368. https://doi.org/10.1016/j.tins.2022.02.006.

Benros ME, Nielsen PR, Nordentoft M et al. (2011). Autoimmune diseases and severe infections as risk factors for schizophrenia: a 30-year population-based register study. Am J Psychiatry 168: 1303–1310. https://doi.org/10.1176/appi.ajp.2011.11030516.

Benros ME, Waltoft BL, Nordentoft M et al. (2013). Autoimmune diseases and severe infections as risk factors for mood disorders: a nationwide study. JAMA Psychiatry 70: 812–820. https://doi.org/10.1001/jamapsychiatry.2013.1111.

Bilinska K, von Bartheld CS, Butowt R (2021). Expression of the A CE2 virus entry protein in the nervus terminalis reveals the potential for an alternative route to brain infection in COVID-19. Front Cell Neurosci 15: 674123. https://doi.org/10.3389/fncel.2021.674123.

Bjerkhaug AU, Granslo HN, Cavanagh JP et al. (2023). Dual inhibition of complement C5 and CD14 attenuates inflammation in a cord blood model. Pediatr Res. https://doi.org/10.1038/s41390-023-02489-2.

Blasi E, Barluzzi R, Mazzolla R et al. (1995). Role of nitric oxide and melanogenesis in the accomplishment of anticryptococcal activity by the BV-2 microglial cell line. J Neuroimmunol 58: 111–116. https://doi.org/10.1016/0165-5728(95)00016-u.

Bleharski JR, Kiessler V, Buonsanti C et al. (2003). A role for triggering receptor expressed on myeloid cells-1 in host defense during the early-induced and adaptive phases of the immune response. J Immunol 170: 3812–3818. https://doi.org/10.4049/jimmunol.170.7.3812.

Boulanger LM (2009). Immune proteins in brain development and synaptic plasticity. Neuron 64: 93–109. https://doi.org/10.1016/j.neuron.2009.09.001.

Braniste V, Al-Asmakh M, Kowal C et al. (2014). The gut microbiota influences blood-brain barrier permeability in mice. Sci Transl Med 6: 263ra158. https://doi.org/10.1126/scitranslmed.3009759.

Braun L, Ghebrehiwet B, Cossart P (2000). gC1q-R/p32, a C1q-binding protein, is a receptor for the InlB invasion protein of Listeria monocytogenes. EMBO J 19: 1458–1466. https://doi.org/10.1093/emboj/19.7.1458.

Brienze VMS, André JC, Liso E et al. (2021). Cryptococcal immune reconstitution inflammatory syndrome: from blood and cerebrospinal fluid biomarkers to treatment approaches. Life 11: 95. https://doi.org/10.3390/life11020095.

Butowt R, von Bartheld CS (2022). The route of SARS-CoV-2 to brain infection: have we been barking up the wrong tree? Mol Neurodegener 17: 20. https://doi.org/10.1186/s13024-022-00529-9.

Campbell GR, To, R.K, Spector SA (2019). TREM-1 protects HIV-1-infected macrophages from apoptosis through maintenance of mitochondrial function. MBio 10. https://doi.org/10.1128/mBio.02638-19e02638-19.

Campbell GR, Rawat P, To, R.K et al. (2023). HIV-1 Tat upregulates TREM1 expression in human microglia. J Immunol 211: 429–442. https://doi.org/10.4049/jimmunol.2300152.

Castellano P, Prevedel L, Eugenin EA (2017). HIV-infected macrophages and microglia that survive acute infection become viral reservoirs by a mechanism involving Bim. Sci Rep 7: 12866. https://doi.org/10.1038/s41598-017-12758-w.

Chaplin DD (2010). Overview of the immune response. J Allergy Clin Immunol 125: S3–23. https://doi.org/10.1016/j.jaci.2009.12.980.

Chapman TR, Barrientos RM, Ahrendsen JT et al. (2012). Aging and infection reduce expression of specific brain-derived neurotrophic factor mRNAs in hippocampus. Neurobiol Aging 33: 832e1–14. https://doi.org/10.1016/j.neurobiolaging.2011.07.015.

Chaudhry UA, Dore S (2009). Cytosolic prostaglandin E synthase: expression patterns in control and Alzheimer's disease brains. Am J Alzheimers Dis Other Demen24: 46–51. https://doi.org/10.1177/1533317508323655.

Chen Z, Zhong D, Li G (2019). The role of microglia in viral encephalitis: a review. J Neuroinflammation16: 76. https://doi.org/10.1186/s12974-019-1443-2.

Chen Y, Li C, Sun D et al. (2021). Quantitative analysis reveals internalisation of Cryptococcus neoformans by brain endothelial cells in vivo. Cell Microbiol 23: e13330. https://doi.org/10.1111/cmi.13330.

Cohen MS, Chen YQ, McCauley M et al. (2011). Prevention of HIV-1 infection with early antiretroviral therapy. N Engl J Med 365: 493–505. https://doi.org/10.1056/NEJMoa1105243.

Colombo AC, Rodrigues ML (2015). Fungal colonization of the brain: anatomopathological aspects of neurological cryptococcosis. Anais Acad Brasil Ci 87: 1293–1309. https://doi.org/10.1590/0001-3765201520140704.

Colonna M, Butovsky O (2017). Microglia function in the central nervous system during health and neurodegeneration. Annu Rev Immunol 35: 441–468. https://doi.org/10.1146/annurev-immunol-051116-052358.

Conrady CD, Zheng M, van Rooijen N et al. (2013). Microglia and a functional type I IFN pathway are required to counter HSV-1–driven brain lateral ventricle enlargement and encephalitis. J Immunol 190: 2807–2817. https://doi.org/10.4049/jimmunol.1203265.

Cortese GP, Barrientos RM, Maier SF et al. (2011). Aging and a peripheral immune challenge interact to reduce mature brain-derived neurotrophic factor and activation of TrkB, PLCgamma1, and ERK in hippocampal synaptoneurosomes. J Neurosci 31: 4274–4279. https://doi.org/10.1523/JNEUROSCI.5818-10.2011.

Costello DA, Lynch MA (2013). Toll-like receptor 3 activation modulates hippocampal network excitability, via glial production of interferon-β. Hippocampus 23: 696–707. https://doi.org/10.1002/hipo.22129.

Cruchaga C, Kauwe JSK, Harari O et al. (2013). GWAS of cerebrospinal fluid tau levels identifies novel risk variants for Alzheimer's disease. Neuron 78: 256–268. https://doi.org/10.1016/j.neuron.2013.02.026.

Cuadros MA, Sepulveda MR, Martin-Oliva D et al. (2022). Microglia and microglia-like cells: similar but different. Front Cell Neurosci 16: 816439. https://doi.org/10.3389/fncel.2022.816439.

da Cunha Colombo ER, Mora DJ, Silva-Vergara ML (2011). Immune reconstitution inflammatory syndrome (IRIS) associated with Cryptococcus neoformans infection in AIDS patients. Mycoses 54: e178–e182. https://doi.org/10.1111/j.1439-0507.2010.01870.x.

Dando SJ, Mackay-Sim A, Norton R et al. (2014). Pathogens penetrating the central nervous system: infection pathways and the cellular and molecular mechanisms of invasion. Clin Microbiol Rev 27: 691–726. https://doi.org/10.1128/CMR.00118-13.

Dantzer R (2018). Neuroimmune interactions: from the brain to the immune system and vice versa. Physiol Rev 98: 477–504. https://doi.org/10.1152/physrev.00039.2016.

Dantzer R, O'Connor JC, Freund GG et al. (2008). From inflammation to sickness and depression: when the immune system subjugates the brain. Nat Rev Neurosci 9: 46–56. https://doi.org/10.1038/nrn2297.

Darbinian N, Popov Y, Khalili K et al. (2008). Creation of a bi-directional protein transduction system for suppression of HIV-1 expression by p27SJ. Antiviral Res 79: 136–141. https://doi.org/10.1016/j.antiviral.2007.11.006.

Darbinian-Sarkissian N, Darbinyan A, Otte J et al. (2006). p27SJ, a novel protein in St John's Wort, that suppresses expression of HIV-1 genome. Gene Ther 13: 288–295. https://doi.org/10.1038/sj.gt.3302649.

Davis JA, Horn DL, Marr KA et al. (2009). Central nervous system involvement in cryptococcal infection in individuals after solid organ transplantation or with AIDS. Transpl Infect Dis 11: 432–437. https://doi.org/10.1111/j.1399-3062.2009.00424.x.

Dheer A, Jain V, Kushwah N et al. (2018). Temporal and spatial changes in glial cells during chronic hypobaric hypoxia: role in neurodegeneration. Neuroscience 383: 235–246. https://doi.org/10.1016/j.neuroscience.2018.04.026.

Dian S, Ganiem AR, van Laarhoven A (2021). Central nervous system tuberculosis. Curr Opin Neurol 34: 396–402. https://doi.org/10.1097/WCO.0000000000000920.

Diesselberg C, Ribes S, Seele J et al. (2018). Activin A increases phagocytosis of *Escherichia coli* K1 by primary murine microglial cells activated by toll-like receptor agonists. J Neuroinflammation 15: 175. https://doi.org/10.1186/s12974-018-1209-2.

Ding C, Festa RA, Chen Y-L et al. (2013). Cryptococcus neoformans copper detoxification machinery is critical for fungal virulence. Cell Host Microbe 13: 265–276. https://doi.org/10.1016/j.chom.2013.02.002.

Ding S, Yang J, Feng X et al. (2021). Interactions between fungal hyaluronic acid and host CD44 promote internalization by recruiting host autophagy proteins to forming phagosomes. iScience 24: 102192. https://doi.org/10.1016/j.isci.2021.102192.

Djukic M, Onken ML, Schütze S et al. (2014). Vitamin d deficiency reduces the immune response, phagocytosis rate, and intracellular killing rate of microglial cells. Infect Immun 82: 2585–2594. https://doi.org/10.1128/IAI.01814-14.

Dufresne SF, Cole DC, Denning DW et al. (2017). Serious fungal infections in Canada. Eur J Clin Microbiol Infect Dis 36: 987–992. https://doi.org/10.1007/s10096-017-2922-y.

Düsedau HP, Steffen J, Figueiredo CA et al. (2023). Influenza a virus (H1N1) infection induces microglial activation and temporal dysbalance in glutamatergic synaptic transmission. MBio 12: e01776-21. https://doi.org/10.1128/mBio.01776-21.

Enlow W, Bordeleau M, Piret J et al. (2021). Microglia are involved in phagocytosis and extracellular digestion during Zika virus encephalitis in young adult immunodeficient mice. J Neuroinflammation 18: 178. https://doi.org/10.1186/s12974-021-02221-z.

Erickson MA, Rhea EM, Knopp RC et al. (2021). Interactions of SARS-CoV-2 with the blood–brain barrier. Int J Mol Sci 22: 2681. https://doi.org/10.3390/ijms22052681.

Ernst T, Jiang CS, Nakama H et al. (2010). Lower brain glutamate is associated with cognitive deficits in HIV patients: a new mechanism for HIV-associated neurocognitive disorder. J Magn Reson Imaging 32: 1045–1053. https://doi.org/10.1002/jmri.22366.

Eschke M, Piehler D, Schulze B et al. (2015). A novel experimental model of Cryptococcus neoformans-related immune reconstitution inflammatory syndrome (IRIS) provides insights into pathogenesis. Eur J Immunol 45: 3339–3350. https://doi.org/10.1002/eji.201545689.

Fazakerley JK, Walker R (2003). Virus demyelination. J Neurovirol 9: 148–164. https://doi.org/10.1080/13550280390194046.

Fekete R, Cserép C, Lénárt N et al. (2018). Microglia control the spread of neurotropic virus infection via P2Y12 signalling and recruit monocytes through P2Y12-independent mechanisms. Acta Neuropathol 136: 461–482. https://doi.org/10.1007/s00401-018-1885-0.

Feng Y, Zeng X, Li W-H et al. (2014). The pathogenesis of optic neuritis caused by Angiostrongylus cantonensis in BALB/c mice. Parasit Vectors 7: 339. https://doi.org/10.1186/1756-3305-7-339.

Feng F, Feng Y, Liu Z et al. (2015). Effects of albendazole combined with TSII-A (a Chinese herb compound) on optic neuritis caused by Angiostrongylus cantonensis in BALB/c mice. Parasit Vectors 8: 606. https://doi.org/10.1186/s13071-015-1214-6.

Feng Y, Feng F, Zheng C et al. (2019). Tanshinone IIA attenuates demyelination and promotes remyelination in A. cantonensis-infected BALB/c mice. Int J Biol Sci 15: 2211–2223. https://doi.org/10.7150/ijbs.35266.

Fernández-Castañeda A, Lu P, Geraghty AC et al. (2022). Mild respiratory SARS-CoV-2 infection can cause multi-lineage cellular dysregulation and myelin loss in the brain. bioRxiv. https://doi.org/10.1101/2022.01.07.475453.

Ferrini F, De Koninck Y (2013). Microglia control neuronal network excitability via BDNF signalling. Neural Plast 2013: 429815. https://doi.org/10.1155/2013/429815.

Field R, Campion S, Warren C et al. (2010). Systemic challenge with the TLR3 agonist poly I:C induces amplified IFNalpha/beta and IL-1beta responses in the diseased brain and exacerbates chronic neurodegeneration. Brain Behav Immun 24: 996–1007. https://doi.org/10.1016/j.bbi.2010.04.004.

Filgueira L, Larionov A, Lannes N (2021). The influence of virus infection on microglia and accelerated brain aging. Cells 10: 1836. https://doi.org/10.3390/cells10071836.

Filippone C, Legros V, Jeannin P et al. (2020). Arboviruses and muscle disorders: from disease to cell biology. Viruses 12: 616. https://doi.org/10.3390/v12060616.

Fontes-Dantas FL, Fernandes GG, Gutman EG et al. (2023). SARS-CoV-2 spike protein induces TLR4-mediated long-term cognitive dysfunction recapitulating post-COVID-19 syndrome in mice. Cell Rep 42. https://doi.org/10.1016/j.celrep.2023.112189.

Franco-Bocanegra DK, McAuley C, Nicoll JAR et al. (2019). Molecular mechanisms of microglial motility: changes in ageing and Alzheimer's disease. Cells 8: 639. https://doi.org/10.3390/cells8060639.

Frank MG, Barrientos RM, Hein AM et al. (2010). IL-1RA blocks E. coli-induced suppression of Arc and long-term memory in aged F344xBN F1 rats. Brain Behav Immun 24: 254–262. https://doi.org/10.1016/j.bbi.2009.10.005.

Gaier ED, Eipper BA, Mains RE (2013). Copper signaling in the mammalian nervous system: synaptic effects. J Neurosci Res 91: 2–19. https://doi.org/10.1002/jnr.23143.

Garber C, Soung A, Vollmer LL et al. (2019). T cells promote microglia-mediated synaptic elimination and cognitive dysfunction during recovery from neuropathogenic flaviviruses. Nat Neurosci 22: 1276–1288. https://doi.org/10.1038/s41593-019-0427-y.

Garden GA (2002). Microglia in human immunodeficiency virus-associated neurodegeneration. Glia 40: 240–251. https://doi.org/10.1002/glia.10155.

Geyer S, Jacobs M, Hsu N-J (2019). Immunity against bacterial infection of the central nervous system: an astrocyte perspective. Front Mol Neurosci 12: 57. https://doi.org/10.3389/fnmol.2019.00057.

Ginhoux F, Lim S, Hoeffel G et al. (2013). Origin and differentiation of microglia. Front Cell Neurosci 7: 45. https://doi.org/10.3389/fncel.2013.00045.

Goldmann T, Zeller N, Raasch J et al. (2015). USP18 lack in microglia causes destructive interferonopathy of the mouse brain. EMBO J 34: 1612–1629. https://doi.org/10.15252/embj.201490791.

Goldmann T, Blank T, Prinz M (2016). Fine-tuning of type I IFN-signaling in microglia—implications for homeostasis, CNS autoimmunity and interferonopathies. Curr Opin Neurobiol 36: 38–42. https://doi.org/10.1016/j.conb.2015.09.003.

Gonçalves de Andrade E, Šimončičová E, Carrier M et al. (2021). Microglia fighting for neurological and mental health: on the central nervous system frontline of COVID-19 pandemic. Front Cell Neurosci 15: 12. https://doi.org/10.3389/fncel.2021.647378.

Gong Y, Chowdhury P, Nagesh PKB et al. (2020a). Novel elvitegravir nanoformulation for drug delivery across the blood-brain barrier to achieve HIV-1 suppression in the CNS macrophages. Sci Rep 10: 3835. https://doi.org/10.1038/s41598-020-60684-1.

Gong Y, Zhi K, Nagesh PKB et al. (2020b). An elvitegravir nanoformulation crosses the blood–brain barrier and suppresses HIV-1 replication in microglia. Viruses 12: 564. https://doi.org/10.3390/v12050564.

Gonzalez H, Podany A, Al-Harthi L et al. (2021). The far-reaching HAND of cART: cART effects on astrocytes. J Neuroimmune Pharmacol 16: 144–158. https://doi.org/10.1007/s11481-020-09907-w.

Góralska K, Blaszkowska J, Dzikowiec M (2018). Neuroinfections caused by fungi. Infection 46: 443–459. https://doi.org/10.1007/s15010-018-1152-2.

Gorlé N, Bauwens E, Haesebrouck F et al. (2020). Helicobacter and the potential role in neurological disorders: there is more than *Helicobacter pylori*. Front Immunol 11: 584165. https://doi.org/10.3389/fimmu.2020.584165.

Govic YL, Demey B, Cassereau J et al. (2022). Pathogens infecting the central nervous system. PLoS Pathog 18: e1010234. https://doi.org/10.1371/journal.ppat.1010234.

Granholm A-C (2023). Long-term effects of SARS-CoV-2 in the brain: clinical consequences and molecular mechanisms. J Clin Med 12: 3190. https://doi.org/10.3390/jcm12093190.

Grant A, Ponia SS, Tripathi S et al. (2016). Zika virus targets human STAT2 to inhibit type I interferon signaling. Cell Host Microbe 19: 882–890. https://doi.org/10.1016/j.chom.2016.05.009.

Gratuze M, Leyns CEG, Holtzman DM (2018). New insights into the role of TREM2 in Alzheimer's disease. Mol Neurodegener 13: 66. https://doi.org/10.1186/s13024-018-0298-9.

Griffin DE, Wesselingh SL, McArthur JC (1994). Elevated central nervous system prostaglandins in human immunodeficiency virus-associated dementia. Ann Neurol 35: 592–597. https://doi.org/10.1002/ana.410350513.

Hammer SM, Squires KE, Hughes MD et al. (1997). A controlled trial of two nucleoside analogues plus indinavir in persons with human immunodeficiency virus infection and CD4 cell counts of 200 per cubic millimeter or less. AIDS Clinical Trials Group 320 study team. N Engl J Med 337: 725–733. https://doi.org/10.1056/NEJM199709113371101.

Hare D, Ayton S, Bush A et al. (2013). A delicate balance: iron metabolism and diseases of the brain. Front Aging Neurosci 5: 34. https://doi.org/10.3389/fnagi.2013.00034.

Hasbun R (2022). Progress and challenges in bacterial meningitis: a review. JAMA 328: 2147–2154. https://doi.org/10.1001/jama.2022.20521.

Hatton CF, Duncan CJA (2019). Microglia are essential to protective antiviral immunity: lessons from mouse models of viral encephalitis. Front Immunol 10: 2656. https://doi.org/10.3389/fimmu.2019.02656.

Hayashida E, Ling ZL, Ashhurst TM et al. (2019). Zika virus encephalitis in immunocompetent mice is dominated by innate immune cells and does not require T or B cells. J Neuroinflammation 16: 177. https://doi.org/10.1186/s12974-019-1566-5.

Hofer U (2013). Fungal pathogenes is: good copper, bad copper. Nat Rev Microbiol 11: 299. https://doi.org/10.1038/nrmicro3013.

Hole C, Wormley FL (2016). Innate host defenses against Cryptococcus neoformans. J Microbiol 54: 202–211. https://doi.org/10.1007/s12275-016-5625-7.

Hong S, Beja-Glasser VF, Nfonoyim BM et al. (2016). Complement and microglia mediate early synapse loss in Alzheimer mouse models. Science (New York, NY) vol. 352, 712–716. https://doi.org/10.1126/science.aad8373.

Hoogland ICM, Yik J, Westhoff D et al. (2022). Microglial cell response in α7 nicotinic acetylcholine receptor-deficient mice after systemic infection with Escherichia coli. J Neuroinflammation 19: 94. https://doi.org/10.1186/s12974-022-02452-8.

Hosseini S, Korte M (2023). How viral infections cause neuronal dysfunction: a focus on the role of microglia and astrocytes. Biochem Soc Trans 51: 259–274. https://doi.org/10.1042/BST20220771.

Hosseini S, Wilk E, Michaelsen-Preusse K et al. (2018). Long-term neuroinflammation induced by influenza a virus infection and the impact on hippocampal neuron morphology and function. J Neurosci 38: 3060–3080. https://doi.org/10.1523/JNEUROSCI.1740-17.2018.

Hosseini S, Michaelsen-Preusse K, Schughart K et al. (2021). Long-term consequen ce of non-neurotropic H3N2 influenza a virus infection for the progression of Alzheimer's disease symptoms. Front Cell Neurosci 15:643650. https://doi.org/10.3389/fncel.2021.643650.

Hou F, Bi F, Jiao R et al. (2020). Gender differences of depression and anxiety among social media users during the COVID-19 outbreak in China: a cross-sectional study. BMC Public Health 20: 1648. https://doi.org/10.1186/s12889-020-09738-7.

Hu B, Guo H, Zhou P et al. (2021). Characteristics of SARS-CoV-2 and COVID-19. Nat Rev Microbiol 19: 141–154. https://doi.org/10.1038/s41579-020-00459-7.

Huang SH, Jong AY (2001). Cellular mechanisms of microbial proteins contributing to invasion of the blood-brain barrier. Cell Microbiol 3: 277–287. https://doi.org/10.1046/j.1462-5822.2001.00116.x.

Huang Z, Nair M (2017). A CRISPR/Cas9 guidance RNA screen platform for HIV provirus disruption and HIV/AIDS gene therapy in astrocytes. Sci Rep 7: 5955. https://doi.org/10.1038/s41598-017-06269-x.

Huang X, Guo M, Zhang Y et al. (2023). Microglial IL-1RA ameliorates brain injury after ischemic stroke by inhibiting astrocytic CXCL1-mediated neutrophil recruitment and microvessel occlusion. Glia 71: 1607–1625. https://doi.org/10.1002/glia.24359.

Jacob J, Steel A, Lin Z et al. (2021). Clinical efficacy and safety of Albendazole and other Benzimidazole Anthelmintics for rat lungworm disease (neuroangiostrongyliasis): a systematic analysis of clinical reports and animal studies. Clin Infect Dis 74: 1293–1302. https://doi.org/10.1093/cid/ciab730.

Janova H, Böttcher C, Holtman IR et al. (2016). CD14 is a key organizer of microglial responses to CNS infection and injury. Glia 64: 635–649. https://doi.org/10.1002/glia.22955.

Jensen BK, Monnerie H, Mannell MV et al. (2015). Altered oligodendrocyte maturation and myelin maintenance: the role of anti-retrovirals in HIV-associated neurocognitive disorders. J Neuropathol Exp Neurol 74: 1093–1118. https://doi.org/10.1097/NEN.0000000000000255.

Jensen BK, Roth LM, Grinspan JB et al. (2019). White matter loss and oligodendrocyte dysfunction in HIV: a consequence of the infection, the antiretroviral therapy or both? Brain Res 1724: 146397. https://doi.org/10.1016/j.brainres.2019.146397.

Johnston SA, May RC (2013). Cryptococcus interactions with macrophages: evasion and manipulation of the phagosome by a fungal pathogen. Cell Microbiol 15: 403–411. https://doi.org/10.1111/cmi.12067.

Jong A, Wu C-H, Chen H-M et al. (2007). Identification and characterization of CPS1 as a hyaluronic acid synthase contributing to the pathogenesis of Cryptococcus neoformans infection. Eukaryot Cell 6: 1486–1496. https://doi.org/10.1128/ec.00120-07.

Jong A, Wu C-H, Gonzales-Gomez I et al. (2012). Hyaluronic acid receptor CD44 deficiency is associated with decreased Cryptococcus neoformans brain infection. J Biol Chem 287: 15298–15306. https://doi.org/10.1074/jbc.M112.353375.

Ju J, Su Y, Zhou Y et al. (2022). The SARS-CoV-2 enve lope protein disrupts barrier function in an in vitro human blood-brain barrier model. Front Cell Neurosci 16: 897564. https://doi.org/10.3389/fncel.2022.897564.

Jurga AM, Paleczna M, Kuter KZ (2020). Overview of general and discriminating markers of differential microglia phenotypes. Front Cell Neurosci 14: 198. https://doi.org/10.3389/fncel.2020.00198.

Jurgens HA, Amancherla K, Johnson RW (2012). Influenza infection induces neuroinflammation, alters hippocampal neuron morphology, and impairs cognition in adult mice. J Neurosci 32: 3958–3968. https://doi.org/10.1523/JNEUROSCI.6389-11.2012.

Kannan M, Sil S, Oladapo A et al. (2023). HIV-1 Tat-mediated microglial ferroptosis involves the miR-204–ACSL4 signaling axis. Redox Biol 62: 102689. https://doi.org/10.1016/j.redox.2023.102689.

Kawakita S, Mandal K, Mou L et al. (2022). Organ-on-a-chip models of the blood–brain barrier: recent advances and future prospects. Small 18: 2201401. https://doi.org/10.1002/smll.202201401.

Kawasaki T, Kawai T (2014). Toll-like rec eptor signaling pathways. Front Immunol 5: 461. https://doi.org/10.3389/fimmu.2014.00461.

Khan F, Sharma P, Pandey S et al. (2021). COVID-19-associated Guillain-Barre syndrome: postinfectious alone or neuroinvasive too? J Med Virol 93: 6045–6049. https://doi.org/10.1002/jmv.27159.

Khaw YM, Aggarwal N, Barclay WE et al. (2020). Th1-dependent Cryptococcus-associated immune reconstitution inflammatory syndrome model with brain damage. Front Immunol 11: 529219. https://doi.org/10.3389/fimmu.2020.529219.

Kim KS (2016). Human meningitis-associated Escherichia coli. EcoSal Plus 7. https://doi.org/10.1128/ecosalplus.ESP-0015–2015.

Kim KJ, Elliott SJ, Di Cello F et al. (2003). The K1 capsule modulates trafficking of E. coli-containing vacuoles and enhances intracellular bacterial survival in human brain microvascular endothelial cells. Cell Microbiol 5: 245–252. https://doi.org/10.1046/j.1462-5822.2003.t01-1-00271.x.

Kitisin T, Muangkaew W, Sukphopetch P (2022). Infections of Cryptococcus species induce degeneration of dopaminergic neurons and accumulation of α-Synuclein in Caenorhabditis elegans. Front Cell Infect Microbiol 12: 1039336. https://doi.org/10.3389/fcimb.2022.1039336.

Knyazev E, Nersisyan S, Tonevitsky A (2021). Endocytosis and transcytosis of SARS-CoV-2 across the intestinal epithelium and other tissue barriers. Front Immunol 12: 636966. https://doi.org/10.3389/fimmu.2021.636966.

Kobayashi SD, Malachowa N, DeLeo FR (2018). Neutrophils and bacterial immune evasion. J Innate Immun 10: 432–441. https://doi.org/10.1159/000487756.

Koyuncu OO, Hogue IB, Enquist LW (2013). Virus infections in the nervous system. Cell Host Microbe 13: 379–393. https://doi.org/10.1016/j.chom.2013.03.010.

Kumar A, Hou S, Airo AM et al. (2016). Zika virus inhibits type-I interferon production and downstream signaling. EMBO Rep 17: 1766–1775. https://doi.org/10.15252/embr.201642627.

Lee SC, Kress Y, Zhao ML et al. (1995). Cryptococcus neoformans survive and replicate in human microglia. Lab Invest 73: 871–879.

Lee M-H, Perl DP, Nair G et al. (2021). Microvascular injury in the brains of patients with Covid-19. N Engl J Med 384: 481–483. https://doi.org/10.1056/NEJMc2033369.

Lei F, Cui N, Zhou C et al. (2020). CSF1R inhibition by a small-molecule inhibitor is not microglia specific; affecting hematopoiesis and the function of macrophages. Proc Natl Acad Sci U S A 117: 23336–23338. https://doi.org/10.1073/pnas.1922788117.

Li G-H, Anderson C, Jaeger L et al. (2015). Cell-to-cell contact facilitates HIV transmission from lymphocytes to astrocytes via CXCR4. AIDS (London, England) 29: 755–766. https://doi.org/10.1097/QAD.0000000000000605.

Li G-H, Henderson L, Nath A (2016). Astrocytes as an HIV reservoir: mechanism of HIV infection. Curr HIV Res 14: 373–381. https://doi.org/10.2174/1570162x14666161006121455.

Lin KY, Chen KM, Lan KP et al. (2010). Alterations of myelin proteins in inflammatory demyelination of BALB/c mice caused by Angiostrongylus cantonensis. Vet Parasitol 171: 74–80. https://doi.org/10.1016/j.vetpar.2010.03.019.

Lisi L, Pizzoferrato M, Miscioscia FT et al. (2017). Interactions between integrase inhibitors and human arginase 1. J Neurochem 142: 153–159. https://doi.org/10.1111/jnc.14039.

Liu P-J, Chen Y-S, Lin H-H et al. (2013). Induction of mouse melioidosis with meningitis by CD11b+ phagocytic cells harboring intracellular B. pseudomallei as a Trojan horse. PLoS Negl Trop Dis 7: e2363. https://doi.org/10.1371/journal.pntd.0002363.

Liu P, Wang X, Yang Q et al. (2022). Collaborative action of microglia and astrocytes mediates neutrophil recruitment to the CNS to defend against *Escherichia coli* K1 infection. Int J Mol Sci 23: 6540. https://doi.org/10.3390/ijms23126540.

Lotz SK, Blackhurst BM, Reagin KL et al. (2021). Microbial infections are a risk factor for neurodegenerative diseases. Front Cell Neurosci 15: 691136. https://doi.org/10.3389/fncel.2021.691136.

Lukiw WJ, Pogue A, Hill JM (2022). SARS-CoV-2 infectivity and neurological targets in the brain. Cell Mol Neurobiol 42: 217–224. https://doi.org/10.1007/s10571-020-00947-7.

Lum F-M, Low DKS, Fan Y et al. (2017). Zika virus infects human fetal brain microglia and induces inflammation. Clin Infect Dis 64: 914–920. https://doi.org/10.1093/cid/ciw878.

Lun J, Li Y, Gao X et al. (2023). Kynurenic acid blunts A1 astrocyte activation against neurodegeneration in HIV-associated neurocognitive disorders. J Neuroinflammation 20: 87. https://doi.org/10.1186/s12974-023-02771-4.

Mamik MK, Hui E, Branton WG et al. (2017). HIV-1 viral protein R activates NLRP3 inflammasome in microglia: implications for HIV-1 associated neuroinflammation. J Neuroimmune Pharmacol 12: 233–248. https://doi.org/10.1007/s11481-016-9708-3.

Manet C, Mansuroglu Z, Conquet L et al. (2022). Zika virus infection of mature neurons from immunocompetent mice generates a disease-associated microglia and a tauopathy-like phenotype in link with a delayed interferon beta response. J Neuroinflammation 19: 307. https://doi.org/10.1186/s12974-022-02668-8.

Marie C, Petri WA (2022). Angiostrongyliasis—infectious diseases. *MSD Manual Professional Edition*, available at: https://www.msdmanuals.com/professional/infectious-diseases/nematodes-roundworms/angiostrongyliasis (accessed 22 March 2023).

Marques CP, Cheeran MC-J, Palmquist JM et al. (2008). Prolonged microglial cell activation and lymphocyte infiltration following experimental herpes Encephalitis1. J Immunol 181: 6417–6426. https://doi.org/10.4049/jimmunol.181.9.6417.

Marrodan M, Alessandro L, Farez MF et al. (2019). The role of infections in multiple sclerosis. Mult Scler Int 25: 891–901. https://doi.org/10.1177/1352458518823940.

Maruvada R, Zhu L, Pearce D et al. (2012). Cryptococcus neoformans phospholipase B1 activates host cell Rac1 for traversal across the blood–brain barrier. Cell Microbiol 14: 1544–1553. https://doi.org/10.1111/j.1462-5822.2012.01819.x.

Masuda T, Amann L, Monaco G et al. (2022). Specification of CNS macrophage subsets occurs postnatally in defined niches. Nature 604: 740–748. https://doi.org/10.1038/s41586-022-04596-2.

Matsumoto T, Imagama S, Hirano K et al. (2012). CD44 expression in astrocytes and microglia is associated with ALS progression in a mouse model. Neurosci Lett 520: 115–120. https://doi.org/10.1016/j.neulet.2012.05.048.

Minhas PS, Latif-Hernandez A, McReynolds MR et al. (2021). Restoring metabolism of myeloid cells reverses cognitive decline in ageing. Nature 590: 122–128. https://doi.org/10.1038/s41586-020-03160-0.

Mohamed SH, Nyazika TK, Ssebambulidde K et al. (2022). Fungal CNS infections in Africa: the neuroimmunology of Cryptococcal meningitis. Front Immunol 13: 804674. https://doi.org/10.3389/fimmu.2022.804674.

Mohamed SH, Fu MS, Hain S et al. (2023). Microglia are not protective against cryptococcal meningitis. Nat Commun 14: 7202. https://doi.org/10.1038/s41467-023-43061-0.

Monje M, Iwasaki A (2022). The neurobiology of long COVID. Neuron. https://doi.org/10.1016/j.neuron.2022.10.006.

Moxon ER, Kroll JS (1990). The role of bacterial polysaccharide capsules as virulence factors. Curr Top Microbiol Immunol 150: 65–85. https://doi.org/10.1007/978-3-642-74694-9_4.

Murta V, Villarreal A, Ramos AJ (2020). Severe acute respiratory syndrome coronavirus 2 impact on the central nervous system: are astrocytes and microglia main players or merely bystanders? ASN Neuro 12: . https://doi.org/10.1177/1759091420954960.

Murthy JMK, Sundaram C (2014). Fungal infections of the central nervous system. Handb Clin Neurol 121: 1383–1401. https://doi.org/10.1016/B978-0-7020-4088-7.00095-X.

Nakajima H, Takayama A, Fujiki Y et al. (2015). Refractory Cryptococcus neoformans meningoencephalitis in an immunocompetent patient: paradoxical antifungal therapy-induced clinical deterioration related to an immune response to Cryptococcal organisms. Case Rep Neurol 7: 204–208. https://doi.org/10.1159/000440948.

Nakamura K, Kinjo T, Saijo S et al. (2007). Dectin-1 is not required for the host defense to Cryptococcus neoformans. Microbiol Immunol 51: 1115–1119. https://doi.org/10.1111/j.1348-0421.2007.tb04007.x.

Nanda A, Vura NVRK, Gravenstein S (2020). COVID-19 in older adults. Aging Clin Exp Res 32: 1199–1202. https://doi.org/10.1007/s40520-020-01581-5.

Nath A, Johnson TP (2022). Mechanisms of viral persistence in the brain and therapeutic approaches. FEBS J 289: 2145–2161. https://doi.org/10.1111/febs.15871.

Neal LM, Xing E, Xu J et al. (2017). CD4+ T cells orchestrate lethal immune pathology despite fungal clearance during Cryptococcus neoformans meningoencephalitis. MBio 8: e01415–e01417. https://doi.org/10.1128/mBio.01415-17.

Nedelcovych MT, Tenora L, Kim B-H et al. (2017). N-(Pivaloyloxy)alkoxy-carbonyl prodrugs of the glutamine antagonist 6-Diazo-5-oxo-l-norleucine (DON) as a potential treatment for HIV associated neurocognitive disorders. J Med Chem 60: 7186–7198. https://doi.org/10.1021/acs.jmedchem.7b00966.

Nedelcovych MT, Kim B-H, Zhu X et al. (2019). Glutamine antagonist JHU083 normalizes aberrant glutamate production and cognitive deficits in the EcoHIV murine model of HIV-associated neurocognitive disorders. J Neuroimmune Pharmacol 14: 391–400. https://doi.org/10.1007/s11481-019-09859-w.

Nersesjan V, Fonsmark L, Christensen RHB et al. (2022). Neuropsychiatric and cognitive outcomes in patients 6 months after COVID-19 requiring hospitalization compared with matched control patients hospitalized for non-COVID-19 illness. JAMA Psychiatry 79: 486–497. https://doi.org/10.1001/jamapsychiatry.2022.0284.

Nimmerjahn A, Kirchhoff F, Helmchen F (2005). Resting microglial cells are highly dynamic surveillants of brain parenchyma in vivo. Science (New York, NY) 308: 1314–1318. https://doi.org/10.1126/science.1110647.

Okorji UP, Olajide OA (2014). A semi-synthetic derivative of artemisinin, artesunate inhibits prostaglandin E2 production in LPS/IFNγ-activated BV2 microglia. Bioorg Med Chem 22: 4726–4734. https://doi.org/10.1016/j.bmc.2014.07.007.

Olszewski MA, Noverr MC, Chen G-H et al. (2004). Urease expression by Cryptococcus neoformans promotes microvascular sequestration, thereby enhancing central nervous system invasion. Am J Pathol 164: 1761–1771. https://doi.org/10.1016/S0002-9440(10)63734-0.

Palella FJ, Delaney KM, Moorman AC et al. (1998). Declining morbidity and mortality among patients with advanced human immunodeficiency virus infection. HIV outpatient study investigators. N Engl J Med 338: 853–860. https://doi.org/10.1056/NEJM199803263381301.

Panackal AA, Wuest SC, Lin Y-C et al. (2015). Paradoxical immune responses in non-HIV Cryptococcal meningitis. PLoS Pathog 11: e1004884. https://doi.org/10.1371/journal.ppat.1004884.

Panackal AA, Komori M, Kosa P et al. (2017). Spinal Arachnoiditis as a complication of Cryptococcal meningoencephalitis in non-HIV previously healthy adults. Clin Infect Dis 64: 275–283. https://doi.org/10.1093/cid/ciw739.

Parkhurst CN, Yang G, Ninan I et al. (2013). Microglia promote learning-dependent synapse formation through brain-derived neurotrophic factor. Cell 155: 1596–1609. https://doi.org/10.1016/j.cell.2013.11.030.

Paul BD, Lemle MD, Komaroff AL et al. (2021). Redox imbalance links COVID-19 and myalgic encephalomyelitis/chronic fatigue syndrome. Proc Natl Acad Sci U S A 118: e2024358118. https://doi.org/10.1073/pnas.2024358118.

Petersen M, Nägele FL, Mayer C et al. (2022). Brain imaging and neuropsychological assessment of individuals recovered from a mild to moderate SARS-CoV-2 infection. medRxiv. https://doi.org/10.1101/2022.07.08.22277420.

Pilotto A, Padovani A (2020). Reply to the letter 'COVID-19-associated encephalopathy and cytokine-mediated neuroinflammation'. Ann Neurol 88: 861–862. https://doi.org/10.1002/ana.25856.

Pires BRB, Silva RCMC, Ferreira GM et al. (2018). NF-kappaB: two sides of the same coin. Genes 9: 24. https://doi.org/10.3390/genes9010024.

Piret J, Carbonneau J, Rhéaume C et al. (2018). Predominant role of IPS-1 over TRIF adaptor proteins in early innate immune response against Zika virus in mice. J Gen Virol 99: 209–218. https://doi.org/10.1099/jgv.0.000992.

Poloni TE, Medici V, Moretti M et al. (2021). COVID-19-related neuropathology and microglial activation in elderly with and without dementia. Brain Pathol 31: e12997. https://doi.org/10.1111/bpa.12997.

Potter MC, Figuera-Losada M, Rojas C et al. (2013). Targeting the glutamatergic system for the treatment of HIV-associated neurocognitive disorders. J Neuroimmune Pharmacol 8: 594–607. https://doi.org/10.1007/s11481-013-9442-z.

Qiu Y, Davis MJ, Dayrit JK et al. (2012). Immune modulation mediated by cryptococcal laccase promotes pulmonary growth and brain dissemination of virulent *Cryptococcus neoformans* in mice. PLoS One 7: e47853. https://doi.org/10.1371/journal.pone.0047853Edited by Valery, C.

Quincozes-Santos A, Santos CL, de Souza Almeida RR et al. (2021). Gliotoxicity and glioprotection: the dual role of glial cells. Mol Neurobiol 58: 6577–6592. https://doi.org/10.1007/s12035-021-02574-9.

Raber J, Rhea EM, Banks WA (2023). The effects of viruses on insulin sensitivity and blood–brain barrier function. Int J Mol Sci 24: 2377. https://doi.org/10.3390/ijms24032377.

Ranki A, Nyberg M, Ovod V et al. (1995). Abundant expression of HIV Nef and Rev proteins in brain astrocytes in vivo is associated with dementia. AIDS (London, England) 9: 1001–1008. https://doi.org/10.1097/00002030-199509000-00004.

Raymond AD, Diaz P, Chevelon S et al. (2016). Microglia-derived HIV Nef+ exosome impairment of the blood–brain barrier is treatable by nanomedicine-based delivery of Nef peptides. J Neurovirol 22: 129–139. https://doi.org/10.1007/s13365-015-0397-0.

Reddy GKK, Padmavathi AR, Nancharaiah YV (2022). Fungal infections: pathogenesis, antifungals and alternate treatment approaches. Curr Res Microb Sci 3: 100137. https://doi.org/10.1016/j.crmicr.2022.100137.

Renz-Polster H, Tremblay M-E, Bienzle D et al. (2022). The pathobiology of myalgic encephalomyelitis/chronic fatigue syndrome: the case for neuroglial failure. Front Cell Neurosci 16: 888232. https://doi.org/10.3389/fncel.2022.888232.

Rhea EM, Logsdon AF, Hansen KM et al. (2021). The S1 protein of SARS-CoV-2 crosses the blood–brain barrier in mice. Nat Neurosci 24: 368–378. https://doi.org/10.1038/s41593-020-00771-8.

Ribes S, Ebert S, Czesnik D et al. (2009). Toll-like receptor prestimulation increases phagocytosis of Escherichia coli DH5alpha and Escherichia coli K1 strains by murine microglial cells. Infect Immun 77: 557–564. https://doi.org/10.1128/IAI.00903-08.

Ribes S, Adam N, Ebert S et al. (2010). The viral TLR3 agonist poly(I:C) stimulates phagocytosis and intracellular killing of Escherichia coli by microglial cells. Neurosci Lett 482: 17–20. https://doi.org/10.1016/j.neulet.2010.06.078.

Ribes S, Regen T, Meister T et al. (2013). Resistance of the brain to Escherichia coli K1 infection depends on MyD88 signaling and the contribution of neutrophils and monocytes. Infect Immun 81: 1810–1819. https://doi.org/10.1128/IAI.01349-12.

Ribes S, Arcilla C, Ott M et al. (2020). Pre-treatment with the viral toll-like receptor 3 agonist poly(I:C) modulates innate immunity and protects neutropenic mice infected intracerebrally with Escherichia coli. J Neuroinflammation 17: 24. https://doi.org/10.1186/s12974-020-1700-4.

Rossi SL, Tesh RB, Azar SR et al. (2016). Characterization of a novel murine model to study Zika virus. Am J Trop Med Hyg 94: 1362–1369. https://doi.org/10.4269/ajtmh.16-0111.

Samudyata N, Oliveira AO, Malwade S et al. (2022). SARS-CoV-2 promotes microglial synapse elimination in human brain organoids. Mol Psychiatry. https://doi.org/10.1038/s41380-022-01786-2.

Sánchez-Garibay C, Salinas-Lara C, Gómez-López MA et al. (2022). Mycobacterium tuberculosis infection induces BCSFB disruption but no BBB disruption in vivo: implications in the pathophysiology of tuberculous meningitis. Int J Mol Sci 23: 6436. https://doi.org/10.3390/ijms23126436.

Sawada R, Nakano-Doi A, Matsuyama T et al. (2020). CD44 expression in stem cells and niche microglia/macrophages following ischemic stroke. Stem Cell Investig 7: 4. https://doi.org/10.21037/sci.2020.02.02.

Saylor D, Dickens AM, Sacktor N et al. (2016). HIV-associated neurocognitive disorder—pathogenesis and prospects for treatment. Nat Rev Neurol 12: 234–248. https://doi.org/10.1038/nrneurol.2016.27.

Schachtele SJ, Hu S, Little MR et al. (2010). Herpes simplex virus induces neural oxidative damage via microglial cell Toll-like receptor-2. J Neuroinflammation 7: 35. https://doi.org/10.1186/1742-2094-7-35.

Schafer DP, Lehrman EK, Kautzman AG et al. (2012). Microglia sculpt postnatal neural circuits in an activity and complement-dependent manner. Neuron 74: 691–705. https://doi.org/10.1016/j.neuron.2012.03.026.

Schnittman SM, Lane HC, Greenhouse J et al. (1990). Preferential infection of CD4+ memory T cells by human immunodeficiency virus type 1: evidence for a role in the selective T-cell functional defects observed in infected individuals. Proc Natl Acad Sci U S A 87: 6058–6062.

Schwartz S, Kontoyiannis DP, Harrison T et al. (2018). Advances in the diagnosis and treatment of fungal infections of the CNS. Lancet Neurol 17: 362–372. https://doi.org/10.1016/S1474-4422(18)30030-9.

Schweighardt B, Atwood WJ (2001). Virus receptors in the human central nervous system. J Neurovirol 7: 187–195. https://doi.org/10.1080/13550280152403236.

Shi ZW, Chen Y, Ogoke KM et al. (2022). Cryptococcal immune reconstitution inflammatory syndrome: from clinical studies to animal experiments. Microorganisms 10: 2419. https://doi.org/10.3390/microorganisms10122419.

Singh A, Kutscher HL, Bulmahn JC et al. (2020). Laser ablation for pharmaceutical nanoformulations: multi-drug nanoencapsulation and theranostics for HIV. Nanomed Nanotechnol Biol Med 25: 102172. https://doi.org/10.1016/j.nano.2020.102172.

Singh H, Koury J, Kaul M (2021). Innate immune sensing of viruses and its consequences for the central nervous system. Viruses 13: 170. https://doi.org/10.3390/v13020170.

Song Z, Feng J, Zhang Q et al. (2021). Tanshinone IIA protects against cerebral ischemia reperfusion injury by regulating microglial activation and polarization via NF-κB pathway. Front Pharmacol 12: 641848. https://doi.org/10.3389/fphar.2021.641848.

Steenbergen SM, Lee Y-C, Vann WF et al. (2006). Separate pathways for O acetylation of polymeric and monomeric sialic acids and identification of sialyl O-acetyl esterase in Escherichia coli K1. J Bacteriol 188: 6195–6206. https://doi.org/10.1128/JB.00466-06.

Stefano GB (2021). Historical insight into infections and disorders associated with neurological and psychiatric sequelae similar to long COVID. Med Sci Monit 27: e931447-1–e931447-4. https://doi.org/10.12659/MSM.931447.

Suh H-S, Zhao M-L, Choi N et al. (2009). TLR3 and TLR4 are innate antiviral immune receptors in human microglia: role of IRF3 in modulating antiviral and inflammatory response in the CNS. Virology 392: 246–259. https://doi.org/10.1016/j.virol.2009.07.001.

Surnar B, Shah AS, Park M et al. (2021). Brain-accumulating nanoparticles for assisting astrocytes to reduce human immunodeficiency virus and drug abuse-induced neuroinflammation and oxidative stress. ACS Nano 15: 15741–15753. https://doi.org/10.1021/acsnano.0c09553.

Swank Z, Senussi Y, Manickas-Hill Z, Yu XG, Li JZ, Alter G, Walt DR (2023). Persistent circulating severe acute respiratory syndrome coronavirus 2 spike is associated with post-acute coronavirus disease 2019 sequelae. Clin Infect Dis 76: e487–e490. https://doi.org/10.1093/cid/ciac722.

Swanson PA, McGavern DB (2015). Viral diseases of the central nervous system. Curr Opin Virol 11: 44–54. https://doi.org/10.1016/j.coviro.2014.12.009.

Takahashi K, Wesselingh SL, Griffin DE et al. (1996). Localization of HIV-1 in human brain using polymerase chain reaction/in situ hybridization and immunocytochemistry. Ann Neurol 39: 705–711. https://doi.org/10.1002/ana.410390606.

Takoutsing BD, Ooi SZY, Egu CB et al. (2023). Management and outcomes of intracranial fungal infections in children and adults in Africa: a scoping review protocol. BMJ Open 13: e065943. https://doi.org/10.1136/bmjopen-2022-065943.

Tanaka N, Cortese GP, Barrientos RM et al. (2018). Aging and an immune challenge interact to produce prolonged, but not permanent, reductions in hippocampal L-LTP and mBDNF in a rodent model with features of delirium. eNeuro 5. https://doi.org/10.1523/ENEURO.0009-18.2018.

Taslem Mourosi J, Anwar S, Hosen MJ (2022). The sex and gender dimensions of COVID-19: a narrative review of the potential underlying factors. Infect Genet Evol 103: 105338. https://doi.org/10.1016/j.meegid.2022.105338.

Tavčar P, Potokar M, Kolenc M et al. (2021). Neurotropic viruses, astrocytes, and COVID-19. Front Cell Neurosci 15: 662578. https://doi.org/10.3389/fncel.2021.662578.

Thygesen C, Ilkjær L, Kempf SJ et al. (2018). Diverse protein profiles in CNS myeloid cells and CNS tissue from lipopolysaccharide- and vehicle-injected APPSWE/PS1ΔE9 transgenic mice implicate Cathepsin Z in Alzheimer's disease. Front Cell Neurosci 12: 397. https://doi.org/10.3389/fncel.2018.00397.

Town T, Jeng D, Alexopoulou L et al. (2006). Microglia recognize double-stranded RNA via TLR3. J Immunol 176: 3804–3812. https://doi.org/10.4049/jimmunol.176.6.3804.

Traylen CM, Patel HR, Fondaw W et al. (2011). Virus reactivation: a panoramic view in human infections. Future Virol 6: 451–463. https://doi.org/10.2217/fvl.11.21.

Tremblay M-E, Madore C, Bordeleau M et al. (2020). Neuropathobiology o f COVID-19the role for glia. Front Cell Neurosci 14: 592214. https://doi.org/10.3389/fncel.2020.592214.

Tseng Y, Wang S-W, Kim KS et al. (2012). NlpI facilitates deposition of C4bp on Escherichia coli by blocking classical complement-mediated killing, which results in high-level bacteremia. Infect Immun 80: 3669–3678. https://doi.org/10.1128/IAI.00320-12.

Utz SG, See P, Mildenberger W et al. (2020). Early fate defines microglia and non-parenchymal brain macrophage development. Cell 181: 557–573.e18. https://doi.org/10.1016/j.cell.2020.03.021.

Uyar O, Laflamme N, Piret J et al. (2020). An early microglial response is needed to efficiently control herpes simplex virus encephalitis. J Virol 94. https://doi.org/10.1128/jvi.01428-20. American Society for Microbiology.

Uyar O, Dominguez JM, Bordeleau M et al. (2022). Single-cell transcriptomics of the ventral posterolateral nucleus-enriched thalamic regions from HSV-1-infected mice reveal a novel microglia/microglia-like transcriptional response. J Neuroinflammation 19: 81. https://doi.org/10.1186/s12974-022-02437-7.

Valenza M, Steardo L, Steardo L et al. (2021). Systemic inflammation and astrocyte reactivity in the neuropsychiatric sequelae of COVID-19: focus on Autism Spectrum disorders. Front Cell Neurosci 15: 748136. https://doi.org/10.3389/fncel.2021.748136.

van Leeuwen LM, van der Kuip M, Youssef SA et al. (2014). Modeling tuberculous meningitis in zebrafish using Mycobacterium marinum. Dis Model Mech 7: 1111–1122. https://doi.org/10.1242/dmm.015453.

Vasek MJ, Garber C, Dorsey D et al. (2016). A complement–microglial axis drives synapse loss during virus-induced memory impairment. Nature 534: 538–543. https://doi.org/10.1038/nature18283.

Vázquez-Santiago FJ, Noel RJ, Porter JT et al. (2014). Glutamate metabolism and HIV-associated neurocognitive disorders. J Neurovirol 20: 315–331. https://doi.org/10.1007/s13365-014-0258-2.

Villeda SA, Luo J, Mosher KI et al. (2011). The ageing systemic milieu negatively regulates neurogenesis and cognitive function. Nature 477: 90–94. https://doi.org/10.1038/nature10357.

Vu K, Tham R, Uhrig JP et al. (2014). Invasion of the central nervous system by Cryptococcus neoformans requires a secreted fungal metalloprotease. MBio 5: e01101–e01114. https://doi.org/10.1128/mBio.01101-14.

Walsh JG, Reinke SN, Mamik MK et al. (2014). Rapid inflammasome activation in microglia contributes to brain disease in HIV/AIDS. Retrovirology 11: 35. https://doi.org/10.1186/1742-4690-11-35.

Walsh NM, Wuthrich M, Wang H et al. (2017). Characterization of C-type lectins reveals an unexpectedly limited interaction between Cryptococcus neoformans spores and Dectin-1. PLoS One 12: e0173866. https://doi.org/10.1371/journal.pone.0173866.

Wang T, Town T, Alexopoulou L et al. (2004). Toll-like receptor 3 mediates West Nile virus entry into the brain causing lethal encephalitis. Nat Med 10: 1366–1373. https://doi.org/10.1038/nm1140.

Wang Y, Liu M, Lu Q et al. (2020). Global prevalence and burden of HIV-associated neurocognitive disorder: a meta-analysis. Neurology 95: e2610–e2621. https://doi.org/10.1212/WNL.0000000000010752.

Watkins BA, Dorn HH, Kelly WB et al. (1990). Specific tropism of HIV-1 for microglial cells in primary human brain cultures. Science (New York, NY) 249: 549–553. https://doi.org/10.1126/science.2200125.

WHO (2019). Global tuberculosis report 2019. Available at: https://www.who.int/publications-detail-redirect/9789241565714 (accessed 23 April 2023).

WHO (2022). WHO Director-General's opening remarks at the media briefing on COVID-19 - 11 March 2020. Available at: https://www.who.int/director-general/speeches/detail/who-director-general-s-opening-remarks-at-the-media-briefing-on-covid-19---11-march-2020 (accessed 1 December 2022).

Williams AE, Blakemore WF (1990). Pathogenesis of meningitis caused by Streptococcus suis type 2. J Infect Dis162: 474–481. https://doi.org/10.1093/infdis/162.2.474.

Woehrl B, Klein M, Grandgirard D et al. (2011). Bacterial meningitis: current therapy and possible future treatment options. Expert Rev Anti Infect Ther 9: 1053–1065. https://doi.org/10.1586/eri.11.129.

Wu Y, Du S, Bimler LH et al. (2023). Toll-like receptor 4 and CD11b expressed on microglia coordinate eradication of Candida albicans cerebral mycosis. Cell Rep 42. https://doi.org/10.1016/j.celrep.2023.113240.

Xu J, Ganguly A, Zhao J et al. (2021). CCR2 signaling promotes brain infiltration of inflammatory monocytes and contributes to neuropathology during Cryptococcal meningoencephalitis. MBio 12: e0107621. https://doi.org/10.1128/mBio.01076-21.

Xu J, Goughenour K, Underwood WR et al. (2023). Immunological analysis of Cryptococcal meningoencephalitis in a murine model. Methods Mol Biol 2667: 71–86. https://doi.org/10.1007/978-1-0716-3199-7_5.

Yagami T, Koma H, Yamamoto Y (2016). Pathophysiological roles of cyclooxygenases and prostaglandins in the central nervous system. Mol Neurobiol 53: 4754–4771. https://doi.org/10.1007/s12035-015-9355-3.

Yan J, Yang X, Han D et al. (2016). Tanshinone IIA attenuates experimental autoimmune encephalomyelitis in rats. Mol Med Rep 14: 1601–1609. https://doi.org/10.3892/mmr.2016.5431.

Yan L, Jayaram M, Chithanathan K et al. (2021). Sex-specific microglial activation and SARS-CoV-2 receptor expression induced by chronic unpredictable stress. Front Cell Neurosci 15.

Yang R, Wang J, Wang F et al. (2023). Blood-brain barrier integrity damage in bacterial meningitis: the underlying link, mechanisms, and therapeutic targets. Int J Mol Sci 24: 2852. https://doi.org/10.3390/ijms24032852.

Ye F, Alvarez-Carbonell D, Nguyen K et al. (2022). Recruitment of the CoREST transcription repressor complexes by nerve growth factor IB-like receptor (Nurr1/NR4A2) mediates silencing of HIV in microglial cells. PLoS Pathog 18: e1010110. https://doi.org/10.1371/journal.ppat.1010110.

Yilmaz A, Price RW, Gisslén M (2012). Antiretroviral drug treatment of CNS HIV-1 infection. J Antimicrob Chemother 67: 299–311. https://doi.org/10.1093/jac/dkr492.

Yuan Z, Fan X, Staitieh B et al. (2017). HIV-related proteins prolong macrophage survival through induction of triggering receptor expressed on myeloid cells-1. Sci Rep 7: 42028. https://doi.org/10.1038/srep42028.

Zhao J, Wohlford-Lenane C, Zhao J et al. (2012). Intranasal treatment with poly(I·C) protects aged mice from lethal respiratory virus infections. J Virol 86: 11416–11424. https://doi.org/10.1128/JVI.01410-12.

Zhou Q, Gault RA, Kozel TR et al. (2006). Immunomodulation with CD40 stimulation and interleukin-2 protects mice from disseminated cryptococcosis. Infect Immun 74: 2161–2168. https://doi.org/10.1128/IAI.74.4.2161-2168.2006.

Zhou Q, Gault RA, Kozel TR et al. (2007). Protection from direct cerebral cryptococcus infection by interferon-gamma-dependent activation of microglial cells. J Immunol 178: 5753–5761. https://doi.org/10.4049/jimmunol.178.9.5753.

Handbook of Clinical Neurology, Vol. 210 (3rd series)
Neuroglia in Neurologic and Psychiatric Disorders, Part II
A. Verkhratsky, L.D. de Witte, E. Aronica, and E.M. Hol, Editors
https://doi.org/10.1016/B978-0-443-19102-2.00017-X

Chapter 14

Physiology and pathophysiology of the retinal neuroglia

ANTJE GROSCHE[1*], JENS GROSCHE[2], AND ALEXEI VERKHRATSKY[3,4,5]

[1]*Department of Physiological Genomics, Ludwig-Maximilians-Universität München, München, Germany*

[2]*Effigos AG, Leipzig, Germany*

[3]*Faculty of Biology, Medicine and Health, University of Manchester, Manchester, United Kingdom*

[4]*Department of Neurosciences, University of the Basque Country UPV/EHU and CIBERNED, Leioa, Bizkaia, Spain*

[5]*IKERBASQUE, Basque Foundation for Science, Bilbao, Spain*

Abstract

Neuroglia of the retina are represented by Müller glia, parenchymal astrocytes, microglia and oligodendrocytes mainly associated with the optic nerve. Müller glia are the most numerous glia, endowed with multiple homeostatic functions and indispensable for the retinal morphofunctional organization. Müller cells integrate retinal neurons into individual functional units (known as retinal columns) and act as a living light guide, transmitting photons to photoreceptors. In pathology, retinal neuroglia undergo complex changes, which include upregulation of neuroprotection, reactive gliosis, and functional asthenia. The balance between all these changes defines the progression and outcome of retinal disorders.

THE RETINA AND THE VISUAL SYSTEM

The retina is the outpost of the central nervous system (CNS) that lines the inner surface of the posterior pole of the eye. It is specialized for detecting photons and generating the primary input to the visual system. The retina is a thin (in the human eye it is 250 μm thick) multilayered tissue (Fig. 14.1; Boulton and Dayhaw-Barker, 2001; Bharti et al., 2011; Reichenbach and Bringmann, 2012; Campbell et al., 2017; Cowan et al., 2020). Photoreceptors are the main sensory element of the retina; they detect photons and convert light energy into the neuronal activity. Photoreceptors are connected to complex neuronal circuits within the retina that represent the primary information processing unit of the visual system. The axons of retinal ganglion cells form the optic nerve that transmits information to the visual centers in the brain. The human retina comprises two layers of the tissue: the neuroretina (which contains retinal neurons, resident tissue macrophages, and neuroglia) and an underlying monolayer of retinal pigment epithelial cells. These two layers are separated by the fluid-filled subretinal space. The inner (apical) side of retinal pigment epithelial cells faces the photoreceptors, while the basal side rests on the basement membrane known as Bruch's membrane that faces the capillaries of the choroid. The choriocapillaries provide metabolic support and blood supply to the retina (Reichenbach and Bringmann, 2012). The retinal pigment epithelium forms the outer retinal blood–brain barrier, sealed by tight junctions. Together, the retinal pigment epithelium and Bruch's membrane with its selective permeability control the transport of ions, energy substrates, and other molecules from

CELLULAR ORGANIZATION OF THE RETINA

The retinal tissue is composed of several cell types (Fig. 14.1). The primary sensory cells of the vertebrate retina are the photoreceptor cells, rods, and cones, which detect photons through light-sensitive proteins of the opsin family, the C-opsins. These opsins are highly expressed in photoreceptor cells, but are also present in some retinal ganglion cells, horizontal cells, and Müller glial cells. In particular, these cells possess several photopigments including melanopsin, encephalopsin, and neuropsin (Sand et al., 2012; Rios et al., 2019; Marchese et al., 2022). Incidentally, opsins are expressed in several regions of the brain (Hang et al., 2016). They contribute to the regulation of many vital functions, including, for example, the regulation of daily rhythms according to the light-dark cycle or the synthesis of melatonin.

The human neuroretina has seven layers (Fig. 14.1). Neuronal cell bodies are localized in the outer nuclear layer (ONL), the inner nuclear layer (INL), and the ganglion cell layer (GCL). The synapses form a neuropil that occupies the outer plexiform layer (OPL) and the inner plexiform layer (IPL). Axons of retinal ganglion cells constitute the nerve fiber layer (NFL). These ganglion cell axons

*Correspondence to: Antje Grosche, Department of Physiological Genomics, Ludwig-Maximilians-Universität München, München, Germany. E-mail: antje.grosche@bmc.med.lmu.de

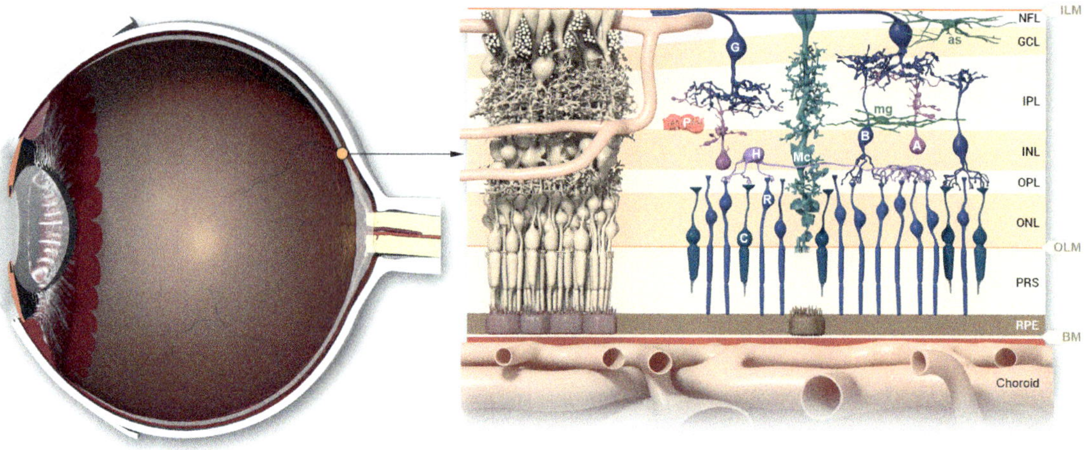

Fig. 14.1. The retina as a light-sensitive extension of the central nervous system in the eye. (*Left*) Gross anatomy of the eye. (*Right*) Cellular architecture of the retina and associated tissues. *ILM*, inner limiting membrane; *GCL*, ganglion cell layer; *IPL*, inner plexiform layer; *INL*, inner nuclear layer; *OPL*, outer plexiform layer; *ONL*, outer nuclear layer; *OLM*, outer limiting membrane; *PRS*, photoreceptor segments; *RPE*, retinal pigment epithelium; *BM*, Bruch's membrane. *Abbreviations: Mc*, Müller cell; *A*, amacrine cell; *B*, bipolar cell; *G*, ganglion cell; *H*, horizontal cell; *P*, rod; *C*, cone; *mg*, microglia; *as*, astrocyte; *P*, pericytes.

converge to form the optic nerve. Finally, the light-sensitive parts of rods and cones, the inner and outer segments, are localized in the outermost photoreceptor segment layer (PRS). The inner surface of the neuroretina is covered by the basement membrane, which, together with the associated Müller cell endfeet, forms the inner limiting membrane (ILM), a structural analog to the *glia limitans* in the brain and spinal cord. The outer boundary of the retina is formed by the outer limiting membrane (OLM), which separates the retina from the subretinal space (Reichenbach and Bringmann, 2012).

About 70 different morphotypes of retinal neurons have been distinguished (Kolb et al., 2001). The main types of retinal neurons (Fig. 14.1; Masland, 2012) are represented by rods and cones (the primary or first-order sensory neurons), bipolar cells, which are the second-order neurons, and retinal ganglion cells, the third-order neurons of the retinal circuit. Retinal interneurons are represented by horizontal and amacrine cells, which contribute to the processing of visual information. A special subtype of amacrine cells, known as the interplexiform cells, is involved in information processing between the two plexiform (synaptic) layers (Dowling, 1979; Jiang and Shen, 2010). Retinal neuroglia are Müller glia, astrocytes, oligodendrocytes, and microglia (Fig. 14.1; Reichenbach and Bringmann, 2020). The retina also contains cells of blood vessels such as endotheliocytes, pericytes, and smooth muscle cells. These cell types contribute to the retinal active milieu (Semyanov and Verkhratsky, 2021) and function. With regard to astrocytes presentation, retinas are divided into vascularized retina, which contains astrocytes, and avascular retina, which does not; in the avascular retina, Müller glia are the only type of astroglial cells (Reichenbach, 1987). Many vertebrates (e.g., birds or some marsupials (Buttery et al., 1990; Ruggeri et al., 2010)) possess an avascular retina; the human retina belongs to the vascular type (Campbell et al., 2017). Oligodendrocytes are present in the retina of fish, birds, and rabbits, but are absent in most mammals and primates (Bringmann et al., 2006). Finally, microglial cells, the brain tissue macrophages, are found in the retina of all vertebrates (Fan et al., 2022).

RETINAL ASTROGLIA: MÜLLER CELLS AND ASTROCYTES

Astroglia: General remarks

Astroglia represent an extended class of neuroglial cells of the CNS, whose main task is to maintain the homeostasis of the nervous tissue in health and upon environmental challenges; astroglial cells are also a fundamental element of the defense system of the CNS (Verkhratsky and Nedergaard, 2016; Verkhratsky and Butt, 2023). Astroglia cover many types of parenchymal astrocytes (including protoplasmic, fibrous, velate, marginal, interlaminar, etc.), radial astrocytes (Bergmann glia, Müller glia, tanycytes, radial stem astrocytes, etc.) and ependymoglia lining the brain ventricles and the central canal of the spinal cord (Verkhratsky et al., 2019).

Astroglia of the visual system are represented by radial retinal Müller glia, protoplasmic astrocytes residing in the retina of some vertebrates, and fibrous astrocytes localized in the optic nerve. Of note, only Müller cells develop form retinal progenitor cells; all other astrocytes migrate to the retina through the optic nerve in the pre- and postnatal period (Provis, 2001). Several other minor subpopulations of neuroglial cells are peripapillary radial glial cells in the avian retina (Stanke et al., 2010; Kim et al., 2011), perivascular glial cells in the mouse retina (Trost et al., 2014), and so named diacytes that share properties of both astrocytes and oligodendrocytes in birds, dogs, and nonhuman primates (Rompani and Cepko, 2010; Cebulla et al., 2012) have been described in the last decades.

Müller cells: The morpho-functional integrators of the retina

Müller glial cells, named so after Heinrich Müller are the main type of astroglia of the retinal tissue. The very first images of radial glial cells in the retina were, in fact, made by Samuel Moritz Pappenheim in 1842 (Fig. 14.2; Pappenheim, 1842). In 1851, the same cells were studied and depicted by Heinrich Müller (Fig. 14.2; Müller, 1851). Some years later, in 1859, Max Johann Sigismund Schultze published a detailed description of the structure of the retina (Schulze, 1859), including radial glial cells (Fig. 14.2), which he named *fibras Muellerianas*, in honor of Heinrich Müller (Schulze, 1859), see also (Chvatal and Verkhratsky, 2018). A remarkable advance in our knowledge of morphology, physiology, and pathophysiology of Müller glia was achieved by studies of Andreas Rechienbach and his colleagues starting in 1980 and continuing today (Fig. 14.2; Reichenbach and Bringmann, 2010, 2012, 2020).

As mentioned before, Müller glia are the major astroglial element of the retina (Reichenbach and Bringmann, 2013). These cells participate in a plethora of physiologic functions and are fundamental in diseases, in which Müller cells undergo complex metamorphoses similar to the reactive astrogliosis (Reichenbach and Bringmann, 2010, 2013, 2020). The physiologic functions of Müller glia are impressively wide (Fig. 14.3). These cells establish contacts and interact with all cellular elements of the retina and provide a morphologic focal point for the assembly of neurons into independent functional units, as explained in detail later. Like many types of astroglia in other regions of the CNS, Müller glia in the retina support the (inner) blood-retinal barrier, regulate retinal blood flow, and provide an interface between the circulation and the retinal parenchyma, contributing to the transport of nutrients, waste, water, and ions. In particular, Müller glia are the central element of the extracellular ionostasis, being central to the local and spatially distant K^+ buffering, water movements, and acid-base homeostasis. Müller glial cells are endowed with morphologic plasticity and regulate the volume of the extracellular space by swelling and shrinking. Müller glia act as unique biologic light guides that transmit light through the thickness of the retina from the ILM to the photoreceptors, which is critical for visual function. Müller cells contact and enwrap synapses to provide local support and act as a synaptic cradle; in addition, they recycle neurotransmitters and provide neurotransmitter precursors to neurons. On top of that, Müller glia contribute to the regeneration of the chromophores of cone photopigments. Finally, Müller glia secrete neuroprotective factors that support both photoreceptor and neuronal survival, but also closely interact with microglia, thereby contributing to the defense of the retina against pathology. Damage and death of Müller cells critically affect the retinal integrity and function. The demise of Müller glia

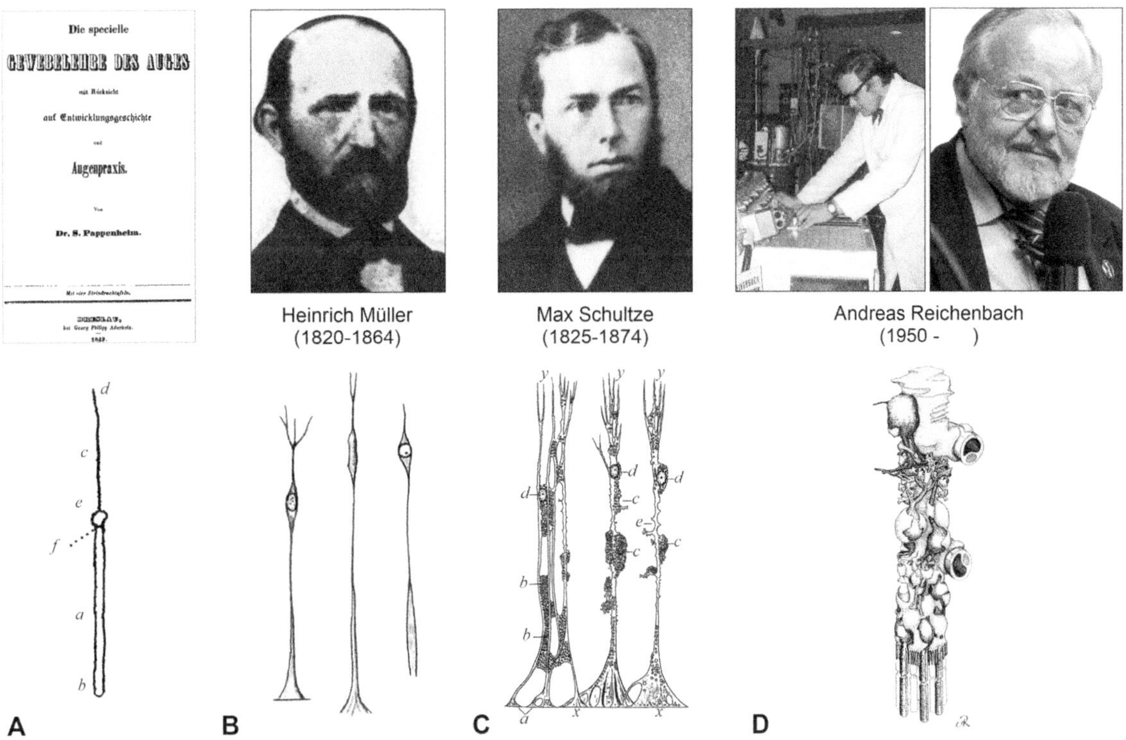

Fig. 14.2. History of Müller glia. (A) The very first image of radial astrocytes of the retina or Muller glial cells made by Samuel Pappenheim; (a) the inner part, (b) part of the retina, (e) an outer part with a large spherical base changing into the fiber (c) and into the ending (d), and the frontispiece of his book *Die specielle gewebelehre des auges mit rücksicht auf entwicklungsgeschichte und augenpraxis* (Pappenheim, 1842). (B) Heinrich Müller and his drawings of radial astrocytes of the retina (Müller, 1851). (C) Radial astrocytes of the retina by Max Schultze (Schulze, 1859). (a) The processes of Müller fibers passing into the outer nuclear layer (s), inner membrane (x), and the holes in membrana limitans interna; (b) the holes in the network of fine membranes for optic nerves; (c) so-called molecular network; (d) the nuclei in Müller fibers; (e) the space with the cell nuclei of the inner nuclear layer. (D) Andreas Reichenbach and his drawing of the Müller glia.

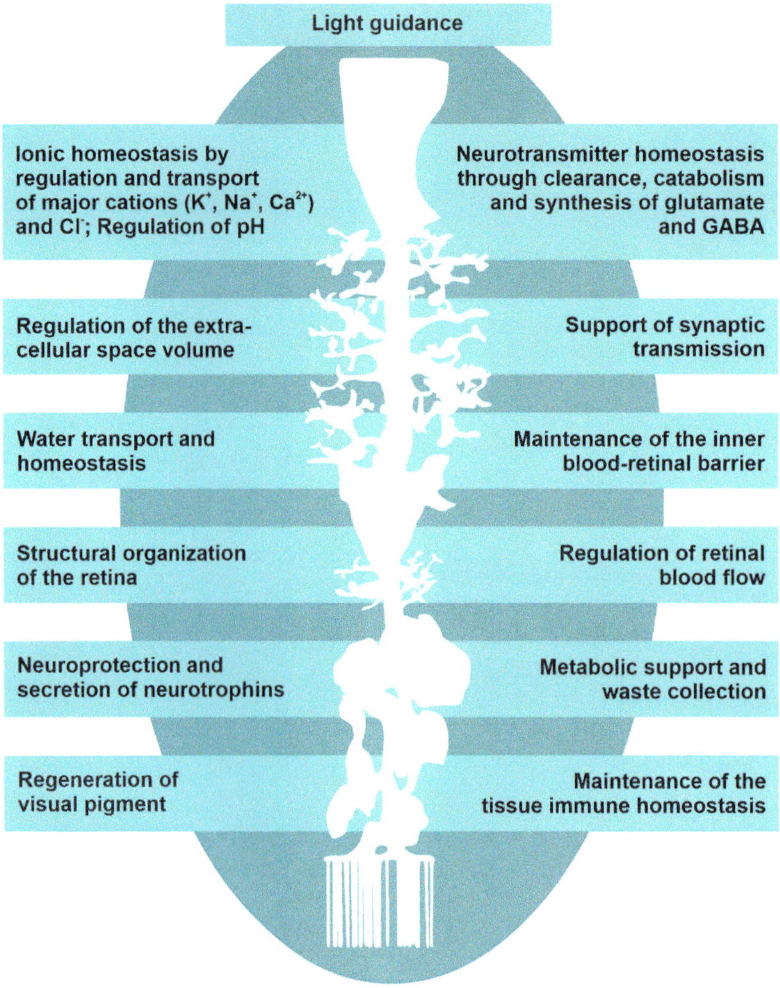

Fig. 14.3. Main functions of Müller glia. (See text for explanations.)

leads to neuronal damage, photoreceptor death, breakdown of the blood-retinal barrier, and ultimately retinal degeneration (Shen et al., 2012; Byrne et al., 2013; Shen et al., 2014).

Müller glia: Numbers and morphology

Müller glial cells are the most numerous glia of the retina. In the turtle, the retina contains 54,000 Müller cells and not a single astrocyte. In the rabbit, there are 4.2 million Müller cells and a few thousand protoplasmic astrocytes. Similarly, the human retina contains 4–5 million Müller cells and much fewer astrocytes (Robinson and Dreher, 1990; Reichenbach et al., 1991; Reichenbach and Bringmann, 2010). The density of Müller cells is quite similar across many species varying between 8000 and 12,000/mm^2. In general, the density of Müller cells is higher in the center (up to 30,000/mm^2 in Rhesus monkeys and 23,000/mm^2 in tree shrews) and lower (5000–10,000 mm^2) in the periphery of the retina (Chao et al., 1997).

Müller glial cells are classified as radial astrocytes (Verkhratsky and Butt, 2023), although they have an idiosyncratic morphology distinguishing them from radial glia and other radial astrocytes, such as tanycytes or Bergmann glial cells. Müller cells are highly specialized radial astrocytes that form the columnar cytoarchitecture of the retina, acting as pillars that integrate functionally interlinked groups of neurons and photoreceptors. It should be noted, however, that the term "pillar" can be easily misunderstood and can be misleading. In fact, Müller cells are not rigid pillars, but are softer than neurons, serving more as a smooth embedding and putative shock absorber for the latter (Lu et al., 2006). Müller glial somata are located in the central line of the inner nuclear, which is also very central in relation to the entire retinal tissue (Figs. 14.1 and 14.4). Two processes (sometimes referred to as trunks) extend from the somata of Müller glial cells in opposite directions (Fig. 14.4). The outer stem process extends to the outer limiting membrane projecting between photoreceptor somata in the outer nuclear layer—the associated Müller glial microvilli even reach into the subretinal space where they intermingle with photoreceptor inner segments. The inner stem process projects toward the vitreous chamber and forms a chalice-like endfoot resting on the basement membrane. Side branches protrude from both stem processes and project into the plexiform layers where they wrap synapses or into the nuclear layers where they cover the somata of neurons with velate sheaths (a kind of a comb-like structure) (Reichenbach, 1989; Reichenbach and Bringmann, 2010).

Müller glial cells, which penetrate the entire thickness of the retina, integrate retinal neurons into the columnar structures (Fig. 14.5),

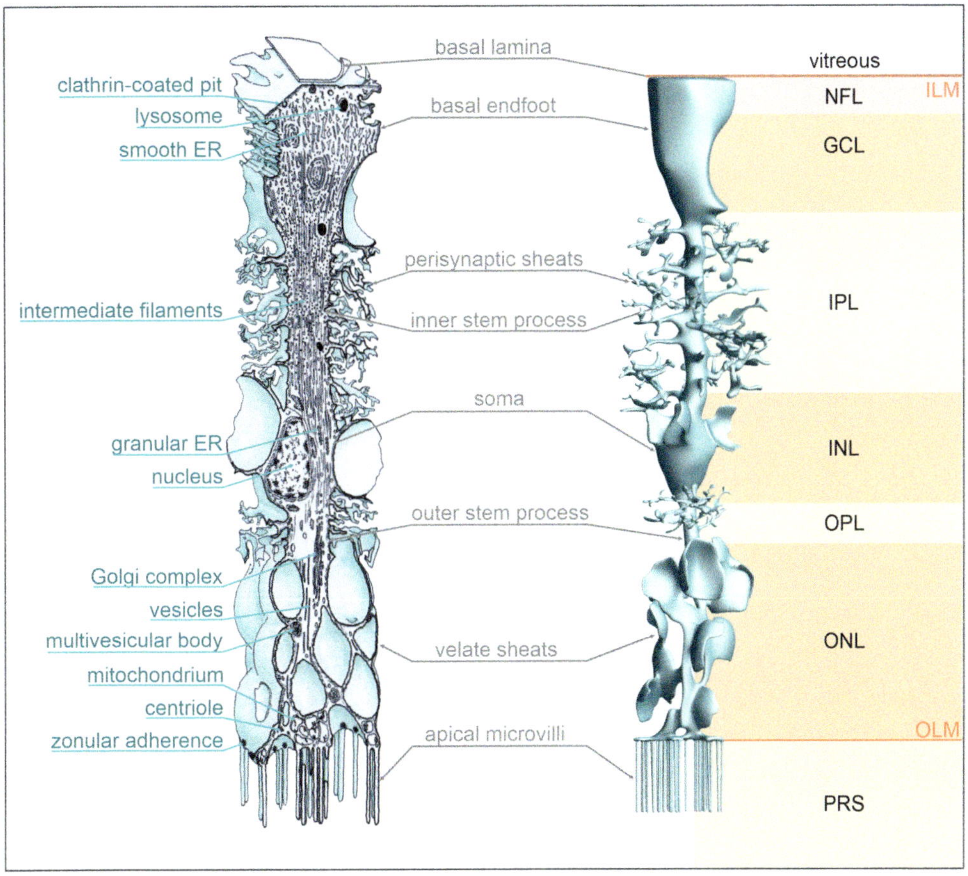

Fig. 14.4. Anatomy of the Müller glial cell. (*Left*) Localization of subcellular structures reconstructed from electron microscopic images of Müller cells in the avascular rabbit retina. (*Right*) 3D model of a Müller cell delineating the main processes and side branches. *Abbreviations: ILM*, inner limiting membrane; *NFL*, nerve fiber layer; *GCL*, ganglion cell layer; *IPL*, inner plexiform layer; *INL*, inner nuclear layer; *OPL*, outer plexiform layer; *ONL*, outer nuclear layer; *OLM*, outer limiting membrane; *PRS*, photoreceptor segments.

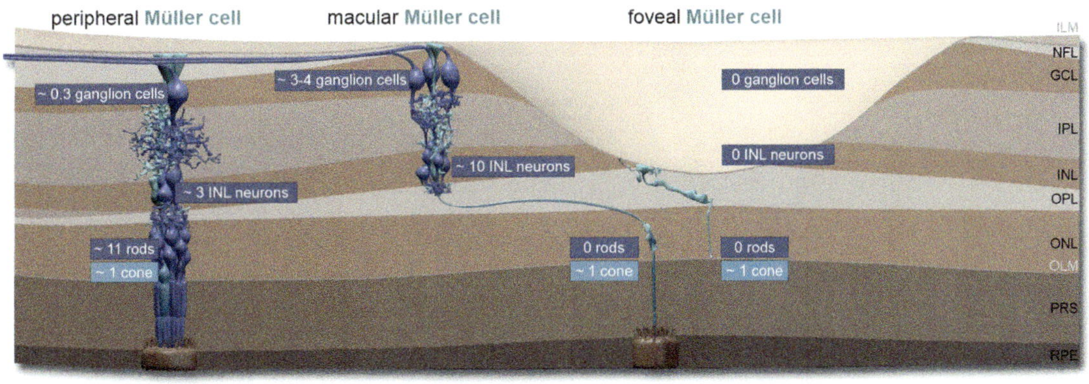

Fig. 14.5. Regional heterogeneity of Müller cells and composition of the smallest functional Müller cell-based retinal columns in the human retina, based on data presented in Bringmann et al. (2018), Kar et al. (2024), Syrbe et al. (2018). *Abbreviations: ILM*, inner limiting membrane; *NFL*, nerve fiber layer; *GCL*, ganglion cell layer; *IPL*, inner plexiform layer; *INL*, inner nuclear layer; *OPL*, outer plexiform layer; *ONL*, outer nuclear layer; *OLM*, outer limiting membrane; *PRS*, photoreceptor segments; *RPE*, retinal pigment epithelium.

and tile retinal space with minimal overlap (Wang et al., 2017b). On average, a single mammalian Müller cell contacts to and brings together ~15 neurons. The Müller cell-based retinal column is the smallest and relatively independent functional unit of the retina that performs forward processing of the primary visual information forward information processing (Reichenbach and Robinson, 1995; Reichenbach and Bringmann, 2010, 2020). A single Müller cell represents the core of such a functional columnar unit, around which neurons are assembled. In the rabbit, for example, every individual columnar unit is made from 1 Müller cell, 1 cone, 10 rods, 2 bipolar cells, and 1 amacrine cell. In humans, the retinal column contains up to 10 rods, 1 cone, 4–6 inner nuclear layer neurons, and anything between 0.3 (in the peripheral retina) and 2.5 (in the fovea) ganglion cells (Bringmann et al., 2018; Reichenbach and Bringmann, 2020). The number of rods linked to an individual Müller cell varies between species and correlates with the life style. In the diurnal tree shrew a single Müller cell is connected to ~0.1 rods, whereas in nocturnal mice up to 27 rods connect to one Müller glia (Reichenbach et al., 1995; Jeon et al., 1998). Rods and cones are also differentially distributed throughout the retina. In primates, for example, the fovea contains only cones, whereas rods dominate the peripheral retina. The association between Müller cells and photoreceptors changes accordingly. In the macaques fovea, a single Müller cell establishes full contact with a single cone and partially covers up to 6 neighboring cones, thus creating an overlap between receptive fields (Burris et al., 2002).

Müller glial cells are highly polarized, in an apical-basal manner, which is reflected by their ultrastructure (Wang et al., 2017b). The inner process contains smooth endoplasmic reticulum and glycogen granules, whereas the outer process contains the Golgi complex, multivesicular bodies, and microtubules. Secretory vesicles with a diameter of 80–100 nm were also visualized in the endfoot of a Müller cell (Reichenbach and Bringmann, 2010; Demais et al., 2022), but also in their microvilli extending into the subretinal space a high accumulation of marker proteins for small extracellular vesicles was described (Demais et al., 2022). Noteworthy, the heterogeneity of Müller cell morphology varies significantly depending on the retinal area in which they reside, especially in human retinae with the macula and fovea as central morphologic specializations for accurate and color vision (Fig. 14.5). Unlike the straight morphology of peripheral Müller cells, macular Müller cells exhibit a z-shape (Fig. 14.5). The formation of the fovea centralis results from the outward movement of the inner retinal layers. This outward shift also affects the macular Müller cells, causing their cell bodies and inner processes to move outward as well. In contrast, their outer stem processes ensheath and follow the axons of photoreceptors through the Henle fiber layer from the fovea to more peripheral areas, where the photoreceptors then form synapses with bipolar and horizontal cells. Finally, the morphology of foveal Müller cells (which number ~25–35 in a human fovea centralis (Syrbe et al., 2018)), located at the very center of the fovea centralis, is quite peculiar and has been a topic of controversial discussion. However, with the development of increasingly sophisticated microscopy approaches including 3D reconstructions from serial EM images, some clarity is emerging. Thus, foveal Müller cells do not extend to parafoveal areas; instead, their inner stem processes and endfeet cover the floor of the foveal pit. Consequently, these unique Müller cells interact solely with cone photoreceptors, but not with synapses, inner retinal neurons, astrocytes, microglia, and blood vessels as the fovea is avascular (Bringmann et al., 2018; Syrbe et al., 2018; Hammer et al., 2020; Reichenbach and Bringmann, 2020; Kar et al., 2024).

Physiology of Müller glia

Similarly to other astroglial cells Müller glia are electrically non-excitable, i.e., they cannot generate regenerative action potentials. Again similarly to other astroglial cells (Verkhratsky et al., 2020) excitability of Müller glia is mediated by coordinated and tightly controlled fluctuation of intercellular ions and second messengers regulated by the activity of a wide range of ion channels and neurotransmitter receptors (Fig. 14.6).

High resting K^+ permeability and membrane potential

Membrane physiology of the Müller glia is defined by a high and predominant expression of several types of K^+ channels active throughout the whole physiologic range of membrane potentials,

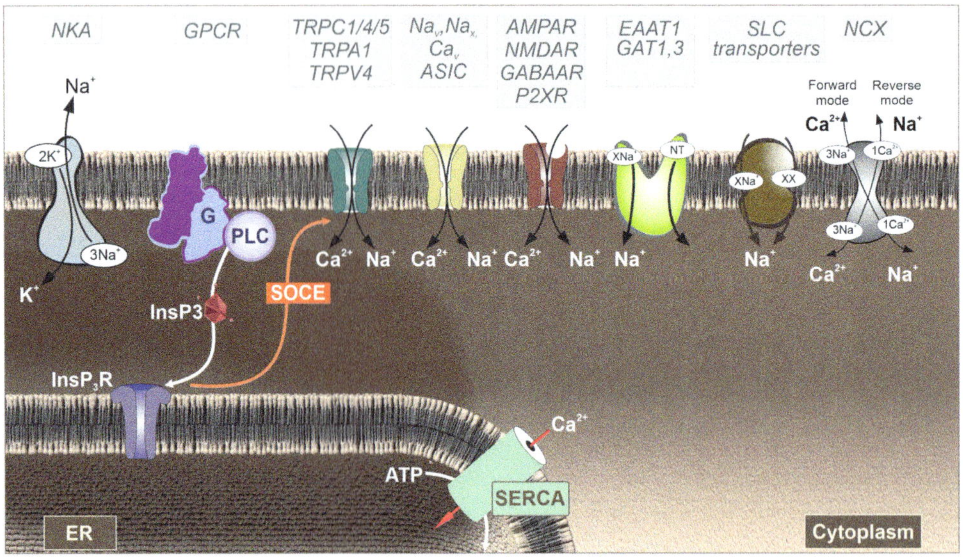

Fig. 14.6. Major types of neurotransmitter receptors and transporters expressed by Müller cells. *Abbreviations: NKA*, Na^+-K^+ ATPase; *GPCR*, G-protein-couple receptors; *TRPC*, transient receptor potential canonical channels; *TRPA*, transient receptor potential ankyrin channels; *TRPV*, transient receptor potential vanilloid channels; Na_v, voltage-dependent Na^+ channels; Ca_v, voltage-dependent Ca^{2+} channels; *ASIC*, acid-sensitive ion channels; *AMPAR*, α-amino-3-hydroxy-5-methyl-4-isoxazolepropionic acid receptors; *NMDAR*, *N*-methyl-D-aspartate receptors; *P2XR*, ionotropic purinoceptors; *GABAAR*, $GABA_A$ receptors; *EAAT1*, excitatory amino acid transporter 1; *GAT*, GABA transporters; *SLC*, solute carrier transporters; *NCXC*, Na^+-Ca^{2+} exchanger.

which stipulates exceptionally high resting K^+ permeability of the plasmalemma (Newman, 1985a). Other resting ion permeability is quite low: Cl^- permeability is negligible, whereas the ratio of K^+ to Na^+ membrane permeability (measured in frog Müller cells) is ~490:1 (Newman, 1985a). In consequence, the membrane of the Müller glial cells behaves almost like a K^+ electrode, and the resting membrane potential is very negative being set around −80 mV, which is close to the K^+ equilibrium potential (Witkovsky et al., 1985; Kofuji et al., 2000).

Potassium channels

The main component of the K^+ permeability of the plasmalemma of Müller cells is represented by several types of inwardly rectifying K^+ channels. Müller glia express weakly rectifying homomeric $K_{ir}4.1$ channels and heteromeric $K_{ir}4.1/K_{ir}5.1$, and strongly rectifying $K_{ir}2.1$ channels (Ishii et al., 1997; Kofuji et al., 2002). At the single channel level, the unitary conductance of K_{ir} channels varies between 17 and 28 pS depending on K^+ transmembrane gradients in different experiments (Newman, 1993; Kusaka and Puro, 1997; Bringmann et al., 1999; Rojas and Orkand, 1999). Homomeric $K_{ir}4.1$ channels are predominantly localized in Müller cells endfeet, as well as in perivascular membranes and microvilli. This is congruent with the main function of these channels in spatial K^+ buffering to mediate K^+ extrusion from the cells into bigger fluid-filled compartments including the blood circulation, the vitreous, or the subretinal space. In contrast, heteromeric $K_{ir}4.1/K_{ir}5.1$ channels are more or less evenly distributed throughout the Müller cells (Ishii et al., 2003). The $K_{ir}2.1$ channels are mainly localized to the side branches of inner stem processes that wrap synapses and the Müller cell endfeet (Ulbricht et al., 2008). About 90% of the resting K^+ permeability of the Müller cell membrane is mediated by $K_{ir}4.1$ channels (Kofuji et al., 2000). Müller glia also express big conductance (BK) Ca^{2+}-dependent K^+ channels (Bringmann et al., 1999), fast K_A channels (Pannicke et al., 2005a), delayed rectifying K^+ channels, ATP-dependent K^+ channels and two-pore domain TASK channels (Eaton et al., 2002; Skatchkov et al., 2006).

Aquaporin 4 water channels

The family of water channels or aquaporins includes 13 members classified as AQP0–AQP12 (Agre et al., 2002; Nagelhus and Ottersen, 2013). In the human retina, transcripts of all AQP channels were detected (Tenckhoff et al., 2005). In Müller glia, the expression of AQP4 dominates. The AQP4 channels are concentrated in perisynaptic and perivascular processes and in the cellular compartments facing the vitreous (Connors and Kofuji, 2006). These AQP4 channels are often colocalized and operate together with inward rectifying $K_{ir}4.1$ channels to coordinate K^+ and water fluxes and to respond to various osmotic challenges (Nagelhus et al., 1999; Bringmann et al., 2006). In addition to AQP4 Müller cells possess AQP6 channels, which are rather selectively concentrated in the membranes surrounding the ribbon synapses in the outer plexiform layer (Iandiev et al., 2011). The AQP6 channels are mainly permeable to glycerol and urea and may not contribute to the water homeostasis of the retina.

Sodium and calcium channels

In contrast to the majority of astroglial cells, Müller glia from all species so far studied (salamander, toad, rat, rabbit, guinea pig, human) express voltage-gated Ca^{2+} and Na^+ channels. In general, however, the voltage-gated Ca^{2+} channels are quite small in amplitude (several tens of nA), and often Ba^{2+} ions (which create larger currents through Ca^{2+} channels) are used for Ca^{2+} current recordings (Newman, 1985b; Puro et al., 1996; Bringmann et al., 2000; Linnertz et al., 2011). Voltage-clamp experiments revealed two types of Ca^{2+} currents, namely, low-voltage activated or T-type and high-voltage activated or L-type. This reflects the coexistence of two distinct types of Ca^{2+} channels encoded by α_{1D}, α_{1D}, α_2, and β_3 subunits (L-type channels) and α_{1G} subunit that forms T-type channels (Puro et al., 1996; Linnertz et al., 2011). Müller glial cells also express functional store-operated Ca^{2+} entry conductance mediated by TRPC1 and ORAI channels (Molnar et al., 2016; Sukkar et al., 2023). Of note, the expression of ORAI1 channels was found to increase in the mouse model of retinal degeneration with a protective effect on photoreceptors (Sukkar et al., 2023). Müller cells also express $Na_v1.6$ sodium channels, which mediate fast, tetrodotoxin (TTX)-sensitive Na^+ current (Francke et al., 1996; Reichelt et al., 1997a; Linnertz et al., 2011). Some Müller glia also express epithelial Na^+ channels of the ENaC family, which may contribute to the regulation of interstitial Na^+ concentration and cell volume regulation in response to osmotic challenges (Deliyanti et al., 2014).

Receptors to neurotransmitters

The most expressed and functionally relevant receptors in Müller glia are glutamate receptors, γ-aminobutyric acid (GABA) receptors, and purinoceptors (Biedermann et al., 2004; Bringmann et al., 2013) and Fig. 14.6. Glutamate is the main excitatory transmitter in the retina, mediating synaptic communications between photoreceptors, bipolar, and ganglion cells (Reichenbach and Bringmann, 2012). Both AMPA (assembled from GluR2/4 subunits) and NMDA (made from NR1, NR2A, and NR2B) ionotropic glutamate receptors are operational in Müller glial cells (Grunder et al., 2000; Vitanova, 2012). Ionotropic glutamate receptors, however, were detected only in cultured, but not in freshly isolated rat Müller glia, thus raising the possibility of in vitro artifact (Pannicke et al., 2005b; Reichenbach and Bringmann, 2010). To the contrary, both freshly isolated and cultured Müller glia express metabotropic glutamate receptors of all three groups, with mGluR1 and mGluR5 mediating Ca^{2+} signaling through stimulating of Ca^{2+} release from the endoplasmic reticulum (ER) Ca^{2+} store (Reichenbach and Bringmann, 2010).

GABA is the major inhibitory transmitter involved in the communication of subsets of horizontal, amacrine, bipolar, and ganglion cells and is therefore key to intraretinal information processing. In isolated human retinal Müller cells, extracellular GABA evoked a fast transient Na^+-independent current and a sustained Na^+-dependent current mediated by GABA transporters (Biedermann et al., 2004). It was found that $GABA_A$ receptors are unevenly distributed across the Müller cell membrane with especially high presence in their endfeet, soma, and outer stem process membranes, suggesting that these cells may sense retinal GABAergic neuronal signal transmission (Biedermann et al., 2004).

Müller glia also express several types of purinoceptors that sense ATP and adenosine released from neuronal and glial retinal elements (Perez et al., 1986; Wurm et al., 2011; Lohr et al., 2014). Purinergic signaling also contributes to neuronal–glial communications both in physiologic and pathologic contexts (Housley et al., 2009; Sanderson et al., 2014). Adenosine A_1, A_{2A}, and A_{2B} receptors as well as $P2Y_{1,2,4,6}$ metabotropic purinoceptors are coupled with Ca^{2+} release form the ER and consequent Ca^{2+} signaling (Weick et al., 2005; Pannicke et al., 2005a; Wurm et al., 2009a, 2010). A predominant role in coordinating glial function in response to neuronal activity is played by $P2Y_1$ and A_1 receptors—both key elements in the glutamatergic-purinergic signaling cascade active in Müller cell volume regulation (Wurm et al., 2008a, 2009a,b, 2010; Grosche et al., 2013).

The main, and seemingly the only, type of functional ionotropic purinoceptor in Müller glia is represented by the $P2X_7$ receptor, known for its low sensitivity to ATP and ability to dilate upon prolonged or excessive stimulation (North and Verkhratsky, 2006; Sluyter, 2017). The $P2X_7$ receptors and $P2X_7$ receptor-mediated cationic currents were found in the Müller glia in humans, macaques, and chickens (Pannicke et al., 2000, 2005a; Francke et al., 2002; Reichenbach and Bringmann, 2010) although there is no clear indication for its expression in murine Müller cells (Kaczmarek-Hajek et al., 2018). Activation of astroglial $P2X_7$ receptors not only generates cationic current but is also connected to several intracellular signaling cascades (Franke et al., 2012); in Müller glia, such activation is linked to the opening of K^+ channels and calcium release from thapsigargin-sensitive intracellular stores (Pannicke et al., 2000; Bringmann et al., 2001).

Photoreceptors in Müller glia

The nonvisual opsins, which are sensitive to detect blue light with a wavelength of around 480 nm (Guido et al., 2022), are expressed in different types of neurons and glial cells. These opsins contribute to the regulation of circadian rhythms and various body and brain systemic and organ-wide functions. Müller glial cells express encephalopsin (Opn3) and neuropsin (Opn5) which, when activated by blue light, trigger Ca^{2+} signaling (Rios et al., 2019; Marchese et al., 2022). This may also partially explain aspects of the distinct two-phase light response observed in Müller cells of the isolated guinea pig or rat retina or in the human MIO-M1 Müller cell line (Newman, 2005; Rillich et al., 2009; Hollborn et al., 2011).

Membrane transporters

Homeostatic control of the molecular environment of the central nervous system is the fundamental function of astroglial cells. In particular, astrocytes in all regions of the brain and the spinal cord regulate the interstitial ionic composition (function known as ionostasis) and control the extracellular presence and homeostasis of major neurotransmitters such as glutamate, GABA, adenosine, and monoamines (Verkhratsky and Nedergaard, 2018). These homeostatic functions of astrocytes are mediated through an extended complement of solute carrier (SLC) transporters, which use the energy stored in ionic gradients for transmembrane translocation of various molecules (Verkhratsky and Rose, 2020; Pizzagalli et al., 2021). The absolute majority of astrocytic SLC transporters are Na^+ dependent, and hence dynamic cytosolic Na^+ signaling plays a primary role in the regulation of astroglial homeostatic response (Rose and Verkhratsky, 2016, 2024); at the same time, the properties of retinal glia Na^+ signaling remain to be explored. Ionostasis is supported mainly through K^+ clearance and buffering (mediated by Na^+-K^+ ATPase and $K_{ir}4.1$ channels), though Cl^- movements, mediated by $GABA_A$ ionotropic receptors, and by Na^+-Ca^{2+} exchanger NCX (Macaulay and Zeuthen, 2012; Untiet et al., 2023; Rose and Verkhratsky, 2024).

The EAAT1/SLC1A3 (also known as glutamate-aspartate transporter or GLAST) is the main glutamate transporter operational in Müller cells (Lehre et al., 1997; Rauen et al., 1998; Rauen, 2000). The EAAT1 is a Na^+-dependent transporter with the stoichiometry of 3 Na^+:1 H^+ (in) 1 K^+ (out); the transporter also contains Cl^- channel which is not coupled with glutamate transport (Zerangue and Kavanaugh, 1996; Owe et al., 2006; Chen et al., 2021). Genetic removal of EAAT1 from Müller cells significantly increases the vulnerability of the retina to ischemia (Harada et al., 1998). Müller glia also contribute to the clearance of GABA by specific transporters, GAT1/SLC6A1 and GAT3/SLC6A11 (Biedermann et al., 2002). GABA transport is unique to mammalian Müller cells and not observed in other species (Marc, 1992). Finally, Müller cells express cystine-glutamate antiporter Sxc^- critical for the production of glutathione, a key player in the retinal anti-oxidative defense (Mysona et al., 2009; Oliveira et al., 2010).

Expression of all three known vesicular glutamate transporters (VGLUTs) was demonstrated in Müller cells. While VGLUT1/Slc17a7 is mainly expressed by retinal neurons, VGLUT2/Slc17a6 is expressed in Müller cells at the same level as in neurons at both mRNA and protein levels. VGLUT3/Slc17a8 shows the lowest overall mRNA expression (while the protein could not be detected) but appears to be specifically present in Müller cells. All VGLUTs, which function as glutamate/proton exchangers (Martineau et al., 2017), allow cells to load glutamate into exocytotic vesicles—a function that is apparently not restricted to synaptic vesicles in neurons.

Functions of Müller glia

Müller glia as light guides

The most idiosyncratic function of Müller cells is their ability to pass the light through the retinal tissue. In the vertebrates the retina is anatomically inverted with respect to the light source: the light has to cross the whole thickness of the retina before hitting the photoreceptors. Such positioning places the latter in close proximity to the choriocapillaris that provide an interrupted supply of oxygen and glucose to highly energy-demanding retinal photoreceptors and neurons. Müller glia act as a living light guide that delivers photons to the photoreceptors with minimal scattering and dispersion (Fig. 14.7, Franze et al., 2007; Agte et al., 2011). The endfoot of the Müller glial cell facing the vitreal surface of the retina is shaped as a funnel. This structure

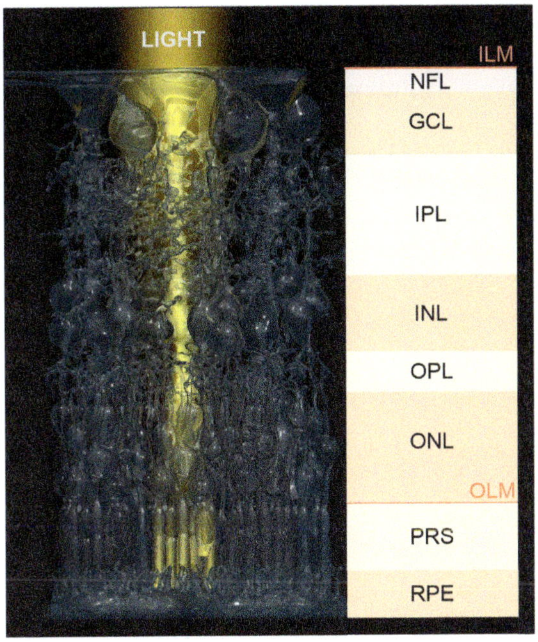

Fig. 14.7. Müller cells as living light guides in the inverse vertebrate retina. *Abbreviations: ILM*, inner limiting membrane; *NFL*, nerve fiber layer; *GCL*, ganglion cell layer; *IPL*, inner plexiform layer; *INL*, inner nuclear layer; *OPL*, outer plexiform layer; *ONL*, outer nuclear layer; *OLM*, outer limiting membrane; *PRS*, photoreceptor segments; *RPE*, retinal pigment epithelium.

effectively collects photons, which subsequently travel through the Müller cell processes directly to the photoreceptors with which these glial cells are connected (Franze et al., 2007). Moreover, the light-guiding properties of Müller cells are wavelength-dependent: they channel the green-red part of the visible spectrum onto cones and deliver the blue-purple part onto nearby rods, thus enhancing the visual acuity (Labin and Ribak, 2010). The light-guiding function of the glia is exceptionally effective: Müller cells occupy ~10% of the retinal tissue and yet they conduct almost all photons through the retina, thus ensuring extremely high sensitivity of the visual system.

Ionostasis: Müller cells support K^+ buffering and water homeostasis

As mentioned before, control over ionostasis is one of the most fundamental functions of astroglia. In particular tight regulation of K^+ concentration in the extracellular space is critical for the excitability of neurons, as K^+ ions define the resting membrane potential thus affecting voltage-gated ion channels and SLC transporters (Rasmussen et al., 2020). Astrocytic regulation of extracellular K^+, known as K^+ buffering, is a complex process involving K^+ clearance by Na^+-K^+ ATPase and/or passive diffusion through $K_{ir}4.1$ channels, K^+ redistribution through astrocytes or astrocytic syncytia, and diffusion of K^+ back to the extracellular space to restore of neuronal ionic gradients (Larsen et al., 2014; MacAulay, 2020; Barros, 2022; Verkhratsky and Semyanov, 2022). In the retina, Müller cells are key elements in the control of extracellular K^+. Neuronal activity associated with photoreception and primary information processing elevates extracellular K^+ concentration by ~1 mM in the plexiform layer (which contains most of the synapses), whereas in the subretinal space the K^+ concentration decreases by 2–3 mM due to K^+ influx mediated hyperpolarization of photoreceptors (Karwoski et al., 1985, 1989). These imbalances in K^+ across the retina are redressed by Müller glial cells by generating the transcellular potassium currents, a process known as K^+ siphoning when Müller cells take up K^+ at the plexiform layer and release K^+ into subretinal space, vitreal fluid, or blood (Fig. 14.8, Newman et al., 1984; Newman and Reichenbach, 1996). It was suggested that the movement of K^+-associated charges through the Müller glia contributes to the electroretinogram (Xu and Karwoski, 1997).

Ion movements drive transcellular water flux following the osmotic gradient and entering/leaving the cells primarily by AQP4 water channels (Bringmann et al., 2006). To protect against cellular edema, Müller cells are equipped with a complex volume-regulatory signaling cascades (Wurm et al., 2008a, 2009a,b, 2010; Grosche et al., 2013) that can be activated either by neurotransmitters such as glutamate or ATP or by growth factors such as vascular endothelial growth factor (Wurm et al., 2008b; Bruckner et al., 2012), heparin-binding epidermal growth

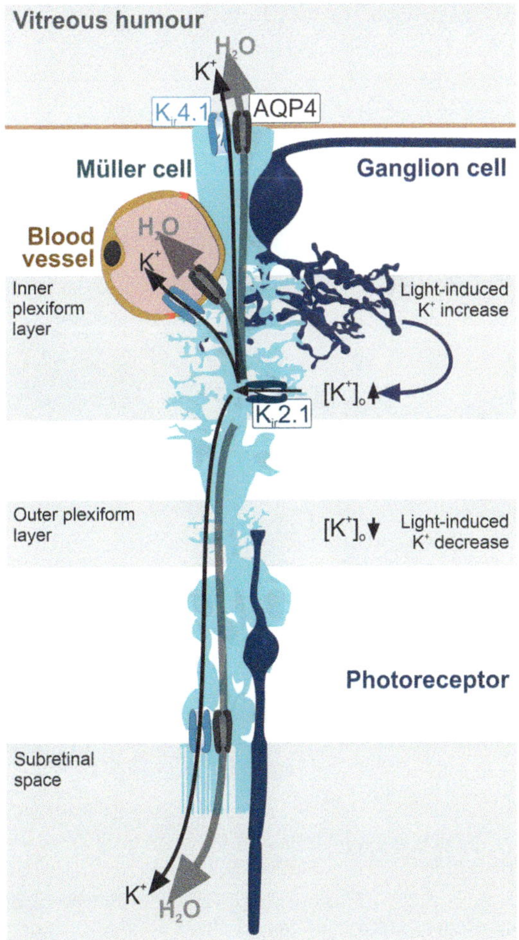

Fig. 14.8. Spatial buffering of K^+ in the retina mediated by Müller cells. (See text for explanations.)

factor (Weuste et al., 2006), platelet-derived growth factor (Diaz-Lezama et al., 2021), or erythropoietin (Krugel et al., 2010). The source of these multiple factors is associated with active neurons, but also can reflect an autocrine signaling from the Müller cells themselves. For instance, Müller cells are capable of releasing glutamate in a Ca^{2+}-dependent manner (Slezak et al., 2012), whereas ATP release from Müller cells seems to be mediated primarily by hemichannels (Bruckner et al., 2012; Voigt et al., 2015; Wagner et al., 2017).

Support of synaptic transmission

Astroglial support of synaptic transmission is manifold. At least half of all synapses in the CNS are closely associated with astroglial processes, which form a multipartite synaptic cradle (Araque et al., 1999; Nedergaard and Verkhratsky, 2012). Perisynaptic astrocytic membranes are densely packed with transporters that provide comprehensive support of neurotransmission for synaptogenesis and synaptic maturation to synaptic maintenance and extinction (Verkhratsky and Nedergaard, 2014). In the retina Müller glial processes cover synapses and contribute to the most fundamental aspects of synaptic transmission (Fig. 14.9). In particular, Müller cells are indispensable for excitatory and inhibitory transmission because they are the host of the glutamate(GABA)-glutamine shuttle, which supports glutamatergic and GABAergic neurotransmission through the removal of glutamate and subsequent conversion of the latter to glutamine, which ultimately is delivered back to neurons for final conversion into glutamate and GABA (Bringmann et al., 2013; Andersen et al., 2022).

Müller cells as key element of the retinal neurovascular unit

Müller glia form the core of the retinal neurovascular unit, performing functions similar to astrocytes in other parts of the CNS (Fig. 14.10, Newman and Reichenbach, 1996; Subirada et al., 2018). In particular, (i) Müller cells are perfectly equipped to sense neuronal activity at the molecular level, through the expression of receptors for all the neurotransmitters present in the retina (as discussed previously). Müller glia generate calcium signals in response to neuronal activity driven by light stimuli (Rillich et al., 2009; Biesecker et al., 2016), and (iii) they form perivascular processes to contact the vessels of the superficial, intermediate, and deep plexuses of the retinal vasculature (Fig. 14.1). The retina is protected by two distinct components of the blood-retinal barrier (BRB). The outer BRB (oBRB) consists of the retinal pigment epithelium. The inner BRB (iBRB) secured by the tight junctions of the endotheliocytes of the retinal capillaries is also supported by Müller cells and pericytes at the level of the intermediate and deep plexi, whereas astrocytes, together with Müller cells and pericytes, form the superficial plexus (Cunha-Vaz et al., 2011; Diaz-Coranguez et al., 2017; Kugler et al., 2021; Grimes et al., 2024). Astrocytes, in particular, are crucial as structural templates for the superficial plexus that develops first during retinal angiogenesis, primarily through the vascular endothelial growth factor (VEGF) and hypoxia-inducible factor (HIF) pathways, following ganglion cell templates (O'Sullivan et al., 2017; Paisley and Kay, 2021). The outgrowth of the intermediate and deep plexus capillaries occurs

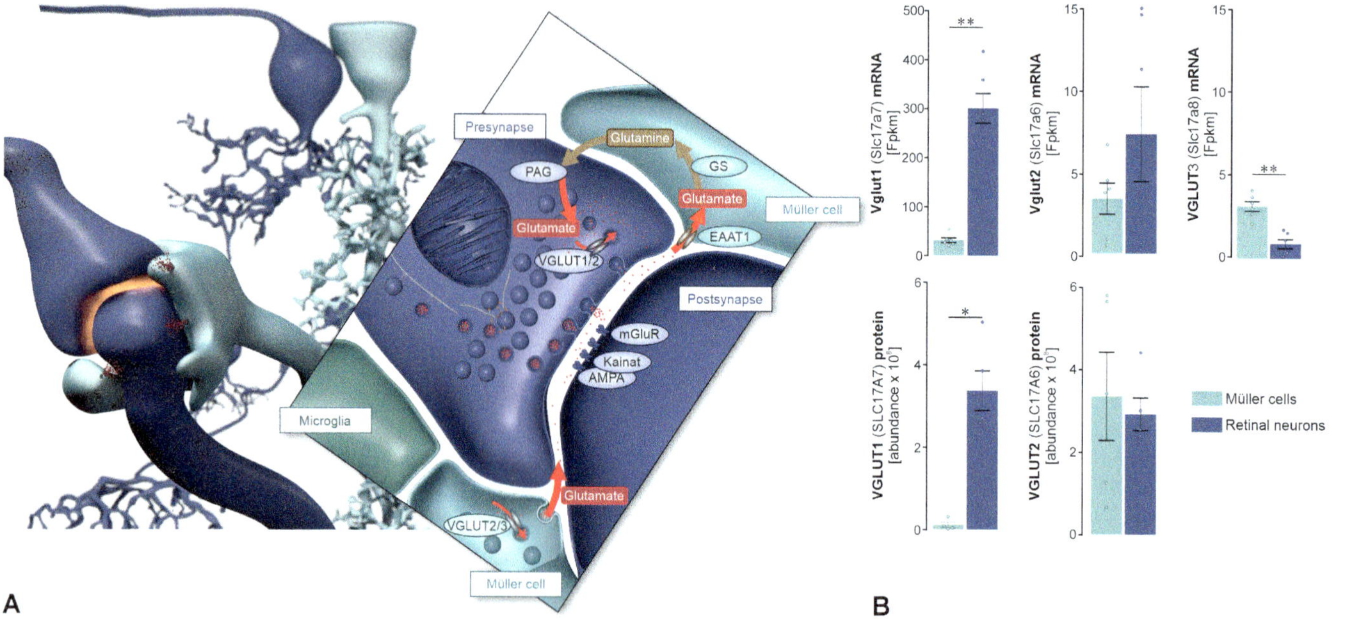

Fig. 14.9. Müller glia and the glutamatergic synapse. (A) Müller cells remove glutamate from the synaptic cleft and inactivate it to glutamine, but can also release glutamate via calcium-dependent exocytosis. *GLAST*, glutamate-aspartate transporter; *GS*, glutamine synthetase; *PAG*, phosphate-activated glutaminase; *VGLUT*, vesicular glutamate transporter. (B) Müller cells are equipped with vesicular glutamate transporters to enable Ca^{2+}-dependent exocytotic glutamate release as demonstrated by cell type-specific expression analysis of purified murine Müller cells cross compared to retinal neurons. Expression was determined at the mRNA level (data extracted from PMID: 38273366) via RNA sequencing and at the protein level via tandem mass spectrometry (own unpublished data). Bars represent mean ± SEM from 4 to 6 biologic replicates. $^{*}P < 0.05$; $^{**}P < 0.01$.

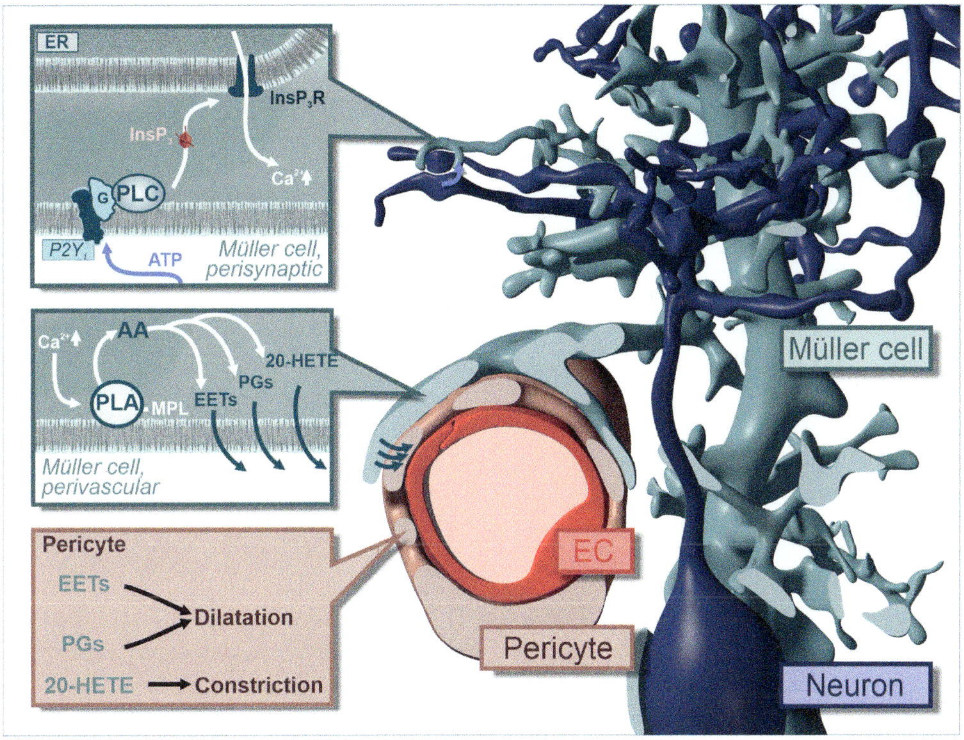

Fig. 14.10. The role of Müller cells in neurovascular coupling. *Top inset*: When neurons in the retina co-release ATP with glutamate at synapses, it activates purinergic receptors ($P2Y_1$) on Müller cells perisynaptic processes. This triggers the production of inositol triphosphate ($InsP_3$) and causes the release of calcium ions (Ca^{2+}) from internal stores within the glial cells. *Middle inset*: The increase in Ca^{2+} activates phospholipase A2 (PLA), which then converts membrane phospholipids (MPL) into arachidonic acid (AA). Arachidonic acid is subsequently metabolized into several compounds: the vasodilators prostaglandin E2 (PGEs) and epoxyeicosatrienoic acids (EETs), as well as the vasoconstrictor 20-hydroxy-eicosatetraenoic acid (20-HETE). *Bottom inset*: These vasomodulators reach contractile pericytes when released by perivascular processes of Müller cells. *$InsP_3$*, inositol-1,4,5-trisphosphate; *$InsP_3R$*, inositol-1,4,5-trisphosphate receptor. Drawn after Newman EA (2015). Glial cell regulation of neuronal activity and blood flow in the retina by release of gliotransmitters. Philos Trans R Soc Lond B Biol Sci, 370.

later. This suggests that Müller cells and astrocytes may work together to influence the development of the neurovascular unit and regulate its function. Patchy ablation of Müller cells in mice resulted in rapid and severe remodeling of the retinal vasculature (Shen et al., 2012; Byrne et al., 2013) precisely at the locations where Müller cells were missing, similar to what was observed after local depletion of cortical astrocytes (Mills et al., 2022).

In addition to maintaining the iBRB, Müller cells, also contribute to the neurovascular coupling (Fig. 14.10). As mentioned earlier, sensory stimulation of the retina induces Ca^{2+} signaling in Müller cells. A functional consequence of this is the induction of functional hyperemia represented by dilatation of retinal arteries, arterioles, and capillaries (but not venules) in all three retinal plexuses (Garhofer et al., 2004; Biesecker et al., 2016; Nippert and Newman, 2020). This vasoregulation is likely to be mediated by the release of ATP from retinal neurons, stimulation of glial $P2Y_1$ receptors with subsequent $InsP_3$-mediated Ca^{2+} release from intracellular stores (Newman, 2005, 2013), with $InsP_3$ receptor type 2 being the major player (Lipp et al., 2009; Wagner et al., 2016). Stimulated Müller cells produce arachidonic acid (AA) through Ca^{2+}-dependent activation of phospholipase A2 (PLA2) and release vasoactive substrates, including epoxyeicosatrienoic acids (EETs), prostaglandin E2 (PGE2), and the vasoconstrictor 20-hydroxyeicosatetraenoic acid (20-HETE) (Metea and Newman, 2007; Newman, 2013), which act on pericytes of the capillaries and/or smooth muscle cells of arteries/arterioles (Fig. 14.10).

Metabolic support of retinal neurons by Müller cells

Müller cells have a consistent metabolic profile across species, characterized by high taurine, medium to high glutamine, medium to low glutamate, and medium levels of glutamine synthetase activity in healthy conditions (Pfeiffer et al., 2020). No clear metabolic subclasses of Müller cells were identified in healthy retinas despite genetic heterogeneity (Roesch et al., 2008). In diseased retinas, Müller glia exhibit different metabolic phenotypes that coincide with morphologic hypertrophy and upregulation of GFAP; The metabolic heterogeneity progresses chaotically during retinal degeneration (Jones et al., 2011; Pfeiffer et al., 2016).

Based mainly on results obtained in primary Müller cell cultures, it was concluded that Müller cells generate ATP primarily by glycolysis when glucose and oxygen are available, with a low rate of oxygen consumption, and that lactate released by Müller glial cells is metabolized by photoreceptors (Poitry-Yamate et al., 1995; Winkler et al., 2000). This metabolic behavior was thought to conserve oxygen for retinal neurons under normal conditions. In addition, cultured retinal Müller cells exhibit resistance to anoxia or glucose deprivation, which helps to explain their lower susceptibility to ischemia or hypoglycemia compared to neurons.

Recent research in the in vivo models challenged this hypothesis, indicating that photoreceptors are the primary sites of aerobic glycolysis in the retina (Lindsay et al., 2014) that provides lactate to Müller glia as an energy source (Kanow et al., 2017). A similar shift of paradigm also applies to astrocytes at large; it seems that the majority of energy needed for astroglial function derives from oxidative phosphorylation and not from glycolysis as was believed previously (Barros, 2022; Verkhratsky and Semyanov, 2022). Müller cells have low levels of hexokinase but high levels of enzymes associated with the tricarboxylic acid (TCA) cycle, anaplerosis, and cataplerosis (Rueda et al., 2016). Furthermore, blocking glycolysis in Müller cells in vivo did not affect their survival (Shen et al., 2021). These findings, together with studies of the resilience of Müller cells to hypoxia (Jassim and Inman, 2019; Jassim et al., 2022), suggest that they may not be as dependent on glycolysis as previously thought. Indeed, when COX10, a farnesyltransferase required for the assembly of cytochrome c oxidase (COXIV), was specifically knocked out from Müller cells to impair their electron transport chain and mitochondrial ATP synthesis, a severe reduction in retinal glycogen stores as well as a shift in retinal metabolism toward increased glycolysis was observed, compromising the physiologic response of both photoreceptors and Müller cells (Nsiah and Inman, 2022). Glycogen, the storage form of glucose, is present in Müller glia and astrocytes in the retina. Low glycogen levels in COX10 knockout MG suggest that internal glycogen stores are used to produce glucose for increased glycolysis (Pfeiffer-Guglielmi et al., 2005; Hurley et al., 2015). Of note, genetic deletion of COX10 in Müller glia increased oxidative stress and led to impaired photoreceptor function (e.g., as determined by a reduced rod-driven the a-wave in electroretinogram recordings), highlighting the close metabolic relationship between Müller cells and photoreceptors.

Mechanical homeostasis of the retinal tissue

As mentioned before, physiologic activity in the retina is associated with substantial movement of ions and water between various cells resulting in regular volume fluctuations of extra and intracellular space. Bipolar and ganglion cells, for example, swell to which Müller glial cells adapt by reshaping their processes (Slezak et al., 2012). Mechanically, Müller cells are twice as soft as neurons and hence are capable of fast and pronounced morphologic plasticity. In essence, Müller cells act as shock absorbers (Reichenbach and Bringmann, 2010, 2013). Their viscoelastic properties allow them to operate as a Voigt element composed of a spring (elastic part) and a dissipative dashpot (viscous part). In stress conditions, when Müller cells upregulate the expression of intermediate filaments, their biomechanic features change toward a stiffer phenotype (Lu et al., 2011).

Another intriguing feature of Müller cells is that they are contractile in vitro (Kaplan et al., 2023) and in vivo (MacDonald et al., 2015). Retinas from which Müller cells were ablated, developed a tearing defect known as retinoschisis (MacDonald et al., 2015), demonstrating that retinal tissue without functional Müller glia has reduced resistance to tensile stress. Laser ablation of inner Müller cell processes revealed that these cells are under tension within the tissue, suggesting that they act as a mechanical support system, similar to springs, that helps hold the neural retina together. A central role for this Müller glial function most likely depends on the aforementioned intermediate filament network (e.g., vimentin, GFAP), its organization (e.g., through the cytolinker epiplakin), and its interaction with other components of the cytoskeleton, including actin and the complex cell–cell adhesion machinery (Kaplan et al., 2023). In the human retina, contractile Müller cells may contribute to the unique architecture of the macula by shaping the slope of the foveal pit (Bringmann et al., 2018; Bringmann, 2019; Kaplan et al., 2023). This hypothesis would explain the massive rupture/edema formation observed in the retinas of patients with macular telangiectasia type 2, in which macular Müller cells degenerate (Powner et al., 2013).

In addition, Müller cells respond to changes in tissue tension (Lindqvist et al., 2010; Pillai and Franze, 2024). Mechanical stretches, which can occur during rapid eye movements or trauma (especially in glaucoma patients) (Fortune, 2019), induce Ca^{2+} signaling in Müller cells. These Ca^{2+} transients are likely mediated by mechanosensitive channels such as transient receptor potential vanilloid 4 (TRPV4) and TRPC1 (Jo et al., 2022). These Ca^{2+} elevations initiate early stress responses that allow cells to adapt to mechanical signals (Lindqvist et al., 2010).

Neuroprotection

Müller cells are fundamental for the protection of retinal neurons both in physiologic and pathologic contexts through many functions already discussed earlier (e.g., neurotransmitter clearance and spatial K^+ buffering to avoid neuronal hyperexcitability, coordination of neurovascular coupling and the maintenance of the iBRB, mechanical and metabolic support, to name but a few). Therefore, it is not surprising that ablation of Müller glia induces retinal dysplasia and photoreceptor death (Dubois-Dauphin et al., 2000; Shen et al., 2012; Byrne et al., 2013). Notably, coculturing Müller glia with neurons greatly increased the resistance of the latter to exogenous glutamate. Such neuroprotective effect was partially mediated by neurotrophins released from Müller glia (Taylor et al., 2003). Glia-derived neurotrophins, such as ciliary neurotrophic factor (CNTF), glial cell line-derived neurotrophic factor (GDNF), leukemia inhibitory factor (LIF), and pigment epithelium-derived factor (PEDF), are also instrumental for protecting photoreceptors and retinal ganglion cells (Bringmann et al., 2009; Wang et al., 2011; Kumar and Shamsuddin, 2012; von Toerne et al., 2014; Ruzafa et al., 2018). For example, CNTF supports photoreceptor regeneration and retinal ganglion cells axonal protection, while GDNF and LIF protect photoreceptors and aid Müller cell survival, especially in diabetic retinopathy. PEDF, produced by retinal pigment epithelial cells and Müller glia, protects photoreceptors from degeneration by reducing intracellular Ca^{2+} signaling and activating calpain. Osteopontin (SPP1), clusterin (CLU), and basigin (BSG) were identified as additional potential neuroprotective agents secreted by Müller cells (Ruzafa et al., 2018). In vitro studies demonstrated that SPP1 and BSG significantly promote the survival of retinal ganglion cells. These findings suggest that the neuroprotective effects of the Müller cell secretome are multifaceted, with SPP1 and BSG playing crucial roles in enhancing retinal ganglion cell survival.

Another mode of neuroprotection mediated by Müller glia is the production and secretion of antioxidants, which is essential due to the high metabolic demands of the retina critical for vision but associated with high levels of oxidative stress in the tissue (Yu and Cringle, 2005). Much of the research on oxidative stress in the retina has overlooked the significant antioxidant role of Müller glia. For instance, glutamate transporters, abundantly expressed by Müller glia, avert excitotoxic retinal damage by facilitating the synthesis of glutathione, a tripeptide formed from glutamate, cysteine, and glycine (Reichelt et al., 1997b). This tripeptide is a major retinal antioxidant released by Müller glia in response to hypoxia and oxidative stress and plays a crucial role in controlling harmful oxidative species in the neural retina (Pow and Crook, 1995). Glutathione prevents retinal ganglion cell degeneration in experimental glaucoma models (Harada et al., 2007).

In response to oxidative stress, Müller cells increase the expression of the transcription factor nuclear factor erythroid 2-related factor 2 (Nrf2), which regulates over 100 antioxidant genes (Inoue et al., 2017; Navneet et al., 2019) and promotes the release of glutathione. Inhibition of Nrf2 in cultured Müller glia significantly reduces antioxidant gene expression and exacerbates oxidative stress, whereas NRF2 activation mitigates this stress (Xu et al., 2014). Nfr2 binds to antioxidant response elements (ARE) in DNA regulating genes involved in cellular stress responses, including key enzymes such as NAD(P)H dehydrogenase (NQO-1), heme-oxygenase 1 (HO-1), superoxide dismutase 2 (SOD2), catalase (CAT), and glutathione peroxidase 1 (GPX1) (Navneet et al., 2019). Suppression of antioxidants, such as observed in Nrf2 knockout diabetic mice, leads to retinal malfunction mediated by increased oxidative stress (Xu et al., 2014). Müller cells also express several enzymes involved in oxidative defense, including glutathione S-transferase mu 1 (GSTM1), glutathione S-transferase mu 5 (GSTM5), and peroxiredoxin 6 (PRDX6), which are critical for retinal neuroprotection (Grosche et al., 2016). For instance, PRDX6 maintains cellular redox balance by neutralizing harmful peroxides and other oxidative molecules (Pacifici et al., 2014) through its dual activity as glutathione peroxidase and phospholipase A2 (Fisher, 2011). High levels of PRDX6 in Müller glia suggest a significant antioxidant role, potentially limiting damage from reactive oxygen species and offering neuroprotection in retinal diseases such as glaucoma and macular degeneration.

Müller cells shape the retinal immune homeostasis

Microglial cells, the tissue macrophages of the CNS and retina, are widely recognized as the primary mediators of retinal immune responses (Karlstetter et al., 2015) and are arguably the first to respond to tissue pathology (Zhang et al., 2010; Liu et al., 2011; Mages et al., 2019; Paolicelli et al., 2022). However, a substantial body of evidence suggests that an effective immune response of the retina depends on close interactions between microglia and Müller cells (Wang et al., 2011; Mages et al., 2019; Di Pierdomenico et al., 2020). Müller cells amplify the immune modulatory responses in the retina through a positive feedback loop (Wang et al., 2011). They express receptors for several cytokines, including interleukin-1β (IL-1β) and tumor necrosis factor-α (TNF-α), secreted by microglia and can in turn release cytokines such as interleukin-6 (IL-6) to modulate immune responses, including microglial activity (Yoshida et al., 2001; Schmalen et al., 2021). Such molecular setup places Müller glia at the central position to balance pro- and anti-inflammatory factors and promote protective inflammation (Krishnan and Chatterjee, 2012). For instance, Müller glia enhance the secretion of neuroprotective factors GDNF, LIF (Wang et al., 2011) or diazepam binding inhibitor (DBI) in response to tissue stress (Mages et al., 2019; Demais et al., 2022). DBI is the putative endogenous ligand of translocator protein 18 kDA (TSPO) expressed by Müller cells (Hector et al., 2024), which is highly upregulated in reactive microglia and its stimulation by exogenous ligands such as XBD173 is neuroprotective in retinal pathology models (Scholz et al., 2015; Mages et al., 2019).

Müller cells also produce and secrete components of the complement cascade, which bridges the innate and adaptive immune responses. They can stimulate complement (e.g., by releasing complement component C3) or downregulate its activity by secreting the complement factor H (Pauly et al., 2019; Zauhar et al., 2022). Microglia, which are highly sensitive to cytokines and interleukins and are the only retinal cell types expressing high levels of receptors for the opsonins generated during complement activation, including C3a or C5a (Pauly et al., 2019; Zauhar et al., 2022). Thus, C3 secretion by Müller cells could assist microglia in sculpting retinal circuitry by synaptic pruning, which is mediated in a complement-dependent manner (Stevens et al., 2007; Schafer et al., 2012; Stevens and Johnson, 2021).

This multifaceted interplay mediated by secreted factors, together with the close spatial association of microglia with Müller cell processes in the plexiform layers of the retina (Wang et al., 2011), supports the concept that these two cell types are critical for coordinated retinal immune homeostasis. Since Müller cells span all retinal layers, they could serve as a highway for microglia to guide them to the lesion site. A molecular correlate for this chemoattractive action of Müller cells may be the upregulation of CCL2 as has been demonstrated in a model of retinal light damage (Rutar et al., 2011). In addition, changes in physical cell–cell interactions in the form of Müller cells acting as an adhesive scaffold to facilitate microglial migration have been suggested by in vitro experiments (Tassi et al., 2006).

Finally, Müller cells do have the capacity to phagocytose cell debris, for example, dead photoreceptors or shed photoreceptor discs (Long et al., 1986; Francke et al., 2001; Sakami et al., 2019) and potentially even to serve as atypical antigen-presenting cells (Mano et al., 1991; Schmalen et al., 2021). Thus, it is tempting to speculate that Müller cells may promote and coordinate the activity of alloreactive naïve or memory T cells (Kambayashi and Laufer, 2014)—another potentially intriguing but certainly understudied Müller cell function, given the recent demonstration of a critical role for regulatory T cells in maintaining retinal homeostasis with age and preventing age-related neurodegeneration (Conedera et al., 2023; Llorian-Salvador et al., 2024).

Retinal astrocytes

Astrocytes proper are found only in mammalian species with vascularized retinas, as well as in some fishes including eel. Nonvascularized neural retinas (lower vertebrates, birds, echidnas, and horses) do not contain astrocytes (Reichenbach and Bringmann, 2012). Morphologically retinal astrocytes are classified into bipolar astrocytes with processes associated with nerve fiber bundles, and perivascular astrocytes (Reichenbach and Bringmann, 2015, 2020). The processes of astrocytes are in contact with the vessels of the superficial vascular plexus and the somata and axons of retinal ganglion cells. Astrocytes provide some local homeostatic support and contribute to the formation of the blood–brain barrier through endfeet plastering of blood vessels.

OLIGODENDROGLIA OF THE RETINA

In the vertebrate retina, oligodendrocytes myelinate axons of retinal ganglion cells in the extraocular part of the optic nerve. The axons, however, are nonmyelinated in the retinal nerve fiber layer and in the intraocular part of the optic nerve. In contrast to other regions of the CNS, Müller glial cells, an astroglial cell type, produce myelin to ensheath large ganglion cell axons (Quesada et al., 2011).

MICROGLIA OF THE RETINA

The retina as an outpost of the CNS, is located behind the blood–brain barrier and hence harbors several types of immune cells, such as dendritic cells or perivascular macrophages. The microglia, however, are the main type of retinal tissue macrophages. Microglia originate from the fetal yolk sac macrophages (Ginhoux and Prinz, 2015) that enter the retina from the ciliary body vasculature and from both subretinal space and hyaloid vessels of the vitreous (Santos et al., 2008; Rashid et al., 2019). After birth, microglia are mainly found in the ganglion cell layer, in the inner and outer plexiform layers, and around blood vessels (Reichenbach and Bringmann, 2020). Surveilling microglia of the retina, similar to microglia in other parts of the CNS (Kettenmann et al., 2011), are characterized by highly dynamic ramified processes with which they scan the retinal parenchyma belonging to their individual territorial domains (Fontainhas et al., 2011). This surveilling phenotype has been linked to important constitutive microglial functions such as clearing apoptotic cells and cellular debris (Kreutzberg, 1996; Ransohoff and Perry, 2009), pruning malfunctional or redundant synapses (Wake et al., 2009; Kettenmann et al., 2013), providing trophic factors (Elkabes et al., 1996; Walton et al., 2006), and regulating synaptic function and plasticity (Bessis et al., 2007; Schafer et al., 2012). In pathology, microglia undergo reactive microgliosis characterized by multiple context-specific states (Paolicelli et al., 2022).

PATHOPHYSIOLOGY OF RETINAL NEUROGLIA

Neuroglia contribute to the pathophysiology of all neurologic diseases through multiple and complex mechanisms. Conceptually glial responses to pathology include reactive gliosis, multiple primary and secondary astrocytopathies, glial atrophy and functional asthenia, gliodegeneration, and death (Verkhratsky et al., 2023). In eye diseases, which often leads to blindness, glial cells act as active participants in neuroprotection and inflammation. Eye disorders include glaucoma and age-related macular degeneration, in which degenerative and inflammatory processes prevail. Neuroglial cells are also involved in the pathogenesis of retinal vascular disorders such as diabetic retinopathy, neovascular age-related macular degeneration, and retinopathy of prematurity. In addition to Müller glia, retinal astrocytes, and microglia, the cells of the retinal pigment epithelium (which originate from neuroepithelium and hence are formally classified as astroglia (Verkhratsky and Butt, 2023)) contribute to eye diseases including macular degeneration, diabetic retinopathy, retinitis pigmentosa, and Stargardt disease.

The advent of more sophisticated methods to investigate cell type-specific expression profiles at mRNA and protein levels, revealed a potential direct contribution of Müller cells to hereditary retinal diseases such as retinitis pigmentosa associated with Usher syndrome (Nagel-Wolfrum et al., 2023) or diverse recessive retinal disorders from severe Leber congenital amaurosis to isolated macular dystrophy associated with pathogenic variants in crumbs homolog 1 (CRB1) (Quinn et al., 2019; Mairot et al., 2021). There is also evidence that a specific functional metabolic deficit of macular Müller cells leads to macular telangiectasia type 2 (Powner et al., 2013; Zucker et al., 2020). Accordingly, in recent years pathologic potential of Müller cells and retinal glia in general have been revisited and are increasingly being investigated as potential targets for novel therapies for retinal diseases that are not amenable to gene correction therapies, e.g., by dampening pro-inflammatory neuroglial responses or by directly targeting a disease-causing mechanism initiated in the neuroglia themselves.

Macular telangiectasia type 2

When first described by Donald Gass in 1977 (Gass and Blodi, 1993; Oransky, 2005), macular telangiectasia type 2 (MacTel2) was thought to be an acquired condition, but observations in multigenerational families suggest that it may follow an autosomal dominant inheritance pattern with a variable penetrance of 0.38 and expressivity (Wu, 2023). The retinas of these patients showed Müller glia loss in the MacTel zone, defined as the distance from the foveal center to the temporal edge of the optic disc (approximately 3 mm), including the macula, fovea and adjacent parafoveal regions in the horizontal dimension. The vertical dimension is approximately 80% of this distance (approximately 2.5 mm) (Heeren et al., 2020). Consistent with this, yellow macular pigment was absent in the area corresponding to Müller glial cell loss (Cherepanoff et al., 2012; Powner et al., 2018). In addition, subretinal debris, reduced outer segment phagosomes, and ultrastructural abnormalities in the retinal pigment epithelium were observed. These abnormalities were not restricted to the MacTel area but were present throughout the central and peripheral retina. In MacTel2 eyes, the vascular density of both the superficial capillary plexus and the deep capillary plexus is reduced compared to non-MacTel2 eyes. In the early stages of the disease, vessels in the deep capillary plexus become dilated and telangiectatic, occasionally causing radial hemorrhages in the Henle layer, leading to sudden vision loss (Au et al., 2018).

Identification of causative genes for MacTel2 has been challenging due to its low penetrance and late onset (Parmalee et al., 2010, 2012). Genome-wide association studies have identified three susceptibility loci associated with MacTel2, linked to serine/glycine metabolism and vascular caliber (Scerri et al., 2017). In addition, whole exome sequencing has shown that functional variants in the phosphoglycerate dehydrogenase (PHGDH) gene, which encodes the rate-limiting enzyme in serine biosynthesis critical for production of D-serine, are responsible for 3.2% of MacTel2 cases (Eade et al., 2021), whereas gain-of-function variants in the serine palmitoyltransferase long chain subunit 1 (SPTLC1) and subunit 2 (SPTLC2) genes, which have been implicated in both MacTel2 and the rare neurologic disease hereditary sensory-autoimmune neuropathy type 1 (HSAN-1), account for $<0.1\%$ of MacTel2 cases (Gantner et al., 2019). Importantly, patients with functional variants in PHGDH, SPTLC1 or SPTLC2 have low serum serine levels and elevated 1-deoxysphingolipids in common and metabolomic analyses of MacTel2 patients without these genetic variants also show low serum serine and elevated 1-deoxysphingolipids (Gantner et al., 2019; Bonelli et al., 2020; Eade et al., 2021).

Serine is essential for several metabolic pathways that maintain retinal homeostasis, particularly as a precursor for sphingolipid synthesis and free radical scavenging. The retina relies on both circulating serine and de novo synthesis by retinal glial cells and the retinal pigment epithelium (Wu, 2023). When serine levels are low, serine palmitoyltransferase uses alanine instead of serine during sphingolipid synthesis to produce L-deoxysphingolipids, which have been shown to induce photoreceptor cell death in retinal organoids (Gantner et al., 2019). Mice fed a diet deficient in glycine and serine had low levels of circulating serine and significant photopic and flicker electroretinographic deficits (Gantner et al., 2019). Electron microscopy-based connectomics of a MacTel2 eye revealed structural mitochondrial changes in all retinal cell types (Zucker et al., 2020). However, Müller glia are particularly sensitive to disturbances in serine metabolism. In vitro studies in Müller cells have shown that inhibition of de novo serine synthesis impairs mitochondrial function and causes cell death, particularly under oxidative stress (Zhang et al., 2018). At the same time studies using induced pluripotent stem cell (iPSC)-differentiated retinal pigment epithelium cells from MacTel2 patients have shown low serine levels and mitochondrial malfunction independent of serine reduction or elevated 1-deoxysphingolipids (Eade et al., 2023).

Therefore, although alterations in retinal serine metabolism appear to be a major driver of the pathology, the exact interplay between retinal cell types and the specific role of Müller cells in the pathology requires further investigation.

Glaucoma

Glaucoma is a class of progressive eye diseases often associated with the rise of intraocular pressure that causes the gradual damage of the optic nerve and retinal ganglion cells (Almasieh et al., 2012; Weinreb et al., 2014; Jayaram et al., 2023). The intraocular pressure is defined by the balance between the production of aqueous humor by the ciliary body and its drainage through the trabecular meshwork and uveoscleral outflow pathway (Weinreb et al., 2014; Braunger et al., 2015). Dysregulation of either of these due to vascular pathology, trauma, or steroids triggers glaucoma. An increase in the intraocular pressure damages both the optic nerve and the retinal tissue and triggers the reactive response as well as loss of function of retinal glia. Neuroglia in particular contribute to retinal inflammation, metabolic failure, oxidative stress, excitotoxicity, and neuronal damage (Seitz et al., 2013; García-Bermúdez et al., 2021; Salkar et al., 2024). In particular, Müller glial cells undergo reactive remodeling manifested by considerable upregulation of expression of intermediate filaments including glial fibrillary acidic protein (GFAP), nestin, and vimentin. These changes are associated with alterations in the mechanical properties of Müller cells making them stiffer (Xue et al., 2006; Lu et al., 2011).

Increased intraocular pressure also triggers reactive remodeling of retinal astrocytes and microglia (Hernandez, 2000; Tamm et al., 2017; Zeng and Shi, 2018); the latter together with reactive Müller glia secrete pro-inflammatory factors including tumor necrosis factor-α (TNF-α), which can be either beneficial or detrimental for the survival of retinal ganglion cells (Tezel, 2008). Reactive Müller cells also release ciliary neurotrophic factor (*CNTF*) and leukemia inhibitory factor (*LIF*), which mediate neuroprotection (Shinozaki and Koizumi, 2021). Neurodegenerative changes in the retina partially also reflect the loss of homeostatic function of both Müller cells and retinal astrocytes manifested, for instance, by a decreased glutamate clearance accelerating excitotoxicity and death of retinal ganglion cells, the central step in the pathogenesis of glaucoma (Vorwerk et al., 2000).

Similarly, astrocytes populating the optic nerve head (where axons of ganglion cells are exiting the retina) demonstrate a prominent reactive response with an increase in GFAP expression, hypertrophy, and remodeling of the extracellular matrix (Chong and Martin, 2015; Schneider and Fuchshofer, 2016). Reactive astrocytes localized near lamina cribrosa are characterized by the loss of thick primary processes and the emergence of very long GFAP-negative processes projecting into axon bundles (Lye-Barthel et al., 2013). These astrocytes were also reported to upregulate the expression of endothelin-1 and inducible nitric oxide synthase (NOS-2) (Liu and Neufeld, 2000; Salvatore and Vingolo, 2010), which may contribute to aberrant regulation of the vasculature tone. In addition reactive astrocytes of the optic nerve head increase synthesis of Nf-κB and complement, which may contribute to the tissue damage (Nikolskaya et al., 2009). Reactive astrocytes are also characterized by malfunctional mitochondria (Ju et al., 2008), which limits astrocytic energy production and hence negatively impacts their homeostatic and protective capacity.

Glaucoma also causes reactive microgliosis (Karlstetter et al., 2015). Of note, reactive microglia are seen in both the ocular hypertensive eye and the healthy contralateral eye (Rojas et al., 2014).

Reactive microgliosis emerges early in the pathogenesis of glaucoma and the degree of microglial reactivity positively correlates with the damage to the optic nerve head (Wei et al., 2019). Reactive microglyocytes secrete cytokines and chemokines including transforming growth factor-β2 (TGF-β2) and TNF-α, which interact with other retinal cells (Kaczmarek-Hajek et al., 2018). In addition, reactive microglia upregulate the expression and release of matrix metalloproteinases (MMP-1, -2, -3, -14), which remodel and ultimately damage the extracellular matrix (Yuan and Neufeld, 2001) and promote loss of synapses through excess pruning of malfunctional synapses in a complement-dependent manner (Trapp et al., 2007; Wake et al., 2009). Accordingly, anticomplement therapies are discussed for the treatment of glaucoma (Harder et al., 2017; Kim et al., 2022).

Age-related macular degeneration

Age-related macular degeneration (AMD) affects ~4%–12% of people over 60 years of age (with a steep increase above 80 years of age) and is the leading pathology resulting in blindness in the elderly (Deng et al., 2022). Pathophysiologically, AMD is considered to be a primary disease of the retinal pigment epithelium and the associated Bruch's membrane (Wong et al., 2022)—atrophy of the retinal pigment epithelium with loss of its homeostatic functions is a leading cause of retinal degeneration in AMD (Thomas et al., 2021). Histopathologically, the disease is characterized by the appearance of the so-called drusen (deposits of lipoprotein), loss of choriocapillary coverage of Bruch's membrane, and loss of rods (Crabb et al., 2002; Curcio et al., 2024). The etiology of the macular degeneration remains enigmatic. The main risk factor is age that comes with changes in the retinal immune homeostasis (Chen et al., 2019); although some genetic predispositions associated with the complement system (Klein et al., 2005; Black and Clark, 2016; Zauhar et al., 2022; He et al., 2024), as well as environmental burdens such as nutrition and smoking have certain impact (Fritsche et al., 2014). AMD is classified into the dry (80%–90% of cases) and more severe wet (also known as exudative or neovascular AMD accounting for up to 20% of cases (Vyawahare and Shinde, 2022)) forms of the disease. The pathophysiology of dry AMD is led by the atrophy of choriocapillaris, the retinal pigment epithelium, and photoreceptors (Curcio et al., 2024). The wet form of AMD often is driven by choroidal neovascularization. The choroidal vessels penetrate Bruch's membrane and infiltrate the subretinal space, causing hemorrhage as these vessels are leaky, inflammation, and retinal detachment.

Neuroglial changes in AMD involve disruption of Müller glia, reactive astrogliosis, and even glial phagocytosis of ganglion neurons (Telegina et al., 2018; Menon et al., 2019). Exploiting the power of scRNA sequencing on the human retina, the activity pattern of 585 genes in the vicinity of the 34 AMD risk loci identified by AMD gene association studies has been studied and it was found that the majority of these genes were associated with Müller glia and astrocytes (Menon et al., 2019). This pattern was also seen for the top AMD-associated genes, such as CFI, TIMP3, VEGFA, and COL4A3, which are the genes closest to the leading GWAS variants, clearly suggesting that changes in Müller glia due to their response to tissue stress may play a key role in disease progression. Others even define an AMD Müller cell phenotype based on scRNA sequencing data that is distinct from classical gliotic Müller cells (Orozco et al., 2023) even though a clear upregulation of GFAP, a classical marker of Müller cell gliosis has been associated with AMD progression and especially with sites of drusen formation (Wu et al., 2003). These AMD-associated changes in Müller cells included increased expression of the drusen components clusterin/CLU and CRYAB (Crabb et al., 2002; Umeda et al., 2005) in AMD samples. Clusterin is a known substrate of the protease HTRA1, a leading candidate gene associated with the onset and progression of AMD (Tom et al., 2020; Thee et al., 2022). CRYAB is a widely expressed small heat shock protein that is abundant in the soft drusen associated with AMD (Crabb et al., 2002). While these findings from transcriptomic approaches suggest that the Müller cell contribution to AMD pathology is certainly much more complex than initially thought, these findings require further functional validation. However, they also highlight the need for increased efforts to understand the role of Müller cells in AMD pathology (Navneet et al., 2024), to enable the development of novel therapies for AMD potentially targeting Müller cells. For example, one such attempt to target the expression of a truncated version of complement factor F to Müller cells showed promising results in a preclinical animal model (Biber et al., 2024), which is one of many reasons to encourage future research into AMD treatment options in this direction (Navneet et al., 2024).

Similarly, microglia get reactive in both human patients and animal models of the disease (Rashid et al., 2019). Moreover, in the context of AMD, microglia were found to migrate into the subretinal space. The injection of microglia in this space in mice damaged the RPE and triggered neovascularization, mimicking the pathogenesis of the disease (Ma et al., 2009).

Retinitis pigmentosa

Retinitis pigmentosa covers a wide range of inherited eye diseases caused my mutations of ~150 different genes linked to various functions. Photoreceptor degeneration, which results in progressive blindness represents the common pathogenetic feature. Both reactivity of Müller cells and astrocytes manifested, for example, by an increased level of GFAP and vimentin are characteristic signs (Roesch et al., 2012). Müller cells also upregulate the expression of genes responsible for the maintenance of photoreceptors and energy production, indicating that reactive Müller cells are neuroprotective (Tomita et al., 2021). Computational molecular phenotyping (CMP) facilitates small molecule fingerprinting and thus the evaluation of metabolite levels in each retinal cell type. Applied to the retina of transgenic P347L rabbits (a model of human autosomal dominant retinitis pigmentosa (Kondo et al., 2009), this methodology revealed distinct variations in metabolite levels among Müller cells. Specifically, neighboring Müller cells exhibited differences in taurine, glutamate, glutamine, glutathione, glutamine synthetase, and CRALBP (Pfeiffer et al., 2020). These variations showed no correlation between metabolites, indicating that the changes are functionally chaotic rather than merely heterogeneous. The inability of any clustering algorithm to classify Müller cells into a single category in the TgP347L retina formally proves a pronounced metabolic variability present in the degenerating retina. It is discussed that although retinal degeneration is the trigger, Müller cell metabolic alterations do not form a coherent response to the microenvironment. While glutamine synthetase is thought to be the primary enzyme for converting glutamate to glutamine in the retina, alternative pathways seem to emerge in the degenerating retina. This long-term remodeling leads to a loss of Müller cell homeostasis, function, and identity, which could have a negative impact on therapeutic interventions targeting neurons alone—especially since neurons also undergo significant metabolic changes that cannot be easily reversed even if therapy is successful (Ayten et al., 2024).

Recent research implies that Müller cells are not only responding to the RP-associated photoreceptor or retinal pigment epithelium death because of disease-causing mutations specific to photoreceptors or retinal pigment epithelium, but the advent of cell type-specific unbiased omics approaches is leading to the identification of old and new genes that are highly and sometimes very specifically expressed in Müller cells that are associated with retinitis pigmentosa. One such example is the *USH1C* gene, which encodes harmonin (Nagel-Wolfrum et al., 2023). Mutations in this gene cause Usher syndrome (USH), a clinically and genetically heterogeneous disease that is the most common cause of hereditary deaf-blindness in humans (Friedman et al., 2011). Usher syndrome is classified as a retinal ciliopathy due to abnormalities in primary cilia including those of photoreceptor cells. Of the three clinical subtypes (USH1, USH2, and USH3), USH1 caused by mutations in *USH1C* is the most severe. It is characterized by severe congenital hearing loss or deafness, balance problems, and early onset of progressive retinal degeneration, specifically autosomal recessive retinitis pigmentosa, before puberty (Fuster-Garcia et al., 2021). The scaffold protein harmonin (USH1C) is expressed in the human retina through different splice variants that localize to multiple subcellular compartments of photoreceptor cells and, importantly, at significant levels also in Müller glial cells. Defects in USH1C/harmonin are thought to affect its functions in these compartments, leading to progressive retinal degeneration and visual impairment in USH1C patients. Understanding these cellular malfunctions may reveal novel targets for treatment, with Müller cells as a promising new target (Nagel-Wolfrum et al., 2023). Progression of retinitis pigmentosa also involves reactive microgliosis; while microglial secretion of various cytokines and increased production of reactive oxygen species may contribute to photoreceptors death (Rashid et al., 2019).

Diabetic retinopathy

Diabetes affects the nervous system in many ways, causing several types of neuropathy, of which diabetic retinopathy (DR) is one of the most serious (Antonetti et al., 2012; Stitt et al., 2016; Hammes, 2018; Semeraro et al., 2019). It is the remodeling of retinal vessels including vasoregression (i.e., formation of avascular capillaries) and impairment of the retinal vascularization that is at the heart of DR, resulting in infarcts in the nerve fiber layer known as "cotton wool" spots and retinal hypoxia (Coorey et al., 2012). Another early pathomorphologic hallmark of DR is the thickening of the retinal vascular basement membrane caused by the hyperglycemia-dependent increase in the production of fibronectin and collagen (Lechner et al., 2017) and a loss of pericytes contributing to the breakdown of the BRB. Even before microvascular remodeling, changes in other nonvascular retinal cell types, including Müller cells, can lead to edema formation and neurodegeneration (Sohn et al., 2016; Hammes, 2018).

Not surprisingly, DR is also associated with a complex reactive remodeling of Müller cells and retinal astrocytes, key players in the neurovascular unit of the retina (Hu et al., 2014), and their malfunction in DR contributes to the progressive decline in retinal function (Grigsby et al., 2014; Lechner et al., 2017; Ghaseminejad et al., 2020; Pfaller et al., 2024). In preclinical models, Müller cells gliosis is observed prior to vassoregression, neurodegeneration, and microgliosis (Hammes, 2018). Müller glia in the diabetic retina display pronounced reactivity, with a significant increase in GFAP and nestin expression, as well as an increase in secretion of *vascular endothelial growth factor (VEGF)* and other neuroprotective agents (Lechner et al., 2017). One such protective factor seems to be IL-33. Knockout of IL-33 in a mouse model of DR promoted inflammation and neurodegeneration (Augustine et al., 2023). Moreover, a negative effect on the glutamate metabolism, neurotransmitter recycling, and neurotrophin secretion by Müller cells was observed.

Müller cells can also initiate and propagate inflammatory signals in DR. For example, they release ATP, which activates $P2X_7$ purinoceptors that are abundantly expressed by microglia (Kaczmarek-Hajek et al., 2018), contributing to neuroinflammation and retinal damage driven by the release of TNF-α and IL-1β by microglia (Abcouwer, 2017). As in AMD, it is likely that Müller cells contribute to the dysregulated complement cascade observed in DR (Huang et al., 2018; Mandava et al., 2020; Shahulhameed et al., 2020), as they produce and secrete key components of the alternative and classical pathways, as well as the complement C3 (Pauly et al., 2019; Zauhar et al., 2022). As discussed earlier, Müller cells can thus induce and enhance reactive microgliosis as microglia are the major retinal cell population expressing receptors for complement activation products. This Müller cell–microglial interaction is therefore likely to be a route by which neuroinflammation unfolds in DR. Therefore, targeting microgliosis has been discussed as a potential treatment option in DR (Arroba and Valverde, 2017; Wang et al., 2017a).

It has been demonstrated that metabolic dysregulation and mitochondrial malfunction in Müller cells significantly impact their homeostatic functions, e.g., the maintenance of the retinal ion and volume homeostasis (Pannicke et al., 2006; Reichenbach et al., 2007; Krugel et al., 2011; Pfaller et al., 2024). These alterations might also contribute to the breakdown of the blood-retinal barrier accelerating the accumulation of intraretinal fluid and the formation of diabetic macular edema (DME), which can develop at any stage of DR and is the leading cause of vision loss in DR. Among various therapeutic approaches, intravitreal administration of glucocorticoids such as triamcinolone acetonide (TA) has been effective in reducing cytotoxic edema in DR (Brooks et al., 2004). Notably, treating Müller cell-ablated mouse models with TA, a selective glucocorticoid receptor agonist, mitigates effects such as photoreceptor degeneration, vascular leakage, and neovascularization, all typical of DR (Shen et al., 2014). TA's effectiveness in reducing vascular leakage, inhibiting VEGF secretion (Itakura et al., 2006), and preventing Müller cell swelling underscores (Wurm et al., 2008a) its therapeutic potential in DR. This effect of TA seems to be primarily mediated through acting on the glucocorticoid receptor (GR) that in the retina is expressed predominantly in Müller cells (Gallina et al., 2014; Pfaller et al., 2024). However, probably due to persistently high blood corticosteroid levels that trigger a negative feedback loop, GR expression in Müller cells of the diabetic retina in a mouse model is downregulated (Pfaller et al., 2024). As an increase in systemically circulating corticosteroids has also been described in DR patients (Nakao and Yokoyama, 1976; Chiodini et al., 2007), a similar aberrant regulation of GR in the retina may limit the efficacy of intravitreal corticosteroids in the treatment of DR. In a gene therapeutic approach applied to a DR mouse model, a beneficial effect of overexpressing GR in Müller cells (instead of injecting more of its ligand) was demonstrated (Pfaller et al., 2024), suggesting a possible new therapeutic concept for DR therapy. Indeed, also another gene therapy approach based on the overexpression of cellular retinaldehyde binding protein (RLBP1), a Müller cell protein that acts as a soluble retinoid carrier essential for proper photoreceptor function in the visual cycle, has shown impressive beneficial effects on neuronal function and survival in a preclinical study (Niu et al., 2021). In light of all these findings, the search for tools to improve Müller glial homeostasis functions in the context of DR could be a game changer for the development of long-term efficient treatment options in DR, as they would interfere with early steps in the progression of the pathology, i.e., the remodeling of the neurovascular unit.

Retinopathy of prematurity

Retinopathy of prematurity (ROP) is a leading cause of visual impairment in children born prematurely. When a premature baby is given high levels of oxygen to support their underdeveloped lungs, the normal vascular development in the retina is disrupted, resulting in an avascular peripheral retina. Upon returning to normal air, this avascular area experiences hypoxia, triggering abnormal blood vessel growth on the retinal surface. Although current treatments aim to address this abnormal angiogenesis, many children still suffer from vision problems, indicating that normal neuronal development might also be affected (O'Connor et al., 2002). Abnormal vascularization causes hypoxia and death of neurons and retinal astrocytes (Chan-Ling and Stone, 1992; Downie et al., 2007). Moreover, this pathology is characterized by prominent reactivity of Müller cells and astrocytes as well as reactive microgliosis; all contributing to retinal inflammation (Fletcher et al., 2010).

Since Müller cells, together with astrocytes, produce angiogenic growth factors, such as VEGF, and act as a scaffold for the developing vasculature in the embryonic retina (Stone et al., 1995; Dorrell and Friedlander, 2006), it was tested whether modulation of the Müller cell response in preclinical models of ROP could improve the outcome of the pathology. A driver of vascular remodeling and leaky vessel formation, as in many other retinal neovascularization-related pathologies, is VEGF. In a rat model of ROP, downregulation of VEGF specifically in Müller cells led to a reduction in aberrant vessel formation (Jiang et al., 2014) supporting a central role of Müller cells in the pathology. This hypothesis is further underpinned by findings regarding the role of norrin in the retina (Ohlmann and Tamm, 2012; Kassumeh et al., 2021). Norrin is specifically expressed in Müller cells and binds to Frizzled-4 receptors, activating the Wnt/β-catenin pathway. This interaction is crucial for the formation of retinal blood vessels during development. Norrin also promotes vessel regrowth and protects retinal ganglion cells from injury (Seitz et al., 2010; Boesl et al., 2020). Therefore, the norrin and/or the Wnt/β-catenin signaling pathway is a potential target for treating ROP and other ischemic retinopathies (Ohlmann and Tamm, 2012). Finally, it is also likely that ROP involves a coordinated action of Müller cells and microglia, as the latter are involved in fine-tuning retinal angiogenesis (Checchin et al., 2006).

Endophthalmitis

Eye infection with bacteria, most commonly with *Staphylococcus epidermidis* and *Staphylococcus aureus*, is one of the most serious complications of ocular surgery or intravitreal injections (Callegan et al., 2007). Bacterial inflammation and retinal innate immunity involve neuroglia, which produce a number of complex responses (Kumar et al., 2013). Müller glia respond to infection by reactive gliosis, probably triggered by pathogen activation of toll-like receptors (TLRs). In particular, simulation of Müller cell-resident TLR initiates synthesis and secretion of interleukins (IL-1β, IL-6, IL-8), and TNF-α (Kumar and Shamsuddin, 2012). There is also indirect evidence that Müller can perform phagocytosis and possibly remove bacteria infecting the retina (Mano and Puro, 1990). Moreover, stimulation of TRL2 receptors in Müller cells was shown to release antimicrobial peptide LL-37 (Shamsuddin and Kumar, 2011). The bactericide effect of Müller glia and reactive retinal microglia can also be mediated through TNF-α, NO, and superoxide (Stevenson et al., 2010; Kumar et al., 2013).

CONCLUSIONS

Retinal neuroglia are central to its physiologic function and pathologic remodeling. Müller glial cells are the most abundant neuroglia in the retina. Müller cells define retinal columnar cytoarchitecture, act as living light guides, provide homeostatic support, immune modulation, and neuroprotection. In pathology, retinal neuroglia undergo complex changes including upregulation of neuroprotection, reactive gliosis, and functional asthenia. The balance between all these changes determines the progression and outcome of retinal diseases.

ACKNOWLEDGMENTS

We extend our heartfelt gratitude to all colleagues and collaborators who supported our journey in the exceptionally fascinating world of retinal glia research, with a special focus on Müller cells. A major thank you goes to Andreas Reichenbach and Andreas Bringmann, whose significant research contributions and enthusiasm for Müller cells have inspired and advanced this field, motivating new generations of Müller cell researchers. Part of this work was funded by the German Research Foundation (DFG) to A.G. (GR4403/1-1; GR 4403/2-1; GR 4403/5-1, GR 4403/7-1; GR 4403/7-2; GR 4403/9-1) and the ProRetina Foundation to A.G. (Pro-Re/Seed/Grosche.1-2014; Pro-Re/Seed/Kaplan-Grosche.8-2019).

References

Abcouwer SF (2017). Muller cell-microglia cross talk drives neuroinflammation in diabetic retinopathy. Diabetes 66: 261–263.

Agre P, King LS, Yasui M et al. (2002). Aquaporin water channels—from atomic structure to clinical medicine. J Physiol 542: 3–16.

Agte S, Junek S, Matthias S et al. (2011). Muller glial cell-provided cellular light guidance through the vital guinea-pig retina. Biophys J 101: 2611–2619.

Almasieh M, Wilson AM, Morquette B et al. (2012). The molecular basis of retinal ganglion cell death in glaucoma. Prog Retin Eye Res 31: 152–181.

Andersen JV, Schousboe A, Verkhratsky A (2022). Astrocyte energy and neurotransmitter metabolism in Alzheimer's disease: integration of the glutamate/GABA-glutamine cycle. Prog Neurobiol 217: 102331.

Antonetti DA, Klein R, Gardner TW (2012). Diabetic retinopathy. N Engl J Med 366: 1227–1239.

Araque A, Parpura V, Sanzgiri RP et al. (1999). Tripartite synapses: glia, the unacknowledged partner. Trends Neurosci 22: 208–215.

Arroba AI, Valverde AM (2017). Modulation of microglia in the retina: new insights into diabetic retinopathy. Acta Diabetol 54: 527–533.

Au A, Hou K, Baumal CR et al. (2018). Radial hemorrhage in Henle layer in macular telangiectasia type 2. JAMA Ophthalmol 136: 1182–1185.

Augustine J, Pavlou S, Harkin K et al. (2023). IL-33 regulates Muller cell-mediated retinal inflammation and neurodegeneration in diabetic retinopathy. Dis Model Mech 16:dmm050174.

Ayten M, Diaz-Lezama N, Ghanawi H et al. (2024). Metabolic plasticity in a Pde6b(STOP/STOP) retinitis pigmentosa mouse model following rescue. Mol Metab 88: 101994.

Barros LF (2022). How expensive is the astrocyte? J Cereb Blood Flow Metab 42: 738–745.

Bessis A, Bechade C, Bernard D et al. (2007). Microglial control of neuronal death and synaptic properties. Glia 55: 233–238.

Bharti K, Miller SS, Arnheiter H (2011). The new paradigm: retinal pigment epithelium cells generated from embryonic or induced pluripotent stem cells. Pigment Cell Melanoma Res 24: 21–34.

Biber J, Jabri Y, Glanzer S et al. (2024). Gliosis-dependent expression of complement factor H truncated variants attenuates retinal neurodegeneration following ischemic injury. J Neuroinflammation 21: 56.

Biedermann B, Bringmann A, Reichenbach A (2002). High-affinity GABA uptake in retinal glial (Muller) cells of the guinea pig: electrophysiological characterization, immunohistochemical localization, and modeling of efficiency. Glia 39: 217–228.

Biedermann B, Bringmann A, Franze K et al. (2004). $GABA_A$ receptors in Muller glial cells of the human retina. Glia 46: 302–310.

Biesecker KR, Srienc AI, Shimoda AM et al. (2016). Glial cell calcium signaling mediates capillary regulation of blood flow in the retina. J Neurosci 36: 9435–9445.

Black JR, Clark SJ (2016). Age-related macular degeneration: genome-wide association studies to translation. Genet Med 18: 283–289.

Boesl F, Drexler K, Muller B et al. (2020). Endogenous Wnt/β-catenin signaling in Muller cells protects retinal ganglion cells from excitotoxic damage. Mol Vis 26: 135–149.

Bonelli R, Woods SM, Ansell BRE et al. (2020). Systemic lipid dysregulation is a risk factor for macular neurodegenerative disease. Sci Rep 10: 12165.

Boulton M, Dayhaw-Barker P (2001). The role of the retinal pigment epithelium: topographical variation and ageing changes. Eye (Lond) 15: 384–389.

Braunger BM, Fuchshofer R, Tamm ER (2015). The aqueous humor outflow pathways in glaucoma: a unifying concept of disease mechanisms and causative treatment. Eur J Pharm Biopharm 95: 173–181.

Bringmann A (2019). Structure and function of the bird fovea. Anat Histol Embryol 48: 177–200.

Bringmann A, Biedermann B, Reichenbach A (1999). Expression of potassium channels during postnatal differentiation of rabbit Muller glial cells. Eur J Neurosci 11: 2883–2896.

Bringmann A, Schopf S, Reichenbach A (2000). Developmental regulation of calcium channel-mediated currents in retinal glial (Muller) cells. J Neurophysiol 84: 2975–2983.

Bringmann A, Pannicke T, Moll V et al. (2001). Upregulation of $P2X_7$ receptor currents in Muller glial cells during proliferative vitreoretinopathy. Invest Ophthalmol Vis Sci 42: 860–867.

Bringmann A, Pannicke T, Grosche J et al. (2006). Muller cells in the healthy and diseased retina. Prog Retin Eye Res 25: 397–424.

Bringmann A, Iandiev I, Pannicke T et al. (2009). Cellular signaling and factors involved in Muller cell gliosis: neuroprotective and detrimental effects. Prog Retin Eye Res 28: 423–451.

Bringmann A, Grosche A, Pannicke T et al. (2013). GABA and glutamate uptake and metabolism in retinal glial (Muller) cells. Front Endocrinol (Lausanne) 4: 48.

Bringmann A, Syrbe S, Gorner K et al. (2018). The primate fovea: structure, function and development. Prog Retin Eye Res 66: 49–84.

Brooks Jr HL, Caballero Jr S, Newell CK et al. (2004). Vitreous levels of vascular endothelial growth factor and stromal-derived factor 1 in patients with diabetic retinopathy and cystoid macular edema before and after intraocular injection of triamcinolone. Arch Ophthalmol 122: 1801–1807.

Bruckner E, Grosche A, Pannicke T et al. (2012). Mechanisms of VEGF- and glutamate-induced inhibition of osmotic swelling of murine retinal glial (Muller) cells: indications for the involvement of vesicular glutamate release and connexin-mediated ATP release. Neurochem Res 37: 268–278.

Burris C, Klug K, Ngo IT et al. (2002). How Muller glial cells in macaque fovea coat and isolate the synaptic terminals of cone photoreceptors. J Comp Neurol 453: 100–111.

Buttery RG, Haight JR, Bell K (1990). Vascular and avascular retinae in mammals. A funduscopic and fluorescein angiographic study. Brain Behav Evol 35: 156–175.

Byrne LC, Khalid F, Lee T et al. (2013). AAV-mediated, optogenetic ablation of Muller glia leads to structural and functional changes in the mouse retina. PLoS One 8: e76075.

Callegan MC, Gilmore MS, Gregory M et al. (2007). Bacterial endophthalmitis: therapeutic challenges and host-pathogen interactions. Prog Retin Eye Res 26: 189–203.

Campbell JP, Zhang M, Hwang TS et al. (2017). Detailed vascular anatomy of the human retina by projection-resolved optical coherence tomography angiography. Sci Rep 7: 42201.

Cebulla CM, Zelinka CP, Scott MA et al. (2012). A chick model of retinal detachment: cone rich and novel. PLoS One 7: e44257.

Chan-Ling T, Stone J (1992). Degeneration of astrocytes in feline retinopathy of prematurity causes failure of the blood-retinal barrier. Invest Ophthalmol Vis Sci 33: 2148–2159.

Chao TI, Grosche J, Friedrich KJ et al. (1997). Comparative studies on mammalian Muller (retinal glial) cells. J Neurocytol 26: 439–454.

Checchin D, Sennlaub F, Levavasseur E et al. (2006). Potential role of microglia in retinal blood vessel formation. Invest Ophthalmol Vis Sci 47: 3595–3602.

Chen M, Luo C, Zhao J et al. (2019). Immune regulation in the aging retina. Prog Retin Eye Res 69: 159–172.

Chen I, Pant S, Wu Q et al. (2021). Glutamate transporters have a chloride channel with two hydrophobic gates. Nature 591: 327–331.

Cherepanoff S, Killingsworth MC, Zhu M et al. (2012). Ultrastructural and clinical evidence of subretinal debris accumulation in type 2 macular telangiectasia. Br J Ophthalmol 96: 1404–1409.

Chiodini I, Adda G, Scillitani A et al. (2007). Cortisol secretion in patients with type 2 diabetes: relationship with chronic complications. Diabetes Care 30: 83–88.

Chong RS, Martin KR (2015). Glial cell interactions and glaucoma. Curr Opin Ophthalmol 26: 73–77.

Chvatal A, Verkhratsky A (2018). Early history of neuroglial research: personalities. Neuroglia 1: 245–281.

Conedera FM, Runnels JM, Stein JV et al. (2023). Assessing the role of T cells in response to retinal injury to uncover new therapeutic targets for the treatment of retinal degeneration. J Neuroinflammation 20: 206.

Connors NC, Kofuji P (2006). Potassium channel $K_{ir}4.1$ macromolecular complex in retinal glial cells. Glia 53: 124–131.

Coorey NJ, Shen W, Chung SH et al. (2012). The role of glia in retinal vascular disease. Clin Exp Optom 95: 266–281.

Cowan CS, Renner M, De Gennaro M et al. (2020). Cell types of the human retina and its organoids at single-cell resolution. Cell 182: 1623–1640.e1634.

Crabb JW, Miyagi M, Gu X et al. (2002). Drusen proteome analysis: an approach to the etiology of age-related macular degeneration. Proc Natl Acad Sci U S A 99: 14682–14687.

Cunha-Vaz J, Bernardes R, Lobo C (2011). Blood-retinal barrier. Eur J Ophthalmol 21: S3–S9.

Curcio CA, Kar D, Owsley C et al. (2024). Age-related macular degeneration, a mathematically tractable disease. Invest Ophthalmol Vis Sci 65: 4.

Deliyanti D, Armani R, Casely D et al. (2014). Retinal vasculopathy is reduced by dietary salt restriction: involvement of glia, ENaCα, and the renin-angiotensin-aldosterone system. Arterioscler Thromb Vasc Biol 34: 2033–2041.

Demais V, Pohl A, Wunderlich KA et al. (2022). Release of VAMP5-positive extracellular vesicles by retinal Muller glia in vivo. J Extracell Vesicles 11: e12254.

Deng Y, Qiao L, Du M et al. (2022). Age-related macular degeneration: epidemiology, genetics, pathophysiology, diagnosis, and targeted therapy. Genes Dis 9: 62–79.

Di Pierdomenico J, Martinez-Vacas A, Hernandez-Munoz D et al. (2020). Coordinated intervention of microglial and Muller cells in light-induced retinal degeneration. Invest Ophthalmol Vis Sci 61: 47.

Diaz-Coranguez M, Ramos C, Antonetti DA (2017). The inner blood-retinal barrier: cellular basis and development. Vision Res 139: 123–137.

Diaz-Lezama N, Wolf A, Koch S et al. (2021). PDGF receptor α signaling is key for Muller cell homeostasis functions. Int J Mol Sci 22: 1700.

Dorrell MI, Friedlander M (2006). Mechanisms of endothelial cell guidance and vascular patterning in the developing mouse retina. Prog Retin Eye Res 25: 277–295.

Dowling JE (1979). A new retinal neurone—the interplexiform cell. Trends Neurosci 2: 189–191.

Downie LE, Pianta MJ, Vingrys AJ et al. (2007). Neuronal and glial cell changes are determined by retinal vascularization in retinopathy of prematurity. J Comp Neurol 504: 404–417.

Dubois-Dauphin M, Poitry-Yamate C, de Bilbao F et al. (2000). Early postnatal Muller cell death leads to retinal but not optic nerve degeneration in NSE-Hu-Bcl-2 transgenic mice. Neuroscience 95: 9–21.

Eade K, Gantner ML, Hostyk JA et al. (2021). Serine biosynthesis defect due to haploinsufficiency of PHGDH causes retinal disease. Nat Metab 3: 366–377.

Eade KT, Ansell BRE, Giles S et al. (2023). iPSC-derived retinal pigmented epithelial cells from patients with macular telangiectasia show decreased mitochondrial function. J Clin Invest 133: e163771.

Eaton MJ, Skatchkov SN, Brune A et al. (2002). SURI and $K_{ir}6.1$ subunits of K_{ATP}-channels are co-localized in retinal glial (Muller) cells. Neuroreport 13: 57–60.

Elkabes S, DiCicco-Bloom EM, Black IB (1996). Brain microglia/macrophages express neurotrophins that selectively regulate microglial proliferation and function. J Neurosci 16: 2508–2521.

Fan W, Huang W, Chen J et al. (2022). Retinal microglia: functions and diseases. Immunology 166: 268–286.

Fisher AB (2011). Peroxiredoxin 6: a bifunctional enzyme with glutathione peroxidase and phospholipase A(2) activities. Antioxid Redox Signal 15: 831–844.

Fletcher EL, Downie LE, Hatzopoulos K et al. (2010). The significance of neuronal and glial cell changes in the rat retina during oxygen-induced retinopathy. Doc Ophthalmol 120: 67–86.

Fontainhas AM, Wang M, Liang KJ et al. (2011). Microglial morphology and dynamic behavior is regulated by ionotropic glutamatergic and GABAergic neurotransmission. PLoS One 6: e15973.

Fortune B (2019). Pulling and tugging on the retina: mechanical impact of glaucoma beyond the optic nerve head. Invest Ophthalmol Vis Sci 60: 26–35.

Francke M, Pannicke T, Biedermann B et al. (1996). Sodium current amplitude increases dramatically in human retinal glial cells during diseases of the eye. Eur J Neurosci 8: 2662–2670.

Francke M, Makarov F, Kacza J et al. (2001). Retinal pigment epithelium melanin granules are phagocytozed by Muller glial cells in experimental retinal detachment. J Neurocytol 30: 131–136.

Francke M, Weick M, Pannicke T et al. (2002). Upregulation of extracellular ATP-induced Muller cell responses in a dispase model of proliferative vitreoretinopathy. Invest Ophthalmol Vis Sci 43: 870–881.

Franke H, Verkhratsky A, Burnstock G et al. (2012). Pathophysiology of astroglial purinergic signalling. Purinergic Signal 8: 629–657.

Franze K, Grosche J, Skatchkov SN et al. (2007). Muller cells are living optical fibers in the vertebrate retina. Proc Natl Acad Sci U S A 104: 8287–8292.

Friedman TB, Schultz JM, Ahmed ZM et al. (2011). Usher syndrome: hearing loss with vision loss. Adv Otorhinolaryngol 70: 56–65.

Fritsche LG, Fariss RN, Stambolian D et al. (2014). Age-related macular degeneration: genetics and biology coming together. Annu Rev Genomics Hum Genet 15: 151–171.

Fuster-Garcia C, Garcia-Bohorquez B, Rodriguez-Munoz A et al. (2021). Usher syndrome genetics of a human ciliopathy. Int J Mol Sci 22: 6723.

Gallina D, Zelinka C, Fischer AJ (2014). Glucocorticoid receptors in the retina, Muller glia and the formation of Muller glia-derived progenitors. Development 141: 3340–3351.

Gantner ML, Eade K, Wallace M et al. (2019). Serine and lipid metabolism in macular disease and peripheral neuropathy. N Engl J Med 381: 1422–1433.

García-Bermúdez MY, Freude KK, Mouhammad ZA et al. (2021). Glial cells in glaucomafriends, foes, and potential therapeutic targets. Front Neurol 12: 624983.

Garhofer G, Zawinka C, Resch H et al. (2004). Reduced response of retinal vessel diameters to flicker stimulation in patients with diabetes. Br J Ophthalmol 88: 887–891.

Gass JD, Blodi BA (1993). Idiopathic juxtafoveolar retinal telangiectasis. Update of classification and follow-up study. Ophthalmology 100: 1536–1546.

Ghaseminejad F, Kaplan L, Pfaller AM et al. (2020). The role of Muller cell glucocorticoid signaling in diabetic retinopathy. Graefes Arch Clin Exp Ophthalmol 258: 221–230.

Ginhoux F, Prinz M (2015). Origin of microglia: current concepts and past controversies. Cold Spring Harb Perspect Biol 7: a020537.

Grigsby JG, Cardona SM, Pouw CE et al. (2014). The role of microglia in diabetic retinopathy. J Ophthalmol 2014: 705783.

Grimes WN, Berson DM, Sabnis A et al. (2024). The retina's neurovascular unit: Muller glial sheaths and neuronal contacts. bioRxiv. https://doi.org/10.1101/2024.04.30.591885.

Grosche A, Pannicke T, Chen J et al. (2013). Disruption of endogenous purinergic signaling inhibits vascular endothelial growth factor- and glutamate-induced osmotic volume regulation of Muller glial cells in knockout mice. Ophthalmic Res 50: 209–214.

Grosche A, Hauser A, Lepper MF et al. (2016). The proteome of native adult Muller glial cells from murine retina. Mol Cell Proteomics 15: 462–480.

Grunder T, Kohler K, Kaletta A et al. (2000). The distribution and developmental regulation of NMDA receptor subunit proteins in the outer and inner retina of the rat. J Neurobiol 44: 333–342.

Guido ME, Marchese NA, Rios MN et al. (2022). Non-visual opsins and novel photo-detectors in the vertebrate inner retina mediate light responses within the blue spectrum region. Cell Mol Neurobiol 42: 59–83.

Hammer DX, Agrawal A, Villanueva R et al. (2020). Label-free adaptive optics imaging of human retinal macrophage distribution and dynamics. Proc Natl Acad Sci U S A 117: 30661–30669.

Hammes HP (2018). Diabetic retinopathy: hyperglycaemia, oxidative stress and beyond. Diabetologia 61: 29–38.

Hang CY, Kitahashi T, Parhar IS (2016). Neuronal organization of deep brain opsin photoreceptors in adult teleosts. Front Neuroanat 10: 48.

Harada T, Harada C, Watanabe M et al. (1998). Functions of the two glutamate transporters GLAST and GLT-1 in the retina. Proc Natl Acad Sci U S A 95: 4663–4666.

Harada T, Harada C, Nakamura K et al. (2007). The potential role of glutamate transporters in the pathogenesis of normal tension glaucoma. J Clin Invest 117: 1763–1770.

Harder JM, Braine CE, Williams PA et al. (2017). Early immune responses are independent of RGC dysfunction in glaucoma with complement component C3 being protective. Proc Natl Acad Sci U S A 114: E3839–E3848.

He W, Han X, Ong JS et al. (2024). Genome-wide meta-analysis identifies risk loci and improves disease prediction of age-related macular degeneration. Ophthalmology 131: 16–29.

Hector M, Langmann T, Wolf A (2024). Translocator protein (18 kDa) (Tspo) in the retina and implications for ocular diseases. Prog Retin Eye Res 100: 101249.

Heeren TFC, Chew EY, Clemons T et al. (2020). Macular telangiectasia type 2: visual acuity, disease end stage, and the MacTel Area: MacTel Project Report Number 8. Ophthalmology 127: 1539–1548.

Hernandez MR (2000). The optic nerve head in glaucoma: role of astrocytes in tissue remodeling. Prog Retin Eye Res 19: 297–321.

Hollborn M, Ulbricht E, Rillich K et al. (2011). The human Muller cell line MIO-M1 expresses opsins. Mol Vis 17: 2738–2750.

Housley GD, Bringmann A, Reichenbach A (2009). Purinergic signaling in special senses. Trends Neurosci 32: 128–141.

Hu J, Popp R, Fromel T et al. (2014). Muller glia cells regulate Notch signaling and retinal angiogenesis via the generation of 19,20-dihydroxydocosapentaenoic acid. J Exp Med 211: 281–295.

Huang C, Fisher KP, Hammer SS et al. (2018). Plasma exosomes contribute to microvascular damage in diabetic retinopathy by activating the classical complement pathway. Diabetes 67: 1639–1649.

Hurley JB, Lindsay KJ, Du J (2015). Glucose, lactate, and shuttling of metabolites in vertebrate retinas. J Neurosci Res 93: 1079–1092.

Iandiev I, Dukic-Stefanovic S, Hollborn M et al. (2011). Immunolocalization of aquaporin-6 in the rat retina. Neurosci Lett 490: 130–134.

Inoue Y, Shimazawa M, Noda Y et al. (2017). RS9, a novel Nrf2 activator, attenuates light-induced death of cells of photoreceptor cells and Muller glia cells. J Neurochem 141: 750–765.

Ishii M, Horio Y, Tada Y et al. (1997). Expression and clustered distribution of an inwardly rectifying potassium channel, KAB-2/Kir4.1, on mammalian retinal Muller cell membrane: their regulation by insulin and laminin signals. J Neurosci 17: 7725–7735.

Ishii M, Fujita A, Iwai K et al. (2003). Differential expression and distribution of $K_{ir}5.1$ and $K_{ir}4.1$ inwardly rectifying K^+ channels in retina. Am J Physiol Cell Physiol 285: C260–C267.

Itakura H, Akiyama H, Hagimura N et al. (2006). Triamcinolone acetonide suppresses interleukin-1 beta-mediated increase in vascular endothelial growth factor expression in cultured rat Muller cells. Graefes Arch Clin Exp Ophthalmol 244: 226–231.

Jassim AH, Inman DM (2019). Evidence of hypoxic glial cells in a model of ocular hypertension. Invest Ophthalmol Vis Sci 60: 1–15.

Jassim AH, Nsiah NY, Inman DM (2022). Ocular hypertension results in hypoxia within glia and neurons throughout the visual projection. Antioxidants (Basel) 11: 888.

Jayaram H, Kolko M, Friedman DS et al. (2023). Glaucoma: now and beyond. Lancet 402: 1788–1801.

Jeon CJ, Strettoi E, Masland RH (1998). The major cell populations of the mouse retina. J Neurosci 18: 8936–8946.

Jiang Z, Shen W (2010). Role of neurotransmitter receptors in mediating light-evoked responses in retinal interplexiform cells. J Neurophysiol 103: 924–933.

Jiang Y, Wang H, Culp D et al. (2014). Targeting Muller cell-derived VEGF164 to reduce intravitreal neovascularization in the rat model of retinopathy of prematurity. Invest Ophthalmol Vis Sci 55: 824–831.

Jo AO, Lakk M, Rudzitis CN et al. (2022). TRPV4 and TRPC1 channels mediate the response to tensile strain in mouse Muller cells. Cell Calcium 104: 102588.

Jones BW, Kondo M, Terasaki H et al. (2011). Retinal remodeling in the Tg P347L rabbit, a large-eye model of retinal degeneration. J Comp Neurol 519: 2713–2733.

Ju WK, Kim KY, Lindsey JD et al. (2008). Intraocular pressure elevation induces mitochondrial fission and triggers OPA1 release in glaucomatous optic nerve. Invest Ophthalmol Vis Sci 49: 4903–4911.

Kaczmarek-Hajek K, Zhang J, Kopp R et al. (2018). Re-evaluation of neuronal $P2X_7$ expression using novel mouse models and a $P2X_7$-specific nanobody. Elife 7: e36217.

Kambayashi T, Laufer TM (2014). Atypical MHC class II-expressing antigen-presenting cells: can anything replace a dendritic cell? Nat Rev Immunol 14: 719–730.

Kanow MA, Giarmarco MM, Jankowski CS et al. (2017). Biochemical adaptations of the retina and retinal pigment epithelium support a metabolic ecosystem in the vertebrate eye. Elife 6: e28899.

Kaplan L, Drexler C, Pfaller AM et al. (2023). Retinal regions shape human and murine Muller cell proteome profile and functionality. Glia 71: 391–414.

Kar D, Singireddy R, Kim YJ et al. (2024). Unusual morphology of foveal Muller glia in an adult human born pre-term. Front Cell Neurosci 18: 1409405.

Karlstetter M, Scholz R, Rutar M et al. (2015). Retinal microglia: just bystander or target for therapy? Prog Retin Eye Res 45: 30–57.

Karwoski CJ, Newman EA, Shimazaki H et al. (1985). Light-evoked increases in extracellular K^+ in the plexiform layers of amphibian retinas. J Gen Physiol 86: 189–213.

Karwoski CJ, Lu HK, Newman EA (1989). Spatial buffering of light-evoked potassium increases by retinal Muller (glial) cells. Science 244: 578–580.

Kassumeh S, Weber GR, Nobl M et al. (2021). The neuroprotective role of Wnt signaling in the retina. Neural Regen Res 16: 1524–1528.

Kettenmann H, Hanisch UK, Noda M et al. (2011). Physiology of microglia. Physiol Rev 91: 461–553.

Kettenmann H, Kirchhoff F, Verkhratsky A (2013). Microglia: new roles for the synaptic stripper. Neuron 77: 10–18.

Kim JY, Sohn HJ, Seo JH (2011). Characterization of the antigenic phenotype of alphaB-crystallin-expressing peripapillary glial cells in the developing chick retina. Anat Cell Biol 44: 35–40.

Kim BJ, Liu T, Mastellos DC et al. (2022). Emerging opportunities for C3 inhibition in the eye. Semin Immunol 59: 101633.

Klein RJ, Zeiss C, Chew EY et al. (2005). Complement factor H polymorphism in age-related macular degeneration. Science 308: 385–389.

Kofuji P, Ceelen P, Zahs KR et al. (2000). Genetic inactivation of an inwardly rectifying potassium channel (Kir4.1 subunit) in mice: phenotypic impact in retina. J Neurosci 20: 5733–5740.

Kofuji P, Biedermann B, Siddharthan V et al. (2002). K_{ir} potassium channel subunit expression in retinal glial cells: implications for spatial potassium buffering. Glia 39: 292–303.

Kolb H, Nelson R, Ahnelt P et al. (2001). Cellular organization of the vertebrate retina. Prog Brain Res 131: 3–26.

Kondo M, Sakai T, Komeima K et al. (2009). Generation of a transgenic rabbit model of retinal degeneration. Invest Ophthalmol Vis Sci 50: 1371–1377.

Kreutzberg GW (1996). Microglia: a sensor for pathological events in the CNS. Trends Neurosci 19: 312–318.

Krishnan G, Chatterjee N (2012). Endocannabinoids alleviate proinflammatory conditions by modulating innate immune response in muller glia during inflammation. Glia 60: 1629–1645.

Krugel K, Wurm A, Linnertz R et al. (2010). Erythropoietin inhibits osmotic swelling of retinal glial cells by Janus kinase- and extracellular signal-regulated kinases1/2-mediated release of vascular endothelial growth factor. Neuroscience 165: 1147–1158.

Krugel K, Wurm A, Pannicke T et al. (2011). Involvement of oxidative stress and mitochondrial dysfunction in the osmotic swelling of retinal glial cells from diabetic rats. Exp Eye Res 92: 87–93.

Kugler EC, Greenwood J, MacDonald RB (2021). The "neuro-glial-vascular" unit: the role of glia in neurovascular unit formation and dysfunction. Front Cell Dev Biol 9: 732820.

Kumar A, Shamsuddin N (2012). Retinal Muller glia initiate innate response to infectious stimuli via toll-like receptor signaling. PLoS One 7: e29830.

Kumar A, Pandey RK, Miller LJ et al. (2013). Muller glia in retinal innate immunity: a perspective on their roles in endophthalmitis. Crit Rev Immunol 33: 119–135.

Kusaka S, Puro DG (1997). Intracellular ATP activates inwardly rectifying K^+ channels in human and monkey retinal Muller (glial) cells. J Physiol 500: 593–604.

Labin AM, Ribak EN (2010). Retinal glial cells enhance human vision acuity. Phys Rev Lett 104: 158102.

Larsen BR, Assentoft M, Cotrina ML et al. (2014). Contributions of the Na^+/K^+-ATPase, NKCC1, and $K_{ir}4.1$ to hippocampal K^+ clearance and volume responses. Glia 62: 608–622.

Lechner J, O'Leary OE, Stitt AW (2017). The pathology associated with diabetic retinopathy. Vision Res 139: 7–14.

Lehre KP, Davanger S, Danbolt NC (1997). Localization of the glutamate transporter protein GLAST in rat retina. Brain Res 744: 129–137.

Lindqvist N, Liu Q, Zajadacz J et al. (2010). Retinal glial (Muller) cells: sensing and responding to tissue stretch. Invest Ophthalmol Vis Sci 51: 1683–1690.

Lindsay KJ, Du J, Sloat SR et al. (2014). Pyruvate kinase and aspartate-glutamate carrier distributions reveal key metabolic links between neurons and glia in retina. Proc Natl Acad Sci U S A 111: 15579–15584.

Linnertz R, Wurm A, Pannicke T et al. (2011). Activation of voltage-gated Na^+ and Ca^{2+} channels is required for glutamate release from retinal glial cells implicated in cell volume regulation. Neuroscience 188: 23–34.

Lipp S, Wurm A, Pannicke T et al. (2009). Calcium responses mediated by type 2 IP3-receptors are required for osmotic volume regulation of retinal glial cells in mice. Neurosci Lett 457: 85–88.

Liu B, Neufeld AH (2000). Expression of nitric oxide synthase-2 (NOS-2) in reactive astrocytes of the human glaucomatous optic nerve head. Glia 30: 178–186.

Liu W, Tang Y, Feng J (2011). Cross talk between activation of microglia and astrocytes in pathological conditions in the central nervous system. Life Sci 89: 141–146.

Llorian-Salvador M, de Fuente AG, McMurran CE et al. (2024). Regulatory T cells limit age-associated retinal inflammation and neurodegeneration. Mol Neurodegener 19: 32.

Lohr C, Grosche A, Reichenbach A et al. (2014). Purinergic neuron-glia interactions in sensory systems. Pflugers Arch 466: 1859–1872.

Long KO, Fisher SK, Fariss RN et al. (1986). Disc shedding and autophagy in the cone-dominant ground squirrel retina. Exp Eye Res 43: 193–205.

Lu YB, Franze K, Seifert G et al. (2006). Viscoelastic properties of individual glial cells and neurons in the CNS. Proc Natl Acad Sci U S A 103: 17759–17764.

Lu YB, Iandiev I, Hollborn M et al. (2011). Reactive glial cells: increased stiffness correlates with increased intermediate filament expression. FASEB J 25: 624–631.

Lye-Barthel M, Sun D, Jakobs TC (2013). Morphology of astrocytes in a glaucomatous optic nerve. Invest Ophthalmol Vis Sci 54: 909–917.

Ma W, Zhao L, Fontainhas AM et al. (2009). Microglia in the mouse retina alter the structure and function of retinal pigmented epithelial cells: a potential cellular interaction relevant to AMD. PLoS One 4: e7945.

MacAulay N (2020). Molecular mechanisms of K^+ clearance and extracellular space shrinkage-Glia cells as the stars. Glia 68: 2192–2211.

Macaulay N, Zeuthen T (2012). Glial K^+ clearance and cell swelling: key roles for cotransporters and pumps. Neurochem Res 37: 2299–2309.

MacDonald RB, Randlett O, Oswald J et al. (2015). Muller glia provide essential tensile strength to the developing retina. J Cell Biol 210: 1075–1083.

Mages K, Grassmann F, Jagle H et al. (2019). The agonistic TSPO ligand XBD173 attenuates the glial response thereby protecting inner retinal neurons in a murine model of retinal ischemia. J Neuroinflammation 16: 43.

Mairot K, Smirnov V, Bocquet B et al. (2021). CRB1-related retinal dystrophies in a cohort of 50 patientsa reappraisal in the light of specific Muller cell and photoreceptor CRB1 isoforms. Int J Mol Sci 22: 12642.

Mandava N, Tirado-Gonzalez V, Geiger MD et al. (2020). Complement activation in the vitreous of patients with proliferative diabetic retinopathy. Invest Ophthalmol Vis Sci 61: 39.

Mano T, Puro DG (1990). Phagocytosis by human retinal glial cells in culture. Invest Ophthalmol Vis Sci 31: 1047–1055.

Mano T, Tokuda N, Puro DG (1991). Interferon-gamma induces the expression of major histocompatibility antigens by human retinal glial cells. Exp Eye Res 53: 603–607.

Marc RE (1992). Structural organization of GABAergic circuitry in ectotherm retinas. Prog Brain Res 90: 61–92.

Marchese NA, Rios MN, Guido ME (2022). The intrinsic blue light responses of avian Muller glial cells imply calcium release from internal stores. ASN Neuro 14. https://doi.org/10.1177/17590914221076698.

Martineau M, Guzman RE, Fahlke C et al. (2017). VGLUT1 functions as a glutamate/proton exchanger with chloride channel activity in hippocampal glutamatergic synapses. Nat Commun 8: 2279.

Masland RH (2012). The neuronal organization of the retina. Neuron 76: 266–280.

Menon M, Mohammadi S, Davila-Velderrain J et al. (2019). Single-cell transcriptomic atlas of the human retina identifies cell types associated with age-related macular degeneration. Nat Commun 10: 4902.

Metea MR, Newman EA (2007). Signalling within the neurovascular unit in the mammalian retina. Exp Physiol 92: 635–640.

Mills 3rd WA, Woo AM, Jiang S et al. (2022). Astrocyte plasticity in mice ensures continued endfoot coverage of cerebral blood vessels following injury and declines with age. Nat Commun 13: 1794.

Molnar T, Yarishkin O, Iuso A et al. (2016). Store-operated calcium entry in Muller glia is controlled by synergistic activation of TRPC and orai channels. J Neurosci 36: 3184–3198.

Müller H (1851). Zur Histologie der Netzhaut. Z Wiss Zool 3: 234–237.

Mysona B, Dun Y, Duplantier J et al. (2009). Effects of hyperglycemia and oxidative stress on the glutamate transporters GLAST and system xc- in mouse retinal Muller glial cells. Cell Tissue Res 335: 477–488.

Nagelhus EA, Ottersen OP (2013). Physiological roles of aquaporin-4 in brain. Physiol Rev 93: 1543–1562.

Nagelhus EA, Horio Y, Inanobe A et al. (1999). Immunogold evidence suggests that coupling of K^+ siphoning and water transport in rat retinal Muller cells is mediated by a coenrichment of $K_{ir}4.1$ and AQP4 in specific membrane domains. Glia 26: 47–54.

Nagel-Wolfrum K, Fadl BR, Becker MM et al. (2023). Expression and subcellular localization of USH1C/harmonin in human retina provides insights into pathomechanisms and therapy. Hum Mol Genet 32: 431–449.

Nakao T, Yokoyama S (1976). Clinical findings in Kawasaki disease (muco-cutaneous lymph node syndrome). Nihon Rinsho 34: 234–237.

Navneet S, Cui X, Zhao J et al. (2019). Excess homocysteine upregulates the NRF2-antioxidant pathway in retinal Muller glial cells. Exp Eye Res 178: 228–237.

Navneet S, Wilson K, Rohrer B (2024). Muller Glial cells in the macula: their activation and cell-cell interactions in age-related macular degeneration. Invest Ophthalmol Vis Sci 65: 42.

Nedergaard M, Verkhratsky A (2012). Artifact versus reality—how astrocytes contribute to synaptic events. Glia 60: 1013–1023.

Newman EA (1985a). Membrane physiology of retinal glial (Muller) cells. J Neurosci 5: 2225–2239.

Newman EA (1985b). Voltage-dependent calcium and potassium channels in retinal glial cells. Nature 317: 809–811.

Newman EA (1993). Inward-rectifying potassium channels in retinal glial (Muller) cells. J Neurosci 13: 3333–3345.

Newman EA (2005). Calcium increases in retinal glial cells evoked by light-induced neuronal activity. J Neurosci 25: 5502–5510.

Newman EA (2013). Functional hyperemia and mechanisms of neurovascular coupling in the retinal vasculature. J Cereb Blood Flow Metab 33: 1685–1695.

Newman E, Reichenbach A (1996). The Muller cell: a functional element of the retina. Trends Neurosci 19: 307–312.

Newman EA, Frambach DA, Odette LL (1984). Control of extracellular potassium levels by retinal glial cell K^+ siphoning. Science 225: 1174–1175.

Nikolskaya T, Nikolsky Y, Serebryiskaya T et al. (2009). Network analysis of human glaucomatous optic nerve head astrocytes. BMC Med Genomics 2: 24.

Nippert AR, Newman EA (2020). Regulation of blood flow in diabetic retinopathy. Vis Neurosci 37: E004.

Niu T, Fang J, Shi X et al. (2021). Pathogenesis study based on high-throughput single-cell sequencing analysis reveals novel transcriptional landscape and heterogeneity of retinal cells in type 2 diabetic mice. Diabetes 70: 1185–1197.

North RA, Verkhratsky A (2006). Purinergic transmission in the central nervous system. Pflugers Arch 452: 479–485.

Nsiah NY, Inman DM (2022). Destabilizing COXIV in Muller glia increases retinal glycolysis and alters scotopic electroretinogram. Cells 11. https://doi.org/10.3390/cells11233756.

O'Connor AR, Stephenson T, Johnson A et al. (2002). Long-term ophthalmic outcome of low birth weight children with and without retinopathy of prematurity. Pediatrics 109: 12–18.

Ohlmann A, Tamm ER (2012). Norrin: molecular and functional properties of an angiogenic and neuroprotective growth factor. Prog Retin Eye Res 31: 243–257.

Oliveira KR, Herculano AM, Crespo-Lopez ME et al. (2010). Pharmacological characterization of glutamate Na^+-independent transport in retinal cell cultures: implications in the glutathione metabolism. Neurochem Int 56: 59–66.

Oransky I (2005). J. Donald M. Gass. Lancet 365: 1302.

Orozco LD, Owen LA, Hofmann J et al. (2023). A systems biology approach uncovers novel disease mechanisms in age-related macular degeneration. Cell Genom 3: 100302.

O'Sullivan ML, Punal VM, Kerstein PC et al. (2017). Astrocytes follow ganglion cell axons to establish an angiogenic template during retinal development. Glia 65: 1697–1716.

Owe SG, Marcaggi P, Attwell D (2006). The ionic stoichiometry of the GLAST glutamate transporter in salamander retinal glia. J Physiol 577: 591–599.

Pacifici F, Arriga R, Sorice GP et al. (2014). Peroxiredoxin 6, a novel player in the pathogenesis of diabetes. Diabetes 63: 3210–3220.

Paisley CE, Kay JN (2021). Seeing stars: development and function of retinal astrocytes. Dev Biol 478: 144–154.

Pannicke T, Fischer W, Biedermann B et al. (2000). $P2X_7$ receptors in Muller glial cells from the human retina. J Neurosci 20: 5965–5972.

Pannicke T, Biedermann B, Uckermann O et al. (2005a). Physiological properties of retinal Muller glial cells from the cynomolgus monkey, Macaca fascicularis—a comparison to human Muller cells. Vision Res 45: 1781–1791.

Pannicke T, Uckermann O, Iandiev I et al. (2005b). Altered membrane physiology in Muller glial cells after transient ischemia of the rat retina. Glia 50: 1–11.
Pannicke T, Iandiev I, Wurm A et al. (2006). Diabetes alters osmotic swelling characteristics and membrane conductance of glial cells in rat retina. Diabetes 55: 633–639.
Paolicelli RC, Sierra A, Stevens B et al. (2022). Microglia states and nomenclature: a field at its crossroads. Neuron 110: 3458–3483.
Pappenheim S (1842). Die specielle Gewebelehre des Auges mit Rücksicht auf Entwicklungsgeschichte und Augenpraxis, Georg Philipp Aderholz. (Original Publication).
Parmalee NL, Schubert C, Merriam JE et al. (2010). Analysis of candidate genes for macular telangiectasia type 2. Mol Vis 16: 2718–2726.
Parmalee NL, Schubert C, Figueroa M et al. (2012). Identification of a potential susceptibility locus for macular telangiectasia type 2. PLoS One 7: e24268.
Pauly D, Agarwal D, Dana N et al. (2019). Cell-type-specific complement expression in the healthy and diseased retina. Cell Rep 29: 2835–2848.e2834.
Perez MT, Ehinger BE, Lindstrom K et al. (1986). Release of endogenous and radioactive purines from the rabbit retina. Brain Res 398: 106–112.
Pfaller AM, Kaplan L, Carido M et al. (2024). The glucocorticoid receptor as a master regulator of the Muller cell response to diabetic conditions in mice. J Neuroinflammation 21: 33.
Pfeiffer RL, Marc RE, Kondo M et al. (2016). Muller cell metabolic chaos during retinal degeneration. Exp Eye Res 150: 62–70.
Pfeiffer RL, Marc RE, Jones BW (2020). Muller cell metabolic signatures: evolutionary conservation and disruption in disease. Trends Endocrinol Metab 31: 320–329.
Pfeiffer-Guglielmi B, Francke M, Reichenbach A et al. (2005). Glycogen phosphorylase isozyme pattern in mammalian retinal Muller (glial) cells and in astrocytes of retina and optic nerve. Glia 49: 84–95.
Pillai EK, Franze K (2024). Mechanics in the nervous system: from development to disease. Neuron 112: 342–361.
Pizzagalli MD, Bensimon A, Superti-Furga G (2021). A guide to plasma membrane solute carrier proteins. FEBS J 288: 2784–2835.
Poitry-Yamate CL, Poitry S, Tsacopoulos M (1995). Lactate released by Muller glial cells is metabolized by photoreceptors from mammalian retina. J Neurosci 15: 5179–5191.
Pow DV, Crook DK (1995). Immunocytochemical evidence for the presence of high levels of reduced glutathione in radial glial cells and horizontal cells in the rabbit retina. Neurosci Lett 193: 25–28.
Powner MB, Gillies MC, Zhu M et al. (2013). Loss of Muller's cells and photoreceptors in macular telangiectasia type 2. Ophthalmology 120: 2344–2352.
Powner MB, Woods SM, Zhu M et al. (2018). Fundus-wide subretinal and pigment epithelial abnormalities in macular telangiectasia type 2. Retina 38: S105–S113.
Provis JM (2001). Development of the primate retinal vasculature. Prog Retin Eye Res 20: 799–821.
Puro DG, Hwang JJ, Kwon O-J et al. (1996). Characterization of an L-type calcium channel expressed by human retinal Müller (glial) cells. Mol Brain Res 37: 41–48.
Quesada A, Aguilera Y, Caparros R et al. (2011). Myelin oligodendrocyte-specific protein is expressed in Muller cells of myelinated vertebrate retinas. J Neurosci Res 89: 674–688.
Quinn PM, Mulder AA, Henrique Alves C et al. (2019). Loss of CRB2 in Muller glial cells modifies a CRB1-associated retinitis pigmentosa phenotype into a Leber congenital amaurosis phenotype. Hum Mol Genet 28: 105–123.
Ransohoff RM, Perry VH (2009). Microglial physiology: unique stimuli, specialized responses. Annu Rev Immunol 27: 119–145.
Rashid K, Akhtar-Schaefer I, Langmann T (2019). Microglia in retinal degeneration. Front Immunol 10: 1975.
Rasmussen R, O'Donnell J, Ding F et al. (2020). Interstitial ions: a key regulator of state-dependent neural activity? Prog Neurobiol 193: 101802.
Rauen T (2000). Diversity of glutamate transporter expression and function in the mammalian retina. Amino Acids 19: 53–62.
Rauen T, Taylor WR, Kuhlbrodt K et al. (1998). High-affinity glutamate transporters in the rat retina: a major role of the glial glutamate transporter GLAST-1 in transmitter clearance. Cell Tissue Res 291: 19–31.
Reichelt W, Hernandez M, Damian RT et al. (1997a). Voltage- and GABA-evoked currents from Muller glial cells of the baboon retina. Neuroreport 8: 541–544.
Reichelt W, Stabel-Burow J, Pannicke T et al. (1997b). The glutathione level of retinal Muller glial cells is dependent on the high-affinity sodium-dependent uptake of glutamate. Neuroscience 77: 1213–1224.
Reichenbach A (1987). Quantitative and qualitative morphology of rabbit retinal glia. A light microscopical study on cells both in situ and isolated by papaine. J Hirnforsch 28: 213–220.
Reichenbach A (1989). Attempt to classify glial cells by means of their process specialization using the rabbit retinal Muller cell as an example of cytotopographic specialization of glial cells. Glia 2: 250–259.
Reichenbach A, Bringmann A (2010). Müller cells in the healthy and diseased retina, Springer. (Original Publication).
Reichenbach A, Bringmann A (2012). Retina: neuroanatomy and physiology. In: DW Pfaff (Ed.), Neuroscience in the 21st century. Springer557–628.
Reichenbach A, Bringmann A (2013). New functions of Muller cells. Glia 61: 651–678.
Reichenbach A, Bringmann A (2015). Retinal glia, Morgan & Claypool Life Sciences. (Original Publication).
Reichenbach A, Bringmann A (2020). Glia of the human retina. Glia 68: 768–796.
Reichenbach A, Robinson SR (1995). Phylogenetic constraints on retinal organisation and development. Prog Retin Eye Res 15: 139–171.
Reichenbach A, Schnitzer J, Friedrich A et al. (1991). Development of the rabbit retina: II. Muller cells. J Comp Neurol 311: 33–44.
Reichenbach A, Fromter C, Engelmann R et al. (1995). Muller glial cells of the tree shrew retina. J Comp Neurol 360: 257–270.
Reichenbach A, Wurm A, Pannicke T et al. (2007). Muller cells as players in retinal degeneration and edema. Graefes Arch Clin Exp Ophthalmol 245: 627–636.
Rillich K, Gentsch J, Reichenbach A et al. (2009). Light stimulation evokes two different calcium responses in Muller glial cells of the guinea pig retina. Eur J Neurosci 29: 1165–1176.
Rios MN, Marchese NA, Guido ME (2019). Expression of non-visual opsins Opn3 and Opn5 in the developing inner retinal cells of birds. Light-responses in Muller glial cells. Front Cell Neurosci *13*: 376.
Robinson SR, Dreher Z (1990). Muller cells in adult rabbit retinae: morphology, distribution and implications for function and development. J Comp Neurol 292: 178–192.
Roesch K, Jadhav AP, Trimarchi JM et al. (2008). The transcriptome of retinal Muller glial cells. J Comp Neurol 509: 225–238.
Roesch K, Stadler MB, Cepko CL (2012). Gene expression changes within Muller glial cells in retinitis pigmentosa. Mol Vis 18: 1197–1214.
Rojas L, Orkand RK (1999). K^+ channel density increases selectively in the endfoot of retinal glial cells during development of Rana catesbiana. Glia 25: 199–203.

Rojas B, Gallego BI, Ramírez AI et al. (2014). Microglia in mouse retina contralateral to experimental glaucoma exhibit multiple signs of activation in all retinal layers. J Neuroinflammation 11: 1–24.
Rompani SB, Cepko CL (2010). A common progenitor for retinal astrocytes and oligodendrocytes. J Neurosci 30: 4970–4980.
Rose CR, Verkhratsky A (2016). Principles of sodium homeostasis and sodium signalling in astroglia. Glia 64: 1611–1627.
Rose CR, Verkhratsky A (2024). Sodium homeostasis and signalling: the core and the hub of astrocyte function. Cell Calcium 117: 102817.
Rueda EM, Johnson Jr JE, Giddabasappa A et al. (2016). The cellular and compartmental profile of mouse retinal glycolysis, tricarboxylic acid cycle, oxidative phosphorylation, and ~P transferring kinases. Mol Vis 22: 847–885.
Ruggeri M, Major Jr JC, McKeown C et al. (2010). Retinal structure of birds of prey revealed by ultra-high resolution spectral-domain optical coherence tomography. Invest Ophthalmol Vis Sci 51: 5789–5795.
Rutar M, Natoli R, Valter K et al. (2011). Early focal expression of the chemokine Ccl2 by Muller cells during exposure to damage-inducing bright continuous light. Invest Ophthalmol Vis Sci 52: 2379–2388.
Ruzafa N, Pereiro X, Lepper MF et al. (2018). A proteomics approach to identify candidate proteins secreted by Muller glia that protect ganglion cells in the retina. Proteomics 18: e1700321.
Sakami S, Imanishi Y, Palczewski K (2019). Muller glia phagocytose dead photoreceptor cells in a mouse model of retinal degenerative disease. FASEB J 33: 3680–3692.
Salkar A, Wall RV, Basavarajappa D et al. (2024). Glial cell activation and immune responses in glaucoma: a systematic review of human postmortem studies of the retina and optic nerve. Aging Dis 15: 2069–2083.
Salvatore S, Vingolo EM (2010). Endothelin-1 role in human eye: a review. J Ophthalmol 2010: 354645.
Sand A, Schmidt TM, Kofuji P (2012). Diverse types of ganglion cell photoreceptors in the mammalian retina. Prog Retin Eye Res 31: 287–302.
Sanderson J, Dartt DA, Trinkaus-Randall V et al. (2014). Purines in the eye: recent evidence for the physiological and pathological role of purines in the RPE, retinal neurons, astrocytes, Muller cells, lens, trabecular meshwork, cornea and lacrimal gland. Exp Eye Res 127: 270–279.
Santos AM, Calvente R, Tassi M et al. (2008). Embryonic and postnatal development of microglial cells in the mouse retina. J Comp Neurol 506: 224–239.
Scerri TS, Quaglieri A, Cai C et al. (2017). Genome-wide analyses identify common variants associated with macular telangiectasia type 2. Nat Genet 49: 559–567.
Schafer DP, Lehrman EK, Kautzman AG et al. (2012). Microglia sculpt postnatal neural circuits in an activity and complement-dependent manner. Neuron 74: 691–705.
Schmalen A, Lorenz L, Grosche A et al. (2021). Proteomic phenotyping of stimulated Muller cells uncovers profound pro-inflammatory signaling and antigen-presenting capacity. Front Pharmacol 12: 771571.
Schneider M, Fuchshofer R (2016). The role of astrocytes in optic nerve head fibrosis in glaucoma. Exp Eye Res 142: 49–55.
Scholz R, Caramoy A, Bhuckory MB et al. (2015). Targeting translocator protein (18 kDa) (TSPO) dampens pro-inflammatory microglia reactivity in the retina and protects from degeneration. J Neuroinflammation 12: 201.
Schulze M (1859). Observationes de retinae structura penitiori, Published lecture at the Univeesity of Bonn. (Original Publication).
Seitz R, Hackl S, Seibuchner T et al. (2010). Norrin mediates neuroprotective effects on retinal ganglion cells via activation of the Wnt/β-catenin signaling pathway and the induction of neuroprotective growth factors in Muller cells. J Neurosci 30: 5998–6010.
Seitz R, Ohlmann A, Tamm ER (2013). The role of Müller glia and microglia in glaucoma. Cell Tissue Res 353: 339–345.
Semeraro F, Morescalchi F, Cancarini A et al. (2019). Diabetic retinopathy, a vascular and inflammatory disease: therapeutic implications. Diabetes Metab 45: 517–527.
Semyanov A, Verkhratsky A (2021). Astrocytic processes: from tripartite synapses to the active milieu. Trends Neurosci 44: 781–792.
Shahulhameed S, Vishwakarma S, Chhablani J et al. (2020). A systematic investigation on complement pathway activation in diabetic retinopathy. Front Immunol 11: 154.
Shamsuddin N, Kumar A (2011). TLR2 mediates the innate response of retinal Muller glia to Staphylococcus aureus. J Immunol 186: 7089–7097.
Shen W, Fruttiger M, Zhu L et al. (2012). Conditional Mullercell ablation causes independent neuronal and vascular pathologies in a novel transgenic model. J Neurosci 32: 15715–15727.
Shen W, Lee SR, Araujo J et al. (2014). Effect of glucocorticoids on neuronal and vascular pathology in a transgenic model of selective Muller cell ablation. Glia 62: 1110–1124.
Shen W, Lee SR, Mathai AE et al. (2021). Effect of selectively knocking down key metabolic genes in Muller glia on photoreceptor health. Glia 69: 1966–1986.
Shinozaki Y, Koizumi S (2021). Potential roles of astrocytes and Müller cells in the pathogenesis of glaucoma. J Pharmacol Sci 145: 262–267.
Skatchkov SN, Eaton MJ, Shuba YM et al. (2006). Tandem-pore domain potassium channels are functionally expressed in retinal (Muller) glial cells. Glia 53: 266–276.
Slezak M, Grosche A, Niemiec A et al. (2012). Relevance of exocytotic glutamate release from retinal glia. Neuron 74: 504–516.
Sluyter R (2017). The $P2X_7$ receptor. Adv Exp Med Biol 1051: 17–53.
Sohn EH, van Dijk HW, Jiao C et al. (2016). Retinal neurodegeneration may precede microvascular changes characteristic of diabetic retinopathy in diabetes mellitus. Proc Natl Acad Sci U S A 113: E2655–E2664.
Stanke J, Moose HE, El-Hodiri HM et al. (2010). Comparative study of Pax2 expression in glial cells in the retina and optic nerve of birds and mammals. J Comp Neurol 518: 2316–2333.
Stevens B, Johnson MB (2021). The complement cascade repurposed in the brain. Nat Rev Immunol 21: 624–625.
Stevens B, Allen NJ, Vazquez LE et al. (2007). The classical complement cascade mediates CNS synapse elimination. Cell 131: 1164–1178.
Stevenson L, Matesanz N, Colhoun L et al. (2010). Reduced nitro-oxidative stress and neural cell death suggests a protective role for microglial cells in TNFalpha-/- mice in ischemic retinopathy. Invest Ophthalmol Vis Sci 51: 3291–3299.
Stitt AW, Curtis TM, Chen M et al. (2016). The progress in understanding and treatment of diabetic retinopathy. Prog Retin Eye Res 51: 156–186.
Stone J, Itin A, Alon T et al. (1995). Development of retinal vasculature is mediated by hypoxia-induced vascular endothelial growth factor (VEGF) expression by neuroglia. J Neurosci 15: 4738–4747.
Subirada PV, Paz MC, Ridano ME et al. (2018). A journey into the retina: Muller glia commanding survival and death. Eur J Neurosci 47: 1429–1443.
Sukkar B, Oktay L, Sahaboglu A et al. (2023). Inhibition of altered Orai1 channels in Muller cells protects photoreceptors in retinal degeneration. Glia 71: 2511–2526.

Syrbe S, Kuhrt H, Gartner U et al. (2018). Muller glial cells of the primate foveola: an electron microscopical study. Exp Eye Res 167: 110–117.

Tamm ER, Ethier CR, The Lasker/IRRF Initiative on Astrocytes and Glaucomatous Neurodegeneration Participants (2017). Biological aspects of axonal damage in glaucoma: a brief review. Exp Eye Res 157: 5–12.

Tassi M, Calvente R, Marin-Teva JL et al. (2006). Behavior of in vitro cultured ameboid microglial cells migrating on Muller cell end-feet in the quail embryo retina. Glia 54: 376–393.

Taylor S, Srinivasan B, Wordinger RJ et al. (2003). Glutamate stimulates neurotrophin expression in cultured Muller cells. Brain Res Mol Brain Res 111: 189–197.

Telegina DV, Kozhevnikova OS, Kolosova NG (2018). Changes in retinal glial cells with age and during development of age-related macular degeneration. Biochemistry (Mosc) 83: 1009–1017.

Tenckhoff S, Hollborn M, Kohen L et al. (2005). Diversity of aquaporin mRNA expressed by rat and human retinas. Neuroreport 16: 53–56.

Tezel G (2008). TNF-α signaling in glaucomatous neurodegeneration. Prog Brain Res 173: 409–421.

Thee EF, Colijn JM, Cougnard-Gregoire A et al. (2022). The phenotypic course of age-related macular degeneration for ARMS2/HTRA1: the EYE-RISK consortium. Ophthalmology 129: 752–764.

Thomas CJ, Mirza RG, Gill MK (2021). Age-related macular degeneration. Med Clin North Am 105: 473–491.

Tom I, Pham VC, Katschke Jr KJ et al. (2020). Development of a therapeutic anti-HtrA1 antibody and the identification of DKK3 as a pharmacodynamic biomarker in geographic atrophy. Proc Natl Acad Sci U S A 117: 9952–9963.

Tomita Y, Qiu C, Bull E et al. (2021). Müller glial responses compensate for degenerating photoreceptors in retinitis pigmentosa. Exp Mol Med 53: 1748–1758.

Trapp BD, Wujek JR, Criste GA et al. (2007). Evidence for synaptic stripping by cortical microglia. Glia 55: 360–368.

Trost A, Schroedl F, Marschallinger J et al. (2014). Characterization of dsRed2-positive cells in the doublecortin-dsRed2 transgenic adult rat retina. Histochem Cell Biol 142: 601–617.

Ulbricht E, Pannicke T, Hollborn M et al. (2008). Proliferative gliosis causes mislocation and inactivation of inwardly rectifying K^+ (K_{ir}) channels in rabbit retinal glial cells. Exp Eye Res 86: 305–313.

Umeda S, Suzuki MT, Okamoto H et al. (2005). Molecular composition of drusen and possible involvement of anti-retinal autoimmunity in two different forms of macular degeneration in cynomolgus monkey (Macaca fascicularis). FASEB J 19: 1683–1685.

Untiet V, Beinlich FRM, Kusk P et al. (2023). Astrocytic chloride is brain state dependent and modulates inhibitory neurotransmission in mice. Nat Commun 14: 1871.

Verkhratsky A, Butt A (2023). Neuroglia: function and pathology, Academic Press, Elsevier. (Original Publication).

Verkhratsky A, Nedergaard M (2014). Astroglial cradle in the life of the synapse. Philos Trans R Soc Lond B Biol Sci 369: 20130595.

Verkhratsky A, Nedergaard M (2016). The homeostatic astroglia emerges from evolutionary specialization of neural cells. Philos Trans R Soc Lond B Biol Sci 371: 20150428.

Verkhratsky A, Nedergaard M (2018). Physiology of astroglia. Physiol Rev 98: 239–389.

Verkhratsky A, Rose CR (2020). Na^+-dependent transporters: the backbone of astroglial homeostatic function. Cell Calcium 85: 102136.

Verkhratsky A, Semyanov A (2022). The great astroglial metabolic revolution: Mitochondria fuel astrocyte homeostatic support and neuroprotection. Cell Calcium 104: 102583.

Verkhratsky A, Parpura V, Vardjan N et al. (2019). Physiology of astroglia. Adv Exp Med Biol 1175: 45–91.

Verkhratsky A, Semyanov A, Zorec R (2020). Physiology of astroglial excitability. Function (Oxf) 1: zqaa016.

Verkhratsky A, Butt A, Li B et al. (2023). Astrocytes in human central nervous system diseases: a frontier for new therapies. Signal Transduct Target Ther 8: 396.

Vitanova LA (2012). NR2C and NR2D subunits of NMDA receptors in frog and turtle retina. Acta Histochem 114: 813–818.

Voigt J, Grosche A, Vogler S et al. (2015). Nonvesicular release of ATP from rat retinal glial (Muller) cells is differentially mediated in response to osmotic stress and glutamate. Neurochem Res 40: 651–660.

von Toerne C, Menzler J, Ly A et al. (2014). Identification of a novel neurotrophic factor from primary retinal Muller cells using stable isotope labeling by amino acids in cell culture (SILAC). Mol Cell Proteomics 13: 2371–2381.

Vorwerk CK, Naskar R, Schuettauf F et al. (2000). Depression of retinal glutamate transporter function leads to elevated intravitreal glutamate levels and ganglion cell death. Invest Ophthalmol Vis Sci 41: 3615–3621.

Vyawahare H, Shinde P (2022). Age-related macular degeneration: epidemiology, pathophysiology, diagnosis, and treatment. Cureus 14: e29583.

Wagner L, Pannicke T, Frommherz I et al. (2016). Effects of IP_3R2 receptor deletion in the ischemic mouse retina. Neurochem Res 41: 677–686.

Wagner L, Pannicke T, Rupprecht V et al. (2017). Suppression of SNARE-dependent exocytosis in retinal glial cells and its effect on ischemia-induced neurodegeneration. Glia 65: 1059–1071.

Wake H, Moorhouse AJ, Jinno S et al. (2009). Resting microglia directly monitor the functional state of synapses in vivo and determine the fate of ischemic terminals. J Neurosci 29: 3974–3980.

Walton NM, Sutter BM, Laywell ED et al. (2006). Microglia instruct subventricular zone neurogenesis. Glia 54: 815–825.

Wang M, Ma W, Zhao L et al. (2011). Adaptive Muller cell responses to microglial activation mediate neuroprotection and coordinate inflammation in the retina. J Neuroinflammation 8: 173.

Wang J, Lin J, Kaiser U et al. (2017a). Absence of macrophage migration inhibitory factor reduces proliferative retinopathy in a mouse model. Acta Diabetol 54: 383–392.

Wang J, O'Sullivan ML, Mukherjee D et al. (2017b). Anatomy and spatial organization of Muller glia in mouse retina. J Comp Neurol 525: 1759–1777.

Wei X, Cho KS, Thee EF et al. (2019). Neuroinflammation and microglia in glaucoma: time for a paradigm shift. J Neurosci Res 97: 70–76.

Weick M, Wiedemann P, Reichenbach A et al. (2005). Resensitization of P2Y receptors by growth factor-mediated activation of the phosphatidylinositol-3 kinase in retinal glial cells. Invest Ophthalmol Vis Sci 46: 1525–1532.

Weinreb RN, Aung T, Medeiros FA (2014). The pathophysiology and treatment of glaucoma: a review. JAMA 311: 1901–1911.

Weuste M, Wurm A, Iandiev I et al. (2006). HB-EGF: increase in the ischemic rat retina and inhibition of osmotic glial cell swelling. Biochem Biophys Res Commun 347: 310–318.

Winkler BS, Arnold MJ, Brassell MA et al. (2000). Energy metabolism in human retinal Muller cells. Invest Ophthalmol Vis Sci 41: 3183–3190.

Witkovsky P, Stone S, Ripps H (1985). Pharmacological modification of the light-induced responses of Muller (glial) cells in the amphibian retina. Brain Res 328: 111–120.

Wong JHC, Ma JYW, Jobling AI et al. (2022). Exploring the pathogenesis of age-related macular degeneration: a review of the interplay between retinal pigment epithelium dysfunction and the innate immune system. Front Neurosci 16: 1009599.
Wu L (2023). Unraveling the mysteries of macular telangiectasia 2: the intersection of philanthropy, multimodal imaging and molecular genetics. The 2022 founders lecture of the pan American vitreoretinal society. Int J Retina Vitreous 9: 69.
Wu KH, Madigan MC, Billson FA et al. (2003). Differential expression of GFAP in early v late AMD: a quantitative analysis. Br J Ophthalmol 87: 1159–1166.
Wurm A, Iandiev I, Hollborn M et al. (2008a). Purinergic receptor activation inhibits osmotic glial cell swelling in the diabetic rat retina. Exp Eye Res 87: 385–393.
Wurm A, Pannicke T, Wiedemann P et al. (2008b). Glial cell-derived glutamate mediates autocrine cell volume regulation in the retina: activation by VEGF. J Neurochem 104: 386–399.
Wurm A, Erdmann I, Bringmann A et al. (2009a). Expression and function of P2Y receptors on Muller cells of the postnatal rat retina. Glia 57: 1680–1690.
Wurm A, Lipp S, Pannicke T et al. (2009b). Involvement of A_1 adenosine receptors in osmotic volume regulation of retinal glial cells in mice. Mol Vis 15: 1858–1867.
Wurm A, Lipp S, Pannicke T et al. (2010). Endogenous purinergic signaling is required for osmotic volume regulation of retinal glial cells. J Neurochem 112: 1261–1272.
Wurm A, Pannicke T, Iandiev I et al. (2011). Purinergic signaling involved in Muller cell function in the mammalian retina. Prog Retin Eye Res 30: 324–342.
Xu X, Karwoski CJ (1997). The origin of slow PIII in frog retina: current source density analysis in the eyecup and isolated retina. Vis Neurosci 14: 827–833.
Xu Z, Wei Y, Gong J et al. (2014). NRF2 plays a protective role in diabetic retinopathy in mice. Diabetologia 57: 204–213.
Xue LP, Lu J, Cao Q et al. (2006). Müller glial cells express nestin coupled with glial fibrillary acidic protein in experimentally induced glaucoma in the rat retina. Neuroscience 139: 723–732.
Yoshida S, Sotozono C, Ikeda T et al. (2001). Interleukin-6 (IL-6) production by cytokine-stimulated human Muller cells. Curr Eye Res 22: 341–347.
Yu DY, Cringle SJ (2005). Retinal degeneration and local oxygen metabolism. Exp Eye Res 80: 745–751.
Yuan L, Neufeld AH (2001). Activated microglia in the human glaucomatous optic nerve head. J Neurosci Res 64: 523–532.
Zauhar R, Biber J, Jabri Y et al. (2022). As in real estate, location matters: cellular expression of complement varies between macular and peripheral regions of the retina and supporting tissues. Front Immunol 13: 895519.
Zeng HL, Shi JM (2018). The role of microglia in the progression of glaucomatous neurodegeneration- a review. Int J Ophthalmol 11: 143–149.
Zerangue N, Kavanaugh MP (1996). Flux coupling in a neuronal gluta mate transporter. Nature 383: 634–637.
Zhang D, Hu X, Qian L et al. (2010). Astrogliosis in CNS pathologies: is there a role for microglia? Mol Neurobiol 41: 232–241.
Zhang T, Gillies MC, Madigan MC et al. (2018). Disruption of de novo serine synthesis in Muller cells induced mitochondrial dysfunction and aggravated oxidative damage. Mol Neurobiol 55: 7025–7037.
Zucker CL, Bernstein PS, Schalek RL et al. (2020). A connectomics approach to understanding a retinal disease. Proc Natl Acad Sci U S A 117: 18780–18787.

Part II

Neuroglia in psychiatric disorders

Handbook of Clinical Neurology, Vol. 210 (3rd series)
Neuroglia in Neurologic and Psychiatric Disorders, Part II
A. Verkhratsky, L.D. de Witte, E. Aronica, and E.M. Hol, Editors
https://doi.org/10.1016/B978-0-443-19102-2.00028-4

Chapter 15

From womb to world: The interplay between maternal immune activation, neuroglia, and neurodevelopment

DANIELE MATTEI[1], DILANSU GUNEYKAYA[2], BILGE UGURSU[3], AND ALICE BUONFIGLIOLI[4*]

[1]*Nash Family Department of Neuroscience, Icahn School of Medicine at Mount Sinai, Friedman Brain Institute, New York, NY, United States*

[2]*Department of Neurobiology, Harvard Medical School, Boston, MA, United States*

[3]*Department of Psychoneuroimmunology, Max-Delbrück-Center for Molecular Medicine in the Helmholtz Association, Berlin, Germany*

[4]*Department of Psychiatry, Icahn School of Medicine at Mount Sinai, New York, NY, United States*

Abstract

This chapter introduces and discusses maternal immune activation (MIA) as a contributing factor in increasing the risk of neurodevelopmental disorders, particularly in relation to its interactions with neuroglia.

Here we first provide an overview of the neuroglia—astroglia, oligodendroglia, microglia, and radial glial cells—and their important role during early brain development and in adulthood. We then present and discuss MIA, followed by a critical overview of inflammatory molecules and temporal stages associated to maternal inflammation during pregnancy. We provide an overview of animal and human models used to mimic and study MIA. Furthermore, we review the possible interaction between MIA and neuroglia, focusing on the current advances in both modeling and therapeutics. Additionally, we discuss and provide preliminary and interesting insights into the most recent pandemic, COVID-19, and how the infection may be associated to MIA and increased risk for neurodevelopmental disorders.

Finally, we provide a critical overview of challenges and future opportunities to study how MIA may contribute to higher risk of developing neurodevelopmental disorders.

THE DYNAMIC ROLE OF NEUROGLIA IN BRAIN DEVELOPMENT AND ADULTHOOD

Neuroglia have recently been recognized as indispensable players throughout brain development and adulthood, with growing evidence highlighting their involvement in disease progression (Rahman et al., 2022). There are four types of neuroglia—astroglia, oligodendroglia, microglia, and radial glial cells (RGCs) (the latter only present in prenatal brain)—each performing diverse functions from supporting and guiding neuronal networks, to maintaining synapse plasticity and the innate immune response (Verkhratsky and Butt, 2023a). Their contribution to shaping and maintaining the nervous system is essential to understanding the brain microenvironment, ultimately unfolding disease mechanisms.

Astroglia account for 20%–40% of the neuroglial cell population in the brain (Verkhratsky and Butt, 2023b); they are essential for the formation of functional synapses during development and crucial for brain plasticity throughout life (Miller, 2018). Astroglia is a heterogenous class of cells, which include radial astrocytes, marginal astrocytes, glia-like stem cells, parenchymal astrocytes, ependymoglia, and tanycytes. Astroglia display brain region specificity in providing neurons with metabolic and trophic support for the release of growth factors and cytokines as well as regulating synaptic homeostasis and blood–brain stability (Molofsky et al., 2012; Allen and Eroglu, 2017; Verkhratsky and Butt, 2023c).

Oligodendroglia (OLs) are responsible for forming and maintaining the myelin sheaths, which are essential for neuronal signaling. Oligodendroglia include oligodendrocyte precursor cells (OPCs) and myelinating OLs, which account for 50%–75% of all the neuroglia. Only a fraction of OPCs differentiate into mature OLs to initiate the myelination process (Simons and Nave, 2015); however, many remain throughout life and whose function is still unknown (Yi et al., 2023). In adulthood, oligodendroglia continue to maintain myelin homeostasis and participate in remyelination to preserve neural circuits (Richardson et al., 2006; Munyeshyaka and Fields, 2022).

*Correspondence to: Alice Buonfiglioli, Department of Psychiatry, Icahn School of Medicine at Mount Sinai, New York, NY, United States. Tel: +1-3322078542, E-mail: alice.buonfiglioli@mssm.edu

Microglia are the resident immune cells of the central nervous system (CNS). They account for approximately 10% of the cell population, and their role varies depending on the context. Microglia regulate synaptic pruning, which is a process that eliminates excessive number of synapses, perform phagocytosis, and modulate the innate immune response while constantly surveying the CNS microenvironment (Bilbo and Stevens, 2017; Salter and Stevens, 2017). Human (Oosterhof et al., 2019) and animal (Erblich et al., 2011) studies have shown that a lack of microglia or perturbation in microglial numbers and reactive states can have severe consequences for the brain development, suggesting their crucial role as neuroglia during neurodevelopment.

RGCs emerge as pivotal players in orchestrating the formation of the neural architecture in prenatal brain (Miranda-Negrón and García-Arrarás, 2022). RGCs are characterized by their unique bipolar morphology, with processes extending radially from the ventricular to the pial surface during early brain development (Lu et al., 2015). RGC processes provide a scaffold along which migrating neurons reach the specific cortical layers to form the laminar organization essential for functional neural circuitry (Malatesta and Götz, 2013). RGCs undergo symmetric and asymmetric neurogenic divisions to contribute to generating neurons, astroglia, and oligodendroglia (Götz and Huttner, 2005). In addition, RGCs participate in the maintenance of the blood–brain barrier—a vascular structure that controls the molecular and cellular exchange between the brain and the periphery—by secreting vascular endothelial growth factor (VEGF) and retinoic acid (Alvarez et al., 2013). RGCs have also been shown to interact with astroglia and microglia and to coordinate brain development such as neurogenesis, neuronal migration, and synaptic pruning. These interactions are mediated through signaling pathways and cellular communications, ensuring neural circuit formation and the maintenance of brain homeostasis (Allen and Lyons, 2018).

Brain development involves diverse cell types, among which the interaction between RGCs and microglia emerges as a critical axis, orchestrating not only neural circuit formation, maintenance, and repair (Ueno et al., 2013) but also immune maturation (Barger et al., 2019). During development, RGCs release cytokines, growth factors, and chemokines, which can modulate microglial development and function (Antony et al., 2011). In a recent study, the ITGB8/TGFβ1 signaling has been proposed to be an important mediator for the radial glia-microglia interaction and to regulate microglial maturation during embryonic stages (McKinsey et al., 2023). Microglial maturation occurs in concert with CNS developmental stages. Disruptions in microglial maturational milestones can lead to deregulated neuroimmune interactions and ultimately disrupted neurologic functions in adulthood (Schafer and Stevens, 2015).

Together, neuroglia influence neurogenesis, neuronal migration, synaptogenesis, myelination, and inflammation in physiology. External factors, such as infection during brain development, can alter their homeostatic states, resulting in an increased risk for neurodevelopmental disorders (NDDs).

MATERNAL IMMUNE ACTIVATION

Maternal immune activation (MIA) refers to the activation of a mother's immune system during pregnancy, often due to infection, inflammation, or other triggers (Fig. 15.1). This immune

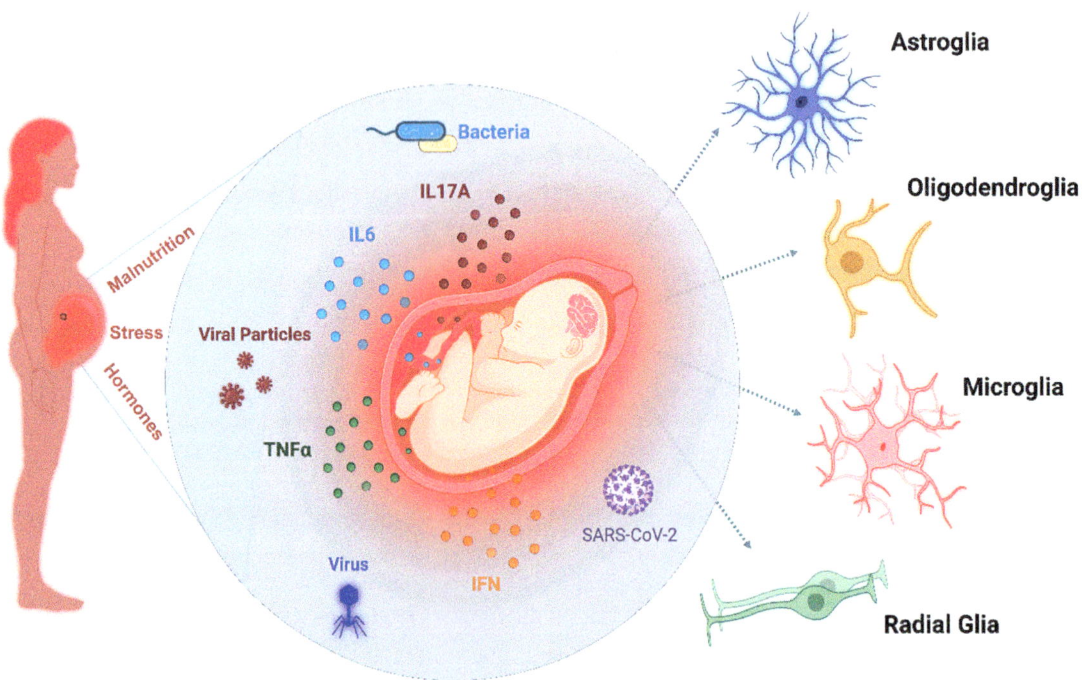

Fig. 15.1. Schematic cartoon of maternal immune activation. Maternal immune activation is represented by maternal cytokine amount changes, induced by any infection or stressor during pregnancy, that affect the fetal brain development and the neuroglial functions.

response can potentially affect the fetal brain development and increase the risk for neurologic disorders (PrabhuDas et al., 2021).

The link between MIA and NDDs was first recognized and proposed in epidemiological studies of influenza A infection. Various neuropsychiatric symptoms were observed after the 1385 German Flu, the 1889 and 1892 Russian Flu, and the Spanish Flu of 1918 (Stefano, 2021). Mental illness was also observed after infectious diseases caused by other pathogens, such as HIV, Zika virus, and Group B *Streptococcus* (Tucci et al., 2017).

In 1971, Andres Nahmias introduced the "TORCH" paradigm to define initially four congenital infections: TOxoplasmosis, Rubella, CYtomegalovirus, and Herpes simplex virus (Nahmias et al., 1971). Later on, more pathogens, like parvovirus B19, hepatitis B virus (Neu et al., 2015), and Zika virus (Coyne and Lazear, 2016), were included in this classification. According to the TORCH paradigm, infections with these pathogens usually cause mild illness in the infected mother, potential vertical transmission to the fetus (through the placenta), and fetus abnormalities (Stegmann and Carey, 2002). However, increasing clinical and scientific studies have shown that the fetus abnormalities go beyond this TORCH paradigm, observing a link between fetal exposure to non-TORCH pathogens and development of neurologic disorders, without necessarily a direct infection of the fetus (Shi et al., 2005). This led to the proposal of the MIA hypothesis consisting of the maternal in utero inflammation that can alter fetal neurodevelopment through placental transmission of inflammatory molecules from the mother to the fetus (Bergdolt and Dunaevsky, 2019). Today, MIA during gestation is considered an environmental risk factor for the development of NDDs (Kim et al., 2019; Han et al., 2021), such as autism spectrum disorder (ASD), schizophrenia (SCZ), attention-deficit/hyperactivity disorder, and learning disorders such as dyslexia (Brown et al., 2004; Babulas et al., 2006; Atladóttir et al., 2010; Brown and Meyer, 2018).

At present, while some infections, such as *Toxoplasma gondii*, Rubella, and Zika virus (Estes and McAllister, 2016), have been shown to potentially cause NDDs, other infections, such as influenza virus, are less clear.

A meta-analysis of 15 case studies found that common bacterial infections during pregnancy, including genitourinary and skin infections, increased the offspring's risk of developing ASD by 13% (Jiang et al., 2016). In addition, genitourinary infections increased the risk of developing attention-deficit/hyperactivity disorder by 26%–33%, while viral respiratory infections during pregnancy have a threefold increased risk of having a child with attention-deficit/hyperactivity disorder (Silva et al., 2014).

Maternal immune response can be also associated with infections and other noninfectious conditions, such as malnutrition, stress, and hormone disbalance, which can act synergistically with MIA in modulating the risk for NDDs (Meyer, 2019).

INFLAMMATION AND CYTOKINES ASSOCIATED WITH MIA

At present, MIA is rather represented by fluctuations in specific cytokine levels in the maternal serum and how they affect neurodevelopment and predisposition to psychiatric disorders.

All epidemiological human studies dispose of maternal and neonatal plasma that can be tested for cytokine amount along with behavioral and cognitive studies performed on the progeny (for schematic illustration, see Fig. 15.2).

Increased concentrations of proinflammatory cytokines in the maternal serum have been found in the progeny that developed psychosis in later life (Allswede et al., 2020). Elevated

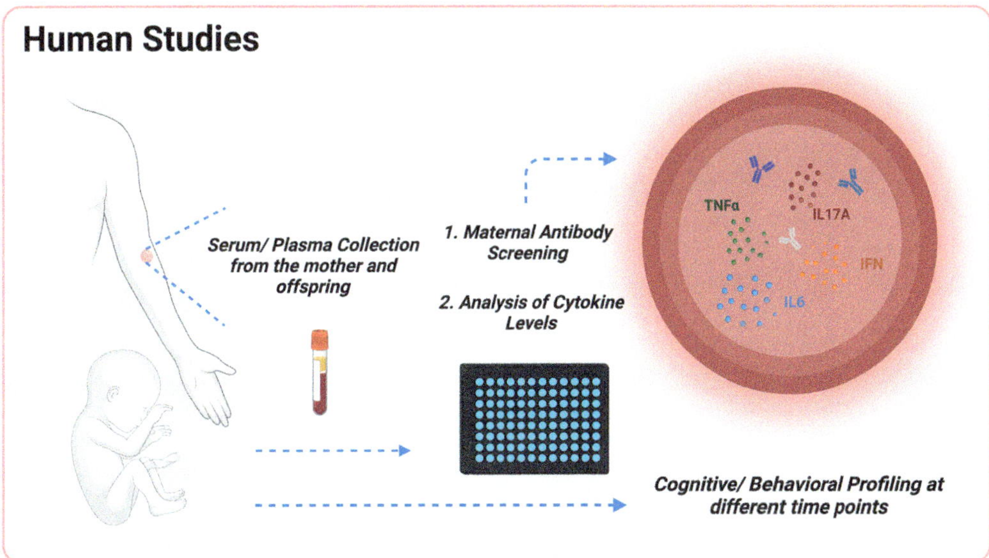

Fig. 15.2. Schematic illustration of epidemiological human studies of MIA. Pregnant women and subsequently the offspring are tested for cytokine (TNF-α, IL-17A, IFNs, IL-6) amount by collection of serum/plasma. Antibody screening is also performed. The offspring is also tested, at the right age and at different time points, for behavioral and cognitive deficits.

concentrations of IFN-γ, IL-4, and IL-5 cytokines have been reported in mid-gestation-maternal serum and correlated with a 50% increased risk of ASD, regardless of the type and the grade of intellectual disability (Goines et al., 2011). Some observed elevated cytokine levels of GM-CSF, IFN-γ, IL-1α, and IL-6 being highly associated with ASD feature intellectual disability (Jones et al., 2017). Other studies characterized additional cytokines, such as IL-1β, IL-6, IL12p70, IL-17A, TNF-α, IFN-γ, and IL-8 in the maternal serum, which can be linked to the offspring's cognitive ability and executive functions (Dozmorov et al., 2018). Several studies using magnetic resonance imaging coupled to behavioral readouts further support the link between high levels of IL-6 in the maternal serum and the offspring's brain structural and functional abnormalities (Graham et al., 2018; Rudolph et al., 2018; Rasmussen et al., 2019).

The ratio of cytokine amounts in the maternal plasma, and subsequent progeny exposure to it, appears to affect the brain activity. Goldstein and colleagues showed that a higher level ratio of TNF-α:IL-10 and dysregulated IL-6 amounts in the maternal serum have an impact on the response to stress in adult (45-year-old) individuals. Additionally, they found a sex-dependent effect in the hypothalamic and hippocampal activity in response to negative stressors, which appear to be regulated by the time of exposure (Goldstein et al., 2021). However, their low number of cases and the weak statistical power is a limiting factor to the MIA sex-based effect theory.

TEMPORAL STAGE OF INFECTION AND CYTOKINE EXPOSURE

The gestational timing at which the infection and immune response occurs is believed to be a regulator of the MIA impact on neurodevelopment.

While Zika virus and *Cytomegalovirus* infections provide a risk of developing NDDs regardless of the gestational time window, other viral infections have been shown to have a greater risk at specific stages of pregnancy. Rubella infection poses a higher risk during the first trimester, while Parvovirus and *Treponema pallidum* poses a higher risk during the second trimester. Lastly, *T. gondii* and herpes simplex virus appear to pose a higher risk toward offspring NDDs during the third trimester and perinatal stage, respectively (Devaraju et al., 2023).

An epidemiological study of a large cohort in California suggested that the incidence of SCZ in the offspring was threefold higher in children of mothers who were infected by the influenza virus in the first half of gestation and sevenfold higher during the first trimester (Brown et al., 2004). Conversely, a study in Denmark, supported by subsequent studies, suggested a higher risk of developing SCZ in the offspring when the influenza infection happened during the second half of gestation (Barr et al., 1990; Brown and Derkits, 2010). The same conflicting outcome has been observed for ASD. While a Danish study in children of women hospitalized for viral and bacterial infections during first and second trimester, respectively, showed an increased risk of ASD (Atladóttir et al., 2010), a Swedish study showed that hospitalization during pregnancy itself increased the risk of ASD with no association to the infection type or gestational timing (Lee et al., 2015). A third study encompassing a smaller case number of Californian individuals suggested an increased risk for ASD associated particularly with bacterial infection during the second and third trimesters (Zerbo et al., 2015).

These conflicting epidemiological studies, together with the fact that these studies did not investigate potential biological pathways, sparked the development of MIA animal and human in vitro and in vivo models to fill this gap and better investigate the effect of MIA on fetal brain development.

MODELING MIA: ANIMAL AND HUMAN

Human epidemiological studies along with preclinical models indicate that MIA may prime for altered neurodevelopment in susceptible individuals, with possible neuropsychiatric manifestations later in life. In this section, we will review the most prominent preclinical in vivo and in vitro MIA models (also listed in Table 15.1 with additional studies), summarizing their translational utility in understanding the role of MIA as a potential risk factor for neuropsychiatric conditions of neurodevelopmental origin (Fig. 15.3).

To date, several animal models have been developed and used to investigate how MIA can alter neurodevelopment in the offspring (Mattei et al., 2015). Such preclinical models have thus far confirmed that MIA results in altered neurodevelopmental trajectories, whereby the offspring presents brain morphologic, cellular, and molecular abnormalities highly reminiscent of several NDDs (Knuesel et al., 2014; Mattei et al., 2015; Reisinger et al., 2015; Estes and McAllister, 2016; Guma et al., 2019; Hall et al., 2023). These abnormalities are accompanied by behavioral deficits that are reminiscent of typical neuropsychiatric symptoms in NDDs, such as decreased social behavior, hyperlocomotion, and cognitive and sensorimotor gating deficits (Knuesel et al., 2014; Mattei et al., 2015; Estes and McAllister, 2016).

Influenza A-mediated MIA model

Although limited and conflicting, there is evidence that influenza infections during pregnancy may lead to NDDs in the offspring (Fung et al., 2022).

To study the potential link between prenatal influenza infection and neuropsychiatric disorders, scientists inoculated pregnant dams with influenza A (Shi et al., 2003; Fatemi et al., 2008; Otero et al., 2024). Early studies demonstrated that prenatal influenza causes changes in the offspring's brain gene expression, including genes related to ASD and SCZ (Fatemi et al., 2008). Such alterations are accompanied by aberrant cortical layering and brain atrophy, as well increased glial fibrillary acidic protein immunoreactivity and behavioral abnormalities reminiscent of various NDDs. Brain morphologic, neurochemical, and cellular defects are observed from fetal and early postnatal stages and in adolescent (postnatal day (P)35) and adult (P56) offspring (Egorova et al., 2024; Fatemi et al., 2002; Fatemi et al., 2008; Fung et al., 2022; Otero et al., 2024; Shi et al., 2003). Importantly,

Table 15.1

In vitro and in vivo models to study MIA.

MIA model	Species—cell target	Effect	References
Influenza A	Mice—in vivo	Maternal Th17 cells Abnormal cortical layers Reactive microglia	Shi et al. (2005), Fatemi et al. (2008), Otero et al. (2024)
Zika virus	Human—iNeurons, iAstro, iMG, HBO	Zika virus can infect all cell types and be detrimental to NPCs	Muffat et al. (2018)
Poly(I:C)	Mice—in vivo		
	Human—iNeurons	Impaired neuronal excitability and synaptic transmission	Ritchie et al. (2018)
LPS	Mice—in vivo	Aberrant cortical organization and increased hippocampal synaptic density Altered GABAergic signaling Fetal brain inflammatory response and altered gene expression	Oskvig et al. (2012), Fernández de Cossío et al. (2021), Gillespie et al. (2024)
IL-6	Human—HBO	Increased MHCI and autism-related gene expression in radial glia	Sarieva et al. (2023)
	Human—iNeurons	Altered mitochondrial respiration	Kathuria et al. (2022)
	Human—iMG	Microglial transcriptomic signature like SCZ geneset	Couch et al. (2023)
IL-17A	Mice—in vivo	ASD-like behavior and cortical disorganization Abnormal expression of synaptic genes Reactive microglia and altered microglial localization	Choi et al. (2016), Gumusoglu et al. (2020), Sasaki et al. (2020)
	Human—iNPCs	Increased neuronal differentiation and synaptic development	Gomes et al. (2022)
IFN-γ	Human—iNeurons	Change expression of genes associated to ASD and SCZ and altered neuronal morphology Altered MHCI and synapsin protein expression Altered dendritic spine density	Warre-Cornish et al. (2020), Kathuria et al. (2022), Pavlinek et al. (2022)
	Human—COiMg	Altered expression of genes associated with ASD and neuronal reduction	Buonfiglioli et al. (2024)

ASD, autism spectrum disorder; *COiMg*, cerebral organoids with integrated microglia; *GABA*, gamma-aminobutyric acid; *HBO*, human brain organoids; *iAstro*, human iPSC-derived astroglia; *iMG*, human iPSC-derived microglia; *iNeurons*, human iPSC-derived neurons; *iNPC*, human-iPSC-derived neuroprogenitor cells; *IFN-β*, interferon-beta; *IL-17a*, interleukin 17A; *IL-1β*, interleukin 1-beta; *IFN-γ*, interferon-gamma; *IL-6*, interleukin 6; *LPS*, lipopolysaccharide; *MHCI*, major histocompatibility complex I; *Poly(I:C)*, polyinosinic:polycytidylic acid; *SCZ*, schizophrenia; *Th17*, T helper 17 cells, *TNF-α*, tumor necrosis factor-alpha.

brain morphologic changes in an influenza-mediated MIA offspring often involve regions relevant to NDDs, such as the hippocampus, frontal, and cingulate cortex (Egorova et al., 2024).

The effect of prenatal influenza on offspring brain development appears to be indirect, as the virus is not detectable in the fetal brain (Shi et al., 2005). In this regard, a recent study demonstrated how Influenza A activates maternal T helper 17 (TH17) lymphocytes in a dose- and time-dependent manner (Otero et al., 2024). This activation appears to mediate the indirect effect of the maternal viral immune response on fetal neurodevelopment. The authors of the study found reactive microglia and boarder-associated macrophages in the fetal brain. Additionally, this fetal brain immune activation was accompanied by abnormalities in cortical layering (Otero et al., 2024). It is important to mention that MIA-mediated reactive fetal microglia has been shown to result in aberrant cortical layering in both rodents (Cunningham et al., 2013). In this context, microglia regulate the number of neuronal precursor cells (NPCs) in the subventricular zone and in the developing cortex (Cunningham et al., 2013). Fetal brain immune activation following MIA can possibly further influence radial glia, which are important for proper NPC migration during neurodevelopment (McEwan et al., 2023).

The poly(I:C)-mediated MIA model

An alternative to study the impact of maternal viral immune response on offspring neurodevelopment is the injection of the viral mimic polyinosinic:polycytidylic acid (Poly(I:C)). The latter is a synthetic double-stranded RNA that acts as an agonist for the toll-like receptor 3 (TLR3), eliciting an antiviral immune response. Injection of Poly(I:C) in pregnant dams has been shown to perturb offspring's neurodevelopment, resulting in behavioral aberrations manifested in adulthood (Meyer, 2014; Mattei et al., 2015; Brown and Meyer, 2018).

The offspring of Poly(I:C)-exposed dams consistently develop abnormalities in social behavior, cognitive deficits, and decreased sensorimotor gating (Meyer, 2014). The behavioral deficits presented by the offspring appear to derive from an altered neurodevelopmental course, resulting in adult brain morphologic, cellular, and neurochemical changes (Knuesel et al., 2014; Mattei et al., 2015; Guma et al., 2019; Mueller et al., 2021; Hall et al., 2023). The specific nature of the abnormalities presented by the Poly(I:C) offspring depends on the gestational timing of MIA induction. For simplicity, we report the general brain morphologic, cellular, and molecular abnormalities along with the behavioral phenotypes reported following Poly(I:C)-MIA for

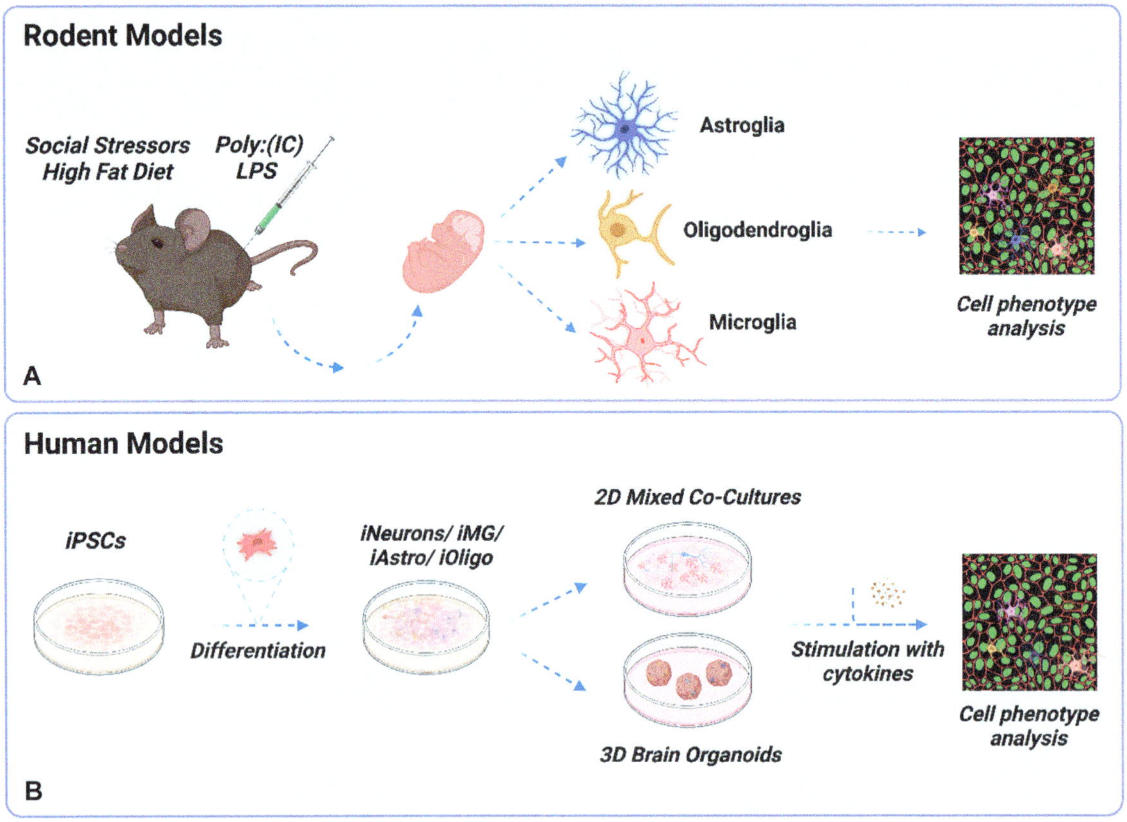

Fig. 15.3. Simplified schematic illustration of the human and murine models to study MIA. (A) Rodents are injected with Poly(I:C), LPS, exposed to stressors or high-fat diet during pregnancy, and the offspring is characterized in terms of cell composition and functionality. (B) iPSC technology allows the differentiation into neurons (iNeurons), microglia (iMG), astroglia (iAstro), and oligodendroglia (iOligo) as monoculture or tridimensional brain organoids. To model MIA, mono or 3D multicultures are stimulated with cytokines, and cell phenotype and function are characterized.

most studied gestational MIA time points. Poly(I:C) offsprings display changes in basal neurotransmitter levels, with increased dopamine and its metabolites in the globus pallidus and prefrontal cortex (PFC) and decreased serotonin in the hippocampus (Hip) and nucleus accumbens (NAc) (Winter et al., 2009). A subsequent study revealed that the cognitive deficits, decreased social interaction, and sensorimotor gating deficits displayed by Poly(I:C) offspring were accompanied by increased dopamine release in the NAc and reduced spontaneous firing rate of dopaminergic neurons (Luchicchi et al., 2016). Zuckerman et al. showed altered dopaminergic function also accompanied by morphologic changes in the hippocampi of Poly(I:C) offspring (Zuckerman et al., 2003). Furthermore, impaired prefrontal excitation-inhibition balance has been demonstrated, along with changes in NMDA receptors number and distribution in the PFC and Hip of Poly(I:C) offspring (Hao et al., 2019; Zhang et al., 2023). In fact, Poly(I:C) offspring also present altered cortical GABAergic interneuron distribution and maturation, as found in many NDDs (Gillespie et al., 2024). Two studies further found impaired Hip-PFC neural synchrony in Poly(I:C) MIA offspring, which is thought to contribute to decreased sensorimotor gating (Dickerson and Bilkey, 2013; Dickerson et al., 2014). Moreover, MIA resulted in decreased adult hippocampal neurogenesis associated with cognitive and sensorimotor gating deficits in offspring (Couch et al., 2021). The reduction in neurogenesis is accompanied by an aberrant hippocampal microglial phenotype, and immunomodulatory treatment of adult Poly(I:C) offspring has been shown to rescue neurogenesis along with the behavioral deficits (Mattei et al., 2014; Couch et al., 2021).

Importantly, the behavioral and cellular abnormalities reported in Poly(I:C) offspring are accompanied by long-lasting alterations in brain DNA methylation pattern as well as brain transcriptional changes, which can be passed on to subsequent generations (Richetto et al., 2017; Weber-Stadlbauer et al., 2017).

The Poly(I:C) MIA model has been extended to nonhuman primates, which is the closest model to human pregnancy, sharing many similarities in the maternal-fetal interface, placental and pregnancy physiology as well as brain developmental course (Ryan and Bauman, 2022). These studies confirmed that MIA can alter neurodevelopmental trajectories also in monkeys. Cognitive deficits along with abnormalities in social behavior have been described in nonhuman primate MIA offspring. There is also evidence of abnormal dendritic arborization in prefrontal cortices following altered brain morphologic features (Ryan and Bauman, 2022). A recent study found that nonhuman primate offspring, born to Poly(I:C) exposed pregnant females, presented with altered cognitive development, decreased frontal cortical gray matter and overall decreased brain volume (Vlasova et al.,

2021). The authors further found that the decrease in gray matter and volume persisted until the final time point of 45 months screened in the study, where a decrease in frontal white matter was also observed (Vlasova et al., 2021).

Lipopolysaccharide (LPS)-mediated MIA model

Prenatal bacterial infections have also been linked to NDDs in human epidemiological studies, with some contradictory results (Khandaker et al., 2013). To study how the maternal immune response to bacterial challenges can impact neurodevelopment, scientists induce MIA through the injection of lipopolysaccharide (LPS), a membrane component of Gram-negative bacteria that binds to TLR4 receptors, eliciting an antibacterial and proinflammatory immune response (Akira et al., 2001).

Maternal LPS has been shown to induce a fetal brain inflammatory response and alter the fetal brain gene expression (Oskvig et al., 2012; Simões et al., 2018). Similar to the Poly(I:C) model, LPS-induced MIA results in an offspring presenting developmental and adult behavioral abnormalities accompanied by brain structural, cellular, and molecular abnormalities (Simões et al., 2018; Fernández de Cossío et al., 2021; Carbone et al., 2023; Lin et al., 2023). Maternal LPS has been shown to alter fetal brain transcriptional programs related to interneuron migration, cellular stress, and various developmental genes (Simões et al., 2018). This transcriptional deregulation was followed by altered social and exploratory behavior in juvenile and adult rats. Several studies have shown that maternal LPS can induce GABAergic abnormalities in the offspring, as well as altered serotonergic functions in PFC (Lin et al., 2023; Gillespie et al., 2024). LPS offspring also display decreased dopamine D2 receptor availability in the PFC, along with increased serotonergic 5-HT2A receptors. Furthermore, LPS offspring displayed increased hippocampal synaptic density along with decreased mRNA and protein levels of synaptic proteins synaptophysin and postsynaptic density-95 (PSD95) and aberrant cortical columnar organization (Aria et al., 2020; Fernández de Cossío et al., 2021; Zhang et al., 2023). Behavioral consequences of maternal LPS include cognitive impairments in learning and memory, sociability deficits, decreased sensorimotor gaiting as well as increased anxiety and depressive-like behaviors (Borrell et al., 2002; Romero et al., 2007; Hao et al., 2010; Depino, 2015; Lee et al., 2021). In summary, although less studied than maternal Poly(I:C), LPS-induced MIA causes similar alterations in neurodevelopmental trajectories, resulting in adolescent and adult animals displaying aberrant behavioral phenotypes related to neuropsychiatric disorders.

Human-induced pluripotent stem cell (iPSC) technology

Human studies rely heavily on epidemiological data to understand the impact of MIA on brain development but do not give insights into the biological pathway and molecular mechanism. Animal models are used to bridge this gap; however, it is known that rodents and humans differ considerably in terms of neurodevelopment as well as the response to environmental triggers. To address this, researchers recently turned to human-induced pluripotent stem cell (iPSC) technology, a valuable tool to enable the in vitro study of MIA in a human context.

This technology was introduced first in 2006 and consists of reprogramming human skin fibroblasts, by using four different transcription factors (Oct3/4, Sox2,c-Myc, and Klf4), into pluripotent stem cells which can then be differentiated into any cell type of the body (Takahashi and Yamanaka, 2006). The iPSC technology has recently advanced to not only differentiate somatic cells into one cell type, originating monocultures, but also to create 3D cultures for the study of a whole organ, called organoids. Today, iPSC models, both in 2D and 3D cultures, are used in vitro to elucidate the molecular mechanism of MIA by either the use of viral and bacterial components or through direct exposure to specific cytokines. 2D cultures enabled scientists to study the effect of MIA on a single cell type, such as neuro-progenitors and mature neurons. Treatment of iPSC-derived neurons (iNeurons) with Poly(I:C) resulted in impaired neuronal excitability and altered synaptic activity, both observed in NDDs (Ritchie et al., 2018). Specifically, the Poly(I:C)-induced TLR3 activation reduced significantly the peak sodium currents, excitatory postsynaptic potential (mEPSC) frequency and amplitude, as well as surface AMPA receptor expression (Ritchie et al., 2018).

Neuronal functions were tested also in response to single cytokines, such as IL-6, IFN-γ, or IL-17A (Warre-Cornish et al., 2020; Gomes et al., 2022; Kathuria et al., 2022; Pavlinek et al., 2022). Treatment of iNeurons with IL-6 resulted in altered mitochondrial respiration (Kathuria et al., 2022). Exposure of iNeurons with IFN-γ has been shown to change the expression of genes linked to NDDs and alter neuronal morphology (Warre-Cornish et al., 2020), dendritic spine density (Kathuria et al., 2022) as well as major histocompatibility complex I (MHCI) and synapsin protein expression (Pavlinek et al., 2022). Treatment with IL-17A was shown to influence the neuronal differentiation of human iPSC-derived NPCs, which may disturb normal neuronal and synaptic development contributing to MIA outcomes (Gomes et al., 2022). Interestingly, acute IL-6 exposure had no significant impact on iPSC-derived NPCs but rather on iPSC-derived microglia (iMG) (Couch et al., 2023). In fact, neuro-progenitors do not express IL-6R and therefore do not respond to such stimulation. On the other hand, iMG, which express IL-6R, when treated with IL-6, increased motility and inflammatory capacities and show a transcriptomic profile similar to microglia derived from postmortem schizophrenic patients (Couch et al., 2023). This study highlights the fact that coculture models and the different cell interactions are essential to fully comprehend the MIA model and its link to NDDs.

Thus, research is now focusing on more complex systems, such as multicellular brain organoids (Lancaster et al., 2013). Zika virus exposure of organoids and neural monocultures has enabled to observe and compare how the virus infects each cell type and affects them, together and separately (Muffat et al., 2018). By using monocultures, Muffat and colleagues observed that the virus has the ability to infect all neural cells, including NPCs, astrocytes, and microglia, but it is only cytotoxic to NPCs. Interestingly, by using a brain organoid system, they were able to show that microglia appear to be the primary targets of the viral initiation of the infection, which then propagates and results in neuronal loss (Muffat et al., 2018).

Dorsal forebrain organoids were used to model MIA by exposing them to IL-6 (Sarieva et al., 2023). Since the receptor IL-6R in early stages of brain development is expressed exclusively by microglia and being these organoids devoid of these immune cells, Sarieva and colleagues engineered hyper-IL-6, a chimeric protein consisting of the soluble IL-6R and IL-6 connected by an oligopeptide linker. By exposing the organoids to the hyper-IL-6, they found that these organoids express the molecular machinery necessary for responding to hyper-IL-6 by activating signal transducer and activator of transcription (STAT) signaling and subsequently upregulating MHCI genes found to be differentially expressed in ASD tissue (Sarieva et al., 2023). In this human in vitro model, RDGs appeared to be the cell with the highest number of differentially expressed genes followed by abnormal cortical layering, indicating that these cells are the most vulnerable to IL-6 exposure (Sarieva et al., 2023). Recently, microglia-containing cerebral organoids have been developed, which allows for a more comprehensive study of neurodevelopmental insults (Ormel et al., 2018; Fagerlund et al., 2021; Cakir et al., 2022; Buonfiglioli et al., 2024). In one study, it was found that acute exposure of brain organoids with IFN-γ causes a reduction in NPCs and neuronal numbers assessed via immunofluorescence followed by transcriptional deregulation of genes associated with ASD (Buonfiglioli et al., 2024). Interestingly, when comparing IFN-γ-treated organoids with and without integrated microglia, they observed a different response in terms of differentially expressed genes and protein expression, suggesting the importance of looking at microglia in the response to that cytokine (Buonfiglioli et al., 2024).

Double-hit MIA models

Similar to the contrasting human studies, where a link between MIA and offspring neuropsychiatric manifestations is not always found, also not all MIA exposed offspring display behavioral abnormalities (Mueller et al., 2021). The Poly(I:C) model has been a helpful tool in revealing that MIA-exposed offspring can be stratified according to its susceptibility to MIA, as it is most likely the case in human MIA (Mueller et al., 2021). The current hypothesis posits that in humans, MIA acts synergistically with other environmental and genetic factors to increase the offsprings' risk of NDDs (Blomström et al., 2016; Davis et al., 2016). This hypothesis has been tested in double-hit models to account for other known etiological risk factors that increase the risk for NDDs. MIA can induce neurodevelopmental changes posing a latent increased risk toward cognitive, social, and sensory processing deficits. Double-hit MIA models explore how pubertal stressors and specific genetic backgrounds can unmask and exacerbate the effects of MIA in exposed offspring.

To date, two studies have explored the interaction between mutations in the gene disrupted in schizophrenia 1 (DISC1) and MIA. DISC1 mutations have been linked to human NDDs, and DISC1 mutant mice display aberrant neurodevelopment and behavioral deficits similar to the MIA models, including decreased sensorimotor gating and impaired cognitive and social functions (Thomson et al., 2013). Abazyan et al. studied the interaction between MIA and DISC1 loss of function in an inducible human mutant DISC1 mouse line (mhDISC1) (Abazyan et al., 2010). The authors found that there was a significant interaction between genotype and MIA with significant effects on adult offspring behavioral and neuroanatomic abnormalities. mhDISC1-MIA offspring displayed a depressive-like phenotype along with decreased sociability and increased anxiety. These behavioral phenotypes were accompanied by decreased hippocampal serotonergic neurotransmission as well as altered spine density in granule cell dendrites (Abazyan et al., 2010). Moreover, compared to MIA and mhDISC1 alone, the mhDISC1-MIA offspring also presented with decreased gray matter volume in the amygdala and periaqueductal gray (Abazyan et al., 2010). Lipina et al. investigated the interaction between two DISC1 gene point-mutations (L100P and Q31L) and Poly(I:C) mediated MIA. The authors found that Disc1-L100P+/− mice displayed higher fetal brain IL-6 upon MIA exposure. Moreover, there was a significant interaction between genotype and MIA that exacerbated deficits in social motivation, latent inhibition, and sensorimotor gating (Lipina et al., 2013).

Peripubertal emotional and social stress is being studied for its link to the development of neuropsychiatric diseases (Gomes and Grace, 2017; Lin and Guo, 2024). To study the interaction between MIA and peripubertal stress, several studies focused on adolescent MIA offspring exposed to various stress paradigms. In a seminal study, Giovanoli et al. combined Poly(I:C)-mediated MIA with various peripubertal stressors, including electric foot shock, restraint stress, and swimming stress (Giovanoli et al., 2013). The synergistic effect of MIA and peripubertal stressors exacerbated sensorimotor gating deficits and resulted in stimulant-hypersensitivity, a sign of altered dopaminergic signaling. Giovanoli and colleagues further found that MIA and pubertal stress synergistically resulted in enhanced hippocampal dopamine levels (Giovanoli et al., 2013). Deslauriers et al. combined MIA with pubertal restraint stress, confirming a synergistic effect on the development of sensorimotor gating deficits (Deslauriers et al., 2013). These deficits were accompanied by increased cortical and striatal dopaminergic D2 receptor and decreased GABAergic interneuron marker GAD67 mRNA and protein levels. Furthermore, abnormalities in hippocampal GABAergic system have also been found when adolescent MIA offspring is exposed to various unpredictable stressors (Giovanoli et al., 2014). Specifically, decreased parvalbumin-positive cells were detected in the ventral dentate gyrus along with increased reelin-positive cells in the CA1 and CA3 hippocampal regions of adult MIA offspring exposed to pubertal stressors (Giovanoli et al., 2014).

MATERNAL IMMUNE ACTIVATION EFFECT ON NEUROGLIA

As mentioned above, preclinical models revealed that MIA is able to induce fetal brain changes via inflammatory mediators such as cytokines (Smith et al., 2007; Pratt et al., 2013; Schaafsma et al., 2017). Preclinical MIA models further implicated IL-6 as a potential propagator of the maternal-to-fetal brain inflammatory response (Smith et al., 2007). Alteration in fetal brain immune functions can have profound consequences for neurodevelopment, as the brain immune system is directly involved in neurodevelopmental mechanisms (Mattei and Notter, 2020; Mastenbroek et al., 2024). In this context, neuroglia are hypothesized to be involved in the fetal immune response and the subsequent altered neurodevelopmental trajectories. Neuroglia are active participants of neurodevelopment, regulating NPC numbers and migration as well as axonal and dendritic sprouting, synaptogenesis, and myelination (Allen and Lyons, 2018; Santos and Fields, 2021; Mastenbroek et al., 2024).

Neuroglia can be impacted during and following MIA, which in turn can have detrimental consequences on neurodevelopmental trajectories. Understanding the role of neuroglia in MIA-mediated NDDs can significantly advance our understanding of their etiology and potentially provide pharmacological targets for NDDs. Below we are going to summarize preclinical evidence of neuroglial involvement in MIA-mediated neurodevelopmental disruptions.

MIA and astroglia

Similar to microglia, astroglia play a pivotal role in brain development. In rodents, they appear after embryonic day 12 (E12), derived from NPCs and radial glia, and they participate in brain circuit refinement (Clarke and Barres, 2013; Vivi and Di Benedetto, 2024). Astroglia support the formation of synapses as well as the pruning of unnecessary synaptic connections, and hence, together with microglial cells, guide the development of brain circuits (Clarke and Barres, 2013). The consequences of MIA on astroglia have been less studied as compared to microglia; however, preclinical data presented so far indicate that astroglia is also affected by the MIA-induced fetal brain immune activation. Increased glial fibrillary acidic protein and S100β protein levels in cortex and hippocampi of LPS-induced MIA offspring are an indication of astrogliosis (Cai et al., 2000; de Souza et al., 2015). Hao et al. further showed that hippocampal glial fibrillary acidic protein expression in LPS-MIA offspring increases with age and is accompanied by cognitive deficits and decreased synaptophysin expression (Hao et al., 2010). Astrogliosis and altered astrocytic morphology was also demonstrated in the amygdala and in the internal and external capsules of LPS-MIA offspring and was paralleled by increased proinflammatory cytokine expression and microgliosis (Rousset et al., 2006; O'Loughlin et al., 2017). Interestingly, studies conducted on Poly(I:C)-MIA offspring did not find any signs of astrogliosis as determined by glial fibrillary acidic protein immunoreactivity (Giovanoli et al., 2015, 2016; Paylor et al., 2016). It is, however, essential to point out that in order to assess the outcome of MIA on astroglia, more in-depth analyses are required. Current evidence is limited to immunostaining and morphologic analyses. More thorough transcriptional and proteomic analyses may provide more insights into the impact of MIA on astroglia.

MIA and oligodendroglia

Oligodendroglia (OLs) are the neuroglial cells responsible for axonal myelination in the developing and adult CNS (Fletcher et al., 2021). OLs are derived from OPCs that originate from radial glia in the subventricular zone around E12 (Fletcher et al., 2021). The OLs' myelination pattern, in terms of myelin sheet thickness, composition, and spacing along axons, modulates neuronal firing and contributes to neural synchronization. This level of brain plasticity provided by OLs is of high relevance during critical periods of brain development for the proper establishment of neuronal connectivity and circuitry function (Fletcher et al., 2021). Given the relevance of myelin in developmental and adult brain plasticity and circuitry maintenance, it is pivotal to investigate whether MIA alters the development and function of OLs. Both OPCs and mature OLs have a degree of immune competence, as they can express proinflammatory cytokines and chemokines, and they can process and present antigens and express immune receptors such as TLR3 (Boccazzi et al., 2022). Moreover, there is an extensive crosstalk between microglia, OPCs, and OLs, where microglia-derived signaling molecules such as cytokines and insulin-like growth factor-1 (IGF1) regulate OPC numbers and maturation (Santos and Fields, 2021). Given the fetal brain immune response to MIA, and the associated microgliosis, it is plausible to infer that OPCs and OLs might be influenced either directly or indirectly by MIA. Although there is at present no study investigating a direct response of OPCs and OLs to MIA, recent work by Boccazzi et al. showed that an injection of interleukin-1β in newborn mice between P1 and P5 results in a proinflammatory response in both OLs and OPCs (Boccazzi et al., 2021). An early study showed that MIA led to a decrease in myelin basic protein (MBP) mRNA and protein levels in hippocampi of juvenile offspring, and overall decreased myelination assessed via electron microscopy. Moreover, axonal diameter was significantly smaller in hippocampi of juvenile MIA offspring (Makinodan et al., 2008). MIA has also been shown to decrease the number of oligodendroglia in the developing spinal cord which persisted through P14 and was associated with increased white matter in the ventral spinal cord (Anderson et al., 2022).

Zhang et al. further demonstrated that MIA alters the expression of oligodendrocyte genes such as SOX10 and myelin-associated glycoprotein (MAG) in the mPFC and hippocampi of adult offspring (Zhang et al., 2020). This is indicative of a potential long-term consequence of MIA on the gene expression of OLs. Finally, an in vivo study using voxel-based analysis of white matter integrity through fractional anisotropy (FA) discovered widespread white matter abnormalities in MIA offspring (Li et al., 2010). In this study, Li et al. found that MIA at gestational day 9 resulted in adult offspring presenting lower FA in the amygdala, cerebral peduncle and right fimbria as well as in the anterior cingulate and ventral striatum (Li et al., 2010). These results suggest a potential long-term decreased myelination in specific brain regions following MIA. Interestingly, a recent study conducted on pregnant women found that maternal IL-6 levels were inversely associated with FA in newborns' uncinate fasciculus, the main frontolimbic tract

(Sanders et al., 2024). Maternal IL-6 was also found to be inversely associated with offspring cognitive development at 12 months (Sanders et al., 2024).

As mentioned earlier, preclinical studies indicate that IL-6 has a major role in fetal brain alterations upon MIA (Smith et al., 2007; Pratt et al., 2013).

MIA and microglia

Microglial cells are the primary immune-competent cells of myeloid origin in the CNS, and they are the first cells to invade the developing brain in early stages (embryonic day 8.5–9.5 in mice and 4.5–5.5 weeks in humans), prior to the appearance of astroglia and oligodendroglia (Mastenbroek et al., 2024). Preclinical studies have demonstrated that MIA induces a microglial response in the fetal brain, which participates to neurodevelopmental disruptions (Pratt et al., 2013; Schaafsma et al., 2017; Ozaki et al., 2020; Ostrem et al., 2024). Microglia appear to be necessary for the fetal brain response to MIA, and postnatal modulation of reactive microgliosis can prevent the development of MIA-induced behavioral phenotypes (Xia et al., 2020; Ostrem et al., 2024). In a seminal study, Cunningham et al. showed that microglia colonize neurogenic niches to regulate the number of migrating NPCs in both rodents and nonhuman primates, through active phagocytosis of NPCs (Cunningham et al., 2013). They further observed reactive fetal microglia induced by MIA in pregnant rats, leading to increased phagocytosis of NPCs and aberrant cortical layering (Cunningham et al., 2013). Subsequent preclinical studies further showed that microglia guide the extension of dopaminergic neurons and patterning of somatostatin and parvalbumin interneurons, whereby MIA, through reactive microglia, results in altered extension of dopaminergic axons and aberrant interneuron patterning in the embryonic forebrain (Mastenbroek et al., 2024). Microglia also regulate developmental myelination and continue to support circuitry maturation in postnatal development along with regulating synaptic density through active pruning and the induction of dendritic spine sprouting through active contacts (Santos and Fields, 2021; Mastenbroek et al., 2024). Intriguingly, preclinical studies have found that MIA-induced changes in microglial phenotype persist from pubertal stages until adulthood, possibly participating in the sustained behavioral deficits characterizing adult MIA offspring (Mattei et al., 2017, 2024; Zhao et al., 2019; Cieślik et al., 2020; Fernández de Cossío et al., 2021; Traetta and Tremblay, 2022). MIA can change the transcriptome of adult hippocampal microglia, accompanied by decreased phagocytic functions and behavioral abnormalities (Mattei et al., 2017). Interestingly, the nature of adult MIA microglial transcriptional deregulation depends on time of sample collection, indicating that the intrinsic microglial circadian regulation is altered in adult MIA offspring (Mattei et al., 2024). Modulating microglial activity through immunomodulatory compounds such as minocycline is able to attenuate aberrant microglial phenotypes and rescue the behavioral deficits in adolescent and adult MIA offspring (Mattei et al., 2014, 2015, 2017; Xia et al., 2020; Ostrem et al., 2024).

The studies summarized above highlight the importance of microglial cells in understanding MIA-induced neurodevelopmental alterations and indicate that microglial modulation is a potential future pharmacological target for the treatment of neuropsychiatric diseases.

THE COVID-19 PANDEMIC

Since November 2019, a novel viral pandemic, called coronavirus disease 2019 (COVID-19) and caused by severe acute respiratory syndrome coronavirus 2 (SARS-CoV-2), placed pregnant women in a high-risk group. Meta-analysis studies reported that pregnant women show a more severe COVID-19, manifesting less of the common symptoms such as fever, cough, and dyspnea, but are at higher risk of admission to the intensive care unit, preterm birth, and subsequent admittance into neonatal unit (Allotey et al., 2020). Vertical transmission of the virus is still a great deal of controversy as there is no clinical proof (Wang et al., 2020; Al-kuraishy et al., 2023). Cytokine profiling on pregnant women infected with SARS-CoV-2 revealed an increased expression of inflammatory cytokines, such as IL-2, IL-6, IL-8, IL-10, and IL-17, in the first trimester and IFN-γ in the third trimester (Cavalcante et al., 2021; Tanacan et al., 2021; Egloff et al., 2023; Gigase et al., 2024). However, the MIA effect caused by COVID-19 is unclear and under investigation.

Elevated plasma levels of IL-6, IP-10, and IL-8 were observed in infected mothers and their infants (Taglauer et al., 2022). Furthermore, neonates born from mothers exposed to SARS-CoV-2 exhibited a higher percentage of innate immune cells and cytokine levels in plasma derived from neonatal cord blood, potentially resulting from a response to the maternal cytokine storm and response to the infection (Gee et al., 2021). Another meta-analysis showed no clear evidence of exacerbated cytokine response to COVID-19 in pregnant mothers and their newborns (Jain et al., 2023). However, they observed slightly dysregulated levels of CXCL10 and IFN-γ that together with and that study limitations, such as sample collection timing and limited cohort study, would suggest a more cautious approach to interpret the data (Jain et al., 2023). Given the fact that COVID-19 is a relatively recent pandemic, very few studies investigated a potential link between COVID-19 maternal infection, and neurodevelopmental impairments and present results are unclear.

Few studies suggest no overall neurodevelopmental deficit in the first 6–12 months of age when neonates are born during a COVID-19 maternal infection (Hessami et al., 2022; Shuffrey et al., 2022). However, Hessami and colleagues identified a significant increase in the odds of developing fine motor impairment. A cohort study of SARS-CoV-2 in utero exposure found preliminary evidence that maternal infection may be associated with NDDs (Edlow et al., 2022). A follow-up study from the same group revealed that SARS-CoV-2 in utero exposure is associated with a greater risk for neurodevelopmental diagnosis of male offspring at 12 months of age (Edlow et al., 2023). A longitudinal cohort study in United States and Brazil observed suboptimal neuromotor development in infants prenatally exposed to COVID-19 (Fajardo Martinez et al., 2023). A national prospective study in

Kuwait found neurodevelopmental delays in neonates born from COVID-19 infected mothers (Ayed et al., 2021).

The pandemic has also caused increased stress levels in pregnant women (Sirait et al., 2024), potentially compounding MIA effects. Preliminary magnetic resonance imaging studies show that maternal distress is linked to aberrant amygdala structural and functional connectivity in infants of 3 months of age. They also suggest that social support moderates this association (Manning et al., 2022). Another study observed an impaired fetal brain growth and delayed cerebral cortical gyrification in neonates born from COVID-19 infected mothers (Lu et al., 2022). Besides the potential effects of maternal COVID-19, Rajagopalan et al. found that abnormalities in fetal brainstem structure, with increased volume, and altered global functional brain activity can be observed in the offspring of mothers in distress due to pandemic (Rajagopalan et al., 2022). This highlights other concomitant factors that may potentially act synergistically with maternal COVID-19 to alter neurodevelopmental trajectories.

Longitudinal monitoring, larger cohorts, and long-term surveillance while accounting for important covariates, such as gestational time window of the infection, severity of maternal symptoms, and cesarean sections or treatments, will be necessary to reliably estimate or refute the hypothesis of COVID-19-MIA--mediated neurodevelopmental disruption.

CONCLUSIONS, SPECULATIONS, AND FUTURE DIRECTIONS

The susceptibility to infections is strongly linked to MIA, but a genetic component and secondary stressors also play a role, especially in neuropsychiatric disorders with late onset like SCZ (Nudel et al., 2019). It was mentioned above that some studies are focusing on modeling MIA via double-hit models. Future research should be focusing more on those studies, in which the MIA hit is later on accompanied by a secondary stressor model. The recently introduced use of iPSC-derived human models can be further exploited in this regard.

Given the fact that NDDs are sex specific, how fetal sex may influence the immunologic response to MIA should be taken into account. For example, ASD is prevalent in males, and several papers suggested that genetics and hormonal and environmental factors may play a role (Werling and Geschwind, 2013; Ribeiro and MacDonald, 2020; Guneykaya et al., 2023). In addition, recent studies about COVID-19 (Bordt et al., 2021; Edlow et al., 2023) indicate sex-specific differences in immune system development and response (Goldstein et al., 2021), potentially leading to differential susceptibility to NDDs. Understanding the vulnerabilities and the differences in immune responses associated with each sex in response to prenatal immune challenges could be crucial for further studies. Giving insight in this regard may be a good option to study MIA.

The gut–brain axis is another area of interest in understanding MIA. Maternal infections can alter the maternal and fetal gut microbiota, influencing brain development through immune and metabolic pathways (Mesa et al., 2020; Suprunowicz et al., 2024). Animal models of MIA already showed how the gut microbiota of the offspring is affected during maternal inflammation (Morais et al., 2018; Lee et al., 2021; Li et al., 2021). Similarly, the maternal gut microbiota during pregnancy can influence the development of the fetal brain, potentially contributing to the risk of NDDs. Maternal diet-induced dysbiosis during pregnancy can significantly increase the risk of developing NDDs (Di Gesù et al., 2021). The microbiome is a potential risk factor not to be excluded, considering its role in regulating the system in the brain (Ratsika et al., 2023). Thus, understanding the role of the gut–brain axis in MIA's effects on brain development is a promising area for future research.

Epidemiological studies control for and adjust their analysis based on the use of antiinflammatory drugs to ensure that the MIA effects are not confounded by drug use. The use of antiinflammatory drugs is another environmental factor to take into consideration not only as one that could blunt MIA effect but also to prevent it, by targeting the activated immune system. One example is the already mentioned minocycline that can rescue NDD-associated symptoms in rodent models (Mattei et al., 2014).

IL-6, used to model MIA, induces RGC proliferation in the cortex of male offspring and human organoid models (Sarieva et al., 2023). Radial glia play an important role in developing nervous system, particularly contributing to the formation of cortical layers (Miranda-Negrón and García-Arrarás, 2022). RGCs have been shown to be promising targets, and further investigating their contribution to the MIA would give more insight into fetal brain development.

References

Abazyan B, Nomura J, Kannan G et al. (2010). Prenatal interaction of mutant DISC1 and immune activation produces adult psychopathology. Biol Psychiatry 68: 1172–1181. https://doi.org/10.1016/j.biopsych.2010.09.022.

Akira S, Takeda K, Kaisho T (2001). Toll-like receptors: critical proteins linking innate and acquired immunity. Nat Immunol 2: 675–680. https://doi.org/10.1038/90609.

Al-kuraishy HM, Al-Gareeb AI, Albezrah NKA et al. (2023). Pregnancy and COVID-19: high or low risk of vertical transmission. Clin Exp Med 23: 957–967. https://doi.org/10.1007/s10238-022-00907-z.

Allen NJ, Eroglu C (2017). Cell biology of astrocyte-synapse interactions. Neuron 96: 697–708. https://doi.org/10.1016/j.neuron.2017.09.056.

Allen NJ, Lyons DA (2018). Glia as architects of central nervous system formation and function. Science 362: 181–185. https://doi.org/10.1126/science.aat0473.

Allotey J, Stallings E, Bonet M et al. (2020). Clinical manifestations, risk factors, and maternal and perinatal outcomes of coronavirus disease 2019 in pregnancy: living systematic review and meta-analysis. BMJ 370: m3320. https://doi.org/10.1136/bmj.m3320.

Allswede DM, Yolken RH, Buka SL et al. (2020). Cytokine concentrations throughout pregnancy and risk for psychosis in adult offspring: a longitudinal case-control study. Lancet Psychiatry 7: 254–261. https://doi.org/10.1016/S2215-0366(20)30006-7.

Alvarez JI, Katayama T, Prat A (2013). Glial influence on the blood brain barrier. Glia 61: 1939–1958. https://doi.org/10.1002/glia.22575.

Anderson RC, O'Keeffe GW, McDermott KW (2022). Characterisation of the consequences of maternal immune activation on distinct cell

populations in the developing rat spinal cord. J Anat 241: 938–950. https://doi.org/10.1111/joa.13726.
Antony JM, Paquin A, Nutt SL et al. (2011). Endogenous microglia regulate development of embryonic cortical precursor cells. J Neurosci Res 89: 286–298. https://doi.org/10.1002/jnr.22533.
Aria F, Bonini SA, Cattaneo V et al. (2020). Brain structural and functional alterations in mice prenatally exposed to LPS are only partially rescued by anti-inflammatory treatment. Brain Sci 10. https://doi.org/10.3390/brainsci10090620.
Atladóttir HO, Thorsen P, Østergaard L et al. (2010). Maternal infection requiring hospitalization during pregnancy and autism spectrum disorders. J Autism Dev Disord 40: 1423–1430. https://doi.org/10.1007/s10803-010-1006-y.
Ayed M, Embaireeg A, Kartam M et al. (2021). Neurodevelopmental outcomes of infants secondary to in utero exposure to maternal SARS-CoV-2 infection: a national prospective study in Kuwait. MedRxiv. https://doi.org/10.1101/2021.11.12.21266291. 2021.2011.2012.21266291.
Babulas V, Factor-Litvak P, Goetz R et al. (2006). Prenatal exposure to maternal genital and reproductive infections and adult schizophrenia. Am J Psychiatry 163: 927–929. https://doi.org/10.1176/ajp.2006.163.5.927.
Barger N, Keiter J, Kreutz A et al. (2019). Microglia: an intrinsic component of the proliferative zones in the fetal Rhesus monkey (Macaca mulatta) cerebral cortex. Cereb Cortex 29: 2782–2796. https://doi.org/10.1093/cercor/bhy145.
Barr CE, Mednick SA, Munk-Jorgensen P (1990). Exposure to influenza epidemics during gestation and adult schizophrenia. A 40-year study. Arch Gen Psychiatry 47: 869–874. https://doi.org/10.1001/archpsyc.1990.01810210077012.
Bergdolt L, Dunaevsky A (2019). Brain changes in a maternal immune activation model of neurodevelopmental brain disorders. Prog Neurobiol 175: 1–19. https://doi.org/10.1016/j.pneurobio.2018.12.002.
Bilbo S, Stevens B (2017). Microglia: the brain's first responders. Cerebrum 2017: cer-14-17.
Blomström Å, Karlsson H, Gardner R et al. (2016). Associations between maternal infection during pregnancy, childhood infections, and the risk of subsequent psychotic disorder–A swedish cohort study of nearly 2 million individuals. Schizophr Bull 42: 125–133. https://doi.org/10.1093/schbul/sbv112.
Boccazzi M, Raffaele S, Fumagalli M (2022). Not only myelination: the immune-inflammatory functions of oligodendrocytes. Neural Regen Res 17: 2661–2663. https://doi.org/10.4103/1673-5374.342678.
Boccazzi M, Van Steenwinckel J, Schang AL et al. (2021). The immune-inflammatory response of oligodendrocytes in a murine model of preterm white matter injury: the role of TLR3 activation. Cell Death Dis 12: 166. https://doi.org/10.1038/s41419-021-03446-9.
Bordt EA, Shook LL, Atyeo C et al. (2021). Maternal SARS-CoV-2 infection elicits sexually dimorphic placental immune responses. Sci Transl Med 13: eabi7428. https://doi.org/10.1126/scitranslmed.abi7428.
Borrell J, Vela JM, Arévalo-Martin A et al. (2002). Prenatal immune challenge disrupts sensorimotor gating in adult rats. Implications for the etiopathogenesis of schizophrenia. Neuropsychopharmacology 26: 204–215. https://doi.org/10.1016/s0893-133x(01)00360-8.
Brown AS, Begg MD, Gravenstein S et al. (2004). Serologic evidence of prenatal influenza in the etiology of schizophrenia. Arch Gen Psychiatry 61: 774–780. https://doi.org/10.1001/archpsyc.61.8.774.
Brown AS, Derkits EJ (2010). Prenatal infection and schizophrenia: a review of epidemiologic and translational studies. Am J Psychiatry 167: 261–280. https://doi.org/10.1176/appi.ajp.2009.09030361.
Brown AS, Meyer U (2018). Maternal immune activation and neuropsychiatric illness: a translational research perspective. Am J Psychiatry 175: 1073–1083. https://doi.org/10.1176/appi.ajp.2018.17121311.
Buonfiglioli A, Kübler R, Missall R et al. (2024). A microglia-containing cerebral organoid model to study early life immune challenges. BioRxiv. https://doi.org/10.1101/2024.05.24.595814.
Cai Z, Pan ZL, Pang Y et al. (2000). Cytokine induction in fetal rat brains and brain injury in neonatal rats after maternal lipopolysaccharide administration. Pediatr Res 47: 64–72. https://doi.org/10.1203/00006450-200001000-00013.
Cakir B, Tanaka Y, Kiral FR et al. (2022). Expression of the transcription factor PU.1 induces the generation of microglia-like cells in human cortical organoids. Nat Commun 13: 430. https://doi.org/10.1038/s41467-022-28043-y.
Carbone E, Buzzelli V, Manduca A et al. (2023). Maternal immune activation induced by prenatal lipopolysaccharide exposure leads to long-lasting autistic-like social, cognitive and immune alterations in male Wistar rats. Int J Mol Sci 24. https://doi.org/10.3390/ijms24043920.
Cavalcante MB, Cavalcante C, Sarno M et al. (2021). Maternal immune responses and obstetrical outcomes of pregnant women with COVID-19 and possible health risks of offspring. J Reprod Immunol 143: 103250. https://doi.org/10.1016/j.jri.2020.103250.
Choi GB, Yim YS, Wong H et al. (2016). The maternal interleukin-17a pathway in mice promotes autism-like phenotypes in offspring. Science 351: 933–939. https://doi.org/10.1126/science.aad0314.
Cieślik M, Gąssowska-Dobrowolska M, Jęśko H et al. (2020). Maternal immune activation induces neuroinflammation and cortical synaptic deficits in the adolescent rat offspring. Int J Mol Sci 21. https://doi.org/10.3390/ijms21114097.
Clarke LE, Barres BA (2013). Emerging roles of astrocytes in neural circuit development. Nat Rev Neurosci 14: 311–321. https://doi.org/10.1038/nrn3484.
Couch ACM, Berger T, Hanger B et al. (2021). Maternal immune activation primes deficiencies in adult hippocampal neurogenesis. Brain Behav Immun 97: 410–422. https://doi.org/10.1016/j.bbi.2021.07.021.
Couch ACM, Solomon S, Duarte RRR et al. (2023). Acute IL-6 exposure triggers canonical IL6Ra signaling in hiPSC microglia, but not neural progenitor cells. Brain Behav Immun 110: 43–59. https://doi.org/10.1016/j.bbi.2023.02.007.
Coyne CB, Lazear HM (2016). Zika virus—reigniting the TORCH. Nat Rev Microbiol 14: 707–715. https://doi.org/10.1038/nrmicro.2016.125.
Cunningham CL, Martínez-Cerdeño V, Noctor SC (2013). Microglia regulate the number of neural precursor cells in the developing cerebral cortex. J Neurosci 33: 4216–4233. https://doi.org/10.1523/jneurosci.3441-12.2013.
Davis J, Eyre H, Jacka FN et al. (2016). A review of vulnerability and risks for schizophrenia: beyond the two hit hypothesis. Neurosci Biobehav Rev 65: 185–194. https://doi.org/10.1016/j.neubiorev.2016.03.017.
de Souza DF, Wartchow KM, Lunardi PS et al. (2015). Changes in astroglial markers in a maternal immune activation model of schizophrenia in Wistar rats are dependent on sex. Front Cell Neurosci 9: 489. https://doi.org/10.3389/fncel.2015.00489.
Depino AM (2015). Early prenatal exposure to LPS results in anxiety- and depression-related behaviors in adulthood. Neuroscience 299: 56–65. https://doi.org/10.1016/j.neuroscience.2015.04.065.
Deslauriers J, Larouche A, Sarret P et al. (2013). Combination of prenatal immune challenge and restraint stress affects prepulse inhibition and dopaminergic/GABAergic markers. Prog

Neuropsychopharmacol Biol Psychiatry 45: 156–164. https://doi.org/10.1016/j.pnpbp.2013.05.006.
Devaraju M, Li A, Ha S et al. (2023). Beyond TORCH: a narrative review of the impact of antenatal and perinatal infections on the risk of disability. Neurosci Biobehav Rev 153: 105390. https://doi.org/10.1016/j.neubiorev.2023.105390.
Di Gesù CM, Matz LM, Buffington SA (2021). Diet-induced dysbiosis of the maternal gut microbiome in early life programming of neurodevelopmental disorders. Neurosci Res 168: 3–19. https://doi.org/10.1016/j.neures.2021.05.003.
Dickerson DD, Bilkey DK (2013). Aberrant neural synchrony in the maternal immune activation model: using translatable measures to explore targeted interventions. Front Behav Neurosci 7: 217. https://doi.org/10.3389/fnbeh.2013.00217.
Dickerson DD, Overeem KA, Wolff AR et al. (2014). Association of aberrant neural synchrony and altered GAD67 expression following exposure to maternal immune activation, a risk factor for schizophrenia. Transl Psychiatry 4: e418. https://doi.org/10.1038/tp.2014.64.
Dozmorov MG, Bilbo SD, Kollins SH et al. (2018). Associations between maternal cytokine levels during gestation and measures of child cognitive abilities and executive functioning. Brain Behav Immun 70: 390–397. https://doi.org/10.1016/j.bbi.2018.03.029.
Edlow AG, Castro VM, Shook LL et al. (2023). Sex-specific neurodevelopmental outcomes among offspring of mothers with SARS-CoV-2 infection during pregnancy. JAMA Netw Open 6: e234415. https://doi.org/10.1001/jamanetworkopen.2023.4415.
Edlow AG, Castro VM, Shook LL et al. (2022). Neurodevelopmental outcomes at 1 year in infants of mothers who tested positive for SARS-CoV-2 during pregnancy. JAMA Netw Open 5: e2215787. https://doi.org/10.1001/jamanetworkopen.2022.15787.
Egloff C, Roques P, Picone O (2023). Impact of COVID-19 on pregnant women's health: consequences in obstetrics two years after the pandemic. J Reprod Immunol 158: 103981. https://doi.org/10.1016/j.jri.2023.103981.
Egorova M, Egorov V, Zabrodskaya Y (2024). Maternal influenza and offspring neurodevelopment. Curr Issues Mol Biol 46: 355–366. https://doi.org/10.3390/cimb46010023.
Erblich B, Zhu L, Etgen AM et al. (2011). Absence of colony stimulation factor-1 receptor results in loss of microglia, disrupted brain development and olfactory deficits. PloS One 6: e26317. https://doi.org/10.1371/journal.pone.0026317.
Estes ML, McAllister AK (2016). Maternal immune activation: implications for neuropsychiatric disorders. Science 353: 772–777. https://doi.org/10.1126/science.aag3194.
Fagerlund I, Dougalis A, Shakirzyanova A et al. (2021). Microglia-like cells promote neuronal functions in cerebral organoids. Cells 11. https://doi.org/10.3390/cells11010124.
Fajardo Martinez V, Zhang D, Paiola S et al. (2023). Neuromotor repertoires in infants exposed to maternal COVID-19 during pregnancy: a cohort study. BMJ Open 13: e069194. https://doi.org/10.1136/bmjopen-2022-069194.
Fatemi SH, Emamian ES, Sidwell RW et al. (2002). Human influenza viral infection in utero alters glial fibrillary acidic protein immunoreactivity in the developing brains of neonatal mice. Mol Psychiatry 7: 633–640. https://doi.org/10.1038/sj.mp.4001046.
Fatemi SH, Reutiman TJ, Folsom TD et al. (2008). Maternal infection leads to abnormal gene regulation and brain atrophy in mouse offspring: implications for genesis of neurodevelopmental disorders. Schizophr Res 99: 56–70. https://doi.org/10.1016/j.schres.2007.11.018.
Fernández de Cossío L, Lacabanne C, Bordeleau M et al. (2021). Lipopolysaccharide-induced maternal immune activation modulates microglial CX3CR1 protein expression and morphological phenotype in the hippocampus and dentate gyrus, resulting in cognitive inflexibility during late adolescence. Brain Behav Immun 97: 440–454. https://doi.org/10.1016/j.bbi.2021.07.025.
Fletcher JL, Makowiecki K, Cullen CL et al. (2021). Oligodendrogenesis and myelination regulate cortical development, plasticity and circuit function. Semin Cell Dev Biol 118: 14–23. https://doi.org/10.1016/j.semcdb.2021.03.017.
Fung SG, Fakhraei R, Condran G et al. (2022). Neuropsychiatric outcomes in offspring after fetal exposure to maternal influenza infection during pregnancy: a systematic review. Reprod Toxicol 113: 155–169. https://doi.org/10.1016/j.reprotox.2022.09.002.
Gee S, Chandiramani M, Seow J et al. (2021). The legacy of maternal SARS-CoV-2 infection on the immunology of the neonate. Nat Immunol 22: 1490–1502. https://doi.org/10.1038/s41590-021-01049-2.
Gigase FAJ, Jessel RH, Kaplowitz E et al. (2024). SARS-CoV-2 infection, inflammation and birth outcomes in a prospective NYC pregnancy cohort. J Reprod Immunol 163: 104243. https://doi.org/10.1016/j.jri.2024.104243.
Gillespie B, Panthi S, Sundram S et al. (2024). The impact of maternal immune activation on GABAergic interneuron development: a systematic review of rodent studies and their translational implications. Neurosci Biobehav Rev 156: 105488. https://doi.org/10.1016/j.neubiorev.2023.105488.
Giovanoli S, Engler H, Engler A et al. (2013). Stress in puberty unmasks latent neuropathological consequences of prenatal immune activation in mice. Science 339: 1095–1099. https://doi.org/10.1126/science.1228261.
Giovanoli S, Notter T, Richetto J et al. (2015). Late prenatal immune activation causes hippocampal deficits in the absence of persistent inflammation across aging. J Neuroinflammation 12: 221. https://doi.org/10.1186/s12974-015-0437-y.
Giovanoli S, Weber L, Meyer U (2014). Single and combined effects of prenatal immune activation and peripubertal stress on parvalbumin and reelin expression in the hippocampal formation. Brain Behav Immun 40: 48–54. https://doi.org/10.1016/j.bbi.2014.04.005.
Giovanoli S, Weber-Stadlbauer U, Schedlowski M et al. (2016). Prenatal immune activation causes hippocampal synaptic deficits in the absence of overt microglia anomalies. Brain Behav Immun 55: 25–38. https://doi.org/10.1016/j.bbi.2015.09.015.
Goines PE, Croen LA, Braunschweig D et al. (2011). Increased midgestational IFN-γ, IL-4 and IL-5 in women bearing a child with autism: a case-control study. Mol Autism 2: 13. https://doi.org/10.1186/2040-2392-2-13.
Goldstein JM, Cohen JE, Mareckova K et al. (2021). Impact of prenatal maternal cytokine exposure on sex differences in brain circuitry regulating stress in offspring 45 years later. Proc Natl Acad Sci U S A 118: e2014464118. https://doi.org/10.1073/pnas.2014464118.
Gomes AKS, Dantas RM, Yokota BY et al. (2022). Interleukin-17a induces neuronal differentiation of induced-pluripotent stem cell-derived neural progenitors from autistic and control subjects. Front Neurosci 16: 828646. https://doi.org/10.3389/fnins.2022.828646.
Gomes FV, Grace AA (2017). Adolescent stress as a driving factor for schizophrenia development—a basic science perspective. Schizophr Bull 43: 486–489. https://doi.org/10.1093/schbul/sbx033.
Götz M, Huttner WB (2005). The cell biology of neurogenesis. Nat Rev Mol Cell Biol 6: 777–788. https://doi.org/10.1038/nrm1739.

Graham AM, Rasmussen JM, Rudolph MD et al. (2018). Maternal systemic Interleukin-6 during pregnancy is associated with newborn amygdala phenotypes and subsequent behavior at 2 years of age. Biol Psychiatry 83: 109–119. https://doi.org/10.1016/j.biopsych.2017.05.027.

Guma E, Plitman E, Chakravarty MM (2019). The role of maternal immune activation in altering the neurodevelopmental trajectories of offspring: a translational review of neuroimaging studies with implications for autism spectrum disorder and schizophrenia. Neurosci Biobehav Rev 104: 141–157. https://doi.org/10.1016/j.neubiorev.2019.06.020.

Gumusoglu SB, Hing BWQ, Chilukuri ASS et al. (2020). Correction: chronic maternal interleukin-17 and autism-related cortical gene expression, neurobiology, and behavior. Neuropsychopharmacology 45: 1588. https://doi.org/10.1038/s41386-020-0701-4.

Guneykaya D, Ugursu B, Logiacco F et al. (2023). Sex-specific microglia state in the neuroligin-4 knock-out mouse model of autism spectrum disorder. Brain Behav Immun 111: 61–75. https://doi.org/10.1016/j.bbi.2023.03.023.

Hall MB, Willis DE, Rodriguez EL et al. (2023). Maternal immune activation as an epidemiological risk factor for neurodevelopmental disorders: considerations of timing, severity, individual differences, and sex in human and rodent studies. Front Neurosci 17: 1135559. https://doi.org/10.3389/fnins.2023.1135559.

Han VX, Patel S, Jones HF et al. (2021). Maternal acute and chronic inflammation in pregnancy is associated with common neurodevelopmental disorders: a systematic review. Transl Psychiatry 11: 71. https://doi.org/10.1038/s41398-021-01198-w.

Hao LY, Hao XQ, Li SH et al. (2010). Prenatal exposure to lipopolysaccharide results in cognitive deficits in age-increasing offspring rats. Neuroscience 166: 763–770. https://doi.org/10.1016/j.neuroscience.2010.01.006.

Hao K, Su X, Luo B et al. (2019). Prenatal immune activation induces age-related alterations in rat offspring: effects upon NMDA receptors and behaviors. Behav Brain Res 370: 111946. https://doi.org/10.1016/j.bbr.2019.111946.

Hessami K, Norooznezhad AH, Monteiro S et al. (2022). COVID-19 pandemic and infant neurodevelopmental impairment: a systematic review and meta-analysis. JAMA Netw Open 5: e2238941. https://doi.org/10.1001/jamanetworkopen.2022.38941.

Jain S, Allen IE, Song D et al. (2023). Cytokine responses to SARS-COV2 infection in mother-infant dyads: a systematic review and meta-analysis. Front Pediatr 11: 1277697. https://doi.org/10.3389/fped.2023.1277697.

Jiang HY, Xu LL, Shao L et al. (2016). Maternal infection during pregnancy and risk of autism spectrum disorders: a systematic review and meta-analysis. Brain Behav Immun 58: 165–172. https://doi.org/10.1016/j.bbi.2016.06.005.

Jones KL, Croen LA, Yoshida CK et al. (2017). Autism with intellectual disability is associated with increased levels of maternal cytokines and chemokines during gestation. Mol Psychiatry 22: 273–279. https://doi.org/10.1038/mp.2016.77.

Kathuria A, Lopez-Lengowski K, Roffman JL et al. (2022). Distinct effects of interleukin-6 and interferon-γ on differentiating human cortical neurons. Brain Behav Immun 103: 97–108. https://doi.org/10.1016/j.bbi.2022.04.007.

Khandaker GM, Zimbron J, Lewis G et al. (2013). Prenatal maternal infection, neurodevelopment and adult schizophrenia: a systematic review of population-based studies. Psychol Med 43: 239–257. https://doi.org/10.1017/s0033291712000736.

Kim JY, Son MJ, Son CY et al. (2019). Environmental risk factors and biomarkers for autism spectrum disorder: an umbrella review of the evidence. Lancet Psychiatry 6: 590–600. https://doi.org/10.1016/s2215-0366(19)30181-6.

Knuesel I, Chicha L, Britschgi M et al. (2014). Maternal immune activation and abnormal brain development across CNS disorders. Nat Rev Neurol 10: 643–660. https://doi.org/10.1038/nrneurol.2014.187.

Lancaster MA, Renner M, Martin CA et al. (2013). Cerebral organoids model human brain development and microcephaly. Nature 501: 373–379. https://doi.org/10.1038/nature12517.

Lee GA, Lin YK, Lai JH et al. (2021). Maternal immune activation causes social behavior deficits and hypomyelination in male rat offspring with an autism-like microbiota profile. Brain Sci 11. https://doi.org/10.3390/brainsci11081085.

Lee BK, Magnusson C, Gardner RM et al. (2015). Maternal hospitalization with infection during pregnancy and risk of autism spectrum disorders. Brain Behav Immun 44: 100–105. https://doi.org/10.1016/j.bbi.2014.09.001.

Li W, Chen M, Feng X et al. (2021). Maternal immune activation alters adult behavior, intestinal integrity, gut microbiota and the gut inflammation. Brain Behav 11: e02133. https://doi.org/10.1002/brb3.2133.

Li Q, Cheung C, Wei R et al. (2010). Voxel-based analysis of postnatal white matter microstructure in mice exposed to immune challenge in early or late pregnancy. Neuroimage 52: 1–8. https://doi.org/10.1016/j.neuroimage.2010.04.015.

Lin J, Guo W (2024). The research on risk factors for Adolescents' mental health. Behav Sci (Basel) 14. https://doi.org/10.3390/bs14040263.

Lin F, Wang X, Luo R et al. (2023). Maternal LPS exposure enhances the 5-HT level in the prefrontal cortex of autism-like young offspring. Brain Sci 13. https://doi.org/10.3390/brainsci13060958.

Lipina TV, Zai C, Hlousek D et al. (2013). Maternal immune activation during gestation interacts with Disc1 point mutation to exacerbate schizophrenia-related behaviors in mice. J Neurosci 33: 7654–7666. https://doi.org/10.1523/jneurosci.0091-13.2013.

Lu YC, Andescavage N, Wu Y et al. (2022). Maternal psychological distress during the COVID-19 pandemic and structural changes of the human fetal brain. Commun Med (Lond) 2: 47. https://doi.org/10.1038/s43856-022-00111-w.

Lu X, Duan M, Song L et al. (2015). Morphological changes of radial glial cells during mouse embryonic development. Brain Res 1599: 57–66. https://doi.org/10.1016/j.brainres.2014.12.039.

Luchicchi A, Lecca S, Melis M et al. (2016). Maternal immune activation disrupts dopamine system in the offspring. Int J Neuropsychopharmacol 19. https://doi.org/10.1093/ijnp/pyw007.

Makinodan M, Tatsumi K, Manabe T et al. (2008). Maternal immune activation in mice delays myelination and axonal development in the hippocampus of the offspring. J Neurosci Res 86: 2190–2200. https://doi.org/10.1002/jnr.21673.

Malatesta P, Götz M (2013). Radial glia—from boring cables to stem cell stars. Development 140: 483–486. https://doi.org/10.1242/dev.085852.

Manning KY, Long X, Watts D et al. (2022). Prenatal maternal distress during the COVID-19 pandemic and associations with infant brain connectivity. Biol Psychiatry 92: 701–708. https://doi.org/10.1016/j.biopsych.2022.05.011.

Mastenbroek LJM, Kooistra SM, Eggen BJL et al. (2024). The role of microglia in early neurodevelopment and the effects of maternal immune activation. Semin Immunopathol 46: 1. https://doi.org/10.1007/s00281-024-01017-6.

Mattei D, Djodari-Irani A, Hadar R et al. (2014). Minocycline rescues decrease in neurogenesis, increase in microglia cytokines and deficits in sensorimotor gating in an animal model of schizophrenia. Brain Behav Immun 38: 175–184. https://doi.org/10.1016/j.bbi.2014.01.019.

Mattei D, Ivanov A, Ferrai C et al. (2017). Maternal immune activation results in complex microglial transcriptome signature in the adult offspring that is reversed by minocycline treatment. Transl Psychiatry 7: e1120. https://doi.org/10.1038/tp.2017.80.

Mattei D, Ivanov A, Hammer J et al. (2024). Microglia undergo molecular and functional adaptations to dark and light phases in male laboratory mice. Brain Behav Immun 120: 571–583. https://doi.org/10.1016/j.bbi.2024.07.007.

Mattei D, Notter T (2020). Basic concept of microglia biology and neuroinflammation in relation to psychiatry. Curr Top Behav Neurosci 44: 9–34. https://doi.org/10.1007/7854_2018_83.

Mattei D, Schweibold R, Wolf SA (2015). Brain in flames—animal models of psychosis: utility and limitations. Neuropsychiatr Dis Treat 11: 1313–1329. https://doi.org/10.2147/ndt.S65564.

McEwan F, Glazier JD, Hager R (2023). The impact of maternal immune activation on embryonic brain development. Front Neurosci 17: 1146710. https://doi.org/10.3389/fnins.2023.1146710.

McKinsey GL, Santander N, Zhang X et al. (2023). Radial glia promote microglial development through integrin α(V)β(8)-TGFβ1 signaling. BioRxiv. https://doi.org/10.1101/2023.07.13.548459.

Mesa MD, Loureiro B, Iglesia I et al. (2020). The evolving microbiome from pregnancy to early infancy: a comprehensive review. Nutrients 12: 133.

Meyer U (2014). Prenatal poly(i:C) exposure and other developmental immune activation models in rodent systems. Biol Psychiatry 75: 307–315. https://doi.org/10.1016/j.biopsych.2013.07.011.

Meyer U (2019). Neurodevelopmental resilience and susceptibility to maternal immune activation. Trends Neurosci 42: 793–806. https://doi.org/10.1016/j.tins.2019.08.001.

Miller SJ (2018). Astrocyte heterogeneity in the adult central nervous system. Front Cell Neurosci 12: 401. https://doi.org/10.3389/fncel.2018.00401.

Miranda-Negrón Y, García-Arrarás JE (2022). Radial glia and radial glia-like cells: their role in neurogenesis and regeneration. Front Neurosci 16: 1006037. https://doi.org/10.3389/fnins.2022.1006037.

Molofsky AV, Krencik R, Ullian EM et al. (2012). Astrocytes and disease: a neurodevelopmental perspective. Genes Dev 26: 891–907. https://doi.org/10.1101/gad.188326.112.

Morais LH, Felice D, Golubeva AV et al. (2018). Strain differences in the susceptibility to the gut-brain axis and neurobehavioural alterations induced by maternal immune activation in mice. Behav Pharmacol 29: 181–198. https://doi.org/10.1097/fbp.0000000000000374.

Mueller FS, Scarborough J, Schalbetter SM et al. (2021). Behavioral, neuroanatomical, and molecular correlates of resilience and susceptibility to maternal immune activation. Mol Psychiatry 26: 396–410. https://doi.org/10.1038/s41380-020-00952-8.

Muffat J, Li Y, Omer A et al. (2018). Human induced pluripotent stem cell-derived glial cells and neural progenitors display divergent responses to Zika and dengue infections. Proc Natl Acad Sci U S A 115: 7117–7122. https://doi.org/10.1073/pnas.1719266115.

Munyeshyaka M, Fields RD (2022). Oligodendroglia are emerging players in several forms of learning and memory. Commun Biol 5: 1148. https://doi.org/10.1038/s42003-022-04116-y.

Nahmias AJ, Walls KW, Stewart JA et al. (1971). The ToRCH complex-perinatal infections associated with toxoplasma and rubella, cytomegol- and herpes simplex viruses. Pediatr Res 5: 405–406. https://doi.org/10.1203/00006450-197108000-00144.

Neu N, Duchon J, Zachariah P (2015). TORCH infections. Clin Perinatol 42: 77–103viiihttps://doi.org/10.1016/j.clp.2014.11.001.

Nudel R, Wang Y, Appadurai V et al. (2019). A large-scale genomic investigation of susceptibility to infection and its association with mental disorders in the Danish population. Transl Psychiatry 9: 283. https://doi.org/10.1038/s41398-019-0622-3.

O'Loughlin E, Pakan JMP, Yilmazer-Hanke D et al. (2017). Acute in utero exposure to lipopolysaccharide induces inflammation in the pre- and postnatal brain and alters the glial cytoarchitecture in the developing amygdala. J Neuroinflammation 14: 212. https://doi.org/10.1186/s12974-017-0981-8.

Oosterhof N, Chang IJ, Karimiani EG et al. (2019). Homozygous mutations in CSF1R cause a pediatric-onset leukoencephalopathy and can result in congenital absence of microglia. Am J Hum Genet 104: 936–947. https://doi.org/10.1016/j.ajhg.2019.03.010.

Ormel PR, Vieira de Sá R, van Bodegraven EJ et al. (2018). Microglia innately develop within cerebral organoids. Nat Commun 9: 4167. https://doi.org/10.1038/s41467-018-06684-2.

Oskvig DB, Elkahloun AG, Johnson KR et al. (2012). Maternal immune activation by LPS selectively alters specific gene expression profiles of interneuron migration and oxidative stress in the fetus without triggering a fetal immune response. Brain Behav Immun 26: 623–634. https://doi.org/10.1016/j.bbi.2012.01.015.

Ostrem BEL, Domínguez-Iturza N, Stogsdill JA et al. (2024). Fetal brain response to maternal inflammation requires microglia. Development 151. https://doi.org/10.1242/dev.202252.

Otero AM, Connolly MG, Gonzalez-Ricon RJ et al. (2024). Influenza A virus during pregnancy disrupts maternal intestinal immunity and fetal cortical development in a dose- and time-dependent manner. Mol Psychiatry. https://doi.org/10.1038/s41380-024-02648-9.

Ozaki K, Kato D, Ikegami A et al. (2020). Maternal immune activation induces sustained changes in fetal microglia motility. Sci Rep 10: 21378. https://doi.org/10.1038/s41598-020-78294-2.

Pavlinek A, Matuleviciute R, Sichlinger L et al. (2022). Interferon-γ exposure of human iPSC-derived neurons alters major histocompatibility complex I and synapsin protein expression. Front Psych 13: 836217. https://doi.org/10.3389/fpsyt.2022.836217.

Paylor JW, Lins BR, Greba Q et al. (2016). Developmental disruption of perineuronal nets in the medial prefrontal cortex after maternal immune activation. Sci Rep 6: 37580. https://doi.org/10.1038/srep37580.

PrabhuDas M, Piper JM, Jean-Philippe P et al. (2021). Immune regulation, maternal infection, vaccination, and pregnancy outcome. J Womens Health (Larchmt) 30: 199–206. https://doi.org/10.1089/jwh.2020.8854.

Pratt L, Ni L, Ponzio NM et al. (2013). Maternal inflammation promotes fetal microglial activation and increased cholinergic expression in the fetal basal forebrain: role of interleukin-6. Pediatr Res 74: 393–401. https://doi.org/10.1038/pr.2013.126.

Rahman MM, Islam MR, Yamin M et al. (2022). Emerging role of neuron-glia in neurological disorders: at a glance. Oxid Med Cell Longev 2022: 3201644. https://doi.org/10.1155/2022/3201644.

Rajagopalan V, Reynolds WT, Zepeda J et al. (2022). Impact of COVID-19 related maternal stress on fetal brain development: a multimodal MRI study. MedRxiv. https://doi.org/10.1101/2022.10.26.22281575.

Rasmussen JM, Graham AM, Entringer S et al. (2019). Maternal Interleukin-6 concentration during pregnancy is associated with variation in frontolimbic white matter and cognitive development in early life. Neuroimage 185: 825–835. https://doi.org/10.1016/j.neuroimage.2018.04.020.

Ratsika A, Cruz Pereira JS, Lynch CMK et al. (2023). Microbiota-immune-brain interactions: a lifespan perspective. Curr Opin Neurobiol 78: 102652. https://doi.org/10.1016/j.conb.2022.102652.

Reisinger S, Khan D, Kong E et al. (2015). The poly(I:C)-induced maternal immune activation model in preclinical neuropsychiatric drug discovery. Pharmacol Ther 149: 213–226. https://doi.org/10.1016/j.pharmthera.2015.01.001.

Ribeiro MC, MacDonald JL (2020). Sex differences in Mecp2-mutant Rett syndrome model mice and the impact of cellular mosaicism in phenotype development. Brain Res 1729: 146644. https://doi.org/10.1016/j.brainres.2019.146644.

Richardson WD, Kessaris N, Pringle N (2006). Oligodendrocyte wars. Nat Rev Neurosci 7: 11–18. https://doi.org/10.1038/nrn1826.

Richetto J, Massart R, Weber-Stadlbauer U et al. (2017). Genome-wide DNA methylation changes in a mouse model of infection-mediated neurodevelopmental disorders. Biol Psychiatry 81: 265–276. https://doi.org/10.1016/j.biopsych.2016.08.010.

Ritchie L, Tate R, Chamberlain LH et al. (2018). Toll-like receptor 3 activation impairs excitability and synaptic activity via TRIF signalling in immature rat and human neurons. Neuropharmacology 135: 1–10. https://doi.org/10.1016/j.neuropharm.2018.02.025.

Romero E, Ali C, Molina-Holgado E et al. (2007). Neurobehavioral and immunological consequences of prenatal immune activation in rats. Influence of antipsychotics. Neuropsychopharmacology 32: 1791–1804. https://doi.org/10.1038/sj.npp.1301292.

Rousset CI, Chalon S, Cantagrel S et al. (2006). Maternal exposure to LPS induces hypomyelination in the internal capsule and programmed cell death in the deep gray matter in newborn rats. Pediatr Res 59: 428–433. https://doi.org/10.1203/01.pdr.0000199905.08848.55.

Rudolph MD, Graham AM, Feczko E et al. (2018). Maternal IL-6 during pregnancy can be estimated from newborn brain connectivity and predicts future working memory in offspring. Nat Neurosci 21: 765–772. https://doi.org/10.1038/s41593-018-0128-y.

Ryan AM, Bauman MD (2022). Primate models as a translational tool for understanding prenatal origins of neurodevelopmental disorders associated with maternal infection. Biol Psychiatry Cogn Neurosci Neuroimaging 7: 510–523. https://doi.org/10.1016/j.bpsc.2022.02.012.

Salter MW, Stevens B (2017). Microglia emerge as central players in brain disease. Nat Med 23: 1018–1027. https://doi.org/10.1038/nm.4397.

Sanders AFP, Tirado B, Seider NA et al. (2024). Prenatal exposure to maternal disadvantage-related inflammatory biomarkers: associations with neonatal white matter microstructure. Transl Psychiatry 14: 72. https://doi.org/10.1038/s41398-024-02782-6.

Santos EN, Fields RD (2021). Regulation of myelination by microglia. Sci Adv 7: eabk1131. https://doi.org/10.1126/sciadv.abk1131.

Sarieva K, Kagermeier T, Khakipoor S et al. (2023). Human brain organoid model of maternal immune activation identifies radial glia cells as selectively vulnerable. Mol Psychiatry. https://doi.org/10.1038/s41380-023-01997-1.

Sasaki T, Tome S, Takei Y (2020). Intraventricular IL-17A administration activates microglia and alters their localization in the mouse embryo cerebral cortex. Mol Brain 13: 93. https://doi.org/10.1186/s13041-020-00635-z.

Schaafsma W, Basterra LB, Jacobs S et al. (2017). Maternal inflammation induces immune activation of fetal microglia and leads to disrupted microglia immune responses, behavior, and learning performance in adulthood. Neurobiol Dis 106: 291–300. https://doi.org/10.1016/j.nbd.2017.07.017.

Schafer DP, Stevens B (2015). Microglia function in central nervous system development and plasticity. Cold Spring Harb Perspect Biol 7: a020545. https://doi.org/10.1101/cshperspect.a020545.

Shi L, Fatemi SH, Sidwell RW et al. (2003). Maternal influenza infection causes marked behavioral and pharmacological changes in the offspring. J Neurosci 23: 297–302. https://doi.org/10.1523/jneurosci.23-01-00297.2003.

Shi L, Tu N, Patterson PH (2005). Maternal influenza infection is likely to alter fetal brain development indirectly: the virus is not detected in the fetus. Int J Dev Neurosci 23: 299–305. https://doi.org/10.1016/j.ijdevneu.2004.05.005.

Shuffrey LC, Firestein MR, Kyle MH et al. (2022). Association of Birth during the COVID-19 pandemic with neurodevelopmental status at 6 months in infants with and without in utero exposure to maternal SARS-CoV-2 infection. JAMA Pediatr 176: e215563. https://doi.org/10.1001/jamapediatrics.2021.5563.

Silva D, Colvin L, Hagemann E et al. (2014). Environmental risk factors by gender associated with attention-deficit/hyperactivity disorder. Pediatrics 133: e14–e22. https://doi.org/10.1542/peds.2013-1434.

Simões LR, Sangiogo G, Tashiro MH et al. (2018). Maternal immune activation induced by lipopolysaccharide triggers immune response in pregnant mother and fetus, and induces behavioral impairment in adult rats. J Psychiatr Res 100: 71–83. https://doi.org/10.1016/j.jpsychires.2018.02.007.

Simons M, Nave KA (2015). Oligodendrocytes: myelination and axonal support. Cold Spring Harb Perspect Biol 8: a020479. https://doi.org/10.1101/cshperspect.a020479.

Sirait SR, Sinaga BY, Tarigan AP et al. (2024). Factors associated with cognitive impairment and the quality-of-life among COVID-19 survivors working as healthcare workers. Narra J 4: e658. https://doi.org/10.52225/narra.v4i1.658.

Smith SE, Li J, Garbett K et al. (2007). Maternal immune activation alters fetal brain development through interleukin-6. J Neurosci 27: 10695–10702. https://doi.org/10.1523/jneurosci.2178-07.2007.

Stefano GB (2021). Historical insight into infections and disorders associated with neurological and psychiatric sequelae similar to long COVID. Med Sci Monit 27: e931447. https://doi.org/10.12659/msm.931447.

Stegmann BJ, Carey JC (2002). TORCH infections. Toxoplasmosis, other (syphilis, varicella-zoster, parvovirus B19), rubella, cytomegalovirus (CMV), and herpes infections. Curr Womens Health Rep 2: 253–258.

Suprunowicz M, Tomaszek N, Urbaniak A et al. (2024). Between dysbiosis, maternal immune activation and autism: is there a common pathway? Nutrients 16: 549.

Taglauer ES, Dhole Y, Boateng J et al. (2022). Evaluation of maternal-infant dyad inflammatory cytokines in pregnancies affected by maternal SARS-CoV-2 infection in early and late gestation. J Perinatol 42: 1319–1327. https://doi.org/10.1038/s41372-022-01391-9.

Takahashi K, Yamanaka S (2006). Induction of pluripotent stem cells from mouse embryonic and adult fibroblast cultures by defined factors. Cell 126: 663–676. https://doi.org/10.1016/j.cell.2006.07.024.

Tanacan A, Yazihan N, Erol SA et al. (2021). The impact of COVID-19 infection on the cytokine profile of pregnant women: a prospective case-control study. Cytokine 140: 155431. https://doi.org/10.1016/j.cyto.2021.155431.

Thomson PA, Malavasi EL, Grünewald E et al. (2013). DISC1 genetics, biology and psychiatric illness. Front Biol (Beijing) 8: 1–31. https://doi.org/10.1007/s11515-012-1254-7.

Traetta ME, Tremblay M-È (2022). Prenatal inflammation shapes microglial immune response into adulthood. Trends Immunol 43: 953–955. https://doi.org/10.1016/j.it.2022.10.009.

Tucci V, Moukaddam N, Meadows J et al. (2017). The forgotten plague: psychiatric manifestations of Ebola, Zika, and emerging infectious diseases. J Glob Infect 9. https://journals.lww.com/jgid/fulltext/2017/09040/the_forgotten_plague__psychiatric_manifestations.6.aspx.

Ueno M, Fujita Y, Tanaka T et al. (2013). Layer V cortical neurons require microglial support for survival during postnatal development. Nat Neurosci 16: 543–551. https://doi.org/10.1038/nn.3358.

Verkhratsky A, Butt AM (2023a). Chapter 2—The concept of neuroglia: definition, classification and evolution. In: A Verkhratsky, AM Butt (Eds.), Neuroglia. Academic Press, pp. 53–88. https://doi.org/10.1016/B978-0-12-821565-4.00003-1.

Verkhratsky A, Butt AM (2023b). Chapter 3—Astroglial physiology. In: A Verkhratsky, AM Butt (Eds.), Neuroglia. Academic Press, pp. 89–197. https://doi.org/10.1016/B978-0-12-821565-4.00009-2.

Verkhratsky A, Butt AM (2023c). Chapter 4—Astroglial functions. In: A Verkhratsky, AM Butt (Eds.), Neuroglia. Academic Press, pp. 199–294. https://doi.org/10.1016/B978-0-12-821565-4.00008-0.

Vivi E, Di Benedetto B (2024). Brain stars take the lead during critical periods of early postnatal brain development: relevance of astrocytes in health and mental disorders. Mol Psychiatry. https://doi.org/10.1038/s41380-024-02534-4.

Vlasova RM, Iosif AM, Ryan AM et al. (2021). Maternal immune activation during pregnancy alters postnatal brain growth and cognitive development in nonhuman primate offspring. J Neurosci 41: 9971–9987. https://doi.org/10.1523/jneurosci.0378-21.2021.

Wang Y, Chen L, Wu T et al. (2020). Impact of Covid-19 in pregnancy on mother's psychological status and infant's neurobehavioral development: a longitudinal cohort study in China. BMC Med 18: 347. https://doi.org/10.1186/s12916-020-01825-1.

Warre-Cornish K, Perfect L, Nagy R et al. (2020). Interferon-γ signaling in human iPSC-derived neurons recapitulates neurodevelopmental disorder phenotypes. Sci Adv 6: eaay9506. https://doi.org/10.1126/sciadv.aay9506.

Weber-Stadlbauer U, Richetto J, Labouesse MA et al. (2017). Transgenerational transmission and modification of pathological traits induced by prenatal immune activation. Mol Psychiatry 22: 102–112. https://doi.org/10.1038/mp.2016.41.

Werling DM, Geschwind DH (2013). Sex differences in autism spectrum disorders. Curr Opin Neurol 26: 146–153. https://doi.org/10.1097/WCO.0b013e32835ee548.

Winter C, Djodari-Irani A, Sohr R et al. (2009). Prenatal immune activation leads to multiple changes in basal neurotransmitter levels in the adult brain: implications for brain disorders of neurodevelopmental origin such as schizophrenia. Int J Neuropsychopharmacol 12: 513–524. https://doi.org/10.1017/s1461145708009206.

Xia Y, Zhang Z, Lin W et al. (2020). Modulating microglia activation prevents maternal immune activation induced schizophrenia-relevant behavior phenotypes via arginase 1 in the dentate gyrus. Neuropsychopharmacology 45: 1896–1908. https://doi.org/10.1038/s41386-020-0743-7.

Yi C, Verkhratsky A, Niu J (2023). Pathological potential of oligodendrocyte precursor cells: terra incognita. Trends Neurosci 46: 581–596. https://doi.org/10.1016/j.tins.2023.04.003.

Zerbo O, Qian Y, Yoshida C et al. (2015). Maternal infection during pregnancy and autism spectrum disorders. J Autism Dev Disord 45: 4015–4025. https://doi.org/10.1007/s10803-013-2016-3.

Zhang XF, Chen T, Yan A et al. (2020). Poly(I:C) challenge alters brain expression of oligodendroglia-related genes of adult progeny in a mouse model of maternal immune activation. Front Mol Neurosci 13: 115. https://doi.org/10.3389/fnmol.2020.00115.

Zhang L, Lin C, Zhu J et al. (2023). Restoring prefrontal cortical excitation-inhibition balance with cannabidiol ameliorates neurobehavioral abnormalities in a mouse model of neurodevelopmental disorders. Neuropharmacology 240: 109715. https://doi.org/10.1016/j.neuropharm.2023.109715.

Zhang YM, Wei RM, Zhang MY et al. (2023). Resveratrol ameliorates maternal immune activation-associated cognitive impairment in adult male offspring by relieving inflammation and improving synaptic dysfunction. Front Behav Neurosci 17: 1271653. https://doi.org/10.3389/fnbeh.2023.1271653.

Zhao Q, Wang Q, Wang J et al. (2019). Maternal immune activation-induced PPARγ-dependent dysfunction of microglia associated with neurogenic impairment and aberrant postnatal behaviors in offspring. Neurobiol Dis 125: 1–13. https://doi.org/10.1016/j.nbd.2019.01.005.

Zuckerman L, Rehavi M, Nachman R et al. (2003). Immune activation during pregnancy in rats leads to a postpubertal emergence of disrupted latent inhibition, dopaminergic hyperfunction, and altered limbic morphology in the offspring: a novel neurodevelopmental model of schizophrenia. Neuropsychopharmacology 28: 1778–1789. https://doi.org/10.1038/sj.npp.1300248.

Handbook of Clinical Neurology, Vol. 210 (3rd series)
Neuroglia in Neurologic and Psychiatric Disorders, Part II
A. Verkhratsky, L.D. de Witte, E. Aronica, and E.M. Hol, Editors
https://doi.org/10.1016/B978-0-443-19102-2.00010-7

Chapter 16

Neuroglia in mood disorders

GIJSJE J.L.J. SNIJDERS* AND FREDERIEKE A.J. GIGASE

Department of Psychiatry, Icahn School of Medicine at Mount Sinai, New York, NY, United States

Abstract

Multiple lines of evidence indicate that mood disorders, such as major depressive and bipolar disorder, are associated with abnormalities in neuroglial cells. This chapter discusses the existing literature investigating the potential role of astrocytes, oligodendrocytes, and microglia in mood pathology. We will describe evidence from in vivo imaging, postmortem, animal models based on (stress) paradigms that mimic depressive-like behavior, and biomarker studies in blood and cerebrospinal fluid in patients with mood disorders. The effect of medication used in the treatment of mood disorders, such as antidepressants and lithium, on glial function is discussed. Lastly, we highlight the most relevant findings about potential deficiencies in glia–glia crosstalk in mood disorders. Overall, decreased astrocyte and oligodendrocyte density and expression and microglial changes in homeostatic functions have frequently been put forward in MDD pathology. Studies of BD report similar findings to some extent; however, the evidence is less well established. Together, these findings are suggestive of reduced glial cell function leading to potential white matter abnormalities, glutamate dysregulation, disrupted neuronal functioning, and neurotransmission. However, more research is required to better understand the exact mechanisms underlying glial cell contributions to mood disorder development.

INTRODUCTION

Mood disorders, such as major depressive disorder (MDD) and bipolar disorder (BD), are debilitating mental disorders that are still poorly understood. Elucidating the cellular pathology is of vital importance to understand the neurobiologic mechanisms that underlie these diseases and to identify targets for treatment. However, studying the key underlying mechanisms is a tremendous challenge since MDD and BD are complex disorders with different subtypes expressing heterogeneous symptoms in individuals (see Section Mood disorders). Both disorders are caused by the interaction of multiple genes and environmental factors. Twin studies have found that the estimated heritability of BD is approximately 60%–80% (Craddock and Sklar, 2013) and for MDD, 30%–50% (Flint and Kendler, 2014). Recent genome-wide association studies have identified over a 100 single nucleotide polymorphisms (SNPs) that each result in a small increase in the risk of developing mood disorders (Stahl et al., 2019; Wu et al., 2021). Mood disorders are polygenic traits arising from the influence of multiple loci with small individual effects. The polygenic traits of mood disorders may also be affected by several environmental factors, including mechanisms involved in circadian rhythm, neuronal development, immunologic- and metabolic aspects, hormonal imbalance, socioeconomic status, substance abuse, the internalizing of behavioral problems, chronic health problems, and environmental stress, including trauma and other stressful life events (Vinberg et al., 2013).

In recent years, many studies have attempted to unravel the key pathophysiologic mechanisms associated with mood disorders. The most widely accepted theories are summarized as follows: (1) The monoaminergic hypothesis postulates that mood or depressive symptoms are due to a deficit or imbalance in the monoaminergic system, which includes serotonergic, dopaminergic, and/or noradrenergic neurotransmission. The neurotransmitter system that has been strongly associated with the pathophysiology of mood pathology is serotonin (5-HT, 5 hydroxytryptamine) (Tissot, 1975); (2) Genetic studies have implied that glutamatergic neurotransmission, calcium signaling, hormonal and epigenetic regulation, and immune mechanisms are disturbed (Harrison, 2016; Stahl et al., 2019; Wu et al., 2021); (3) Neurochemical and structural alterations in mesolimbic and corticolimbic neural circuitry that regulate mood and behavior, including the prefrontal cortex, amygdala, nucleus accumbens and hippocampus have been reported to be the cause of depressive symptoms (Koo and Wohleb, 2021); (4) Brain imaging studies have pointed toward abnormalities in white matter integrity (Birur et al., 2017); (5) Neuroendocrine studies have indicated overactivity of the hypothalamic pituitary adrenal (HPA) axis in

*Correspondence to: G.J.L.J. Snijders, Department of Psychiatry, Icahn School of Medicine at Mount Sinai, New York, United States. Tel: +31628685410, E-mail: gijsjesnijders@gmail.com

depression. Interestingly, there is also research suggesting that the effects of the serotonergic system and the HPA axis overlap. This is indicated by the fact that serotonin, while regulating mood, also affects the activity of the HPA axis (Du and Pang, 2015); (6) Environmental factors, such as traumatic events and chronic stress, can increase the risk of developing mood disorders. These factors can interact with genetic vulnerabilities, and can contribute to dysregulation of neurotransmitters, neuroplasticity and the HPA axis (Mariani et al., 2021); (7) Chronic inflammation has been implicated in mood disorders. Increased levels of cytokines have been observed in individuals with mood disorders. Inflammation can impact neuronal functioning (Benedetti et al., 2020); (8) Reproductive-related hormonal changes play a role in increasing the risk of mood symptoms premenstrual, postpartum, and in the perimenopausal period (Antonelli et al., 2022); (9) There is evidence of seasonality in mood disorders, with emerging evidence that climatic conditions may trigger depressive or manic symptoms or episodes (Geoffroy et al., 2014; Øverland et al., 2019). Nevertheless, the cellular mechanisms that underlie and explain these findings remain unclear. In this chapter, we will discuss the association of mood disorders and abnormalities in the brain's cellular composition, namely the glial cells (Fig. 16.1).

In this chapter, we will discuss the association of mood disorders and abnormalities in the brain's cellular composition, namely the neuroglia (Fig. 16.1). Previous studies have demonstrated that neuroglial cells play an important role in neurodevelopment, connectivity, brain metabolism, the regulation of monoamines, formation, and maintenance of neuronal circuits and contribute to neurotransmission and clearance of both extracellular ions and neurotransmitters. In short, glial cells affect crucial central nervous system (CNS) functions and neurodevelopmental processes that might be disrupted in mood disorders as described earlier (Elsayed and Magistretti, 2015; Jäkel and Dimou, 2017). Several models have been applied to study glial cells in mood disorders, including imaging studies, animal models, postmortem brain studies, and biomarker studies. In vivo neuroimaging is a valuable tool to examine glial alterations in mood disorders. Lately, numerous novel imaging methods have been applied, including diffusion tensor imaging (DTI), functional magnetic resonance imaging (fMRI), proton magnetic spectroscopy (1H-MRS), and positron emission tomography (PET) imaging (Ding et al., 2022). Appropriate animal models and experimental studies are important as well. However, due to the heterogeneity of mood disorders, reproducing symptoms in models poses a challenge. Current models are based on manipulating either the environment to which rodents are exposed or the biological underpinnings. These manipulations can alter specific behavioral and biologic outcomes that can be related to different symptomatic and pathophysiologic dimensions of mood disorders but cannot fully replicate the complexity of depression in humans and display substantial shortcomings. Most models currently rely on responses to stress, such as chronic stress, maternal deprivation, social defeat paradigms, or actions of known antidepressants (Planchez et al., 2019). We only discuss results of animal models

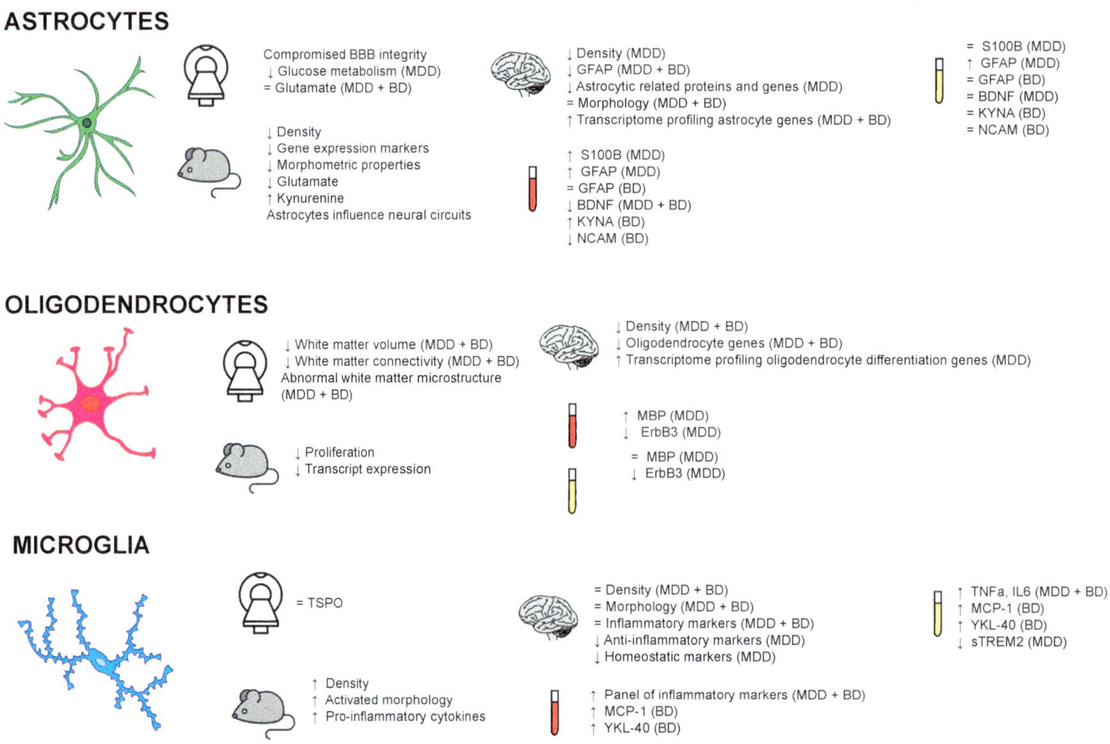

Fig. 16.1. Schematic representation of main finding, in mood disorders. Most robust findings across the different cell types (astrocytes, oligodendrocytes, and microglia) from imaging studies, animal models, postmortem studies, and biomarker blood and cerebrospinal fluid data. *BD*, bipolar disorder; *BDNF*, brain derived neurotrophic factor; *ErbB3*, Erb-B2 receptor tyrosine kinase 3; *GFAP*, glial fibrillary acidic protein; *IL-6*, Interleukin-6; *MCP-1*; *KYNA*, kynurenic acid; monocyte chemoattractant protein-1; *MBP*, myelin basic protein; *MDD*, major depressive disorder; *NCAM*, neuronal cell adhesion molecule. YKL-40: Chitinase-3-like protein TNFa: tumor necrosis factor alpha; *TSPO*, translocator protein. 1 = no differences are observed, or data are inconclusive. ↑: increase, ↓: decrease.

for depressive-like behaviors (Li et al., 2023). Recently, there is a promising novel murine animal model of BD that exhibits mania-like symptoms (Li et al., 2023), but this model is not yet established in the field. Postmortem brain studies provide the main possibility to study psychiatric pathologies at the level of individual cell type and have investigated structural cellular changes in mood disorders. Transcriptome studies have allowed the identification of distinct cell types based on cellular population-specific genes, including glial cell types (Zhang et al., 2014; Darmanis et al., 2015). Studies based on humans, especially postmortem studies, also have many limitations, including individual differences, multiple causes of death, and the postmortem interval affecting cellular integrity. Also, postmortem studies show a large variety in terms of research question, evaluated brain region, sample size, methodology, and outcome measurement. Additionally, samples from postmortem brain material are scarce and often reflect later stages of the disease (Lewis, 2002). Numerous studies examined expression levels of specific markers for glial cells in the blood. The rationale for investigating blood is that changes in the periphery may reflect similar changes in the brain or could directly affect brain function. However, whether this is indeed the case is a complicated issue, as the blood–brain barrier (BBB) separates the CNS compartment from the periphery. We will discuss findings of biomarkers for glial cells in relation to mood disorder symptomatology and staging, however, direct functional translation of these findings is questionable. We evaluated whether the observed glial biomarker changes in the blood reflect similar changes in the brain by examining glial biomarkers in the cerebrospinal fluid (CSF). It has been suggested that currently known treatments for mood disorders attenuate reactive gliosis, enhance glial proliferation, and modify glial physiology and morphology (Öngür et al., 2014). Therefore, we discuss the effects of antidepressants and lithium on glial cell function. Glial cells also influence each other, and deficiencies in this glia–glia crosstalk might contribute to diseases. Lastly, we highlight the most relevant findings in this potential crosstalk deficiency in the context of mood disorders.

MOOD DISORDERS

The term mood disorders includes both BD and unipolar depression (MDD); however, the symptomatology as described in the DSM is distinct. MDD is characterized by a low or depressed mood and might include inability to experience pleasurable activities (anhedonia), sleep disturbances, psychomotor retardation or agitation, fatigue, feelings of worthlessness or guilt, impaired thinking or concentration, change of appetite or weight, and suicidal thoughts. The overall lifetime prevalence of MDD is estimated at 10%–15% (Kessler and Bromet, 2013). The longitudinal course of MDD is heterogeneous, with tremendous variation in remission and chronicity between patients. A combination of psychotherapy and pharmaceutical interventions (with antidepressants and mood stabilizers) is often recommended. More treatment-resistant patients may benefit from electroconvulsive therapy (Fang, 2019). The prevalence of MDD in females is about twice that in males. Recent evidence suggests that treatment strategies should be tailored to sex, since a one-size-fits-all strategy has not proven effective for meeting the therapeutic needs (Albert, 2015). Patients diagnosed with BD experience both manic and depressive episodes. A manic episode is characterized by an elevated, expansive, or irritable mood and increased energy or activity. Other symptoms include inflated self-esteem or grandiosity, decreased need for sleep, excessive talking, flight of ideas, distractibility, an increase in goal-directed activity, psychomotor agitation, and hedonistic tendencies. Bipolar disorder I is the classic and most severe form, characterized by one or more manic episodes or mixed episodes (in which standards are met for both depression and mania). Bipolar disorder II is characterized by one or more major depressive episodes accompanied by at least one hypomanic episode, i.e., an episode with manic symptoms but without associated functional impairment. The key difference between bipolar disorder I and bipolar disorder II is that subtype II consists of hypomanic but not manic episodes. Cyclothymic disorder is a chronic, fluctuating mood disturbance characterized by numerous hypomanic and depressive moods, of which none meet the formal criteria for a full depressive or manic episode with a duration of minimal 2 years. The term bipolar disorder not otherwise specified (NOS) is used for patients who clearly have certain symptoms of BD but do not meet the criteria for one of the subtypes described earlier. The lifetime prevalence of bipolar disorder I and II is estimated at 1%–2% in the general population, while the overall prevalence of bipolar spectrum disorders is estimated to be around 5% (Goodwin and Jamison, 2007). In most cases, there is a full recovery between mood episodes. Treatment is often a combination of psychotherapy and pharmaceutic interventions such as mood stabilizers. Lithium has been used the longest among these mood stabilizers and is still considered a top choice for treatment because of its proven effectiveness. Mood disorders are a significant contributor to the global burden of disease and a leading cause of disability worldwide. The incidence of death by suicide is high in mood disorders (Bachmann, 2018). Insights into disease mechanisms and therapeutics are therefore urgently needed.

TOTAL GLIA

Three major glial cell types are present in the CNS: astrocytes, oligodendrocytes, and microglia. We will first describe studies that did not differentiate between the three glial cell types. Next, we will outline findings for astrocytes, oligodendrocytes, and microglia, respectively. Findings from imaging, animal models, postmortem, and biomarker studies in blood and CSF in mood disorder patients are summarized, and the most robust findings are highlighted. Meta-analyses and systematic reviews were preferred to provide the most robust evidence.

Imaging

Imaging studies indirectly offer insights into neuroglial abnormalities through various imaging techniques. Magnetic resonance imaging (MRI) provides structural information about the brain, and MRI studies have found decreased glial cell density and reduced glial volume in certain brain regions, such as the prefrontal cortex and hippocampus, in individuals with MDD and BD (Kempton et al., 2011; Harrison et al., 2020). fMRI primarily measures changes in blood oxygen level-dependent signals associated with neural activity. fMRI studies have revealed disrupted functional connectivity in brain regions involved in mood regulation in patients with mood disorders (Czéh and Nagy, 2018; Schumer et al., 2023), suggesting impaired neuroglial modulation of emotional processing. Magnetic resonance spectroscopy (MRS) is a technique that measures brain metabolites. Studies using MRS have shown altered levels of glia-specific metabolites, such as myo-inositol, in patients with mood disorders (Silverstone et al., 2005). Myo-inositol is one of the most abundant metabolites in the human brain located mainly in glial cells and functions as an osmolyte. It should be noted that these techniques allow only indirect measurement of glial functioning in the live brain.

Animal models

Animal models of chronic stress, social defeat, and (maternal) deprivation have been utilized to study the involvement of glial cells in depression (Sanacora and Banasr, 2013; Czéh and Simon, 2021). The direction of effect regarding total glial cell numbers was inconsistent across these models, which led to the conclusion that distinct glial cell types, such as astrocytes, oligodendrocytes, and microglia, may be affected differently in depression (Czéh et al., 2006, 2007; Banasr and Duman, 2007; Rajkowska and Miguel-Hidalgo, 2007; Banasr et al., 2007; Vazquez-Matias et al., 2023). It has been suggested that besides stress, other factors such as excess glucocorticoids, altered gene expression of neurotrophic factors and glial transporters, and changes in extracellular levels of neurotransmitters released by neurons may modify these glial cell numbers. Also, widespread differences between brain regions were observed in animal models with more pronounced glial changes in the (stress-sensitive) corticolimbic regions (Wegener and Neigh, 2021). However, these findings might be biased since these brain regions have been investigated more often. Experiments in animal models with depressive-like phenotypes showed that altered cytogenesis between brain regions might be related to differences in the observed cell densities (Rajkowska and Miguel-Hidalgo, 2007).

Human postmortem

Human postmortem cell counting and morphology studies have reported decreased glial cell density and number in certain brain regions, including the dorsolateral prefrontal cortex and the amygdala in MDD compared to controls (Cotter et al., 2001, 2002; Bowley et al., 2002; Gittins and Harrison, 2011). However, studies did not find clear differences in other brain regions such as the anterior cingulate cortex and hippocampus (Rajkowska et al., 1999; Rajkowska and Miguel-Hidalgo, 2007; Khundakar et al., 2011; Cobb et al., 2013; Zhou et al., 2019).

For BD, a few studies reported a decrease in total glial density in cortical regions (Ongür et al., 1998; Rajkowska et al., 2001; Brauch et al., 2006) compared to controls (Gigante et al., 2011). However, most studies indicate no change in glia number, density of size across different brain regions (Gigase et al., 2019).

In summary, neuropathology data in mood disorders suggest that cellular alterations do exist in the frontal cortex of the mood disorder brain. In line with earlier findings from animal models, a recent systematic review suggested that astrocytes, oligodendrocytes, and microglia seem to have specific patterns of changes in major psychiatric diseases (Liu et al., 2022).

Biomarkers in blood and CSF

In recent years, an increasing amount of novel glial-specific markers has been found to measure in the blood and CSF of patients with psychiatric diseases. For example, glial cell line-derived neurotrophic factor (GDNF) is a protein that is primarily secreted by glial cells and plays a crucial role in the functioning of neurons (Amidfar et al., 2021). A meta-analysis of studies of mood disorders revealed that the blood GDNF levels of MDD patients in a remission state were decreased compared to those of controls. In BD patients, findings of blood GDNF levels were inconsistent (Lin and Tseng, 2015). More recent studies have suggested that blood GDNF levels may be promising biomarkers for both BD and MDD in remission or depressive states (Idemoto et al., 2021; Zinchuk et al., 2022). Findings regarding GDNF levels in CSF in mood disorder patients, on the other hand, are not convincing (Mousten et al., 2022). Also, GDNF levels in the blood and CSF did not significantly correlate (Gadad et al., 2021). How GDNF contributes to the pathogenesis of MDD, or BD specifically is not known.

Summary

In general, the association between abnormalities in glial cells and mood disorders is complex and not fully understood. A tendency toward reduction in total glial cells has been reported, yet results are conflicting and not always replicated. Previous studies have suggested looking more into detail at the specific glial cell subtypes and the crosstalk among them to identify potential linking mechanisms.

ASTROCYTES

Astrocytes are involved in neurodevelopment, regulate synaptic transmission, provide physical and metabolic support to neurons, mediate glutamate reuptake, and aid in the maintenance of the BBB (Verkhratsky et al., 2023). In response to brain injury, astrocytes mount an evolutionary conserved defensive response known as reactive astrogliosis (Escartin et al., 2021). This serves as a defense mechanism to provide neuroprotection. For decades, astrogliosis has been described as a hallmark for mood disorder pathology. Disturbances in this process, as evidenced by increased or reduced astrogliosis, have been implicated in mood symptoms. Additionally, other forms of astrogliopathology exist, such as astrocytic atrophy and asthenia, which are prominent in psychiatric diseases (Verkhratsky et al., 2023). Astrocytes have been hypothesized to play a role in mood disorders (Rajkowska and Stockmeier, 2013), since abnormal glutamate and calcium signaling are implicated in mood pathology (de Ceglia et al., 2023; González-Arias et al., 2023). Additionally, dysregulation of homeostatic functions of astrocytes, such as regulation of circadian rhythm and metabolism, is implicated in mood disorders (Vadnie and McClung, 2017).

Imaging

Astrocytes are crucial for maintaining the integrity of the BBB (Knox et al., 2022). MRI studies have indicated compromised BBB integrity in individuals with mood disorders (Zhao et al., 2022), potentially reflecting astrocyte dysfunction. Additionally, astrocytes carry out an important function for the neuronal microenvironment, including regulation of glucose metabolism and glutamate uptake. The uptake of glucose could be measured using in vivo PET imaging. Reduced glucose metabolism in the prefrontal cortex, amygdala, thalamus, and other cortical regions is replicated in PET imaging studies in patients with mood disorders (Videbech, 2000; Chen et al., 2011). These findings suggest that metabolic changes are likely to be a consequence of altered astrocyte functioning. 1-HMRS studies examined *Glx* (a composite measure of glutamate and glutamine) levels in the context of mood disorders. Previous meta-analytic evidence suggested reduced levels of *Glx* in the medial frontal cortex of MDD patients (Moriguchi et al., 2019). However, a recent meta-analysis on studies using 1H-MRS on different clinical populations revealed no difference in the *Glx* levels in MDD and BD compared to the control group (Gudmundson et al., 2023). Other PET imaging studies investigated tracer binding to the metabotropic glutamate receptor (*mGlur5*), which is a key component of the glutamatergic system, also expressed by astrocyte cells. These studies showed inconsistent results (Esterlis et al., 2018).

Animal models

Animal studies have consistently found an association between decreased density, morphometric properties, and expression of specific astrocyte markers with depression and chronic stress (Elsayed and Magistretti, 2015). However, findings from chronic stress animal models imply that astrocytes not only change astrocyte density, morphology, and physiologic characteristics but also influence neural circuits via synapse structure and formation (Zhou et al., 2019). The underlying physiologic mechanisms whereby astrocytes modulate neurons are not completely understood in mood disorders. Studies have revealed potential mechanisms through which astrocytes and neurons interact, such as through the release of gliotransmitters (Harada et al., 2015). Also, indirect measurements associated with astrocyte functions, including glutamine and kynurenine levels, were found to be altered in a systematic analysis of metabolic changes in animal models of depression (Pu et al., 2021). Moreover, astrocytic atrophy common for stress-induced models is associated with down-regulation of plasmalemma-actin linker ezrin. This has been shown to control terminal astrocytic processes, so-called leaflets, which are mainly responsible for synaptic coverage (Lin et al., 2023).

Human postmortem

For MDD, atrophy and reduction in number of astrocytes as well as a reduction in various astrocytic proteins, such as glial fibrillary intermediate protein (GFAP), S100B, and ALDH1L1, have been described (Kim et al., 2018; Enache et al., 2019). Specifically, reduced astrocyte numbers in the amygdala, hippocampus, cerebellum, thalamus, caudate nucleus, and prefrontal cortex of MDD patients have been found (Fatemi et al., 2004; Altshuler et al., 2010; Gos et al., 2013; Wesseling et al., 2014; Bernstein et al., 2015; Cobb et al., 2016; O'Leary et al., 2021). Studies have suggested that lower astrocyte densities and decreased GFAP expression are associated with an early onset of depression (Rajkowska and Miguel-Hidalgo, 2007). A few studies reported no differences in astrocyte density and astrocytic proteins and genes in MDD compared to controls (Damadzic et al., 2001; Malchow et al., 2015). The results from studies that have examined the morphology of astrocytes in human postmortem tissue samples of subjects with depression have been mixed (Kim et al., 2018) and no straightforward conclusions can be drawn. Gene and protein expression of other astrocyte function-related proteins, including glutamine synthetase, glutamate transporters, and gap junction proteins such as connexins, are downregulated in depression (Blacker et al., 2020). Also, aquaporin 4, which influences many astrocytic functions, such as maintenance of the BBB, glutamate turnover, and synaptic plasticity, has lower expression levels in MDD (Rajkowska and Stockmeier, 2013). In contrast to these findings, an upregulation of astrocyte specific genes, such as *ALDH1L1* and *S100B* were reported on the gene level in the postmortem prefrontal cortex of MDD patients (Zhang et al., 2021). A transcriptome study reported lower expression of genes involved in astrocyte migration in postmortem prefrontal cortex of MDD participants compared to controls (Pantazatos et al., 2016). Astrocytes were consistent cell associates of depression in a large study ($n > 23{,}723$) that linked anatomic, functional, cellular, transcriptional, and genetic levels of analysis (Anderson et al., 2020). These postmortem studies together suggest a role for astrocytes in MDD, but the direction of effect is unclear.

The evidence in BD is less consistent (Gigase et al., 2019; Dai et al., 2022). Astrocyte density and numbers were found to be unchanged in most assessed brain regions. Mixed results were shown for the hippocampus. Lower levels of the traditional astrocyte marker, GFAP, have been found in the frontal cortex and hippocampus in patients with BD; however, studies show inconsistencies (Giridharan et al., 2020). Other astrocytic-related proteins, such as glutamine synthetase and glutamate transporters, were not altered in BD (Toro et al., 2006; Bernstein et al., 2015; Blacker et al., 2020). Upregulation of gap junction proteins, such as glial connexins, has been proposed as a potential mechanism that may cause synaptic imbalances responsible for the pathophysiology of mania (Mitterauer, 2011), but this concept has not been proven. To the best of our knowledge, no studies have investigated Aquaporin-4 in BD so far. Transcriptome studies of postmortem material in BD observed an upregulation in the expression profiles of cortical astrocyte genes (Ramaker et al., 2017), but not in subcortical regions.

Biomarkers in blood and CSF

S100B

S100B has extensively been discussed in the literature as a potential biomarker that plays a role in the pathogenesis of mood disorders. S100B is a calcium-binding protein that may act as a growth and/or differentiation factor for astrocytes but also neurons. It is located on and can be actively released from astrocytes and oligodendrocytes in the human brain. One might consider S100B as a glial marker protein. A meta-analysis reported an increase in S100B in MDD subjects compared to controls (Schroeter et al., 2011), suggesting that MDD may be associated with astrogliosis or brain damage (Schroeter et al., 2013). Few studies that have assessed S100B levels in CSF reported inconsistent results (Kroksmark and Vinberg, 2018; Gadad et al., 2021).

GFAP

Another well-known astrocyte-specific marker that has been measured in the blood and CSF, also in mood disorder patients, is GFAP. GFAP is the prominent intermediate filament protein in astrocytes. During processes such as CNS trauma, ischemia, and neurodegeneration, astrocytes assume a reactive phenotype (astrogliosis), which is marked by the upregulation of GFAP (Abdelhak et al., 2022). Increased GFAP expression levels were found in the blood and CSF of MDD patients, suggesting astrogliosis, but not in BD (Michel et al., 2021; Steinacker et al., 2021).

Neurotrophic factors

Astrocytes are a source of neurotrophic factors, and it has been suggested that changes in these factors may be a mechanism through which astrocytes influence mood disorders. Neurotrophic factors can promote neurogenesis, gliogenesis, and synaptic structure remodeling. The most investigated growth factor in the pathogenesis of mood disorders is brain-derived neurotrophic factor (BDNF). BDNF is produced in the endoplasmic reticulum and secreted from dense core vesicles. BDNF is abundantly expressed in the brain and periphery. BDNF promotes the growth, maturation, differentiation, and maintenance of neurons. BDNF is a critical regulator of the formation and plasticity of neuronal networks. BDNF is also involved in glutamatergic, serotonergic, and GABAergic neurotransmission. Meta-analytic evidence has repeatedly indicated that serum or plasma levels of BDNF are decreased in mood disorders (Fernandes et al., 2015; Pisoni et al., 2018; Dombi et al., 2022). BDNF levels were also significantly lower in the CSF of MDD patients compared to controls in a recent meta-analysis (Mousten et al., 2022). There is a lack of studies assessing BDNF in the CSF of patients with BD. The findings of BDNF in postmortem brain tissue in mood disorders are inconclusive (Ray et al., 2011; Reinhart et al., 2015).

Other astrocytic markers

Other astrocyte-specific markers, such as kynurenic acid (KYNA) and neuronal cell adhesion molecule (NCAM), are less well investigated. KYNA is mainly produced by astrocytes and is an *N*-methyl-D-aspartate (NMDA) antagonist with neuroprotective properties. NCAM is a synaptic membrane glycoprotein expressed in astrocytes but also neurons. BD subjects with ongoing depressive symptoms displayed a tendency of decreased plasma KYNA concentrations (Sellgren et al., 2019). Increased levels of NCAM were found in BD patients in a depressive state, but changes were not related to MDD or remissive states (Wakabayashi et al., 2008). However, the markers NCAM and KYNA were both unchanged in CSF (Sellgren et al., 2019; Mousten et al., 2022), calling the specificity of these markers as proxy for ongoing processes in the CNS into question.

Summary

To conclude, in mood disorders, and particularly MDD, a lower expression of astrocyte markers and lower densities of astrocytes labeled for these markers have been reported, suggesting a loss of astrocytes in this mental illness. Based on the current morphometric findings, it is unlikely that astrogliosis is implicated in the mood disorder brain (Rajkowska and Stockmeier, 2013; Mechawar and Savitz, 2016). Many other hypotheses have been proposed as linking mechanisms between mood disorder symptomatology and the decrease in astrocytes, such as an altered astrocyte-microglia crosstalk (Yang et al., 2020) or neuron-astrocyte signaling (Yao et al., 2023), synaptic dysfunction (Rial et al., 2015), changes in Ca+ signaling (Okubo, 2020), imbalance between excitatory and inhibitory neurotransmission (Zhou et al., 2019), potassium imbalance (O'Leary and Mechawar, 2021), and gut microbiota (Minuti et al., 2022). However, the exact link is not clarified yet.

OLIGODENDROCYTES

Oligodendrocytes are a type of glial cells in the CNS that play a crucial role in myelinating axons, which is essential for proper neuronal function (Nave and Ehrenreich, 2014). Oligodendrocytes show overlap with other glial cell functions and are also involved in neuronal signaling to nearby neurons, synaptic function, and neuronal development, repair, and survival (Duncan et al., 2021). Oligodendrocytes originate from their progenitor's oligodendrocyte precursor cells (OPCs), which have also been implicated in psychiatric pathologies including depression (Nagy et al., 2020; Yi et al., 2023). Additionally, it was found that in patients with MDD, as well as in a stress-induced depression mouse model, the density of OPC's is reduced (Birey et al., 2015). Here, we focus on oligodendrocytes.

Imaging

Imaging studies, such as MRI, have provided evidence of white matter abnormalities in mood disorders, including reduced white matter volume and disrupted white matter connectivity (Tham et al., 2011). Oligodendrocytes play a crucial role in maintaining the integrity of white matter tracts of the brain. DTI studies documented abnormal white matter microstructure in MDD and BD patients (Favre et al., 2019; van Velzen et al., 2019).

Animal models

Animal studies that used chemically induced lesions or genetic manipulations that cause oligodendrocyte disruption and demyelination showed behavioral consequences that resemble depression-related symptoms. However, demyelination in the brain could confound behavioral tasks, making it difficult to discern mood changes (Takahashi et al., 2011). Also, animal models based on chronic stress paradigms showed cellular and molecular changes in corticolimbic structures that uncover the involvement of oligodendrocytes (Lucassen et al., 2014). Rodents exposed to chronic stress showed a reduction in the proliferation of oligodendrocytes in the cortex and amygdala but also a downregulation of oligodendrocyte transcripts in the corticolimbic brain regions. These changes were reversed by antidepressant treatments, while nonstressed rodents displayed no changes after treatment. Recent reviews have suggested that the paranodal regions of oligodendrocytes may represent sites of dysregulation. The paranode region of oligodendrocytes is a critical site of interaction between myelin and axons. For example, the lack of CNP1, a paranodal-associated gene product, can impact the function of the amygdala, which is a critical region within the broader corticolimbic circuitry of mood regulation. It has been suggested that this paranode region, being the interface between axons and the myelin sheath, may be selectively vulnerable to stress (Edgar and Sibille, 2012).

Human postmortem

Human postmortem studies have reported decreased oligodendrocyte density in MDD and patterns of downregulation of oligodendrocyte-related genes in subjects with MDD compared to control subjects, primarily in the frontal cortex. No significant changes were observed in several subcortical brain regions (Tkachev et al., 2003; Tham et al., 2011; Öngür et al., 2014; Mechawar and Savitz, 2016). Whole transcriptome brain expression profiling in the dorsolateral prefrontal cortex of MDD (and suicide) is suggestive of reduced expression of genes involved in oligodendrocyte differentiation (Pantazatos et al., 2016).

Although evidence in BD is mixed, some reports suggest a similar decrease in oligodendrocyte density and oligodendrocyte-related genes, such as *Olig2, MOG,* and *MBP,* in BD patients compared to controls. These reductions are mainly described in the frontal cortex (McIntosh et al., 2009; Mahon et al., 2010; Gigase et al., 2019). A review on transcriptome studies in BD concluded that genes expressed in oligodendrocytes are downregulated (Konradi et al., 2012).

Biomarkers in blood and CSF

Oligodendrocyte specific markers

Myelin basic protein (MBP) is a structural hydrophilic protein involved in maintaining the structural integrity of myelin and reflecting the degree of axon damage and thus a presumed marker of white matter degeneration. Serum levels of MBP have been measured in MDD patients, and increased levels of MBP were found in the group with recurrent depression. The diagnostic value of MBP was also assessed and indicates that MPB contributes to the development of (recurrent) depression (Levchuk et al., 2023). MBP was also measured in the CSF across different mood disorder cohorts, but no significant differences in MBP were found (Jakobsson et al., 2014). Another oligodendrocyte-specific marker that has been investigated is ErbB3. ErbB3 is important for neuroplasticity due to its involvement in oligodendroglia differentiation and myelination. ErbB3 was identified as a candidate biomarker associated with depressive state in patients affected by BD (Le-Niculescu et al., 2009). Lower levels have been found in blood and CSF from MDD patients (Milanesi et al., 2012).

Summary

These findings altogether led to the support of a hypothesis model of dysconnectivity due to reduced oligodendrocyte function via white matter tracts between the frontal cortex and emotion regulation regions in mood pathology. However, the causality between the potentially decreased oligodendrocyte components, white matter abnormalities, and the phenotype of mood disorder is unknown (Edgar and Sibille, 2012).

MICROGLIA

Microglia are the resident immune cells of the brain and are important for the initiation and control of inflammation in the CNS (Biswas, 2023). Evidence suggests that immune-related mechanisms are involved in mood disorders (Rosenblat et al., 2014). In addition to immune functions, it is becoming increasingly clear that microglia are involved in neurodevelopment, neuronal functioning, survival and neurogenesis, synaptic pruning, and control of homeostasis (Paolicelli et al., 2022). Moreover, they contribute to the metabolism of tryptophan and thereby influence serotonergic neurotransmission. A dysfunction of microglia is therefore hypothesized to be the connection between the immune system and the neurobiologic changes that underlie mood disorders (Watkins et al., 2014).

Imaging

For a long time, studies focused mostly on signs of reactive microgliosis in the pathogenesis of mood disorders (Beumer et al., 2012) since PET imaging studies showed that translocator protein (TSPO), a marker of neuroinflammation, was increased in patients with mood disorders (Eggerstorfer et al., 2022). A correlation between the degree of reactive microglia and severity of depressive symptoms has been described as well (Setiawan et al., 2015). However, recent literature suggests that increased TSPO might be more associated with suicidality than the mood disorder itself (Yan et al., 2021). Also, the cellular specificity of TSPO for microglia in human diseases has been the topic of intense debate for the past years (Guilarte et al., 2022; Nutma et al., 2023).

Animal models

Animal models have explored the role of specific molecules important for microglial functioning. For example, the inhibition of the P2X7 receptor or CX3CR1-deficient mice has shown antidepressant-like effects in various animal models, suggesting that targeting microglial receptors and signaling pathways might present potential therapeutic strategies for depression (Ribeiro et al., 2019).

Various animal models subjected to stress paradigms, such as the chronic unpredictable stress and social defeat stress model, reported sustained reactive microgliosis that exhibited high levels of proinflammatory cytokines in different brain regions, while inhibition of microglia alleviated depressive-type behaviors (Wang et al., 2022; Afridi and Suk, 2023). Moreover, these animal models have demonstrated that stress selectively increases the number of microglia in certain stress-sensitive brain regions and also causes a marked transition of microglia from a ramified-resting state to a rounded nonresting state (Tynan, 2014). More evidence indicating the involvement of reactive microglia comes from minocycline studies. Minocycline is an antiinflammatory tetracycline that inhibits reactive microgliosis. Minocycline treatment does not have antidepressive behavioral effects in naive mice; however, it elicits significant antidepressive effects in the rat model of chronic stress (Rahimian et al., 2022a). Even though a strong correlation between microglia reactivity and depression has been observed in animal models, determining whether microglia play a significant role in depression remains challenging. A few hypotheses have been postulated in recent years based on microglia reactivity in animal models that are thought to be causing the mood disorder symptomatology: dysfunctional interactions between neurons and microglia (Kwon et al., 2017), stress-causing inhibition of neurogenesis that leads to depressive-like behavior (Bassett et al., 2021), activation of the NLRP3 inflammasome (Leonard and Maes, 2012), and mediation of the kynurenine pathway (Garrison et al., 2018). Most of the animal research has been done using stress as a paradigm to mimic depression-like behavior in rodents. However, stress itself is also capable of triggering significant production of inflammatory cytokines (Golovatscka et al., 2012). Therefore, findings in these animal models for depression are hard to directly translate to a depressed state in humans.

Human postmortem

Reactive microgliosis has been studied in human postmortem material in several ways: (1) microglia density, (2) microglia morphology, (3) expression levels of microglia reactivity markers, such as HLA-DRA, QUIN, *IL6*, and *IL1β*, and (4) (anti) inflammatory responses of microglia to specific stressors, such as lipopolysaccharides (LPS) and dexamethasone. Most postmortem studies did not report a difference in density of microglial cells in MDD and BD compared to nonpsychiatric subjects (Enache et al., 2019; Gigase et al., 2019; Harrison et al., 2020; Snijders et al., 2021). Morphologic alterations that point to an increased microglia reactive state in mood disorders were described in some studies, but not all (Sneeboer et al., 2019; Harrison et al., 2020; Snijders et al., 2021). Overall, an upregulation of immune activation markers or differences in the response of microglia to stressors was not found in mood disorder patients compared to controls (Sneeboer et al., 2019; Snijders et al., 2021). Studies focusing solely on suicide found more evidence for reactive microgliosis (Suzuki et al., 2019; Gonçalves de Andrade et al., 2022). Recently, it also seems clear that the extraordinary heterogeneity and diversity of microglial cells requires the reconsideration of the outdated dogma of their different reactive states being linked to one morphology (Vidal-Itriago et al., 2022). Additionally, most studies investigated different brain regions, making validation impossible (Liu et al., 2022). Recent studies on the sex specificity of transcriptomic differences in MDD suggest that in females with MDD there is reduced reactive microgliosis, while the opposite is true for males (Maitra et al., 2023). In BD, microglia gene-enriched modules were downregulated (Gandal et al., 2018; Zandi et al., 2022), providing additional support for the involvement of microglia in BD.

Recently, changes in homeostatic properties of microglial cells have been the topic of interest in mood disorders (Brisch et al., 2022). Studies that characterize the phenotype and function of microglial cells in MDD found a significant upregulation of homeostatic microglia-specific markers, such as TMEM119, P2RY12, and CX3CR1, and a downregulation of antiinflammatory markers such as CD163 (Böttcher et al., 2020; Snijders et al., 2021). Interestingly, RNA sequencing of microglia isolated from the cortical gray matter also showed less abundant expression of antiinflammatory microglia genes and enhanced homeostatic gene expression (Scheepstra et al., 2023). Also, whole exome gene and exon expression in the dorsolateral prefrontal cortex of MDD subjects showed markedly lower expression of genes associated with microglia cell functions (chemokine activity, regulation of cytokine biosynthetic processes) in depression, suggesting that altered microglia-mediated synaptic pruning may be present in MDD (Pantazatos et al., 2016).

Biomarkers

Inflammatory markers

Microglia involvement in immune activation has been suggested based on blood analyses of circulating inflammatory markers, including cytokines, chemokines, and cytokine receptors. In addition to microglia, reactive astrocytes can also produce these inflammatory markers in the brain. Several meta-analyses listed increased serum levels of c–c chemokine ligand 2, interleukin (IL)-4, IL-6, IL-10, IL-12, IL-13, IL-18, tumor necrosis factor (TNF)-α, sIL-2 (R), sIL-6R, and sTNF-R1 as the most consistent findings in patients with mood disorders (Goldsmith et al., 2016). However, previous studies on inflammatory markers have rarely accounted for potential confounders, such as body mass index (BMI), smoking, and ongoing medication use. Moreover, the interpretation of the findings is hampered by the fact that elevated inflammatory markers in the blood cannot be presumed to reflect inflammatory processes in the CNS. Only increased levels of IL-6 and TNF-α were replicated in the CSF (Enache et al., 2019).

Other microglia specific markers

In addition to inflammatory markers, other candidate markers primarily secreted by microglia have been measured in serum and CSF in mood disorders. Monocyte chemoattractant protein 1 (MCP-1; also called CCL-2) and YKL-40 (also called chitinase-3-like protein 1) are primarily expressed by microglia (and astrocytes) during neuroinflammation and were found to be increased in CSF in patients with BD. Serum levels correlated with levels in the CSF (Jakobsson et al., 2015). Other examples that are (primarily) expressed by microglia and are involved in reactive microgliosis or inflammation are soluble TREM2 (sTREM2), soluble CD44 (sCD44), and matrix metalloproteinases (MMPS). These factors have been measured in the blood or CSF of patients with mood disorders during different stages of disease or treatment (Ventorp et al., 2016; Tateishi et al., 2023). Only sTREM2 was found to be significantly reduced in the CSF of patients with MDD in more than one study (Pomara et al., 2021; Teipel et al., 2021).

Summary

As extensively described in recent literature (Zhu et al., 2023), microglia are so versatile that they not only provide immune surveillance but participate in many other functions such as brain development, synaptic plasticity, neuronal synchronization, priming, durotaxis, and homeostasis. Research has only just started to unravel the broad range of microglial functions. Most recent findings combined suggest an immune suppressed microglia phenotype in mood disorders, which is probably more pronounced in females compared to males. Other microglial functions are also dysregulated in mood disorders such as homeostasis, phagocytosis, neurogenesis, or neuronal repair; however, these mechanisms are not fully understood yet (Priller and Prinz, 2019; Wang et al., 2022).

MOOD DISORDER TREATMENTS

Antidepressants

Most research on the effects of antidepressants has been focused on their actions on neurons and neurotransmitters; however, emerging evidence has suggested that antidepressants may also directly affect glial cells. MRI and fMRI studies have shown that certain antidepressants can restore glial cell density, volume, glial activity, and connectivity in individuals with mood disorders (Czéh and Di Benedetto, 2013; Mariani et al., 2022). Postmortem studies involve analyzing neuroglial changes in brain tissue from individuals who had been diagnosed with mood disorders and were treated with antidepressants prior to death. These types of studies are very limited, and results are not convincing (Kandratavicius et al., 2015). In humans, particularly, most studies have been performed in blood and show decreased inflammatory cytokine levels after antidepressant treatment (Liu et al., 2019). However, most of the current evidence has come from studies utilizing animal and in vitro models. Clinical antidepressants, including selective serotonin reuptake inhibitors (SSRI's) and tricyclic antidepressants (TCA's), exert inhibitory effects in in vivo studies using animal models on the expression of inflammatory mediators, including cytokines, as well as both microgliosis and astrogliosis in the inflamed CNS. In vitro studies using mixed glial cells or reactive microglia have demonstrated that various types of antidepressants diminish glial generation of inflammatory molecules (Nicolai et al., 2023). The antiinflammatory effects of antidepressants on microglia may reduce the detrimental effects on synaptic pruning. It has also been suggested that the modulation of neuroinflammation and oxidative stress indirectly protects oligodendrocytes from damage and promotes myelin repair. Overall, a decrease in inflammatory markers is not observed in cell studies or animal models without prior evidence of classical reactive gliosis. Nevertheless, it revealed conflicting views that antidepressants might sometimes increase the inflammatory load in the brain (Wang et al., 2022). These differences may depend on regional brain specificity and heterogeneity among depression models and patients, but also on factors such as type of antidepressant, dosage, and duration of treatment. To conclude, the mechanisms through which antidepressants affect glial cells have still to be elucidated.

Besides the effects of antidepressants on neuroinflammation, various other important potential effects of antidepressants have been described in the literature for individual glial cell types, as depicted in Table 16.1.

Lithium

Lithium is a well-known medication used in the treatment of particularly BD. Although the exact mechanism of action is not fully understood, lithium has been found to have various effects on glial cells and astrocytes in particular (Nayak et al., 2021). Overall, it has been suggested that lithium exerts antiinflammatory properties and reduces the production of proinflammatory cytokines, which could

Table 16.1

Overview of impact of antidepressants on glial cells (astrocytes, oligodendrocytes, and microglia) based on current literature

Cell type	Types of action
Astrocytes	Regulation of astrocyte function by increasing expression of astrocyte specific markers and enhancing astrocyte-mediated clearance of neurotransmitters from the synaptic cleft (Allaman et al., 2011)
	Regulation of astrocyte-mediated glutamate clearance to influence glutamate neurotransmission (Czéh and Di Benedetto, 2013)
	Regulation of the release of neurotrophic factors from astrocytes, which can promote neuronal survival and synaptic plasticity (Kajitani et al., 2012; Marathe et al., 2018)
Oligodendrocytes	Promotion of maturation and myelination of oligodendrocytes to improve conduction of electrical signals and enhance neuronal communication (Chao et al., 2020)
	Regulation of survival and function of oligodendrocytes by promoting expression of myelin-related genes to impact maintenance and integrity of myelin (Zhu et al., 2023)
Microglia	Modulation of microglial function to promote neurogenesis and enhance synaptic plasticity (Zhang et al., 2018)
	Enhancement of microglial phagocytic activity, potentially aiding in the clearance of cellular debris to promote brain homeostasis (Park et al., 2021)
	Modulation of neurotransmission to trigger specific signaling pathways (Turkin et al., 2021)

Table 16.2

Overview of the impact of lithium on glial cells (astrocytes, oligodendrocytes, and microglia) based on current literature

Cell type	Types of action
Astrocytes	Promoting survival of astrocytes and preventing cell death (Zhou et al., 2017)
	Modulation of astrocytic glutamate transporters, leading to increased uptake and clearance of glutamate from the synaptic cleft (Dixon and Hokin, 1998)
	Regulation of the release of neurotrophic factors from astrocytes to contribute to neuroprotection (Emamghoreishi et al., 2015)
Oligodendrocytes	Promotion of survival and differentiation of oligodendrocyte precursor cells (Dong et al., 2015)
	Upregulation of expression of myelin-related genes, potentially enhancing the production and integrity of myelin (Fang et al., 2016)
	Enhance remyelination to protect loss or damage of myelin (Makoukji et al., 2012)
	Inhibit GSK-3 activity, which enhances white matter integrity (Benedetti et al., 2013)
Microglia	Promotion of microglial phagocytosis (Li et al., 2019)
	Neuroprotection by preventing ATP cell death (Gubert et al., 2021)

indirectly protect the glial cells from inflammation-mediated damage (Nassar and Azab, 2014). PET imaging studies have shown that chronic lithium treatment can increase astrocyte density or activity in certain brain regions (Rajkowska et al., 2016). This was also demonstrated in animal models for depression (Verkhratsky et al., 2023). In vitro studies have found that lithium can reduce reactive microglia (Dong et al., 2014). See Table 16.2 for an explanation of the described different potential types of action.

DEFICIENCIES IN GLIA–GLIA CROSSTALK AND MOOD DISORDERS

The underlying mechanisms by which glial cells are orchestrated and regulated by each other are quite important to understand the significance of these events during the development of mood symptoms. Therefore, we summarized the recent research findings in this field. Studies have suggested that the interplay between glial cell types is complex and have shown that glial cells are involved in each other's development, proliferation, and maturation. Impaired glia–glia crosstalk has been implicated in the pathology of various CNS disorders, including schizophrenia, autism spectrum disorder, Alzheimer disease, and multiple sclerosis (Bernaus et al., 2020). The glia–glia crosstalk in mood disorders is less well researched. In a mouse model of depressive-like symptoms, it was found that microglia induced oligodendrocyte proliferation by signaling to the adhesion G-protein-coupled receptor GRP56 (which is downregulated in MD patients), thereby improving remyelination, suggesting a role for microglia-oligodendrocyte cross-talk in MD pathology (Rahimian et al., 2022b). Similarly, interaction between microglia ligands and oligodendrocyte receptors was found to be different between MDD patients and controls, with most interactions being increased in cases compared to controls (Rahimian et al., 2022b). Moreover, a study showed that astrocyte specific connexin Cx30 expression, as well as major connexin-interacting proteins, were reduced on oligodendrocytes in MD patients, implicating changes in astrocyte-oligodendrocyte coupling in MD pathology (Tanti et al., 2019). Astrocyte-microglia crosstalk has been implicated in homeostasis as well as in processes including synaptogenesis and the immune response. Microglia can produce proinflammatory cytokines, which in turn may trigger astrogliosis. Reactive astrocytes may release neurotrophic factors and proinflammatory molecules that lead to increased cytokine release by microglia, synaptogenesis, and increased BBB permeability (Wang et al., 2020a). Due to the potential role of these cells in mood disorders, disruption of the astrocyte-microglia crosstalk has been implicated in mood disorders; however, additional research is needed to better understand the potential role of glia–glia interactions in mood disorder pathology (Yang et al., 2020).

CONCLUSIONS AND FUTURE DIRECTIONS

In summary, existing evidence suggests that glial cellular alterations do exist in mood disorder brains, yet results are somewhat conflicting and are limited by sample size, lack of replication, differences in study design, and varying models, making comparison of findings complex. Decreased astrocyte and oligodendrocyte density and expression and microglial changes in homeostatic functions have frequently been put forward in MDD pathology. These findings are suggestive of reduced glial cell function leading to potential white matter abnormalities, glutamate dysregulation, disrupted neuronal functioning, and neurotransmission. Studies of BD report similar findings to some extent; however, the evidence is less well established. Overall, there are numerous reports substantiating glial reductions in different corticolimbic brain regions, while others state otherwise. Current reviews are still lacking convincing theories explaining how the glial abnormalities develop and how exactly they contribute to the emotional and cognitive disturbances observed in mood disorders. Also, the actual mechanisms and associations between potential glial biomarkers and mood disorders are unclear. As an important negative result, it can be concluded that gliosis is not an evident feature of mood disorders. The evidence from imaging, animal models, postmortem, and biomarker studies is not sufficient to suggest an inflammatory process as a driving force in the pathogenesis of mood disorders.

More research is required to better understand the mechanism underlying glial cell contributions to mood disorder development. This chapter brings to light issues in validation and replication of glial abnormalities in mood disorders, particularly regarding neuropathologic findings. Surprisingly few studies have attempted to replicate previous studies, precluding substantial meta-analyses. Also, we should note that there are comparable numbers of negative findings in the literature. Probably, most studies are limited by the small sample size. We suggest replicating and validating current findings using convergent approaches in the same individual. It is also important to note that the measures might not be specific to mood disorders and might be influenced by many other factors such as brain injury, neurodegeneration, and aging, cause of death (suicide, for example), comorbidities, and medication use. Studies perform adjustments for these confounding factors to varying degrees. It is a tremendous challenge to identify an organic component that can be directly linked to the pathogenesis of mood disorders. This might be complicated by the fact that psychiatry is suffering from the contested validity of the current diagnostic system. Mood disorders are clinically heterogeneous, and mood disorder patients present a wide range of symptoms, making it difficult to detect one and the same "organic" component. Analysis of smaller subpopulations with similar specific mood disorder phenotypes might be preferable in this context. Glial cells were initially suggested to reflect responses to relevant environmental factors; however, recent novel genetic analyses revealed that a proportion of genetic risk might exert their impacts through glial cells, highlighting a more casual role of glial subtypes in the pathogenesis of mood disorders (Gandal et al., 2018; de P Lopes et al., 2022). Current models cannot capture the in vivo situation of mood symptomatology in the human body fully, since they did not capture the genetic diversity of the patients (Alciati et al., 2022; Wang et al., 2020a,b). Recent advances in technology showed how complex genetic variants lead to the pathogenesis of disease with the use of human-induced pluripotent stem cells (hIPSCs) from neuropsychiatric patients (Hong et al., 2023). However, most iPSCs of mood disorder patients were differentiated into neurons (Alciati et al., 2022). Only one study used glial progenitor cells from BD patients and showed that BD astrocytes were functionally less supportive of neuronal activity (Vadodaria et al., 2021). In addition, hIPSCs could be applied to generate cerebral organoids with disease-specific genetic backgrounds (Kathuria et al., 2020). Single-cell or nuclei RNA sequencing, another technologic advance in the past decade, enables precise classification of cell subtypes based on their transcriptome and allows identification of previously unknown cell types. This might offer further possibilities for the discovery of cell-specific molecular and mood disorder-relevant disease mechanisms (Antunes and Martins-de-Souza, 2023; Walter et al., 2023). These high resolution approaches revealed previously undetectable changes in specific cell types (such as immature OPCs and homeostatic microglia) in the context of complex phenotypes and heterogeneous tissues, as described in a few studies performed in MDD (Böttcher et al., 2020; Nagy et al., 2020; Zeng et al., 2024).

References

Abdelhak A, Foschi M, Abu-Rumeileh S et al. (2022). Blood GFAP as an emerging biomarker in brain and spinal cord disorders. Nat Rev Neurol 18: 158–172.

Afridi R, Suk K (2023). Microglial responses to stress-induced depression: causes and consequences. Cells 12: 1521.

Albert PR (2015). Why is depression more prevalent in women? J Psychiatry Neurosci 40: 219–221.

Alciati A, Reggiani A, Caldirola D et al. (2022). Human-induced pluripotent stem cell technology: toward the future of personalized psychiatry. J Pers Med 12: 1340.

Allaman I, Fiumelli H, Magistretti PJ et al. (2011). Fluoxetine regulates the expression of neurotrophic/growth factors and glucose metabolism in astrocytes. Psychopharmacology (Berl) 216: 75–84.

Altshuler LL, Abulseoud OA, Foland-Ross L et al. (2010). Amygdala astrocyte reduction in subjects with major depressive disorder but not bipolar disorder. Bipolar Disord 12: 541–549.

Amidfar M, Réus GZ, de Moura AB et al. (2021). The role of neurotrophic factors in pathophysiology of major depressive disorder. Adv Exp Med Biol 1305: 257–272.

Anderson KM, Collins MA, Kong R et al. (2020). Convergent molecular, cellular, and cortical neuroimaging signatures of major depressive disorder. Proc Natl Acad Sci U S A 117: 25138–25149.

Antonelli A, Giannini A, Chedraui P et al. (2022). Mood disorders and hormonal status across women's life: a narrative review. Gynecol Endocrinol 38: 1019–1027.

Antunes ASLM, Martins-de-Souza D (2023). Single-cell RNA sequencing and its applications in the study of psychiatric disorders. Biol Psychiatry Glob Open Sci 3: 329–339.

Bachmann S (2018). Epidemiology of suicide and the psychiatric perspective. Int J Environ Res Public Health 15: 1425.

Banasr M, Duman RS (2007). Regulation of neurogenesis and gliogenesis by stress and antidepressant treatment. CNS Neurol Disord Drug Targets 6: 311–320.

Banasr M, Valentine GW, Li X-Y et al. (2007). Chronic unpredictable stress decreases cell proliferation in the cerebral cortex of the adult rat. Biol Psychiatry 62: 496–504.

Bassett B, Subramaniyam S, Fan Y et al. (2021). Minocycline alleviates depression-like symptoms by rescuing decrease in neurogenesis in dorsal hippocampus via blocking microglia activation/phagocytosis. Brain Behav Immun 91: 519–530.

Benedetti F, Bollettini I, Barberi I et al. (2013). Lithium and GSK3-β promoter gene variants influence white matter microstructure in bipolar disorder. Neuropsychopharmacology 38: 313–327.

Benedetti F, Aggio V, Pratesi ML et al. (2020). Neuroinflammation in bipolar depression. Front Psychiatry 11: 71.

Bernaus A, Blanco S, Sevilla A (2020). Glia crosstalk in neuroinflammatory diseases. Front Cell Neurosci 14: 209.

Bernstein H-G, Meyer-Lotz G, Dobrowolny H et al. (2015). Reduced density of glutamine synthetase immunoreactive astrocytes in different cortical areas in major depression but not in bipolar I disorder. Front Cell Neurosci 9: 273.

Beumer W, Gibney SM, Drexhage RC et al. (2012). The immune theory of psychiatric diseases: a key role for activated microglia and circulating monocytes. J Leukoc Biol 92: 959–975.

Birey F, Kloc M, Chavali M et al. (2015). Genetic and stress-induced loss of NG2 glia triggers emergence of depressive-like behaviors through reduced secretion of FGF2. Neuron 88: 941–956. https://doi.org/10.1016/j.neuron.2015.10.046.

Birur B, Kraguljac NV, Shelton RC et al. (2017). Brain structure, function, and neurochemistry in schizophrenia and bipolar disorder—a systematic review of the magnetic resonance neuroimaging literature. NPJ Schizophr 3: 15.

Biswas K (2023). Microglia mediated neuroinflammation in neurodegenerative diseases: a review on the cell signaling pathways involved in microglial activation. J Neuroimmunol 383: 578180.

Blacker CJ, Millischer V, Webb LM et al. (2020). EAAT2 as a research target in bipolar disorder and unipolar depression: a systematic review. Mol Neuropsychiatry 5: 44–59.

Böttcher C, Fernández-Zapata C, Snijders GJL et al. (2020). Single-cell mass cytometry of microglia in major depressive disorder reveals a non-inflammatory phenotype with increased homeostatic marker expression. Transl Psychiatry 10: 310.

Bowley MP, Drevets WC, Ongür D et al. (2002). Low glial numbers in the amygdala in major depressive disorder. Biol Psychiatry 52: 404–412.

Brauch RA, Adnan El-Masri M, Parker Jr JC et al. (2006). Glial cell number and neuron/glial cell ratios in postmortem brains of bipolar individuals. J Affect Disord 91: 87–90.

Brisch R, Wojtylak S, Saniotis A et al. (2022). The role of microglia in neuropsychiatric disorders and suicide. Eur Arch Psychiatry Clin Neurosci 272: 929–945.

Chao F-L, Zhang Y, Zhang L et al. (2020). Fluoxetine promotes hippocampal oligodendrocyte maturation and delays learning and memory decline in APP/PS1 mice. Front Aging Neurosci 12: 627362.

Chen Q, Liu W, Li H et al. (2011). Molecular imaging in patients with mood disorders: a review of PET findings. Eur J Nucl Med Mol Imaging 38: 1367–1380.

Cobb JA, Simpson J, Mahajan GJ et al. (2013). Hippocampal volume and total cell numbers in major depressive disorder. J Psychiatr Res 47: 299–306.

Cobb JA, O'Neill K, Milner J et al. (2016). Density of GFAP-immunoreactive astrocytes is decreased in left hippocampi in major depressive disorder. Neuroscience 316: 209–220.

Cotter D, Mackay D, Landau S et al. (2001). Reduced glial cell density and neuronal size in the anterior cingulate cortex in major depressive disorder. Arch Gen Psychiatry 58: 545–553.

Cotter D, Mackay D, Chana G et al. (2002). Reduced neuronal size and glial cell density in area 9 of the dorsolateral prefrontal cortex in subjects with major depressive disorder. Cereb Cortex 12: 386–394.

Craddock N, Sklar P (2013). Genetics of bipolar disorder. Lancet 381: 1654–1662.

Czéh B, Di Benedetto B (2013). Antidepressants act directly on astrocytes: evidences and functional consequences. Eur Neuropsychopharmacol 23: 171–185.

Czéh B, Nagy SA (2018). Clinical findings documenting cellular and molecular abnormalities of glia in depressive disorders. Front Mol Neurosci 11: 56.

Czéh B, Simon M (2021). Benefits of animal models to understand the pathophysiology of depressive disorders. Prog Neuropsychopharmacol Biol Psychiatry 106: 110049.

Czéh B, Simon M, Schmelting B et al. (2006). Astroglial plasticity in the hippocampus is affected by chronic psychosocial stress and concomitant fluoxetine treatment. Neuropsychopharmacology 31: 1616–1626.

Czéh B, Müller-Keuker JIH, Rygula R et al. (2007). Chronic social stress inhibits cell proliferation in the adult medial prefrontal cortex: hemispheric asymmetry and reversal by fluoxetine treatment. Neuropsychopharmacology 32: 1490–1503.

Dai N, Jones BDM, Husain MI (2022). Astrocytes in the neuropathology of bipolar disorder: review of current evidence. Brain Sci 12: 1513.

Damadzic R, Bigelow LB, Krimer LS et al. (2001). A quantitative immunohistochemical study of astrocytes in the entorhinal cortex in schizophrenia, bipolar disorder and major depression: absence of significant astrocytosis. Brain Res Bull 55: 611–618.

Darmanis S, Sloan SA, Zhang Y et al. (2015). A survey of human brain transcriptome diversity at the single cell level. Proc Natl Acad Sci U S A 112: 7285–7290.

de Ceglia R, Ledonne A, Litvin DG et al. (2023). Specialized astrocytes mediate glutamatergic gliotransmission in the CNS. Nature 622: 120–129.

de P Lopes K, Snijders GJL, Humphrey J et al. (2022). Genetic analysis of the human microglial transcriptome across brain regions, aging and disease pathologies. Nat Genet 54: 4–17.

Ding J, Liao W, Zhu D (2022). Advanced neuroimaging methods in brain disorders, Frontiers Media SA.

Dixon JF, Hokin LE (1998). Lithium acutely inhibits and chronically up-regulates and stabilizes glutamate uptake by presynaptic nerve endings in mouse cerebral cortex. Proc Natl Acad Sci U S A 95: 8363–8368.

Dombi ZB, Szendi I, Burnet PWJ (2022). Brain derived neurotrophic factor and cognitive dysfunction in the schizophrenia-bipolar spectrum: a systematic review and meta-analysis. Front Psych 13: 827322.

Dong H, Zhang X, Dai X et al. (2014). Lithium ameliorates lipopolysaccharide-induced microglial activation via inhibition of toll-like receptor 4 expression by activating the PI3K/Akt/FoxO1 pathway. J Neuroinflammation 11: 140.

Dong B-T, Tu G-J, Han Y-X et al. (2015). Lithium enhanced cell proliferation and differentiation of mesenchymal stem cells to neural cells in rat spinal cord. Int J Clin Exp Pathol 8: 2473–2483.

Du X, Pang TY (2015). Is dysregulation of the HPA-Axis a core pathophysiology mediating co-morbid depression in neurodegenerative diseases? Front Psych 6: 32.

Duncan GJ, Simkins TJ, Emery B (2021). Neuron-oligodendrocyte interactions in the structure and integrity of axons. Front Cell Dev Biol 9: 653101.

Edgar N, Sibille E (2012). A putative functional role for oligodendrocytes in mood regulation. Transl Psychiatry 2: e109.

Eggerstorfer B, Kim J-H, Cumming P et al. (2022). Meta-analysis of molecular imaging of translocator protein in major depression. Front Mol Neurosci 15: 981442.

Elsayed M, Magistretti PJ (2015). A new outlook on mental illnesses: glial involvement beyond the glue. Front Cell Neurosci 9: 468.

Emamghoreishi M, Keshavarz M, Nekooeian AA (2015). Acute and chronic effects of lithium on BDNF and GDNF mRNA and protein levels in rat primary neuronal, astroglial and neuroastroglia cultures. Iran J Basic Med Sci 18: 240–246.

Enache D, Pariante CM, Mondelli V (2019). Markers of central inflammation in major depressive disorder: a systematic review and meta-analysis of studies examining cerebrospinal fluid, positron emission tomography and post-mortem brain tissue. Brain Behav Immun 81: 24–40.

Escartin C, Galea E, Lakatos A et al. (2021). Reactive astrocyte nomenclature, definitions, and future directions. Nat Neurosci 24: 312–325. https://doi.org/10.1038/s41593-020-00783-4.

Esterlis I, Holmes SE, Sharma P et al. (2018). Metabotropic glutamatergic receptor 5 and stress disorders: knowledge gained from receptor imaging studies. Biol Psychiatry 84: 95–105.

Fang Y (2019). Depressive disorders: mechanisms, measurement and management, Springer Nature.

Fang X-Y, Zhang W-M, Zhang C-F et al. (2016). Lithium accelerates functional motor recovery by improving remyelination of regenerating axons following ventral root avulsion and reimplantation. Neuroscience 329: 213–225.

Fatemi SH, Laurence JA, Araghi-Niknam M et al. (2004). Glial fibrillary acidic protein is reduced in cerebellum of subjects with major depression, but not schizophrenia. Schizophr Res 69: 317–323.

Favre P, Pauling M, Stout J et al. (2019). Widespread white matter microstructural abnormalities in bipolar disorder: evidence from

mega- and meta-analyses across 3033 individuals. Neuropsychopharmacology 44: 2285–2293.
Fernandes BS, Molendijk ML, Köhler CA et al. (2015). Peripheral brain-derived neurotrophic factor (BDNF) as a biomarker in bipolar disorder: a meta-analysis of 52 studies. BMC Med 13: 289.
Flint J, Kendler KS (2014). The genetics of major depression. Neuron 81: 1214.
Gadad BS, Vargas-Medrano J, Ramos EI et al. (2021). Altered levels of interleukins and neurotrophic growth factors in mood disorders and suicidality: an analysis from periphery to central nervous system. Transl Psychiatry 11: 341.
Gandal MJ, Zhang P, Hadjimichael E et al. (2018). Transcriptome-wide isoform-level dysregulation in ASD, schizophrenia, and bipolar disorder. Science 362: eaat8127.
Garrison AM, Parrott JM, Tuñon A et al. (2018). Kynurenine pathway metabolic balance influences microglia activity: targeting kynurenine monooxygenase to dampen neuroinflammation. Psychoneuroendocrinology 94: 1–10.
Geoffroy PA, Bellivier F, Scott J et al. (2014). Seasonality and bipolar disorder: a systematic review, from admission rates to seasonality of symptoms. J Affect Disord 168: 210–223.
Gigante AD, Young LT, Yatham LN et al. (2011). Morphometric post-mortem studies in bipolar disorder: possible association with oxidative stress and apoptosis. Int J Neuropsychopharmacol 14: 1075–1089.
Gigase FAJ, Snijders GJLJ, Boks MP et al. (2019). Neurons and glial cells in bipolar disorder: a systematic review of postmortem brain studies of cell number and size. Neurosci Biobehav Rev 103: 150–162.
Giridharan VV, Sayana P, Pinjari OF et al. (2020). Postmortem evidence of brain inflammatory markers in bipolar disorder: a systematic review. Mol Psychiatry 25: 94–113.
Gittins RA, Harrison PJ (2011). A morphometric study of glia and neurons in the anterior cingulate cortex in mood disorder. J Affect Disord 133: 328–332.
Goldsmith DR, Rapaport MH, Miller BJ (2016). A meta-analysis of blood cytokine network alterations in psychiatric patients: comparisons between schizophrenia, bipolar disorder and depression. Mol Psychiatry 21: 1696–1709.
Golovatscka V, Ennes H, Mayer EA et al. (2012). Chronic stress-induced changes in pro-inflammatory cytokines and spinal glia markers in the rat: a time course study. Neuroimmunomodulation 19: 367–376.
Gonçalves de Andrade E, González Ibáñez F, Tremblay M-È (2022). Microglia as a hub for suicide neuropathology: future investigation and prevention targets. Front Cell Neurosci 16: 839396.
González-Arias C, Sánchez-Ruiz A, Esparza J et al. (2023). Dysfunctional serotonergic neuron-astrocyte signaling in depressive-like states. Mol Psychiatry 28: 3856–3873.
Goodwin FK, Jamison KR (2007). Manic-depressive illness: bipolar disorders and recurrent depression, Oxford University Press, USA.
Gos T, Schroeter ML, Lessel W et al. (2013). S100B-immunopositive astrocytes and oligodendrocytes in the hippocampus are differentially afflicted in unipolar and bipolar depression: a postmortem study. J Psychiatr Res 47: 1694–1699.
Gubert C, Andrejew R, Figueiro F et al. (2021). Lithium-induced neuroprotective activity in neuronal and microglial cells: a purinergic perspective. Psychiatry Res 295: 113562.
Gudmundson AT, Koo A, Virovka A et al. (2023). Meta-analysis and open-source database for in vivo brain magnetic resonance spectroscopy in health and disease. Anal Biochem 676: 115227.
Guilarte TR, Rodichkin AN, McGlothan JL et al. (2022). Imaging neuroinflammation with TSPO: a new perspective on the cellular sources and subcellular localization. Pharmacol Ther 234: 108048.
Harada K, Kamiya T, Tsuboi T (2015). Gliotransmitter release from astrocytes: functional, developmental, and pathological implications in the brain. Front Neurosci 9: 499.
Harrison PJ (2016). Molecular neurobiological clues to the pathogenesis of bipolar disorder. Curr Opin Neurobiol 36: 1–6.
Harrison PJ, Colbourne L, Harrison CH (2020). The neuropathology of bipolar disorder: systematic review and meta-analysis. Mol Psychiatry 25: 1787–1808.
Hong Y, Yang Q, Song H et al. (2023). Opportunities and limitations for studying neuropsychiatric disorders using patient-derived induced pluripotent stem cells. Mol Psychiatry 28: 1430–1439.
Idemoto K, Niitsu T, Hata T et al. (2021). Serum levels of glial cell line-derived neurotrophic factor as a biomarker for mood disorders and lithium response. Psychiatry Res 301: 113967.
Jäkel S, Dimou L (2017). Glial cells and their function in the adult brain: a journey through the history of their ablation. Front Cell Neurosci 11: 235525.
Jakobsson J, Bjerke M, Ekman CJ et al. (2014). Elevated concentrations of neurofilament light chain in the cerebrospinal fluid of bipolar disorder patients. Neuropsychopharmacology 39: 2349–2356.
Jakobsson J, Bjerke M, Sahebi S et al. (2015). Monocyte and microglial activation in patients with mood-stabilized bipolar disorder. J Psychiatry Neurosci 40: 250–258.
Kajitani N, Hisaoka-Nakashima K, Morioka N et al. (2012). Antidepressant acts on astrocytes leading to an increase in the expression of neurotrophic/growth factors: differential regulation of FGF-2 by noradrenaline. PLoS One 7: e51197.
Kandratavicius L, Peixoto-Santos JE, Monteiro MR et al. (2015). Mesial temporal lobe epilepsy with psychiatric comorbidities: a place for differential neuroinflammatory interplay. J Neuroinflammation 12: 1–12.
Kathuria A, Lopez-Lengowski K, Vater M et al. (2020). Transcriptome analysis and functional characterization of cerebral organoids in bipolar disorder. Genome Med 12: 34.
Kempton MJ, Salvador Z, Munafò MR et al. (2011). Structural neuroimaging studies in major depressive disorder. Meta-analysis and comparison with bipolar disorder. Arch Gen Psychiatry 68: 675–690.
Kessler RC, Bromet EJ (2013). The epidemiology of depression across cultures. Annu Rev Public Health 34: 119–138.
Khundakar A, Morris C, Oakley A et al. (2011). A morphometric examination of neuronal and glial cell pathology in the orbitofrontal cortex in late-life depression. Int Psychogeriatr 23: 132–140.
Kim R, Healey KL, Sepulveda-Orengo MT et al. (2018). Astroglial correlates of neuropsychiatric disease: from astrocytopathy to astrogliosis. Prog Neuropsychopharmacol Biol Psychiatry 87: 126–146.
Knox EG, Aburto MR, Clarke G et al. (2022). The blood-brain barrier in aging and neurodegeneration. Mol Psychiatry 27: 2659–2673.
Konradi C, Sillivan SE, Clay HB (2012). Mitochondria, oligodendrocytes and inflammation in bipolar disorder: evidence from transcriptome studies points to intriguing parallels with multiple sclerosis. Neurobiol Dis 45: 37–47.
Koo JW, Wohleb ES (2021). How stress shapes neuroimmune function: implications for the neurobiology of psychiatric disorders. Biol Psychiatry 90: 74–84.
Kroksmark H, Vinberg M (2018). Does S100B have a potential role in affective disorders? A literature review. Nord J Psychiatry 72: 462–470.

Kwon S-H, Han J-K, Choi M et al. (2017). Dysfunction of microglial STAT3 alleviates depressive behavior via neuron–microglia interactions. Neuropsychopharmacology 42: 2072–2086.
Le-Niculescu H, Kurian SM, Yehyawi N et al. (2009). Identifying blood biomarkers for mood disorders using convergent functional genomics. Mol Psychiatry 14: 156–174.
Leonard B, Maes M (2012). Mechanistic explanations how cell-mediated immune activation, inflammation and oxidative and nitrosative stress pathways and their sequels and concomitants play a role in the pathophysiology of unipolar depression. Neurosci Biobehav Rev 36: 764–785.
Levchuk LA, Roschina OV, Mikhalitskaya EV et al. (2023). Serum levels of S100B protein and myelin basic protein as a potential biomarkers of recurrent depressive disorders. J Pers Med 13: 1423.
Lewis DA (2002). The human brain revisited: opportunities and challenges in postmortem studies of psychiatric disorders. Neuropsychopharmacology 26: 143–154.
Li R, Liu Z, Wu X et al. (2019). Lithium chloride promoted hematoma resolution after intracerebral hemorrhage through GSK-3β-mediated pathways-dependent microglia phagocytosis and M2-phenotype differentiation, angiogenesis and neurogenesis in a rat model. Brain Res Bull 152: 117–127.
Li X, Chen B, Zhang D et al. (2023). A novel murine model of mania. Mol Psychiatry 28: 3044–3054. https://doi.org/10.1038/s41380-023-02037-8.
Lin P-Y, Tseng P-T (2015). Decreased glial cell line-derived neurotrophic factor levels in patients with depression: a meta-analytic study. J Psychiatr Res 63: 20–27.
Lin SS, Zhou B, Chen BJ et al. (2023). Electroacupuncture prevents astrocyte atrophy to alleviate depression. Cell Death Dis14: 343. https://doi.org/10.1038/s41419-023-05839-4.
Liu JJ, Wei YB, Strawbridge R et al. (2019). Peripheral cytokine levels and response to antidepressant treatment in depression: a systematic review and meta-analysis. Mol Psychiatry 25: 339–350.
Liu S-H, Du Y, Chen L et al. (2022). Glial cell abnormalities in major psychiatric diseases: a systematic review of postmortem brain studies. Mol Neurobiol 59: 1665–1692.
Lucassen PJ, Pruessner J, Sousa N et al. (2014). Neuropathology of stress. Acta Neuropathol 127: 109–135.
Mahon K, Burdick KE, Szeszko PR (2010). A role for white matter abnormalities in the pathophysiology of bipolar disorder. Neurosci Biobehav Rev 34: 533–554.
Maitra M, Mitsuhashi H, Rahimian R et al. (2023). Cell type specific transcriptomic differences in depression show similar patterns between males and females but implicate distinct cell types and genes. Nat Commun 14: 1–18.
Makoukji J, Belle M, Meffre D et al. (2012). Lithium enhances remyelination of peripheral nerves. Proc Natl Acad Sci U S A 109: 3973–3978.
Malchow B, Strocka S, Frank F et al. (2015). Stereological investigation of the posterior hippocampus in affective disorders. J Neural Transm 122: 1019–1033.
Marathe SV, D'almeida PL, Virmani G et al. (2018). Effects of monoamines and antidepressants on astrocyte physiology: implications for monoamine hypothesis of depression. J Exp Neurosci 12: 1179069518789149.
Mariani N, Cattane N, Pariante C et al. (2021). Gene expression studies in depression development and treatment: an overview of the underlying molecular mechanisms and biological processes to identify biomarkers. Transl Psychiatry 11: 354.
Mariani N, Everson J, Pariante CM et al. (2022). Modulation of microglial activation by antidepressants. J Psychopharmacol 36: 131–150.
McIntosh AM, Hall J, Lymer GKS et al. (2009). Genetic risk for white matter abnormalities in bipolar disorder. Int Rev Psychiatry 21: 387–393.
Mechawar N, Savitz J (2016). Neuropathology of mood disorders: do we see the stigmata of inflammation? Transl Psychiatry 6: e946.
Michel M, Fiebich BL, Kuzior H et al. (2021). Increased GFAP concentrations in the cerebrospinal fluid of patients with unipolar depression. Transl Psychiatry 11: 308.
Milanesi E, Minelli A, Cattane N et al. (2012). ErbB3 mRNA leukocyte levels as a biomarker for major depressive disorder. BMC Psychiatry 12: 145.
Minuti A, Brufani F, Menculini G et al. (2022). The complex relationship between gut microbiota dysregulation and mood disorders: a narrative review. Curr Res Neurobiol 3: 100044.
Mitterauer BJ (2011). Downregulation and upregulation of glial connexins may cause synaptic imbalances responsible for the pathophysiology of bipolar disorder. CNS Neurosci Ther 17: 281–293.
Moriguchi S, Takamiya A, Noda Y et al. (2019). Glutamatergic neurometabolite levels in major depressive disorder: a systematic review and meta-analysis of proton magnetic resonance spectroscopy studies. Mol Psychiatry 24: 952–964.
Mousten IV, Sørensen NV, Christensen RHB et al. (2022). Cerebrospinal fluid biomarkers in patients with unipolar depression compared with healthy control individuals: a systematic review and meta-analysis. JAMA Psychiatry 79: 571–581.
Nagy C, Maitra M, Tanti A et al. (2020). Single-nucleus transcriptomics of the prefrontal cortex in major depressive disorder implicates oligodendrocyte precursor cells and excitatory neurons. Nat Neurosci 23: 771–781.
Nassar A, Azab AN (2014). Effects of lithium on inflammation. ACS Chem Nerosci 5: 451–458.
Nave K-A, Ehrenreich H (2014). Myelination and oligodendrocyte functions in psychiatric diseases. JAMA Psychiatry 71: 582–584.
Nayak R, Rosh I, Kustanovich I et al. (2021). Mood stabilizers in psychiatric disorders and mechanisms learnt from in vitro model systems. Int J Mol Sci 22: 9315.
Nicolai L, Nettesheim P, de Witte LD et al. (2023). Antidepressants as a potential candidate to reduce microglia activation in neurodegenerative diseases. A systematic review and meta-analysis of preclinical studies. J Affect Disord Rep 11: 100465.
Nutma E, Fancy N, Weinert M et al. (2023). Translocator protein is a marker of activated microglia in rodent models but not human neurodegenerative diseases. Nat Commun 14: 5247.
O'Leary LA, Mechawar N (2021). Implication of cerebral astrocytes in major depression: a review of fine neuroanatomical evidence in humans. Glia 69: 2077–2099.
O'Leary LA, Belliveau C, Davoli MA et al. (2021). Widespread decrease of cerebral vimentin-immunoreactive astrocytes in depressed suicides. Front Psych 12: 640963.
Okubo Y (2020). Astrocytic Ca2+ signaling mediated by the endoplasmic reticulum in health and disease. J Pharmacol Sci 144: 83–88.
Ongür D, Drevets WC, Price JL (1998). Glial reduction in the subgenual prefrontal cortex in mood disorders. Proc Natl Acad Sci U S A 95: 13290–13295.
Öngür D, Bechtholt AJ, Carlezon Jr WA et al. (2014). Glial abnormalities in mood disorders. Harv Rev Psychiatry 22: 334–337.
Øverland S, Woicik W, Sikora L et al. (2019). Seasonality and symptoms of depression: a systematic review of the literature. Epidemiol Psychiatr Sci 29: e31.

Pantazatos SP, Huang Y-Y, Rosoklija GB et al. (2016). Whole-transcriptome brain expression and exon-usage profiling in major depression and suicide: evidence for altered glial, endothelial and ATPase activity. Mol Psychiatry 22: 760–773.

Paolicelli RC, Sierra A, Stevens B et al. (2022). Microglia states and nomenclature: a field at its crossroads. Neuron 110: 3458–3483.

Park SH, Lee Y-S, Yang H-J et al. (2021). Fluoxetine potentiates phagocytosis and autophagy in microglia. Front Pharmacol 12: 770610.

Pisoni A, Strawbridge R, Hodsoll J et al. (2018). Growth factor proteins and treatment-resistant depression: a place on the path to precision. Front Psych 9: 386.

Planchez B, Surget A, Belzung C (2019). Animal models of major depression: drawbacks and challenges. J Neural Transm 126: 1383–1408.

Pomara N, Bruno D, Plaska CR et al. (2021). Evidence of upregulation of the cholinergic anti-inflammatory pathway in late-life depression. J Affect Disord 286: 275–281.

Priller J, Prinz M (2019). Targeting microglia in brain disorders. Science 365: 32–33.

Pu J, Liu Y, Gui S et al. (2021). Metabolomic changes in animal models of depression: a systematic analysis. Mol Psychiatry 26: 7328–7336.

Rahimian R, Belliveau C, Chen R et al. (2022a). Microglial inflammatory-metabolic pathways and their potential therapeutic implication in major depressive disorder. Front Psych 13: 871997.

Rahimian R, Perlman K, Canonne C et al. (2022b). Targeting microglia-oligodendrocyte crosstalk in neurodegenerative and psychiatric disorders. Drug Discov Today 27: 2562–2573.

Rajkowska G, Miguel-Hidalgo JJ (2007). Gliogenesis and glial pathology in depression. CNS Neurol Disord Drug Targets 6: 219–233.

Rajkowska G, Stockmeier CA (2013). Astrocyte pathology in major depressive disorder: insights from human postmortem brain tissue. Curr Drug Targets 14: 1225–1236.

Rajkowska G, Miguel-Hidalgo JJ, Wei J et al. (1999). Morphometric evidence for neuronal and glial prefrontal cell pathology in major depression. Biol Psychiatry 45: 1085–1098.

Rajkowska G, Halaris A, Selemon LD (2001). Reductions in neuronal and glial density characterize the dorsolateral prefrontal cortex in bipolar disorder. Biol Psychiatry 49: 741–752.

Rajkowska G, Clarke G, Mahajan G et al. (2016). Differential effect of lithium on cell number in the hippocampus and prefrontal cortex in adult mice: a stereological study. Bipolar Disord 18: 41–51.

Ramaker RC, Bowling KM, Lasseigne BN et al. (2017). Post-mortem molecular profiling of three psychiatric disorders. Genome Med 9: 72.

Ray MT, Weickert CS, Wyatt E et al. (2011). Decreased BDNF, trkB-TK+ and GAD67 mRNA expression in the hippocampus of individuals with schizophrenia and mood disorders. J Psychiatry Neurosci 36: 195–203.

Reinhart V, Bove SE, Volfson D et al. (2015). Evaluation of TrkB and BDNF transcripts in prefrontal cortex, hippocampus, and striatum from subjects with schizophrenia, bipolar disorder, and major depressive disorder. Neurobiol Dis 77: 220–227.

Rial D, Lemos C, Pinheiro H et al. (2015). Depression as a glial-based synaptic dysfunction. Front Cell Neurosci 9: 521.

Ribeiro DE, Roncalho AL, Glaser T et al. (2019). P2X7 receptor signaling in stress and depression. Int J Mol Sci 20: 2778.

Rosenblat JD, Cha DS, Mansur RB et al. (2014). Inflamed moods: a review of the interactions between inflammation and mood disorders. Prog Neuropsychopharmacol Biol Psychiatry 53: 23–34.

Sanacora G, Banasr M (2013). From pathophysiology to novel antidepressant drugs: glial contributions to the pathology and treatment of mood disorders. Biol Psychiatry 73: 1172–1179.

Scheepstra KWF, Mizee MR, van Scheppingen J et al. (2023). Microglia transcriptional profiling in major depressive disorder shows inhibition of cortical gray matter microglia. Biol Psychiatry 94: 619–629.

Schroeter ML, Steiner J, Mueller K (2011). Glial pathology is modified by age in mood disorders–a systematic meta-analysis of serum S100B in vivo studies. J Affect Disord 134: 32–38.

Schroeter ML, Sacher J, Steiner J et al. (2013). Serum S100B represents a new biomarker for mood disorders. Curr Drug Targets 14: 1237–1248.

Schumer MC, Chase HW, Rozovsky R et al. (2023). Prefrontal, parietal, and limbic condition-dependent differences in bipolar disorder: a large-scale meta-analysis of functional neuroimaging studies. Mol Psychiatry 28: 2826–2838.

Sellgren CM, Gracias J, Jungholm O et al. (2019). Peripheral and central levels of kynurenic acid in bipolar disorder subjects and healthy controls. Transl Psychiatry 9: 37.

Setiawan E, Wilson AA, Mizrahi R et al. (2015). Role of translocator protein density, a marker of neuroinflammation, in the brain during major depressive episodes. JAMA Psychiatry 72: 268–275.

Silverstone PH, McGrath BM, Kim H (2005). Bipolar disorder and myo-inositol: a review of the magnetic resonance spectroscopy findings. Bipolar Disord 7: 1–10.

Sneeboer MAM, Snijders GJLJ, Berdowski WM et al. (2019). Microglia in post-mortem brain tissue of patients with bipolar disorder are not immune activated. Transl Psychiatry 9: 153.

Snijders GJLJ, Sneeboer MAM, Fernández-Andreu A et al. (2021). Distinct non-inflammatory signature of microglia in post-mortem brain tissue of patients with major depressive disorder. Mol Psychiatry 26: 3336–3349.

Stahl EA, Breen G, Forstner AJ et al. (2019). Genome-wide association study identifies 30 loci associated with bipolar disorder. Nat Genet 51: 793–803.

Steinacker P, Al Shweiki MR, Oeckl P et al. (2021). Glial fibrillary acidic protein as blood biomarker for differential diagnosis and severity of major depressive disorder. J Psychiatr Res 144: 54–58.

Suzuki H, Ohgidani M, Kuwano N et al. (2019). Suicide and microglia: recent findings and future perspectives based on human studies. Front Cell Neurosci 13: 31.

Takahashi N, Sakurai T, Davis KL et al. (2011). Linking oligodendrocyte and myelin dysfunction to neurocircuitry abnormalities in schizophrenia. Prog Neurobiol 93: 13–24.

Tanti A, Lutz P-E, Kim J et al. (2019). Evidence of decreased gap junction coupling between astrocytes and oligodendrocytes in the anterior cingulate cortex of depressed suicides. Neuropsychopharmacology 44: 2099–2111.

Tateishi H, Matsushima J, Kunitake H et al. (2023). Serum soluble triggering receptor expressed on myeloid cells-2 was not altered by rTMS in patients with treatment-resistant depression. Neuropsychopharmacol Rep 43: 222–227.

Teipel S, Bruno D, Plaska CR et al. (2021). Association of CSF sTREM2, a marker of microglia activation, with cholinergic basal forebrain volume in major depressive disorder. J Affect Disord 293: 429–434.

Tham MW, Woon PS, Sum MY et al. (2011). White matter abnormalities in major depression: evidence from post-mortem, neuroimaging and genetic studies. J Affect Disord 132: 26–36.

Tissot R (1975). The common pathophysiology of monaminergic psychoses: a new hypothesis. Neuropsychobiology 1: 243–260.

Tkachev D, Mimmack ML, Ryan MM et al. (2003). Oligodendrocyte dysfunction in schizophrenia and bipolar disorder. Lancet 362: 798–805.

Toro CT, Hallak JEC, Dunham JS et al. (2006). Glial fibrillary acidic protein and glutamine synthetase in subregions of prefrontal cortex in schizophrenia and mood disorder. Neurosci Lett 404: 276–281.

Turkin A, Tuchina O, Klempin F (2021). Microglia function on precursor cells in the adult hippocampus and their responsiveness to serotonin signaling. Front Cell Dev Biol 9: 665739.

Tynan R (2014) The impact of stress on glia: a characterisation of the chronic stress-induced density and morphological alterations observed in astrocytes and microglia. Doctoral dissertation, University of Newcastle.

Vadnie CA, McClung CA (2017). Circadian rhythm disturbances in mood disorders: insights into the role of the suprachiasmatic nucleus. Neural Plast 2017: 1504507.

Vadodaria KC, Mendes APD, Mei A et al. (2021). Altered neuronal support and inflammatory response in bipolar disorder patient-derived astrocytes. Stem Cell Rep 16: 825–835.

van Velzen LS, Kelly S, Isaev D et al. (2019). White matter disturbances in major depressive disorder: a coordinated analysis across 20 international cohorts in the ENIGMA MDD working group. Mol Psychiatry 25: 1511–1525.

Vazquez-Matias DA, de Vries EFJ, Dierckx RAJO et al. (2023). PET imaging of animal models with depressive-like phenotypes. Eur J Nucl Med Mol Imaging 50: 1564–1584.

Ventorp F, Barzilay R, Erhardt S et al. (2016). The CD44 ligand hyaluronic acid is elevated in the cerebrospinal fluid of suicide attempters and is associated with increased blood–brain barrier permeability. J Affect Disord 193: 349–354.

Verkhratsky A, Butt A, Li B et al. (2023). Astrocytes in human central nervous system diseases: a frontier for new therapies. Signal Transduct Target Ther 8: 396.

Vidal-Itriago A, Radford RAW, Aramideh JA et al. (2022). Microglia morphophysiological diversity and its implications for the CNS. Front Immunol 13: 997786.

Videbech P (2000). PET measurements of brain glucose metabolism and blood flow in major depressive disorder: a critical review. Acta Psychiatr Scand 101: 11–20.

Vinberg M, Miskowiak K, Kessing LV (2013). Risk markers for affective disorder, a seven-years follow up study of a twin cohort at low and high risk for affective disorder. J Psychiatr Res 47: 565–571.

Wakabayashi Y, Uchida S, Funato H et al. (2008). State-dependent changes in the expression levels of NCAM-140 and L1 in the peripheral blood cells of bipolar disorders, but not in the major depressive disorders. Prog Neuropsychopharmacol Biol Psychiatry 32: 1199–1205.

Walter TJ, Suter RK, Ayad NG (2023). An overview of human single-cell RNA sequencing studies in neurobiological disease. Neurobiol Dis 184: 106201.

Wang J, Qin J, Wang P et al. (2020a). Molecular mechanisms of glial cells related signaling pathways involved in the neuroinflammatory response of depression. Mediators Inflamm 2020: 3497920.

Wang M, Zhang L, Gage FH (2020b). Modeling neuropsychiatric disorders using human induced pluripotent stem cells. Protein Cell 11: 45–59.

Wang H, He Y, Sun Z et al. (2022). Microglia in depression: an overview of microglia in the pathogenesis and treatment of depression. J Neuroinflammation 19: 132.

Watkins CC, Sawa A, Pomper MG (2014). Glia and immune cell signaling in bipolar disorder: insights from neuropharmacology and molecular imaging to clinical application. Transl Psychiatry 4: e350.

Wegener AJ, Neigh GN (2021). Animal models of anxiety and depression: incorporating the underlying mechanisms of sex differences in macroglia biology. Front Behav Neurosci 15: 780190.

Wesseling H, Gottschalk MG, Bahn S (2014). Targeted multiplexed selected reaction monitoring analysis evaluates protein expression changes of molecular risk factors for major psychiatric disorders. Int J Neuropsychopharmacol 18: yu015.

Wu W, Howard D, Sibille E et al. (2021). Differential and spatial expression meta-analysis of genes identified in genome-wide association studies of depression. Transl Psychiatry 11: 8.

Yan W-J, Jiang C-L, Su W-J (2021). Life in the flame: inflammation sounds the alarm for suicide risk. Brain Behav Immun Health 14: 100250.

Yang L, Zhou Y, Jia H et al. (2020). Affective immunology: the crosstalk between microglia and astrocytes plays key role? Front Immunol 11: 1818.

Yao J, Chen C, Guo Y et al. (2023). A review of research on the association between neuron–astrocyte signaling processes and depressive symptoms. Int J Mol Sci 24: 6985.

Yi C, Verkhratsky A, Niu J (2023). Pathological potential of oligodendrocyte precursor cells: terra incognita. Trends Neurosci 46: 581–596. https://doi.org/10.1016/j.tins.2023.04.003.

Zandi PP, Jaffe AE, Goes FS et al. (2022). Amygdala and anterior cingulate transcriptomes from individuals with bipolar disorder reveal downregulated neuroimmune and synaptic pathways. Nat Neurosci 25: 381–389.

Zeng L, Fujita M, GaoZ et al. (2024). A single-nucleus transcriptome-wide association study implicates novel genes in depression pathogenesis. BiolPsychiatry 96: 34–43.

Zhang Y, Chen K, Sloan SA et al. (2014). An RNA-sequencing transcriptome and splicing database of glia, neurons, and vascular cells of the cerebral cortex. J Neurosci 34: 11929–11947.

Zhang L, Zhang J, You Z (2018). Switching of the microglial activation phenotype is a possible treatment for depression disorder. Front Cell Neurosci 12: 306.

Zhang L, Verwer RWH, Zhao J et al. (2021). Changes in glial gene expression in the prefrontal cortex in relation to major depressive disorder, suicide and psychotic features. J Affect Disord 295: 893–903.

Zhao NO, Topolski N, Tusconi M et al. (2022). Blood-brain barrier dysfunction in bipolar disorder: molecular mechanisms and clinical implications. Brain Behav Immun Health 21: 100441.

Zhou K, Xie C, Wickström M et al. (2017). Lithium protects hippocampal progenitors, cognitive performance and hypothalamus-pituitary function after irradiation to the juvenile rat brain. Oncotarget 8: 34111–34127.

Zhou X, Xiao Q, Xie L et al. (2019). Astrocyte, a promising target for mood disorder interventions. Front Mol Neurosci 12: 136.

Zhu H, Guan A, Liu J et al. (2023). Noteworthy perspectives on microglia in neuropsychiatric disorders. J Neuroinflammation 20: 223.

Zinchuk MS, Guekht AB, Druzhkova TA et al. (2022). Glial cell line-derived neurotrophic factor (GDNF) in blood serum and lacrimal fluid of patients with a current depressive episode. J Affect Disord 318: 409–413.

Handbook of Clinical Neurology, Vol. 210 (3rd series)
Neuroglia in Neurologic and Psychiatric Disorders, Part II
A. Verkhratsky, L.D. de Witte, E. Aronica, and E.M. Hol, Editors
https://doi.org/10.1016/B978-0-443-19102-2.00006-5

Chapter 17

Neuroglia in autism spectrum disorders

VIVI M. HEINE[1,2*] AND STEPHANIE DOOVES[1,2]

[1]*Department of Child and Adolescence Psychiatry, Emma Center for Personalized Medicine, Amsterdam Neuroscience, Emma Children's Hospital, Amsterdam UMC Location Vrije Universiteit Amsterdam, Amsterdam, The Netherlands*

[2]*Department of Complex Trait Genetics, Center for Neurogenomics and Cognitive Research, Vrije Universiteit Amsterdam, Amsterdam Neuroscience, Amsterdam, The Netherlands*

Abstract

Autism spectrum disorder (ASD) is characterized by difficulties in social interaction, communication, and repetitive behavior, typically diagnosed during early childhood and attributed to altered neuronal network connectivity. Several genetic and environmental risk factors contribute to ASD, including pre- or early life immune activation, which can trigger microglial and astroglial reactivity, impacting early neurodevelopment. In ASD, astrocytes show altered glutamate metabolism, directly influencing neuronal network activity, while microglia display impaired synaptic pruning, an essential developmental process for the refinement of neuronal connections. Additionally, reduced myelination in specific cortical and subcortical regions may affect brain connectivity in ASD, with white matter integrity correlating with the severity of the disorder, suggesting an important role for oligodendrocytes and myelin in ASD. This chapter provides an overview of current literature on the role of neuroglia cells in ASD, with a focus on immune activation, glutamate signaling, synaptic pruning, and myelination.

ASD DIAGNOSIS AND SYMPTOMS

Autism was first described in 1943 by Leo Kanner (Kanner, 1943; Harris, 2018). His observations involved a group of children who displayed challenges in social interaction, communication, and repetitive behaviors. These early findings laid the foundation for the diagnostic criteria of autism that is now regarded as autism spectrum disorder (ASD), encompassing several related conditions such as Asperger syndrome and pervasive developmental disorder not otherwise specified (PDD-NOS). ASD symptoms can vary widely, as outlined in the Diagnostic and Statistical Manual of Mental Disorders, Fifth Edition (DSM-5) (American Psychiatric Association, 2013). The symptoms associated with ASD encompass deficits in social and emotional reciprocity, nonverbal communication, establishing and maintaining relationships, repetitive speech or movements, adherence to routines, restricted interests, and sensory reactivity. To receive an ASD diagnosis, a person should exhibit persistent deficits in all three areas of social communication and interaction, as well as demonstrate two types of repetitive behaviors (American Psychiatric Association, 2013). ASD affects individuals worldwide, with an estimated global prevalence of 1 in 100 children (Zeidan et al., 2022). It transcends cultures and regions, typically becoming apparent in children between the ages of 1 and 4 years, but could also become apparent later in life. Although more commonly diagnosed in males, females may exhibit different symptom presentations or be underdiagnosed due to potential gender bias (Lai et al., 2015). It is worth noting that increased awareness and improved screening efforts have contributed to a greater recognition of ASD in recent decades (Zeidan et al., 2022).

The neurobiologic mechanisms underlying ASD are not fully understood, but there is evidence suggesting that changes in brain connectivity play a role. Neuroimaging studies have shown that individuals with ASD tend to have larger brain volumes compared to neurotypical individuals in early life (Hazlett et al., 2017). Moreover, differences have been reported in specific brain regions, such as the amygdala, hippocampus, and various areas of the cortex (Kim et al., 2010; Boedhoe et al., 2020). Increased growth rate of the striatum in individuals with ASD is associated with repetitive behavior (Langen et al., 2014). It is hypothesized that these

*Correspondence to: Vivi M. Heine, Department of Child and Adolescence Psychiatry, Emma Center for Personalized Medicine, Amsterdam Neuroscience, Emma Children's Hospital, Amsterdam UMC Location Vrije Universiteit Amsterdam, Amsterdam, The Netherlands; Department of Complex Trait Genetics, Center for Neurogenomics and Cognitive Research, Vrije Universiteit Amsterdam, Amsterdam Neuroscience, Amsterdam, The Netherlands. Tel: +31-20-5986695, E-mail: vm.heine@amsterdamumc.nl

alterations in neuronal networks may result from disturbances in neuroglial support, synaptogenesis (formation of synaptic connections), excitatory/inhibitory balance, or disrupted neuronal differentiation (Fetit et al., 2021). Specifically, changes in neurotransmitter levels or abnormalities in the receptors for GABA, glutamate, dopamine, and serotonin are thought to be involved in the manifestation of ASD (Cellot and Cherubini, 2014; Fetit et al., 2021). This hypothesis is supported by the observation that medications targeting these neurotransmitter systems can alleviate ASD symptoms in some cases (Lamy and Erickson, 2018). Despite significant progress in research, the precise disease mechanisms underlying ASD remain unclear. Further studies are necessary to gain a deeper understanding of the neurobiologic processes involved in ASD and to identify potential therapeutic targets.

Understanding the underlying mechanisms of ASD is crucial for advancing diagnostic methods and developing effective interventions to improve the lives of individuals with ASD and their families. While the cause of the majority of ASD cases remains unknown (idiopathic), emerging evidence suggests that a combination of genetic and environmental factors, as well as gene–environment interactions, may contribute to the increased risk of developing an ASD. In this chapter, we will explore research findings and hypotheses that shed light on the involvement of neuroglial cells in the pathomechanisms of ASD (Fig. 17.1).

GENETIC FACTORS ASSOCIATED WITH ASD

Approximately 15% of ASD cases have a clear genetic cause, which can involve specific gene mutations, chromosomal rearrangements, or copy number variations (CNVs). GWAS studies conducted on different cohorts of individuals with ASD have identified several common genetic variants and revealed specific

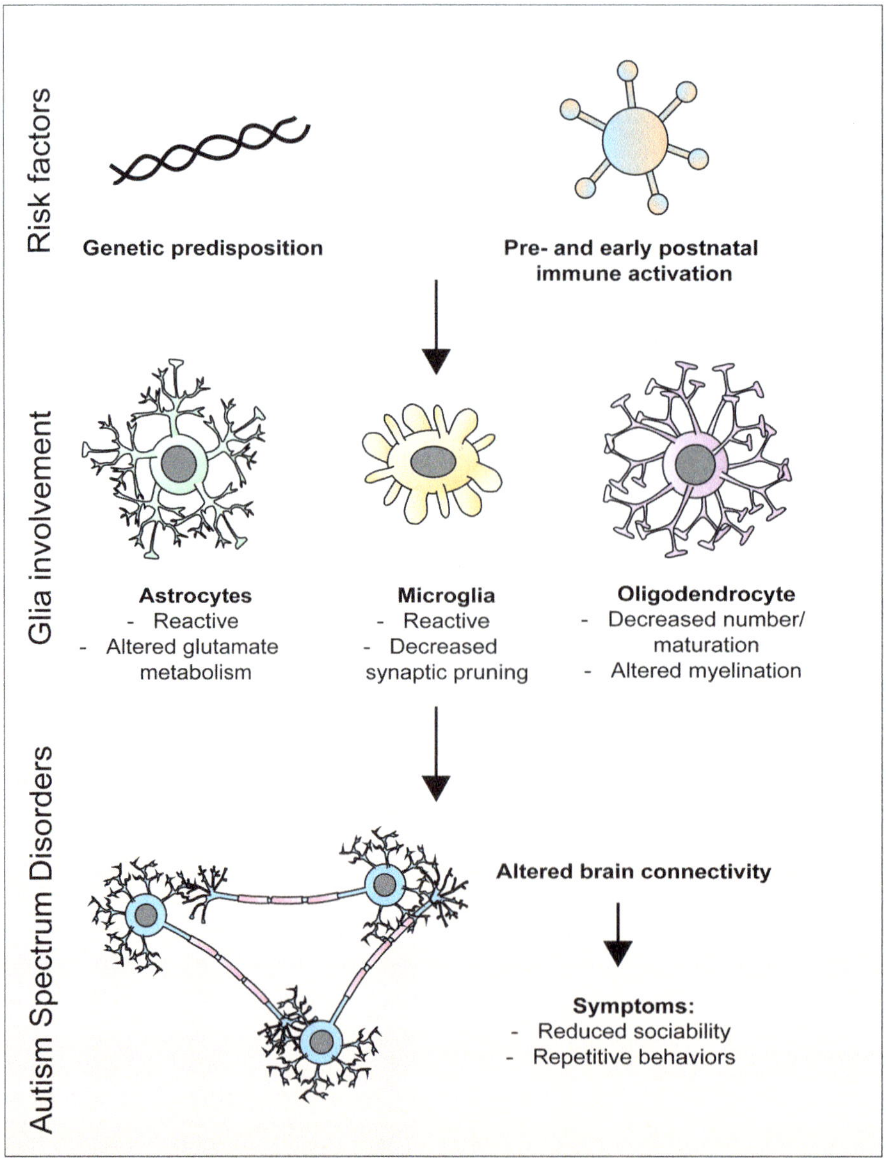

Fig. 17.1. The role of neuroglial cells in the etiology of ASD.

regions of the genome that are statistically linked to an increased risk of ASD (Autism Spectrum Disorders Working Group of The Psychiatric Genomics C, 2017; Grove et al., 2019; Iakoucheva et al., 2019). One large whole-exome sequencing study (>35,000 total samples) identified 100 ASD risk genes, and gene set analysis revealed an enrichment for genes associated with oligodendrocyte and astrocyte lineage cells together with mature excitatory and inhibitory neurons (Satterstrom et al., 2020). Notably, rare de novo CNVs in *IP3R2,* a gene involved in astrocyte reactivity, are associated with ASD and were shown to cause ASD-like behavior in mice (Wang et al., 2021). Multiple neurodevelopmental disorders with monogenic causes can present with ASDs, accounting for about 5% of ASD cases (Ziats et al., 2021). Here we will focus on three well-studied syndromic neurodevelopmental disorders with a high prevalence of ASD: Fragile X, Tuberous Sclerosis Complex, and Rett syndrome. Fragile X syndrome, caused by an expansion in the *FMR1* gene, is the most common syndrome causing ASD. Fragile X patients often experience intellectual disability and may exhibit hyperactivity and seizures (Salcedo-Arellano et al., 2020). Tuberous Sclerosis Complex (TSC), resulting from mutations in the *TSC1* or *TSC2* genes, is another syndrome associated with an increased incidence of ASD (Northrup et al., 2021), as well as Rett syndrome, caused by *MECP2* mutations. Rett affects females and can manifest with ASD symptoms, along with distinctive repetitive behaviors and other neurologic impairments (Vidal et al., 2019). Although these syndromes often present with additional neurologic abnormalities, studying the underlying cellular mechanisms can still offer insights into the development of ASD symptoms. By investigating the cellular mechanisms involved in these disorders, we can gain a better understanding of how an ASD manifests and potentially identify common pathways or targets for therapeutic interventions.

PRENATAL INFLAMMATION IS A RISK FACTOR FOR ASD

Various environmental factors have been associated with the development of ASD. While these findings may be influenced by complex gene–environment interactions, epidemiologic studies consistently indicate a strong association between maternal immune activation and the risk of developing ASD (Knuesel et al., 2014). This association extends beyond a specific environmental factor and includes risk factors such as maternal infections, obesity, and diabetes; alterations in gut microbiota during pregnancy and early life; and prenatal exposure to specific medications, heavy metals, or toxins (Kwon et al., 2022). For instance, maternal obesity triggers chronic low-grade inflammation characterized by elevated levels of proinflammatory molecules (Kwon et al., 2022). Preterm birth is another risk factor for ASD, which is often accompanied by perinatal infections and hypoxia (van Tilborg et al., 2016). Interestingly, maternal immune activation has also been identified as a risk factor for other psychiatric conditions, including schizophrenia and hyperactivity (Knuesel et al., 2014; Kwon et al., 2022), indicating a potential shared underlying mechanism. When the maternal immune system is activated, specific cytokines can cross the placenta and reach the developing fetal brain, potentially impacting neurodevelopment.

Circulating cytokines have the potential to induce reactivity of neuroglial cells in the brain, thereby contributing to the pathogenesis of ASD. Studies have shown that alterations in the gut microbiota, both during pregnancy and early life, can lead to an increased risk of ASD. Dysbiosis of the gut microbiota can disrupt the normal immune response balance, leading to inflammation and alterations in the permeability of the gastrointestinal wall. This increased permeability allows neurotoxins and inflammatory molecules to pass from the gut into the bloodstream (Roe, 2022). Positron emission tomography (PET) imaging studies have demonstrated reactive microglial cells in young adults with ASD (Petrelli et al., 2016). Additionally, postmortem tissue analyses have revealed an increase in reactive microglia and astrocytes within ASD brains (Vargas et al., 2005; Morgan et al., 2010, 2014; Fetit et al., 2021; Wamsley et al., 2023). Both astrocytes and microglia produce the cytokine CCL2, which has been found to be elevated in the brains and blood of individuals with ASD (Matta et al., 2019; Ye et al., 2021). Furthermore, elevated levels of glial fibrillary acidic protein (GFAP), a marker of reactive astrocytes, have been observed in ASD brains (Fetit et al., 2021; Wamsley et al., 2023). Gene expression analyses demonstrated increased astrocyte reactivity and downregulation of membrane transport across the blood–brain barrier (BBB) throughout the cortex in individuals with ASD (Gandal et al., 2022). In astrocytes and microglia specifically, upregulation of inflammatory response genes and increased cytokine-mediated signaling, notably interferon-gamma and inflammatory JAK/STAT/NFκB signaling pathways, has been observed (Wamsley et al., 2023). The collective findings in individuals with ASD, including reactivity of microglia and astrocytes and the altered cytokine signaling in neuroglial cells, suggest a potential contribution to the immune and inflammatory dysregulation observed in ASD.

Model systems have provided further evidence supporting the involvement of maternal immune activation in the development of ASD, even though mouse models cannot fully capture the cognitive and communicative abilities seen in humans. For example, pregnant mice treated with immune response-inducing agents, such as PolyI:C, produce offspring displaying ASD-like symptoms like deficits in social interaction (Lammert and Lukens, 2019). These models help to study how immune activation affects neurodevelopment and provide insights into potential mechanisms underlying the association between inflammation and ASD. For example, IL-6 has been identified as a crucial cytokine in the induction of ASD-like abnormalities following maternal inflammation. IL-6 is released in response to infection or inflammation and can cross the placental barrier, directly affecting IL-6 receptors in the developing brain (Smith et al., 2007; Mandal et al., 2011). A significant role for IL-6 is further supported by a study using human induced pluripotent stem cells (iPSCs) from idiopathic ASD patients. The authors showed that ASD iPSC-derived astrocytes reduce synaptic connectivity when cocultured with control or ASD neurons, an effect mediated by increased secretion of IL-6 (Russo et al., 2018). Besides an effect on neurodevelopment, IL-6 has also been implicated in altering the differentiation of Th17 cells, a subset of T-helper cells involved in immune responses (Mandal et al., 2011; Nadeem et al., 2020). Interestingly, Th17 cells and their main cytokine IL-17 have been associated with ASD, and both IL-6 receptor and IL-17 levels are increased in serum individuals with ASD (Enstrom et al., 2008; Mandal et al., 2011; Nadeem et al., 2020). Besides maternal immune activation, early postnatal inflammation has been hypothesized to play a causal role in ASD. Indeed, a mouse model of early allergic inflammation demonstrated altered microglia morphology and impaired synaptic pruning, which correlated with an increase in excitatory synapses and ASD-like behavior (Saitoh et al., 2021). Interestingly, other studies have reported sex differences in the response of offspring to maternal immune activation, with a larger effect in male offspring. For example, after maternal immune activation, male offspring showed more behavioral abnormalities than female offspring (Carlezon Jr. et al., 2019; Haida et al., 2019). Additionally, only male offspring showed a decreased expression of anti-inflammatory factors (Carlezon Jr. et al., 2019) and a decreased expression of the microglial chemokine receptor CX3CR1, which is involved in synaptic pruning (Fernandez de Cossio et al., 2017). These studies suggest that gender differences in immune reactions may underlie the higher prevalence of ASD in males. Despite limitations to mouse and human in vitro models as tools in ASD research, these can provide basic insights into underlying mechanisms.

The growing body of evidence emphasizes the fundamental significance of inflammation, particularly maternal immune activation, in the etiology of ASD. It underscores the intricate interplay between immune responses, brain development, and neuroinflammation in the pathogenesis of ASD. Further research in this field is necessary to unravel the precise mechanisms and identify potential therapeutic targets for mitigating the impact of maternal immune activation on neurodevelopment and reducing the risk of ASD.

THE ROLE OF NEUROGLIA IN ASD-ASSOCIATED CHANGES IN NEURONS

Postmortem analysis of ASD patients' brain tissue revealed changes in the astrocytic genes *AQP4* and *CX43*, which are crucial for blood–brain barrier functions and cell-to-cell communication (Fatemi et al., 2008), supporting the involvement of neuroglial dysfunction in ASD. The importance of astrocytes in neuronal development and neurotransmitter metabolism, along with their secretion of various factors, means that astrocytic dysfunction can significantly impact neuronal network functionality. An important role for neuroglial dysfunction in ASD-like behaviors, such as hyperactive behavior and alterations in social interaction in rodents, is confirmed by different model systems. Disruption of cortical astrocytic Ca^{2+} signaling (Luo et al., 2023) or deficiencies in NPAS3, a critical transcription factor for astrogenesis, have induced ASD-like behaviors in mice (Li et al., 2022). Additionally, a rat model of ASD induced through prenatal exposure to valproic acid exhibited ASD-like behavior and an altered GABA/glutamate ratio, which could be restored by enhancing astrocytic CREB/BDNF signaling and EAAT2/SLC1A2 expression (Zohny et al., 2023). When the astrocyte-specific transporter EAAT2/SLC1A2, responsible for the majority of glutamate uptake, is deficient, mice present increased neuronal excitation and excessive repetitive behavior (Aida et al., 2015). Glutamate is the main excitatory neurotransmitter, and neurons are dependent on astrocytic production of the glutamate precursor glutamine and for fast clearance of glutamate from the synaptic cleft (Schousboe et al., 2014; Mahmoud et al., 2019). Excessive amount of extracellular glutamate can cause neuronal excitotoxicity, and glutamate metabolism is therefore strictly regulated (Mahmoud et al., 2019). Glutamine synthetase (GS), exclusively expressed in astrocytes, has also been implicated in ASD, supported by elevated GS levels in the blood plasma of individuals with ASD (Hamed et al., 2018). A role for glutamate metabolism in ASD is also shown in different syndromes with a high ASD prevalence. Mouse models with an astrocyte-specific deletion of *Tsc1* showed decreased abundance of EAAT2/SLC1A2 and EAAT1/SLC1A3 and a reduction of glutamate transporter currents (Wong et al., 2003). In an astrocyte-specific Fragile X mouse model, *Slc1a2* expression and glutamate uptake were reduced, which corresponded to increased neuronal excitability (Higashimori et al., 2016). Rett syndrome astrocytes showed abnormal glutamate clearance and increased GS protein levels (Okabe et al., 2012), which were also observed in TSC brain tumors (Buccoliero et al., 2016). Besides their role in glutamate metabolism, astrocytes contribute to ASD-like behaviors in mice through the release of ATP and astrocyte-derived neuromodulators L-serine and lactate, which have essential roles in memory formation (Wang et al., 2021). Further supported by human in vitro studies, iPSC studies showed an increased generation of astrocytes at the expense of neuronal differentiation in Rett syndrome patient cultures (Andoh-Noda et al., 2015), although other studies found opposite results (Kim et al., 2019; Hong et al., 2023). Data from mouse and iPSC models of ASD syndromes showed that affected astrocytes disrupt neuronal network functioning by impacting neuronal development, neurite outgrowth, and/or the excitation/inhibition balance (Williams et al., 2014; Wang et al., 2016; Dooves et al., 2021; Ehinger et al., 2021; Caldwell et al., 2022). Together, these studies suggest a significant role for astrocytes in ASD-related behaviors.

Many ASD mouse models demonstrate a critical role for affected microglia-mediated synaptic pruning. Microglia refine the neuronal network by pruning excess synapses, a process that is mediated by the complement system and the microglia CX3CR1 receptor (Faust et al., 2021). Knockout mouse models of microglial protein TREM2, responsible for synapse phagocytosis, show altered sociability and repetitive, with TREM2 levels correlating negatively with the severity of ASD symptoms (Filipello et al., 2018). A common feature in many ASD mouse models (for example, TMEM59 or TREM2 knockout mice) is a transient decrease in microglia, followed by a synaptic pruning defect, which is strongly associated with social deficits (Zhan et al., 2014; Fernandez de Cossio et al., 2017; Kim et al., 2017; Filipello et al., 2018; Meng et al., 2022). Increased translation in microglia led to ASD-like behavior in male mice specifically, which was associated with an increased microglial density and size, enhanced phagocytic capacity despite reduced motility and synapse engulfment of microglia, and a high synapse density and E/I ratio in cortical neurons (Xu et al., 2020). Although not much is known about pruning defects in TSC, one study reports pruning defects in a TSC mouse model, although this was hypothesized to be mediated by neuronal autophagy rather than microglia (Tang et al., 2014). In Fragile X, loss of the FMRP protein leads to a hyperabundance of immature spines. It is thought that this hyperabundance is caused by a failure in synaptic pruning through altered neuron–glia communications (Patel et al., 2014; Jawaid et al., 2018; Khayachi et al., 2018; Vita et al., 2021). In Rett syndrome mice, microglia showed excessive pruning, although not through cell autonomous factors (Schafer et al., 2016). Ablation of CX3CR1, blocking the neuron–microglia interactions necessary for synaptic pruning, prolonged the lifespan and symptoms of Rett mice (Horiuchi et al., 2017). These studies suggest that in these ASD syndromes, microglia exacerbate but are probably not causing aberrant neuronal network functioning. Overall, microglia seem to play a pivotal role in abnormal neuronal network refinement in ASD and may provide a link between inflammation and brain connectivity.

MYELIN ALTERATIONS UNDERLYING BRAIN CONNECTIVITY

The connectivity of the brain is highly dependent on fast, accurate signal transduction. An essential element in signal transduction is the myelin sheath around neuronal processes produced by oligodendrocytes. Imaging studies on individuals with ASD have consistently shown changes in white matter areas, with affected myelin integrity in several cortical and subcortical regions (Deoni et al., 2015; Chen et al., 2022). These studies report on either increased or decreased white matter density in individuals with ASD. Depending on factors such as brain regions, age, or symptom severity, affected individuals may exhibit diverse changes, which could explain these apparent disparities. Interestingly, a reduction in WM density and impaired structural integrity of the corpus callosum are consistently found in ASD cases (Galvez-Contreras et al., 2020). Involvement of hypomyelination in ASD is confirmed by individuals who are affected by neurodevelopmental disorders that are often diagnosed with ASD. Fragile X patients show altered development of white matter tracts compared to controls (Swanson et al., 2018). TSC patients show alterations in oligodendrocyte number and myelin content and a lower amount of MBP, which correlates with the presence of ASD in affected individuals (Muhlebner et al., 2020). Also, TSC infants that are diagnosed with ASD later showed a poorer white matter microstructure on MRI compared to TSC infants that did not develop ASD (Prohl et al., 2019), indicating a potential association between diagnosis and impaired white matter integrity in TSC. In line with these findings, increased symptom severity is found to be correlated with reduced myelin content in the brain (Deoni et al., 2015) and altered myelin basic protein (MBP) levels in the serum of individuals with ASD (Han et al., 2021). This is further supported by postmortem studies showing that ASD brains present a lower number of oligodendrocytes, altered gene expression associated with oligodendrocyte lineage cells, aberrant myelination thickness, and a decrease in the number of myelinated axons (Morgan et al., 2014; Fetit et al., 2021). Molecular studies of human ASD brain tissue have demonstrated that oligodendrocytes exhibit upregulation of genes associated with inflammatory and energy stress responses, as well as histone-modifying enzymes. Additionally, there is a downregulation of genes associated with cadherin signaling, transport along microtubules, and myelination (Wamsley et al., 2023). Various human studies investigating individuals affected by ASD have reported changes in myelin and oligodendrocytes. Although the specific nature and extent of these changes can vary among individuals and brain regions, these findings suggest a potential involvement of myelin-related mechanisms in the pathophysiology of ASD.

Animal models for ASD, including models for syndromic disorders, could provide clues about underlying and involved mechanisms, as these often present hypomyelination and changes in oligodendrocyte lineage cells (Wei et al., 2016; Graciarena et al., 2018; Silva et al., 2019; Ju et al., 2021). Animal models of TSC (Carson et al., 2015; Jiang et al., 2016; Gruber et al., 2021) and Fragile X (Pacey et al., 2013; Lucas et al., 2021) present affected oligodendrocyte maturation and myelin generation. Although most ASD animal studies report hypomyelination, some studies find increased expression of myelin proteins (Lee et al., 2019), an increased number of (early) oligodendrocytes (Khanbabaei et al., 2019; Lee et al., 2019), and/or increased myelination (Khanbabaei et al., 2019). Rat oligodendrocytes showed an increase in myelin genes after knockdown of MeCP2, the causal gene of Rett syndrome (Sharma et al., 2015), but in contrast, the level of myelin-related proteins was decreased in MeCP2 knockout mice (Nguyen et al., 2013) and postmortem brain tissue (Petazzi et al., 2023). As indicated earlier, these opposing findings might be due to the use of different models, technique, or a focus on different brain areas. Interestingly, the BTBR mouse model, an inbred mouse presenting ASD-like behaviors (Wei et al., 2016), demonstrates enhanced sociability (Yang et al., 2011) and increased myelination (Makinodan et al., 2017) following cohabitation with a different mouse line known for its sociability. This suggests the significance of early socialization in ASD-related contexts. Furthermore, considering that prenatal inflammation and different complications associated with preterm birth are risk factors for ASD, studying models for these conditions could provide important clues about the impact of immune activation on myelination processes associated with ASD. Models for maternal immune activation, such as mouse, rat, and nonhuman primate models, demonstrated alterations in oligodendrocyte gene expression (Zhang et al., 2020; Page et al., 2021) and hypomyelination (Lee et al., 2021). Notably, the period from 24 to 32 weeks of gestation is an especially critical developmental period for oligodendrocytes, as at this stage many oligodendrocyte progenitor cells (OPCs) start differentiating toward mature oligodendrocytes. Therefore, inflammation or hypoxic insults during this period can disrupt oligodendrocyte maturation and have a major impact on white matter development (van Tilborg et al., 2016). In a rat model of diffuse white matter injury, it was demonstrated that the combination of fetal inflammation and postnatal hypoxia leads to delayed cortical myelination, reactive microglia, and astrogliosis. Notably, these rats show signs of ASD-like behavior, including reduced social play and increased repetitive grooming (van Tilborg et al., 2018), highlighting the relevance of this model in studying mechanisms in ASD-related features. A number of studies suggest that the effect of (maternal) immune activation on myelination is mediated through growth factor IGF1. As discussed earlier, maternal immune activation increases IL-6 levels, which in turn can reduce the amount of placental IGF1 release (Steinman, 2020). Interestingly, studies showed that serum IGF1 levels are lower in boys than girls (Yuksel et al., 2011), and lower in the cerebrospinal fluid of individuals with ASD compared to controls (Riikonen et al., 2006). IGF1 is important for the differentiation of OPCs to functional oligodendrocytes (Simons and Nave, 2015), and low levels of IGF1 are associated with deficient myelination and disorganization of neuronal circuits (Steinman, 2015). This is further supported by investigations in the *Shank-3*-deficient mouse model of ASD, in which IGF-1 treatment improved neuronal functioning and decreased stereotypic behaviors (Bozdagi et al., 2013), suggesting prospects for IGF-1 supplementation for ASD. Together, investigating animal models of ASD offers valuable insights into the underlying mechanisms associated with white matter changes found in individuals affected with ASD.

In conclusion, research indicates that neuroglial cells play a significant role in the complex mechanisms underlying ASD, emphasizing their involvement in the disorder. As our understanding of the risk factors for ASD continues to grow, the development of new model systems that better mimic the complex genetic, environmental, and neurodevelopmental aspects of ASD becomes crucial. These model systems will enable researchers to deepen their understanding of the underlying mechanisms of ASD and potentially identify novel therapeutic targets or interventions for the disorder.

References

Aida T, Yoshida J, Nomura M et al. (2015). Astroglial glutamate transporter deficiency increases synaptic excitability and leads to pathological repetitive behaviors in mice. Neuropsychopharmacology 40: 1569–1579.

American Psychiatric Association (2013). Diagnostic and statistical manual of mental disorders, American Psychiatric Publishing, Inc., Washington, DC.

Andoh-Noda T, Akamatsu W, Miyake K et al. (2015). Differentiation of multipotent neural stem cells derived from Rett syndrome patients is biased toward the astrocytic lineage. Mol Brain 8: 31.

Autism Spectrum Disorders Working Group of The Psychiatric Genomics C (2017). Meta-analysis of GWAS of over 16,000 individuals with autism spectrum disorder highlights a novel locus at 10q24.32 and a significant overlap with schizophrenia. Mol Autism 8: 21.

Boedhoe PSW, van Rooij D, Hoogman M et al. (2020). Subcortical brain volume, regional cortical thickness, and cortical surface area across disorders: findings from the ENIGMA ADHD, ASD, and OCD working groups. Am J Psychiatry 177: 834–843.

Bozdagi O, Tavassoli T, Buxbaum JD (2013). Insulin-like growth factor-1 rescues synaptic and motor deficits in a mouse model of autism and developmental delay. Mol Autism 4: 9.

Buccoliero AM, Caporalini C, Giordano F et al. (2016). Subependymal giant cell astrocytoma: a lesion with activated mTOR pathway and constant expression of glutamine synthetase. Clin Neuropathol 35: 295–301.

Caldwell ALM, Sancho L, Deng J et al. (2022). Aberrant astrocyte protein secretion contributes to altered neuronal development in multiple models of neurodevelopmental disorders. Nat Neurosci 25: 1163–1178.

Carlezon Jr WA, Kim W, Missig G et al. (2019). Maternal and early postnatal immune activation produce sex-specific effects on autism-like behaviors and neuroimmune function in mice. Sci Rep 9: 16928.

Carson RP, Kelm ND, West KL et al. (2015). Hypomyelination following deletion of Tsc2 in oligodendrocyte precursors. Ann Clin Transl Neurol 2: 1041–1054.

Cellot G, Cherubini E (2014). GABAergic signaling as therapeutic target for autism spectrum disorders. Front Pediatr 2: 70.

Chen B, Linke A, Olson L et al. (2022). Cortical myelination in toddlers and preschoolers with autism spectrum disorder. Dev Neurobiol 82: 261–274.

Deoni SC, Zinkstok JR, Daly E et al. (2015). White-matter relaxation time and myelin water fraction differences in young adults with autism. Psychol Med 45: 795–805.

Dooves S, van Velthoven AJH, Suciati LG et al. (2021). Neuron-glia interactions in tuberous sclerosis complex affect the synaptic balance in 2D and organoid cultures. Cells 10: 134.

Ehinger Y, Matagne V, Cunin V et al. (2021). Analysis of Astroglial Secretomic profile in the Mecp2-deficient male mouse model of Rett syndrome. Int J Mol Sci 22: 4316.

Enstrom A, Onore C, Hertz-Picciotto I et al. (2008). Detection of IL-17 and IL-23 in plasma samples of children with autism. Am J Biochem Biotechnol 4: 114–120.

Fatemi SH, Folsom TD, Reutiman TJ et al. (2008). Expression of astrocytic markers aquaporin 4 and connexin 43 is altered in brains of subjects with autism. Synapse 62: 501–507.

Faust TE, Gunner G, Schafer DP (2021). Mechanisms governing activity-dependent synaptic pruning in the developing mammalian CNS. Nat Rev Neurosci 22: 657–673.

Fernandez de Cossio L, Guzman A, van der Veldt S et al. (2017). Prenatal infection leads to ASD-like behavior and altered synaptic pruning in the mouse offspring. Brain Behav Immun 63: 88–98.

Fetit R, Hillary RF, Price DJ et al. (2021). The neuropathology of autism: a systematic review of post-mortem studies of autism and related disorders. Neurosci Biobehav Rev 129: 35–62.

Filipello F, Morini R, Corradini I et al. (2018). The microglial innate immune receptor TREM2 is required for synapse elimination and Normal brain connectivity. Immunity 48: 979–991.e978.

Galvez-Contreras AY, Zarate-Lopez D, Torres-Chavez AL et al. (2020). Role of oligodendrocytes and myelin in the pathophysiology of autism Spectrum disorder. Brain Sci 10: 951.

Gandal MJ, Haney JR, Wamsley B et al. (2022). Broad transcriptomic dysregulation occurs across the cerebral cortex in ASD. Nature 611: 532–539.

Graciarena M, Seiffe A, Nait-Oumesmar B et al. (2018). Hypomyelination and oligodendroglial alterations in a mouse model of autism Spectrum disorder. Front Cell Neurosci 12: 517.

Grove J, Ripke S, Als TD et al. (2019). Identification of common genetic risk variants for autism spectrum disorder. Nat Genet 51: 431–444.

Gruber VE, Lang J, Endmayr V et al. (2021). Impaired myelin production due to an intrinsic failure of oligodendrocytes in mTORpathies. Neuropathol Appl Neurobiol 47: 812–825.

Haida O, Al Sagheer T, Balbous A et al. (2019). Sex-dependent behavioral deficits and neuropathology in a maternal immune activation model of autism. Transl Psychiatry 9: 124.

Hamed NO, Al-Ayadhi L, Osman MA et al. (2018). Understanding the roles of glutamine synthetase, glutaminase, and glutamate decarboxylase autoantibodies in imbalanced excitatory/inhibitory neurotransmission as etiological mechanisms of autism. Psychiatry Clin Neurosci 72: 362–373.

Han Y, Xiong W, Liu J et al. (2021). Associations of serum cytokine levels and Interleukin-6-572C/G polymorphism with myelin damage in Chinese children with autism spectrum disorder. Neuroscience 465: 95–104.

Harris J (2018). Leo Kanner and autism: a 75-year perspective. Int Rev Psychiatry 30: 3–17.

Hazlett HC, Gu H, Munsell BC et al. (2017). Early brain development in infants at high risk for autism spectrum disorder. Nature 542: 348–351.

Higashimori H, Schin CS, Chiang MS et al. (2016). Selective deletion of Astroglial FMRP dysregulates glutamate transporter GLT1 and contributes to fragile X syndrome phenotypes in vivo. J Neurosci 36: 7079–7094.

Hong H, Yoon SB, Park JE et al. (2023). MeCP2 dysfunction prevents proper BMP signaling and neural progenitor expansion in brain organoid. Ann Clin Transl Neurol 10: 1170–1185.

Horiuchi M, Smith L, Maezawa I et al. (2017). CX(3)CR1 ablation ameliorates motor and respiratory dysfunctions and improves survival of a Rett syndrome mouse model. Brain Behav Immun 60: 106–116.

Iakoucheva LM, Muotri AR, Sebat J (2019). Getting to the cores of autism. Cell 178: 1287–1298.

Jawaid S, Kidd GJ, Wang J et al. (2018). Alterations in CA1 hippocampal synapses in a mouse model of fragile X syndrome. Glia 66: 789–800.

Jiang M, Liu L, He X et al. (2016). Regulation of PERK-eIF2alpha signalling by tuberous sclerosis complex-1 controls homoeostasis and survival of myelinating oligodendrocytes. Nat Commun 7: 12185.

Ju J, Yang X, Jiang J et al. (2021). Structural and lipidomic alterations of striatal myelin in 16p11.2 deletion mouse model of autism Spectrum disorder. Front Cell Neurosci 15: 718720.

Kanner L (1943). Autistic disturbances of affective contact. Nervous Child 2: 217–250.

Khanbabaei M, Hughes E, Ellegood J et al. (2019). Precocious myelination in a mouse model of autism. Transl Psychiatry 9: 251.

Khayachi A, Gwizdek C, Poupon G et al. (2018). Sumoylation regulates FMRP-mediated dendritic spine elimination and maturation. Nat Commun 9: 757.

Kim JE, Lyoo IK, Estes AM et al. (2010). Laterobasal amygdalar enlargement in 6- to 7-year-old children with autism spectrum disorder. Arch Gen Psychiatry 67: 1187–1197.

Kim HJ, Cho MH, Shim WH et al. (2017). Deficient autophagy in microglia impairs synaptic pruning and causes social behavioral defects. Mol Psychiatry 22: 1576–1584.

Kim JJ, Savas JN, Miller MT et al. (2019). Proteomic analyses reveal misregulation of LIN28 expression and delayed timing of glial differentiation in human iPS cells with MECP2 loss-of-function. PLoS One 14: e0212553.

Knuesel I, Chicha L, Britschgi M et al. (2014). Maternal immune activation and abnormal brain development across CNS disorders. Nat Rev Neurol 10: 643–660.

Kwon HK, Choi GB, Huh JR (2022). Maternal inflammation and its ramifications on fetal neurodevelopment. Trends Immunol 43: 230–244.

Lai MC, Lombardo MV, Auyeung B et al. (2015). Sex/gender differences and autism: setting the scene for future research. J Am Acad Child Adolesc Psychiatry 54: 11–24.

Lammert CR, Lukens JR (2019). Modeling autism-related disorders in mice with maternal immune activation (MIA). Methods Mol Biol 1960: 227–236.

Lamy M, Erickson CA (2018). Pharmacological management of behavioral disturbances in children and adolescents with autism spectrum disorders. Curr Probl Pediatr Adolesc Health Care 48: 250–264.

Langen M, Bos D, Noordermeer SD et al. (2014). Changes in the development of striatum are involved in repetitive behavior in autism. Biol Psychiatry 76: 405–411.

Lee H, Thacker S, Sarn N et al. (2019). Constitutional mislocalization of Pten drives precocious maturation in oligodendrocytes and aberrant myelination in model of autism spectrum disorder. Transl Psychiatry 9: 13.

Lee GA, Lin YK, Lai JH et al. (2021). Maternal immune +activation causes social behavior deficits and hypomyelination in male rat offspring with an autism-like microbiota profile. Brain Sci 11: 1085.

Li Y, Fan T, Li X et al. (2022). Npas3 deficiency impairs cortical astrogenesis and induces autistic-like behaviors. Cell Rep 40: 111289.

Lucas A, Poleg S, Klug A et al. (2021). Myelination deficits in the auditory brainstem of a mouse model of fragile X syndrome. Front Neurosci 15: 772943.

Luo R, Zhou B, Liao P et al. (2023). Disrupting cortical astrocyte ca(2+) signaling in developing brain induces social deficits and depressive-like behaviors. Glia 71: 1592–1606.

Mahmoud S, Gharagozloo M, Simard C et al. (2019). Astrocytes maintain glutamate homeostasis in the CNS by controlling the balance between glutamate uptake and release. Cells 8: 184.

Makinodan M, Okumura K, Ikawa D et al. (2017). Effects of cross-rearing with social peers on myelination in the medial prefrontal cortex of a mouse model with autism spectrum disorder. Heliyon 3: e00468.

Mandal M, Marzouk AC, Donnelly R et al. (2011). Maternal immune stimulation during pregnancy affects adaptive immunity in offspring to promote development of TH17 cells. Brain Behav Immun 25: 863–871.

Matta SM, Hill-Yardin EL, Crack PJ (2019). The influence of neuroinflammation in autism spectrum disorder. Brain Behav Immun 79: 75–90.

Meng J, Han L, Zheng N et al. (2022). Microglial Tmem59 deficiency impairs phagocytosis of synapse and leads to autism-like behaviors in mice. J Neurosci 42: 4958–4979.

Morgan JT, Chana G, Pardo CA et al. (2010). Microglial activation and increased microglial density observed in the dorsolateral prefrontal cortex in autism. Biol Psychiatry 68: 368–376.

Morgan JT, Barger N, Amaral DG et al. (2014). Stereological study of amygdala glial populations in adolescents and adults with autism spectrum disorder. PloS One 9: e110356.

Muhlebner A, van Scheppingen J, de Neef A et al. (2020). Myelin pathology beyond white matter in tuberous sclerosis complex (TSC) cortical tubers. J Neuropathol Exp Neurol 79: 1054–1064.

Nadeem A, Ahmad SF, Attia SM et al. (2020). Dysregulation in IL-6 receptors is associated with upregulated IL-17A related signaling in CD4+ T cells of children with autism. Prog Neuropsychopharmacol Biol Psychiatry 97: 109783.

Nguyen MV, Felice CA, Du F et al. (2013). Oligodendrocyte lineage cells contribute unique features to Rett syndrome neuropathology. J Neurosci 33: 18764–18774.

Northrup H, Aronow ME, Bebin EM et al. (2021). Updated international tuberous sclerosis complex diagnostic criteria and surveillance and management recommendations. Pediatr Neurol 123: 50–66.

Okabe Y, Takahashi T, Mitsumasu C et al. (2012). Alterations of gene expression and glutamate clearance in astrocytes derived from an MeCP2-null mouse model of Rett syndrome. PloS One 7: e35354.

Pacey LK, Xuan IC, Guan S et al. (2013). Delayed myelination in a mouse model of fragile X syndrome. Hum Mol Genet 22: 3920–3930.

Page NF, Gandal MJ, Estes ML et al. (2021). Alterations in retrotransposition, synaptic connectivity, and myelination implicated by transcriptomic changes following maternal immune activation in nonhuman primates. Biol Psychiatry 89: 896–910.

Patel AB, Loerwald KW, Huber KM et al. (2014). Postsynaptic FMRP promotes the pruning of cell-to-cell connections among pyramidal neurons in the L5A neocortical network. J Neurosci 34: 3413–3418.

Petazzi P, Jorge-Torres OC, Gomez A et al. (2023). Global impairment of immediate-early genes expression in Rett syndrome models and patients linked to myelination defects. Int J Mol Sci 24: 1453.

Petrelli F, Pucci L, Bezzi P (2016). Astrocytes and microglia and their potential link with autism spectrum disorders. Front Cell Neurosci 10: 21.

Prohl AK, Scherrer B, Tomas-Fernandez X et al. (2019). Early white matter development is abnormal in tuberous sclerosis complex patients who develop autism spectrum disorder. J Neurodev Disord 11: 36.

Riikonen R, Makkonen I, Vanhala R et al. (2006). Cerebrospinal fluid insulin-like growth factors IGF-1 and IGF-2 in infantile autism. Dev Med Child Neurol 48: 751–755.

Roe K (2022). Autism Spectrum disorder initiation by inflammation-facilitated neurotoxin transport. Neurochem Res 47: 1150–1165.

Russo FB, Freitas BC, Pignatari GC et al. (2018). Modeling the interplay between neurons and astrocytes in autism using human induced pluripotent stem cells. Biol Psychiatry 83: 569–578.

Saitoh BY, Tanaka E, Yamamoto N et al. (2021). Early postnatal allergic airway inflammation induces dystrophic microglia leading to excitatory postsynaptic surplus and autism-like behavior. Brain Behav Immun 95: 362–380.

Salcedo-Arellano MJ, Dufour B, McLennan Y et al. (2020). Fragile X syndrome and associated disorders: clinical aspects and pathology. Neurobiol Dis 136: 104740.

Satterstrom FK, Kosmicki JA, Wang J et al. (2020). Large-scale exome sequencing study implicates both developmental and functional changes in the neurobiology of autism. Cell 180: e523.

Schafer DP, Heller CT, Gunner G et al. (2016). Microglia contribute to circuit defects in Mecp2 null mice independent of microglia-specific loss of Mecp2 expression. ELife 5: e15224.

Schousboe A, Scafidi S, Bak LK et al. (2014). Glutamate metabolism in the brain focusing on astrocytes. Adv Neurobiol 11: 13–30.

Sharma K, Singh J, Pillai PP et al. (2015). Involvement of MeCP2 in regulation of myelin-related gene expression in cultured rat oligodendrocytes. J Mol Neurosci 57: 176–184.

Silva AI, Haddon JE, Ahmed Syed Y et al. (2019). Cyfip1 haploinsufficient rats show white matter changes, myelin thinning, abnormal oligodendrocytes and behavioural inflexibility. Nat Commun 10: 3455.

Simons M, Nave KA (2015). Oligodendrocytes: myelination and axonal support. Cold Spring Harb Perspect Biol 8: a020479.

Smith SE, Li J, Garbett K et al. (2007). Maternal immune activation alters fetal brain development through interleukin-6. J Neurosci 27: 10695–10702.

Steinman G (2015). Plausible etiology of brain dysconnectivity in autism—review and prospectus. Med Hypotheses 85: 405–407.

Steinman G (2020). The putative etiology and prevention of autism. Prog Mol Biol Transl Sci 173: 1–34.

Swanson MR, Wolff JJ, Shen MD et al. (2018). Development of white matter circuitry in infants with fragile X syndrome. JAMA Psychiatry 75: 505–513.

Tang G, Gudsnuk K, Kuo SH et al. (2014). Loss of mTOR-dependent macroautophagy causes autistic-like synaptic pruning deficits. Neuron 83: 1131–1143.

van Tilborg E, Heijnen CJ, Benders MJ et al. (2016). Impaired oligodendrocyte maturation in preterm infants: potential therapeutic targets. Prog Neurobiol 136: 28–49.

van Tilborg E, Achterberg EJM, van Kammen CM et al. (2018). Combined fetal inflammation and postnatal hypoxia causes myelin deficits and autism-like behavior in a rat model of diffuse white matter injury. Glia 66: 78–93.

Vargas DL, Nascimbene C, Krishnan C et al. (2005). Neuroglial activation and neuroinflammation in the brain of patients with autism. Ann Neurol 57: 67–81.

Vidal S, Xiol C, Pascual-Alonso A et al. (2019). Genetic landscape of Rett syndrome spectrum: improvements and challenges. Int J Mol Sci 20.

Vita DJ, Meier CJ, Broadie K (2021). Neuronal fragile X mental retardation protein activates glial insulin receptor mediated PDF-Tri neuron developmental clearance. Nat Commun 12: 1160.

Wamsley B, Bicks L, Cheng Y et al. (2023). Molecular cascades and cell-type specific signatures in ASD revealed by single cell genomics. bioRxiv. https://doi.org/10.1101/2023.03.10.530869.

Wang L, Wang Y, Zhou S et al. (2016). Imbalance between glutamate and GABA in Fmr1 knockout astrocytes influences neuronal development. Genes (Basel) 7: 45.

Wang Q, Kong Y, Wu DY et al. (2021). Impaired calcium signaling in astrocytes modulates autism spectrum disorder-like behaviors in mice. Nat Commun 12: 3321.

Wei H, Ma Y, Liu J et al. (2016). Proteomic analysis of cortical brain tissue from the BTBR mouse model of autism: evidence for changes in STOP and myelin-related proteins. Neuroscience 312: 26–34.

Williams EC, Zhong X, Mohamed A et al. (2014). Mutant astrocytes differentiated from Rett syndrome patients-specific iPSCs have adverse effects on wild-type neurons. Hum Mol Genet 23: 2968–2980.

Wong M, Ess KC, Uhlmann EJ et al. (2003). Impaired glial glutamate transport in a mouse tuberous sclerosis epilepsy model. Ann Neurol 54: 251–256.

Xu ZX, Kim GH, Tan JW et al. (2020). Elevated protein synthesis in microglia causes autism-like synaptic and behavioral aberrations. Nat Commun 11: 1797.

Yang M, Perry K, Weber MD et al. (2011). Social peers rescue autism-relevant sociability deficits in adolescent mice. Autism Res 4: 17–27.

Ye J, Wang H, Cui L et al. (2021). The progress of chemokines and chemokine receptors in autism spectrum disorders. Brain Res Bull 174: 268–280.

Yuksel B, Ozbek MN, Mungan NO et al. (2011). Serum IGF-1 and IGFBP-3 levels in healthy children between 0 and 6 years of age. J Clin Res Pediatr Endocrinol 3: 84–88.

Zeidan J, Fombonne E, Scorah J et al. (2022). Global prevalence of autism: a systematic review update. Autism Res 15: 778–790.

Zhan Y, Paolicelli RC, Sforazzini F et al. (2014). Deficient neuron-microglia signaling results in impaired functional brain connectivity and social behavior. Nat Neurosci 17: 400–406.

Zhang XF, Chen T, Yan A et al. (2020). Poly(I:C) challenge alters brain expression of oligodendroglia-related genes of adult progeny in a mouse model of maternal immune activation. Front Mol Neurosci 13: 115.

Ziats CA, Patterson WG, Friez M (2021). Syndromic autism revisited: review of the literature and lessons learned. Pediatr Neurol 114: 21–25.

Zohny SM, Habib MZ, Mohamad MI et al. (2023). Memantine/aripiprazole combination alleviates cognitive dysfunction in valproic acid rat model of autism: hippocampal CREB/BDNF signaling and glutamate homeostasis. Neurotherapeutics 20: 464–483.

Handbook of Clinical Neurology, Vol. 210 (3rd series)
Neuroglia in Neurologic and Psychiatric Disorders, Part II
A. Verkhratsky, L.D. de Witte, E. Aronica, and E.M. Hol, Editors
https://doi.org/10.1016/B978-0-443-19102-2.00019-3

Chapter 18

Neuroglia in eating disorders (obesity, Prader-Willi syndrome and anorexia nervosa)

FELIPE CORREA-DA-SILVA[1,2,3] AND CHUN-XIA YI[1,2,3*]

[1]*Department of Endocrinology and Metabolism, Amsterdam Gastroenterology Endocrinology and Metabolism, Amsterdam University Medical Center (UMC), University of Amsterdam, Amsterdam, The Netherlands*

[2]*Laboratory of Endocrinology, Amsterdam University Medical Center (UMC), University of Amsterdam, Amsterdam, The Netherlands*

[3]*Netherlands Institute for Neuroscience, Amsterdam, The Netherlands*

Abstract

The hypothalamus is widely recognized as one of the most extensively studied brain regions involved in the central regulation of energy homeostasis. Within the hypothalamus, peptidergic neurons play a crucial role in monitoring peripheral concentrations of metabolites and hormones, and they finely adjust the sensing of these factors, leading to the activation of either anorexigenic (appetite-suppressing) or orexigenic (appetite-stimulating) pathways. While cortical innervation of the hypothalamus does influence these processes, it is generally considered of secondary importance. Eating-related disorders, such as obesity and anorexia nervosa, are strongly associated with imbalances in energy intake and expenditure. The phenotypes of these disorders can be attributed to dysfunctions in the hypothalamus. Traditionally, it has been believed that hypothalamic dysfunction in these disorders primarily stems from defects in neural pathways. However, recent evidence challenges this perception, highlighting the active participation of neuroglial cells in shaping both physiologic and behavioral characteristics. This review aims to provide an overview of the latest insights into glial biology in three specific eating disorders: obesity, Prader-Willi syndrome, and anorexia. In these disorders, neural dysfunction coincides with glial malfunction, suggesting that neuroglia actively contribute to the development and progression of various neurologic disorders. These findings underscore the importance of glial cells and open up potential new avenues for therapeutic interventions.

INTRODUCTION[1]

Maintenance of body regulation relies on genetic, physiologic, and environmental components (Waterson and Horvath, 2015). Specialized neuronal networks that integrate feeding behavior with energy expenditure underpin body mass stability (Roh et al., 2016). These neural circuits sense and respond to central- and peripheral-born signals, leading to adaptive adjustments in feeding behavior and fuel flux in metabolically relevant organs (Roh et al., 2016). Adjustments in energy homeostasis combine a plethora of behaviors in response to contextual cues (i.e., food seeking) with physiologic adaptations (i.e., fluctuations in insulin levels) (Oomura and Kita, 1981; McLachlan et al., 2022). Evidence from both animal models and humans demonstrates that disruption of such neurocircuitry has direct consequences on body weight and glucose homeostasis (Farooqi et al., 2006). Nonetheless, dysfunction of these neural pathways is a hallmark of eating disorders, such as anorexia nervosa and obesity.

Over the last few decades, advances in understanding the role of the central nervous system in energy homeostasis have been made through the identification of the anatomic and molecular identities of neuronal populations engaged in energy homeostasis (Stachniak et al., 2014; Saucisse et al., 2021). Several mechanisms have been postulated to correlate the dysregulated metabolism with impaired neural networks. Glial cells are the main support system for neurons and promote and maintain a healthy microenvironment for optimal neuronal function. Extensive research has demonstrated the role of glial cells in sensing and

[1]Abbreviations used in the chapter are listed at the end of the chapter before References section.

*Correspondence to: Chun-Xia Yi, Department of Endocrinology and Metabolism, Amsterdam Gastroenterology Endocrinology and Metabolism, Amsterdam University Medical Center (UMC), University of Amsterdam, Amsterdam, The Netherlands; Laboratory of Endocrinology, Amsterdam University Medical Center (UMC), University of Amsterdam, Amsterdam, The Netherlands. Tel: +31-020-5669111, E-mail: c.yi@amsterdamumc.nl

integrating a multitude of peripheral and central cues, particularly microglia and astrocytes (Jais and Bruning, 2017). Deviations in the function and structure of these cells lead to neuronal malfunction, ultimately culminating in metabolic disorders (Bocarsly et al., 2015; Garcia-Caceres et al., 2016).

This review focuses on the role of glial cells in the pathophysiology of eating disorders. First, we provide an overview of the control of feeding behavior, highlighting the role of the hypothalamus and other (sub)cortical areas. Next, we provide an overview of the role of glial cells in eating disorders. Finally, we briefly discuss the similarities between these pathologies and the importance of glial cells in therapeutic approaches.

HOMEOSTATIC AND NON-HOMEOSTATIC CONTROL OF FEEDING BEHAVIOR

Energy homeostasis is tightly regulated and results from the monitoring of microenvironmental cues and the activation of orexigenic and anorexigenic neural pathways (Waterson and Horvath, 2015). The hypothalamus is a highly heterogeneous central nervous system (CNS) structure that harbors neural populations responsible for a variety of neuroendocrine functions, including feeding behavior. In a simplified manner, neurons situated in the mediobasal hypothalamus coordinate food intake and energy expenditure by coupling periphery-born signals, central inputs, and fluctuations in key metabolites such as glucose (Meek et al., 2016). Neurons located in the infundibular nucleus (the arcuate nucleus in rodents) are the primary responders to these environmental cues, and metabolic signals propagate to a variety of distinct neuroendocrine populations throughout the hypothalamus (Campbell et al., 2017).

The major neurons in the control of energy homeostasis orchestrate metabolism by sensing dietary and endocrine mediators. The receptors for energy homeostasis-controlling hormones are broadly expressed in the central nervous system and are enriched in the hypothalamus (Koch et al., 2010). The hormonal control of metabolism is multilayered and depends on the temporal and tissue-dependent actions of a great variety of hormones. Considerable attention has been paid to insulin (of pancreatic origin), leptin (of adipose tissue origin), and ghrelin (of stomach origin) (Clement et al., 1998; Marzullo et al., 2004; Kalsbeek et al., 2020; Abdelhak et al., 2022). All of these interfere with the production of hypothalamic neuropeptides that result in food intake and energy expenditure rate and facilitate the synchronization of peripheral metabolic status with the central nervous system. Notably, glial cells also express functional receptors for these hormones, indicating that their action extrapolates to neural modulation (Garcia-Caceres et al., 2016; Gao et al., 2018; Duquenne et al., 2021; Milanova et al., 2022).

Furthermore, food intake is influenced by non-homeostatic components, including the hedonic and other cognitive aspects of feeding. A great diversity of neurocircuitries is engaged with these non-homoeostatic features of feeding and is composed of cortical structures and mesolimbic dopaminergic systems (Petrovich et al., 2007; Gaykema et al., 2014). The mechanisms by which these neurons modulate feeding include sensory-specific satiety, motivational cues, reward systems, and behavioral adjustments to food availability or nutritional composition (Clarke et al., 2018). All of these systems are of great complexity and can be influenced by other biologic variables, such as gender and age (Haase et al., 2011). Of importance, there is reciprocal and constant communication between the hypothalamic and non-homeostatic circuitries.

Hypothalamic neurons are in close association with neuroglial cells, and dynamic neuron–glia interactions enable proper information processing, and consequently, behavior (Yi et al., 2017). Recent evidence has demonstrated the role of neuroglia in metabolic integration, with special attention to microglia, astrocytes, and tanycytes—a unique type of glial cells located primarily in the wall of the third ventricle. Ultimately, exposure to metabolic challenges leads to the impairment of glial function and contributes to abnormal eating and impaired glucose homeostasis, among other metabolic disruptions. Such challenges include, but are not limited to, hypercaloric diets, dysregulation of circadian rhythms, and abrupt body weight fluctuations (Lechuga-Sancho et al., 2006; Milanova et al., 2019). Like neurons, microglia, astrocytes, and tanycytes specialize in nutrient sensing and adapt to stimuli from the CNS. It is noteworthy that, despite the role of astrocytes and microglia as protagonists in metabolic dysregulation, recent insights have demonstrated a role for oligodendrocytes in such physiologic disruptions (Djogo et al., 2016; Yoo et al., 2020).

OBESITY

Obesity is defined as abnormal or excessive fat. Obesity is an important health challenge, as it significantly increases the risk of life-threatening comorbidities, such as atherosclerosis, myocardial infarction, and type 2 diabetes mellitus (T2DM) (Must et al., 1999). T2DM is one of the most common comorbidities of obesity and it is characterized by insulin resistance and consequent poor glycemic control.

The progression and development of the obese phenotype are rooted in both genetic and environmental factors, which are intertwined with feeding regulation (Farooqi et al., 2006; Xie et al., 2022). Hypothalamic feeding centers suffer neurocircuitry disorganization upon obesogenic cues, with numeric and functional loss of anorexigenic neuronal populations (in special proopiomelanocortin- and oxytocin-expressing neurons) (Challis et al., 2004; Lawson, 2017). These findings have been extensively recapitulated in diverse animal models and these losses are associated with sustained microglial activity and gliosis (Kalin et al., 2015). To date, few studies have explored hypothalamic anomalies in obese humans owing to technical and availability challenges. Our current knowledge points to a great overlap between obesity-associated hypothalamic dysfunction in rodents and humans with a strong glial reactivity signature (Kalin et al., 2015). Neuroimaging studies have identified reactive gliosis and hypothalamic tissue damage associated with increased inflammatory markers in obese patients (Thaler et al., 2012). Furthermore, systemic low-grade inflammation found in obesity has been linked to reduced integrity of cortical areas in the control of the hedonic aspect of feeding in humans (O'Brien et al., 2017). Here, we summarize the cellular and temporal bases of glial biology under obesogenic conditions.

Tanycytes

The infundibular (arcuate) nucleus is positioned in an anatomically privileged site, with higher permeability of the blood–brain barrier. This is an essential feature of food intake regulation, allowing neurons to be exposed to fluctuations in peripheral concentrations of nutrients, metabolites, and hormones (Lhomme et al., 2021). Tanycytes are specialized radial cells that are key components of this neurovascular structure. Typically, their cell bodies are lined in the ventricular wall associated with circumventricular organs (third ventricle for the mediobasal hypothalamus), but their processes extend to the gray matter (Langlet, 2014). Thus, tanycytic morphologic features underpin their role as hypothalamic gatekeepers by conveying peripheral metabolic cues to the neural populations. Signal transduction is based on the transport of selective blood-borne molecules (Balland et al., 2014). A large range of metabolites is sensed by tanycytes, and interestingly, they are capable of fatty acid storage in lipid-rich organelles called lipid droplets, especially in the dorsal part of the third ventricle wall (Maya-Monteiro et al., 2021). Tanycytic lipid droplets accumulate in the hypothalamus of diet-induced obese mice, suggesting a role for these cells in buffering excessive lipid flux and the associated lipotoxicity (Maya-Monteiro et al., 2021). It is noteworthy that high-fat feeding leads to the loss of morphologic integrity of tanycytes in the median eminence, a process that appears to be transient (Ramalho et al., 2018). Whether lipid droplets also accumulate in the human obese hypothalamus is still unknown; however, in diabetic patients, there is a reduction in their content (Maya-Monteiro et al., 2021). These findings suggest that metabolic sensing performed by tanycytes is dynamic and is subject to the global metabolic status of the individual. Moreover, special attention has been given to the role of tanycytes in leptin transport and sensitivity, and many point to a potential therapeutic value in manipulating tanycytic leptin transport in obesity. The extent to which it is possible to apply these findings to human biology is yet to be determined.

Microglia

Microglia are the resident immune cells of the central nervous system and serve critical homeostatic functions such as synaptic pruning, circuit remodeling, and tissue clearance. These cells have a characteristic morphology, with long processes that are in constant contraction and expansion, allowing them to scan their microenvironment (Nimmerjahn, 2012). Upon neuronal injury, microglia no longer display a homeostatic signature and respond through the production of cytokines and other neurotoxic intermediates. Obesity is associated with a chronic low-grade inflammatory state that affects both the peripheral and central nervous systems. Research in animal models has shown concomitant hypothalamic neuronal loss with reactive microgliosis, indicating that inflammatory mediators are a major mechanistic node in neural impairment in obesity (Yi et al., 2017). Interestingly, inhibition of hypothalamic neuroinflammation prevents local neuronal loss and is associated with the attenuation of inflammatory markers and improvement of peripheral insulin sensitivity (Andre et al., 2017).

Of multifactorial etiology, an increase in the hypothalamic inflammatory state is mainly triggered by dietary components and endocrine imbalance. This is linked to metabolic disturbances, including, but not restricted to, body weight gain, ectopic fat accumulation, insulin resistance, and dyslipidemia (Lehrke and Lazar, 2004). Hypothalamic inflammation has been linked to the development and progression of obesity and dysregulation of glucose homeostasis (Posey et al., 2009). Moreover, hypothalamic inflammation is observed in rodents within 24h of the introduction of an obesogenic diet, in contrast to peripheral inflammation, which appears to be a consequence of obesity. It is noteworthy that this initial inflammatory response is transient, with normalization of the expression levels of pro-inflammatory cytokines within the days following the introduction of an obesogenic diet. Persistent hypothalamic reactive microgliosis in rodents is observed at least 4 weeks after the introduction of an obesogenic diet (Thaler et al., 2012). To date, it has been impossible to pinpoint chronic hypothalamic microglial reactivity as a cause or consequence of metabolic disruptions. Nevertheless, growing evidence has shown that microglia are key components of neuronal health, and continuous microglial malfunction is undoubtedly associated with the pathophysiology of obesity. Therapeutic approaches to obesity and its comorbidities with anti-inflammatory properties have been shown to ameliorate dysregulated glycemia and, to some extent, excess body weight (Jing et al., 2018).

Microglia-associated changes are not restricted to the hypothalamic nuclei, although disturbances in these neurons are directly related to the dysregulation of feeding behavior and energy expenditure. In chronically obese rodents, increased reactive microgliosis is observed in the cortical areas and is correlated with obesity comorbidities, such as cognitive decline (Bocarsly et al., 2015). Finally, translational research has confirmed obesity-associated, low-grade inflammation at the peripheral and CNS levels.

Hypothalamic inflammation precedes body weight gain, and microglial reactivity in obese rodents is primed by dietary components (1) and gut- and adipose tissue-derived hormones (2), rather than adiposity per se.

The role of dietary components in reactive microgliosis in obesity

Excessive intake of dietary components (i.e., sugars and fatty acids) promotes obesity, which is associated with drastic changes in metabolically relevant organs including the brain (Koopman et al., 2013). The use of dietary interventions in combination with other behavioral approaches has been proven to reverse or ameliorate the pathologic features of obesity and its comorbidities (Ruddick-Collins et al., 2022). Two major classes of macronutrients have been implicated in the establishment and progression of inflammatory changes associated with obesity.

Plasma free fatty acid levels are elevated in obesity, and the lipotoxic effects of saturated fatty acids have emerged as a major mechanistic link between excessive body weight and the development of metabolic syndrome (Golay and Bobbioni, 1997). The first description of the induction of the inflammatory profile in the hypothalamus of rodents upon exposure to a hyperlipidic diet was dated back to 2005 (De Souza et al., 2005), and microarray analysis of the hypothalamus of normal and hyperlipidic diet-fed animals was performed, and a significant increase in pro-inflammatory cytokines was observed. Neuroinflammation-related transcripts include cytokines and interleukins (i.e., tumor necrosis factor-alpha and interleukin 6), and their expression is orchestrated by the nuclear factor-κB (NF-κB) signaling pathway (De Souza et al., 2005). Since then, numerous independent research groups have reached similar conclusions, and the concept that fatty acids are inducers of the inflammatory response has emerged. Accumulating evidence has indicated a microglia-derived neurotoxic environment with an apparent predilection that affects anorexigenic neurons (Valdearcos et al., 2014). Importantly, palmitate (a saturated fatty acid) accumulates in the cerebrospinal fluid of patients with obesity (Melo et al., 2020). The mechanistic hypothesis in this scenario is that saturated fatty acids are able to bind Toll-like receptor 4 (TLR4) proteins in myeloid cells, both in the peripheral and central nervous systems (Milanski et al., 2009). TLR4 is an innate immune receptor that has been highly conserved throughout evolution and specializes in recognizing lipopolysaccharides (LPS), a microbial component. Because of their similar molecular structures, saturated fatty acids were deemed to act as TLR4 nonmicrobial agonists and, therefore, to induce classical pro-inflammatory immunity (Rocha et al., 2016). In the periphery, the exclusive presence of TLR4 in myeloid cells is sufficient to cause diet-induced insulin resistance and liver dysfunction (Razolli et al., 2015). These results support the notion that dietary fats are key orchestrators of microglial inflammation in a TLR4-dependent manner. However, recent studies have indicated that lipid-induced inflammation in myeloid cells is not through their action as a direct agonist of TLR4; rather, saturated fatty acids promote molecular and metabolic remodeling of myeloid cells and prime TLR4-dependent responses (Lancaster et al., 2018). These results do not exclude the role of microglial lipid-induced inflammation but challenge the prevailing view of participation and the importance of TLR4. These data indicate that the sustained inflammatory response observed in obese animals requires continuous exposure to dietary fat.

Notably, dietary carbohydrates are also necessary for the neuroinflammatory response prompted by microglia. Epidemiologic studies have extensively linked dietary sugar intake to obesity, and it is intimately associated with the development of T2DM (Veit et al., 2022). The combination of a high-sugar, high-fat diet is more representative of the human obesogenic diet profile, and further research is needed to gain detailed mechanistic insights into the role of dietary sugars in hypothalamic cellular reorganization in obesity. Interestingly, the combined overconsumption of fat and sugar leads to increased hypothalamic neuronal glycation end-products, which trigger microgliosis and an inflammatory response (Gao et al., 2017). The introduction of a diet with low carbohydrate and high-fat content fails to induce glycation end-product accumulation in hypothalamic neurons and microgliosis (Gao et al., 2017). These results indicate that fat consumption is not a deterministic factor in diet-induced microglial malfunction.

Metabolic-relevant hormones and microglial immunity

Leptin is an adipokine that promotes an anorexigenic state. In humans, leptin deficiency owing to genetic defects leads to early-onset obesity and infertility (Clement et al., 1998). Rodents with nonfunctional leptin or its receptor also display severe metabolic disturbances, including obesity and glycemic imbalance (Zhang et al., 1994). Leptin is a known immunomodulator, and its signaling leads to the production of pro-inflammatory cytokines in a diverse range of immune cells (Lord, 2006; Lafrance et al., 2010). Leptin-deficient mice fed a standard chow diet had a lower number of microglia in the arcuate nucleus of the hypothalamus than wild-type control mice. Moreover, the microglial processes in mutant mice are less branched, indicating impaired immunosurveillance capacity. When fed a high-fat diet, these mice display increased microglial numbers, comparable to diet-induced obese mice (Gao et al., 2014). Interestingly, the mouse strain with a global deficiency of the leptin receptor had comparable microglial numbers in a standard chow diet, but with significantly decreased phagocytic capacity. Likewise, depletion of the leptin receptor, specifically in myeloid cells, downregulates hypothalamic microglial phagocytic capacity and defective hypothalamic neurocircuitry (Gao et al., 2018). It is well known that obese subjects are hyperleptinemic due to adipose tissue hyperplasia but concomitantly present molecular resistance to its signal in neurons (Knight et al., 2010), which hampers the use of leptin as an anti-obesity therapy.

Insulin is a hormone produced in the pancreatic islets, with the main function of glycemic control. Virtually every tissue is sensitive to insulin, and its impact on global metabolism is coordinated by its actions on adipose tissue, liver, skeletal muscle, and the brain (Jensen et al., 1989). Obese humans exhibit hyperinsulinemia, but similar to leptin, its receptor also displays molecular resistance to its signals (Ferrannini et al., 2005). Insulin is a potent anorexigenic factor, and growing evidence points to its immunoregulatory role. Interestingly, under metabolic stress, peripheral macrophages lacking insulin signaling acquire an anti-inflammatory phenotype (Kubota et al., 2018). Further research is needed to elucidate the role of insulin in immunity, as in vitro studies have pointed to an anti-inflammatory role for insulin. This is also recapitulated in microglia in culture, in which insulin-treated microglia suppress pro-inflammatory cytokine production (Brabazon et al., 2018). The apparent paradoxical findings indicate that the immunomodulatory roles of insulin might be time-dependent or concentration-dependent; hence, plasma and central concentrations might differ significantly. To date, only one study in mice has depleted insulin receptors, specifically in microglia. This study showed that a lack of microglial insulin receptor signaling promotes disruption of microglial morphology, irrespective of diet (Milanova et al., 2022). The consequences on global metabolism and other cells in the hypothalamus seem to be sex-dependent, but overall, there are mild impacts. Further research is necessary to elucidate the role of insulin in regulating microglial immunity in obesity.

Ghrelin is the only circulating orexigenic hormone produced by the stomach and gastrointestinal tract (Kojima et al., 1999). It promotes feeding behavior by activating appetite-modifying neurons in the arcuate nucleus of the hypothalamus, and its actions are mediated by G protein-coupled growth hormone secretagogue receptor 1a (GSH-R1a) (Gnanapavan et al., 2002). Notably, recent evidence has demonstrated that ghrelin is rapidly sensed by appetite-curbing neurons (Schaeffer et al., 2013). Furthermore, the local blood flow velocity in the median eminence of the hypothalamus directly affects ghrelin-stimulated food intake (Romano et al., 2023). Therefore, changes in vascular permeability in the hypothalamus that are associated with obesity (discussed later) have a direct impact on ghrelin-associated changes in the feeding circuitry. Pharmacologic ghrelin treatment in rodents leads to fat mass gain, partially due to elevated food consumption (Tschop et al., 2000), and ghrelin-deficient mouse models show improved peripheral insulin sensitivity (Zigman et al., 2005). Moreover, growing evidence has shown that ghrelin acts on immune cells to suppress inflammation and induce an anti-inflammatory profile (Correa da Silva et al., 2019). Ghrelin-treated microglia display an anti-inflammatory and pro-resolution immune response, supporting the notion that hypothalamic microglia may display an anti-inflammatory phenotype upon ghrelin exposure (Liu et al., 2019). Interestingly, the plasma levels of ghrelin in obese subjects were lower than those in lean-matched controls (Tschop et al., 2001). Furthermore, an inverse relationship has been established between plasma ghrelin levels, insulin resistance, and T2DM. Altogether, these data indicate that there is a reasonable potential for targeting ghrelin-GHS-R systems to treat metabolic disorders in humans, and ameliorated reactive microgliosis may be an important mechanistic node in this process.

Astrocytes

Hypothalamic astrocytes play major roles in nutrient and endocrine sensing, neurovascular coupling, and synaptic modulation. Astrocytic proliferation and hypertrophy are well-recognized features of central nervous system pathologies (Escartin et al., 2021). Moreover, reactive astrocytes exhibit marked upregulation of structural proteins such as glial fibrillary acid protein (GFAP) and vimentin (Galou et al., 1996). In contrast, there was a reduction in the number of astrocytes in the hypothalamus of poorly controlled diabetic rats. Growing evidence also indicates that impairment of astrocytic function leads to exaggerated body weight gain in rodents fed an obesogenic diet. Among the key nodes in astrocyte function, it is possible to highlight its responsiveness to metabolism-regulating hormones (i.e., insulin and leptin) and key components of cellular metabolism (Kim et al., 2014; Garcia-Caceres et al., 2016). Altogether, this evidence suggests that endocrine mediators and adiposity are major regulators of hypothalamic function. Notably, reactive astrogliosis is obesity, which is deemed to be reversible after weight loss in mice (Berkseth et al., 2014). Importantly, reactive gliosis was also observed in the hypothalamus of obese subjects, indicating that these observations in murine models are recapitulated in humans (Thaler et al., 2012).

Obesity-induced astrocytic dysfunction can be summarized into three pillars: (1) perturbations in tissue microarchitecture, (2) functional impairment of hypothalamic neurons, and (3) production of neurotoxic/inflammatory mediators.

Reactive astrocytes promote hypothalamic tissue remodeling in obesity

The distinct morphologic features of astrocytes allow them to have close contact with neurons, providing them with trophic support. Moreover, astrocytes surround endothelial cells that compose the blood–brain barrier (BBB) and, therefore, have privileged access to periphery-derived metabolites and hormones (Obermeier et al., 2013).

A chronic high-fat diet promotes increased permeability of the BBB, and this intense vascular remodeling is dependent on angiogenic factors released by astrocytes. In particular, vascular endothelial growth factor (VEGF) appears to be one of the major players in this process, alongside hyperleptinemia (Gruber et al., 2021). Interestingly, a shift in the cytoarchitecture of the BBB impairs the access of hypothalamic arcuate nucleus neurons to blood vessels in mice (Horvath et al., 2010). These findings indicate that gliovascular remodeling after exposure to an obesogenic diet is more permissive to the entry of neurotoxic- and metabolic-derived factors from the periphery to the central nervous system. Moreover, key neuronal populations controlling metabolism are deprived of proper nutritional and oxygen support under chronic obesogenic diet conditions. Dysregulated cellular metabolism and continuous hypoxia lead to the functional and numeric loss of these neural populations.

Astroglia modulates neuronal activity and feeding behavior

Astrocytes are homeostatic keepers of the neural microenvironment that provide optimal metabolic and trophic support for neuronal functions. This translates into the transfer of metabolites from the bloodstream, production and release of growth and neuroprotective factors, and influences the formation, function, and plasticity of synapses (Ventura and Harris, 1999). Increased markers of neuronal injury and synaptic disorganization have been reported in the hypothalamus of obese rodents, in parallel with reactive gliosis (Horvath et al., 2010; Thaler et al., 2012). It is noteworthy that these perturbations are not solely due to astrocytic dysfunction; the reactive phenotype limits the functional and structural support that astrocytes would provide to neurons. Furthermore, reactive astrocytes release neurotransmitters and modulate neural functions (Lee et al., 2010). This is particularly observed in the lateral hypothalamic area (LHA), which harbors a variety of neuronal populations that control energy homeostasis through control of feeding behavior and innervation of peripheral metabolically relevant organs, such as adipose tissue. In diet-induced obesity, astrocytes from the LHA have been reported to release gamma-aminobutyric acid (GABA), an inhibitory neurotransmitter, inhibiting the function of surrounding neurons responsive to this signal (Sa et al., 2022). Both genetic and pharmacologic manipulation of astrocytic GABA synthesis reduce astrogliosis in the LHA, accompanied by a reduction in body weight in high-fat, diet-fed rodents (Sa et al., 2022). Furthermore, a recent report demonstrated that astrocytic control of extracellular glutamate in the hypothalamus is impaired in obesity (Herrera Moro Chao et al., 2022). This directly affects neuronal function, with direct implications for energy homeostasis and dysregulated glucose metabolism. Similarly, astrocytes in the orbitofrontal cortex have impaired glutamate transport, which eventually leads to synaptic disorganization of neurons in the control of the hedonic components of feeding (Lau et al., 2021).

Astrocyte-derived inflammatory mediators in overnutrition

Resident immune cells of the central nervous system produce neurotoxic/inflammatory mediators following neuronal injury. The main cellular subset implicated in the production and release of pro-inflammatory cytokines and neurotoxic molecules is the microglia. Hypothalamic neuroinflammation is maintained by microglial features, as well as through the inflammatory reactivity of an astrocytic subset (Buckman et al., 2015). Astroglia also produce and release inflammatory mediators during overnutrition, to a lesser extent than microglia. Inflammatory astrocytes lose the ability to promote synapse formation and exert neurotoxic effects. The astrocytic inflammatory response is orchestrated by the NF-κB pathway, and animals with selective genetic ablation of NF-κB components in astrocytes do not display an astrocytic reactive phenotype or characteristic hyperphagia and weight gain in obesity (Zhang et al., 2008; Douglass et al., 2017; Jin et al., 2020).

Oligodendrocytes

In metabolic diseases, the role of myelinating-oligodendrocytes and oligodendrocyte progenitor cells (OPCs) is virtually underexplored. Most reports linking obesity with these cells establish impairment of myelinating capacity associated with the obese phenotype, both in the hypothalamic and extra-hypothalamic areas (Sena et al., 1985; Huang et al., 2019; Bouhrara et al., 2021). However, little is known about the effect of oligodendrocytes on global metabolic regulation. Notably, oligodendrocytes respond to leptin, and hypothalamic OPCs are key components for proper hypothalamic leptin signaling (Udagawa et al., 2006; Djogo et al., 2016). Genetic and pharmacologic ablation of OPCs in the hypothalamic median eminence led to weight gain, suggesting a fundamental role for these cells in body weight regulation. However, oligodendrocytes also support neuronal functions in addition to myelin production. In particular, nutritional support for neurons from oligodendrocytes has been documented and is proposed to be disrupted in the obese hypothalamus (Ou et al., 2019). Further mechanistic insights are needed, but initial evidence in mice seems to corroborate the impact of hypothalamic oligodendrocytes as coordinators of neuronal health and function (Trevisiol and Nave, 2015). Thus, dysregulation of oligodendrocyte biology is implicated in the malfunctioning of hypothalamic neuropeptidergic systems. Importantly, these findings have yet to be confirmed in humans.

PRADER-WILLI SYNDROME

Prader-Willi syndrome (PWS) is a rare congenital neurologic disorder caused by the loss of expression of protein-coding genes and non-coding RNAs located on chromosome 15 q11-q13.1 (Holm et al., 1993). The clinical features of PWS include a multitude of endocrine abnormalities. Typically, there are delays in motor and linguistic milestones, intellectual disabilities, hypogonadism, hyperphagia, and gross obesity (Cassidy et al., 2012). This broad symptomatology is explained by the genetic basis of the pathology; however, it is widely recognized that the most prevalent clinical manifestations are of hypothalamic etiology (i.e., hyperphagia, obesity, and hypogonadism) (Correa-da-Silva et al., 2021). Recent neuroimaging findings show reduced hypothalamic volume compared to matched controls, including obese individuals (Brown et al., 2022), and neuronal loss has been extensively reported in postmortem and fresh specimens of PWS hypothalami (Swaab et al., 1995; Bochukova et al., 2018). Secondary to hypothalamic dysfunction, defects in the cortical circuitry, and imbalances in the periphery have also been reported (Shapira et al., 2005). No animal model can fully recapitulate the PWS phenotype, limiting the pool of viable research tools available to further understand the cellular basis of the disease (Correa-da-Silva et al., 2021). Remarkably, paradoxical phenotypes, such as leanness, can be observed in animal models based on the global or conditional deletion of singular genes of the PWS critical genomic region (Ding et al., 2008). In addition, human research material is difficult to access because of the low prevalence of this disorder, which poses a great challenge in the PWS research field. This is aggravated by the fact that there is a clear genotype–phenotype association in PWS, in which different extensions of chromosomal loss and imprinting defects lead to gene silencing instead of physical loss of the chromosome, which has a direct effect on the severity of symptoms (Costa et al., 2019). Importantly, PWS is uncurable, and symptom management is the only therapeutic approach currently available. Despite substantial reports of neurodegeneration (both in the hypothalamic and extra-hypothalamic areas), little is known about the contribution of glial cells to PWS pathophysiology. Correspondingly, there is considerable overlap between neuronal loss in PWS and nongenetic obesity (Swaab et al., 1995; Goldstone et al., 2002); however, whether this is due to the genetic components of the disease or a consequence of obesity remains to be further elucidated. Of note, *Magel2*-null and *Snord116*-null mice (both genes in the PWS critical genomic region) have defects in the hypothalamic circuitry in the control of energy homeostasis (Mercer et al., 2013; Polex-Wolf et al., 2018). Further investigation is needed to strengthen our understanding of glial biology in PWS, and the elucidation of pending questions can eventually lead to therapeutic approaches to this disease. Unexplored topics, such as the cellular specificity of PWS-related genes, trigger the observed inflammatory response, and whether these changes are restrained to the hypothalamus are key milestones to be achieved. Here, we aimed to explore the current knowledge on the (potential) role of glia in PWS-associated hyperphagia and body weight imbalance, in parallel with nongenetic obesity.

Microglia and astrocytes

As previously discussed, these two cell types are the key homeostatic keepers of neurons. Whether spatial and temporal microglial and astroglial outcomes in PWS are comparable to those in non-syndromic obesity remain to be elucidated. Plasma levels of pro-inflammatory cytokines are increased in PWS but do not differ from those in obese subjects, which suggests a similar peripheral inflammatory profile between these two groups (Haqq et al., 2011). Recent evidence from transcriptomic data suggests astrocytic loss and classical reactive microgliosis in the hypothalami of patients with PWS (Bochukova et al., 2018). This study demonstrated the reduction of hypothalamic trophic mediators (i.e., brain-derived neurotrophic factor), which are essential for neuronal survival. Associated with this, reduced astrocytic cell numbers reflect lesser glial support for neighboring neurons. Finally, a robust increase in microglia-associated transcripts suggests that this cell type plays a key role in PWS-associated neuronal dysfunction (Bochukova et al., 2018). Thus, a speculative interpretation of these data points to similar, but not identical, mechanisms underlying neuronal loss in PWS and obesity. Recent findings of our laboratory indicate similar findings, with classical microglial reactivity features in PWS hypothalami associated with a lower number of specific neuronal populations (Correa-da-Silva et al., 2024). Curiously, we evaluated different PWS subgenotypes (PWS Type 1 and Type 2 deletions, based on the extent of chromosomal loss) and found distinct microglial and astrocytic phenotypes. PWS Type 1 has highly fragmented and poorly phagocytic microglia, an indication of cellular senescence, whereas PWS Type 2 hypothalami present morphologically intact and apparently functional microglia (Correa-da-Silva et al., 2024). Both subgenotypes do not have differences in anthropometric parameters, but PWS Type 1 subjects have a more severe phenotype, especially in social cognition-related features (Butler et al., 2004). These findings indicate that despite similar outcomes in neural cell death both subgenotypes seem to have distinct cellular and molecular mechanisms for this effect. Thus, further investigation of the glial role in PWS pathophysiology has the potential to create targeted therapeutic approaches for different subsets of PWS.

Oligodendrocytes

To our knowledge, there are no reports, which directly access the contribution of this cell type in PWS despite the evidence of white matter thinning in cortical structures of PWS patients, already observed during childhood (Lukoshe et al., 2013). These findings could be interpreted as either brain atrophy or delayed neural development and potentially affect cortico-hypothalamic myelinated tracts in the control of social cognition and metabolism. Interestingly, RNA single nuclei-sequence analysis in the hypothalamus of *Magel2* knockout animals revealed that oligodendrocyte-specific transcripts were highly modulated compared to controls (Yu et al., 2022). As previously stated, the loss of hypothalamic neurons is a hallmark of PWS, and diverse neuronal populations in distinct hypothalamic nuclei are affected. Oligodendrocyte dysfunction and consequent reduction of their support to hypothalamic neurons can be one of the underlying mechanisms behind this broad neuronal loss, in a similar way to what is proposed for nongenetic obesity.

ANOREXIA NERVOSA

Anorexia nervosa (AN) is an eating disorder characterized by restrictive nutritional intake relative to energy requirements, which leads to excessive weight loss. AN patients fear body weight gain and present a distorted self-image that limits their perception of the seriousness of the disease (Treasure et al., 2015). AN is strongly linked to mood and anxiety disorders, and it has multisystemic complications that include, but are not limited to, bone, cardiac, endocrine, and gastrointestinal changes (Treasure et al., 2015). Complications worsen with disease progression and are associated with a lower body mass index (Mehler et al., 2018).

Experimental animal models of AN can be achieved by either activity-based anorexia (this model combines maximum ad libitum food intake of 3 h daily combined with free access to a running wheel) or dehydration (animals are offered a 2.5% NaCl solution leading to reduced body weight due to dehydration) (Klenotich and Dulawa, 2012; Reyes-Ortega et al., 2022). Both have limitations but are considered to recapitulate the major features of human pathology and trigger brain alterations through the promotion of a negative energy balance. It has been widely recognized that there is evidence of brain atrophy in the acute underweight stage of AN. Neuroimaging studies have shown cortical thinning and a reduction in both gray and white matter volumes (King et al., 2018; Seitz et al., 2018). These changes are attributed to changes in neurons (i.e., reduced synapses, smaller neurons, and accumulation of neurotoxic particles in neuronal cell bodies), but these changes are uncoupled from differences in oligodendrocyte density in a rat AN model (Frintrop et al., 2018). These neurologic changes are restored upon body weight gain, suggesting dynamic remodeling of circuitries and cells related to the AN phenotype.

Microglia

A considerable amount of evidence indicates the preservation of immune function in AN, in contrast to the pattern observed in malnutrition (Slotwinska and Slotwinski, 2017). Interestingly, pro-inflammatory cytokines are known anorectic mediators, and increased serum levels of cytokines have been reported in patients with AN compared to healthy controls (Kapas and Krueger, 1992; Fantino and Wieteska, 1993; Kahl et al., 2004). Limited information regarding microglial cells in AN pathophysiology is available, especially for those derived from human material.

Experimental models of anorexia display increased microglial presence in different cortical areas, such as the hippocampus and prefrontal cortex (Ragu-Varman et al., 2019; Reyes-Ortega et al., 2020). These microglia are less ramified and are associated with augmented expression of pro-inflammatory cytokines, which is consistent with a classic reactive profile in microglia (Ragu-Varman et al., 2019; Reyes-Ortega et al., 2020). This reactive state has implications for neuronal health and affects astrocytic functions. Such reactive microgliosis is aligned with the behavioral features of animal anorexia models, such as memory performance and cognitive impairment (Paulukat et al., 2016).

Astrocytes

There is a significant reduction in GFAP-expressing astrocytes in the corpus callosum and cerebral cortex of experimental anorectic rats compared with matched controls (Reyes-Haro et al., 2016; Reyes-Ortega et al., 2020). These changes in cortical astrocytes are reversible upon normalization of feeding, but not in the corpus callosum, possibly due to differences in cellular turnover in the gray and white matter (Reyes-Haro et al., 2015). Interestingly, (partial) body weight restoration normalizes astrocytic count (Frintrop et al., 2019). When rodents are separated into resistant or prone to anorexia groups, it is possible to detect increased astrocytic density in the prefrontal cortex of AN-prone animals compared to resistant or control animals (Reyes-Ortega et al., 2020; Hurley et al., 2022). The prefrontal cortex integrates information on food perception and therefore contributes to food-seeking behavior (Petrovich et al., 2007).

One possible explanation for the diminished astrocytic density in experimental models of anorexia relies on the metabolic support these cells provide to the neurons. To provide optimal neuronal function, astrocytes perform metabolic shuttle reactions with neurons, typically known as the glutamate/GABA-glutamine cycle (Hertz, 2013). Here, astrocytes remove neurotransmitters from the synaptic site and convert them into glutamine, which is then incorporated into neurons. This serves the purpose of preventing toxicity due to excessive neurotransmitter content and provides metabolic support to neurons once glutamine is easily metabolized in mitochondrial respiration (Hertz, 2013). Anorexia reduces the expression of astrocytic enzymes associated with the glutamate-glutamine cycle in the prefrontal cortex, possibly to promote glutamine catabolism and therefore cope with the negative energy balance (Reyes-Ortega et al., 2022).

Finally, emerging evidence supports astrocytic dysfunction in patients with AN. Functional imaging studies have also reported reduced brain volume in AN subjects compared to controls (King et al., 2018), and it is reasonable to speculate that the lessening of astrocytes contributes to this. Notably, the use of serum GFAP as a biomarker for different neuropathologic conditions has been proposed and validated (Abdelhak et al., 2022). A recent report showed increased serum GFAP levels in AN subjects compared to controls, and partial weight restoration decreased it (Hellerhoff et al., 2021). These findings need to be further validated in different cohorts, and whether they reflect central astrocytic profiles also needs to be determined. In contrast, a case report indicated prominent gliosis in the periventricular regions related to dopaminergic mesolimbic pathways (Umeda et al., 2021). These regions are linked to emotion and cognition and might indicate dysfunction of the hedonic component of feeding.

Taken together, these findings point to an emerging role for microglia and astrocytes in AN pathology. Glial changes appear to be reversible after body weight restoration, and the main mechanism underlying glial dysfunction is starvation (Frintrop et al., 2019). Curiously, hypothalamic nuclei have not been the focus of research in the field of AN, despite their central role in feeding coordination.

CONCLUSION

The regulation of energy homeostasis involves a complex and multilayered interplay between neurons and glial cells. The major regulatory system in the central nervous system that controls metabolism is the hypothalamus, which harbors numerous neuronal populations engaging in feeding behavior, glycemic control, and energy expenditure. This review focused on eating-related disorders related to long-term positive or negative homeostasis, obesity/PWS, and anorexia nervosa. Although they do not necessarily have a hypothalamic etiology, hypothalamic dysfunction is a central node in their development and progression. In brief, appetite-curbing neurons are severely affected, which consequently disrupts the normal feeding and energy expenditure patterns. These eating disorders considerably impair physical health and can evolve into disabling and deadly conditions.

Glial cells are classically considered to be only a support system for neurons; however, emerging evidence demonstrates the active role of these cells in virtually all homeostatic systems. Thus, glial malfunction has been implicated in physiologic and behavioral impairments. A considerable portion of the research discussed here is based on experimental animals, which cannot fully mimic human conditions. Nonetheless, current translational research demonstrates a considerable overlap in glia-related cellular and molecular alterations. In general, the roles of oligodendrocytes and tanycytes in the pathology of metabolic disorders are underexplored. In contrast, many studies have evaluated the roles of astrocytes and microglia. Both cell types provide trophic and structural support to neurons, and their malfunctioning is a key mechanism of neural dysfunction. Astrocytes are overrepresented in the hypothalamus of obese subjects but are not necessarily linked to a positive energy balance; hence, PWS subjects have diminished astrocytic cell counts. Interestingly, PWS has been proposed to be a genetic starvation model; hence, their satiety-related pathways are bluntly impaired. AN subjects also experience starvation, and this disorder is associated with a reduced number of astrocytes. Finally, microglia assume a pro-inflammatory state in the three disorders discussed here. This reactive microgliosis is triggered by dietary and endocrine components, but it is reasonable to speculate that neuronal dysfunction and death also lead to increased local immunity and, consequently, microglial reactivity.

Nevertheless, all the findings discussed here indicate that glial biology needs to be considered in therapeutic approaches for feeding-related disorders. The apparent reversibility of these glial changes, in contrast to those observed in neurons that appear to be permanently lost, reinforces the need to equate glia in future therapies. This is especially evident in positive energy balance disorders, but it also seems plausible in AN, based on empiric evidence obtained from other mood disorders. Pharmacologic therapies targeting obesity and T2DM can either be based on or have off-target anti-inflammatory outcomes. Collectively, this review covers the latest data on the proposed eating-related disorders involving glial cells, which possess several mechanistic targets for future therapies.

ABBREVIATIONS

AN, anorexia nervosa; CNS, central nervous system; OPC, oligodendrocyte progenitor cell; T2DM, type 2 diabetes mellitus.

References

Abdelhak A, Foschi M, Abu-Rumeileh S et al. (2022). Blood GFAP as an emerging biomarker in brain and spinal cord disorders. Nat Rev Neurol 18: 158–172.

Andre C, Guzman-Quevedo O, Rey C et al. (2017). Inhibiting microglia expansion prevents diet-induced hypothalamic and peripheral inflammation. Diabetes 66: 908–919.

Balland E, Dam J, Langlet F et al. (2014). Hypothalamic tanycytes are an ERK-gated conduit for leptin into the brain. Cell Metab 19: 293–301.

Berkseth KE, Guyenet SJ, Melhorn SJ et al. (2014). Hypothalamic gliosis associated with high-fat diet feeding is reversible in mice: a combined immunohistochemical and magnetic resonance imaging study. Endocrinology 155: 2858–2867.

Bocarsly ME, Fasolino M, Kane GA et al. (2015). Obesity diminishes synaptic markers, alters microglial morphology, and impairs cognitive function. Proc Natl Acad Sci U S A 112: 15731–15736.

Bochukova EG, Lawler K, Croizier S et al. (2018). A transcriptomic signature of the hypothalamic response to fasting and BDNF deficiency in Prader-Willi syndrome. Cell Rep 22: 3401–3408.

Bouhrara M, Khattar N, Elango P et al. (2021). Evidence of association between obesity and lower cerebral myelin content in cognitively unimpaired adults. Int J Obes (Lond) 45: 850–859.

Brabazon F, Bermudez S, Shaughness M et al. (2018). The effects of insulin on the inflammatory activity of BV2 microglia. PLoS One 13: e0201878.

Brown SSG, Manning KE, Fletcher P et al. (2022). In vivo neuroimaging evidence of hypothalamic alteration in Prader-Willi syndrome. Brain Commun 4: fcac229.

Buckman LB, Thompson MM, Lippert RN et al. (2015). Evidence for a novel functional role of astrocytes in the acute homeostatic response to high-fat diet intake in mice. Mol Metab 4: 58–63.

Butler MG, Bittel DC, Kibiryeva N et al. (2004). Behavioral differences among subjects with Prader-Willi syndrome and type I or type II deletion and maternal disomy. Pediatrics 113: 565–573.

Campbell JN, Macosko EZ, Fenselau H et al. (2017). A molecular census of arcuate hypothalamus and median eminence cell types. Nat Neurosci 20: 484–496.

Cassidy SB, Schwartz S, Miller JL et al. (2012). Prader-Willi syndrome. Genet Med 14: 10–26.

Challis BG, Coll AP, Yeo GS et al. (2004). Mice lacking pro-opiomelanocortin are sensitive to high-fat feeding but respond normally to the acute anorectic effects of peptide-YY(3-36). Proc Natl Acad Sci U S A 101: 4695–4700.

Clarke RE, Verdejo-Garcia A, Andrews ZB (2018). The role of corticostriatal-hypothalamic neural circuits in feeding behaviour: implications for obesity. J Neurochem 147: 715–729.

Clement K, Vaisse C, Lahlou N et al. (1998). A mutation in the human leptin receptor gene causes obesity and pituitary dysfunction. Nature 392: 398–401.

Correa da Silva F, Aguiar C, Pereira JAS et al. (2019). Ghrelin effects on mitochondrial fitness modulates macrophage function. Free Radic Biol Med 145: 61–66.

Correa-da-Silva F, Fliers E, Swaab DF et al. (2021). Hypothalamic neuropeptides and neurocircuitries in Prader-Willi syndrome. J Neuroendocrinol 33: e12994.

Correa-da-Silva F, Carter J, Wang XY et al. (2024). Microglial phagolysosome dysfunction and altered neural communication amplify phenotypic severity in Prader-Willi syndrome with larger deletion. Acta Neuropathol 147: 64.

Costa RA, Ferreira IR, Cintra HA et al. (2019). Genotype-phenotype relationships and endocrine findings in Prader-Willi syndrome. Front Endocrinol (Lausanne) 10: 864.

De Souza CT, Araujo EP, Bordin S et al. (2005). Consumption of a fat-rich diet activates a proinflammatory response and induces insulin resistance in the hypothalamus. Endocrinology 146: 4192–4199.

Ding F, Li HH, Zhang S et al. (2008). SnoRNA Snord116 (Pwcr1/MBII-85) deletion causes growth deficiency and hyperphagia in mice. PLoS One 3: e1709.

Djogo T, Robins SC, Schneider S et al. (2016). Adult NG2-glia are required for median eminence-mediated leptin sensing and body weight control. Cell Metab 23: 797–810.

Douglass JD, Dorfman MD, Fasnacht R et al. (2017). Astrocyte IKKbeta/NF-kappaB signaling is required for diet-induced obesity and hypothalamic inflammation. Mol Metab 6: 366–373.

Duquenne M, Folgueira C, Bourouh C et al. (2021). Leptin brain entry via a tanycytic LepR-EGFR shuttle controls lipid metabolism and pancreas function. Nat Metab 3: 1071–1090.

Escartin C, Galea E, Lakatos A et al. (2021). Reactive astrocyte nomenclature, definitions, and future directions. Nat Neurosci 24: 312–325.

Fantino M, Wieteska L (1993). Evidence for a direct central anorectic effect of tumor-necrosis-factor-alpha in the rat. Physiol Behav 53: 477–483.

Farooqi IS, Drop S, Clements A et al. (2006). Heterozygosity for a POMC-null mutation and increased obesity risk in humans. Diabetes 55: 2549–2553.

Ferrannini E, Gastaldelli A, Miyazaki Y et al. (2005). β-Cell function in subjects spanning the range from normal glucose tolerance to overt diabetes: a new analysis. J Clin Endocrinol Metab 90: 493–500.

Frintrop L, Liesbrock J, Paulukat L et al. (2018). Reduced astrocyte density underlying brain volume reduction in activity-based anorexia rats. World J Biol Psychiatry 19: 225–235.

Frintrop L, Trinh S, Liesbrock J et al. (2019). The reduction of astrocytes and brain volume loss in anorexia nervosa-the impact of starvation and refeeding in a rodent model. Transl Psychiatry 9: 159.

Galou M, Colucci-Guyon E, Ensergueix D et al. (1996). Disrupted glial fibrillary acidic protein network in astrocytes from vimentin knockout mice. J Cell Biol 133: 853–863.

Gao Y, Ottaway N, Schriever SC et al. (2014). Hormones and diet, but not body weight, control hypothalamic microglial activity. Glia 62: 17–25.

Gao Y, Bielohuby M, Fleming T et al. (2017). Dietary sugars, not lipids, drive hypothalamic inflammation. Mol Metab 6: 897–908.

Gao Y, Vidal-Itriago A, Milanova I et al. (2018). Deficiency of leptin receptor in myeloid cells disrupts hypothalamic metabolic circuits and causes body weight increase. Mol Metab 7: 155–160.

Garcia-Caceres C, Quarta C, Varela L et al. (2016). Astrocytic insulin signaling couples brain glucose uptake with nutrient availability. Cell 166: 867–880.

Gaykema RP, Nguyen XM, Boehret JM et al. (2014). Characterization of excitatory and inhibitory neuron activation in the mouse medial prefrontal cortex following palatable food ingestion and food driven exploratory behavior. Front Neuroanat 8: 60.

Gnanapavan S, Kola B, Bustin SA et al. (2002). The tissue distribution of the mRNA of ghrelin and subtypes of its receptor, GHS-R, in humans. J Clin Endocrinol Metab 87: 2988.

Golay A, Bobbioni E (1997). The role of dietary fat in obesity. Int J Obes Relat Metab Disord 21: S2–11.

Goldstone AP, Unmehopa UA, Bloom SR et al. (2002). Hypothalamic NPY and agouti-related protein are increased in human illness but not in Prader-Willi syndrome and other obese subjects. J Clin Endocrinol Metab 87: 927–937.

Gruber T, Pan C, Contreras RE et al. (2021). Obesity-associated hyperleptinemia alters the gliovascular interface of the hypothalamus to promote hypertension. Cell Metab 33: 1155–1170. e1110.

Haase L, Green E, Murphy C (2011). Males and females show differential brain activation to taste when hungry and sated in gustatory and reward areas. Appetite 57: 421–434.

Haqq AM, Muehlbauer MJ, Newgard CB et al. (2011). The metabolic phenotype of Prader-Willi syndrome (PWS) in childhood: heightened insulin sensitivity relative to body mass index. J Clin Endocrinol Metab 96: E225–E232.

Hellerhoff I, King JA, Tam FI et al. (2021). Differential longitudinal changes of neuronal and glial damage markers in anorexia nervosa after partial weight restoration. Transl Psychiatry 11: 86.

Herrera Moro Chao D, Kirchner MK, Pham C et al. (2022). Hypothalamic astrocytes control systemic glucose metabolism and energy balance. Cell Metab 34: 1532–1547.e1536.

Hertz L (2013). The glutamate-glutamine (GABA) cycle: importance of late postnatal development and potential reciprocal interactions between biosynthesis and degradation. Front Endocrinol (Lausanne) 4: 59.

Holm VA, Cassidy SB, Butler MG et al. (1993). Prader-Willi syndrome: consensus diagnostic criteria. Pediatrics 91: 398–402.

Horvath TL, Sarman B, Garcia-Caceres C et al. (2010). Synaptic input organization of the melanocortin system predicts diet-induced hypothalamic reactive gliosis and obesity. Proc Natl Acad Sci U S A 107: 14875–14880.

Huang HT, Tsai SF, Wu HT et al. (2019). Chronic exposure to high fat diet triggers myelin disruption and interleukin-33 upregulation in hypothalamus. BMC Neurosci 20: 33.

Hurley MM, Collica SC, Aston SA et al. (2022). Adolescent female rats prone to the activity based anorexia (ABA) paradigm have altered hedonic responses and cortical astrocyte density compared to resistant animals. Appetite 168: 105666.

Jais A, Bruning JC (2017). Hypothalamic inflammation in obesity and metabolic disease. J Clin Invest 127: 24–32.

Jensen MD, Caruso M, Heiling V et al. (1989). Insulin regulation of lipolysis in nondiabetic and IDDM subjects. Diabetes 38: 1595–1601.

Jin S, Kim KK, Park BS et al. (2020). Function of astrocyte MyD88 in high-fat-diet-induced hypothalamic inflammation. J Neuroinflammation 17: 195.

Jing Y, Wu F, Li D et al. (2018). Metformin improves obesity-associated inflammation by altering macrophages polarization. Mol Cell Endocrinol 461: 256–264.

Kahl KG, Kruse N, Rieckmann P et al. (2004). Cytokine mRNA expression patterns in the disease course of female adolescents with anorexia nervosa. Psychoneuroendocrinology 29: 13–20.

Kalin S, Heppner FL, Bechmann I et al. (2015). Hypothalamic innate immune reaction in obesity. Nat Rev Endocrinol 11: 339–351.

Kalsbeek MJ, Wolff SE, Korpel NL et al. (2020). The impact of antidiabetic treatment on human hypothalamic infundibular neurons and microglia. JCI Insight 5: e133868.

Kapas L, Krueger JM (1992). Tumor necrosis factor-beta induces sleep, fever, and anorexia. Am J Physiol 263: R703–R707.

Kim JG, Suyama S, Koch M et al. (2014). Leptin signaling in astrocytes regulates hypothalamic neuronal circuits and feeding. Nat Neurosci 17: 908–910.

King JA, Frank GKW, Thompson PM et al. (2018). Structural neuroimaging of anorexia nervosa: future directions in the quest for mechanisms underlying dynamic alterations. Biol Psychiatry 83: 224–234.

Klenotich SJ, Dulawa SC (2012). The activity-based anorexia mouse model. Methods Mol Biol 829: 377–393.

Knight ZA, Hannan KS, Greenberg ML et al. (2010). Hyperleptinemia is required for the development of leptin resistance. PLoS One 5: e11376.

Koch C, Augustine RA, Steger J et al. (2010). Leptin rapidly improves glucose homeostasis in obese mice by increasing hypothalamic insulin sensitivity. J Neurosci 30: 16180–16187.

Kojima M, Hosoda H, Date Y et al. (1999). Ghrelin is a growth-hormone-releasing acylated peptide from stomach. Nature 402: 656–660.

Koopman KE, Booij J, Fliers E et al. (2013). Diet-induced changes in the lean brain: hypercaloric high-fat-high-sugar snacking decreases serotonin transporters in the human hypothalamic region. Mol Metab 2: 417–422.

Kubota T, Inoue M, Kubota N et al. (2018). Downregulation of macrophage Irs2 by hyperinsulinemia impairs IL-4-indeuced M2a-subtype macrophage activation in obesity. Nat Commun 9: 4863.

Lafrance V, Inoue W, Kan B et al. (2010). Leptin modulates cell morphology and cytokine release in microglia. Brain Behav Immun 24: 358–365.

Lancaster GI, Langley KG, Berglund NA et al. (2018). Evidence that TLR4 is not a receptor for saturated fatty acids but mediates lipid-induced inflammation by reprogramming macrophage metabolism. Cell Metab 27: 1096–1110.e1095.

Langlet F (2014). Tanycytes: a gateway to the metabolic hypothalamus. J Neuroendocrinol 26: 753–760.

Lau BK, Murphy-Royal C, Kaur M et al. (2021). Obesity-induced astrocyte dysfunction impairs heterosynaptic plasticity in the orbitofrontal cortex. Cell Rep 36: 109563.

Lawson EA (2017). The effects of oxytocin on eating behaviour and metabolism in humans. Nat Rev Endocrinol 13: 700–709.

Lechuga-Sancho AM, Arroba AI, Frago LM et al. (2006). Reduction in the number of astrocytes and their projections is associated with increased synaptic protein density in the hypothalamus of poorly controlled diabetic rats. Endocrinology 147: 5314–5324.

Lee S, Yoon BE, Berglund K et al. (2010). Channel-mediated tonic GABA release from glia. Science 330: 790–796.

Lehrke M, Lazar MA (2004). Inflamed about obesity. Nat Med 10: 126–127.

Lhomme T, Clasadonte J, Imbernon M et al. (2021). Tanycytic networks mediate energy balance by feeding lactate to glucose-insensitive POMC neurons. J Clin Invest 131: e140521.

Liu F, Li Z, He X et al. (2019). Ghrelin attenuates neuroinflammation and demyelination in experimental autoimmune encephalomyelitis involving NLRP3 inflammasome signaling pathway and pyroptosis. Front Pharmacol 10: 1320.

Lord GM (2006). Leptin as a proinflammatory cytokine. Contrib Nephrol 151: 151–164.

Lukoshe A, White T, Schmidt MN et al. (2013). Divergent structural brain abnormalities between different genetic subtypes of children with Prader-Willi syndrome. J Neurodev Disord 5: 31.

Marzullo P, Verti B, Savia G et al. (2004). The relationship between active ghrelin levels and human obesity involves alterations in resting energy expenditure. J Clin Endocrinol Metab 89: 936–939.

Maya-Monteiro CM, Correa-da-Silva F, Hofmann SS et al. (2021). Lipid droplets accumulate in the hypothalamus of mice and humans with and without metabolic diseases. Neuroendocrinology 111: 263–272.

McLachlan IG, Kramer TS, Dua M et al. (2022). Diverse states and stimuli tune olfactory receptor expression levels to modulate food-seeking behavior. Elife 11. https://doi.org/10.7554/eLife.79557.

Meek TH, Nelson JT, Matsen ME et al. (2016). Functional identification of a neurocircuit regulating blood glucose. Proc Natl Acad Sci U S A 113: E2073–E2082.

Mehler PS, Blalock DV, Walden K et al. (2018). Medical findings in 1,026 consecutive adult inpatient-residential eating disordered patients. Int J Eat Disord 51: 305–313.

Melo HM, Seixas da Silva GDS, Sant'Ana MR et al. (2020). Palmitate is increased in the cerebrospinal fluid of humans with obesity and induces memory impairment in mice via pro-inflammatory TNF-alpha. Cell Rep 30: 2180–2194.e2188.

Mercer RE, Michaelson SD, Chee MJ et al. (2013). Magel2 is required for leptin-mediated depolarization of POMC neurons in the hypothalamic arcuate nucleus in mice. PLoS Genet 9: e1003207.

Milanova IV, Kalsbeek MJT, Wang XL et al. (2019). Diet-induced obesity disturbs microglial immunometabolism in a time-of-day manner. Front Endocrinol (Lausanne) 10: 424.

Milanova IV, Korpel NL, Correa-da-Silva F et al. (2022). Loss of microglial insulin receptor leads to sex-dependent metabolic disorders in obese mice. Int J Mol Sci 23: 2933.

Milanski M, Degasperi G, Coope A et al. (2009). Saturated fatty acids produce an inflammatory response predominantly through the activation of TLR4 signaling in hypothalamus: implications for the pathogenesis of obesity. J Neurosci 29: 359–370.

Must A, Spadano J, Coakley EH et al. (1999). The disease burden associated with overweight and obesity. JAMA 282: 1523–1529.

Nimmerjahn A (2012). Two-photon imaging of microglia in the mouse cortex in vivo. Cold Spring Harb Protoc 2012. https://doi.org/10.1101/pdb.prot069294.

Obermeier B, Daneman R, Ransohoff RM (2013). Development, maintenance and disruption of the blood-brain barrier. Nat Med 19: 1584–1596.

O'Brien PD, Hinder LM, Callaghan BC et al. (2017). Neurological consequences of obesity. Lancet Neurol 16: 465–477.

Oomura Y, Kita H (1981). Insulin acting as a modulator of feeding through the hypothalamus. Diabetologia 20: 290–298.

Ou Z, Ma Y, Sun Y et al. (2019). A GPR17-cAMP-lactate signaling axis in oligodendrocytes regulates whole-body metabolism. Cell Rep 26: 2984–2997.e2984.

Paulukat L, Frintrop L, Liesbrock J et al. (2016). Memory impairment is associated with the loss of regular oestrous cycle and plasma oestradiol levels in an activity-based anorexia animal model. World J Biol Psychiatry 17: 274–284.

Petrovich GD, Ross CA, Holland PC et al. (2007). Medial prefrontal cortex is necessary for an appetitive contextual conditioned stimulus to promote eating in sated rats. J Neurosci 27: 6436–6441.

Polex-Wolf J, Lam BY, Larder R et al. (2018). Hypothalamic loss of Snord116 recapitulates the hyperphagia of Prader-Willi syndrome. J Clin Invest 128: 960–969.

Posey KA, Clegg DJ, Printz RL et al. (2009). Hypothalamic proinflammatory lipid accumulation, inflammation, and insulin resistance in rats fed a high-fat diet. Am J Physiol Endocrinol Metab 296: E1003–E1012.

Ragu-Varman D, Macedo-Mendoza M, Labrada-Moncada FE et al. (2019). Anorexia increases microglial density and cytokine expression in the hippocampus of young female rats. Behav Brain Res 363: 118–125.

Ramalho AF, Bombassaro B, Dragano NR et al. (2018). Dietary fats promote functional and structural changes in the median eminence blood/spinal fluid interface-the protective role for BDNF. J Neuroinflammation 15: 10.

Razolli DS, Moraes JC, Morari J et al. (2015). TLR4 expression in bone marrow-derived cells is both necessary and sufficient to produce the insulin resistance phenotype in diet-induced obesity. Endocrinology 156: 103–113.

Reyes-Haro D, Labrada-Moncada FE, Miledi R et al. (2015). Dehydration-induced anorexia reduces astrocyte density in the rat Corpus callosum. Neural Plast 2015: 474917.

Reyes-Haro D, Labrada-Moncada FE, Varman DR et al. (2016). Anorexia reduces GFAP+ cell density in the rat Hippocampus. Neural Plast 2016: 2426413.

Reyes-Ortega P, Ragu Varman D, Rodriguez VM et al. (2020). Anorexia induces a microglial associated pro-inflammatory environment and correlates with neurodegeneration in the prefrontal cortex of young female rats. Behav Brain Res 392: 112606.

Reyes-Ortega P, Soria-Ortiz MB, Rodriguez VM et al. (2022). Anorexia disrupts glutamate-glutamine homeostasis associated with astroglia in the prefrontal cortex of young female rats. Behav Brain Res 420: 113715.

Rocha DM, Caldas AP, Oliveira LL et al. (2016). Saturated fatty acids trigger TLR4-mediated inflammatory response. Atherosclerosis 244: 211–215.

Roh E, Song DK, Kim MS (2016). Emerging role of the brain in the homeostatic regulation of energy and glucose metabolism. Exp Mol Med 48: e216.

Romano N, Lafont C, Campos P et al. (2023). Median eminence blood flow influences food intake by regulating ghrelin access to the metabolic brain. JCI Insight 8: e165763.

Ruddick-Collins LC, Morgan PJ, Fyfe CL et al. (2022). Timing of daily calorie loading affects appetite and hunger responses without changes in energy metabolism in healthy subjects with obesity. Cell Metab 34: 1472–1485.e1476.

Sa M, Park MG, Lee CJ (2022). Role of hypothalamic reactive astrocytes in diet-induced obesity. Mol Cells 45: 65–75.

Saucisse N, Mazier W, Simon V et al. (2021). Functional heterogeneity of POMC neurons relies on mTORC1 signaling. Cell Rep 37: 109800.

Schaeffer M, Langlet F, Lafont C et al. (2013). Rapid sensing of circulating ghrelin by hypothalamic appetite-modifying neurons. Proc Natl Acad Sci U S A 110: 1512–1517.

Seitz J, Konrad K, Herpertz-Dahlmann B (2018). Extend, pathomechanism and clinical consequences of brain volume changes in anorexia nervosa. Curr Neuropharmacol 16: 1164–1173.

Sena A, Sarlieve LL, Rebel G (1985). Brain myelin of genetically obese mice. J Neurol Sci 68: 233–243.

Shapira NA, Lessig MC, He AG et al. (2005). Satiety dysfunction in Prader-Willi syndrome demonstrated by fMRI. J Neurol Neurosurg Psychiatry 76: 260–262.

Slotwinska SM, Slotwinski R (2017). Immune disorders in anorexia. Cent Eur J Immunol 42: 294–300.

Stachniak TJ, Ghosh A, Sternson SM (2014). Chemogenetic synaptic silencing of neural circuits localizes a hypothalamus–>midbrain pathway for feeding behavior. Neuron 82: 797–808.

Swaab DF, Purba JS, Hofman MA (1995). Alterations in the hypothalamic paraventricular nucleus and its oxytocin neurons (putative satiety cells) in Prader-Willi syndrome: a study of five cases. J Clin Endocrinol Metab 80: 573–579.

Thaler JP, Yi CX, Schur EA et al. (2012). Obesity is associated with hypothalamic injury in rodents and humans. J Clin Invest 122: 153–162.

Treasure J, Zipfel S, Micali N et al. (2015). Anorexia nervosa. Nat Rev Dis Primers 1: 15074.

Trevisiol A, Nave KA (2015). Brain energy metabolism: conserved functions of glycolytic glial cells. Cell Metab 22: 361–363.

Tschop M, Smiley DL, Heiman ML (2000). Ghrelin induces adiposity in rodents. Nature 407: 908–913.

Tschop M, Weyer C, Tataranni PA et al. (2001). Circulating ghrelin levels are decreased in human obesity. Diabetes 50: 707–709.

Udagawa J, Nimura M, Otani H (2006). Leptin affects oligodendroglial development in the mouse embryonic cerebral cortex. Neuro Endocrinol Lett 27: 177–182.

Umeda K, Kawakami I, Ikeda K et al. (2021). Case report of anorexia nervosa showing periventricular gliosis at autopsy. Neuropathology 41: 127–132.

Valdearcos M, Robblee MM, Benjamin DI et al. (2014). Microglia dictate the impact of saturated fat consumption on hypothalamic inflammation and neuronal function. Cell Rep 9: 2124–2138.

Veit M, van Asten R, Olie A et al. (2022). The role of dietary sugars, overweight, and obesity in type 2 diabetes mellitus: a narrative review. Eur J Clin Nutr 76: 1497–1501.

Ventura R, Harris KM (1999). Three-dimensional relationships between hippocampal synapses and astrocytes. J Neurosci 19: 6897–6906.

Waterson MJ, Horvath TL (2015). Neuronal regulation of energy homeostasis: beyond the hypothalamus and feeding. Cell Metab 22: 962–970.

Xie Z, Sun Y, Ye Y et al. (2022). Randomized controlled trial for time-restricted eating in healthy volunteers without obesity. Nat Commun 13: 1003.

Yi CX, Walter M, Gao Y et al. (2017). TNFalpha drives mitochondrial stress in POMC neurons in obesity. Nat Commun 8: 15143.

Yoo S, Cha D, Kim S et al. (2020). Tanycyte ablation in the arcuate nucleus and median eminence increases obesity susceptibility by increasing body fat content in male mice. Glia 68: 1987–2000.

Yu H, Rubinstein M, Low MJ (2022). Developmental single-cell transcriptomics of hypothalamic POMC neurons reveal the genetic trajectories of multiple neuropeptidergic phenotypes. Elife 11.

Zhang Y, Proenca R, Maffei M et al. (1994). Positional cloning of the mouse obese gene and its human homologue. Nature 372: 425–432.

Zhang X, Zhang G, Zhang H et al. (2008). Hypothalamic IKKbeta/NF-kappaB and ER stress link overnutrition to energy imbalance and obesity. Cell 135: 61–73.

Zigman JM, Nakano Y, Coppari R et al. (2005). Mice lacking ghrelin receptors resist the development of diet-induced obesity. J Clin Invest 115: 3564–3572.

Handbook of Clinical Neurology, Vol. 210 (3rd series)
Neuroglia in Neurologic and Psychiatric Disorders, Part II
A. Verkhratsky, L.D. de Witte, E. Aronica, and E.M. Hol, Editors
https://doi.org/10.1016/B978-0-443-19102-2.00005-3

Chapter 19

Neuroglia in Tourette syndrome and obsessive-compulsive disorder

LUCIANA R. FRICK*

Departments of Neurology and Medicine, Neuroscience Graduate Program, Jacobs School of Medicine & Biomedical Sciences, State University of New York at Buffalo, Clinical and Translational Research Center, Buffalo, NY, United States

Abstract

In recent years, neuroglia have drawn the attention of researchers in the fields of neurology and psychiatry. Besides their well-known functions providing support to neurons, myelinating axons, and clearing up debris, a constantly growing of evidence indicates that glial cells are key contributors to the pathophysiology of neuropsychiatric disorders. Alterations in microglia, astrocytes, and oligodendrocytes have been described in Tourette syndrome (TS) and obsessive-compulsive disorder (OCD). The sudden onset of tics and OCD-like symptoms after infection in children (Pediatric Autoimmune Neuropsychiatric Disorder Associated with Streptococcal Infections) suggests a connection with the immune system; in fact, neuroinflammation has been reported. Many imaging studies revealed abnormal myelination in the brain of TS and OCD patients, highlighting the implication of oligodendroglia in the connectivity alterations. Moreover, animal models have unveiled a cell-autonomous role of microglia and astrocytes in the etiology of pathologic grooming, which links these glial cells to the related disorder trichotillomania. This chapter reviews the state of the art and current gaps in the literature, proposing possible pathomechanisms and future research directions.

GILLES DE LA TOURETTE'S SYNDROME

Tourette syndrome (TS) is a neurodevelopmental disorder with an estimated prevalence of 0.1%–1% (Tinker et al., 2022). It is characterized by motor and vocal tics that start in childhood usually at age 5–7 years old. These tics are sudden, unwanted, and uncontrolled, worsen in the early teenage years, and often improve during the transition into adulthood. The causes of TS are unknown, but it is well established that both genetic and environmental factors are involved. A few rare but highly penetrant mutations in the *SLITRK1, HDC,* and *CNTNAP2* genes have been found in single families (Fernandez et al., 2018). A diversity of other factors influences the onset of TS, among which infection and autoimmunity have been reported. Several studies have reported peripheral abnormalities in immune cell populations and cytokine profiles in TS (reviewed in Murphy et al., 2010). This leads to a belief that TS may be accompanied by inflammation of the nervous system. In fact, there is evidence that supports this hypothesis. The literature on neuroinflammation, microglia, and astrocytes, as well as oligodendroglia and myelin in this disorder, is reviewed below.

Microglia dysfunction in TS

Two clinical studies suggest microglia may become reactive in TS patients. Reactive microgliosis can be estimated by measuring the levels of the 18 kDa translocator protein (TSPO) by positron emission tomography (PET) using the tracer (*R*)-[^{11}C]PK11195. TSPO is a protein of the outer mitochondrial membrane that is upregulated in reactive microglia, but its expression is not restricted to this cell type. For this reason, increased binding of TSPO tracers can reflect processes beyond microglia reactivity; thus findings obtained using this approach must be interpreted cautiously (Guilarte et al., 2022). That being said, increased (*R*)-[^{11}C]PK11195 tracer binding was found in the bilateral caudate nuclei of children with TS, although the control group consisted of healthy adults (mean ages: 11.0 ± 3.0 vs 28.7 ± 7.9 years) (Kumar et al., 2015). Another study analyzed postmortem brains from adult patients with severe refractory TS and found an increased number of CD45$^+$ cells in the caudate by immunohistochemistry. CD45 is expressed by brain resident microglia but also by infiltrating monocytes, and this study could not distinguish between CD45^{high} and CD45^{low} cell populations. In fact, mRNA expression of the monocyte chemoattractant protein-1

*Correspondence to: Luciana R. Frick, PhD, Departments of Neurology and Medicine, Neuroscience Graduate Program, Jacobs School of Medicine & Biomedical Sciences, The State University of New York at Buffalo, Clinical and Translational Research Center, 875 Ellicott Street, 6th Floor, Room C6079, Buffalo, NY 14203, United States. Tel: +1-716-878-3314, Fax: +1-716-859-4850, E-mail: lfrick@buffalo.edu

(MCP-1/CCL2), a key chemokine that regulates migration and infiltration of monocytes/macrophages, is reportedly upregulated in the basal ganglia of TS patients (Morer et al., 2010; Alshammery et al., 2022). Nevertheless, another study showed that IBA1$^+$ cells in the caudate present abnormal morphology with significantly reduced cell body size and retracted ramifications (Lennington et al., 2016). Further evidence of reactive microgliosis is the upregulated expression of genes related to the inflammatory response and to the development and trafficking of immune cells, such as proinflammatory cytokines, chemokines, and their receptors, in the basal ganglia of TS brains (Lennington et al., 2016).

In laboratory animals, administration of two different substances that induce stereotypic behaviors, namely 1-(2,5-dimethoxy-4-iodophenyl)-2-aminopropane and 3,3′-iminodipropionitrile (IDPN), causes overexpression of markers associated with reactive microgliosis such as IBA1, CD11b, interleukins IL-1β and IL-6, and tumor necrosis factor (TNF)-α in the striatum (Long et al., 2019; Zhao et al., 2020; Xi et al., 2022; Zhongling et al., 2022). However, these animal models lack construct validity. Based on the observations of mutations in the L-histidine decarboxylase (*HDC*) gene in patients with TS (Ercan-Sencicek et al., 2010; Fernandez et al., 2012), the *HDC* knockout (KO) mouse model was developed to have construct, face, and predictive validity. HDC-KO mice exhibit stereotypic behavior after exposure to stress or amphetamines, which are attenuated by haloperidol treatment, and recapitulate the neurochemical abnormalities seen in patients with TS (Baldan et al., 2014; Xu et al., 2015b). These animals also show microglial abnormalities in the striatum and hypothalamus, with a reduced density of ramifications. While the total number of IBA1$^+$ cells was normal, striatal microglia positive for insulin-like growth factor 1 (IGF-1) are reduced in HDC-KO mice. More interestingly, HDC-KO microglia are more susceptible to bacterial lipopolysaccharide (LPS) challenge than wild-type microglia, exhibiting a pronounced reactivity and the induction of the proinflammatory cytokines IL-1β and TNF-α in the striatum (Frick et al., 2016). These observations are similar to those reported in the postmortem studies discussed earlier (Lennington et al., 2016), and suggest that reactive microgliosis in TS may be caused by gene X environment interaction. This is also consistent with the finding of higher (*R*)-[^{11}C]PK11195 binding, indicative of reactive microgliosis, in a subset of TS children with streptococcal infection (discussed later). Thus, the current literature suggests that microglia may be involved in the pathophysiology of TS, possibly in association with its classic inflammatory function. Postmortem and imaging studies of TS patients support this hypothesis, and the HDC-KO model emphasizes the importance of environmental factors in the immune aspects of the disease.

Although the neurobiologic bases of TS are not fully understood, a growing body of evidence suggests that excessive dopamine in the striatum may excite the thalamocortical circuits, producing tics. Interestingly, a link between microglia and dopamine has been established: microglia cells engulf and modulate the outgrowth of dopaminergic axons in the striatum (Squarzoni et al., 2014). Therefore, it is plausible that dysfunctional microglia may contribute to dopamine dysregulation in TS. In line with this idea, administration of IDPN, which induces stereotypic behaviors in rats, was reported to promote a shift from protective to inflammatory microglia reducing expression of dopamine-associated markers in neurons (Wang et al., 2023). Furthermore, haloperidol, a potent antagonist of the dopamine D2 receptor, alleviated microglia reactivity induced by IDPN (Zhongling et al., 2022). Conversely, microglial cells are known to express dopamine receptors and to be regulated by dopamine (Thomas Broome et al., 2020); thus reactive microgliosis may be a direct consequence of dopamine dysregulation.

Another interesting hypothesis is that impairment of the neuroprotective function of microglia may lead to neuronal death in TS. Postmortem analysis of TS brains revealed loss of interneurons in the basal ganglia (Kalanithi et al., 2005; Kataoka et al., 2010; Lennington et al., 2016). HDC-KO striata have reduced the number of protective IGF-1$^+$ microglia (Frick et al., 2016), which have been shown to promote neuronal survival in other contexts (Ziv et al., 2006; Kohman et al., 2012; Ueno et al., 2013). Given that ablation of parvalbumin-expressing fast-spiking or cholinergic interneurons triggers Tourette-like symptoms in mice (Xu et al., 2015a, 2016), it is tempting to speculate that defective neuroprotection mediated by IGF-1$^+$ microglia may result in striatal interneuron loss thus causing repetitive behaviors. This exciting possibility has not been yet investigated.

Are astrocytes involved in TS?

Evidence linking astroglia with TS is scarce. The *SLC1A3* gene encodes the main glutamate transporter in astrocytes, excitatory amino acid transporter 1 (EAAT1). Its expression is restricted to the astrocyte lineage. One study found a missense mutation in a highly conserved residue (E219D) of the *SLC1A3* gene in a small number of individuals with TS (Adamczyk et al., 2011). In HEK293 cells, this E219D variant exhibits higher glutamate intake and enrichment in the membrane fraction (Adamczyk et al., 2011). However, another study found no evidence of an association between TS and *SLC1A3* (Laurin et al., 2009). Larger genetic or preclinical studies are needed to confirm or rule out the involvement of this gene in TS. But it is also possible that many genes, each of small effect, contribute to the polygenic nature of this disorder. In support of this, a gene-set analysis clustered the largest available TS genome-wide association data by cellular function and found a connection with a set of genes involved in astrocyte function (de Leeuw et al., 2015). This gene set is associated with astrocyte carbohydrate metabolism, specifically with a subset of genes involved in glycolysis and glutamate metabolism through which astrocytes support synaptic function. Untreated patients with TS reportedly have significant reductions in thalamic and striatal concentrations of glutamine and glutamate, as well as lower glutamine/glutamate ratio in comparison to healthy controls (Kanaan et al., 2017). Moreover, striatal glutamine levels were negatively correlated with tic severity, further suggesting a role of astrocytes in TS. It is plausible that alterations in the astrocytic-neuronal coupling systems that normally maintain the balance between excitatory and inhibitory neurotransmission contribute to the pathophysiology of TS in a similar way as in obsessive-compulsive disorder (OCD) (discussed later). However, this needs to be investigated.

Oligodendroglia, myelin and TS

The clinical evidence on oligodendroglia, the myelinating cells of the brain, in TS is limited and results are inconclusive. One study reported that children with TS had higher circulating levels of S100B, a glial-specific protein expressed in the brain by myelinating oligodendrocytes (and astrocytes), than the reference group (van Passel et al., 2001). On the contrary, two genetic studies found no association between oligodendroglia-specific genes and TS: negative results were obtained on microsatellite polymorphisms in the gene encoding the myelin oligodendrocyte glycoprotein (MOG) and single nucleotide polymorphisms in the *OLIG2* gene (Huang et al., 2004; Stewart et al., 2007). Regardless, structural magnetic resonance imaging (MRI) studies suggest that TS may be accompanied by changes in the brain's white matter, which is composed of myelinated axons. For example, two small studies of boys with TS showed a larger proportion of white matter in the frontal lobe (Fredericksen et al., 2002; Hong et al., 2002), while another study found no difference in whole-brain white matter volume (Roessner et al., 2011). Conversely, both reduced and increased white matter volumes were also found in many other brain regions (Muller-Vahl et al., 2009). A larger multicenter study comprising 103 children and adolescents with TS and 103 well-matched children without tics demonstrated a consistent lower white matter volume bilaterally in orbital prefrontal cortex and medial prefrontal cortex (mPFC) (Greene et al., 2017). In addition, decreases in white matter volume in the right frontal pole were inversely related to tic severity in Liu et al. (2013).

Unfortunately, it is difficult to reach a conclusion regarding white matter abnormalities in TS based on volumetric analysis, due to such limited and contradictory evidence. Another cautionary note is that increased/decreased white matter volume does not necessarily mean increased/decreased myelination. Other factors such as inflammation or axonal damage may account for volume differences. Similarly, normal white matter volume does not always mean unaffected myelin, since it may still be structurally disorganized. Diffusion tensor imaging (DTI) analysis can shed light on the microstructure of white matter; in fact, several studies have found microstructural abnormalities in patients with TS, which seem to occur independently of comorbidities. For instance, reduced fractional anisotropy indexes were reported in the corpus callosum of children and adults with TS (Plessen et al., 2006; Neuner et al., 2010). A meta-analysis of 8 datasets including 168 patients with TS and 163 healthy controls found a robust decrease in fractional anisotropy in the corpus callosum and right inferior longitudinal fasciculus (Yang et al., 2021). Moreover, fractional anisotropy reductions in the corpus callosum and the nucleus accumbens of patients with TS correlate with increased tics severity (Draganski et al., 2010; Neuner et al., 2011). Lower values of fractional anisotropy may represent a decrease in the number of axons, a reduction of their degree of myelination, or a disruption of their structural organization within the white matter. Remarkably, the areas with low fractional anisotropy overlapped with those with high radial diffusivity (Neuner et al., 2010). Since the latter is believed to be more sensitive to myelin than axons, the researchers attributed the observed differences to less myelination or compromised oligodendroglia integrity. In confirmation of this, fractional anisotropy reductions and radial diffusivity increases were also observed in white matter tracts in the cortico-striato-thalamo-cortical circuit, the neural substrate for TS (Wen et al., 2016). Furthermore, lower fractional anisotropy values and higher radial diffusivity values in white matter regions were correlated with more severe tics, supporting the hypothesis of reduced or abnormal myelination being involved in TS. On the contrary, another study found elevated fractional anisotropy and diminished radial diffusivity in tracts connecting the cortex with the striatum and the thalamus of TS patients (Worbe et al., 2015). Thus, it is also plausible that the cortico-striato-thalamic white matter connections have higher axonal density and/or dysmyelination, perhaps with pathologic enlargement of myelin sheaths.

Animal studies can help understand the nature of diffusivity alterations. Whole-brain ex vivo DTI showed reduced fractional anisotropy in the dorsal striatum of HDC-KO mice (Jindachomthong et al., 2022). But what does a lower striatal fractional anisotropy mean? Unlike the clinical studies previously discussed, the HDC-KO mouse model of TS helped address this question. RNAseq analysis of the striatum revealed an increase in genes associated with axon ensheathment and myelination. When clustered by cell type, upregulated transcripts were found to be specific to oligodendrocytes. Overexpression of myelin basic protein (MBP) and myelin-associated oligodendrocyte basic protein was confirmed by immunoblot analysis and immunohistochemistry. Altogether, these pieces of evidence suggest that TS brains may have more but less organized myelin maybe caused by oligodendroglia dysfunction, which could be directly modulated by dysregulated neurotransmitters (Karadottir and Attwell, 2007). Among them, it is worth mentioning that histamine negatively regulates oligodendrocyte differentiation and remyelination via the H3 receptor, which is elevated in the striatum of HDC-KO mice (Rapanelli et al., 2016; Chen et al., 2017). Conversely, neuronal hyperactivation can stimulate axonal myelination, whereas reducing their activity has the opposite effect (Gibson et al., 2014; Mitew et al., 2018). Thus, it is possible that elevated neural activity observed in the striatum of HDC-KO mice (Baldan et al., 2014) may lead to increased but disordered myelination (Jindachomthong et al., 2022). This is particularly important given that hypermyelination can be pathologic (Macklin, 2010). More research in this area is needed to understand the precise role of oligodendroglia in this disorder.

OBSESSIVE-COMPULSIVE DISORDER

OCD affects 1%–3% of the population and is characterized by uncontrollable reoccurring thoughts (obsessions) and repetitive behaviors (compulsions). The obsessive-compulsive spectrum comprises disorders such as TS (reviewed earlier) and trichotillomania (hair pulling). Although the neurobiologic basis of OCD remains to be elucidated, one theory is the hyperactivation of the corticostriatal circuit due to a higher excitation-to-inhibition ratio (E/I) caused by increased glutamatergic transmission and/or reduced GABAergic inhibition (Pittenger et al., 2011; Dougherty et al., 2018). As will be reviewed below, a growing body of evidence implicates neuroglia in OCD. Furthermore, genetic manipulations of microglia and astrocytes in mice are able to trigger pathologic grooming in a cell-autonomous manner, strongly suggesting glial dysfunction as a pathomechanism for OCD.

OCD as a microgliopathy

The first cue that microglia may be involved in OCD comes from a serendipitous finding from mice lacking the HoxB8 gene. HoxB8 is a transcription factor with a homeobox DNA-binding domain involved in development. Knockout mice for this gene exhibit compulsive grooming and excessive hair removal to the point of developing skin lesions, in a similar way to humans with the OCD-spectrum disorder trichotillomania (Greer and Capecchi, 2002). Surprisingly, in the brain Hoxb8 expression is restricted to $CD11b^+$ cells, a marker specific to resident microglia and infiltrating myeloid cells (Chen et al., 2010). It was further demonstrated that $HoxB8^+$ cells are also $CX3CR1^+$ (the fractalkine receptor) and account for about one-third of total brain microglia (De et al., 2018). While the "canonical" (non-HoxB8) microglia progenitors arise at embryonic day E7.5 during yolk sac hematopoiesis and colonize the brain at E9.5, "noncanonical" $Hoxb8^+$ progenitors appear to be generated during the second wave of yolk sac hematopoiesis and infiltrate the E12.5 brain. This novel lineage also contributes to brain-resident macrophages: about two-thirds of border-associated macrophages in the subdural meninges, perivascular spaces, and choroid plexus are $HoxB8^+$ (Van Deren et al., 2022). In the adult brain, $HoxB8^+$ parenchymal microglia are unevenly distributed, being abundant in the brain cortex and the dorsal striatum, coincident with the "OCD circuit" regions. Ablation of the Hoxb8 lineage microglia by crossing $HoxB8^{IRES\text{-}Cre}$ X $CSFR1^{flox}$ mutants can recapitulate the pathologic grooming seen in the knockout mice (Trankner et al., 2019), whereas transplantation of wildtype bone barrow into HoxB8 mutant mice rescues the OCD-like phenotype (Chen et al., 2010).

The linkage between microglia and OCD is not restricted to evidence collected from the HoxB8 model. Granulin and its precursor form, progranulin, are proteins implicated in development, inflammation, cell proliferation, and protein homeostasis. Subjects carrying mutations of the *PGRN* gene were found to be more likely to have obsessive-compulsive behaviors such as simple stereotypies, repetitive checking, grooming/personal cleanliness, and collecting. *PGRN* single nucleotide polymorphisms have also been associated with more severe symptoms in OCD patients who experienced early-life trauma (Wang et al., 2020). In the CNS, progranulin is highly expressed by microglia, where it prevents excessive reactivity caused by aging or neurotoxins such as MPTP (1-methyl-4-phenyl-1,2,3,6-tetrahydropyridine), consequently reducing neuronal loss (Martens et al., 2012; Lui et al., 2016). Constitutive GRN-KO mice were reported to exhibit excessive grooming, possibly caused by overreactive microglia via the nuclear factor NFκB–TNFα pathway and complement hyperactivation (Lui et al., 2016; Krabbe et al., 2017). Finally, ablation of the *GRN* gene specifically in microglia is sufficient to recapitulate the excessive grooming and other OCD-related behaviors such as nesting and marble burying further reinforcing the connection of these glial cells to this disease (Krabbe et al., 2017).

Clinical data further strengthen the findings from animal models of pathologic grooming and repetitive behaviors. Using the TSPO radioligand $[^{18}F]$FEPPA, reactive microgliosis was suggested in adults with OCD by PET, particularly in brain regions associated with the pathophysiology of the disease: the orbitofrontal and anterior cingulate cortices, thalamus, ventral striatum, and the dorsal putamen (Attwells et al., 2017). Additionally, the distress associated with preventing compulsive behaviors correlated with the PET tracer binding in the orbitofrontal cortex (Attwells et al., 2017). In line with these findings, reactive microgliosis is a promising therapeutic target for OCD. A pilot, open-label study suggested that minocycline—a tetracycline antibiotic known to suppress reactive microgliosis—may reduce the Yale-Brown Obsessive Compulsive Scale (Y-BOCS) score in adult patients with early onset and with primary hoarding receiving fluoxetine (Rodriguez et al., 2010). A second study that was randomized and double-blinded, comprising a larger number of patients with moderate to severe OCD, confirmed that minocycline in combination with fluvoxamine can further reduce the Y-BOCS score in comparison to the placebo (fluvoxamine alone) group, and achieved shorter times to partial response, as well as higher remission and complete response rates (Esalatmanesh et al., 2016). Similar results were observed with other antiinflammatory agents such as celecoxib and *N*-acetylcysteine in small clinical trials (Meyer, 2021). Although this point still requires systematic investigation in large cohorts, it supports the idea of microglia playing a crucial role in the pathophysiology of OCD.

How are microglia implicated in OCD? The literature suggests that abnormal pruning of neuronal synapses may be involved. Synaptic pruning by microglia is critical during development and more recently it has been proposed, when dysfunctional, as a convergent mechanism that contributes to diverse neuropathologic conditions (Scott-Hewitt et al., 2023). Hoxb8 microglia have a conserved function in synaptic pruning, and Hoxb8 mutant mice exhibit excessive dendritic spines in the corticostriatal circuit, which can be a pathogenic mechanism directly associated with microglial dysfunction that leads to neuronal defects (Nagarajan et al., 2018). In GRN-KO mice, excessive pruning of inhibitory synapses in the ventral thalamus and the nucleus accumbens by overreactive microglia was associated with neuronal hyperexcitability in the OCD circuit (Lui et al., 2016), further supporting a role for these glial cells by contributing to the increased the E/I ratio.

Astrocytes, glutamate, and pathologic grooming

Clinical evidence linking OCD with astroglial abnormalities is limited to one RNAseq study of postmortem samples. Gene-set data analysis revealed an increased representation of astrocyte cell fractions across the orbitofrontal cortex, caudate, and nucleus accumbens regions of OCD brains (Piantadosi et al., 2021). This finding is interesting and promising, warranting further investigation. More evidence connecting astrocytes with OCD comes from animal models.

As discussed before, glutamatergic dysregulation is known to be essential in OCD, and astrocytes seem to be implicated in this neurochemical abnormality. The excitatory amino acid transporter 2 (EAAT2), also known as solute carrier family 1 member 2 (SLC1A2) and glutamate transporter 1 (GLT-1), is the major glutamate transporters expressed predominantly in astroglial cells and is responsible for 90% of total glutamate uptake. Although clinical evidence linking mutations in the *SLC1A2* gene with OCD is weak (Mattheisen et al., 2015), preclinical studies suggest that its mouse homolog gene *GLT1* may play a role. Full GLT1-KO is lethal, but when this gene is conditionally ablated in astrocytes, mice develop pathologic repetitive behaviors such as excessive self-grooming leading to skin lesions (Aida et al., 2015; Jia et al., 2021). Concomitantly, these mice display altered synaptic transmission efficacy, with corticostriatal glutamatergic hyperactivity after prolonged repetitive stimulation (Aida et al., 2015). It is worth mentioning that astrocyte-specific GLT1-KO mice also exhibit tic-like head movements (Aida et al., 2015); its connection with TS remains to be investigated.

Similar results were observed in knockout mice for the scaffolding SAP90/PSD95-associated protein 3 (SAPAP3), which also displays excessive grooming and is used to study the pathophysiology of OCD (Welch et al., 2007). Several human studies have tried to connect variations of this gene to OCD with confounding results, but it is possible that SAPAP3 could be associated with TS or specifically with grooming disorders such as pathologic nail biting, skin picking, and trichotillomania (Bienvenu et al., 2009; Zuchner et al., 2009; Boardman et al., 2011; Crane et al., 2011; Naaz et al., 2022). SAPAP3 is highly expressed in postsynaptic excitatory synapses, but recently it has also been shown to be expressed by striatal astrocytes, where it interacts with glutamate uptake including SLC1A2 (Soto et al., 2023). More interestingly, targeted delivery of SAPAP3 specifically to astrocytes in the striatum of knockout mice significantly ameliorated grooming behavior and reduced facial lesions. Hyperactivation of striatal neurons was also rescued upon astrocytic SAPAP3 restoration, further supporting a role for these glial cells in OCD pathophysiology (Soto et al., 2023).

Another paper linked astrocytes in the mPFC with OCD-like behaviors (Petrelli et al., 2023). Deletion of the vesicular monoamine transporter 2 in cortical astrocytes increased the number and duration of grooming bouts. Interestingly, this is accompanied by a reduction of dopamine levels in the mPFC and increased glutamatergic synaptic transmission to medium spiny neurons (MSN) in the dorsomedial striatum, thus connecting astrocyte-mediated dysregulation of the corticostriatal pathway in repetitive behaviors. Finally, the reduction of Ca^{2+}-dependent signaling in striatal astrocytes was reported to increase repetitive self-grooming behaviors by reducing tonic inhibition of MSN in mice (Yu et al., 2018). This reinforces the idea that increased E/I ratio due to increased glutamatergic and/or reduced GABAergic inhibition may be causative of OCD and other neuropsychiatric disorders, with the contribution of both neurons and astrocytes.

Oligodendroglia and myelination in OCD

Abnormal myelination in OCD is suggested by many imaging studies of using different methodologies, such as DTI and MRI. Most of them comprise a small number of subjects, and some have mixed samples with demographic and clinical variability, generating confounding results. Differences in white matter volumes and diffusivity parameters such as fractional anisotropy have been reported in OCD patients, but an extensive review of the literature unveiled contradictions in the directionality of the changes, being reportedly higher, lower, or unchanged (Fontenelle et al., 2009; Piras et al., 2013). It was postulated that many factors account for the inconsistencies in the findings, namely treatment status (naïve, responsive, refractory), age at onset and evaluation, symptom heterogeneity and severity, comorbidity with other diseases, family history, and duration of the disorder (Fontenelle et al., 2009). For this reason, only large multicentered studies and meta-analyses are reviewed later.

A voxel-based morphometry study comprising 3 continents, 6 imaging centers, 412 adult OCD patients, and 368 healthy controls demonstrated that frontal white matter volumes remained significantly smaller after multiple corrections (de Wit et al., 2014). None of the other differences reported previously survived pooled statistical analysis. Meta-analysis of eight brain voxel-based studies using coordinate activation likelihood estimation found decreased fractional anisotropy in the splenium of the corpus callosum and increased fractional anisotropy in its body. Additionally, ROI-based meta-analysis of six publications showed a reduction of fractional anisotropy in the anterior limb of the internal capsule, which remarkably carries fibers connecting the caudate nucleus to the putamen, the cortex with the striatum, and the thalamus, all associated with OCD (Piras et al., 2013).

Another meta-analysis of white matter microstructural changes that included 23 DTI studies, 707 OCD patients, and 674 controls confirmed with confidence a reduction of the fractional anisotropy in the corpus callosum, and extended it to other brain regions such as the cerebellum (Tao et al., 2023). Considering that the changes in other diffusivity parameters overlap with the area of consistently lower fractional anisotropy, it can be inferred that myelin microarchitecture is compromised in the corpus callosum of OCD patients (Bora et al., 2011; Gan et al., 2017). This DTI meta-analysis also revealed that the method used, i.e., tract-based spatial statistics and voxel-based analysis, can introduce biases in the results generated. Another conclusion is that, despite the relatively high number of subjects evaluated in this meta-analysis, there is not enough data published on different subgroups of OCD. This is an important caveat, especially in light of the findings suggesting that microstructural alterations of white matter may be associated with symptoms' dimensions (Koch et al., 2012; Lazaro et al., 2014; Bollettini et al., 2018). This also helps understand the origin of data heterogeneity, as proposed before (Fontenelle et al., 2009). It may be safely concluded that abnormal myelination has been demonstrated confidently in only a few regions of the brain, nevertheless, other alterations have been observed in mixed conditions and should not be ruled out.

Genetic studies also linked myelin and oligodendroglia-specific genes with OCD. Biased transmission of one particular allele of tetranucleotide repeat $(TAAA)_n$ located in the 3′-UTR of the *MOG* gene was observed in adult OCD patients (34.9 ± 10.7 years old) with early onset (14.7 ± 9.0 years old) (Zai et al., 2004). This allele was also significantly associated with high Y-BOCS severity scores and with both obsession and compulsion subscales. Linkage disequilibrium of intronic C to T base-exchange polymorphisms was also described (Zai et al., 2004). Among these, the genotype C/C of MOG(C10991T) was subsequently associated with lower total white matter volume in pediatric cases with a mean age of 11.7 ± 2.9 years old (Zai et al., 2021). However, another study found that adult OCD patients (30.63 ± 5.65 years old) carrying the Val/Val genotype of the MOG(G511C)/(Val142Leu) polymorphism had significantly larger total white matter volumes (Atmaca et al., 2010) than the other genotypes carriers. Finally, a more recent paper found a significant relationship between the $(TAAA)_n$ MOG microsatellite marker and increased total white matter volume in a small group of pediatric OCD cases, with ages ranging from 7 to 18 years old (Zai et al., 2023).

There are two considerations here: first, it is not known what each of those gene variations entail, i.e., whether and how would MOG expression and function by affected, and ultimately how that would impact on myelination. Second, all these studies were limited by relatively small sample; the authors argued that there was not enough statistical power to examine white matter volume in specific areas, which could dissect regional contributions to the opposing directions of changes found in whole brains. In addition, these differences could be also explained by the age of disease onset, since correlations with fractional anisotropy have been described, for example, in the corpus callosum and the thalamus with opposite directions (Rosso et al., 2014; Li et al., 2021).

Remarkably, methylation profiles of the *MOG* gene have been connected to baseline severity, treatment effect, and responder status (Nissen et al., 2016). All these factors have been related to white matter microstructural abnormalities in OCD. For instance, higher Y-BOCS significantly correlate with lower fractional anisotropy values in the major long fiber tracts; the rostrum of the corpus callosum was particularly identified as a subregion of change (Saito et al., 2008; Garibotto et al., 2010). Conversely, positive correlations between symptom severity and fractional anisotropy were reported in various white matter tracts and brain regions of relevance to OCD such as the striatum, cingulate, and orbitofrontal area (Zarei et al., 2011; Maleki et al., 2020). It is important to remark that these mixed observations may be confounded by variables such as the neurodevelopmental stage (discussed earlier) and the treatment status. In fact, Benedetti and colleagues showed an effect of treatment on fractional anisotropy and radial diffusivity in comparison with drug-naïve OCD patients (Benedetti et al., 2013). Longitudinal studies provided evidence of posttreatment white matter volume and diffusivity changes, suggesting that pharmacologic and cognitive behavioral therapy may modulate myelin plasticity (Yoo et al., 2007; Lazaro et al., 2009; Fan et al., 2012; Zhong et al., 2019). White matter volumes were also shown to differ between responders and nonresponders to cognitive behavioral therapy (Hashimoto et al., 2014; Cao et al., 2021). Similarly, a study comparing OCD patients with resistance to pharmacologic treatment versus responders to first-line antidepressants revealed fractional anisotropy differences in the nucleus accumbens (Li et al., 2014). Thus, parameters that are commonly considered nuisance covariates may actually help to understand the role of MOG in the pathophysiology of OCD. Perhaps genetic and epigenetic MOG profiles could even be useful predictors of treatment response or help personalize the drug of choice for individuals with OCD. This speculation warrants further investigation.

Notably, an observation from a mouse model of multiple sclerosis provides a connection between demyelination induced by targeting MOG and OCD-like behavioral anomalies. MOG (35–55)-induced experimental autoimmune encephalomyelitis was shown to be accompanied by OCD-resembling behaviors such as increased grooming, marble burying, and nestlet shredding (Kant et al., 2018). Thus, it is still intriguing if the polymorphisms discovered in the *MOG* gene and/or its epigenetic modifications could underlie the abnormal myelination described in patients. The involvement of MOG in the etiology of OCD remains to be unraveled.

Single nucleotide polymorphisms have also been discovered in the *OLIG2* gene of patients with OCD (Stewart et al., 2007). Interestingly, these SNPs were not observed in patients with TS comorbidity, suggesting a disease-specific role. Further studies have associated variants of the *OLIG2* gene with the sex and age of OCD onset, and with the occurrence of intrusive thoughts (Zhang et al., 2015; Vieira-Fonseca et al., 2019). Since OLIG2 encodes a transcription factor well-known for determining oligodendrocyte differentiation, it is also plausible that oligodendrogenesis may play a role in the pathophysiology of OCD. It is plausible that oligodendroglia dysfunction may cause microstructural abnormalities in the myelination of the OCD circuit, ultimately contributing to brain connectivity alterations and disrupted inhibitory control in patients.

THE PARTICULAR CASE OF PANDAS/PANS

Pediatric Autoimmune Neuropsychiatric Disorder Associated with Streptococcal Infections (PANDAS), or more generally Pediatric Acute-onset Neuropsychiatric Syndrome (PANS), is characterized by dramatic and sudden onset of neuropsychiatric symptoms in children, specifically OCD and/or tics. Because of its onset, PANDAS has been proposed to arise from an autoimmune response triggered by group A beta-hemolytic streptococcal (GABHS) infection, with autoreactive antibodies that affect the basal ganglia functioning. The antigens recognized by these autoantibodies remain elusive despite recent progress in the identification of protein/cellular targets (Frick et al., 2018; Xu et al., 2021). Levels of anti-MBP IgA were reported to be significantly lower in children diagnosed with PANDAS than in healthy control subjects, which could potentially increase the vulnerability of patients to develop an autoimmune response (Kawikova et al., 2010; Singh et al., 2014). Although one small study comprising 14 PANDAS patients and 14 age- and sex-matched controls found no significant group differences in any DTI measures (fractional anisotropy, mean diffusivity, radial diffusivity, and axial diffusivity), a more recent paper described an overall increased median diffusivity in deep gray matter (e.g., the thalamus, basal ganglia, and amygdala) of 34 PANS children in comparison to 64 healthy controls (Cabrera et al., 2019; Zheng et al., 2020). These alterations may imply abnormal oligodendroglia and myelination, but this has not been demonstrated yet.

Because of its etiology, it would be expected that PANDAS/PANS is accompanied by neuroinflammation. Consistently, TSPO receptor binding was found to be increased bilaterally in the caudate and lentiform nucleus of PANDAS patients compared to controls, suggesting an underlying reactive microglia-mediated neuroinflammation (Kumar et al., 2015). In line with this idea, whole exome and whole genome sequencing analysis identified mutations in a set of genes related to microglia in children with PANS (Trifiletti et al., 2022). Moreover, reactive microgliosis seems to be more spread throughout the basal ganglia of PANDAS patients than in noninfectious TS age-matched cases: higher binding was observed not only in the caudate but also in the bilateral lentiform nucleus (Kumar et al., 2015). This finding further supports a role for the environment influence on the individual genetics as mentioned earlier. Levels of glial fibrillary acidic protein secreted into the CSF or blood may be used to estimate astrocyte reactivity, but no differences were found in the serum of children with severe PANDAS/PANS (Johnson et al., 2021). Although this observation suggests astrocytes might be normal, more studies are needed to confirm it. It is important to note that microglia have been proposed to be more sensitive to pathogens than astrocytes and might become reactive chronologically first (Garland et al., 2022). For instance, astrocytes do not express the Toll-like receptor (TLR4), but microglia do. Thus, they are not directly stimulated by LPS, a component of the cell wall of Gram-negative bacteria such as GABHS. Instead, the reactive astrogliosis may be instigated by reactive microglia and be propagated via inflammatory cytokines. This might explain the absence of astrogliosis reported in PANDAS/PANS.

FINAL NOTES

As thoroughly reviewed earlier, the literature available suggests a role for glial cells in TS and OCD. Although some aspects of glial abnormalities in both disorders are similar, it is possible that some mechanisms are specific, as evidenced by the differences reported between TS, OCD, and PANDAS/PANS. This novel discovery is exciting given that treatment is limited and understanding neuroglial dysfunction may lead to the development of novel therapeutic approaches. There are still some points that need to be addressed, as discussed later.

First, it is not clear how defective microglia-mediated neuroprotection and inflammatory reactive microgliosis are related to TS. The latter could be a consequence of neuronal abnormalities or caused by environmental factors (i.e., infection). Alternatively, impaired microglial support during early development could contribute to the onset of the disease. Second, both malfunctional microglia and astrocytes were shown to be causative of pathologic grooming in laboratory animals; however, the connection of these glial cells with the etiology of OCD in humans is weak and needs to be demonstrated. Third, myelin imaging studies in TS and OCD have provided inconsistent results, which might be due to the intrinsic limitations of DTI analysis and interpretation. Preclinical models would be useful to investigate the nature of oligodendroglia alterations and their impact on brain circuits associated with these disorders.

As mentioned earlier, microglia and astrocytes communicate with each other to coordinate their activities as well as with the neurons to support them. This crosstalk has been investigated in neurodegenerative conditions, such as Alzheimer, Parkinson, and other diseases (Liddelow et al., 2017; Garland et al., 2022). Moreover, astrocytes and microglia can also regulate oligodendrogenesis as well as the production of myelin in development and in pathologic conditions (extensively reviewed in Traiffort et al., 2020). The interaction of different glial cell types in TS and OCD remains to be investigated.

References

Adamczyk A, Gause CD, Sattler R et al. (2011). Genetic and functional studies of a missense variant in a glutamate transporter, SLC1A3, in Tourette syndrome. Psychiatr Genet 21: 90–97.

Aida T, Yoshida J, Nomura M et al. (2015). Astroglial glutamate transporter deficiency increases synaptic excitability and leads to pathological repetitive behaviors in mice. Neuropsychopharmacology 40: 1569–1579.

Alshammery S, Patel S, Jones HF et al. (2022). Common targetable inflammatory pathways in brain transcriptome of autism spectrum disorders and Tourette syndrome. Front Neurosci 16: 999346.

Atmaca M, Onalan E, Yildirim H et al. (2010). The association of myelin oligodendrocyte glycoprotein gene and white matter volume in obsessive-compulsive disorder. J Affect Disord 124: 309–313.

Attwells S, Setiawan E, Wilson AA et al. (2017). Inflammation in the neurocircuitry of obsessive-compulsive disorder. JAMA Psychiatry 74: 833–840.

Baldan LC, Williams KA, Gallezot JD et al. (2014). Histidine decarboxylase deficiency causes Tourette syndrome: parallel findings in humans and mice. Neuron 81: 77–90.

Benedetti F, Giacosa C, Radaelli D et al. (2013). Widespread changes of white matter microstructure in obsessive-compulsive disorder: effect of drug status. Eur Neuropsychopharmacol 23: 581–593.

Bienvenu OJ, Wang Y, Shugart YY et al. (2009). Sapap3 and pathological grooming in humans: results from the OCD collaborative genetics study. Am J Med Genet B Neuropsychiatr Genet 150B: 710–720.

Boardman L, van der Merwe L, Lochner C et al. (2011). Investigating SAPAP3 variants in the etiology of obsessive-compulsive disorder and trichotillomania in the South African white population. Compr Psychiatry 52: 181–187.

Bollettini I, Mazza MG, Muzzarelli L et al. (2018). White matter alterations associate with onset symptom dimension in obsessive-compulsive disorder. Psychiatry Clin Neurosci 72: 13–27.

Bora E, Harrison BJ, Fornito A et al. (2011). White matter microstructure in patients with obsessive-compulsive disorder. J Psychiatry Neurosci 36: 42–46.

Cabrera B, Romero-Rebollar C, Jimenez-Angeles L et al. (2019). Neuroanatomical features and its usefulness in classification of patients with PANDAS. CNS Spectr 24: 533–543.

Cao R, Yang X, Luo J et al. (2021). The effects of cognitive behavioral therapy on the whole brain structural connectome in unmedicated patients with obsessive-compulsive disorder. Prog Neuropsychopharmacol Biol Psychiatry 104: 110037.

Chen SK, Tvrdik P, Peden E et al. (2010). Hematopoietic origin of pathological grooming in Hoxb8 mutant mice. Cell 141: 775–785.

Chen Y, Zhen W, Guo T et al. (2017). Histamine receptor 3 negatively regulates oligodendrocyte differentiation and remyelination. PLoS One 12: e0189380.

Crane J, Fagerness J, Osiecki L et al. (2011). Family-based genetic association study of DLGAP3 in Tourette syndrome. Am J Med Genet B Neuropsychiatr Genet 156B: 108–114.

de Leeuw C, Goudriaan A, Smit AB et al. (2015). Involvement of astrocyte metabolic coupling in Tourette syndrome pathogenesis. Eur J Hum Genet 23: 1519–1522.

de Wit SJ, Alonso P, Schweren L et al. (2014). Multicenter voxel-based morphometry mega-analysis of structural brain scans in obsessive-compulsive disorder. Am J Psychiatry 171: 340–349.

De S, Van Deren D, Peden E et al. (2018). Two distinct ontogenies confer heterogeneity to mouse brain microglia. Development 145.

Dougherty DD, Brennan BP, Stewart SE et al. (2018). Neuroscientifically informed formulation and treatment planning for patients with obsessive-compulsive disorder: a review. JAMA Psychiatry 75: 1081–1087.

Draganski B, Martino D, Cavanna AE et al. (2010). Multispectral brain morphometry in Tourette syndrome persisting into adulthood. Brain 133: 3661–3675.

Ercan-Sencicek AG, Stillman AA, Ghosh AK et al. (2010). L-Histidine decarboxylase and Tourette's syndrome. N Engl J Med 362: 1901–1908.

Esalatmanesh S, Abrishami Z, Zeinoddini A et al. (2016). Minocycline combination therapy with fluvoxamine in moderate-to-severe obsessive-compulsive disorder: a placebo-controlled, double-blind, randomized trial. Psychiatry Clin Neurosci 70: 517–526.

Fan Q, Yan X, Wang J et al. (2012). Abnormalities of white matter microstructure in unmedicated obsessive-compulsive disorder and changes after medication. PLoS One 7: e35889.

Fernandez TV, Sanders SJ, Yurkiewicz IR et al. (2012). Rare copy number variants in Tourette syndrome disrupt genes in histaminergic pathways and overlap with autism. Biol Psychiatry 71: 392–402.

Fernandez TV, State MW, Pittenger C (2018). Tourette disorder and other tic disorders. Handb Clin Neurol 147: 343–354.

Fontenelle LF, Harrison BJ, Yucel M et al. (2009). Is there evidence of brain white-matter abnormalities in obsessive-compulsive disorder? A narrative review. Top Magn Reson Imaging 20: 291–298.

Fredericksen KA, Cutting LE, Kates WR et al. (2002). Disproportionate increases of white matter in right frontal lobe in Tourette syndrome. Neurology 58: 85–89.

Frick L, Rapanelli M, Abbasi E et al. (2016). Histamine regulation of microglia: gene-environment interaction in the regulation of central nervous system inflammation. Brain Behav Immun 57: 326–337.

Frick LR, Rapanelli M, Jindachomthong K et al. (2018). Differential binding of antibodies in PANDAS patients to cholinergic interneurons in the striatum. Brain Behav Immun 69: 304–311.

Gan J, Zhong M, Fan J et al. (2017). Abnormal white matter structural connectivity in adults with obsessive-compulsive disorder. Transl Psychiatry 7: e1062.

Garibotto V, Scifo P, Gorini A et al. (2010). Disorganization of anatomical connectivity in obsessive compulsive disorder: a multi-parameter diffusion tensor imaging study in a subpopulation of patients. Neurobiol Dis 37: 468–476.

Garland EF, Hartnell IJ, Boche D (2022). Microglia and astrocyte function and communication: what do we know in humans? Front Neurosci 16: 824888.

Gibson EM, Purger D, Mount CW et al. (2014). Neuronal activity promotes oligodendrogenesis and adaptive myelination in the mammalian brain. Science 344: 1252304.

Greene DJ, Williams Iii AC, Koller JM et al. (2017). Brain structure in pediatric Tourette syndrome. Mol Psychiatry 22: 972–980.

Greer JM, Capecchi MR (2002). Hoxb8 is required for normal grooming behavior in mice. Neuron 33: 23–34.

Guilarte TR, Rodichkin AN, McGlothan JL et al. (2022). Imaging neuroinflammation with TSPO: a new perspective on the cellular sources and subcellular localization. Pharmacol Ther 234: 108048.

Hashimoto N, Nakaaki S, Kawaguchi A et al. (2014). Brain structural abnormalities in behavior therapy-resistant obsessive-compulsive disorder revealed by voxel-based morphometry. Neuropsychiatr Dis Treat 10: 1987–1996.

Hong KE, Ock SM, Kang MH et al. (2002). The segmented regional volumes of the cerebrum and cerebellum in boys with Tourette syndrome. J Korean Med Sci 17: 530–536.

Huang Y, Li T, Wang Y et al. (2004). Linkage disequilibrium analysis of polymorphisms in the gene for myelin oligodendrocyte glycoprotein in Tourette's syndrome patients from a Chinese sample. Am J Med Genet B Neuropsychiatr Genet 124B: 76–80.

Jia YF, Wininger K, Peyton L et al. (2021). Astrocytic glutamate transporter 1 (GLT1) deficient mice exhibit repetitive behaviors. Behav Brain Res 396: 112906.

Jindachomthong K, Yang C, Huang Y et al. (2022). White matter abnormalities in the Hdc knockout mouse, a model of tic and OCD pathophysiology. Front Mol Neurosci 15: 1037481.

Johnson M, Fernell E, Gillberg C et al. (2021). No neurochemical evidence of neuronal injury or glial activation in children with paediatric acute-onset neuropsychiatric syndrome. An explorative pilot study. World J Biol Psychiatry 22: 800–804.

Kalanithi PS, Zheng W, Kataoka Y et al. (2005). Altered parvalbumin-positive neuron distribution in basal ganglia of individuals with Tourette syndrome. Proc Natl Acad Sci USA 102: 13307–13312.

Kanaan AS, Gerasch S, Garcia-Garcia I et al. (2017). Pathological glutamatergic neurotransmission in Gilles de la Tourette syndrome. Brain 140: 218–234.

Kant R, Pasi S, Surolia A (2018). Auto-reactive Th17-cells trigger obsessive-compulsive-disorder like behavior in mice with experimental autoimmune encephalomyelitis. Front Immunol 9: 2508.

Karadottir R, Attwell D (2007). Neurotransmitter receptors in the life and death of oligodendrocytes. Neuroscience 145: 1426–1438.

Kataoka Y, Kalanithi PS, Grantz H et al. (2010). Decreased number of parvalbumin and cholinergic interneurons in the striatum of individuals with Tourette syndrome. J Comp Neurol 518: 277–291.

Kawikova I, Grady BP, Tobiasova Z et al. (2010). Children with Tourette's syndrome may suffer immunoglobulin A dysgammaglobulinemia: preliminary report. Biol Psychiatry 67: 679–683.

Koch K, Wagner G, Schachtzabel C et al. (2012). White matter structure and symptom dimensions in obsessive-compulsive disorder. J Psychiatr Res 46: 264–270.

Kohman RA, DeYoung EK, Bhattacharya TK et al. (2012). Wheel running attenuates microglia proliferation and increases expression of a proneurogenic phenotype in the hippocampus of aged mice. Brain Behav Immun 26: 803–810.

Krabbe G, Minami SS, Etchegaray JI et al. (2017). Microglial NFkappaB-TNFalpha hyperactivation induces obsessive-compulsive behavior in mouse models of progranulin-deficient frontotemporal dementia. Proc Natl Acad Sci USA 114: 5029–5034.

Kumar A, Williams MT, Chugani HT (2015). Evaluation of basal ganglia and thalamic inflammation in children with pediatric autoimmune neuropsychiatric disorders associated with streptococcal infection and Tourette syndrome: a positron emission tomographic (PET) study using 11C-[R]-PK11195. J Child Neurol 30: 749–756.

Laurin N, Wigg KG, Feng Y et al. (2009). Chromosome 5 and Gilles de la Tourette syndrome: linkage in a large pedigree and association study of six candidates in the region. Am J Med Genet B Neuropsychiatr Genet 150B: 95–103.

Lazaro L, Bargallo N, Castro-Fornieles J et al. (2009). Brain changes in children and adolescents with obsessive-compulsive disorder before and after treatment: a voxel-based morphometric MRI study. Psychiatry Res 172: 140–146.

Lazaro L, Calvo A, Ortiz AG et al. (2014). Microstructural brain abnormalities and symptom dimensions in child and adolescent patients with obsessive-compulsive disorder: a diffusion tensor imaging study. Depress Anxiety 31: 1007–1017.

Lennington JB, Coppola G, Kataoka-Sasaki Y et al. (2016). Transcriptome analysis of the human striatum in Tourette syndrome. Biol Psychiatry 79: 372–382.

Li Z, Ji W, Li D et al. (2014). Microstructural abnormality in left nucleus accumbens predicts dysfunctional beliefs in treatment-resistant obsessive-compulsive disorder. Med Sci Monit 20: 2275–2282.

Li Q, Zhao Y, Huang Z et al. (2021). Microstructural white matter abnormalities in pediatric and adult obsessive-compulsive disorder: a systematic review and meta-analysis. Brain Behav 11: e01975.

Liddelow SA, Guttenplan KA, Clarke LE et al. (2017). Neurotoxic reactive astrocytes are induced by activated microglia. Nature 541: 481–487.

Liu Y, Miao W, Wang J et al. (2013). Structural abnormalities in early Tourette syndrome children: a combined voxel-based morphometry and tract-based spatial statistics study. PLoS One 8: e76105.

Long H, Wang C, Ruan J et al. (2019). Gastrodin attenuates neuroinflammation in DOI-induce Tourette syndrome in rats. J Biochem Mol Toxicol 33: e22302.

Lui H, Zhang J, Makinson SR et al. (2016). Progranulin deficiency promotes circuit-specific synaptic pruning by microglia via complement activation. Cell 165: 921–935.

Macklin WB (2010). The myelin brake: when enough is enough. Sci Signal 3: pe32.

Maleki S, Chye Y, Zhang X et al. (2020). Neural correlates of symptom severity in obsessive-compulsive disorder using magnetization transfer and diffusion tensor imaging. Psychiatry Res Neuroimaging 298: 111046.

Martens LH, Zhang J, Barmada SJ et al. (2012). Progranulin deficiency promotes neuroinflammation and neuron loss following toxin-induced injury. J Clin Invest 122: 3955–3959.

Mattheisen M, Samuels JF, Wang Y et al. (2015). Genome-wide association study in obsessive-compulsive disorder: results from the OCGAS. Mol Psychiatry 20: 337–344.

Meyer J (2021). Inflammation, obsessive-compulsive disorder, and related disorders. Curr Top Behav Neurosci 49: 31–53.

Mitew S, Gobius I, Fenlon LR et al. (2018). Pharmacogenetic stimulation of neuronal activity increases myelination in an axon-specific manner. Nat Commun 9: 306.

Morer A, Chae W, Henegariu O et al. (2010). Elevated expression of MCP-1, IL-2 and PTPR-N in basal ganglia of Tourette syndrome cases. Brain Behav Immun 24: 1069–1073.

Muller-Vahl KR, Kaufmann J, Grosskreutz J et al. (2009). Prefrontal and anterior cingulate cortex abnormalities in Tourette syndrome: evidence from voxel-based morphometry and magnetization transfer imaging. BMC Neurosci 10: 47.

Murphy TK, Kurlan R, Leckman J (2010). The immunobiology of Tourette's disorder, pediatric autoimmune neuropsychiatric disorders associated with Streptococcus, and related disorders: a way forward. J Child Adolesc Psychopharmacol 20: 317–331.

Naaz S, Balachander S, Srinivasa Murthy N et al. (2022). Association of SAPAP3 allelic variants with symptom dimensions and pharmacological treatment response in obsessive-compulsive disorder. Exp Clin Psychopharmacol 30: 106–112.

Nagarajan N, Jones BW, West PJ et al. (2018). Corticostriatal circuit defects in Hoxb8 mutant mice. Mol Psychiatry 23: 1868–1877.

Neuner I, Kupriyanova Y, Stocker T et al. (2010). White-matter abnormalities in Tourette syndrome extend beyond motor pathways. Neuroimage 51: 1184–1193.

Neuner I, Kupriyanova Y, Stocker T et al. (2011). Microstructure assessment of grey matter nuclei in adult Tourette patients by diffusion tensor imaging. Neurosci Lett 487: 22–26.

Nissen JB, Hansen CS, Starnawska A et al. (2016). DNA methylation at the neonatal State and at the time of diagnosis: preliminary support for an association with the estrogen receptor 1, gamma-aminobutyric acid B receptor 1, and myelin oligodendrocyte glycoprotein in female adolescent patients with OCD. Front Psych 7: 35.

Petrelli F, Zehnder T, Laugeray A et al. (2023). Disruption of astrocyte-dependent dopamine control in the developing medial prefrontal cortex leads to excessive grooming in mice. Biol Psychiatry 93: 966–975.

Piantadosi SC, McClain LL, Klei L et al. (2021). Transcriptome alterations are enriched for synapse-associated genes in the striatum of subjects with obsessive-compulsive disorder. Transl Psychiatry 11: 171.

Piras F, Piras F, Caltagirone C et al. (2013). Brain circuitries of obsessive compulsive disorder: a systematic review and meta-analysis of diffusion tensor imaging studies. Neurosci Biobehav Rev 37: 2856–2877.

Pittenger C, Bloch MH, Williams K (2011). Glutamate abnormalities in obsessive compulsive disorder: neurobiology, pathophysiology, and treatment. Pharmacol Ther 132: 314–332.

Plessen KJ, Gruner R, Lundervold A et al. (2006). Reduced white matter connectivity in the corpus callosum of children with Tourette syndrome. J Child Psychol Psychiatry 47: 1013–1022.

Rapanelli M, Frick LR, Horn KD et al. (2016). The histamine H3 receptor differentially modulates mitogen-activated protein kinase (MAPK) and Akt signaling in striatonigral and striatopallidal neurons. J Biol Chem 291: 21042–21052.

Rodriguez CI, Bender Jr J, Marcus SM et al. (2010). Minocycline augmentation of pharmacotherapy in obsessive-compulsive disorder: an open-label trial. J Clin Psychiatry 71: 1247–1249.

Roessner V, Overlack S, Schmidt-Samoa C et al. (2011). Increased putamen and callosal motor subregion in treatment-naive boys with Tourette syndrome indicates changes in the bihemispheric motor network. J Child Psychol Psychiatry 52: 306–314.

Rosso IM, Olson EA, Britton JC et al. (2014). Brain white matter integrity and association with age at onset in pediatric obsessive-compulsive disorder. Biol Mood Anxiety Disord 4: 13.

Saito Y, Nobuhara K, Okugawa G et al. (2008). Corpus callosum in patients with obsessive-compulsive disorder: diffusion-tensor imaging study. Radiology 246: 536–542.

Scott-Hewitt N, Huang Y, Stevens B (2023). Convergent mechanisms of microglia-mediated synaptic dysfunction contribute to diverse neuropathological conditions. Ann N Y Acad Sci.

Singh K, Chang C, Gershwin ME (2014). IgA deficiency and autoimmunity. Autoimmun Rev 13: 163–177.

Soto JS, Jami-Alahmadi Y, Chacon J et al. (2023). Astrocyte-neuron subproteomes and obsessive-compulsive disorder mechanisms. Nature 616: 764–773.

Squarzoni P, Oller G, Hoeffel G et al. (2014). Microglia modulate wiring of the embryonic forebrain. Cell Rep 8: 1271–1279.

Stewart SE, Platko J, Fagerness J et al. (2007). A genetic family-based association study of OLIG2 in obsessive-compulsive disorder. Arch Gen Psychiatry 64: 209–214.

Tao Q, Dang J, Niu X et al. (2023). White matter microstructural abnormalities and gray matter volume alterations in obsessive-compulsive disorder: a coordinate-based meta-analysis. J Affect Disord 320: 751–761.

Thomas Broome S, Louangaphay K, Keay KA et al. (2020). Dopamine: an immune transmitter. Neural Regen Res 15: 2173–2185.

Tinker SC, Bitsko RH, Danielson ML et al. (2022). Estimating the number of people with Tourette syndrome and persistent tic disorder in the United States. Psychiatry Res 314: 114684.

Traiffort E, Kassoussi A, Zahaf A et al. (2020). Astrocytes and microglia as major players of myelin production in normal and pathological conditions. Front Cell Neurosci 14: 79.

Trankner D, Boulet A, Peden E et al. (2019). A microglia sublineage protects from sex-linked anxiety symptoms and obsessive compulsion. Cell Rep 29: 791–799.e3.

Trifiletti R, Lachman HM, Manusama O et al. (2022). Identification of ultra-rare genetic variants in pediatric acute onset neuropsychiatric syndrome (PANS) by exome and whole genome sequencing. Sci Rep 12: 11106.

Ueno M, Fujita Y, Tanaka T et al. (2013). Layer V cortical neurons require microglial support for survival during postnatal development. Nat Neurosci 16: 543–551.

Van Deren DA, De S, Xu B et al. (2022). Defining the Hoxb8 cell lineage during murine definitive hematopoiesis. Development 149.

van Passel R, Schlooz WA, Lamers KJ et al. (2001). S100B protein, glia and Gilles de la Tourette syndrome. Eur J Paediatr Neurol 5: 15–19.

Vieira-Fonseca T, Fontenelle LF, Kohlrausch FB (2019). OLIG2 gene polymorphisms are associated with nasty, unpleasant and uncontrollable thoughts in obsessive-compulsive disorder. J Clin Neurosci 70: 202–207.

Wang P, Zhao Q, Xu T et al. (2020). Interaction between PGRN gene and the early trauma on clinical characteristics in patients with obsessive-compulsive disorder. J Affect Disord 263: 134–140.

Wang X, Liu X, Chen L et al. (2023). The inflammatory injury in the striatal microglia-dopaminergic-neuron crosstalk involved in Tourette syndrome development. Front Immunol 14: 1178113.

Welch JM, Lu J, Rodriguiz RM et al. (2007). Cortico-striatal synaptic defects and OCD-like behaviours in Sapap3-mutant mice. Nature 448: 894–900.

Wen H, Liu Y, Wang J et al. (2016). Combining tract- and atlas-based analysis reveals microstructural abnormalities in early Tourette syndrome children. Hum Brain Mapp 37: 1903–1919.

Worbe Y, Marrakchi-Kacem L, Lecomte S et al. (2015). Altered structural connectivity of cortico-striato-pallido-thalamic networks in Gilles de la Tourette syndrome. Brain 138: 472–482.

Xi L, Ji X, Ji W et al. (2022). Jing-an oral liquid alleviates Tourette syndrome via the NMDAR/MAPK/CREB pathway in vivo and in vitro. Pharm Biol 60: 1790–1800.

Xu M, Kobets A, Du JC et al. (2015a). Targeted ablation of cholinergic interneurons in the dorsolateral striatum produces behavioral manifestations of Tourette syndrome. Proc Natl Acad Sci USA 112: 893–898.

Xu M, Li L, Ohtsu H et al. (2015b). Histidine decarboxylase knockout mice, a genetic model of Tourette syndrome, show repetitive grooming after induced fear. Neurosci Lett 595: 50–53.

Xu M, Li L, Pittenger C (2016). Ablation of fast-spiking interneurons in the dorsal striatum, recapitulating abnormalities seen post-mortem in Tourette syndrome, produces anxiety and elevated grooming. Neuroscience 324: 321–329.

Xu J, Liu RJ, Fahey S et al. (2021). Antibodies from children with PANDAS bind specifically to striatal cholinergic interneurons and alter their activity. Am J Psychiatry 178: 48–64.

Yang C, Yao L, Liu N et al. (2021). Microstructural abnormalities of white matter across Tourette syndrome: a voxel-based meta-analysis of fractional anisotropy. Front Neurol 12: 659250.

Yoo SY, Jang JH, Shin YW et al. (2007). White matter abnormalities in drug-naive patients with obsessive-compulsive disorder: a diffusion tensor study before and after citalopram treatment. Acta Psychiatr Scand 116: 211–219.

Yu X, Taylor AMW, Nagai J et al. (2018). Reducing astrocyte calcium signaling in vivo alters striatal microcircuits and causes repetitive behavior. Neuron 99: 1170–1187.e9.

Zai G, Bezchlibnyk YB, Richter MA et al. (2004). Myelin oligodendrocyte glycoprotein (MOG) gene is associated with obsessive-compulsive disorder. Am J Med Genet B Neuropsychiatr Genet 129B: 64–68.

Zai G, Arnold PD, Richter MA et al. (2021). An association of myelin oligodendrocyte glycoprotein (MOG) gene variants with white matter volume in pediatric obsessive-compulsive disorder. Psychiatry Res Neuroimaging 307: 111231.

Zai G, Zai CC, Arnold PD et al. (2023). White matter volumeand myelin oligodendrocyte glycoprotein (MOG) microsatellites in pediatric obsessive-compulsive disorder. Psychiatr Genet 33: 160–163.

Zarei M, Mataix-Cols D, Heyman I et al. (2011). Changes in gray matter volume and white matter microstructure in adolescents with obsessive-compulsive disorder. Biol Psychiatry 70: 1083–1090.

Zhang X, Liu J, Guo Y et al. (2015). Association study between oligodendrocyte transcription factor 2 gene and Obsessive-compulsive disorder in a Chinese Han population. Depress Anxiety 32: 720–727.

Zhao L, Cheng N, Sun B et al. (2020). Regulatory effects of Ningdong granule on microglia-mediated neuroinflammation in a rat model of Tourette's syndrome. Biosci Trends 14: 271–278.

Zheng J, Frankovich J, McKenna ES et al. (2020). Association of pediatric acute-onset neuropsychiatric syndrome with microstructural differences in brain regions detected via diffusion-weighted magnetic resonance imaging. JAMA Netw Open 3: e204063.

Zhong Z, Yang X, Cao R et al. (2019). Abnormalities of white matter microstructure in unmedicated patients with obsessive-compulsive disorder: changes after cognitive behavioral therapy. Brain Behav 9: e01201.

Zhongling K, Yanhui C, Guofeng C et al. (2022). Neuroinflammation in a rat model of Tourette syndrome. Front Behav Neurosci 16: 710116.

Ziv Y, Ron N, Butovsky O et al. (2006). Immune cells contribute to the maintenance of neurogenesis and spatial learning abilities in adulthood. Nat Neurosci 9: 268–275.

Zuchner S, Wendland JR, Ashley-Koch AE et al. (2009). Multiple rare SAPAP3 missense variants in trichotillomania and OCD. Mol Psychiatry 14: 6–9.

Handbook of Clinical Neurology, Vol. 210 (3rd series)
Neuroglia in Neurologic and Psychiatric Disorders, Part II
A. Verkhratsky, L.D. de Witte, E. Aronica, and E.M. Hol, Editors
https://doi.org/10.1016/B978-0-443-19102-2.00008-9

Chapter 20

Neuroglia in anxiety disorders

ROBIN E. BONOMI, ROBERT PIETRZAK, AND KELLY P. COSGROVE*

Department of Psychiatry, Yale University, New Haven, CT, United States

Abstract

Anxiety disorders are some of the most prevalent in the world and are extraordinarily debilitating to many individuals, costing millions in disability. One of the most disabling is posttraumatic stress disorder (Snijders et al., 2020). Understanding the pathophysiology of these illnesses further and the cell types involved will allow better targeting of treatments. Glial cells, encompassing microglia, astrocytes, and oligodendrocytes, play critical roles in the pathophysiology of PTSD and other anxiety illnesses. Each of these cell types interacts with aspects of neuro-epigenetics, neuroimmune, and neuronal signaling and may contribute to the pathophysiology of anxiety illnesses. This chapter covers the literature on the role of glial cells in the neurobiology and pathology of anxiety disorders, more specifically PTSD. PTSD is one of the most debilitating anxiety disorders and one of the most complicated from a neurobiologic perspective. This chapter also features a discussion surrounding the current state of treatment and some of the hypothesized mechanisms for novel treatments including tetrahydrocannabidiol and 3,4-methylenedioxymethamphetamine. Finally, thoughts on the future directions for precision treatment and pharmacologic development with a focus on neuroglia are undertaken.

INTRODUCTION

Anxiety disorders are among the most prevalent mental illnesses worldwide, and especially so within the United States and Western European countries. When accounting for differences in diagnosis, the prevalence ranges from 2% to 29% (Baxter et al., 2013). In the United States, the Diagnostic and Statistical Manual of Mental Disorders 5 (DSM 5) (Kilpatrick et al., 2013) is the primary text used for classifying, characterizing, and providing diagnostic criteria for mental disorders. Anxiety disorders are characterized as a generalized feeling of nervousness, restlessness, irritability, disturbed sleep, and difficulty in concentrating; an individual must meet three of these criteria for 6 months prior to diagnosis. Specific anxiety disorders can be focused on in an individual area of a person's life (i.e., agoraphobia, claustrophobia). Together, these disorders are a major contributor to disability, as well as an economic burden throughout the world, and may cost the global economy up to $1 trillion per year (Kessler et al., 2005). Recently, posttraumatic stress disorder (PTSD) (Snijders et al., 2020), characterized by intrusive fear and fear-conditioned behaviors with an inability to extinguish fear-based memories, has been reclassified to be included in the anxiety disorder spectrum within the DSM 5 (Davidson et al., 2004; Adamou et al., 2007). PTSD is defined as a condition that occurs 6 or more months after experiencing a significant, life-threatening trauma, either directly, by witnessing it, or by learning that it happened to a loved one that includes major symptomatology of hyperarousal, irritability, flashbacks, nightmares, and avoidance (Kessler et al., 1995, 2005). While approximately 70% of people in the United States report exposure to a traumatic event, the prevalence of PTSD is ~7%–9% (Kilpatrick et al., 2013). Certain individuals may be more prone to develop PTSD, and many anxiety and PTSD phenotypes are influenced by psychosocial stressors. Current first-line treatment includes serotonin reuptake inhibitors (SSRI) and serotonin-norepinephrine reuptake inhibitors (SNRIs) that have only 50% efficacy (Williams et al., 2022). Recently, studies have indicated that the neuroimmune system may play a critical role in the neurobiology of PTSD opening up new avenues for treatment development.

The neurobiology of PTSD is still unclear. Animal models and some clinical studies suggest that PTSD arises from a failure in memory consolidation after trauma. This creates memories that are difficult to extinguish attached to elevated emotions. These memories can lead to difficulty with over-reaction-cued fear responses in the future. Although there is more to learn about the brain's biologic response to trauma, this circuitry is thought to be made up of four major centers of the brain including the hippocampus, insula, amygdala, and prefrontal cortex (PFC). Within this circuitry, neuroglial cells are a key component of stress-related changes contributing

*Correspondence to: Kelly P. Cosgrove, Department of Psychiatry, Yale University, New Haven, CT 06511, United States. Tel: +1-203-737-6969, E-mail: kelly.cosgrove@yale.edu

to neuroplasticity and neuroinflammation, with evidence for microglia, the brain's primary immune cells, changing morphology and function in the corticolimbic brain regions following anxiety and depressive symptoms (Rohan Walker et al., 2013). There is also evidence that repeated stress exposure leads to a dysregulation of neuroimmune cells that may be implicated in PTSD (Beumer et al., 2012). Furthermore, it has been hypothesized that the microglia play a key role in the link between peripheral and central immune responses to stress. Specifically, through reactive microgliosis and recruitment or decruitment as evidenced by neuroimaging studies (PET) labeling of translocator protein (TSPO), a surrogate marker for microglia (Setiawan et al., 2015; Bhatt et al., 2020).

Significant research in the last 20 years has advanced our understanding of the brain after trauma with regard to neurons, neuroglia, and neuroimmune changes. Connections between neuroglial cells and immune modulation will also be discussed. Data from the living human brain in the neurobiology of stress and PTSD has thus far been relatively limited and focused more on postmortem brain tissue examination, therefore, relevant preclinical studies will be discussed to inform the clinical literature. Last, future steps and potential areas for treatment development will be examined.

OVERVIEW OF NEUROGLIA

In the brain, there are multiple cell types with varying roles, particularly within neuroglia, which include microglia, astrocytes, oligodendrocytes, oligodendrocyte progenitor cells (OPC), and radial glia (progenitor cells). Each cell type serves an multiple functions in the overall regulation and support of the neuroimmune system and neuronal signaling. In most cases, neuroglial cells serve dynamic roles, which may be altered in disease states (Allen and Lyons, 2018), contributing to neuroimmune and neuronal dysfunction. Within disease states such as PTSD, microglia and astrocytes are the most well studied with increasing investigation into oligodendrocytes (Machelska and Celik, 2016). Microglia primarily act to release soluble factors for blood–brain barrier stability, neuroimmune response, and vascularization, and to serve as the primary phagocytic body for the brain. Thus, these cells are primarily responsible for synaptic pruning (Tremblay et al., 2011). Astrocytes are cells with complex, often spongiform, morphology, which dynamically regulate neurotransmitter uptake by altering the neurotransmitter concentration available at synapses. Therefore, increases in astrocytes would increase the ensheathment by astrocytes and decrease the diffusion of the transmitter away from the synapse (Allen, 2014; Araque et al., 2014). Astrocytes have been shown to be necessary for plasticity and memory in their ability to induce *N*-methyl-D-aspartate (NMDA)-dependent long-term potentiation within the hippocampus (Adamsky et al., 2018). The oligodendrocytes primarily serve to regulate myelin sheaths surrounding the axon of the neuron to regulate the signal conduction velocity. OPCs and microglia also regulate synaptic function together with astrocytes. The OPCs may aid in controlling postsynaptic NMDA and α-amino-3-hydroxy-5-methyl-4-isoxazolepropionic acid receptor-mediated currents. Thus, these cells function to regulate the synaptic conduction and therefore connectivity within the brain. Radial glia are neuroprogenitor cells, originally arising from the neural tube neuroepithelial cells that have acquired astroglia-like characteristics (Miranda-Negrón and García-Arrarás, 2022). Together, these cells help regulate memory and fear circuitry that may be dysregulated in disorders such as PTSD.

PRECLINICAL ANIMAL MODELS OF STRESS AND ANXIETY

Due to the limitations of studying human brain pathology in vivo, many animal models have been developed to study PTSD and anxiety phenotypes (Campos et al., 2013; Zhang et al., 2019). The models characterize anxiety symptoms of isolation, anhedonia, hypervigilance, and increased fear response. Models have traditionally focused primarily on males, though there is increasing development of stress models for female animals. The primary models include chronic social defeat stress (CSDS) (Erburu et al., 2015; Harris et al., 2018; Van Doeselaar et al., 2021), single prolonged stress (SPS) (Knox et al., 2010; Eagle et al., 2013; Lisieski et al., 2018), and predator scent (Albrechet-Souza and Gilpin, 2019).

The predominant models mentioned earlier serve to highlight different features of PTSD. Notably, CSDS displays more of the depressive characteristics of PTSD including withdrawn animals and anhedonia features, as well as an altered hypothalamic–pituitary axis (Campos et al., 2013; Van Doeselaar et al., 2021). In this model, the animal is repeatedly exposed to stress in the form of an aggressor rodent in the cage with them for periods of time. The SPS model demonstrates a nice correlation with single event trauma as it is comprised of a 3-h trauma including forced swimming, restraint, and exposure to noxious stimuli, and produces a significant portion of the anxiety phenotype for PTSD including hypervigilance, increased arousal, and decreased appetite (Lisieski et al., 2018). The predator scent model is another robust PTSD model wherein a mouse or rat is exposed to the scent of a predator, such as a fox or cat, for a prolonged period of time in their home cage. This potentiates a significant stressful event, especially when repeated, and produces a mixture of depressive and anxious phenotypes of PTSD (Verbitsky et al., 2020). Herein, these models have been used in various studies to uncover brain pathology in stress disorders.

OVERVIEW OF SPECIFIC BRAIN REGIONS INVOLVED IN PTSD

Within the primary fear memory learning circuits involved in PTSD and anxiety, i.e., hippocampus, amygdala, and PFC, neuroglia play integral roles in the progression and subsequent reversal of disease pathology. PTSD is thought to arise from a failure of memory consolidation and reconsolidation after trauma. This process renders the memory inextinguishable from long-term memory and fear responses (Kida, 2019). After trauma occurs and is brought into the brain, the memory is processed through the amygdala to the PFC with concurrent input from the hippocampus. Through the amygdala and hippocampus, a new stable memory is formed, and when it is cued, the memory is brought up and temporarily destabilized. There are two pathways for this memory: (1) to reconsolidate the memory through a CREB-dependent gene expression pathway and continue to hold this memory as it is; (2) the memory can undergo extinction through the PFC and amygdala (Mamiya et al., 2009; Kida, 2019). Therefore, the destabilization of a memory is a pivotal point at which time the conditioned fear response to this memory can be further solidified or rewritten. This process is primarily driven by glutamatergic signaling changes (Duek et al., 2019) as demonstrated by the use of NMDA receptor agonists and nitric oxide antagonists to treat animals after fear extinction conditioning. The results revealed that learning and memory dysfunction could be induced by increasing glutamatergic response and decreasing nitric oxide, which also led to decreases in the expression of glial fibrillary acidic protein, a marker of astrocytes (Li et al., 2022b). From these changes, synaptic pruning and restructuring of neuronal density often occur. The majority of the literature has focused on microglial, oligodendrocytes, and astrocytes and their roles within hippocampus, amygdala, and PFC. Herein, we will focus predominantly on these cells and the evidence that currently exists in the literature to support their role in the neuropathology of PTSD.

MICROGLIA

Microglial cells have demonstrated important roles in the regulation of fear memory learning through their role as prime mediators of synaptic pruning and neuroimmune signaling (Allen and Lyons, 2018). They regulate the synaptic pruning that occurs during normal learning and memory functions through the complement pathway and chemokine signaling, such as CX3CR1 (Cornell et al., 2022). Through their release of multiple soluble proteins, microglia function to activate the complement cascade and signal for destruction of synapses no longer needed. Complement proteins such as C1q and CR3 allow microglia to interact with neurons and mediate neuronal functioning (Hong et al., 2016). Their interactions with brain-derived neurotrophic factors are permissive for brain development and maturation, as well as synaptic plasticity in the mature brain. Dysfunctional microglial signaling can contribute to learning and memory deficits, through molecular and cellular mechanisms in the proinflammatory pathways (Cornell et al., 2022) (Fig. 20.1).

Deficits in synaptic plasticity, specifically the loss of synapses, are present in PTSD, particularly when it is comorbid with depressive disorders. These diseases have demonstrated significant changes in the dendritic spine density (Holmes et al., 2019) and the neuronal restructuring of diseased individuals. PTSD has been described as a "synaptic disconnection syndrome" to highlight the possibility for ketamine as a therapeutic, synaptogenic, agent for PTSD (Krystal et al., 2017) and with the upcoming 3,4-methylenedioxymethamphetamine (MDMA)-assisted therapy (Mitchell et al., 2021). The areas of the brain that are most affected by synaptic loss and restructuring include the hippocampus, nucleus accumbens, amygdala, and caudate (Krystal et al., 2017). It is possible that synaptic loss and compromised microglia, particularly within the corticolimbic connections of the brain are responsible for large portions of PTSD pathophysiology (Koch et al., 2016). This could be tied to a primary role of microglia to physically remove synapses and alter

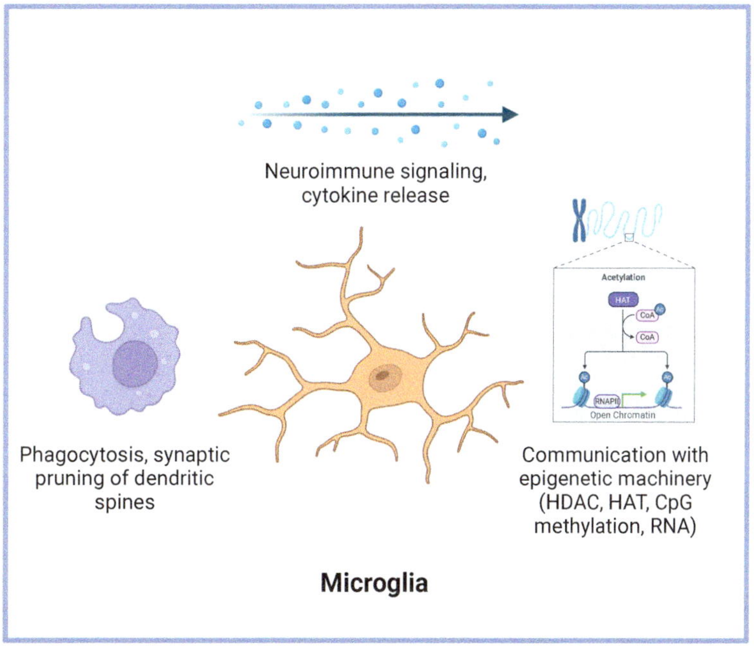

Fig. 20.1. Microglia play important roles in maintaining homeostasis of dendritic spine density through chemokine signaling and phagocytosis, this is one of the primary mechanisms of action for some of the new treatments for PTSD. Studies have demonstrated microglia dysfunction leading to neuroimmune dysfunction and synaptic spine loss in subjects with PTSD. Epigenetic machinery interacts with microglia to regulate genomic transcription and alter cellular plasticity following a traumatic event. When traumatic memories are remembered and repackaged it is hypothesized that microglia become less responsive over time and therefore do not respond with appropriate chemokine signaling to successfully repackage a memory with less fear response. This dysfunction may also play a role in a reduced neuroimmune response in individuals with PTSD.

neuronal activity (Wohleb, 2016). Therefore, understanding the role that microglia play in altering synapse structure may help to better understand and treat the ensuing pathology.

Rats exposed to the predator scent stress model demonstrated reductions in the chemokine receptor 1 (CX3CR1) within the hippocampus of animals that were more behaviorally disrupted (PTSD-like phenotype) as compared to those that were not. These animals displayed increased ionized calcium binding adaptor molecule (IBA1) as well, indicating changes in the microglial reactivity in the hippocampus (Nahum et al., 2022). The ratio of microglia to immunocytes and the sheer number of microglia have increased with foot shock-induced PTSD, particularly within the hippocampus (Li et al., 2021). Some of these preclinical studies have indicated that poststress hippocampal-dependent cognitive decline may be correlated with blood–brain barrier impairment secondary to microglial dysfunction (Ni et al., 2022).

Microglia and epigenetics

It has also been thought that microglia play a key role in the connection between cell signaling and plastic epigenetic changes. Memories that are traumatic and intrusive need to be repackaged within the memory circuitry on a dynamic basis. Much of this repackaging occurs on the epigenetic level. While there are many forms of epigenetics that play a role in brain function and subsequent pathology including DNA methylation (DNAm), long noncoding RNA, small interfering RNA, microRNA, and piwi-interacting RNA, histone acetylation and transcription factor acetylation are the most common (Whittle and Singewald, 2014; Kim et al., 2018; Blacker et al., 2019) and the most well studied in the PTSD literature is DNAm.

One of the primary regulators of microglia is in part the epigenetic machinery of the brain. Histone deacetylases (HDACs) are one primary type of epigenetics thought to play a key role in both glial cell regulation and neuronal regulation in the pathologic response to trauma. It is key to understand these changes within the glial cells to better identify areas for intervention. These enzymes are classically known for their ability to remove acetyl residue from histone core protein tails. Some of the HDACs are critical for cellular differentiation into astroglia lineage rather than neuronal lineage (Prozorovski et al., 2008). Sirtuin1, a class III HDAC, is one of the primary regulators for mouse neural progenitor cells (Prozorovski et al., 2008). Another example of HDAC involvement is through a class I HDAC, HDAC2, wherein binding to glial-derived neurotrophic factor promoter decreases glial-derived neurotrophic factor expression in the nucleus accumbens in rodent models (Cheng et al., 2011). Furthermore, decreasing HDAC2 availability attenuated some of the stress-induced depression-like phenotype seen in these mice. Therefore, it is hypothesized that the response of HDAC2 to glial cell growth and differentiation may play a role in the behavior changes observed after stress (Sun et al., 2019). HDAC6, a class IIB HDAC, typically associated with microtubule deacetylation, has been found to play a prominent role in regulating the phenotype of mutant vs wild-type GFAP in certain neurologic disorders (Melchionda et al., 2013). It is hypothesized that HDAC6 may be important in regulating and supporting glial fibrillary acidic protein production in neuronal cells of the healthy brain, thereby increasing glial cell proliferation (Melchionda et al., 2013). HDACs class IIA (namely HDAC 4 and 5 due to their relative abundance in the brain) demonstrated significant roles in the microglial inflammatory response following a trauma-induced anxiety phenotype in a rodent model (Kamal et al., 2022). Using longitudinal imaging with an HDAC class IIA selective radiotracer, traumatic-brain-injury-induced depression of HDAC class IIA expression correlated with changes in cellular localization of these enzymes and thus changes in histone acetylation levels. These changes have been correlated to the increased microglia inflammatory response following traumatic brain injury as well (Zhang et al., 2008). Furthermore, in trauma models, systemic inflammatory response syndrome is frequently observed after a traumatic injury. The response from injured tissues to release large amounts of inflammatory mediators began to be an independent predictor of mortality after trauma. This lead to the beginning of an understanding of how blockade for endogenous substances might improve clinical outcomes in patients after trauma (Manson et al., 2011). These data also began to demonstrate links to glial changes and reactive microgliosis causing the rise in inflammatory cytokines. Therefore, research began to focus on a possible link between inflammation and prediction of PTSD sequelae. The link from systemic inflammation to neuroimmune signaling is an area of active and ongoing study.

HDAC inhibition, which increases levels of histone core protein acetylation, has demonstrated a modulating effect on the microglia changes posttrauma, which alters the anxiety phenotype (Donat et al., 2017). These effects are thought to be modulated through the PTEN/Akt axis (Wang et al., 2015). The HDAC-associated effects may also be linked to the endocannabinoid effects that are induced during trauma, causing increasing levels of MKP-1 (Eljaschewitsch et al., 2006) thus attenuating damage through activation of CB-1 receptors. Published data suggest that microglia are the primary mediators of endocannabinoid signaling with high expression of CB-1 and CB-2 receptors within the brain, specifically within the brainstem (Finn et al., 2021).

Epigenome-wide association studies (Snijders et al., 2020) examining individuals with PTSD as compared to trauma-exposed healthy control subjects found the gene with the highest impact to be MAD1L1, a primary gene involved in the mitotic spindle assembly used in cellular replication (Snijders et al., 2020). Another EWAS study also demonstrated a positive correlation between DNAm in individuals with PTSD at cg19534438 in the gene G0S2 (Logue et al., 2020). These are further evidence that epigenetics and DNAm changes may play a significant role in the pathology of PTSD. In pediatric subjects with PTSD, methylome-wide studies found methylation differences from healthy controls on nine genes including those in glucocorticoid functioning, with the most predominant in the OLFM3 (a microglial marker) gene in the hippocampus (Ensink et al., 2021), which correlated with PTSD symptoms. This is particularly important because the molecular pathways highlighted in this study point to changes in the stress response, neuroplasticity, and changes in the microglial cell response.

Microglia and the neuroimmune system

More recently, studies have demonstrated links between the immune system and mental health. PTSD and major depressive disorder (MDD) are both associated with immune system dysfunction. This has primarily been demonstrated through relationships between clinical symptoms and peripheral immune markers in individuals managing these disorders as compared to healthy controls. Many of these studies have focused on primary proinflammatory or antiinflammatory cytokines released by the white blood cells in the body during an immune response. One of the most widely studied examples is C-reactive protein (CRP), which is a systemic inflammatory marker and indicator of immune reactivity. Increased levels of systemic CRP were linked to trauma exposure in human subjects and correlated to an increased risk of developing PTSD and worsening PTSD symptoms (Eraly et al., 2014; O'donovan et al., 2017). In addition to CRP, many trauma-exposed individuals with PTSD symptoms have increased levels of IL-6 and IL-1B as compared to trauma-exposed individuals without PTSD and healthy controls (Baker et al., 2001; Lerman et al., 2016). Though individuals with PTSD have increased in some specific systemic inflammatory markers, it is not clear whether these inflammatory changes are consequential to PTSD or influence a person's risk of developing PTSD. Additionally, these findings are not consistent as some studies have demonstrated lower levels of IL-6 and CRP in patients with PTSD. More information on this area is currently needed (Hori and Kim, 2019).

The positron emission tomography radiotracer [^{11}C]PBR28, which binds to 18-kDa TSPO a microglial biomarker, has been used to probe the neuroimmune system in many psychiatric disorders including PTSD (Bhatt et al., 2020). Interestingly, and contrary to what was hypothesized, this microglial marker was significantly lower in those with PTSD as compared to the levels of healthy controls (Bhatt et al., 2020). Additionally, elevated peripheral inflammatory markers such as CRP were also correlated with lower prefrontal–limbic microglia levels. This pivotal study in human subjects demonstrated a quantitative link between a marker of microglial and PTSD. This study posits that while initially the neuroimmune system may be reactive to, e.g., trauma, over time, with chronic stress, there may be a subsequent downregulation in individuals with PTSD. There may be increased systemic inflammation with a suppressed neuroimmune profile.

OLIGODENDROCYTES

The myelination of axons within the central nervous system is necessary for cognition, sensation, and motor function. Myelination functions as an insulation for the neuronal system allowing electrical signals to flow between neuronal axons with less impedance. Many neurodegenerative and psychiatric illnesses are connected to changes in myelination and thus deficits of oligodendrocytes, the primary cells responsible for myelination in the CNS. While some studies have demonstrated this link, it is difficult to fully elucidate structural changes and causation due to the limited ability to study human neurobiology in vivo. Therefore, animal models offer a useful construct for studying this correlation. A full comprehensive review of the myelin changes in animal models of anxiety has been published by Antontseva et al. (2020), which will be briefly reviewed here as well (Antontseva et al., 2020) (Fig. 20.2).

Oligodendrocytes in preclinical models

Chronic unpredictable stress has been widely used to model a depression-like phenotype with anhedonia, agitation, psychomotor retardation, and sleep disturbances (Van Rijn et al., 2017). Following this stress, rodents have a significant decrease in myelination levels in several brain regions (Xiao et al., 2018). There are some benefits and some ability to reverse these changes using physical exercise in rodents, whereas antidepressants, such as fluoxetine, did not demonstrate any ability to rescue these changes (Xiao et al., 2018). Another model, CSDS, causes animals to exhibit more social avoidance behavior or absence of stress effects, thus producing a bifurcation of the phenotype into susceptible and resilient animals. Within the susceptible mice after 10 days, there was a significant decrease in the myelination levels of the nucleus accumbens and the medial PFC (Bonnefil et al., 2019). However, some of these changes appeared to be regionally specific as no changes were seen in the ventral hippocampus as opposed to many myelin-related genes downregulated in the medial and basolateral nuclei of the amygdala. The social isolation stress model in rodents produces a mix of depressive and anxiety phenotypes, similar to that of PTSD (Ieraci et al., 2016; Mumtaz et al., 2018). After 8 weeks of isolation, animals had social withdrawal behaviors and thinner myelin sheaths in the PFC, however, with no difference in the numbers of oligodendrocytes when compared to the controls. However, these animals did demonstrate greater histone acetylation and therefore a lower portion of heterochromatin, pointing to more immature cells (Liu et al., 2012). These alterations were correlated with the downregulation of two myelin-related genes myelin basic protein and myelin-associated oligodendrocyte basic protein, in the PFC. However, none of these changes were seen in other brain regions outside of the PFC, such as the nucleus accumbens, cingulate cortex, and cerebellum (Liu et al., 2012).

Another notable finding is that the structural integrity of neurons is closely linked to the myelin structure and thus the oligodendrocyte function within a specific region of the brain. Within the ventrolateral thalamus, a primary switchboard for neurocircuitry pathways, structural atrophy can be visualized after extreme stress in rats. This change indicates significant neuronal and glial cell death in this area (Yoshii et al., 2017) likely including both microglia and oligodendrocytes. Similar findings have been seen in the hippocampus following SPS in rats, up to 14 days poststress, which could be correlated with decreased performance on hippocampal-dependent cognitive tasks (Han et al., 2015). Together these findings support that myelinated axon density plays a crucial role in stress-related memory and functional changes.

Oligodendrocytes in clinical PTSD

Clinically, myelination changes and oligodendrocyte function have also been linked to PTSD symptomatology. An increased level of

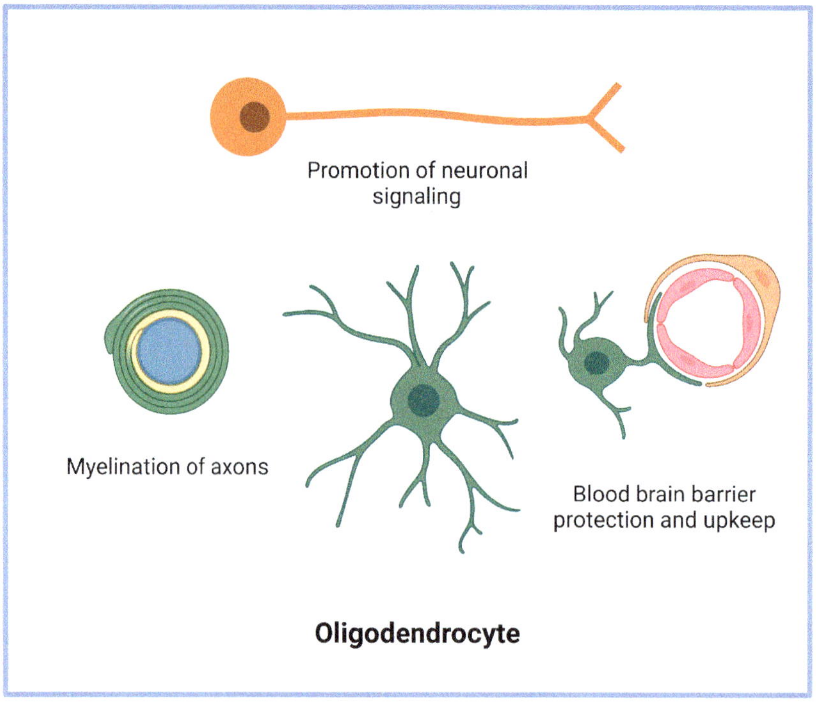

Fig. 20.2. Oligodendrocytes are well known as the myelinators of the central nervous system. However, their link to disease and the link to myelin-associated dysfunction in psychiatric illness is becoming increasingly well known. Breakdown in myelin due to oligodendrocyte suppression is correlated with worsening PTSD, particularly in cases with traumatic brain injury. Furthermore, the resilience and neuronal response to stress relies heavily on the structural integrity of its myelin. Stress may trigger a breakdown in the neuronal signaling and microglial environment thus altering the integrity of myelin sheaths and oligodendrocyte function. This may feedback to further slowed and dysfunctional limbic pathways involved in the pathophysiology of PTSD.

oligodendrocyte density (as quantified by myelin water fraction in magnetic resonance imaging) was positively correlated with PTSD symptoms across all regions of interest within the brain (Jak et al., 2020). Translational studies have also confirmed oligodendrocyte changes and myelin basic protein content in areas of the brain such as hippocampal dentate gyrus, which were positively correlated with stress-induced avoidance behaviors in rats as measured by avoidance, startle, and fear learning behavior. Furthermore, in a group of trauma-exposed veterans, the hippocampal oligodendrocyte data match the rats (Long et al., 2021). In adults with histories of childhood abuse, there are substantial differences in axonal integrity and myelination of small-diameter axon fibers, which also correlate to decreased expression of myelin-related genes (Lutz et al., 2014). Other results point to decreased numbers of mature oligodendrocytes in the ventromedial PFC white matter in individuals who died by suicide as compared to healthy controls (Tanti et al., 2018). These studies indicate that anxiety and depressive phenotypes in adults are frequently linked with myelin-related abnormalities in their CNS.

In a study of first-episode treatment naïve generalized anxiety disorder (GAD) patients without comorbid disorders as compared to healthy controls, the anxiety disorder patients demonstrated significantly higher white matter microstructure in the amygdala as seen by elevated fractional anisotropy (Zhang et al., 2013). This is in line with other studies that have demonstrated a reduction of integrity in the uncinate fasciculus (connecting amygdala and frontal cortex) (Hettema et al., 2012; Tromp et al., 2012). Together, these findings suggest that the dysregulation of myelination in this key frontolimbic pathway may play a large part in the phenotype of generalized anxiety patients.

ASTROCYTES

As stated, PTSD pathology can stem from either mental stress trauma and/or physical trauma such as traumatic brain injury (TBI). Within both etiologies of PTSD, astroglia play a major regulatory role in determining the phenotype. Typically, animal models have focused on the mental stress pathway of PTSD. A recent review detailed the role of astrocytes in both traumatic brain injury and nontraumatic psychological stress PTSD. Mild TBI has demonstrated a significant correlation to PTSD in prior studies and analyses (Yurgil et al., 2014). Because astrocytes clear glutamate within the CNS and support the glutamate (GABA)-glutamine shuttle, they are necessary for protection against neuronal excitotoxicity (Verkhratsky and Nedergaard, 2018). Thus, these cells might play an important role in the formation of symptoms of PTSD such as hyperarousal and heightened anxiety after a traumatic event (Li et al., 2022a). These preclinical data suggest that there may be strong links between astrocyte function and stress phenotype induced after a TBI, more so than that of a mental stress trauma. This is likely due in large part to the links between astrocytes and endothelial vascularization, as well as the structural integrity of the tissue (Verkhratsky et al., 2017; Allen and Lyons, 2018; Verkhratsky and Nedergaard, 2018; Li et al., 2022a) (Fig. 20.3).

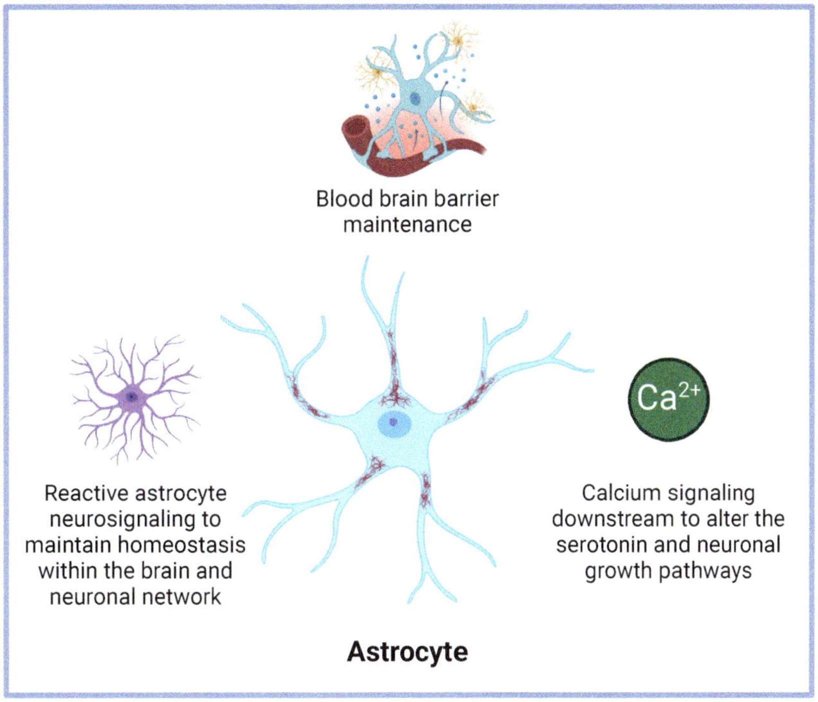

Fig. 20.3. Astrocytes are important for the maintenance of the blood–brain barrier, an area that has demonstrated a breakdown in psychiatric illness, including PTSD. Additionally, they maintain homeostasis of the nervous tissue, which also has demonstrated links to individuals with PTSD. Lower density of astrocytes and astrocyte-related proteins have been noted in human subjects with PTSD, as well as in animal models. Critically, astrocytes may be one of the primary targets for a current mainstay of treatment, SSRI. Astrocytes may contribute to the changes in neurotransmitter levels in individuals with anxiety and PTSD, thus are acted upon directly by many of the primary treatments. The crucial blood–brain barrier allows for an intact neuroimmune system, astrocytes may contribute to a dysfunctional neuroimmune system through a breakdown in the blood–brain barrier following trauma.

Specifically, astrocyte-related proteins have demonstrated dynamic decreases in areas related to PTSD. Within the specific neuronal circuitry underlying PTSD pathology, astrocytes modulate a large part of the behavior changes. Attention and behavioral flexibility have demonstrated significant changes when astrocytes are altered using in vivo probes such as optogenetics or chemogenetics (Lyon and Allen, 2021). In many examples, the corticolimbic brain regions display both atrophic astrocytes and decreased glial fibrillary acidic protein. Individuals who have died by suicide have robust changes in glial fibrillary acidic protein expression across multiple cortical and subcortical brain regions (Torres-Platas et al., 2016). On autopsy, individuals with higher depressive symptoms had lower density of glial fibrillary acidic protein-positive astrocytes and vimentin-positive astrocytes in the dorsomedial PFC, caudate nucleus, and medial thalamus (Verkhratsky and Nedergaard, 2018) at the time of death. Vimentin is a type III intermediate filament and closely related to glial fibrillary acidic protein and is a marker of astrocyte function. It is found to be significantly lower within astrocytes of individuals who died by suicide as compared to healthy controls (O'leary et al., 2021). Vimentin is also expressed in vascular endothelial cells and thus an exploration into the possible link between astrocytic expression and vascularization has been explored, particularly due to the changes in CD31 (mark of endothelial differentiation) marker. These astrocytic changes correlate with preclinical models and changes seen in astrocytic structure following stress-enhanced fear learning behaviors, particularly after a TBI (Lopez-Rodriguez et al., 2015; O'leary et al., 2021). Another study recently demonstrated that astrocytes within the basolateral amygdala may be one of the main drivers of the fear memory but not extinction memory learning in a foot-shock paradigm for rodents (Suthard et al., 2023). The mechanics of this are not well understood but currently believed to be due to astrocyte modulation of calcium channel dependent signaling cascades.

Astroglia in clinical PTSD

Sleep disturbance is one of the most troubling and common symptoms among individuals with PTSD and is reported by up to 70% of patients with PTSD (Babson and Feldner, 2010). Astrocytes are critical to the sleep regulatory cells within the hypothalamic–pituitary–adrenal axis. It is possible that astrocytes may be a link to stress-induced insomnia that is seen within the PTSD phenotype (Birch and Vanderheyden, 2022). PTSD has also been associated with significant activation of the nucleotide-binding domain and leucine-rich repeat protein 3 (NLRP3) inflammasome, which contributes to the behavioral and cognitive impairments seen after trauma (Ji et al., 2023) through depression of astrocyte mitochondrial function. Disrupted sleep is also linked to the IL-6 cytokine alterations in gene transcription that include the circadian rhythm genes within the brain, including astrocytes (Halassa et al., 2009). Some of the fear memory learning has been ameliorated through the use of leptin as a possible treatment through reversal of the astrocyte mitochondrial inhibition (Ji et al., 2023).

These data also demonstrate that monoamine oxidase B (MAO-B) levels may act as a surrogate peripheral biomarker for astrocyte function as this protein is expressed in reactive astrocytes and circulates systemically. Studies have demonstrated a genetic polymorphism and changes in platelet MAO-B activity in PTSD individuals exposed to trauma as compared to those without PTSD (Pivac et al., 2007). This notable finding may allow for future use of a peripheral biomarker to better understand brain pathophysiology and astrocyte function. Another method for probing this system includes the use of novel imaging technologies. In vivo imaging with positron emission tomography using receptor or enzyme-selective radiotracers allows measurement of these proteins in the living human brain. One such study with the MAO-B selective radioligand [^{11}C]SL25.1188 (Gill et al., 2022) demonstrated lower MAO-B volume of distribution (brain outcome measure that is a proxy for protein levels) in regions of the brain including the ventral striatum and medial PFC potentially indicating lower levels of astrocytes in people with PTSD compared to healthy controls (Gill et al., 2022). Additional studies demonstrated hypometabolic brain regions, as demonstrated by reduced radiofluorinated glucose volume of distribution, which correlate to areas of elevated MAO-B (Fowler et al., 1999). These two pieces of information together combine to further demonstrate functional deficits, evidenced by reduced glucose metabolism, in brain regions that are associated with loss of astrocyte function, as evidenced by reduced MAO-B in the brains of individuals with PTSD vs healthy controls. Furthermore, there was an inverse relationship between the glucose distribution and the normalized MAO-B values within the medial temporal cortex but a reversed relationship within the lateral temporal cortex (Fowler et al., 1999). This indicates that there may be regional specificity to some of these changes playing important roles in the pathophysiology of stress-related phenotypes.

Murine models have also demonstrated a reduction in astrocyte volume by more than 50% in the PFC after chronic stress (Tynan et al., 2013) and in the hippocampus (CA1, CA3) in the restraint stress model (Kassem et al., 2013). In the stress-enhanced fear learning model of PTSD, animals display increased IL-1B and decreased IL-1 receptors after stress, which are predominantly driven by astrocytic changes (Li et al., 2022a). Animals exposed to the SPS model of rodent PTSD-like behaviors exhibited reductions in the extracellular matrix secreted by astrocytes and reduction in astrocyte volume, which in some cases could be rescued by intraperitoneal injection of FGF2 (Xia et al., 2013).

Astroglia in neuroimmune response

Along with microglia, astrocytes also play an instrumental role in navigating the neuroimmune response to trauma. Astrocytes work closely with the inflammatory system to provide a fundamental homeostatic mechanism and maintain synaptic transmission and plasticity through secretion of important molecules. And, within synaptogenesis, there is a close relationship between neurons and astrocytes such that the neurons can influence astrocyte maturity while astrocytes provide necessary support to the neuronal synapse development (Li et al., 2022a). Astrocytes are responsible for synthesizing and secreting neurotrophic factors and inflammatory factors, both pro- and anti-, to regulate the environment of nervous tissue. Molecules stimulated by astrocytes are responsible for activation of inflammatory pathways such as nuclear factor kB (Gorina et al., 2011; Lerman et al., 2016) and tumor necrosis factor alpha, as well as cyclooxygenase 1 and matrix metalloproteinase 9 (Hoarau et al., 2011). Therefore, there is substantial evidence that astrocytes are involved in the pathophysiology of PTSD, specifically in the remodeling of fear memories and other stress disorders (Verkhratsky et al., 2017; Li et al., 2020). Astrocytic repression of the connexons (CX30 and CX43) is linked with increased methylation at specific sites, i.e., H3K9me3, and is correlated with depression and suicide risk (Nagy et al., 2017).

Taken together, data regarding astrocytic connections to methylation in connexons suggest that astrocytes may play a substantial role in propagating stress-related neuropathology following a traumatic event. This further opens up the opportunity for astrocyte-targeted pharmacologic treatments. Interestingly, one of the CX30 abnormalities manifesting in gap junction dysfunction linked with astrocyte disruption is ameliorated by selective SSRI treatments (Duek et al., 2019; Williams et al., 2022; Li et al., 2022a). This is one reason that these treatments are a mainstay of PTSD treatment in the current pharmacologic regimens.

PHARMACOLOGIC TREATMENTS FOR PTSD

Current pharmacologic treatments for PTSD and anxiety disorders are primarily based on symptom management, including first-line treatment with SSRIs and SNRIs (Williams et al., 2022), with additional treatments ranging from alpha-1 antagonists for nightmares or mood stabilizers for anger. There are not yet targeted pharmacologic treatments for PTSD. The psychotherapeutic approaches are effective but can be challenging to access and require time and personnel-intensive resources. These treatments include cognitive behavioral therapy, prolonged exposure therapy, eye movement desensitization and reprocessing therapy, and trauma-focused psychotherapies. Pharmacologic and nonpharmacologic treatments are often combined for maximum benefit to treat the patient. There are now promising new studies using riluzole (glutamatergic modulator), MDMA, and ketamine, though the agents have not yet been approved by the FDA for use in PTSD (Al Jowf et al., 2023).

Recently, hypotheses include the mechanisms behind novel treatments such as MDMA or the use of tetrahydrocannabinols (Garzia et al., 2009) and cannabidiol to ameliorate PTSD (García-Gutiérrez et al., 2020; Xie et al., 2024) symptoms have centered on the role of glial cells. Microglia may be a primary mediator for the restructuring that occurs following treatment with these novel compounds. These studies suggest that the endocannabinoid system (Lopez-Rodriguez et al., 2015) and microglial-mediated neuroimmune system may play a role in treatment response. It is possible that changes in chemokine signaling from microglia and astrocytes are allowing the therapy to be more effective following these treatments.

Additionally, THC has offered a new way of thinking about symptom management for sleep and anxiety in PTSD. Many individuals have found that access to THC has significantly improved their symptom burden (Abizaid et al., 2019). However, more research is needed to fully understand the role that THC plays in the neuropathology of PTSD and the overall safety and efficacy profile of these substances.

FUTURE STEPS FOR UNDERSTANDING NEUROBIOLOGY OF ANXIETY

In addition to future neurobiologic research, it is important to consider the phenotypic heterogeneity of PTSD and related disorders. According to the DSM-5, PTSD is comprised of four separate symptom clusters—intrusions, avoidance, negative alterations in cognition and mood, and alterations in arousal and reactivity. A burgeoning body of confirmatory factor analytic (CFA) studies, however, has suggested that the phenotypic expression of PTSD may be more nuanced and characterized by as many as eight symptom dimensions. For example, a recent CFA study (Duek et al., 2022) found that PTSD symptoms were best characterized by separate dimensions of internally generated intrusions (e.g., intrusive memories), externally generated intrusions (e.g., emotional reactivity to trauma cues), avoidance, negative affect (e.g., persistent negative emotional state), anhedonia (e.g., detachment), externalizing behaviors (e.g., reckless behavior), anxious arousal (e.g., hypervigilance), and dysphoric arousal (e.g., concentration difficulties). In addition to within-disorder heterogeneity, there is also emerging evidence of transdiagnostic symptom dimensions of common mood and anxiety disorders (Pietrzak et al., 2015; Ruggero et al., 2019; Stanton et al., 2020). For example, a study of US veterans with chronic PTSD found that symptoms of PTSD, MDD, and GAD were comprised of separate loss (e.g., dysphoria), threat (e.g., anxious arousal and reexperiencing), and somatic anxiety (e.g., physiologic activation) symptoms (Pietrzak et al., 2015).

While most neurobiologic studies typically employ case–control designs, greater consideration of the dimensional and transdiagnostic expression of heterogeneous disorders such as PTSD—both with respect to recruitment, phenotyping, and data analysis—may help identify neurobiologic underpinnings of component aspects of these disorders. Such approaches, which align with the National Institute of Mental Health Research Domain Criteria (RDoC) initiative, could employ dimensional recruitment strategies that include the full spectrum of symptom severity (e.g., no/minimal, mild/subthreshold, moderate, and severe). This would allow researchers to examine how neurobiologic measures relate to overall symptom severity, as well as subdimensions of PTSD and related anxiety phenotypes (e.g., eight-factor model). For example, a recent PET imaging study found that higher cortical metabotropic glutamate receptor 5 (mGluR5) availability was associated with greater severity of avoidance symptoms of PTSD (Holmes et al., 2017). In summary, greater consideration of the state-of-the-science regarding the phenotypic modeling of symptoms of PTSD and related disorders is critical to advancing knowledge regarding the neurobiology, as well as more targeted, precision medicine-based treatments for multifaceted aspects of these disorders.

CONCLUSION

Within the central nervous system, each class fo neuroglia plays an important role in the regulation of neuronal signaling. Through the many pathways including epigenetics, genomic transcription, and immune signaling, these cells are dynamically regulated in times of stress and trauma. The regulation and subsequent dysregulation that can occur within microglia, oligodendrocytes, and astrocytes play important roles in the anxiety phenotypes that are seen in mental health disorders following stress or trauma. Anxiety and depressive phenotypes are often mixed within a PTSD spectrum and depict the great heterogeneity that is often seen within this spectrum of disorders. Each of these cell types is important in the development of anxiety and PTSD symptoms. Understanding specific relationships between glial cells, the interactions with neurons, and PTSD symptomatology will offer insights into novel pharmacotherapy development for these disorders. Notably, more data are arising demonstrating immune dysfunction in PTSD and the links to microglia and astrocyte dysfunction. This may help to better understand and direct treatment methodologies. Additionally, studies demonstrating the efficacy of psychedelics in PTSD and anxiety symptoms offer promise for alternative therapy mechanisms such as increasing synaptic density. Improving the understanding of histone deacetylases and their interactions with microglia may also offer novel treatments such as histone deacetylase inhibitors, some of which are already used clinically in oncology and antiseizure treatments. PTSD is particularly hard to treat as a psychiatric illness due to its variety of symptoms and phenotypes. Ideally, there will be multiple treatment methods available and clinicians may choose the most appropriate treatment for the symptoms or biomarkers present in a given patient. Through this work, these goals may be attainable in the future.

ACKNOWLEDGMENTS

We are thankful for our funding from the National Institute of Mental Health grants R01MH110674 and T32MH019961, the Veterans Affairs National Center for PTSD, and the Thomas P. Detre Award for Neurosciences.

REFERENCES

Abizaid A, Merali Z, Anisman H (2019). Cannabis: a potential efficacious intervention for PTSD or simply snake oil? J Psychiatry Neurosci 44: 75–78.

Adamou M, Puchalska S, Plummer W et al. (2007). Valproate in the treatment of PTSD: systematic review and meta analysis. Curr Med Res Opin 23: 1285–1291.

Adamsky A, Kol A, Kreisel T et al. (2018). Astrocytic activation generates de novo neuronal potentiation and memory enhancement. Cell 174: 59–71.e14.

Al Jowf GI, Ahmed ZT, Reijnders RA et al. (2023). To predict, prevent, and manage post-traumatic stress disorder (PTSD): a review of pathophysiology, treatment, and biomarkers. Int J Mol Sci 24: 5238. https://doi.org/10.3390/ijms24065238. PMID: 36982313; PMCID: PMC10049301.

Albrecht-Souza L, Gilpin NW (2019). The predator odor avoidance model of post-traumatic stress disorder in rats. Behav Pharmacol 30: 105–114.

Allen NJ (2014). Astrocyte regulation of synaptic behavior. Annu Rev Cell Dev Biol 30: 439–463.

Allen NJ, Lyons DA (2018). Glia as architects of central nervous system formation and function. Science 362: 181–185.

Antontseva E, Bondar N, Reshetnikov V et al. (2020). The effects of chronic stress on brain myelination in humans and in various rodent models. Neuroscience 441: 226–238.

Araque A, Carmignoto G, Haydon PG et al. (2014). Gliotransmitters travel in time and space. Neuron 81: 728–739.

Babson KA, Feldner MT (2010). Temporal relations between sleep problems and both traumatic event exposure and PTSD: a critical review of the empirical literature. J Anxiety Disord 24: 1–15.

Baker DG, Ekhator NN, Kasckow JW et al. (2001). Plasma and cerebrospinal fluid interleukin-6 concentrations in posttraumatic stress disorder. Neuroimmunomodulation 9: 209–217.

Baxter AJ, Scott KM, Vos T et al. (2013). Global prevalence of anxiety disorders: a systematic review and meta-regression. Psychol Med 43: 897–910.

Beumer W, Gibney SM, Drexhage RC et al. (2012). The immune theory of psychiatric diseases: a key role for activated microglia and circulating monocytes. J Leukoc Biol 92: 959–975.

Bhatt S, Hillmer AT, Girgenti MJ et al. (2020). PTSD is associated with neuroimmune suppression: evidence from PET imaging and postmortem transcriptomic studies. Nat Commun 11: 2360.

Birch JN, Vanderheyden WM (2022). The molecular relationship between stress and insomnia. Adv Biol (Weinh) 6: e2101203.

Blacker CJ, Frye MA, Morava E et al. (2019). A review of epigenetics of PTSD in comorbid psychiatric conditions. Genes 10: 140.

Bonnefil V, Dietz K, Amatruda M et al. (2019). Region-specific myelin differences define behavioral consequences of chronic social defeat stress in mice. Elife 8: e40855.

Campos AC, Fogaça MV, Aguiar DC et al. (2013). Animal models of anxiety disorders and stress. Braz J Psychiatry 35: S101–S111.

Cheng J, Uchida M, Zhang W et al. (2011). Role of salt-induced kinase 1 in androgen neuroprotection against cerebral ischemia. J Cereb Blood Flow Metab 31: 339–350.

Cornell J, Salinas S, Huang HY et al. (2022). Microglia regulation of synaptic plasticity and learning and memory. Neural Regen Res 17: 705–716.

Davidson JR, Stein DJ, Shalev AY et al. (2004). Posttraumatic stress disorder: acquisition, recognition, course, and treatment. J Neuropsychiatry Clin Neurosci 16: 135–147.

Donat C, Scott G, Gentleman S et al. (2017). Microglial activation in traumatic brain injury. Front Aging Neurosci 9: 208.

Duek O, Kelmendi B, Pietrzak RH et al. (2019). Augmenting the treatment of PTSD with ketamine—a review. Curr Treat Options Psychiatry 6: 143–153.

Duek O, Spiller TR, Rubenstein A et al. (2022). Exploration of a novel model of intrusive symptoms in posttraumatic stress disorder among US veterans. JAMA Netw Open 5: e223555.

Eagle AL, Fitzpatrick CJ, Perrine SA (2013). Single prolonged stress impairs social and object novelty recognition in rats. Behav Brain Res 256: . https://doi.org/10.1016/j.bbr.2013.09.014.

Eljaschewitsch E, Witting A, Mawrin C et al. (2006). The endocannabinoid anandamide protects neurons during CNS inflammation by induction of MKP-1 in microglial cells. Neuron 49: 67–79.

Ensink JBM, Keding TJ, Henneman P et al. (2021). Differential DNA methylation is associated with hippocampal abnormalities in pediatric posttraumatic stress disorder. Biol Psychiatry Cogn Neurosci Neuroimaging 6: 1063–1070.

Eraly SA, Nievergelt CM, Maihofer AX et al. (2014). Assessment of plasma C-reactive protein as a biomarker of posttraumatic stress disorder risk. JAMA Psychiatry 71: 423–431.

Erburu M, Muñoz-Cobo I, Domínguez-Andrés J et al. (2015). Chronic stress and antidepressant induced changes in Hdac5 and Sirt2 affect synaptic plasticity. Eur Neuropsychopharmacol 25: 2036–2048.

Finn DP, Haroutounian S, Hohmann AG et al. (2021). Cannabinoids, the endocannabinoid system, and pain: a review of preclinical studies. Pain 162: S5–s25.

Fowler JS, Volkow ND, Cilento R et al. (1999). Comparison of brain glucose metabolism and monoamine oxidase B (MAO B) in traumatic brain injury. Clinical Positron Imaging 2: 71–79.

García-Gutiérrez MS, Navarrete F, Gasparyan A et al. (2020). Cannabidiol: a potential new alternative for the treatment of anxiety, depression, and psychotic disorders. Biomolecules 10: 1575. https://doi.org/10.3390/biom10111575. PMID: 33228239; PMCID: PMC7699613.

Garzia L, Andolfo I, Cusanelli E et al. (2009). MicroRNA-199b-5p impairs cancer stem cells through negative regulation of HES1 in medulloblastoma. PLoS One 4: e4998. https://doi.org/10.1371/journal.pone.0004998. Epub 2009 Mar 24. PMID: 19308264; PMCID: PMC2656623.

Gill T, Watling SE, Richardson JD et al. (2022). Imaging of astrocytes in posttraumatic stress disorder: a PET study with the monoamine oxidase B radioligand [^{11}C]SL25.1188. Eur Neuropsychopharmacol 54: 54–61.

Gorina R, Font-Nieves M, Márquez-Kisinousky L et al. (2011). Astrocyte TLR4 activation induces a proinflammatory environment through the interplay between MyD88-dependent NFκB signaling, MAPK, and Jak1/Stat1 pathways. Glia 59: 242–255.

Halassa MM, Florian C, Fellin T et al. (2009). Astrocytic modulation of sleep homeostasis and cognitive consequences of sleep loss. Neuron 61: 213–219.

Han F, Xiao B, Wen L (2015). Loss of glial cells of the hippocampus in a rat model of post-traumatic stress disorder. Neurochem Res 40: 942–951.

Harris AZ, Atsak P, Bretton ZH et al. (2018). A novel method for chronic social defeat stress in female mice. Neuropsychopharmacology 43: 1276–1283.

Hettema JM, Kettenmann B, Ahluwalia V et al. (2012). Pilot multimodal twin imaging study of generalized anxiety disorder. Depress Anxiety 29: 202–209.

Hoarau J-J, Krejbich-Trotot P, Jaffar-Bandjee M-C et al. (2011). Activation and control of CNS innate immune responses in health and diseases: a balancing act finely tuned by neuroimmune regulators (NIReg). CNS Neurol Disord Drug Targets 10: 25–43.

Holmes SE, Girgenti MJ, Davis MT et al. (2017). Altered metabotropic glutamate receptor 5 markers in PTSD: in vivo and postmortem evidence. Proc Natl Acad Sci USA 114: 8390–8395.

Holmes SE, Scheinost D, Finnema SJ et al. (2019). Lower synaptic density is associated with depression severity and network alterations. Nat Commun 10: 1529. https://doi.org/10.1038/s41467-019-09562-7. PMID: 30948709; PMCID: PMC6449365.

Hong S, Beja-Glasser VF, Nfonoyim BM et al. (2016). Complement and microglia mediate early synapse loss in Alzheimer mouse models. Science 352: 712–716.

Hori H, Kim Y (2019). Inflammation and post-traumatic stress disorder. Psychiatry Clin Neurosci 73: 143–153.

Ieraci A, Mallei A, Popoli M (2016). Social isolation stress induces anxious-depressive-like behavior and alterations of neuroplasticity-related genes in adult male mice. Neural Plast 2016: 6212983. https://doi.org/10.1155/2016/6212983. Epub 2016 Jan 6. PMID: 26881124; PMCID: PMC4736811.

Jak AJ, Jurick S, Hoffman S et al. (2020). PTSD, but not history of mTBI, is associated with altered myelin in combat-exposed Iraq and Afghanistan veterans. Clin Neuropsychol 34: 1070–1087.

Ji M, Gong W, Wang S et al. (2023). Leptin attenuates fear memory by inhibiting astrocytic NLRP3 inflammasome in post-traumatic stress disorder model. Neurochem Res 48: 1180–1190. https://doi.org/10.1007/s11064-022-03655-4. Epub 2022 Jun 24. PMID: 35750877.

Kamal SR, Potukutchi S, Gelovani DJ et al. (2022). Spatial and temporal dynamics of HDACs class IIa following mild traumatic brain injury in adult rats. Mol Psychiatry 27: 1683–1693. https://doi.org/10.1038/s41380-021-01369-7. PMID: 35027678.

Kassem MS, Lagopoulos J, Stait-Gardner T et al. (2013). Stress-induced grey matter loss determined by MRI is primarily due to loss of dendrites and their synapses. Mol Neurobiol 47: 645–661.

Kessler RC, Sonnega A, Bromet E et al. (1995). Posttraumatic stress disorder in the National Comorbidity Survey. Arch Gen Psychiatry 52: 1048–1060.

Kessler RC, Chiu WT, Demler O et al. (2005). Prevalence, severity, and comorbidity of 12-month DSM-IV disorders in the National Comorbidity Survey Replication. Arch Gen Psychiatry 62: 617–627.

Kida S (2019). Reconsolidation/destabilization, extinction and forgetting of fear memory as therapeutic targets for PTSD. Psychopharmacology (Berl) 236: 49–57.

Kilpatrick DG, Resnick HS, Milanak ME et al. (2013). National estimates of exposure to traumatic events and PTSD prevalence using DSM-IV and DSM-5 criteria. J Trauma Stress 26: 537–547.

Kim GS, Smith AK, Nievergelt CM et al. (2018). Neuroepigenetics of post-traumatic stress disorder. Prog Mol Biol Transl Sci 158: 227–253.

Knox D, Perrine SA, George SA et al. (2010). Single prolonged stress decreases glutamate, glutamine, and creatine concentrations in the rat medial prefrontal cortex. Neurosci Lett 480: 16–20.

Koch SB, Van Zuiden M, Nawijn L et al. (2016). Aberrant resting-state brain activity in posttraumatic stress disorder: a meta-analysis and systematic review. Depress Anxiety 33: 592–605.

Krystal JH, Abdallah CG, Averill LA et al. (2017). Synaptic loss and the pathophysiology of PTSD: implications for ketamine as a prototype novel therapeutic. Curr Psychiatry Rep 19: 74.

Lerman I, Davis BA, Bertram TM et al. (2016). Posttraumatic stress disorder influences the nociceptive and intrathecal cytokine response to a painful stimulus in combat veterans. Psychoneuroendocrinology 73: 99–108.

Li Y, Li L, Wu J et al. (2020). Activation of astrocytes in hippocampus decreases fear memory through adenosine A1 receptors. eLife 9: e57155.

Li S, Liao Y, Dong Y et al. (2021). Microglial deletion and inhibition alleviate behavior of post-traumatic stress disorder in mice. J Neuroinflammation 18: 7.

Li B, Zhang D, Verkhratsky A (2022a). Astrocytes in post-traumatic stress disorder. Neurosci Bull 38: 953–965.

Li H, Tofigh AM, Amirfakhraei A et al. (2022b). Modulation of astrocyte activity and improvement of oxidative stress through blockage of NO/NMDAR pathway improve posttraumatic stress disorder (PTSD)-like behavior induced by social isolation stress. Brain Behav 12: e2620.

Lisieski MJ, Eagle AL, Conti AC et al. (2018). Single-prolonged stress: a review of two decades of progress in a rodent model of post-traumatic stress disorder. Front Psych 9: 196.

Liu J, Dietz K, Deloyht JM et al. (2012). Impaired adult myelination in the prefrontal cortex of socially isolated mice. Nat Neurosci 15: 1621–1623.

Logue MW, Miller MW, Wolf EJ et al. (2020). An epigenome-wide association study of posttraumatic stress disorder in US veterans implicates several new DNA methylation loci. Clin Epigenetics 12: 46.

Long KLP, Chao LL, Kazama Y et al. (2021). Regional gray matter oligodendrocyte- and myelin-related measures are associated with differential susceptibility to stress-induced behavior in rats and humans. Transl Psychiatry 11: 631.

Lopez-Rodriguez AB, Acaz-Fonseca E, Viveros MP et al. (2015). Changes in cannabinoid receptors, aquaporin 4 and vimentin expression after traumatic brain injury in adolescent male mice. Association with edema and neurological deficit. PLoS One 10: e0128782.

Lutz MI, Milenkovic I, Regelsberger G et al. (2014). Distinct patterns of sirtuin expression during progression of Alzheimer's disease. Neuromolecular Med 16: 405–414.

Lyon KA, Allen NJ (2021). From synapses to circuits, astrocytes regulate behavior. Front Neural Circuits 15: 786293.

Machelska H, Celik MÖ (2016). Recent advances in understanding neuropathic pain: glia, sex differences, and epigenetics. F1000Research 5: 2743. https://doi.org/10.12688/f1000research.9621.1. PMID: 28105313; PMCID: PMC5224690.

Mamiya N, Fukushima H, Suzuki A et al. (2009). Brain region-specific gene expression activation required for reconsolidation and extinction of contextual fear memory. J Neurosci 29: 402–413.

Manson J, Thiemermann C, Brohi K (2011). Trauma alarmins as activators of damage-induced inflammation. Br J Surg 99: 12–20.

Melchionda L, Fang M, Wang H et al. (2013). Adult-onset Alexander disease, associated with a mutation in an alternative GFAP transcript, may be phenotypically modulated by a non-neutral HDAC6 variant. Orphanet J Rare Dis 8: 66.

Miranda-Negrón Y, García-Arrarás JE (2022). Radial glia and radial glia-like cells: their role in neurogenesis and regeneration. Front Neurosci 16: 1006037.

Mitchell JM, Bogenschutz M, Lilienstein A et al. (2021). MDMA-assisted therapy for severe PTSD: a randomized, double-blind, placebo-controlled phase 3 study. Nat Med 27: 1025–1033.

Mumtaz F, Khan MI, Zubair M et al. (2018). Neurobiology and consequences of social isolation stress in animal model—a comprehensive review. Biomed Pharmacother 105: 1205–1222.

Nagy C, Torres-Platas SG, Mechawar N et al. (2017). Repression of astrocytic connexins in cortical and subcortical brain regions and prefrontal enrichment of H3K9me3 in depression and suicide. Int J Neuropsychopharmacol 20: 50–57.

Nahum K, Todder D, Zohar J et al. (2022). The role of microglia in the (mal)adaptive response to traumatic experience in an animal model of PTSD. Int J Mol Sci 23: 7185. https://doi.org/10.3390/ijms23137185. PMID: 35806185; PMCID: PMC9266429.

Ni K, Zhu J, Xu X et al. (2022). Hippocampal activated microglia may contribute to blood-brain barrier impairment and cognitive dysfunction in post-traumatic stress disorder-like rats. J Mol Neurosci 72: 975–982.

O'donovan A, Ahmadian AJ, Neylan TC et al. (2017). Current posttraumatic stress disorder and exaggerated threat sensitivity associated with elevated inflammation in the Mind Your Heart Study. Brain Behav Immun 60: 198–205.

O'leary LA, Belliveau C, Davoli MA et al. (2021). Widespread decrease of cerebral vimentin-immunoreactive astrocytes in depressed suicides. Front Psychiatry 12: 640963.

Pietrzak RH, Rosenheck RA, Cramer JA et al. (2015). Elucidating the transdiagnostic dimensional structure of trauma-related psychopathology: findings from VA cooperative study 504—risperidone treatment for military service related chronic post traumatic stress disorder. J Affect Disord 172: 331–336.

Pivac N, Knezevic J, Kozaric-Kovacic D et al. (2007). Monoamine oxidase (MAO) intron 13 polymorphism and platelet MAO-B activity in combat-related posttraumatic stress disorder. J Affect Disord 103: 131–138.
Prozorovski T, Schulze-Topphoff U, Glumm R et al. (2008). Sirt1 contributes critically to the redox-dependent fate of neural progenitors. Nat Cell Biol 10: 385–394.
Rohan Walker F, Nilsson M, Jones K (2013). Acute and chronic stress-induced disturbances of microglial plasticity, phenotype and function. Curr Drug Targets 14: 1262–1276.
Ruggero CJ, Kotov R, Hopwood CJ et al. (2019). Integrating the hierarchical taxonomy of psychopathology (HiTOP) into clinical practice. J Consult Clin Psychol 87: 1069–1084.
Setiawan E, Wilson AA, Mizrahi R et al. (2015). Role of translocator protein density, a marker of neuroinflammation, in the brain during major depressive episodes. JAMA Psychiatry 72: 268–275.
Snijders C, Maihofer AX, Ratanatharathorn A et al. (2020). Longitudinal epigenome-wide association studies of three male military cohorts reveal multiple CpG sites associated with post-traumatic stress disorder. Clin Epigenetics 12: 11.
Stanton K, McDonnell CG, Hayden EP et al. (2020). Transdiagnostic approaches to psychopathology measurement: recommendations for measure selection, data analysis, and participant recruitment. J Abnorm Psychol 129: 21–28.
Sun X-Y, Zheng T, Yang X et al. (2019). HDAC2 hyperexpression alters hippocampal neuronal transcription and microglial activity in neuroinflammation-induced cognitive dysfunction. J Neuroinflammation 16: 249.
Suthard RL, Senne RA, Buzharsky MD et al. (2023). Basolateral amygdala astrocytes are engaged by the acquisition and expression of a contextual fear memory. J Neurosci 43: 4997–5013.
Tanti A, Kim J, Wakid M et al. (2018). Child abuse associates with an imbalance of oligodendrocyte-lineage cells in ventromedial prefrontal white matter. Mol Psychiatry 23: 2018–2028. https://doi.org/10.1038/mp.2017.231. Epub 2017 Nov 21. PMID: 29158585.
Torres-Platas S, Nagy C, Wakid M et al. (2016). Glial fibrillary acidic protein is differentially expressed across cortical and subcortical regions in healthy brains and downregulated in the thalamus and caudate nucleus of depressed suicides. Mol Psychiatry 21: 509–515.
Tremblay M-È, Stevens B, Sierra A et al. (2011). The role of microglia in the healthy brain. J Neurosci 31: 16064–16069.
Tromp DP, Grupe DW, Oathes DJ et al. (2012). Reduced structural connectivity of a major frontolimbic pathway in generalized anxiety disorder. Arch Gen Psychiatry 69: 925–934.
Tynan RJ, Beynon SB, Hinwood M et al. (2013). Chronic stress-induced disruption of the astrocyte network is driven by structural atrophy and not loss of astrocytes. Acta Neuropathol 126: 75–91.
Van Doeselaar L, Yang H, Bordes J et al. (2021). Chronic social defeat stress in female mice leads to sex-specific behavioral and neuroendocrine effects. Stress 24: 168–180.
Van Rijn E, Carter N, Mcmurtrie H et al. (2017). Sleep does not cause false memories on a story-based test of suggestibility. Conscious Cogn 52: 39–46.
Verbitsky A, Dopfel D, Zhang N (2020). Rodent models of post-traumatic stress disorder: behavioral assessment. Transl Psychiatry 10: 132. https://doi.org/10.1038/s41398-020-0806-x.
Verkhratsky A, Nedergaard M (2018). Physiology of astroglia. Physiol Rev 98: 239–389.
Verkhratsky A, Zorec R, Parpura V (2017). Stratification of astrocytes in healthy and diseased brain. Brain Pathol 27: 629–644.
Wang G, Shi Y, Jiang X et al. (2015). HDAC inhibition prevents white matter injury by modulating microglia/macrophage polarization through the GSK3β/PTEN/Akt axis. Proc Natl Acad Sci USA 112: 2853–2858.
Whittle N, Singewald N (2014). HDAC inhibitors as cognitive enhancers in fear, anxiety and trauma therapy: where do we stand? Biochem Soc Trans 42: 569–581.
Williams T, Phillips NJ, Stein DJ et al. (2022). Pharmacotherapy for post traumatic stress disorder (PTSD). Cochrane Database Syst Rev 3: CD002795.
Wohleb ES (2016). Neuron-microglia interactions in mental health disorders: "for better, and for worse". Front Immunol 7: 544.
Xia L, Zhai M, Wang L et al. (2013). FGF2 blocks PTSD symptoms via an astrocyte-based mechanism. Behav Brain Res 256: 472–480.
Xiao Q, Wang F, Luo Y et al. (2018). Exercise protects myelinated fibers of white matter in a rat model of depression. J Comp Neurol 526: 537–549.
Xie G, Gao X, Guo Q et al. (2024). Cannabidiol ameliorates PTSD-like symptoms by inhibiting neuroinflammation through its action on CB2 receptors in the brain of male mice. Brain Behav Immun 119: 945–964.
Yoshii T, Oishi N, Ikoma K et al. (2017). Brain atrophy in the visual cortex and thalamus induced by severe stress in animal model. Sci Rep 7: 12731.
Yurgil KA, Barkauskas DA, Vasterling JJ et al. (2014). Association between traumatic brain injury and risk of posttraumatic stress disorder in active-duty marines. JAMA Psychiatry 71: 149–157.
Zhang B, West EJ, Van KC et al. (2008). HDAC inhibitor increases histone H3 acetylation and reduces microglia inflammatory response following traumatic brain injury in rats. Brain Res 1226: 181–191.
Zhang Y, Li L, Yu R et al. (2013). White matter integrity alterations in first episode, treatment-naive generalized anxiety disorder. J Affect Disord 148: 196–201.
Zhang L, Hu XZ, Li H et al. (2019). Updates in PTSD animal models characterization. Methods Mol Biol 2011: 331–344.

Handbook of Clinical Neurology, Vol. 210 (3rd series)
Neuroglia in Neurologic and Psychiatric Disorders, Part II
A. Verkhratsky, L.D. de Witte, E. Aronica, and E.M. Hol, Editors
https://doi.org/10.1016/B978-0-443-19102-2.00014-4

Chapter 21

Neuroglia in substance use disorders

EMILY M. CASTRO[1], SHAHRDAD LOTFIPOUR[1,2,3], AND FRANCES M. LESLIE[1*]

[1]*Department of Pharmaceutical Sciences, School of Pharmacy and Pharmaceutical Sciences, University of California, Irvine, Irvine, CA, United States*

[2]*Department of Emergency Medicine, School of Medicine, University of California, Irvine, Irvine, CA, United States*

[3]*Department of Pathology and Laboratory Medicine, School of Medicine, University of California, Irvine, Irvine, CA, United States*

Abstract

Substance use disorders (SUD) remain a major public health concern in which individuals are unable to control their use of substances despite significant harm and negative consequences. Drugs of abuse dysregulate major brain and behavioral functions. Glial cells, primarily microglia and astrocytes, play a crucial role in these drug-induced molecular and behavioral changes. This review explores preclinical and clinical studies of how neuroglia and their associated neuroinflammatory responses contribute to SUD and reward-related properties. We evaluate preclinical and clinical evidence for targeting neuroglia as therapeutic interventions. In addition, we evaluate the literature on the gut microbiome and its role in SUD. Clinical treatments are most effective for reducing drug cravings, and some have yielded promising results in other measures of drug use. N-Acetylcysteine, through modulation of cysteine-glutamate antiporter of glial cells, shows encouraging results across a variety of drug classes. Neuroglia and gut microbiome interactions are important factors to consider with regard to SUD and could lead to novel therapeutic avenues. Age- and sex-dependent properties of neuroglia, gut microbiome, and drug use behaviors are important areas in need of further investigation.

INTRODUCTION

Substance use disorder (SUD) is a medical condition in which a person is unable to control their use of drugs and other substances despite substantial harm and negative consequences. It often impairs an individual's ability to meet obligations, and disrupts various aspects of life including family, work, and individual health (Hasin et al., 2013). Although neuronal alterations and dysregulation are heavily implicated in substance use, drugs of abuse also impact glial cells such as microglia and astrocytes. They trigger an immune response mediated by glial cells that influence neuronal function directly through physical connections (Corty and Freeman, 2013) or indirectly through chemical communication (Guo et al., 2007; York et al., 2018). Neuroglia contribute to drug-induced molecular and synaptic changes in neurons and consequently impact behavioral responses to these drugs. In this review, we discuss current knowledge of glial-mediated mechanisms involved in substance use and evaluate clinical evidence for targeting glia in therapeutic interventions. We also describe the influence of the gut microbiome in the effects of drugs of abuse on glial cells.

NEUROGLIA

Neuroglia are nonneuronal cells in the central nervous system (CNS) that are crucial for proper neuronal function and neuroimmune response. Beyond a supportive role in homeostasis, glia actively contribute to proper neuronal functions such as synaptic plasticity and neurotransmitter modulation. Due to the central role of astrocytes and microglia in the neuroimmune response, most drug abuse studies focus on these two cell types. Astrocytes support structural and metabolic functions for neurons including homeostasis of the extracellular environment via neurotransmitters, releasing signaling molecules, and regulating synapse function (Sofroniew and Vinters, 2010). Microglia are highly mobile immunocompetent cells that respond to external threats, detect pathogens and engage in phagocytosis (Graeber, 2010). Astrocytes and microglia are the principal immune regulators within the brain, surveilling the environment and initiating immune signals via cytokine and chemokine release after biologic insult or trauma. In this inflammatory state, astrocytes, and microglia are described as *reactive*. This state is accompanied by morphologic, transcriptional, and functional

*Correspondence to: Frances M. Leslie, Ph.D., Department of Pharmaceutical Sciences, School of Pharmacy and Pharmaceutical Sciences, University of California, Irvine, 367 Med Surge II, Irvine, CA 92697, United States. Tel: +1-949-824-6699, Fax: +1-949-824-4855, E-mail: fmleslie@uci.edu

changes. Reactive states are further differentiated by proinflammatory vs antiinflammatory characteristics with context-dependent functions (Escartin et al., 2021; Paolicelli et al., 2022). These properties are determined by the nature of the released chemokines and cytokines; many proinflammatory cytokines are from the tumor necrosis factor (TNF), gamma-interferon (IFNγ), and interleukin (IL) families. Some cytokines such as IL-10, IL-13, IFN-α, and TGF-β are largely considered to be antiinflammatory. Astrocytes can directly modulate synaptic signaling with their perisynaptic astroglial processes (PAPs) which interact with neuronal presynaptic nerve terminals and postsynaptic dendritic specializations, known as the "tripartite synapse" (Volterra and Meldolesi, 2005; Kruyer et al., 2023). The scope of glial functions in normal health and development are still emerging and represent a critical area of study for conditions such as SUDs.

NEUROGLIAL MECHANISMS UNDERLYING SUBSTANCE USE

Drugs of abuse activate reward-related circuitry within the brain and trigger positively-valenced emotions with reinforcing effects. Following repeated exposure and use, drugs eventually impair normal cognitive and executive function leading to loss of control in drug-related behaviors (Volkow and Morales, 2015). Studies largely focus on the ventral tegmental area (VTA) to nucleus accumbens (NAc) pathway due to its role in dopamine (DA) mediated reward processing (Wise and Robble, 2020). Due to this key role, DA-related neuronal signaling and receptor expression have been extensively investigated as mechanisms underlying substance use and SUD (previously reviewed in Volkow and Morales, 2015). However, all drugs of abuse alter the function of glial cells which promote drug-induced changes in the brain and behavior (Miguel-Hidalgo, 2009; Coller and Hutchinson, 2012; Lacagnina et al., 2017; Linker et al., 2019). Drugs of abuse have a few overlapping mechanisms of action such as binding to toll-like receptors (TLRs), a receptor found on both microglia and astrocytes, and induction of proinflammatory cytokines. Given the active maturation of the brain and reward-related DA circuitry during adolescence, the primary time of onset of drug use (Yuan et al., 2015), we will evaluate age effects in the glial mediated activity (Reynolds and Flores, 2021).

Opioids

Neuroglia express all opioid receptor subtypes and can be altered by endogenous and exogenous opioid receptor ligands (Gomez-Flores and Weber, 1998; Bidlack, 2000; Liang et al., 2016; Zare et al., 2023). Opioid-induced patterns in gliosis in reward-related regions such as the VTA and NAc are dependent on length of exposure and influence subsequent behavior (Watkins et al., 2009; Schwarz et al., 2011; Yoon et al., 2012). Acute morphine exposure increases the mRNA expression of both microglial CD11b and astrocytic glial fibrillary acidic protein (GFAP) in the NAc (Schwarz et al., 2011), while chronic exposure increases microglial CD11b and astrocytic GFAP density in the VTA (Hutchinson et al., 2007). GFAP and CD11b are considered specific markers for astrocytes and microglia respectively and upregulated expression is associated with inflammatory responses in the CNS. Chronic morphine-induced upregulation of GFAP in the VTA is abolished when treated with the opioid receptor antagonist, naltrexone (Beitner-Johnson et al., 1993). Opioids induce proinflammatory neuroimmune signaling pathways largely (Schwarz et al., 2011; Coller and Hutchinson, 2012). Toll-like receptor 4 (TLR-4), which is heavily involved in inflammatory intracellular signaling pathways, strongly influences the reinforcing and motivational properties of opioids such as morphine and heroin (Hutchinson et al., 2012; Wang et al., 2012; Theberge et al., 2013). Morphine can bind to microglial TLR4 directly and enhance TLR4 signaling and microglial activation within the NAc of adolescent rats (Schwarz and Bilbo, 2013). Of note, robust morphine-induced reinstatement of conditioned place preference (CPP) unique to adolescent, but not adult, rats was paralleled by with upregulated CD11b and microglial TLR-4 expression (Schwarz and Bilbo, 2013). In adults, opioid-induced reactive microgliosis in the VTA reduces DA signaling and DA-dependent rewards behaviors via microglia–brain–derived neurotrophic factor (BDNF) signaling pathway (Taylor et al., 2015a,b, 2016). Thus, direct microglia-induced dysregulation in DA signaling contributes to impaired reward behaviors involved in opioid use disorders. Cue-induced heroin seeking behavior engages two distinct astrocyte subpopulations, where synaptic proximity of astroglial processes in the NAc core increases in one and increases surface expression of astroglial glutamate transporter-1 (GLT-1) in another (Kruyer et al., 2022). Synaptic proximity of astrocytes directly regulates excitatory synapses and neurotransmitter transmission (Lehre and Rusakov, 2002; Kruyer et al., 2023), whereas GLT-1 is highly expressed in astrocytes and removes excess glutamate (Rothstein et al., 1994; Sutherland et al., 1996). Such distinct mechanisms within glial subpopulations, as well as age, are critical to consider in SUD treatment.

Opioid use disorder treatments

Pioglitazone, a peroxisome proliferator-activated receptor-gamma (PPARγ) agonist, inhibits the expression of cytokines and reduces activation of microglia and astrocytes (Kielian and Drew, 2003; Qiu and Li, 2015). Pioglitazone is commonly used to treat diabetes due to the primary function of PPARγ as a fatty acid sensor and lipoprotein metabolism regulator (Wahli and Michalik, 2012; Lefterova et al., 2014). PPARγ is highly expressed in neurons within reward-related regions, importantly within the VTA and NAc (Moreno et al., 2004; Gofflot et al., 2007; Sarruf et al., 2009; Margolis et al., 2012). Pioglitazone reduces morphine withdrawal, heroin-induced reinstatement, heroin self-administration, and heroin-induced DA release in rodents (de Guglielmo et al., 2014, 2015, 2017). However, clinical trials have been largely unsuccessful, although this may be due to insufficient enrollment (Schroeder et al., 2018) Ibudilast (AKA, AV-411), a nonselective nucleotide phosphodiesterases (PDEs) inhibitor, reduces reactive gliosis (Suzumura et al., 1999), is a TLR-4 antagonist (Wu and Wang, 2020), and is considered to be neuroprotective in Alzheimer disease (Schwenkgrub et al., 2017) (see Fig. 21.1). Coadministration of morphine with ibudilast reduces glial reactivity (Suzumura et al., 1999) and diminishes morphine-induced DA transmission within

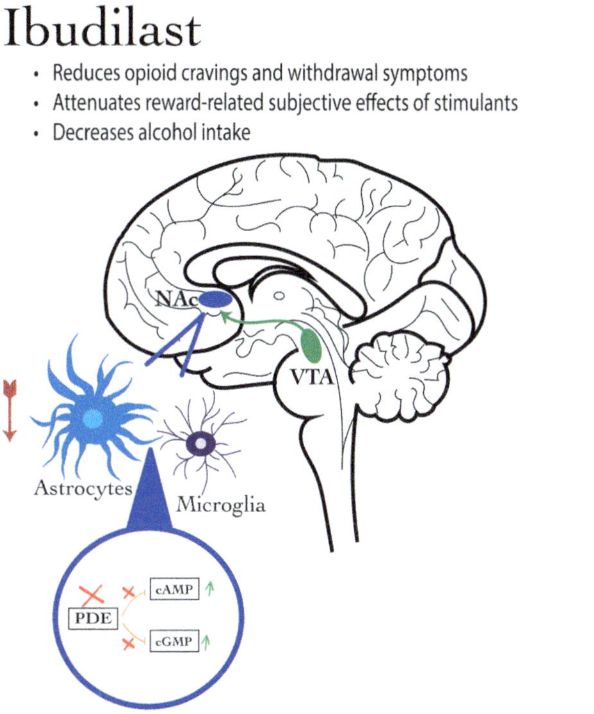

Fig. 21.1. Major preclinical/clinical findings for Ibudilast and impacts on glia. *cAMP*, cyclic adenosine monophosphate; cGMP, cyclic guanosine monophosphate; *NAc*, nucleus accumbens; *PDE*, phosphodiesterase; *VTA*, ventral tegmental area.

the NAc, as well as the somatic signs of opioid withdrawal (Ledeboer et al., 2007) in rodents. Clinically, ibudilast decreases the subjective ratings of severe withdrawal symptoms in opioid-dependent human populations (Cooper et al., 2016, 2017). Ibudilast attenuates opioid-induced positive subjective responses, opioid responding, and drug cravings in humans with opioid use disorder (Metz et al., 2017). Propentofylline, another PDE inhibitor, reduces glial-induced TNF-α release, inhibits microglial and astrocyte reactivity, and increases expression of astroglial GLT-1 (Raghavendra et al., 2004; Tawfik et al., 2008; Norsted Gregory et al., 2013). Propentofylline reduces the development of morphine CPP (Narita et al., 2006) and attenuates morphine tolerance in rodents (Eidson and Murphy, 2013). Propentofylline shows promise clinically, and is safe since it has been used in Asia for dementia; however, it is often poorly absorbed systemically and thus difficult to pursue further investigation in clinical trials for SUD (Sweitzer and De Leo, 2011). Minocycline is an antibiotic with antiinflammatory and neuroprotective properties known to inhibit reactive microgliosis and proliferation, and reduces cytokine release and phagocytosis (Tikka and Koistinaho, 2001; Yong et al., 2004; Hashimoto et al., 2007; Linker et al., 2020; Šimončičová et al., 2022) (see Fig. 21.2). It enhances opioid analgesia while suppressing other opioid-induced effects such as respiratory depression in rats (Hutchinson et al., 2008). Minocycline also blocks the maintenance of morphine CPP and the reinstatement of morphine-seeking behavior in rats (Arezoomandan and Haghparast, 2016). Broad-spectrum antibiotics have also been implicated in morphine tolerance in preclinical trials (Kang et al., 2017), but have not been directly

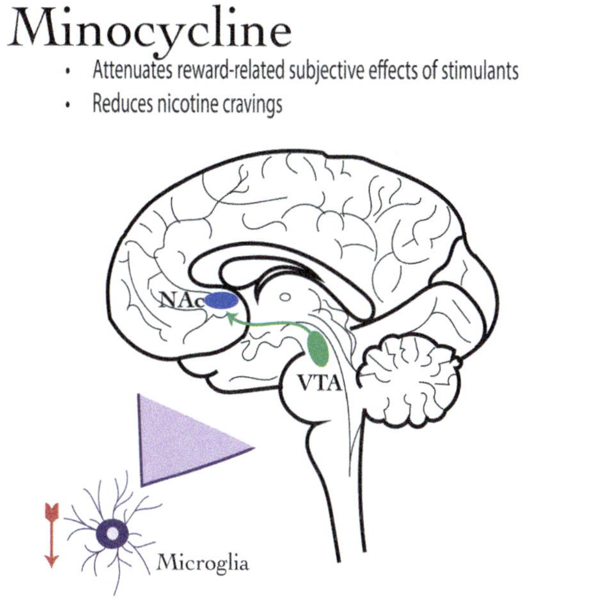

Fig. 21.2. Major preclinical/clinical findings for minocycline and impacts on glia.

implicated in drug seeking behavior suggesting a unique role for minocycline in suppressing microglia. Clinical studies support the enhancing effects of minocycline with morphine in pain management, but has not shown significant effects on tolerance or craving in human patients with opioid use disorders (Arout et al., 2019). Future studies may evaluate the potential effects of minocycline in the development of opioid tolerance. Taken together, ibudilast shows promise in ameliorating opioid use and withdrawal symptoms and in humans (Cooper et al., 2016, 2017; Metz et al., 2017). Given the success in preclinical studies when treatments are administered before drug exposure or coadministered, future clinical studies may evaluate the efficacy of these treatments in the development of tolerance (see Table 21.1).

Stimulants

Many psychostimulants induce reactive glia (Coller and Hutchinson, 2012; Lacagnina et al., 2017) and increase IBA1 and GFAP expression, markers for microglia, and astrocytes, respectively (Fattore et al., 2002; Thomas et al., 2004; Liao et al., 2016). Acute cocaine and methamphetamine exposure upregulates the GFAP expression within the NAc, a critical region for reward processing and DA release in rodents (Armstrong et al., 2004; Narita et al., 2006; Cearley et al., 2011). The colony-stimulating factor 1 (CSF1R) inhibitor, PLX3397, depletes microglia and reduces cocaine behavioral sensitization, suggesting a critical role for microglia in cocaine-induced behaviors (da Silva et al., 2021). Methamphetamine triggers astrogliosis and microgliosis via intracellular sigma-1 receptors (Zhang et al., 2015; Fernandes et al., 2016), upregulating IL-6 and downregulating TNFα (Wang et al., 2018a,b). Cocaine releases largely proinflammatory cytokines such as IL-1β, IL-6, TNFα, and toll-like receptor-2 (TLR2) (Ahmed et al., 2005; Piechota et al., 2010; Northcutt et al., 2015; Lewitus et al., 2016; Liao et al., 2016). Of particular interest, microglial

Table 21.1

Overview of treatments and preclinical/clinical findings for opioids.

Treatment	Model	Treatment dose	Drug behavior	Findings	Reference
Pioglitazone	Mice	Morphine tolerance (9 days; IP; 10 mg/kg)	Morphine tolerance (9 days; IP; 10 mg/kg)	Diminished the development of tolerance to morphine measured by thermal analgesia	de Guglielmo et al. (2014)
	Mice and rats	30 mg/kg 60 mg/kg (oral)	Heroin self-administration and heroin-induced DA release	Acute treatment (60 mg/kg) reduced heroin self-administration. Chronic treatment reduced heroin-induced DA release in the NAc	de Guglielmo et al. (2015)
	Mice	10 mL/kg (oral)	Precipitated morphine withdrawal (6 days; IP; 10 mg/kg; followed by naloxone)	Reduced development of precipitated withdrawal	de Guglielmo et al. (2017)
Ibudilast	Humans	20 mg 40 mg (oral; 2 weeks)	Opioid withdrawal symptoms	All doses decrease several somatic withdrawal symptoms	Cooper et al. (2016)
	Humans	50 mg (oral; 5–6 days)	Morphine and heroin craving, reinforcement, and analgesia	Decreased ratings of drug liking, heroin craving, and enhanced analgesia	Cooper et al. (2017)
	Humans	50 mg (oral; 7 days; twice daily)	Oxycodone administration and heroin craving	Ibudilast reduced subjective effects and responding to oxycodone, and reduced heroin craving	Metz et al. (2017)
Propentofylline	Mice	3 mg/kg (IP)	Morphine conditioned place preference	Prevented morphine conditioned place preference with associated reduction in reactive astrogliosis	Narita et al. (2006)
	Rats	100 fmol (microinfusions into ventrolateral periaqueductal gray)	Morphine tolerance	Attenuates the development of morphine tolerance with associated reductions in morphine-induced microglial and reactive astrogliosis	Eidson and Murphy (2013)
Minocycline	Rats	5 mg/kg (oral)	Morphine-induced respiratory depression and conditioned place preference	Coadministration of minocycline and morphine attenuated morphine-induced respiratory depression and condition place preference while enhancing morphine analgesia	Hutchinson et al. (2008)
	Rats	1 μg/0.5 μL 5 μg/0.5 μL 10 μg/0.5 μL (microinfusions into the nucleus accumbens during extinction)	Morphine conditioned place preference reinstatement	Enhances morphine extinction and decreased morphine conditioned place preference reinstatement	Arezoomandan and Haghparast (2016)
	Humans	200 mg/day (oral; 1, 8, or 15 days)	Opioid tolerance and hyperalgesia	No impact on opioid tolerance or opioid-induced hyperalgesia	Arout et al. (2019)

TLR4 plays a critical role in cocaine-induced increase in extracellular DA in the NAc, as well as cocaine CPP and self-administration (Northcutt et al., 2015). Cocaine decreases VTA GABA neuronal activity via GABA release from astrocytes (Yang et al., 2023). Psychostimulants are also known to alter glutamate homeostasis and disrupt normal neurotransmitter release (Cadet and Bisagno, 2014; Scofield et al., 2016). Methamphetamine promotes extracellular glutamate release via astroglial GLT-1 (Rothstein et al., 1994; Flores et al., 1998; Pereira et al., 2012; Parsegian and See, 2014; Scofield et al., 2016). Chronic cocaine exposure reduces the GFAP expression in the NAc (Scofield et al., 2016), while withdrawal downregulates the GLT1 expression (Fischer-Smith et al., 2012). Thus, the regulatory role of glia in psychostimulant-induced glutamate dysregulationmay be a critical consideration in treatment options.

Stimulant use disorder treatments

Minocycline, by inhibiting microglial activation, has shown promising effects in methamphetamine studies. Preclinical studies show administration of minocycline before methamphetamine exposure may reduce DA release and be protective in drug-induced behavioral measures such as self-administration (Zhang et al., 2006; Hashimoto et al., 2007; Mizoguchi et al., 2008; Snider et al., 2013). Minocycline also prevents the development of cocaine sensitization (Chen et al., 2009), suppresses cocaine CPP, and disrupts upregulation of extracellular DA (Northcutt et al., 2015). In humans, minocycline attenuates the subjective rewarding effects of dextroamphetamine, an isomer of amphetamine (Sofuoglu et al., 2011). The PDE inhibitor, ibudilast, is another drug that has been evaluated as a therapeutic target for psychostimulant use. Ibudilast reduces cocaine and methamphetamine sensitization (Beardsley et al., 2010; Snider et al., 2012; Poland et al., 2016), attenuates self-administration (Snider et al., 2013; Mu et al., 2021), and reinstatement in animal models (Beardsley et al., 2010; Mu et al., 2021). In humans, ibudilast also reduces subjective effects of methamphetamine (Worley et al., 2016; Birath et al., 2017). Minocycline and ibudilast are most effective in reducing the development of dependence and may be feasible treatments to support subjective withdrawal symptoms.

N-Acetylcysteine (N-Acet), a drug that acts at the cysteine-glutamate antiporter of glial cells, shows encouraging outcomes for treatment of cocaine use disorder (see Fig. 21.3). In preclinical studies, N-Acet blocks cocaine-induced glutamate release in the NAc (Baker et al., 2003), and decreases plasticity-dependent drug seeking behavior (Madayag et al., 2007) and cocaine-induced reinstatement (Murray et al., 2012; Jastrzębska et al., 2016). Critically, chronic N-Acet contributes to maintained reductions in cocaine seeking behavior (Reichel et al., 2011). N-Acet also reduces cocaine intake in rats and reverses cocaine-induced decreases the GLT-1 expression in the NAc and dorsolateral striatum (Ducret et al., 2016). Thus, cocaine-induced decreases in the GLT-1 expression lead to an overall upregulation of glutamate levels in the NAc and may suggest a role for astrocytes in this process. Restoring GLT-1 expression and reducing overall glutamate levels using N-Acet presents a viable therapeutic avenue for cocaine treatments

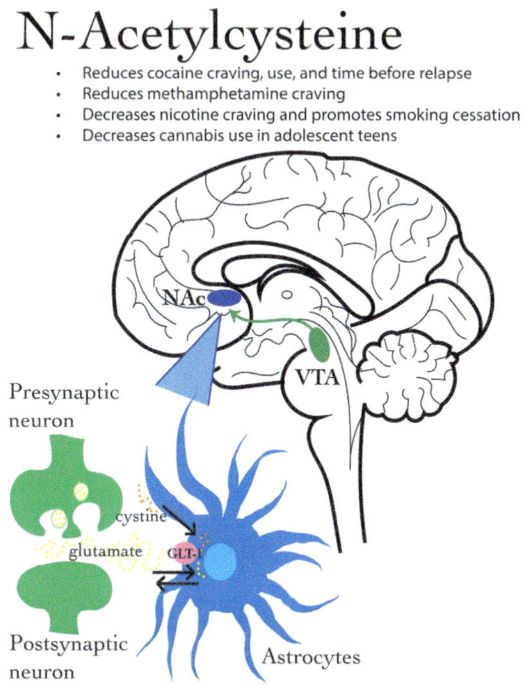

Fig. 21.3. Major preclinical/clinical findings for N-Acet and associated glial-mediated effects. *GLT-1*, glutamate transporter-1; *NAc*, nucleus accumbens; *VTA*, ventral tegmental area.

(Woodcock et al., 2021). In a human treatment study, high doses (2400 mg/day) of N-Acet reduced cocaine use and craving and had the highest treatment retention rate of patients (Mardikian et al., 2007; Amen et al., 2011). Additional clinical studies have shown that N-Acet reduces time to relapse (LaRowe et al., 2013) and cocaine seeking behavior (Woodcock et al., 2021). This dose also reduces cravings for methamphetamine in humans (Mousavi et al., 2015). Indeed, N-Acet has shown overall highly promising results in treating drug cravings (Duailibi et al., 2017; Nocito Echevarria et al., 2017) (see Table 21.2).

Alcohol

Largely proinflammatory glial mechanisms have been implicated in alcohol use. Chronic alcohol exposure upregulates glial markers, microglial CD11b, and astrocytic GFAP, eventually inducing apoptosis via increased caspase-3 activity, that is dependent on TLR4 signaling in mice (Alfonso-Loeches et al., 2010). TLR4 activation in the VTA is associated with binge drinking in alcohol-preferring rats (Liu et al., 2011; June et al., 2015). Critically, there are sex differences to consider in glial mediated effects in ethanol exposure. Greater ethanol-induced astrogliosis and less neuronal microtubule-associated protein 2 (MAP2) were found in female mice when compared to male mice given the same ethanol treatment (Alfonso-Loeches et al., 2013). Adolescent exposure to alcohol upregulates GFAP (Fernandes et al., 2018) which persists into adulthood with marked effects in the hippocampus of female rats as compared to males (Nwachukwu et al., 2022). Postmortem tissue from human patients diagnosed with alcohol dependence has shown reduced astrocytic immunoreactivity in the

Table 21.2

Overview of treatments and preclinical/clinical findings for stimulants.

Treatment	Model	Treatment dose	Drug behavior	Findings	Reference
Ibudilast	Rats	2.5 mg/kg 7.5 mg/kg (IP; twice daily for 2 days, once on test day)	Methamphetamine reinstatement and seeking behavior	Reduced prime and stress-induced reinstatement	Beardsley et al. (2010)
	Mice	1.8–13 mg/kg (IP; twice daily for 7 days)	Methamphetamine behavioral sensitization	Decreased methamphetamine-induced locomotor activity following acute and chronic exposure	Snider et al. (2012)
	Rats	15 mg/kg (IP; twice daily for 3 days, once on test day)	Cocaine behavioral sensitization	Reversed sensitized cocaine-induced locomotor behaviors	Poland et al. (2016)
	Rats	1 mg/kg 7.5 mg/kg 10 mg/kg (IP; twice daily for 3 days)	Methamphetamine self-administration	Reduced methamphetamine self-administration	Snider et al. (2013)
	Rats	1 mg/kg 3 mg/kg 10 mg/kg (IP)	Cocaine self-administration, prime, and stress-induced reinstatement	Ibudilast (3 and 10 mg/kg) decreased cocaine self-administration, total intake, prime, and stress-induced reinstatement	Mu et al. (2021)
	Humans	40 mg 100 mg (oral; twice daily for 1 week)	Methamphetamine subjective effects in dependent users	Attenuates reward subjective effects of methamphetamine	Worley et al. (2016)
	Humans	20 mg 50 mg (oral; 5 days)	Attention in early abstinence from methamphetamine dependence	Ibudilast (50 mg) improved attentional state during early abstinence from methamphetamine	Birath et al. (2017)
Minocycline	Mice	40 mg/kg (IP)	Methamphetamine-induced behavioral sensitization	Diminishes the development of methamphetamine-induced locomotor sensitization with parallel reduction in dopamine release	Zhang et al. (2006)
	Monkeys	200 mg (SC; twice daily)	Methamphetamine administration with PET imaging	Treatment of minocycline (before or after drug administration) attenuates methamphetamine-induced reduction of dopamine transporters	Hashimoto et al. (2007)
	Mice	20–60 mg/kg: IP; 7 days)	Methamphetamine-induced behavioral sensitization and recognition memory	Posttreatment of minocycline (40 mg/kg) attenuated methamphetamine-induced memory impairment. Coadministration of methamphetamine and minocycline abolished behavioral sensitization	Mizoguchi et al. (2008)
	Rats	10 mg/kg 30 mg/kg 60 mg/kg (IP: 3 days)	Methamphetamine self-administration	Reduced methamphetamine self-administration	Snider et al. (2013)
	Mice	80 mg/kg 40 mg/kg (IP)	Cocaine-induced locomotor behavioral sensitization	Prevents the development of behavioral cocaine sensitization	Chen et al. (2009)
	Rats	25 mg/kg (oral)	Cocaine conditioned place preference	Suppresses cocaine conditioned place preference	Northcutt et al. (2015)
	Humans	200 mg/kg (oral; 5 days)	Subjective responses to dextroamphetamine	Decreased subjective rewarding effects of dextroamphetamine	Sofuoglu et al. (2011)

N-Acetylcysteine (N-Acet)	Humans	1200 mg 2400 mg 3600 mg (oral; daily; 4 weeks)	Cocaine dependence	Retention rates best for higher doses. Patients who completed the 4 week trial on N-Acet terminated cocaine use or significantly reduced use	Mardikian et al. (2007)
	Rats and Humans	60 mg/kg (IP; 7 days)	Cocaine reinstatement	Reduced cocaine reinstatement in rats	Amen et al. (2011)
		1200–2400 mg (oral; daily; 4 days)	Cocaine dependence	All doses reduced cocaine craving	
	Humans	1200 mg 2400 mg (oral; 8 weeks)	Cocaine dependence	N-Acet (2400 mg) reduced cocaine craving and had longer time before relapse	LaRowe et al. (2013)
	Humans	1200 mg 3600 mg (oral; daily; 16 days)	Cocaine-seeking behavior	Attenuated cocaine-primed increases cocaine seeking with reduced glutamate levels in critical reward-related region, anterior cingulate cortex	Woodcock et al. (2021)
	Humans	1200 mg (oral; daily; 8 weeks)	Methamphetamine dependence	Suppressed methamphetamine craving	Mousavi et al. (2015)
	Rats	60 mg/kg (IP; 11 days)	Cocaine-seeking behaviors	Administering N-Acet before cocaine exposure blocks cocaine seeking behaviors and reduced long-term reductions of glutamate	Madayag et al. (2007)
	Rats	100 mg/kg (IP; 10 days)	Reinstatement of cocaine seeking	Repeated exposure of N-Acet (100 mg/kg) reduced cocaine reinstatement	Jastrzębska et al. (2016)
	Rats	30 mg/kg 60 mg/kg 90 mg/kg (IP)	Cocaine seeking and reinstatement	Decreased cocaine seeking and prevented cocaine reinstatement	Murray et al. (2012)
	Rats	60 mg/kg 100 mg/kg (IP; daily; 12 days)	Cocaine withdrawal and reinstatement	Administration of N-Acet during extinction from cocaine self-administration decreased cue and cue plus cocaine-induced reinstatement	Reichel et al. (2011)
	Rats	60 mg/kg (IP; daily; 23 days)	Cocaine intake	Decreases cocaine intake after development of abstinence following punishment	Ducret et al. (2016)

prefrontal cortex and hippocampus (Miguel-Hidalgo et al., 2002), and increased microglial immunoreactivity in the cingulate cortex (He and Crews, 2008). Comprehensive evaluation of animal and postmortem human alcoholic brain tissue reveals an increased expression of high mobility group box 1 (HMGB1), an endogenous proinflammatory cytokine, that activates TLR4 signaling (Crews et al., 2013). HMGB1-mediated sensitization of microglia may result in neuroimmune sensitization (Zou and Crews, 2014) that contributes to the impulsivity and negative affect seen in alcohol use disorder (Perry, 2014; Coleman and Crews, 2018).

Alcohol use disorder treatments

Minocycline reduces alcohol self-administration and reinstatement of alcohol CPP in preclinical studies (Plane et al., 2010; Agrawal et al., 2011; Gajbhiye et al., 2017; Harper et al., 2018). However, minocycline has not shown the same promise in human studies (Petrakis et al., 2019). Ceftriaxone, an antibiotic and glial GLT-1 modulator, has been effective in attenuating cue-primed reinstatement and relapse-like ethanol-drinking behavior in alcohol-preferring rats (Qrunfleh et al., 2013; Rao and Sari, 2014; Weiland et al., 2015; Stennett et al., 2017) but has not been evaluated in humans. Pioglitazone decreases ethanol consumption, self-administration, and CPP in preclinical studies (Stopponi et al., 2011, 2013; Blednov et al., 2015; Fotio et al., 2021), but was unsuccessful in clinical trials (Schwandt et al., 2020). Ibudilast reduces ethanol drinking by 50% and relapse in alcohol preferring rats (Bell et al., 2015). Human participants taking ibudilast showed reduced alcohol craving, decreased cue-response and attenuation of some subjective effects of alcohol such as "wanting" and "liking" (Ray et al., 2017). Over a 2-week trial, ibudilast reduced odds of heavy drinking by 45% and decreased alcohol cue-induced functional connectivity in frontostriatal circuits (Burnette et al., 2021; Grodin et al., 2021). Examination of neurometabolites in individuals with alcohol use disorder treated with ibudilast confirmed increases in antiinflammatory markers that were predictive of reduced drinking behaviors (Grodin et al., 2022). Thus, ibudilast presents the most promising potential in reducing alcohol preference and relapse (see Table 21.3).

Nicotine

Nicotine suppresses reactive microglia and promotes an antiinflammatory role in the adult brain (Liu et al., 2018; Noda and Kobayashi, 2011). It inhibits microglial proliferation and microgliosis (Guan et al., 2015; Linker et al., 2020), and suppresses lipopolysaccharide (LPS)-induced microglial reactivity and TLR4 expression (Kim et al., 2014; Li et al., 2016). Nicotine also directly impacts morphologic and functional properties of astrocytes in an nAChR dependent manner (Aryal et al., 2021). Nicotine-induced downregulation of microglia and its antiinflammatory effects are mediated by the α7 nicotinic acetylcholine receptor (nAChR) (Tsoyi et al., 2011; Shen and Yakel, 2012; Egea et al., 2015; Noda and Kobayashi, 2011). A brain imaging study in active adult human smokers has demonstrated decreased microglial reactivity compared to nonsmokers (Brody et al., 2017). In contrast to what has been found in adults, nicotine increases IBA1 expression and induces a proinflammatory microglial state in adolescent rats (Linker et al., 2020). The underlying mechanism involves activation of NAc D2 receptors and fractalkine signaling cascades, leading to enhancement of subsequent cocaine self-administration (Linker et al., 2020). Distinct glial dynamics in drug-induced behaviors during adolescence highlight the importance of considering age when developing treatments.

Nicotine use treatments

In preclinical studies, ceftriaxone reduced nicotine acquisition, intake, withdrawal, and nicotine CPP (Alajaji et al., 2013; Sari et al., 2016; Philogene-Khalid et al., 2017) but has not been evaluated in human trials. Ceftriaxone highly modulates the GLT-1 expression and highlights this as a potential mechanism to consider in future treatments. Pioglitazone and minocycline reduce nicotine craving in humans, but have shown limited potential for smoking cessation (Sofuoglu et al., 2009; Jones et al., 2017). In contrast, N-Acet has shown promise as a viable treatment option, reducing nicotine self-administration (Ramirez-Niño et al., 2013), cue-induced reinstatement (Namba et al., 2020), and withdrawal in adult rodent models (Goenaga et al., 2020). Chronic treatment with N-Acet induces long-lasting reductions in cue-induced nicotine seeking behavior and decreases in extinction responding in rats (Moro et al., 2018; Powell et al., 2019). In humans, N-Acet can decrease nicotine craving, smoking, and increase positive affect during withdrawal (Grant et al., 2013; Froeliger et al., 2015). However, clinical randomized controlled trials were most successful when combined with psychotherapy and/or other nicotine cessation medications such as varenicline and bupropion (McClure et al., 2015; Prado et al., 2015; Machado et al., 2020). Taken together, this suggests adjunctive care with N-Acet may be the optimal treatment course for nicotine use disorder (see Table 21.4).

Cannabis

Astrocytes and microglia express CB1 and CB2 receptors, the latter of which is known to dampen the peripheral inflammatory responses (Herring and Kaminski, 1999; Járai et al., 1999; Stella, 2010). Acute exposure to cannabinoids inhibits inflammatory cytokines and chemokines (Tanasescu and Constantinescu, 2010; McCoy, 2016). Human neuroimaging of long-term cannabis users showed upregulated glutamate metabolites and diminished glial function (Watts et al., 2020). In a cohort of users who started during adolescence, cannabis causes glial dysregulation independent of glutamate changes within the hippocampus, a region heavily implicated in memory (Blest-Hopley et al., 2020). Preclinical studies largely highlight astrocytes in endocannabinoid signaling and subsequent memory impairments following cannabis exposure (Navarrete and Araque, 2008, 2010; Navarrete et al., 2014). Of note, delta-9-tetrahydrocannabinol (THC) exposure during adolescence upregulated the astrocytic marker GFAP in both sexes, but downregulated CB1 receptor expression in female adults only (Lopez-Rodriguez et al., 2014). Detrimental effects following THC exposure during adolescence

Table 21.3

Overview of treatments and preclinical/clinical findings for alcohol.

Treatment	Model	Treatment dose	Drug behavior	Findings	Reference
Pioglitazone	Mice	10 mg/kg 30 mg/kg (oral)	Oral ethanol consumption (15%)	Pioglitazone (30 mg/kg) reduced ethanol intake and preference	Blednov et al. (2015)
	Rat	10 mg/kg 30 mg/kg (oral)	Ethanol drinking (10%), relapse-like behavior, and precipitated withdrawal	Pioglitazone (30 mg/kg) decreased ethanol intake, reinstatement, and self-administration	Stopponi et al. (2011)
	Rats	10 mg/kg 30 mg/kg (IP)	Ethanol intake and relapse-like behavior (10%)	All doses reduced alcohol intake and relapse	Stopponi et al. (2013)
	Rats	2.5 μg/μl 5 μg/μl 10 μg/μl (microinfusions into reward-related brain regions)	Ethanol drinking and seeking behavior (10%)	Decreased voluntary ethanol drinking, self-administration, and seeking	Fotio et al. (2021)
Ibudilast	Rats	3 mg/kg 6 mg/kg 9 mg/kg 12 mg/kg 18 mg/kg (SC; 4 days)	Voluntary ethanol consumption (15%)	Ibudilast (9 mg/kg) decreased ethanol intake by approximately 50%	Bell et al. (2015)
	Humans	50 mg (oral; twice daily; 7 days)	Primary and secondary subjective measures of response to alcohol	Reduced craving and improved mood following stress and alcohol cue exposure	Ray et al. (2017)
	Humans	20 mg on days 1–2 50 mg on days 3–14 (oral; twice a day)	Alcohol cue-induced functional connectivity and use	Decreased alcohol cue-induced connectivity between the ventral striatum and reward-related regions with correlated drinks per day	Burnette et al. (2021)
	Humans	20 mg on days 1–2 50 mg on days 3–14 (oral; twice a day)	Alcohol cue-induced neuroimaging and use	Reduced probability of heavy drinking and lowered cue-induced activation of ventral striatum	Grodin et al. (2021)
	Humans	20 mg on days 1–2 50 mg on days 3–14 (oral; twice a day)	Alcohol-induced neuromarkers and use behavior in individuals with use disorder	Reduced neuroinflammatory markers and correlated with reduced drinking	Grodin et al. (2022)
Minocycline	Mice	10 mL/kg (IP)	Ethanol (10%) voluntary drinking	Reduced total ethanol intake	Agrawal et al. (2011)
	Mice	10 mg/kg 30 mg/kg 50 mg/kg (IP)	Ethanol (20%) conditioned place preference and reinstatement	Attenuated alcohol conditioned place preference and blocked reinstatement	Gajbhiye et al. (2017)
	Humans	100 mg 200 mg (oral; 10 days)	Alcohol administration	No differences in alcohol responses or alcohol-induced craving	Petrakis et al. (2019)
Ceftriaxone	Rats	50 mg/kg 100 mg/kg (IP; 5 days)	Ethanol (15% and 20%) drinking	Reduced ethanol drinking and attenuated relapse with upregulation of GLT-1 in prefrontal cortex and nucleus accumbens	Qrunfleh et al. (2013)
	Rats	100 mg/kg (IP; 5 days)	Ethanol (15% and 30%) drinking	Attenuates ethanol intake with associated upregulation of GLT-1 in prefrontal cortex, amygdala, and nucleus accumbens	Rao and Sari (2014)
	Rats	200 mg/kg (IP; 5 days)	Ethanol (20%) drinking	Decreases ethanol intake that is not associated with GLT-1 expression	Stennett et al. (2017)
	Rats	200 mg/kg (IP; 5 days)	Ethanol (20%) reinstatement	Reduces alcohol cue-induced reinstatement with dampened GABA transmission	Weiland et al. (2015)

Table 21.4

Overview of treatments and preclinical/clinical findings for nicotine.

Treatment	Model	Treatment dose	Drug behavior	Findings	Reference
Pioglitazone	Humans	45 mg (oral; 3 weeks)	Cigarette smoking	Reduced measures of cigarette cravings	Jones et al. (2017)
Minocycline	Humans	200 mg (oral; 4 days)	Nicotine self-administration (IV)	Did not alter nicotine intake but significantly reduced cravings	Sofuoglu et al. (2009)
N-Acetylcysteine (N-Acet)	Rats	30 mg/kg 60 mg/kg 90 mg/kg (IP; daily; acute and 14 days)	Nicotine self-administration and reinstatement	Acute N-Acet treatment reduced nicotine self-administration and cue-induced reinstatement. Chronic N-Acet decreased nicotine self-administration	Ramirez-Niño et al. (2013)
	Rats	100 mg/kg (IP; daily; 5 days)	Nicotine reinstatement	Inhibits cue-induced nicotine seeking with associated upregulation of GLT-1 expression and downregulation of TNFα in nucleus accumbens	Namba et al. (2020)
	Rats	30 mg/kg 60 mg/kg 100 mg/kg (IP)	Nicotine-seeking behavior	N-Acet (100 mg/kg) reduced cue-induced reinstatement	Moro et al. (2018)
	Rats	100 mg/kg (IP; daily;15 days)	Nicotine reinstatement and extinction	Treatment during extinction reduced reinstatement of nicotine seeking	Powell et al. (2019)
	Humans	1200 mg (twice daily; 3 days)	Nicotine withdrawal symptoms and smoking abstinence	Patients treated with N-Acet maintained abstinence and had decreased cravings along with restructured corticostriatal connectivity	Froeliger et al. (2015)
	Humans	1200 mg (twice daily; 4 weeks)	Nicotine dependence	Coadministration of varenicline and N-Acet reduced the number of cigarettes per day	McClure et al. (2015)
	Humans	3 g (oral; daily; 12 weeks)	Nicotine craving and reward	Lowered daily number of cigarettes and increased number of successful cessations	Prado et al. (2015)
	Humans	1800 mg (oral; twice daily; 12 weeks)	Smoking cessation	N-Acet and psychotherapy improved smoking cessation and long-term abstinence	Machado et al. (2020)
Ceftriaxone	Mice	200 mg/kg (IP; twice daily; 4 days)	Nicotine conditioned place preference and reinstatement	Reduced nicotine-induced reinstatement of conditioned place preference	Alajaji et al. (2013)
	Rats	200 mg/kg (IP; daily)	Nicotine conditioned place preference	Reduced development of conditioned place preference	Philogene-Khalid et al. (2017)
	Rats	100 mg/kg (IP; daily; 7 days)	Nicotine intake	Reduced nicotine intake by approximately 70% and upregulated GLT-1 expression in the prefrontal cortex and nucleus accumbens	Sari et al. (2016)

have also been associated with increased expression of glutamate receptor subunits GluA1 and GluA2 in male rats only (Zamberletti et al., 2016). Sex differences and age of exposure in cannabis use and future treatment will be critical to consider given differences in metabolism and pharmacodynamics observed in adolescent females (Ruiz et al., 2021).

Cannabis use treatments

Preclinical and clinical trials in glial modulators and cannabis use are limited (see Table 21.5). N-Acet is effective in reducing craving and abstinence in cannabis use disorder in clinical trials (Gray et al., 2012; Sharma et al., 2022). Interestingly, this treatment has shown more promise in adolescent patients than adults (McClure et al., 2014a,b; Gray et al., 2017). This further emphasizes the need to expand on sex and age differences in cannabis use and treatments.

GUT MICROBIOME

The enteric nervous system (ENS), a major component of the autonomic nervous system, comprises more than 100 million afferent sensory-motor neurons lining the gastrointestinal tract (GI tract). Described as the "second brain," the enteric circuitry processes chemical and electrical input signals and functions independently of the CNS (Rao and Gershon, 2016). Bidirectional biochemical signaling between the GI tract and CNS is referred to as the "gut-brain axis (GBA)." This communication involves immune, hormonal, and neural connections between the brain and gut. Broadly described, the GBA communicates through afferent signaling via the vagus nerve, neuroendocrine signaling via the hypothalamic–pituitary–adrenal axis, enteroendocrine signaling via enteroendocrine cells, gut-derived neurotransmitters, and production of microbial metabolites (Sudo et al., 2004; Mayer, 2011).

The gut microbiome refers to the microorganisms (bacteria, viruses, and eukaryotic microbes) present in the human digestive tract. It is composed of approximately 10 million genes (Kelly et al., 2015; Skosnik and Cortes-Briones, 2016). The variability within an individual's genome, in addition to microbial genome present within the gut, can have profound influences on the brain and behavior. The gut microbiota release chemical signals that interact and directly modulate the bidirectional GBA communication responsible for proper gastrointestinal function and maintenance (Cryan and Dinan, 2012; Carabotti et al., 2015). Functions of gut microbiome have also emerged beyond homeostasis and may be crucial in understanding other medical conditions. Characterizing the richness and diversity of the bacterial phyla is complicated and creates challenges in finding disease causing disruptions, or dysbiosis. Directionality of potential microbiome-derived mechanisms in diseased states is difficult to establish as diseases can also alter the microbiome. Beyond diet and disease, the gut microbiome has direct contributions to the development, homeostasis, and function of astrocytes and microglia (Heiss and Olofsson, 2019). Drugs of abuse can also disrupt gut microbiota and induce an immune response via release of cytokines (Petra et al., 2015). Thus, the gut microbiome has emerged as a critical factor to consider in GBA derived mechanisms underlying substance use.

Gut microbiome and neuroglia

Gut microbiota modulate glial function in the brain. The gut microbiome has direct contributions to microglial homeostasis and development as germ-free mice have altered microglial morphology, density, immature phenotype, and diminished inflammatory response (Erny et al., 2015; Castillo-Ruiz et al., 2018). The gut likely communicates directly with the CNS via the vagus nerve to influence microglia (Forsythe et al., 2014). Furthermore, microbiota-induced "leaky gut," with resulting increased inflammatory cytokines and short chain fatty acids in the circulatory system and BBB permeability, indirectly influences astrocytes and microglia in the brain (Banks, 2009; Mayer et al., 2014; Gorky and Schwaber, 2016; Meckel and Kiraly, 2019). Given this interconnected network, gut microbiome and glial contributions to substance use represent a viable avenue to consider for patient treatment.

Gut microbiome and substance use

Given the established communication between the gut microbiome and the brain, it is essential to consider the impact of the gut microbiome on molecular mechanisms involved in SUDs (Angoa-Pérez and Kuhn, 2021). Drugs of abuse disrupt bacterial homeostasis and cause gut dysbiosis (Volpe et al., 2014; Temko et al., 2017; Wang and Roy, 2017; Hillemacher et al., 2018; Hofford et al., 2019). Alcohol abuse in humans reduces abundance of beneficial

Table 21.5

Overview of treatments and preclinical/clinical findings for cannabis.

Treatment	Model	Treatment dose	Drug behavior	Findings	Reference
N-Acetylcysteine (N-Acet)	Humans	1200 mg (oral; twice daily; 8 weeks)	Cannabis cessation	Decreased cannabis use in adolescent teens	Gray et al. (2012)
	Humans	1200 mg (oral; twice daily; 12 weeks)	Cannabis cessation	Improved smoking cessation rates	McClure et al. (2014b)
	Human	1200 (oral; twice daily; 12 weeks)	Cannabis cessation	Not improvement of smoking cessation rates in adults	Gray et al. (2017)

microbiota such as bifidobacterial, lactobacilli, Bacteroidetes, and bacteroidaceae (Kirpich et al., 2008; Mutlu et al., 2012). Chronic exposure to alcohol in mice increases relative abundance of Bacteroidetes and Verrucomicrobia bacteria while decreasing relative abundance of Firmicutes compared to control (Yan et al., 2011). Similarly, cocaine and alcohol use in humans also increase relative abundance of Bacteroidetes and decrease relative abundance of Firmicutes (Volpe et al., 2014). Cocaine use alone causes gut dysbiosis with increased relative abundance of the Bacteroidetes phylum in humans (Volpe et al., 2014). Given the concentration of opioid receptors within the GI tract (Holzer, 2004), opioid-induced dysbiosis is known to cause constipation and specific gut microbiome alterations are still under investigation in the context of substance use. Morphine-induced dysbiosis in mice alters functional gut microbiome profiles after a single exposure (Wang et al., 2018a). Following methamphetamine CPP, rats showed increased relative abundance of Bacillaceae and Ruminococcaceae in methamphetamine-treated animals compared to controls (Ning et al., 2017). Fentanyl use reduces alpha diversity of gut microbiome in female rats with changes in *Prevotella*, and increases alpha diversity in male rats with marked changes in *Verrucomicrobia*, *Ruminococcus*, and *Akkermansia* (Ren and Lotfipour, 2022a,b). Overall, both clinical and preclinical studies reveal drug-induced alterations of the gut microbiome.

Evaluating how alterations in the gut microbiome directly influence drug-related behaviors has relied primarily on animal studies and antibiotic treatments. Knockdown of gut bacterial content through antibiotics heightens sensitivity to cocaine CPP and increases locomotor sensitization in mice (Kiraly et al., 2016). Broad-spectrum antibiotic treatment attenuates the development of tolerance to morphine in mice (Kang et al., 2017). Short-chain fatty acids (SCFAs), a microbial metabolite, influence gut microbiome driven processes by influencing gene expression, blood–brain barrier (BBB) permeability, and cytokine production (Kelly et al., 2015; Sarkar et al., 2016; Schirmer et al., 2016). SCFAs reverse cocaine place preference in mice (Kiraly et al., 2016). Antibiotic treatment increases fentanyl self-administration in male rats which is reduced by SCFA supplementation (Ren and Lotfipour, 2022b).

Altered microbiota have a significant impact on inflammatory markers in the circulatory system. Alcohol dependency in humans increases intestinal permeability, enhances all proinflammatory markers and *Bifidobacterium*, while decreasing antiinflammatory bacterial species *Faecalibacterium prausnitzii* (Leclercq et al., 2014a,b). Further investigation has revealed lipopolysaccharides and peptidoglycans from the gut microbiota enter the systemic circulation and activate Toll-like receptors to stimulate inflammatory pathways (Leclercq et al., 2014a,b). Alcohol and opioid use animal models highlight impaired epithelial barriers in the intestine lead to translocation of gut bacteria into the circulatory system and causes a systemic immune response (Bala et al., 2014; Leclercq et al., 2014a, b; Hillemacher et al., 2018; Yang et al., 2021). This "leaky gut" and subsequent translocated microbiota trigger immune responses via host TLRs, releasing proinflammatory cytokines (Kelly et al., 2005, 2015). These disruptions can be rescued by SCFA supplementation (Erny et al., 2015). Gut dysbiosis in morphine-dependent rats is directly associated with upregulated neuroinflammatory genes and reactive astrocytes (O'Sullivan et al., 2019). Administration of probiotic *Lactobacillus rhamnosus* attenuates cocaine-induced locomotion and both microgliosis and astrogliosis in mice (Chivero et al., 2022).

Recent studies have examined the impact of microbiome depletion on both brain and behavior. Male mice in which the gut microbiome is depleted by antibiotics show decreased morphine CPP and morphine-induced locomotor sensitization (Hofford et al., 2019). Following 7 days of treatment with high-dose morphine, these animals exhibited changes in NAc gene expression, including genes involved in gliogenesis. The behavioral changes, and many transcriptional changes, were reversed by coadministration with SCFAs. In a recent study of male mice, both constitutive and antibiotic-induced depletion of gut microbiota enhanced nicotine-induced activation of DA neurons in the posterior VTA and modified the motivational effects of the drug (Lakosa et al., 2023). Gut microbiome depletion increased astrocyte density within the pVTA and levels of the postsynaptic density protein PSD-95 in non-DA neurons, without affecting GLT-1. Reorganization of microglial subpopulations within VTA was also observed.

CONCLUSIONS

SUD remains a major health concern and glia present promising therapeutic targets. An investigation of mechanisms underlying drug and glial interactions is ongoing. Ibudilast shows promise in reducing opioid withdrawal symptoms, stimulant rewarding effects, and alcohol use clinical trials (see Table 21.6). Minocycline has many positive findings in preclinical studies for many drugs which have not been as successful in clinical trials. In humans, it has decreased cravings for stimulants and nicotine (see Table 21.7). N-Acet has been most successful in preclinical and clinical trials (see Table 21.8). It is critical to consider models of drug initiation and the nature of dependence when developing "successful" therapeutic interventions. Reducing use, cravings, and time to relapse are common measures of efficacy. Patients with stimulant use disorder often require more support with withdrawal symptoms. Thus, reducing cravings or instances of relapse is prioritized. For opioids, we must also consider that use often initiates in a medical setting. Thus, attenuating the development of tolerance or reducing subjective feelings of "reward" is a potential preventative measure to consider. The rise in electronic cigarettes popularity has caused a surge of use amongst adolescents. We have established nicotine-induced glia dynamics differ between adolescence and adulthood (Linker et al., 2020). Therefore, efficacy of glia-mediated treatments may be reevaluated in adolescent populations. In particular, age-related differences in drug-induced glia or glia-mediated treatments are critical to understand given that adolescence is a major developmental window in which exploratory drug use occurs and often coincide with onset of sex differences. There are sex differences observed in drugs use, glial dynamics, and gut microbiome that are critical to evaluate at a clinical level. The relationship between gut microbiome, brain, and substance use is gaining attention and presents a

Table 21.6

Overview of treatments and preclinical/clinical findings for ibudilast.

Model	Treatment dose	Drug behavior	Findings	Reference
Humans	20 mg 40 mg (oral; 2 weeks)	Opioid withdrawal symptoms	All doses decrease several somatic withdrawal symptoms	Cooper et al. (2016)
Humans	50 mg (oral; 5–6 days)	Morphine and heroin craving, reinforcement, and analgesia	Decreased ratings of drug liking, heroin craving, and enhanced analgesia	Cooper et al. (2017)
Rats	2.5 mg/kg 7.5 mg/kg (IP; twice daily for 2 days, once on test day)	Methamphetamine reinstatement and seeking behavior	Reduced prime and stress-induced reinstatement	Beardsley et al. (2010)
Mice	1.8–13 mg/kg (IP; twice daily for 7 days)	Methamphetamine behavioral sensitization	Decreased methamphetamine-induced locomotor activity following acute and chronic exposure	Snider et al. (2012)
Rats	15 mg/kg (IP; twice daily for 3 days, once on test day)	Cocaine behavioral sensitization	Reversed sensitized cocaine-induced locomotor behaviors	Poland et al. (2016)
Rats	1 mg/kg 7.5 mg/kg 10 mg/kg (IP; twice daily for 3 days)	Methamphetamine self-administration	Reduced methamphetamine self-administration	Snider et al. (2013)
Rats	1 mg/kg 3 mg/kg 10 mg/kg (IP)	Cocaine self-administration, prime, and stress-induced reinstatement	Ibudilast (3 mg/kg and 10 mg/kg) decreased cocaine self-administration, total intake, prime, and stress-induced reinstatement	Mu et al. (2021)
Humans	40 mg 100 mg (oral; twice daily for 1 week)	Methamphetamine subjective effects in dependent users	Attenuates reward subjective effects of methamphetamine	Worley et al. (2016)
Humans	20 mg 50 mg (oral; 5 days)	Attention in early abstinence from methamphetamine dependence	Ibudilast (50 mg) improved attentional state during early abstinence from methamphetamine	Birath et al. (2017)
Rats	3 mg/kg 6 mg/kg 9 mg/kg 12 mg/kg 18 mg/kg (SC; 4 days)	Voluntary ethanol consumption (15%)	Ibudilast (9 mg/kg) decreased ethanol intake by approximately 50%	Bell et al. (2015)
Humans	50 mg (oral; twice daily; 7 days)	Primary and secondary subjective measures of response to alcohol	Reduced craving and improved mood following stress and alcohol cue exposure	Ray et al. (2017)
Humans	20 mg on days 1–2 50 mg on days 3–14 (oral; twice a day)	Alcohol cue-induced functional connectivity and use	Decreased alcohol cue-induced connectivity between the ventral striatum and reward-related regions with correlated drinks per day	Burnette et al. (2021)
Humans	20 mg on days 1–2 50 mg on days 3–14 (oral; twice a day)	Alcohol cue-induced neuroimaging and use	Reduced probability of heavy drinking and lowered cue-induced activation of ventral striatum	Grodin et al. (2021)
Humans	20 mg on days 1–2 50 mg on days 3–14 (oral; twice a day)	Alcohol-induced neuromarkers and use behavior in individuals with use disorder.	Reduced neuroinflammatory markers and correlated with reduced drinking	Grodin et al. (2022)

Table 21.7

Overview of treatments and preclinical/clinical findings for minocycline.

Model	Treatment dose	Drug behavior	Findings	Reference
Rats	5 mg/kg (oral)	Morphine-induced respiratory depression and conditioned place preference	Coadministration of minocycline and morphine attenuated morphine-induced respiratory depression and condition place preference while enhancing morphine analgesia	Hutchinson et al. (2008)
Rats	1 μg/0.5 μL 5 μg/0.5 μL 10 μg/0.5 μL (microinfusions into the nucleus accumbens during extinction)	Morphine conditioned place preference reinstatement	Enhances morphine extinction and decreased morphine conditioned place preference reinstatement	Arezoomandan and Haghparast (2016)
Humans	200 mg/day (oral; 1, 8, or 15 days)	Opioid tolerance and hyperalgesia	No impact on opioid tolerance or opioid-induced hyperalgesia	Arout et al. (2019)
Mice	40 mg/kg (IP)	Methamphetamine-induced behavioral sensitization	Diminishes the development of methamphetamine-induced locomotor sensitization with parallel reduction in dopamine release	Zhang et al. (2006)
Monkeys	200 mg (SC; twice daily)	Methamphetamine administration with PET imaging	Treatment of minocycline (before or after drug administration) attenuates methamphetamine-induced reduction of dopamine transporters	Hashimoto et al. (2007)
Mice	20–60 mg/kg (IP; 7 days)	Methamphetamine-induced behavioral sensitization and recognition memory	Posttreatment of minocycline (40 mg/kg) attenuated methamphetamine-induced memory impairment. Coadministration of methamphetamine and minocycline abolished behavioral sensitization	Mizoguchi et al. (2008)
Rats	10 mg/kg 30 mg/kg 60 mg/kg (IP: 3 days)	Methamphetamine self-administration	Reduced methamphetamine self-administration	Snider et al. (2013)
Mice	80 mg/kg 40 mg/kg (IP)	Cocaine-induced locomotor behavioral sensitization	Prevents the development of behavioral cocaine sensitization	Chen et al. (2009)
Rats	25 mg/kg (oral)	Cocaine conditioned place preference	Suppresses cocaine conditioned place preference	Northcutt et al. (2015)
Humans	200 mg/kg (oral; 5 days)	Subjective responses to dextroamphetamine	Decreased subjective rewarding effects of dextroamphetamine	Sofuoglu et al. (2011)
Mice	10 mL/kg (IP)	Ethanol (10%) voluntary drinking	Reduced total ethanol intake	Agrawal et al. (2011)
Mice	10 mg/kg 30 mg/kg 50 mg/kg (IP)	Ethanol (20%) conditioned place preference and reinstatement	Attenuated alcohol conditioned place preference and blocked reinstatement	Gajbhiye et al. (2017)
Humans	100 mg 200 mg (oral; 10 days)	Alcohol administration	No differences in alcohol responses or alcohol-induced craving	Petrakis et al. (2019)
Humans	200 mg (oral; 4 days)	Nicotine self-administration (IV)	Did not alter nicotine intake but significantly reduced cravings	Sofuoglu et al. (2009)

Table 21.8

Overview of treatments and preclinical/clinical findings for N-Acetylcysteine (N-Acet).

Model	Treatment dose	Drug behavior	Findings	Reference
Humans	1200 mg 2400 mg 3600 mg (oral; daily; 4 weeks)	Cocaine dependence	Retention rates best for higher doses. Patients who completed the 4 week trial on N-Acet terminated cocaine use or significantly reduced use	Mardikian et al. (2007)
Rats and Humans	60 mg/kg (IP; 7 days)	Cocaine reinstatement	Reduced cocaine reinstatement in rats	Amen et al. (2011)
	1200–2400 mg (oral; daily; 4 days)	Cocaine dependence	All doses reduced cocaine craving	
Humans	1200 mg 2400 mg (oral; 8 weeks)	Cocaine dependence	N-Acet (2400 mg) reduced cocaine craving and had longer time before relapse	LaRowe et al. (2013)
Humans	1200 mg 3600 mg (oral; daily; 16 days)	Cocaine-seeking behavior	Attenuated cocaine-primed increases cocaine seeking with reduced glutamate levels in critical reward-related region, anterior cingulate cortex	Woodcock et al. (2021)
Humans	1200 mg (oral; daily; 8 weeks)	Methamphetamine dependence	Suppressed methamphetamine craving	Mousavi et al. (2015)
Rats	60 mg/kg (IP; 11 days)	Cocaine-seeking behaviors	Administering N-Acet before cocaine exposure blocks cocaine seeking behaviors and reduced long-term reductions of glutamate	Madayag et al. (2007)
Rats	100 mg/kg (IP; 10 days)	Reinstatement of cocaine seeking	Repeated exposure of N-Acet (100 mg/kg) reduced cocaine reinstatement	Jastrzębska et al. (2016)
Rats	30 mg/kg 60 mg/kg 90 mg/kg (IP)	Cocaine seeking and reinstatement	Decreased cocaine seeking and prevented cocaine reinstatement	Murray et al. (2012)
Rats	60 mg/kg 100 mg/kg (IP; daily; 12 days)	Cocaine withdrawal and reinstatement	Administration of N-Acet during extinction from cocaine self-administration decreased cue and cue plus cocaine-induced reinstatement	Reichel et al. (2011)
Rats	60 mg/kg (IP; daily; 23 days)	Cocaine intake	Decreases cocaine intake after development of abstinence following punishment	Ducret et al. (2016)
Rats	30 mg/kg 60 mg/kg 90 mg/kg (IP; daily; acute and 14 days)	Nicotine self-administration and reinstatement	Acute N-Acet treatment reduced nicotine self-administration and cue-induced reinstatement. Chronic N-Acet decreased nicotine self-administration	Ramirez-Niño et al. (2013)
Rats	100 mg/kg (IP; daily; 5 days)	Nicotine reinstatement	Inhibits cue-induced nicotine seeking with associated upregulation of GLT-1 expression and downregulation of TNFα in nucleus accumbens	Namba et al. (2020)
Rats	30 mg/kg 60 mg/kg 100 mg/kg (IP)	Nicotine-seeking behavior	N-Acet (100 mg/kg) reduced cue-induced reinstatement	Moro et al. (2018)
Rats	100 mg/kg (IP; daily;15 days)	Nicotine reinstatement and extinction	Treatment during extinction reduced reinstatement of nicotine seeking	Powell et al. (2019)

Continued

Table 21.8

Continued

Model	Treatment dose	Drug behavior	Findings	Reference
Humans	1200 mg (twice daily; 3 days)	Nicotine withdrawal symptoms and smoking abstinence	Patients treated with N-Acet maintained abstinence and had decreased cravings along with restructured corticostriatal connectivity	Froeliger et al. (2015)
Humans	1200 mg (twice daily; 4 weeks)	Nicotine dependence	Coadministration of varenicline and N-Acet reduced the number of cigarettes per day	McClure et al. (2015)
Humans	3 g (oral; daily; 12 weeks)	Nicotine craving and reward	Lowered daily number of cigarettes and increased number of successful cessations	Prado et al. (2015)
Humans	1800 mg (oral; twice daily; 12 weeks)	Smoking cessation	N-Acet and psychotherapy improved smoking cessation and long-term abstinence	Machado et al. (2020)
Humans	1200 mg (oral; twice daily; 8 weeks)	Cannabis cessation	Decreased cannabis use in adolescent teens	Gray et al. (2012)
Humans	1200 mg (oral; twice daily; 12 weeks)	Cannabis cessation	Improved smoking cessation rates	McClure et al. (2014a,b)
Human	1200 (oral; twice daily; 12 weeks)	Cannabis cessation	Not improvement of smoking cessation rates in adults	Gray et al. (2017)

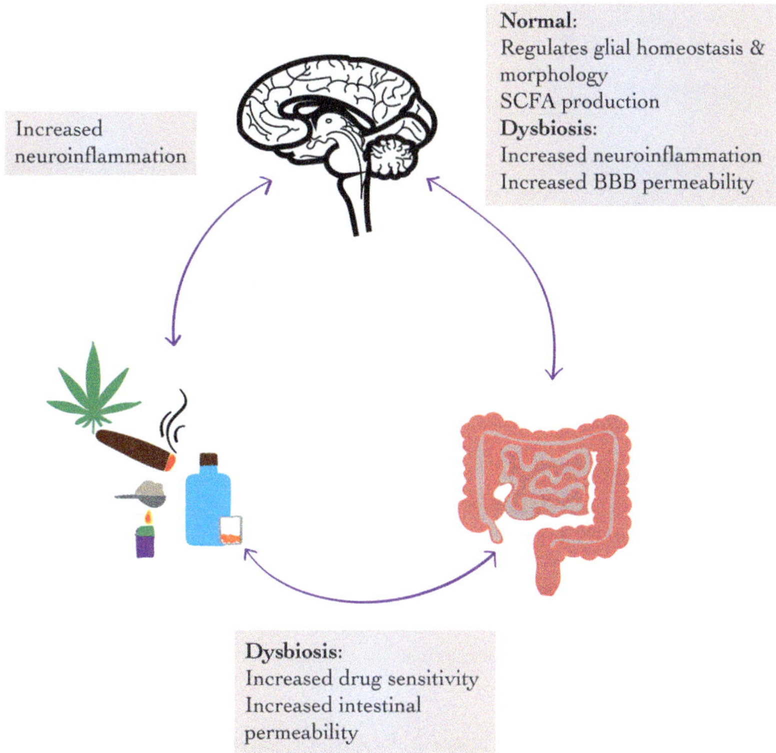

Fig. 21.4. Interactions between drugs of abuse, the brain, and the gut microbiome. *BBB*, blood–brain barrier; *SCFA*, short chain fatty acids.

complex dynamic to consider (see Fig. 21.4). We have described the aforementioned preclinical evidence for the interactions between gut microbiome and substance use with associated glial alterations in reward-related regions. Future clinical studies could explore the potential benefits of gut microbial SCFA supplementation.

References

Agrawal RG, Hewetson A, George CM et al. (2011). Minocycline reduces ethanol drinking. Brain Behav Immun 25: 165–169.

Ahmed SH, Lutjens R, van der Stap LD et al. (2005). Gene expression evidence for remodeling of lateral hypothalamic circuitry in cocaine addiction. Proc Natl Acad Sci U S A 102: 11533–11538.

Alajaji M, Bowers MS, Knackstedt L et al. (2013). Effects of the beta-lactam antibiotic ceftriaxone on nicotine withdrawal and nicotine-induced reinstatement of preference in mice. Psychopharmacology (Berl) 228: 419–426.

Alfonso-Loeches S, Pascual M, Guerri C (2013). Gender differences in alcohol-induced neurotoxicity and brain damage. Toxicology 311: 27–34.

Alfonso-Loeches S, Pascual-Lucas M, Blanco AM et al. (2010). Pivotal role of TLR4 receptors in alcohol-induced neuroinflammation and brain damage. J Neurosci 30: 8285–8295.

Amen SL, Piacentine LB, Ahmad ME et al. (2011). Repeated N-acetyl cysteine reduces cocaine seeking in rodents and craving in cocaine-dependent humans. Neuropsychopharmacology 36: 871–878.

Angoa-Pérez M, Kuhn DM (2021). Evidence for modulation of substance use disorders by the gut microbiome: hidden in plain sight. Pharmacol Rev 73: 571–596.

Arezoomandan R, Haghparast A (2016). Administration of the glial cell modulator, minocycline, in the nucleus accumbens attenuated the maintenance and reinstatement of morphine-seeking behavior. Can J Physiol Pharmacol 94: 257–264.

Armstrong V, Reichel CM, Doti JF et al. (2004). Repeated amphetamine treatment causes a persistent elevation of glial fibrillary acidic protein in the caudate–putamen. Eur J Pharmacol 488: 111–115.

Arout CA, Waters AJ, MacLean RR et al. (2019). Minocycline does not affect experimental pain or addiction-related outcomes in opioid maintained patients. Psychopharmacology (Berl) 236: 2857–2866.

Aryal SP, Fu X, Sandin JN et al. (2021). Nicotine induces morphological and functional changes in astrocytes via nicotinic receptor activity. Glia 69: 2037–2053.

Baker DA, McFarland K, Lake RW et al. (2003). Neuroadaptations in cystine-glutamate exchange underlie cocaine relapse. Nat Neurosci 6: 743–749.

Bala S, Marcos M, Gattu A et al. (2014). Acute binge drinking increases serum endotoxin and bacterial DNA levels in healthy individuals. PLoS One 9: e96864.

Banks WA (2009). The blood-brain barrier in psychoneuroimmunology. Immunol Allergy Clin North Am 29: 223–228.

Beardsley PM, Shelton KL, Hendrick E et al. (2010). The glial cell modulator and phosphodiesterase inhibitor, AV411 (ibudilast), attenuates prime- and stress-induced methamphetamine relapse. Eur J Pharmacol 637: 102–108.

Beitner-Johnson D, Guitart X, Nestler EJ (1993). Glial fibrillary acidic protein and the mesolimbic dopamine system: regulation by chronic morphine and Lewis-Fischer strain differences in the rat ventral tegmental area. J Neurochem 61: 1766–1773.

Bell RL, Lopez MF, Cui C et al. (2015). Ibudilast reduces alcohol drinking in multiple animal models of alcohol-dependence. Addict Biol 20: 38–42.

Bidlack JM (2000). Detection and function of opioid receptors on cells from the immune system. Clin Diagn Lab Immunol 7: 719–723.

Birath JB, Briones M, Amaya S et al. (2017). Ibudilast may improve attention during early abstinence from methamphetamine. Drug Alcohol Depend 178: 386–390.

Blednov YA, Benavidez JM, Black M et al. (2015). Peroxisome proliferator-activated receptors α and γ are linked with alcohol consumption in mice and withdrawal and dependence in humans. Alcohol Clin Exp Res 39: 136–145.

Blest-Hopley G, O'Neill A, Wilson R et al. (2020). Adolescent-onset heavy cannabis use associated with significantly reduced glial but not neuronal markers and glutamate levels in the hippocampus. Addict Biol 25: e12827.

Brody AL, Hubert R, Enoki R et al. (2017). Effect of cigarette smoking on a marker for neuroinflammation: a [11C]DAA1106 Positron Emission Tomography Study. Neuropsychopharmacology 42: 1630–1639.

Burnette EM, Ray LA, Irwin MR et al. (2021). Ibudilast attenuates alcohol cue–elicited frontostriatal functional connectivity in alcohol use disorder. Alcohol Clin Exp Res 45: 2017–2028.

Cadet JL, Bisagno V (2014). Glial-neuronal ensembles: partners in drug addiction-associated synaptic plasticity. Front Pharmacol 5: 1–5.

Carabotti M, Scirocco A, Maselli MA et al. (2015). The gut-brain axis: interactions between enteric microbiota, central and enteric nervous systems. Ann Gastroenterol 28: 203–209.

Castillo-Ruiz A, Mosley M, George AJ et al. (2018). The microbiota influences cell death and microglial colonization in the perinatal mouse brain. Brain Behav Immun 67: 218–229.

Cearley CN, Blindheim K, Sorg BA et al. (2011). Acute cocaine increases interleukin-1β mRNA and immunoreactive cells in the cortex and nucleus accumbens. Neurochem Res 36: 686–692.

Chen H, Uz T, Manev H (2009). Minocycline affects cocaine sensitization in mice. Neurosci Lett 452: 258–261.

Chivero ET, Sil S, Singh S et al. (2022). Protective role of *Lactobacillus rhamnosus* probiotic in reversing cocaine-induced oxidative stress, glial activation and locomotion in mice. J Neuroimmune Pharmacol 17: 62–75.

Coleman LG, Crews FT (2018). Innate immune signaling and alcohol use disorders. In: KA Grant, DM Lovinger (Eds.), The neuropharmacology of alcohol, handbook of experimental pharmacology. Springer International Publishing, Cham, pp. 369–396.

Coller JK, Hutchinson MR (2012). Implications of central immune signaling caused by drugs of abuse: mechanisms, mediators and new therapeutic approaches for prediction and treatment of drug dependence. Pharmacol Ther 134: 219–245.

Cooper ZD, Johnson KW, Pavlicova M et al. (2016). The effects of ibudilast, a glial activation inhibitor, on opioid withdrawal symptoms in opioid-dependent volunteers. Addict Biol 21: 895–903.

Cooper ZD, Johnson KW, Vosburg SK et al. (2017). Effects of ibudilast on oxycodone-induced analgesia and subjective effects in opioid-dependent volunteers. Drug Alcohol Depend 178: 340–347.

Corty MM, Freeman MR (2013). Architects in neural circuit design: glia control neuron numbers and connectivity. J Cell Biol 203: 395–405.

Crews FT, Qin L, Sheedy D et al. (2013). High mobility group box 1/toll-like receptor danger signaling increases brain neuroimmune activation in alcohol dependence. Biol Psychiatry 73: 602–612.

Cryan JF, Dinan TG (2012). Mind-altering microorganisms: the impact of the gut microbiota on brain and behaviour. Nat Rev Neurosci 13: 701–712.

da Silva MCM, Gomes GF, de Barros FH et al. (2021). Inhibition of CSF1R, a receptor involved in microglia viability, alters behavioral and molecular changes induced by cocaine. Sci Rep 11: 15989.

de Guglielmo G, Kallupi M, Scuppa G et al. (2014). Analgesic tolerance to morphine is regulated by PPARγ. Br J Pharmacol 171: 5407–5416.

de Guglielmo G, Melis M, De Luca MA et al. (2015). PPARγ activation attenuates opioid consumption and modulates mesolimbic dopamine transmission. Neuropsychopharmacology 40: 927–937.

de Guglielmo G, Kallupi M, Scuppa G et al. (2017). Pioglitazone attenuates the opioid withdrawal and vulnerability to relapse to heroin seeking in rodents. Psychopharmacology (Berl) 234: 223–234.

Duailibi MS, Cordeiro Q, Brietzke E et al. (2017). N-acetylcysteine in the treatment of craving in substance use disorders: systematic review and meta-analysis. Am J Addict 26: 660–666.

Ducret E, Puaud M, Lacoste J et al. (2016). N-acetylcysteine facilitates self-imposed abstinence after escalation of cocaine intake. Biol Psychiatry 80: 226–234.

Egea J, Buendia I, Parada E et al. (2015). Anti-inflammatory role of microglial alpha7 nAChRs and its role in neuroprotection. Biochem Pharmacol 97: 463–472.

Eidson LN, Murphy AZ (2013). Blockade of toll-like receptor 4 attenuates morphine tolerance and facilitates the pain relieving properties of morphine. J Neurosci 33: 15952–15963.

Erny D, de Angelis ALH, Jaitin D et al. (2015). Host microbiota constantly control maturation and function of microglia in the CNS. Nat Neurosci 18: 965–977.

Escartin C, Galea E, Lakatos A et al. (2021). Reactive astrocyte nomenclature, definitions, and future directions. Nat Neurosci 24: 312–325.

Fattore L, Puddu MC, Picciau S et al. (2002). Astroglial in vivo response to cocaine in mouse dentate gyrus: a quantitative and qualitative analysis by confocal microscopy. Neuroscience 110: 1–6.

Fernandes NC, Sriram U, Gofman L et al. (2016). Methamphetamine alters microglial immune function through P2X7R signaling. J Neuroinflammation 13: 91.

Fernandes LMP, Cartágenes SC, Barros MA et al. (2018). Repeated cycles of binge-like ethanol exposure induce immediate and delayed neurobehavioral changes and hippocampal dysfunction in adolescent female rats. Behav Brain Res 350: 99–108.

Fischer-Smith KD, Houston ACW, Rebec GV (2012). Differential effects of cocaine access and withdrawal on glutamate type 1 transporter expression in rat nucleus accumbens core and shell. Neuroscience 210: 333–339.

Flores C, Rodaros D, Stewart J (1998). Long-lasting induction of astrocytic basic fibroblast growth factor by repeated injections of amphetamine: blockade by concurrent treatment with a glutamate antagonist. J Neurosci 18: 9547–9555.

Forsythe P, Bienenstock J, Kunze WA (2014). Vagal pathways for microbiome-brain-gut axis communication. In: M Lyte, JF Cryan (Eds.), Microbial endocrinology: the microbiota-gut-brain axis in health and disease, advances in experimental medicine and biology. Springer, New York, NY, pp. 115–133.

Fotio Y, Borruto AM, Benvenuti F et al. (2021). Activation of peroxisome proliferator-activated receptor γ reduces alcohol drinking and seeking by modulating multiple mesocorticolimbic regions in rats. Neuropsychopharmacology 46: 360–367.

Froeliger B, McConnell PA, Stankeviciute N et al. (2015). The effects of N-Acetylcysteine on frontostriatal resting-state functional connectivity, withdrawal symptoms and smoking abstinence: a double-blind, placebo-controlled fMRI pilot study. Drug Alcohol Depend 156: 234–242.

Gajbhiye SV, Tripathi RK, Salve B et al. (2017). Evaluation of effect of minocycline on rewarding potential and alcohol relapse in place preference model in mice. Neurosci Lett 649: 28–33.

Goenaga J, Powell GL, Leyrer-Jackson JM et al. (2020). N-acetylcysteine yields sex-specific efficacy for cue-induced reinstatement of nicotine seeking. Addict Biol 25: e12711.

Gofflot F, Chartoire N, Vasseur L et al. (2007). Systematic gene expression mapping clusters nuclear receptors according to their function in the brain. Cell 131: 405–418.

Gomez-Flores R, Weber RJ (1998). Immunomodulation of macrophage functions by opioids. In: H Friedman, JJ Madden, TW Klein (Eds.), Drugs of abuse, immunomodulation, and aids, advances in experimental medicine and biology. Springer US, Boston, MA, pp. 13–19.

Gorky J, Schwaber J (2016). The role of the gut-brain axis in alcohol use disorders. Prog Neuropsychopharmacol Biol Psychiatry 65: 234–241.

Graeber MB (2010). Changing face of microglia. Science 330: 783–788.

Grant JE, Odlaug BL, Chamberlain SR et al. (2013). A randomized, placebo-controlled trial of N-acetylcysteine plus imaginal desensitization for nicotine-dependent pathological gamblers. J Clin Psychiatry 74: 16833.

Gray KM, Carpenter MJ, Baker NL et al. (2012). A double-blind randomized controlled trial of N-acetylcysteine in cannabis-dependent adolescents. AJP 169: 805–812.

Gray KM, Sonne SC, McClure EA et al. (2017). A randomized placebo-controlled trial of N-acetylcysteine for cannabis use disorder in adults. Drug Alcohol Depend 177: 249–257.

Grodin EN, Bujarski S, Towns B et al. (2021). Ibudilast, a neuroimmune modulator, reduces heavy drinking and alcohol cue-elicited neural activation: a randomized trial. Transl Psychiatry 11: 355.

Grodin EN, Nieto SJ, Meredith LR et al. (2022). Effects of ibudilast on central and peripheral markers of inflammation in alcohol use disorder: a randomized clinical trial. Addict Biol 27: e13182.

Guan YZ, Jin XD, Guan LX et al. (2015). Nicotine inhibits microglial proliferation and is neuroprotective in global ischemia rats. Mol Neurobiol 51: 1480–1488.

Guo W, Wang H, Watanabe M et al. (2007). Glial–cytokine–neuronal interactions underlying the mechanisms of persistent pain. J Neurosci 27: 6006–6018.

Harper KM, Knapp DJ, Park MA et al. (2018). Differential effects of single versus repeated minocycline administration—lack of significant interaction with chronic alcohol history. Pharmacol Biochem Behav 168: 33–42.

Hashimoto K, Tsukada H, Nishiyama S et al. (2007). Protective effects of minocycline on the reduction of dopamine transporters in the striatum after administration of methamphetamine: a positron emission tomography study in conscious monkeys. Biol Psychiatry 61: 577–581.

Hasin DS, O'Brien CP, Auriacombe M et al. (2013). DSM-5 criteria for substance use disorders: recommendations and rationale. AJP 170: 834–851.

He J, Crews FT (2008). Increased MCP-1 and microglia in various regions of the human alcoholic brain. Exp Neurol 210: 349–358.

Heiss CN, Olofsson LE (2019). The role of the gut microbiota in development, function and disorders of the central nervous system and the enteric nervous system. J Neuroendocrinol 31: e12684.

Herring AC, Kaminski NE (1999). Cannabinol-mediated inhibition of nuclear factor-kappaB, cAMP response element-binding protein, and interleukin-2 secretion by activated thymocytes. J Pharmacol Exp Ther 291: 1156–1163.

Hillemacher T, Bachmann O, Kahl KG et al. (2018). Alcohol, microbiome, and their effect on psychiatric disorders. Prog Neuro-Psychopharmacol Biol Psychiatry 85: 105–115.

Hofford RS, Russo SJ, Kiraly DD (2019). Neuroimmune mechanisms of psychostimulant and opioid use disorders. Eur J Neurosci 50: 2562–2573.

Holzer P (2004). Opioids and opioid receptors in the enteric nervous system: from a problem in opioid analgesia to a possible new prokinetic therapy in humans. Neurosci Lett 361: 192–195.

Hutchinson MR, Bland ST, Johnson KW et al. (2007). Opioid-induced glial activation: mechanisms of activation and implications for opioid analgesia, dependence, and reward. Sci World J 7: 98–111.

Hutchinson MR, Northcutt AL, Chao LW et al. (2008). Minocycline suppresses morphine-induced respiratory depression, suppresses morphine-induced reward, and enhances systemic morphine-induced analgesia. Brain Behav Immun 22: 1248–1256.

Hutchinson MR, Northcutt AL, Hiranita T et al. (2012). Opioid activation of Toll-like receptor 4 contributes to drug reinforcement. J Neurosci 32: 11187–11200.

Járai Z, Wagner JA, Varga K et al. (1999). Cannabinoid-induced mesenteric vasodilation through an endothelial site distinct from CB1 or CB2 receptors. Proc Natl Acad Sci U S A 96: 14136–14141.

Jastrzębska J, Frankowska M, Filip M et al. (2016). N-acetylcysteine amide (AD4) reduces cocaine-induced reinstatement. Psychopharmacology (Berl) 233: 3437–3448.

Jones JD, Comer SD, Metz VE et al. (2017). Pioglitazone, a PPARγ agonist, reduces nicotine craving in humans, with marginal effects on abuse potential. Pharmacol Biochem Behav 163: 90–100.

June HL, Liu J, Warnock KT et al. (2015). CRF-amplified neuronal TLR4/MCP-1 signaling regulates alcohol self-administration. Neuropsychopharmacology 40: 1549–1559.

Kang M, Mischel R, Bhave S et al. (2017). The effect of gut microbiome on tolerance to morphine mediated antinociception in mice. Sci Rep 7: 42658.

Kelly D, Conway S, Aminov R (2005). Commensal gut bacteria: mechanisms of immune modulation. Trends Immunol 26: 326–333.

Kelly J, Kennedy P, Cryan J et al. (2015). Breaking down the barriersthe gut microbiome, intestinal permeability and stress-related psychiatric disorders. Front Cell Neurosci 91–20.

Kielian T, Drew PD (2003). Effects of peroxisome proliferator-activated receptor-gamma agonists on central nervous system inflammation. J Neurosci Res 71: 315–325.

Kim TH, Kim SJ, Lee SM (2014). Stimulation of the α7 nicotinic acetylcholine receptor protects against sepsis by inhibiting toll-like receptor via phosphoinositide 3-kinase activation. J Infect Dis 209: 1668–1677.

Kiraly DD, Walker DM, Calipari ES et al. (2016). Alterations of the host microbiome affect behavioral responses to cocaine. Sci Rep 6: 35455.

Kirpich IA, Solovieva NV, Leikhter SN et al. (2008). Probiotics restore bowel flora and improve liver enzymes in human alcohol-induced liver injury: a pilot study. Alcohol 42: 675–682.

Kruyer A, Angelis A, Garcia-Keller C et al. (2022). Plasticity in astrocyte subpopulations regulates heroin relapse. Sci Adv 8: eabo7044.

Kruyer A, Kalivas PW, Scofield MD (2023). Astrocyte regulation of synaptic signaling in psychiatric disorders. Neuropsychopharmacology 48: 21–36.

Lacagnina MJ, Rivera PD, Bilbo SD (2017). Glial and neuroimmune mechanisms as critical modulators of drug use and abuse. Neuropsychopharmacology 42: 156–177.

Lakosa A, Rahimian A, Tomasi F et al. (2023). Impact of the gut microbiome on nicotine's motivational effects and glial cells in the ventral tegmental area in male mice. Neuropsychopharmacology 48: 963–974.

LaRowe SD, Kalivas PW, Nicholas JS et al. (2013). A double-blind placebo-controlled trial of N-acetylcysteine in the treatment of cocaine dependence. Am J Addict 22: 443–452.

Leclercq S, De Saeger C, Delzenne N et al. (2014a). Role of inflammatory pathways, blood mononuclear cells, and gut-derived bacterial products in alcohol dependence. Biol Psychiatry 76: 725–733.

Leclercq S, Matamoros S, Cani PD et al. (2014b). Intestinal permeability, gut-bacterial dysbiosis, and behavioral markers of alcohol-dependence severity. Proc Natl Acad Sci U S A 111: E4485–E4493.
Ledeboer A, Hutchinson MR, Watkins LR et al. (2007). Ibudilast (AV-411). A new class therapeutic candidate for neuropathic pain and opioid withdrawal syndromes. Expert Opin Investig Drugs 16: 935–950.
Lefterova MI, Haakonsson AK, Lazar MA et al. (2014). PPARγ and the global map of adipogenesis and beyond. Trends Endocrinol Metab 25: 293–302.
Lehre KP, Rusakov DA (2002). Asymmetry of glia near central synapses favors presynaptically directed glutamate escape. Biophys J 83: 125–134.
Lewitus GM, Konefal SC, Greenhalgh AD et al. (2016). Microglial TNFα suppresses cocaine-induced plasticity and behavioral sensitization. Neuron 90: 483–491.
Li X, Han X, Bao J et al. (2016). Nicotine increases eclampsia-like seizure threshold and attenuates microglial activity in rat hippocampus through the α7 nicotinic acetylcholine receptor. Brain Res 1642: 487–496.
Liang X, Liu R, Chen C et al. (2016). Opioid system modulates the immune function: a review. Transl Perioper Pain Med 1: 5–13.
Liao K, Guo M, Niu F et al. (2016). Cocaine-mediated induction of microglial activation involves the ER stress-TLR2 axis. J Neuroinflammation 13: 33.
Linker KE, Cross SJ, Leslie FM (2019). Glial mechanisms underlying substance use disorders. Eur J Neurosci 50: 2574–2589.
Linker KE, Elabd MG, Tawadrous P et al. (2020). Microglial activation increases cocaine self-administration following adolescent nicotine exposure. Nat Commun 11: 306.
Liu J, Yang AR, Kelly T et al. (2011). Binge alcohol drinking is associated with GABAA α2-regulated Toll-like receptor 4 (TLR4) expression in the central amygdala. Proc Natl Acad Sci U S A 108: 4465–4470.
Liu Q, Liu C, Jiang L et al. (2018). α7 Nicotinic acetylcholine receptor-mediated anti-inflammatory effect in a chronic migraine rat model via the attenuation of glial cell activation. J Pain Res 11: 1129–1140.
Lopez-Rodriguez AB, Llorente-Berzal A, Garcia-Segura LM et al. (2014). Sex-dependent long-term effects of adolescent exposure to THC and/or MDMA on neuroinflammation and serotoninergic and cannabinoid systems in rats. Br J Pharmacol 171: 1435–1447.
Machado RCBR, Vargas HO, Baracat MM et al. (2020). N-acetylcysteine as an adjunctive treatment for smoking cessation: a randomized clinical trial. Braz J Psychiatry 42: 519–526.
Madayag A, Lobner D, Kau KS et al. (2007). Repeated *N*-acetylcysteine administration alters plasticity-dependent effects of cocaine. J Neurosci 27: 13968–13976.
Mardikian PN, LaRowe SD, Hedden S et al. (2007). An open-label trial of N-acetylcysteine for the treatment of cocaine dependence: a pilot study. Prog Neuro-Psychopharmacol Biol Psychiatry 31: 389–394.
Margolis EB, Toy B, Himmels P et al. (2012). Identification of rat ventral tegmental area GABAergic neurons. PLoS One 7: e42365.
Mayer EA (2011). Gut feelings: the emerging biology of gut–brain communication. Nat Rev Neurosci 12. https://doi.org/10.1038/nrn3071.
Mayer EA, Knight R, Mazmanian SK et al. (2014). Gut microbes and the brain: paradigm shift in neuroscience. J Neurosci 34: 15490–15496.
McClure EA, Gipson CD, Malcolm RJ et al. (2014a). Potential role of N-acetylcysteine in the management of substance use disorders. CNS Drugs 28: 95–106.
McClure EA, Sonne SC, Winhusen T et al. (2014b). Achieving Cannabis cessation—evaluating N-acetylcysteine treatment (ACCENT): design and implementation of a multi-site, randomized controlled study in the National Institute on Drug Abuse Clinical Trials Network. Contemp Clin Trials 39: 211–223.
McClure EA, Baker NL, Gipson CD et al. (2015). An open-label pilot trial of N-acetylcysteine and varenicline in adult cigarette smokers. Am J Drug Alcohol Abuse 41: 52–56.
McCoy KL (2016). Interaction between Cannabinoid system and toll-like receptors controls inflammation. Mediators Inflamm 2016: e5831315.
Meckel KR, Kiraly DD (2019). A potential role for the gut microbiome in substance use disorders. Psychopharmacology 236: 1513–1530.
Metz VE, Jones JD, Manubay J et al. (2017). Effects of ibudilast on the subjective, reinforcing, and analgesic effects of oxycodone in recently detoxified adults with opioid dependence. Neuropsychopharmacology 42: 1825–1832.
Miguel-Hidalgo JJ (2009). The role of glial cells in drug abuse. Curr Drug Abuse Rev 2: 76–82.
Miguel-Hidalgo JJ, Wei J, Andrew M et al. (2002). Glia pathology in the prefrontal cortex in alcohol dependence with and without depressive symptoms. Biol Psychiatry 52: 1121–1133.
Mizoguchi H, Takuma K, Fukakusa A et al. (2008). Improvement by minocycline of methamphetamine-induced impairment of recognition memory in mice. Psychopharmacology (Berl) 196: 233–241.
Moreno S, Farioli-Vecchioli S, Cerù MP (2004). Immunolocalization of peroxisome proliferator-activated receptors and retinoid x receptors in the adult rat CNS. Neuroscience 123: 131–145.
Moro F, Orrù A, Marzo CM et al. (2018). mGluR2/3 mediates short-term control of nicotine-seeking by acute systemic N-acetylcysteine. Addict Biol 23: 28–40.
Mousavi SG, Sharbafchi MR, Peykanpour M et al. (2015). Methamphetamine dependence: a double-blind controlled, crossover study. Arch Iran Med 18: 28–33.
Mu L, Liu X, Yu H et al. (2021). Ibudilast attenuates cocaine self-administration and prime- and cue-induced reinstatement of cocaine seeking in rats. Neuropharmacology 201: 108830.
Murray JE, Everitt BJ, Belin D (2012). N-Acetylcysteine reduces early- and late-stage cocaine seeking without affecting cocaine taking in rats. Addict Biol 17: 437–440.
Mutlu EA, Gillevet PM, Rangwala H et al. (2012). Colonic microbiome is altered in alcoholism. Am J Physiol Gastrointest Liver Physiol 302: G966–G978.
Namba MD, Kupchik YM, Spencer SM et al. (2020). Accumbens neuroimmune signaling and dysregulation of astrocytic glutamate transport underlie conditioned nicotine-seeking behavior. Addict Biol 25: e12797.
Narita M, Miyatake M, Narita M et al. (2006). Direct evidence of astrocytic modulation in the development of rewarding effects induced by drugs of abuse. Neuropsychopharmacology 31: 2476–2488.
Navarrete M, Araque A (2008). Endocannabinoids mediate neuron-astrocyte communication. Neuron 57: 883–893.
Navarrete M, Araque A (2010). Endocannabinoids potentiate synaptic transmission through stimulation of astrocytes. Neuron 68: 113–126.
Navarrete M, Díez A, Araque A (2014). Astrocytes in endocannabinoid signalling. Philos Trans R Soc Lond B Biol Sci 369: 20130599.
Ning T, Gong X, Xie L et al. (2017). Gut microbiota analysis in rats with methamphetamine-induced conditioned place preference. Front Microbiol 8: 1–9.
Nocito Echevarria MA, Andrade Reis T, Ruffo Capatti G et al. (2017). N-acetylcysteine for treating cocaine addiction—a systematic review. Psychiatry Res 251: 197–203.
Noda M, Kobayashi AI (2011). Nicotine inhibits activation of microglial proton currents via interactions with α7 acetylcholine receptors. J Physiol Sci 67: 235–245.

Norsted Gregory E, Delaney A, AbdelMoaty S et al. (2013). Pentoxifylline and propentofylline prevent proliferation and activation of the mammalian target of rapamycin and mitogen activated protein kinase in cultured spinal astrocytes. J Neurosci Res 91: 300–312.

Northcutt AL, Hutchinson MR, Wang X et al. (2015). DAT isn't all that: cocaine reward and reinforcement require Toll-like receptor 4 signaling. Mol Psychiatry 20: 1525–1537.

Nwachukwu KN, King DM, Healey KL et al. (2022). Sex-specific effects of adolescent intermittent ethanol exposure-induced dysregulation of hippocampal glial cells in adulthood. Alcohol 100: 31–39.

O'Sullivan SJ, Malahias E, Park J (2019). Single-cell glia and neuron gene expression in the central amygdala in opioid withdrawal suggests inflammation with correlated gut dysbiosis. Front Neurosci 13: 1–14.

Paolicelli RC, Sierra A, Stevens B et al. (2022). Microglia states and nomenclature: a field at its crossroads. Neuron 110: 3458–3483.

Parsegian A, See RE (2014). Dysregulation of dopamine and glutamate release in the prefrontal cortex and nucleus accumbens following methamphetamine self-administration and during reinstatement in rats. Neuropsychopharmacology 39: 811–822.

Pereira FC, Cunha-Oliveira T, Viana SD et al. (2012). Disruption of striatal glutamatergic/GABAergic homeostasis following acute methamphetamine in mice. Neurotoxicol Teratol 34: 522–529.

Perry EC (2014). Inpatient management of acute alcohol withdrawal syndrome. CNS Drugs 28: 401–410.

Petra AI, Panagiotidou S, Hatziagelaki E et al. (2015). Gut-microbiota-brain axis and its effect on neuropsychiatric disorders with suspected immune dysregulation. Clin Ther 37: 984–995.

Petrakis IL, Ralevski E, Gueorguieva R et al. (2019). Targeting neuroinflammation with minocycline in heavy drinkers. Psychopharmacology (Berl) 236: 3013–3021.

Philogene-Khalid H, Simmons SJ, Muschamp JW et al. (2017). Effects of ceftriaxone on conditioned nicotine reward in rats. Behav Pharmacol 28: 485–488.

Piechota M, Korostynski M, Solecki W et al. (2010). The dissection of transcriptional modules regulated by various drugs of abuse in the mouse striatum. Genome Biol 11: R48.

Plane JM, Shen Y, Pleasure DE et al. (2010). Prospects for minocycline neuroprotection. Arch Neurol 67: 1442–1448.

Poland RS, Hahn YK, Knapp PE et al. (2016). Ibudilast attenuates expression of behavioral sensitization to cocaine in male and female rats. Neuropharmacology 109: 281–292.

Powell GL, Leyrer-Jackson JM, Goenaga J et al. (2019). Chronic treatment with N-acetylcysteine decreases extinction responding and reduces cue-induced nicotine-seeking. Physiol Rep 7: e13958.

Prado E, Maes M, Piccoli LG et al. (2015). N-acetylcysteine for therapy-resistant tobacco use disorder: a pilot study. Redox Rep 20: 215–222.

Qiu D, Li X (2015). Pioglitazone inhibits the secretion of proinflammatory cytokines and chemokines in astrocytes stimulated with lipopolysaccharide. Int J Clin Pharmacol Ther 53: 746–752.

Qrunfleh AM, Alazizi A, Sari Y (2013). Ceftriaxone, a beta-lactam antibiotic, attenuates relapse-like ethanol-drinking behavior in alcohol-preferring rats. J Psychopharmacol 27: 541–549.

Raghavendra V, Tanga FY, DeLeo JA (2004). Attenuation of morphine tolerance, withdrawal-induced hyperalgesia, and associated spinal inflammatory immune responses by propentofylline in rats. Neuropsychopharmacology 29: 327–334.

Ramirez-Niño AM, D'Souza MS, Markou A (2013). N-acetylcysteine decreased nicotine self-administration and cue-induced reinstatement of nicotine seeking in rats: comparison with the effects of N-acetylcysteine on food responding and food seeking. Psychopharmacology (Berl) 225: 473–482.

Rao M, Gershon MD (2016). The bowel and beyond: the enteric nervous system in neurological disorders. Nat Rev Gastroenterol Hepatol 13: 517–528.

Rao P, Sari Y (2014). Effectiveness of ceftriaxone treatment in preventing relapse-like drinking behavior following long-term ethanol dependence in P rats. J Addict Res Ther 5: 1000183.

Ray LA, Bujarski S, Shoptaw S et al. (2017). Development of the neuroimmune modulator ibudilast for the treatment of alcoholism: a randomized, placebo-controlled, human laboratory trial. Neuropsychopharmacology 42: 1776–1788.

Reichel CM, Moussawi K, Do PH et al. (2011). Chronic N-acetylcysteine during abstinence or extinction after cocaine self-administration produces enduring reductions in drug seeking. J Pharmacol Exp Ther 337: 487–493.

Ren M, Lotfipour S (2022a). Antibiotic knockdown of gut bacteria sex-dependently enhances intravenous fentanyl self-administration in adult Sprague Dawley rats. IJMS 24: 409.

Ren M, Lotfipour S (2022b). Dose- and sex-dependent bidirectional relationship between intravenous fentanyl self-administration and gut microbiota. Microorganisms 10: 1127.

Reynolds LM, Flores C (2021). Mesocorticolimbic dopamine pathways across adolescence: diversity in development. Front Neural Circuits 15: 1–17.

Rothstein JD, Martin L, Levey AI et al. (1994). Localization of neuronal and glial glutamate transporters. Neuron 13: 713–725.

Ruiz CM, Torrens A, Lallai V et al. (2021). Pharmacokinetic and pharmacodynamic properties of aerosolized ("vaped") THC in adolescent male and female rats. Psychopharmacology (Berl) 238: 3595–3605.

Sari Y, Toalston JE, Rao PSS et al. (2016). Effects of ceftriaxone on ethanol, nicotine or sucrose intake by alcohol-preferring (P) rats and its association with GLT-1 expression. Neuroscience 326: 117–125.

Sarkar A, Lehto SM, Harty S et al. (2016). Psychobiotics and the manipulation of bacteria–gut–brain signals. Trends Neurosci 39: 763–781.

Sarruf DA, Yu F, Nguyen HT et al. (2009). Expression of peroxisome proliferator-activated receptor-γ in key neuronal subsets regulating glucose metabolism and energy homeostasis. Endocrinology 150: 707–712.

Schirmer M, Smeekens SP, Vlamakis H et al. (2016). Linking the human gut microbiome to inflammatory cytokine production capacity. Cell 167: 1125–1136.e8.

Schroeder JR, Phillips KA, Epstein DH et al. (2018). Assessment of pioglitazone and proinflammatory cytokines during buprenorphine taper in patients with opioid use disorder. Psychopharmacology (Berl) 235: 2957–2966.

Schwandt ML, Diazgranados N, Umhau JC et al. (2020). PPARγ activation by pioglitazone does not suppress cravings for alcohol, and is associated with a risk of myopathy in treatment seeking alcohol dependent patients: a randomized controlled proof of principle study. Psychopharmacology (Berl) 237: 2367–2380.

Schwarz JM, Bilbo SD (2013). Adolescent morphine exposure affects long-term microglial function and later-life relapse liability in a model of addiction. J Neurosci 33: 961–971.

Schwarz JM, Hutchinson MR, Bilbo SD (2011). Early-life experience decreases drug-induced reinstatement of morphine CPP in adulthood via microglial-specific epigenetic programming of anti-inflammatory IL-10 expression. J Neurosci 31: 17835–17847.

Schwenkgrub J, Zaremba M, Mirowska-Guzel D et al. (2017). Ibudilast: a non-selective phosphodiesterase inhibitor in brain disorders. Postepy Hig Med Dosw (Online) 71: 137–148.

Scofield MD, Heinsbroek JA, Gipson CD et al. (2016). The nucleus accumbens: mechanisms of addiction across drug classes reflect the importance of glutamate homeostasis. Pharmacol Rev 68: 816–871.

Sharma R, Tikka SK, Bhute AR et al. (2022). N-acetyl cysteine in the treatment of cannabis use disorder: a systematic review of clinical trials. Addict Behav 129: 107283.

Shen J, Yakel JL (2012). Functional α7 nicotinic ACh receptors on astrocytes in rat hippocampal CA1 slices. J Mol Neurosci 48: 14–21.

Šimončičová E, Gonçalves de Andrade E, Vecchiarelli HA et al. (2022). Present and future of microglial pharmacology. Trends Pharmacol Sci 43: 669–685.

Skosnik PD, Cortes-Briones JA (2016). Targeting the ecology within: the role of the gut-brain axis and human microbiota in drug addiction. Med Hypotheses 93: 77–80.

Snider SE, Vunck SA, van den Oord EJCG et al. (2012). The glial cell modulators, ibudilast and its amino analog, AV1013, attenuate methamphetamine locomotor activity and its sensitization in mice. Eur J Pharmacol 679: 75–80.

Snider SE, Hendrick ES, Beardsley PM (2013). Glial cell modulators attenuate methamphetamine self-administration in therat. Eur J Pharmacol 701: 124–130.

Sofroniew MV, Vinters HV (2010). Astrocytes: biology and pathology. Acta Neuropathol 119: 7–35.

Sofuoglu M, Waters AJ, Mooney M et al. (2009). Minocycline reduced craving for cigarettes but did not affect smoking or intravenous nicotine responses in humans. Pharmacol Biochem Behav 92: 135–140.

Sofuoglu M, Mooney M, Kosten T et al. (2011). Minocycline attenuates subjective rewarding effects of dextroamphetamine in humans. Psychopharmacology (Berl) 213: 61–68.

Stella N (2010). Cannabinoid and cannabinoid-like receptors in microglia, astrocytes, and astrocytomas. Glia 58: 1017–1030.

Stennett BA, Frankowski JC, Peris J et al. (2017). Ceftriaxone reduces alcohol intake in outbred rats while upregulating xCT in the nucleus accumbens core. Pharmacol Biochem Behav 159: 18–23.

Stopponi S, Somaini L, Cippitelli A et al. (2011). Activation of nuclear PPARγ receptors by the antidiabetic agent pioglitazone suppresses alcohol drinking and relapse to alcohol seeking. Biol Psychiatry 69: 642–649.

Stopponi S, de Guglielmo G, Somaini L et al. (2013). Activation of PPARγ by pioglitazone potentiates the effects of naltrexone on alcohol drinking and relapse in msP rats. Alcohol Clin Exp Res 37: 1351–1360.

Sudo N, Chida Y, Aiba Y et al. (2004). Postnatal microbial colonization programs the hypothalamic–pituitary–adrenal system for stress response in mice. J Physiol 558: 263–275.

Sutherland ML, Delaney TA, Noebels JL (1996). Glutamate transporter mRNA expression in proliferative zones of the developing and adult murine CNS. J Neurosci 16: 2191–2207.

Suzumura A, Ito A, Yoshikawa M et al. (1999). Ibudilast suppresses TNFα production by glial cells functioning mainly as type III phosphodiesterase inhibitor in the CNS. Brain Res 837: 203–212.

Sweitzer S, De Leo J (2011). Propentofylline: glial modulation, neuroprotection, and alleviation of chronic pain. In: BB Fredholm (Ed.), Methylxanthines, handbook of experimental pharmacology. Springer, Berlin, Heidelberg, pp. 235–250.

Tanasescu R, Constantinescu CS (2010). Cannabinoids and the immune system: an overview. Immunobiology, Special Issue: Cannabinoids and Immunology 215: 588–597.

Tawfik VL, Regan MR, Haenggeli C et al. (2008). Propentofylline-induced astrocyte modulation leads to alterations in glial glutamate promoter activation following spinal nerve transection. Neuroscience 152: 1086–1092.

Taylor AMW, Castonguay A, Taylor AJ et al. (2015a). Microglia disrupt mesolimbic reward circuitry in chronic pain. J Neurosci 35: 8442–8450.

Taylor A, Ghogha A, Cook C et al. (2015b). (251) Opioids and microglia: a murine-derived microglia stable cell line study. J Pain 16: S38.

Taylor AMW, Castonguay A, Ghogha A et al. (2016). Neuroimmune regulation of GABAergic neurons within the ventral tegmental area during withdrawal from chronic morphine. Neuropsychopharmacology 41: 949–959.

Temko JE, Bouhlal S, Farokhnia M et al. (2017). The microbiota, the gut and the brain in eating and alcohol use disorders: a 'Ménage à Trois'? Alcohol Alcohol 52: 403–413.

Theberge FR, Li X, Kambhampati S et al. (2013). Effect of chronic delivery of the Toll-like receptor 4 antagonist (+)-naltrexone on incubation of heroin craving. Biol Psychiatry 73: 729–737.

Thomas DM, Dowgiert J, Geddes TJ et al. (2004). Microglial activation is a pharmacologically specific marker for the neurotoxic amphetamines. Neurosci Lett 367: 349–354.

Tikka TM, Koistinaho JE (2001). Minocycline provides neuroprotection against N-methyl-D-aspartate neurotoxicity by inhibiting microglia1. J Immunol 166: 7527–7533.

Tsoyi K, Jang HJ, Kim JW et al. (2011). Stimulation of alpha7 nicotinic acetylcholine receptor by nicotine attenuates inflammatory response in macrophages and improves survival in experimental model of sepsis through heme oxygenase-1 induction. Antioxid Redox Signal 14: 2057–2070.

Volkow ND, Morales M (2015). The brain on drugs: from reward to addiction. Cell 162: 712–725.

Volpe GE, Ward H, Mwamburi M et al. (2014). Associations of cocaine use and HIV infection with the intestinal microbiota, microbial translocation, and inflammation. J Stud Alcohol Drugs 75: 347–357.

Volterra A, Meldolesi J (2005). Astrocytes, from brain glue to communication elements: the revolution continues. Nat Rev Neurosci 6: 626–640.

Wahli W, Michalik L (2012). PPARs at the crossroads of lipid signaling and inflammation. Trends Endocrinol Metab 23: 351–363.

Wang F, Roy S (2017). Gut homeostasis, microbial dysbiosis, and opioids. Toxicol Pathol 45: 150–156.

Wang B, Chen T, Wang J et al. (2018a). Methamphetamine modulates the production of interleukin-6 and tumor necrosis factor-alpha via the cAMP/PKA/CREB signaling pathway in lipopolysaccharide-activated microglia. Int Immunopharmacol 56: 168–178.

Wang F, Meng J, Zhang L et al. (2018b). Morphine induces changes in the gut microbiome and metabolome in a morphine dependence model. Sci Rep 8: 3596.

Wang X, Loram LC, Ramos K et al. (2012). Morphine activates neuroinflammation in a manner parallel to endotoxin. Proc Natl Acad Sci 109: 6325–6330.

Watkins LR, Hutchinson MR, Rice KC et al. (2009). The "Toll" of opioid-induced glial activation: improving the clinical efficacy of opioids by targeting glia. Trends Pharmacol Sci 30: 581–591.

Watts JJ, Garani R, Da Silva T et al. (2020). Evidence that cannabis exposure, abuse, and dependence are related to glutamate metabolism and glial function in the anterior cingulate cortex: a 1H-magnetic resonance spectroscopy study. Front Psych 11: 1–9.

Weiland A, Garcia S, Knackstedt LA (2015). Ceftriaxone and cefazolin attenuate the cue-primed reinstatement of alcohol-seeking. Front Pharmacol 6: 1–11.

Wise RA, Robble MA (2020). Dopamine and addiction. Annu Rev Psychol 71: 79–106.

Woodcock EA, Lundahl LH, Khatib D et al. (2021). N-acetylcysteine reduces cocaine-seeking behavior and anterior cingulate glutamate/glutamine levels among cocaine-dependent individuals. Addict Biol 26: e12900.

Worley MJ, Heinzerling KG, Roche DJO et al. (2016). Ibudilast attenuates subjective effects of methamphetamine in a placebo-controlled inpatient study. Drug Alcohol Depend 162: 245–250.

Wu NC, Wang JJ (2020). Ibudilast, a phosphodiesterase inhibitor and toll-like receptor-4 antagonist, improves hemorrhagic shock and reperfusion-induced left ventricular dysfunction by reducing myocardial tumor necrosis factor α. Transplant Proc 52: 1869–1874.

Yan AW, Fouts D, Brandl J et al. (2011). Enteric dysbiosis associated with a mouse model of alcoholic liver disease. Hepatology 53: 96–105.

Yang J, Xiong P, Bai L et al. (2021). The association of altered gut microbiota and intestinal mucosal barrier integrity in mice with heroin dependence. Front Nutr 8: 1–14.

Yang J, Chen J, Liu Y et al. (2023). Ventral tegmental area astrocytes modulate cocaine reward by tonically releasing GABA. Neuron 111: 1104–1117.e6.

Yong VW, Wells J, Giuliani F et al. (2004). The promise of minocycline in neurology. Lancet Neurol 3: 744–751.

Yoon SY, Patel D, Dougherty PM (2012). Minocycline blocks lipopolysaccharide induced hyperalgesia by suppression of microglia but not astrocytes. Neuroscience 221: 214–224.

York EM, Bernier LP, MacVicar BA (2018). Microglial modulation of neuronal activity in the healthy brain. Dev Neurobiol 78: 593–603.

Yuan M, Cross SJ, Loughlin SE et al. (2015). Nicotine and the adolescent brain. J Physiol 593: 3397–3412.

Zamberletti E, Gabaglio M, Grilli M et al. (2016). Long-term hippocampal glutamate synapse and astrocyte dysfunctions underlying the altered phenotype induced by adolescent THC treatment in male rats. Pharmacol Res 111: 459–470.

Zare N, Pourhadi M, Vaseghi G et al. (2023). The potential interplay between opioid and the toll-like receptor 4 (TLR-4). Immunopharmacol Immunotoxicol 45: 240–252.

Zhang L, Kitaichi K, Fujimoto Y et al. (2006). Protective effects of minocycline on behavioral changes and neurotoxicity in mice after administration of methamphetamine. Prog Neuro-Psychopharmacol Biol Psychiatry 30: 1381–1393.

Zhang Y, Lv X, Bai Y et al. (2015). Involvement of sigma-1 receptor in astrocyte activation induced by methamphetamine via up-regulation of its own expression. J Neuroinflammation 12: 29.

Zou JY, Crews FT (2014). Release of neuronal HMGB1 by ethanol through decreased HDAC activity activates brain neuroimmune signaling. PLoS ONE 9: e87915.

Handbook of Clinical Neurology, Vol. 210 (3rd series)
Neuroglia in Neurologic and Psychiatric Disorders, Part II
A. Verkhratsky, L.D. de Witte, E. Aronica, and E.M. Hol, Editors
https://doi.org/10.1016/B978-0-443-19102-2.00018-1

Chapter 22

Neuroglia in suicide

LIN ZHANG AND DICK F. SWAAB*

Neuropsychiatric Disorders Lab, Neuroimmunology Group, Netherlands Institute for Neuroscience, An Institute of the Royal Netherlands Academy of Arts and Sciences, Amsterdam, The Netherlands

Abstract

Suicide is the worst outcome for many neuropsychiatric disorders having a high social and economic burden. There is a great need to determine the neurobiologic background of the etiopathogenesis and resilience toward suicide and to find novel pharmacologic strategies to treat suicidal behaviors. Neuroglia have been found to actively participate in the regulation of many cerebral functions, but it is debated whether these cells are structurally or functionally involved in the neuropathology of suicide, or merely follow the changes of comorbid psychiatric disorders. The purpose of this chapter is to review the scattered literature on the involvement of neuroglia in suicide and to describe how these cells might be responsive to the current pharmacologic interventions. We describe the different biological features of neuroglia in relation to suicide and the underlying psychiatric disorders, the molecular commonalities of neuroglial alterations in suicide across different psychiatric disorders, and the evidence for morphologic neuroglia changes in relation to the severity and resilience of suicide. Illuminating the mechanisms by which neuroglia are involved in suicide may ultimately lead to the development of suicide-related biomarkers and novel therapies for suicide prevention.

Suicide is defined as the death ideations and behaviors that intentionally end one's life (WHO, 2023). As a major global health concern, more than 700,000 people across the world die due to suicide, and 20 times more people attempt suicide every year (Turecki and Brent, 2016). One in every four to five people have experienced suicidal thoughts in their lifetime. Suicide is a leading cause of death of all ages. Among adolescent and young populations, in particular, it has been the second leading cause of death from age 10 to 14, and from 25 to 34 (Bachmann, 2018). The harm of suicide is not only limited to lifespan shortening but also a heavy mental load on the people that experienced suicidal behaviors, on the left behind and the whole society (O'Dea and Tucker, 2005). It has been reported that the rate of suicide in family members of suicide victims was twice as high as compared to subjects who died of other causes (Runeson and Åsberg, 2003). This is not due solely to genetics, as copycat behavior also may be involved. Among the individuals exposed to friends and acquaintances who died by suicide, the rate of newly onset psychiatric diseases was shown to be more than five times higher than their matching controls lacking suicide exposure (Brent et al., 1992). In addition, patient suicide is an occupational hazard for psychological and psychiatric practitioners (Chemtob et al., 1989). More than a quarter of psychiatrists considered a change in career path as a result of their patient's death due to suicide (Gibbons et al., 2019). Together, effective suicide prevention is crucial not only in relation to the treatment of mental disorders but also for the maintenance of social functioning.

RISK FACTORS AND TREATMENTS FOR SUICIDE

A series of risk factors can increase the possibility of suicide attempts, including previous suicide attempts, comorbid mental disorders, substance abuse, recent life events, adverse childhood experiences, family history, etc. It is noteworthy to consider that the risk factors for suicide that are indicated differ in developed and developing countries. In general, less than a quarter of suicides worldwide occurred in developed countries, while more than three-quarters occurred in low- and middle-income countries. Pre-existing psychiatric disorders are considered to be the most significant risk for suicide in high-income countries because more than 90% of suicide completers were reported to have a psychiatric diagnosis during their lifetime (Bertolote and Fleischmann, 2002). However, more than 80% of these investigated cases were from North America and highly developed European countries, where psychiatric service is fully

*Correspondence to: Prof. Dick F. Swaab M.D., Ph.D., Netherlands Institute for Neuroscience, An Institute of the Royal Netherlands Academy of Arts and Sciences, Amsterdam, The Netherlands. Tel: +31-20-566-5507, Fax: +31-20-566-6121, E-mail: d.f.swaab@nin.knaw.nl

equipped and easily accessible (Bertolote et al., 2003). In contrast, the profiles of risk factors mentioned in developing countries show some differences from those in developed countries. Female gender, financial problems, and serious illnesses (incurable diseases or chronic pain) are more commonly seen as risk factors in developing countries (Vijayakumar et al., 2005; WHO, 2023). One should note that the uneven distribution of mental healthcare, together with religious, economic, and cultural reasons, can make psychiatric assistance unreachable, especially in developing countries (WHO, 2023). However, even in the United States, a lack of access to mental healthcare holds for half of the people with severe and persistent mental disorders, especially those without insurance, frequently resulting in dramatic deterioration (Drake and Latimer, 2012).

There is clinical consensus that the effects of remitted psychiatric disorders on suicide risks are largely determined by the co-existing mental disorders (Hoertel et al., 2015). However, the diagnostic variability and self-concealment of suicide attempts have confounded suicide prevention by mental healthcare (Hogge and Blankenship, 2020; Jørgensen et al., 2023). In addition, due to the differences in risk factors, approaches to diminish the suicide rate in developed countries may not be appropriate for the current situation in developing countries. Even so, efforts to prevent suicide still showed poor responses in developed countries. According to a complete follow-up of the patient records, more than 5% of the patients with major depressive disorder (MDD), bipolar disorder (BD), or schizophrenia (SCZ) continue to be at risk for suicide even when they are receiving active mental health services (Nordentoft et al., 2011). There are signs that indicated increased risks for suicide regarding antipsychotics or antidepressants. For example, among the selective serotonin reuptake inhibitors, paroxetine treatment has been reported to result in an increased intensity of attempted suicides compared to placebo treatments (Aursnes et al., 2005). As to serotonin-norepinephrine reuptake inhibitors, the prescription of venlafaxine has been consistently linked to a higher risk of suicide (Rubino et al., 2007). In addition, a possible link between benzodiazepine overprescription and increased potential for suicide attempts in males and adolescents is worth discussing (Neutel and Patten, 1997; Niederkrotenthaler et al., 2019). Zolpidem exposure was found to be significantly associated with the risk of suicide or suicide attempt, and this risk increased with the level of zolpidem use, independent of the comorbidity of psychiatric disorders (Sun et al., 2016; Khan et al., 2022). Of note, some cases documented a relationship between abrupt clozapine discontinuation and a subsequent rise in suicide events (Patchan et al., 2015).

To sum up, a possible solution is to develop therapeutic strategies grounded in the molecular neuropathology of suicide, which should be investigated independent of the underlying mental disorders (Zhao et al., 2019; Zhang et al., 2022a).

NEUROGLIA BIOLOGY PROVIDES MOLECULAR FEATURES OF SUICIDE

Neuroglia consist of non-neuronal cells in the brain and spinal cord. In the central nervous system, they include astroglia, microglia, oligodendroglia, and ependymal cells. The major functions of neuroglia were for a long time thought to be physical support, but these cells are also actively involved in metabolism, oxygen supplies to neurons, and removal of pathogens and dead neurons. They also play important roles in neurotransmission and synaptic functions. Postmortem studies examining neuroglia in individuals who died by suicide have revealed significant alterations in cell density and morphology, and functional phenotypes specific to biologic mechanisms.

Astroglia in suicide

Astroglia are the primary homeostatic cells of the central nervous system, essential for normal neuronal development and function, metabolism, and response to mood regulation and neuroinflammation. Astroglial alterations are implicated in the pathophysiology of suicide. From epigenetics, morphology, quantity, functional phenotypes, and confounder aspects, postmortem studies have shown functional astroglial alterations in cortical and subcortical regions of suicide completers.

Investigations on genome-wide DNA methylation associated with astroglial markers have indicated significant differences in the methylation patterns in the prefrontal cortex of depressed suicide victims compared to nonpsychiatric controls (Nagy et al., 2015). A morphometric analysis after Golgi staining further displayed that fibrous astroglia in the anterior cingulate cortex (ACC) of depressed suicides had significantly larger cell bodies, as well as longer, more ramified processes than those measured in controls (Torres-Platas et al., 2011). Kekesi et al. (2012) observed a decreased expression of the glial fibrillary acidic protein (GFAP) in the prefrontal cortex of depressed suicide victims. Another study also reported, in a group of subjects, the majority of whom died by suicide, that GFAP expression was reduced in the locus coeruleus (Chandley et al., 2013). Cobb et al. (2016) further showed in suicide completers, and not in natural death cases, that the expression of $GFAP^+$ astroglia in the hippocampal subregions was negatively correlated to the duration of comorbid depressive illness. However, we found that the overall expression of $GFAP^+$ astroglia in the hypothalamic paraventricular nucleus was much higher in patients with mood disorders who died by suicide than in patients with mood disorders who died naturally, and in nonpsychiatric controls (Zhang et al., 2022b). These immunohistochemical data indicate heterogeneous alterations in astroglia across cortical and subcortical brain regions in relation to suicide.

In postmortem studies, altered astroglial expression can be represented by specific astroglia phenotypes. In the noradrenergic locus coeruleus, excitatory amino acid transporters 1 and 2 (EAAT1 and EAAT2) showed reduced expressions in depressed patients (Chandley et al., 2013). However, the evidence shown by our previous study has displayed that this decrease may be largely due to the high proportion of suicide completers involved in Chandley's collection (Zhao et al., 2016). In addition, microarray analyses showed reduced expression of connexin genes in the neocortex, mediodorsal thalamus, and caudate nucleus of suicide completers (Ernst et al., 2011; Nagy et al., 2017). This could be associated with reduced connectivity between astroglia-oligodendroglia gap junctions due to decreased connexin 30 coupling to oligodendroglia- and myelin-specific connexins (Tanti et al., 2019). Similarly, these areas also showed reduced densities of vimentin$^+$ astroglia in depressed suicides relative to controls (O'Leary et al., 2021). In vitro studies have proven that membrane protein connexins are involved in glutamate release, while EAATs are crucial for glutamate uptake (Magdy et al., 2021). The above data implied that astroglia could be candidate targets in mediating the abnormalities in glutamatergic neurotransmission in suicide (Zhao et al., 2016).

Microglia in suicide

Microglia originate from myeloid precursors in the yolk sac and are regarded as the resident macrophages of the central nervous system (Mosser et al., 2017). The functions of microglia include participating in the formation of new synapses, pruning of weak synapses, induction of neuronal apoptosis, and phagocytosing apoptotic neurons (Brisch et al., 2022). Microglia critically contribute to the psychopathology of suicide through a series of biologic mechanisms.

Steiner et al. (2006) were the first to observe highly elevated microglial cell numbers in the ACC and mediodorsal thalamus of two schizophrenic patients who died of suicide during acute psychosis. Later, significant microgliosis was reported in the dorsolateral prefrontal cortex (DLPFC) as well, in suicidal patients with schizophrenia and depression (Steiner et al., 2008). Torres-Platas et al. (2014) further showed in the ACC that the ratio of primed over ramified microglia was significantly increased in depressed suicides compared to nonpsychiatric controls.

Studies also showed altered microglia densities in relation to functional abnormalities via molecular phenotypes. Among patients with bipolar disorder (BD), the densities of microglia immunoreactive for the purinergic receptor P2Y12 (P2RY12) or major histocompatibility complex II (MHC-II) in the hippocampus of suicide completers were significantly higher than their nonsuicidal counterparts as well as nonpsychiatric controls (Naggan et al., 2023). Interestingly, their densities were negatively correlated with the proportion of cells that double-labeled lymphocyte activation gene 3 (LAG3), a negative regulator of microglia activation, indicating that reduced expression of microglial LAG3 enabled the proliferation of microglia in suicide (Naggan et al., 2023). Similar differences were also evident in the anterior midcingulate cortex between suicidal and nonsuicidal BD patients, showing that suicide completers displayed a higher ionized calcium-binding adapter molecule 1 immunoreactive (iba1$^+$) microglia density than nonsuicidal subjects (Petrasch-Parwez et al., 2020). This difference was not shown in schizophrenic patients following accomplished suicide (Petrasch-Parwez et al., 2020).

There is a significant association between the presence of neuroinflammation and suicidal behaviors (Vasupanrajit et al., 2022). In general, quinolinic acid (QUIN), kynurenic acid (KYN), and L-kynurenine are neuroactive kynurenine pathway metabolites that accumulate in inflammatory neurologic diseases (Heyes et al., 1993). Plasma KYN levels were shown to be negatively associated with suicidal ideations, one of the most serious symptoms of MDD (Setoyama et al., 2016). In the central nervous system, QUIN is synthesized by activated microglia, which acts as an endogenous *N*-methyl-D-aspartate glutamate receptor agonist with neurotoxic properties. Steiner et al. (2011) found an increased microglial QUIN expression in the subgenual and supracallosal parts, but not in the pregenual part of the ACC in postmortem brains of suicide cases with severe depression. They further reported reduced expression of QUIN$^+$ microglia in the hippocampus of depressed patients who died by suicide (Busse et al., 2015), again indicating the involvement of disturbed glutamatergic neurotransmission and mononuclear phagocyte system activation in suicide pathophysiology. In addition, transcriptome sequencing data from the DLPFC showed different expression patterns that were significantly enriched in microglia and endothelial-related genes between depressed patients who died of suicide and nonpsychiatric controls, indicating the participation of cell death and apoptosis (Zeng et al., 2020). However, one does not know from this comparison whether the alterations are related to depression or suicide.

Using positron emission tomography, researchers have measured the availability of translocator protein (TSPO) *in vivo*, which showed significant elevation in the ACC of patients with major depression, resulting from their suicidal thoughts (Holmes et al., 2018). Since TSPO is upregulated predominately in microglia, this study indicates the involvement of neuroinflammation in the ACC of severe major depressive episodes with suicidal thoughts.

Oligodendroglia in suicide

The main function of oligodendroglia is to provide support and insulation to axons within the central nervous system (CNS). Inflammatory mechanisms in the CNS may be active during suicidal behaviors, along with oligodendroglia dysfunction and altered glutamate neurotransmission (Sha et al., 2023).

Using oligonucleotide microarrays, Klempan et al. (2009b) first claimed pronounced misexpression of genes central to astroglia/oligodendroglia function and GABAergic signaling in the ventral prefrontal cortex of suicide victims with and without major depression. They subsequently examined the brain expression and genetic/epigenetic regulation of KH domain containing RNA-binding gene (QKI), an oligodendroglia-specific RNA-binding protein important for myelination, and found reductions in transcriptional and translational levels of QKI mRNA in the hippocampus, amygdala, and orbitofrontal cortex of depressed suicide victims, supporting the growing body of evidence for a role of myelination-related deficits in the etiology of suicide (Klempan et al., 2009a). In addition, Pantazatos et al. (2017) also identified a lower expression of genes involved in oligodendroglia differentiation in the DLPFC in both suicide and depression. There is a strong indication that a history of early life adversity, instead of suicide, might be the key explanation for cell-type-specific changes in DNA methylation of oligodendroglia genes and a global impairment of the myelin-related transcriptional program (Lutz et al., 2017). On the other hand, Chandley et al. (2022) disclosed that oligodendrocyte pathology in depressed suicide victims is widespread in the brain and not restricted to regions commonly associated with depression psychopathology. Interestingly, Zhou et al. (2023) found that oligodendroglia precursor cells contributed most to cell-type-specific gene expression alterations and methylated sites in suicidal behaviors. More studies are needed to disclose the exact involvement of oligodendroglia in suicide pathology and clarify whether they are involved in early life adversities leading to suicide.

PITFALLS IN SUICIDE RESEARCH

Animal models of suicide

Suicide is a complex of suicidal thoughts, planning, preparations, and execution, which, so far, cannot be properly simulated in animal models. There are no animal models of suicide. Current animal stress models can neither reproduce the entire process of suicide nor imitate the intention in suicide mechanisms (Preti, 2011). The major function of animal experiments in suicide research at present is to help understand certain pathophysiologic pathways to the underlying neuropsychiatric disorders, and suicide-trait-related behaviors and pharmacologic means to improve our ability to manage the risk of suicide (Malkesman et al., 2009; Gould et al., 2017).

In addition, one should keep in mind the discrepancies in neurochemistry, neuroanatomy, and neuropharmacology between animal studies and human research. For example, our recent study showed a sharply reduced number of neuropeptide Y-immunoreactive (NPY^+) neurons, which perform anxiolytic and antidepressant actions in mood regulation, in the infundibular nucleus of depressed patients who died by suicide and other death causes (Zhang et al., 2024). However, an increased expression of NPY^+ fibers was shown in a mouse model of chronic stress, without expression of NPY in the soma of mouse arcuate nucleus (infundibular nucleus in humans) as neurobiologic basics (Yam et al., 2017). Evidence obtained from animal models may miss key molecules and, thus, may not always be representative of human data with similar pathophysiology.

Human postmortem brain samples in suicide studies

So far, in vivo human brain neuroimaging studies have proven to be able to prospectively predict a treatment response of psychiatric disorders (Poirot et al., 2024), but it cannot yet foresee therapeutic outcomes, such as suicide or natural death. Neuropathologic studies are so far fundamental to provide suicide-related biomarkers in functionally involved key regions. However, suicide is a serious confounder in many postmortem studies on neuropsychiatric and neurodegenerative diseases (Zhao et al., 2019; Zhang et al., 2022a). Evidence showed that patients with psychiatric disorders who died by suicide or natural death may have very different neurobiologic alterations when compared to nonpsychiatric control subjects (Zhao et al., 2016, 2018). The often-used comparison of human brain samples of depressed patients who died by suicide with nonpsychiatric controls has produced alterations that can either be caused by suicide or by the underlying psychiatric disorder. In addition, a direct comparison between suicide and nonpsychiatric controls, without the endorsement of a psychiatric diagnosis, will ignore the confounding role of the underlying psychiatric disorders in suicide pathology. For example, reduced expression of $GFAP^+$ astroglia in the thalamus and caudate nucleus and an upregulated microglia expression in the ACC were reported in depressed suicides compared to the controls, while one could not conclude whether these alterations were due to suicide or depression (Torres-Platas et al., 2014, 2016). Furthermore, when completed suicide and natural death cases are mixed into the comparisons between psychiatric illnesses and their matched controls, one also cannot tell whether significant changes were due to psychiatric disorders, or have been confounded by the suicide cases (Catts et al., 2014).

Thus, we suggest that suicide research should, as a standard, at least include nonpsychiatric control subjects, patients with a psychiatric disorder who died by natural death, and patients with the same psychiatric disorder who completed suicide. In the following context, we will illustrate the value of such a protocol with our findings concerning neuroglial involvements in suicide.

NOVEL NEUROGLIAL CHANGES IN SUICIDE

Suicide and psychiatric disorders: Commonalities and differences in neuroglia

The development of suicidality is often presumed to be a pathogenic extension of its underlying psychiatric disorders, such as MDD, BD, and SCZ. Indeed, suicide goes together with distinct transcriptional alterations in the human brains relative to underlying psychiatric disorders. Scheepstra et al. (2023, 2024) have reported microglia transcriptional profiling in the occipital cortex and corpus callosum and indicated an immune-suppressed microglia phenotype that was present in MDD, either with or without suicidality. Our recent studies also revealed, in the DLPFC of individuals with MDD, that the transcripts of astroglial S100 calcium-binding protein B (S100B) and oligodendrocytic myelin basic protein (MBP) were increased compared to nonpsychiatric controls; this elevation did not differentiate between patients who died by suicide and patients who died by natural death (Zhang et al., 2021). These examples showed that MDD might affect microglial homeostasis or oligodendrocytic myelination irrespective of natural death or suicide.

However, more studies have observed biological differences between existing psychiatric disorders and suicidality. One pattern observed is that the alteration appeared only in psychiatric disorders, and not in suicide. For instance, aldehyde dehydrogenase 1 family member L1 (ALDH1L1), an astroglial marker involved in dopaminergic metabolism, only showed an elevated expression in the DLPFC of depressed patients who died naturally, compared to the control subjects (Zhang et al., 2021), and a similar alteration occurred in individuals with SCZ (Zhang et al., 2020a). In addition, increased glutamate transmission, involving higher expression of EAAT2 and glutamine synthetase (GS), was found in the DLPFC of nonsuicidal, and not in suicidal depressed patients compared to the controls (Zhao et al., 2016). Concerning microglia, we found only depressed patients who died naturally showed altered expression of cluster of differentiation molecule 11B (CD11B) and P2RY12 in the ACC, and TSPO in the DLPFC, compared to the controls (Zhang et al., 2021). In BD, the transcripts of CD68 and P2RY12 were reduced in the DLPFC of patients who died naturally compared to nonpsychiatric controls and suicide completers (Zhang et al., 2020b). In BD and SCZ, a reduced expression of triggering receptor expressed on myeloid cell 2 (TREM2) was found only in natural deaths, and not in suicide completers (Zhang et al., 2020a,b).

Another pattern observed is that neuroglia alterations merely occurred in suicide completers, not in nonsuicidal patients. For example, the expression of $GFAP^+$ astroglia in the hypothalamic paraventricular nucleus only showed increases in depressed patients who died by suicide, compared to those who died naturally (Zhang et al., 2022b). In patients with MDD, the expression of CD68 in the DLPFC was higher than controls only in suicide completers, whereas the elevation was not found in nonsuicidal BD patients (Zhang et al., 2020b, 2021). These examples showed that suicide-related molecular alterations can well be independent of comorbid psychiatric disorders.

Neuroglia changes in suicide can be similar in different psychiatric disorders

Suicide can be accompanied by similar neurobiologic features in different psychiatric diseases. Our previous studies found that in the prefrontal cortex of suicidal individuals with psychiatric disorders, e.g., MDD, BD, and SCZ, the transcriptional levels of P2RY12 were elevated compared to nonsuicidal patients (Zhang et al., 2020a,b, 2021), which is consistent with a disturbed purine metabolism in the PFC as observed in suicide victims (Pantazatos et al., 2017). Similar findings were later reported by Punzi et al. (2022), that expressed genes that altered purinergic signaling in microglia were associated with the violent manners of suicide, across different psychiatric disorders.

Psychiatric suicides and nonpsychiatric suicides may have shared or independent neuroglia changes

More than 77% of global suicides occurred in low- and middle-income countries, where psychiatric services were not well organized compared to high-income countries (WHO, 2023). The imbalanced mental health support has resulted in a huge number of suicides without psychiatric diagnosis. Therefore, clinical documents could not provide a reliable history of underlying psychiatric illness.

The strong and intense ideations that drive a person to end his/her own life can be shared with comorbid psychiatric disorders, irrespective of comorbid psychiatric disorders, or even absent psychiatric diagnosis. Therefore, we hypothesized that psychiatric suicides and nonpsychiatric suicides have molecular commonalities. To investigate this possibility, brain samples from donors who died by legal euthanasia, a group of people with strong and intense suicidality, were introduced in neurobiologic research. According to the clinically documented mental states, these donors could be subdivided into subjects with or without psychiatric disorders. The latter often had cancer. By comparing them with natural death cases with or without psychiatric illnesses, we obtained specific suicide-related biomarkers in psychiatric suicides and nonpsychiatric suicides, and also obtained the molecular features that co-existed (Zhang, 2023). First, purinergic microglia were found to represent an example of such a molecular commonality. We found that the expression of P2RY12 in the hypothalamic supraoptic nucleus showed increases in (1) controls with legal euthanasia versus controls with natural deaths, and (2) mood disorders (MD) with legal euthanasia or suicide versus MDs with natural deaths (Zhang et al., 2022b).

Second, suicide-related biomarkers are not always identical between psychiatric suicides and nonpsychiatric suicides. In the entorhinal and prefrontal cortex of depressed patients who died by suicide or legal euthanasia, a significant amount of microglia with fragmented processes (Fig. 22.1) was observed, which was absent in nonpsychiatric control subjects who received legal euthanasia (Zhang et al., 2022b).

Together, we distinguished two patterns of suicide-related biomarkers: (i) suicide-specific markers that occur across different psychiatric disorders that end in suicide and are also present in nonpsychiatric suicides. We aim to identify a wide spectrum of

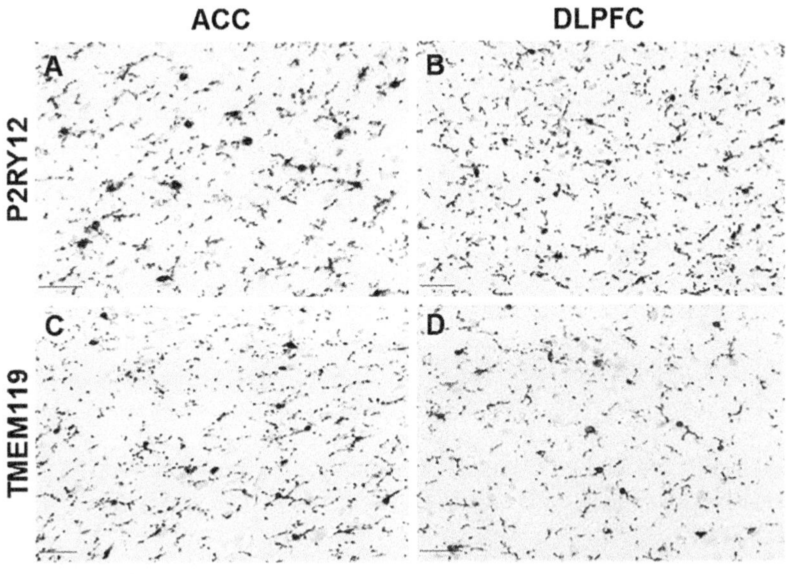

Fig. 22.1. $P2RY12^+$ (A and B) or $TMEM119^+$ (C and D) microglia showed smaller cell sizes and fragmented processes in the anterior cingulate cortex (ACC) and dorsolateral prefrontal cortex (DLPFC) of donors who died by suicide or legal euthanasia. Scale bar = 50 μm. From Lin Zhang's studies, with permission.

molecular commonalities across different domains, which provide potential suicide-specific therapeutic targets. (ii) Suicide risk markers that are limited to a particular psychiatric disorder. This has special value for understanding how suicidality originates in specific psychiatric disorders and, therefore, provides specific treatment options based on diagnosed disorders.

Microglia are involved in resilience to suicide

To translate the suicide-related molecules into clinically relevant treatment targets, one should validate which ones are resilient to suicide. Resilience is a positive adaptation to adversity, requiring the presence of both adversity and positive adaptation (Bhatnagar, 2021). People who are resilient to suicide are those who experienced suicidal intentions or behaviors but bounced back from their suicidality. In order to select such people, we proposed the following grouping: (1) Predisposition to suicide: Donors with diagnosed psychiatric disorders, who did not experience suicidal ideations throughout their lifespans. These patients have a higher risk for suicide than nonpsychiatric control subjects. (2) Suicide onset: A group of people with psychiatric disorders who completed suicide. (3) Resilience to suicide: People who had been diagnosed with psychiatric disorders and experienced suicidality (ideations/attempts) but died of natural causes. By the comparisons between these three groups, we can tell whether a suicide-related biomarker might be a suicide target in potential, a trait marker that indicates the pathologic alterations in association with previous suicidalities, or a state marker that acts as a dynamic monitor of the suicide.

We interpreted, from the alterations we observed in morphometric analysis that microglial changes were associated with completed suicide, resilience to suicide, and the presence of previous suicidality. In the entorhinal cortex of depressed patients, the size of iba1^{+} microglia was smaller in suicide completers than in patients who did not have suicidal ideations throughout their lifetime. When patients experienced suicidal ideations or attempted suicide, and survived until they had a natural death, the iba1^{+} microglia became larger and returned to their original sizes. Correspondingly, we showed that the fragments of P2RY12^{+} microglia in the entorhinal cortex were larger in suicide completers than in nonsuicidal patients and suicide survivors. On the other hand, the density of transmembrane protein 119 immunoreactive (TMEM119^{+}) microglial fragments was increased in patients who died by suicide compared to the nonsuicidals, but this increase did not alter even when patients died naturally. Alterations shown by these morphologic variables indicated that the presence of more fragmented microglia was part of the pathophysiology of suicide. A previous study has shown that the presence of suicidality could accelerate the aging of immune cells (Schiweck et al., 2020). Together with our findings, this indicates that microglial homeostasis would be a promising target in therapeutic development. More importantly, the significance of studying suicide resilience further confirms the insufficiency of using a two-group comparison (depressed suicides vs nonpsychiatric controls) to explore suicide biomarkers, because it could not tell whether a biomarker could be a therapeutic target, a trait marker, or a state marker.

NEUROGLIA AND ANTISUICIDAL THERAPIES

Neuroglia can be targeted for therapeutic approaches. Based on studies providing human data, we described this possibility based upon three general antisuicidal treatments: cognitive-behavioral therapy (CBT), ketamine, and lithium.

CBT has been widely applied, and its efficacy in suicide prevention has been proven across ages and following various adverse life triggers, including self-injury, post-traumatic stress disorder, sexual assault, and cognitive degeneration (Zhang et al., 2022a). Among adults, CBT is effective in preventing suicide attempts for patients with depression and/or a history of recent suicide attempts (Brown et al., 2005). Using positron emission tomography (PET), a recent study has examined the microglia marker translocator protein total distribution volume (TSPO VT) in the brains of individuals with MDD before and after CBT. It is shown that TSPO VT was significantly elevated in neocortical gray matter, frontal cortex, temporal cortex, and hippocampus in MDD relative to the control subjects before therapy, while after CBT, TSPO VT was significantly normalized (Li et al., 2018). Another study further indicated significantly higher TSPO VT in patients with MDD compared with nonpsychiatric controls in the ACC, which was especially pronounced in patients with suicidal thinking (Holmes et al., 2018). In addition, studies measured the myelin oligodendrocyte glycoprotein in plasma peripheral blood mononuclear cells in patients with comorbid diagnoses of MDD and relapsing–remitting multiple sclerosis, with evidence showing that the level of depression and MOG-stimulated IFN-γ production both declined significantly over the 16-week CBT treatment (Mohr et al., 2001).

Ketamine has become an effective drug in suicide interventions. Many studies have confirmed that ketamine can rapidly alleviate suicidal thoughts regardless of the administered frequency (single/repeated) or mode (intravenous/intranasal/oral) (Grunebaum et al., 2017, 2018; Beaudequin et al., 2020; Domany and Mccullumsmith, 2022). By treating HMC3 human microglia with ketamine, Ho et al. (2019) found that ketamine regulated the type 1 interferon pathway in human microglia through the signal transducer and activation of transcription 3 (STAT-3), which played a major role in the immune response.

Lithium is a mood stabilizer for managing mood disorders such as mania and bipolar disorder. Observational studies and small randomized controlled trials have disclosed the idea that taking lithium is associated with a reduction in these symptoms (Smith and Cipriani, 2017). Astroglia are direct cellular targets of lithium treatment. Yu et al. (2011) have shown that lithium treatment may exert its cytoprotective and mood-stabilizing effects by inducing expression of fasciculation and elongation protein zeta-1 in astroglia, which is necessary for normal axonal bundling and elongation of axon bundles. Similar neuroprotective effects were also revealed by Nishino et al. (2012), who showed that lithium reduced the release of an intracellular brain-derived neurotrophic factor in human astrocytoma cells. In addition, lithium may increase the size of the hippocampus in bipolar patients by promoting the generation of new neurons and microglia (Jahandideh et al., 2023). Using human-induced pluripotent stem cell-derived microglia, it was demonstrated that lithium inhibited neuroinflammation through the kynurenine pathway in the microglia compartment of the human brain (Göttert et al., 2022). Animal studies have shown that the use of lithium could increase stem cell differentiation into oligodendroglia and improve re-myelination (Ghosouri et al., 2023). Correspondingly, the level of oligodendrocyte transcription factor 2 in the peripheral blood of BD patients with lithium treatments was higher than their counterparts who are not taking lithium (Ferensztajn-Rochowiak et al., 2016).

Since evidence from human tissues that correlates antisuicidal therapies and glial activities was scarce, more studies are expected to disclose whether neuroglial cells can indeed be effective target cells with antisuicidal properties.

CONCLUSIONS

As our understanding of the neuropathologic and homeostatic functions of these heterogeneous cell types expands, neuroglia have emerged as prominent targets for research on suicide. This chapter summarized our current knowledge on the molecular alterations in neuroglia in the pathopsychological processes of suicide. This topic is rapidly developing into a novel research line on suicide. Postmortem studies offered invaluable insights into neuroglia in the development, onset, and resilience of suicidal behavior, albeit still with certain insufficiencies, as discussed. It is proposed that by normalizing the aberrant expression of neuroglial genes in relation to suicide, we may be able to devise novel therapeutic strategies to ameliorate suicidal thoughts/attempts.

References

Aursnes I, Tvete IF, Gaasemyr J et al. (2005). Suicide attempts in clinical trials with paroxetine randomised against placebo. BMC Med 3: 1–5.

Bachmann S (2018). Epidemiology of suicide and the psychiatric perspective. Int J Environ Res Public Health 15: 1425.

Beaudequin D, Can AT, Dutton M et al. (2020). Predicting therapeutic response to oral ketamine for chronic suicidal ideation: a Bayesian network for clinical decision support. BMC Psychiatry 20: 1–15.

Bertolote JM, Fleischmann A (2002). Suicide and psychiatric diagnosis: a worldwide perspective. World Psychiatry 1: 181.

Bertolote JM, Fleischmann A, De Leo D et al. (2003). Suicide and mental disorders: do we know enough? Br J Psychiatry 183: 382–383.

Bhatnagar S (2021). Rethinking stress resilience. Trends Neurosci 44: 936–945.

Brent DA, Perper J, Moritz G et al. (1992). Psychiatric effects of exposure to suicide among the friends and acquaintances of adolescent suicide victims. J Am Acad Child Adolesc Psychiatry 31: 629–639.

Brisch R, Wojtylak S, Saniotis A et al. (2022). The role of microglia in neuropsychiatric disorders and suicide. Eur Arch Psychiatry Clin Neurosci 272: 929–945.

Brown GK, Ten Have T, Henriques GR et al. (2005). Cognitive therapy for the prevention of suicide attempts: a randomized controlled trial. JAMA 294: 563–570.

Busse M, Busse S, Myint AM et al. (2015). Decreased quinolinic acid in the hippocampus of depressive patients: evidence for local anti-inflammatory and neuroprotective responses? Eur Arch Psychiatry Clin Neurosci 265: 321–329.

Catts VS, Wong J, Fillman SG et al. (2014). Increased expression of astrocyte markers in schizophrenia: association with neuroinflammation. Aust N Z J Psychiatry 48: 722–734.

Chandley MJ, Szebeni K, Szebeni A et al. (2013). Gene expression deficits in pontine locus coeruleus astrocytes in men with major depressive disorder. J Psychiatry Neurosci 38: 276–284.

Chandley MJ, Szebeni A, Szebeni K et al. (2022). Markers of elevated oxidative stress in oligodendrocytes captured from the brainstem and occipital cortex in major depressive disorder and suicide. Prog Neuro-Psychopharmacol Biol Psychiatry 117: 110559.

Chemtob CM, Bauer GB, Hamada RS et al. (1989). Patient suicide: occupational hazard for psychologists and psychiatrists. Prof Psychol Res Pract 20: 294.

Cobb JA, O'Neill K, Milner J et al. (2016). Density of GFAP-immunoreactive astrocytes is decreased in left hippocampi in major depressive disorder. Neuroscience 316: 209–220.

Domany Y, Mccullumsmith CB (2022). Single, fixed-dose intranasal ketamine for alleviation of acute suicidal ideation. An emergency department, trans-diagnostic approach: A randomized, double-blind, placebo-controlled, proof-of-concept trial. Arch Suicide Res 26: 1250–1265.

Drake RE, Latimer E (2012). Lessons learned in developing community mental health care in North America. World Psychiatry 11: 47–51.

Ernst C, Nagy C, Kim S et al. (2011). Dysfunction of astrocyte connexins 30 and 43 in dorsal lateral prefrontal cortex of suicide completers. Biol Psychiatry 70: 312–319.

Ferensztajn-Rochowiak E, Tarnowski M, Samochowiec J et al. (2016). Increased mRNA expression of peripheral glial cell markers in bipolar disorder: the effect of long-term lithium treatment. Eur Neuropsychopharmacol 26: 1516–1521.

Ghosouri S, Soleimani M, Bakhtiari M et al. (2023). Evaluation of in vivo lithium chloride effects as a GSK3-β inhibitor on human adipose derived stem cells differentiation into oligodendrocytes and re-myelination in an animal model of multiple sclerosis. Mol Biol Rep 50: 1617–1625.

Gibbons R, Brand F, Carbonnier A et al. (2019). Effects of patient suicide on psychiatrists: survey of experiences and support required. BJPsych Bulletin 43: 236–241.

Göttert R, Fidzinski P, Kraus L et al. (2022). Lithium inhibits tryptophan catabolism via the inflammation-induced kynurenine pathway in human microglia. Glia 70: 558–571.

Gould T, Georgiou P, Brenner L et al. (2017). Animal models to improve our understanding and treatment of suicidal behavior. Transl Psychiatry 7: e1092.

Grunebaum MF, Ellis SP, Keilp JG et al. (2017). Ketamine versus midazolam in bipolar depression with suicidal thoughts: a pilot midazolam-controlled randomized clinical trial. Bipolar Disord 19: 176–183.

Grunebaum MF, Galfalvy HC, Choo T-H et al. (2018). Ketamine for rapid reduction of suicidal thoughts in major depression: a midazolam-controlled randomized clinical trial. Am J Psychiatry 175: 327–335.

Heyes MP, Saito K, Major EO et al. (1993). A mechanism of quinolinic acid formation by brain in inflammatory neurological disease: attenuation of synthesis from L-tryptophan by 6-chlorotryptophan and 4-chloro-3-hydroxyanthranilate. Brain 116: 1425–1450.

Ho M-F, Zhang C, Zhang L et al. (2019). Ketamine and active ketamine metabolites regulate STAT3 and the type I interferon pathway in human microglia: molecular mechanisms linked to the antidepressant effects of ketamine. Front Pharmacol 10: 1302.

Hoertel N, Franco S, Wall MM et al. (2015). Mental disorders and risk of suicide attempt: a national prospective study. Mol Psychiatry 20: 718–726.

Hogge I, Blankenship P (2020). Self-concealment and suicidality: mediating roles of unmet interpersonal needs and attitudes toward help-seeking. J Clin Psychol 76: 1893–1903.

Holmes SE, Hinz R, Conen S et al. (2018). Elevated translocator protein in anterior cingulate in major depression and a role for inflammation in suicidal thinking: a positron emission tomography study. Biol Psychiatry 83: 61–69.

Jahandideh M, Ebrahimi E, Farzaei MH et al. (2023). The effect of chronic lithium treatment on hippocampal progenitor cells: transcriptomic analysis and systems pharmacology. Brain Behav 13: e3215.

Jørgensen TSH, Osler M, Jorgensen MB et al. (2023). Mapping diagnostic trajectories from the first hospital diagnosis of a psychiatric disorder: a Danish nationwide cohort study using sequence analysis. Lancet Psychiatry 10: 12–20.

Kekesi KA, Juhasz G, Simor A et al. (2012). Altered functional protein networks in the prefrontal cortex and amygdala of victims of suicide. PLoS One 7: e50532.

Khan H, Garg A, Agarwal NB et al. (2022). Zolpidem use and risk of suicide: a systematic review and meta-analysis. Psychiatry Res 316: 114777.

Klempan TA, Ernst C, Deleva V et al. (2009a). Characterization of QKI gene expression, genetics, and epigenetics in suicide victims with major depressive disorder. Biol Psychiatry 66: 824–831.

Klempan TA, Sequeira A, Canetti L et al. (2009b). Altered expression of genes involved in ATP biosynthesis and GABAergic neurotransmission in the ventral prefrontal cortex of suicides with and without major depression. Mol Psychiatry 14: 175–189.

Li H, Sagar AP, Kéri S (2018). Translocator protein (18 kDa TSPO) binding, a marker of microglia, is reduced in major depression during cognitive-behavioral therapy. Prog Neuro-Psychopharmacol Biol Psychiatry 83: 1–7.

Lutz P-E, Tanti A, Gasecka A et al. (2017). Association of a history of child abuse with impaired myelination in the anterior cingulate cortex: convergent epigenetic, transcriptional, and morphological evidence. Am J Psychiatry 174: 1185–1194.

Magdy S, Gamal M, Samir NF et al. (2021). IκB kinase inhibition remodeled connexins, pannexin-1, and excitatory amino-acid transporters expressions to promote neuroprotection of galantamine and morphine. J Cell Physiol 236: 7516–7532.

Malkesman O, Pine DS, Tragon T et al. (2009). Animal models of suicide-trait-related behaviors. Trends Pharmacol Sci 30: 165–173.

Mohr DC, Goodkin DE, Islar J et al. (2001). Treatment of depression is associated with suppression of nonspecific and antigen-specific TH1 responses in multiple sclerosis. Arch Neurol 58: 1081–1086.

Mosser C-A, Baptista S, Arnoux I et al. (2017). Microglia in CNS development: shaping the brain for the future. Prog Neurobiol 149: 1–20.

Naggan L, Robinson E, Dinur E et al. (2023). Suicide in bipolar disorder patients is associated with hippocampal microglia activation and reduction of lymphocytes-activation gene 3 (LAG3) microglial checkpoint expression. Brain Behav Immun 110: 185–194.

Nagy C, Suderman M, Yang J et al. (2015). Astrocytic abnormalities and global DNA methylation patterns in depression and suicide. Mol Psychiatry 20: 320–328.

Nagy C, Torres-Platas SG, Mechawar N et al. (2017). Repression of astrocytic connexins in cortical and subcortical brain regions and prefrontal enrichment of H3K9me3 in depression and suicide. Int J Neuropsychopharmacol 20: 50–57.

Neutel CI, Patten SB (1997). Risk of suicide attempts after benzodiazepine and/or antidepressant use. Ann Epidemiol 7: 568–574.

Niederkrotenthaler T, Stack S, Till B et al. (2019). Association of increased youth suicides in the United States with the release of 13 reasons why. JAMA Psychiatry 76: 933–940.

Nishino S, Ohtomo K, Numata Y et al. (2012). Divergent effects of lithium and sodium valproate on brain-derived neurotrophic factor (BDNF) production in human astrocytoma cells at therapeutic concentrations. Prog Neuro-Psychopharmacol Biol Psychiatry 39: 17–22.

Nordentoft M, Mortensen PB, Pedersen CB (2011). Absolute risk of suicide after first hospital contact in mental disorder. Arch Gen Psychiatry 68: 1058–1064.

O'Dea D, Tucker S (2005). The cost of suicide to society, Ministry of Health, Wellington.

O'Leary LA, Belliveau C, Davoli MA et al. (2021). Widespread decrease of cerebral vimentin-immunoreactive astrocytes in depressed suicides. Front Psych 12: 640963.

Pantazatos SP, Huang Y, Rosoklija GB et al. (2017). Whole-transcriptome brain expression and exon-usage profiling in major depression and suicide: evidence for altered glial, endothelial and ATPase activity. Mol Psychiatry 22: 760–773.

Patchan KM, Richardson C, Vyas G et al. (2015). The risk of suicide after clozapine discontinuation: cause for concern. Ann Clin Psychiatry 27: 253.

Petrasch-Parwez E, Schöbel A, Benali A et al. (2020). Lateralization of increased density of Iba1-immunopositive microglial cells in the anterior midcingulate cortex of schizophrenia and bipolar disorder. Eur Arch Psychiatry Clin Neurosci 270: 819–828.

Poirot MG, Ruhe HG, Mutsaerts H-JM et al. (2024). Treatment response prediction in major depressive disorder using multimodal MRI and clinical data: secondary analysis of a randomized clinical trial. Am J Psychiatry. appi. ajp. 20230206.

Preti A (2011). Animal model and neurobiology of suicide. Prog Neuro-Psychopharmacol Biol Psychiatry 35: 818–830.

Punzi G, Ursini G, Chen Q et al. (2022). Genetics and brain transcriptomics of completed suicide. Am J Psychiatry 179: 226–241.

Rubino A, Roskell N, Tennis P et al. (2007). Risk of suicide during treatment with venlafaxine, citalopram, fluoxetine, and dothiepin: retrospective cohort study. BMJ 334: 242.

Runeson B, Åsberg M (2003). Family history of suicide among suicide victims. Am J Psychiatry 160: 1525–1526.

Scheepstra KW, Mizee MR, Van Scheppingen J et al. (2023). Microglia transcriptional profiling in major depressive disorder shows inhibition of cortical gray matter microglia. Biol Psychiatry 94: 619–629.

Scheepstra K, Mizee M, Wever D et al. (2024). Reporting psychiatric disease characteristics in post-mortem-and biological research. Neurosci Insights 19. https://doi.org/10.1177/26331055241252632.

Schiweck C, Valles-Colomer M, Arolt V et al. (2020). Depression and suicidality: a link to premature T helper cell aging and increased Th17 cells. Brain Behav Immun 87: 603–609.

Setoyama D, Kato TA, Hashimoto R et al. (2016). Plasma metabolites predict severity of depression and suicidal ideation in psychiatric patients-a multicenter pilot analysis. PLoS One 11: e0165267.

Sha Q, Fu Z, Escobar Galvis ML et al. (2023). Integrative transcriptome-and DNA methylation analysis of brain tissue from the temporal pole in suicide decedents and their controls. Mol Psychiatry 29: 134–145.

Smith KA, Cipriani A (2017). Lithium and suicide in mood disorders: updated meta-review of the scientific literature. Bipolar Disord 19: 575–586.

Steiner J, Mawrin C, Ziegeler A et al. (2006). Distribution of HLA-DR-positive microglia in schizophrenia reflects impaired cerebral lateralization. Acta Neuropathol 112: 305–316.

Steiner J, Bielau H, Brisch R et al. (2008). Immunological aspects in the neurobiology of suicide: elevated microglial density in schizophrenia and depression is associated with suicide. J Psychiatr Res 42: 151–157.

Steiner J, Walter M, Gos T et al. (2011). Severe depression is associated with increased microglial quinolinic acid in subregions of the anterior cingulate gyrus: evidence for an immune-modulated glutamatergic neurotransmission? J Neuroinflammation 8: 94.

Sun Y, Lin C-C, Lu C-J et al. (2016). Association between zolpidem and suicide: a nationwide population-based case-control study. Mayo Clinic Proceedings. Elsevier308–315.

Tanti A, Lutz P-E, Kim J et al. (2019). Evidence of decreased gap junction coupling between astrocytes and oligodendrocytes in the anterior cingulate cortex of depressed suicides. Neuropsychopharmacology 44: 2099–2111.

Torres-Platas SG, Hercher C, Davoli MA et al. (2011). Astrocytic hypertrophy in anterior cingulate white matter of depressed suicides. Neuropsychopharmacology 36: 2650–2658.

Torres-Platas SG, Cruceanu C, Chen GG et al. (2014). Evidence for increased microglial priming and macrophage recruitment in the dorsal anterior cingulate white matter of depressed suicides. Brain Behav Immun 42: 50–59.

Torres-Platas S, Nagy C, Wakid M et al. (2016). Glial fibrillary acidic protein is differentially expressed across cortical and subcortical regions in healthy brains and downregulated in the thalamus and caudate nucleus of depressed suicides. Mol Psychiatry 21: 509–515.

Turecki G, Brent DA (2016). Suicide and suicidal behaviour. The Lancet 387: 1227–1239.

Vasupanrajit A, Jirakran K, Tunvirachaisakul C et al. (2022). Inflammation and nitro-oxidative stress in current suicidal attempts and current suicidal ideation: a systematic review and meta-analysis. Mol Psychiatry 27: 1350–1361.

Vijayakumar L, John S, Pirkis J et al. (2005). Suicide in developing countries (2) risk factors. Crisis 26: 112–119.

WHO (2023). *Suicide* [Online]. Available: https://www.who.int/newsroom/fact-sheets/detail/suicide [Accessed 2023].

Yam K-Y, Ruigrok S, Ziko I et al. (2017). Ghrelin and hypothalamic NPY/AgRP expression in mice are affected by chronic early-life stress exposure in a sex-specific manner. Psychoneuroendocrinology 86: 73–77.

Yu Z, Ono C, Kim HB et al. (2011). Four mood stabilizers commonly induce FEZ1 expression in human astrocytes. Bipolar Disord 13: 486–499.

Zeng D, He S, Ma C et al. (2020). Network-based approach to identify molecular signatures in the brains of depressed suicides. Psychiatry Res 294: 113513.

Zhang L (2023). Molecular pathology of suicide: a postmortem study.

Zhang L, Verwer RW, Lucassen PJ et al. (2020a). Prefrontal cortex alterations in glia gene expression in schizophrenia with and without suicide. J Psychiatr Res 121: 31–38.

Zhang L, Verwer RW, Lucassen PJ et al. (2020b). Sex difference in glia gene expression in the dorsolateral prefrontal cortex in bipolar disorder: relation to psychotic features. J Psychiatr Res 125: 66–74.

Zhang L, Verwer RW, Zhao J et al. (2021). Changes in glial gene expression in the prefrontal cortex in relation to major depressive disorder, suicide and psychotic features. J Affect Disord 295: 893–903.

Zhang L, Lucassen PJ, Salta E et al. (2022a). Hippocampal neuropathology in suicide: gaps in our knowledge and opportunities for a breakthrough. Neurosci Biobehav Rev 132: 542–552.

Zhang L, Verwer RW, Van Heerikhuize J et al. (2022b). Stress-associated purinergic receptors code for fatal suicidality in the hippocampal-hypothalamic-prefrontal circuit. bioRxiv 2022.11. 22.516142.

Zhang L, Verwer RW, Van Heerikhuize J et al. (2024). Progesterone receptor distribution in the human hypothalamus and its association with suicide. Acta Neuropathol Commun 12: 16.

Zhao J, Verwer R, Van Wamelen D et al. (2016). Prefrontal changes in the glutamate-glutamine cycle and neuronal/glial glutamate transporters in depression with and without suicide. J Psychiatr Res 82: 8–15.

Zhao J, Verwer R, Gao S-F et al. (2018). Prefrontal alterations in Gabaergic and glutamatergic gene expression in relation to depression and suicide. J Psychiatr Res 102: 261–274.

Zhao J, Lucassen PJ, Swaab DF (2019). Suicide is a confounder in postmortem studies on depression. Biol Psychiatry 86: e37–e40.

Zhou Y, Xiong L, Chen J et al. (2023). Integrative analyses of scRNA-seq, bulk mRNA-seq, and DNA methylation profiling in depressed suicide brain tissues. Int J Neuropsychopharmacol 26: 840–855.

Index

NB: Page numbers in *italics* refer to figures and tables.